Ara Monadjem, Peter J. Taylor, Christiane Denys, Fenton P.D. Cotterill
Rodents of Sub-Saharan Africa

Ara Monadjem, Peter J. Taylor,
Christiane Denys, Fenton P.D. Cotterill

Rodents of Sub-Saharan Africa

A biogeographic and taxonomic synthesis

DE GRUYTER

Authors

Prof. Dr. Ara Monadjem
University of Swaziland (UNISWA)
Dept. of Biological Sciences
Private Bag 4
Kwaluseni, Swaziland
E-Mail: ara@uniswa.sz

Prof. Christiane Denys
Museum National d'Histoire Naturelle
Dept. Systematique et Evolution
USM 601-UMR 5202 OSEB
55 Rue Buffon, CP 51
75231 Paris Cedex 05, France
E-Mail: denys@mnhn.fr

Prof. Dr. Peter J. Taylor
South African Research Chair on Biodiversity
Value & Change
University of Venda
School of Mathematical & Natural Sciences
Private Bag X5050
Thohoyandou, Limpopo Province
0950, South Africa
E-Mail: peter.taylor.univen@gmail.com

Dr. Fenton P.D. Cotterill
Stellenbosch University
Dept. of Botany and Zoology - AEON
X1 Private Bag
Matieland 7602, South Africa
E-Mail: fcotterill@gmail.com

ISBN 978-3-11-030166-3
e-ISBN (PDF) 978-3-11-030191-5
e-ISBN (EPUB) 978-3-11-038923-4

Library of Congress Cataloging-in-Publication Data
A CIP catalog record for this book has been applied for at the Library of Congress.

Bibliographic information published by the Deutsche Nationalbibliothek
The Deutsche Nationalbibliothek lists this publication in the Deutsche Nationalbibliografie; detailed bibliographic data are available on the Internet at http://dnb.dnb.de.

Coverimage: Ground squirrel; Nico Smit/iStock/Thinkstock
Typesetting: Michael Peschke, Berlin
Printing: CPI books GmbH, Leck
♾ Printed on acid free paper
Printed in Germany

www.degruyter.com

Foreword

This is an excellent book for anyone interested in rodents. The authors should be applauded for providing the most up-to-date information on a highly diverse group of mammals in one of the most important continents of the world for mammal diversity. The book is not only for taxonomists and will be of immense help to all kinds of research from zoonosis, pest management, conservation and climate change, because knowing the species is the foundation of any ecological and environmental research. Understanding rodent speciation through the application of relatively new genetic tools has helped emphasise the importance of rodents in understanding biodiversity and evolution. The maps and identification tools will be an invaluable resource for expert taxonomists, research professionals and amateurs alike. Despite the considerable work that has been done and reflected in this book on understanding African rodents, the field continues to rapidly change, and this book has made every effort to include very recent discoveries whilst delivering a high quality reference in a very short production time.

Prof. Steven Belmain, Natural Resources Institute, University of Greenwich, UK

Acknowledgements

It has become a cliché to acknowledge the role of numerous players in the development of a book, whose input was deemed essential to the book's successful publication. This book is no different. As a "synthesis", this book owes its very existence to the works of previous small mammal researchers and enthusiasts. In particular, the distribution maps are drawn from tens of thousands of vouchered specimen records deposited in dozens of museums scattered around the globe. These humble museum specimens, and the persons that collected them, curated them and documented them are the real reason this book exists. This book is dedicated to all the museum workers who have toiled under difficult, and sometimes impossible, circumstances to ensure their collections not only survive, but continue to grow. Some of these institutions are mentioned below and the rest are listed in the "General Methods and Layout of Book" section. All the photographs appearing in the book have been acknowledged in the captions. We are sincerely grateful to all these people for making their images freely available to us. Bill Branch, Josef Bryja, Radim Sumbera and Darcy Ogada went out of their way to help locate suitable images for this book; without their generous support this book would have been significantly poorer.

Ara Monadjem
Dr. Teresa Kearney (Ditsong National Museum of Natural History, South Africa), Dr. Leigh Richards (Durban Natural Science Museum, South Africa), Simon Musila (National Museums of Kenya, Kenya) and Seth Eiseb (National Museum of Namibia) granted us permission to examine their small mammal collections and went out of their way to make our visits to their institutions as comfortable and productive as possible. I would like to thank my Head of Department Dr. Themb'a A. Mahlaba, himself a well-respected African small mammal expert, who has whole-heartedly supported my involvement with the writing of this book, including sanctioning several lengthy museum excursions. The rest of the lecturers in the Department of Biological Sciences at the University of Swaziland are sincerely thanked for making whatever sacrifices necessary to allow me to conduct the research needed for this project. Fezile Mtsetfwa is thanked for carefully checking all the references. Prof. Steven Belmain kindly reviewed the book and wrote the forward to the book. Dr. Bob McCleery, University of Florida, has provided valuable suggestions and input over the years. My wife, Dr. Sara Padidar, has had to put up with the "usual nonsense" associated with living with a field biologist, including long stints away in the field and getting up at ridiculous hours of the morning; but never with a single word (spoken or unspoken) of disapproval. Thank you! To my boys: Keyan, Damian and Liam, I hope that at least some of my passion for the study of wild animals has rubbed off.

Christiane Denys
I wish to thank the staff of the Muséum national d'histoire naturelle (MNHN) (Mammalogy) that helped during the early process of collection consultation and allowed SEM images to be taken of many specimens (Michel Tranier, Jacques Cuisin). The late Francis Petter was a key promoter of this book. He participated in the earliest phase of the identification key and donated some of his personal photographs. We thank the different curators of collections that loaned to us some specimens, images or gave us access to holotypes of very rare species (Dr. Renate Angermann, Zoologische Museum Berlin, Germany; Dr. Fritz Dieterlen, Stuttgart State Museum of Natural History,

Germany; Dr. Leigh Richards, Durban Natural Science Museum, South Africa; Wim Wendelen, Museum for Central Africa, Belgium; Dr. Paula Jenkins, Natural History Museum, London, UK; Drs. Julian C. Kerbis and Bruce Patterson, Field Museum of Natural History, Chicago, USA). Most of the SEM pictures were taken thanks to a BQR grant from the MNHN (1992-1993) and UMR OSEB (2005, 2011). Dr. Philippe Grandcolas, Dr. Guillaume Lecointre and Mrs Séverine Deruyter (MNHN) provided administrative assistance for the visits of Ara Monadjem and Peter Taylor to the MNHN as invited professors. The late Mrs Laroche, Mrs Chancogne and Dr. Rausch are thanked for assisting with the taking of the SEM images. Guy Rabache is warmly thanked for the preparation of dental plates, some of the skull pictures, the scanning of negatives and of photographs.

Peter Taylor
I would like to thank the South African National Research Foundation (NRF) and the University of Venda for financial support and for study leave to complete this project through several study visits to museum collections in South Africa and France and to Swaziland. I am most grateful to all the museum curators who generously provided advice and access to collections, particularly Teresa Kearney (Ditsong National Museum of Natural History), Leigh Richards (Durban Natural Science Museum) and mammal curatorial staff of the Paris Museum of Natural History. Colleagues from the "Ratzooman", "Ecorat" and "Stoprats" projects provided enlightening discussions and important insights into the economic importance of rodents and the need for improved taxonomy and field guides in order to optimally implement ecologically-based rodent management (EBRM) in Africa. As always, I am incredibly grateful to my wife Frances and four children (Ashleigh, Robyn, Lauren and Benjamin) for unstinting support and understanding.

Fenton P.D. Cotterill
Natural History Museum of Zimbabwe (Bulawayo) and Associated Mammal Surveys: Since the early 20th century, the commitments of many scientists and technicians collected the many rodent specimens preserved in the NMZB (including A. Ndlovu, K. Mkwananzi, L. Sango-Moyo, N. Sango-Moyo, F. M. Masiyandima, J. Chimombe and A. N. Sango). Special mention is due to R.H.N. Smithers who made so much possible, and W.F.H. Ansell and J. L. P. Lobão Tello who focused on pioneering small mammal surveys in Zambia and Mozambique, respectively. Participants, and high school expeditions - namely the Exploration Societies of the Falcon College, Girls College, Peterhouse and the Rhodesian Schools (and latterly Operation Raleigh) aided the collection of many specimens. For over three decades, the leaders of these trips, notably John Stakesby Lewis, organised these pedagogical expeditions in association with the Natural History Museum, almost annually and often into remote regions. Overall, it is important to underscore the efforts of Reay Smithers for directing small mammal collections for nearly four decades. By far the greatest part of the Smithers' Legacy lies in the collaboration of the Rhodesian Museums with the Smithsonian Institution African surveys (driven by H.W. Setzer and carried out by many committed fieldworkers) made possible the remarkable discoveries of the Botswana Mammal Survey, and associated expeditions in Mozambique and Zimbabwe (see Schmidt et al., 2008).

Schmidt D.F., C.A. Ludwig and M.D. Carleton. 2008. The Smithsonian Institution African Mammal Project (1961–1972) : An Annotated Gazetteer of Collecting Localities and Summary of Its Taxonomic and Geographic Scope. Smithsonian Contributions to Zoology 628: 1-320

Table of Contents

Introduction

Rationale and Scope of the Book

Our aim is to provide an accessible, accurate and updated identification, distribution and taxonomic reference for rodent biologists, conservationists, public health workers, epidemiologists and agriculturalists for all 89 genera and 463 currently recognized species.

This is NOT a re-hash of already published taxonomic or ecological reviews (Musser & Carleton, 2005; Happold, 2013). It is NOT a complete review of the biology of all Sub-Saharan African rodents, nor does it pretend to contain a complete bibliography (Musser & Carleton, 2005 and Happold, 2013 can be consulted for this) although we have included all the recent and key biogeographical and taxonomic literature. Furthermore, we have not attempted to compile, let alone present, all the scientific names that have been described over the past few hundred years. For a source of synonomies, readers are referred to the key publications of Allen (1939), Ansell (1989) and Musser & Carleton (2005). The emphasis of this book is taxonomic and biogeographic, focusing on both classical works and the recent literature which is documenting an extraordinary renaissance in species discovery and molecular phylogenetic synthesis, often confirming the results of original and classical studies from the early twentieth century which were too often obscured by the lumpers and pheneticists of the middle twentieth century (see commentary by Groves & Grubb, 2011; Groves, 2013). Aided by new collections and the reduced cost and increased ease of molecular techniques, this accelerated pace of taxonomic discovery has resulted in many changes to rodent taxonomy at the genus and species level and the status quo has changed considerably since the synthesis of Musser & Carleton (2005). To take one example, in the gerbils two genera recognized by Musser & Carleton (2005) have been synonymised within other genera and new species have been described, e.g., within *Gerbilliscus* (Granjon et al., 2012). To take another example, 17 species of Otomyinae were recognised by Happold (2013) but we present 31 in this book.

In compiling this work, we have relied heavily on a few recent publications that deserve to be singled out. Firstly, Mammals of Africa, the rodent volume edited by Happold (2013), is an amazing source of information, particularly on the ecology, behaviour and reproductive biology of these creatures. We found this volume to be indispensable to us in compiling our own book. We must emphasize, again, that we do not view our book as being in competition with Happold (2013), but rather complementary to it. Another recently published book that we refer to regularly is Granjon & Duplantier (2009) that deals with the rodents of the Sahelo-Sudanian zone of West Africa. The extensive work of West African rodent biologists (often published in French) is not as well-known as it should be in Anglophone Africa, but is exhaustively synthesised in this book and, despite having been written in French, it should be on the shelf of every African rodent biologist.

This book seeks to update and elucidate distribution and taxonomy by considering original museum collection specimens, or studies that refer to them, and the recent spate of publications using molecular and chromosomal methods to uncover cryptic species and clarify relationships within and between genera. By considering or re-considering museum collections and traditional reviews of these, in the light of modern molecular studies, we have uncovered anomalies and interesting novelties which have escaped earlier syntheses based often only on uncritical preceding reviews. The effects of taxonomic lumping, the polytypic (biological) species concept and uncritical phenetic analyses based only on quantitative and not qualitative characters has been to obscure much of the diversity which exists and which had already been recognized by earlier workers e.g. see *Rhabdomys bechuanae* which was clearly diagnosed by Hill & Carter (1941), but then synonymised with *R. pumilio* for more than half a century before molecular analysis confirmed its distinctness (Du Toit et al., 2012). Whilst this process of discovery is still under way, we seek to capture the emerging consensus of rodent taxonomy and classification and thus help to identify gaps for further study. Defining the real geographic limits of species defined on molecular or chromosomal grounds remains a challenge.

Examples of overlooked anomalies include *Dasymys* cf *foxi* from plateaux of Nigeria east of Jos and including the Cameroon Volcanic Line extension (these populations mentioned in Happold, 1987 are totally ignored in recent reviews including Happold, 2013) and similarly, totally overlooked forest populations of *Heliosciurus* cf *rufobrachium* from south of the Congo River (and north of the range of *H. gambianus*). Many more such examples exist and are highlighted in the relevant species accounts.

With the plethora of molecular phylogenetic and phylogeographical studies, there is great demand for accurate data on earliest fossil occurrences of taxa in order to calibrate molecular clocks. The discovery of new fossil data on African rodents is continuing apace (e.g. see Winkler et al., 2010). For this reason, the next section (page 5) synthesizes classical and recent studies to provide fossil data in a useful and accessible summary; at the same time offering important evolutionary insights. In addition we provide in the genus accounts a summary of first fossil occurrences as well as the temporal range for each genus in northern, eastern and southern Africa.

The importance of rodents to agriculture and human health and the resulting risks to food security and human health are being increasingly acknowledged and investigated. Many multidisciplinary projects have been carried out in Africa in recent years, particularly in West, East and Southern Africa, and we provide a succinct description of publications arising from such studies on page 19. Such applied studies, as well as pure biological studies, all depend critically on accurate knowledge of the distribution and habitats of different rodent species. For the first time, we provide accurate point distribution maps for 463 species of African rodents based on museum specimen records (many verified by one of the authors) as well as from reputable studies based themselves on such collections (such as Hollister, 1919; Hill & Carter, 1941). A review of the conservation status of Sub-Saharan African rodents is provided in on page 22.

A problem encountered when conducting any field investigations of rodents in Africa concerns the lack of coherent field identification keys to genera or species. Such keys are dispersed in the literature and are usually restricted to a particular region or subset of characters. In this book, we attempt to provide keys based on external and skull characters for nearly all the rodent species covered in this book. The few genera that we were unable to provide keys for include a large proportion of cryptic species that simply cannot be identified based on traditional morphological features. Instead of attempting to simplify the situation by presenting keys that clearly would not work effectively, we have chosen to not present keys for these groups. Our keys are purely functional in nature, and we do not claim that they underscore evolutionary relationships. Furthermore, these keys should be viewed as a guide to preliminary identification; for confirmation of the identification, we suggest that the detailed genus and species accounts be consulted. We have occasionally been forced to use geographical distributions in distinguishing species; this is generally undesirable since distributional limits of most African rodents are still relatively poorly known and new records of range extension are continuously being published. Therefore, basing identification on geographical location may lead to mis-identifications. However, we have attempted to use distribution as a last resort, and typically when the distributions separated by large geographical distances or when the species are restricted to different mountain top habitats. Lastly, these keys should be viewed as a work in progress. There may be errors in some keys and we may have overlooked certain diagnostic features in others. We are eager to hear from field and museum workers alike as to the usefulness of these keys, and would welcome any suggestions for improving them.

Evolution of Fossil Rodents in Africa

Sub-Saharan Africa has yielded a rodent fossil record of importance, especially in East and Southern Africa. Many geographical and temporal gaps can be identified in this record such as western Central Africa or the Early to Middle Oligocene period (Table 1, Figure 1). A succession of different radiations may explain the modern diversity in this group in Africa and these events must be related to the tectonic and climatic evolution of the sub-continent through the Cenozoic. In fact, from the Late Cretaceous to the Middle Miocene, the African plate (including part of modern Africa, Arabia and part of Asia) was isolated from Eurasia, North America and the Atlantic Ocean has started to open creating a barrier between the South American and African continents (Cox, 2000). However, some exchanges occurred between Africa and South Asia during the Paleogene (middle to late Eocene) and allowed some groups of primates and nearly all groups of hystricognath rodents to establish and diversify (Dawson et al., 2003; Marivaux et al., 2005). This allowed endemic African rodent families to diversify under warm climatic conditions. During the Oligo-Miocene boundary (23.8 MYA), the Tethys seaway that was isolating Africa and Eurasia start to close gradually allowing some faunal interchanges between Pakistan and Africa through Arabia and Anatolia (Koufos et al., 2005). This allowed elephants, artiodactyls, and primates to enter Eurasia at different periods (Bernor et al., 1987). The bulk of the migration events are recorded between 20.7 and 17.0 MYA, mostly around 18.5 MYA ("Burdigalian interchange"). Concerning the rodents, this episode allowed Pedetidae to enter Asia and *Democricetodon, Megacricetodon* (belonging to Cricetidae extinct sub-families) to arrive in Africa. Due to short-term sea level oscillations of the Tethys, temporal landbridges were active at different times and exchanges still continue especially between North Africa, Arabia and Pakistan between 17 to 10.5 MYA. It is during this period that the muroid rodent *Potwarmus* reached Libya from Pakistan (18 MYA) and the true murid *Progonomys* colonised Africa from Eurasia (12-10 MYA). The exchanges continued sporadically through the paleo Saharo-Sindian region (or "province" sensu Brandy et al., 1980) until the late Miocene (6.1 MYA) between Europe and North Africa. The orogenic uplift of the Himalaya chain around 21 MYA but with a strong uplift at 8 MYA, considerably modified the climate by the development of the monsoon (Partridge et al., 1995). The connections between north-east Africa and east-central Africa are not very clear due to the lack of fossils at this critical time. In East Africa rifting began in the Middle Miocene resulting in the formation of the Rift Valley. The formation of the Rift Valley, the development of the mega-Lake Chad, and the evolution and spread of C3 grasses during the late Miocene times may explain the progressive differentiation into biogeographic regions of the rodent faunas (Denys et al., 1985; Denys, 1999; de Menocal, 2000; Cerling et al., 1977).

Rodents belong to one of the earliest modern mammalian radiations and appear in Eurasia as early as the Paleocene. No rodent of this age has been found in Africa; the oldest African record dates from the Early Eocene (Winkler et al., 2010). The first rodent radiation started during the Lower

Eocene with the Ctenohystrica and the Anomaluropmorpha which belong to basal clades of the order Rodentia (Huchon et al., 2007; Blanga-Kanfi et al., 2009). Despite the opening of the Atlantic Ocean during Cretaceous times some African hystricognath rodents are nearly identical to ones observed in South America, resulting in much debate over the dispersal routes of these rodents e.g. via North America or across the Atlantic (Wood, 1985; Lavocat, 1985; Marivaux et al., 2002). However the molecular data all converge towards the monophyly of the Hystricognathi and with Ctenodactylidae as sister taxon and basal members of a clade named Ctenohystrica (Blanga-Koffi et al., 2009). The discovery of new fossils in South-East Asia has shed new light on the origins of the modern African endemic families, especially the discovery of *Laonastes* (Jenckins et al., 2005; Marivaux et al., 2011) as the only known living representative of Diatomyinae an extinct Asiatic fossil taxon. Recent molecular analyses confirm that the dichotomy *Ctenodactylus-Laonastes* dates from 44.7 MYA ± 3.5 MYA and indicates that these families diverged by vicariance. Furthermore, the Pedetidae are not related to Diatomyidae (Huchon et al., 2007). During the Eocene an extinct family, the Zegdoumyidae, appears in Algeria (Vianey-Liaud & Jaeger, 1996; Marivaux et al., 2011) and may be related to the modern Anomaluridae also occurring for the first times in the Middle to Late Eocene in Algeria (Jaeger et al., 1985), in Egypt (Sallam et al., 2010) and later in Kenya and Uganda during the Miocene.

The isolation of Africa during the Middle and Upper Eocene and the Oligocene has led to the diversification of various endemic families of hystricognath rodents from the Asiatic ancestral Ctenohystrica stock including the Phiomorpha (Petromuridae, Bathyergidae), the Anomaluromorpha and the tribe Xerini (Sciuromorpha). Later, the opening of terrestrial connections between Eurasia and Africa between Middle and Upper Miocene allowed the Muridae to enter Africa, diversify and constitute the bulk of the present-day rodent diversity of Africa.

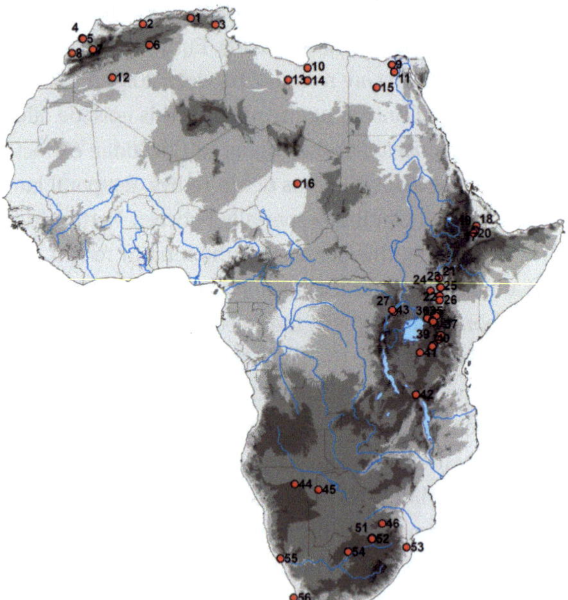

Figure 1.1 Map showing the distribution of sites where rodent fossils have been excavated in Africa. The sites have been numbered and the legend appears below

Number	Site	Country	Number	Site	Country
1	Amama	Morocco	29	Tabarin	Kenya
2	Bou Hanifia	Algeria	30	Ngorora	Kenya
3	Chambi	Tunisia	31	Lukeino	Kenya
4	Lissafsa	Morocco	32	Maboko	Kenya
5	Al Oughlam	Morocco	33	Meswa Bridge	Kenya
6	Nementcha	Algeria	34	Songhor	Kenya
7	Beni Mellal	Morocco	35	Fort Ternan	Kenya
8	Jebel Irhoud	Morocco	36	Koru	Kenya
9	Wadi El Natrun	Egypt	37	Lainyamok	Kenya
10	Sahabi	Libya	38	Peninj	Tanzania
11	Fayum	Egypt	39	Olduvai Gorge	Tanzania
12	Glib Zegdou	Algeria	40	Laetoli	Tanzania
13	Zallah Oasis	Libya	41	Manonga	Tanzania
14	Jebel Zelten	Lybia	42	Songwe	Tanzania
15	Sheik Abdallah	Egypt	43	Lusso Beds	DRC
16	Kossom Bougoudi	Chad	44	Otavi (Berg Aukas, Harasib)	Namibia
17	Hadar	Ethiopia	45	Ngamiland	Botswana
18	Chorora	Ethiopia	46	Makapansgat	South Africa
19	Aramis	Ethiopia	47	Gladysvale	South Africa
20	Adu Asa	Ethiopia	48	Swartkrans	South Africa
21	Omo	Ethiopia	49	Plovers Lake	South Africa
22	Koobi Fora	Kenya	50	Bolt'sFarm/Waypoint160	South Africa
23	East Turkana	Kenya	51	Kromdraai	South Africa
24	Lokone	Kenya	52	Sterkfontein	South Africa
25	Lothagham	Kenya	53	Border Cave	South Africa
26	Kanapoi	Kenya	54	Taung	South Africa
27	Kakara	Uganda	55	Arrisdrift	Namibia
28	Muruyur	Kenya	56	Langebaanweg	South Africa

Table 1. Stratigraphic position of the known African sites that have yielded fossil rodents.

Age (MYA)	Period	Kenya-Uganda	Tanzania	Ethiopia	Egypt/North Africa	Namibia	South Africa
0-1	Upper Pleistocene Holocene	Isenya, Kapthurin	Olduvai Bed IV,	Porcupine cave		Berg AukasII	Swartkrans 3, Florisbad
1-2.5	Lower, Middle Pleistocene	Koobi Fora rodents	Olduvai beds I,II,III, Peninj	Melka Kunturé, Omo E,F,J,K,L			Swratkrans1,2, Kromdraai, Elandsfontein
2.5-5.5	Lower & Middle Pliocene	Kanapoi, Tabarin Kakara	Laetoli, Manonga	Hadar, Kanapoi, Omo B,C,D, Aramis	Aghoiuri, lac Ich-keul,Amama3, Ain Brimba, Al Oughlam	Jägersquelle I (1,3), Nosib I, IIBerg Aukas I(8,54)	Waypoint160, Bolts Farm, MKP, STK2, ST4
5.5-8	Upper/Late Miocene	Lothagham, Lemudong'o		Adu Asa	Sahabi,Wadi Natron, Amama2, Lissafsa		Langebaanweg
8-10.5	Upper Miocene	Lukeino - Mpesida				Berg Aukas I (31,90)	
10.5-12	Middle Miocene	Ngorora		Chorora	Bou Hanifia Sheik Abdallah	Harasib 3a	
12-14.5	Middle Miocene	Fort Ternan, Serek, Kapsibor			Beni Mellal, Azdal,-Testour	Berg Aukas I (1,47,63)	
14.5-16	Middle Miocene	Nyakach, Ombo,Maboko, Majiwa, Muruyur			Jebel Zelten	Arrisdrift	
16-18.5	Middle Miocene					Langental, Plant 4,	
18.5-23	Lower Miocene	Songhor, Chamtwara,Legetet, Koru, Napak, Bukwa, Meswa Bridge	Mbeya region		Gebel Zelten	Lüderitz, Elisabeth-feld	
23-28	Upper Oligocene	Lokone Losodok	Songwe				
28-33.9	Early, Lower /Middle Oligocene				Dor el Talh, Zallah Oasis		
33.9-40	Upper Eocene				Bir l Ater, Fayum		
40-50	Middle Eocene				Glib Zegdou, Chambi		
55-60	Paleocene						

Ctenohystrica diversification in Africa

During the Upper Eocene and Lower Oligocene in Algeria and in Egypt an extinct family of hystricognath rodent diversifies: the Phiomyidae which are the sister clade of modern Thryonomyidae (Lopez Antonanzas et al., 2004). The first genus to occur is *Protophiomys* in Algeria (Jaeger et al., 1985), Egypt (Sallam et al., 2009), and possibly Namibia (Pickford et al., 2008). The Phiomyidae are found during the Early Oligocene in south-western Tanzania (Stevens et al., 2006, 2009), Central Libya (Coster et al., 2012), Late Oligocene in Kenya (Marivaux et al., 2012) and by the Early Miocene are recovered through the whole African continent where they are represented by different sub-families of rodents both found in East and Southern Africa (Diamantomyinae in Namibia and Phiomyine, Myophiomyinae, Kenyamyinae in East Africa) (Lavocat, 1973; Holroyd & Stevens, 2009; Winkler et al., 2005). The early modern representatives of Thryonomyidae (*Apodecter, Paraphiomys* and *Epiphiomys*) are known in the early Miocene sites of Kenya and Uganda (around 24-18 MYA) while *Paraulacodus* (considered as the ancestor of modern *Thryonomys*) appears in the Middle Miocene sites of Kenya (around 14MYA) (Winkler et al., 2010). *Paraphiomys* is found until the Late Miocene in Namibia (Mein et al., 2000). *Paraulacodus* is also recovered in Asia later (Flynn & Winkler, 1994). Early modern *Thryonomys* representatives occur during the Late Miocene in Ethiopia (Wesselman et al., 2009) (Table 2). The replacement of *Paraulacaudus* by *Thryonomys* occurred at 5.3-5.7 MYA in East Africa (Mein & Pickford, 2006). A comprehensive phylogenetical analysis of the whole modern and fossil Thryonomyidae at the scale of their Asiatic, Arabic and African distributions led to the conclusion that *Epiphiomys corundoni* from the Lower Miocene of Kenya is the most primitive species of the family while *Gaudeamus aegyptius* from the Upper Eocene of Fayoum is its oldest representative (Lopez Antonanzas et al., 2004). However, according to Sallam et al. (2011), *Gaudeamus* belongs to the Gaudeamuridae, a sister taxon of Hystricidae. Furthermore, the thryonomyids appear to originate in Africa and to have reached the Arabian Peninsula in the early Miocene and Pakistan and Greece during Middle Miocene. According to Mein & Pickford (2006), two lineages of phiomyids differing by the number of lophs in the molars diverge from the Early Miocene stock. *Apodecter* survived in East Africa until 5.8 MYA and until the Lower Pliocene in South Africa while true *Paraphiomys* disappeared around 16 MYA (Mein & Pickford, 2006).

The other endemic hystricognath African family the Petromuridae appears with fossils attributed to the modern *Petromus* in the Late Miocene of Kenya around 6 MYA (Mein and Pickford, 2006), in the Early Pliocene of Namibia (Pickford et al., 1994), and South Africa (Senegas, 2004). Some *incertae sedis* genera not yet clearly placed in a phylogenetic tree may belong to the same initial hystricognath radiation of rodents that constitutes the major part of the fossil communities during the Eocene-Oligocene times through Africa.

During the Miocene, the rodent faunas of Africa experienced the diversification of most of the modern families. Two other families of hystricognath, the mole-rats and the porcupines are known as early as the Middle and Late Miocene, respectively, in East Africa. Mole-rats (Bathyergidae) are endemics of tropical Africa and have diversified both in East and Southern Africa. Five extinct genera have been attributed to Bathyerginae (*Bathyergoides, Geofossor, Richardus, Proheliophobius, Gypsorychus*) and are described from the Lower and Middle Miocene (Lavocat, 1973; Denys & Jaeger, 1992). Modern representatives of *Bathyergus, Georychus, Cryptomys, Heliophobius* and *Heterocephalus* are recognized only since the Early and Middle Pliocene (Winkler et al., 2010) (Table 2).

Hystricidae originate from central Europe and eastern Mediterranean sites at around 10.5 MYA and seem to enter East and Southern Africa in the Late Miocene. The oldest record of *Hystrix* is in Carriere Marceau (Algeria) at around 9 -10 MYA. In East Africa it is the Lower Nawata site (Kenya) (Winkler 2003) at about 7.4 MYA. *Atherurus*, the other genus present in Africa is described as early as 11MYA in Egypt (Mein & Pickford, 2010) and in Lemudong'o, Kenya (Hlusko 2007). During the Lower Pliocene an extinct large porcupine (*Xenohystrix*) was recorded in most of Southern and East African sites and various species of extinct *Hystrix* were described (Denys 2011, Winkler et al., 2010).

Ctenodactylidae fossils have been discovered in early Miocene in North Africa. They are well represented by five extinct genera *Metasyimys, Sayimys, Africanomys, Irhoudia* and *Testouromys* (Winkler et al., 2010) in this region. They have never been found in any of the Miocene sites of East and Southern Africa. Today this family is restricted to dry and arid environments and perhaps the tropical forest belt has acted as a barrier preventing their dispersal into East and Southern Africa. The first modern *Ctenodactylus* occurs in an upper Pleistocene cave of Morocco (Bouzouggar et al., 2007) and *Massoutiera* in an Holocene cave of Central Sahara (Lopez-Antonanzas, 2010). A phylogeny including modern and fossil Ctenodactylidae was proposed by Lopez-Antonanzas & Knoll (2011).

Table 2. First occurrence of modern genera of the Ctenohystrica and Anomaluromorpha.

Family/Genus	FAD (MYA)	Site	Region	reference
Bathyergidae				
Bathyergus	5	Langebaanweg	South Africa	Denys (1998)
Cryptomys	5	Langebaanweg	South Africa	Denys (1998)
Heliophobius	0.3	Isenya	Kenya	Denys & Brugal (1989)
Heterocephalus	4.3	Kakesio Beds	Tanzania	Denys (2011)
Hystricidae				
Hystrix	7.44	Lothagham	Kenya	Winkler (2003)
Atherurus	11-10	Sheik Abdallah	Egypt	Pickford et al. (2008)
Thryonomyidae				
Thryonomys	6	Lemudong'o	Kenya	Manthi (2007)
Petromuridae				
Petromus	6.1-5.8	Lukeino	Kenya	Mein & Pickford (2006)
Pedetidae				
Pedetes	2	Taung, Lainyamok	South Africa, Kenya	Pickford & Mein (2011)
Anomaluridae				
Anomalurus	13-14	Muruyur Beds	Kenya	Winkler (2002)
Zenkerella	16-18	Songhor	Kenya	Lavocat (1973)
Ctenodactylidae				
Ctenodactylus	130.000 years BP	Taforalt	Morocco	Bouzouggar et al., 2007
Massoutiera	10.000 years BP	Ti-n-Thora	Libya	Lopez-Antonanzas, 2010

Anomaluromorpha diversification

Zegdoumyidae have been recovered in the Early to Middle Eocene of Algeria, Egypt and possibly Namibia and are considered as the earliest offshoots of a large clade constituted of highly diversified Anomaluridae plus the modern Anomaluridae group and possibly the Pedetidae. Their origins have to be searched among Paleogene Myodonta or North American Sciuravidae or a more primitive unknown African lineage (Marivaux et al., 2011).

With the exception of anomalurid fossils from Middle Eocene sites in Algeria (*Nementchamys*), Egypt (*Kabirmys, Shazurus*), and Myanmar (*Pondaugimys*) very few records of this family are known from the Upper Eocene-Lower Oligocene. Lavocat (1973) described *Paranomalurus* and the earliest modern *Zenkerella* from Early Miocene sites of Kenya and Uganda and the first *Anomalurus* from the Middle Miocene of Kenya (Table 2). Anomaluridae have never been recorded from the Southern African region (Winkler et al., 2010). A phylogeny and biogeographical scenario including both modern and fossil anomaluroid rodents is proposed by Sallam et al. (2010).

The Springhares or Pedetidae are first known from the Early Miocene of Kenya (23 MYA, Meswa Bridge) by *Oldrichpedetes praecursor* (Winkler, 1992; Pickford & Mein, 2011). This genus has been recovered throughout Africa up until the Early Pliocene. Three other extinct genera named *Parapedetes, Megapedetes, Rusingapedetes* have also been described from East, Southern and North Africa (Lavocat 1973). Their relationships with Diatomyidae have been questioned by Marivaux et al. (2004) but it seems clear that they belong to two distinct families and that the Pedetidae differentiated in Africa (Pickford & Mein, 2011). According to Pickford & Mein (2011) *Propedetes* discovered in the Upper Miocene of Namibia and Kenya, in the Lower Pliocene of Southern and East Africa would be the closest to the modern *Pedetes* whose early representatives date from the Middle Pliocene (Table 2).

Sciuromorpha diversification

The African lineage of Sciuromorpha essentially belongs to the subfamily Xerinae, which are today found in Africa and Asia. The oldest Sciuridae recorded in Africa dates from the Early Miocene from East Africa and was represented by the extinct genus *Vulcanisciurus* (Lavocat, 1973). During the Late Miocene, the presence of an extinct giant sciurid *Kubwaxerus* is first noted from East Africa as well as the occurrence of the first representatives of modern genera. *Xerus* appears in the Upper Miocene of Ethiopia whereas the first *Paraxerus* is recorded in Lemudong'o (Kenya) (Table 3). In South Africa, the first squirrels are represented by *Heteroxerus* (also described from Eurasia) and *Paraxerus* in Namibia during the Late Miocene (Mein et al., 2000a). During the Plio-cene-Pleistocene times, extinct species of sciurids have been described like *Paraxerus meini* or *Xerus janenschi* from Laetoli , Tanzania, 3.7 -3.5 MYA (Denys, 2011), *Xerus daamsi* in Chad (Denys et al., 2003) which is closely related to the modern *X. erythropus* and *X. rutilus*, and *Xerus* cf *inauris* at Olduvai Bed I (Denys, 1989a) All these taxa are closely related to modern Xerini species. However, the earliest squirrel attributable to a modern species is *Xerus erythropus* from the Omo deposits, 3.03-2.45 MYA (Wesselman, 1984).

The Gliridae experienced an important radiation in North Africa during the Late Miocene with European affinities while in Southern Africa the oldest presence of the family is attested in

Namibia in the Late Miocene with the presence of an extinct genus *Otaviglis* Mein et al. (2000a). Modern *Graphiurus* representatives are found later during the Late Pliocene in South Africa and Early Pleistocene in East Africa (Winkler et al., 2010) (Table 3).

Table 3. First occurrence of modern genera of the Sciuromorpha.

Family/Genus	FAD (MYA)	site	region	ref
Sciuridae				
Heliosciurus	4.5-4.4	Tabarin	Kenya	Winkler (1990, 2002)
Paraxerus	6	Lemudong'o	Kenya	Manthi (2007)
Xerus	11	Chorora	Ethiopia	Geraads (2001)
Gliridae				
Graphiurus	3.3	Makapansgat	South Africa	Pocock (1987)

Myomorpha diversification

The bulk of the rodent radiation in Africa is constituted by the Muroidea which are today the most abundant group of rodents. This taxon is represented by the Spalacidae, Nesomyidae, and Muridae in Africa, with the Cricetidae occurring in North Africa outside of Sub-Saharan Africa. The genus *Lophiomys*, which was previously included in the Cricetidae (Musser & Carleton, 2005), has convincingly been shown (on molecular grounds) to belong to a subfamily of the Muridae (Jansa & Weksler, 2004; Schenk et al., 2013). The muroids have been considered as originating from Pakistan and more generally Asia and have developed different endemic lineages in Africa which are distinct from those on other continents.

Spalacidae

Today, this family comprises fossorial and subterranean rodents known mainly in Asia with the exception of the Tachyoryctinae or root rats from Africa. The Spalacidae diversified during the Lower and Middle Miocene in the Siwaliks (Pakistan) and probably entered Africa during the Burdigalian interchange episode (around 18-20 MYA). The first representative of this subfamily is *Pronakalimys* from the Middle Miocene of Kenya. It is then replaced by *Nakalimys* and the Early Pliocene modern *Tachyoryctes* in Kenya and Ethiopia (Table 4). The only Spalacidae fossils encountered in Southern Africa date from the Late Miocene of North Namibia are *Nakalimys* and *Harasibomys* (Mein et al., 2000a). The subfamily is no longer present in Southern Africa and never occurred in Northern Africa.

Nesomyidae

The earliest muroid rodent fossils from East Africa belong to an extinct taxon called the Afrocricetodontinae (Lavocat, 1973). They occur for the first time in the Early Miocene of Kenya, Uganda and Namibia, and are represented by three genera *Afrocricetodon, Notocricetodon* and *Protarsomys*; these are interpreted as early representatives of the Nesomyidae. In the recent classification of Musser & Carleton (2005), the Nesomyidae are considered to include the subfamilies of Nesomyinae together with Cricetomyinae, Dendromurinae, Otavyminae, Mystromyinae, Delamyinae and Petromyscinae, all of which are endemic to Africa. Their relationships with Democricetodontinae and Myocricetodontinae (fossil Cricetinae) are not clear and have also been discussed. Since the latter taxa are supposed to have entered Africa only 20.7-17.0 MYA (Koufos et al., 2005), this would make them contemporaneous with the Afrocricetodontinae. The sites of Songhor and Rusinga date from this period which implies either very fast diversification or an earlier bridge at around 23 MYA. The Afrocricetodontinae probably went extinct around the Middle Miocene. They were replaced in East and Southern Africa by Myocricetodontinae (*Dakkamys, Mioharimys, Myocricetodon*) and Megacricetodontinae (Winkler et al., 2010). According to Mein et al. (2004) *Mioharimys* would be the ancestor of the unique modern representative of the tribe *Mystromys*, which is endemic to South Africa. Other Cricetodontinae genera such as *Democricetodon* and *Afaromys*, and an extinct subfamily Namybinae also occur in East and Southern Africa from the Middle to Late Miocene.

The earliest modern Cricetomyinae to appear is *Saccostomus* from Ch'orora Ethiopia and Harasib Namibia at around 9.5-10.0 MYA (Mein et al., 2004). The genus is well represented in East Africa by various species during the Pliocene-Pleistocene while other members of the Cricetomyinae are absent from the fossil record (*Beamys*) or only found in the Upper Pleistocene (*Cricetomys*) (Table 4).

Dendromurinae are described for the first time from the Middle Miocene of Pakistan (Siwaliks) by Lindsay (1988). During the same period rodents with Dendromurinae-like molars appear in East Africa (*Ternania, Mabokomys, Dakkamys*) as well as in North Africa 9.5-10.0 MYA (*Senoussimys*) and Arabia (Abu Dhabi) (Mein et al., 2004; de Bruijn & Whybrow, 1984; Ameur, 1984; Geraads, 2001; Aguilar et al., 1984). The odest recovered modern *Steatomys* and *Dendromus* date from the Late Miocene sites of Kenya, Ethiopia and Namibia (Table 4). The oldest *Malacothrix* appears at around 4 MYA in South Africa (Winkler et al., 2010).

The Otavimyinae, created by Mein et al. (2004) and thought to be closely related to the Dendromurinae, are exclusively found in Southern Africa and comprise two extinct genera: *Otavimys* and *Boltimys*. Their relationships are not yet clearly understood.

The Mystromyinae, exclusively found in Southern Africa, first appear in the Early Pliocene represented by two genera: the extinct *Proodontomys* and *Mystromys* (Winkler et al., 2010).

Similarly, Petromyscinae were also restricted to Southern Africa. They occur for the first time during the Late Miocene of Namibia and are known by two extinct genera (*Stenodontomys* and *Harimyscus*). *Stenodontomys* was also recorded in South Africa up until the early Pliocene (Denys, 1991).

No fossil representative of the East African endemic *Delanymys* (Delanymyinae) has yet been uncovered.

Table 4. First occurrences of modern genera (Dipodidae, Spalacidae, Nesomyidae) of the Myomorpha.

	FAD (MYA)	site	region	ref
Dipodidae				
Jaculus	3-2.5	Omo F	Ethiopia	Wesselman (1984)
Spalacidae				
Tachyoryctes	5.7-4.8	Adu-Asa	Ethiopia	Wesselman et al. (2009)
Nesomyidae				
Cricetomyinae				
Cricetomys	0.3	Kapthurin	Kenya	Denys (1999)
Saccostomus	8-11	Harasib3a, Chorora	Namibia, Ethiopia	Mein et al. (2004); Geraads (2001)
Dendromurinae				
Dendromus	8.8- 8-10	Sidi Salem, Harasib3a	North Africa, Namibia	Ameur (1988) ; Mein et al. (2004)
Malacothrix	4	Bolt's Farm	South Africa	Senegas (2000)
Steatomys	8-10	Harasib3a, Chorora	Namibia, Ethiopia	Mein et al. (2004); Geraads (2001)
Mystromyinae				
Mystromys	5	Langebaanweg	South Africa	Pocock (1987)

Muridae

The bulk of the African Murinae radiation started with *Potwarmus* 19-18 MYA in the Siwaliks (Pakistan) and 18-15 MYA in Libya (Jebel Zelten). This taxon is considered the ancestor of the murid lineage (*Antemus-Progonomys*) that appeared 14-12 MYA. It has also been considered as closely related to the Deomyinae as well as the Nesomyidae subfamily the Dendromurinae. Rodents with Dendromurinae-like molars have been recorded from Pakistan (Lindsay et al., 1988), but Denys et al. (1985) have demonstrated that this subfamily is not monophyletic and multiple exchanges or vicariance episodes may explain this complex radiation.

Lophiomyinae

Due to its peculiar molar and skull morphology, *Lophiomys* has been described as an unique modern representative of its own subfamily Lophiomyinae. It appears for the first time during the Late Miocene in Spain with the genus *Protolophiomys* found later in the Middle Pliocene in Algeria and Morrocco (Winkler et al., 2010). Only two fossil *Lophiomys* are reported from Ethiopia at 5.7 MYA (Wesselman et al., 2009) and Morocco at about the same time (Geraads 1998) (Table 4). Until recently the origins of this subfamily were not yet clear and it was previously thought to belong in the family Cricetidae; some extinct cricetids such as *Democricetodon* or *Megacricetodon* were thought to share close affinities (Tong & Jaeger, 1993). However, recent molecular analyses clearly show *Lophiomys* to belong to the Muridae (Jansa & Weksler, 2004; Schenk et al., 2013).

Myocricetodontinae and Gerbillinae

During Middle to Late Miocene, the North African rodent faunas are characterized by the domination of the Myocricetodontinae an extinct family that has been suggested as being the sister taxon of the Gerbillinae and of the Ctenodactylidae (Jaeger, 1977; Flynn et al., 2003; Benami, 1997) while at the same time very rare representatives of this very diverse family are described from Namibia and Ethiopia. According to Tong (1989) the Myocricetodontinae may have been the ancestors of Gerbillinae and their differentiation would have occurred in North Africa during the Late Miocene. More recent discoveries have also found early Gerbillinae "dental morphs" in various Middle and Upper Miocene rodents of Spain, North Africa and Asia. Currently, there is a consensus that the earliest gerbil in tropical Africa is *Abudhabia* from Kenya (Winkler, 2003; Manthi, 2007; Mein & Pickford, 2006) during the Upper Miocene. This genus was originally described from Pakistan and Afghanistan (Flynn et al., 2003). Mein et al. (2000b) describe the extinct genus *Mioharimys* in Harasib (Namibia) as a new genus of Myocricetodontinae. One of the earliest modern Gerbillinae is *Gerbilliscus* sp. from the Late Miocene in Ethiopia (Wesselman et al., 2009) or from Kanapoi (4 MYA) (Denys, 1987). Denys (2011) described three fossil *Gerbilliscus* species from Laetoli, Tanzania: the extinct *Gerbilliscus satimani* (3.7-3.5 MYA), which is closely related to the modern *G. leucogaster*; *G. winkleri* (2.7-2.5 MYA); and *G.* cf *inclusus* (2.7-2.5 MYA). Other Gerbillinae are known from Pliocene-Pleistocene deposits of East Africa like *Gerbilliscus petteri* (Olduvai Bed I, Denys, 1989) or *Gerbillus* sp. from the Omo (Wesselman 2004) and *Gerbilliscus* and *Desmodillus* from Southern African caves (Senegas, 2000). One representative of the genus *Gerbillus* in Tanzania at 1.7 MYA (Denys, 1989) and in the Omo member F (2.08 MYA) (Wesselman, 1984); this would be the earliest incursion of this North African genus in tropical Africa.

Cricetodontinae

Cricetodontinae appear during the Late Miocene in North Africa but some of them also occur in East and Southern Africa. This group of extinct rodents is known by one genus *Democriceton* in tropical regions.

Namybiminae

There is only one genus in this extinct subfamily, created by Mein et al (2000b), to name small molars displaying longitudinal crests and a t4 on the upper first molar as in Petromyscinae but distinguished by the presence of accessory cusplets on the lower molars. This lineage occurs from the Late Miocene to the Pliocene and is thought to be derived from the Myocricetodontinae complex (Mein et al., 2000).

Deomyinae

From 11-9 MYA, early Deomyinae appear in the fossil record in Namibia and in East Africa, and are represented by two extinct genera: *Preacomys* and *Tectonomys*. Three species of *Preacomys* have been described from Namibia: *Preacomys* cf *kikiae*, *P. karsticus*, *P. griffini* (Mein et al., 2004). *Preacomys griffini* bears some morphological resemblances to *Acomys mabele* of Langebaanweg as well as with modern *A. subspinosus*. The two other species have an elongated t1 that has some similarities with modern *A. spinosissimus*. It is not clear whether *Tectonomys* represents a synonym of *Preacomys* or belongs to the murines. The earliest *Acomys* to occur is in Late Miocene of East Africa (Manthi 2007) as well as in the South-western Cape, South Africa (Denys, 1990c). Other Deomyinae genera such *Deomys* and *Lophuromys* have not yet been recorded in the fossil record which may be due to their forest habitat. The fourth member of the subfamily *Uranomys* was uncovered in the Pliocene-Pleistocene caves of Angola (Winkler et al., 2010; Table 5).

Murinae

This subfamily includes the most diverse (and abundant) group of rodents alive today. The initial radiation took place in the Siwaliks (Pakistan) with *Antemus chinjiensis* at 13.8 MYA. The latter taxon was extinct at 12.5 MYA (Barry, 1995) and was replaced by *Progonomys* that appears at the same sites as *Hipparion* 10.5-11.5 MYA in North Africa (Bernor et al., 1987). *Progonomys* occurs for the first time in the Siwaliks in Pakistan at 12.1 MYA (Barry, 1995). Different species of *Progonomys* have been described both in Europe (Jacobs, 1978) and North Africa but has not been described from any of the Sub-Saharan African Miocene sites. The oldest true Murinae to occur in the tropical African record are from the Upper Miocene sites of Kenya, Ethiopia and Namibia; they are mainly representatives of the Arvicanthini tribe. The earliest murines to occur in East Africa are cf *Stenocephalemys* and *Parapelomys* in Chorora (late Miocene) Geraads (2001). In Southern Africa the oldest Murinae of Harasib 3a have been attributed to *Aethomys* sp. by Mein et al. (2004) and the same genus is recorded in Lemudong'o (Manthi, 2007) and in Langebaanweg (Denys, 1999) (Table 5).

Parapelomys was recorded 7.4-5.7 MYA in the Siwaliks (Jacobs, 1978; Barry, 1995). In the Pliocene of Afghanistan, Sen et al. (1979) described typical 'African rodents' such as *Arvicanthis* and *Pelomys*. The first *Arvicanthis* within Africa comes from Uganda (Mein, 2004) whereas the earliest *Lemniscomys* is from Ethiopia 5.6 MYA and Kenya 6 MYA (Wesselman et al., 2009; Manthi 2007) (Table 5).

Other extinct genera of the tribe Arvicanthini (sensu Lecompte et al., 2008) are found both in the Late Miocene and Early Pliocene of Egypt, Tanzania, and Kenya (*Saidomys, Lukeinomys*) whose affinities with modern representatives are not clear. In South Africa the genus *Euryotomys* has been considered as the ancestor of the tribe Otomyini (Pocock, 1976; Senegas & Avery, 1998). The true early Otomyinae representatives occur at Makapansgat at around 3.3 MYA (table 5). This subfamily diversified in Southern Africa during the Pliocene and is recorded for the first time in the Lower Pleistocene of Lusso Beds (Zaire) and later entered Tanzania and Ethiopia (Denys, 2003).

The earliest Praomyini are found in Laetoli, Hadar, Omo and Makapansgat and they already are modern genera; these include *Mastomys, Praomys, Oenomys,* and *Zelotomys* (Denys, 1999)(Table 5). During the Middle and Upper Pliocene other members of the Arvicanthini diversified such as *Thallomys, Rhabdomys,* and *Pelomys*. However, a strong turnover appears to occur in the murid faunas at around 2.4 MYA in the rift Valley (Wesselman, 1995; Reed, 2011).

The earliest *Mus* recorded in the fossil record of Africa come from the Early Pliocene of North Africa (*Mus ique*) and is later found in Southern Africa. The origins of the genus are found in Asia where old *Mus* have been described (Barry, 1995).

Table 5. First occurrences of modern genera (Muridae) of the Myomorpha.

Family/tribe/genus	FAD (MYA)	Site	Region	Ref
Lophiomyinae				
Lophiomys	5.7-4.8	Adu-Asa	Ethiopia	Wesselman et al. (2009)
Deomyinae				
Acomys	6	Lemudong'o	Kenya	Manthi (2007)
Uranomys	Plio-Pleistocene	Angola	South Africa	Pickford et al. (1992)
Gerbillinae				
Desmodillus	5	Langebaanweg	South Africa	Pocock (1976), Denys et al. (in prep.)
Gerbilliscus	4	Kanapoi	East Africa	Denys (1999)
Gerbillurus	2.5	Jägerquelle, Nosib	Namibia	Senut et al. (1992)
Gerbillus	3-2.3	Al Oughlam Omo F	North Africa Ethiopia	Geraads (1995); Wesselman (1984)
Murinae				
Arvicanthini				
Aethomys	6	Harasib, Lemudong'o	Namibia, Kenya	Mein et al. (2004); Manthi (2007)
Arvicanthis	6	Lemudongo	East Africa	Manthi (2007)
Dasymys	4 - 5	Bolts Farm	South Africa	Senegas (2000)
Grammomys	3.3	Makapansgat	South Africa	Pocock (1976)
Lemniscomys	6	Lemudon'go	East Africa	Manthi (2007)
Mastomys	6	Lemudon'go	East Africa	Manthi (2007)
Oenomys	3.3	Hadar	East Africa	Sabatier (1982)
Pelomys	5.3	Amama3	North Africa	Ameur (1988)
Rhabdomys	3.3	Makapansgat	South Africa	Pocock (1976)
Thallomys	3.7	Laetoli	East Africa	Denys (1987, 2011)
Praomyini				
Praomys	3.3	Hadar, Makapansgat	Ethiopia, South Africa	Sabatier (1978)
Zelotomys	3.3	Makapansgat	South Africa	Pocock (1976)
Otomyini				
Otomys	3.3	Makapansgat	South Africa	Pocock (1976)
Myotomys	3.3	Makapansgat	South Africa	Pocock (1976)
Murini				
Mus	5.5	Lissafsa	North Africa	Geraads (1998)

Conclusions

The history of rodent diversification in Sub-Saharan Africa is still poorly known and the fossil record is relatively scarce. In summary, three different periods can be identified: the early diversification during the isolation of Africa (Eocene-Oligocene), an important turnover after the Burdigalian interchange (Miocene) and the appearance of the modern faunas (Late Miocene-Early Pliocene).

Concerning Anomaluromorpha and Ctenohystrica clades, it seems clear that they originate from North Africa and their roots have to been searched for in Eurasia during the Paleocene. In consequence the modern diversity of these taxa seems considerably lower today compared to their Eocene-Oligocene counterparts. This is specially the case for Thryonomyidae with one extant genus compared with six extinct genera grouped under the name Phiomyidae/Gaudeamuridae. The only group within this clade that diversified was the family Bathyergidae with now seven genera compared with four extinct ones. This group comprises only fossorial rodents (mole rats) with some that have developed a high degree of sociality (Jarvis, 1981). This may have allowed the family to survive changes in climate and invasions by other rodents after the 18.5 MYA interchange (Table 6).

The Lower Miocene faunas of East and Southern Africa were dominated by Phiomorpha and sciurids, and some "muroids" whose affinities are difficult to assess. These rodents are quite rare and their fossil record relatively scarce. During the Middle and Late Miocene one sees the progressive appearance of the modern endemic African rodent fauna and a beginning of specialization between different regions. This regionalisation of the rodent faunas was exacerbated during the Pliocene-Pleistocene due to the rifting of East Africa (Denys et al., 1985) and global climate cooling at 2.5 MYA (Wesselman, 1995; Reed & Geraads, 2011). However, during the Late Miocene, North, East and Southern African faunas are dominated by different early representatives of the Dendromurinae, Deomyinae and Arvicanthini. This suggests a beginning of differentiation of these three regions. Exchanges between Arabia and Africa however persist until the Middle Pliocene in Ethiopia and appear to only have stopped by the opening of the red sea (Denys et al., 1985). Little is known of past rodent faunas in West and Central Africa; this is due to the fact that fossils of tropical forest rodents are rarely preserved. Hence the knowledge of the evolutionary history of fossil rodents in Africa is still heavily biased towards savanna and arid-zone groups.

Table 6. Main evolutionary patterns of diversification of African rodents through time.

Date (MYA)	Climatic events	Tectonic events	Faunal events
1-2.4			extinction some fossil rodents
2.4	Northern Hemisphere ice sheet glaciation onset	Rift valley formation	early modern *Homo*
3		closure Panama isthmus	modern genera diversification
5-3		Great escarpment in South Africa	early Praomyini,
5.6		opening Red sea, closure Mediterranean,	early *Mus*

7-8	Asian Monsoon intensifies	major rapid uplift Tibetan plateau	Hominidae appear
8-6			C3 plants spread, early Arvicanthini, Turolian turnover
11			*Hipparion* and *Progonomys* event, early Nesomyidae
16	mid-Miocene climatic optimum		Myocricetodontinae in Kenya, Namibia
18		uplift of the East and Southern Africa superswell	*Potwarmus* in Libya
18.5			Burdigalian interchange (Proboscidean datum)
20		Doming in Western and Kenya Rift Systems	
23-24		Red Sea rifting, Andean Uplift, Tibetan plateau Uplift	early Sciuridae in East Africa
33.9	First Antarctic ice sheet		first cetaceans, Phiomyidae diversification in Fayum
49-33		early doming, volcanism, faulting in Afar region	
50-40		India-Asia contact	Earliest higher primates
52	Early Eocene climatic optimum		early rodents in Africa (Zegdoumyinae)
52-54.5	Late Paleocene thermal maximum (PETM)	North Atlantic Rifting & Volcanism	
55-60	Paleocene		Earliest modern primates
65	Cretaceous-Tertiary boundary	Meteoritic impact	K-T extinction

Importance of Rodents

Introduction

Out of some 463 rodent species in Africa, 77 are known to damage crops but only between 12 and 20 are known to be significant crop pests (Singleton et al., 2010). Even fewer are known to be invasive aliens but these include the infamous house mouse (*Mus musculus*) and house rat (*Rattus rattus*) which have been listed among the IUCN's 100 world's worst invaders (Lowe et al., 2000; www.issg.org/database). Similarly, only a few rodent species are responsible for harbouring some 60 known zoonotic diseases worldwide (Meerburg et al., 2009), although the impact of these diseases is considerable, particularly in densely populated rural and urban settlements in Africa (Gratz, 1997; Begon, 2003; Taylor et al., 2008a; Meerburg et al., 2009). On the other hand, rodents have important beneficial effects in natural ecosystems and agro-ecosystems, e.g. as bioengineers and as prey for reptilian, avian and mammalian predators. For example, the seed dispersal behaviour of *Cricetomys kivuensis* in montane forests has recently been demonstrated (Nyiramana et al., 2011).

Medical science has benefitted enormously from laboratory-bred rats and mice. In many parts of the world rodents are an important protein source for human communities (Singleton et al., 2010) and this is particularly true throughout much of Central and West African savannas where rodents are the second most important component of bush meat after ungulates (Granjon & Duplantier, 2009) or third after primates in forest habitats (Fa et al., 2006). In parts of southern Africa, such as Malawi, certain rodents are also considered delicacies (Denys pers comm.). In West Africa, such as in Benin, cane rats (*Thryonomys swinderianus*) are commercially farmed and are delicacies on most restaurant menus; an estimated 16,000 cane rats are managed in 500 farms in Benin alone (Granjon & Duplantier, 2009). Phylogeographic studies of commensal rats and mice have provided useful insights into historical human movements (Van Vuuren & Chown, 2007; Matisoo-Smith & Robins, 2009; Searle et al., 2009a,b; Tollenaere et al., 2010; Kaleme et al., 2011). And a relatively recent project in Tanzania (APOPO) has exploited the acute olfactory sense of *Cricetomys* species to train them to detect mines buried under soil.

In some cases, native species have expanded their ranges within Africa or their populations may periodically attain outbreak proportions in agricultural areas, making them economically important pest species. Examples include several gerbil species in the Sahel of West Africa (Bâ et al., 2006) and drier regions of southern Africa (De Graaff, 1981), the Nile rat (*Arvicanthis*) and multimammate mice (*Mastomys*) in the savannas of Africa (Leirs, 1995; Leirs et al., 1996, 1989; Sluydts et al., 2009).

Whilst rodent control in Africa and worldwide was until recently dominated by research into rodenticides, a new and integrated approach advocates ecologically-based rodent management (EBRM) (Singleton et al., 1999). EBRM has been widely accepted in South-East Asia, Australia and parts of Africa and has been the centre of several multinational, internationally-funded projects in eastern, southern and western Africa, such as STAPLERAT (cordis.europa.eu/data/PROJ_FP5), RATZOOMAN (projects.nri.org/ratzooman), ECORAT (projects.nri.org/ecorat/), STOPRATS (projects.nri.org/stoprats), CHANCIRA (agence-nationale-recherche.fr) and ENEMI (agence-nationale-recherche.fr).

It is beyond the scope of this brief chapter to completely review the vast literature on the subject of EBRM. Rather, the aim of this chapter is to direct readers to excellent sources of literature as well as online resources related to the projects mentioned above. For starters, two monographs published by the Australian Centre for International Agricultural Research (ACIAR) in Canberra (Singleton et al., 1999; 2003) are available online and contain a number of important chapters on rodent biology and management.

Zoonotic Diseases

More than 60 zoonotic diseases (diseases transmitted from animals to humans) can be transmitted to humans from rodents via direct and indirect means (see the comprehensive global review of Meerberg et al., 2009). Gratz (1997) reviewed rodent-borne diseases in Africa, stressing the dearth of data for most countries. Plague remains a major cause of mortality in certain countries where it is endemic (e.g. DRC, Madagascar, Mozambique, Tanzania and Uganda). The 1269 cases of plague reported in Africa during 1994 comprised 43% of all cases globally and contributed 50% of global mortality due to the disease (Gratz, 1997). Granjon & Duplantier (2009) reviewed rodent-borne diseases in West Africa transmitted via three modes of transmission, bites, faeces and urine, and intermediate vectors. A further mode of transmission is the human consumption of rodents (ter Meulen et al., 1996). Whilst rat bites are frequently reported by residents of slums or rural dwellings (e.g. 15% of 84 residents of a Durban slum: Taylor et al., 2008a), this is not a common path of disease transmission although bites may become infected by secondary bacterial infections (Granjon & Duplantier, 2009). Rodents may be involved in disease transmission in different ways either directly or indirectly. Diseases transmitted by urine, faeces and saliva include leptospirosis, salmonella, Lassa fever, Lujo fever and other haemorrhagic fevers such as Hantaviruses. Diseases spread by vectors include plague, murine and scrub typhus, Leishmaniasis, borreliosis, toxoplasmosis, Congo-Crimean haemorrhagic fever, intestinal schistosomiasis and Rift Valley fever. Some of these diseases may be initially transmitted via rodents to human but then lead to human-to-human transmission (becoming "communicable"). The conjunction of rodent consumption and population displacements during civil war has also clearly been established as a cause of outbreaks of Lassa fever in West Africa (Lalis et al., 2012).

Few rodent species are implicated in the transmission of rodent-borne diseases, but these include alien invaders (*R. rattus*, *R. norvegicus* and *M. musculus*), as well as indigenous species such as multimammate mice (*Mastomys* spp.), pygmy mice (*Mus* spp.), Nile rats (*Arvicanthis niloticus*), giant rats (*Cricetomys* spp.) and some gerbils (Gerbillinae). Granjon and Duplantier (2009) provide a comprehensive review of the life cycle and details of particular rodent reservoirs and associated vectors for the major rodent-borne diseases known in Africa. Knowledge of the identity and biology of rodents is critical to understanding the epidemiology of rodent-borne diseases, particularly due to the prevalence of cryptic rodent species which may vary in their roles as disease vectors. For example, in South Africa, *Mastomys coucha* is more susceptible to experimental plague infection than *M. natalensis* and thus more implicated in plague epidemiology (Isaacson 1975, Isaacson et al., 1981; Arntzen et al., 1991; Venturi et al., 2004). In areas of Lassa fever prevalence in West Africa, where three cryptic species of *Mastomys* co-occur, only *M. natalensis* is important for transmission of the disease (Lecompte et al., 2006; Fichet-Calvet et al., 2007). In plague, the chain of transmission is complex and involves different species of rodents in different endemic areas. The aetiological agent (the bacterium, *Yersinia pestis*) is maintained in certain sylvatic species which are resistant to infection. It is then transmitted via fleas to more commensal species such as multimammate mice (*Mastomys* spp.) or house rats (*Rattus* spp.) which may share habitats in peri-urban environments, thus bringing infected fleas in close proximity to humans.

The RatZooMan project (projects.nri.org/ratzooman/) "used a multidisciplinary approach to further understanding of emerging infectious diseases affecting people and their livelihoods". The project focused on three rodent-borne diseases (plague, leptospirosis and toxoplasmosis) in south-eastern Africa. Readers are referred to the project website which provides access to all project documents, a useful bibliography on rodent-bornes diseases in Africa and globally, as well the option to subscribe to the RatZooMan Network on rodent-borne diseases.

Crop Damage

Fiedler (1988; 1994), Leirs (1999), Makundi et al. (1999), Granjon & Duplantier (2009) and Singleton et al. (2010) provide reviews of agricultural impacts of rodents in Africa. Granjon and

Duplantier (2009) provide a useful synthesis of African pest rodent reproduction and population dynamics, pre and post-harvest damage to agriculture (with focus on West Africa) and chemical and ecological principles in rodent control. Impacts on food security may be massive, e.g. up to 80% of the potential harvest may be lost during rodent outbreak years in East Africa (Mwanjabe et al., 2002). At densities of up to 140 rodents per hectare, losses of maize in the field amounted to 17% to 80% (Mulungu et al., 2007). In Tanzania, it was estimated that rodents caused losses to maize production of 5-15%, amounting to 412,500 tonnes per annum, sufficient to feed two million people (Makundi et al., 1999).

Singleton et al. (2010) listed details of 30 rodent outbreaks documented between 1925 and 2005 across Africa, associated with murine species such as gerbils (*Gerbilliscus*, *Gerbillus*, *Taterillus*), multimammate mice (*Mastomys*), Nile rats (*Arvicanthis niloticus*) and striped mice (*Rhabdomys pumilio*). In these species, the extent of damage varies considerably between outbreak and non-outbreak years. Outbreaks have been shown to be correlated with rainfall, particularly when this follows multiple dry years and where it occurs late in the season (Leirs, 1995; Leirs et al., 1996). Regression-based models have been developed to forecast rodent outbreaks in *M. natalensis* in East Africa (Leirs et al., 1996) and in *M. erythroleucus* (Hubert & Adam, 1983) and *A. niloticus* (Poulet, 1985) in West Africa. Based on the model of Leirs et al. (1996) a bioeconomic model was developed to model the costs and benefits of varying the timing and duration of control, e.g., rodenticide treatment (Skonhoft et al., 2006).

Larger non-murine species such as cane rats and giant rats may cause extensive damage to agriculture due to their large body size, but damage does not vary as much from year to year.

Several internationally-funded, multidisciplinary studies conducted over the past two decades have documented rodent population dynamics and agricultural impacts in several African countries. These include the StapleRat (cordis.europa.eu/data/PROJ_FP5; Corti et al., 2005), EcoRat (projects.nri.org/ecorat; Belmain et al., 2008; Katakweba et al., 2012; Massawe et al., 2011; Monadjem et al., 2011; Mulungu et al., 2011; Taylor et al., 2012) and StopRats (projects.nri.org/stoprats) projects focusing on southeastern Africa, and the CHANCIRA (agence-nationale-recherche.fr) and ENEMI (agence-nationale-recherche.fr) projects in West Africa. Information and project reports from the various projects are available from the above-cited websites. An extensive bibliography on agricultural impacts of rodents in Africa is available from the EcoRat website.

Conservation of African rodents

Introduction

With 463 species, rodents are by far the most diverse order of mammals in Africa, accounting for almost half of all the species on the continent. African rodents occupy a wide range of terrestrial environments and are present in all the habitats in the region from dry desert to humid rainforest, and from lowland swamps to high-altitude montane grasslands. A few species (such as *Colomys goslingi* and *Nilopegamys plumbeus*) have even adapted to living in aquatic situations. As such, rodents are exposed to a diverse array of threats and the populations of certain species are dwindling, with a few species even on the brink of extinction (Schlitter, 1989). However, compared with other groups of mammals (in particular carnivores, elephants and ungulates), rodent populations are relatively secure and most species have been able to adapt to the changing African landscape and the concomitant threats posed by it.

The IUCN Red Data process

Identifying threatened species is not an easy task. Currently, the IUCN Red Data list (www.iucnredlist.org) is the internationally recognized authority for identifying and listing threatened species of animals and plants. The process of identifying threatened species is relatively simple to understand, but not necessarily simple to apply. Each species, in this case of rodent, is screened by relevant experts to determine whether its population is secure and likely to persist into the future or not. A species with a dwindling or restricted population is then assessed to see whether it is likely to go extinct. This process has been quantified and a detailed account is provided in IUCN (2001) which is freely available as a download off the IUCN website. However, there is rarely sufficient (good quality) information available for rodents, especially for rare and elusive species with small distributional ranges, for an unbiased (and therefore repeatable) assessment.

The currently recognized IUCN threat categories are listed in Table 7 (IUCN, 2001). As can be seen, all evaluated species are placed in a category ranging from Least Concern (the species is thought to be safe and unlikely to go extinct in the near future) to Extinct, in ever-increasing threat categories. Where insufficient information exists to make a reliable assessment (for example in situations where the species is known by just a single specimen, often collected decades ago), it is listed as Data Deficient. Since new species are being described continuously, some species may not yet have been assessed and are listed as Not Evaluated. Species falling into one of the three categories of Critically Endangered, Endangered and Vulnerable are traditionally considered threatened (IUCN, 2001).

Table 7. A list of the IUCN threat categories (based on version 3.1), the recognized abbreviations in brackets, and a description of what conditions are necessary for a species to be listed under that category (IUCN, 2001).

Category	Description
Extinct (EX)	No reasonable doubt that the last individual has died
Extinct in the Wild (EW)	Only known to survive in captivity or naturalised populations outside its natural range
Critically Endangered (CR)	Facing an extremely high risk of extinction in the wild
Endangered (EN)	Facing a very high risk of extinction in the wild
Vulnerable (VU)	Facing a high risk of extinction in the wild
Near Threatened (NT)	Not currently threatened, but close to qualifying for a threatened category in the near future
Least Concern (LC)	Typically widespread and/or abundant with currently no signs of being included in any of the threatened or near threatened categories
Data Deficient (DD)	Inadequate information to make a direct, or indirect, assessment of its risk of extinction
Not Evaluated (NE)	Not yet evaluated against the IUCN criteria

The IUCN assessments are conducted on global populations of a species, and the list maintained by this organization pertains to the global status of a species. Most African countries also maintain national lists and assessments but these are often not updated taxonomically or do not follow recent literature. In such cases, a slightly modified version of the international criteria is applied. National lists are important since most conservation action is taken at this level. However, due to the random nature of species distributions with respect to international boundaries, a species may be assessed as highly threatened in a particular country (where it might occur only marginally), whereas it is not considered threatened on a global scale.

Threatened rodents of Sub-Saharan Africa

The conservation status of each Sub-Saharan African rodent was obtained from the IUCN website (www.iucnredlist.org; downloaded 3 Jan 2014) and summary statistics are presented in Table 8. The status of each threatened species is presented in Table 9.

Table 8. The number of species of Sub-Saharan African rodents in the various IUCN threat categories (as of Jan 2014).

IUCN Category	Number of species
Extinct	0
Extinct in the Wild	0
Critically Endangered	3
Endangered	21
Vulnerable	15
Near Threatened	5
Least Concern	267
Data Deficient	52
Not Evaluated	100

Some species have not been found in the wild for the 50 or 100 years and may even be extinct (for example: *Nilopegamys plumbeus, Leimacomys buettneri, Dendroprionomys rousselotti*).

Attention should be drawn to a number of points. Firstly, a large proportion of the species have not yet been evaluated, due to the fact that they have only recently been described or elevated to species rank. Secondly, a significant percentage of species (14%) are so poorly known that they are listed as Data Deficient. Thirdly, no Sub-Saharan African rodents are listed as Extinct or Extinct in the Wild. Finally, most rodent species are listed as Least Concern and are in no imminent danger of going extinct. Hence, compared with many other mammalian groups, rodents are faring relatively well. In fact some species are considered pests of one sort or another, requiring management of their populations (see previous Section). Approximately 10% of the species are listed as Critically Endangered, Endangered or Vulnerable. Despite this being a small proportion, it still adds up to 39 species which is greater in absolute terms than the numbers of threatened carnivores or primates on mainland Africa.

All three Critically Endangered species (see Table 9) share the following characteristics: 1) they occupy highly restricted ranges on mountain-tops, typically only a few km^2 in extent; and 2) they are not known to occur in any protected area and their natural habitat is under extreme pressure. The majority of the Endangered and Vulnerable species are similarly range-restricted, occurring on single mountains or within a very small area. Exceptions to this rule include *Mystromys albicaudatus* (EN) which is relatively widespread on the plateau of South Africa and Lesotho, but occurs at low densities (Dean, 1978; Avenant, 2011), and *Hylomyscus baeri* (EN) which occurs across the Upper Guinea rainforest zone but has only been recorded at a few, widely spaced sites.

Table 9. A list of the threatened (Critically Endangered, Endangered and Vulnerable) rodents occurring in Sub-Saharan Africa

Family	Species	IUCN status	Distribution
Muridae	*Hylomyscus grandis*	CR	Mt Oku, Cameroon
Muridae	*Nilopegamys plumbeus*	CR	Gojam, Ethiopia
Nesomyidae	*Dendromus kahuziensis*	CR	Mt Kahuza, DRC
Muridae	*Lophuromys dieterleni*	EN	Mt Oku, Cameroon
Muridae	*Lophuromys eisentrauti*	EN	Mt Lefo, Cameroon
Muridae	*Lophuromys rahmi*	EN	Kivu Mts, DRC, Rwanda
Muridae	*Dasymys montanus*	EN	Ruwenzori Mtns, Uganda
Muridae	*Desmomys yaldeni*	EN	Sheko Forest, Ethiopia
Muridae	*Grammomys gigas*	EN	Mt Kenya, Kenya
Muridae	*Hybomys badius*	EN	Mt Cameroon, Cameroon
Muridae	*Hybomys basilii*	EN	Bioko , Equatorial Guinea
Muridae	*Hylomyscus baeri*	EN	Upper Guinea forest
Muridae	*Lamottemys okuensis*	EN	Mt Oku, Cameroon
Muridae	*Praomys hartwigi*	EN	Mt Oku , Cameroon
Muridae	*Praomys morio*	EN	Mt Cameroon, Cameroon
Muridae	*Praomys obscurus*	EN	Gotel Mts, Nigeria
Muridae	*Otomys barbouri*	EN	Mt Elgon,Uganda,Kenya
Muridae	*Otomys burtoni*	EN	Mt Cameroon, Cameroon
Muridae	*Otomys jacksoni*	EN	Mt Elgon, Uganda
Muridae	*Otomys uzungwensis*	EN	Uzungwe Mts, Tanzania
Nesomyidae	*Mystromys albicaudatus*	EN	South Africa, Lesotho
Sciuridae	*Paraxerus vincenti*	EN	Mt Namuli, Mozambique
Spalacidae	*Tachyoryctes macrocephalus*	EN	High Southern Plateau, Ethiopia
Spalacidae	*Tachyoryctes splendens*	EN	Ethiopia, Somalia, NW Kenya
Bathyergidae	*Fukomys kafuensis*	VU	Zambia
Muridae	*Lophuromys medicaudatus*	VU	Kivu highlands, DRC, Rwanda
Muridae	*Lophuromys melanonyx*	VU	SC Ethiopia
Muridae	*Grammomys minnae*	VU	S Ethiopia
Muridae	*Hybomys lunaris*	VU	E Ruwenzori, Uganda
Muridae	*Lemniscomys mittendorfi*	VU	Mt Oku, Cameroon
Muridae	*Mastomys awashensis*	VU	Upper Awash Valley, Ethiopia
Muridae	*Praomys degraaffii*	VU	Albertine Rift Mtns, Burundi, Rwanda, Uganda
Muridae	*Thamnomys kempi*	VU	N Kivu province, DRC
Muridae	*Thamnomys venustus*	VU	E Ruwenzori, Uganda, DRC
Muridae	*Otomys dartmouthi*	VU	E Ruwenzori, Uganda
Muridae	*Otomys lacustris*	VU	N Malawi,SWKenya, SC Tanzania
Muridae	*Otomys occidentalis*	VU	Mt Oku, Gotel Mtns, Cameroon Nigeria
Nesomyidae	*Delanymys brooksi*	VU	SW Uganda, DRC, Rwanda, Burundi
Nesomyidae	*Dendromus oreas*	VU	Mt Cameroon, Cameroon

Major threats facing African rodents

The threats facing African rodents are not unique to the group, and affect many other components of biodiversity on the continent. Habitat loss probably constitutes the most important threat to African rodents. This impact is exacerbated by the association between vertebrate (including mammal) species richness and human population; areas with high diversity of species also support high densities of humans (Fjeldsa & Burgess, 2008). Logging of forests, slash-and-burn agriculture, afforestation of montane grasslands (typically with exotic plants such as *Pinus* and *Eucalyptus*), expansion of small-scale and commercial agriculture, rapid growth of urban areas, overgrazing of pastures are among the many drivers of habitat loss.

Over-harvesting of organisms is a well-known threat, and extends to rodents as well. The larger species, including porcupines, squirrels, cane rats and springhares are heavily sought after in certain communities with the obvious consequence of reductions in numbers of the rodents. This threat is not currently thought to be endangering any rodent species across its Sub-Saharan range, but certainly at finer scales, rodent numbers may be reduced to local extinction.

One of these habitat threats is worth discussing in a little more detail as it has only recently been recognised as a threat to small mammals. Grazing by ungulates, both domestic and wild, affects plant cover; overgrazing often leading to the loss of grass layer. Many grassland- savanna rodents, especially the smaller rats and mice, typically require good grass cover for hiding from predators, particularly raptors and medium-sized mammalian carnivores. As ungulate populations increase in density, such as on communal pastoral lands (livestock) or in protected areas (wild game), rodents decline or even disappear from these areas. Africa's priceless and wonderful protected area network is mostly geared towards tourists that want to see large game; hence these reserves are managed for maximizing ungulate numbers, often at the detriment of rodents and other smaller mammals.

Climate change is a complicated process that is still poorly understood, particularly its effect on biota. No doubt, some species will benefit from increases in temperature and/or changes in rainfall patterns, and others will be negatively impacted. Mountain top species (which includes some of the most threatened rodents, see above) are likely to fall into the latter group. This is because as temperatures increase, suitable montane and moorland habitat shifts up the mountain to higher elevations; at some point even tall mountains will "run out" of altitude and as a result the habitat itself disappears leaving the species stranded in unfamiliar territory.

General Methods and Layout of Book

Methods

The data on which the distribution maps in this book are based were obtained from three sources: museum specimens examined by us; literature records published by various authors; and online databases of museum records. Museum specimens play a pivotal role in archiving and describing biodiversity, hence our focus on them; see Monadjem et al. (2010) for a detailed discussion on the incredible value of museum specimens.

Our area of interest is Sub-Saharan Africa, in other words, Africa south of the Sahara desert. We chose to exclude the Sahara and North Africa for two simple reasons. Firstly this vast area has a very high diversity of rodents which would have made this bulky book even bulkier. Secondly, the affinities of North African rodents are more with Europe and Arabia than with Sub-Saharan Africa. We have defined our region politically to include Sudan, Chad, Niger and Mali, all of which have true desert in their far northern regions.

Between us, we have visited numerous museums and examined tens of thousands of specimen records, including from the following institutions (we thank the curators who allowed us access to these collections): Museum national d'Histoire naturelle, Paris (Dr. Christiane Denys); Durban Natural Science Museum, Durban (Prof Peter Taylor was curator 1989-2010, followed by Dr. Leigh Richards); Natural History Museum of Zimbabwe, Bulawayo (Dr. Woody Cotterill was curator 1992-2004); Ditsong Museum, Pretoria (formally known as the Transvaal Museum, Dr. Teresa Kearney); National Museum of Namibia, Windhoek (Seth Eiseb); National Museums of Kenya, Nairobi (Simon Musila); Royal Museum for Central Africa, Tervuren (Wim Wendelen).

Literature records came from numerous published sources, the most important of which are listed here (an asterisk denotes a regional synthesis for rodents or major subgroup of rodents): *Ansell (1957, 1960, 1966, 1969, 1973, 1974, 1978); *Ansell & Dowsett (1988); Ba et al. (2006); Barome et al. (2000); Bastos et al. (2011); Bates (1988); Ben Faleh et al. (2012b); Beolchini & Corti (2004); Boratynski et al. (2012); Brito et al. (2010); Britton-Davidian (2012); Brouat et al. (2009); Bryja et al. (2010, 2014); Burda (2000); Carleton & Robbins (1985); Carleton & Stanley (2005); Carleton & Stanley (2012); Carleton et al. (2006); Colangelo et al. (2005, 2007, 2013); *Crawford-Cabral (1998); *De Graaff (1981); *Delany (1975); Denys & Aniskine (2012); Denys et al. (2006, 2007, 2009a,b, 2011b) ; Dieterlen (1975, 1976a,b, 1983, 1991, 2009); Dieterlen & Nikolaus (1985) ; Dobigny et al. (2001a,b, 2002a,c, 2005, 2008, 2013); Dudu et al. (1989); Eisentraut (1968a,b, 1969, 1970) ; Fadda et al. (2001a,b) ; Faulkes et al. (2004, 2010, 2011); Gautun & Bellier (1970) ; Gautun & Petter (1972) ; Gautun et al. (1986) ; Gordon (1978); *Granjon & Duplantier (2009); Granjon et al. (2002a,b, 2004, 2012); Grubb (1982); *Grubb et al. (1998); *Happold (1987, 2013); *Hill & Carter (1941); Holden (1996, 2005); Holden & Levine (2009); *Hollister (1919); Hutterer & Dieterlen (1984, 1986); Hutterer et al. (1992); Jarvis & Sherman (2002); Jones (1971); Julliot et al. (1998) ; Kan Kouassi et al. (2008); Kennis et al. (2011); Kingdon (1974); Krystufek (2008); *Kuhn (1965); Lamb et al. (2014); Lavrenchenko et al. (1998a,b, 1999, 2007, 2011); Lecompte et al. (2008) ; Linzey et al. (2003); Mboumba et al. (2011); Misonne (1965); Missoup et al. (2009, 2012); *Monadjem (1998a,b); Mouline et al. (2008); Mullin et al. (2002, 2004a,b, 2005); Musser & Carleton (2005); Nicholas et al. (2003a, 2005, 2006, 2008a,b,c, 2009, 2010a, 2011, 2012a); *Pearch et al. (2001); Perez del Val et al. (1995); Petter (1981); *Rautenbach (1982); Rautenbach & Schlitter (1978); *Robbins & Van der Straeten (1996); Roche & Petter (1968); Roche et al. (1984); *Rosevear (1969); *Sanderson (1940); Schlitter et al. (1985); Schunke & Hutterer (2004, 2005, 2007); *Setzer (1956); *Skinner & Chimimba (2005); *Smithers (1971); *Smithers & Labao Tello (1976); *Smithers & Wilson (1979); Sokolov et al. (1993); Stanley & Goodman (2011); Stanley et al. (1998, 2000, 2005a,b, 2007a,b,c); Swanepoel & Schlitter (1978); *Swynnerton & Hayman (1951); *Taylor (1998); Taylor et al. (1995, 2009a,b, 2011); Van Daele et al. (1999, 2013); Van der Merwe & Avenant (2010); Van der Straeten (1980a,b, 1999, 2008); Van der Straeten & Dieterlen (1983, 1987); Van der Straeten & Dudu (1990); Van der Straeten & Kerbis Peterhans (1999); Van der Straeten & Verheyen (1979a,b, 1983); Van der Straeten et al. (2003); Venturi et al. (2004); Verheyen (1964); Verheyen et al. (1996b, 2000, 2002, 2003, 2007, 2011); Veyrunes et al. (2004, 2005, 2006); Volobouev et al. (2002a,b, 2007a,b); Williams et al. (1983); *Yalden & Largen (1992); *Yalden et al. (1976, 1996).

Finally, we extracted a large number of records from two very important datasets that are available online: the Smithsonian's National Museum of Natural History, Washington D.C., and the Royal Museum of Central Africa, Tervuren.

From these various sources, we compiled a total of 21,602 unique locality-species records (Figure 1.2) that were included in our distributional database; and maps were plotted based on these records using ArcGIS. For most species, distribution maps were based on all the records in our database. However, the limits of certain species are currently unclear and some of these can only be identified on cytogenetic or molecular grounds (e.g. many species of *Hylomyscus, Praomys, Mus* and *Taterillus*). In such circumstances, we have only presented verified records (i.e. with known karyotype or which have been sequenced).

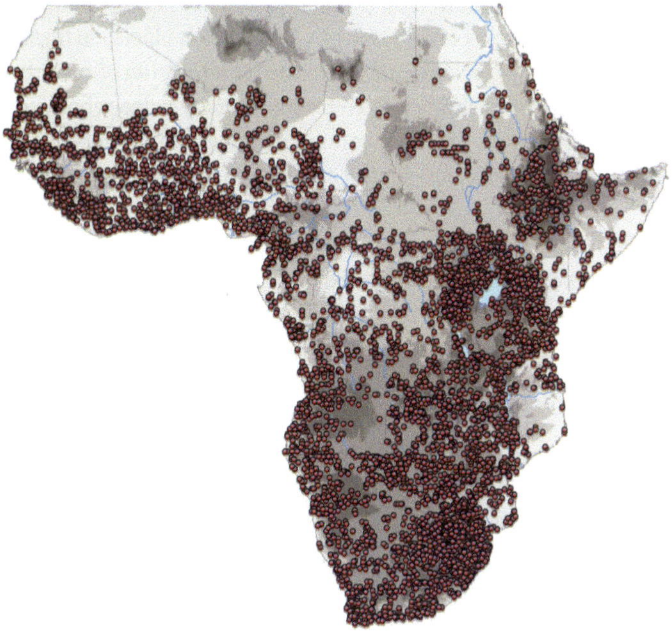

Figure 1.2 Map showing the collecting localities (sites) on which the species distribution maps are based.

Layout of book

This book starts with a brief introductory chapter that deals with a number of unrelated but hopefully interesting topics including fossil rodents, the ecology of managing pest rodents and the conservation status of Africa rodents in Sub-Saharan Africa. The remainder of the book is dedicated to the genus and species accounts. Due to the fact that species of rodents within a particular genus tend to share many morphological (external and cranium) features and to display similar ecologies, we have chosen to focus our accounts around the taxonomic level of the genus. As a result, the genus accounts are more detailed and bulkier than the species accounts.

For each genus we present images of both the cranium (skull) and the upper and lower molar rows. We hope that these images will assist understanding our descriptions of the main morphological characters provided in the genus (and occasionally in the species) descriptions, particularly for palaeontologists, anatomists or mammalogists (such as those studying diet of predators). Cranial pictures were taken with a Nikon D3000 camera, mostly of specimens from the MNHN collection in Paris. All images of molars were taken in Paris (MNHN) using two different Scanning Electron Microscopes. A first set of images were taken using a Jeol JSM 25 from 1988 to 1991 by M. Laroche and C. Chancogne at the SEM service of Palaeontology laboratory. Then a second set of images were taken in 2013 and 2014 using a Hitachi SU3500 and the facilities of the Microscopy platform (Plateau technique de Microscopie Electronique et de Microanalyse (PMEM) Dr C. Rausch) of the MNHN. Most of the specimens (71) used for SEM pictures come from the MNHN-ZM-AC (Zoology Mammals) collections.

A typical genus account begins with a "Diagnosis" where the important characters that define the genus are presented. This is followed by "Cranial Characters" and "Dentition" which describe in greater detail the skull and teeth of these rodents. Emphasis is placed on these features because they have traditionally been of great importance to the identification of rodents. We have placed less emphasis on describing pelage (fur) for two reasons. Firstly pelage characters are beautifully described in the recent Mammals of Africa volume dealing with rodents (Happold, 2013). Secondly, pelage is often not very useful for identification of rodents; where it is, it has been described under "Diagnosis".

The next section "Habitat and Biogeography" outlines the distribution of the genus and the habitats which its species occupy. This is followed by "Fossil Record" which presents known fossil records of this genus from Sub-Saharan Africa. The section on "Biology and Ecology" is purposefully kept very brief as this information is fully synthesized in Happold (2013). Finally, "Systematic Notes" discusses the latest developments within the genus, emphasizing recent molecular studies (mostly post-2008).

A list of the species that we recognise follows the genus account which in turn is followed by a key to their identification. The species accounts are typically shorter and less detailed than the genus account, and they include only four sections. The first is "Diagnosis" which describes the features that distinguish the species from other similar species. "Taxonomic Issues" highlights any taxonomic problems with the species such as whether cryptic species may be present or whether it may be conspecific with a closely related taxon. Where known, the diploid number of chromosomes (2n) and fundamental number (FN) are provided. The "Distribution" section includes a short description of the range of the species in Sub-Saharan Africa and its preferred habitat, followed by a map plotting known records. Each map has been zoomed in to the range of the species so as to provide as much detail as possible. A digital elevation model is presented as the background to the map and the main rivers and lakes are also shown. These may be viewed by some as unnecessary clutter, however, mountains (i.e. high elevations) and large rivers probably form the most important barriers to rodents in our region and many species boundaries closely follow these geological features. Finally, under "Measurements" a table presents the standard museum measurements of each species. These measurements are often very useful for field identification of rodents. Where possible, we have taken these measurements from the labels of specimens that we have examined or collected, or measured them ourselves. For species with small sample sizes, we have opted to present measurements based on literature, particularly from Happold (2013) which contains measurements for many species that hitherto remained unknown.

A key to the 14 rodent families occurring in Sub-Saharan Africa is presented and further keys to the genera within each family precede each genus account. Where possible, keys to the species within a genus are also presented.

Skull characters

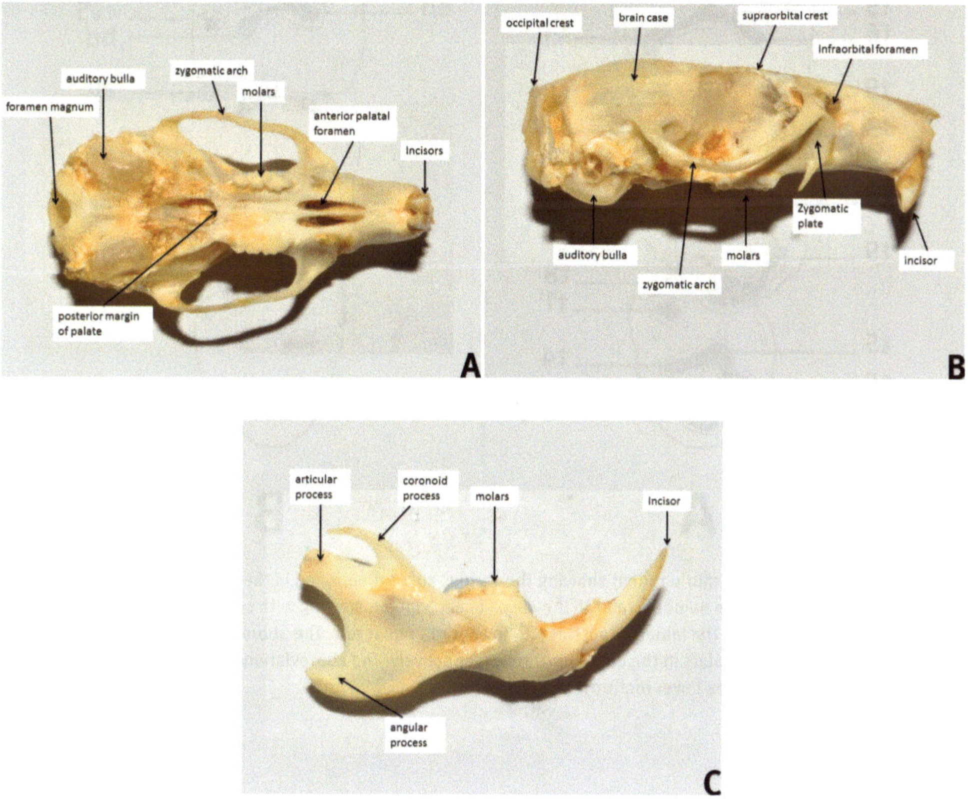

Figure 1.3 a-c Cranium and jaws of a typical murid rodent: (A) ventral view of the cranium; (B) lateral view of the cranium; (C) lateral view of the lower jaw (mandible).

For the molar rows, cusp nomenclature (see Figure 1.4) follows Denys & Tranier (1992).

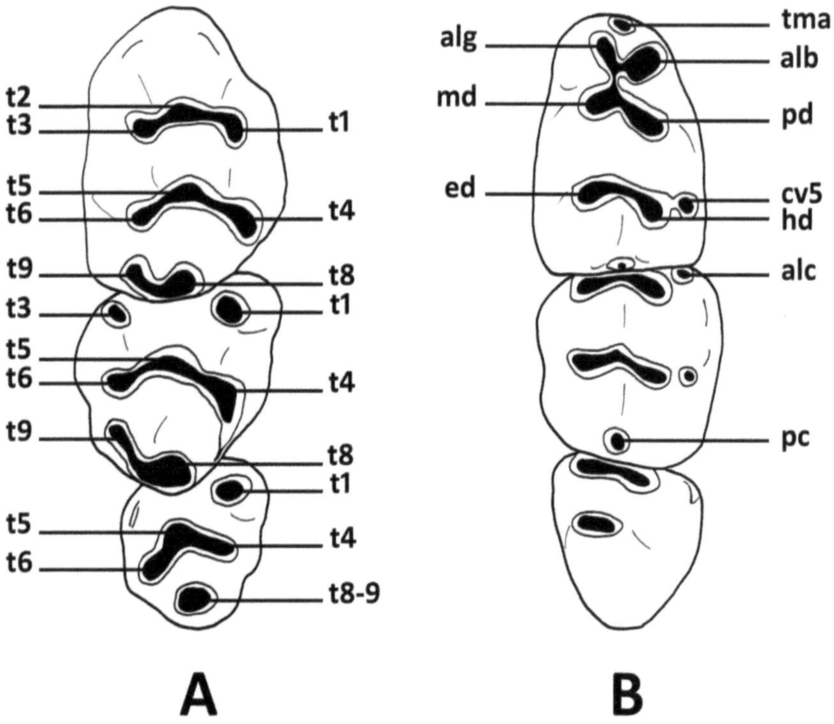

Figure 1.4 Molars of a typical murid rodent showing the various cusps mentioned in the genus and species accounts: (A) upper right molar row (the numbers refer to the cusps of each molar, starting with cusp 1 on the lingual, or inside, and ending with 9 on the labial, or outside; (B) lower right molar row. The abbreviations M^1, M^2, M^3 refer to the first, second and third molars in the upper toothrow, respectively. The abbreviations M_1, M_2, M_3 refer to the first, second and third molars in the lower toothrow, respectively.

Species accounts

Order: Rodentia

Key to Rodent Families of Africa

1	Undersurface of tail with obvious large scales at its base; flight membrane present (except *Zenkerella*); strictly arboreal	Anomaluridae
	Undersurface of tail without obvious scales at its base	2
2	Body covered in spines; typically large (mass > 2 kg; HB > 500 mm)	Hystricidae
	Body not covered in spines; small to large	3
3	Bipedal with enormous hindlimbs in proportion to forelimbs (at least five times as long)	4
	Not bipedal with hindlimbs similar in length to forelimbs	5
4	Large (mass > 1 kg; HB > 300 mm)	Pedetidae
	Small (mass < 100 g; HB < 200 mm)	Dipodidae
5	Body large (mass > 1.5 kg; HB > 400 mm) and stocky with short tail (30-75% of HB)	6
	Body not as above	7
6	Body beaver-shaped with long, shaggy fur and longer tail (ca. 70-75% of HB); upper incisors ungrooved; feral populations in Kenya only	Myocastoridae
	Body with shorter fur and short tail (30-40% of HB); upper incisors with deep longitudinal grooves	Thryonomyidae
7	Fossorial with mole-like appearance; short tail (< 50% of HB), small eyes, short ears	8
	Not fossorial	9
8	Hard palate extends well beyond posterior margin of last upper molar	Bathyergidae
	Hard palate does not extend beyond posterior margin of last upper molar	Spalacidae
9	Guinea pig-shaped with stocky body, soft fur, short tail (30-45% of HB); strictly rupiculous	Ctenodactylidae
	Not as above	10
10	Three upper cheekteeth; typically rat-like with long, sparsely furred tail	13
	Four upper cheekteeth; typically squirrel- or dormouse-like in appearance	11
11	Tail sparsely covered in long hairs; strictly rupicolous; restricted to arid Namibia (marginally into South Africa and Angola)	Petromuridae
	Tail bushy	12
12	Smaller (mass < 100 g; HB < 160 mm); dormouse-like typically with grey pelage; uniform dark-coloured tail (with or without white tip); skull with interorbital constriction; postorbital process absent; nocturnal	Gliridae
	Larger (mass > 40 g; HB > 100 mm); squirrel-like with pelage not uniform grey; tail typically grizzled or banded; skull without interorbital constriction; postorbital process well-developed; diurnal	Sciuridae
13	First lobe of M^1 with three cusps (t1 present)*	Muridae
	First lobe of M^1 with two cusps (t1 absent)	Nesomyidae

*Except for the Gerbillinae, *Deomys* and *Leimacomys* (there are in fact no unique morphological features that distinguish these two families, although they are clearly separable on molecular grounds).

Family: Gliridae

This family is represented by 15 species in Sub-Saharan Africa. The original name for the family (Myoxidae) was ruled invalid by the Commission on Zoological Nomenclature (Holden, 2005).

Genus Account

Graphiurus Smuts, 1832 (Dormice)

Diagnosis

This diverse genus is small-sized (mostly 15-85 g). *Graphiurus* has the distinctively shaped body resembling that of a tiny squirrel with dense, soft fur on its body, a relatively long, bushy tail and large eyes. Although the pelage varies considerably between species, it is always soft and usually some shade of grey without obvious markings except on the face or tip of tail. Fur colour ranges from light grey to bright rufous dorsally and pale grey or white ventrally. The ears are relatively small and rounded, and a black eye ring is often present and in at least one species the face is boldly patterned in black and white (*G. ocularis*). There are five digits on the hindfoot and four digits on the forefoot, all with sharp and curved claws. The tail is mostly the same length (or slightly shorter) than the body length, and typically very bushy. Its colour is usually the same as that on the dorsal body and in some species ending in a black or white tip. Where known, there are three or four pairs of nipples (2 + 1 = 6 or 2 + 2 = 8) placed along the thorax and abdomen.

(a) (b) (c)

Figure 2.1 a-c Photographs of *Graphiurus*, from left to top right to bottom right: *Graphiurus* sp. (Bill Branch), *G. kelleni* (James Harvey) and *G. nagtglasii* (Jan Decher)."

Cranial Characters

The skull is small in size (mostly 21-38 mm in greatest length). It is typically flat, curving down posteriorly towards the occipitals, but is particularly flattened in *G. platyops* and *G. rupicola*. The palate ends at the posterior margin of the third molars. The zygomatic arch is long, narrow and distinctive in the structure of the zygomatic plate in which the dorsal and ventral arms are of equal size, forming a relatively wide, oval infraorbital foramen. The supraorbital foramen and postorbital process are absent. The anterior palatal foramina are short and narrow but widening posteriorly. The bullae are relatively large and slightly inflated.

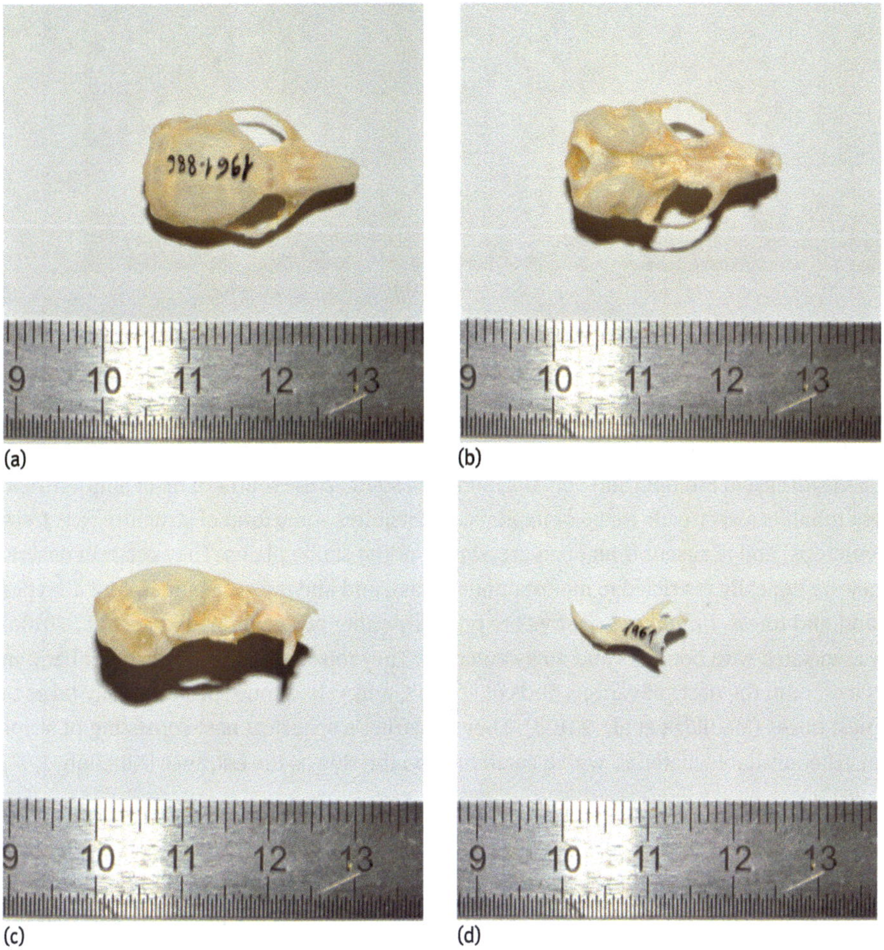

Figure 2.2 a-d Cranium of *Graphiurus murinus* (MNHN ZM-MO- 1961-886) (from top left to bottom right: dorsal view, ventral view, lateral view and lower jaw).

Dentition

Graphiurus has ungrooved incisors which are usually orthodont or slightly opisthodont, but pro-odont in *G. crassicaudatus,* with a V-shaped cutting edge (in anterior view). The dental formula is: 1013/1013 (Rosevear, 1969), with the premolar P^4 being larger than (or equal to) the third molar (except in *G. ocularis* in which it is reduced), but smaller than the first and second molars. The presence of a premolar immediately distinguishes *Graphiurus* from any of the murid or nesomyid genera. The cheekteeth are brachyodont with low transverse ridges forming shallow occlusal surfaces.

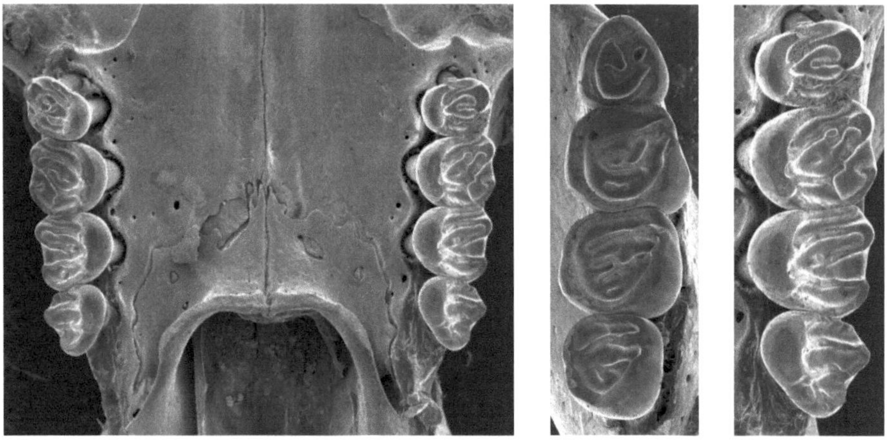

(a) (b) (c)

Figure 2.3 Cheek teeth of *Graphiurus lorraineus* (MNHN-CG 2011-925) from Cameroon (A: upper cheek teeth; B: lower right toothrow; C: upper left toothrow).

Habitat & Biogeography

Graphiurus has a wide, almost unbroken, distribution across Sub-Saharan Africa. It occurs from southern Senegal east to Ethiopia and Somalia, and then south to the southern tip of South Africa. This rodent inhabits a very wide range of habitats, but requires some form of structure (e.g. trees or rocky outcrops) and is absent from open grasslands or flat sandy plains. They occur in deserts (where they are typically restricted to mountainous areas), arid and moist savannas, and all types of woodland and forest. *Graphiurus* species are typically either arboreal (Madikiza et al., 2010a) or closely associated with boulders and rocky outcrops. They roost in hollow trees, under bark, in rocky crevices, or in the nests of various birds or spiders, and at least one species readily takes to artificial nest boxes (Madikiza et al., 2010b). They construct a spherical nest consisting of strips of grass or other available material, with a small hole on the side as the entrance (Kingdon, 1974; Happold, 2013).

Fossil Record

The oldest *Graphiurus* fossil is from 3.3 MYA, Makapansgat, South Africa (Pocock, 1987). In East Africa, it has been reported from the Early and Middle Pleistocene; in Southern Africa from the Late Pliocene to the Late Pleistocene (Winkler et al, 2010).

Biology & Ecology

Only a few species of *Graphiurus* have been studied in detail, and for most species knowledge has not progressed beyond anecdotal observations. The diet of most species is thought to be omnivorous consisting of insects, fruits, seeds and nuts (Webb & Skinner, 1995; Happold, 2013) and may include eggs and small birds (Kingdon, 1974), but only two species have been studied in any detail. The large *G. ocularis* appears to specialise on vertebrates (Channing, 1984), whilst stomach content analysis of *G. murinus* indicates geographical variability in diet with insects, seeds and fruit all being important components (Rowe-Rowe, 1986; Wirminghaus & Perrin, 1992). Its spatial ecology is also poorly known. Home ranges of *G. ocularis* were relatively large at 1.1-3.8 ha (Van Hensbergen & Channing, 1989), but were less than 0.5 ha in a capture-mark-recapture study of *G. murinus* (Madikiza et al., 2011). Members of this genus are well known for using torpor (Webb & Skinner, 1996; McKechnie & Mzilikazi, 2011) as well as true hibernation (Perrin & Ridgard, 1999; Mzilikazi et al., 2012). At least one species (*G. kelleni*) has a rich vocal repertoire including ultrasonic sounds (Hutterer & Peters, 2001). Where known, litter size is 2-6 but mostly 3-4 (de Graaff, 1981; Webb & Skinner, 1995; Happold, 2013). Gestation period has been estimated by Kingdon (1974) as 24 days. In *G. ocularis* females are thought to produce two litters per year, 6-8 weeks apart (Channing, 1984), whereas in *G. murinus* at one site in South Africa, breeding appeared to be aseasonal (Wirminghaus & Perrin, 1992). This species is thought to exhibit a promiscuous mating system (Madikiza et al., 2011).

Systematic Notes

Graphiurus belongs to the family Gliridae which has had an unstable nomenclatural history; the family has previously been named Myoxidae, and as well as Muscardinidae. This history has been reviewed by Walhert et al. (1993) and Holden (2005), and the name Gliridae should now, hopefully, stick for this group. The suggestion that the family is paraphyletic, and that *Graphiurus* is not a glirid (Vianey-Liaud & Jaeger, 1996), is not supported by molecular data (Bentz & Montgelard, 1999; Montgelard et al., 2002, 2003; Nunome et al., 2007). However, *Graphiurus* is unusual in showing hystricomorphy in the zygomasseteric architecture, in contrast to other glirid genera which show the myomorph condition (Hautier et al., 2008). Yet, the Gliridae and Sciuridae (which exhibit sciuromorphy) form a well-supported monophyletic clade (Blanga-Kanfi et al., 2009) and the myomorph condition is therefore an example of convergent evolution. *Graphiurus* is placed in its own monotypic subfamily Graphiurinae and is considered basal to the Gliridae both on morphological (Wahlert et al., 1993) and molecular grounds (Montgelard et al., 2003), and is restricted to Sub-Saharan Africa, which it probably colonised via Europe about 8-10 million years ago (Montgelard et al., 2003). Four subgenera have been described within this genus: *Graphiurus, Aethoglis, Claviglis* and *Gliriscus* (Pavlinov & Potapova, 2003) but their validity cannot be asserted at present because a phylogeny for the entire genus is not yet available. Holden (2005), based on Pavlinov & Potapova's (2003) morphological analysis, recognised the first three of these subgenera but not *Gliriscus*. The number of species recognised within *Graphiurus* has varied considerably over the past few decades. For example, five species were recognised by Misonne (1974), six species by Genest-Villard (1978), but 14 by Holden (2005). Much of this instability rests on uncertainty within the *murinus* group (Holden, 1996), from which a new species was described recently *G. walterverheyeni* (Holden & Levine, 2009). Reviews of the genus, or species groups, have been conducted either regionally (Schlitter et al., 1985) or continentally (Genest-Villard, 1978; Holden, 1996; Holden & Levine, 2009).

The diploid number of chromosomes in various *Graphiurus* species is as follows:
Graphiurus nagtglasii: 2n = 40, FN = 66 (Tranier & Gautun, 1979; Tranier & Dosso, 1979)
G. murinus: 2n = 70 (Tranier & Gautun, 1979)
G. kelleni: 2n = 70 (Dobigny et al., 2002a)

G. murinus: 2n = 46, FN = 92 (Krystufek et al., 2004)
G. microtis: 2n = 52 (Chitaukali et al., 2001)

Species List

Fifteen species are described within the genus *Graphiurus*, all restricted to Sub-Saharan Africa.
Graphiurus angolensis de Winton, 1897
Graphiurus christyi Dollman, 1914
Graphiurus crassicaudatus (Jentink, 1888)
Graphiurus johnstoni (Thomas, 1898)
Graphiurus kelleni (Peuvens, 1890)
Graphiurus lorraineus Dollman, 1910
Graphiurus microtis (Noack, 1887)
Graphiurus monardi (St. Leger, 1936)
Graphiurus murinus (Desmerest, 1822)
Graphiurus nagtglasii Jentink, 1888
Graphiurus ocularis (Smith, 1829)
Graphiurus platyops Thomas, 1897
Graphiurus rupicola (Thomas & Hinton, 1925)
Graphiurus surdus Dollman, 1912
Graphiurus walterverheyeni Holden & Levine, 2009

Key to Species

1	Size larger (typically GLS > 33 mm)	2
	Size smaller (typically GLS < 32 mm)	4
2	Conspicuous black facial mask; restricted to the southern parts of South Africa	*G. ocularis*
	Facial mask inconspicuous or absent; does not occur in South Africa	3
3	Tail white tipped; restricted distribution in north-eastern Angola, and neighbouring regions of the DRC and Zambia	*G. monardi*
	Tail not white tipped; restricted to West Africa and northern-western parts of Central Africa	*G. nagtglasii*
4	Nasals parallel-sided; dorsal pelage rufous or with rufous hue; upper incisors pro-odont or orthoodont	5
	Nasals broaden out distally; pelage may brown or rufous; upper incisors orthodont or opisthodont	6
5	Upper incisors pro-odont; size larger (GLS > 24 mm); restricted to West Africa and western parts of Central Africa	*G. crassicaudatus*
	Upper incisors orthodont; size much smaller (GLS < 23 mm); known from two localities on the left bank of the Congo River	*G. walterverheyeni*
6	Size smaller (typically GLS < 25 mm, occasionally up to 26.2 mm)	7
	Size larger (typically GLS > 26 mm, occasionally down to 25.2 mm)	9
7	Dorsal and ventral pelage not clearly demarcated; typical inhabitant of rainforest	*G. lorraineus*
	Dorsal and ventral pelage clearly demarcated; typical inhabitant of savanna	8
8	Restricted to southern Malawi	*G. johnstoni*
	Not restricted to southern Malawi	*G. kelleni*
9	Inhabitant of lowland rainforest	10
	Inhabitant of savanna/woodland or montane/temperate habitats	11
10	Pelage usually rufous; tail typically longer > 84 mm without white tipped hairs along its entire length (not frosted in appearance); facial mask more conspicuous	*G. christyi*
	Pelage grey-brown; tail typically shorter < 84 mm with white tipped hairs along its entire length giving a frosted appearance; facial mask inconspicuous	*G. surdus*
11	Braincase appears flattened in lateral view; inhabits rocky habitats	12
	Braincase not particularly flattened; inhabits savanna/woodland habitats	13
12	Restricted to Angola, Namibia and marginally into extreme western parts of South Africa	*G. rupicola*
	Widely distributed in the eastern parts of Southern Africa	*G. platyops*
13	Restricted to miombo woodlands of Angola and neighbouring western Zambia	*G. angolensis*
	Occurs elsewhere in Southern and East Africa	14
14	Dorsal and ventral pelage not clearly demarcated; occurs in montane or temperate habitats	*G. murinus*
	Dorsal and ventral pelage clearly demarcated; occurs in savanna/woodland habitats	*G. microtis*

Species Accounts

Graphiurus angolensis de Winton, 1897

Diagnosis

This species is similar is external and skull characteristics to *Graphiurus platyops* (with which it does not overlap in distribution) and *G. rupicola* (with which it occurs sympatrically in Angola). *Graphiurus rupicola* is larger (typically, GLS > 30 mm and HF > 20 mm), has a flattened skull, and occurs in rocky situations (woodland in *G. angolensis*). *Graphiurus microtis* is smaller (typically, GLS < 29 mm, HF < 18 mm), but measurements can overlap making a definitive identification difficult; however the two species apparently only overlap in north-western Zambia.

Taxonomic Issues

The relationship between *G. angolensis* and *G. microtis* is confused and has not yet been assessed.

Distribution

This species is restricted to the savannas of Angola and western Zambia.

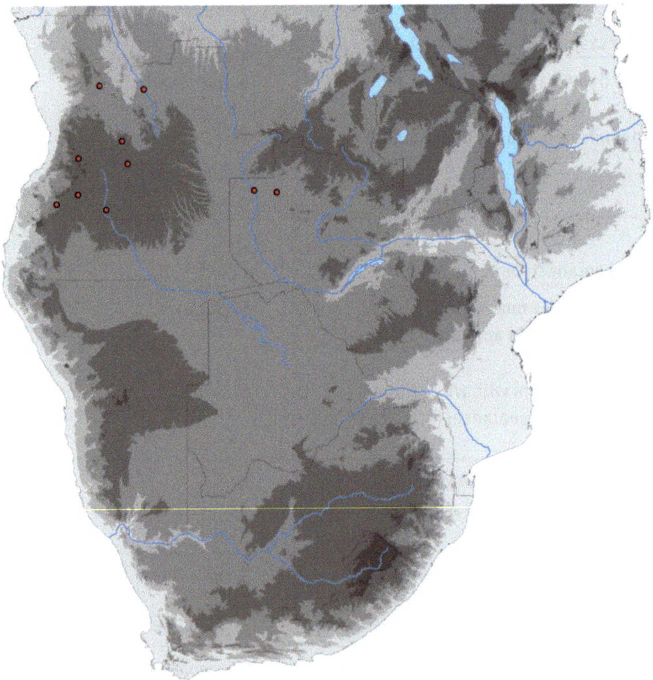

Figure 2.4 Distribution map

Measurements

External and cranial measurements (mm) and mass (g) for *Graphiurus angolensis*, sexes combined

	Mean	Min	Max	N
Mass	-	-	-	-
HB*	104.6	96	118	14
Tail*	69.4	60	77	14
HF*	19.3	18	20	14
Ear*	-	-	-	-
GLS*	29.3	28.3	30.5	14

*Hill & Carter (1941)

Graphiurus christyi Dollman, 1914

Diagnosis

This medium-sized *Graphiurus* with reddish-brown pelage and obvious supraorbital ridges can be confused with the following species with which it shares its range: *Graphiurus lorraineus* (smaller, GLS < 26 mm), *G. crassicaudatus* (pro-odont upper incisors, parallel-sided nasals), *G. surdus* (grey-brown pelage, tail typically shorter < 84 mm with white tipped hairs along its entire length giving a frosted appearance), *G. walterverheyeni* (much smaller, GLS < 23 mm).

Taxonomic Issues

None.

Distribution

This species has a patchy distribution in the Congo basin (right bank of the Congo River) and southern Cameroon.

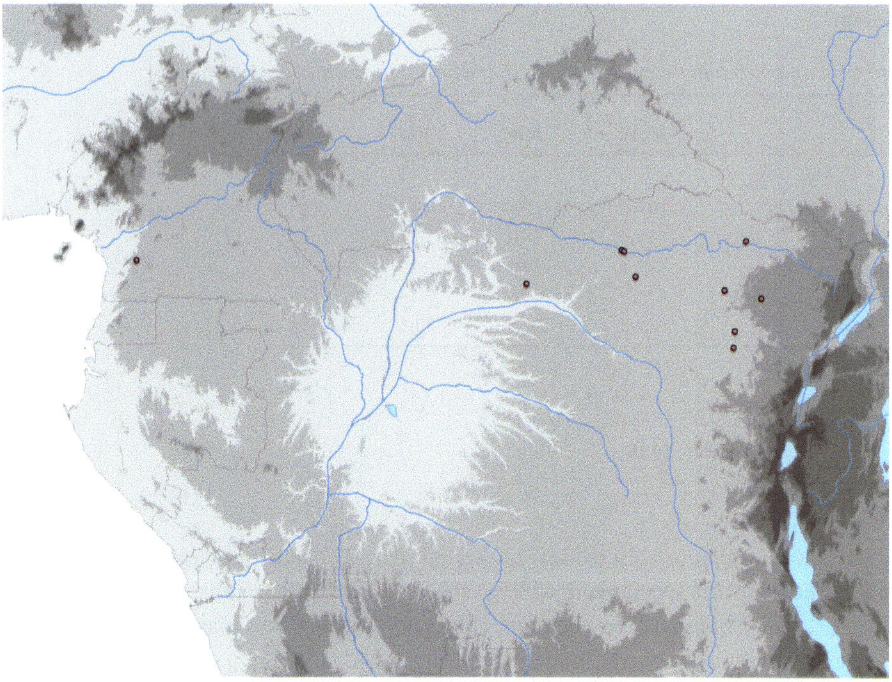

Figure 2.5 Distribution map

Measurements

External and cranial measurements (mm) and mass (g) for *Graphiurus christyi*, sexes combined

	Mean	Min	Max	N
Mass*	29.0	25	33	6
HB*	96.9	86	107	30
Tail*	80.2	73	95	27
HF*	18.1	16	20	31
Ear*	14.2	12	17	27
GLS*	28.0	26.7	29.7	25

*Holden (1996)

Graphiurus crassicaudatus (Jentink, 1888)

Diagnosis

This is a relatively distinct species with its rufous pelage, pro-odont upper incisors, parallel-sided nasals and strong supraorbital ridges. However, it may be confused with the following species within its range: *G. lorraineus* (smaller, GLS < 26 mm, upper incisors orthodont, nasals broaden out distally), *G. surdus* (upper incisors orthodont, pelage grey-brown without any rufous, nasals broaden out distally) and *G. christyi* (upper incisors orthodont, nasals broaden out distally). It is significantly larger than *G. walterverheyeni* (GLS < 23 mm) with which it is allopatric.

Taxonomic Issues

The phylogenetic relationships of this species are not known.

Distribution

This species is widely distributed in the forest zone of West Africa, and extending marginally into Central Africa in southern Cameroon.

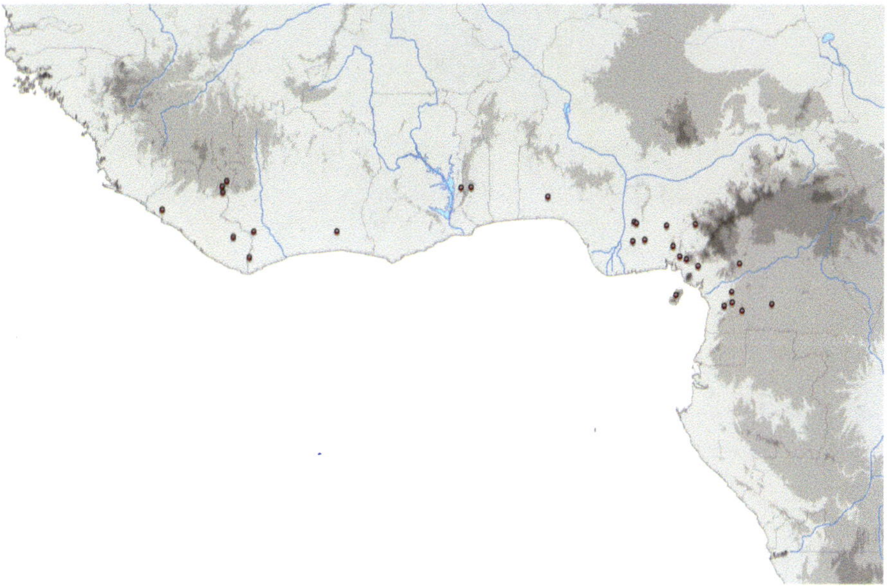

Figure 2.6 Distribution map

Measurements

External and cranial measurements (mm) and mass (g) for *Graphiurus crassicaudatus*, sexes combined

	Mean	Min	Max	N
Mass*	23.6	19	29	8
HB*	88.9	77	100	26
Tail*	58.4	53	70	15
HF*	16.4	13	19	26
Ear*	12.0	7	15	24
GLS*	25.6	24.1	27.8	29

*Holden & Levine (2009)

Graphiurus johnstoni (Thomas, 1898)

Diagnosis

The differences between *G. johnstoni* and *G. kelleni* have not been clearly stated and therefore these two species cannot be distinguished at present.

Taxonomic Issues

This species was previously listed as *G. kelleni* (Holden, 1993), but was elevated to species level based on slight differences in cranial characteristics (Holden, 2005).

Distribution

This species is known from just three localities in southern Malawi (Happold & Happold, 1989a, 1997).

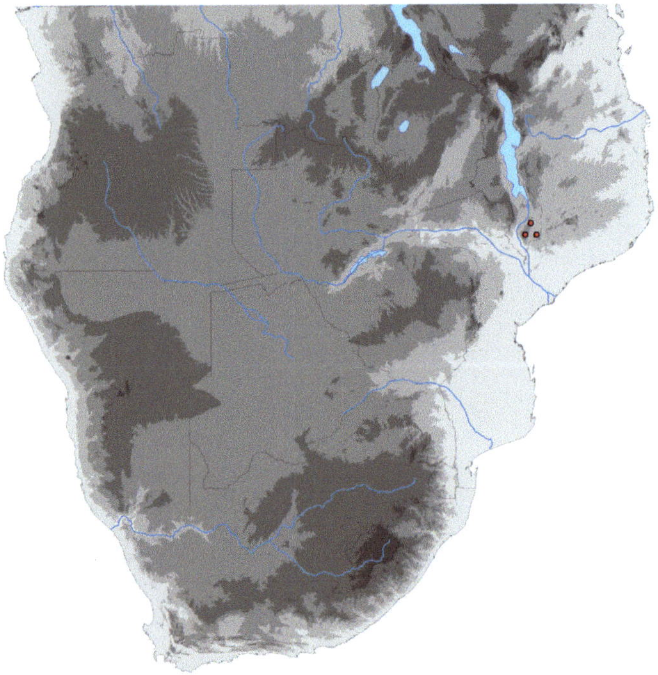

Figure 2.7 Distribution map

Measurements

External and cranial measurements (mm) and mass (g) for *Graphiurus johnstoni*, sexes combined

	Mean	Min	Max	N
Mass*	-	-	-	-
HB*	74.3	69	84	3
Tail*	68.5	65	75.5	3
HF*	16.0	15	17	4
Ear*	11.8	11	12	4
GLS*	23.6	23.3	23.9	2

*Happold (2013)

Graphiurus kelleni (Peuvens, 1890)

Diagnosis

This small *Graphiurus* is relatively easily distinguished from most congeners by its diminutive size and savanna habitat. *Graphiurus microtis* is larger (GLS > 25 mm) and has a prominent white tip to the tail (in *G. kelleni* tip of tail may or may not be white). *Graphiurus murinus* is also larger (GLS > 25 mm), and typically occupies a different habitat (montane or temperate habitats). It is not clear what distinguishes *G. kelleni* from *G. johnstoni*. The diminutive *G. walterverheyeni* is a rainforest species that does not overlap in distribution with *G. kelleni*.

Taxonomic Issues

The phylogenetic relationships of this species are not known.

Distribution

This species is widely distributed throughout the savanna zone, but absent in the southern savannas of Botswana, Namibia and South Africa.

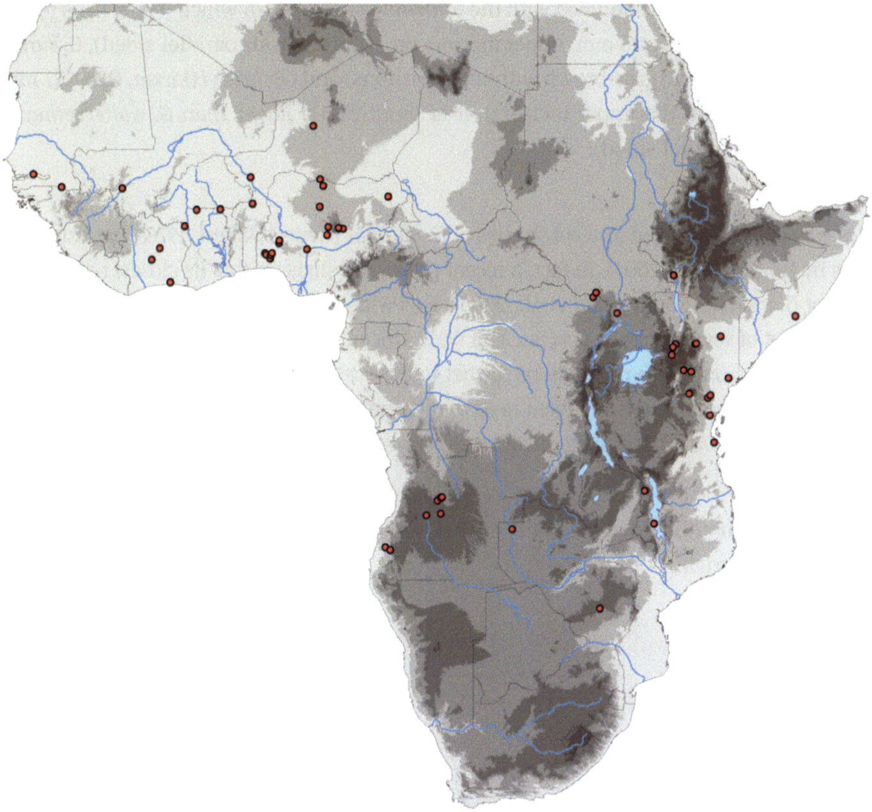

Figure 2.8 Distribution map

Measurements

External and cranial measurements (mm) and mass (g) for *Graphiurus kelleni*, sexes combined

	Mean	Min	Max	N
Mass*	19.1	10.9	23.5	8
HB	76.3	66	87	10
Tail	75.9	60	100	8
HF	14.7	13	17	10
Ear	12.8	11	15	8
GLS	23.1	20.7	25.0	14

*Happold (2013)

Graphiurus lorraineus Dollman, 1910

Diagnosis

This forest zone species can be confused with the following species with which it shares its range: *G. crassicaudatus* (larger, GLS > 26 mm, upper incisors pro-odont, nasals parallel-sided), *G. surdus* (larger, GLS > 26.5 mm, pelage grey-brown without any rufous) and *G. christyi* (larger, GLS > 27 mm, nasals extend well beyond the upper incisors). It is significantly larger than *G. walterverheyeni* (GLS < 23 mm, nasals parallel-sided).

Taxonomic Issues

This species was previously listed as a subspecies of *G. murinus* (Genest-Villard, 1978), but listed as a valid species by Holden (1993, 2005). In fact, a recent phylogeny suggests that its closest relative is *G. kelleni*, and not *G. murinus* (Montgelard et al., 2003). This widely distributed forest species may well be shown to consist of several cryptic species. The West African taxon *spurelli* is separated by a large gap (western Nigeria) from those populations further to the east, and may represent a valid species. This species (or species complex) is in urgent need of revision. The taxon *haedulus* has been recorded in montane habitats (> 1700 m) in Cameroon and may represent a different species (Happold, 2013).

Distribution

This species is widely distributed in the forest zone of tropical Africa, extending south into Zambia, but absent from East Africa.

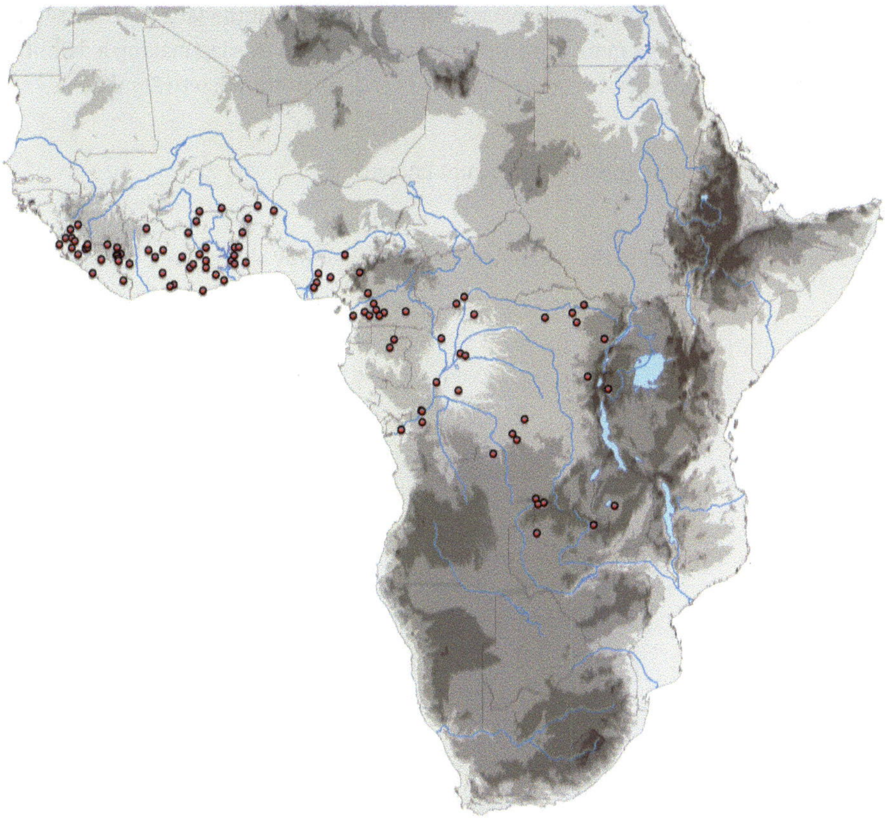

Figure 2.9 Distribution map

Measurements

External and cranial measurements (mm) and mass (g) for *Graphiurus lorraineus*, sexes combined

	Mean	Min	Max	N
Mass*	16.8	12	24	5
HB*	80.5	70	93	57
Tail*	66.8	45	77	50
HF*	16.5	13	19	57
Ear*	12.4	9	15	27
GLS*	24.5	22.7	26.2	47

*Holden & Levine (2009)

Graphiurus microtis (Noack, 1887)

Diagnosis

This species is not easily distinguished from the following savanna species: *G. kelleni* and *G. johnstoni* (both species are smaller, GLS < 25 mm, and lack a conspicuous white tip to the tail); *G. murinus* (dorsal and ventral pelage not clearly demarcated, typically occupies montane or temperate habitats). *Graphiurus microtis* and *G. angolensis* are very similar but the latter species is typically larger (GLS > 29 mm, HF > 18 mm), however, measurements can overlap making a definitive identification difficult; the two species apparently only overlap in north-western Zambia.

Taxonomic Issues

The phylogenetic relationships of this species are not known.

Distribution

This species is widely distributed in the savanna zone, but apparently absent from West Africa.

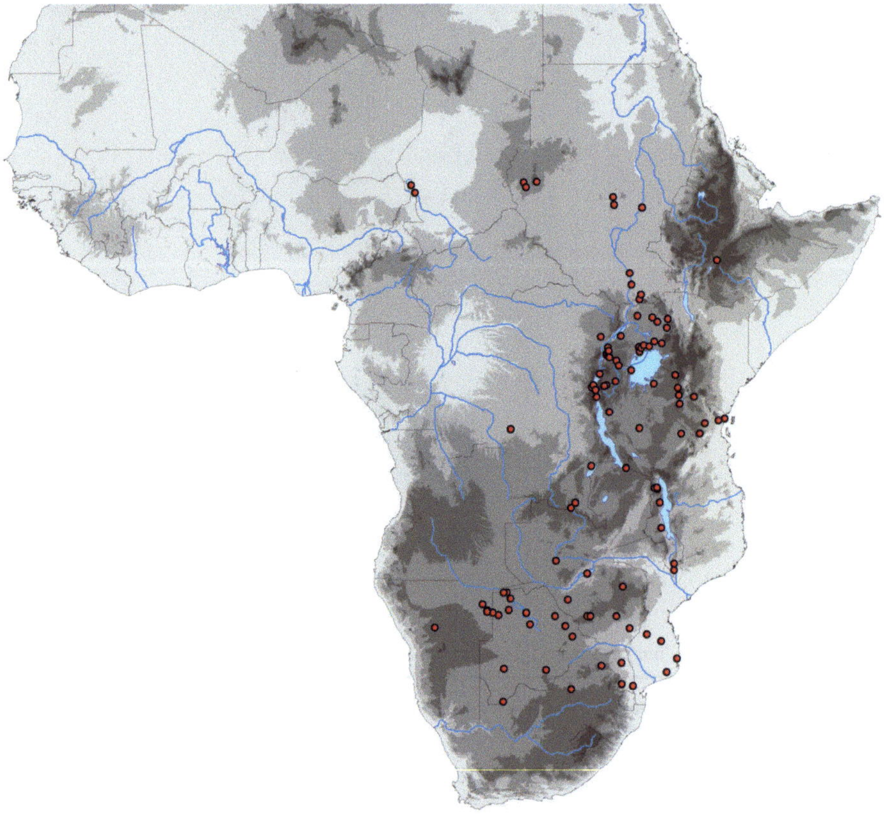

Figure 2.10 Distribution map

Measurements

External and cranial measurements (mm) and mass (g) for *Graphiurus microtis*, sexes combined

	Mean	Min	Max	N
Mass*	29.5	17.6	42.5	21
HB*	98.8	75	115	33
Tail*	75.2	62	86	28
HF*	16.9	14	20	33
Ear*	15.5	13	21	31
GLS*	27.4	25.5	29.1	28

*Happold (2013)

Graphiurus monardi (St. Leger, 1936)

Diagnosis

This is one of three very large *Graphiurus* species (the other two being *G. nagtglasii* and *G. ocularis*), which are easily separated from all other congeners by their large size (GLS > 31 mm, HF > 21 mm). It can be distinguished from *G. ocularis* by the lack of a very conspicuous facial mask and distribution (*G. ocularis* is restricted to the southern parts of South Africa), and from *G. nagtglasii* by its white-tipped tail, longer fur and distribution (*G. nagtglasii* does not have a white-tipped tail and occurs in West and Central African rainforest).

Taxonomic Issues

The relationship of this species to other *Graphiurus* is not known, although Happold (2013) speculates that it is not closely related to *G. nagtglasii*.

Distribution

This species is narrowly distributed in miombo woodland, in north-eastern Angola and north-western Zambia.

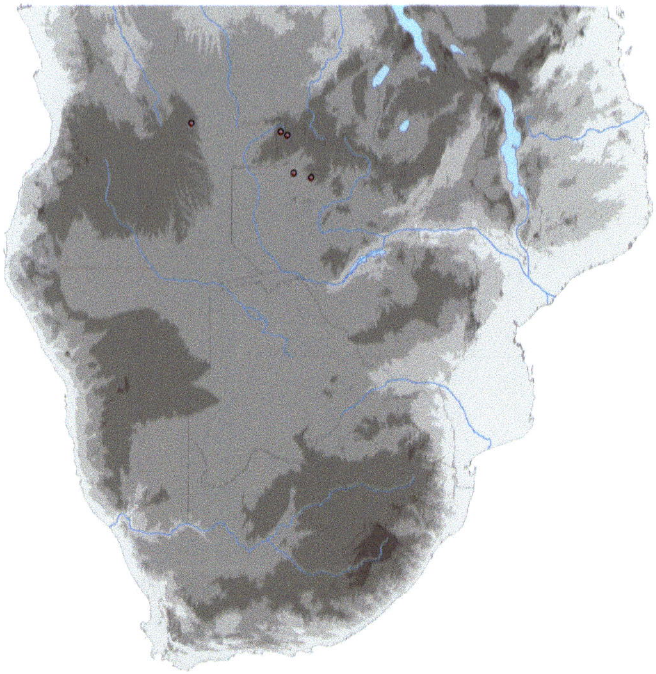

Figure 2.11 Distribution map

Measurements

External and cranial measurements (mm) and mass (g) for *Graphiurus monardi*, sexes combined

	Mean	Min	Max	N
Mass*	-	-	-	-
HB*	143.4	130	160	5
Tail*	90.0	90	90	3
HF*	21.6	21	22	5
Ear	-	-	-	-
GLS*	34.0	32.0	36.6	5

*Hill & Carter (1941)

Graphiurus murinus (Desmerest, 1822)

Diagnosis

This species is not easily distinguished from the following savanna species with which it may marginally overlap (although it is associated with montane and temperate regions rather than savannas): *G. kelleni* and *G. johnstoni* (both species are smaller, GLS < 25 mm, and lack a conspicuous white tip to the tail); *G. microtis* (dorsal and ventral pelage clearly demarcated, typically occupies savanna habitats, avoiding montane or temperate zones); *G. angolensis* is larger (GLS > 29 mm), however, measurements can overlap making a definitive identification difficult; the two species apparently do not overlap in distribution. *Graphiurus lorraineus* is smaller (typically, GLS < 26 mm) and is restricted to the forest zone; the two species may co-occur in the Albertine Rift, where they are altitudinally stratified (*G. murinus* > 1000 m, *G. lorraineus* < 1000 m).

Taxonomic Issues

The taxonomy of this species and its phylogenetic relationships are poorly understood, and it may prove to be a species complex.

Distribution

This species is widely distributed in montane or temperate regions of Southern and East Africa.

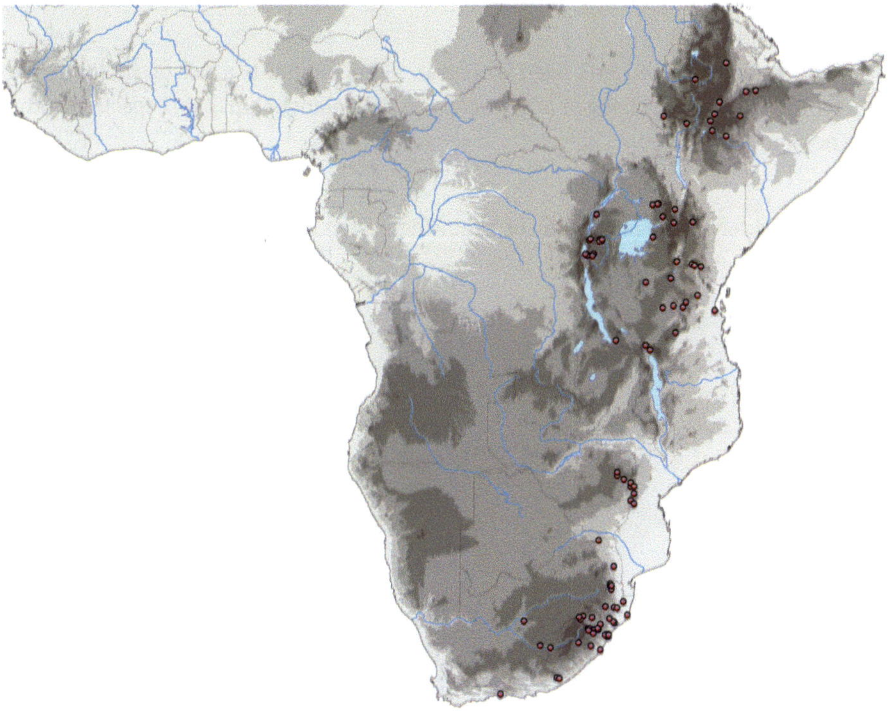

Figure 2.12 Distribution map

Measurements

External and cranial measurements (mm) and mass (g) for *Graphiurus murinus*, sexes combined

	Mean	Min	Max	N
Mass*	28.8	19	41	10
HB	90.1	75	115	23
Tail	77.2	50	97	22
HF	17.7	15	20	15
Ear	13.3	10	19	22
GLS**	26.4	25.2	28.8	19

*Taylor (1998); **Happold (2013)

Graphiurus nagtglasii Jentink, 1888

Diagnosis

This is one of three very large *Graphiurus* species (the other two being *G. monardi* and *G. ocularis*), which are easily separated from all other congeners by their large size (GLS > 31 mm, HF > 21 mm). It can be distinguished from *G. ocularis* by the lack of a very conspicuous facial mask and distribution (*G. ocularis* is restricted to the southern parts of South Africa), and from *G. monardi* by its tail tip which is never white and its distribution (*G. monardi* has a white-tipped tail and occurs in miombo woodlands of north-eastern Angola and western Zambia).

Taxonomic Issues

This species was previously known as *G. hueti* (Rosevear, 1969), but due to a lack of a holotype and an uncertain type locality it was changed to *G. nagtglasii*. The relationship of this species to other *Graphiurus* is not known.

Distribution

This species occurs widely in the forest zone of West Africa extending east to the right bank of the Congo River.

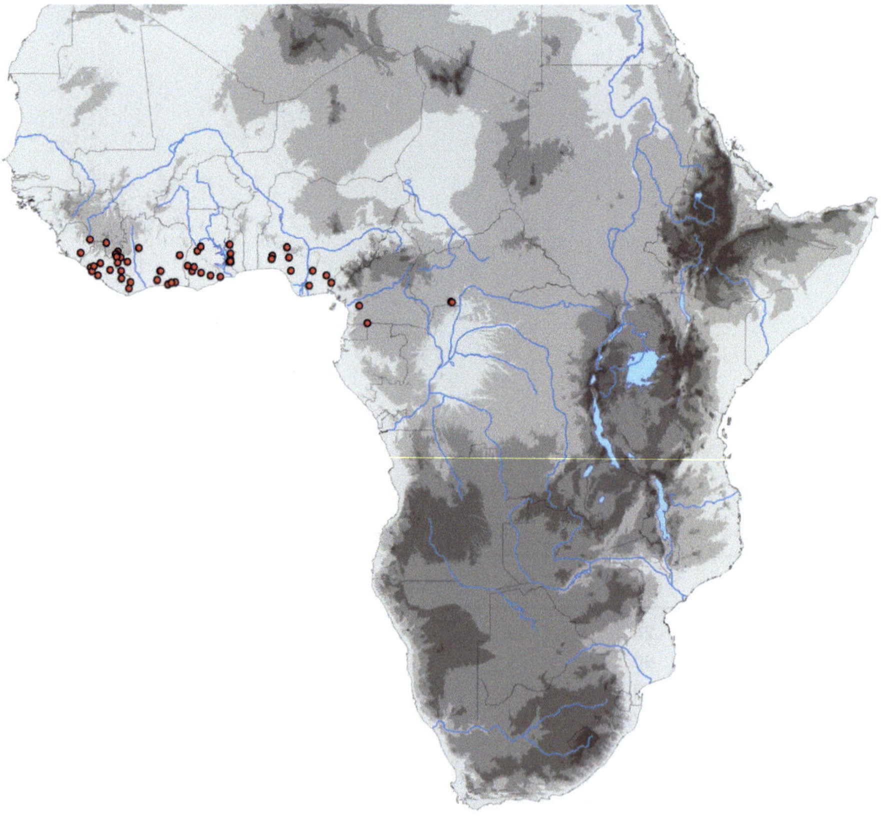

Figure 2.13 Distribution map

Measurements

External and cranial measurements (mm) and mass (g) for *Graphiurus nagtglasii*, sexes combined

	Mean	Min	Max	N
Mass	70.2	62	77	6
HB	132.4	108	160	18
Tail	126.0	102	160	13
HF	27.3	24	31	18
Ear	19.4	16	22	15
GLS	36.3	32.9	39.9	20

Graphiurus ocularis (Smith, 1829)

Diagnosis

This is one of three very large *Graphiurus* species (the other two being *G. nagtglasii* and *G. monardi*), which are easily separated from all other congeners by their large size (GLS > 31 mm, HF > 21 mm). The characteristic feature of *Graphiurus ocularis* is its facial mask which is very conspicuous and extends posterior the ears. It does not share its range with either *G. nagtglasii* (which does not have a white tip to the tail, the facial mask is less pronounced, and inhabits rainforest) or *G. monardi* (facial mask is less pronounced, miombo woodlands of north-east Angola and western Zambia).

Taxonomic Issues

Graphiurus ocularis is apparently closely related to *G. platyops* (Montgelard et al., 2003), but *G. rupicola, G. monardi* and *G. nagtglasii* were not included in the latter study. A complete molecular phylogeny for the genus *Graphiurus* is long overdue.

Distribution

This species is restricted to the southern parts of South Africa. Two records from north of the Gariep River (De Graaff, 1981) could not be traced and are not plotted here.

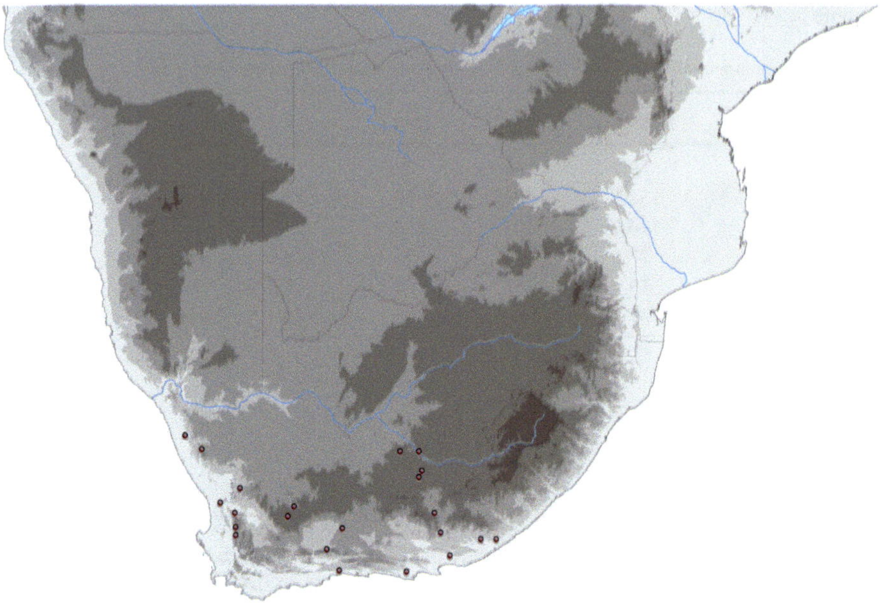

Figure 2.14 Distribution map

Measurements

External and cranial measurements (mm) and mass (g) for *Graphiurus ocularis*, sexes combined

	Mean	Min	Max	N
Mass	78.3	49	105	3
HB	135.6	118	152	8
Tail	114.6	78	149	7
HF	23.6	20	25	7
Ear	19.4	15	23	8
GLS	35.4	33.6	37.1	6

Graphiurus platyops Thomas, 1897

Diagnosis
Due to its flattened skull and rupiculous habits (features shared with the allopatric *G. rupicola*), this species is unlikely to be confused with any other *Graphiurus* within its range.

Taxonomic Issues
This taxon was considered a valid species by Roberts (1951), but listed as a subspecies of *G. platyops* by Genest-Villard (1978). It was elevated back to species level by Holden (2005).

Distribution
This species is patchily distributed in the north-eastern savannas of southern Africa.

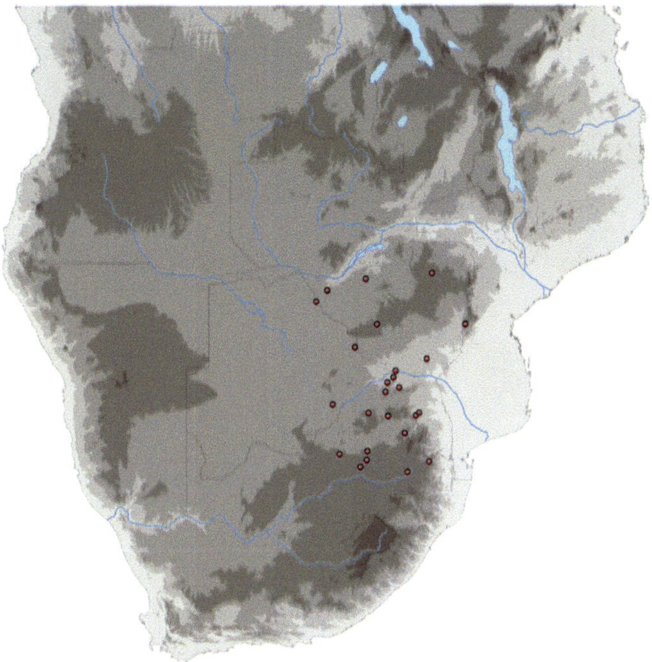

Figure 2.15 Distribution map

Measurements

External and cranial measurements (mm) and mass (g) for *Graphiurus platyops*, sexes combined

	Mean	Min	Max	N
Mass	41.6	29	65	8
HB	106.1	85	120	17
Tail	73.0	48	98	17
HF	20.4	19	23	15
Ear	15.3	13	18	16
GLS	29.5	26.6	31.9	14

Graphiurus rupicola (Thomas & Hinton, 1925)

Diagnosis

Due to its flattened skull and rupiculous habits (features shared with the allopatric *G. platyops*), this species is unlikely to be confused with any other *Graphiurus* within its range, except *G. angolensis* which is smaller (typically, GLS < 30 mm and HF < 20 mm), does not have a flattened skull, and occurs in woodland. *Graphiurus microtis* is significantly smaller (GLS < 29 mm) and does not have a flattened skull.

Taxonomic Issues

This taxon was considered a valid species by Roberts (1951), but listed as a subspecies of *G. platyops* by Genest-Villard (1978). It was elevated back to species level by Holden (1993, 2005).

Distribution
This species occurs in a narrow belt along the escarpment of Namibia and marginally into South Africa.

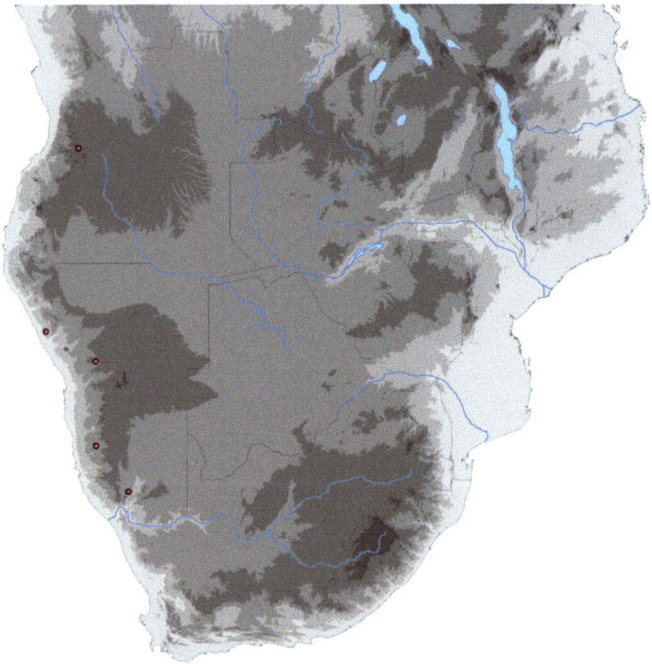

Figure 2.16 Distribution map

Measurements

External and cranial measurements (mm) and mass (g) for *Graphiurus rupicola*, sexes combined

	Mean	Min	Max	N
Mass	37.5	26	63	7
HB	103.9	93	120	8
Tail	93.6	74	100	8
HF	22.0	21	25	7
Ear	18.2	16	20	6
GLS	31.0	28.3	32.9	4

Graphiurus surdus Dollman, 1912

Diagnosis
This forest zone species can be confused with the following species with which it shares its range: *G. crassicaudatus* (pelage conspicuously with rufous, upper incisors pro-odont, nasals parallel-sided), *G. lorraineus* (smaller, GLS < 26.5, pelage often with some rufous) and *G. christyi* (pelage often with some rufous, tail longer > 85 mm without white tipped hairs along its entire length). It is significantly larger than *G. walterverheyeni* (GLS < 23 mm, nasals parallel-sided).

Taxonomic Issues

This species was previously listed as a subspecies of *G. murinus*, but given specific status by Holden (1993, 2005). Its relationship with other species of *Graphiurus* is not known, and a molecular analysis is long overdue.

Distribution

This species has a patchy distribution in the Lower Guinea forest zone with two outliers in the Congo basin.

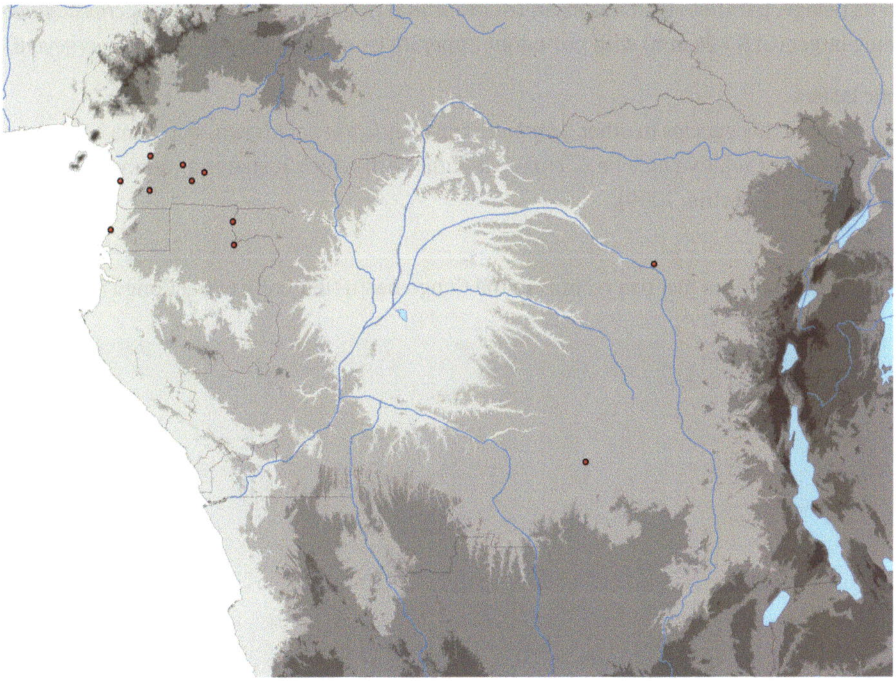

Figure 2.17 Distribution map

Measurements

External and cranial measurements (mm) and mass (g) for *Graphiurus surdus*, sexes combined

	Mean	Min	Max	N
Mass*	25.0	18	34	7
HB*	98.5	87	110	10
Tail*	71.8	65	82	8
HF*	20.4	18	22	14
Ear*	12.5	9	15	13
GLS*	27.7	26.5	29.4	11

*Holden & Levine (2009)

Graphiurus walterverheyeni Holden & Levine, 2009

Diagnosis

This forest zone species is relatively distinct due to its diminutive size and parallel-sided nasals, but could be confused with the following species with which it shares its range: *G. lorraineus* (larger, GLS > 23 mm, nasals broaden distally), *G. christyi* (larger, GLS > 26.5 mm, HF > 16 mm, tail longer > 85 mm, nasals broaden distally), *G. surdus* (larger, GLS > 26.5 mm, HB > 18 mm, nasals broaden distally, pelage grey-brown). It closely resembles *Graphiurus crassicaudatus* in having rufous in its pelage, parallel-sided nasals and supraorbital ridges, but differs from *G. crassicaudatus* by being larger (GLS > 24 mm) with pro-odont upper incisors (orthodont in *G. walterverheyeni*).

Taxonomic Issues

The relationship of this species to other *Graphiurus* species is not yet known, but it has been speculated that its closest relative may be *G. crassicaudatus* with which it shares a number of cranial characters (Holden & Levine, 2009).

Distribution

This species is known from just two records in the Congo basin (left bank of the Congo River).

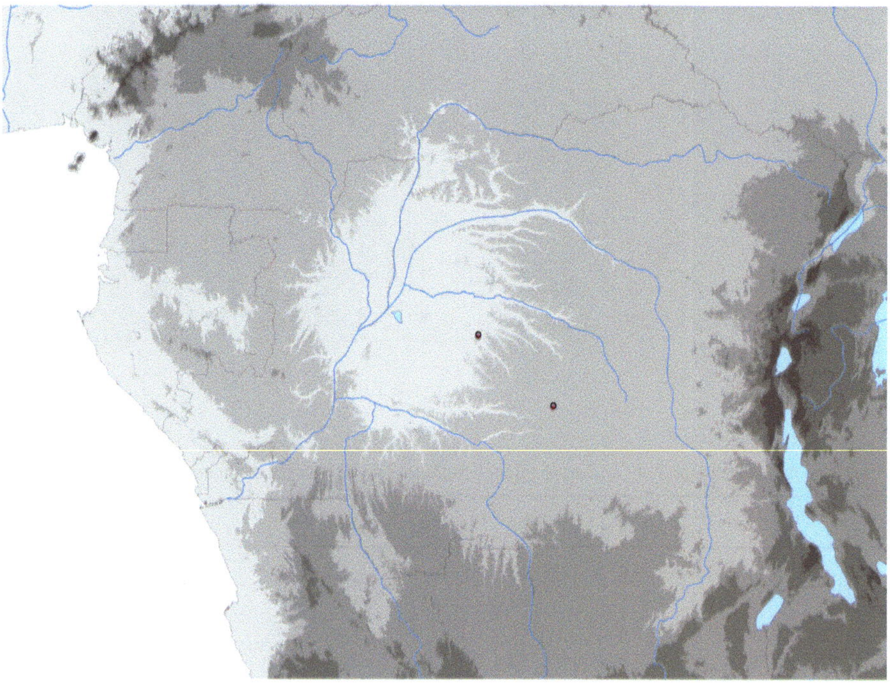

Figure 2.18 Distribution map

Measurements

External and cranial measurements (mm) and mass (g) for *Graphiurus walterverheyeni*, sexes combined

	Mean	Min	Max	N
Mass*	-	-	-	-
HB*	72.0	-	-	1
Tail*	64.0	-	-	1
HF*	14.5	14	15	2
Ear*	11.3	11	11.5	2
GLS*	22.7	22.5	22.9	2

*Holden & Levine (2009)

Family: Sciuridae

This family includes the familiar squirrels and is represented by 36 species in Sub-Saharan Africa.

Key to Genera*

1	Bony palate extends well beyond the end of the molar toothrow; masseteric tubercule well developed; predominantly ground-dwelling	*Xerus*
	Bony palate ending at, or extending marginally beyond, the molars; masseteric tubercule reduced or absent; ground-dwelling or arboreal	2
2	Size tiny (HB < 75 mm, GSL < 25 mm)	*Myosciurus*
	Size larger (mostly far larger)	3
3	Bony palate extending marginally beyond molars; predominantly terrestrial	*Epixerus*
	Bony palate ending at the molars; predominantly arboreal	4
4	Size very large (GSL > 58 mm; HF > 55 mm)	5
	Size smaller (GSL < 56 mm; HF < 54 mm)	6
5	Pelage uniformly dark brown; tail uniformly dark brown; no sharp boundary between dorsal pelage and ventral pelage	*Allosciurus*
	Pelage speckled; tail banded with black and white rings; ventral surface almost hairless	*Protoxerus*
6	Masseteric tubercule small but clearly present; 4 teeth in upper toothrow (except *H. ruwenzorii* which has 5); no stripes or bands on flanks	*Heliosciurus*
	Masseteric tubercule small and insignificant; 5 teeth in upper toothrow; stripes or bands present in some species	7
7	Cheek teeth flat-crowned in adults; tail typically held above back	*Funisciurus*
	Cheek teeth cuspidate in adults; tail not held above back	*Paraxerus*

*The exotic species *Sciurus carolinensis* has been introduced to Cape Town, South Africa where it is the only species of squirrel within its range

Genus Account

Allosciurus Conisbee, 1953 (Slender-tailed Squirrels)

Diagnosis

This monotypic genus is readily distinguished from all other African squirrels by its large size (about 400-500 g), uniform dark brown pelage, and very long, slender tail. It is typically squirrel-shaped with a slender body, relatively long legs and very long tail. The pelage is short, soft and uniform dark brown without any obvious flecking or grizzling, being the same dorsally and, occasionally, slightly paler ventrally. There is never an orange wash on the limbs, chest and face, and a lateral band is absent. The ears are relatively long. There are five digits on the hindfoot and four digits on the forefoot, all with sharp, curved claws. The tail is significantly longer than the body (ca. 125% of HB) and not bushy; individual hairs are far shorter than in *Epixerus* and *Protoxerus*, dark brown at the base with black tips giving the tail a distinctly slender and black appearance. There are four pairs of nipples (1 + 1 + 1 + 1 = 8) placed along the chest and abdomen.

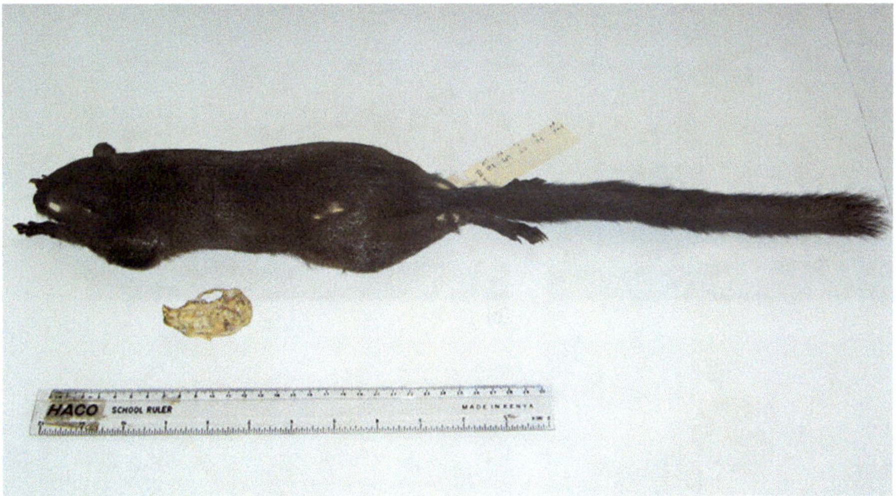

Figure 2.19 Photograph of *Allosciurus aubinnii* museum specimen (NMK 306) (Ara Monadjem).

Cranial Characters

The skull is large (about 59 mm in greatest length) and characteristically domed, curving down both anteriorly and posteriorly. In contrast to *Epixerus, Protoxerus* and *Heliosciurus*, the rostrum is obviously down-curved. The palate extends to the posterior margin of the third molars, or just beyond. The zygomatic arch is long and narrow, thickening in the middle. A supraorbital notch is present, being open to the exterior (and not an enclosed foramen), and the post-orbital process is well developed. The anterior palatal foramina are short and narrow. The bullae are relatively large, being slightly longer than the molar toothrow.

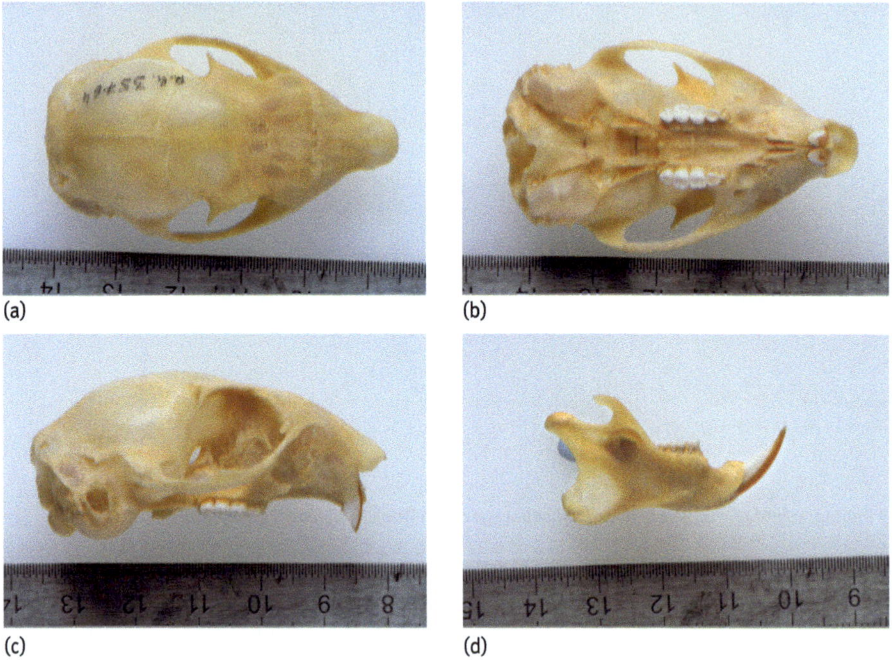

Figure 2.20 a-d Cranium of *Allosciurus aubinnii* (RMCA RG 35764) (from top left to bottom right: dorsal view, ventral view, lateral view and lower jaw).

Dentition

Allosciurus has large, ungrooved, orthodont (or slightly pro-odont) incisors. The dental formula is: 1023/1013 (Rosevear, 1969). The first upper premolar is reduced to a peg, resulting in four functional cheekteeth being present in each half jaw. The second premolar and molars are all roughly the same size, each with four transverse ridges.

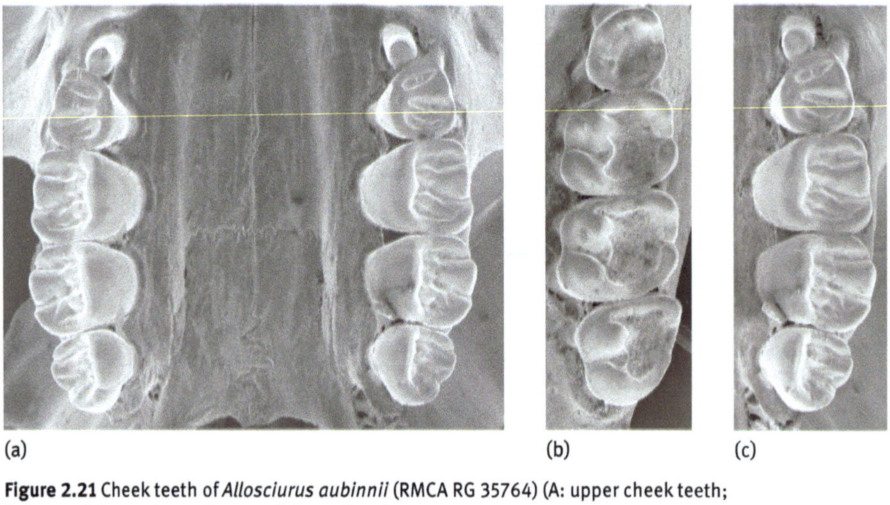

Figure 2.21 Cheek teeth of *Allosciurus aubinnii* (RMCA RG 35764) (A: upper cheek teeth; B: lower right toothrow; C: upper left toothrow).

Habitat & Biogeography

Allosciurus has a narrow distribution, being endemic to the Upper Guinea rainforest zone of West Africa. It occurs from west of the Volta River, Ghana, westward through Ivory Coast, Liberia to the Gola Forest on the Sierra Leone-Liberia border, and marginally into south-eastern Guinea. It is a lowland rainforest specialist, and said to be associated with *Raphia* palms, on which the species may feed (Rosevear, 1969).

Fossil Record

There are currently no known fossils of *Allosciurus* (Winkler et al., 2010).

Biology & Ecology

The biology of this species is practically unknown, and purely based on anecdotal observations, most of which were reported either by Cansdale (in 1938) or by Büttifoker (in the 1880s), and summarised by Rosevear (1969). A dedicated field study on the ecology of this species is long overdue.

Systematic Notes

The status of this monospecific genus has been under dispute. It was originally described in the genus *Myrsilus*, but this name was already occupied by a hemipteran (Rosevear, 1969), and was then assigned to *Allosciurus*. Thorington & Hoffmann (2005), placed it in the genus *Protoxerus*, but this overlooks the many unique features of *Allosciurus aubinnii*, including its uniformly dark-coloured pelage, dense ventral fur, very long and dark tail, greatly curved skull and rostrum, and the presence of an additional (greatly reduced) premolar anterior to the last premolar (see Rosevear, 1969 for a full account of the differences). We have thus assigned it to a separate genus from *Protoxerus*. Two subspecies have been described (*A. a. aubinnii* from Ghana and Ivory Coast, and *A. a. salae* from Liberia, presumably divided by the Sassandra River in western Ivory Coast), but the phylogenetic and systematic relationships of these have not been studied.

Species List

One species described within the genus *Allosciurus*, restricted to West Africa.
Allosciurus aubinnii (Gray, 1873)

Species Account

Allosciurus aubinnii (Gray, 1873)

Diagnosis

See genus account.

Taxonomic Issues

See genus account.

Distribution

This species is restricted to the Upper Guinea Forest of West Africa.

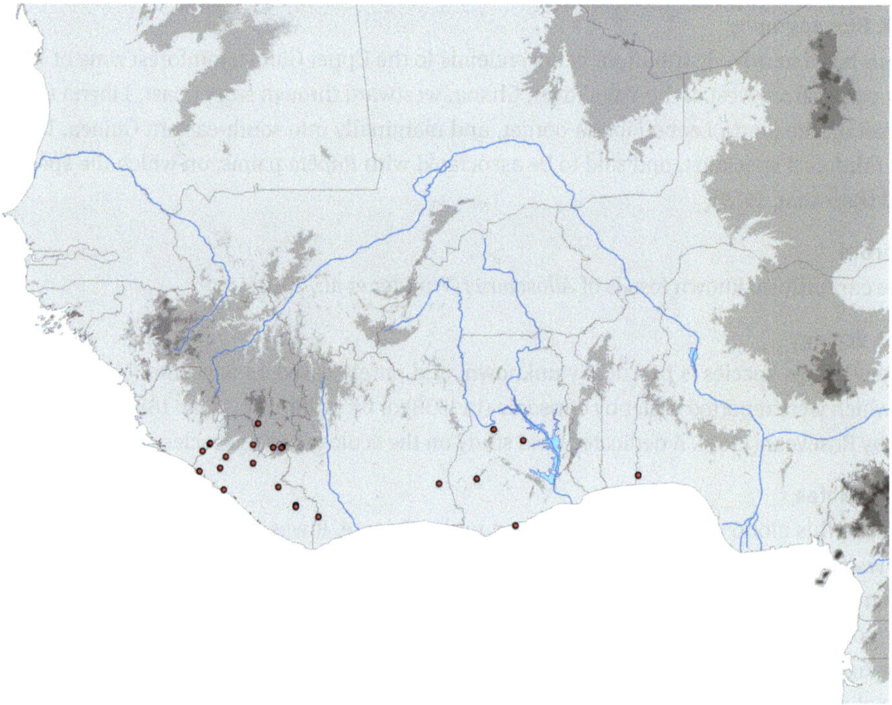

Figure 2.22 Distribution map

Measurements

External and cranial measurements (mm) and mass (g) for *Allosciurus aubinnii*, sexes combined

	Mean	Min	Max	N
Mass	470.7	429	498	3
HB	253	240	267	3
Tail	295	273	310	3
HF	54.5	52	57	3
Ear	20.1	19	21	3
GLS	59.2	58.5	60.1	3

Genus Account

Epixerus Thomas, 1909 (Western Palm Squirrels)

Diagnosis

This monotypic genus is characteristically large (Mass ca. 500-650 g), second only to *Protoxerus*, from which it can be distinguished by its rich orange-brown coloured forelimbs and its terrestrial tendencies. It is typically squirrel-shaped with a slender body, relatively long legs and long tail. The pelage is short, soft and a rich dark brown flecked with white dorsally, and sparse and pale orange-brown ventrally. There is an orange wash on the limbs, chest and face. A lateral band is never present. The ears are relatively long. There are five digits on the hindfoot and four digits on the forefoot, all with sharp, curved claws. The tail is subequal to body length and bushy; individual hairs are long, with red, black and grey bands giving the tail an indistinct black and white banded appearance dorsally. There are four pairs of nipples (1 + 1 + 1 + 1 = 8) placed along the chest and abdomen.

Figure 2.23 Photograph of *Epixerus ebii* museum specimen (NMK 432) (Ara Monadjem).

Cranial Characters

The skull is large (about 68 mm in greatest length) but, compared with *Protoxerus*, relatively delicate. It is flat dorsally, but sloping down posteriorly towards the occipital. The rostrum is relatively long, and the palate extends posteriorly well beyond the third molars. The zygomatic arch is long and narrow. A supraorbital foramen is present, and the post-orbital process is well developed. The anterior palatal foramina are short and narrow. The bullae are relatively small, being the same length as the molar toothrow.

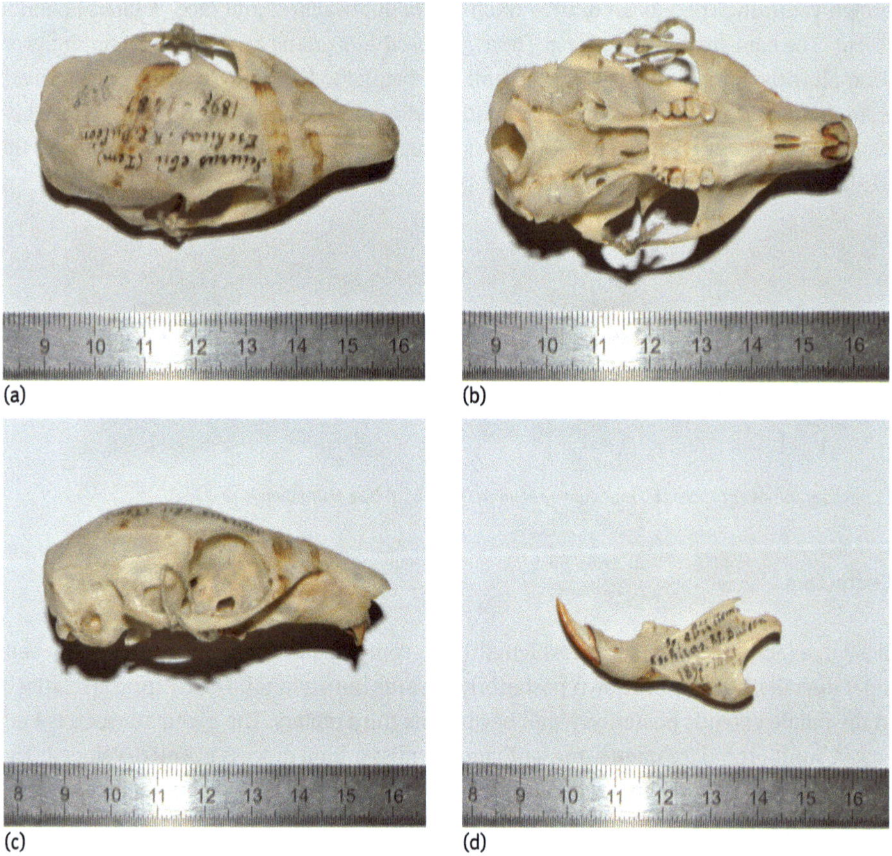

Figure 2.24 a-d Cranium of *Epixerus ebii* (MNHN-ZM-MO 1897-1481) (from top left to bottom right: dorsal view, ventral view, lateral view and lower jaw).

Dentition

Epixerus has large, ungrooved, orthodont (or slightly opisthodont), orange incisors. The dental formula is: 1013/1013 (Rosevear, 1969), resulting in four cheekteeth being present in each half jaw. The premolar and molars are all roughly the same size with four transverse ridges and a longitudinal ridge on the inner side of each molar.

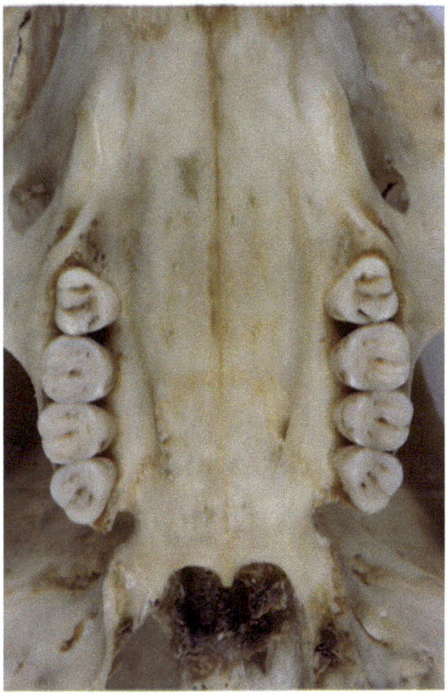

Figure 2.25 Upper cheek teeth of *Epixerus ebii* (MNHN-ZM-MO 1876-2036).

Habitat & Biogeography

Epixerus has a relatively narrow distribution, occurring in two widely isolated pockets: Upper Guinea rainforest from west of the Volta, Ghana, though Ivory Coast, Liberia and Sierra Leone; and in the Lower Guinea rainforest south of the Sanaga, Cameroon through Gabon and Congo. It is closely associated with lowland rainforest, but has been recorded up to 1000 m above sea level in Sierra Leone (Grubb et al., 1998). In stark contrast to *Protoxerus*, this species is strictly terrestrial, foraging on the ground and only climbing up a tree to reach its nest (in a tree hollow usually about 8 m above the ground) (Emmons, 1980).

Fossil Record

There are currently no known fossils of *Epixerus* (Winkler et al., 2010), but *Kubwaxerus pattersoni*, described from the Late Miocene, Kenya (Cifelli et al., 1986) may be related to it.

Biology & Ecology

The biology of this species is poorly known. It feeds on hard shelled nuts and fruits (Emmons, 1980). Home range size for three radio-tracked individuals was 14-22 ha, and activity was highest in the mornings (Emmons, 1980). Litter size, based on a sample size of one, is two (Emmons, 1979a).

Systematic Notes

Based on a molecular phylogeny, the closest relative of *Epixerus* is *Protoxerus*, and both are placed in the tribe Protoxerini with all other Sub-Saharan African squirrels, except *Xerus* (Mercer & Roth, 2003). Currently only a single species is recognised, *Epixerus ebii*, which occurs in two isolated populations separated by more than a thousand kilometres. Three subspecies have been described: *E. e. jonesi* in Liberia and Sierra Leone; *E. e. ebii* in Ivory Coast and Ghana; and *E. e.*

wilsoni in Cameroon, Gabon, Equatorial Guinea and Congo (Happold, 2013). Genetic structure in this species is worthwhile examining, as we suspect that at least the population to the south of the Sanaga River, Cameroon, may represent a distinct species.

Species List

One species described within the genus *Epixerus*, restricted to Sub-Saharan Africa.
Epixerus ebii (Temminck, 1853)

Species Account

***Epixerus ebii* (Temminck, 1853)**

Diagnosis

See genus account.

Taxonomic Issues

See genus account.

Distribution

This species is patchily distributed in the Upper and Lower Guinea forests of West and Central Africa, but absent from Togo and Nigeria.

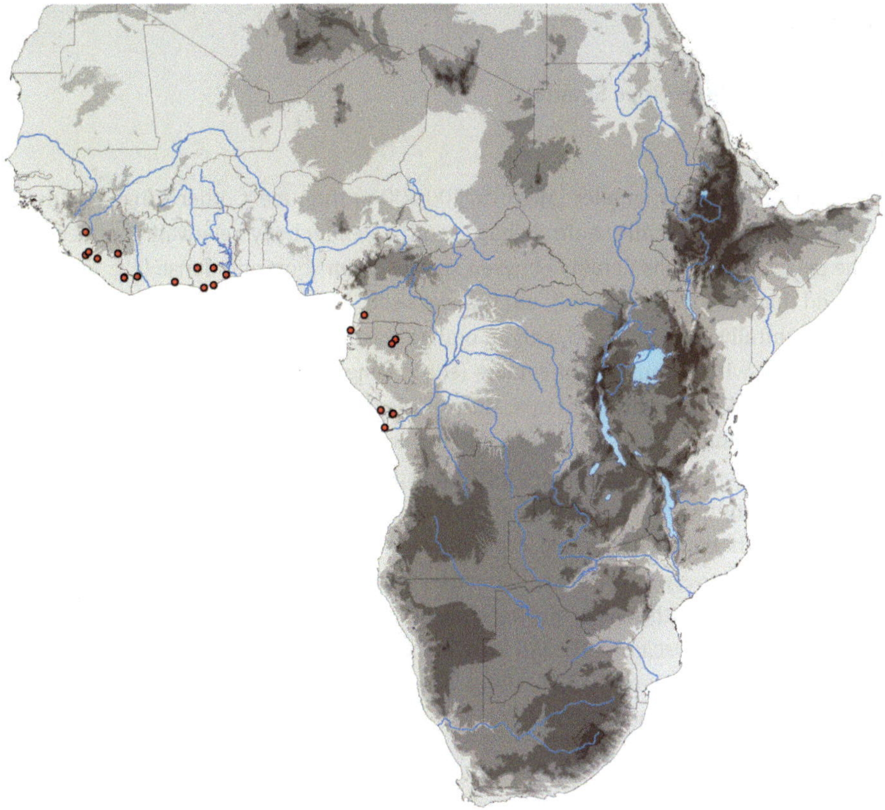

Figure 2.26 Distribution map

Measurements

External and cranial measurements (mm) and mass (g) for *Epixerus ebii*, sexes combined

	Mean	Min	Max	N
Mass	470.3	346	543	4
HB	263.8	240	280	4
Tail	283.8	260	320	4
HF	65.1	62.3	68.3	4
Ear	19.4	18	20	4
GLS	68.4	64.9	70.5	4

Genus Account

Funisciurus Trouessart, 1880 (Rope Squirrels)

Diagnosis

This diverse genus is typically smaller (Mass ca. 100-300 g) than most other African squirrels. Members of this genus are typically squirrel-shaped with a slender body, relatively short legs, and long tail. The pelage varies considerably between species, from short and dense (e.g. *F. anerythrus*), to long and thick (e.g. *F. carruthersi*), but is always soft. The fur is often a bright brown dorsally, and paler ventrally often suffused with an orange wash (e.g. *F. pyrropus* and *F. leucogenys*) or is dull green-olive without any bright colours (e.g. *F. congicus* and *F. carruthersi*). Individual hairs typically have dark bases with light tips. Most species have one or more longitudinal stripes or bands on the back, which may be very bold and distinct (e.g. *F. isabella* and *F. lemniscatus*), or faded (*F. anerythrus*) or absent (*F. carruthersi*). The ears are moderate in size and rounded. There are five digits on the hindfoot and four digits on the forefoot, all with sharp and curved claws. The tail is variable in length, being shorter (e.g. *F. pyrropus*) or equal (e.g. *F. substriatus*) to the length of the body, and typically very bushy. Its colour varies between species and individuals within a species, but is usually dark with some white flecking, but not generally forming alternating bands of black and white as in *Heliosciurus*. The tail is frequently held forwards over the back and neck, and is flicked when the animal is excited or calling. There are typically two pairs of nipples (0 + 0 + 1 + 1 = 4) placed along the abdomen, but in *F. leucogenys*, there are four pairs of nipples (2 + 2 = 4).

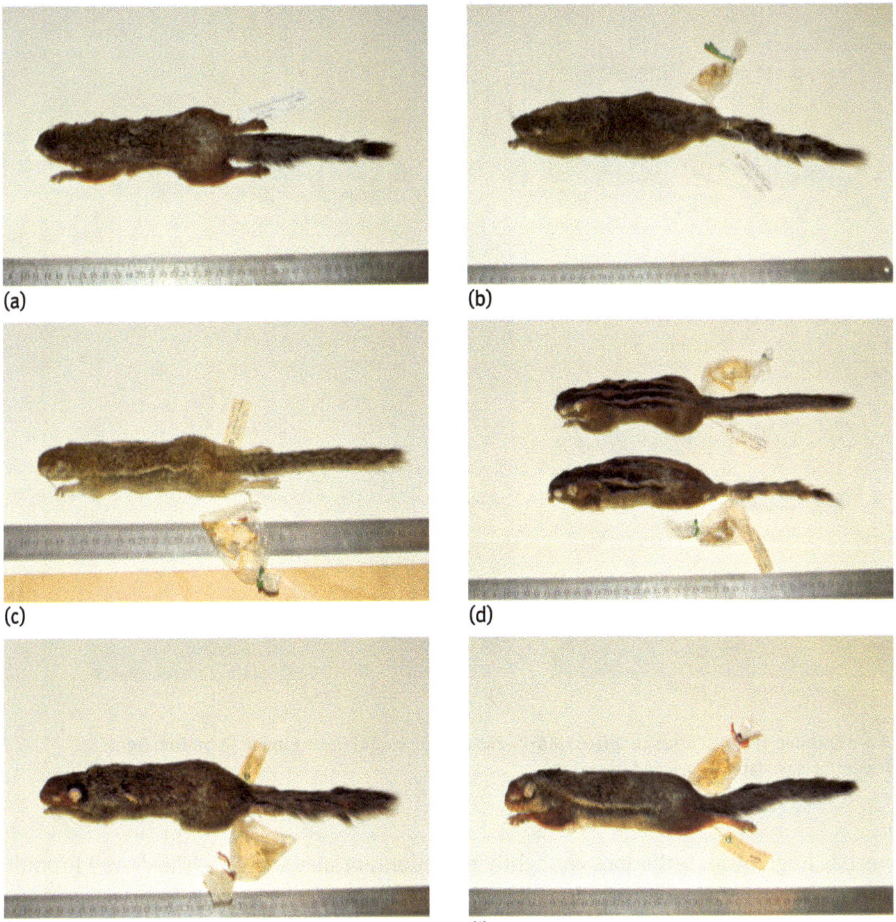

(a) (b) (c) (d) (e) (f)

Figure 2.27 a-f Photograph of *Funisciurus* museum specimens (MNHN), from top left to bottom right: *F. anerythrus, F. carruthersi, F. congicus, F. isabella & F. lemniscatus, F. leucogenys*, and *F. pyrropus* (Ara Monadjem)

Cranial Characters

The skull is small to medium in size (mostly 38-52 mm in greatest length) but less robust than *Heliosciurus* with which it overlaps in greatest length. It is domed, curving down posteriorly towards the occipitals. The rostrum slopes down but is relatively flat. The palate ends at, or slightly before, the posterior margin of the third molars. The zygomatic arch is long and narrow. The supraorbital foramen is absent, but the post-orbital process is well developed. The anterior palatal foramina are short and narrow. The bullae are relatively large, being longer than the molar toothrow.

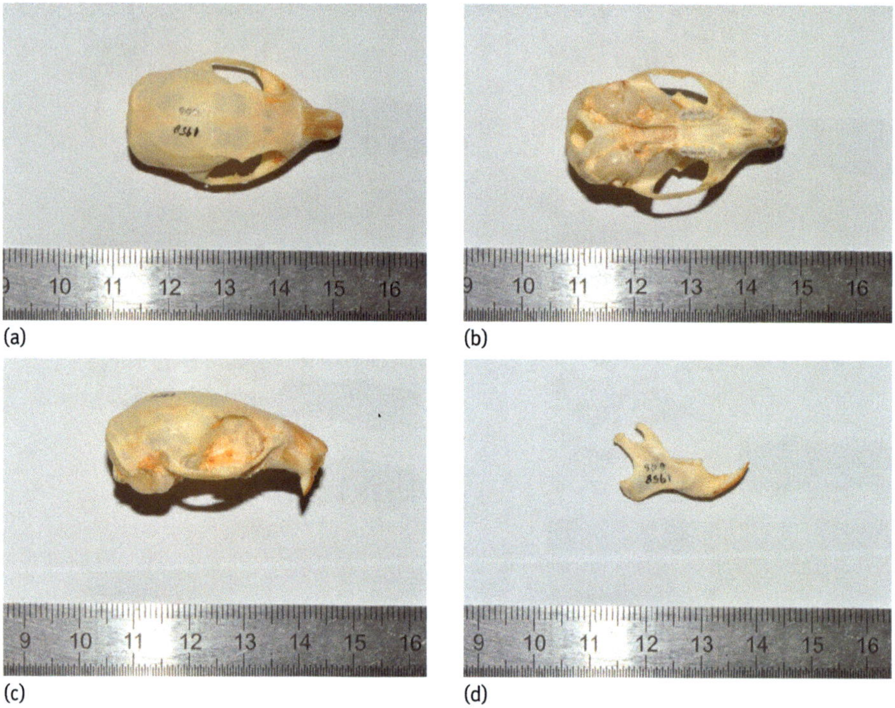

Figure 2.28 a-d Cranium of *Funisciurus congicus* (MNHN-ZM-MO 1958-606) (from top left to bottom right: dorsal view, ventral view, lateral view and lower jaw).

Dentition

Funisciurus has ungrooved, orthodont or slightly pro-odont, orange incisors. The dental formula is: 1023/1013 (Rosevear, 1969), with the first premolar being reduced in size resulting in four functional cheekteeth being present in each half jaw. The second premolar and molars are all roughly the same size with four transverse ridges and a longitudinal ridge on the inner side of each molar.

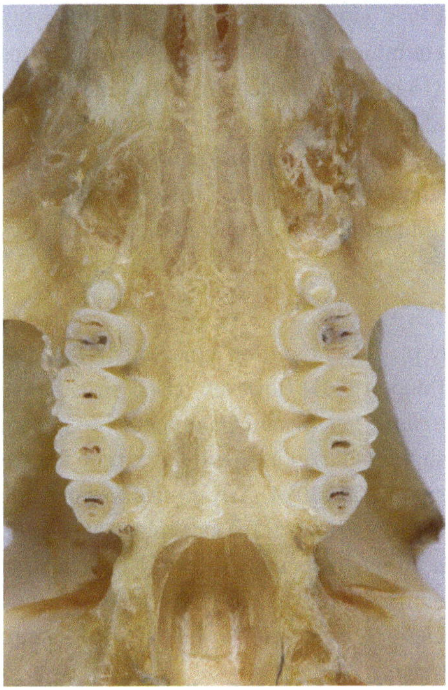

Figure 2.29 Upper cheek teeth of *Funisciurus pyrropus* (MNHN-ZM-MO 1965-104).

Habitat & Biogeography

Funisciurus has a wide distribution across the rainforest belt of tropical Africa. It occurs from The Gambia in the west through the extreme south of Senegal to Guinea, Sierra Leone and Liberia, and then east to southern Nigeria and Cameroon to the Lower Guinea forest and Congo basin. One species (*F. congicus*) continues south to southern Angola and northern Namibia in southern Africa. Only two species (*F. pyrropus* and *F. carruthersi*) occur marginally east of the Albertine Rift. It is closely associated with rainforest, but may also occur in a variety of secondary and degraded forests, plantations and agricultural fields on the edges of forest (Emmons, 1980; Angelici & Luiselli, 2005). One species (*F. anerythrus*) specialises in occupying seasonally inundated shores of rivers and islands within larger tropical rivers. *Funisciurus* occurs from sea level up to 2,800m (*F. carruthersi*). Habitat use varies between species with some being primarily terrestrial or associated with the undergrowth (e.g. *F. pyrropus* and *F. lemniscatus*) or occurring in mid-stratum (e.g. *F. anerythrus* and *F. isabella*), but the forest canopy is avoided or used less than lower strata (Emmons, 1980; Gautier-Hion et al., 1980; Brugiere et al., 2005). They roost in burrows (e.g. *F. pyrropus*) or more frequently in nests constructed out of leaves and situated from ground level (in roots of trees) to several metres above ground (Rosevear, 1969; Kingdon, 1974; Emmons, 1980).

Fossil Record

There are currently no known fossils of *Funisciurus* (Winkler et al., 2010).

Biology & Ecology

A comprehensive study of the ecology of nine forest squirrels was conducted in Gabon (Emmons, 1980). Up to four *Funisciurus* species co-occur together at a single site, differing in body size and habitat use (Emmons, 1980), and up to five may possibly co-occur in southern Cameroon. *Funis-*

ciurus species are mostly frugivorous, but also take leaves, gum, flowers and insects in varying proportions (Emmons, 1980). Home range has been determined by radio-tracking for two species (*F. pyrropus* and *F. lemniscatus*), and is typically small (1-5 ha), and overlapping that of conspecific individuals (Emmons, 1980; Happold, 2013). The reproductive biology of *Funisciurus* species is not well known. Litter size is 1-3, but mostly 1-2, and the gestation and lactation periods of 52 and 50 days are known for only *F. congicus* (Emmons, 1979a; Viljoen & du Toit, 1985). In the latter species, offspring reached 90% of adult mass after 100 days, and the interval between births was 60 days (Viljoen & du Toit, 1985).

Systematic Notes

On morphological grounds, this genus was previously placed in the tribe Funambulini together with the Indian *Funambulus* and African *Paraxerus* (Moore, 1959) However, based on a molecular phylogeny, the closest relative of the genus *Funisciurus* is *Paraxerus*, and these two genera together with *Protoxerus* and *Heliosciurus* form a well supported monophyletic clade (tribe Protoxerini) with *Funambulus* not included in this group (Steppan et al., 2004). Ten species are currently recognised in this genus, and species limits are still not fully resolved. For example, *F. duchaillui* was originally described as a subspecies of *F. isabella* until convincingly shown to be a distinct species by Brugiere et al. (2005). The taxon *poolii* (Jentink, 1906) was listed as a full species by Hoffmann et al. (1993), but then relegated to a subspecies of *F. congicus* (Thorington & Hoffmann, 2005). A molecular phylogeny of the genus is much needed.

Species List

Ten species described within the genus *Funisciurus*, all restricted to Sub-Saharan Africa.
Funisciurus anerythrus (Thomas, 1890)
Funisciurus bayonii (Bocage, 1890)
Funisciurus carruthersi Thomas, 1906
Funisciurus congicus (Kuhl, 1820)
Funisciurus duchaillui Sanborn, 1953
Funisciurus isabella (Gray, 1862)
Funisciurus lemniscatus (Le Conte, 1857)
Funisciurus leucogenys (Waterhouse, 1842)
Funisciurus pyrropus (F. Cuvier, 1833)
Funisciurus substriatus de Winton, 1899

Key to Species

1	Flank stripe absent, pelage long and soft, endemic to the Albertine Rift	*F. carruthersi*
	Flank stripe present (conspicuous or inconspicuous)	2
2	Two dark (blackish) flank stripes on each side, bordering a pale stripe	3
	A single pale flank stripe, without conspicuous black border stripes, which may or may not be broken into spots and with or without a single dark border stripe above or below (but not both)	5
3	Larger (GLS > 46 mm), restricted to a narrow region south of the Ogooue River (Gabon)	*F. duchaillui*
	Smaller (GLS < 45 mm)	4
4	Mid-dorsal "stripe" (region between the two inner black flank stripes) dark brown; inner black flank stripe not extending forward to the ears	*H. lemniscatus*
	Mid-dorsal "stripe" (region between the two inner black flank stripes) pale yellow-brown; inner black flank stripe extending forward to the ears	*H. isabella*
5	Bright orange-red or red on sides of face	6
	No red on sides of face (although ventral pelage, extending to throat, may be orange-brown)	7
6	Bright red on limbs and side of face, conspicuous white flank stripe	*F. pyrropus*
	Bright red not extending onto limbs (on side of face only); flank stripe broken into yellow spots	*F. leucogenys*
7	Smaller (GLS < 40 mm); dorsal pelage conspicuously pale yellow-brown	*F. congicus*
	Larger (GLS > 41 mm); dorsal pelage darker	8
8	Flank stripe inconspicuous; restricted to north-east Angola and neighbouring DRC	F. bayonii
	Flank stripe more conspicuous; occurring north of Angola	9
9	Larger (GLS > 46 mm); ventral pelage usually orange or yellow (occasionally grey)	*F. anerythrus*
	Smaller (GLS < 46 mm); ventral pelage never orange or yellow (usually white)	*F. substriatus*

Species Accounts

Funisciurus anerythrus (Thomas, 1890)

Diagnosis

This species is most likely to be confused with *F. pyrropus* from which it can be distinguished by the lack of red on limbs and side of face (in *F. pyrropus*, limbs and side of face reddish), and the presence of a pale brown stripe on its flanks (flank stripe is white in *F. pyrropus*). *Funisciurus leucogenys* differs by having the flank stripe broken into spots and the presence of red on face and cheeks; *F. bayonii* is smaller (GLS < 44 mm), has an indistinct flank stripe, and does not overlap in distribution with *F. anerythrus*; *F. substriatus* closely resembles *F. anerythrus* but distributions do not overlap. This species is the only one in the genus that is restricted to swamp forest.

Taxonomic Issues

Based on external characteristics, this species appears to be closely related to *F. pyrropus*, *F. bayonii* and *F. substriatus*, and the distinction between them is not always clear. A review of this group is necessary.

Distribution

This species occurs widely in the forest zone of tropical Africa but is absent from eastern Nigeria to south-western Cameroon.

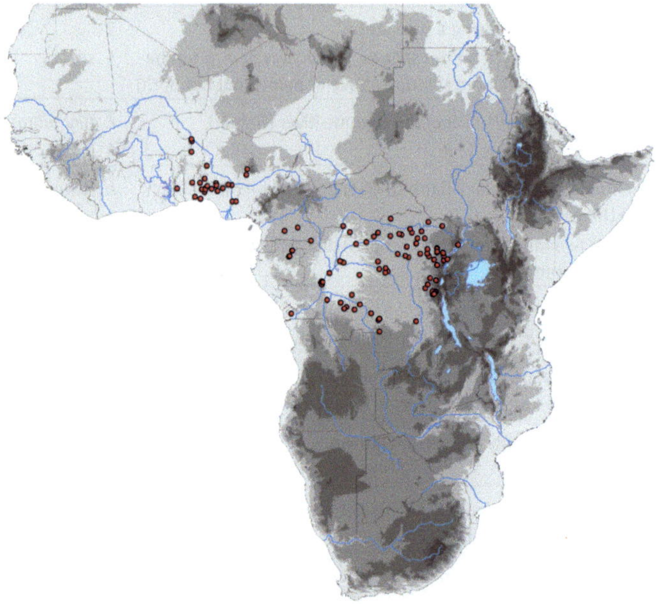

Figure 2.30 Distribution map

Measurements

External and cranial measurements (mm) and mass (g) for *Funisciurus anerythrus*, sexes combined

	Mean	Min	Max	N
Mass*	219.0	190	240	14
HB	179.5	144	197	10
Tail	161.4	143	200	10
HF	43.0	40	48	4
Ear	17.2	12	19	9
GLS	46.1	43.3	47.9	11

*Happold (2013)

Funisciurus bayonii (Bocage, 1890)

Diagnosis

This species is most likely to be confused with *F. anerythrus* from which it can be distinguished by its indistinct flank stripe, smaller size (GLS < 44 mm), and non-overlapping distribution. It lacks any red on the face, distinguishing it from *F. pyrropus* and *F. leucogenys* (the latter has its flank stripe broken into spots). *Funisciurus substriatus* is similar in size and appearance but their distributions are widely spaced apart (*F. substriatus* is restricted to West Africa).

Taxonomic Issues

The relationship of *F. bayonii* to other *Funisciurus* species is not known.

Distribution

This species has a restricted distribution in south-western DRC and north-eastern Angola.

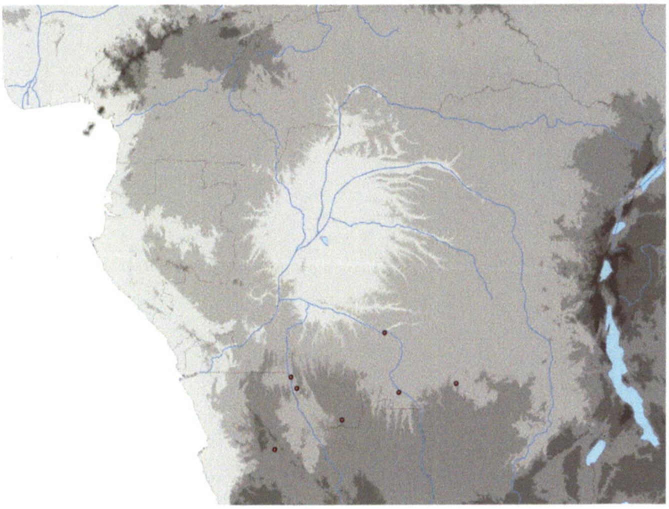

Figure 2.31 Distribution map

Measurements

External and cranial measurements (mm) and mass (g) for *Funisciurus bayonii*, sexes combined

	Mean	Min	Max	N
Mass	-	-	-	-
HB*	178.5	167	190	2
Tail*	174.0	170	178	2
HF*	41.5	41	42	2
Ear	-	-	-	-
GLS**	42.9	41.0	44.6	9

*Happold (2013), **Hayman (1963)

Funisciurus carruthersi Thomas, 1906

Diagnosis

This species is unlikely to be confused with any species of *Funisciurus* due to its long, soft fur and lack of a flank stripe. *Paraxerus poensis* is similar in appearance but smaller (GLS < 44 mm, HF < 42 mm), and does not overlap in distribution. *Heliosciurus ruwenzori* is similar in size but has a tail banded black and white (in *F. carruthersi* tail is dark brown with inconspicuous bands of pale brown).

Taxonomic Issues

It is not clear whether this species belongs in the genus *Funisciurus* or *Paraxerus*, sharing features common to both taxa (Happold, 2013).

Distribution

This species is restricted to the Albertine Rift.

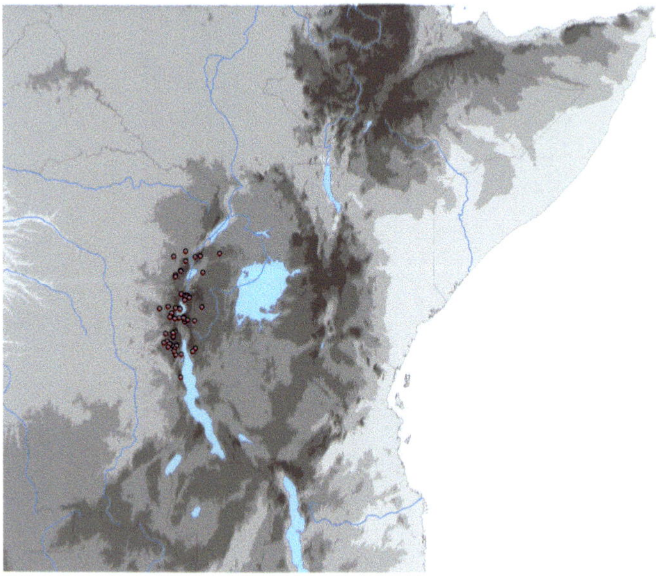

Figure 2.32 Distribution map

Measurements

External and cranial measurements (mm) and mass (g) for *Funisciurus carruthersi*, sexes combined

	Mean	Min	Max	N
Mass*	222.0	200	280	4
HB*	203.0	187	236	7
Tail*	201.0	180	231	6
HF*	48.0	46	50	7
Ear*	19.0	17	20	7
GLS*	48.7	46.4	50.2	7

*Happold (2013)

Funisciurus congicus (Kuhl, 1820)

Diagnosis

This species should not be confused with any others in the genus *Funisciurus* due to its small size (GLS < 40 mm, HF < 42 mm) and straw-brown dorsal pelage. *Funisciurus bayonii* is larger (GLS > 41 mm) and darker in colour.

Taxonomic Issues

None.

Distribution

This species occurs widely in the Congo basin (left bank of the Congo River), south through western Angola and into north-western Namibia.

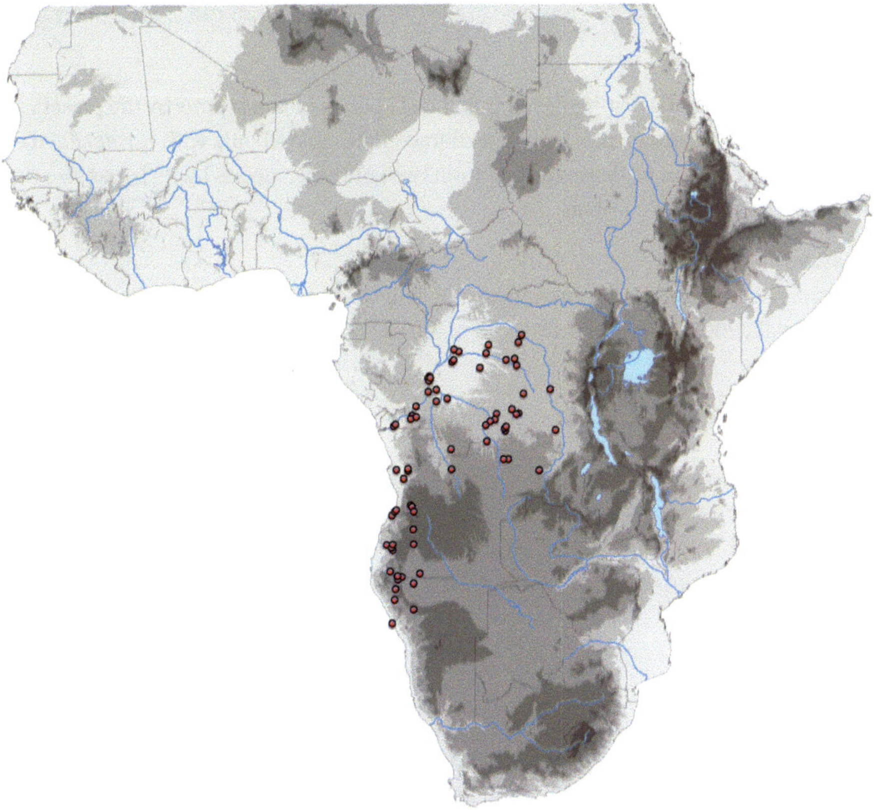

Figure 2.33 Distribution map

Measurements

External and cranial measurements (mm) and mass (g) for *Funisciurus congicus*, sexes combined

	Mean	Min	Max	N
Mass	111.3	100	122	3
HB	150.4	137	165	18
Tail	164.2	130	193	18
HF	38.6	34	42	16
Ear	15.6	11	18	17
GLS	37.9	36.6	39.4	15

Funisciurus duchaillui Sanborn, 1953

Diagnosis

Funisciurus duchaillui is readily distinguished from most species of *Funisciurus* by the presence of two dark flank stripes bordering a pale stripe (a characteristic shared only with *F. isabella* and *F. lemniscatus*). This species can be distinguished from both of these species by its larger size (in *F. isabella* and *F. lemniscatus* GLS < 45 mm).

Taxonomic Issues

This taxon was previously considered a synonym of *F. isabella*, but recently elevated to full species status (Brugiere et al., 2005). It occurs only south of the Ogooue River.

Distribution

This species has a narrow distribution in southern Gabon on the left bank of the Ogooue River.

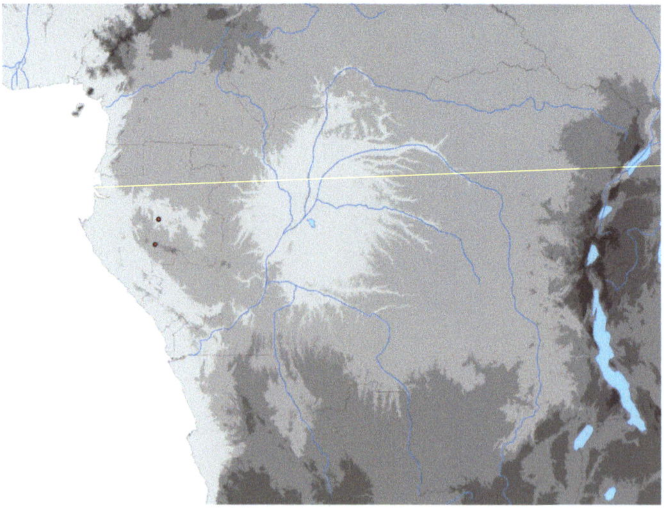

Figure 2.34 Distribution map

Measurements

External and cranial measurements (mm) and mass (g) for *Funisciurus duchaillui*, sexes combined

	Mean	Min	Max	N
Mass*	200.0	180	220	4
HB*	197.0	185	212	4
Tail*	210.0	190	230	3
HF*	45.0	42	50	4
Ear	-	-	-	-
GLS*	47.2	46.9	48.3	3

*Brugiere et al. (2005); min and max values from Happold (2013)

Funisciurus isabella (Gray, 1862)

Diagnosis

Funisciurus isabella is readily distinguished from most species of *Funisciurus* by the presence of two dark flank stripes bordering a pale stripe (a characteristic shared only with *F. lemniscatus* and *F. duchaillui*). This species can be distinguished from *F. lemniscatus* by dorsal pelage (in *F. isabella* the dorsal pelage between the two inner dark stripes is the same pale colour as between the inner and outer flank stripes; in *F. lemniscatus* the dorsal pelage between the two inner dark stripes is brown and contrasts with the pale yellow stripe bordered by the inner and outer flank stripes), and the fact that the inner dark flank stripe extends further forward to behind the ears in *F. isabella* (it terminates posterior to the ears in *F. lemniscatus*). Habitat may also be useful in separating these two species, with *F. lemniscatus* preferring the ground and open understorey of mature forest; whereas *F. isabella* typically occurs in dense thickets, often in secondary or disturbed forest (Emmons, 1980). *Funisciurus duchaillui* is larger (GLS > 46 mm) and does not co-occur with *F. isabella* (Brugiere et al., 2005).

Taxonomic Issues

The taxon *duchaillui* (considered a valid species here) was formerly a synonym of *F. isabella*.

Distribution

This species occurs widely in the Lower Guinea Forest on the right bank of the Ogooue River with an isolated record from Brazzaville, Congo.

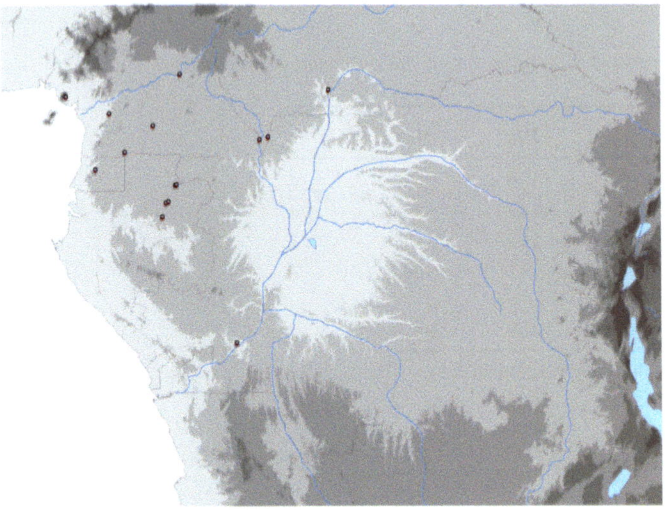

Figure 2.35 Distribution map

Measurements

External and cranial measurements (mm) and mass (g) for *Funisciurus isabella*, sexes combined

	Mean	Min	Max	N
Mass*	110.0	90	116	17
HB	151.3	140	178	6
Tail	176.0	145	199	6
HF	38.0	36	41	6
Ear	13.2	12	14	6
GLS	40.9	39.7	42.0	5

*Emmons (1980); min and max values from Happold (2013)

Funisciurus lemniscatus (Le Conte, 1857)

Diagnosis

Funisciurus lemniscatus is readily distinguished from most species of *Funisciurus* by the presence of two dark flank stripes bordering a pale stripe (a characteristic shared only with *F. isabella* and *F. duchaillui*). This species can be distinguished from *F. isabella* by dorsal pelage (in *F. lemniscatus* the dorsal pelage between the two inner dark stripes is brown and contrasts with the pale yellow stripe bordered by the inner and outer flank stripes; in *F. isabella* the dorsal pelage between the two inner dark stripes is the same pale colour as between the inner and outer flank stripes), and the fact that the inner dark flank stripe extends further forward to behind the ears in *F. isabella* (it terminates posterior to the ears in *F. lemniscatus*). Habitat may also be useful in separating these two species, with *F. lemniscatus* preferring the ground and open understorey of mature forest; whereas *F. isabella* typically occurs in dense thickets, often in secondary or disturbed forest (Emmons, 1980). *Funisciurus duchaillui* is larger (GLS > 46 mm).

Taxonomic Issues

This species occurs in two discrete populations separated by the Ogooue River which are currently recognised as subspecies (*F. l. lemniscatus* north of the river, and *F. l. mayumbicus* south of the river) (Happold, 2013) which may provide to be distinct species.

Distribution

This species occurs widely in the Lower Guinea Forest zone of Central Africa.

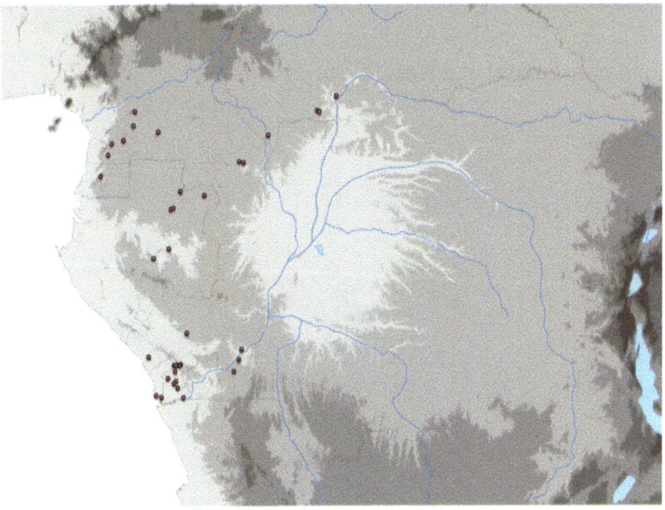

Figure 2.36 Distribution map

Measurements

External and cranial measurements (mm) and mass (g) for *Funisciurus lemniscatus*, sexes combined

	Mean	Min	Max	N
Mass*	141.0	123	158	46
HB	150.2	130	168	6
Tail	169.8	143	200	6
HF	38.8	35	42	6
Ear	13.3	9	16	6
GLS	40.9	38.2	42.6	6

*Emmons (1980), min and max values from Happold (2013)

Funisciurus leucogenys (Waterhouse, 1842)

Diagnosis

This species should be readily distinguishable by the combination of its flank stripe which is broken into spots and the presence of a black postauricular patch. It is most likely to be confused with *F. pyrropus* (both have red on side of face) from which it can be distinguished by lack of red on limbs (in *F. pyrropus*, limbs reddish), and flank stripe broken into spots; furthermore, their distributions overlap only very marginally (eastern Ghana and Togo in the west and Central African Republic in the east). *F. bayonii* is smaller (GLS < 44 mm), has an indistinct flank stripe, and does not overlap in distribution with *F. leucogenys*; *F. substriatus* overlaps in distribution (in Togo and Benin) but is smaller (GLS < 46 mm, HF < 45 mm), lacks the black postauricular patch and does not have red on face.

Taxonomic Issues

The relationship of this species to other *Funisciurus* is not known.

Distribution

This species has a narrow distribution in Central and West Africa.

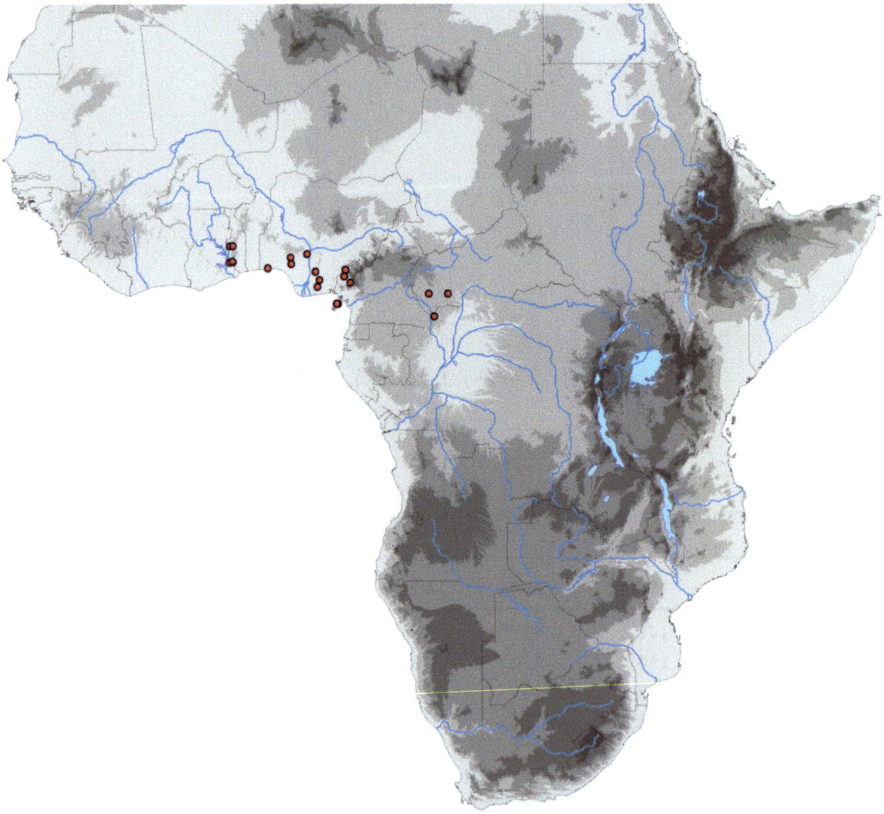

Figure 2.37 Distribution map

Measurements

External and cranial measurements (mm) and mass (g) for *Funisciurus leucogenys*, sexes combined

	Mean	Min	Max	N
Mass*	237.3	171	298	12
HB	211.5	205	218	2
Tail	184.0	168	200	2
HF	46.5	45	48	2
Ear	14.5	14	15	2
GLS	52.3	52.1	52.5	2

*Happold (2013)

Funisciurus pyrropus (F. Cuvier, 1833)

Diagnosis

This species is most likely to be confused with *F. leucogenys* from which it can be distinguished by the lack of red on limbs (in *F. leucogenys*, only face is reddish), and the presence of a white stripe on its flanks (flank stripe is broken into yellow spots in *F. leucogenys*). *Funisciurus anerythrus* and *F. substriatus* differ by having a less distinct flank stripe and the absence of conspicuous red on the face and limbs. *Funisciurus bayonii* is smaller (GLS < 44 mm), has an indistinct flank stripe, and overlaps only marginally in northern Angola and south-western DRC.

Taxonomic Issues

The taxonomy of this species complex is confused and requires urgent revision. At present nine subspecies are recognised (Happold, 2013), and some of these may well prove to be distinct species.

Distribution

This species occurs widely in the forest zone of tropical Africa, extending marginally into Guinea Woodland to the north.

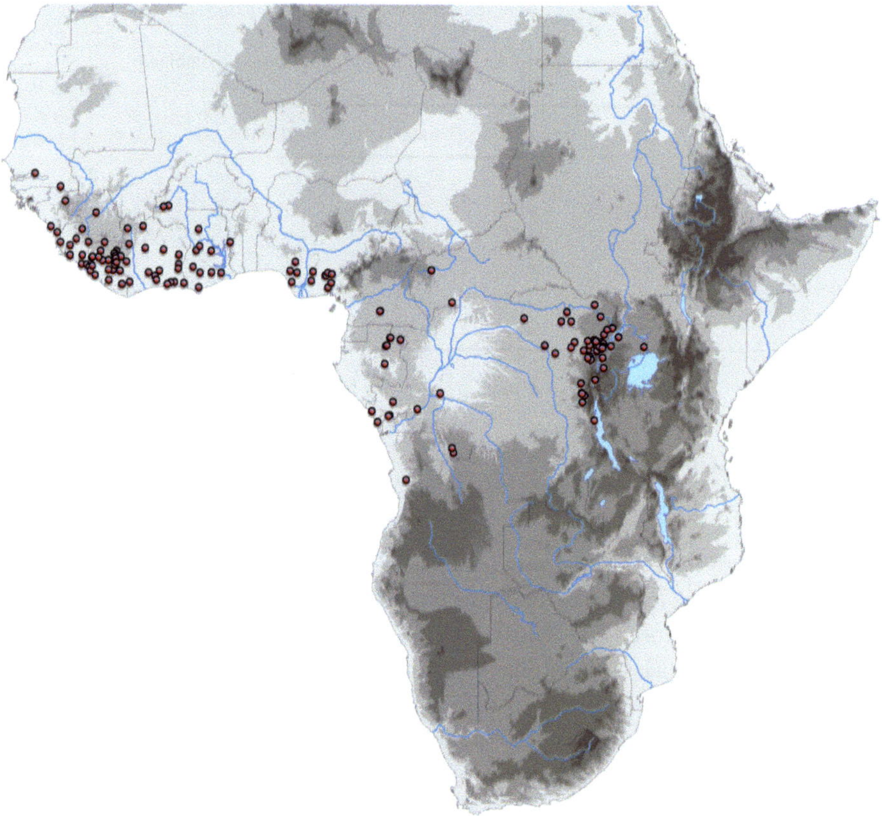

Figure 2.38 Distribution map

Measurements

External and cranial measurements (mm) and mass (g) for *Funisciurus pyrropus*, sexes combined

	Mean	Min	Max	N
Mass	241.6	187	289	8
HB	198.8	183	215	12
Tail	178.0	154	210	12
HF	46.2	41	52	12
Ear	17.8	117	20	12
GLS	49.0	46.7	51.9	13

Funisciurus substriatus **de Winton, 1899**

Diagnosis

This species is most likely to be confused with *F. pyrropus* and *F. anerythrus*. It can be distinguished from the former by its smaller size (GLS < 46 mm), the absence of red on the face and limbs (although there may be a limited amount of reddish-brown on the flanks and legs, it is absent from the face), and an indistinct flank stripe (in *F. pyrropus* flank stripe is very conspicuous); and from the latter by the lack of orange on the ventral pelage (in *F. anerythrus* ventral is usually orange or yellow). *Funisciurus leucogenys* differs by having the flank stripe broken into spots and the presence of red on face and cheeks; *F. bayonii* is similar in size but distributions are widely separated.

Taxonomic Issues

The relationship of this species to other *Funisciurus* (in particular *F. anerythrus* and *F. pyrropus*) is not known.

Distribution

This species has a restricted distribution in the eastern parts of West Africa.

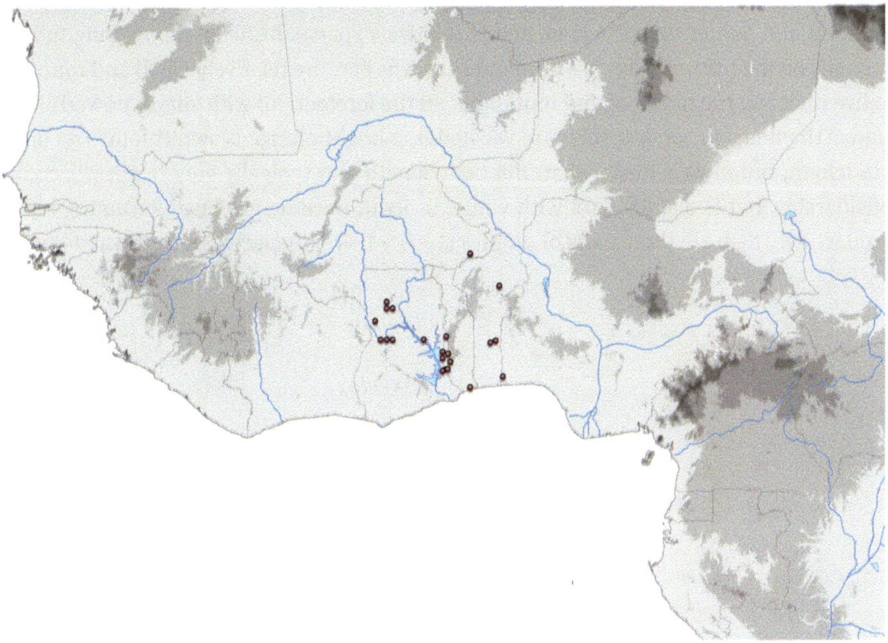

Figure 2.39 Distribution map

Measurements

External and cranial measurements (mm) and mass (g) for *Funisciurus substriatus*, sexes combined

	Mean	Min	Max	N
Mass*	142.5	115	187	28
HB*	171.6	-	-	28
Tail*	154.7	130	178	28
HF*	43.1	40	45	33
Ear*	17.5	13	19	33
GLS*	44.2	42.5	45.3	10

*Happold (2013); HB calculated by substracting Tail from Total Length

Genus Account

Heliosciurus Trouessart, 1880 (Sun Squirrels)

Diagnosis

This diverse genus is typically medium-sized (mostly 150-390 g) compared with other African squirrels. Members of this genus are squirrel-shaped with a slender body, relatively long legs, and long tail. The pelage varies considerably between species and regions, for example being shorter in *H. gambianus* and longer in *H. rufobrachium*, and is harsher than in *Funisciurus* but softer than *Protoxerus*. The fur varies from a bright brown to a dull grey dorsally, but is always flecked or speckled with white or pale shades of brown or grey. The ventral fur varies extensively but is usually uniform and lighter than the back and is relatively sparse. In *Heliosciurus*, longitudinal stripes or bands on the back or sides are never present. The ears are relatively small and rounded. There are five digits on the hindfoot and four digits on the forefoot, all with sharp, powerful and curved claws. The limbs in some species are washed in a bright orange brown colour. The tail is variable in length, but always longer than the body length and typically very bushy. Its colour varies considerably, but is usually dark with white, or light, flecking forming alternating bands of black and white. There are three pairs of nipples (1 + 0 + 1 + 1 = 6) placed along the abdomen.

Figure 2.40 a-e Photographs of *Heliosciurus undulatus* (Bill Stanley) (top) and *Heliosciurus* museum specimens (MNHN), from middle left to bottom right: *H. gambianus* (two specimens), *H. punctatus*, *H. rufobrachium* and *H. undulatus* (Ara Monadjem).

Cranial Characters

The skull is medium-size (mostly 38-56 mm in greatest length) and robust for its size (more so than that of *Funisciurus* of similar size). It is domed, curving down posteriorly towards the occipitals. The rostrum slopes down but is relatively flat. The palate ends at the posterior margin of the third molars. The zygomatic arch is long and narrow, but thickening somewhat in the middle. The supraorbital foramen is absent, but the post-orbital process is well developed. The anterior palatal foramina are short and narrow but relatively longer than in *Funisciurus*. The bullae are moderately large, being only slightly longer than the molar toothrow.

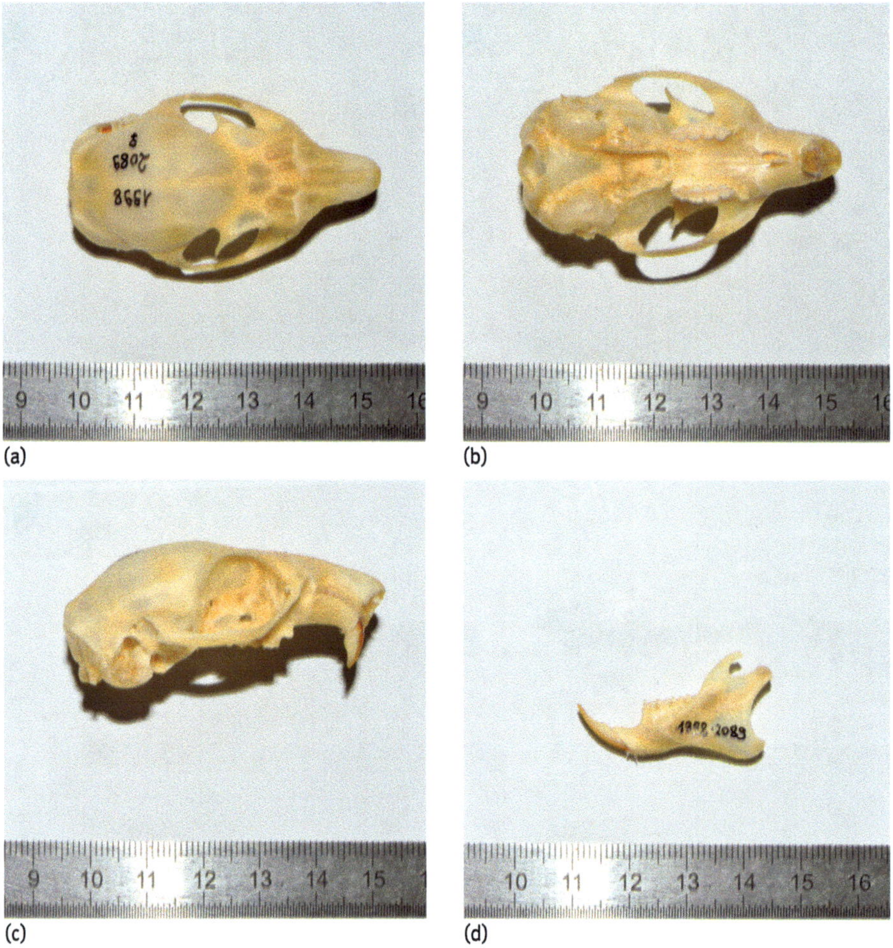

Figure 2.41 a-d Cranium of *Heliosciurus rufobrachium* MNHN- ZM-MO 1998-2089 (from top left to bottom right: dorsal view, ventral view, lateral view and lower jaw).

Dentition

Heliosciurus has ungrooved, orthodont, orange incisors. The dental formula is: 1013/1013 (Rosevear, 1969), resulting in four cheekteeth being present in each half jaw. The premolar and molars are all roughly the same size with four transverse ridges and a longitudinal ridge on the inner side of each molar.

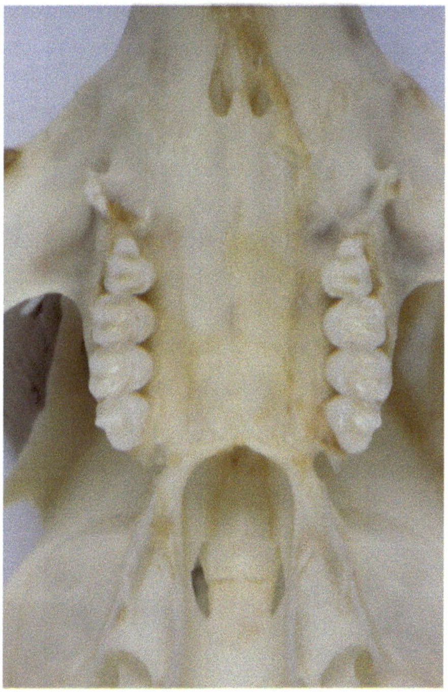

Figure 2.42 Upper cheek teeth of *Heliosciurus punctatus* (MNHN-ZM-MO 1959-723).

Habitat & Biogeography

Heliosciurus has a wide distribution across tropical Africa. It occurs from the extreme north of Senegal, west to Ethiopia and then south to northern Angola, eastern Zimbabwe and southern Mozambique. Habitat use varies between the species, with the majority of species associated with some form of forest or woodland. *Heliosciurus rufobrachium* inhabits lowland rainforest, but may also occur in a variety of secondary and degraded forests, plantations and agricultural fields on the edges of forest (Emmons, 1980). Two species, *H. mutabilis* and *H. ruwenzorii*, are primarily inhabitats of montane forest although the former species also occurs in riparian forest in lowland savanna and in coastal forest (Kingdon, 1974; Skinner & Chimimba, 2005). The remaining species are mostly woodland or savanna species, but penetrating forest where settlements have opened up fields and plantations (Rosevear, 1969). Members of this genus occur from sea level up to 2,700m (*H. ruwenzorii*)(Kingdon, 1974). All *Heliosciurus* species are arboreal, rarely descending to the ground. For example, *H. rufobrachium* typically forage in the middle or upper strata of the forest (Emmons, 1980). They roost in hollows on the trunk of trees (Rosevear, 1969; Kingdon, 1974; Emmons, 1980).

Fossil Record

The oldest, and only, *Heliosciurus* fossil is from the Early Pliocene 4.4 MYA, Kenya (Winkler et al, 2010).

Biology & Ecology

A comprehensive study of the ecology of nine forest squirrels was conducted in Gabon, of which one was a *Heliosciurus* (Emmons, 1980). Similar ecological studies of other *Heliosciurus* species have not yet been conducted. In at least one species (*H. rufobrachium*), male and female pairs seem to forage together (Emmons, 1980). *Heliosciurus* is mostly frugivorous, feeding on fruits and leaves, but also taking insects and a variety of vertebrate prey (Smithers & Wilson, 1979; Emmons, 1980). Litter size is 1-3, but mostly 1-2 (Emmons, 1979a; Hayssen, 2008).

Systematic Notes

Based on a molecular phylogeny, the closest relatives of the genus *Heliosciurus* are members of a clade that include *Funisciurus*, *Protoxerus* and *Paraxerus*; these four genera form a well-supported monophyletic clade and, together with *Myosciurus*, form the tribe Protoxerini (Steppan et al., 2004). This differs from the traditional phylogeny based on morphological characters (Moore, 1959). Six species are currently recognised in this genus, but species limits are still not fully resolved. For example, the two widely separately populations of *H. gambianus* may refer to distinct species. A molecular phylogeny of the genus is long overdue.

Species List

Six species described within the genus *Heliosciurus*, all restricted to Sub-Saharan Africa.
Heliosciurus gambianus (Ogilby, 1835)
Heliosciurus mutabilis (Peters, 1852)
Heliosciurus punctatus (Temminck, 1853)
Heliosciurus rufobrachium (Waterhouse, 1842)
Heliosciurus ruwenzorii (Schwann, 1904)
Heliosciurus undulatus (True, 1892)

Key to Species

1	Red or reddish-brown on limbs, contrasting with duller dorsal pelage	*H. rufobrachium*
	Limbs not red and not contrasting with dorsal pelage	2
2	Conspicuous mid-ventral white stripe (flanked by cream or grey ventral pelage)	*H. ruwenzorii*
	No mid-ventral white stripe (although ventral pelage may be white)	3
3	Dorsal pelage rich brown or rufous-brown; restricted to southern Africa	*H. mutabilis*
	Dorsal pelage duller, usually grey or grey-brown; marginally present in southern Africa	4
4	Larger (GLS > 51 mm); restricted to eastern regions of Kenya and Tanzania	H. undulatus
	Smaller (GLS < 51 mm); absent from eastern regions of East Africa	5
5	Dorsal pelage lighter grey; occurs in savannas	*H. gambianus*
	Dorsal pelage darker brown-grey; occurs in forests	*H. punctatus*

Species Accounts

Heliosciurus gambianus (Ogilby, 1835)

Diagnosis

This species is differentiated from most of its congeners by its lighter grey-brown pelage, typically lacking any bright colours. *Heliosciurus rufobrachium* has red or reddish-brown limbs; *H. mutabilis* and *H. punctatus* have darker pelage; *H. undulatus* is larger (GLS > 51 mm), and distributions do not overlap; *H. ruwenzorii* has pure white stripe on ventral pelage bordered by off-white or cream (no white stripe in *H. gambianus*), and distributions do not overlap.

Taxonomic Issues

This species exhibits significant geographical variation in pelage characteristics and at least four subspecies are currently recognised (Happold, 2013), at least some of which may be shown to be valid species. The southern and northern populations are isolated by the forest belt.

Distribution

This species is widely distributed in the moister savannas of tropical Africa, avoiding rainforest and arid regions.

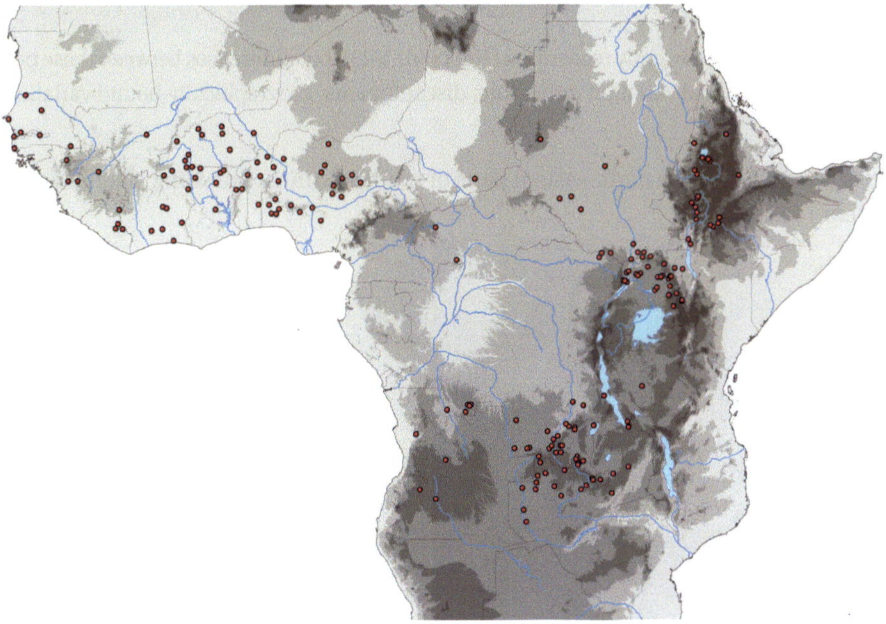

Figure 2.43 Distribution map

Measurements

External and cranial measurements (mm) and mass (g) for *Heliosciurus gambianus*, sexes combined

	Mean	Min	Max	N
Mass	245.6	157	325	8
HB	205.5	188	230	10
Tail	248.0	200	305	10
HF	50.2	45	54	10
Ear	14.8	10	18	10
GLS	47.4	42.5	51.6	17

Heliosciurus mutabilis (Peters, 1852)

Diagnosis

This species is differentiated from its congeners by its rich brown (rufous-brown) dorsal pelage. It is the only species of *Heliosciurus* within its range. *Heliosciurus gambianus* and *H. undulatus* have a duller grey-brown dorsal pelage.

Taxonomic Issues

At least five subspecies are currently recognised (Happold, 2013), but differences between these populations may represent clinal variation rather than discontinuous, geographically-bound variation.

Distribution

This species is restricted to the north-eastern parts of southern Africa, extending marginally into south-western Tanzania.

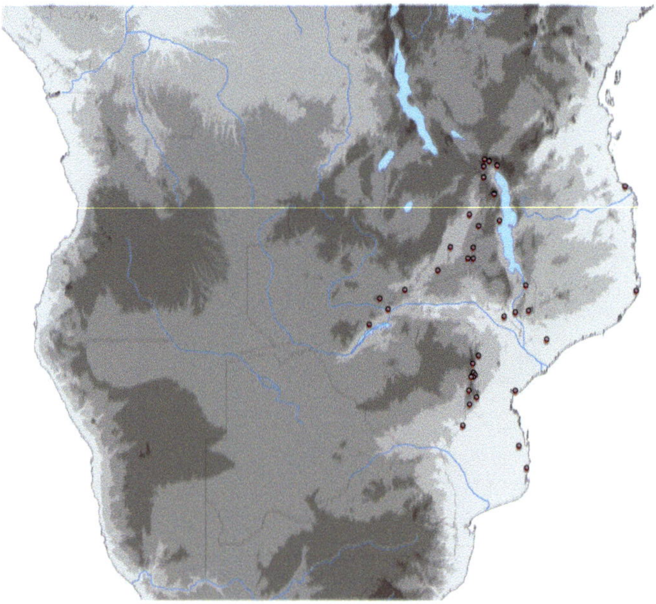

Figure 2.44 Distribution map

Measurements

External and cranial measurements (mm) and mass (g) for *Heliosciurus mutabilis*, sexes combined

	Mean	Min	Max	N
Mass	356	-	-	1
HB	225.9	205	243.7	6
Tail	263.1	230	291	6
HF	55.0	48	60	5
Ear	17.5	15	20	5
GLS	54.2	50.5	56.3	9

Heliosciurus punctatus (Temminck, 1853)

Diagnosis

Only three species of *Heliosciurus* (*H. gambianus, H. punctatus* and *H. rufobrachium*) occur in West Africa. *Heliosciurus punctatus* can be differentiated from *H. rufobrachium* by the lack of red on its limbs, and from *H. gambianus* by its darker ventral pelage (in *H. gambianus*, ventral pelage white or off-white, grey-brown in *H. punctatus*) and non-overlapping distribution (*H. gambianus* occurs in savanna habitats to the north and *H. punctatus* in forests to the south).

Taxonomic Issues

This species was previously included as a subspecies of *H. gambianus*.

Distribution

This species is restricted to the Upper Guinea Forest of West Africa.

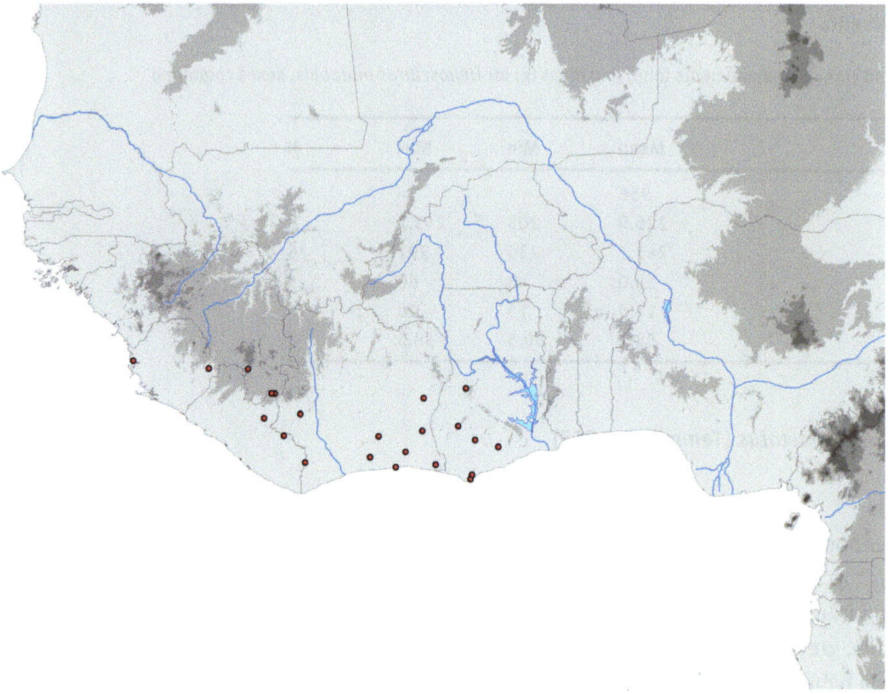

Figure 2.45 Distribution map

Measurements

External and cranial measurements (mm) and mass (g) for *Heliosciurus punctatus*, sexes combined

	Mean	Min	Max	N
Mass	145.5	119	163	4
HB	186.7	155	206	7
Tail	227.3	202	270	7
HF	46.7	44	51	7
Ear	15.2	13	17	5
GLS	45.0	41.3	48.0	6

Heliosciurus rufobrachium (Waterhouse, 1842)

Diagnosis

This species is differentiated from its congeners by the presence of bright red or reddish-brown on its limbs which contrasts with its brown dorsal pelage (although the intensity of the red varies considerably, for example being almost absent in the subspecies *H. r. isabellinus* from Nigeria and Togo). *Heliosciurus mutabilis* may have reddish hues to its dorsal pelage, but limbs are the same colour as the back, and ranges do not overlap.

Taxonomic Issues

This species is in urgent need of revision. At least 14 subspecies are currently recognised (Happold, 2013), which show considerable variation in pelage colour.

Distribution

This species is widespread in the forest zone of tropical Africa, extending marginally into western Kenya.

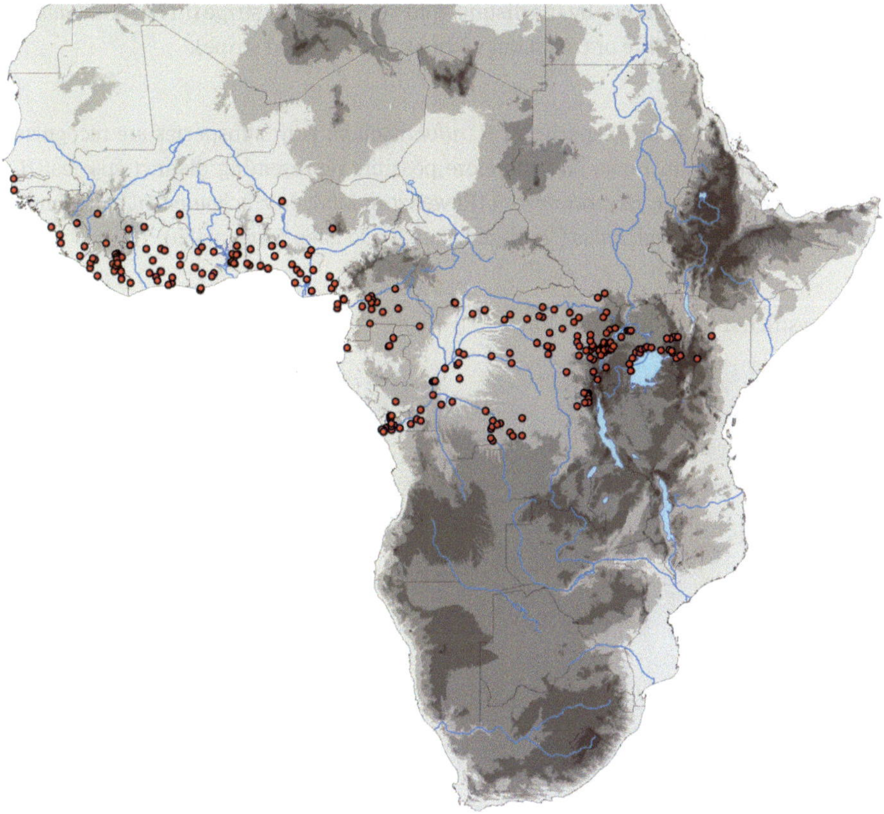

Figure 2.46 Distribution map

Measurements

External and cranial measurements (mm) and mass (g) for *Heliosciurus rufobrachium*, sexes combined

	Mean	Min	Max	N
Mass	329.3	250	388	8
HB	227.7	193	256	21
Tail	267.3	203	320	22
HF	53.9	47	60	22
Ear	15.7	10	19	10
GLS	52.1	46.3	56.8	14

Heliosciurus ruwenzorii (Schwann, 1904)

Diagnosis

This species is immediately recognisable by its distinctive white mid-ventral stripe flanked by cream or buff, the presence of a third upper premolar, and its restricted distribution (Albertine Rift endemic where it occupies montane habitats). It can be distinguished from *Heliosciurus rufobrachium* by the lack of red on its limbs (in *H. rufobrachium* the limbs are bright red or reddish-brown), and from *H. gambianus* by the distinct mid-ventral white stripe (in *H. gambianus* the entire ventral pelage may be white or cream but without a mid-ventral stripe).

Taxonomic Issues

It was previously regarded as a subspecies of *H. rufobrachium*. Four subspecies are recognised, each restricted to a mountain range and therefore possibly representing distinct species: *H. r. ituriensis* (west of Lake Albert); *H. r. ruwenzorii* (Ruwenzorii Mts); *H. r. schoutedeni* (surrounding Lakes Edward and Kivu); *H. r. vulcanius* (south of Lake Kivu to Lake Tanganyika).

Distribution

This species is restricted to the Albertine Rift.

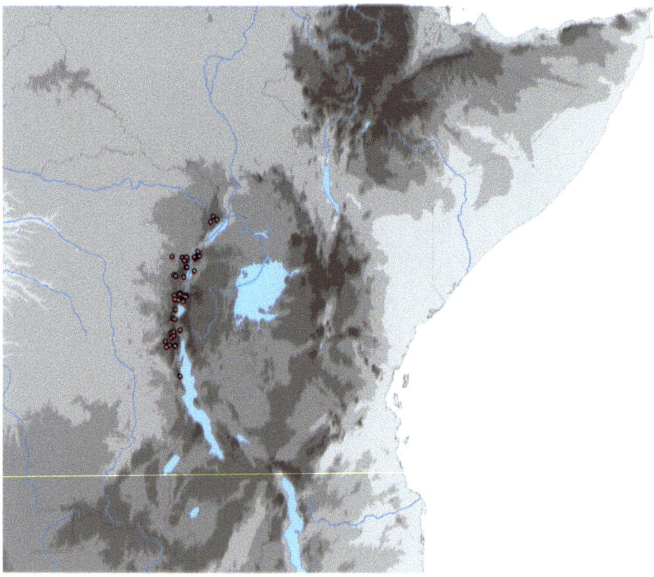

Figure 2.47 Distribution map

Measurements

External and cranial measurements (mm) and mass (g) for *Heliosciurus ruwenzorii*, sexes combined

	Mean	Min	Max	N
Mass	286.0	249	318	8
HB	217.0	192	242	19
Tail	240.0	220	267	19
HF	54.9	51	59	18
Ear	17.7	15	20	15
GLS	51.3	48.8	54.4	24

Heliosciurus undulatus (True, 1892)

Diagnosis

This is the only species of *Heliosciurus* within its range (eastern regions of Kenya and Tanzania). It can be distinguished from *H. rufobrachium* by the lack of red on the limbs. *Heliosciurus gambianus* is smaller (GLS < 51 mm), and *H. mutabilis* has a darker dorsal pelage.

Taxonomic Issues

This taxon was previously listed as a subspecies of *H. rufobrachium*. Despite its relatively small distribution, there is considerable geographic variation in pelage, much of which may be clinal in nature (Grubb, 1982).

Distribution

This species is restricted to coastal East Africa with inland records from around Mt Kilimanjaro and the Eastern Arc mountains of Tanzania.

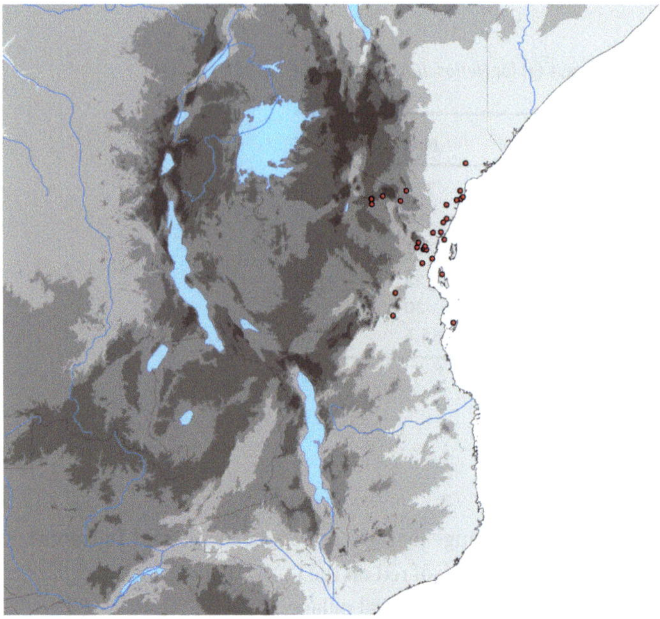

Figure 2.48 Distribution map

Measurements

External and cranial measurements (mm) and mass (g) for *Heliosciurus undulatus*, sexes combined

	Mean	Min	Max	N
Mass	305.8	270	348	4
HB	221.5	190	237	12
Tail	282.6	232	345	12
HF	54.1	50	63	10
Ear	17.0	13	20	10
GLS	53.6	51.0	55.5	12

Genus Account

Myosciurus (Le Conte, 1857) (African Pygmy Squirrels)

Diagnosis

This monotypic genus is immediately recognisable by its diminutive size (about 15-20 g) compared with other African squirrels. The small size makes is rather atypical for a squirrel, superficially appearing more similar to a dormouse (Rosevear, 1969). The pelage is relatively long and soft, being a dark brown with a hue of red or orange dorsally, and paler ventrally. The ears are relatively large and conspicuous due to their ring of white hair. There is an orange eye ring present. There are five digits on the hindfoot and four digits on the forefoot, all elongated and with sharp claws. The feet are typically rotated clockwise, an adaptation allowing the squirrel to hang onto bark with equal pull on the four toes (Emmons, 1979b). The tail is marginally shorter than the body length, and cylindrical rather than bushy. It is reddish brown at the base, becoming browner towards the tip. Based on a limited sample size, there are either one or two pairs of nipples (0 + 0 + 1 + 1 = 4 or 0 + 0 + 0 + 1 = 2) placed along the abdomen (Gharaibeh & Jones, 1996).

Figure 2.49 Photograph of *Myosciurus pumilio* museum specimen (FHNH 48925) (Julian Peterhans Kerbis).

Cranial Characters

The skull is very small (about 20-22 mm in greatest length) and delicate for a squirrel. It is high and domed, curving down posteriorly towards the occipitals and sloping down anteriorly to the rostrum. The rostrum is relatively long and narrow. The palate ends well beyond the posterior margin of the third molars. The zygomatic arch is long and narrow, and thinnest in the middle, and in contrast to other African squirrels, attaching approximately right angles to the rostrum (very obvious in ventral view). The supraorbital foramen is absent, and there is no post-orbital process. The anterior palatal foramina are short and very narrow. The bullae are large, being far longer than the molar toothrow.

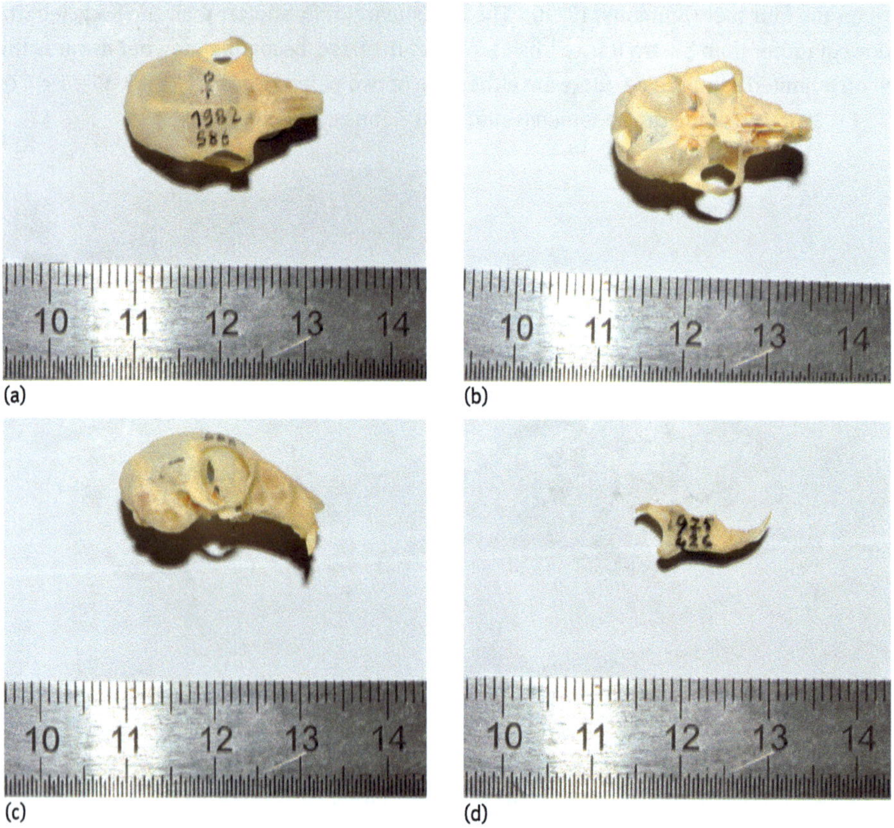

Figure 2.50 a-d Cranium of *Myosciurus pumilio* (MNHN-ZM-MO 1975-426) (from top left to bottom right: dorsal view, ventral view, lateral view and lower jaw).

Dentition

Myosciurus has ungrooved, orthodont or slightly pro-odont, yellow incisors. The dental formula is: 1013/1013 (Rosevear, 1969), with the premolar being smaller than the three molars. The molars have faint transverse ridges.

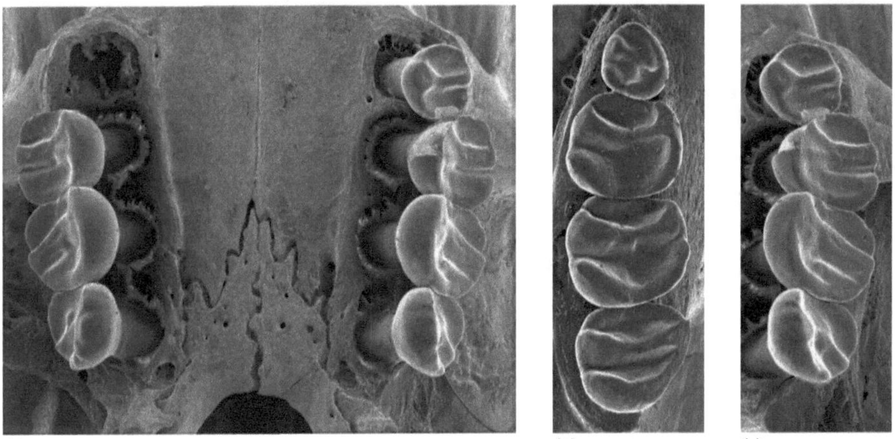

Figure 2.51 Cheek teeth of *Myosciurus pumilio* (MNHN-ZM-MO 1965-148) (A: upper cheek teeth; B: lower left toothrow; C: upper left toothrow).

Habitat & Biogeography

Myosciurus has a limited distribution in central Africa, where it is restricted to the Lower Guinea forest belt. It occurs only in Cameroon, Gabon, Equatorial Guinea and the north-west of Congo (Gharaibeh & Jones, 1996). It is an inhabitant of rainforest where it can occur in primary and secondary forest, as well as plantations (Emmons, 1980). It is predominantly arboreal, rarely descending to the ground, and mostly occupying the lower and middle strata but also venturing into the canopy (Emmons, 1980).

Fossil Record

There are currently no known fossils of *Myosciurus* (Winkler et al., 2010).

Biology & Ecology

The only ecological study conducted on this species was by Emmons (1980) in Gabon. *Myosciurus* is omnivorous, taking fruit, bark chips and insects in approximately similar proportions (Emmons, 1980). A single female had a litter size of 2 (Emmons, 1979a).

Systematic Notes

Myosciurus does not have any close relationships with pygmy squirrels from other parts of the world (e.g. *Nannosciurus* from south-east Asia), but was previously placed in the tribe Funambulini together with *Funisciurus*, *Paraxerus* and the Indian *Funambulus* (Moore, 1959; Heaney, 1985; Thorington et al., 1997). However, a recent molecular phylogeny of the Sciuridae places *Myosciurus* within the tribe Protoxerini which includes *Protoxerus*, *Epixerus*, *Heliosciurus*, *Funisciurus* and *Paraxerus*, but not the Asian *Funambulus* (Mercer & Roth, 2004); *Myosciurus* is basal to this clade of African squirrels.

Species List

One species described within the genus *Myosciurus*, restricted to Sub-Saharan Africa.
Myosciurus pumilio (Le Conte, 1857)

Species Account

Myosciurus pumilio (Le Conte, 1857)

Diagnosis

This tiny squirrel (GLS < 23 mm, HF < 23 mm) cannot be confused with any other sciurid in Africa. The next smallest Sub-Saharan African squirrel is *Paraxerus alexandri* which is considerably larger (GLS > 28 mm) and has a conspicuous flank stripe.

Taxonomic Issues

None.

Distribution

This species is restricted to Central African rainforests.

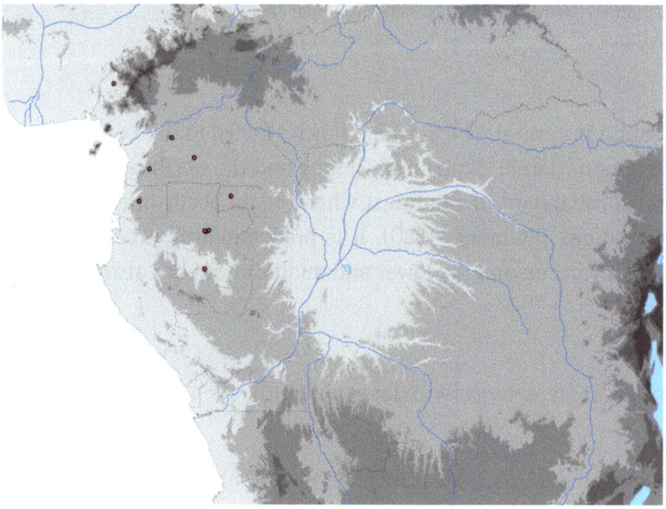

Figure 2.52 Distribution map

Measurements

External and cranial measurements (mm) and mass (g) for *Myosciurus pumilio*, sexes combined

	Mean	Min	Max	N
Mass*	16.5	12	20	6
HB	70.0	60	75	4
Tail	73.3	65	80	3
HF	19.9	18	22	4
Ear	8.5	7	10	4
GLS	22.3	-	-	1

*Emmons (1980), min and max values from Happold (2013)

Genus Account

Paraxerus Forsyth Major, 1893 (Bush Squirrels)

Diagnosis

This diverse genus ranges from small to medium-sized (mostly 50-500 g), with *P. lucifer* and *P. palliatus* approaching the size of some of the larger African squirrels. *Paraxerus* is typically squirrel-shaped with a slender body, relatively long legs, and long tail. The pelage varies considerably between species, being typically longer and softer in montane species; it is exceptionally soft in *P. boehmi*. Fur colour ranges from drab brown (e.g. *P. cepapi*) to bright red (e.g. *P. palliatus* and *P. lucifer*) and dull olive green (*P. poensis*), and melanistic individuals have also been reported (Monks, 1988). Lateral stripes or bands are frequently present and well developed in some species (e.g. *P. alexandri* and *P. boehmi*). The ears are relatively small and rounded, and an eye ring or cheek stripe is frequently present. There are five digits on the hindfoot and four digits on the forefoot, all with sharp and curved claws. The tail is variable in length, being longer (e.g. *P. alexandri* and *P. boehmi*) or subequal to the length of the body, and typically very bushy. Its colour varies between species and individuals within a species, but is often speckles or grizzles (but banding is usually absent or indistinct), whilst in some species it is bright orange-red (*P. palliatus* and *P. lucifer*). Where known, there are three pairs of nipples (0 + 1 + 1 + 1 = 6) placed along the thorax and abdomen.

Figure 2.53 a-f Photograph of *Paraxerus cepapi* (Sara Padidar) (top left) and of museum specimens (NMK and MNHN), from top right to bottom right: *P. boehmi* (two upper specimens) & *P. alexandri* (two lower specimens), and *P. flavovittis*, *P. ochraceus*, *P. palliatus*, and *P. poensis*.

Cranial Characters

The skull is small to medium in size (mostly 30-55 mm in greatest length). It is domed, curving down posteriorly towards the occipitals and anteriorly towards the rostrum. The palate ends slightly beyond the posterior margin of the third molars. The zygomatic arch is long and narrow. The supraorbital foramen is absent, but the post-orbital process is well developed. The anterior palatal foramina are short and narrow. The bullae are moderate in size.

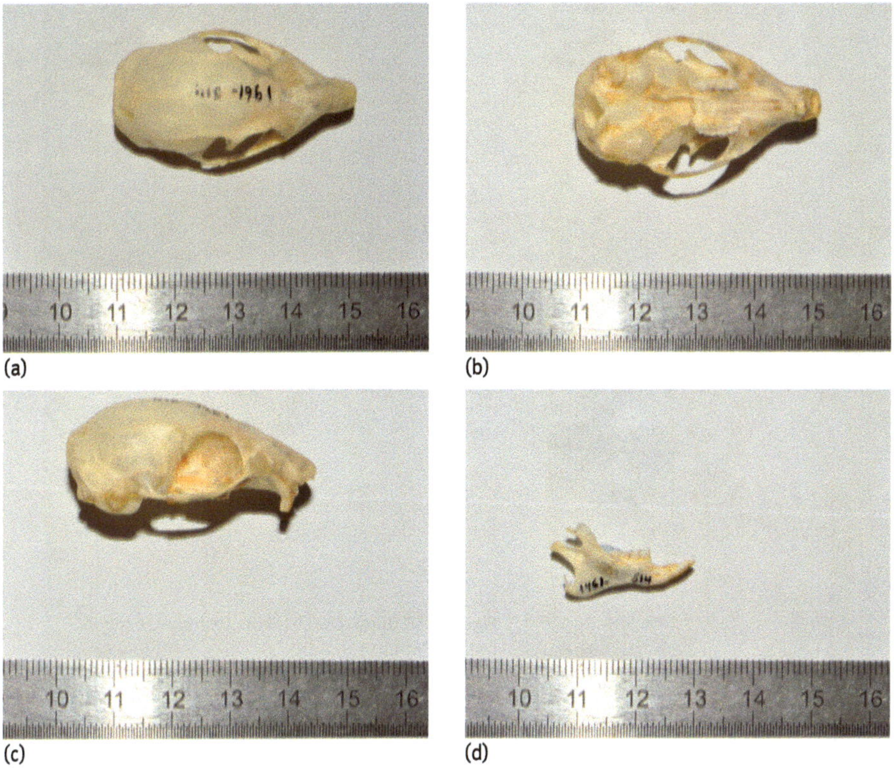

Figure 2.54 a-d Cranium of *Paraxerus poensis* (MNHN-ZM-MO 1961-514) (from top left to bottom right: dorsal view, ventral view, lateral view and lower jaw).

Dentition

Paraxerus has ungrooved, orthodont, orange incisors. The dental formula is: 1023/1013 (Rosevear, 1969), with the first premolar being reduced to a peg resulting in four functional cheekteeth being present in each half jaw. The second premolar and molars are all roughly the same size with transverse ridges that are raised on the either side of each molar.

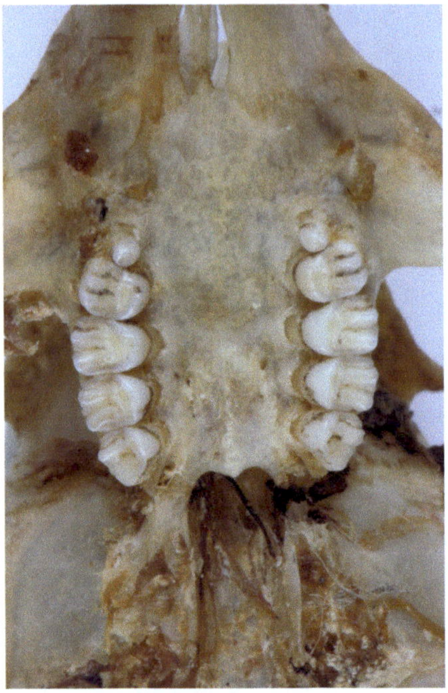

Figure 2.55 Upper cheek teeth of *Paraxerus cepapi* (MNHN-ZM-MO 1969-68).

Habitat & Biogeography

Paraxerus has a wide distribution across Sub-Saharan Africa, mostly confined to tropical latitudes. It occurs from Sierra Leone and Guinea in the west to Ethiopia and Somalia in the east, and south to Namibia, Botswana and South Africa. Unlike in *Funisciurus* there is usually only one species of *Paraxerus* present at a site, and where multiple species co-occur these are separated by habitat. There is only one species present in Central and West Africa (*P. poensis*), and south-western Africa (*P. cepapi*). The centre of diversity for this genus is in East Africa. *Paraxerus* species occur in a variety of different wooded habitats such as rainforest (e.g. *P. poensis, P. alexandri* and *P. boehmi*), coastal and riparian forest (*P. palliatus*), montane forest (*P. lucifer,* and *P. vexillaris*), dry savanna and woodland (*P. cepapi,* and *P. ochraceus*). This genus occurs from sea level up to 2300m (*P. boehmi* and *P. lucifer*) in East Africa (Kingdon, 1974) and up to 2700m in Cameroon (*P. cooperi*) (Eisentraut, 1975) The forest species (e.g. *P. poensis*) are typically arboreal, and may restrict their activity to middle and higher strata (Emmons, 1980). By contrast, the savanna species (e.g. *P. cepapi*) regularly descend to the ground to move between trees as well as for foraging (Kingdon, 1974). They roost in hollow tree trunks which they line with leaves (Kingdon, 1974; Viljoen, 1983a; Skinner & Chimimba, 2005).

Fossil Record

The oldest *Paraxerus* fossil is from 6 MYA, Kenya (Manthi, 2007). In East Africa, it has been reported from the Late Miocene to the Late Pliocene; in Southern Africa from the Late Miocene to the Late Pleistocene (Winkler et al, 2010).

Biology & Ecology

Only a few species of *Paraxerus* have been studied in detail, and for many species knowledge has not progressed beyond anecdotal observations. The species that have been studied are mostly frugivorous, but also taking insects and eggs (Emmons, 1980; Viljoen, 1983a), and even possibly vertebrates (Stanley & Collett, 2009). One species (*P. cepapi*) may be a disperser of the marula tree *Sclerocarya birrea* (Midgley et al., 2012). Home range has been reported for two species (*P. cepapi* and *P. palliatus*), and is typically small (1-5 ha) and may, or may not, overlap that of neighbouring individuals (Viljoen, 1986). The reproductive biology of most *Paraxerus* species is not well known. Where known, litter size is 1-3 but mostly 1-2, and the gestation and lactation periods are 55-61 and 24-45 days (Viljoen, 1977a,b; Emmons, 1979a). In *P. cepapi*, the interval between births was 62 days (Viljoen, 1977b).

Systematic Notes

On morphological grounds, this genus was previously placed in the tribe Funambulini together with the Indian *Funambulus* and African *Funisciurus*. Based on a recent molecular phylogeny, the closest relative of *Paraxerus* is *Funisciurus*, and these two genera together with *Protoxerus* and *Heliosciurus* form a well supported monophyletic clade (Steppan et al., 2004), with *Funambulus* not falling within this group (Mercer & Roth, 2003). These four African genera, together with *Myosciurus* form the tribe Protoxerini. The relationships within the genus are not yet understood, and a molecular phylogeny for this genus (along with its closest relative *Funisciurus*) is required to shed light on this. Eleven species are currently recognised in this genus, and species limits are still not fully resolved. For example, *P. palliatus* shows great geographical variation in its size and colouring, and numerous subspecies have been described. At least some of these are likely to represent distinct species.

Species List

Eleven species are described within the genus *Paraxerus*, all restricted to Sub-Saharan Africa.
Paraxerus alexandri (Thomas & Wroughton, 1907)
Paraxerus boehmi (Reichenow, 1886)
Paraxerus cepapi (A. Smith, 1836)
Paraxerus cooperi Hayman, 1950
Paraxerus flavovittis (Peters, 1852)
Paraxerus lucifer (Thomas, 1897)
Paraxerus ochraceus (Huet, 1880)
Paraxerus palliatus (Peters, 1852)
Paraxerus poensis (A. Smith, 1830)
Paraxerus vexillarius (Kershaw, 1923)
Paraxerus vincenti Hayman, 1950

Key to Species

1	Conspicuous white flank stripe present with black border below and/or above	2
	Flank stripe absent (or inconspicuous pale stripe without black border)	4
2	Smaller (GLS < 37 mm, HF < 35 mm); restricted to forests of the eastern Congo basin and Albertine Rift	3
	Larger (GLS > 38 mm, HF > 35 mm); restricted to savannas of East Africa	*P. flavovittis*
3	Smaller (GLS < 33 mm); white flank stripe with a black border above; ears with white hairs	*P. alexandri*
	Larger (GLS > 33 mm); white flank stripe with a black border above and below; ears without white hairs	*P. boehmi*
4	Dorsal and/or ventral pelage bright rufous or orange (sometimes confined to the limbs)	5
	No bright rufous or orange anywhere in the pelage	8
5	Ventral pelage bright rufous or orange	6
	Ventral pelage grey (dorsal pelage, limbs and flanks may be rufous)	7
6	Restricted to montane forest at Mt Namuli (northern Mozambique)	*P. vincenti*
	Occurs in forests elsewhere in Southern and East Africa	*P. palliatus*
7	Dorsal pelage bright rufous	*P. lucifer*
	Bright rufous restricted to the limbs (and sometimes the tail)	*P. vexillarius*
8	Occurs in forests in West and/or Central Africa	9
	Occurs in savannas of Southern and/or East Africa	10
9	Smaller (GLS < 40 mm, HF < 36 mm); no rufous on limbs	*P. poensis*
	Larger (GLS > 43 mm, HF > 38 mm); rufous wash on limbs; restricted to south-western Cameroon mountains	*P. cooperi*
10	Dorsal pelage with black tipped hairs; mostly restricted to Southern Africa	*P. cepapi*
	Dorsal pelage with white tipped hairs; restricted to East Africa	*P. ochraceus*

Species Accounts

Paraxerus alexandri (Thomas & Wroughton, 1907)

Diagnosis

Only three species of *Paraxerus* have conspicuous white flank stripes extending from the shoulders to the rump (*P. alexandri, P. boehmi* and *P. flavovittis*). *Paraxerus alexandri* can be distinguished from *P. boehmi* by its smaller size, absence of a black border below the white flank stripe, and white hairs on the ears (in *P. boehmi*, GLS > 33 mm, HF > 28 mm, the white flank stripe is bordered above and below by black stripes, and the ears without white hairs). *Paraxerus flavovittis* can be distinguished by its much larger size (GLS > 38 mm, HF > 35 mm), a single black border below the white flank stripe (in *P. alexandri*, black border is above white flank stripe) and non-overlapping distribution.

Taxonomic Issues

None.

Distribution

This species has a restricted distribution in the forests of north-eastern DRC and Uganda.

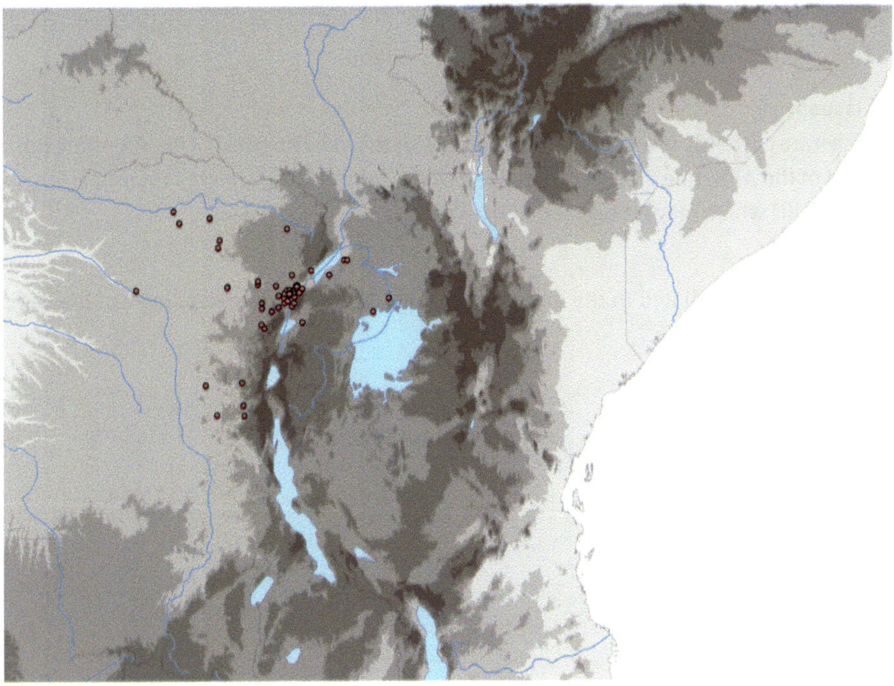

Figure 2.56 Distribution map

Measurements

External and cranial measurements (mm) and mass (g) for *Paraxerus alexandri*, sexes combined

	Mean	Min	Max	N
Mass*	46.0	40	72	9
HB	105.0	96	116	4
Tail	116.0	110	126	4
HF	28.0	26	30	4
Ear	11.5	11	13	4
GLS	29.4	28.7	29.9	3

*Happold (2013)

Paraxerus boehmi (Reichenow, 1886)

Diagnosis

Only three species of *Paraxerus* have conspicuous white flank stripes extending from the shoulders to the rump (*P. alexandri, P. boehmi* and *P. flavovittis*). *Paraxerus boehmi* can be distinguished from *P. alexandri* by its larger size, presence of black borders below and above the white flank stripe, and the lack of white hairs on the ears (in *P. alexandri*, GLS < 33 mm, HF < 28 mm, the white flank stripe is only bordered above by a black stripe, and the ears have white hairs). *Paraxerus fla-*

vovittis can be distinguished by its much larger (GLS > 38 mm, HF > 35 mm), a single black border below the white flank stripe, and non-overlapping distribution.

Taxonomic Issues

Four subspecies are currently recognised (Happold, 2013): *P. b. boehmi* (west of the Albertine Rift), *P. b. emini* (east of the Albertine Rift), *P. b. gazellae* (Sudan), and *P. b. vulcanorum* (montane forests in the Albertine Rift) which may represent distinct species.

Distribution

This species is widely distributed in the forest regions of Central Africa, extending as far east as Uganda.

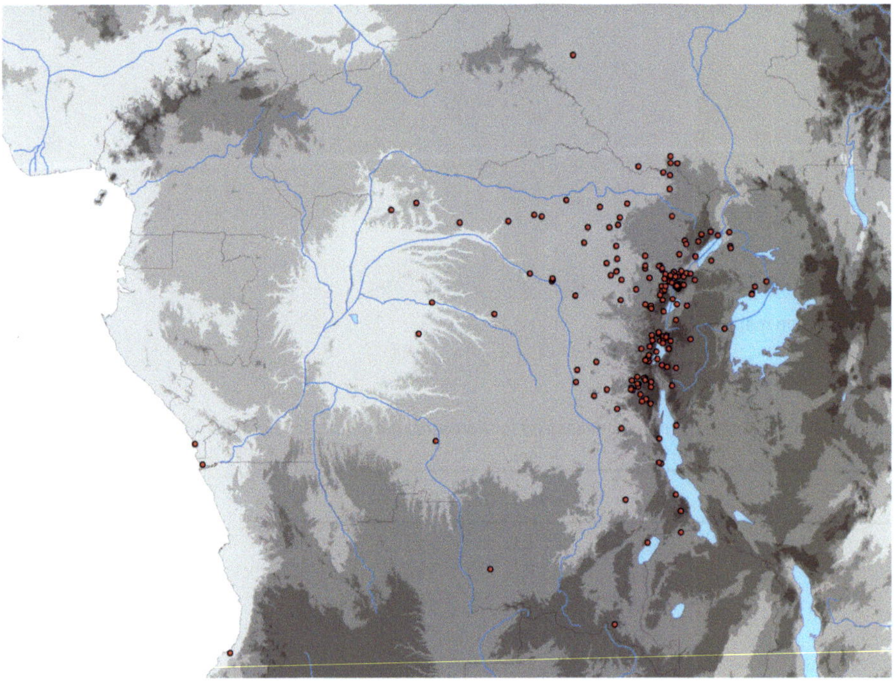

Figure 2.57 Distribution map

Measurements

External and cranial measurements (mm) and mass (g) for *Paraxerus boehmi*, sexes combined

	Mean	Min	Max	N
Mass*	73.0	48	83	15
HB	120.0	118	122	2
Tail	158.5	145	172	2
HF	32.0	31	33	2
Ear	14.5	14	15	2
GLS	35.0	34.5	35.6	4

*Happold (2013)

Paraxerus cepapi (A. Smith, 1836)

Diagnosis

This is the only *Paraxerus* species within most of its range. It could be confused with: *P. flavovittis* (has a white flank stripe which is absent in *P. cepapi*); *P. ochraceus* (has hairs in dorsal pelage with white tips. In *P. cepapi* dorsal hairs are usually black tipped, and distributions do not overlap); *P. palliatus* (larger, GLS > 45 mm, and has bright red, orange or yellow ventral pelage. In *P. cepapi*, ventral pelage is white or dull cream, never bright).

Taxonomic Issues

Nine subspecies are currently recognised, but their phylogenetic relationships are poorly understood.

Distribution

This species is widely distributed in the savannas of southern Africa, extending north into southern DRC and Tanzania.

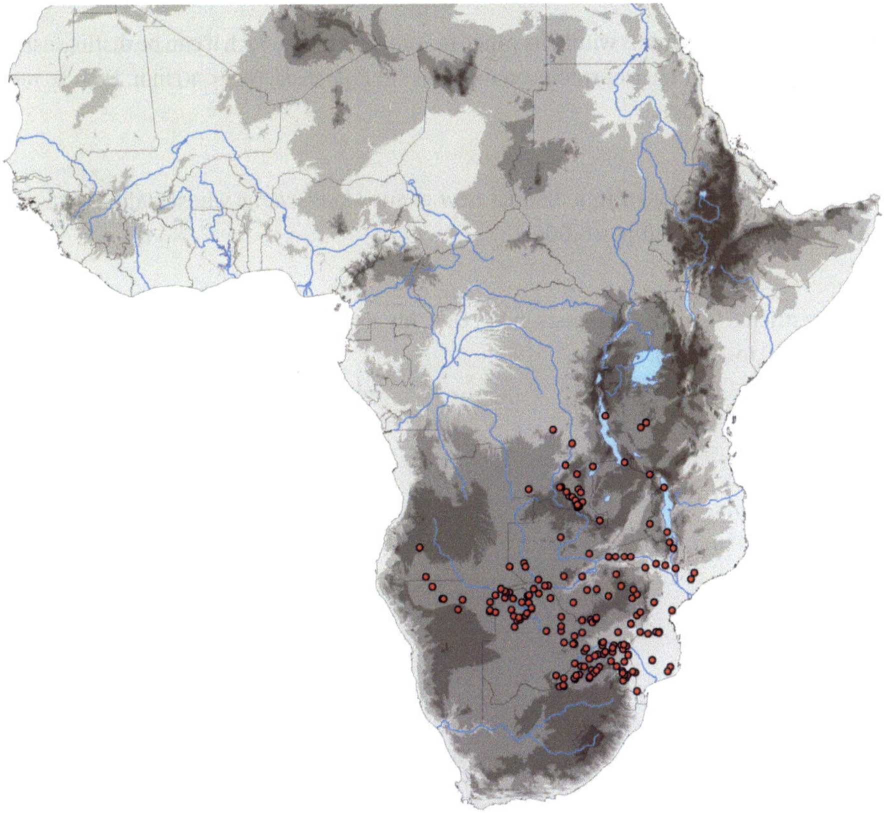

Figure 2.58 Distribution map

Measurements

External and cranial measurements (mm) and mass (g) for *Paraxerus cepapi*, sexes combined

	Mean	Min	Max	N
Mass*	192.3	76	265	52
HB	182.2	137	220	26
Tail	172.7	116	215	26
HF	44.4	40	48	17
Ear	18.4	13	21	24
GLS	42.5	40.0	44.7	13

*Rautenbach (1982)

Paraxerus cooperi Hayman, 1950

Diagnosis
The only other species of *Paraxerus* within its range is *P. poensis* from which it can be distinguished by its larger size and the presence of rufous on its limbs (in *P. poensis*, GLS < 40 mm, HF < 36 mm, and the limbs lack any rufous).

Taxonomic Issues
Eisentraut (1975) placed *P. cooperi* in a distinct new genus *Montisciurus* whose validity was not recognized by Thorington & Hoffman (2005).

Distribution
This species is restricted to montane forests in Mt. Oku, Mt. Kupe and Rumpi Hills of south-eastern Cameroon (from 800 to 2500m).

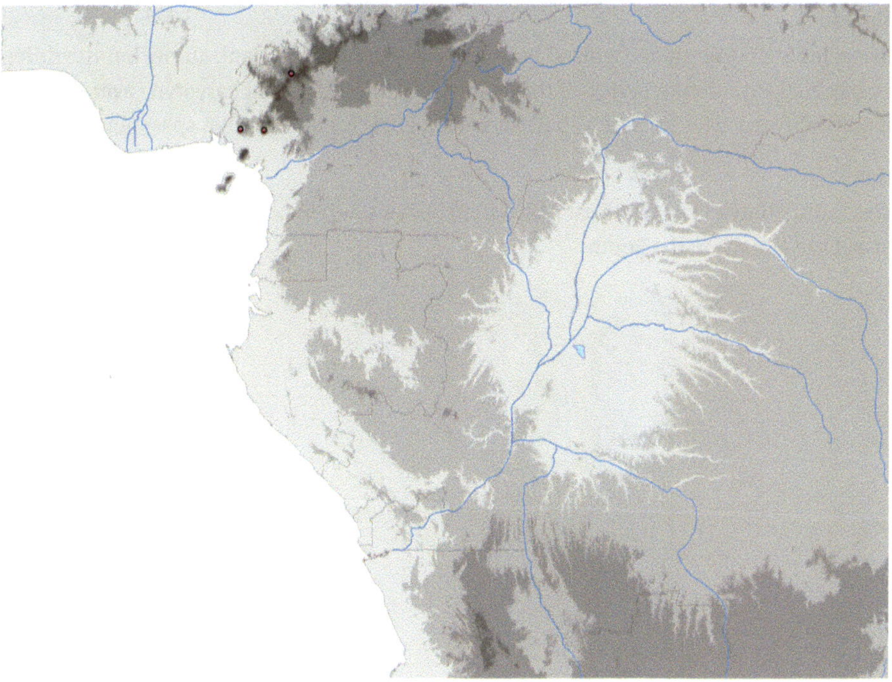

Figure 2.59 Distribution map

Measurements

External and cranial measurements (mm) and mass (g) for *Paraxerus cooperi*, sexes combined

	Mean	Min	Max	N
Mass	-	-	-	-
HB*	201.8	182	212	21
Tail*	183.7	161	200	21
HF*	41.6	38	45	21
Ear*	16.0	15	17	21
GLS*	46.1	44.4	48.4	21

*Eisentraut (1968a)

Paraxerus flavovittis (Peters, 1852)

Diagnosis

Only three species of *Paraxerus* have conspicuous white flank stripes extending from the shoulders to the rump (*P. alexandri, P. boehmi* and *P. flavovittis*). *Paraxerus flavovittis* can be readily distinguished from *P. alexandri* and *P. boehmi* by its much larger size (GLS > 38 mm, HF > 35 mm), the presence of a single black border below white flank stripe, and its non-overlapping distribution.

Taxonomic Issues

Three subspecies are currently recognised based on the extent of the flank stripe, but the differences may be clinal variation. A review of this taxon is required. The first karyotype ever recorded for this genus indicated 2N = 38, aFN = 70 (Denys et al., 2011) for a Tanzanian specimen.

Distribution

This species is distributed in coastal East Africa from Kenya south to northern Mozambique and as far inland as Lake Malawi in the South.

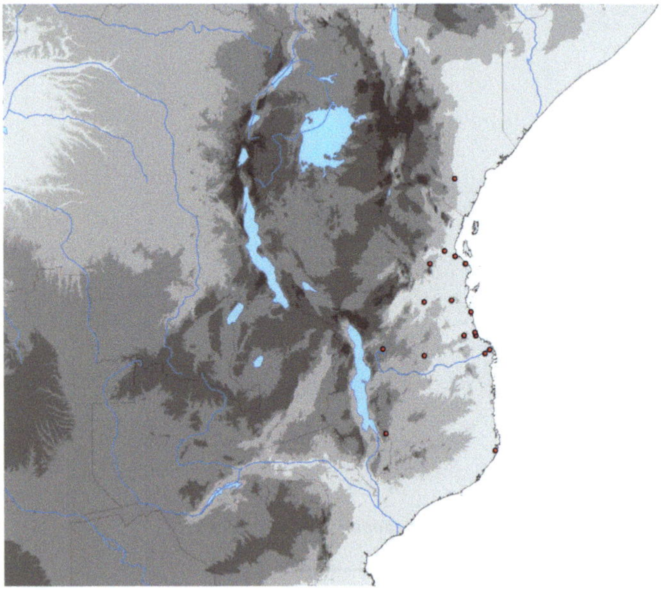

Figure 2.60 Distribution map

Measurements

External and cranial measurements (mm) and mass (g) for *Paraxerus flavovittis*, sexes combined

	Mean	Min	Max	N
Mass	129	-	-	1
HB	169	-	-	1
Tail	191	-	-	1
HF	45	-	-	1
Ear	-	-	-	1
GLS	40.9	40.8	41.1	3

Paraxerus lucifer (Thomas, 1897)

Diagnosis

This is the only large-sized, bright rufous *Paraxerus* species in its range. It co-occurs with *P. cepapi* from which it can easily be distinguished by its larger size and dorsal pelage (in *P. cepapi*, GLS < 45 mm, HF < 48 mm and dorsal pelage is yellow-brown). There are three other *Paraxerus* species with bright rufous pelage which can be distinguished as follows: *P. palliatus* (has rufous ventral pelage, in *P. lucifer* is grey); *P. vexillarius* (bright rufous restricted to the limbs, in *P. lucifer* the dorsal pelage is rufous); and *P. vincenti* (ventral pelage rufous, confined to Mt Namuli, Mozambique).

Taxonomic Issues

The phylogenetic relationships of this species with the other rufous *Paraxerus* is not known.

Distribution

This species is restricted to the mountains of northern Malawi (Nyika Plateau, Misuku Hills) and south-western Tanzania (Poroto mountains).

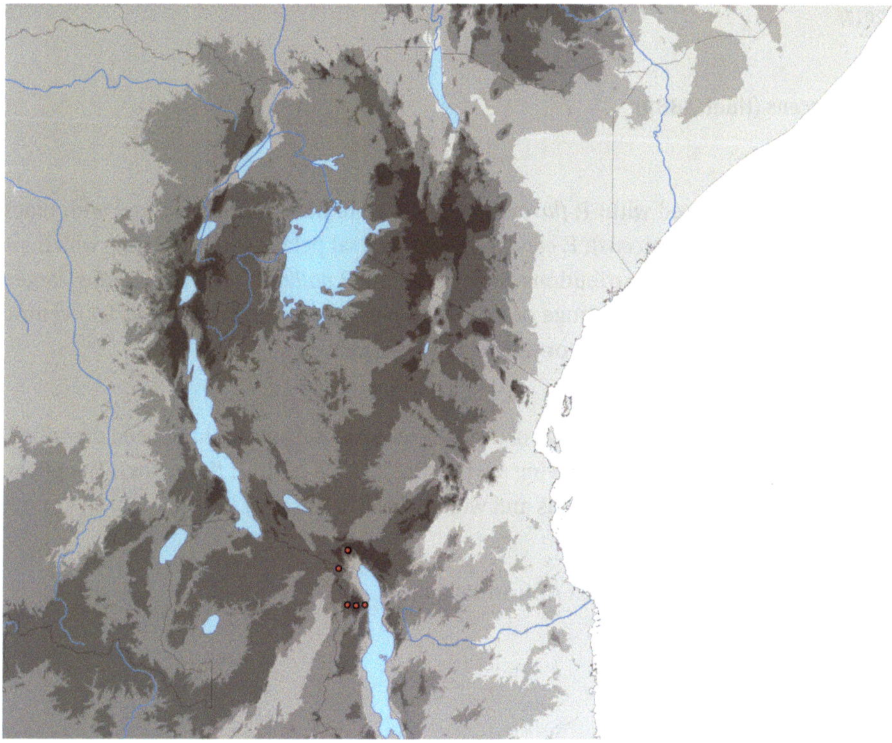

Figure 2.61 Distribution map

Measurements

External and cranial measurements (mm) and mass (g) for *Paraxerus lucifer*, sexes combined

	Mean	Min	Max	N
Mass*	496	300	675	7
HB*	222	201	241	10
Tail*	202	186	218	10
HF*	52.2	48	55	13
Ear*	19.7	15	22	12
GLS*	55.6	53.8	56.4	12

*Happold (2013)

Paraxerus ochraceus (Huet, 1880)

Diagnosis

This species could be confused with: *P. flavovittis* (has a white flank stripe bordered with black below which is absent in *P. ochraceus*); *P. cepapi* (hairs in dorsal pelage with black tips which are white tipped in *P. ochraceus* and distributions do not overlap); *P. palliatus* and *P. vexillarius* (larger, GLS > 45 mm, and with bright red, orange or yellow (*P. palliatus*) or grey (*P. vexillarius*) ventral pelage which in *P. ochraceus* is off-white or yellow, never bright).

Taxonomic Issues

There is considerable variation in the pelage of this species (nine subspecies have been described) with specimens from coastal Tanzania having a white flank stripe (as in *P. flavovittis*). The phylogenetic relationships of these various forms and with *P. flavovittis* are not known.

Distribution

This species is widely distributed in East African savannas.

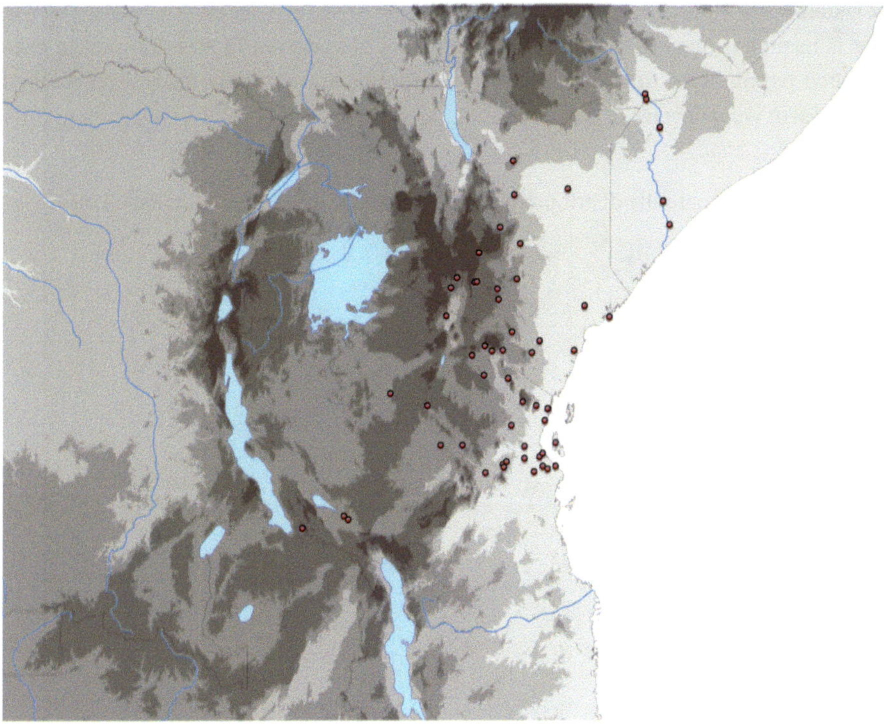

Figure 2.62 Distribution map

Measurements

External and cranial measurements (mm) and mass (g) for *Paraxerus ochraceus*, sexes combined

	Mean	Min	Max	N
Mass	-	-	-	-
HB	160.5	147	199	12
Tail	170.0	131	190	13
HF	38.5	34	45	13
Ear	16.6	11	19	11
GLS	39.8	36.7	42.6	11

Paraxerus palliatus (Peters, 1852)

Diagnosis

This species co-occurs with *P. cepapi*, *P. flavovittis* and *P. ochraceus* from which it can easily be distinguished by its bright rufous dorsal pelage, bright rufous, orange or yellow ventral pelage and (generally) larger size. There are three other *Paraxerus* species with bright rufous pelage which can be distinguished as follows: *P. lucifer* (has grey ventral pelage); *P. vexillarius* (has grey ventral pelage, and bright rufous restricted to the limbs); and *P. vincenti* (confined to Mt Namuli, Mozambique, and distributions do not overlap).

Taxonomic Issues

This species is in urgent need of revision. Large amounts of variation in size and pelage exist between isolated populations, and at least seven subspecies are currently recognised some of which may represent valid species.

Distribution

This species is widely, but patchily distributed in coastal forests of Southern and East Africa.

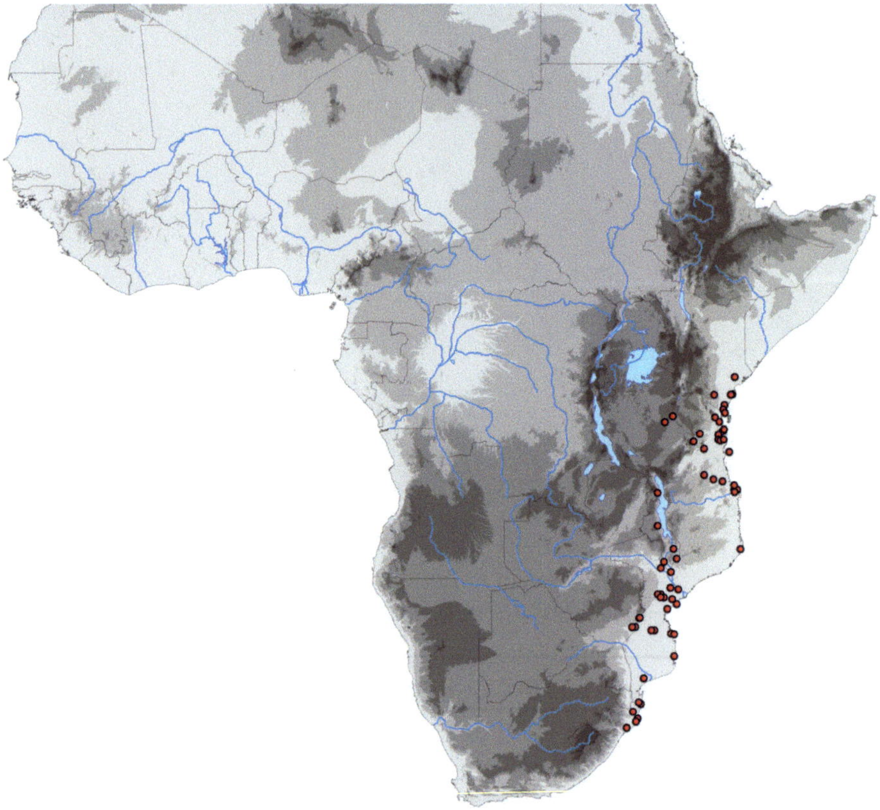

Figure 2.63 Distribution map

Measurements

External and cranial measurements (mm) and mass (g) for *Paraxerus palliatus*, sexes combined. Considerable variation exists between the various populations of this species which is masked by the mean values presented here.

	Mean	Min	Max	N
Mass	219.6	218	220	2
HB	202.2	180	235	19
Tail	200.3	135	240	19
HF	48.3	43	53	16
Ear	18.7	12	22	17
GLS	48.8	43.1	51.7	17

Paraxerus poensis (A. Smith, 1830)

Diagnosis

This relatively drab-coloured species is the only *Paraxerus* in most of its range. It co-occurs in northern DRC with *P. boehmi* which is readily distinguished by its conspicuous white flank stripe. It co-occurs with *P. cooperi* from which it can be distinguished by its smaller size and the absence of rufous on its limbs (in *P. cooperi*, GLS > 43 mm, HF > 40 mm, and the limbs are rufous).

Taxonomic Issues

No subspecies are currently recognised. However, due to the large hiatus in its distribution (being absent from Benin and Nigeria), the West African animals may prove to be distinct from the Central African ones.

Distribution

This species occurs widely in the Upper Guinea Forest of West Africa, and in the Lower Guinea Forest of Central Africa. An isolated population occurs in the far north of the DRC (on the right bank of the Congo River).

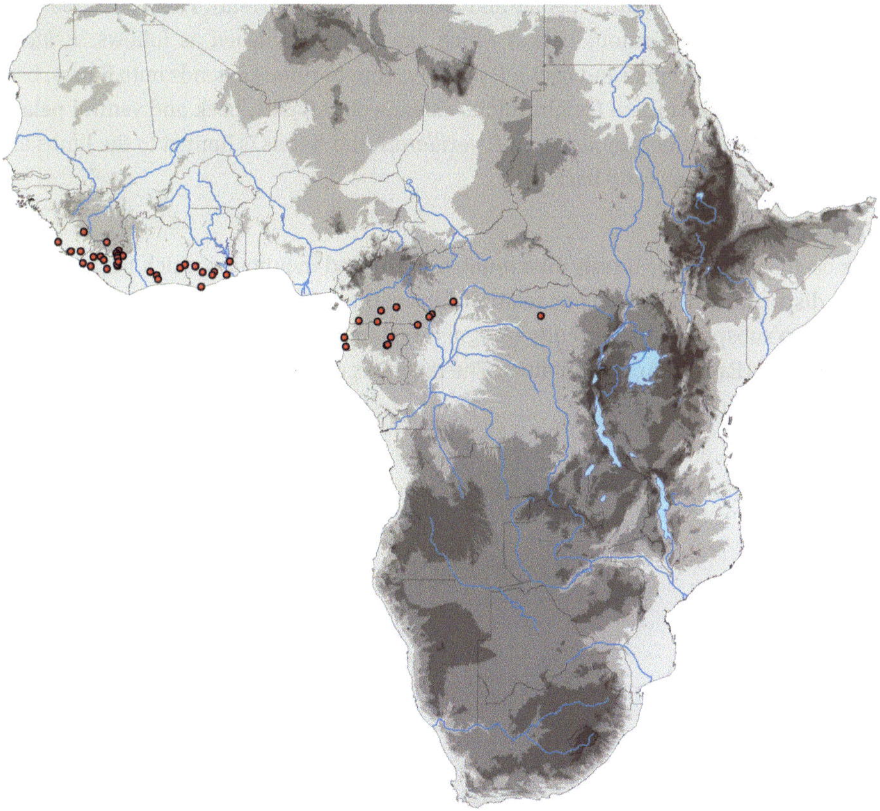

Figure 2.64 Distribution map

Measurements

External and cranial measurements (mm) and mass (g) for *Paraxerus poensis*, sexes combined

	Mean	Min	Max	N
Mass	171.3	114	346	9
HB	157.9	132	179	12
Tail	184.6	106	218	12
HF	40.0	32	41	12
Ear	12.8	10	17	11
GLS	40.5	34.7	43.5	10

Paraxerus vexillarius (Kershaw, 1923)

Diagnosis

This species co-occurs with *P. flavovittis* and *P. ochraceus* from which it can easily be distinguished by its bright rufous limbs, grey ventral pelage and larger size (GLS > 50 mm). There are three other *Paraxerus* species with bright rufous pelage which can be distinguished as follows: *P. lucifer* (bright rufous extends onto flanks and back); *P. palliatus* (bright rufous extends onto flanks, back and ventral pelage); and *P. vincenti* (bright rufous extends onto flanks, back and ventral pelage, and confined to Mt Namuli, Mozambique). In *P. vexillarius*, the rufous is limited to the limbs and does not extend onto the flanks and back.

Taxonomic Issues

This species occurs in montane forests on a number of isolated mountains which may represent barriers to its dispersal. The phylogenetic relationships of these various populations have not yet been studied. Two subspecies are currently recognised (Happold, 2013): *P. v. byatti* (Mt. Kilimanjaro); *P. v. vexillarius* (Eastern Arc mountains). The taxon occurring on Mt Kilimanjaro (*byatti*) is often considered a separate species.

Distribution

This species occurs in the montane forests of the Eastern Arc and Mt Kilimanjaro, Tanzania.

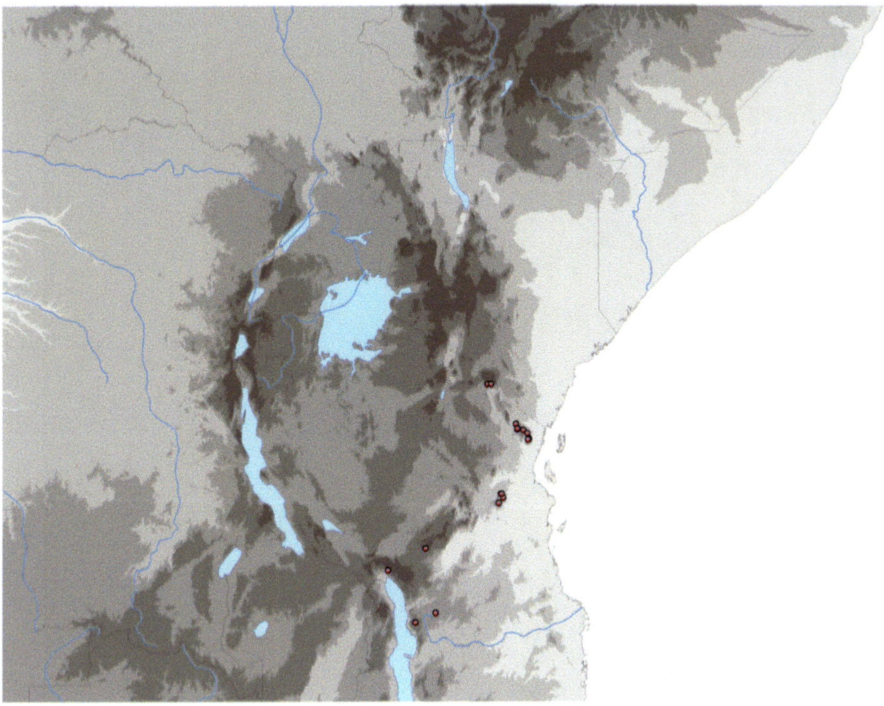

Figure 2.65 Distribution map

Measurements

External and cranial measurements (mm) and mass (g) for *Paraxerus vexillarius*, sexes combined

	Mean	Min	Max	N
Mass	-	-	-	-
HB*	230	190	264	4
Tail*	195	180	210	4
HF*	49.7	45	52	4
Ear*	16.5	15	18	2
GLS*	53.5	53	54	2

*Happold (2013)

Paraxerus vincenti Hayman, 1950

Diagnosis

This is the only *Paraxerus* species occurring on Mt Namuli, northern Mozambique. It is easily distinguished from all other *Paraxerus* (except *P. palliatus*) by its bright rufous dorsal pelage and rufous ventral pelage. It is similar in appearance to *P. palliatus*, but the latter species does not occur at Mt Namuli.

Taxonomic Issues

Its relationship to other *P. palliatus* populations is currently not known.

Distribution

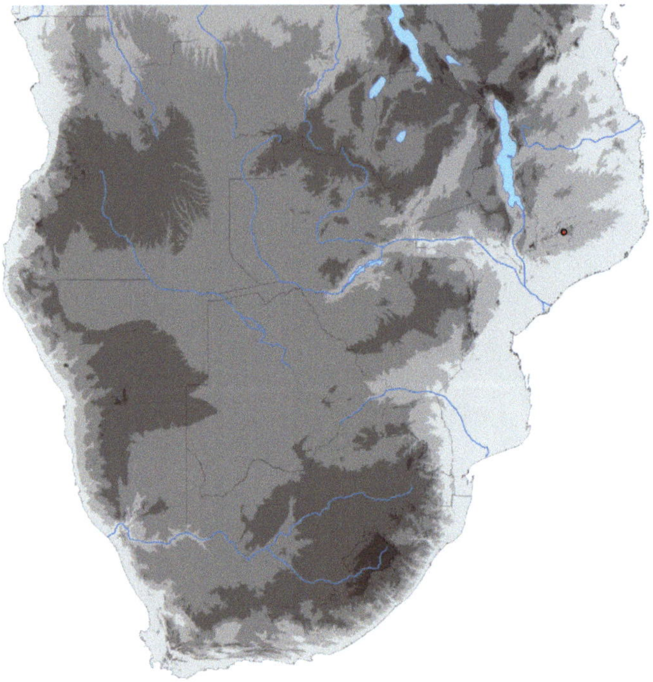

Figure 2.66 Distribution map

Measurements

External and cranial measurements (mm) and mass (g) for *Paraxerus vincenti*, sexes combined

	Mean	Min	Max	N
Mass	-	-	-	-
HB	212	-	-	5
Tail	209	-	-	5
HF	46.6	-	-	5
Ear	21.1	-	-	5
GLS	50.4	-	-	5

Genus Account

Protoxerus Forsyth Major, 1893 (African Giant Squirrels)

Diagnosis

This monotypic genus is the largest (Mass ca. 550-750 g) squirrel in Africa, although *Epixerus* and some *Heliosciurus* overlap with it in size. It is typically squirrel-shaped with a slender body, relatively short legs, largish head and long tail. The pelage is relatively long and coarse (harsher than *Epixerus*) and variably brown and speckled dorsally, and very sparse (almost naked) ventrally. There is occasionally a slight wash of yellow on the hindlimbs, but never on the forelimbs. The hairs on the head are white-tipped, giving it a characteristic speckled appearance. A lateral band or stripe is usually absent, but may occasionally be indistinctly present. Pelage colour is highly variable (Amtmann, 1965). The ears are of moderate length. There are five digits on the hindfoot and four digits on the forefoot, all with powerful, sharp, and curved claws. The tail, unless damaged, is slightly longer than the body (ca. 110% of HB) and bushy with alternating black (broad) and white (narrow) bands. There are four pairs of nipples (1 + 1 + 1 + 1 = 8) placed along the chest and abdomen.

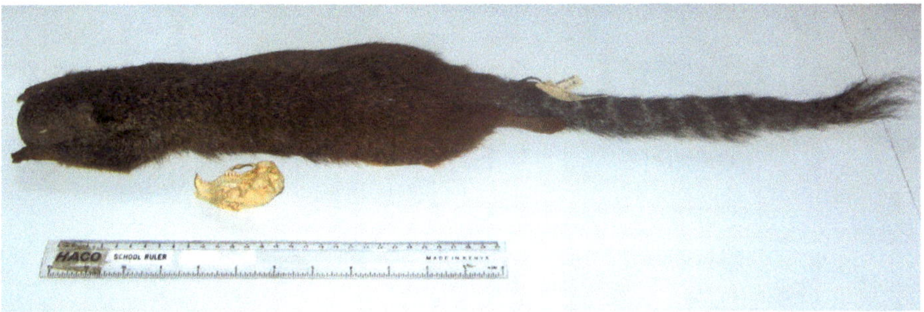

Figure 2.67 Photograph of *Protoxerus stangeri* museum specimen (NMK 1772) (Ara Monadjem).

Cranial Characters

The skull is large (about 68 mm in greatest length) and robust, compared with *Epixerus*. It is flat dorsally, but sloping down posteriorly towards the occipitals. The rostrum is relatively short, and the palate ends at the posterior margin of the third molars. The zygomatic arch is long and expanded in the middle. A supraorbital foramen is present, and the post-orbital process is well developed. The anterior palatal foramina are short and narrow. The bullae are relatively large, being longer than the molar toothrow.

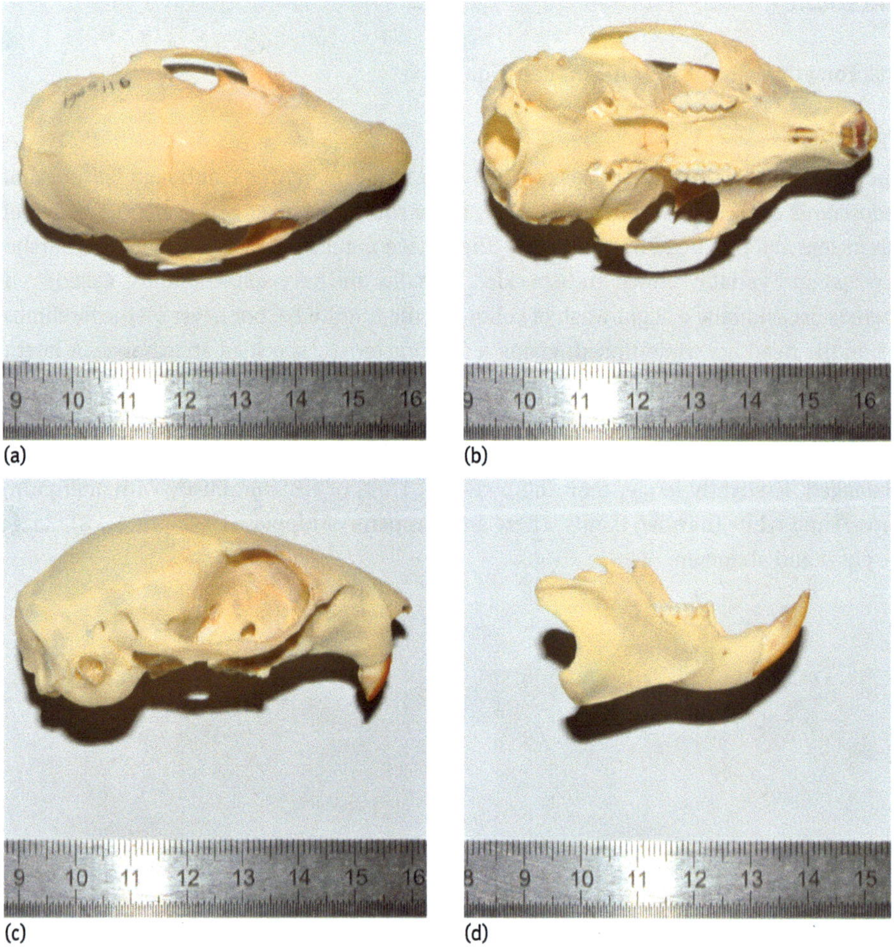

Figure 2.68 a-d Cranium of *Protoxerus stangeri* (MNHN-ZM-MO 1965-116) (from top left to bottom right: dorsal view, ventral view, lateral view and lower jaw).

Dentition

Protoxerus has large, ungrooved, orthodont, orange incisors. The dental formula is: 1013/1013 (Rosevear, 1969), resulting in four cheekteeth being present in each half jaw. The premolar and molars are all roughly the same size with four transverse ridges and a longitudinal ridge on the inner side of each molar.

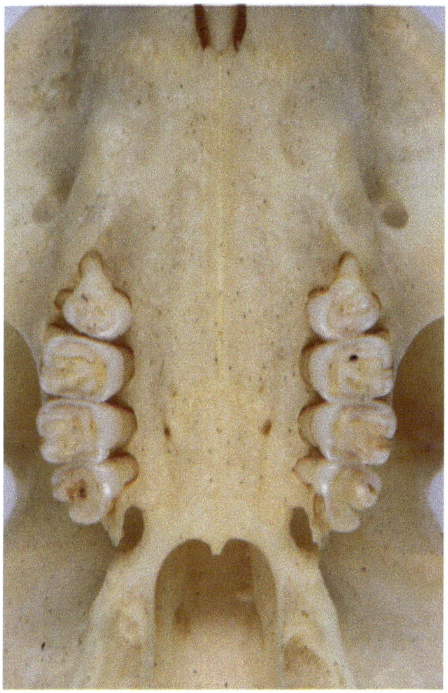

Figure 2.69 Upper cheek teeth of *Protoxerus stangeri* (MNHN-ZM-MO 2001-1282).

Habitat & Biogeography

Protoxerus has a wide distribution across the rainforest belt of tropical Africa. It occurs from Sierra Leone in the west through to Kakamega forest, Kenya in the east, and extends southwards east of the Albertine Rift throughout the northern half of the Democratic Republic of Congo to northern Angola (Hill & Carter, 1941). There is an isolated population on the eastern shores of Lake Tanganyika on Mt Rungwe (Kingdon, 1974). It is closely associated with rainforest, but may also occur in secondary forest and plantations. It occurs from sea level up to 2100m in eastern DRC (Kingdon, 1974). *Protoxerus stangeri* is arboreal, foraging at mid-stratum and in the canopy, very rarely descending to the ground (Emmons, 1980). It roosts in tree hollows, where it also builds its nest, usually high up in the canopy (Kingdon, 1974).

Fossil Record

The oldest putative *Protoxerus* fossil is from the Early Oligocene from the White River Formation of North America (Cifelli et al., 1986), but its relationship to the African *Protoxerus* is unclear. A large African giant squirrel, *Kubwaxerus pattersoni*, described from the Late Miocene, Kenya (Cifelli et al., 1986) may be related to extant *Protoxerus*.

Biology & Ecology

Despite its large size and widespread distribution, the biology of this species is poorly known. It is frequently seen foraging in the company of forest monkeys but the nature of this relationship has not been studied (Emmons, 1980). It feeds primarily on fruits and nuts, especially those with a hard shell, using its powerful incisors to break the shell, but also eats a smaller proportion of leaves (Emmons, 1980). Home range of two sub-adults was 3-5 ha (Happold, 2013). Litter size, based on a sample size of two, is 1-2 (Emmons, 1979a).

Systematic Notes

Based on a molecular phylogeny, the closest relative of *Protoxerus* is *Epixerus* and both are placed in the tribe Protoxerini with all other Sub-Saharan African squirrels, except *Xerus* (Mercer & Roth, 2003). The relationship of *Protoxerus* to *Allosciurus* has not yet been resolved. This genus is currently considered to be monospecific with *Protoxerus stangeri* as the sole representative. Happold (2013) listed 11 subspecies. Molecular studies are required to examine the genetic structure of this widespread species; in particular, the isolated population at Mt Rungwe, Tanzania, and the northern Angolan population.

Species List

One species described within the genus *Protoxerus*, restricted to Sub-Saharan Africa.
Protoxerus stangeri (Waterhouse, 1842)

Species Account

Protoxerus stangeri (Waterhouse, 1842)

Diagnosis

See genus account.

Taxonomic Issues

See genus account.

Distribution

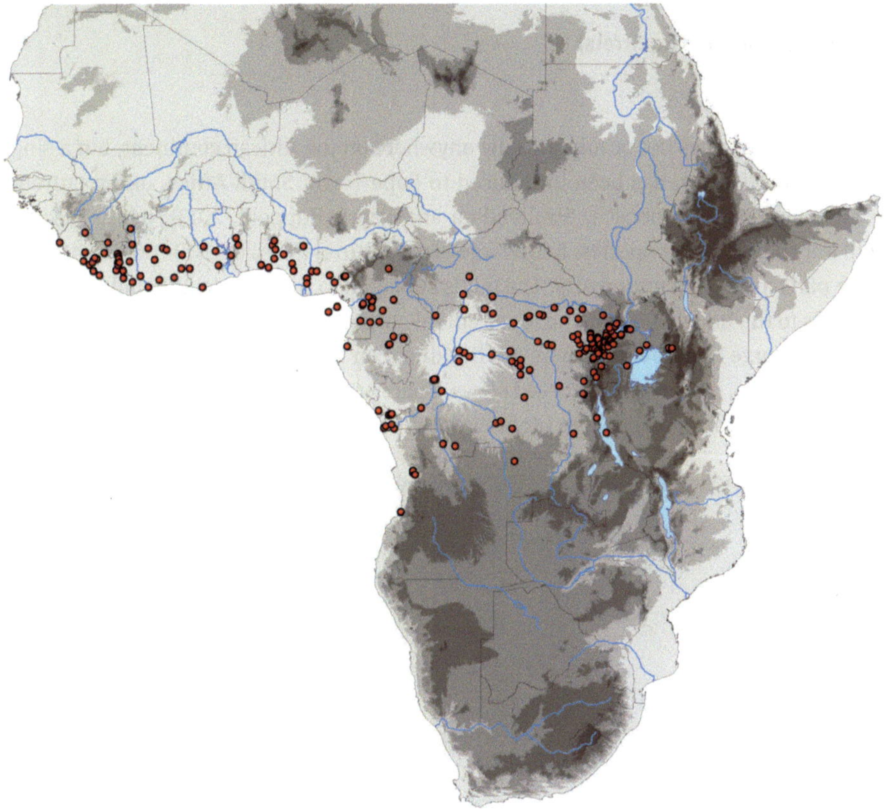

Figure 2.70 Distribution map

Measurements

External and cranial measurements (mm) and mass (g) for *Protoxerus stangeri*, sexes combined

	Mean	Min	Max	N
Mass	616.8	393	716	6
HB	276.7	210	310	12
Tail	316.8	230	394	15
HF	64.9	60	72	15
Ear	21.4	15	25	14
GLS	66.4	62.0	70.5	9

Genus Account

Sciurus Linnaeus, 1758 (Sun Squirrels)

Diagnosis

This very diverse genus does not occur naturally anywhere on the African continent, but a single species, *Sciurus carolinensis*, has been introduced to Cape Town, South Africa, and it currently occurs as far inland as Stellenbosch, Swellendam and Paarl (Happold, 2013); hence this brief account will only describe this species. It is a large-sized squirrel (Mass ca. 300-700 g), with grey-brown upper parts and white underparts. The tail is subequal to the body length and bushy with broad white tips to the hairs. There are no longitudinal stripes or bands on the back or sides. The ears are relatively large and pointed. There are five digits on the hindfoot and four digits on the forefoot, all with sharp, powerful and curved claws. There are four pairs of nipples (1 + 1 + 1 + 1 = 8) placed along the thorax and abdomen (Koprowski, 1994).

Figure 2.71 Photograph of *Sciurus carolinensis* (Alex Sliwa).

Cranial Characters

The skull is large (57-63 mm in greatest length) (Moncrief, 1993) and robust. It is domed, curving down posteriorly towards the occipitals and anteriorly to the rostrum. The palate ends well beyond the posterior margin of the third molars. The zygomatic arch is long and narrow, but thickening somewhat in the middle. The supraorbital foramen is absent and replaced with an open notch. The post-orbital process is sharply pointed downwards. The anterior palatal foramina are short and narrow. The bullae are moderately large, being similar in length to the molar toothrow.

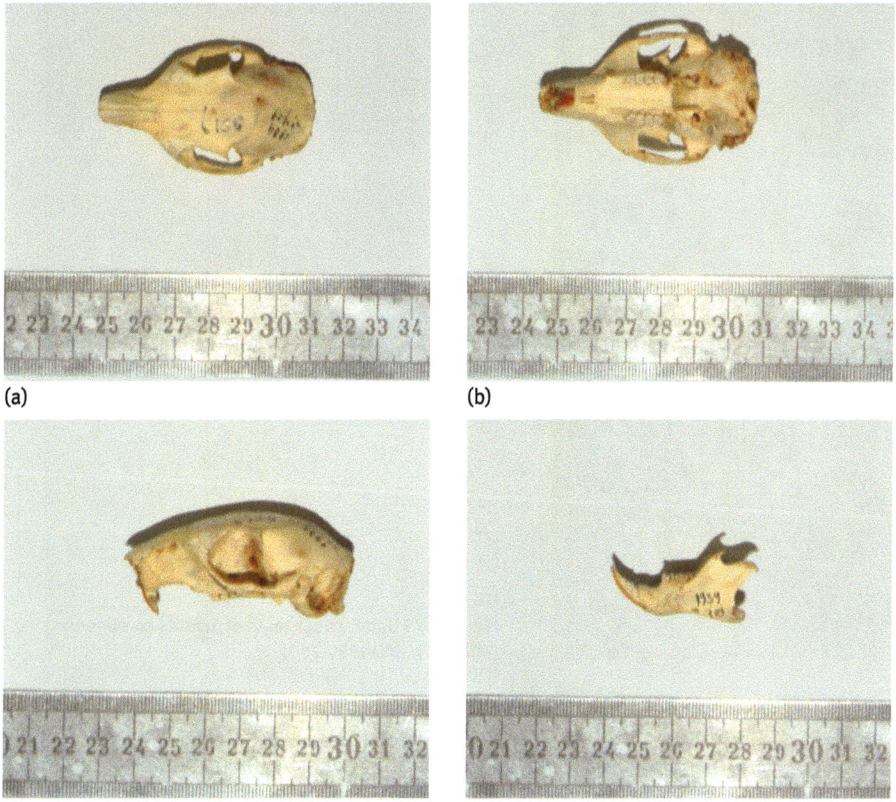

(a) (b) (c) (d)

Figure 2.72 a-d Cranium of *Sciurus carolinensis* (MNHN 1939-709) (from top left to bottom right: dorsal view, ventral view, lateral view and lower jaw).

Dentition

Sciurus carolinensis has ungrooved, opisthodont, orange incisors. The dental formula is: 1023/1013, with the first upper premolar being reduced to a small peg, resulting in four cheekteeth being present in each half jaw. The last premolar is slightly smaller than the molars, with four transverse ridges and a longitudinal ridge on the inner side of each molar.

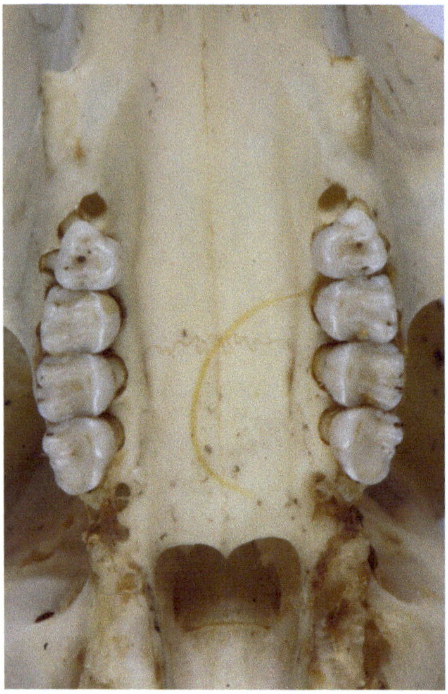

Figure 2.73 Upper cheek teeth of *Sciurus carolinensis* (MNHN-ZM-MO 1939-709).

General

Sciurus carolinensis has been introduced to Cape Town, South Africa, and has spread to neighbouring towns including Paarl and Stellenbosch. It inhabits gardens, parks, and exotic plantations in urban and suburban environments. This species has been introduced world-wide, where it has displaced native species (e.g. *S. vulgaris* in the British Isles), and has therefore received much research attention as a result (e.g. Reynolds, 1985; Kenward, et al., 1998; Sandro, 2008). By comparison with African species, *S. carolinensis* has a large litter size 1-6 (mean of 2.9), with gestation and lactation periods of 44 and 65 days, and two births per year (Hayssen, 2008). A full review of its biology is given in Koprowski (1994). Happold (2013) reviews its biology in South Africa.

Systematic Notes

The genus *Sciurus* is placed in the subfamily Sciurinae, and is not closely related to the African squirrels.

Species List

One introduced species from the genus *Sciurus* occurs in Sub-Saharan Africa.
Sciurus carolinensis Gmelin, 1788

Species Account

Sciurus carolinensis Gmelin, 1788

Diagnosis

See genus account.

Taxonomic Issues

See genus account.

Distribution

This species is restricted to Cape Town and surrounding areas in the extreme south-western parts of South Africa where it is associated with urban areas.

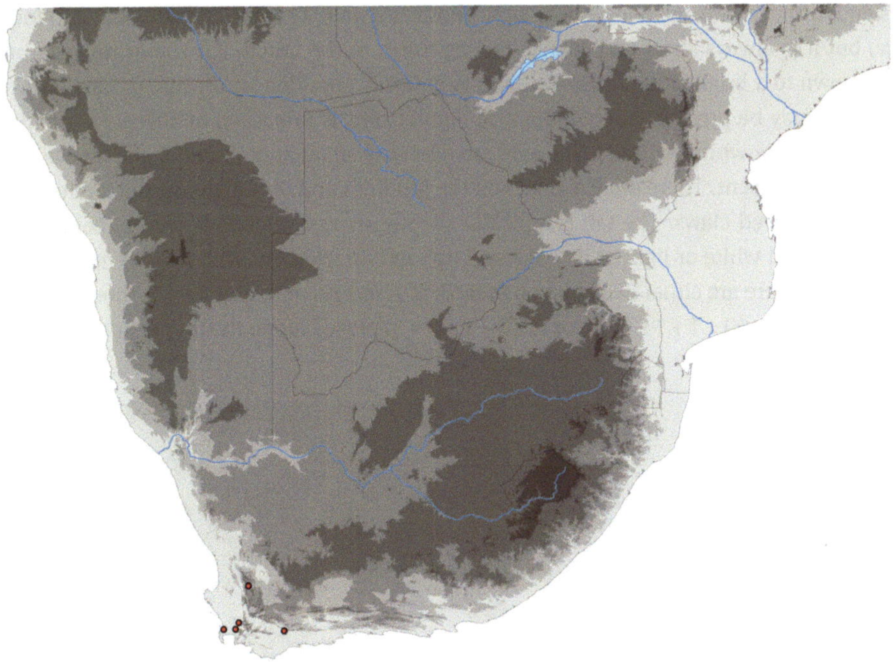

Figure 2.74 Distribution map

Measurements

External and cranial measurements (mm) and mass (g) for *Sciurus carolinensis*, sexes combined

	Mean	Min	Max	N
Mass	579	434	750	256
HB	282	-	-	250
Tail	216	115	269	250
HF	60	51	67	179
Ear	-	25	33	-
GLS	59.0	54.6	62.9	13

*Happold (2013)

Genus Account

Xerus Hemprich & Ehrenberg, 1933 (Ground Squirrels)

Diagnosis

This distinctive genus includes species of medium to large size (Mass ca. 300-500 g). *Xerus* has a long, slender body, with relatively short legs, and bushy tail. There is considerable variation in pelage, but it is always short and coarse, feeling rough to the touch. Fur colour ranges from drab light brown to a brighter sandy brown, and is typically speckled with white or pale brown. A lateral stripe may be present (*X. erythropus*, *X. inauris*, and *X. princeps*), or absent (*X. rutilus*). The hairs are paler or white ventrally. The ears are relatively small and rounded, and an eye ring or cheek stripe is present. There are five digits on the hindfoot and four digits on the forefoot, all with sharp and curved claws. The tail is typically slightly shorter than body length, being dark along its centre and white or light at its edge and tip, and grizzled but without alternating dark and light bands. There are either two pairs (*X. inauris*, *X. princeps*, and *X. rutilus*) or three pairs (*X. erythropus*) of nipples (0 + 1 + 1 + 1 = 6, or 0 + 0 + 1 + 1 = 4) placed along the thorax and abdomen (Moore, 1961).

(a)

(b)

(c)

Figure 2.75 a-c Photograph of *Xerus inauris* (Peter Taylor and Emmanuel Do Linh San) (top two) and of museum specimens (NMK) (below), from top to bottom: *X. erythropus* and *X. rutilus* (Ara Monadjem).

Cranial Characters

The skull is medium in size (mostly 50-62 mm in greatest length). It is domed, curving down posteriorly towards the occipitals and anteriorly towards the rostrum. The palate ends well beyond the posterior margin of the third molars. The zygomatic arch is long and narrow. The supraorbital foramen is absent, but the post-orbital process is narrow and projecting downwards. The anterior palatal foramina are short and narrow. The bullae are large and inflated, being longer than the molar toothrow.

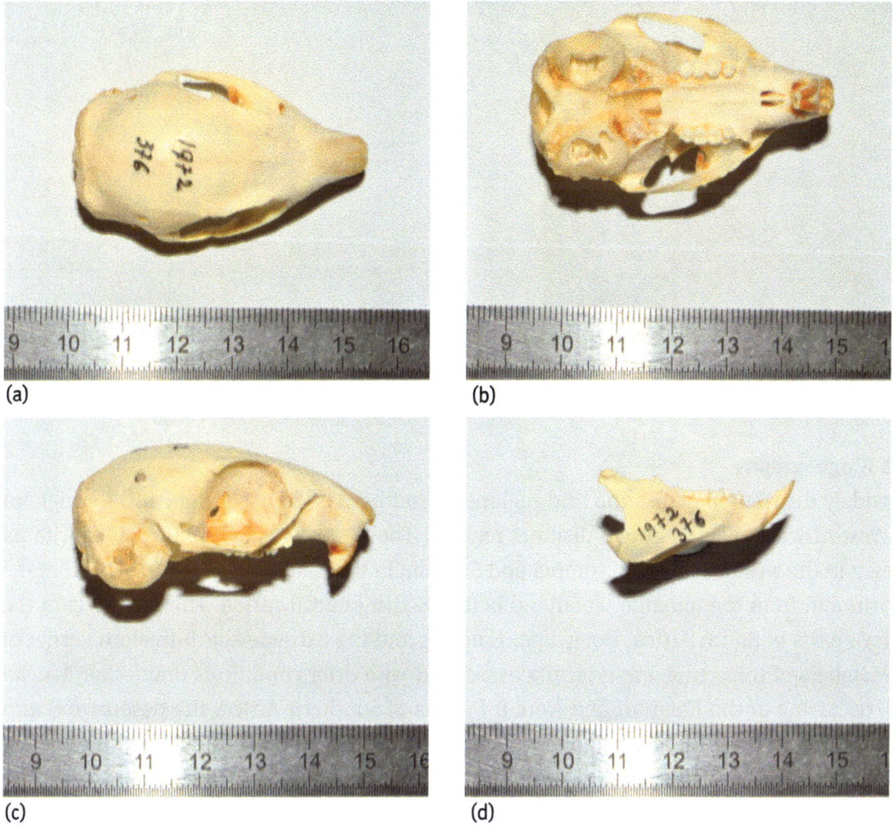

Figure 2.76 a-d Cranium of *Xerus rutilus* (MNHN-ZM-MO 1972-376) (from top left to bottom right: dorsal view, ventral view, lateral view and lower jaw).

Dentition

Xerus has ungrooved, opisthodont, orange incisors. The dental formula is: 1013/1013 or 1023/1013 (Rosevear, 1969). When present, the additional premolar is reduced to a peg resulting in four functional cheekteeth being present in each half jaw. The second premolar and third molar are slightly smaller than the first and second molars. There are well developed transverse ridges and a longitudinal ridge on the inner side of each molar.

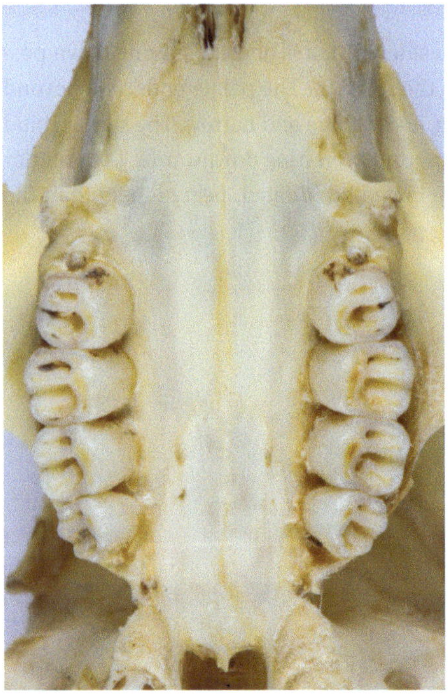

Figure 2.77 Upper cheek teeth of *Xerus erythropus* (MNHN-ZM-MO 1995-3071).

Habitat & Biogeography

Xerus is widely distributed across the arid regions of Southern and East Africa, and throughout tropical West Africa. It occurs in two disjunct regions. The first covers an area from Mauritania and Senegal in the west through to Ethiopia and Somalia in the east. It is largely absent from the Congo basin and from the miombo woodland belt of South-Central Africa. The second is in the arid western parts of South Africa, Botswana, Namibia and the extreme south-western corner of Angola. Members of this genus are typically associated with drier conditions than other African squirrels, occurring in the Kalahari and Namib Deserts of southern Africa, the deserts and arid savannas of the horn of Africa and neighbouring Kenya-Tanzania, and the arid Sahel. One species (*X. erythropus*) extends its range into the rainforest belt, but only occurring in clearings within the forest, not in the forest itself. Two species (*X. erythropus* and *X. inauris*) also occur in dry grasslands, and one in broken, rocky terrain (*X. princeps*) (Herron et al., 2005). Ge et al. (2013) present a biogeographical hypothesis suggesting the colonisation of Africa by the Xerini (*Xerus* and *Atlantoxerus*, see Systematic Notes, below) 13-11 MYA, some 3-9 MYA after the Protoxerini (the remaining African squirrels). *Xerus* is terrestrial, and does not climb trees, resting up in burrows at the base of trees, in termite mounds, piles of rocks or in the ground (Happold, 2013).

Fossil Record

The oldest *Xerus* fossil is from 11 MYA, Ethiopia (Geraads, 2001). In North Africa, it has been reported from the Late Miocene to the Early Pliocene; in Central Africa (Chad) in the Early Pliocene; in Eastern Africa from the Late Miocene to the Late Pliocene; and in Southern Africa from the Late Pleistocene (Winkler et al, 2010).

Biology & Ecology

In contrast to most other African squirrels, members of this genus have been studied in detail, and comprehensive reviews are available for all four species: *X. erythropus* (Herron & Waterman, 2004), *X. inauris* (Skurski & Waterman, 2005), *X. princeps* (Waterman & Herron, 2004), and *X. rutilus* (O'Shea, 1991). Furthermore, numerous studies have been conducted on the morphology, basic biology, behavioural ecology, behaviour, physiology and biogeography of this group (e.g. Waterman, 1995, 1996, 2002; Linn & Key, 1996; O'Shea, 1976; Herron et al., 2005; Michaux et al., 2008; Scantlebury et al., 2008; Fick et al., 2009). All species of *Xerus* are mainly herbivorous, but also taking insects (see further details in the review papers). Home ranges vary with species, sex (larger in males) and environment (larger in drier, harsher environments), mostly 1-14 ha in extent (O'Shea, 1976; Herzig-Straschil, 1978; Linn & Key, 1996; Waterman, 1995). Reproduction has been relatively well studied. Litter size is 2-4 (*X. erythropus*), or 1-3 (the other three species), and the gestation and lactation periods are 42-49 and 45-56 days (Herzig-Straschil, 1978; Herzig-Straschil & Herzig, 1989; Waterman, 1996). Up to 4 litters may be raised in a year (Waterman, 1996).

Systematic Notes

On morphological grounds, this genus was previously been placed in the tribe Xerini together with the North African *Atlantoxerus* and the Asian *Spermophilopsis*. A recent molecular phylogeny supports these relationships (Mercer & Roth, 2003; Steppan et al., 2004). The Xerini are basal to the subfamily Xerinae in which all African squirrels are placed, but also includes the Asian Marmotini (Steppan et al., 2004). The closest relative of *Xerus* is the North African *Atlantoxerus*. The four recognised species of *Xerus* are clearly distinct, despite morphological similarities, especially between *X. inauris* and *X. princeps*. These two species are not distinguishable on external appearance, but occupy different habitats (*X. princeps* prefers rocky, hilly terrain or gravel plains) and exhibit different behaviour (social in *X. inauris* and solitary in *X. princeps*), and are molecularly distinct (Herron et al., 2005). A recent molecular phylogeny for the genus *Xerus* indicates that *X. erythropus* is more distantly related to the other three species, with *X. inauris* and *X. princeps* each other's closest relatives (Herron et al., 2005). There appears to be significant geographically-based genetic structure in populations of *X. inauris* from Namibia and South Africa, however the genetic distances between these populations are relatively small (Herron et al., 2005), suggesting a single species. Six subspecies have are recognised in *X. erythropus*, based on pelage colour (Happold, 2013). This species is highly variable in its pelage, being darker in tropical habitats, and paler in arid regions. Whether this variability is solely due to environmental pressure or evolutionary history has yet to be investigated.

The diploid number of chromosomes in all *Xerus* species has been recorded as: 2n = 38 (Nadler & Hoffmann, 1974; Robinson et al., 1986; Dobigny et al., 2002a).

Species List

Four species are described within the genus *Xerus*, all restricted to Africa.
Xerus erythropus (E. Geoffroy, 1803)
Xerus inauris (Zimmermann, 1780)
Xerus princeps (Thomas, 1929)
Xerus rutilus (Cretzschmar, 1828)

Key to Species

1	Flank stripe absent; restricted to East Africa and the Horn of Africa	*X. rutilus*
	A conspicuous white flank stripe present	2
2	Widespread in the northern savannas, extending south to the equator in East Africa	*X. erythropus*
	Restricted to Southern Africa	3
3	Incisors white; tail hairs with two black bands; widespread in the arid regions of Southern Africa	*X. inauris*
	Incisors orange; tail hairs with three black bands; restricted to the Namib desert and neighbouring arid zone	*X. princeps*

Species Accounts

Xerus erythropus (E. Geoffroy, 1803)

Diagnosis

This is the only *Xerus* species over most of its range, overlapping marginally with *X. rutilus* in Kenya, Ethiopia and Eritrea from which it can readily be distinguished by the presence of a white flank stripe (in *X. rutilus* there is no flank stripe).

Taxonomic Issues

Six subspecies have been described based on extensive variation in pelage, but their taxonomic relationships are not known. Much of this variation in pelage colour may be clinal in nature with specimens from high rainfall areas being darker than those from more arid areas.

Distribution

This species is widely distributed throughout the savannas and forest clearings of West Africa from Senegal in the west and extending eastward to the northern parts of East Africa.

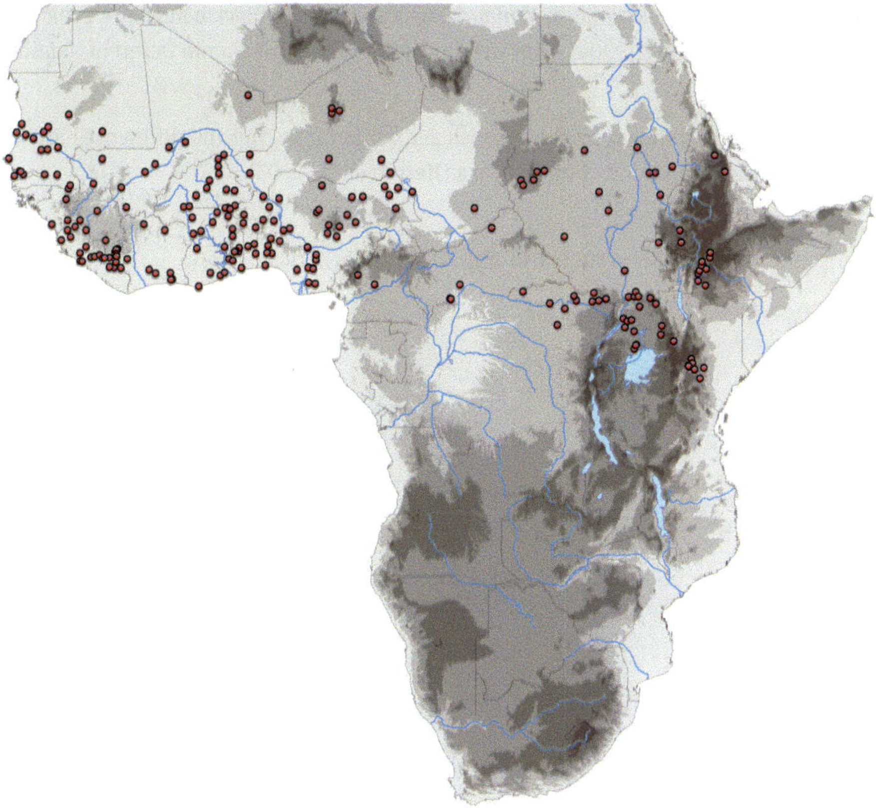

Figure 2.78 Distribution map

Measurements

External and cranial measurements (mm) and mass (g) for *Xerus erythropus*, sexes combined

	Mean	Min	Max	N
Mass	491.6	300	770	14
HB	254.1	210	300	20
Tail	229.5	164	280	20
HF	65.6	56	73	19
Ear	16.9	11	20	19
GLS	59.3	52.9	63.7	18

Xerus inauris (Zimmermann, 1780)

Diagnosis

Only two species of *Xerus* occur in Southern Africa: *X. inauris* and *X. princeps* which can be distinguished based on: the colour of the incisors (white in *X. inauris*, orange in *X. princeps*); the colour of tail hairs (white with two black bands in *X. inauris*, white with three black bands in *X. princeps*); and some dental patterns (Denys 1990a). They overlap marginally in the Namib Desert of Namibia.

Taxonomic Issues

Shallow genetic structure is shown by this species across its range, but the differences are not sufficient to warrant the recognition of further taxa (Herron et al., 2005).

Distribution

This species occurs widely in the western arid zone of southern Africa.

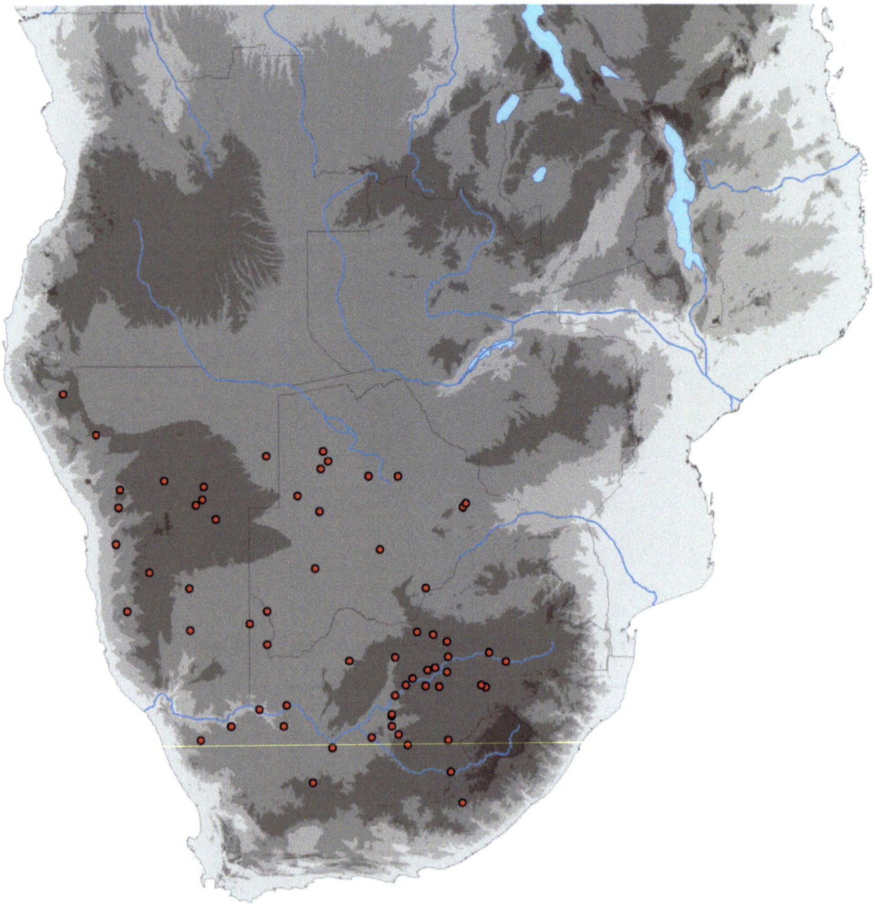

Figure 2.79 Distribution map

Measurements

External and cranial measurements (mm) and mass (g) for *Xerus inauris*, sexes combined

	Mean	Min	Max	N
Mass	592.5	520	720	6
HB	247.4	215	275	20
Tail	213.5	177	257	20
HF	66.4	62	70	17
Ear	10.8	9	12	5
GLS	56.3	52.4	62.2	23

Xerus princeps (Thomas, 1929)

Diagnosis

Only two species of *Xerus* occur in Southern Africa: *X. inauris* and *X. princeps* which can be distinguished based on: the colour of the incisors (white in *X. inauris*, orange in *X. princeps*); the colour of tail hairs (white with two black bands in *X. inauris*, white with three black bands in *X. princeps*); and dental characters (Denys 1990). They overlap marginally in the Namib Desert of Namibia.

Taxonomic Issues

None.

Distribution

This species has a narrow distribution from south-western Angola through western Namibia and marginally into South Africa.

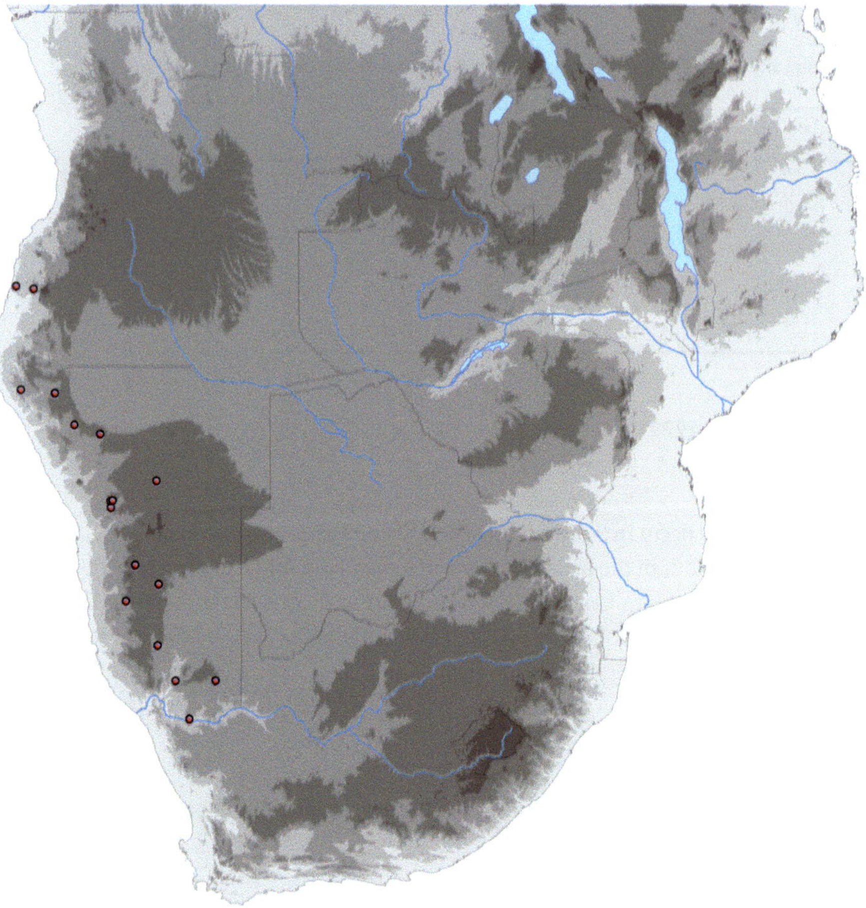

Figure 2.80 Distribution map

Measurements

External and cranial measurements (mm) and mass (g) for *Xerus princeps*, sexes combined

	Mean	Min	Max	N
Mass	612.7	528	700	3
HB	256.8	210	285	13
Tail	250.2	220	285	13
HF	68.6	65	72	10
Ear	13.4	10	15	5
GLS	58.7	56.3	62.2	11

Xerus rutilus (Cretzschmar, 1828)

Diagnosis

This is the only *Xerus* species over most of its range, overlapping marginally with *X. erythropus* in Kenya, Ethiopia and Eritrea from which it can readily be distinguished by the lack of a flank stripe (in *X. rutilus* there is a conspicuous white flank stripe).

Taxonomic Issues

Eight subspecies have been described based on extensive variation in pelage, but their validity is not known. Much of this variation in pelage colour may be clinal in nature with specimens from more arid zones being paler or related to differences in soil colour (Happold, 2013).

Distribution

This species is widely distributed in the arid zone of East Africa and extending into the Horn of Africa.

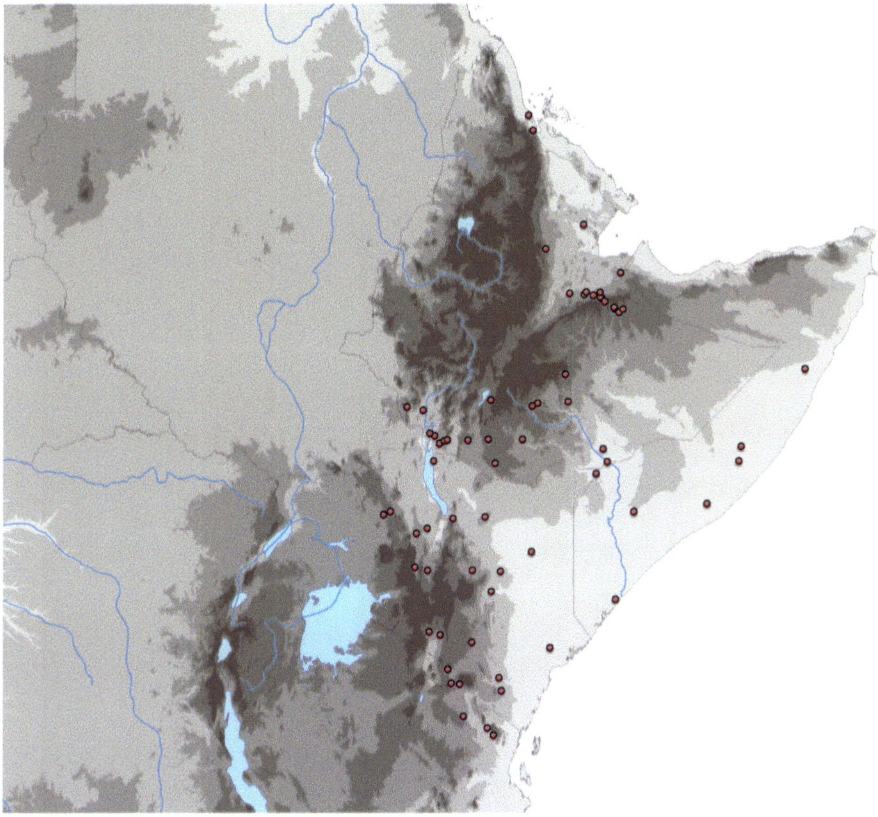

Figure 2.81 Distribution map

Measurements

External and cranial measurements (mm) and mass (g) for *Xerus rutilus*, sexes combined

	Mean	Min	Max	N
Mass	308.8	200	410	8
HB	219.1	155	250	21
Tail	181.2	85	220	21
HF	51.8	48	55	12
Ear	14.6	10	17	16
GLS	52.1	46.8	55.4	12

Family: Myocastoridae

This family is represented by a single exotic species introduced to Sub-Saharan Africa.

Genus Account

Myocastor Kerr, 1792 (Coypu or Nutria)

Diagnosis

This monotypic genus does not occur naturally anywhere on the African continent, but has been introduced, with feral populations only occurring in Kenya, East Africa. It is a very large (mostly > 6 kg in adults), robust beaver-shaped rodent, with long shaggy fur. It has an arched back, short legs and a long, round tail. The pelage is typically dark brown dorsally and ventrally, appearing almost black when it is wet. The tail is shorter than the body length (about 70-75% of HB) and mostly naked. The ears are relatively small and rounded. There are four long digits on each of the forefeet and hindfeet, all with long claws. Digits 2-5 of the hindfeet are webbed. There are four to five pairs of nipples situated dorsolaterally (Woods et al., 1992).

(a) (b)

Figure 2.82 a and b Photograph of *Myocastor coypus* (Paolo Torchio, left and Olivier Hamerlynck, right).

Cranial Characters

The skull is large (102-126 mm in greatest length) and robust (Woods et al., 1992). It is flat dorsally, and the rostrum is long and narrow. The palate ends at, or slightly beyond, the posterior margin of the third molars. The zygomatic arch is robust and deep. The infra-orbital foramen is well developed, and the paraoccipital processes are large and pointed. The anterior palatal foramina are short and narrow. The bullae are relatively small.

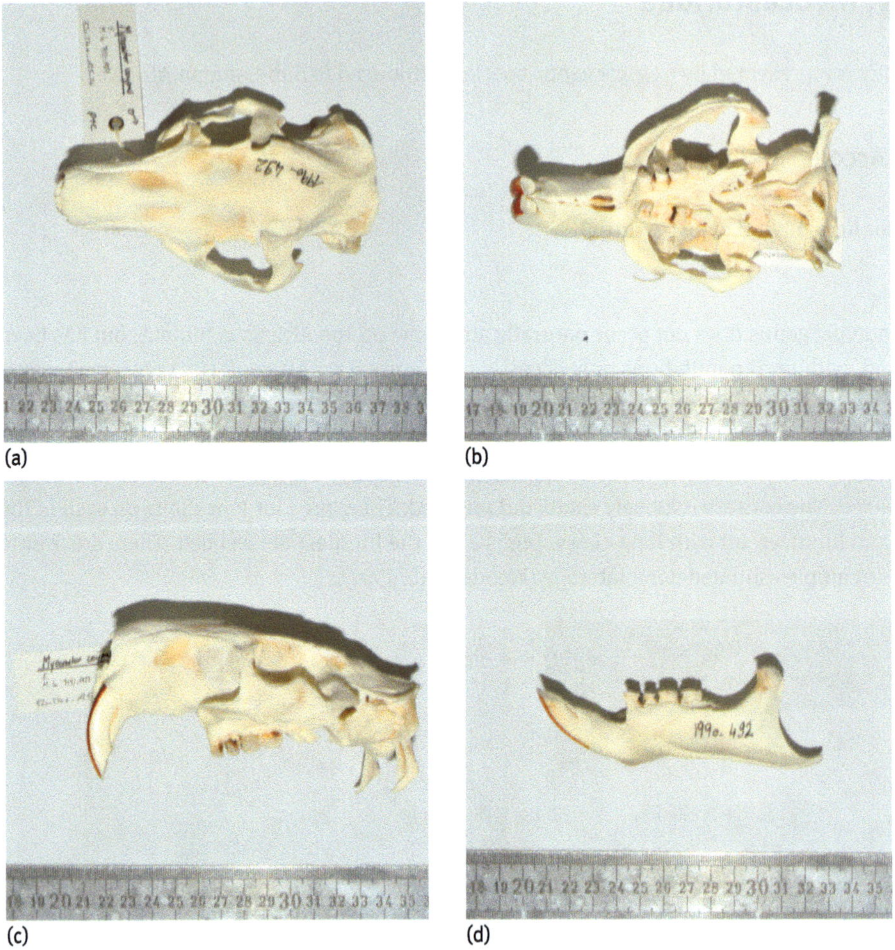

Figure 2.83 a-d Cranium of *Myocastor coypus* (MNHN-ZM-MO 1990-492) (from top left to bottom right: dorsal view, ventral view, lateral view and lower jaw).

Dentition

Myocastor coypus has ungrooved, slighly orthodont, orange incisors. The dental formula is: 1013/1013, resulting in four cheekteeth being present in each half jaw. Characteristically, the toothrow of cheekteeth converge anteriorly. The last premolar and three molars are similar in size, with transverse ridges.

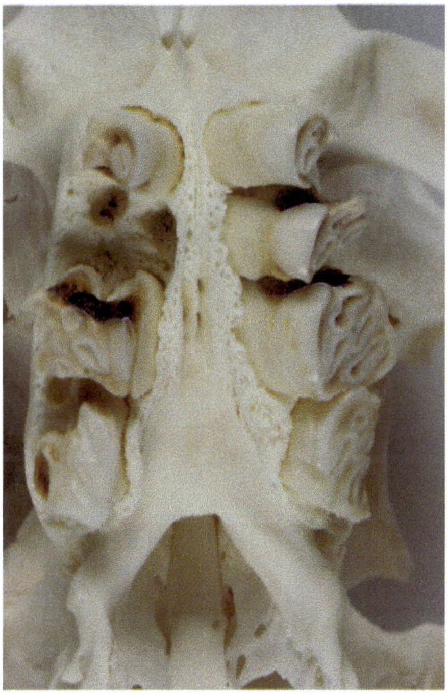

Figure 2.84 Upper cheek teeth of *Myocastor coypus* (MNHN-ZM-MO 1990-492).

Habitat & Biogeography

The biology of this species has been reviewed by Woods et al. (1992). *Myocastor* is native to the Patagonian region of South America, but has been introduced widely across the world (reviewed by Carter & Leonard, 2002). The African introduction has recently been reviewed by Happold (2013), who found no evidence for the existence of feral populations of Coypu in Southern Africa or Tanzania, and the only current populations in existence are in Kenya, east of the Rift Valley. Harper et al. (1990) give an account of the colonisation of Lake Naivasha by Coypu in 1965, the ineffective control of its population by introduced pythons (*Python sebae*), and the final decline and extinction of that population in 1984 for reasons unknown. *Myocastor coypus* has large litter sizes (mean 3-6, but up to 12), with long gestation and lactation periods of ca. 130 and 54 days, respectively (Woods et al., 1992). In good conditions, females can produce a mean of 15 offspring per year (Woods et al., 1992).

Taxonomy

Myocastor belongs to the Caviomorpha, a group of South American rodents which are the sister taxa of African Phiomorpha (Thryonomyidae and Hystricidae) among the Ctenohystrica clade (Blanga-Kanfi et al., 2009).

Species List

One species of *Myocastor* occurs as an introduced species in Kenya, East Africa.
Myocastor coypus (Molina, 1782)

Species Accounts

Myocastor coypus (Molina, 1782)

Diagnosis
See genus account.

Taxonomic Issues
See genus account.

Distribution
This species is currently only known from a few localities in central Kenya.

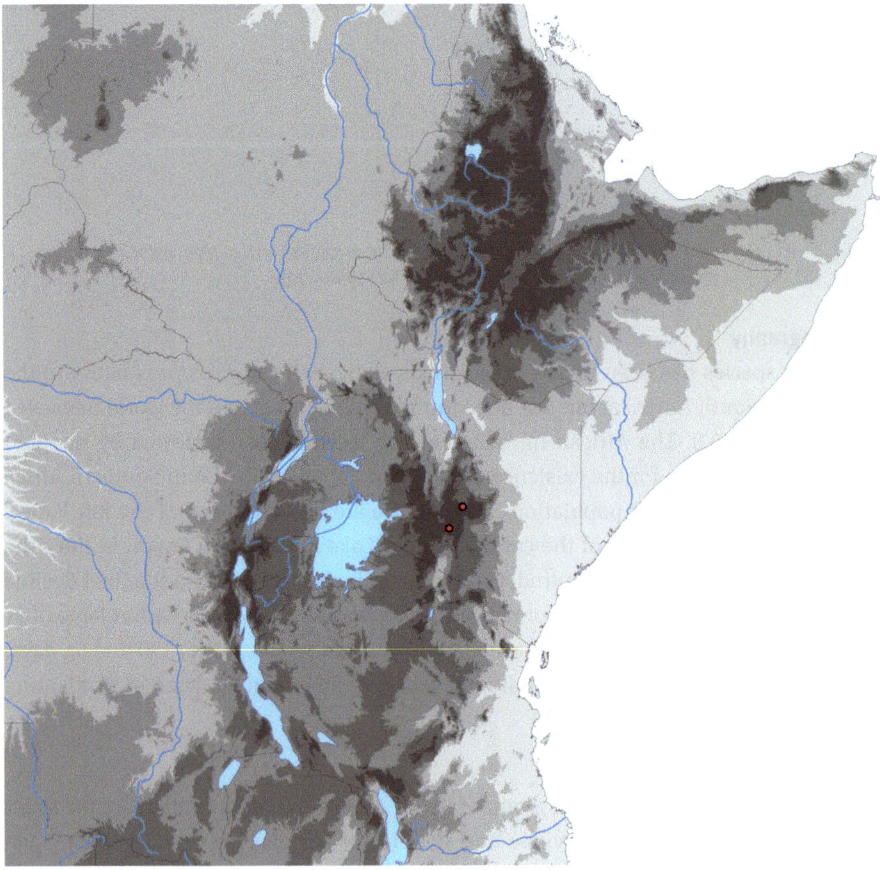

Figure 2.85 Distribution map

Measurements

External and cranial measurements (mm) and mass (g) for *Myocastor coypus*, sexes combined

	Mean	Min	Max	N
Mass*	6400, 6700	-	-	-
HB	557	-	-	1
Tail	435	-	-	1
HF	140	-	-	1
Ear	30	-	-	1
GLS*	114.2	102.2	126.4	-

*Woods et al. (1992). Mass quoted is the mean for females and males, respectively

Family: Ctenodactylidae

This family is represented by three species in Sub-Saharan Africa.

Key to Genera

1	Upper incisors clearly grooved; auditory bullae poorly inflated; restricted to Mauritania and extreme western Mali	*Felovia*
	Upper incisors not grooved (or only faintly grooved); auditory bullae greatly inflated	2
2	Palate extends posteriorly beyond the last molars; present in the central Sahara	*Massoutiera*
	Palate does not extend posteriorly beyond the last molars; restricted to the Horn of Africa	*Pectinator*

Genus Account

Felovia Lataste, 1886 (Felou Gundis)

Diagnosis

This monotypic genus has a highly restricted distribution in Mauritania and Mali, and is the only species of ctenodactylid within its range. It is a medium-size rodent (Mass ca. 180 g), with a characteristically stout body, short limbs and a short, furry tail typical of the family. The pelage is dense, short and soft to the touch, being soft bright orange-brown dorsally, and slightly paler ventrally. The hairs have light grey bases, giving the animal a lighter and brighter appearance compared with *Massoutiera*. The feet are dark brown dorsally with four digits and sharp claws, and a row of stiff hairs at the tips. These hairs are particularly brittle-like on the digits of the hindfeet, hiding the claws. The ears are small and pressed against the head. There is a distinct pale eye ring. The tail is short, covered in feathery hair with a black tip. There are two pairs of nipples (2 + 0 = 4) placed anteriorly and laterally (George, 1978a).

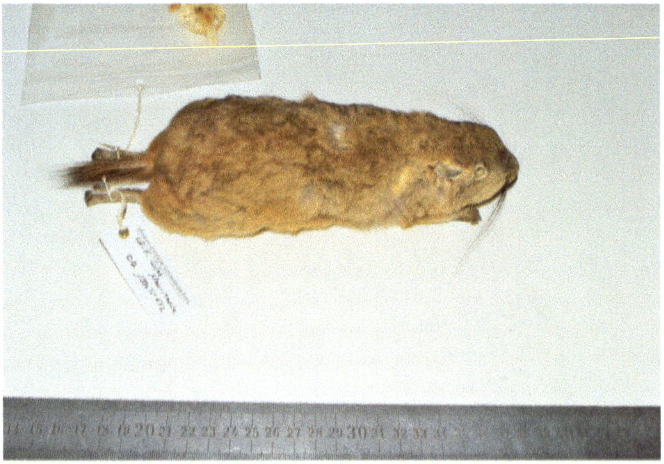

Figure 2.86 Photograph of *Felovia vae* museum specimen (MNHN-ZM-MO 1994-612) (Ara Monadjem).

Cranial Characters

The skull is robust and medium-sized (typically around 45 mm in greatest length in adults). It is broad but relatively flat for its size, and shallowly domed posteriorly. The rostrum is narrow and parallel-sided, but the cranium widens considerably posteriorly with the zygomatic arches at a sharp angle to each other. The nasal bones narrow posteriorly. There is a well-developed process on the vertical arm of the zygomatic arch extending posteriorly into the orbit. The infraorbital foramen is large and elliptical. The anterior palatal foramina are long and widening posteriorly, extending for more than half the length of the diastema. The bullae are relatively large, but not particularly inflated. Like all Ctenodactylidae the mandible is low in its posterior part and lacks the coronoid apophysis.

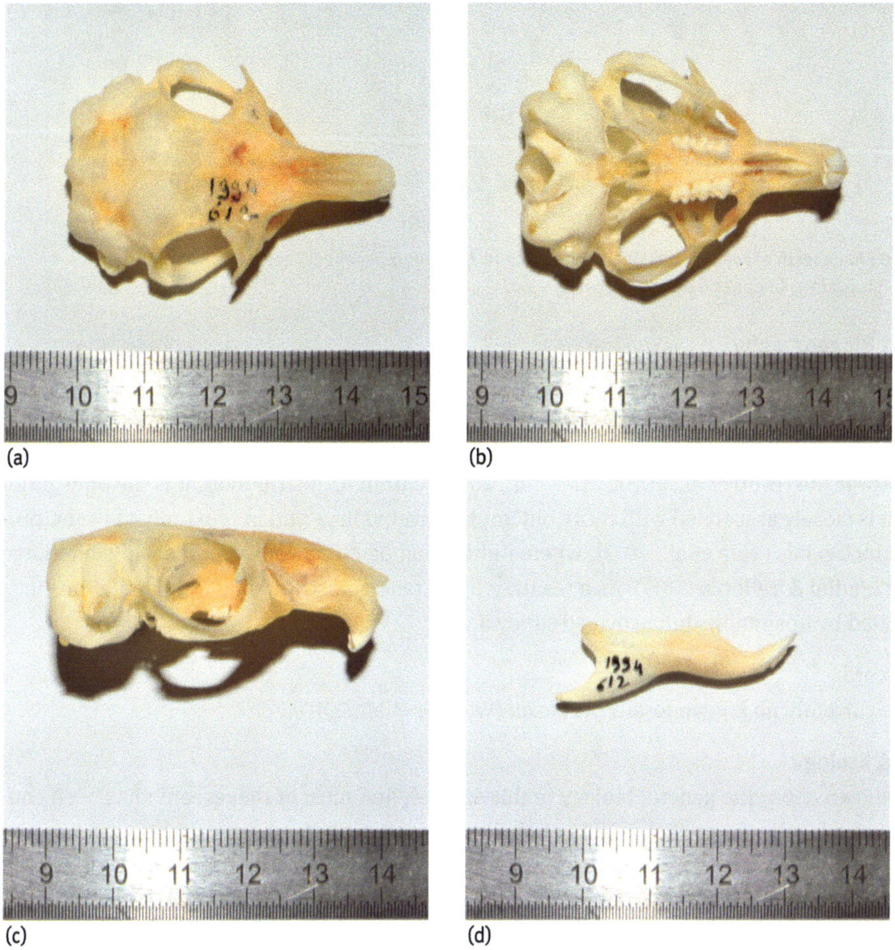

Figure 2.87 a-d Cranium of *Felovia vae* (MNHN-ZM-MO 1994-612) (from top left to bottom right: dorsal view, ventral view, lateral view and lower jaw).

Dentition

Felovia has strong, grooved heavily opisthodont upper incisors. The dental formula is: 1013/1013, but the premolars are shed when the last molars erupt, so that there are usually only three cheekteeth in each half jaw (Rosevear, 1969). The molars are all roughly the same size, simple and uncusped. Each molar comprises of two oval laminae which tend to be more angular than in *Massoutiera*.

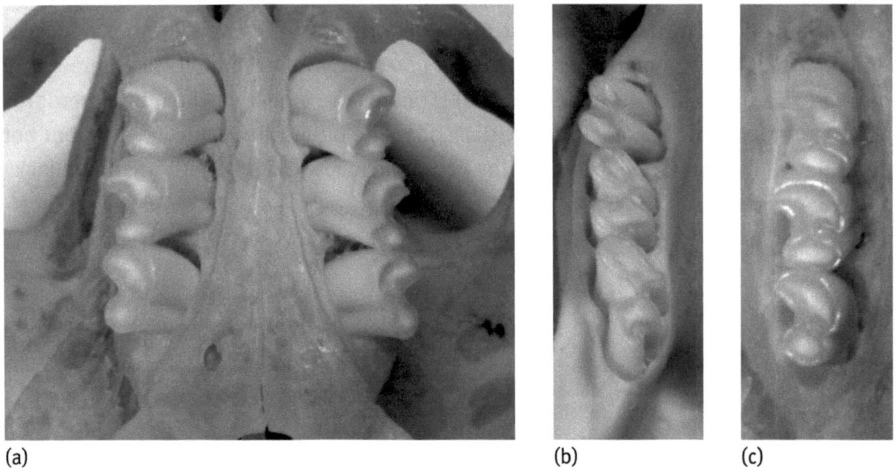

Figure 2.88 Cheek teeth of *Felovia vae* (MNHN-ZM-MO 1994-612) (A: upper cheek teeth; B: lower left toothrow; C: upper right toothrow).

Habitat & Biogeography

The Felou Gundi has a very limited distribution, restricted to a narrow zone of rocky, broken country in north-western Mali and south-central Mauritania. It occurs in the Felou hills along the upper Senegal River in Mali, and various mountains in Mauritania including, Assaba, Afolle, Tagant, Adrar Ata (Brito et al., 2010, Vale et al., 2012). Within its distribution, it is the only gundi known. It is closely associated with rocky outcrops in river valleys and in proximity to water, provided by rock pools (Vale et al., 2012), where individuals perch on canyon walls, cliffs and large boulders (Padial & Telleria, 2009). As a result, it has a patchy distribution, and populations may be separated by unsuitable dune-covered sandy areas.

Fossil Record

There are currently no known fossils of *Felovia* (Winkler et al., 2010).

Biology & Ecology

Little is known about the general biology of this species, and most of the research has been conducted at one locality in the Felou hills of Mali. It is vegetarian, feeding on leaves, grasses and seeds (George, 1974). *Felovia vae* has a distinct and noisy alarm call (George, 1981a). It lives in family groups, which are larger in Mali than in Mauritania (George, 1981; Padial & Telleria, 2009). Based on a limited dataset, breeding appears be seasonal, with births occurring in December-January; litter size is 1 (George, 1978a).

Systematic Notes

The systematic position of the ctenodactylids has been debated for decades (e.g. Graur et al., 1992), but a recent phylogeny based on five nuclear genes places the clade as a sister to the Hystricognathi (Blanga-Kanfi et al., 2009). The closest relative of the genus *Felovia* appears to be *Massout-*

iera (George, 1985; Lopez Antonanzas & Knoll, 2011; Gomes Rodrigues et al., 2012), but a molecular phylogeny for the family has yet to be published. There is currently one species of *Felovia* recognised. However, given the patchy nature of its distribution, it would be interesting to examine the genetic structure of the species. The possibility exists that the Mali population is isolated from those in Mauritania by unsuitable habitat (Vale et al., 2012). The diploid number of chromosomes in *F. vae* has been recorded as: 2n = 36, FN = 72 (Mali, George, 1979).

Species List

One species described within the genus *Felovia*, and restricted to Africa.
Felovia vae (Lataste, 1886)

Species Account

Felovia vae (Lataste, 1886)

Diagnosis

See genus account.

Taxonomic Issues

The southern population in Mali is apparently isolated from populations further north (Vale et al., 2012) and may therefore prove to be distinct.

Distribution

This species has a narrow distribution restricted to Mauritania and marginally into western Mali.

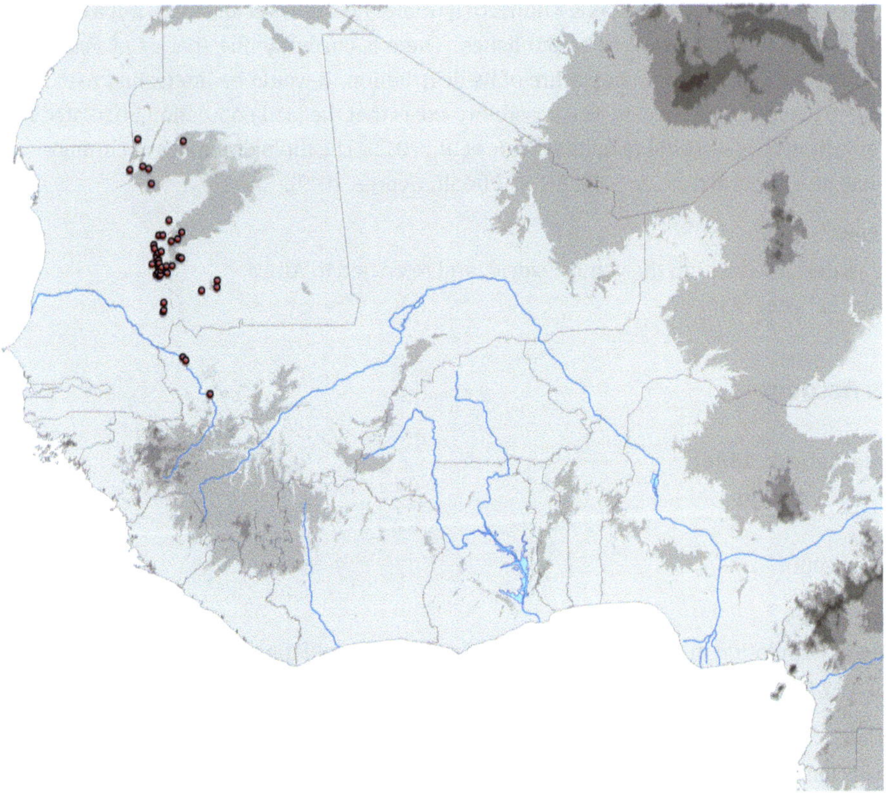

Figure 2.89 Distribution map

Measurements

External and cranial measurements (mm) and mass (g) for *Felovia vae*, sexes combined

	Mean	Min	Max	N
Mass*	185.8	-	-	10
HB	180.7	169	190	3
Tail	57.0	-	-	1
HF	33.3	32	35	3
Ear	15.0	14	16	2
GLS	45.3	42.9	47.2	4

*George (1978a)

Genus Account

Massoutiera Lataste, 1885 (Mzab Gundis)

Diagnosis

This monotypic genus has a relatively widespread distribution in the Sahara, and is the only species of ctenodactylid within its range. It is a medium-size rodent (Mass ca. 200 g), with a characteristically stout body, short limbs and a short, furry tail typical of the family. The pelage is dense, short and soft to the touch, being variable in colour from pale- to orange-brown, and paler ventrally. The hairs have dark grey bases, giving the animal a darker appearance compared with *Felovia*. The feet are whitish dorsally with four digits and sharp claws, and a row of stiff hairs at the tips. These hairs are particularly brittle-like on the digits of the hindfeet, hiding the claws. The ears are small and pressed against the head, with a band of pale-coloured hairs around them which may be indistinct. The eye ring, if present, is indistinct. The tail is short, covered in feathery hair with a dark tip. There are two pairs of nipples (2 + 0 = 4) placed anteriorly and laterally (George, 1978a).

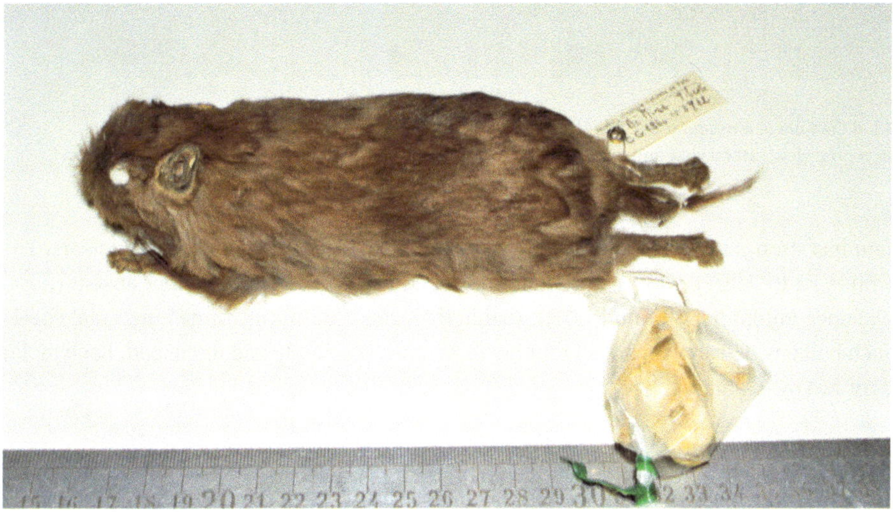

Figure 2.90 Photograph of *Massoutiera mzabi* museum specimen (MNHN-ZM-MO 1960-3812) (Ara Monadjem).

Cranial Characters

The skull is robust and medium-sized (typically around 45 mm in greatest length in adults). It is broad but relatively flat for its size, and shallowly domed. The rostrum is narrow and parallel-sided, but the cranium widens considerably posteriorly with the zygomatic arches at a sharp angle to each other. The nasal bones narrow posteriorly. There is a well-developed supraorbital process on the vertical arm of the zygomatic arch extending posteriorly into the orbit. The infraorbital foramen is large and elliptical. The anterior palatal foramina are long and widening posteriorly, extending for more than half the length of the diastema. The bullae are very large and inflated, extending anteriorly almost to the base of the orbits, and nearly meeting medially. Like all Ctenodactylidae the mandible is low in its posterior part and lacks the coronoid apophysis.

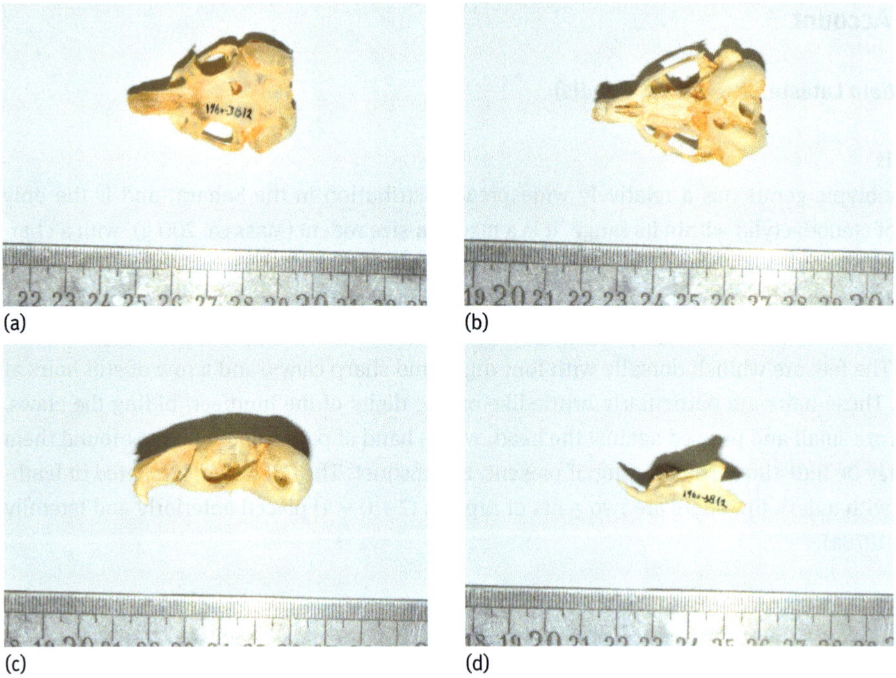

Figure 2.91 a-d Cranium of *Massoutiera mzabi* (MNHN-ZM-MO 1960-3812) (from top left to bottom right: dorsal view, ventral view, lateral view and lower jaw).

Dentition

Massoutiera has strong, ungrooved (or faintly grooved), heavily opisthodont upper incisors. The dental formula is: 1013/1013, with the premolars reduced to tiny pegs or altogether absent (Rosevear, 1969; Lopez Antonanzas & Knoll, 2011), so that there are usually only three functional cheek-teeth in each half jaw. The molars are all roughly the same size, simple and uncusped. Each molar comprises of two oval laminae.

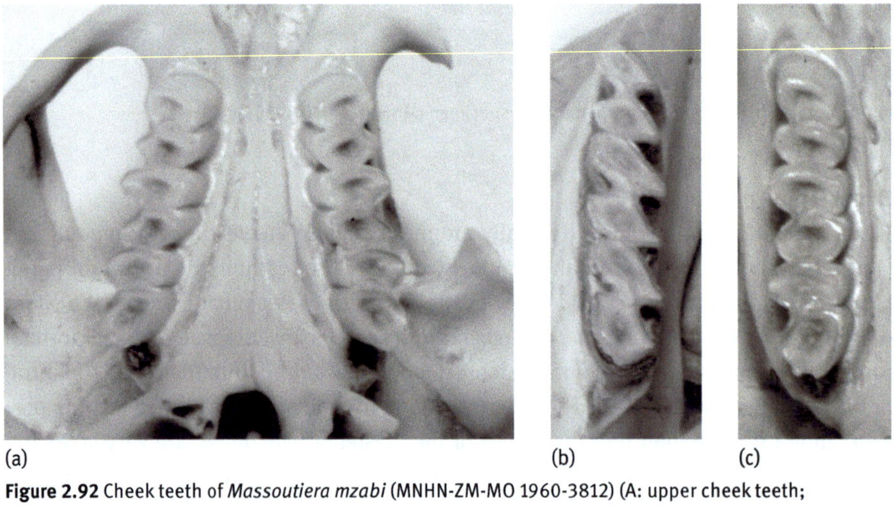

Figure 2.92 Cheek teeth of *Massoutiera mzabi* (MNHN-ZM-MO 1960-3812) (A: upper cheek teeth; B: lower left toothrow; C: upper right toothrow).

Habitat & Biogeography
This species is widely, but patchily distributed in the central Sahara occurring in Algeria, Libya, Mali, Niger and Chad (Gouat et al., 1984; Granjon & Duplantier, 2009; Happold, 2013). Within the region covered by this book, it occurs in three isolated massifs: the Adrar des Iforas, Mali (Dobigny et al., 2001a); Air, Niger (Dobigny et al., 2002a); and Tibesti, Chad. It is closely associated with rocky habitats (Dobigny et al., 2002a; Granjon & Duplantier, 2009), where it shelters in crevices.

Fossil Record
The oldest fossils of *Massoutiera* are from Ti-n Thora East, Libya dated from 9080 BP (Lopez Antonanzas & Knoll, 2010).

Biology & Ecology
Little is known about the general biology of this species in Sub-Saharan Africa, with most of the research having been conducted in Algeria (Gouat et al., 1984, George, 1988). It is vegetarian, feeding primarily on herbs and grasses (George, 1988). *Massoutiera mzabi* is generally silent but may produce a distinct alarm call (George, 1981a). It lives in widely dispersed family groups (George, 1981a). Breeding is seasonal in Algeria, but not known in Sub-Saharan Africa. The litter size is typically two, and a single female offspring took 184 days to reach adult size (George, 1978a).

Systematic Notes
The systematic position of the ctenodactylids has been debated for decades (e.g. Graur et al., 1992), but a recent phylogeny based on five nuclear genes places the clade as a sister to the Hystricognathi (Blanga-Kanfi et al., 2009). The closest relative of the genus *Massoutiera* appears to be *Felovia* (George, 1985; Lopez Antonanzas & Knoll, 2011; Gomes Rodrigues et al., 2012), but a molecular phylogeny for the family has not yet been published. There is currently only one species of *Massoutiera* recognised. However, given the patchy nature of its distribution, with some populations isolated by hundreds of kilometres of unsuitable habitat (e.g. Tibesti, Chad), it would be interesting to examine the genetic structure of the species. The diploid number of chromosomes in *M. mzabi* has been recorded as: 2n = 36, FN = 72 (Algeria, George, 1979; Niger, Dobigny et al., 2002a).

Species List
One species described within the genus *Massoutiera*, restricted to Africa and occurring beyond Sub-Saharan Africa in the Sahara to the north.
Massoutiera mzabi (Lataste, 1881)

Species Account

Massoutiera mzabi (Lataste, 1881)

Diagnosis
See genus account.

Taxonomic Issues
Currently a single species is recognised. The taxa *harterti* and *rothschildi* are considered synonyms of *M. mzabi* (Allen, 1939).

Distribution

This species is patchily distributed through the central Sahara in Adrar des Iforas (Mali), Air (Niger) and Tibesti (Chad) mountain ranges.

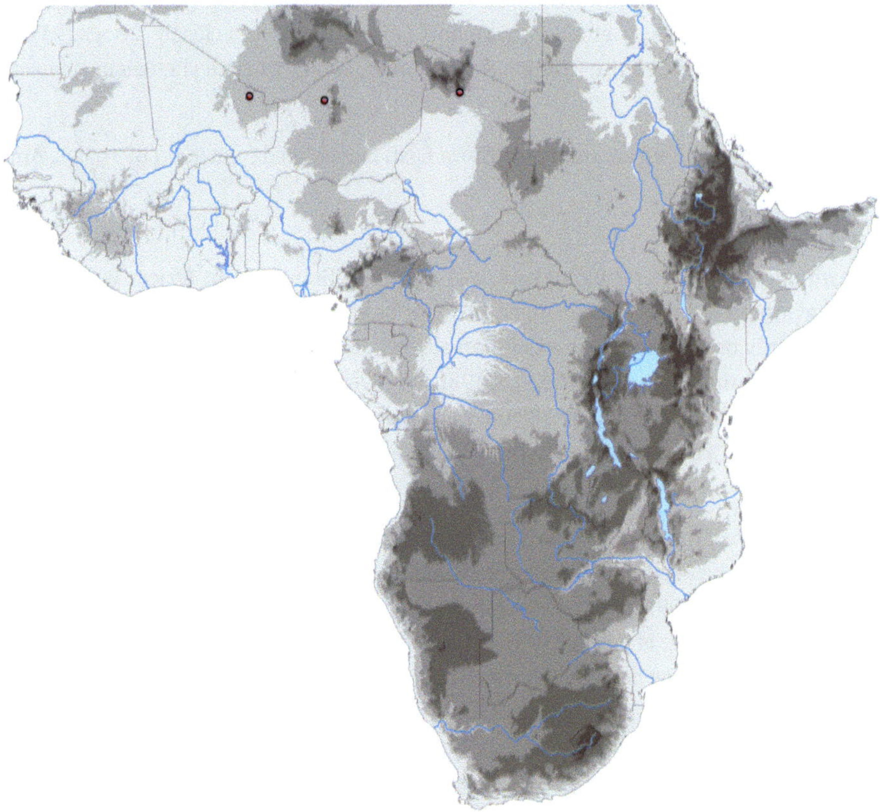

Figure 2.93 Distribution map

Measurements

External and cranial measurements (mm) and mass (g) for *Massoutiera mzabi*, sexes combined

	Mean	Min	Max	N
Mass*	200.0	132	234	7
HB*	176.0	125	210	16
Tail*	52.0	33	85	14
HF*	35.0	33	36	10
Ear*	17.0	15	17	10
GLS*	45.5	39.8	48.5	24

*Happold (2013)

Genus Account

Pectinator Blyth, 1856 (Pectinators)

Diagnosis

This monotypic genus has a narrow and patchy distribution in the horn of Africa where it is the only species of ctenodactylid within its range. It is a medium-size rodent (Mass ca. 180 g), with a characteristically stout body, short limbs and a short, furry tail typical of the family. The pelage is dense, short and soft to the touch. Dorsally, it is sandy coloured, pencilled with black, fading to off-white ventrally. Hairs have light grey or white bases, giving the animal a light appearance. The feet are whitish dorsally with four digits and sharp claws, and a row of stiff hairs at the tips. These hairs are particularly long and brittle-like on the digits of the hindfeet, hiding the claws. The ears are small and pressed against the head, with a band of white hairs around them. The eye ring, if present, is indistinct. The tail is short, covered in feathery hair with a dark or black tip (the ends of which may be tipped white). There are two pairs of nipples (2 + 0 = 4) placed anteriorly and laterally (George, 1978a).

Figure 2.94 Photograph of *Pectinator spekei* (David Hegner).

Cranial Characters

The skull is robust and medium-sized (typically around 45 mm in greatest length in adults). It is broad (but relatively narrower than in *Felovia* or *Massoutiera*), relatively flat for its size, and shallowly domed. The rostrum is narrow and parallel-sided, but the cranium widens considerably posteriorly with the zygomatic arches at a sharp angle to each other. The nasal bones narrow slightly posteriorly. There is a well-developed supraorbital process on the vertical arm of the zygomatic arch extending posteriorly into the orbit. The infraorbital foramen is large and elliptical. The anterior palatal foramina are long and widening posteriorly, extending for more than half the length of the diastema. The bullae are relatively large, but not particularly inflated. Like all Ctenodactylidae the mandible is low in its posterior part and lacks the coronoid apophysis.

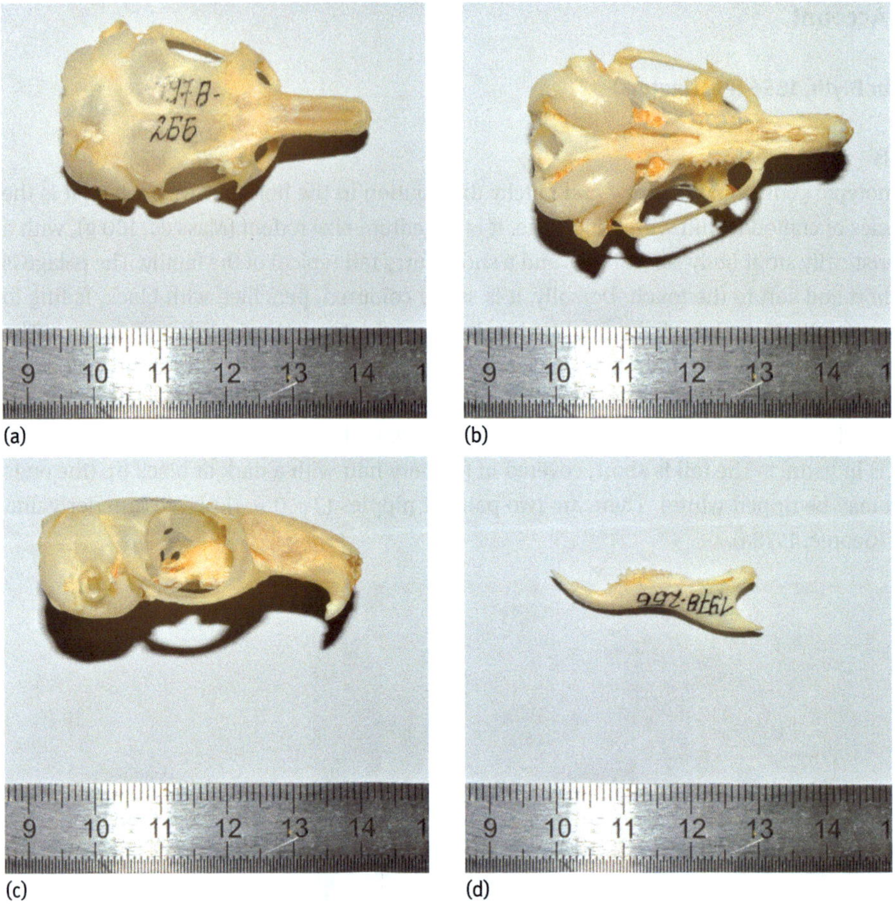

Figure 2.95 a-d Cranium of *Pectinator spekei* (MNHN-ZM-MO 1978-266) (from top left to bottom right: dorsal view, ventral view, lateral view and lower jaw).

Dentition

Pectinator has strong, ungrooved heavily opisthodont upper incisors. The dental formula is: 1013/1013, with the premolar greatly reduced in size (Lopez Antonanzas & Knoll, 2011). The molars are all roughly the same size and uncusped with two connected laminae on each molar created by an infold on the outside and one on the inside that do not meet medially (unlike *Felovia* and *Massoutiera*, where the two laminae are completely separated).

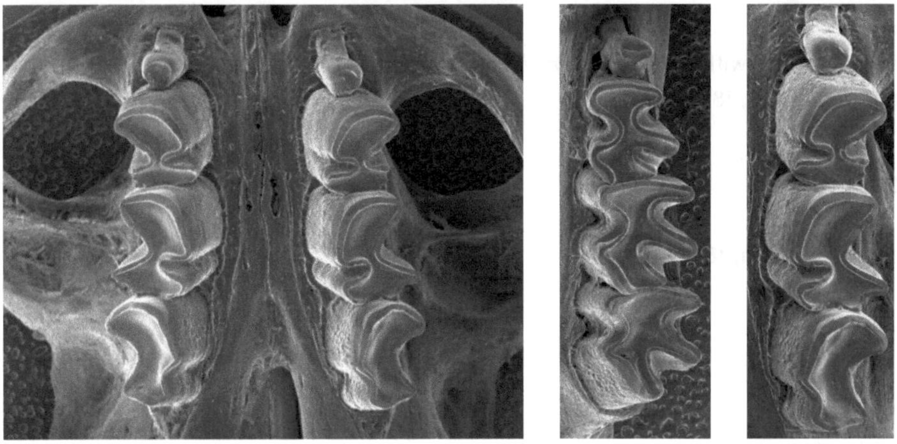

(a) (b) (c)

Figure 2.96 Cheek teeth of *Pectinator spekei* (MNHN-ZM-MO 1981-504) (A: upper cheek teeth; B: lower left toothrow; C: upper left toothrow).

Habitat & Biogeography

This species has a narrow distribution in the horn of Africa, occurring in Eritrea, Djibouti, Ethiopia and Somalia. It has been collected from within 40 km of the Kenyan border (Yalden et al., 1976), but has yet to be reported from that country. It occupies rocky outcrops and cliffs in desert or semi-desert environments, from sea level up to 1200 m above sea level (Yalden et al., 1976).

Fossil Record

There are currently no known fossils of *Pectinator* (Winkler et al., 2010).

Biology & Ecology

Very little is known about the general biology of this species. It is vegetarian, feeding primarily on grasses and leaves of herbs and trees (George, 1974). *Pectinator spekei* gives a distinctive alarm which includes a whistle component (George, 1981a). It lives in large groups at higher densities than any other African gundi, peaking at more than 2 animals/100m^2 (George, 1981a). Based on limited field observations, it would appear that this species exhibits aseasonal breeding, with one, or rarely two, offspring being born (George, 1978a). A single male and female offspring took 174 days each to reach adult size (George, 1978a).

Systematic Notes

The systematic position of the ctenodactylids has been debated for decades (e.g. Graur et al., 1992), but a recent phylogeny based on five nuclear genes places the clade as a sister to the Hystricognathi (Blanga-Kanfi et al., 2009). Based on morphological and dental grounds, the genus *Pectinator* has been placed basal in the subfamily Ctenodactylinae (George, 1985; Lopez Antonanzas & Knoll, 2011; Gomes Rodrigues et al., 2012), but a molecular phylogeny for the family has not yet been published. There is currently one species of *Pectinator* recognised. The diploid number of chromosomes in *P. spekei* has been recorded as: 2n = 40, FN = 80 (Ethiopia, George, 1979).

Species List

One species described within the genus *Pectinator*, endemic to the horn of Africa.
Pectinator spekei Blyth, 1856

Species Account

Pectinator spekei Blyth, 1856

Diagnosis

See genus account.

Taxonomic Issues

See genus account.

Distribution

This species is restricted to the Horn of Africa.

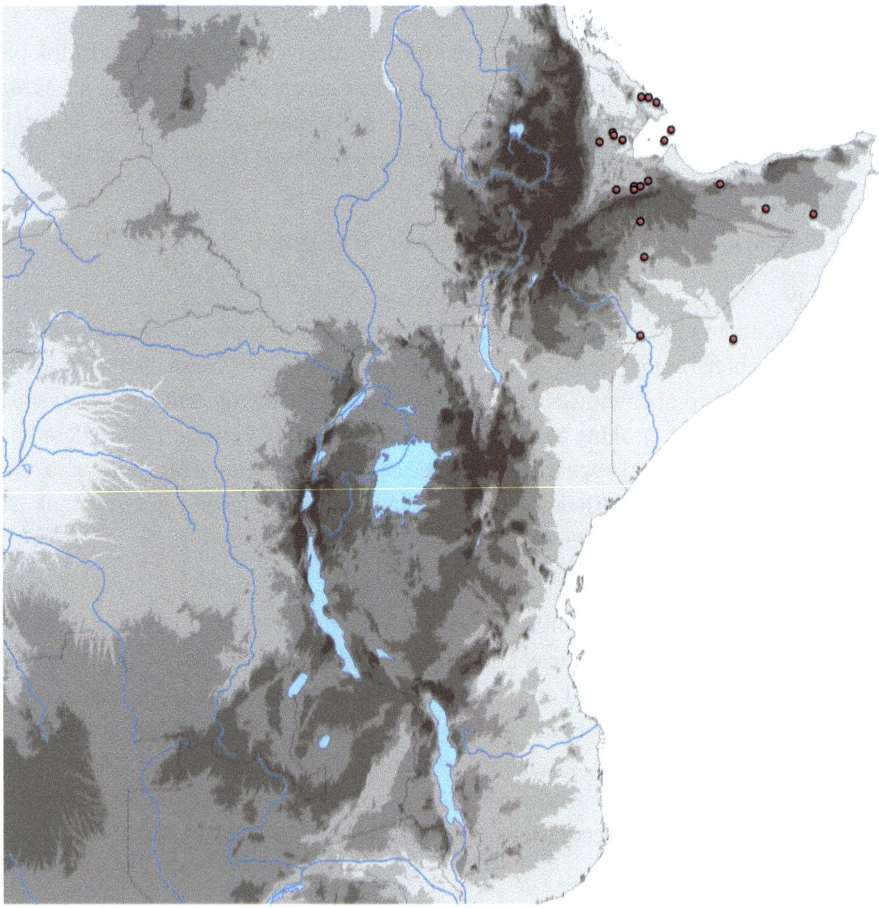

Figure 2.97 Distribution map

Measurements

External and cranial measurements (mm) and mass (g) for *Pectinator spekei*, sexes combined

	Mean	Min	Max	N
Mass*	178.2	-	-	4
HB	175.2	165	190	5
Tail	65.0	48	80	5
HF	35	34	36	5
Ear	18.8	17	20	5
GLS	45.3	43.1	47.6	8

*George (1978a)

Family: Hystricidae

This family is represented by three species in Sub-Saharan Africa.

Key to Genera

1	Body covered in long, cylindrical quills; tail shorter (< 130 mm)	*Hystrix*
	Body covered in shorter, flattened quills; tail longer (> 150 mm)	*Atherurus*

Genus Account

Atherurus F. Cuvier, 1829 (Brush-tailed Porcupines)

Diagnosis

This genus is readily distinguishable by its large size (2.5-3.4 kg), spiny coat of fur and long tail. The quills are longest on the lower back and rump, but do not approach the length of those in *Hystrix*. The dorsal spines are brown and white in colour, giving the animal a grizzled dark brown appearance. The ventral pelage is paler brown or off-white. The ears are small but visible. The legs are short and stout, with four well-developed digits, and a reduced 1st digit. All digits have claws. The tail is long and covered with spines, ending in rattle-quills. There are two pairs of nipples ($2 + 0 = 4$), situated laterally on the thorax near the armpits. This genus includes a single African species, the African brush-tailed porcupine (*Atherurus africanus*), which is easily told apart from members of the genus *Hystrix* by its smaller size, absence of a crest on the back of the neck, a long tail, and its preferred rainforest habitat.

Figure 2.98 Photograph of *Atherurus africanus* (Guy Colborne).

Cranial Characters

Although the skull is large (mostly > 90 mm in length in adults), it is far smaller than that of *Hystrix*, and further differs from that genus in being only gently domed dorsally, and having smaller nasal bones. The nasals are parallel-sized, situated above the rostrum, and do not extend back to cover the dorsal part of the cranium as in *Hystrix*. As a result, the frontal and parietal bones are not greatly reduced in size, and cover most of the dorsal part of the skull. The supraoccipital and sagittal crests are well developed, with the paraoccipital process greatly reduced and not obvious. The zygomatic arch is robust and deeper anteriorly, narrowing posteriorly. The infraorbital foramen is large and oval. The anterior palatal foramina are greatly reduced and not much longer than the width of a cheektooth. The bullae are relatively large in comparison with *Hystrix*.

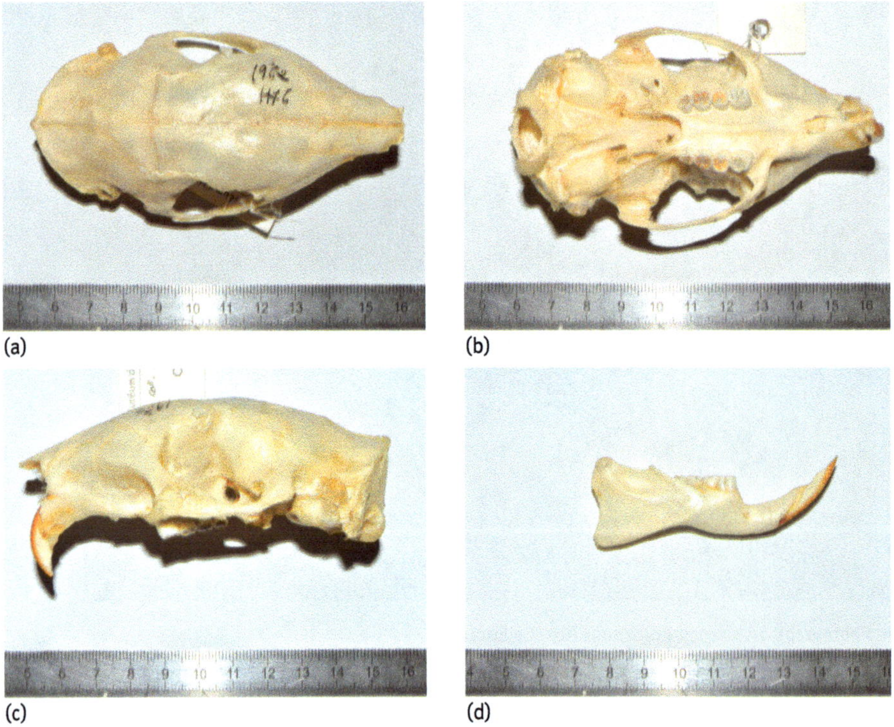

Figure 2.99 a-d Cranium of *Atherurus africanus* (MNHN-ZM-MO 1962-1176) (from top left to bottom right: dorsal view, ventral view, lateral view and lower jaw).

Dentition

Compared with *Hystrix, Atherurus* has relatively small, ungrooved orthodont (or slightly prodont) incisors. The dental formula is: 1013/1013 (Rosevear, 1969), resulting in four cheekteeth being present in each half jaw. The premolar and molars are all roughly the same size and uncusped. A detailed description of the molars of *Atherurus* is given in van Weers (1993).

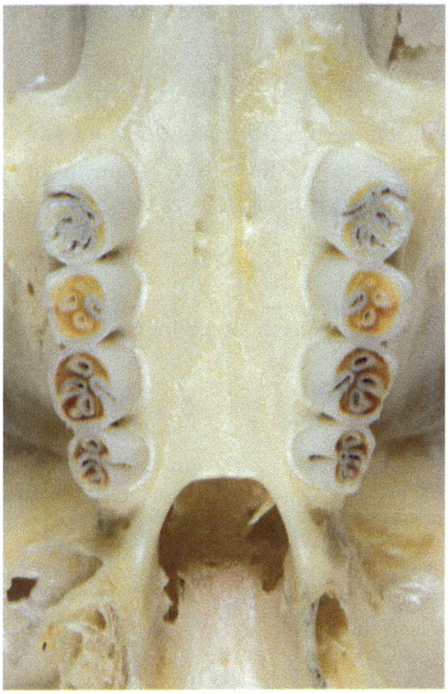

Figure 2.100 Upper cheek teeth of *Atherurus africanus* (MNHN-ZM-MO 1962-1176).

Habitat & Biogeography

The single species of Brush-tailed Porcupine, *Atherurus africanus*, occurs widely in rainforests of tropical Africa, from the Kakamega forest (western Kenya) in the east to The Gambia in the west. They are typically associated with rainforest and may occur from sea level up to 1,500 m above sea level on Mt Cameroon (Rosevear, 1969). Although a forest specialist, this species may venture out into clearings where forest has been cleared in patches e.g. for slash-and-burn agriculture. Under such conditions, it may even become a pest (Rosevear, 1969). The African brush-tailed porcupines typically rest up during the day in rock crevices, hollow trees or fallen logs. The home range of this species is relatively small, 11-22 ha, with large amounts of overlap between different individuals of both sexes (Emmons, 1983).

Fossil Record

The oldest *Atherurus* fossil is from 10-11 MYA, Egypt (Pickford et al., 2008). In North Africa, it has been reported from the Late Miocene only; in East Africa from the Late Miocene to the Early Pliocene (Winkler et al, 2010).

Biology & Ecology

Atherurus africanus is principally vegetarian feeding on leaves, fruits and roots of forest plants, and may also feed on cultivated root crops such as cassava (Rosevear, 1969). The reproductive biology of this species has recently been reviewed by Happold (2013). It is long-lived and relatively slow reproducing for a rodent. Births occur throughout the year. The gestation period is long for its size, 100-120 days (Rahm, 1962). Mean litter size is 1-2, and young are usually suckled for about two months. Offspring attain sexual maturity between 1-2 years of age. Females are capable of having multiple litters per year. This species is greatly prized as bush meat and hunted in large numbers throughout its range (Jori, 1998; MacDonald, 2012).

Systematic Notes

There are currently only two species recognised within the genus: *Atherurus africanus* which occurs widely in tropical Africa, and *A. macrourus* in East Asia. The following are synonyms of *A. africanus*: *armata* Gervais, 1854; *burrowsi* Thomas, 1902; *centralis* Thomas, 1895; *turneri* St. Ledger, 1932.

Species List

Two species described within the genus *Atherurus*, with only one species occurring in Africa, and one occurring beyond in East Asia (Woods & Kilpatrick, 2005).
Atherurus africanus Gray, 1842

Species Account

***Atherurus africanus* Gray, 1842**

Diagnosis

See genus account.

Taxonomic Issues

None.

Distribution

This species is widely distributed throughout the forest zone of tropical Africa.

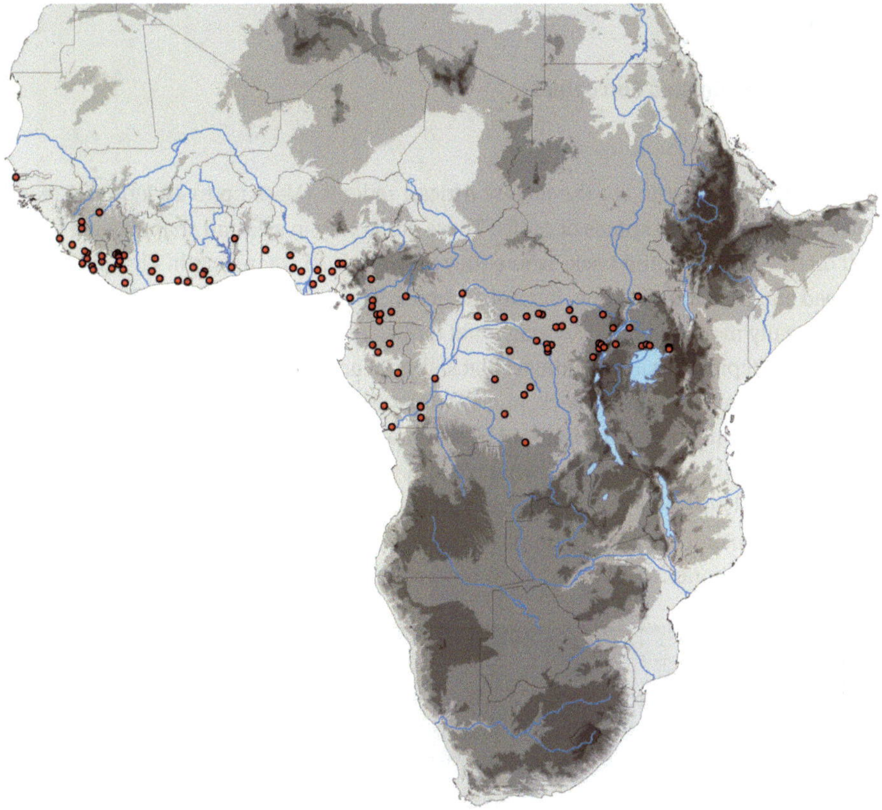

Figure 2.101 Distribution map

Measurements

External and cranial measurements (mm) and mass (g) for *Atherurus africanus*, sexes combined

	Mean	Min	Max	N
Mass*	-	2500	3400	-
HB	521	295	648	5
Tail	203.6	150	255	5
HF	73.0	62	80	5
Ear	35.2	33	37	5
GLS	89.3	63.7	101.0	10

*Rosevear (1969)

Genus Account

Hystrix Linnaeus, 1758 (Crested Porcupines)

Diagnosis

Porcupines of the genus *Hystrix* are the largest African rodents (Mass up to 20 kg), and readily distinguishable by their coat of spines or quills. These quills vary in size and shape, being longest and stoutest on the shoulders and back, where they may reach 30 cm in length. The quills on the back of the head and neck are also long, but not as robust. The quills on the rest of the body are far shorter and generally more flexible. The individual quills are various shades of black and white, although the animal typical has a dark, almost blackish appearance, flecked with white. Their very large size, long quills on the back and neck, and short, indistinct tail differentiate members of this genus from the smaller *Atherurus*. The limbs of *Hystrix* are short and stout, with four clawed digits on each foot. The ears are small and somewhat concealed by the quills. The tail is short, covered in quills, and not easily distinguished in a living animal, being obscured by the long spines on the back. There are two or three pairs of nipples (2 or 3 + 0 = 4 or 6), varying between individuals, placed laterally on the thorax.

Figure 2.102 Photograph of *Hystrix* sp. (Charles Foley).

Cranial Characters

The skull is very large (typically over 140 mm in greatest length in adults) and domed. The nasal bones are characteristically wide and long, covering most of the roof of the cranium. By contrast, the frontal and parietal bones are reduced in size. The rostrum is short but nasal fossae deep and pneumatized. The paraoccipital process is relatively short. The zygomatic arch is short but robust, narrowing posteriorly. The infraorbital foramen is large and oval. The anterior palatal foramina are reduced but visible, extending for far less than half the length of the diastema. The bullae are relatively small.

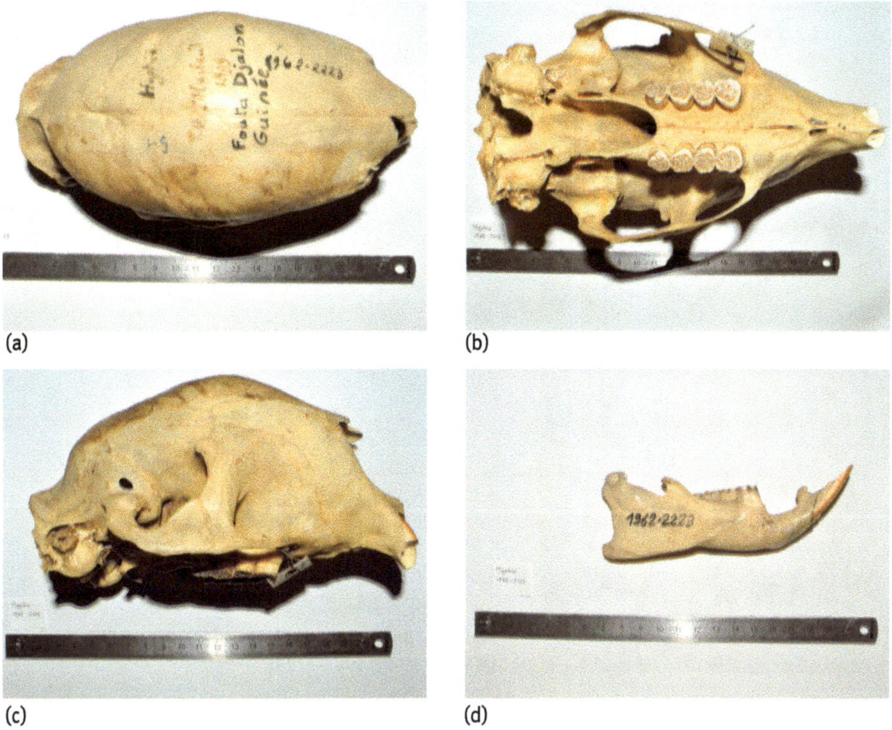

Figure 2.103 a-d Cranium of *Hystrix cristata* (MNHN-ZM-MO 1962-2223) (from top left to bottom right: dorsal view, ventral view, lateral view and lower jaw).

Dentition

Hystrix has large, ungrooved pro-odont incisors. The dental formula is: 1013/1013 (Rosevear, 1969), resulting in four cheekteeth being present in each half jaw. The premolar and molars are all roughly the same size and uncusped. The molar crowns are high and display some crests of enamel; these crests are sometimes interrupted defining rounded islands of enamel in the middle of the molars.

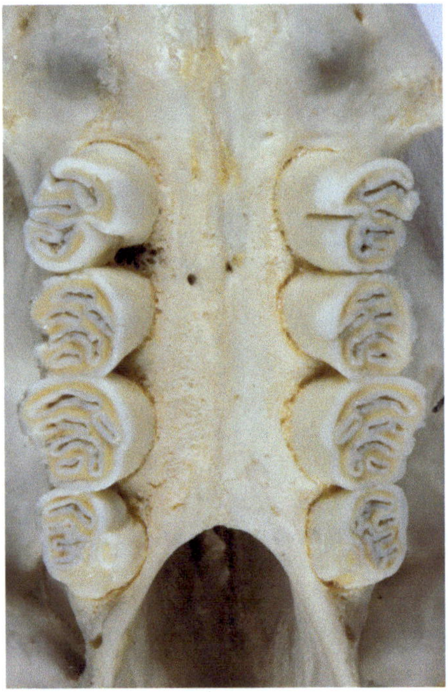

Figure 2.104 Upper cheek teeth of *Hystrix cristata* (MNHN-ZM-MO 1962-2223).

Habitat & Biogeography

Crested porcupines of the genus *Hystrix* occur widely in Sub-Saharan Africa, avoiding the tropical forests of the Congo basin, and the driest parts of the Namib Desert, and the Sahara (Granjon & Duplantier, 2009). *Hystrix* species typically are associated with savannah habitats, but there are a few records from lowland tropical forest in West Africa (Rosevear, 1969). For the most part, the two African species occur in different parts of the continent, but overlap broadly in East Africa. *Hystrix cristata* occurs from Senegal in the west to Ethiopia in the east, and then south through Kenya and Tanzania, whilst *H. africaeaustralis* occurs from South Africa northwards to Kenya.

Crested Porcupines are tolerant of a wide range of habitats, occurring in dry, semi-arid savannas to moist miombo woodlands, and from sea level up to 3500 m on Mt Kilimanjaro (Kingdon, 1974). In Tanzania, where the two species overlap, the northern species (*H. cristata*) has been collected at more sites than *H. africaeaustralis* (Kingdon, 1974); however, since field-based differentiation of these two species is not currently possible, no conclusions can be reached regarding whether they utilise the habitat differently or not. Crested porcupines typically rest up during the day in rock crevices, small caves, or burrows, the latter either dug by the antbear (*Orycteropus afer*) or by the porcupines themselves.

Fossil Record

The oldest *Hystrix* fossil is from 7.44 MYA, Kenya (Winkler, 2003). In North Africa, it has been reported from the Late Miocene to the Middle Pleistocene; in East Africa from the Late Miocene to the Middle Pleistocene; and in Southern Africa from the Early Pliocene to the late Pleistocene (Winkler et al, 2010).

Biology & Ecology

Crested Porcupines feed predominantly on roots and tubers which are dug up from under the ground using their strong incisors. They also feed on fallen fruits, and gnaw bones. Due to the combination of their diet and digging abilities, porcupines can become agricultural pests in farming areas. The extensive literature on the diet of porcupines is reviewed in Happold (2013), as is their reproductive biology. Porcupines are long-lived and slow reproducing. Breeding is seasonal and varies with location. In captivity, the gestation period is 93-94 days, average litter size is 1-3, and young are usually suckled for 100 days or more (Van Aarde, 1985a,b). Offspring attain sexual maturity between 1-2 years of age, during which time annual survival is relatively high for a rodent (Van Aarde, 1987a,b,c). The inter-litter interval in captivity is approximately 1 year (Van Aarde, 1985b).

Systematic Notes

The family Hystricidae is basal to the clade Hystricognathi, which includes the Old World Phiomorpha and New World Caviomorpha (Blanga-Kanfi et al., 2009). There are currently only two species recognised within the genus, *Hystrix cristata* and *H. africaeaustralis*, which are not readily distinguishable on external appearance. The following characters may assist in distinguishing the two species. First, the short spines on the rump (beneath the long quills) are black (with or without a white tip) in *H. cristata*, but pure white in *H. africaeaustralis*. Second, the nasal bones are slightly less extended in *H. africaeaustralis* than in *H. cristata*. Third, the rattle-quills are shorter in *H. cristata* (being generally under 50 mm) than in *H. africaeaustralis* (Rosevear, 1969). The diploid number of chromosomes has been recorded as: 2n = 60, FN = 120 in *H. cristata* (Kenya, George, 1980), as 2n = 66, FN = 114 in *H. africaeaustralis* (George & Weir, 1974).

Species List

Two species described within the genus *Hystrix*, one restricted to Africa, and one occurring beyond in Europe and the Middle East.
Hystrix africaeaustralis Peters, 1852
Hystrix cristata Linneaus, 1758

Key to Species

1	short spines on rump (beneath the long quills) white; 2n = 66	*H. africaeaustralis*
	short spines on rump (beneath the long quills) black, with or without a white tip; 2n = 60	*H. cristata*

Species Accounts

Hystrix africaeaustralis Peters, 1852

Diagnosis

This species cannot easily be distinguished from *H. cristata* in the field, but specimens examined closely can be distinguished by external and cranial features (see genus account).

Taxonomic Issues

None.

Distribution

This species is widely distributed in the savannas of southern Africa, south of the equator, overlapping with *H. cristata* in East Africa.

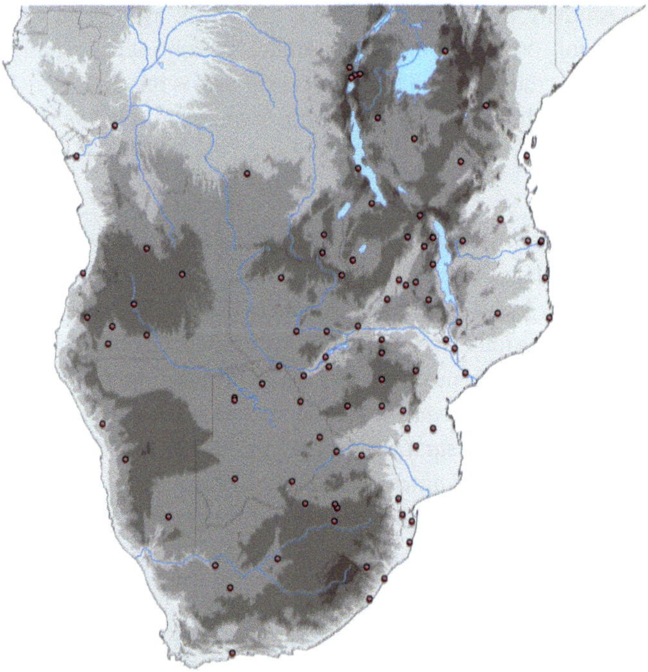

Figure 2.105 Distribution map

Measurements

External and cranial measurements (mm) and mass (g) for *Hystrix africaeaustralis*, sexes combined

	Mean	Min	Max	N
Mass*	12.5	9.1	16.5	4
HB*	686.4	-	-	5
Tail*	73.0	40	105	5
HF*	92.0	83	100	5
Ear*	38.6	30	49	5
GLS	152.1	104.2	160.0	3

*Rautenbach (1982); HB was calculated by substracting the tail length from the total length.

Hystrix cristata Linneaus, 1758

Diagnosis

This species cannot easily be distinguished from *H. africaeaustralis* in the field, but specimens examined closely can be distinguished by external and cranial features (see genus account).

Taxonomic Issues

The East African populations have been considered a separate taxon (*galeata*) in the past, but shown to overlap extensively with populations in other parts of its range (Corbet & Jones, 1965). No subspecies are currently recognised (Happold, 2013).

Distribution

This species is widely distributed in the northern savannas, and overlapping with *H. africaeaustralis* in East Africa.

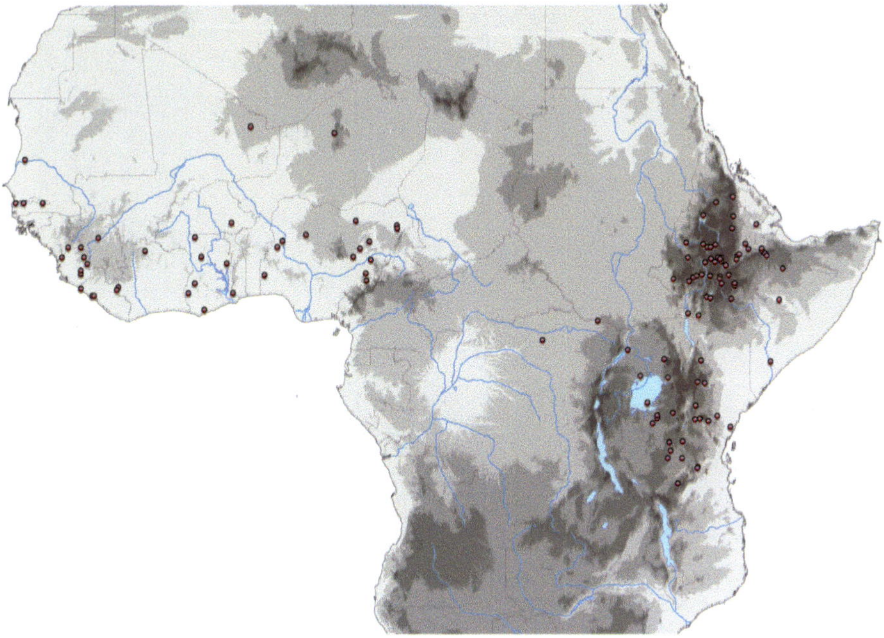

Figure 2.106 Distribution map

Measurements

External and cranial measurements (mm) and mass (g) for *Hystrix cristata*, sexes combined

	Mean	Min	Max	N
Mass	-	-	-	-
HB	770	-	-	1
Tail	120	-	-	1
HF	102	-	-	1
Ear	42	-	-	1
GLS*	158	152	170	-

*Rosevear (1969)

Family: Thryonomyidae

This family is represented by two species in Sub-Saharan Africa.

Genus Account

Thryonomys Fitzinger, 1867 (Cane Rats)

Diagnosis

Their very large size (Mass ca. 1.4-5.2 kg; HB ca. 400-550 mm), coarse pelage, short legs, stocky body, blunt face and relatively short tail (30-40% of HB) readily distinguish members of this genus. Within African rodents, they are only surpassed in size by porcupines of the genus *Hystrix*. The skull is massive for a rodent with a well-developed supraoccipital crest, and a heavy zygomatic arch. The ears are small and partially hidden by the thick fur. The pelage is medium brown and noticeably grizzled dorsally, and white or whitish ventrally. The hindfeet are well furred dorsally with black soles ventrally. There are five digits (the 1st and 5th greatly reduced) on the forefeet and four digits (1st missing, 5th greatly reduced) on the hindfeet. The three well-developed digits all end in a thick, large claw. The number of nipples varies with species, being two pairs (0 + 2 = 4) in *T. gregorianus*, and three pairs (0 + 3 = 6) in *T. swinderianus*, situated laterally on the abdomen.

Figure 2.107 Photograph of *Thryonomys swinderianus* (Christiane Denys)

Cranial Characters

The skull is very large (generally well over 70 mm in greatest length in adults) and robust but with a short rostrum giving the face a blunt appearance. The nasal bones are wide and parallel-sided. Unusually for rodents, the supraoccipital and sagittal crests are well developed and obvious. The paraoccipital process is long and finger-like, extending ventrally from the posterior edge of the cranium. The zygomatic arch is robust, and very deep anteriorly, narrowing posteriorly. The infra-orbital foramen is enormous and elliptical. The anterior palate is deeply concave in lateral view.

The anterior palatal foramina are clearly visible, extending for about half the length of the diastema. The bullae are relatively small.

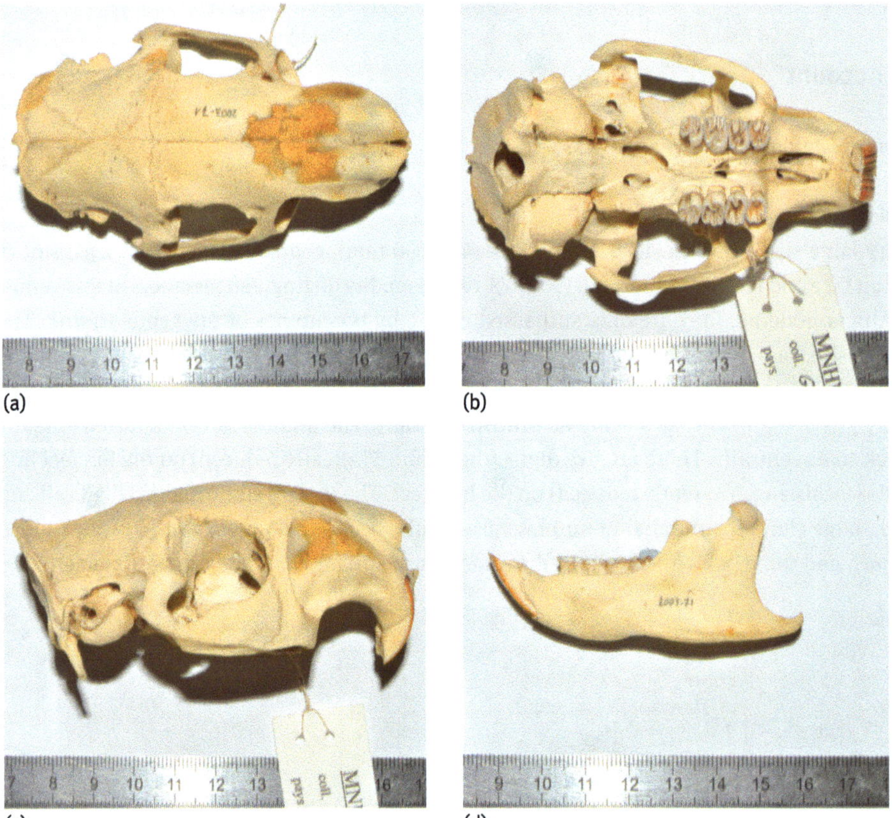

(a) (b) (c) (d)

Figure 2.108 a-d Cranium of *Thryonomys swinderianus* (MNHN-ZM-MO 2003-71) (from top left to bottom right: dorsal view, ventral view, lateral view and lower jaw).

Dentition

Thryonomys has large, orange-coloured opisthodont upper incisors, each with three characteristic longitudinal grooves running on the outer (anterior) surface. The lower incisors are not grooved. The dental formula is: 1013/1013 (van der Merwe, 2007), resulting in four cheekteeth being present in each half jaw. The premolar is the smallest cheektooth, the three molars being roughly the same size. The upper toothrows are parallel. Each molar has a distinct pattern of transverse folds of enamel, with two outer and one inner fold.

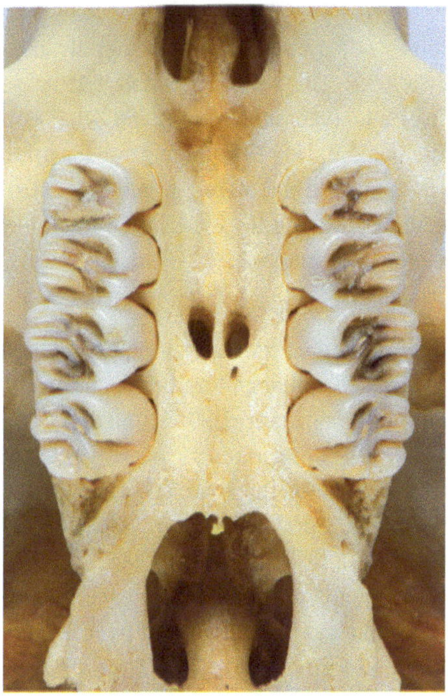

Figure 2.109 Upper cheek teeth of *Thryonomys swinderianus* (MNHN-ZM-MO 1974-297).

Habitat & Biogeography

Cane Rats are widespread throughout the savannas of Sub-Saharan Africa occurring from the Free State and Eastern Cape provinces of South Africa north through to Ethiopia and Angola and west to southern Senegal. They occur throughout the tropical forest zone (such as the Congo basin and the Upper Guinean Forest), although they do not occur within the forest itself, but in clearings and along the edges of the forest. *Thryonomys swinderianus* is more widespread, covering most of the sub-continent, whereas *T. gregorianus* is restricted to East and Southern Africa. The two species occur sympatrically with *T. swinderianus* being more common in drier savannas at lower altitudes and *T. gregorianus* in moist savannas at higher altitudes. Only the latter species has been recorded from Ethiopia.

Both Cane Rat species are closely associated with tall, rank grasslands. Despite its association with moist savannas, the upland species (*T. gregorianus*) appears to utilise drier habitats than the lowland species (*T. swinderianus*). Typical habitat includes good cover in the form of thick, dense grass at the edge of a swamp, wetland, or river. Both species may be semi-aquatic. The presence of cane rats is readily identified by their runways in dense, rank grassland which lead to their feeding sites. Cane Rats may persist in an area even after the grass cover has temporarily been lost e.g. by fire, which happens annually throughout much of their range. Presumably during this time, vulnerability to predation is increased (Kingdon, 1974).

Fossil Record

The family diversified during the Miocene and ranged across Africa and Asia. The earliest *Thryonomys* species is from 6 MYA, Kenya (Manthi, 2007). In East Africa it has been reported from the Late Miocene to the Late Pleistocene; and in Southern Africa from the Late Pleistocene (Winkler et al, 2010).

Biology & Ecology

Cane Rats are exclusively herbivorous, and their diet consists of grasses. They are generally long-lived, and slow reproducing for a rodent (van der Merwe 1999). The gestation period is 155 days (van der Merwe, 1999), and mean litter size ranges between 3-4 (van der Merwe & van Zyl, 2001), with a higher limit of probably around five or six in the wild. Males continue to grow for up to a year whereas females generally stop growing at 10 months (van der Merwe & van Zyl, 2001). Numerous studies have been conducted on the reproduction of this species, especially with a view to raising cane rats for meat production, and recently reviewed by Happold (2013).

Systematic Notes

The closest relative of the Thryonomyidae is the family Petromuridae, which is supported on both morphological and molecular grounds (Woods & Kirkpatrick, 2005; Blanga-Kanfi et al, 2009). The family Thryonomyidae was more diverse in the past (Lopez Antonanzas et al, 2004), but at present is known by just two closely related species, *T. gregorianus* and *T. swinderianus*. There is confusion in the literature regarding the best characters to use in distinguishing these two species, with the two most frequently cited being the arching of the skull in the frontal region and the position of the longitudinal grooves on the upper incisors (Happold, 2013). However, van der Merwe (2007) showed that these characters are not reliable as they overlap significantly in the two species. Three additional characters proposed by van der Merwe (2007) include: 1) dorsal view of the cranium, the frontal bones are triangular-shaped in *T. swinderianus*, and square-shaped in *T. gregorianus*; 2) in lateral view of the cranium, the angle of the anterior (outer) rim of the infraorbital foramen with the base of the zygomatic arch is sharp in *T. swinderianus* and at right angles in *T. gregorianus*; 3) the presence of a bar across the foramen ovale in *T. gregorianus* which is usually absent in *T. swinderianus*. The diploid number of chromosomes has been recorded as: 2n = 44, FN = 86 in *T. swinderianus* (Benin, Civitelli et al., 1996), or 2n = 44, FN = 84 (Senegal, Granjon et al., 1992), and in *T. gregorianus* as 2n = 40, FN = 80 (George, 1980).

Species List

Two species described within the genus *Thryonomys*, both restricted to Sub-Saharan Africa.
Thryonomys gregorianus (Thomas, 1894)
Thryonomys swinderianus (Temminck, 1827)

Key to Species

1	Frontal bones triangular in shape (dorsal view)	*T. swinderianus*
	Frontal bones rectangular or square in shape (dorsal view)	*T. gregorianus*

Species Accounts

Thryonomys gregorianus (Thomas, 1894)

Diagnosis

This species is easily confused with *T. swinderianus*, especially if the cranium is not available for examination, with which it co-occurs in South-central and East Africa. It occupies drier habitats than *T. swinderianus* and in East Africa has a proportionately shorter tail (in *T. swinderianus* tail is about twice the length of HF, in *T. gregorianus* tail is same length as HF) but this relationship does not hold for Southern Africa, where *T. gregorianus* has a longer tail (Skinner & Chimimba, 2005; Happold, 2013).

Taxonomic Issues

No subspecies are currently recognised, however the outlying population in Cameroon appears to be isolated from the rest of the populations in East and Southern Africa. A molecular analysis is needed to establish whether any phylogenetic structure occurs in this species.

Distribution

This species is widely distributed in South-central and East Africa, with isolated records from Cameroon.

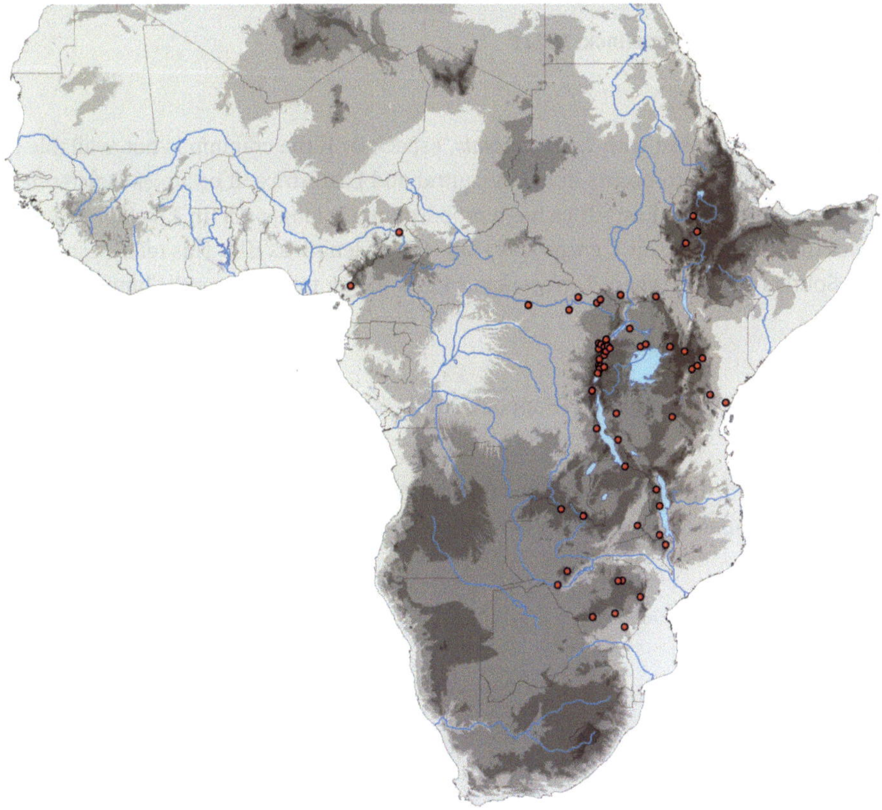

Figure 2.110 Distribution map

Measurements

External and cranial measurements (mm) and mass (g) for *Thryonomys gregorianus*, sexes combined

	Mean	Min	Max	N
Mass*	1900	1400	2400	10
HB	457.0	400	514	2
Tail	103.0	-	-	1
HF	82.0	77	87	2
Ear	27.0	-	-	1
GLS	71.1	68.0	7402	3

*Happold (2013)

Thryonomys swinderianus (Temminck, 1827)

Diagnosis
This species is easily confused with *T. gregorianus*, especially if the cranium is not available for examination, with which it co-occurs in South-central and East Africa. It occupies moister habitats than *T. gregorianus* and in East Africa has a proportionately longer tail (in *T. swinderianus* tail is about twice the length of HF, in *T. gregorianus* tail is same length as HF). This relationship does not hold for Southern Africa, where *T. gregorianus* has a longer tail (Skinner & Chimimba, 2005; Happold, 2013).

Taxonomic Issues
No subspecies are currently recognised (Happold, 2013).

Distribution
This species is widely distributed throughout the moister parts of Sub-Saharan Africa, avoiding only deserts, dry savannas and the interior of rainforest.

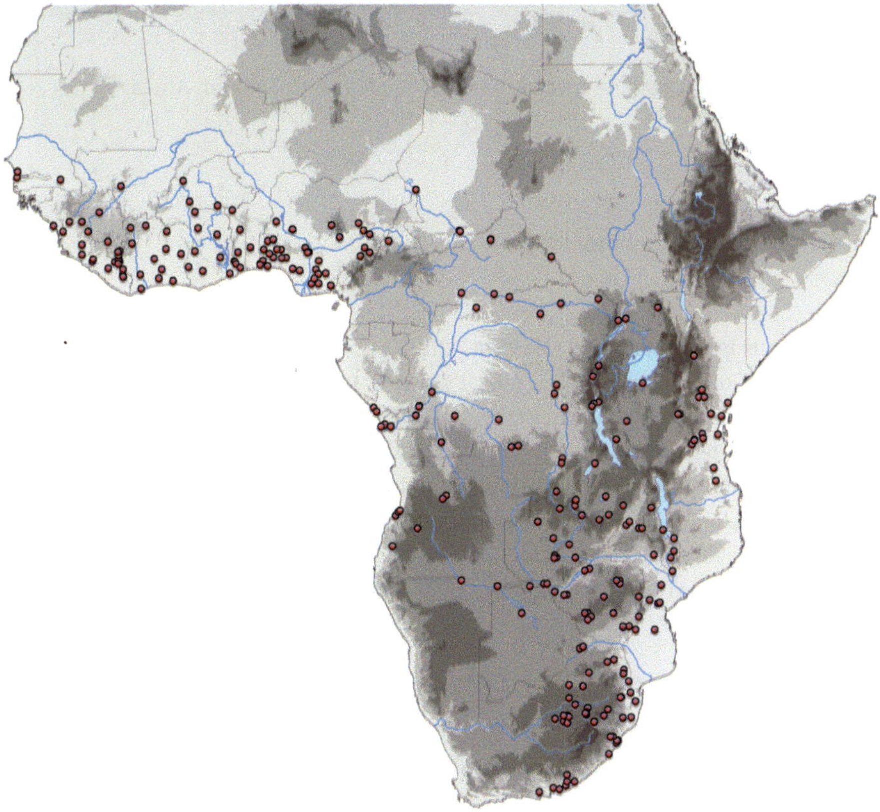

Figure 2.111 Distribution map

Measurements

External and cranial measurements (mm) and mass (g) for *Thryonomys swinderianus*, sexes combined

	Mean	Min	Max	N
Mass	3815.0	2500	5100	3
HB	479.9	356	640	12
Tail	192.9	132	250	16
HF	86.4	65	102	16
Ear	29.9	24	35	15
GLS	91.8	79.5	106.0	14

Family: Petromuridae

This family is represented by a single species that is endemic to Sub-Saharan Africa.

Genus Account

Petromus A. Smith, 1831 (Nokis or Dassie Rats)

Diagnosis

This monotypic genus is medium-sized (Mass ca. 120-170 g), and has a squirrel-like appearance with short limbs and long, furry tail. The dorsal pelage varies geographically from light sandy-brown in the south to dark grey-brown in the north of its distribution, with the ventral pelage being paler. The ears are relatively large and rounded, and a pale eye ring is present. There are five digits on the hindfoot and four digits on the forefoot, all with short claws. The tail is subequal or slightly longer than the length of the body, typically black except at the base which is the same colour as the dorsal pelage, and covered in long hairs giving it a bushy appearance. There are three pairs of nipples (2 + 1 = 6), two placed laterally posterior to the shoulders, and one on the abdomen (Happold, 2013).

(a) (b)

Figure 2.112 a and b Photograph of *Petromus typicus* (Galen Rathbun) (left), Beryl Wilson (right).

Cranial Characters

The skull is medium in size (mostly 38-43 mm in greatest length), and dorsoventrally flattened, with a flat rostrum and occipital region (George & Crowther, 1981). The palate ends well beyond the posterior margin of the third molars, typically forming a V-shape on its posterior edge. The zygomatic arch is robust and deep, narrowing slightly posteriorly. The infraorbital foramen is large. The anterior palatal foramina are long and wide, reaching posterior to the anterior margin of the last premolars. The bullae are relatively large and inflated.

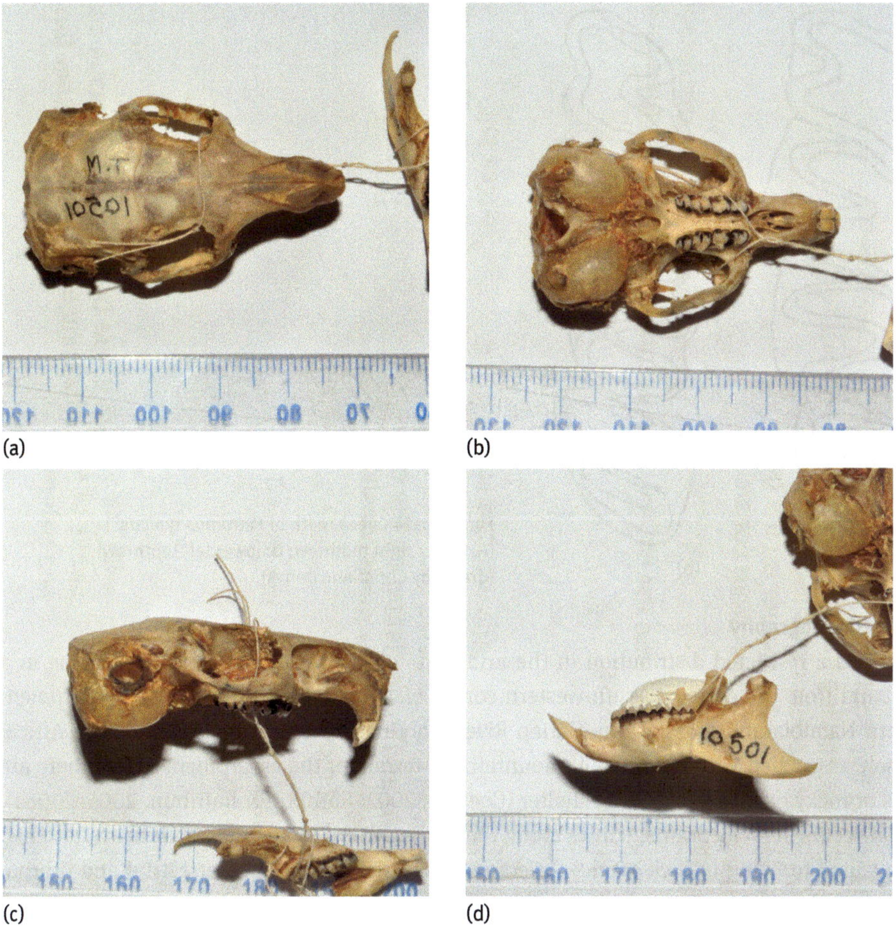

Figure 2.113 a-d Cranium of *Petromus typicus* TM 10501 (from top left to bottom right: dorsal view, ventral view, lateral view and lower jaw).

Dentition

Petromys has ungrooved, opisthodont incisors. The dental formula is: 1013/1013 (Happold, 2013), with the last premolar and three molars being roughly the same size, resulting in four functional cheekteeth being present in each half jaw. These are characteristically hypsodont with deep infolds on the inner and outer sides, giving the appearance of dividing each molar into two separate parts.

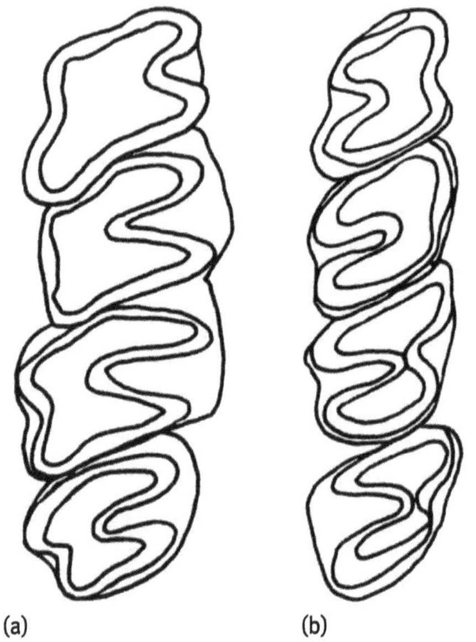

Figure 2.114 Cheek teeth of *Petromus typicus* (A: upper right toothrow; B: lower left toothrow) (drawn by Christiane Denys).

Habitat & Biogeography

Petromus has a restricted distribution in the arid parts of south-western Africa, occurring in a narrow band from the extreme south-western corner of Angola, south through the escarpment of western Namibia and crossing the Gariep River marginally into north-western South Africa. It is closely associated with the broken, mountainous terrain of the escarpment where there are boulders or rocky outcrops providing shelter (Coetzee, 2002; Rathbun & Rathbun, 2005, 2006a), but it also occurs in rocky drainage lines along the Gariep River (George, 1981b). It has a flattened cranium suited to living in rock crevices (George & Crowther, 1981). It is terrestrial and diurnal (Rathbun & Rathbun, 2005).

Fossil Record

The oldest *Petromus* fossil is from 5.8-6.1 MYA, Kenya (Mein & Pickford, 2006). In East Africa, it has been reported from the Late Miocene; in Southern Africa from the Early Pliocene to the Late Pleistocene (Winkler et al, 2010).

Biology & Ecology

The ecology of *Petromus* has been relatively well studied, and reviewed in Happold (2013). It is vegetarian feeding predominantly on dry grass stems (George, 1981b), but also feeding on fresh leaves, fruits, and flowers when available (Rathbun & Rathbun, 2005). It exhibits coprophagy; fibrous grass stems are broken down in the caecum and re-ingested by the animal (Mess & Ade, 2005). It forms monogamous pairs that occupy small overlapping home ranges of 0.03-1.04 ha (Rathbun & Rathbun, 2006b). Litter size is 1-3, but mostly two, the gestation period is about 84 days, and adult size is reached after 9-12 months (Coetzee, 2002; Happold, 2013).

Systematic Notes

Based on a molecular phylogeny, the closest relative of the family Petromuridae is Thryonomyidae (Huchon et al., 2002), forming a monophyletic clade with the Bathyergidae within the Hystricognathi rodents (Blanga-Kanfi et al., 2009). These two families (Petromuridae and Thryonomyidae) are also linked by various morphological characters (Mess, 2007a,b). The genus *Petromus* was previously mis-spelt as *Petromys* and the family as Petromyidae (e.g. Roberts, 1951), but corrected by Wood (1955). *Petromus* is currently known by a single species *P. typicus*. Numerous forms and subspecies have been described (see Roberts, 1951), but none are recognised here pending a molecular phylogenetic analysis of this species. The diploid number of chromosomes in *Petromus typicus* has been recorded as: 2n = 56, FN = 112 (George, 1980).

Species List

One species is described in the genus *Petromus*, restricted south-western Africa.
Petromus typicus A. Smith, 1831

Species Account

Petromus typicus A. Smith, 1831

Diagnosis

See genus account.

Taxonomic Issues

See genus account.

Distribution

This species is restricted to the western parts of Namibia, marginally extending south into South Africa.

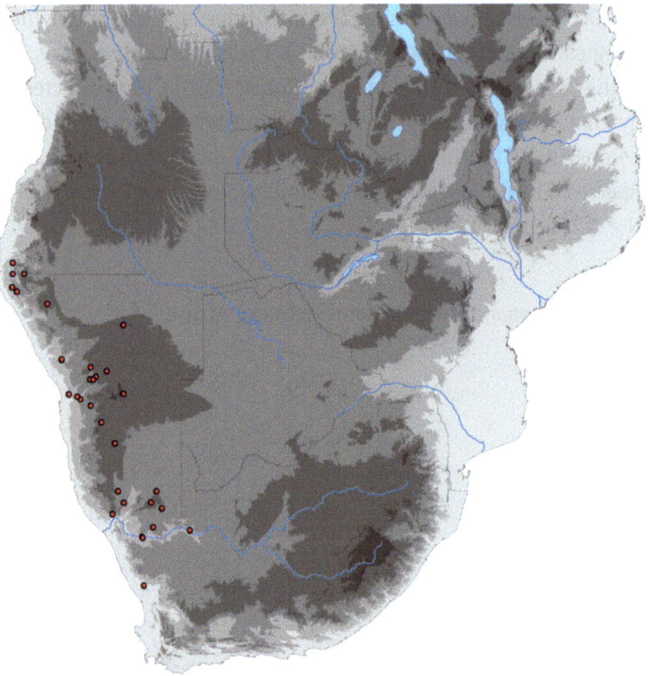

Figure 2.115 Distribution map

Measurements

External and cranial measurements (mm) and mass (g) for *Petromus typicus*, sexes combined

	Mean	Min	Max	N
Mass	146.7	109	176	17
HB	160.6	140	195	22
Tail	143.4	110	176	18
HF	32.5	29	36	20
Ear	13.3	10	16	21
GLS	40.7	37.7	44.3	10

Family: Bathyergidae

This family is represented by 29 species all endemic to Sub-Saharan Africa.

Key to Genera

1	Body entirely naked (with the exception of a few isolated tactile hairs)	*Heterocephalus*
	Body covered in fur	2
2	Digits with elongated claws (used for digging); upper incisors grooved; restricted to arid regions of Southern Africa	*Bathyergus*
	Digits without elongated claws; upper incisors ungrooved; widely distributed in Africa	3
3	Head and dorsal pelage strikingly marked in black, white and russet	*Georychus*
	Not marked as above	4
4	Maximum of 6 cheek teeth in upper toothrow; palate extends only slightly beyond the last molars; the angular process of the lower jaw does not extend posteriorly; predominantly occurring in East Africa	*Heliophobius*
	Only 4 cheek teeth in upper toothrow; palate extends well beyond the last molars; the angular process of the lower jaw extends posteriorly; predominantly in Southern and South-central Africa	5
	2n = 54; restricted to South Africa, Swaziland and southern Zimbabwe	*Cryptomys*
	Chromosome number variable; most diverse in South-central Africa	*Fukomys*

Genus Account

Bathyergus Illiger, 1811 (Dune Mole-rats)

Diagnosis

This genus has a highly restricted distribution in southern Namibia and western South Africa. They are large rodents (Mass ca. 300-900 g), with a characteristically stout body, very short limbs and a short tail typical of this fossorial family. The pelage varies between and within species, being light brown, silvery or black dorsally, and paler ventrally. External ear pinnae are absent. The large body size, grooved upper incisors and enlarged claws on the forelimbs readily distinguish this genus from other mole-rat genera. Three pairs of nipples (2 + 1 = 6) are present.

Figure 2.116 Photograph of *Bathyergus suillus* (John Visser).

Cranial Characters

The skull is robust, large (GLS ca. 45-65 mm), flattened with a narrow rostrum. The zygomatic arches are sturdy and wide. The interorbital constriction is wide. The sagital crest is well-developed. The infraorbital foramen is small. The anterior palatal foramina are very short and end well before the M^1. The bullae are slightly enlarged, but not markedly so.

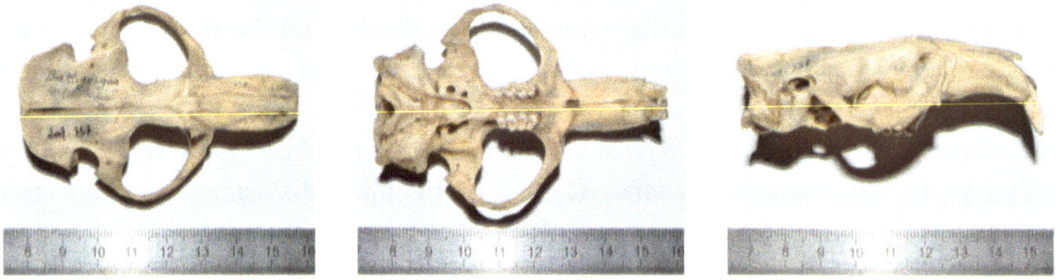

(a) (b) (c)

Figure 2.117 a-c Cranium of *Bathyergus suillus* (MNHN-ZM-MO 2007-357) (from left to right: dorsal view, ventral view and lateral view).

Dentition

Bathyergus has robust, grooved and pro-odont upper incisors. The dental formula is: 1013/1013. The molars are all roughly the same size, simple and uncusped (or with folds in juveniles).

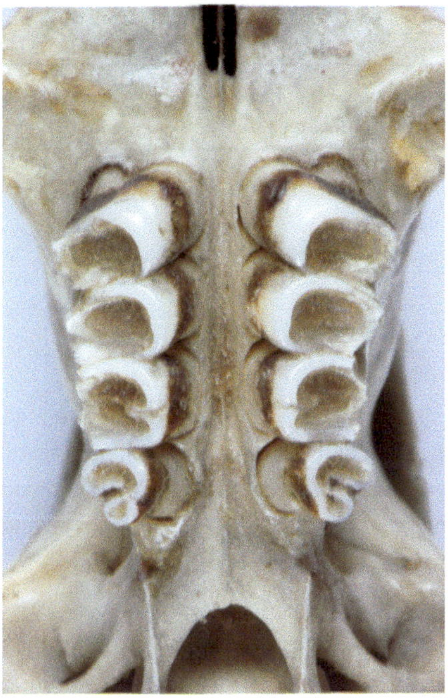

Figure 2.118 Upper cheek teeth of *Bathyergus suillus* (MNHN-ZM-MO 1962-2298).

Habitat & Biogeography

The genus has a very limited distribution, restricted to a narrow belt from the south-western tip of South Africa northwards to south-western Namibia. It is closely associated with coastal sand dunes and alluvial sands along rivers.

Fossil Record

Fossils of this genus are known from the Early Pliocene to the Late Pleistocene from Southern Africa (Winkler et al., 2010).

Biology & Ecology

The ecology of this genus is relatively well known and has been reviewed in Bennett & Faulkes (2000) and Happold (2013). These mole-rats are solitary and territorial, digging their burrows with their large front claws.

Systematic Notes

The genus was originally considered to be monospecific, with the taxon *janetta* listed as a subspecies of *B. suillus*. The two taxa, however, differ on morphological (De Graaff, 1981) and molecular grounds, with a 5.6% divergence between the two species on the cytochrome b gene (Faulkes et al., 2004). The possibility of a hybrid zone in the area of sympatry has been suggested (Van Daele et al., 2007).

Species List

Two species are recognised:
Bathyergus janetta Thomas & Schwann, 1904
Bathyergus suillus (Schreber, 1782)

Key to Species

1	Dorsal pelage with broad, dark band from neck to rump, contrasting with paler flanks	*B. janetta*
	Dorsal pelage without broad, dark band from neck to rump, and not contrasting with flanks	*B. suillus*

Species Accounts

Bathyergus janetta Thomas & Schwann, 1904

Diagnosis

This species is readily distinguished from *B. suillus* (size larger, GLS > 55 mm (males) or > 40 mm (females); dorsal pelage uniform cinnamon-brown, lacking a dark mid-dorsal band; palate extends beyond last upper molar). In this species, the dorsal pelage has a distinct dark, broad dorsal band from neck to rump which contrasts with the paler flanks; the palate does not extend beyond the last upper molar; and GLS < 55 mm (males) or < 50 mm (females).

Taxonomic Issues

Chromosome number 2n = 54, FN = 104 (Nevo et al., 1986; Deuve et al., 2008).

Distribution

This species is restricted to the coastal regions of western South Africa extending marginally into south-western Namibia where it inhabits sandy soils, particularly sand dunes (De Graaff, 1981).

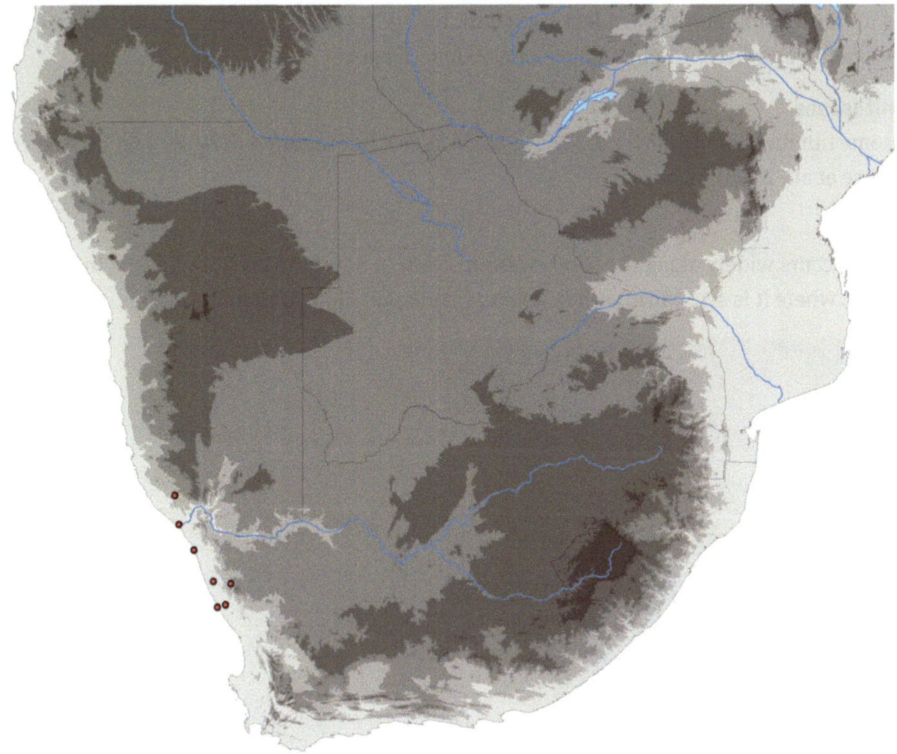

Figure 2.119 Distribution map

Measurements

External and cranial measurements (mm) and mass (g) for *Bathyergus janetta*, sexes combined

	Mean	Min	Max	N
Mass*	345.1	198	500	19
HB**	196.5	170	235	23
Tail**	45.3	38	52	23
HF**	38.7	34	45	23
Ear	-	-	-	-
GLS*	46.3	40.4	54.8	17

*Happold (2013); **De Graaff (1981)

Bathyergus suillus (Schreber, 1782)

Diagnosis

This species is readily distinguished from *B. janetta* (size smaller, GLS < 55 mm (males) or < 50 mm (females); has a distinct dark, broad dorsal band from neck to rump which contrasts with the paler flanks; the palate does not extend beyond the last upper molar). In this species, the dorsal pelage is uniform cinnamon-brown, lacking a dark mid-dorsal band; the palate extends beyond last upper molar; and GLS > 55 mm (males) or > 40 mm (females).

Taxonomic Issues

Chromosome number 2n = 56, FN = 102 (Nevo et al., 1986), but reported to be 2n = 54 by a recent study (Deuve et al., 2008).

Distribution

This species occurs widely along the south-western coast of South Africa from Knysna in the east to Rondawel, where it is sympatric with *B. janetta* (Faulkes et al., 2004), in the west.

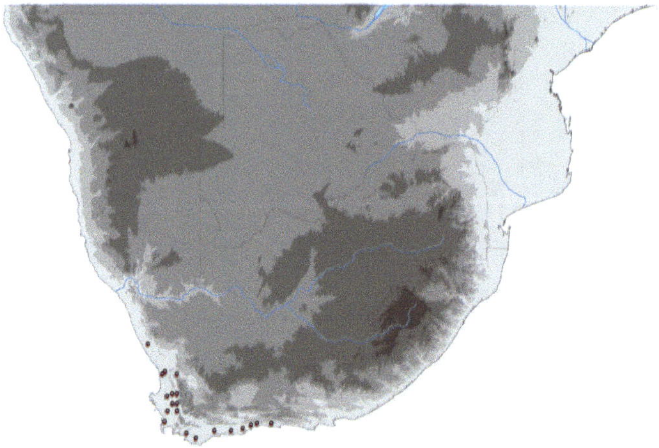

Figure 2.120 Distribution map

Measurements

External and cranial measurements (mm) and mass (g) for *Bathyergus suillus*, sexes combined

	Mean	Min	Max	N
Mass*	771.1	494	2200	465
HB**	267.6	204	330	84
Tail**	48.4	27	70	84
HF**	48.3	42	65	80
Ear	-	-	-	-
GLS*	59.7	39.4	76.7	59

*Happold (2013); **De Graaff (1981)

Genus Account

Cryptomys Gray, 1864 (Mole-rats)

Diagnosis

This genus is restricted to South Africa, Swaziland and southern Zimbabwe. They are small rodents (Mass ca. 50-80 g), with a characteristically stout body, very short limbs and a short tail typical of this fossorial family. The pelage varies between and within species, being grey-brown, cinnamon-brown or dark grey dorsally, and paler ventrally. This genus can be distinguished from other mole-rats within its range by its small size, uniform colour of dorsal pelage and lack of folds on its upper molars. This genus is indistinguishable from *Fukomys* on morphological grounds, but the two genera are highly divergent on molecular grounds (Faulkes et al., 2004; Kock et al., 2006). Three pairs of nipples (2 + 1 = 6) are present.

Figure 2.121 Photograph of *Cryptomys natalensis* (Tim Jackson).

Cranial Characters

The skull is sturdy, small (GLS ca. 30-40 mm), flattened with a narrow rostrum. The zygomatic arches are sturdy and wide. The interorbital constriction is wide. The sagittal crest is poorly developed. The infraorbital foramen is small. The anterior palatal foramina are tiny and end well before the M^1. The bullae are slightly enlarged, but not markedly so.

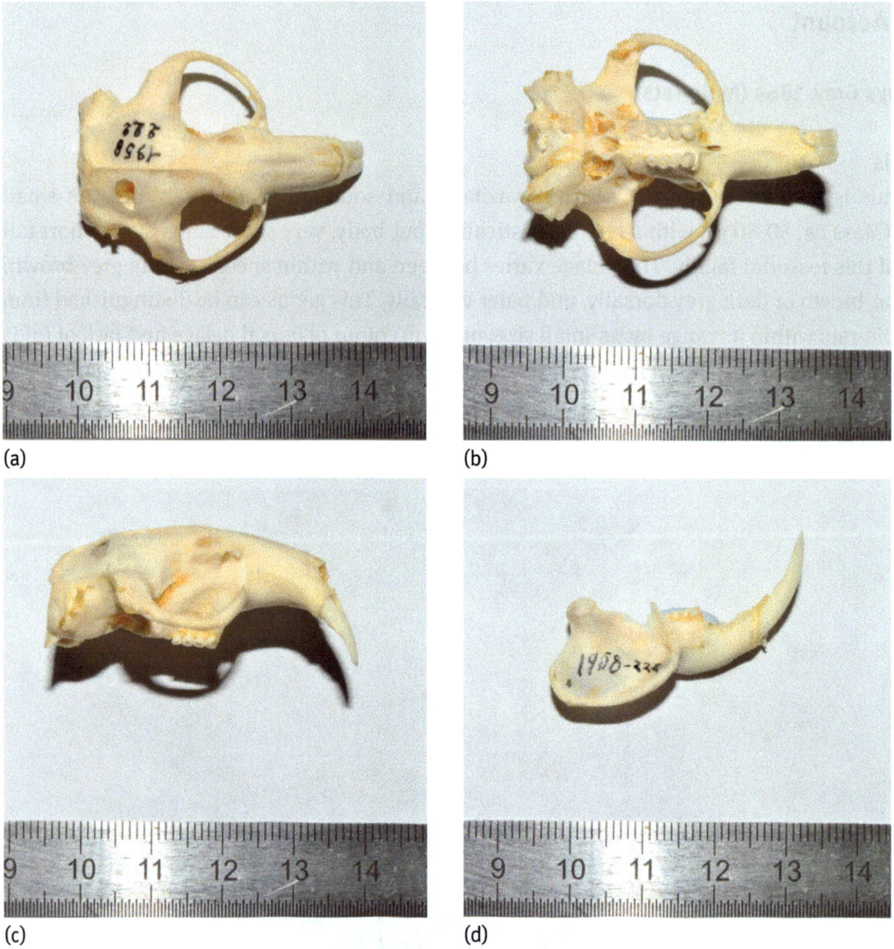

Figure 2.122 a-d Cranium of *Cryptomys hottentotus* (MNHN-ZM-MO 1958-222) (from top left to bottom right: dorsal view, ventral view, lateral view and lower jaw).

Dentition

Cryptomys has robust, ungrooved and pro-odont upper incisors. The dental formula is: 1013/1013. The molars are all roughly the same size, simple and uncusped and without folds.

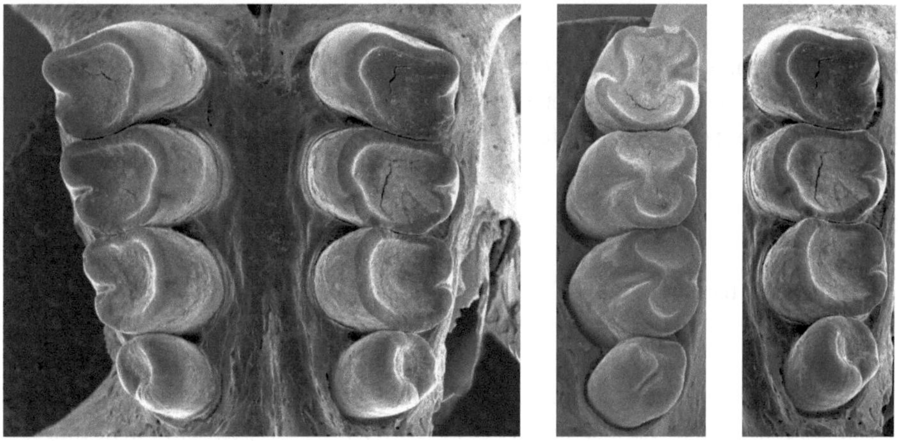

(a) (b) (c)

Figure 2.123 Cheek teeth of *Cryptomys hottentotus* (MNHN-ZM-MO 1988-115) (A: upper cheek teeth; B: lower left toothrow; C: upper left toothrow).

Habitat & Biogeography

The genus is widely but patchily distributed in South Africa, Swaziland and southern Zimbabwe where it is associated with sandy, alluvial or friable soils in which it can dig its burrows (De Graaff, 1981).

Fossil Record

Fossils of this genus are known from the Early Pliocene to the Late Pleistocene from Southern Africa, and Middle Pleistocene in East Africa (Winkler et al., 2010).

Biology & Ecology

The ecology and behaviour of this genus has been intensively studied and is reviewed in Bennett & Faulkes (2000) and Happold (2013). These mole-rats are colonial with a single reproductive pair per colony. They dig their burrows with their large incisors.

Systematic Notes

This genus previously also included all the species now placed in *Fukomys*. The two genera are not obviously distinguishable on morphological grounds, but differ on molecular grounds (Faulkes et al., 2004). The five species recognised here within *Cryptomys* were previously lumped into a single species *C. hottentotus*, but differ significantly by 11-18% on the cytochrome b gene (Faulkes et al., 2004). These five species are also typically allopatric, although *C. mahali* and *C. pretoriae* occur within 20 km of each other in the Pretoria region of South Africa.

Species List

Five species are recognised:
Cryptomys hottentotus (Lesson, 1826)
Cryptomys mahali (Roberts, 1913)
Cryptomys natalensis (Roberts, 1913)
Cryptomys nimrodi (de Winton, 1896)
Cryptomys pretoriae (Roberts, 1913)

Key to Species

The members of this genus are not readily diagnosable on traditional morphological grounds and therefore a key cannot currently be presented. The species are distinct on chromosomal and/or molecular grounds (see species accounts for further details).

Species Accounts

Cryptomys hottentotus (Lesson, 1826)

Diagnosis

This species can currently only be distinguished from other *Cryptomys* species on molecular grounds (Faulkes et al., 2004), but it appears to be the only *Cryptomys* species within its range.

Taxonomic Issues

Chromosome number 2n = 54 (Faulkes et al., 2004).

Distribution

This species occurs widely in western and southern parts of South Africa from Cape Town north to the Gariep River and east to beyond Port Elisabeth. However, the northern and eastern limits of its range are currently unresolved due to confusion with other *Cryptomys* species.

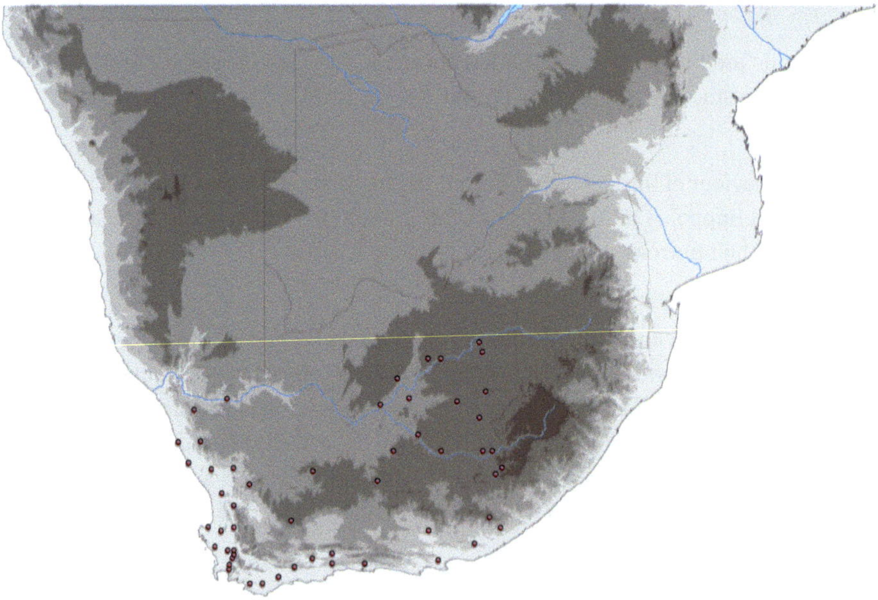

Figure 2.124 Distribution map

Measurements

External and cranial measurements (mm) and mass (g) for *Cryptomys hottentotus*, sexes combined

	Mean	Min	Max	N
Mass	-	-	-	-
HB*	-	110	160	43
Tail*	-	13	29	37
HF*	-	19	28	40
Ear	-	-	-	-
GLS*	-	34.0	42.5	43

*Roberts (1951)

Cryptomys mahali (Roberts, 1913)

Diagnosis

This species can currently only be distinguished from other *Cryptomys* species on molecular grounds (Faulkes et al., 2004).

Taxonomic Issues

Chromosome number 2n = 54 (Faulkes et al., 2004).

Distribution

This species is currently only known from the type locality in Pretoria, South Africa (Faulkes et al., 2004), but may occur more widely in surrounding areas.

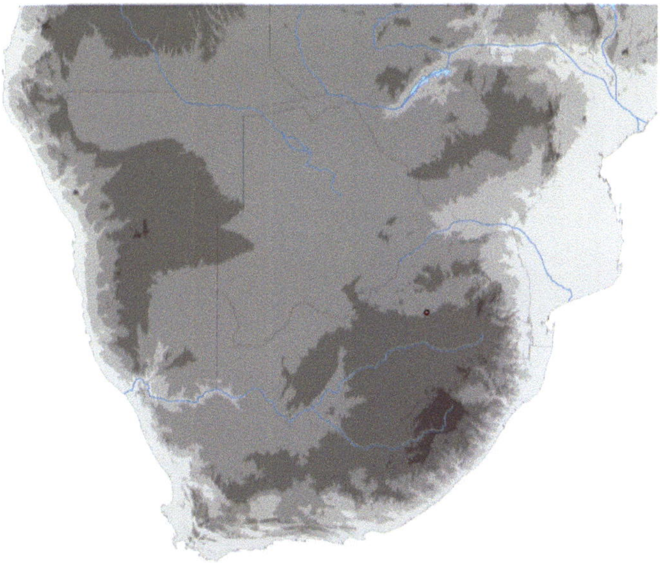

Figure 2.125 Distribution map

Measurements

External and cranial measurements (mm) and mass (g) for *Cryptomys mahali*, sexes combined

	Mean	Min	Max	N
Mass	-	-	-	-
HB*	-	135	156	23
Tail*	-	16	24	23
HF*	-	22	28	23
Ear	-	-	-	-
GLS*	-	38.0	42.0	23

*Roberts (1951)

Cryptomys natalensis (Roberts, 1913)

Diagnosis

This species can currently only be distinguished from other *Cryptomys* species on molecular grounds (Faulkes et al., 2004), but it appears to be the only *Cryptomys* species within its range.

Taxonomic Issues

Chromosome number 2n = 54 (Faulkes et al., 2004).

Distribution

This species occurs widely in the eastern parts of South Africa and Swaziland. The northern limits of its range are unknown and the localities plotted on the map from north-eastern South Africa (northern Kruger National Park) may refer to *C. nimrodi*.

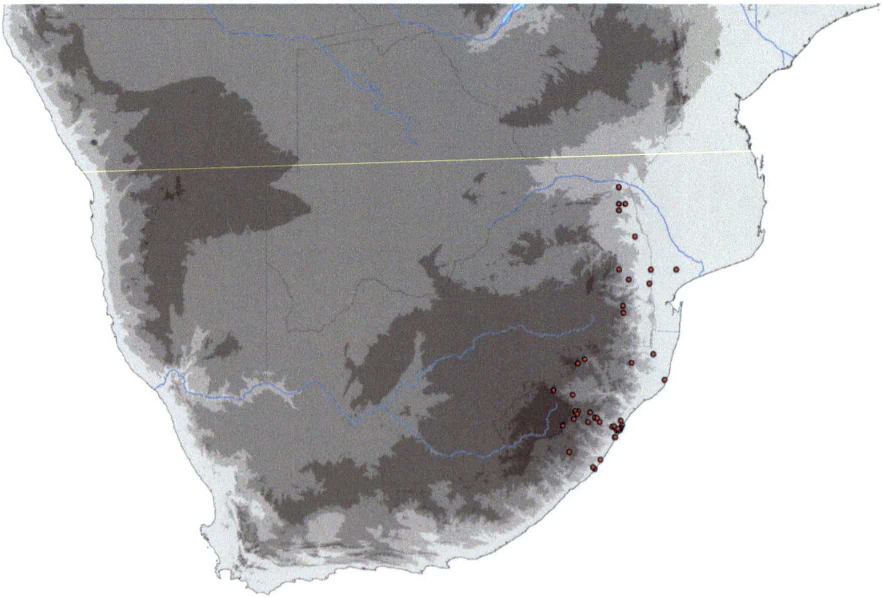

Figure 2.126 Distribution map

Measurements

External and cranial measurements (mm) and mass (g) for *Cryptomys natalensis*, sexes combined

	Mean	Min	Max	N
Mass	79.0	48	110	7
HB	123.1	79	178	28
Tail	13.3	8	20	27
HF	22.6	17	29	20
Ear	-	-	-	-
GLS*	39.7	35.5	42.3	11

*Roberts (1951)

Cryptomys nimrodi (de Winton, 1896)

Diagnosis

This species can currently only be distinguished from other *Cryptomys* species on molecular grounds (Faulkes et al., 2004), but it appears to be the only *Cryptomys* species within its range.

Taxonomic Issues

See genus account.

Distribution

This species appears to be restricted to southern Zimbabwe, but its southern limits are unresolved and it may occur in northern South Africa.

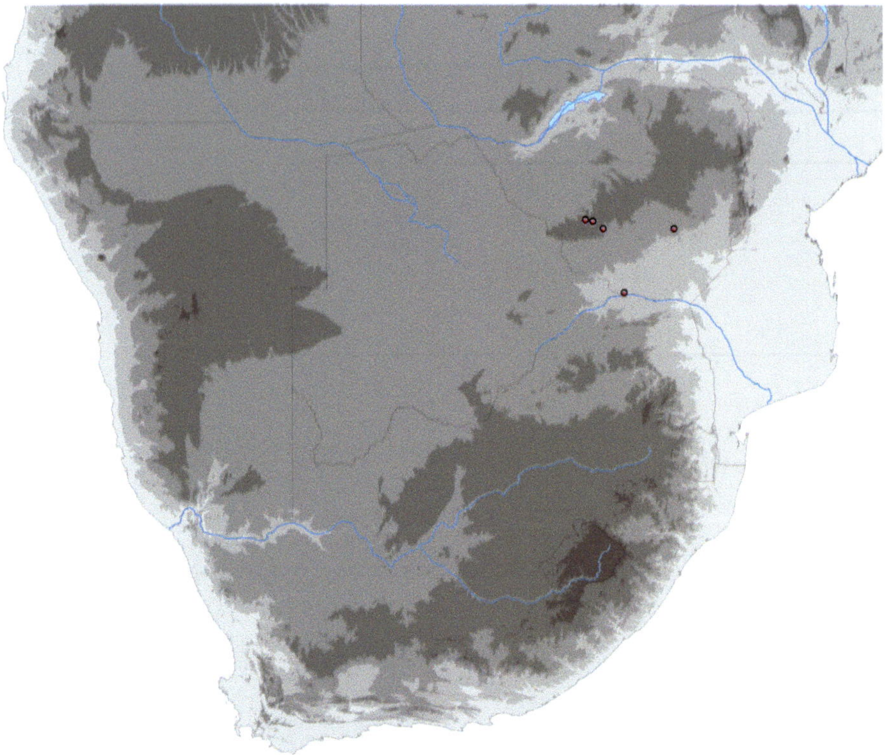

Figure 1.127 Distribution map

Measurements

External and cranial measurements (mm) and mass (g) for *Cryptomys nimrodi*, sexes combined

	Mean	Min	Max	N
Mass	-	-	-	-
HB*	147	-	-	1
Tail	-	-	-	-
HF*	24.5	-	-	1
Ear	-	-	-	-
GLS*	31.0	-	-	1

*de Winton (1896), measurements of the holotype

Cryptomys pretoriae (Roberts, 1913)

Diagnosis

This species can currently only be distinguished from other *Cryptomys* species on molecular grounds (Faulkes et al., 2004).

Taxonomic Issues

Chromosome number 2n = 54 (Faulkes et al., 2004).

Distribution

This species has currently only been confirmed from four localities in the Gauteng province, South Africa (Faulkes et al., 2004). Its distributional limits are currently unresolved but it is likely to extend from Gauteng to the Limpopo River, where there are numerous records of "*Cryptomys hottentotus*" (De Graaff, 1981; Rautenbach, 1982). Its eastern limit is likely to be the escarpment beyond which *C. natalensis* is known to occur (Faulkes et al., 2004).

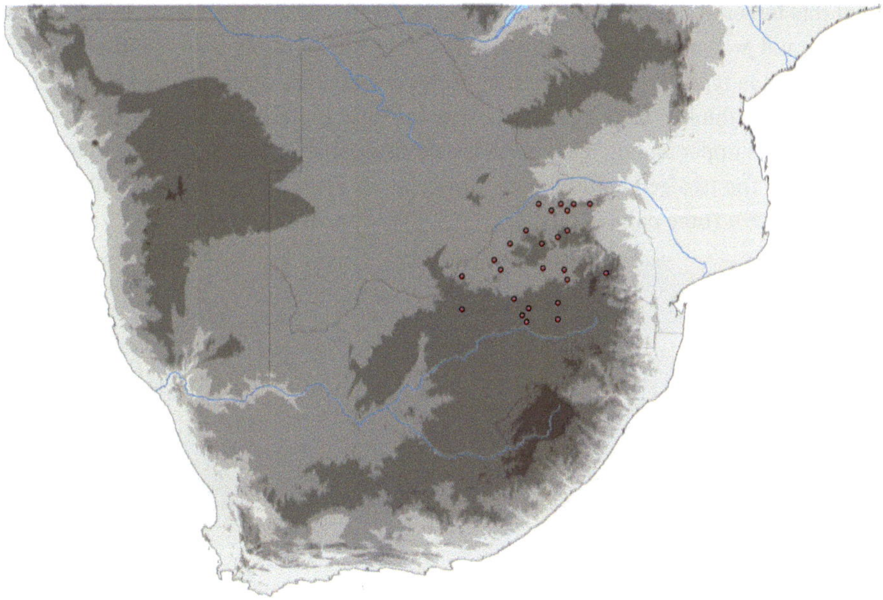

Figure 1.128 Distribution map

Measurements

External and cranial measurements (mm) and mass (g) for *Cryptomys pretoriae*, sexes combined

	Mean	Min	Max	N
Mass	-	-	-	-
HB	-	-	-	-
Tail	-	-	-	-
HF	-	-	-	-
Ear	-	-	-	-
GLS*	36.4	-	-	1

*Roberts (1913)

Genus Account

Fukomys Kock et al., 2006 (Mole-rats)

Diagnosis

This genus is widespread in South-central Africa, extending south into Southern Africa (*F. damarensis* and *F. darlingi*), and north to the northern savannas (*F. ochraceocinereus*) and into West Africa (*F. foxi* and *F. zechi*). They are small to medium-sized rodents (Mass mostly ca. 80-160 g, but *F. mechowi* can attain > 500 g), with a characteristically stout body, very short limbs and a short tail typical of this fossorial family. The pelage varies between and within species, being various shades of brown and grey dorsally often with a conspicuous white patch on top of the head and neck. Ventral pelage is typically paler than that of the dorsal. This genus can be distinguished from other mole-rats within its range by its small size, the (usual) presence of a white head patch, and lack of folds on its upper molars. This genus is indistinguishable from *Cryptomys* on morphological grounds, but the two genera are highly divergent on molecular grounds (Faulkes et al., 2004; Kock et al., 2006). Three pairs of nipples (2 + 1 = 6) are present.

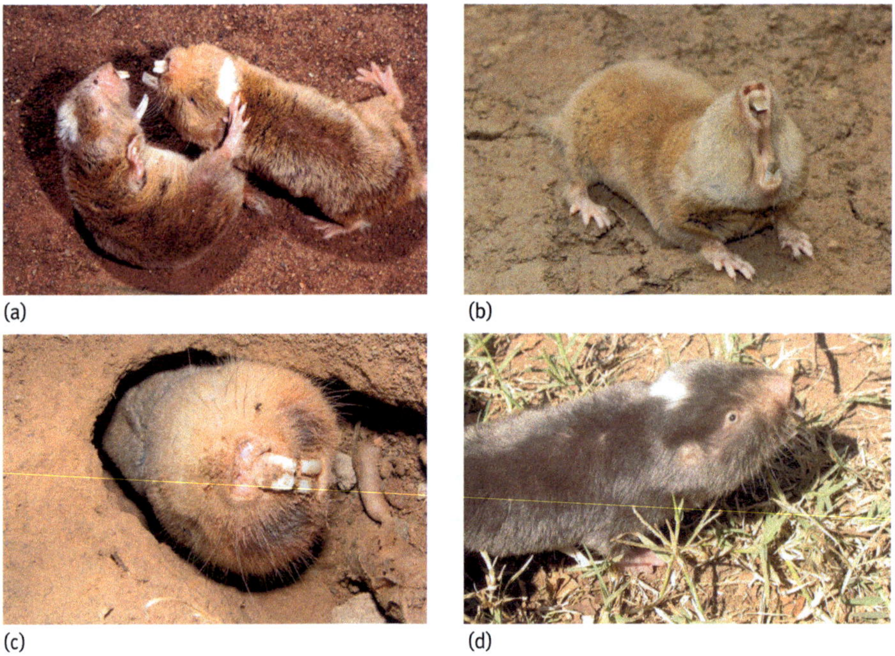

(a) (b) (c) (d)

Figure 2.129 a-d Photographs of *Fukomys*, from top left to bottom right: *F. damarensis* (Tim Jackson), *F. mechowi* (James Harvey), *F. mechowi* (Radim Sumbera) and *F. darlingi* (Radim Sumbera).

Cranial Characters

The skull is sturdy, small (GLS mostly ca. 30-45 mm, but may be > 50 mm in *F. mechowi*), flattened with a narrow rostrum. The zygomatic arches are sturdy and wide. The interorbital constriction is wide. The sagital crest is poorly developed. The infraorbital foramen is small. The anterior palatal foramina are tiny and end well before the M^1. The bullae are slightly enlarged, but not markedly so.

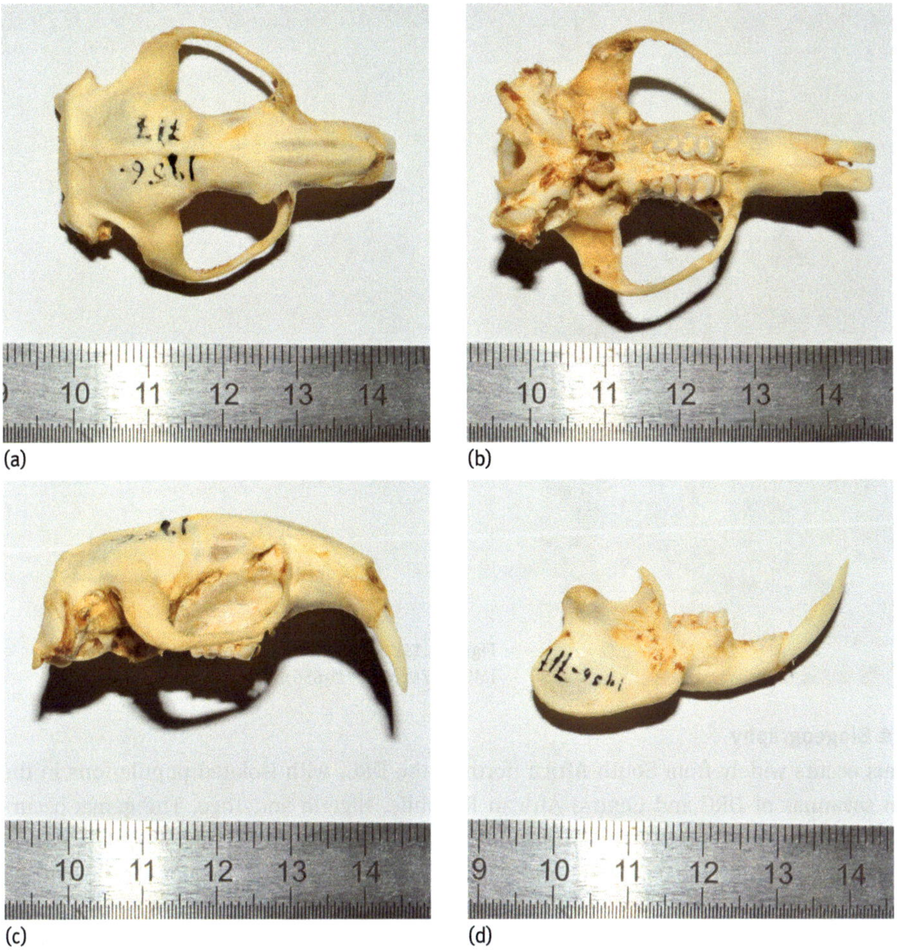

Figure 2.130 a-d Cranium of *Fukomys ochraceocinereus* (MNHN-ZM-MO 1956-717) (from top left to bottom right: dorsal view, ventral view, lateral view and lower jaw).

Dentition

Fukomys resembles *Cryptomys* in having robust, ungrooved and pro-odont upper incisors. The dental formula is: 1013/1013. The molars are all roughly the same size, simple and uncusped and without folds.

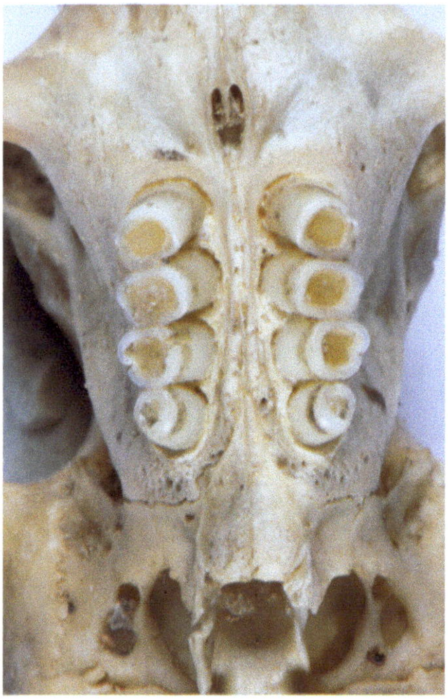

Figure 2.131 Upper cheek teeth of *Fukomys mechowi* (MNHN-ZM-MO 1911-656).

Habitat & Biogeography

The genus occurs widely from South Africa north to the DRC, with isolated populations in the northern savannas of DRC and Central African Republic, Nigeria and Togo. The genus occurs predominantly west of the East African Rift Valley, with the great lakes (such as Lakes Malawi and Tanganyika) forming the eastern boundary of the genus (the genus *Heliophobius* occurring predominantly east of these lakes). The few exceptions include the newly described is *F. ilariae* that is known from a single site on the coastal plains of Somalia (Gippoliti & Amori, 2011) and an undescribed species has been recorded from Hanang in northern Tanzania (Faulkes et al., 2010). *Fukomys* is typically associated with sandy, alluvial or friable soils in savanna habitats where it can dig its burrows.

Fossil Record

No fossils are known of this genus, but see *Cryptomys* account for fossils of that genus (which presumably are not distinguishable from fossils of *Fukomys*).

Biology & Ecology

The ecology of this genus is relatively well known and has been reviewed in Bennett & Faulkes (2000) and Happold (2013). These mole-rats are colonial with a single reproductive pair per colony. They dig their burrows with their large incisors.

Systematic Notes

Members of this genus were previously placed in *Cryptomys* which was recently shown on molecular grounds to comprise two deeply divergent clades (Faulkes et al., 2004), hence justifying the description of the new genus *Fukomys* (Kock et al., 2006). *Fukomys* occurs widely in Sub-Saharan

Africa, but the hotspot of species diversity is situated in south-central Zambia (Van Daele et al., 2007). The species limits within this genus are still mostly unresolved, and several undescribed taxa await description (e.g. Van Daele et al., 2007; Faulkes et al., 2010).

Species List

Fourteen species have currently been described with at least four new species awaiting description:

Fukomys amatus (Wroughton 1907)
Fukomys anselli (Burda et al., 1999)
Fukomys bocagei (de Winton 1897)
Fukomys damarensis (Ogilby, 1838)
Fukomys darlingi (Thomas, 1895)
Fukomys foxi (Thomas, 1911)
Fukomys ilariae Gippoliti & Amori, 2011
Fukomys kafuensis (Burda et al., 1999)
Fukomys mechowi (Peters, 1881)
Fukomys micklemi (Chubb, 1909)
Fukomys ochraceocinereus (Heuglin, 1864)
Fukomys whytei (Thomas, 1897)
Fukomys zechi (Marschie, 1900)
Fukomys vandewoestijneae Van Daele et al., 2013

Key to Species

The members of this genus are not readily diagnosable on traditional morphological grounds and therefore a key cannot currently be presented. The species are distinct on chromosomal and/or molecular grounds (see species accounts for further details).

Species Accounts

Fukomys amatus (Wroughton 1907)

Diagnosis

This species can be differentiated from other *Fukomys* species by its karyotype (see below). The white head patch typical of many *Fukomys* species is absent in *F. amatus*

Taxonomic Issues

This species is most closely related to *F. whytei* (Burda et al., 2005; Faulkes et al., 2010). Chromosome number 2n = 50, FN = 92 (Macholan et al., 1998).

Distribution

This species is known from a few sites in Zambia and the DRC.

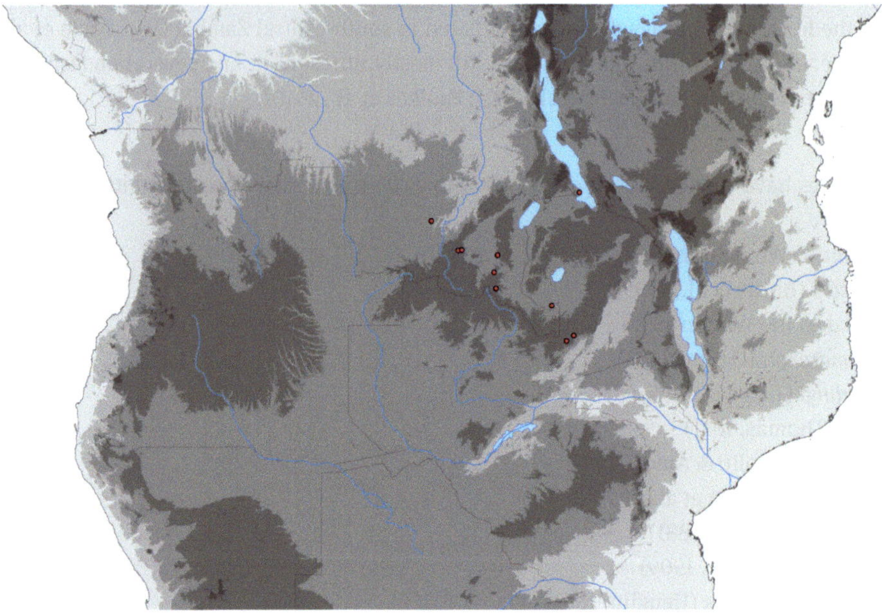

Figure 2.132 Distribution map

Measurements

No measurements could be located for this species.

Fukomys anselli (Burda et al., 1999)

Diagnosis

This species can only be differentiated from other *Fukomys* species on chromosomal and molecular grounds (see below).

Taxonomic Issues

This species is closely related to *F. kafuensis* and *F. micklemi* with only 1.4% divergence on the cytochrome b gene between these three species (Ingram et al., 2004; Van Daele et al., 2007), but each with a distinct karyotype (Burda et al., 1999). Chromosome number 2n = 68, FN = 75-78 (Burda et al., 1999).

Distribution

This species is known from a few scattered sites in Zambia.

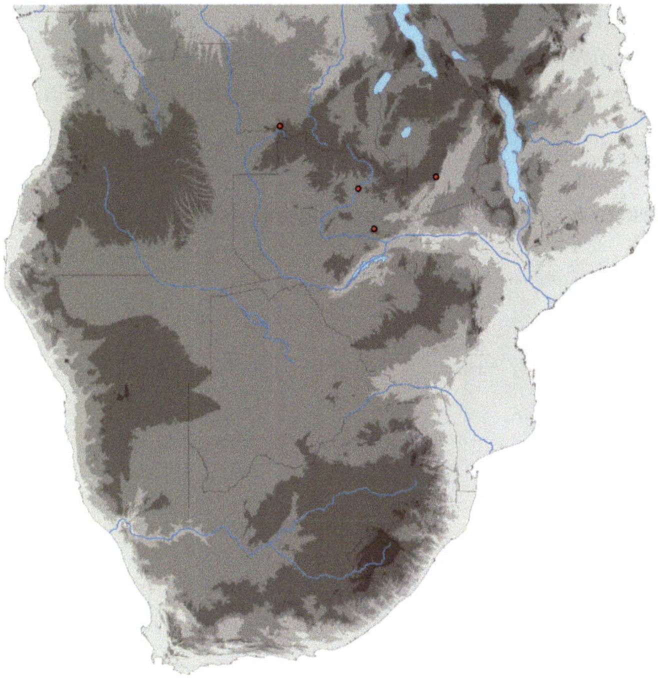

Figure 2.133 Distribution map

Measurements

External and cranial measurements (mm) and mass (g) for *Fukomys anselli*, sexes combined

	Mean	Min	Max	N
Mass*	84.0	65	145	140
HB*	120.2	108	135	50
Tail*	18.1	14	23	50
HF*	23.0	22	26	50
Ear*	0	-	-	-
GLS*	32.9	29.0	38.8	20

*Happold (2013)

Fukomys bocagei (de Winton 1897)

Diagnosis

This species co-occurs with *F. mechowi* (size larger, GLS typically > 45 (males), or > 35 (females); HF > 30 mm).

Taxonomic Issues

The recently described species *F. vandewoestijneae* was previously included in this species; as a result of this split, *F. bocagei* is now an Angolan endemic. Chromosome number 2n = 58, FN = 98 (Van Daele et al., 2013).

Distribution

This species occurs widely in western Angola extending marginally into northern Namibia.

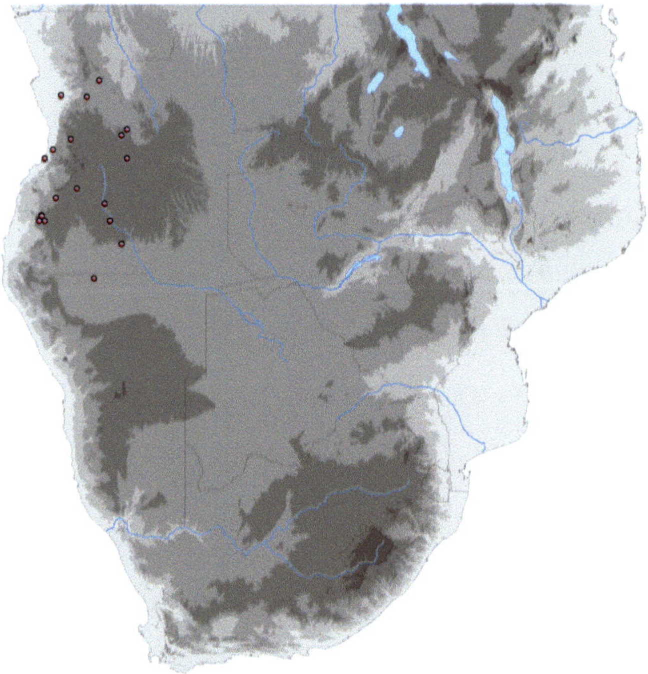

Figure 2.134 Distribution map

Measurements

External and cranial measurements (mm) and mass (g) for *Fukomys bocagei*, sexes combined

	Mean	Min	Max	N
Mass	-	-	-	-
HB*	153.0	141	165	8
Tail*	10.5	6	15	8
HF*	21.0	19	24	8
Ear*	0	-	-	-
GLS**	33.9	29.5	41.7	20

*Happold (2013); **Van Daele et al. (2013)

Fukomys damarensis (Ogilby, 1838)

Diagnosis

This species can be readily distinguished on chromosomal and molecular grounds (Faulkes et al., 2004; Ingram et al., 2004). Throughout most of its range it is also the only species of *Fukomys* (or *Cryptomys*) present.

Taxonomic Issues

Chromosome number 2n = 74 or 78, FN = 92 (Nevo et al., 1986), or 2n = 80 (Deuve et al., 2008). This chromosomal diversity appears to be rather high for intraspecific variation, and may suggest a species complex. Further research on the systematics of this species is warranted.

Distribution

This species occurs widely on Kalahari sands in Botswana, Namibia and Zimbabwe, extending marginally in Zambia and South Africa; it has probably been overlooked in south-eastern Angola.

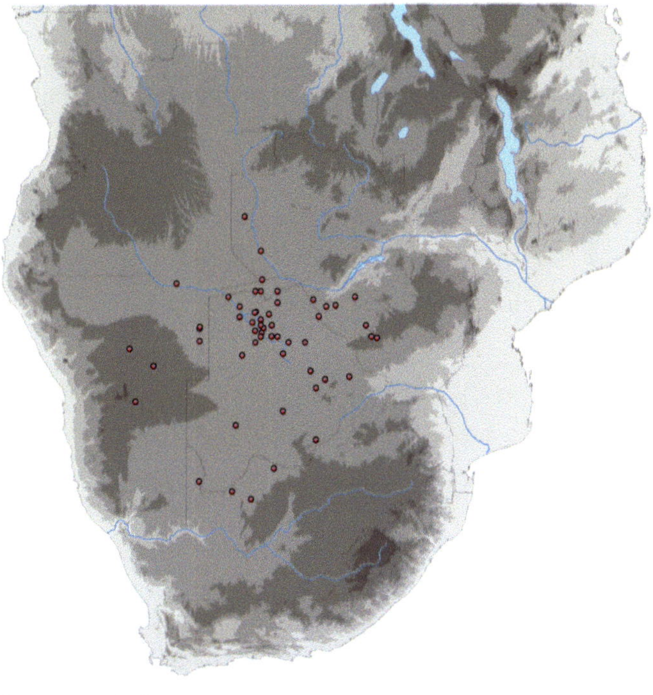

Figure 2.135 Distribution map

Measurements

External and cranial measurements (mm) and mass (g) for *Fukomys damarensis*, sexes combined

	Mean	Min	Max	N
Mass*	136.0	49	234	42
HB*	161.8	141	185	24
Tail*	25.5	23	32	24
HF*	27.0	26	30	24
Ear*	0	-	-	-
GLS*	36.1	31.6	44.1	24

*Happold (2013)

Fukomys darlingi (Thomas, 1895)

Diagnosis
This species can be readily distinguished on chromosomal and molecular grounds (Faulkes et al., 2004; Ingram et al., 2004). Throughout most of its range it is also the only species of *Fukomys* (or *Cryptomys*) present.

Taxonomic Issues
Chromosome number 2n = 54, FN = 80 (Aguilar, 1993).

Distribution
This species is known only from Zimbabwe and neighbouring Mozambique.

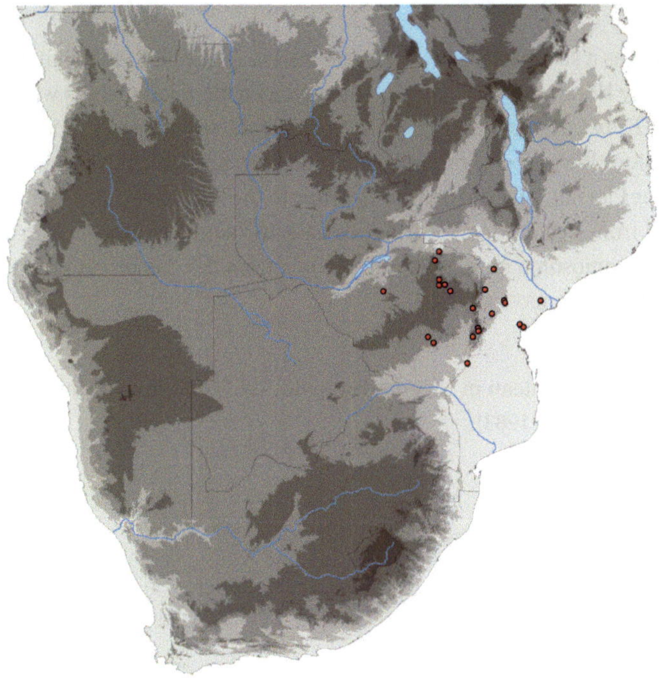

Figure 2.136 Distribution map

Measurements

External and cranial measurements (mm) and mass (g) for *Fukomys darlingi*, sexes combined

	Mean	Min	Max	N
Mass*	76.6	54	92	18
HB*	143.7	125	165	56
Tail*	10.0	8	13	56
HF*	22.7	20	30	56
Ear*	0	-	-	-
GLS*	33.1	30.6	37.9	56

*Happold (2013)

Fukomys foxi (Thomas, 1911)

Diagnosis

This is the only species of *Fukomys* (or in fact of the family Bathyergidae) within its range (but see below). The white head patch is usually present but is highly variable and may occasionally be absent (Williams et al., 1983).

Taxonomic Issues

This species was previously listed as a subspecies of *F. ochraceocinereus* (Happold, 1987), but was recognised as a distinct species by Honeycutt et al. (1991). The phylogenetic relationships within the northern *Fukomys* species (*F. foxi, F. zechi* and *F. ochraceocinereus*) are unknown and this group is in urgent need of revision. We suspect, on biogeographic and morphological grounds, that *F. foxi* (described from the Jos Plateau, Nigeria) may be endemic to the Jos Plateau and that the Cameroon population (recorded from Ngaoundere) is either conspecific with *F. ochraceocinereus* or refers to a separate species. Chromosome number is only known from the Cameroon population: 2n = 66 or 70, FN = 122 or 130 (Williams et al., 1983).

Distribution

This species is only known from the Jos Plateau in Nigeria (Happold, 1987) and from a single site in neighbouring Cameroon (Williams et al., 1983).

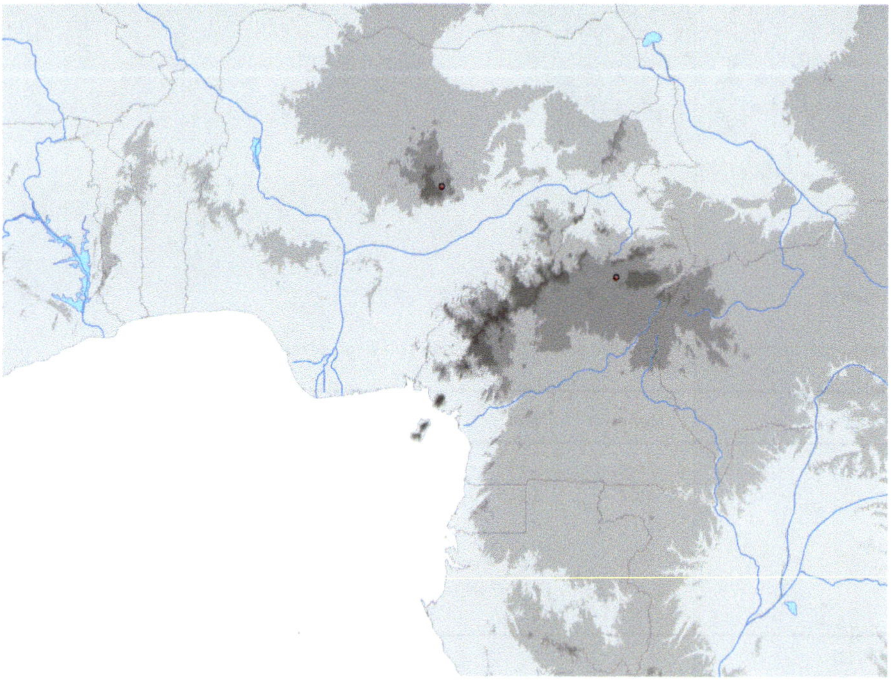

Figure 2.137 Distribution map

Measurements

External and cranial measurements (mm) and mass (g) for *Fukomys foxi*, sexes combined

	Mean	Min	Max	N
Mass	-	-	-	-
HB*	145	135	159	10
Tail*	14	11	17	10
HF*	29	26	31	10
Ear*	0	-	-	10
GLS*	40.7	39.3	43.0	10

*Rosevear (1969) based on a sample of 10 "normal" sized individuals. Measurements are also presented for two larger individuals, HB = 177 mm, Tail = 20 mm; HF = 32 mm; GSL = 46.2 mm. The measurements from the larger individuals fall within the range of values of the Cameroon specimens, HB = 161-191 mm; T = 16-25 mm; HF = 29-36 mm; GSL = 39.7-48.4 mm (Williams et al., 1983).

Fukomys ilariae Gippoliti & Amori, 2011

Diagnosis

This is the only *Fukomys* species occurring within its range. This species also has a unique dorsal pelage pattern with a white band extending from the head and neck posteriorly, dividing into two lateral stripes over the mid-back.

Taxonomic Issues

This species is only known from the holotype which consists of a mounted specimen without the skull (Gippoliti & Amori, 2011). In the absence of the skull it is difficult to distinguish *Fukomys* from *Heliophobius* and the possibility exists that this specimen belongs to the latter genus.

Distribution

This species is only known from the type locality in Somalia (Gippoliti & Amori, 2011)

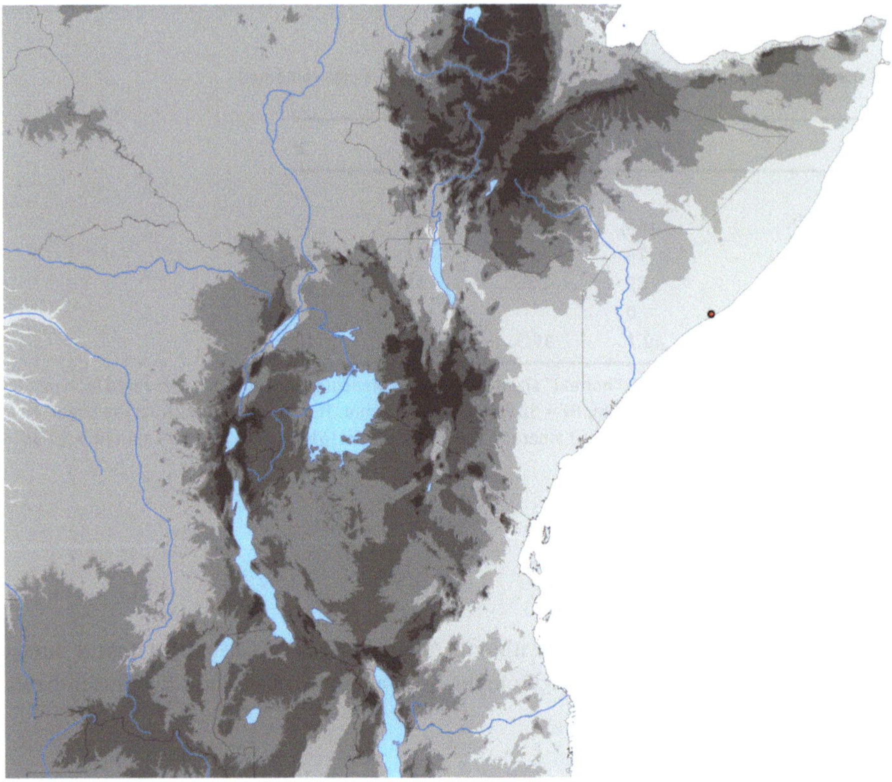

Figure 2.138 Distribution map

Measurements

External and cranial measurements (mm) and mass (g) for *Fukomys ilariae*, sexes combined

	Mean	Min	Max	N
Mass	-	-	-	-
HB*	118	-	-	1
Tail	-	-	-	-
HF	-	-	-	-
Ear	-	-	-	-
GLS	-	-	-	-

*Gippoliti & Amori (2011)

Fukomys kafuensis (Burda et al., 1999)

Diagnosis

This species can only be differentiated from other *Fukomys* species on chromosomal and molecular grounds (see below).

Taxonomic Issues

This species is closely related to *F. anselli* and *F. micklemi* with only 1.4% divergence on the cytochrome b gene between these three species (Ingram et al., 2004; Van Daele et al., 2007), but each with a distinct karyotype (Burda et al., 1999). Chromosome number 2n = 58, FN = 82 (Burda et al., 1999).

Distribution

This species is only known from a few localities in southern Zambia (Burda, 2000).

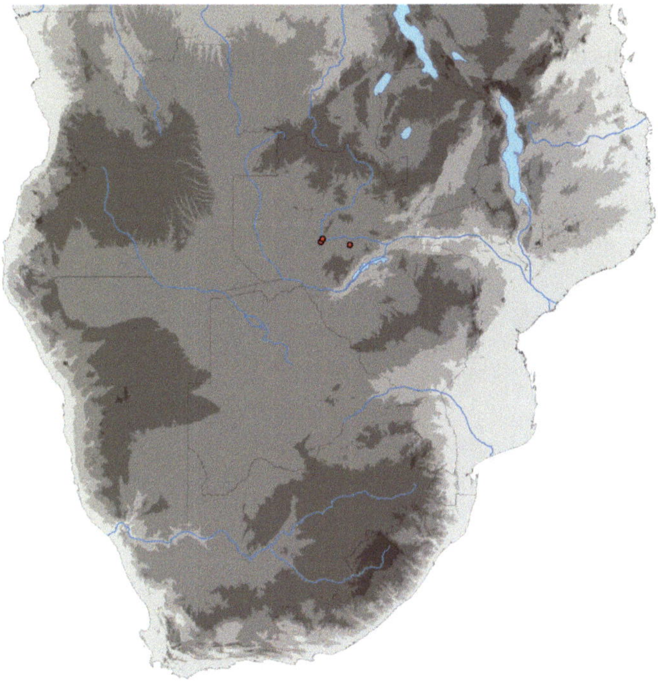

Figure 2.139 Distribution map

Measurements

External and cranial measurements (mm) and mass (g) for *Fukomys kafuensis*, sexes combined

	Mean	Min	Max	N
Mass*	85.0	61	127	12
HB*	107.6	96	129	9
Tail*	16.7	14	20	9
HF*	23.0	19	25	9
Ear*	0	-	-	-
GLS*	32.9	30.3	35.6	10

*Happold (2013)

Fukomys mechowi (Peters, 1881)

Diagnosis

This is the largest species of *Fukomys* which readily differentiates it from other sympatric members of the genus (GLS typically > 45 (males), or > 35 (females); HF > 30 mm).

Taxonomic Issues

Chromosome number 2n = 40, FN = 78 (Van Daele et al., 2013).

Distribution

This species occurs widely in South-central Africa in Zambia and neighbouring DRC and central Angola with an isolated record from Kabinda, north of the Congo River.

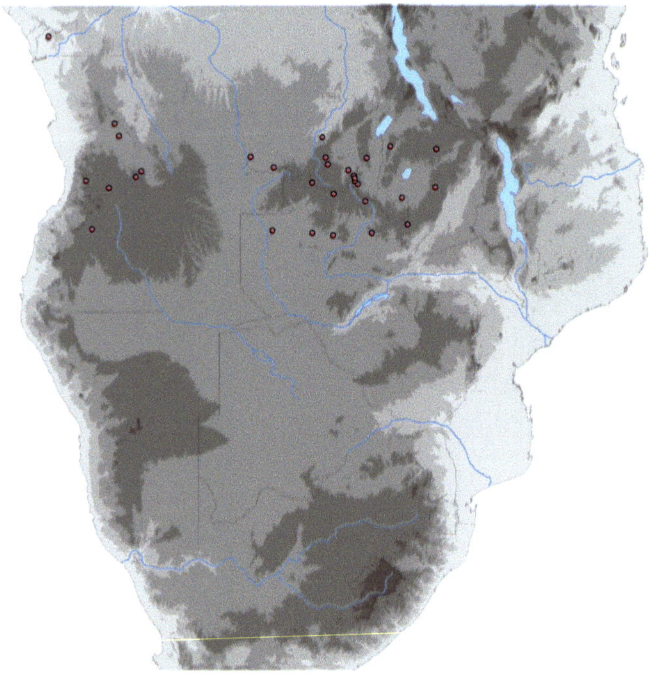

Figure 2.140 Distribution map

Measurements

External and cranial measurements (mm) and mass (g) for *Fukomys mechowi*, sexes combined

	Mean	Min	Max	N
Mass*	307.5	200	560	22
HB*	177.5	135	262	20
Tail*	27.6	23	33	20
HF*	33.8	30	38	20
Ear*	0	-	-	-
GLS**	47.8	38.6	61.8	47

*Happold (2013); **Van Daele et al. (2013)

Fukomys micklemi (Chubb, 1909)

Diagnosis

This species can only be differentiated from other *Fukomys* species on chromosomal and molecular grounds (see below).

Taxonomic Issues

This species is closely related to *F. anselli* and *F. kafuensis* with only 1.4% divergence on the cytochrome b gene between these three species (Ingram et al., 2004; Van Daele et al., 2007), but each with a distinct karyotype (Burda et al., 1999). Chromosome number 2n = 58 (Ingram et al., 2004).

Distribution

This species is only known from western Zambia (Van Daele et al., 2007).

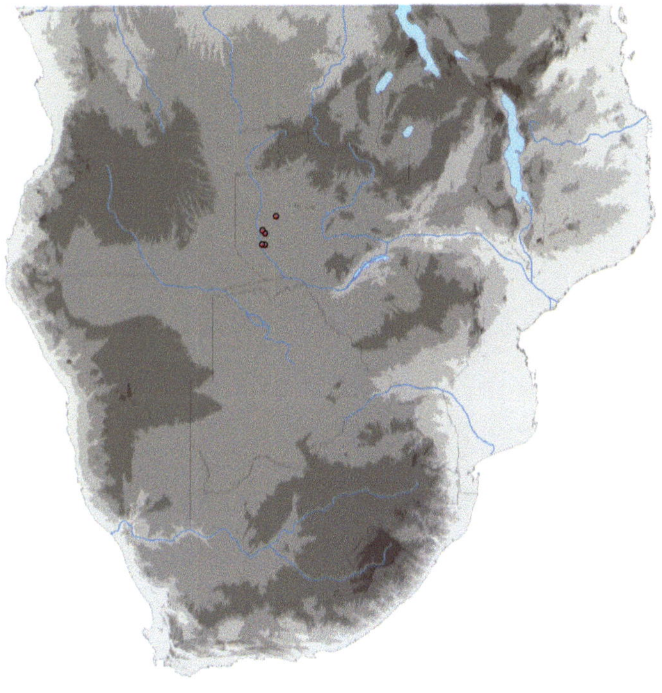

Figure 2.141 Distribution map

Measurements

No measurements could be located for this species.

Fukomys ochraceocinereus (Heuglin, 1864)

Diagnosis

This is the only species of *Fukomys* (or in fact of the family Bathyergidae) within its range (but see below).

Taxonomic Issues

The taxa *foxi* and *zechi* were previously listed as subspecies of *F. ochraceocinereus*. The phylogenetic relationships within these three northern *Fukomys* species are unknown and this group is in urgent need of revision. We suspect, on biogeographic and morphological grounds, that *F. foxi* (described from the Jos Plateau, Nigeria) may be endemic to the Jos Plateau and that the Cameroon population (recorded from Ngaoundere) of *F. foxi* may in fact be conspecific with *F. ochraceocinereus* or refer to a separate species.

Distribution

This species occurs in the northern savannas from northern Uganda and southern Sudan west to Central African Republic. The westerns limits of its distribution have not yet been resolved and if it is conspecific with the populations of *F. foxi* in Ngaoundere, Cameroon, then this species would have a much wider distribution than shown in this map.

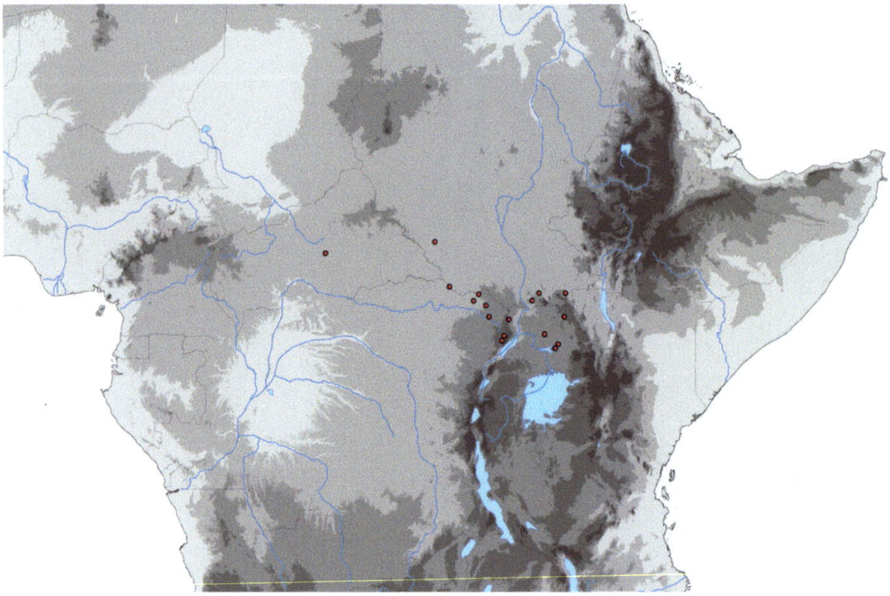

Figure 2.142 Distribution map

Measurements

External and cranial measurements (mm) and mass (g) for *Fukomys ochraceocinereus*, sexes combined

	Mean	Min	Max	N
Mass	-	-	-	-
HB*	201	-	-	4
Tail*	14	-	-	4
HF*	29	-	-	5
Ear	-	-	-	-
GLS*	47.8	-	-	5

*Rosevear (1969)

Fukomys vandewoestijneae Van Daele et al., 2013

Diagnosis

This species co-occurs with *F. mechowi* (size larger, GLS typically > 45 (males), or > 35 (females); HF > 30 mm). The white head patch typical of many *Fukomys* species is absent in *F. vandewoestijneae.*

Taxonomic Issues

This recently described species was previously included in *F. bocagei*, but differs on morphological, chromosomal and molecular grounds (Van Daele et al., 2013). Chromosome number 2n = 44, FN = 76 (Van Daele et al., 2013).

Distribution

This species is only known from north-western Zambia and neighbouring DRC where it is endemic to the Ikelenge Pedicle (Van Daele et al., 2013)

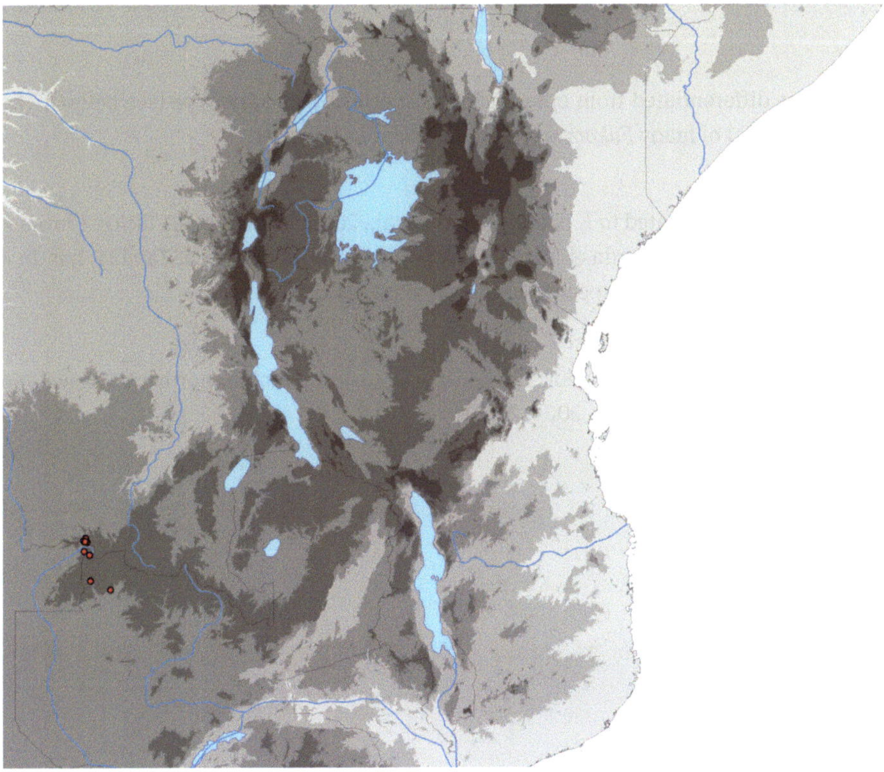

Figure 2.143 Distribution map

Measurements

External and cranial measurements (mm) and mass (g) for *Fukomys vandewoestijneae*, sexes combined

	Mean	Min	Max	N
Mass*	91.6	64	134	5
HB*	125.6	87	150	5
Tail*	18.7	14.9	21.3	5
HF*	22.3	18.5	24.1	5
Ear	-	-	-	-
GLS*	35.8	30.3	41.7	21

*Van Daele et al. (2013)

Fukomys whytei (Thomas, 1897)

Diagnosis

This species can be differentiated from other *Fukomys* species by its karyotype (see below). The white head patch typical of many *Fukomys* species is absent in *F. whytei*.

Taxonomic Issues

This species is most closely related to *F. amatus* (Burda et al., 2005; Faulkes et al., 2010). Chromosome number 2n = 46, FN = 76 (Burda et al., 2005). A population from Kasama, Zambia, has 2n = 64 and may refer to an undescribed species (Ingram et al., 2004).

Distribution

This species is only known from a few localities in north-eastern Zambia, northern Malawi and neighbouring Tanzania (Faulkes et al., 2004, 2010).

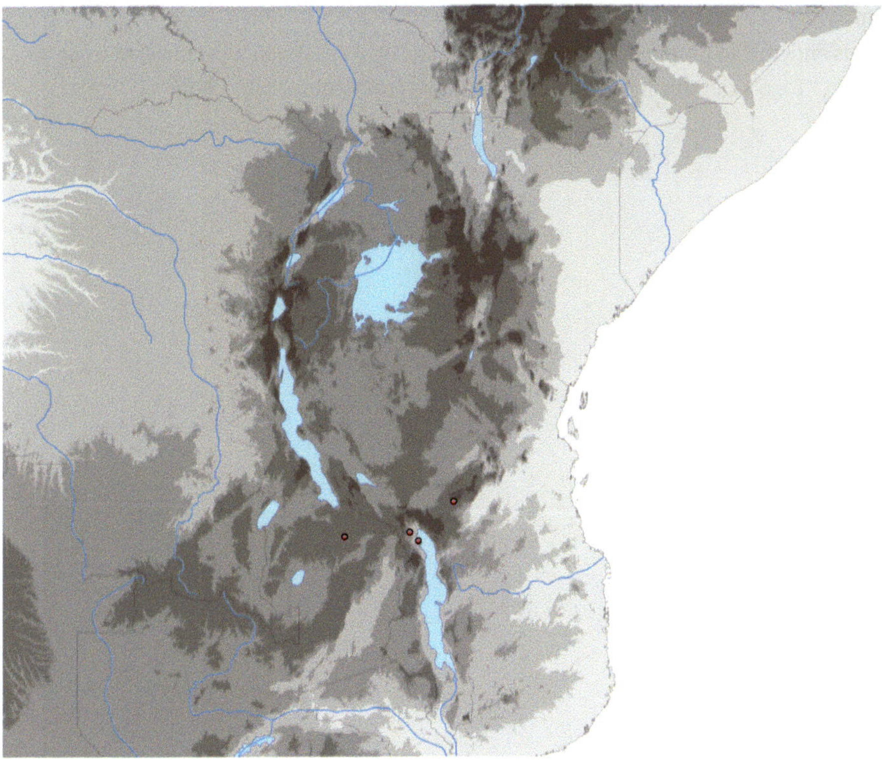

Figure 2.144 Distribution map

Measurements

External and cranial measurements (mm) and mass (g) for *Fukomys whytei*, sexes combined

	Mean	Min	Max	N
Mass*	127.0	-	-	8
HB*	143.0	-	-	8
Tail*	17.5	-	-	8
HF*	23.9	-	-	8
Ear	-	-	-	-
GLS*	37.6	-	-	8

*Burda et al. (2005)

Fukomys zechi (Marschie, 1900)

Diagnosis

This is the only species of *Fukomys* (or in fact of the family Bathyergidae) in its range.

Taxonomic Issues

See genus account.

Distribution

This species is endemic to Ghana (Grubb et al., 1998). The type specimen was collected from "near Kete Krachi" in "Togo", but this locality is now in Ghana (see Grubb et al., 1998).

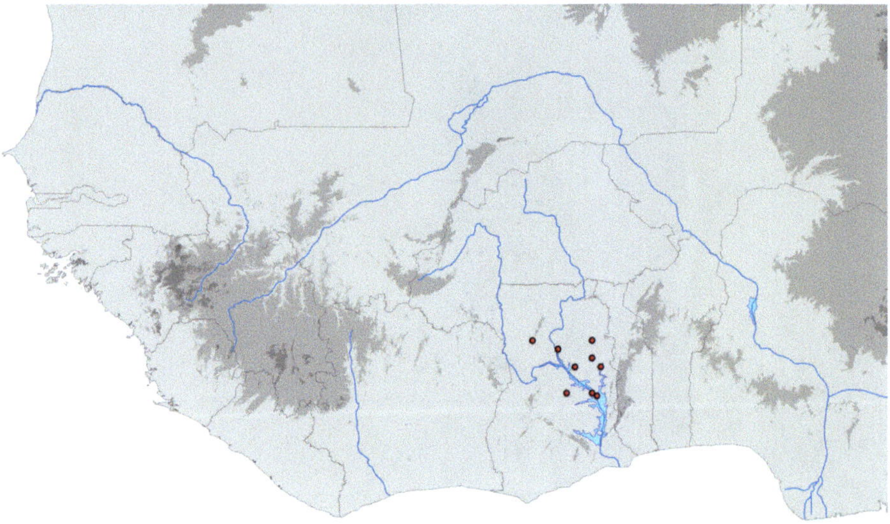

Figure 2.145 Distribution map

Measurements

External and cranial measurements (mm) and mass (g) for *Fukomys zechi*, sexes combined

	Mean	Min	Max	N
Mass*	217	155	283	48
HB**	167	-	-	4
Tail**	15	-	-	4
HF**	27	-	-	4
Ear**	0	-	-	4
GLS**	42.7	-	-	4

*Yeboah & Dakwa (2002); **Rosevear (1969)

Genus Account

Georychus Illiger, 1811 (Cape Mole-rats)

Diagnosis

This monotypic genus is endemic to South Africa. It is a medium-sized rodent (Mass ca. 180 g), with a characteristically stout body, very short limbs and a short tail typical of this fossorial family. The dorsal pelage is distinctively russet in colour, fading to silvery-brown on the rump, with black on the head, except around the eyes and mouth which are white. Ventral pelage is whitish. The brightly coloured pelage, ungrooved upper incisors and molars with folds readily distinguish this genus from other mole-rat genera. Three pairs of nipples (2 + 1 = 6) are present.

Figure 2.146 Photograph of *Georychus capensis* (John Visser).

Cranial Characters

The skull is robust, relatively large (GLS ca. 45-50 mm), flattened with a narrow rostrum. The zygomatic arches are sturdy and wide. The interorbital constriction is wide. The sagittal crest is relatively well-developed. The infraorbital foramen is small. The anterior palatal foramina are very short and end well before the M^1. The palate extends beyond the molar toothrow. The bullae are not particularly enlarged.

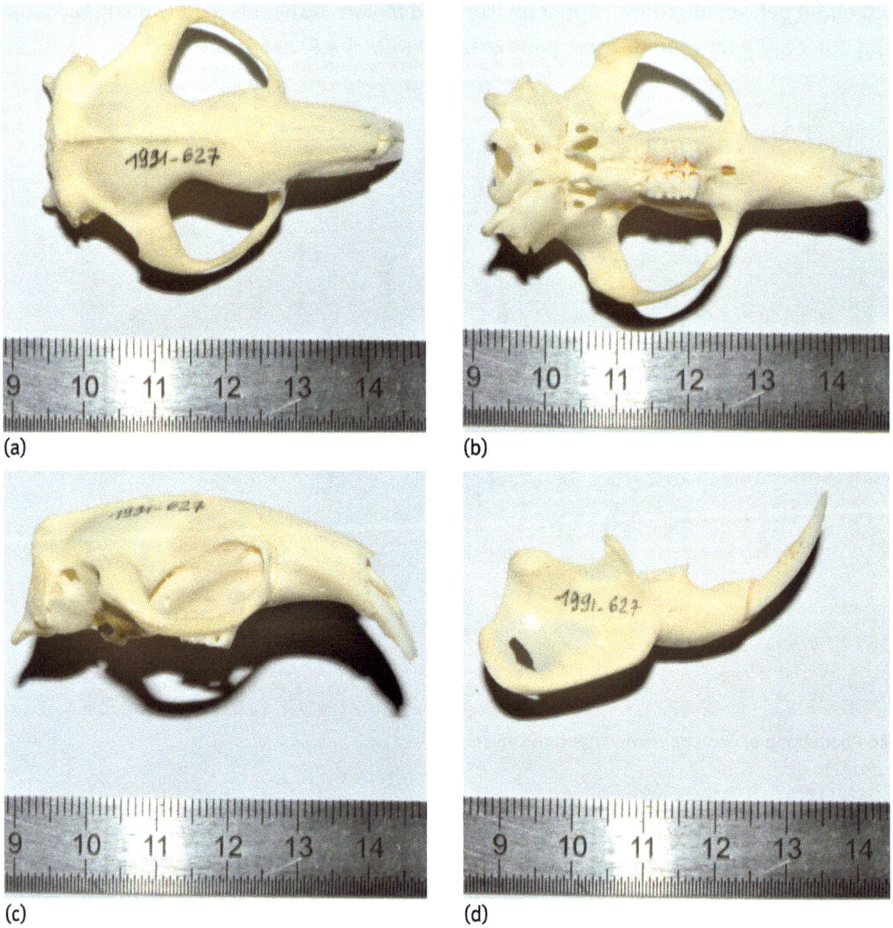

Figure 2.147 a-d Cranium of *Georychus capensis* (MNHN-ZM-MO 1991-627) (from top left to bottom right: dorsal view, ventral view, lateral view and lower jaw).

Dentition

Georychus has robust, ungrooved and pro-odont upper incisors. The dental formula is: 1013/1013. The molars are all roughly the same size, simple and uncusped but with conspicuous folds (Taylor et al., 1985).

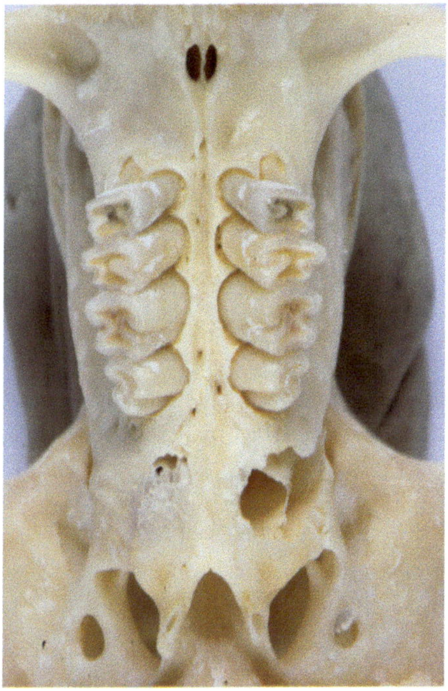

Figure 2.148 Upper cheek teeth of *Georychus capensis* (MNHN-ZM-MO 1988-107).

Habitat & Biogeography

The genus has a very limited distribution, restricted to a narrow belt from the south-western region of South Africa eastwards to Port Elisabeth, with isolated pockets in KwaZulu-Natal and Mpumalanga provinces. It is associated with fynbos, forest and grassland habitats (De Graaff, 1981).

Fossil Record

Fossils of this genus are known from the Plio-Pleistocene to the Late Pleistocene from Southern Africa (Winkler et al., 2010).

Biology & Ecology

The ecology of this genus is not well known and has recently been reviewed in Happold (2013). These mole-rats are solitary and territorial, digging their burrows with their large incisors.

Systematic Notes

This taxon may comprise a species complex with populations in KwaZulu-Natal and Mpumalanga differing considerably on molecular grounds from those in the Western Cape (Honeycutt et al., 1987).

Species List

Currently a single species is recognised:
Georychus capensis (Pallas, 1778)

Species Accounts

Georychus capensis (Pallas, 1778)

Diagnosis
See genus account.

Taxonomic Issues
Chromosome number 2n = 54, FN = 100 (Nevo et al., 1986; Deuve et al., 2008).

Distribution
This South African endemic has a narrow distribution in the Western Cape Province where it occurs along the coast and inland; isolated populations also occur in KwaZulu-Natal and Mpumalanga provinces where the species inhabits high elevation grasslands (De Graaff, 1981).

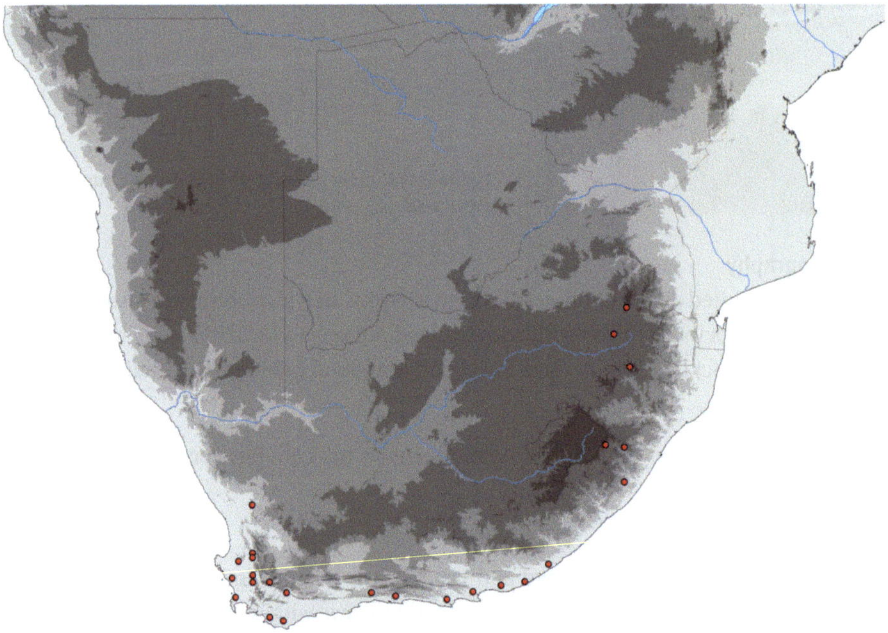

Figure 2.149 Distribution map

Measurements

External and cranial measurements (mm) and mass (g) for *Georychus capensis*, sexes combined

	Mean	Min	Max	N
Mass*	247.6	124	360	7
HB*	179.1	143	204	36
Tail*	27.4	20	40	28
HF*	30.0	25	35	34
Ear*	-	-	-	-
GLS**	46.0	41.0	53.3	41

*De Graaff (1981); **Happold (2013)

Genus Account

Heliophobius Peters, 1846 (Silvery Mole-rats)

Diagnosis

This genus is widely distributed in East Africa, extending into South-Central Africa. They are small to medium-sized rodents (Mass ca. 150 g), with a characteristically stout body, very short limbs and a short tail typical of this fossorial family. The pelage varies between species, but is typically some shade of silvery-grey dorsally and paler ventrally. This genus is distinguished by its long silky pelage (with hairs > 20 mm in length), palate not extending beyond toothrow and the presence of up to six cheekteeth (but typically only four or five are present at a time). *Heliophobius* is morphologically similar to *Fukomys* and *Cryptomys*, but is distinct on chromosomal and molecular grounds (Faulkes et al., 2004; Igram et al., 2004), and typically occurs east of the African Rift Valley (but overlapping in Zambia, Malawi, DRC and very locally in Tanzania. Three pairs of nipples (2 + 1 = 6) are present.

(a) (b) (c)

Figure 2.150 a-c Photographs of *Heliophobius* sp., from top left to bottom left (Radim Sumbera, Jan Skliba and Bill Stanley).

Cranial Characters

The skull is robust, small to medium in size (GLS ca. 40 mm), flattened with a narrow rostrum. The zygomatic arches are sturdy and wide. The interorbital constriction is wide. The sagittal crest is well-developed and the infraorbital foramen is small. The anterior palatal foramina are very short and end well before the M^1. The palate is narrow and does not extend posteriorly beyond the molar toothrow. The bullae are slightly enlarged, but not markedly so. The angular process of the lower jaw does not extend posteriorly (as in *Bathyergus*, and to a lesser extend *Fukomys*/*Cryptomys*).

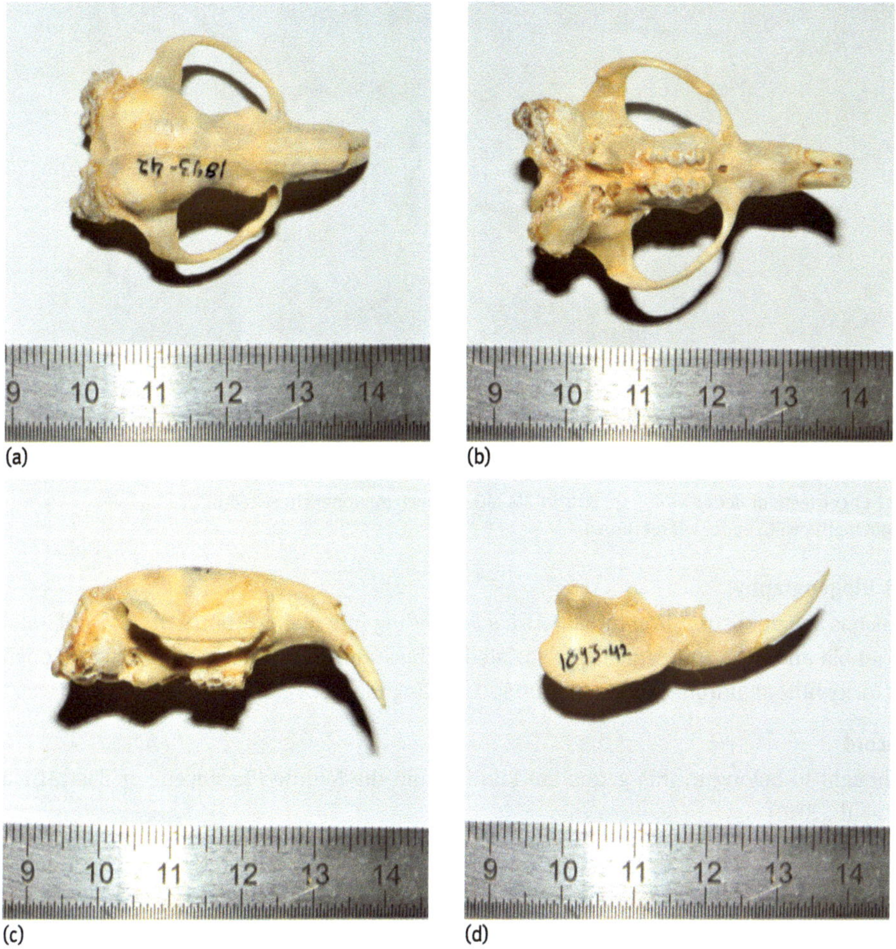

Figure 2.151 a-d Cranium of *Heliophobius* sp. (MNHN-ZM-MO 1893-42) (from top left to bottom right: dorsal view, ventral view, lateral view and lower jaw).

Dentition

Heliophobius has robust, ungrooved and pro-odont upper incisors. There are a maximum of six cheek teeth in the upper toothrow (but the premolars are generally shed before the last molars erupt and hence there are usually only 4-5 cheek teeth present at a time). The molars are all roughly the same size, simple and uncusped (or with folds in juveniles).

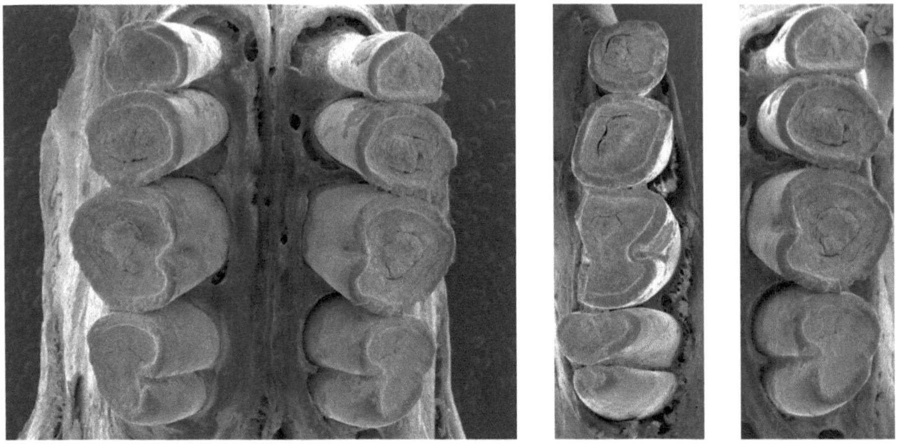

(a) (b) (c)

Figure 2.152 Cheek teeth of *Heliophobius* sp. (MNHN-ZM-MO 1893-41) (A: upper cheek teeth; B: lower right toothrow; C: upper left toothrow).

Habitat & Biogeography

The genus has a wide distribution in East Africa, extending into Malawi, northern Mozambique, eastern Zambia and DRC. It is typically associated with sandy soils in Miombo woodland, open plains or rocky hills at altitudes between 750-1500 m (Kingdon, 1974).

Fossil Record

Fossils thought to belong to this genus are known from the Middle Pleistocene of East Africa (Winkler et al., 2010).

Biology & Ecology

The ecology of this genus is not well known although Sumbera et al. (2003a,b) describe its reproductive biology and burrow formation. These mole-rats are solitary and territorial, digging their burrows with their large incisors.

Systematic Notes

The genus was previously considered to be monospecific (Woods & Kilpatrick, 2005), but with a large number of described races. For example, nine taxa were listed by Ellerman (1940), who also recognised two species: *H. argenteocinereus* and *H. spalax*. Honeycutt et al. (1991) considered the latter species to be no more than age variation and therefore recognised a single species *H. argenteocinereus*. Recent molecular work, based on extensive sampling of populations and races, confirms the existence of multiple distinct species as part of an adaptive radiation that can be linked to tectonic activity that created the East African Rift Valley (Faulkes et al., 2011). These species have only been diagnosed on molecular grounds (Faulkes et al., 2011), and therefore they cannot currently be distinguished on traditional morphological grounds. However, all six species appear to be allopatric suggesting that they could be differentiated based on geographical location. Chromosome number for the genus is 2n = 60 or 62 (see species accounts).

Species List

Six species are currently recognised:
Heliophobius argenteocinereus Peters, 1852
Heliophobius emini Noack, 1894
Heliophobius kapiti (Heller, 1909)
Heliophobius marungensis Noack, 1887
Heliophobius mottuolei (Schouteden, 1913)
Heliophobius spalax (Thomas, 1910)

Key to Species

The members of this genus have recently been elevated to full species status purely on molecular grounds (see Systematic Notes), and therefore a key cannot currently be presented. However, most of these species appear to have allopatric distributions which may assist in their identification (see species accounts for further details).

Species Accounts

Heliophobius argenteocinereus Peters, 1852

Diagnosis

Currently, this species can only be diagnosed on molecular grounds, but it appears to be the only *Heliophobius* present within its range.

Taxonomic Issues

In the designation of this species, Faulkes et al. (2011) did not have specimens at their disposal from the type locality (Tete, north-central Mozambique), and therefore some doubt remains as to their taxonomic conclusions. However, the specimen from Mt Namuli (northern Mozambique) clearly grouped with those from Zambia and Malawi, suggesting that populations from Tete (which lies between Mt Namuli and Zambia/Malawi) are conspecific with them. Chromosome number 2n = 62 (Scharff et al., 2001).

Distribution

This species appears to be restricted to Malawi, south-western Zambia and northern Mozambique (Faulkes et al., 2011).

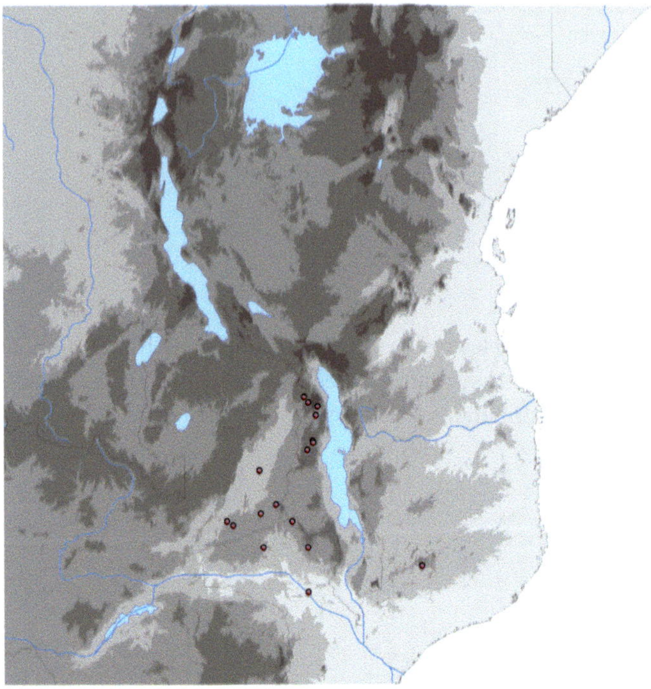

Figure 2.153 Distribution map

Measurements

External and cranial measurements (mm) and mass (g) for *Heliophobius argenteocinereus* sensu lato sexes combined

	Mean	Min	Max	N
Mass*	175.7	51	331	144
HB*	152.1	106	193	126
Tail**	12.6	6	17	126
HF*	29.3	23	36	126
Ear**	0	-	-	-
GLS**	39.5	32.9	45.0	192

*Sumbera (2003a) from Malawi (probably referable to *H. argenteocinereus*);
**Happold (2013) from Tanzania (probably referable to *H. spalax*).

Heliophobius emini Noack, 1894

Diagnosis

Currently, this species can only be diagnosed on molecular grounds, but it appears to be the only *Heliophobius* present within its range. Its distribution appears to overlap that of *H. spalax*, but the two species have not yet been collected at the same locality.

Taxonomic Issues

Chromosome number 2n = 62 (Deuve et al., 2008).

Distribution

This species occurs in southern and central Tanzania (Faulkes et al., 2011).

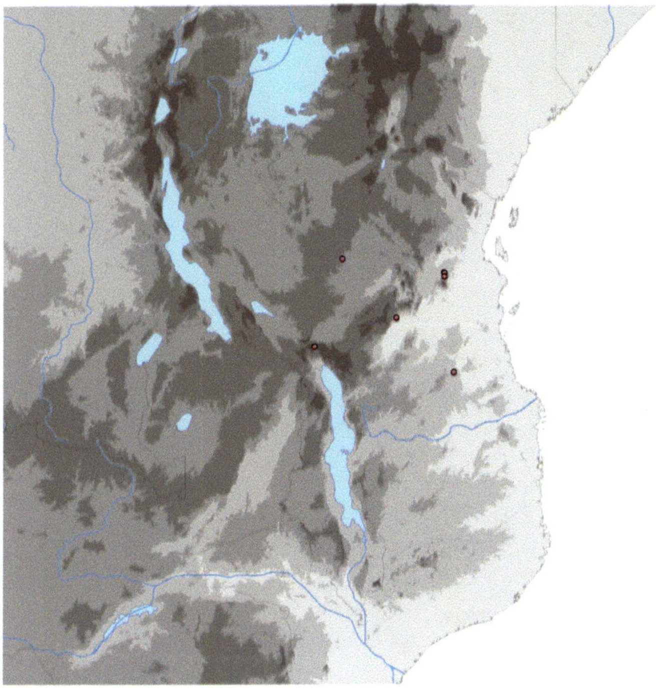

Figure 2.154 Distribution map

Measurements

See measurements for *Heliophobius argenteocinereus*.

Heliophobius kapiti (Heller, 1909)

Diagnosis

Currently, this species can only be diagnosed on molecular grounds, but it appears to be the only *Heliophobius* present within its range.

Taxonomic Issues

Chromosome number 2n = 60 (George, 1979).

Distribution

This species is only known from a small region in south-central Kenya (Faulkes et al., 2011).

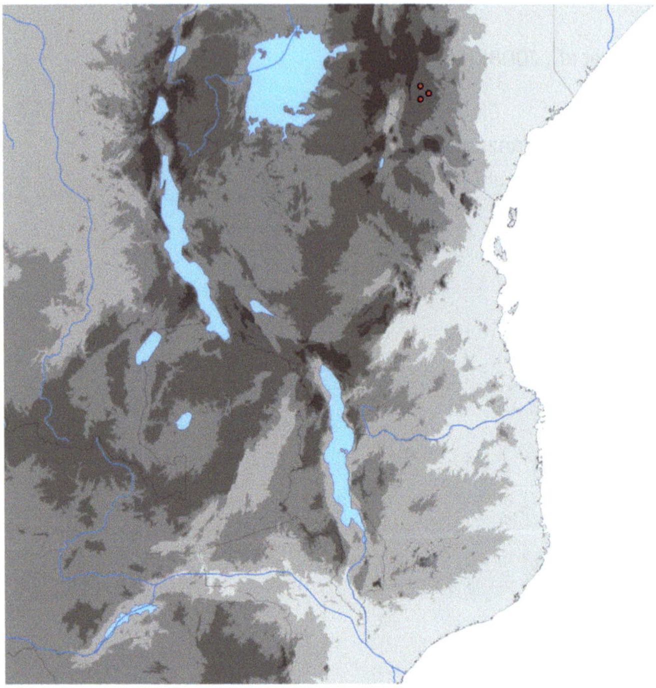

Figure 2.155 Distribution map

Measurements

See measurements for *Heliophobius argenteocinereus*.

Heliophobius marungensis Noack, 1887

Diagnosis

Currently, this species can only be diagnosed on molecular grounds, but it appears to be the only *Heliophobius* present within its range.

Taxonomic Issues

See genus account.

Distribution

This species is known from several widely scattered localities in eastern DRC and western Rwanda from the Albertine Rift in the north, southward along the western margins of Lake Tanganyika (Faulkes et al., 2011).

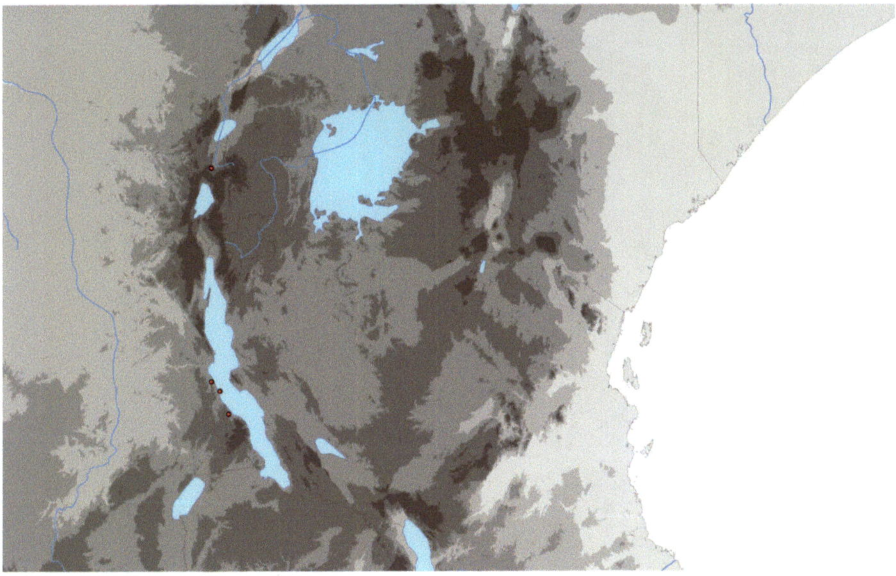

Figure 2.156 Distribution map

Measurements

See measurements for *Heliophobius argenteocinereus*.

Heliophobius mottuolei (Schouteden, 1913)

Diagnosis

Currently, this species can only be diagnosed on molecular grounds, but it appears to be the only *Heliophobius* present within its range.

Taxonomic Issues

See genus account.

Distribution

This species is only known from scattered localities in the southern DRC (Faulkes et al., 2011).

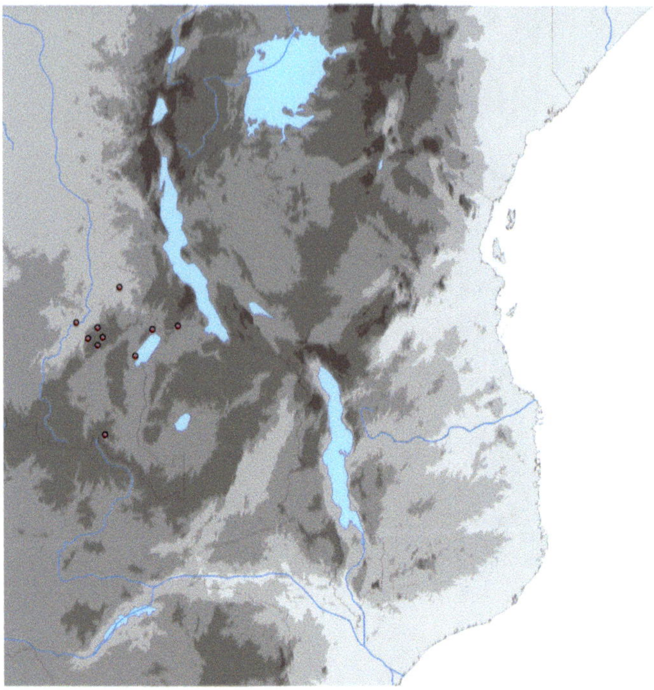

Figure 2.157 Distribution map

Measurements

See measurements for *Heliophobius argenteocinereus*.

Heliophobius spalax (Thomas, 1910)

Diagnosis

Currently, this species can only be diagnosed on molecular grounds, but it appears to be the only *Heliophobius* present within its range. Its distribution appears to overlap that of *H. emini*, but the two species have not yet been collected at the same locality.

Taxonomic Issues

See genus account.

Distribution

This species occurs in the Eastern Arc Mts (Usambara and Uluguru, Ukaguru Mts) and Mbizi Mts in Tanzania (Faulkes et al., 2011).

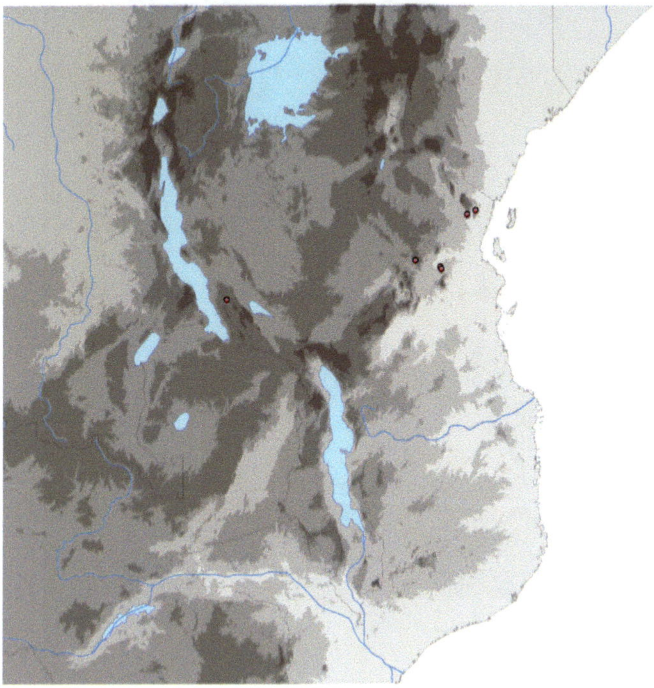

Figure 2.158 Distribution map

Measurements

See measurements for *Heliophobius argenteocinereus*.

Genus Account

Heterocephalus Ruppell, 1842 (Naked Mole-rats)

Diagnosis

This genus is endemic to the Horn of Africa. They are small rodents (Mass ca. 30-80 g), with a characteristically stout body, very short limbs and a short tail typical of this fossorial family. This genus is unique amongst rodents in completely lacking a pelage, with just a few isolated tactile hairs on the face and body. The lack of pelage readily distinguishes this genus from other mole-rat genera; the number of upper cheekteeth is also reduced in *Heterocephalus* (three pairs, compared with four or more in other Bathyergidae). The number of nipples varies considerably (range: 9-15), and nipples are situated in two parallel rows (Sherman et al., 1999).

Figure 2.159 Photograph of *Heterocephalus glaber* (John Visser).

Cranial Characters

The skull is robust, small (GLS ca. 20 mm), flattened with a narrow rostrum. The zygomatic arches are sturdy and wide. The interorbital constriction is wide. The sagittal crest is not well-developed. The infraorbital foramen is small. The anterior palatal foramina are very short and end well before the M^1. The bullae are slightly enlarged, but not markedly so.

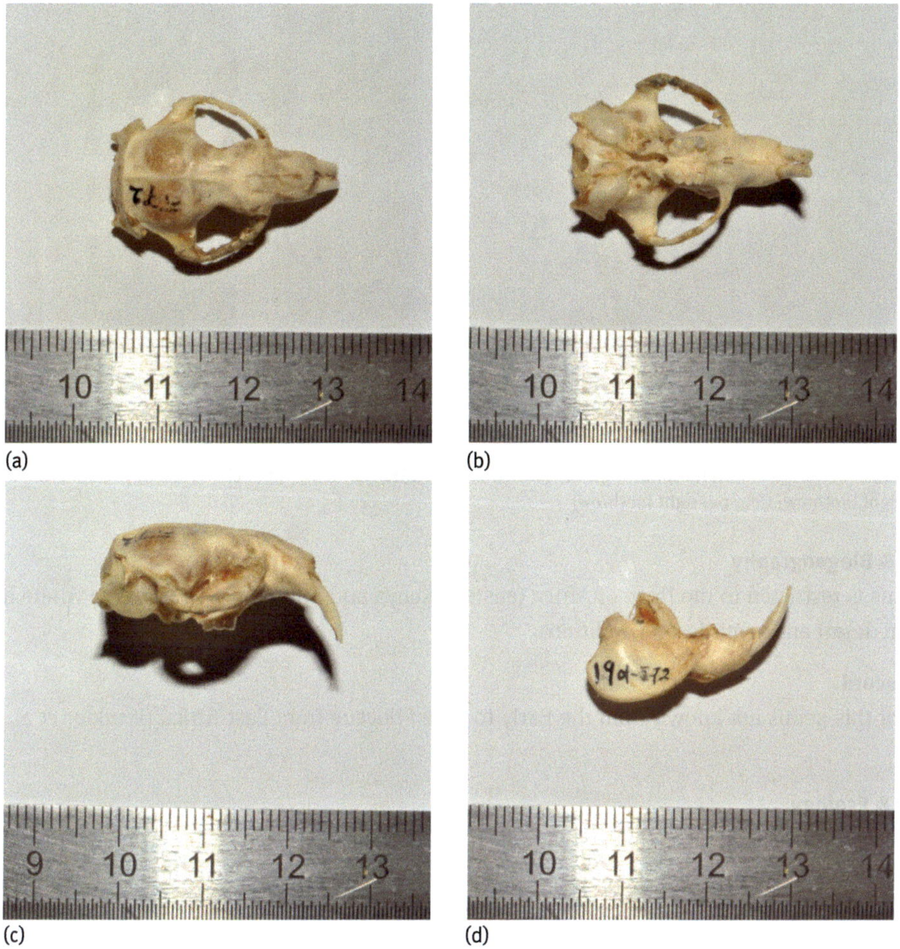

Figure 2.160 a-d Cranium of *Heterocephalus glaber* (MNHN-ZM-MO 1901-572) (from top left to bottom right: dorsal view, ventral view, lateral view and lower jaw).

Dentition

Heterocephalus has robust, ungrooved and pro-odont upper incisors. The dental formula is: 1003/1003. The molars are all roughly the same size, simple and uncusped (or with folds in juveniles).

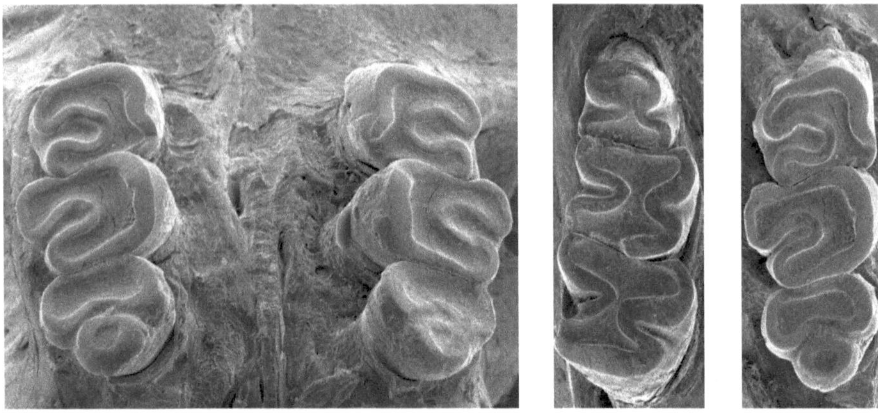

(a) (b) (c)

Figure 2.161 Cheek teeth of *Heterocephalus glaber* (MNHN-ZM-MO 1884-1578) (A: upper cheek teeth; B: lower right toothrow; C: upper right toothrow).

Habitat & Biogeography

The genus is restricted to the Horn of Africa (eastern Kenya and Ethiopia, and Somalia) where it occurs in desert and semi-desert conditions.

Fossil Record

Fossils of this genus are known from the Early to Late Pliocene from East Africa (Winkler et al., 2010).

Biology & Ecology

The ecology and behaviour of this genus is well known and has been reviewed in Bennett & Faulkes (2000) and Jarvis & Sherman (2002). This is the only known eusocial mammal where there is reproductive division of labour, overlap of generations and cooperative care of young (Jarvis, 1981).

Systematic Notes

This genus is currently considered to be monotypic (Musser & Carleton, 2005; Happold, 2013). There are deep genetic divisions between certain populations (Van Daele et al., 1997), however, it has been suggested that this merely represents intraspecific variation (Faulkes et al., 2004).

Species List

One species is recognised:
Heterocephalus glaber Ruppell, 1842

Species Accounts

Heterocephalus glaber Ruppell, 1842

Diagnosis

See genus account.

Taxonomic Issues

Chromosome number 2n = 60 (George, 1979).

Distribution

This species occurs widely in the Horn of Africa occurring from south-eastern Kenya, north to eastern Ethiopia and Somalia (Yalden et al., 1976; Jarvis & Sherman, 2002).

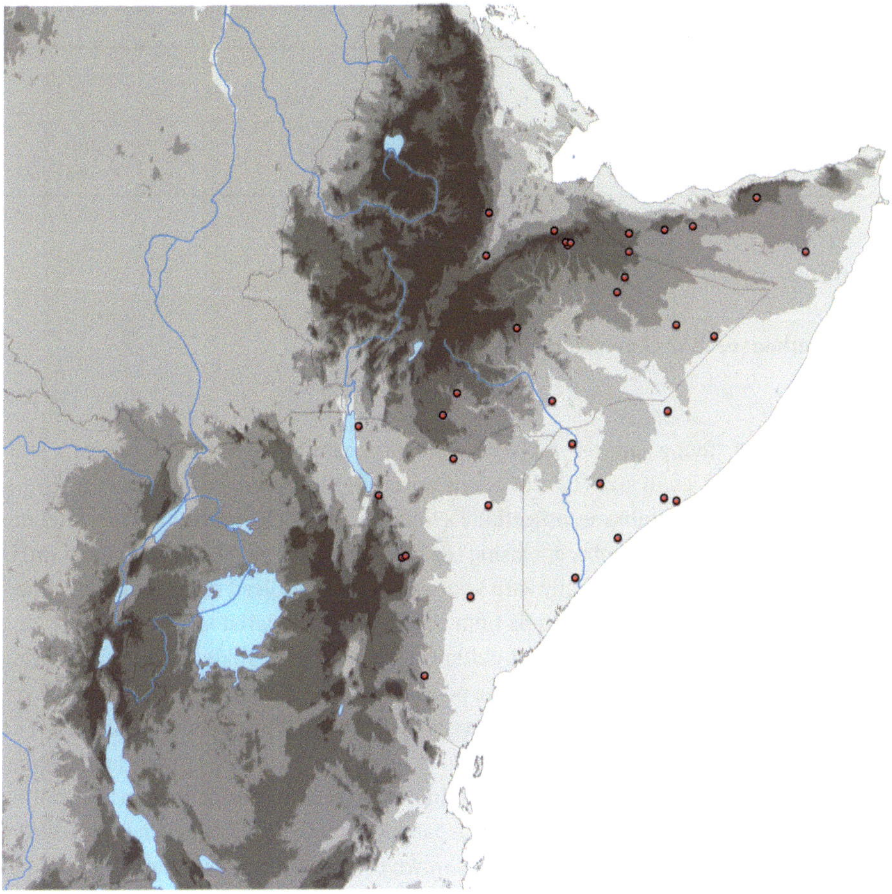

Figure 2.162 Distribution map

Measurements

External and cranial measurements (mm) and mass (g) for *Heterocephalus glaber*, sexes combined

	Mean	Min	Max	N
Mass*	33.9	25	80	715
HB*	83.4	66	110	131
Tail*	42.1	34	50	123
HF*	20.1	18	22	127
Ear	-	-	-	-
GLS*	21.4	16.5	24.4	119

*Happold (2013)

Family: Anomaluridae

This family is represented by seven species all endemic to Sub-Saharan Africa.

Key to Genera

1	Flight membrane (patagium) absent	*Zenkerella*
	Flight membrane (patagium) present	2
2	Size smaller (GLS < 30 mm, HF < 25 mm)	*Idiurus*
	Size larger (GLS > 40 mm, HF > 40 mm)	*Anomarulus*

Genus Account

Anomalurus Waterhouse, 1843 (Anomalures)

Diagnosis

This distinctive genus of "flying squirrel" includes animals of medium to large size (Mass ca. 500-1400 g, three species), and small size (150-200g, one species). *Anomalurus* can be readily distinguished by its size and flight membrane, essentially a fold of skin which is stretched from forelimb to hindlimb and between the hindlimbs, enclosing the tail. This flight membrane (or patagium) is unique among African rodents, shared only with *Idiurus*, but it has evolved independently numerous times in mammals. The patagium extends from the thorax to the wrist and posteriorly along the abdomen to the ankles, and between the hindlimbs, from ankle to ankle, effectively creating a flattening surface which is used for gliding; true powered flight is not possible. A long cartilaginous projection from the elbow (the calcaneum) assists in supporting the patagium anteriorly. The patagium only becomes visible when the limbs are stretched out (i.e. in flight), but is folded along the body when the animal is walking or climbing. The fur is very soft and long, densely coating the body. The dorsal pelage, which extends onto the patagium, varies greatly between and within species, but is usually a shade of grey-brown, olive-brown or slate-brown, and may be grizzled or frosted with white. The ventral pelage is paler or white, but suffused with a rich orange-brown in one species (*A. beecrofti*). The ears are large, mostly naked and rise well above the crown. There are five digits on the hindfoot and four digits on the forefoot, all with sharp, powerful and curved claws. The tail is shorter than body length (between 60-95% of HB) typically narrow at its base, broadening out into a bush towards the tip. The colour varies as in the dorsal and ventral pelage. Characteristic of the family Anomaluridae, two rows of posteriorly-pointed scales are present on the ventral surface of the tail, near its base, which anchor the animal when pressed against the trunk of a tree. There are two pairs of nipples (2 + 0 = 4) usually placed along the thorax (Rosevear, 1969).

Figure 2.163 a and b Photographs of *Anomalurus pusillus* (Ara Monadjem) (top) and *A. beecrofti* museum specimen (MNHN-ZM-MO 1955-659) (Ara Monadjem) (bottom)

Cranial Characters

Depending on species, the skull is medium to large in size (GSL ca. 45 mm in *A. pusillus*, and 50-70 mm in the other species). The rostrum and roof of skull are relatively flat, curving down posteriorly towards the occipitals. The rostrum is relatively short, giving the animal a blunt face. A weak supra-occipital crest may be present. Medially, the palate ends well before the posterior margin of the third molars. The zygomatic arch is long and curved outwards, giving the animal a broad head. There is a large, elliptical infraorbital foramen present. The post-orbital process is absent, or if present, greatly reduced to a small vestige. The anterior palatal foramina are short and narrow. The bullae are relatively large.

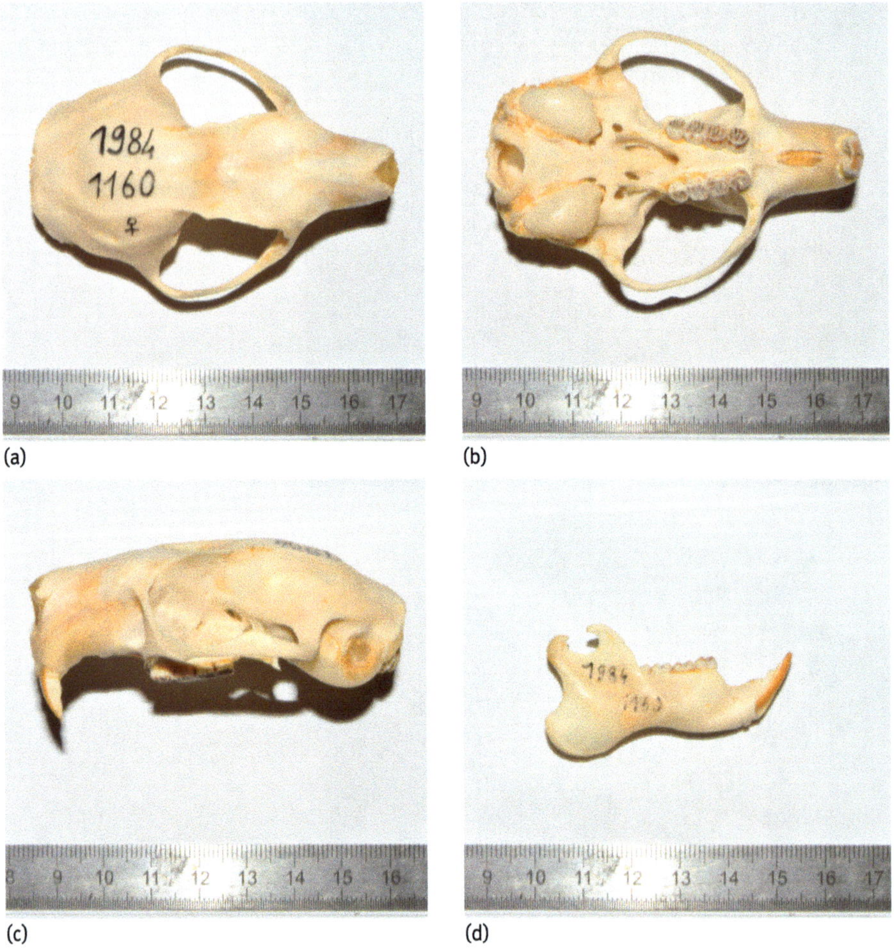

Figure 2.164 a-d Cranium of *Anomalurus peli* (MNHN-ZM-MO 1984-1160) (from top left to bottom right: dorsal view, ventral view, lateral view and lower jaw).

Dentition

Anomalurus has ungrooved, orthodont, orange incisors. The dental formula is: 1013/1013 (Rosevear, 1969). The premolar and three molars are similar in size and flat-crowned, with well-developed transverse ridges.

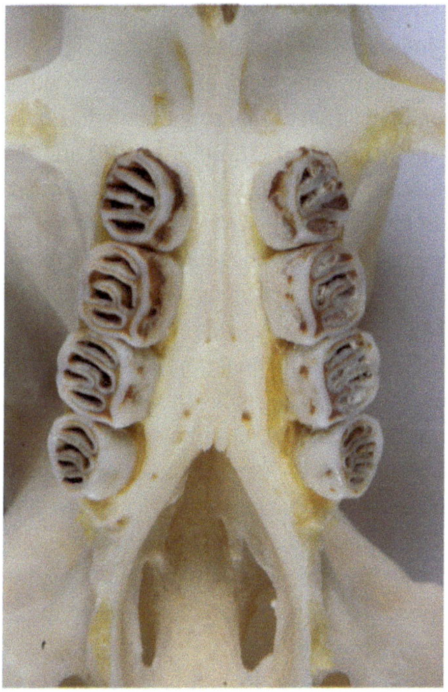

Figure 2.165 Upper cheek teeth of *Anomalurus peli* (MNHN-ZM-MO 1984-1160).

Habitat & Biogeography

Anomalurus is widely distributed across tropical Africa, from southern Senegal, south to Guinea and Sierra Leone, and then east through to Nigeria, southern Cameroon to western Kenya. It occurs throughout the Congo basin, south to central Angola, Zambia, Malawi and Mozambique (north of the Zambezi River), and south-eastern Tanzania. Members of this genus are closely tied to forested habitats, primarily lowland rainforest, but one species (*A. derbianus*) extends into moist woodland, and montane forest, up to 2400 m above sea level (Kingdon, 1974). *Anomalurus* is strictly arboreal and mostly nocturnal, tending to remain in the higher strata of the forest, gliding between trees (rather than descending to the ground). Glide distance has been accurately recorded for two individuals as 4.5m and 40m, with a glide angle of 24° (Corbin & Cordeiro, 2006). It rests up in hollows of trees during the day, clinging to the inside of the trunk (Rosevear, 1969, Kingdon, 1974).

Fossil Record

The oldest, and only known, *Anomalurus* fossil is from 13-14 MYA, Kenya (Winkler, 2002).

Biology & Ecology

The ecology and reproductive biology of *Anomalurus* species has been poorly studied. In fact, the skilful reviews of Rosevear (1969) and Kingdon (1974) are still relevant today, with just a single detailed ecological study since then (Julliot et al., 1998). *Anomalurus* species are vegetarian, eating fruits, leaves, inner bark and sap. Home range of a single female tracked in Gabon was 3.7 ha (Julliot et al., 1998). Litter size is 1-3 (*A. derbianus*), and practically unknown for the other species.

Systematic Notes

The closest relative of the family Anomaluridae is Pedetidae (the Springhares), although this relationship is not particularly close (the two groups having diverged deep in the past); together they form the superfamily Anomaluroidea. The taxonomic position of this superfamily has been much disputed and they have been placed with both the Sciurognathi (Rosevear, 1969) and the Hystricognathi (Horner et al., 2007). However, a recent molecular phylogeny for the order Rodentia shows strong support for the Anomaluroidea within the "mouse group" (Blanga-Kanfi et al., 2009). A molecular phylogeny is not available for the Anomaluridae, so the relationships between *Anomalurus, Idiurus* and *Zenkerella* are only known by reference to morphology, with *Anomalurus* being placed in a separate subfamily (Anomalurinae) from the other two genera (Rosevear, 1969). The species *A. beecrofti* is often placed in its own monotypic genus *Anomalurops* based on slight differences in dental morphology and pelage (e.g. Rosevear, 1969), however, we retain this species within *Anomalurus* until a molecular phylogeny has been developed. Morphological variation (particularly of pelage and skull) of the four species of *Anomalurus* has recently been investigated by Schunke & Hutterer (2004, 2005), who showed considerable variation in these characters for three species (all except *A. pusillus*). For two species (*A. pelii* and *A. derbianus*) variation was geographically based and, for *A. pelii*, discontinuities were strongly associated with environmental barriers (rivers, in this case), rather than being clinal. For this species, which is restricted to the Upper Guinea rainforest, the Bandama and Sassandra rivers coincide with the ranges of the three recognised subspecies. From west to east, these are as follows. *Anomalurus peli auzembergeri* (west of Sassandra), *Anomalurus p. peralbus* (between Sassandra and Bandama), *A p. pelii* (east of Bandama) (Schunke & Hutterer, 2005). In *A. beecrofti*, different populations show the same range of variation, but in different proportions, typical of a polymorphic species. The situation is more complicated for *A. derbianus* and it is likely that at least some populations are specifically distinct. Whether these various populations refer to distinct species or to subspecific variation will require the integration of molecular and morphological analysis, long overdue for this group. Lastly, the species *A. pusillus* is poorly known having been rarely collected and, in West Africa, is known from only two localities (both in Liberia). This West African population is separated by its closes neighbour in southern Cameroon by at least 2000 km, strongly suggesting a distinct species. The diploid number of chromosomes is not known for any *Anomalurus* species.

Species List

Four species are described within the genus *Anomalurus*, all restricted to tropical Africa.

Anomalurus beecrofti Fraser, 1853
Anomalurus derbianus (Gray, 1842)
Anomalurus pelii (Schlegel & Muller, 1845)
Anomalurus pusillus Thomas, 1887

Key to Species

1	Size very larger (GLS > 65 mm, HB > 400 mm), dorsal pelage uniform black	*A. pelii*
	Size smaller (GLS < 65 mm, HB < 400 mm), dorsal pelage not black	2
2	Size smaller (GLS < 50 mm, HB < 250 mm)	*A. pusillus*
	Size larger (GLS > 50 mm, HB > 250 mm)	3
3	Tail with bushy black terminal brush, ventral pelage white, yellow or grey	*A. derbianus*
	Tail not bushy and mostly grey, ventral pelage usually with some orange	*A. beecrofti*

Species Accounts

Anomalurus beecrofti Fraser, 1853

Diagnosis

This is a large anomalure with great variation in the colour and pattern of the fur (Rosevear, 1969; Schunke & Hutterer, 2004). It is larger in absolute size than *Anomalurus pusillus* (GLS < 50 mm, HF < 50 mm). It overlaps in size with *A. derbianus*, from which it can be distinguished by a lack of a bushy, black tail and the presence of bright orange fur on the ventral pelage (which may occasionally be reduced or altogether absent). *Anomalurus pelii* is larger with a distinctive black (dorsal) and white or grey (ventral) pelage.

Taxonomic Issues

Although the pelage exhibits great variability, it is not apparently geographically based (Schunke & Hutterer, 2004), and therefore unlikely to represent different cryptic species. However a molecular analysis of this species has yet to be conducted.

Distribution

This species occurs widely in the forest zone of tropical Africa.

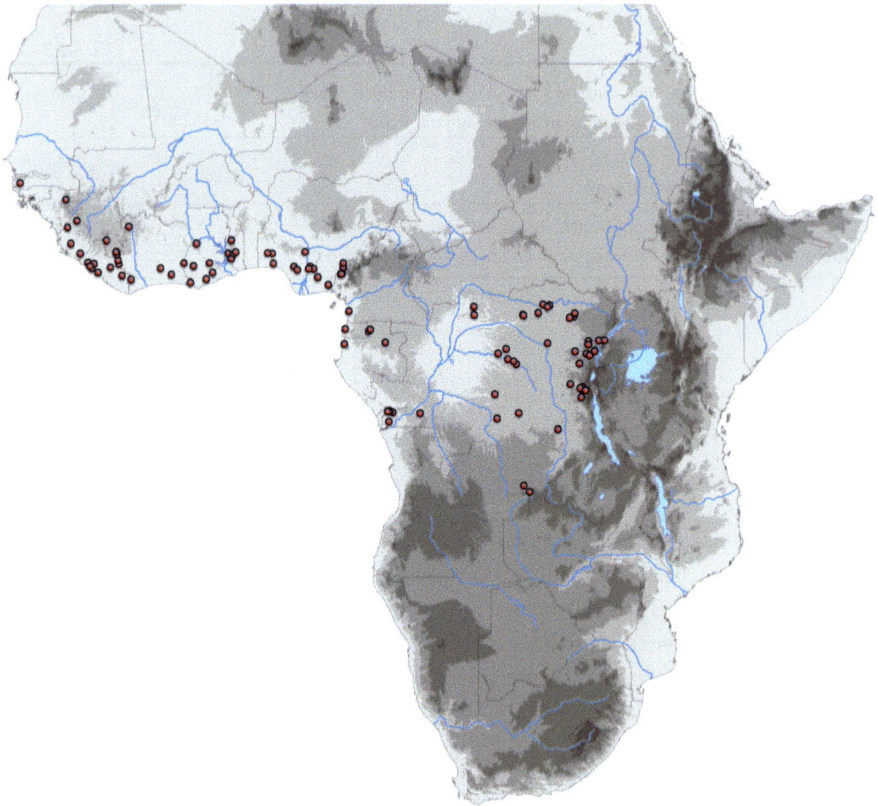

Figure 2.166 Distribution map

Measurements

External and cranial measurements (mm) and mass (g) for *Anomalurus beecrofti*, sexes combined

	Mean	Min	Max	N
Mass	-	-	-	-
HB	301.7	260	340	3
Tail	179.0	161	195	3
HF	57.0	55	60	3
Ear	27.3	26	29	3
GLS	55.8	51.7	60.8	6

Anomalurus derbianus (Gray, 1842)

Diagnosis

This is a large anomalure with great variation in the colour and pattern of the fur (Rosevear, 1969; Schunke & Hutterer, 2004). It is larger in absolute size than *Anomalurus pusillus* (GLS < 50 mm, HF < 50 mm). It overlaps in size with *A. beecrofti*, from which it can be distinguished by the presence of a bushy, black tail and the absence of bright orange fur on the ventral pelage (which is usually present in *A. beecrofti*). *Anomalurus pelii* is larger with a distinctive black (dorsal) and white or grey (ventral) pelage.

Taxonomic Issues

The pelage exhibits great variability, at least some of which it is geographically based (Schunke & Hutterer, 2004), suggesting the presence of cryptic species. However a molecular analysis of this species has yet to be conducted.

Distribution

This species occurs widely in tropical Africa, inhabiting rainforest, gallery forest and woodland.

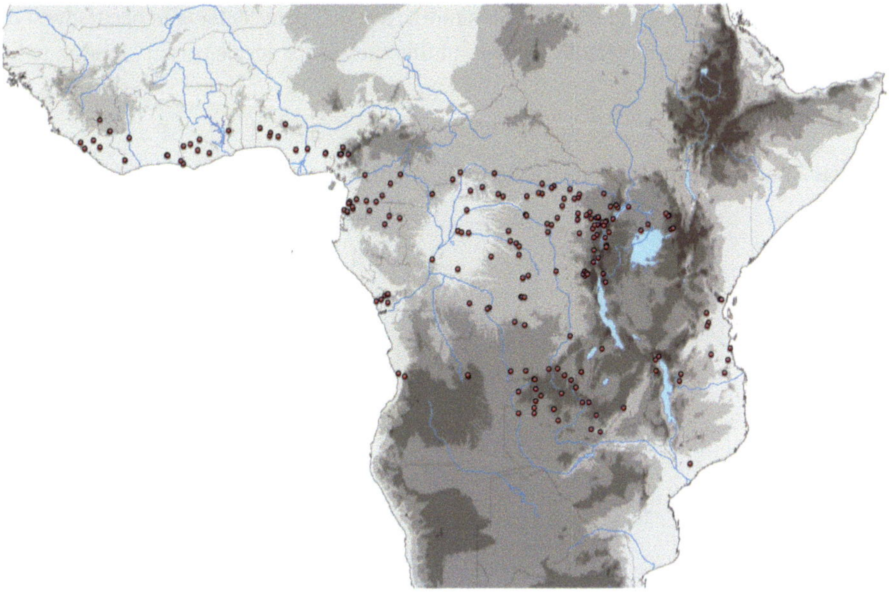

Figure 2.167 Distribution map

Measurements

External and cranial measurements (mm) and mass (g) for *Anomalurus derbianus*, sexes combined

	Mean	Min	Max	N
Mass	760	-	-	1
HB	330.8	275	388	8
Tail	297	255	360	8
HF	61.1	50	70	8
Ear	38.2	36	42	5
GLS	56.0	50.5	60.3	15

Anomalurus pelii (Schlegel & Muller, 1845)

Diagnosis

This is a very large and distinctive anomalure. It is larger in absolute size than all the other species of *Anomalurus*. Furthermore, it can be distinguished by its pitch black dorsal pelage and white (or off-white or grey) ventral pelage.

Taxonomic Issues

The pelage varies geographically, which has been the basis for the recognition of three subspecies (Schunke & Hutterer, 2004); each subspecies separated by a large river from neighbouring subspecies (Schunke & Hutterer, 2005). *A. p. pelii* occurs east of the Bandama River (Ivory Coast); *A. p. peralbus* occurs between the Bandama and Sassandra rivers; *A. p. auzembergeri* occurs west of the Sassandra River. However a molecular analysis of this species (or species group) has yet to be conducted.

Distribution

This species has a narrow distribution restricted to West Africa.

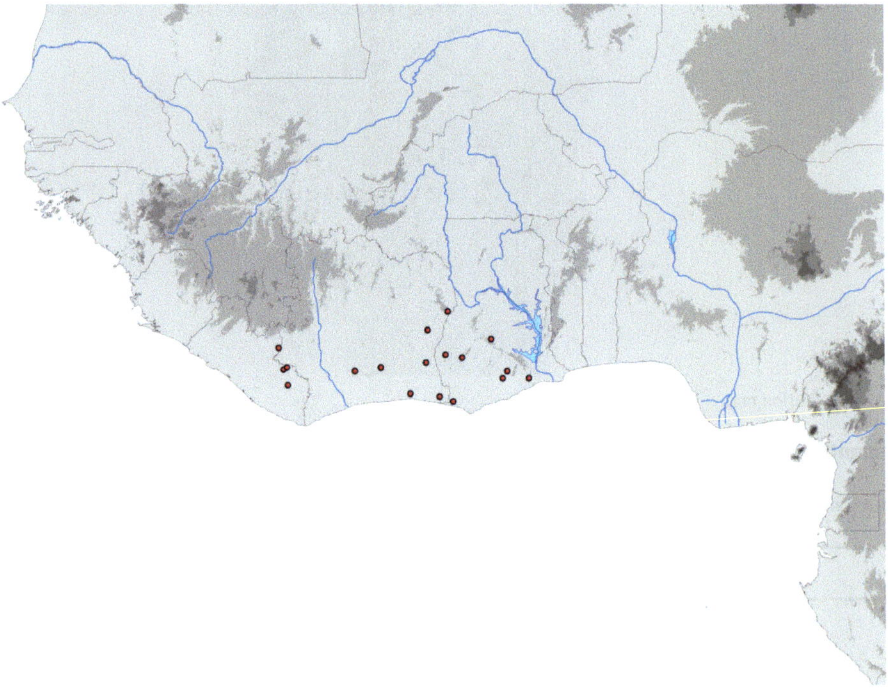

Figure 2.168 Distribution map

Measurements

External and cranial measurements (mm) and mass (g) for *Anomalurus pelii*, sexes combined

	Mean	Min	Max	N
Mass*	1400	-	-	1
HB	430	-	-	1
Tail	520	-	-	1
HF	90	-	-	1
Ear	40	-	-	1
GLS	69.8	67.1	72.5	2

*Happold (2013)

Anomalurus pusillus Thomas, 1887

Diagnosis
This is a medium-sized anomalure, but the smallest in absolute size (GLS < 50 mm, HF < 50 mm) within the genus *Anomalurus*.

Taxonomic Issues
Although the pelage exhibits great variability, it is not apparently geographically based (Schunke & Hutterer, 2005), and therefore unlikely to represent different cryptic species. However a molecular analysis of this species has yet to be conducted.

Distribution
This species occurs primarily in the Congo basin on the right bank of the Congo River, with an isolated population in the Upper Guinea Forest of Liberia.

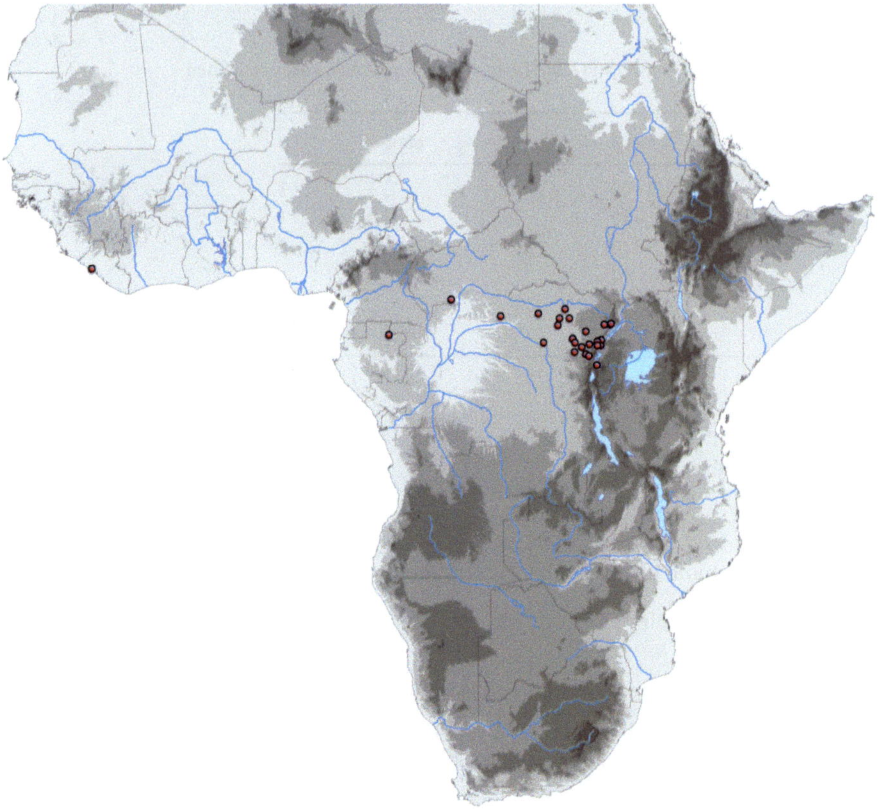

Figure 2.169 Distribution map

Measurements

External and cranial measurements (mm) and mass (g) for *Anomalurus pusillus*, sexes combined

	Mean	Min	Max	N
Mass*	170	-	-	1
HB	235	-	-	1
Tail	200	-	-	1
HF	44	-	-	1
Ear	29	-	-	1
GLS	46.7	-	-	1

*Julliot et al. (1998)

Genus Account

Idiurus Matschie, 1894 (Pygmy Anomalures)

Diagnosis

This distinctive genus of diminutive (Mass ca. 15-30g) "flying squirrel" can be readily distinguished by its small size, long feathery tail, and the presence of a flight membrane, essentially a fold of skin which is stretched from forelimb to hindlimb and between the hindlimbs, enclosing the tail. This flight membrane (or patagium) is unique among African rodents, shared only with *Anomalurus*, but it has evolved independently several times in other mammalian groups. The patagium extends from the thorax to the wrist and posteriorly along the abdomen to the ankles, and between the hindlimbs, from ankle to ankle, effectively creating a flattening surface which is used for gliding; true powered flight is not possible. A long cartilaginous projection from the elbow (the calcaneum) assists in supporting the patagium anteriorly. The patagium only becomes visible when the limbs are stretched out (i.e. in flight), but is folded along the body when the animal is walking or climbing. The fur is very soft and long, densely coating the body. The pelage, which extends onto the patagium, is usually a shade of grey-brown to medium-brown, and is not grizzled or frosted with white as in *Anomalurus*. The ventral pelage is usually the same colour as the dorsal, occasionally slightly paler. The ears are moderately large, mostly naked and rise above the crown. Individual hairs are dark based with light tips. There are five digits on the hindfoot and four digits on the forefoot, all with sharp, powerful and curved claws. The tail is significantly longer than body length (between 130-160% of HB) and distinctively feathery, with sparsely spaced, long, fine hairs distributed along its length. Characteristic of the family Anomaluridae, two rows of posteriorly-pointed scales are present on the ventral surface of the tail, near its base, which anchor the animal when pressed against the trunk of a tree. There are two pairs of nipples (2 + 0 = 4) placed along the thorax (Happold, 2013).

Figure 2.170 Photograph of *Idiurus macrotis* (Natalie Webber)

Cranial Characters

The skull is very small (mostly 21-26 mm). The rostrum and roof of skull are relatively flat, curving down posteriorly towards the occipitals. The rostrum is relatively short, giving the animal a blunt face. The palate ends well before the posterior margin of the third molars, and is characteristically V-shaped at its posterior edge. The zygomatic arch is long, thin and curved outwards, giving the animal a broad head. There is a large, but narrow infraorbital foramen present. The post-orbital process is absent. The anterior palatal foramina are very short and narrow. The bullae are large.

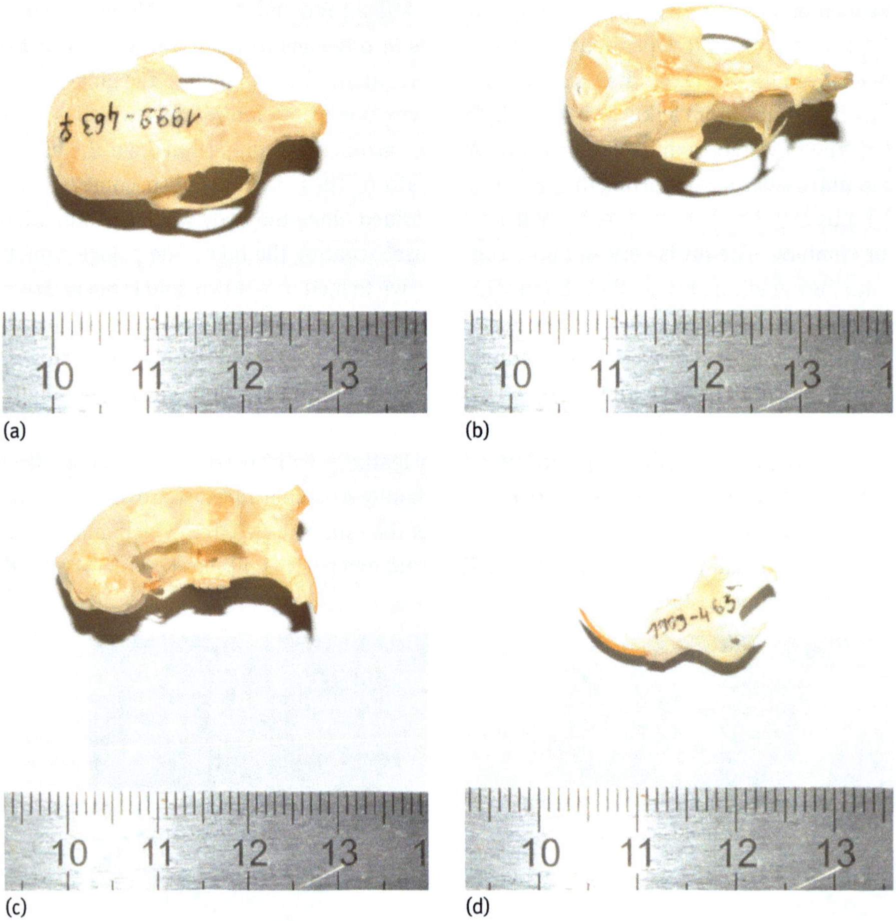

Figure 2.171 a-d Cranium of *Idiurus macrotis* (MNHN-ZM-MO 1999-463) (from top left to bottom right: dorsal view, ventral view, lateral view and lower jaw).

Dentition

Idiurus has ungrooved, orthodont, orange incisors, the posterior surface of which has a well-defined notch. The dental formula is: 1013/1013 (Rosevear, 1969). The premolar and first two molars are similar in size and flat-crowned, with well-developed transverse ridges; the last molar is the smallest of the cheekteeth.

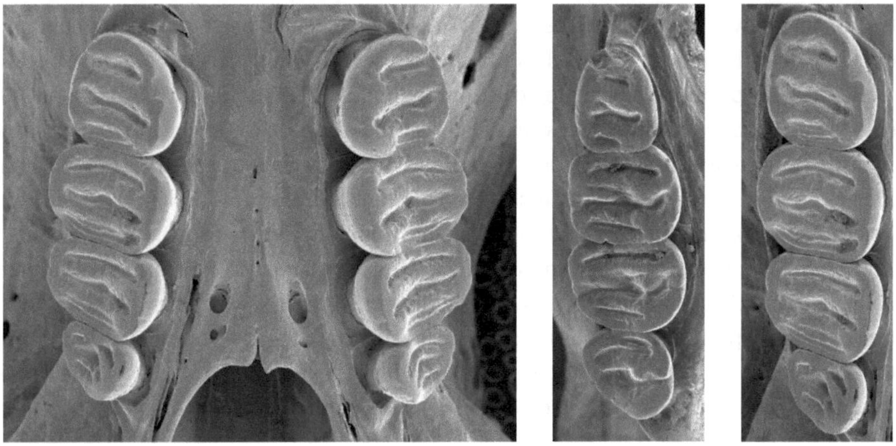

(a) (b) (c)

Figure 2.172 Cheek teeth of *Idiurus macrotis* (MNHN-ZM-MO 1985-220) (A: upper cheek teeth; B: lower right toothrow; C: upper right toothrow).

Habitat & Biogeography

Idiurus is sparsely distributed across the rainforest belt of tropical Africa in three apparently disjunct locations: Upper Guinea rainforest (Sierra Leone east to Ghana); Lower Guinea rainforest (Cameroon, Equatorial Guinea, Gabon and the extreme south-west of the Central African Republic); and eastern DRC. It has not been recorded south of the Congo River. The two species (*I. macrotis* and *I. zenkeri*) co-occur in two of these regions, but only *I. macrotis* occurs in the Upper Guinea rainforest. A further isolated record exists for *I. macrotis* on the north-western shores of Lake Victoria in Tanzania, based on a specimen collected in 1905 (Schunke & Hutterer, 2007). *Idiurus* is arboreal and mostly nocturnal. It rests up in small to large groups in hollows of trees during the day, which may be shared with *Anomalurus* species and bats (Sanderson, 1940, Rosevear, 1969, Jones, 1971, Kingdon, 1974).

Fossil Record

There are currently no known fossils of *Idiurus* (Winkler et al., 2010).

Biology & Ecology

The ecology and reproductive biology of *Idiurus* species are poorly known. They are vegetarian, eating fruits, bark and sap. Home range of a single male tracked in Gabon was 3 ha (Julliot et al., 1998). Litter size is 1 (*I. macrotis*) (Rahm, 1969).

Systematic Notes

The closest relative of the family Anomaluridae is Pedetidae (the Springhares), although this relationship is not particularly close (the two groups having diverged deep in the past); together they form the superfamily Anomaluroidea. The taxonomic position of this superfamily has been much disputed and they have been placed with both the Sciurognathi (Rosevear, 1969) and the Hystricognathi (Horner et al., 2007). However, a recent molecular phylogeny for the order Rodentia shows strong support for the Anomaluroidea within the "mouse group" (Blanga-Kanfi et al., 2009). A molecular phylogeny is not available for the Anomaluridae, so the relationships between the genera are not known; however, *Idiurus* and *Zenkerella* have been placed in the same subfamily Zenkerellinae based on morphological similarities (Rosevear, 1969). The two *Idiurus* species are clearly delineated species on morphological grounds (Schunke & Hutterer, 2007). Both species also show clear geographical differences in skull size and shape. Five such populations can be distinguished in *I. macrotis*: West Africa from Sierra Leone to Ghana; south-eastern Nigeria and north-western Cameroon; southern Cameroon, Equatorial Guinea and Gabon; eastern DRC; north-western Tanzania. *Idiurus zenkeri* showed similar differences between the three populations from which it is known (being absent from West Africa and Tanzania). Schunke & Hutterer (2007) concluded that these differences were not significant enough to warrant subspecies level recognition (let alone species recognition). However, considering the large gaps in distribution, a molecular examination of these populations could possibly reveal hidden diversity within these two species. The diploid number of chromosomes is not known for any *Idiurus* species.

Species List

Two species are described within the genus *Idiurus*, all restricted to tropical Africa.
Idiurus macrotis Miller, 1898
Idiurus zenkeri Matschie, 1894

Key to Species

1	Size larger (GLS > 24 mm, HB > 80 mm, HF > 18 mm)	*I. macrotis*
	Size smaller (GLS < 23 mm, HB < 80 mm, HF < 18 mm)	*I. zenkeri*

Species Accounts

Idiurus macrotis Miller, 1898

Diagnosis

This small anomalurid can only be confused with *I. zenkeri* from which it can be distinguished by its larger non-overlapping size (in *I. zenkeri* GLS < 23 mm, HF < 18 mm), and its relatively shorter ears (Schunke & Hutterer, 2007). *Zenkerella insignis* is larger (GLS > 40 mm) and has no patagium.

Taxonomic Issues

Despite geographically based variation in cranio-dental characters and pelage colouration, no subspecies were recognised by Schunke & Hutterer (2007). A molecular analysis of these populations, however, has not yet been conducted.

Distribution

This species occurs patchily but widely in the forest zone of tropical Africa.

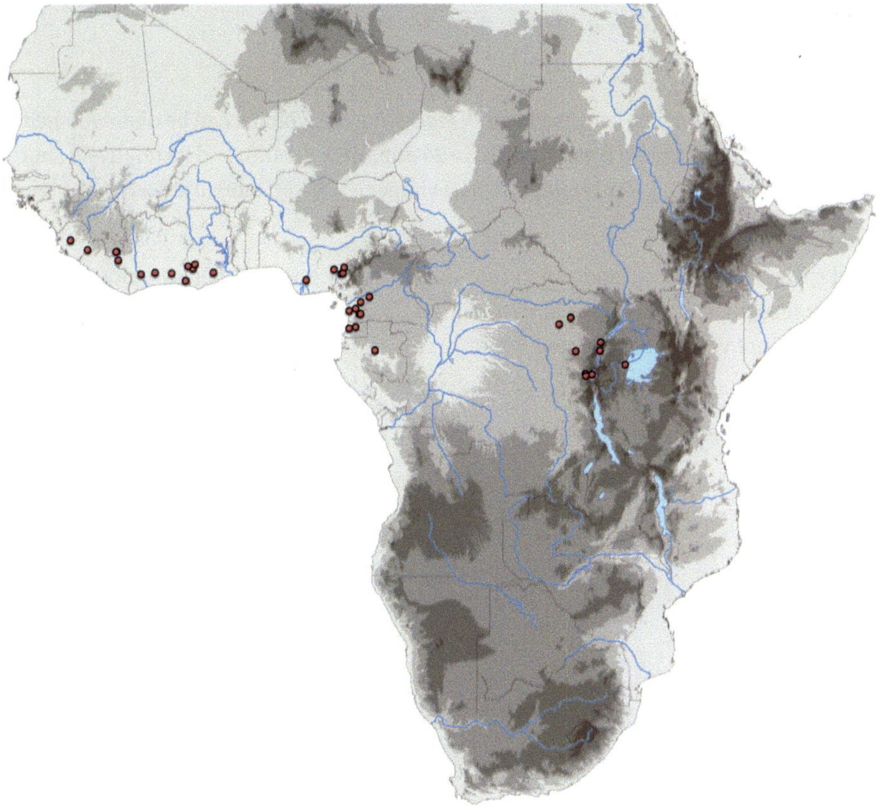

Figure 2.173 Distribution map

Measurements

External and cranial measurements (mm) and mass (g) for *Idiurus macrotis*, sexes combined

	Mean	Min	Max	N
Mass*	28.7	23	40	14
HB	86.6	83	90	5
Tail	127.0	121	132	5
HF	19.4	18	20	5
Ear	15.8	15	17	5
GLS	25.8	24.8	26.8	6

*Schunke & Hutterer (2007)

Idiurus zenkeri Matschie, 1894

Diagnosis

This small anomalurid can only be confused with *I. macrotis* from which it can be distinguished by its smaller non-overlapping size (in *I. macrotis* GLS > 24 mm, HF > 18 mm), and its relatively shorter ears (Schunke & Hutterer, 2007). *Zenkerella insignis* is larger (GLS > 40 mm) and has no patagium.

Taxonomic Issues

Despite geographically based variation in cranio-dental characters and pelage colouration, no subspecies were recognised by Schunke & Hutterer (2007). A molecular analysis of these populations, however, has not yet been conducted.

Distribution

This species is restricted to the forest zone of Central Africa.

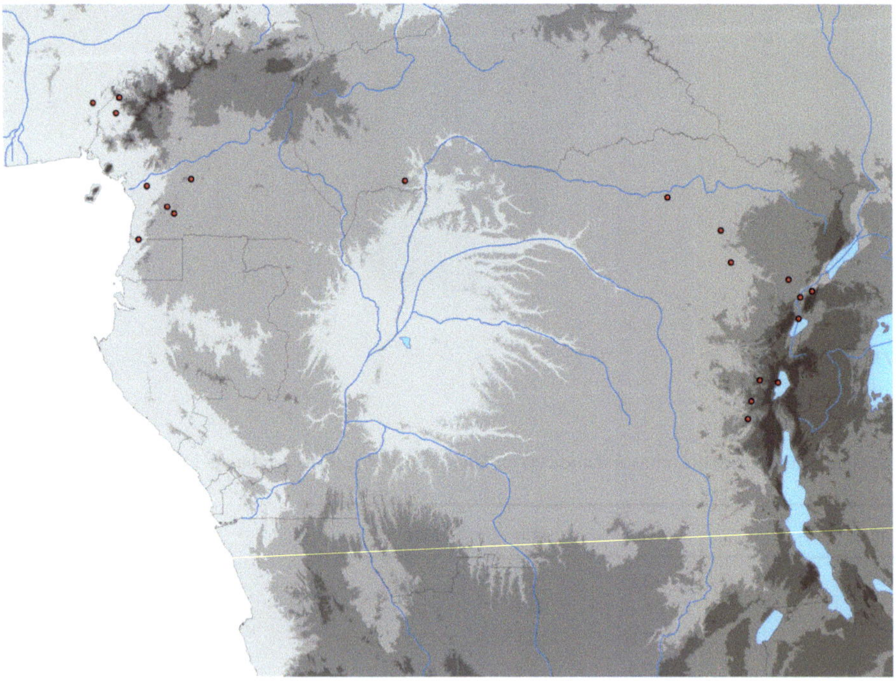

Figure 2.174 Distribution map

Measurements

External and cranial measurements (mm) and mass (g) for *Idiurus zenkeri*, sexes combined

	Mean	Min	Max	N
Mass*	17.4	-	-	1
HB**	72.7	70	79	10
Tail**	96.9	75	108	10
HF**	16.2	15	19	11
Ear**	15.6	9	14	11
GLS**	20.8	19.5	22.6	42

*Happold (2013), **Schunke & Hutterer (2007)

Genus Account

Zenkerella Matschie, 1898 (Cameroon Anomalures)

Diagnosis

This distinctive, but poorly known, monotypic genus is a medium-sized (about 450g) "flying squirrel" and can be distinguished by its long, black-tipped bushy tail, and (uniquely among the anomalures) the absence of a flight membrane. The fur is very soft and long, densely coating the body. The pelage is usually a shade of grey-brown is not grizzled or frosted with white as in *Anomalurus*. The ventral pelage is usually the same colour as the dorsal but paler. The ears are large, mostly naked and rise above the crown. There are five digits on the hindfoot and four digits on the forefoot, all with sharp, powerful and curved claws. The tail is slightly shorter than the body length and distinctively dark brown to black in colour, with the hairs long on the distal portion of the tail creating a bushy effect. Characteristic of the family Anomaluridae, two rows of posteriorly-pointed scales are present on the ventral surface of the tail, near its base, which anchor the animal when pressed against the trunk of a tree.

Figure 2.175 Photograph of *Zenkerella insignis* museum specimen (MNHN-ZM-MO 1970-112) (Ara Monadjem).

Cranial Characters

The skull is medium-sized (about 45 mm in greatest length). The rostrum and roof of skull are relatively flat, curving down posteriorly towards the occipitals. The rostrum is relatively short, giving the animal a blunt face. The palate ends at, or slightly before, the posterior margin of the third molars, and is characteristically V-shaped at its posterior edge. The zygomatic arch is long, thin and curved outwards, giving the animal a broad head. There is a large, but narrow infra-orbital foramen present. The post-orbital process is absent.

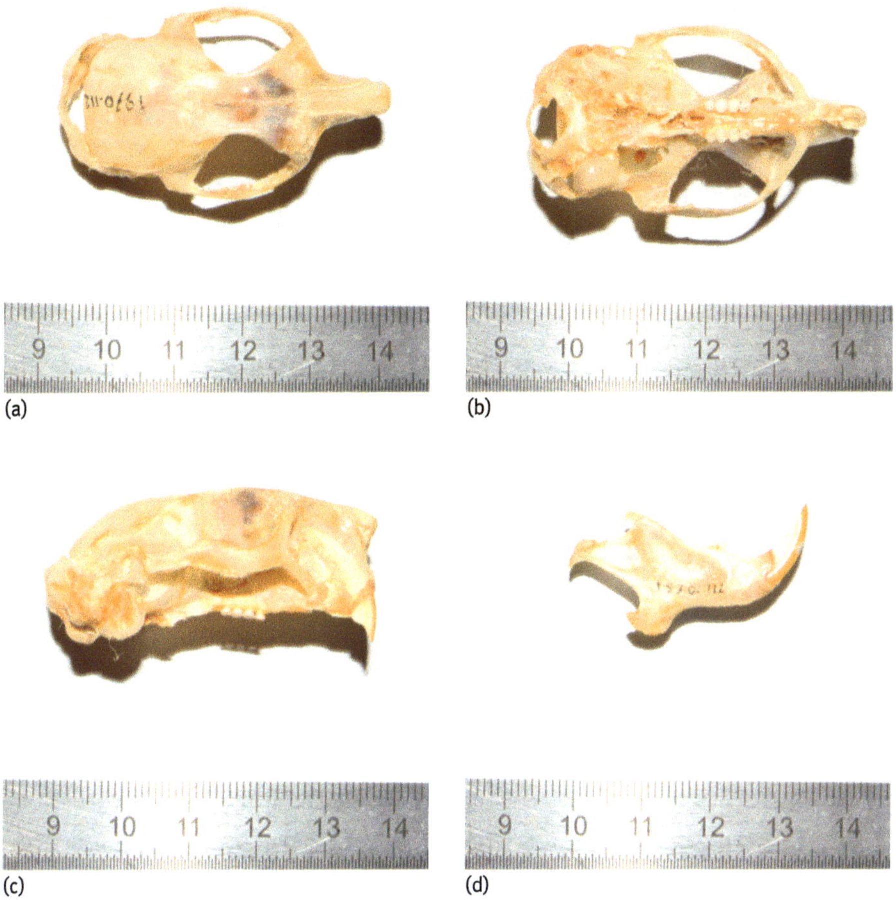

Figure 2.176 a-d Cranium of *Zenkerella insignis* (MNHN-ZM-MO 1970-112) (from top left to bottom right: dorsal view, ventral view, lateral view and lower jaw).

Dentition

Zenkerella has ungrooved, orthodont, orange incisors, the posterior surface of which has a notch (not as well developed as in *Idiurus*). The dental formula is: 1013/1013 (Rosevear, 1969). The premolar and last molar are similar in size and smaller than the first and second molars, with well-developed transverse ridges.

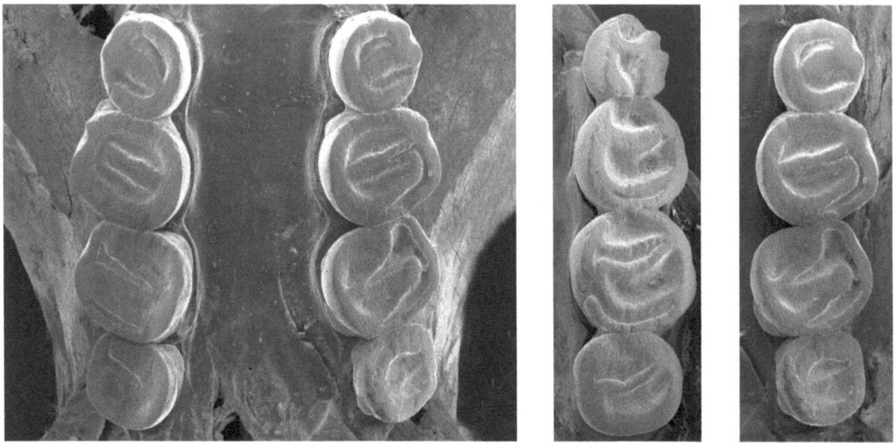
(a) (b) (c)

Figure 2.177 Cheek teeth of *Zenkerella insignis* (MNHN-ZM-MO 1970-112) (A: upper cheek teeth; B: lower right toothrow; C: upper left toothrow).

Habitat & Biogeography

Zenkerella is known by just 11 specimens (Schunke, 2005) from the Lower Guinea rainforests of Central Africa, making it the least-known anomalurid by a long margin. It has been recorded from Cameroon, Congo, Central African Republic, Equatorial Guinea, and from Bioko Island, where it occurs in rainforest and semi-deciduous forest (Perez del Val., 1995). It is arboreal but has been captured on the ground, suggesting that it might be partially terrestrial (Perez del Val., 1995).

Fossil Record

The oldest *Zenkerella* fossils are from 18-17 MYA, Uganda (Pickford & Mein, 2006), Kenya (Lavocat, 1973).

Biology & Ecology

Nothing is known about the biology or ecology of this species. A recent study on the impact of roads, hunting and habitat alteration on nocturnal mammals in Gabon presented quantitative data on *Z. insignis* which were reportedly observed during spot-light transects (Laurance et al., 2008). Considering the paucity of specimens mentioned above, and the fact that *Zenkerella* has not yet been collected from Gabon, we regard these observations in need of confirmation before their conclusions regarding this species are taken seriously. Our view is supported by the fact that an intensive six month survey of anomalures in Gabon failed to locate a single sighting of *Zenkerella* (Juilott et al., 1998).

Systematic Notes

The closest relative of the family Anomaluridae is Pedetidae (the Springhares), although this relationship is not particularly close (the two groups having diverged deep in the past); together they form the superfamily Anomaluroidea. The taxonomic position of this superfamily has been much disputed and they have been placed with both the Sciurognathi (Rosevear, 1969) and the Hystricognathi (Horner et al., 2007). However, a recent molecular phylogeny for the order Rodentia shows strong support for the Anomaluroidea within the "mouse group" (Blanga-Kanfi et al.,

2009). A molecular phylogeny is not available for the Anomaluridae, so the relationships between the genera are not known; however, *Idiurus* and *Zenkerella* have been placed in the same subfamily Zenkerellinae based on morphological similarities (Rosevear, 1969). There are currently no taxonomic issues in the designation of the sole species within this genus. The diploid number of chromosomes is not known for *Zenkerella*.

Species List

One species is described within the genus *Zenkerella*, restricted to tropical Central Africa.
Zenkerella insignis Matschie, 1898

Species Account

Zenkerella insignis Matschie, 1898

Diagnosis

This medium-sized anomalure can easily be distinguished from all other anomalures by the lack of a patagium.

Taxonomic Issues

See genus account.

Distribution

This species has a narrow distribution restricted to the Lower Guinea Forest of Central Africa.

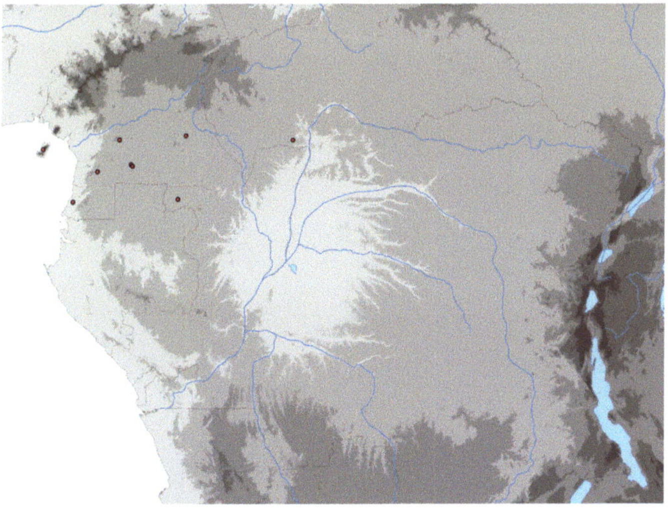

Figure 2.178 Distribution map

Measurements

External and cranial measurements (mm) and mass (g) for *Zenkerella insignis*, sexes combined

	Mean	Min	Max	N
Mass*	460	-	-	1
HB*	202.7	180	225	3
Tail*	172.5	165	180	4
HF*	41.0	40	42	3
Ear*	19.5	14	23	3
GLS*	45.7	44.8	46.3	3

*Happold (2013)

Family: Pedetidae

Genus Account

Pedetes Fitzinger, 1867 (Springhares)

Diagnosis

Their very large size (mostly 2.5-3.5 kg), bipedal gait, large ears and bushy tail readily distinguish members of this genus. Within African rodents, *Pedetes* is only surpassed in size by the Crested Porcupines (*Hystrix*) and Cane Rats (*Thryonomys*), and is the only large bipedal rodent within its range. The pelage is long and soft, sandy or pale brown dorsally, and white or off-white ventrally, with large amounts of geographical variation. The moult of *Pedetes capensis* is presented by Butynski (1982). The ears are elongated giving the impression of that of a hare, and the eyes are large. The hindlimbs are an order of magnitude larger than the forelimbs, with robust hindfeet. There are five clawed digits on the forefeet and four digits on the hindfeet, of which the first is reduced in size and has a sharp claw; the three remaining digits each have a robust triangle-shaped nail. There are two pairs of nipples (2 + 0 = 4) on the thorax.

Figure 2.179 Photograph of *Pedetes capensis* (Judy Hamilton).

Cranial Characters

The skull is very large (generally well over 80 mm in greatest length in adults) and robust but with a shortened rostrum giving the face a blunt appearance. The nasal bones are wide and parallel-sided. The palate ends at the posterior margin of the first molar. The zygomatic arch is robust and very deep anteriorly, narrowing posteriorly. The orbits are large, and the infraorbital foramen is very large (larger than the orbits in depth) and elliptical. The anterior palatal foramina are relatively short and wide. The palate is narrower anteriorly and the molar rows are oblique and divergent. The bullae are large and extended posteriorly to the occipital region.

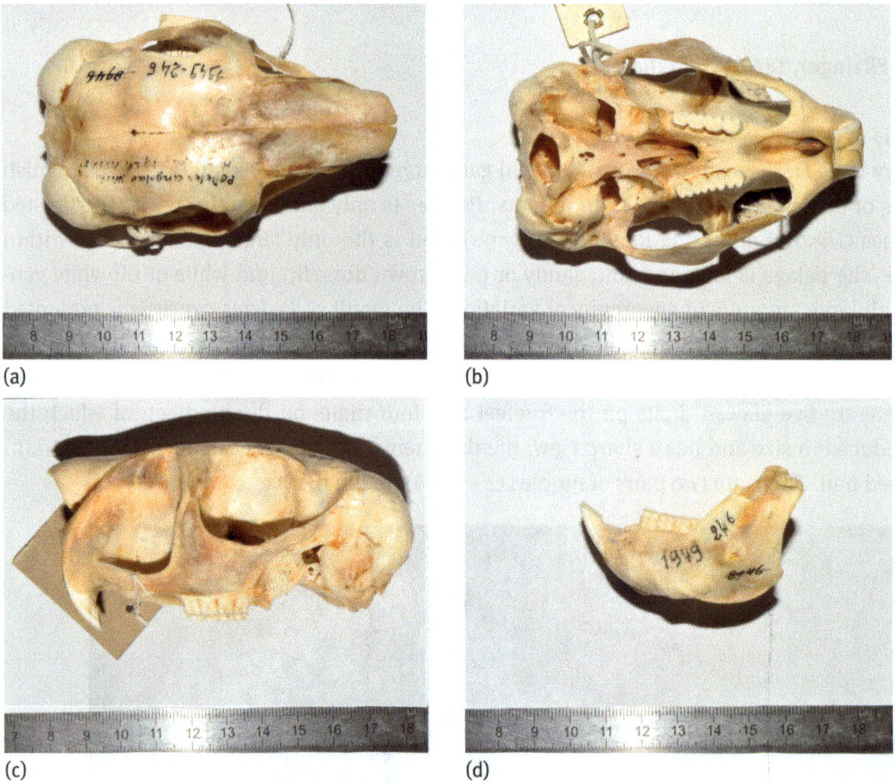

Figure 2.180 a-d Cranium of *Pedetes capensis* (MNHN-ZM-MO 1949-246) (from top left to bottom right: dorsal view, ventral view, lateral view and lower jaw).

Dentition

Pedetes has large, cream-coloured and opisthodont upper incisors. The dental formula is: 1013/1013, with the premolar and molars being approximately equal in size, hypsodont and bilobed, the lobes uniting lingually (Wood, 1965).

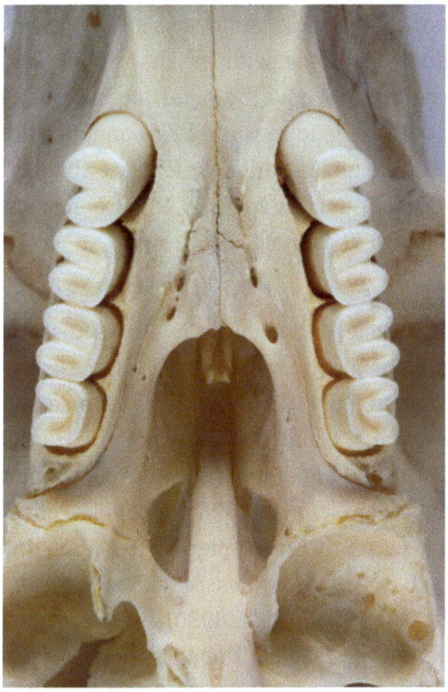

Figure 2.181 Upper cheek teeth of *Pedetes surdaster* (MNHN-ZM-MO 1939-92).

Habitat & Biogeography

Springhares are widespread throughout the arid savannas of Southern (*P. capensis*) and East Africa (*P. surdaster*). The Southern Africa *P. capensis* occurs in South Africa, southern Mozambique, Zimbabwe, Botswana, Namibia, southern Angola, western Zambia and the extreme southern tip of the DRC. *Pedetes surdaster* has been recorded patchily from Tanzania and southern Kenya. The two species are separated by the band of moist miombo woodland dominating the plateau of South-central Africa (Matthee & Robinson, 1997a). *Pedetes* is closely tied to open savanna with deep, sandy substrate, preferring areas with short grass (in particular *Cynodon*), and little woody vegetation (Butynski, 1979; Peinke & Brown, 2006). Seasonally flooded pans (Butynski, 1984) and heavily grazed areas are particularly favoured. *Pedetes* digs its own burrows, consisting of a simple tunnel about 0.8 m in depth, in which it rests up during the day (Butynski, 1979). Hence, a soft substrate is an essential component of its habitat.

Fossil Record

The earliest *Pedetes* is from 2 MYA, South Africa (Pickford & Mein, 2011). In East Africa it has been reported from the Middle Pliocene to the Middle Pleistocene; and in Southern Africa from the Early Pliocene to the Late Pleistocene (Winkler et al, 2010).

Biology & Ecology

The biology and ecology of *Pedetes* is relatively well-known and an excellent review is provided by Happold (2013). *Pedetes* is herbivorous feeding mainly on grasses, including the stem, roots, leaves, as well as corms and tubers of other plants (Peinke & Brown, 2006). It typically remains close to its burrow when foraging (Butynski, 1984), but changes these frequently, remaining in any one burrow for only a few consecutive days (Peinke & Brown, 2005). Its home range varies greatly between different individuals from 0.3 ha up to 28 ha (Peinke & Brown, 2005). *Pedetes* exhibits a slow reproductive rate for a rodent. Modal litter size is one, with less than 1% of pregnancies containing twins, and offspring are generally born throughout the year (Butynski, 1979; Van der Merwe et al., 1980; Peinke & Bernard, 2005), but may tend towards a seasonal peak in some areas during droughts (e.g. Kofron, 1987). The gestation and lactation periods are about 80 days for each, and between one and three births may occur annual (Butynski, 1979). Sexual maturity occurs at about 12 months in males and 1-2 years in females (Happold, 2013).

Systematic Notes

The closest relative of the family Pedetidae is the Anomaluridae, although this relationship is not particularly close (the two groups having diverged deep in the past); together they form the superfamily Anomaluroidea. The taxonomic position of this superfamily has been much disputed and they have been placed with both the Sciurognathi (Rosevear, 1969) and the Hystricognathi (Horner et al., 2007). However, a recent molecular phylogeny for the order Rodentia shows strong support for the Anomaluroidea within the "mouse group" (Blanga-Kanfi et al., 2009). The family Pedetidae contains a single living genus *Pedetes*; this genus is currently represented by two allopatric species: *P. capensis* in southern Africa and *P. surdaster* in East Africa (Matthee & Robinson, 1997a). These two populations have been considered conspecific (e.g. Meester et al., 1986), but are distinct on morphological and molecular grounds (Matthee & Robinson, 1997a,b; Happold, 2013). A large number of subspecies have been described for both species (nine for *P. capensis* and five for *P. surdaster*) however their validity is not certain. *Pedetes capensis* shows only weak phylogenetic structure across its range (Matthee & Robinson, 1997a) suggesting a single, uniform species. In contrast, *P. surdaster* shows strong phylogenetic structure, probably as a result of barriers created by rifting (Matthee & Robinson, 1997a) and a study of geographic variation in this species may identify distinct lineages. The diploid number of chromosomes has been recorded as: 2n = 38 in *P. capensis* (South Africa, Matthee & Robinson, 1997a), and as 2n = 40 in *P. surdaster* (Kenya, George, 1980; Matthee & Robinson, 1997a).

Species List

Two species described within the genus *Pedetes*, both restricted to Sub-Saharan Africa.
Pedetes capensis (Forster, 1778)
Pedetes surdaster Thomas, 1902

Key to Species

1	Restricted to savannas of Southern Africa	*P. capensis*
	Restricted to East Africa (Kenya and Tanzania, only)	*P. surdaster*

Species Accounts

Pedetes capensis (Forster, 1778)

Diagnosis
This species is immediately recognisable by its large size and bipedal gait. It is similar in size and appearance to *P. surdaster* (restricted to East Africa), but the two species do not overlap in distribution. Furthermore, the premaxillary process is more robust in *P. capensis* (breadth ca. 3 mm) and auditory bulla is larger (length ca. 19 mm) than in *P. surdaster* (premaxillary process breadth ca. 1 mm, auditory bulla ca. 13 mm in length)(Happold, 2013).

Taxonomic Issues
See genus account.

Distribution
This species is widely distributed in the more arid regions of southern Africa.

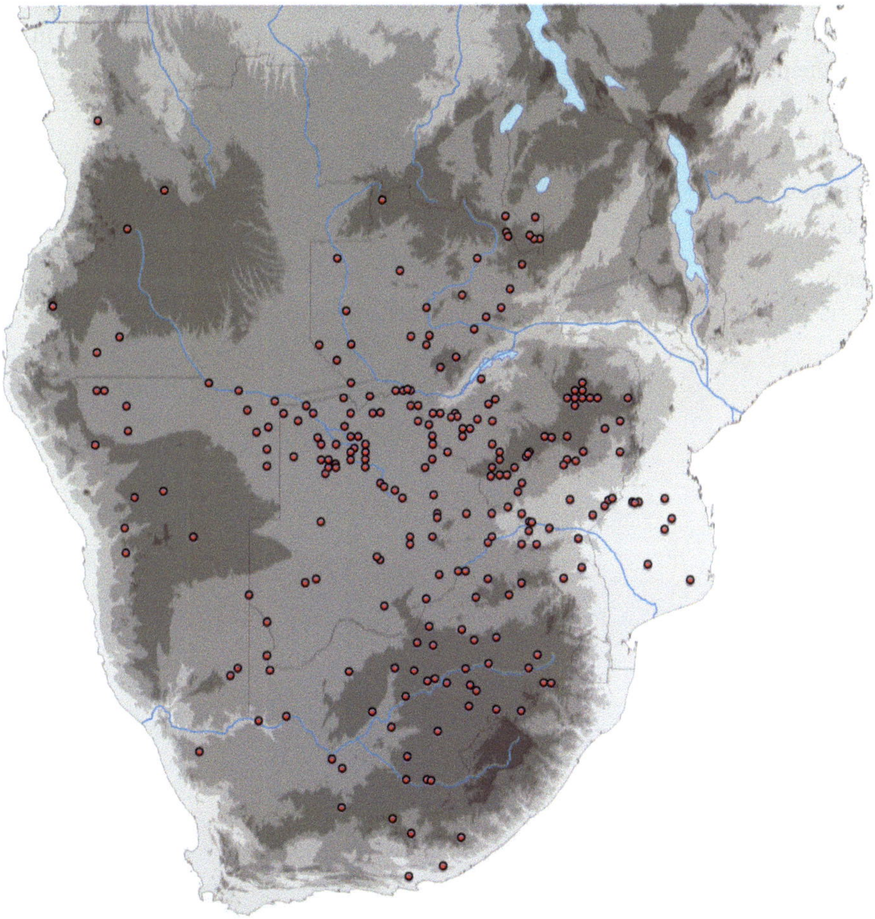

Figure 2.182 Distribution map

Measurements

External and cranial measurements (mm) and mass (g) for *Pedetes capensis*, sexes combined

	Mean	Min	Max	N
Mass	2945.8	2400	3750	12
HB	378.6	320	402	18
Tail	401.2	350	450	19
HF	153.8	140	185	20
Ear	72.4	65	82	19
GLS	88.0	82.6	94.0	12

Pedetes surdaster Thomas, 1902

Diagnosis

This species is immediately recognisable by its large size and bipedal gait. It is similar in size and appearance to *P. capensis* (restricted to Southern Africa), but the two species do not overlap in distribution. Furthermore, the premaxillary process is more robust in *P. capensis* (breadth ca. 3 mm) and auditory bulla is larger (length ca. 19 mm) than in *P. surdaster* (premaxillary process breadth ca. 1 mm, auditory bulla ca. 13 mm in length)(Happold, 2013).

Taxonomic Issues

See genus account.

Distribution

This species is patchily distributed in southern Kenya and Tanzania, extending south and east to Lake Rukwa.

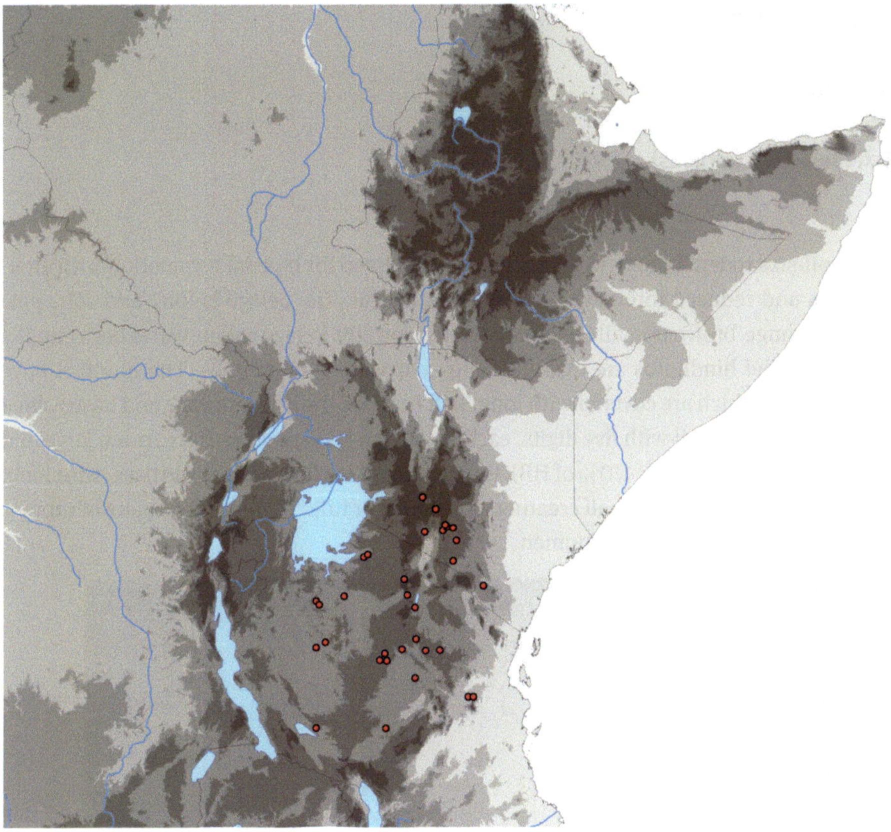

Figure 2.183 Distribution map

Measurements

External and cranial measurements (mm) and mass (g) for *Pedetes surdaster*, sexes combined

	Mean	Min	Max	N
Mass	-	-	-	-
HB	369.3	345	400	7
Tail	431.5	405	475	7
HF	145.3	133	161	7
Ear	71.4	60	78	7
GLS	81.8	68.2	87.8	5

Family: Dipodidae

Genus Account

Jaculus Erxleben, 1777 (Jerboas)

Diagnosis

Jaculus is a small-size rodent (about 30 g), distinctively adapted for bipedal locomotion with enormous hindlimbs and reduced forelimbs, and a very long tail. The pelage is long and soft, pale sandy-brown to orange-brown dorsally and pure white ventrally. Pelage colour varies both clinally and within a site. The hindlimbs are massive compared with the forelimbs. The hindfeet have three very long digits which are covered with long, stiff hairs on the undersurface and have robust claws. The forelimbs are small with five digits, each with a long, sharp claw. The ears are large and rounded. The tail is very long (160-170% of HB), finely covered in short hairs at the base, with long, black and white hairs on the distal half creating a bushy tip or tuft. There are four pairs of nipples (2 + 2 = 8) placed on the thorax and abdomen.

Figure 2.184 Photograph of *Jaculus jaculus* (Ara Monadjem).

Cranial Characters

The skull is small (typically 30-33 mm in greatest length). It is relatively broad and shallowly domed posteriorly. The rostrum is narrow and long. The zygomatic arch is distinctive with a well-developed jugal plate extending vertically along the entire anterior margin of the orbit. The palate ends well beyond the posterior margin of the third molars. The infraorbital foramen is large and elliptical. The anterior palatal foramina are long and relatively wide. The bullae are very large and inflated. The mandibles have a characteristic perforation on the angular process.

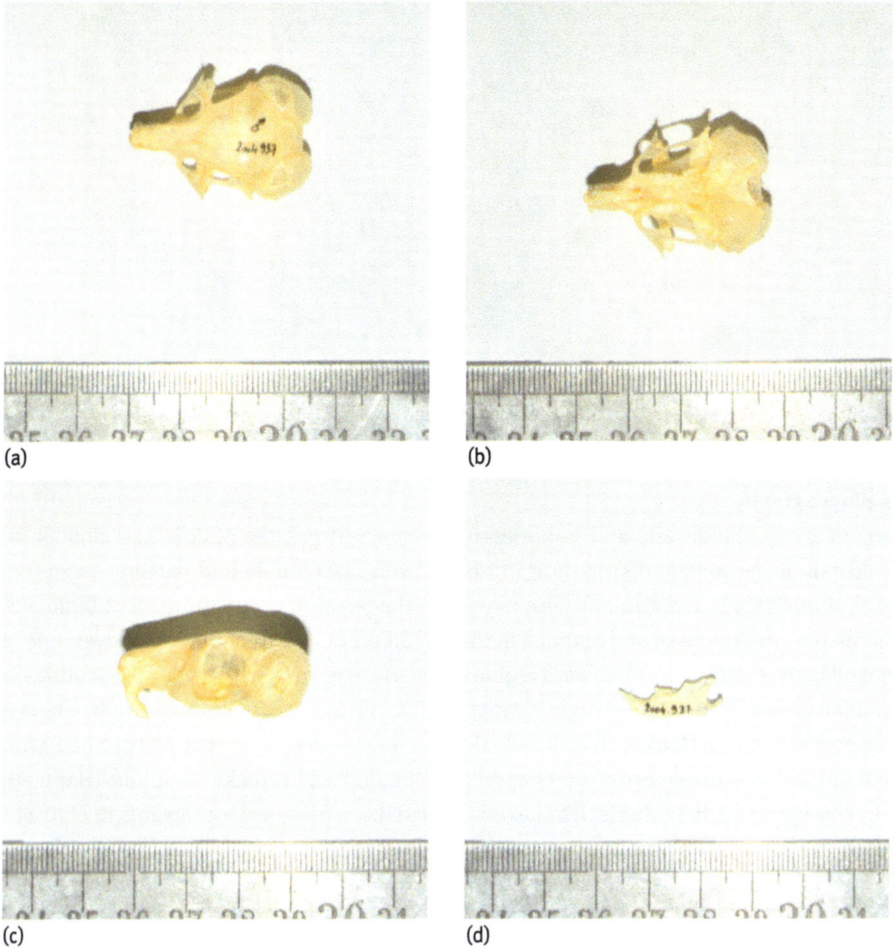

Figure 2.185 a-d Cranium of *Jaculus jaculus* (MNHN-ZM-MO 2004-937) (from top left to bottom right: dorsal view, ventral view, lateral view and lower jaw).

Dentition

Jaculus has long, ungrooved and distinctly opisthodont incisors. The dental formula is: 1003/1003, with only three cheekteeth in each half jaw (Rosevear, 1969). The last molar is slightly smaller than the two preceding it. The molars have transverse ridges created by infolding of the enamel on the inner and outer side, dividing them into two lobes (Happold, 1967).

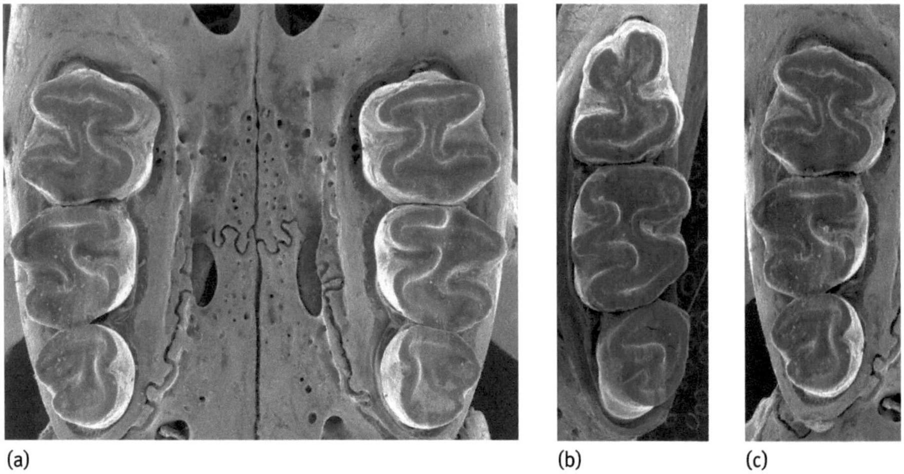

Figure 2.186 Cheek teeth of *Jaculus jaculus* (MNHN-ZM-MO 1978-300) (A: upper cheek teeth; B: lower left toothrow; C: upper right toothrow).

Habitat & Biogeography

Juculus occurs across the deserts and semi-deserts of North Africa, the Middle East and as far west as Pakistan. It has a wide distribution in North Africa, marginally and patchily extending into Sub-Saharan Africa in the arid northern regions of the Sahel. It occurs from Mauritania and Senegal in the west to Ethiopian and Somalia in the east. It occurs north of the Niger River in Mali and Niger, and in the extreme north-eastern region of Nigeria. It is known from various localities in northern Sudan (Setzer, 1956), and a single location in Chad (Granjon & Duplantier, 2009) where it has most probably been overlooked. It is closely tied to sandy, desert or semi-desert habitats with soft, friable soil in which to dig its burrow (Happold, 2013), and avoids rocky situations (Happold, 1967, 1969). This species is hypothesised to have extended its distribution southwards in Mali as a result of desertification in the past few decades (Dobigny et al., 2002a).

Fossil Record

The oldest *Jaculus* fossil is from 2.5-3.0 MYA, Ethiopia (Wesselman, 1984). In North Africa, it has been reported from the Late Pliocene to the Early Pleistocene; in East Africa from the Late Pliocene to the Middle Pleistocene (Winkler et al, 2010).

Biology & Ecology

The biology and ecology of *Jaculus jaculus* in Africa has been recently review by Happold (2013), where it is still poorly known (e.g. little is known about its spatial ecology in this region e.g. see Schropfer & Klenner-Fringes, 1991). It is vegetarian, feeding on leaves, grasses and small seeds. Breeding is seasonal, and litter size is 2-5 and varies during the year (Happold, 1967). The development of young is given in Happold (1970).

Systematic Notes

Within Africa, the genus *Jaculus* has two clearly defined species (*J. jaculus* and *J. orientalis*), and a third taxon (*deserti*) that has only recently had its status (partially) resolved. The taxon *deserti* has had a long history from when it was first described by Loche (1867). It was regarded as a separate species by Ranck (1968), but was subsequently relegated as a synonym of *J. jaculus* (see Holden & Musser, 2005). Granjon & Duplantier (2009) pointed out the existence of two distinct clades in West Africa based on differences in karyotype (2n = 48 and 50). Recent studies have shown that the Cytochrome b sequences of these two taxa differ by more than 10%, suggesting distinctiveness at a species level (Ben Faleh et al., 2010b; Ben Faleh et al., 2012a; Boratynski et al., 2012). These two cryptic species also show differences in cranial morphology but, unusually for African rodents, not in karyology (Ben Faleh et al., 2010a). It is not clear how the two different colour forms or morphs described in the literature, and reviewed by Ben Faleh et al. (2010a,b), relate to these two species; clarification of this would have important implications for field identification of these two species. However, they differ in their biogeographic histories, with *J. deserti* only recently having expanded its range into north-western Africa. The larger *J. orientalis* is readily distinguishable from both *J. jaculus* and *J. deserti* on morphological and molecular-genetic grounds (Ben Faleh, 2012a; Ben Faleh et al., 2012b) but does not occur south of the Mediterranean coast of North Africa (Happold, 2013). The diploid number of chromosomes in *J. jaculus* appears to be variable with two karyotypes having been recorded: 2n = 48 (Niger, Dobigny et al., 2002a; Tunisia, Ben Faleh et al., 2010a) and 2n = 50, FN = 90 (Senegal, Granjon et al., 1992).

Species List

Two species within the genus *Jaculus* extend marginally into Sub-Saharan Africa in the north.
Jaculus jaculus (Linneaus, 1758)
Jaculus deserti Loche, 1867

Key to Species

The two species of *Jaculus* listed here cannot be distinguished on any morphological feature and even share the same karyotype; they can only be identified on molecular grounds (Ben Faleh et al., 2010b; Ben Faleh et al., 2012a). Hence a key cannot currently be presented.

Species Accounts

Jaculus deserti Loche, 1867

Diagnosis

This species can only be differentiated from *J. jaculus* on molecular grounds; morphologically and karyotypically the two species are indistinguishable (Ben Faleh et al., 2010a; Ben Faleh et al., 2012a).

Taxonomic Issues

See genus account.

Distribution

This species occurs widely throughout the northern parts of the Sahel and the Sahara, but has not yet been recorded east of Niger.

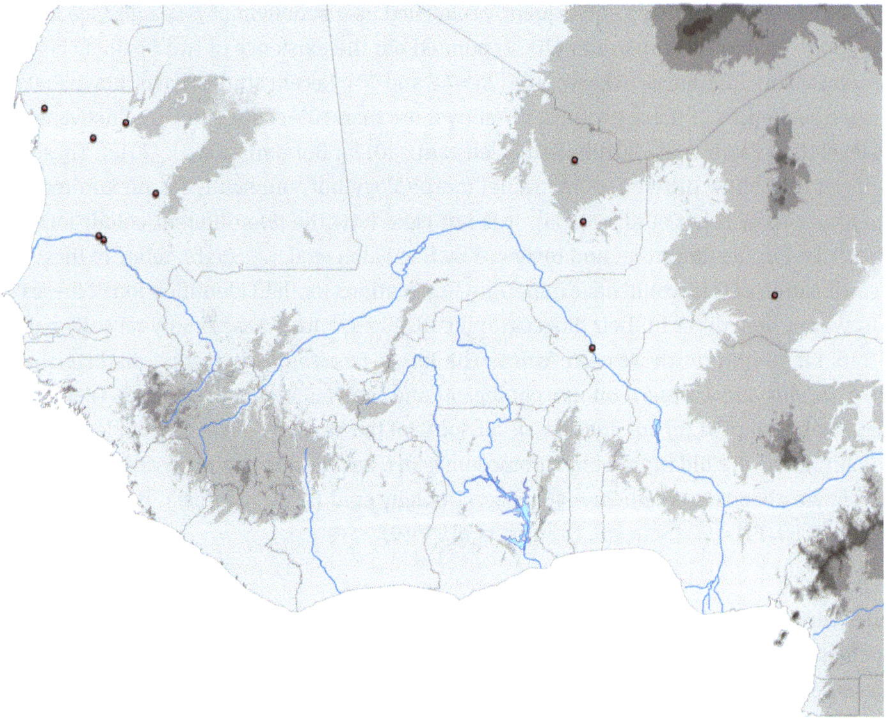

Figure 2.187 Distribution map

Measurements

External and cranial measurements (mm) and mass (g) for *Jaculus deserti*, sexes combined

	Mean	Min	Max	N
Mass	68.6	62	75	9
HB	123.2	119	129	9
Tail	192.9	190	195	9
HF	64.7	62	67	9
Ear	20.3	19	22	9
GLS	-	-	-	-

Jaculus jaculus (Linneaus, 1758)

Diagnosis

This species can only be differentiated from *J. deserti* on molecular grounds; morphologically and karyotypically the two species are indistinguishable (Ben Faleh et al., 2010a; Ben Faleh et al., 2012a).

Taxonomic Issues

See genus account.

Distribution

This species occurs widely throughout the northern parts of the Sahel and the Sahara.

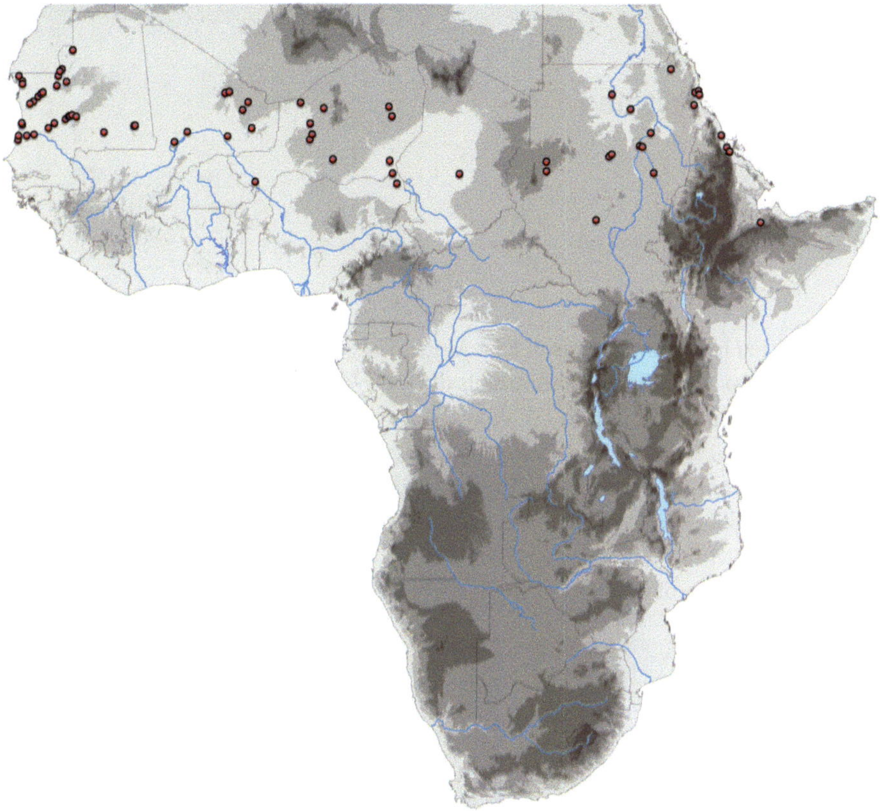

Figure 2.188 Distribution map

Measurements

External and cranial measurements (mm) and mass (g) for *jaculus jaculus*, sexes combined

	Mean	Min	Max	N
Mass	68.5	64	74	26
HB	123.5	119	130	26
Tail	194.4	191	197	26
HF	64.5	60	67	26
Ear	20.3	19	21	26
GLS*	32.3	31.2	33.3	40

*Happold (1967)

Family: Spalacidae

This mostly Eurasian family is represented in Sub-Saharan Africa by a single endemic subfamily and 13 species.

Subfamily: Tachyoryctinae

Genus Account

Tachyoryctes Ruppell, 1835 (Root-rats)

Diagnosis

This genus is restricted to highland areas of East Africa. They are medium to large rodents (Mass ca. 200-600 g, but > 900 g in very large individuals of *T. macrocephalus*), with a characteristically stout body, very short limbs and a short tail typical of fossorial species. In contrast to the other group of African fossorial rodents (family Bathyergidae), this genus has clearly visible (but tiny) ear pinnae, larger eyes, and orange incisors. The tail is also somewhat longer in *Tachyoryctes*. The pelage varies considerably, but is thick and silky, and some shade of light to dark brown, often with a reddish hue; individual hairs have dark grey bases. The ventral pelage is paler than that of the dorsal pelage. Four pairs of nipples (2 + 2 = 8) are present.

(a) (b) (c) (d)

Figure 2.189 a-d Photographs of *Tachyoryctes*, from top left to bottom right: *T. daemon* (Charles Foley, Bill Stanley, Radim Sumbera) and *T. macrocephalus* (Marian Polak).

Cranial Characters

The skull is robust, large (GLS ca. 45-65 mm), flattened with a narrow rostrum. The zygomatic arches are sturdy and wide. The brain case is relatively swollen distinctly narrowing to the interorbital constriction (a feature that readily differentiates *Tachyoryctes* from the Bathyergidae). The sagittal crest is well-developed. The infraorbital foramen is small. The anterior palatal foramina are very short and end well before the M^1. The bullae are slightly enlarged, but not markedly so.

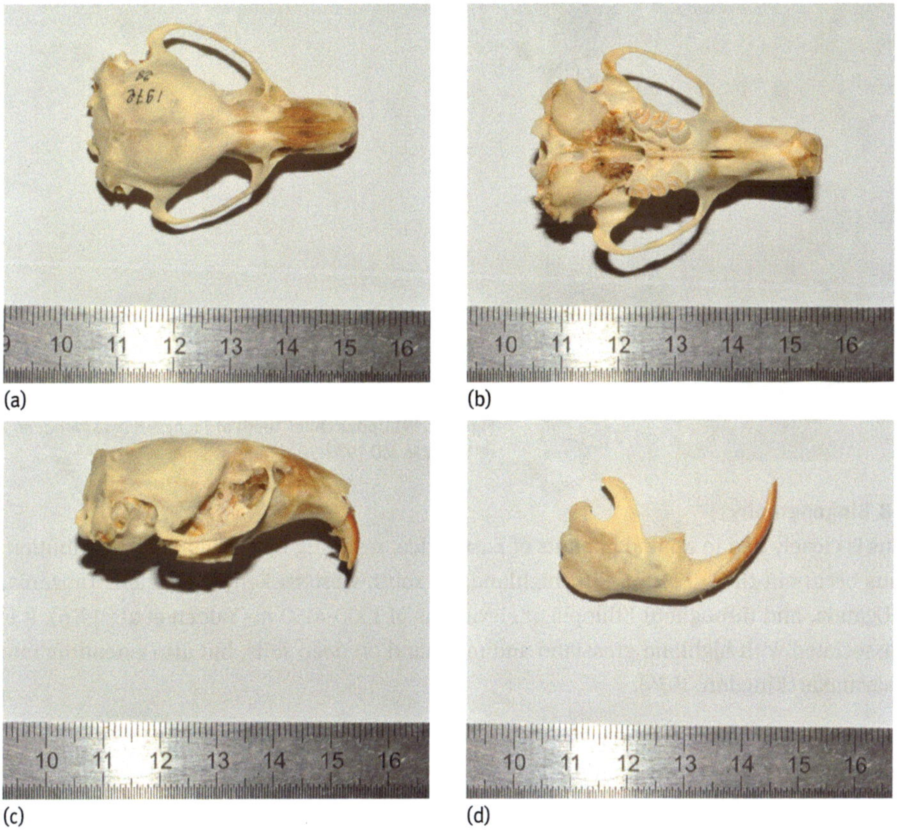

Figure 2.190 a-d Cranium of *Tachyoryctes splendens* (MNHN-ZM-MO 1972-28) (from top left to bottom right: dorsal view, ventral view, lateral view and lower jaw).

Dentition

Tachyoryctes has robust, ungrooved and pro-odont upper incisors; all incisors are conspicuously orange (in contrast to the Bathyeridae which have white incisors). The dental formula is: 1003/1003. The molars are all roughly the same size, simple but with deep folds. The molar rows are not parallel resulting in the palate widening posteriorly; the palate does not extend beyond the last upper molar. The coronoid process of the lower jaw is well developed (in contrast to the Bathyergidae).

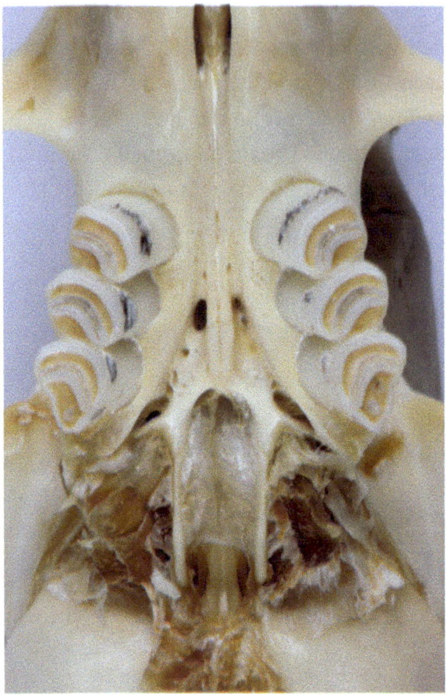

Figure 2.191 Upper cheek teeth of *Tachyoryctes splendens* (MNHN-ZM-MO 1972-30).

Habitat & Biogeography

The genus is closely tied to upland habitats of East Africa, resulting in a fragmented distribution. The genus occurs in the Albertine Rift, highlands of south-western Kenya, northern Tanzania, eastern Uganda, and throughout Ethiopia at elevations of 1300-4150 m (Yalden et al., 1976). It is closely associated with highland grassland and moorland on deep soils, but also extending into upland savannas (Kingdon, 1974).

Fossil Record

Fossils of this genus are known from the Late Miocene to the Late Pleistocene of Ethiopia, East Africa (Winkler et al., 2010).
Three fossil species are known:
T. makooka, Adu Asa (Lopez-Antonanzas & Wesselman, 2013)
T. pliocaenicus, Hadar, (Sabatier, 1982)

Biology & Ecology

Tachyoryctes is solitary and territorial digging burrows with its large incisors. It is predominantly subterranean where it forages for food (predominantly herbs and grasses) by excavating tunnels, but also forages on the surface, and may be a crop pest (Kingdon, 1974). It can occur at high densities: 49 animals were excavated in 48 separate burrows on a plot of 0.72 ha (Rahm, 1971); and 200 individuals per ha (Jarvis, 1973).

Systematic Notes

The genus was considered to comprise two species by Happold (2013), *T. macrocephalus* (endemic to Ethiopia) and *T. splendens* (occurring widely in Ethiopia and East Africa). However, a large number of taxa have been described with eight species recognised by Hollister (1919) and 13 species by Musser & Carleton (2005). However, a recent and extensive morphometric study was only able to distinguish the two species mentioned above (Beolchini & Corti, 2004). This is rather surprising since at least one taxon (*rex*), within the *T. splendens* group, is significantly larger than other taxa (Hollister, 1919). Here we follow Musser & Carleton (2005) in recognising 13 species, pending a full revision of the genus, including molecular data. However, we suspect that several taxa may be conspecific. In particular, we suspect that *T. naivashae, T. daemon, T. storeyi,* and *T. ibeanus* may refer to the same species (the name *ibeanus* taking taxonomic priority), and *T. annectens* may be conspecific with *T. audax* (*annectens* taking taxonomic priority), which would reduce the number of species to nine. See the relevant species account for further details.

Species List

Thirteen species are recognised:
Tachyoryctes ankoliae Thomas, 1909
Tachyoryctes annectens (Thomas, 1891)
Tachyoryctes audax Thomas, 1910
Tachyoryctes daemon Thomas, 1909
Tachyoryctes ibeanus Thomas, 1900
Tachyoryctes macrocephalus (Ruppell, 1842)
Tachyoryctes naivashae Thomas, 1909
Tachyoryctes rex Heller, 1910
Tachyoryctes ruandae Lonnberg & Gyldenstolpe, 1925
Tachyoryctes ruddi Thomas, 1909
Tachyoryctes spalacinus Thomas, 1909
Tachyoryctes splendens (Ruppell, 1835)
Tachyoryctes storeyi Thomas, 1909

Key to Species

1	Distribution restricted to Ethiopia	2
	Distribution does not include Ethiopia	3
2	Size larger (GLS > 60 mm, HB > 230 mm); typically occurs at higher elevations (3000-4150 m); endemic to the Bale Mts and neighbouring areas east of the Rift Valley	*T. macrocephalus*
	Size smaller (GLS < 55 mm, HB < 230 mm); typically occurs at lower elevations (1300-3900 m); widely distributed in Ethiopia on both sizes of the Rift Valley	*T. splendens*
3	Size larger (GLS typically ca. 48 mm or larger); endemic to the higher slopes of the Aberdare Mts or Mt Kenya (typically above 2600 m)	4
	Size smaller (GLS typically < 48 mm); not occurring on the higher slopes of the Aberdare Mts or Mt Kenya	5
4	Size larger (GLS ca. 52 mm); Endemic to the higher slopes of Mt Kenya above the bamboo zone (ca. 2600-3300 m)	*T. rex*
	Size smaller (GLS ca. 48 mm); Endemic to the higher slopes of the Aberdare Mts above 2700 m	*T. audax**
5	Posterior margin of the cranium (in lateral profile) is slanted at an angle (as opposed to being near vertical in other *Tachyoryctes* species)	*T. spalacinus*

	Posterior margin of the cranium (in lateral profile) is not slanted at an angle, being near vertical	6
6	Occurs from Kakamega (western Kenya) westwards to the Albertine Rift	7
	Occurs east of Kakamega in south-central Kenya and Tanzania	*T. naivashae***
7	Size slightly larger (HB ca. 215 mm); endemic to the Albertine Rift	*T. ruandae*
	Size slightly smaller (HB ca. 200 mm or less); occurs mostly to the east of the Albertine Rift	8
8	Size slightly larger (HB ca. 200 mm); parietal ridges never join to form well-developed sagital crest	*T. ankoliae*
	Size slightly smaller (HB ca. 185 mm); parietal ridges join to form well-developed sagital crest	*T. ruddi*

*We suspect that *T. annectens* is conspecific with *T. audax* (in which case the name *annectens* takes priority); **we suspect that *T. storeyi*, *T. daemon* and *T. ibeanus* are all conspecific with *T. naivashae* (in which case the name *ibeanus* takes priority). See relevant species accounts for more information

Species Accounts

Tachyoryctes ankoliae Thomas, 1909

Diagnosis

This species co-occurs with *T. ruddi* and *T. ruandae* (Delany, 1975; Lunde & Sarmiento, 2002) from which it can be distinguished as follows: *T. ruddi* (size on average smaller, HB ca. 185 mm; parietal ridges join to form a distinct sagital crest); *T. ruandae* (size on average larger, HB typically > 210 mm). In this species, the parietal ridges do not join to form a distinct sagital crest, although the frontal ridges are well developed (Thomas, 1909).

Taxonomic Issues

See genus account.

Distribution

This species occurs from the eastern mountains of the Albertine Rift to the northern shores of Lake Victoria as far as the Nile River in the east.

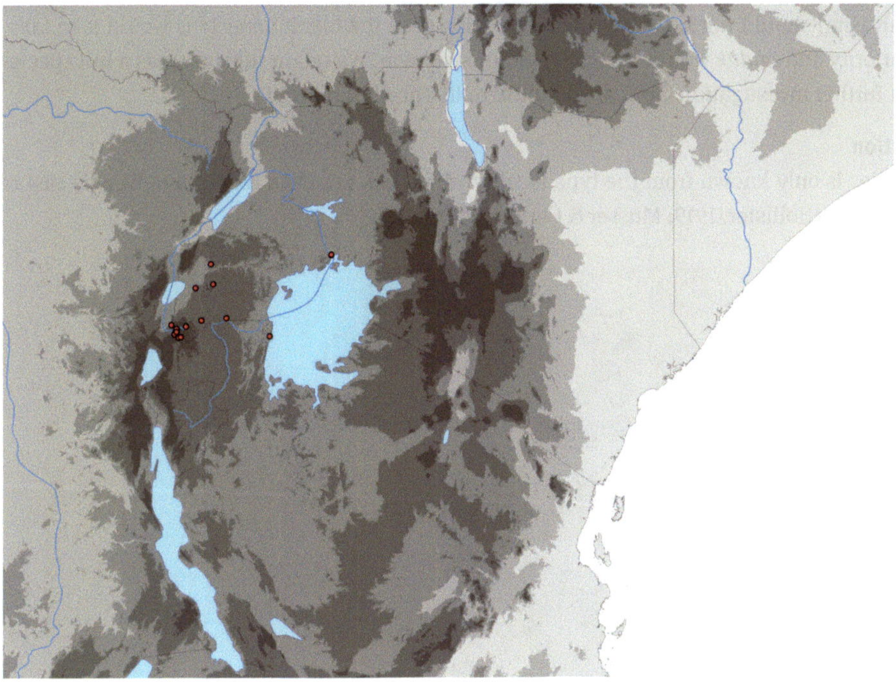

Figure 2.192 Distribution map

Measurements

External and cranial measurements (mm) and mass (g) for *Tachyoryctes ankoliae*, sexes combined

	Mean	Min	Max	N
Mass*	288	-	-	1
HB**	199.0	198	200	2
Tail**	68	-	-	1
HF**	28.0	28	28	2
Ear	-	-	-	-
GLS**	46.2	-	-	2

*Lunde & Sarmiento (2002); **Hollister (1919); GLS combines samples from both these papers

Tachyoryctes annectens (Thomas, 1891)

Diagnosis

This species was described based on its large size which distinguishes it from the smaller *T. naivashae* with which it may co-occur in southern Kenya (Thomas, 1909) (but see below).

Taxonomic Issues

The status of this species is uncertain, particularly since the type locality is not known. The possibility of it being conspecific with the large *T. audax* from the Aberdare Mts (ca. 20-30 km east of

Lake Naivasha) should not be discounted; in such a case the taxon *annectens* would take taxonomic priority and *audax* would become a junior synonym. We retain this taxon as a full species pending further investigations, but are doubtful of its specific validity.

Distribution

This species is only known from the type locality which has not been fixed, but given as east of Lake Naivasha (Hollister, 1919; Musser & Carleton (2005).

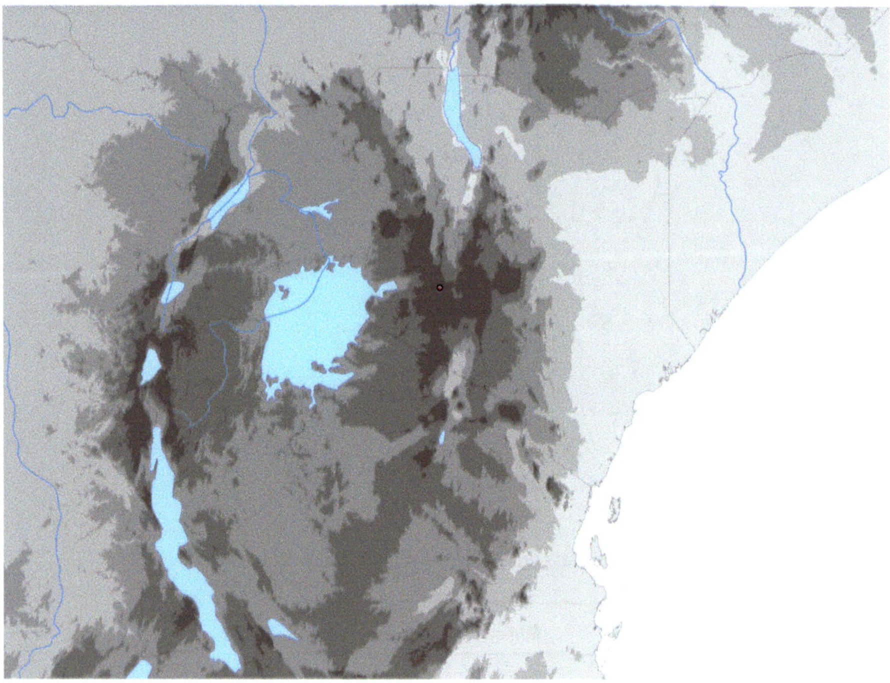

Figure 2.193 Distribution map

Measurements

No measurements could be located for this species, but see *T. audax* account (below).

Tachyoryctes audax Thomas, 1910

Diagnosis

This is the only species of *Tachyoryctes* occurring within its range (the Aberdare Mts). It can be distinguished on size from the smaller *T. naivashae* (GLS < 46 mm) which occurs to the west of the Aberdares.

Taxonomic Issues

This species may be conspecific with *T. annectens* (see latter species account).

Distribution

This species is only known from the higher slopes of the Aberdare Mts, south-central Kenya where it occurs in moorland above 2700 m (Hollister, 1919).

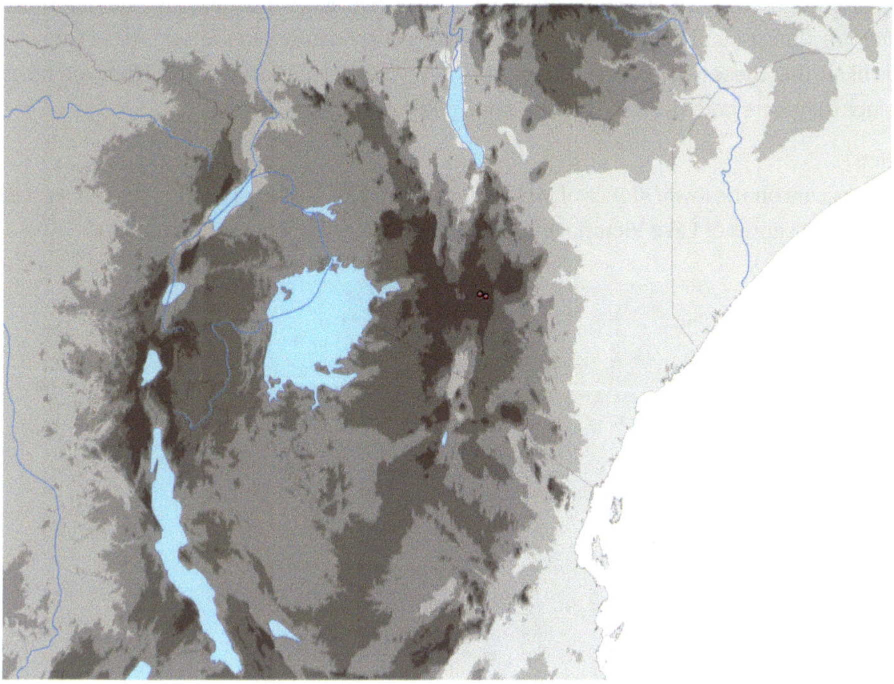

Figure 2.194 Distribution map

Measurements

External and cranial measurements (mm) and mass (g) for *Tachyoryctes audax*, sexes combined

	Mean	Min	Max	N
Mass	-	-	-	-
HB*	209.4	180	225	9
Tail*	67.0	63	78	8
HF*	28.6	28	30	8
Ear	-	-	-	-
GLS*	48.6	44	53	9

*Hollister (1919)

Tachyoryctes daemon Thomas, 1909

Diagnosis

This species appears to be the only *Tachyoryctes* within its range, however, the northern limits of its range have not yet been resolved and it may be shown to co-occur with the similar-sized *T. naivashae* (mean GLS 44.3 mm).

Taxonomic Issues

The character distinguishing this species from *T. naivashae* is supposed to be the smaller size of the latter species (Thomas, 1909). However, Hollister (1919) presented measurements of a large series of *T. naivashae*, collected from the environs of Lake Naivasha, that demonstrate that these two species are almost identical in size. Both also have poorly developed sagital crests (Thomas, 1909). It is likely that these two taxa are conspecific; in such a case it is not clear which taxon takes priority since they were both described in the same publication.

Distribution

This species occurs on the lower slopes of Mt Kilimanjaro and Mt Meru, up to 4000 m, and west to the south-eastern shores of Lake Victoria and the north-eastern shores Lake Tanganyika.

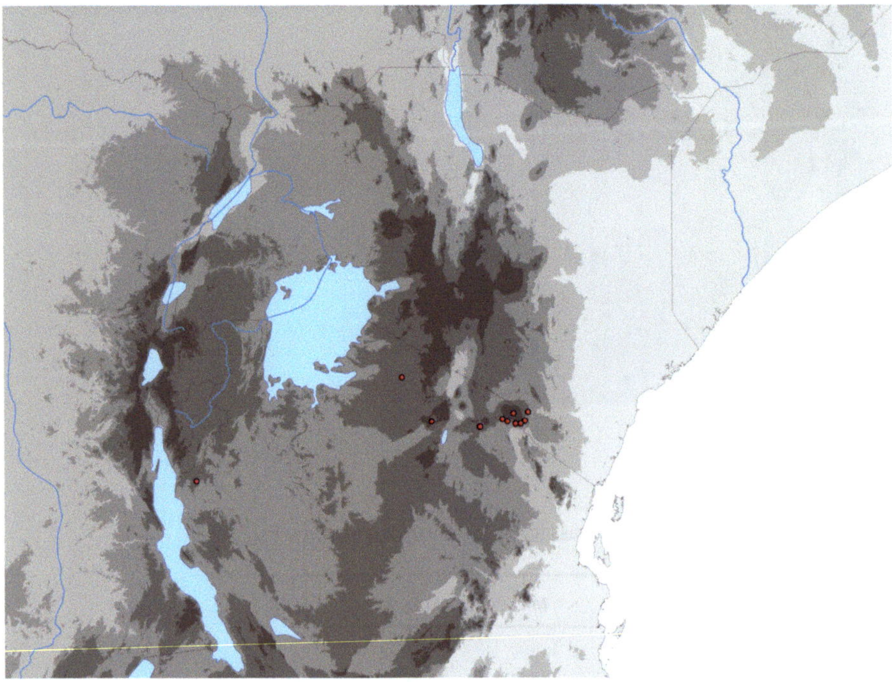

Figure 2.195 Distribution map

Measurements

External and cranial measurements (mm) and mass (g) for *Tachyoryctes daemon*, sexes combined

	Mean	Min	Max	N
Mass	-	-	-	-
HB	-	-	-	-
Tail	-	-	-	-
HF	-	-	-	-
Ear	-	-	-	-
GLS*	44.0	43	45	3

*Hollister (1919)

Tachyoryctes ibeanus Thomas, 1900

Diagnosis

This species is similar in size and general skull characteristics to *T. naivashae* and the two species are not currenlty diagnosable on morphological grounds (see below). This species may also co-occur with *T. spalacinus* but differs in the shape of the posterior margin of the cranium (which is slanted in *T. spalacinus* and near vertical in *T. ibeanus*).

Taxonomic Issues

This species is supposedly larger than *T. naivashae* (Thomas, 1909), but not according to measurements of a large series from Lake Naivasha taken by Hollister (1919) (see *T. naivashae* account). It is currently unclear how these two species differ and we suspect that they may refer to the same species. If this is the case, the name *ibeanus* takes taxonomic priority over *naivashae*.

Distribution

This species is known from a few localities on the plains to the south of the main Kenyan highlands.

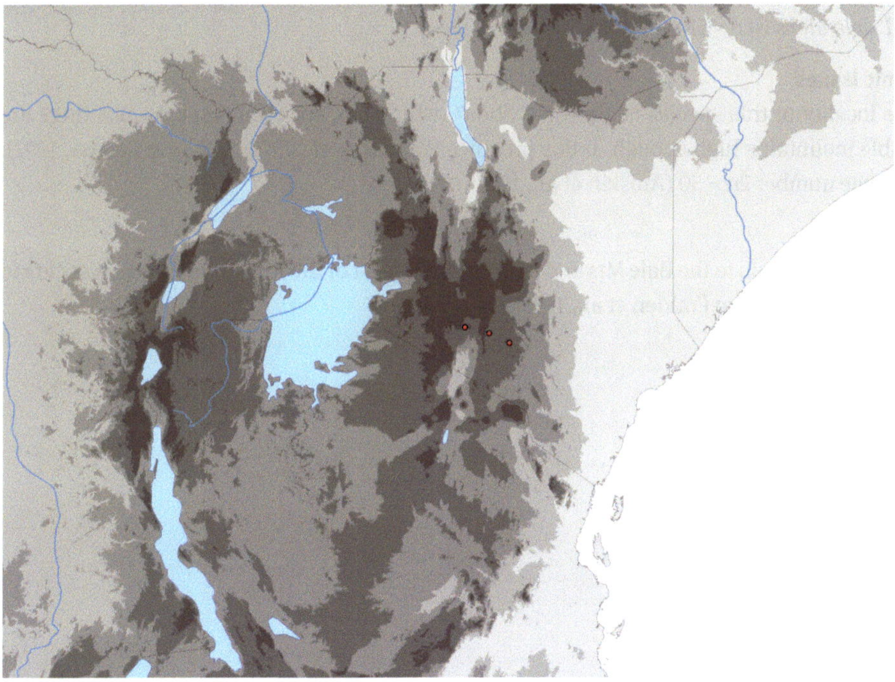

Figure 2.196 Distribution map

Measurements

External and cranial measurements (mm) and mass (g) for *Tachyoryctes ibeanus*, sexes combined

	Mean	Min	Max	N
Mass	-	-	-	-
HB*	195.8	181	220	13
Tail*	61.5	54	73	13
HF*	27.7	26	30	14
Ear	-	-	-	-
GLS*	43.3	40	47	13

*Hollister (1919)

Tachyoryctes macrocephalus (Ruppell, 1842)

Diagnosis

This is one of two *Tachyoryctes* species endemic to Ethiopia; it can be distinguished by its larger size (in *T. splendens* GLS < 55 mm, HB < 230 mm).

Taxonomic Issues

The type locality of this species is given as "Shoa", which has not been located since there are no suitable mountains high enough in this province (Yalden et al., 1976; Yalden & Largen, 1992). Chromosome number 2n = 50 (Aniskin et al., 1997a).

Distribution

This species is endemic to the Bale Mts and surroundings, east of the Rift Valley, where it occurs in moorland at 3000-4150 m (Yalden et al., 1976).

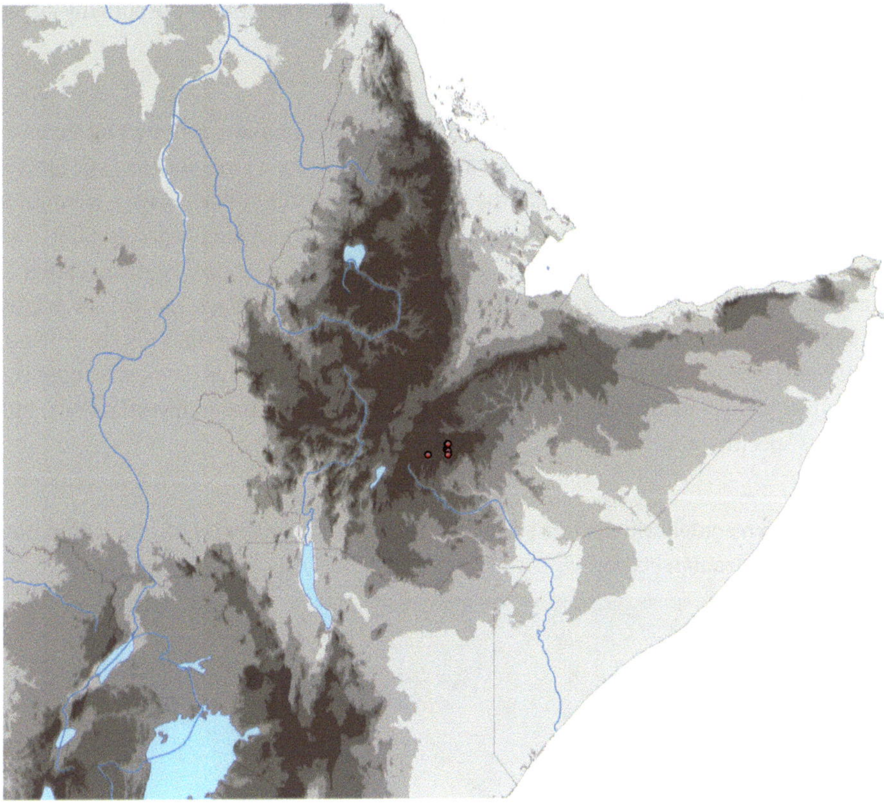

Figure 2.197 Distribution map

Measurements

External and cranial measurements (mm) and mass (g) for *Tachyoryctes macrocephalus*, sexes combined

	Mean	Min	Max	N
Mass	590.0	540	640	2
HB	263.8	230	290	4
Tail	51.3	40	65	4
HF	36.0	35	37	4
Ear	13.3	12	15	4
GLS*	64.1	60.4	69.6	11

*Happold (2013)

Tachyoryctes naivashae Thomas, 1909

Diagnosis

This species is supposed to be readily distinguished by its smaller size (Thomas, 1909), however measurements taken by Hollister (1919) suggest that this is not a particularly small animal (see

Taxonomic Notes). As a result, it cannot be distinguished from *T. daemon* or the purportedly larger *T. storeyi* based on size alone.

Taxonomic Issues

This species was described based on its small size; a GSL of 39.5 mm was reported by Thomas (1909) which indeed would make this a small *Tachyoryctes*. However, a large series of specimens collected from the environs of Lake Naivasha (the type locality for this species) have a mean GLS of 44.3 mm (Hollister, 1919), demonstrating that this is not a particularly small species. Similarly, Beolchini & Corti (2004) showed that this species is not remarkable for its size. We suspect that *T. naivashae* may be conspecific with *T. daemon*, and possibly also with *T. storeyi*. The last taxon falls within the range of *T. naivashae*, albeit on the large side. Finally, this species may also be conspecific with *T. ibeanus* (see the latter species account), in which case the name *ibeanus* would take priority over *naivashae*. We retain this taxon as a full species pending further investigations, but are doubtful of its specific validity.

Distribution

This species has been recorded widely from south-central Kenya on both sides of, and within, the Rift Valley up to the Tanzanian border.

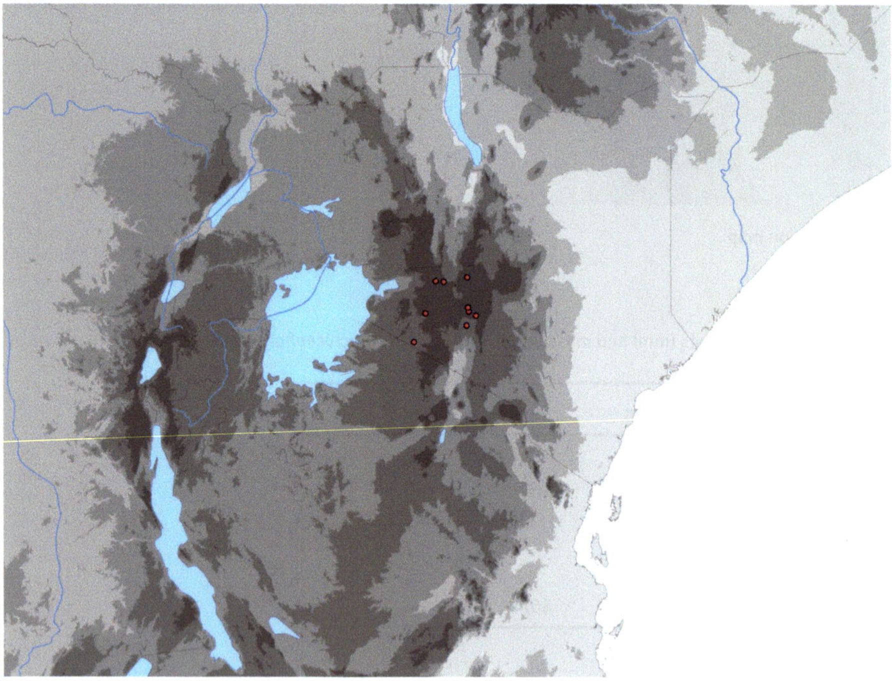

Figure 2.198 Distribution map

Measurements

External and cranial measurements (mm) and mass (g) for *Tachyoryctes naivashae*, sexes combined

	Mean	Min	Max	N
Mass	-	-	-	-
HB*	190.8	187	228	18
Tail*	67.1	53	95	18
HF*	27.7	25	30	18
Ear	-	-	-	-
GLS*	44.3	41.0	46.0	18

*Hollister (1919)

Tachyoryctes rex Heller, 1910

Diagnosis

This species is the largest of the *splendens* group (exceeded only by *T. macrocephalus* from Ethiopia). It can be distinguished by its large size (GLS typically > 50 mm), and is the only *Tachyoryctes* species occurring on Mt Kenya.

Taxonomic Issues

See genus account.

Distribution

This species is restricted to Mt Kenya where it occurs in moorland above the bamboo zone, ca. 2600-3300 m (Hollister, 1919).

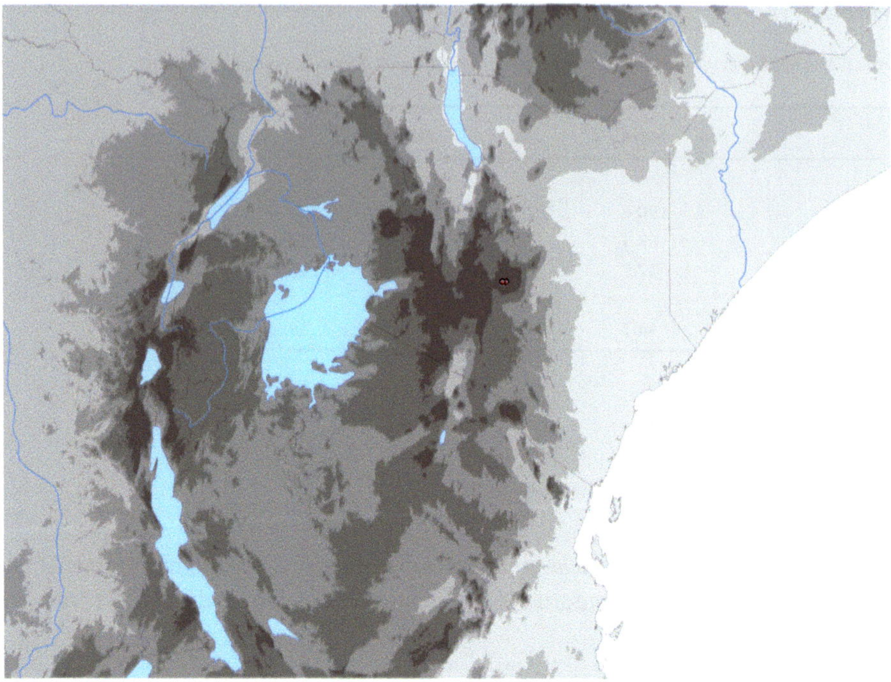

Figure 2.199 Distribution map

Measurements

External and cranial measurements (mm) and mass (g) for *Tachyoryctes rex*, sexes combined

	Mean	Min	Max	N
Mass	-	-	-	-
HB*	237.2	222	268	14
Tail*	66.8	54	80	14
HF*	31.0	29	33	14
Ear	-	-	-	-
GLS*	52.1	47	57	14

*Hollister (1919)

Tachyoryctes ruandae Lonnberg & Gyldenstolpe, 1925

Diagnosis

This species co-occurs with *T. ankoliae* (Delany, 1975) from which it can be distinguished by its larger size (in *T. ankoliae* size on average smaller, HB ca. 200 mm).

Taxonomic Issues

The distinction between *T. ruandae* and *T. ankoliae* is not clear and size is the only criterion by which they can be currently separated, but with significant overlap. Chromosome number 2n = 48, FN = 54, (Musser & Carleton, 2005).

Distribution

This species is restricted to the southern parts of the Albertine Rift.

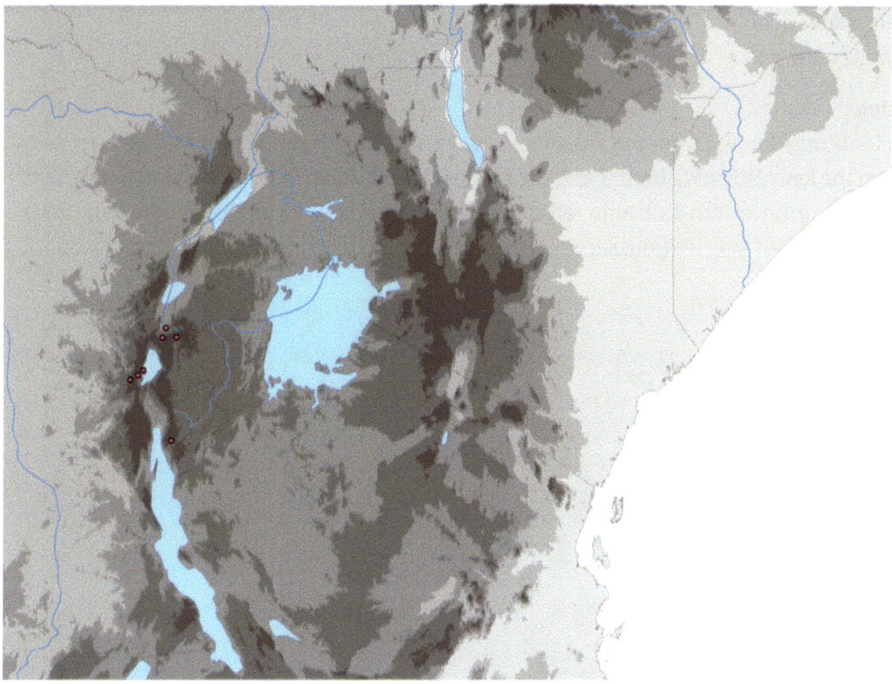

Figure 2.200 Distribution map

Measurements

External and cranial measurements (mm) and mass (g) for *Tachyoryctes ruandae*, sexes combined

	Mean	Min	Max	N
Mass	229.0	205	255	5
HB	215.0	208	222	5
Tail	51.4	49	54	5
HF	31.2	29	33	5
Ear	-	-	-	-
GLS	44.6	43.3	46.5	5

Tachyoryctes ruddi Thomas, 1909

Diagnosis

This species co-occurs with *T. ankoliae* (Lunde & Sarmiento, 2002; Musser & Carleton, 2005) from which it can be distinguished as follows: *T. ankoliae* (size on average larger, HB ca. 200 mm; parietal ridges never join to form a distinct sagital crest). In this species, the parietal ridges join to form a distinct sagital crest (Thomas, 1909).

Taxonomic Issues

See genus account.

Distribution

This species is only known from the Kakamega Forest and neighbouring areas of western Kenya, Mt Elgon on the Kenya-Uganda border and the Kalinzu Forest, Uganda (Lunde & Sarmiento, 2002). A record from north-western Tanzania referred to as *T. daemon* (Swynnerton & Hayman, 1951) is thought to refer to this species (Musser & Carleton, 2005), but has not been plotted on the map.

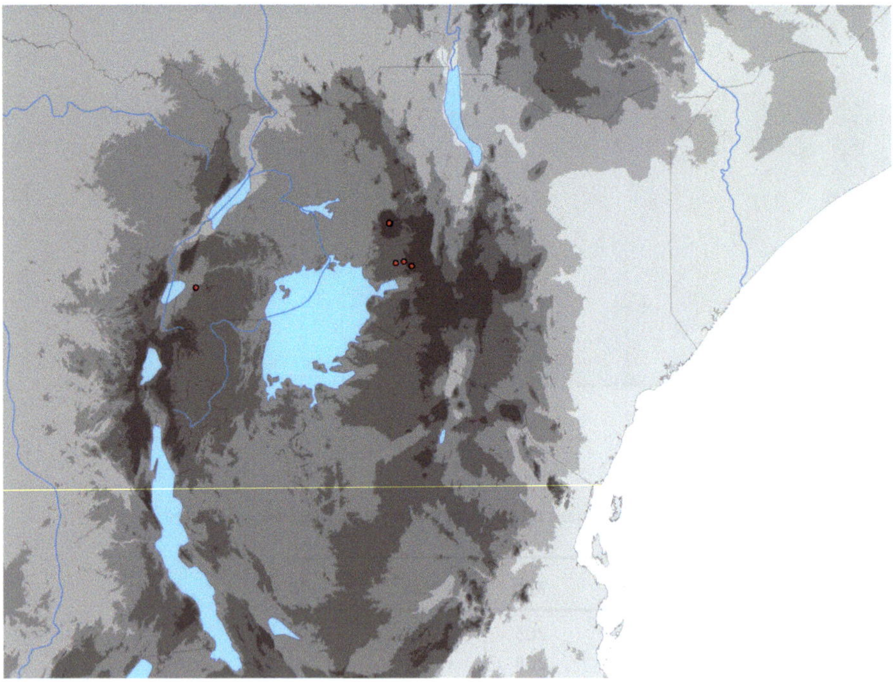

Figure 2.201 Distribution map

Measurements

External and cranial measurements (mm) and mass (g) for *Tachyoryctes ruddi*, sexes combined

	Mean	Min	Max	N
Mass*	215	-	-	1
HB**	184.4	170	190	8
Tail**	70.6	57	78	8
HF**	27.8	26	29	8
Ear	-	-	-	-
GLS**	44.1	41	46	8

*Lunde & Sarmiento (2002); **Hollister (1919)

Tachyoryctes spalacinus Thomas, 1909

Diagnosis

In this species, the latter profile of the posterior margin of the cranium is slanted at an angle (as opposed to being near vertical in other *Tachyoryctes* species) (Thomas, 1909).

Taxonomic Issues

We have not seen the cranium of this species, but the description of Thomas (1909) suggests that this is a diagnostic character for distinguishing *T. spalacinus* from other *Tachyoryctes* species.

Distribution

This species occurs in the plains and foothills south and west of Mt Kenya (and is replaced by *T. rex* on Mt Kenya above 2600 m).

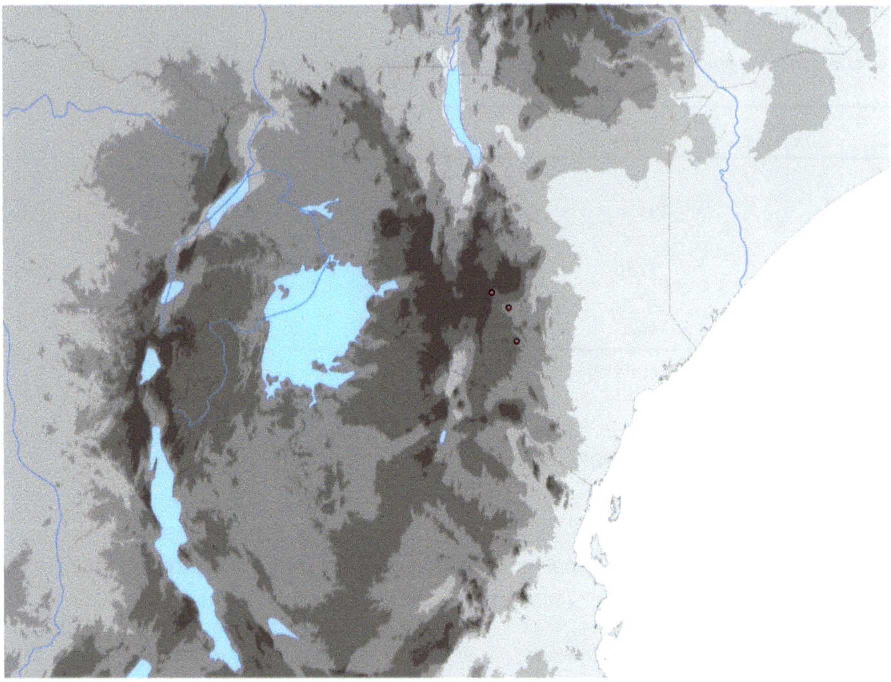

Figure 2.202 Distribution map

Measurements

External and cranial measurements (mm) and mass (g) for *Tachyoryctes spalacinus*, sexes combined

	Mean	Min	Max	N
Mass	-	-	-	-
HB*	228.6	224	235	7
Tail*	65.7	57	80	7
HF*	29.6	28	31	7
Ear	-	-	-	-
GLS*	48.3	46	52	7

*Hollister (1919)

Tachyoryctes splendens (Ruppell, 1835)

Diagnosis

This is one of two *Tachyoryctes* species endemic to Ethiopia; it can be distinguished by its smaller size (in *T. macrocephalus* GLS > 60 mm, HB > 220 mm).

Taxonomic Issues

Chromosome number 2n = 48 (Aniskin et al., 1997a).

Distribution

This species is endemic to Ethiopia where it typically occurs above 2000 m, but occasionally down to 1300 m (Yalden et al., 1976). It is sympatric with *T. macrocephalus* in the Bale Mts.

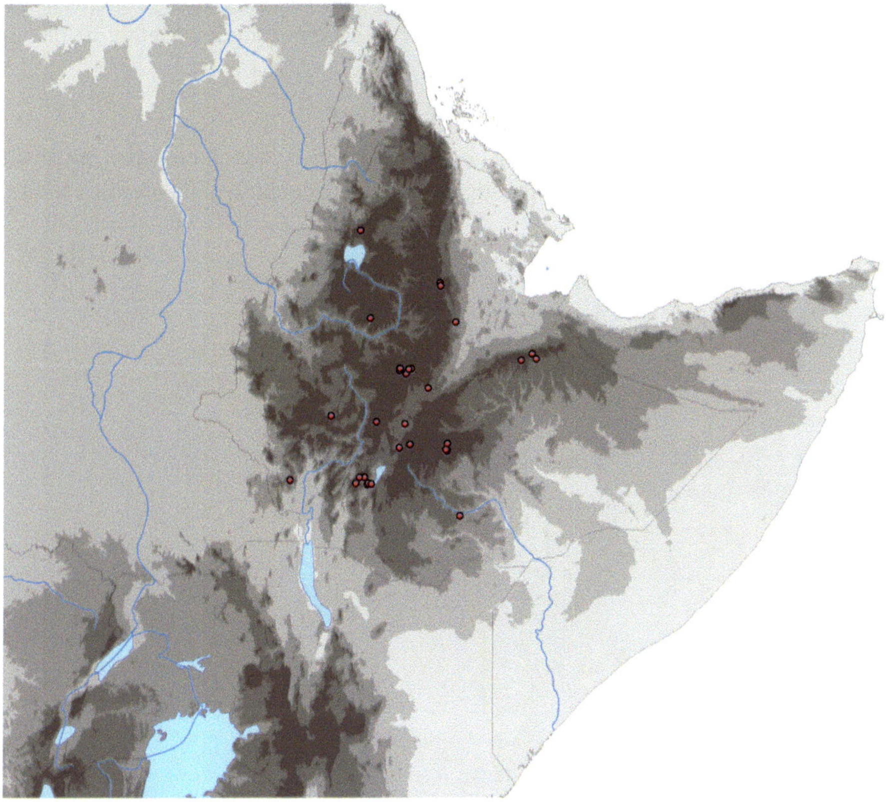

Figure 2.203 Distribution map

Measurements

External and cranial measurements (mm) and mass (g) for *Tachyoryctes splendens*, sexes combined

	Mean	Min	Max	N
Mass	265.7	210	330	7
HB	210.0	190	225	7
Tail	63.3	56	69	6
HF	33.0	31	35	7
Ear	12.9	11	15	7
GLS	50.7	47.2	53.2	7

Tachyoryctes storeyi Thomas, 1909

Diagnosis
This species was described based on its larger skull size, but which actually falls within the range of *T. naivashae* (see the latter species account). These two species cannot currently be distinguished.

Taxonomic Issues
This taxon may be conspecific with *T. naivashae* (see latter species account for details). We retain this taxon as a full species pending further investigations, but are doubtful of its specific validity.

Distribution
This species is only known from the type locality, Lake Elmenteita, Kenya.

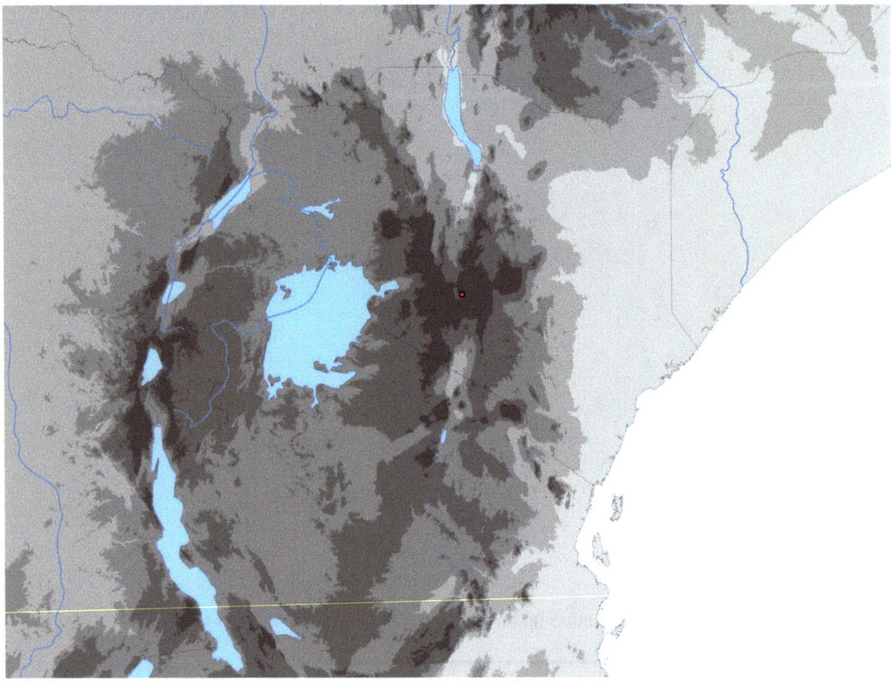

Figure 2.204 Distribution map

Measurements

External and cranial measurements (mm) and mass (g) for *Tachyoryctes storeyi*, sexes combined

	Mean	Min	Max	N
Mass	-	-	-	-
HB	-	-	-	-
Tail	-	-	-	-
HF	-	-	-	-
Ear	-	-	-	-
GLS*	46.5	-	-	1

*Thomas (1909); external measurements were not available

Family: Nesomyidae

This family is endemic to Sub-Saharan Africa and Madagascar. It is represented by five subfamilies and 40 species in Sub-Saharan Africa. The subfamily Nesomyinae is endemic to Madagascar. There are currently no known morphological features that distinguish it from the Muridae, although the two families are clearly distinct on molecular grounds (Jansa & Weksler, 2004).

Key to Subfamilies and Genera

1	Presence of internal cheek pouches (Cricetomyinae)	2
	Internal cheek pouches absent	4
2	Size very large (HB typically > 270 mm)	*Cricetomys*
	Size smaller (HB typically < 180 mm)	3
3	Tail longer (ca. 90% of HB); typically confined to forested habitats	*Beamys*
	Tail far shorter (ca. 45-50% of HB); mostly in savanna habitats	*Saccostomus*
4	Tail ca. twice the length of HB (Delanymyinae)	*Delanymys*
	Tail relatively shorter (160% of HB or shorter)	5
5	Upper incisors not grooved	6
	Upper incisors grooved (except *Prionomys* which is distinguishable from (6) by relatively long tail ca. 130% of HB) (Dendromurinae)	7
6	Tail relatively shorter (ca. 35-40% of HB) (Mystromyinae)	*Mystromys*
	Tail relatively longer (ca. 95-110% of HB) (Petromyscinae)	*Petromyscus*
7	Upper incisors not grooved; Digit 5 of hindfoot not opposable	*Prionomys*
	Upper incisors grooved; Digit 5 of hindfoot opposable	*8*
8	Tail relatively shorter (ca. 60% of HB or shorter)	9
	Tail relatively longer (ca. 80% of HB or longer)	10
9	Hindfoot with 4 digits; prominant dorsal stripes and patches	*Malacothrix*
	Hindfoot with 5 digits; no dorsal stripes or patches	*Steatomys*
10	Size larger (HB ca. >100 mm; GLS > 28 mm)	*Megadendromus*
	Size smaller (HB ca. <90 mm; GLS < 26 mm)	11
11	Digit 1 of hindfoot elongated; no mid-dorsal stripe; known only from type locality (Brazzaville, Congo)	*Dendroprionomys*
	Digit 1 of hindfoot not elongated; mid-dorsal stripe (or stripes) often present; widespread in Sub-Saharan Africa	*Dendromus*

Subfamily: Cricetomyinae

Genus Account

Beamys Thomas, 1909 (Long-tailed Pouched Rat)

Diagnosis

This medium-sized rat (HB ca. 145-155 mm) is characterized by a thick long prehensile bicoloured tail (100% of HB), and well developed cheek pouches. It differs from *Cricetomys* in being distinctly smaller in size and from *Saccostomus* by a relatively longer tail. The dorsal pelage is grey and the belly white. The dentition is highly specialized and shows a typical Cricetomyinae pattern.

Figure 2.205 Photograph of *Beamys hindei* (Bill Stanley).

Cranial Characters

The skull is massively built and the posterior of the skull is rectangular in shape (GLS ca. 35-38 mm). The rostrum is long and robust. The zygomatic arches are thin and parallel-sided. The incisors are orthodont or opisthodont. The anterior palatal foramina are short and end well before the roots of the M^1. The two molar rows are almost parallel, the palate between M^1 being slightly wider than between M^3. The bullae are small (5-6 mm in greatest diameter).

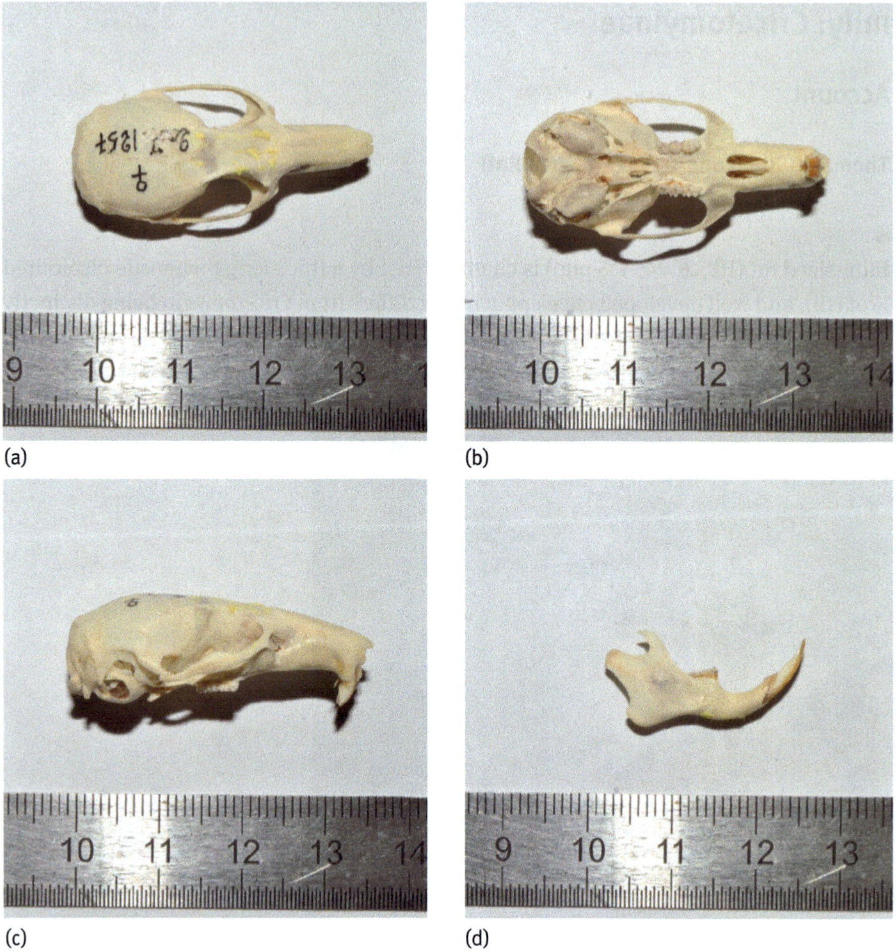

Figure 2.206 a-d Cranium of *Beamys hindei* (MNHN-ZM-MO 2007-1257) (from top left to bottom right: dorsal view, ventral view, lateral view and lower jaw).

Dentition

The cusps on the upper molars are large and typically have three transverse lobes. The first lobe on M^1 comprises two large cusps (t2 and t3), whilst the second and third lobes display a t4 and a t7. Molar M^2, is composed of two lobes with three cusps on each, and has a posterior cingulum. Cusp t3 is present on M^2 and M^3, but cusp t1 is absent. Molar M^3 is the smallest in the toothrow. Molar M_1 is the longest tooth in the lower toothrow and is narrow. It is composed of three lobes, the first two with their 4 cusps fused along their middle by a longitudinal crest and a last lobe with two cusps and a small posterior cingulum. Two conspicuous external supplementary cusps are present on the M_1, and connected to the main cusps. Molar M_2 has a supplementary cusp on the cingulum, two rows of cusps and a posterior cingulum. Molar M_3 has a tiny cingulum and two rows of cusps.

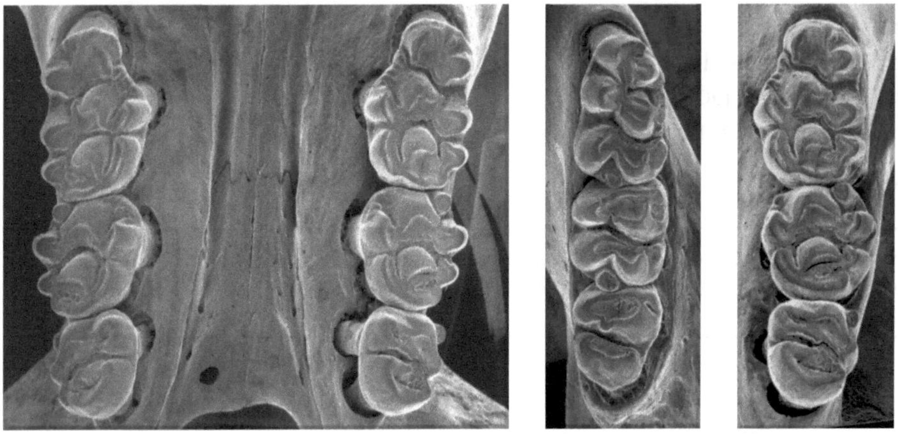

(a) (b) (c)

Figure 2.207 Upper cheek teeth of *Beamys hindei* (TM 16808) (A: upper cheek teeth; B: lower right toothrow; C: upper left toothrow).

Habitat & Biogeography

Beamys hindei inhabits coastal forests (below 2000 m) of East Africa extending from southern Kenya to Tanzania, and possibly into northern Mozambique. By contrast, *B. major* occurs inland to the west of Lake Malawi, extending southward into neighbouring Mozambique.

Fossil Record

None.

Biology & Ecology

This nocturnal solitary rodent is found in low abundance and its biology is poorly known. It is able to climb and stores food collected in its mouth pouches in its burrow, where it rests during the day (Hanney & Morris 1962). Its diet is omnivorous and includes seeds, fruits and insects. The females give birth to 1-5 young after 22-23 days gestation (Egoscue 1972) and appear to be reproductively active during the whole year. It has been recorded in owl pellets (Hanney & Morris 1962), in the scats of an unknown mammalian carnivore (Denys et al., 1999), and in the stomach of a snake (Fitzgibbon et al., 1995).

Systematic Notes

Based on external morphology, and cranio-dental features, *Beamys* was placed with *Cricetomys* and *Saccostomus* into the family Cricetomyidae by Roberts (1951), Petter (1966a), and Ryan (1989). The affinities of the genus were confirmed on molecular grounds, where *Beamys* is sister to *Cricetomys* (Corti et al., 2004a; Jansa & Weksler, 2004) and distant from *Saccostomus* which may belong to a different tribe (Roberts, 1951). Musser & Carleton (2005) listed two species, *B. hindei* and *B. major*, but Happold (2013) recognised these as subspecies of *B. hindei*. These two species have non-overlapping distributions and occupy different habitats (Fitzgibbon et al., 1995). Denys et al. (2011b) reported on new specimens of *B. hindei* from Tanzania with a karyotype distinct (2n = 56, FN = 72) from specimens collected in Kenya (2n = 52) (Fitzgibbon et al., 1995) that could indicate the existence of a third species of the genus. No karyotypes have yet been reported for *B. major*.

Species List

Two species are recognised.
Beamys hindei Thomas, 1909
Beamys major Dollman, 1914

Key to Species

1	Size smaller (typically GLS < 37 mm, HB < 150 mm), occurs in coastal Kenya and Tanzania	*B. hindei*
	Size larger (typically GLS > 37 mm, HB > 140 mm), occurs inland to the west and south of Lake Malawi	*B. major*

Species Accounts

Beamys hindei Thomas, 1909

Diagnosis

This species is similar to *B. major* from which it can be distinguished by its smaller, but overlapping, size (in *B. major*, typically GLS > 37 mm, HB > 140 mm); the distributions of the two species do not overlap.

Taxonomic Issues

The karyotype of animals from Tanzania differs from those in Kenya (see Genus account) and may represent a distinct species. The limits of the two populations are currently not known. However, the Kenyan karyotype was reported from the Arabuko-Sokoke Forest which is well known for its high number of endemic species of birds and mammals. The Usambara populations (northern Tanzania) may group with either the Kenyan or southern Tanzanian ones.

Distribution

This species is restricted to coastal forests in Kenya and Tanzania (and possibly in northern Mozambique).

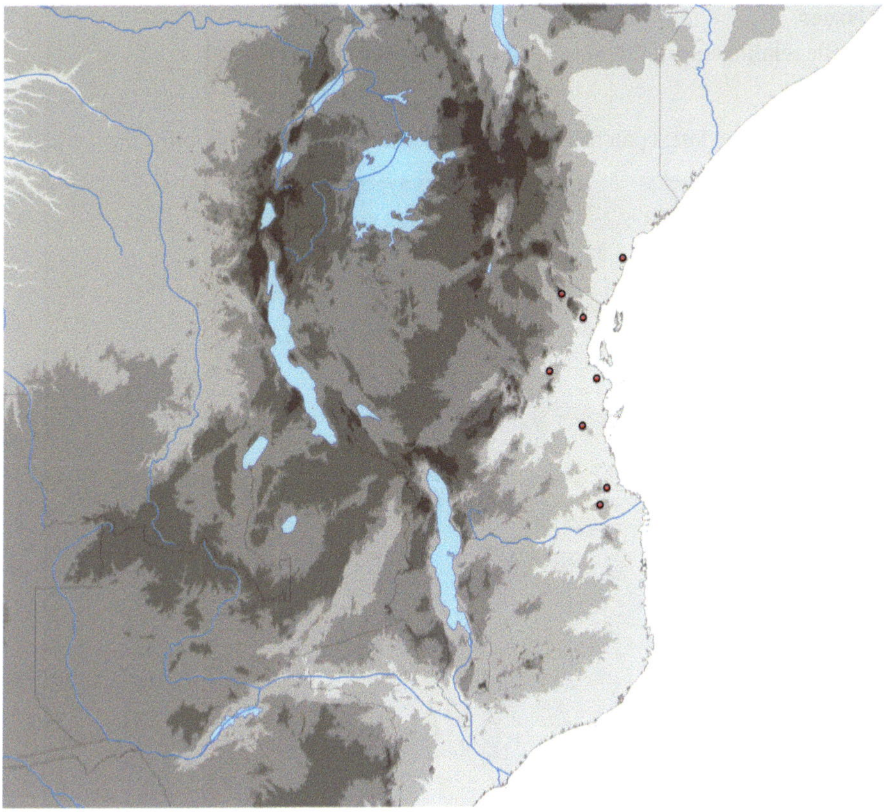

Figure 2.208 Distribution map

Measurements

External and cranial measurements (mm) and mass (g) for *Beamys hindei*, sexes combined

	Mean	Min	Max	N
Mass	60.2	49	76	5
HB	144.1	125	155	7
Tail	114.1	70	127	7
HF	23.1	22	25	7
Ear	22.0	20	23	7
GLS	35.4	34.7	36.7	5

Beamys major Dollman, 1914

Diagnosis

This species is similar to *B. hindei* from which it can be distinguished by its larger, but overlapping, size (in *B. hindei*, typically GLS < 37 mm, HB < 150 mm); the distributions of the two species do not overlap.

Taxonomic Issues

The phylogenetic relationship of this species to *B. hindei* needs to be resolved.

Distribution

This species is restricted to forest patches in Malawi and at Mt Mabu in neighbouring Mozambique.

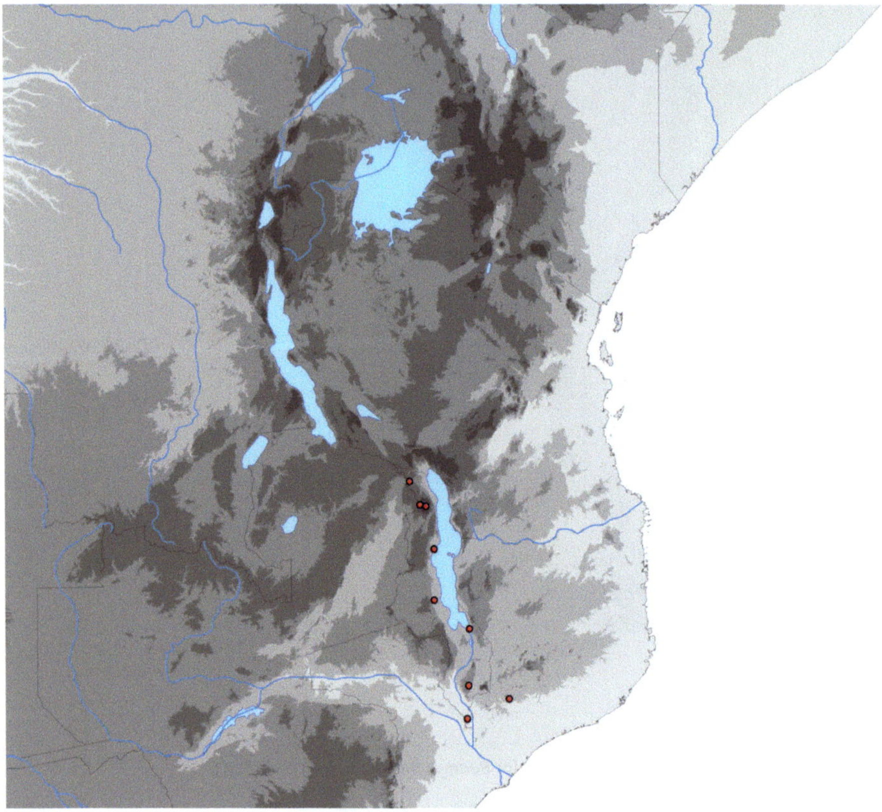

Figure 2.209 Distribution map

Measurements

External and cranial measurements (mm) and mass (g) for *Beamys major*, sexes combined

	Mean	Min	Max	N
Mass	98.5	95	102	2
HB	156.5	136	173	4
Tail	135.0	126	143	4
HF	23.8	21	25	4
Ear	20.0	19	21	4
GLS*	37.7	36.8	38.5	2

*Ansell (1957)

Genus Account

Cricetomys Waterhouse, 1840 (Giant Pouched Rats)

Diagnosis

These very large rodents (HB ca. 320-350 mm) are characterized by a long bi-coloured tail (85-130% of HB), well developed cheek pouches, small feet and large ears. Their large size, long tail and check pouches readily distinguish them from all other African rodents. The narrow skull is massively built and characterized by a large rostrum, small anterior palatal foramina, robust zygomatic arches and small tympanic bullae. The dentition is highly specialized and has a Cricetomyinae pattern with cusps t1, t4, t7 all present on M^1. The dorsal pelage is typically a shade of grey-brown and contrasts with the white ventral pelage. The terminal half of the tail typically is white.

Figure 2.210 Photograph of *Cricetomys ansorgei* (Ara Monadjem).

Cranial Characters

The skull is massive (GLS ca. 65-75mm), and relatively narrow with a large rostrum. The zygomatic arches are thick and oblique. The incisors are opisthodont and ungrooved. The anterior palatal foramina are short and end well before M^1. The interorbital constriction is wide with the fronto-parietal crests more or less developed. The toothrows are parallel. The bullae are relatively small and not inflated (8-10 mm in greatest diameter).

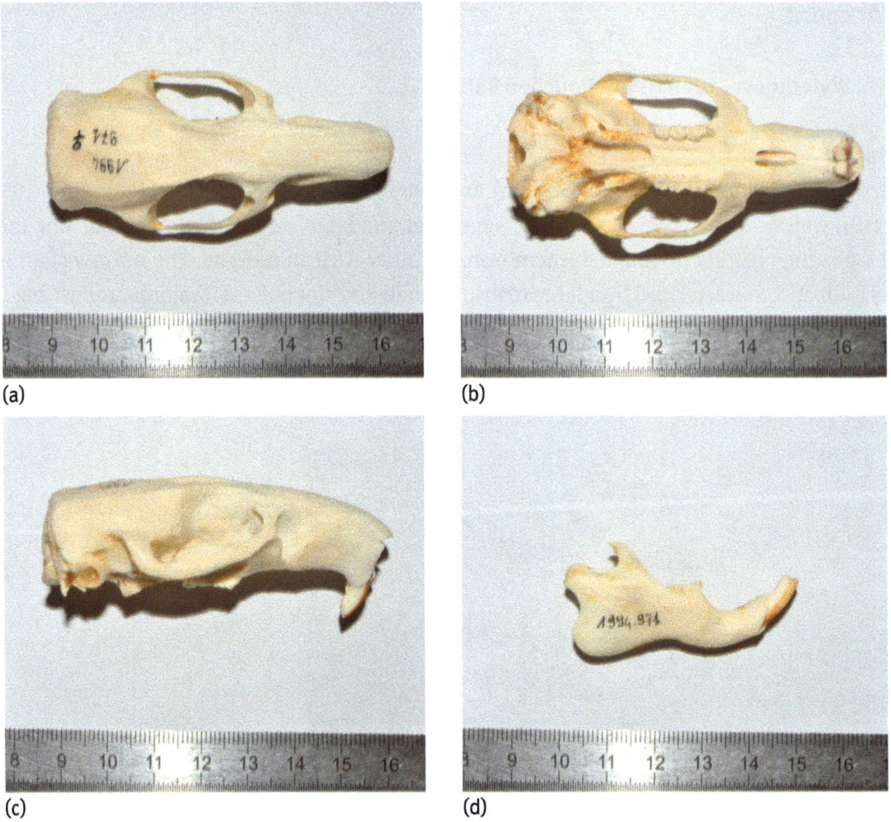

Figure 2.211 a-d Cranium of *Cricetomys gambianus* (MNHN-ZM-MO 1994-971) (from top left to bottom right: dorsal view, ventral view, lateral view and lower jaw).

Dentition

The cusps on the upper molars are high and separate. Molars M^2 and M^3 are similar in size and smaller than M^1. Molar M^1 displays three distorted lobes with different numbers of cusps. The first lobe has two large cusps (t2 and t3) which are separated from the t1 (which is isolated and posterior to the t2). On the second lobe t5 and t6 are large and connected, but separated from the t4 which is posterior to the t5. There is no t7 on the third lobe but there is a large posterior cingulum. Molar M^2 has two lobes, with separated t1 and t4, and a posterior cingulum. Molar M^3 also has two lobes but t1 and t4 are absent. Molar M_1 is the longest tooth in the lower toothrow, and has two separated lobes with cusps that are aligned transversally. There is a small posterior cingulum and two external supplementary cusps. Molars M_2 and M_3 have two rows of cusps and a posterior cingulum.

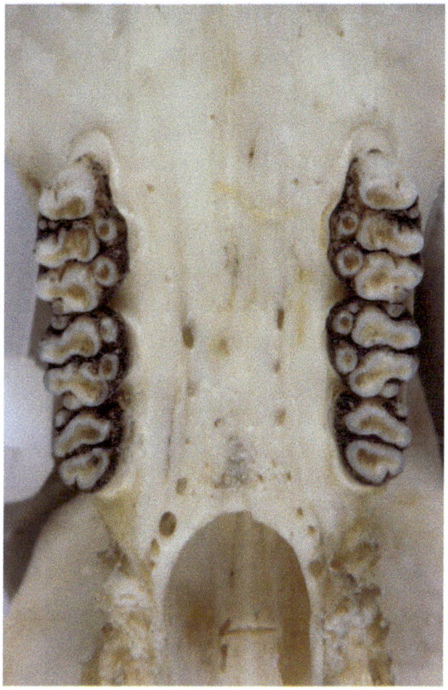

Figure 2.212 Upper cheek teeth of *Cricetomys gambianus* (MNHN-ZM-MO 1934-99).

Habitat & Biogeography

Cricetomys is found in a variety of habitats from lowland rainforest to montane forest up to 2000m as well as in northern and southern savannas. Some species are also closely associated with anthropogenic habitats like cocoa farms, secondary forest, rice fields and in and around homes.

Fossil Record

Fossils are known from the Middle Pleistocene in East Africa (Winkler et al., 2010). Fossils include: *Cricetomys* sp. Recorded from Kapthurin (0.3 Ma) by Denys (1999).

Biology & Ecology

This nocturnal rodent is terrestrial but can also climb. It digs complex burrows or uses old termite mounds or tree roots to rest up in (Ajayi, 1977). This omnivorous vegetarian rodent is a hoarder and transports food in its cheek pouches which it may store in the nest. Palm nuts, seeds, fruit, snails and insects are included in its diet (Rahm & Christiaensen, 1963; Rosevear, 1969; Knight & Eloff, 1987). Females give birth to 1-5 young after a gestation period of 27-42 days (Morris, 1963; Ewer, 1967; Ajayi, 1977). Large nocturnal birds of prey (*Bubo* spp.), eagles as well as mammalian carnivores like mongooses, genets, civets, wild cats are its main predators (De Graaf, 1981; Ray & Sunquist, 2001). Giant rats also serve as bushmeat and are frequently sold in markets (Colyn et al., 1987).

Systematic Notes

The placement of *Cricetomys* within Cricetomyinae was proposed by Roberts (1951) and confirmed by Chaline et al. (1977). Various studies (on dental, morphological and molecular grounds) have confirmed the monophyly of the subfamily and the inclusion of *Cricetomys* closer to *Beamys* than to *Saccostomus*. The taxonomy of the genus is under revision and the number of species is not yet clarified. Ellerman (1941) recognised a single species, but Genest-Villard (1967) separated *C. emini* and *C. gambianus*. This classification was retained by Happold (2013), but Musser & Carleton (2005) elevated *C. ansorgei* and *C. kivuensis* based on morphological grounds. A recent molecular analysis (Olayemi et al., 2012) has confirmed the specific validity of *C. ansorgei, C. emini,* and *C. gambianus* but not of *C. kivuensis*. Moreover the molecular clades indicate the existence of three new species yet to be described, one from the forest region of West Africa and two from the forest regions of Central Africa (left and right bank of Congo River, respectively). Morphological and molecular characters are not closely associated, but rather morphology appears to be linked with habitat, with clear separation between forest and savanna forms (Olayemi et al., 2012). Hence, morphological characters may inappropriate for identification of systematic relationships within *Cricetomys*. This genus is in need of a thorough revision.

Species List

4 species are currently recognized.
Cricetomys ansorgei Thomas, 1904
Cricetomys emini Wroughton, 1910
Cricetomys gambianus Waterhouse, 1840
Cricetomys kivuensis Lonnberg, 1917

Key to Species

At present the various species of *Cricetomys* cannot be confidently distinguished based on morphological grounds, and the distributional limits of the various species are not known. However, the four species recognised here are mostly allopatric and distribution may be used for now to differentiate them until the genus has been fully revised and the species properly diagnosed.

1	Restricted to the Albertine Rift	*C. kivuensis*
	Occurring elsewhere in Africa	2
2	Occurring in savanna habitats of southern and eastern Africa from South Africa north to Angola and southern DRC and east to Kenya	*C. ansorgei*
	Not occurring in the southern African region as defined above	3
3	Occurring in savanna habitats of northern Africa from Guinea to Central African Republic and probably Uganda	*C. gambianus*
	Occurring in rainforest habitat from Guinea and Sierra Leone in the west to Uganda in the east	*C. emini**

**Cricetomys emini* as defined here consists of four separate and allopatric species (see species account)

Species Accounts

Cricetomys ansorgei Thomas, 1904

Diagnosis

This species cannot, at present, be distinguished from other *Cricetomys* species on morphological grounds, but is distinct on molecular grounds (Olayemi et al., 2012). On average, it is the largest species of *Cricetomys*, with GLS significantly larger than that of *C. gambianus* (GLS typically < 70 mm). The forest forms (lumped under *C. emini* here) have softer, shorter fur with dark (but bright) brown dorsal pelage which is crisply delineated from the pure white ventral pelage; in *C. ansorgei*, the dorsal pelage is greyish, rather coarse with longer hairs and without clear delineation of flanks from ventral pelage.

Taxonomic Issues

This taxon has recently been elevated to full species based on morphological (Musser & Carleton, 2005) and molecular (Olayemi et al., 2012) analysis.

Distribution

This species occurs in the savannas of Southern and East Africa, extending into the Congo Basin where the forest has been cleared. The northern limits of this species are unknown, particularly in East Africa where its relationship to *C. gambianus* is not currently known.

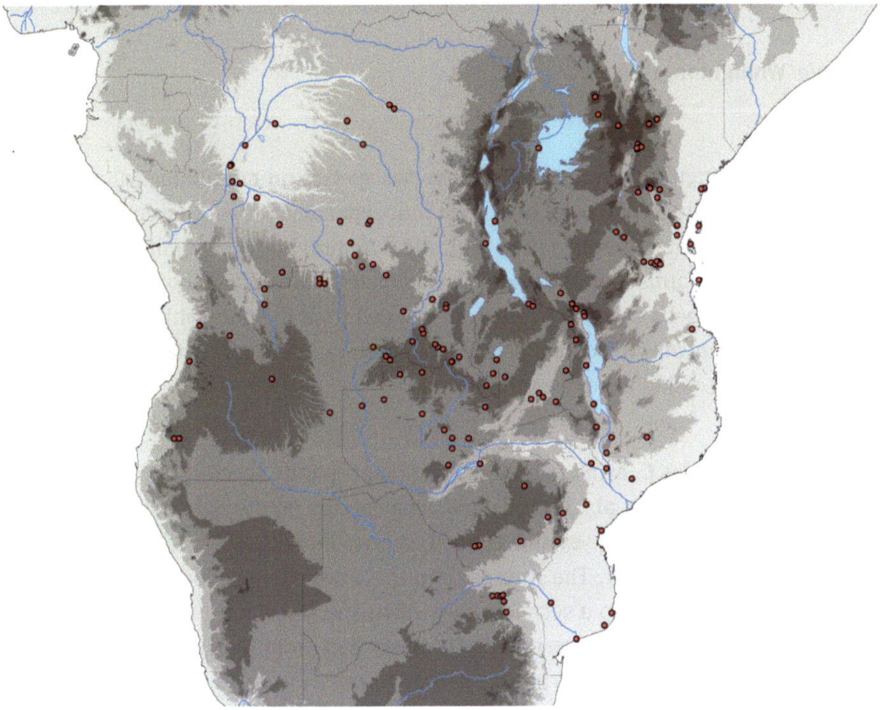

Figure 2.213 Distribution map

Measurements

External and cranial measurements (mm) and mass (g) for *Cricetomys ansorgei*, sexes combined

	Mean	Min	Max	N
Mass*	1240.0	960	2800	50
HB	346.5	288	413	18
Tail	402.2	345	449	17
HF	70.9	68	78	17
Ear	39.8	32	46	17
GLS	75.1	68.4	81.5	21

*Smithers & Wilson (1979)

Cricetomys emini Wroughton, 1910

Diagnosis

This species can be distinguished from other *Cricetomys* species on morphological and molecular grounds, but is actually a composite of at least four separate species (Olayemi et al., 2012). On average, it is smaller than *C. ansorgei* (GLS typically > 70 mm) but with significant overlap in measurements. The forest forms (lumped under *C. emini* here) have softer, shorter fur with dark (but bright) brown dorsal pelage which is crisply delineated from the pure white ventral pelage. In *C. ansorgei* and *C. gambianus* the dorsal pelage is greyish, rather coarse with longer hairs and without clear delineation of flanks from ventral pelage.

Taxonomic Issues

The records presented here as *C. emini* refer to at least four separate species, which do not form a monophyletic group (Olayemi et al., 2012). True *C. emini* occurs on the right bank of the Congo River in the DRC. A separate, and as yet undescribed, species more closely related to *C. ansorgei* occurs on the left bank of the river. The records from the Lower Guinea Forest (Cameroon and Gabon), extending into DRC refer to a second undescribed species. A third undescribed species occurs in the Upper Guinea rainforest west of the Dahomey Gap (from Togo westward to Guinea), but also extends eastward across the gap into western Nigeria.

Distribution

This species occurs widely in the tropical rainforest belt from Guinea in the west to the eastern shores of Lake Victoria in the east. But see Taxonomic Issues (above).

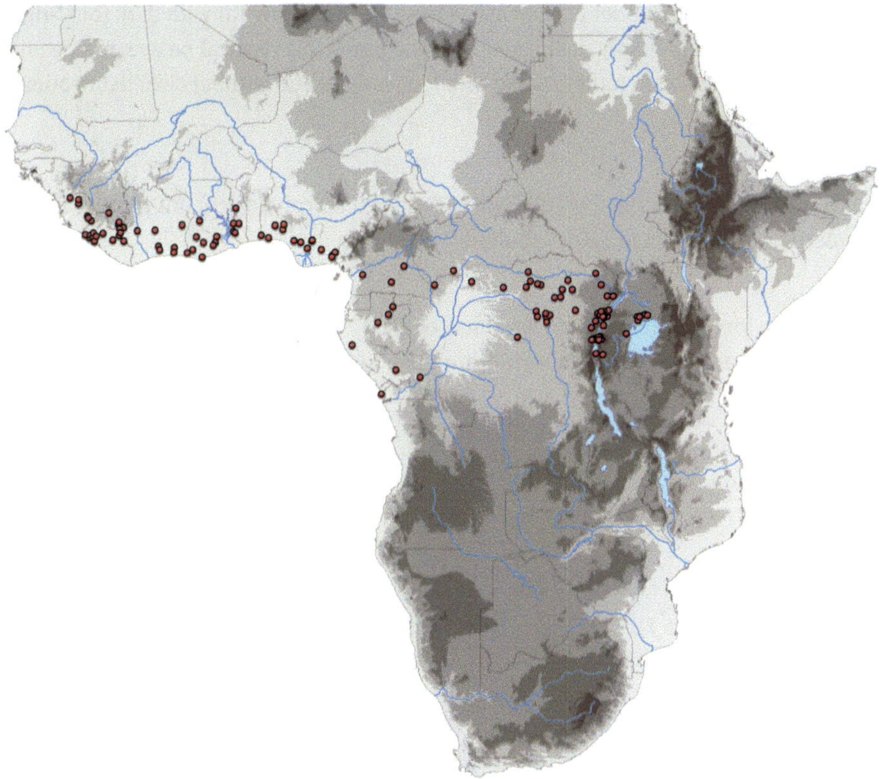

Figure 2.214 Distribution map

Measurements

External and cranial measurements (mm) and mass (g) for *Cricetomys emini*, sexes combined. These measurements reflect the means of several different, but as yet undescribed, species (see Taxonomic Issues).

	Mean	Min	Max	N
Mass	1000	-	-	1
HB	328.5	300	355	10
Tail	381.6	320	429	9
HF	67.7	64	71	10
Ear	38.1	33	45	10
GLS	70.3	59.5	77.2	15

Cricetomys gambianus Waterhouse, 1840

Diagnosis

This species cannot, at present, be distinguished from other *Cricetomys* species on morphological grounds, but is distinct on molecular grounds (Olayemi et al., 2012). On average, it is a small species of *Cricetomys*, with GLS significantly smaller than that of *C. ansorgei* (GLS typically > 70 mm). The forest forms (lumped under *C. emini* here) have softer, shorter fur with dark (but bright) brown dorsal pelage which is crisply delineated from the pure white ventral pelage; in *C. gambianus*, the dorsal pelage is greyish, rather coarse with longer hairs and without clear delineation of flanks from ventral pelage.

Taxonomic Issues

This taxon has recently been elevated to full species based on morphological (Musser & Carleton, 2005) and molecular (Olayemi et al., 2012) analysis.

Distribution

This species occurs widely in the northern savannas of West and Central Africa. Its eastern limits are currently unknown and the records from northern Uganda may refer to *C. gambianus*.

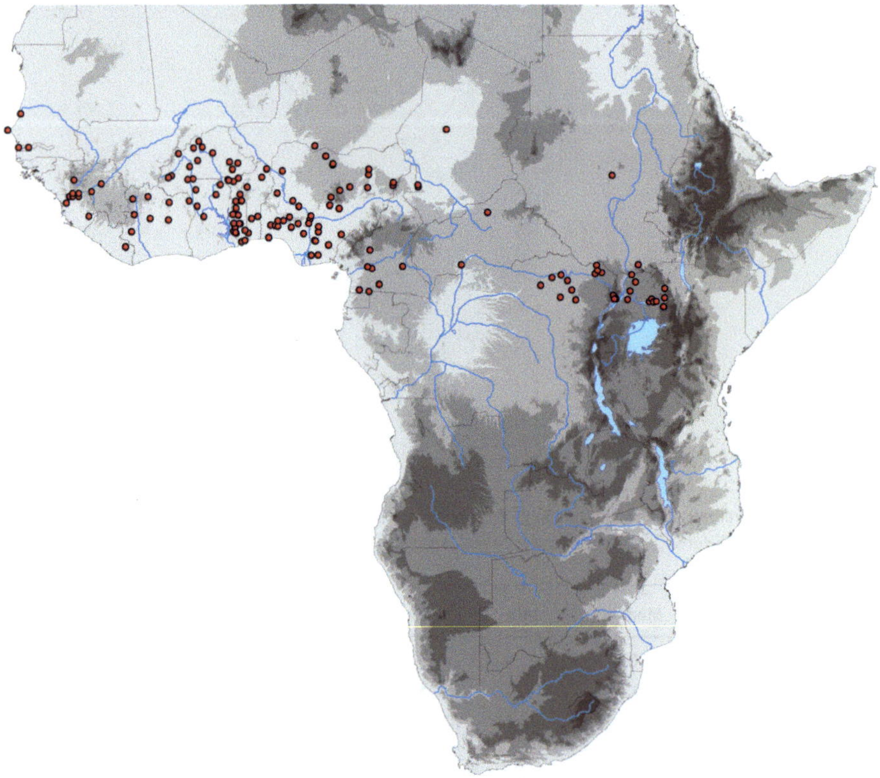

Figure 2.215 Distribution map

Measurements

External and cranial measurements (mm) and mass (g) for *Cricetomys gambianus*, sexes combined

	Mean	Min	Max	N
Mass	998.6	550	1465	7
HB	332.9	284	386	7
Tail	336.8	247	405	6
HF	65.3	62	73	7
Ear	37.7	35	40	7
GLS	66.4	57.6	72.5	10

Cricetomys kivuensis Lonnberg, 1917

Diagnosis

This species cannot, at present, be distinguished from other *Cricetomys* species on either morphological or molecular grounds (Olayemi et al., 2012). The descriptions of this species (e.g. Delany, 1975; Musser & Carleton, 2005) suggest a closer association with the savanna forms in having coarser, longer fur with greyish-brown dorsal pelage which gradually fades on the flanks to the greyish-white ventral pelage; in *C. emini*, the dorsal pelage is dark brown, soft to the touch and shows clear delineation of flanks from ventral pelage. The only identifying feature at present seems to be its distribution which is restricted to the Albertine Rift.

Taxonomic Issues

This taxon has recently been elevated to full species based on morphological analysis (Musser & Carleton, 2005), but not yet on molecular grounds (Olayemi et al., 2012). On morphological grounds, it has been suggested that this species is more similar to *C. ansorgei*/*C. gambianus* than to *C. emini* (Musser & Carleton, 2005) suggesting a closer link with the savanna forms. Furthermore, Delany (1975) who included this taxon under *C. emini* (rather than in *C. gambianus*) also noted that it shared the characters of *C. gambianus* with respect to lack of a clear demarcation between the flanks and ventral pelage. We retain this species for now, but note that it has yet to be properly diagnosed.

Distribution

This species is restricted to the Albertine Rift where it occurs in montane habitats.

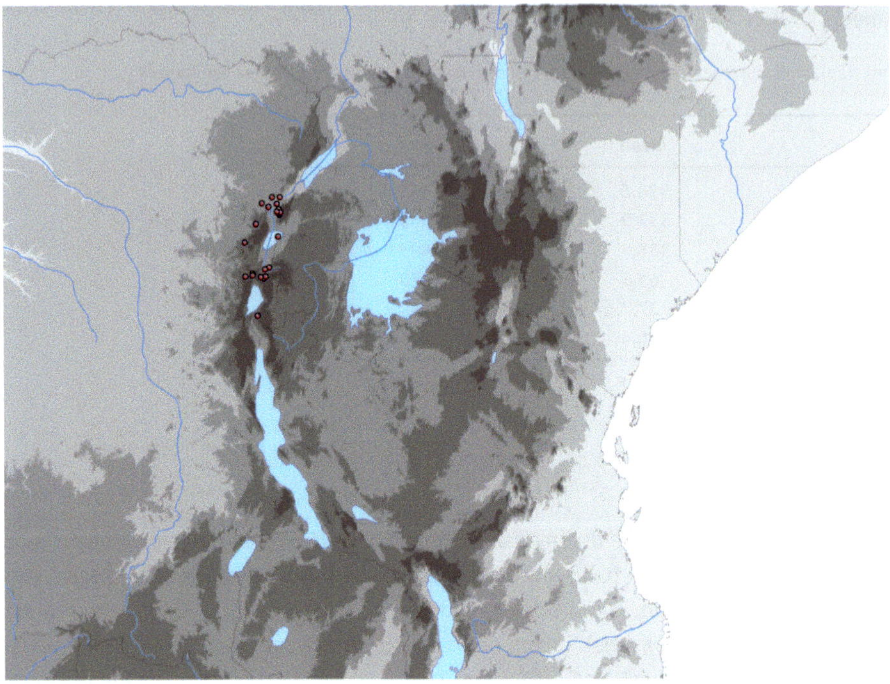

Figure 2.216 Distribution map

Measurements

External and cranial measurements (mm) and mass (g) for *Cricetomys kivuensis*, sexes combined

	Mean	Min	Max	N
Mass	-	-	-	-
HB	320	-	-	1
Tail	342	-	-	1
HF	69	-	-	1
Ear	45	-	-	1
GLS	69.0	-	-	1

Genus Account

Saccostomus Peters, 1846 (Pouched Mice)

Diagnosis

This small to medium sized rat (HB ca. 130-135 mm) is characterized by a short tail (50% of HB), well developed cheek pouches, small feet and large ears. Its dentition is very specialized with a Cricetomyinae pattern. *Saccostomus* is adapted to the woodlands and grasslands of East and Southern Africa.

Figure 2.217 Photograph of *Saccostomus campestris* (Emmanuel Do Linh San).

Cranial Characters

The skull is relatively long (GLS ca. 32 mm) and narrow, with a short rostrum. The zygomatic arches are thick and parallel-sided. The incisors are opisthodont and ungrooved. The anterior palatal foramina are long but do not extend beyond M^1. At the posterior of the skull, the mastoid is inflated and the occipital bone oblique, the condylo-articular surface wide and inflated. The bullae are relatively inflated (4.4-8.6 mm in greatest diameter). The upper molar rows are broadly parallel.

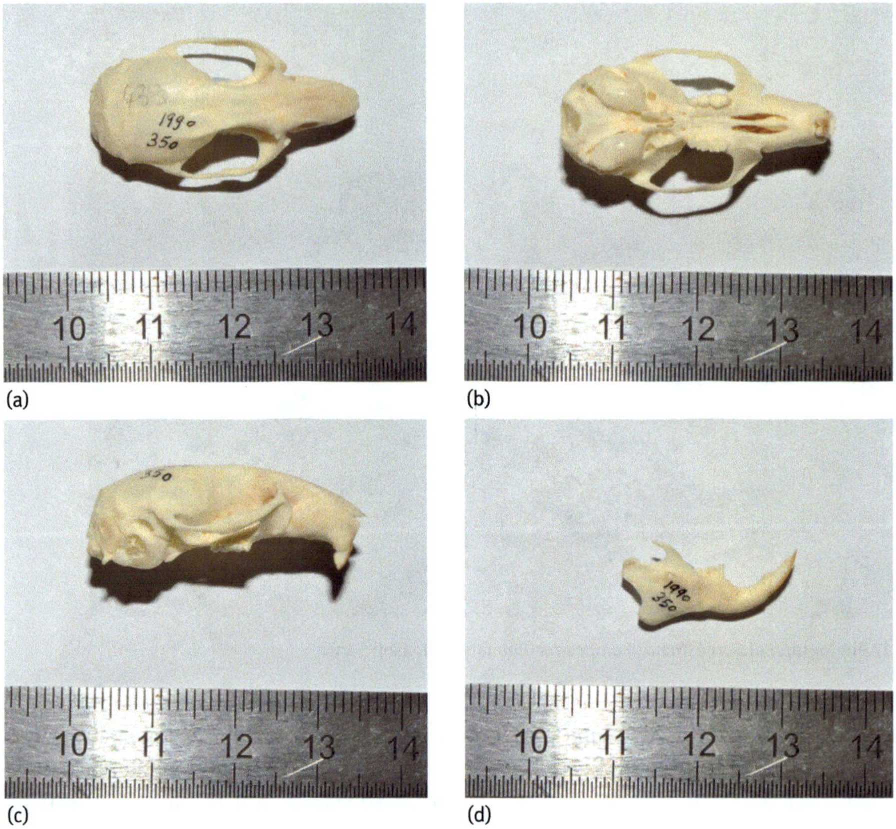

Figure 2.218 a-d Cranium of *Saccostomus campestris* (MNHN-ZM-MO 1990-350) (from top left to bottom right: dorsal view, ventral view, lateral view and lower jaw).

Dentition

Molar M^1 has three lobes with separated cusps. The first lobe has two cusps, the second and third both have three cusps with a relatively well-developed posterior cingulum on the latter lobe. Molar M^2 has two lobes with a posterior cingulum. Molar M^3 is the smallest molar in the toothrow and also has two lobes. Molar M_1 is the longest in the lower toothrow, is narrow and has three lobes. The first has a single cusp which may be connected to the second lobe. The remaining two lobes each have two fused cusps. Molar M_2 has an antero-external cusp, two rows of cusps and a posterior cingulum. Molar M_3 is reduced with a tiny antero-external cingulum and two cusps.

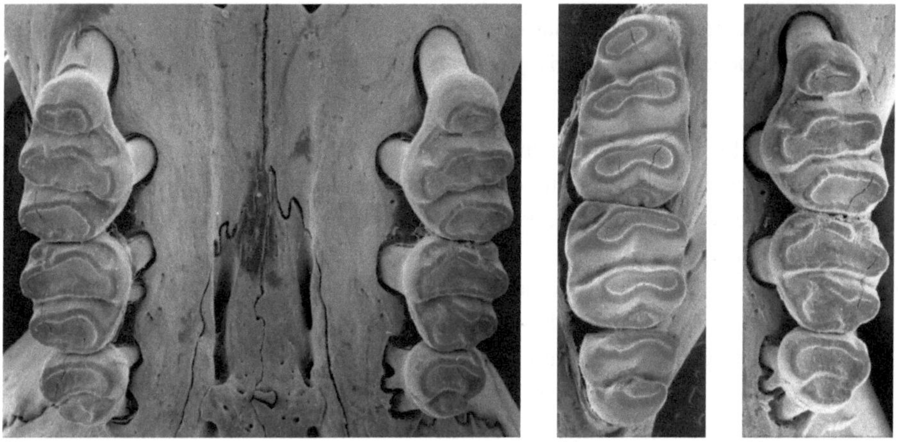

(a) (b) (c)

Figure 2.219 Cheek teeth of *Saccostomus campestris* (MNHN-ZM-MO-1969-56) (A: upper cheek teeth; B: lower left toothrow; C: upper left toothrow).

Habitat & Biogeography

Saccostomus inhabits lowland savannas from Ethiopia to South Africa, not occurring above 1500 m. It typically occurs in a variety of habitats including thickets, acacia and mopane woodlands, dry river beds, sandy habitats and kopjes.

Fossil Record

This genus is first recorded at around 11-9.5 Ma in the Miocene deposits of Chorora (Ethiopia) and Harasib 3a (Namibia). A number of different fossil species have been described in the Plio-Pleistocene.

Saccostomus geraadsi Mein et al. (2004) Harasib 3a, Ch'orora.

Saccostomus cf *geraadsi* Lukeino, Kenya, 6-5 Ma (Mein et al., 2004).

Saccostomus major (Laetolil beds, Laetoli, 3,7 Ma, Tanzania, Denys, 1987a; Ibole, Manonga Valley, Tanzania 4.3 Ma; Winkler, 1997).

Saccostomus cf *mearnsi* (Peninj, Olduvai Bed I, East Turkana Tanzania, 1.7 -1.1 Ma, Denys, 1992).

Saccostomus sp. (Isenya, Olduvai Bed IV, Kenya, Tanzania, Plio-Pleistocene) (Brugal & Denys, 1989; Denys, 1992).

Saccostomus sp. Makapansgat Cave of Hearths (Pleistocene) (De Graaff, 1960).

Saccostomus campestris Gladysvale, Kabwe (Avery, 1995).

Biology & Ecology

This nocturnal ground-dwelling rodent digs complex burrows where it stores food or uses old termite mounds or burrows made by springhares or aardwarks. They are solitary and territorial (Korn, 1987; Keesing & Crawford, 2001). The diet of this genus is mainly composed of seeds that are gathered in the pouches and then deposited in caches. However, termites and other insects and vegetable matter are also consumed in variable proportions (Neal, 1984; Keesing, 1998; Metz & Keesing, 2001). Females give birth to 1-10 young after 20-21 days gestation, mostly during the warm wet season (Hanney, 1965; Smithers & Wilson, 1979; Happold, 2013). Owls such as *Tyto alba, T.*

capensis and *Bubo* spp. are the main predators of this species (De Graaf, 1981), and their remains have been found in *Genetta* scat in Malawi (Denys et al., 1999).

Systematic Notes

Roberts (1951) was the first to provide a diagnosis for the Cricetomyinae and to align *Saccostomus* with *Cricetomys* and *Beamys*. This was later confirmed by molar morphology (Petter, 1966a) and anatomy of internal cheek pouches (Ryan, 1989) as well as molecular analyses (Jansa et al., 1999; Michaux et al., 2001; Michaux & Catzeflis, 2000). *Saccostomus* diverged from the *Beamys-Cricetomys* group approximately 13.8 million years ago (Corti et al., 2004a). There are at least 13 described forms of this genus which were combined into a single species *S. campestris* by Ellerman (1941). This view was followed by authorities until Hubert (1978b) who rehabilitated *S. mearnsi* Heller, 1910 as a valid species based on a karyological study of Ethiopian specimens. Gordon (1986) and Gordon & Rautenbach (1980) reported high karyological variablity for the South African *S. campestris*. Denys (1988) confirmed skull and teeth differences between the two species as well as the high morphological variability in South African specimens. A karyological and molecular study confirmed that the two taxa are distinct species (Corti et al., 2005; Fadda et al., 2001a), having diverged about 7.9 million years ago (Corti et al., 2004a).

Species List

Two species are currently recognized.
Saccostomus campestris Peters, 1846
Saccostomus mearnsi Heller, 1910

Key to Genus

1	Ventral pelage white; upper molar row shorter (< 5 mm); occurs in Southern Africa (extending into the extreme south-west of Tanzania in the Lake Rukwa region)	*S. campestris*
	Ventral pelage gray; upper molar row longer (> 5 mm); occurs in East Africa	*S. mearnsi*

Species Accounts

Saccostomus campestris Peters, 1846

Diagnosis

This species differs from *S. mearnsi* by its white ventral pelage, relatively shorter tail, shorter upper molar row (in *S. mearnsi* upper molar row > 5 mm, Hubert, 1978b), different karyotype (see Corti et al., 2004a, 2005), and non-overlapping distribution.

Taxonomic Issues

This species may represent a species complex with one or more cryptic species having been overlooked (Corti et al., 2004a; Fadda et al., 2001a). There is extreme karyotypic variation in Southern Africa with diploid number ranging from 2n = 28 up to 2n = 50 (Gordon & Rautenbach, 1980; Corti et al., 2005). However, how this variability is related to taxonomy and systematics of the group has not yet been established and it has been suggested that speciation within this group has not yet resulted in distinct species (Corti et al., 2005).

Distribution

This species occurs widely in the savannas of Southern Africa.

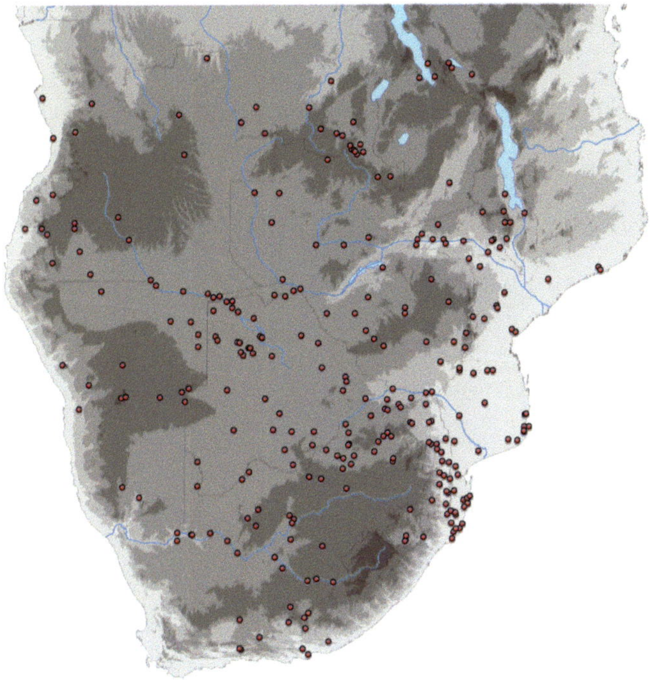

Figure 2.220 Distribution map

Measurements

External and cranial measurements (mm) and mass (g) for *Saccostomus campestris*, sexes combined

	Mean	Min	Max	N
Mass	50.3	26	81	29
HB	135.4	86	173	39
Tail	40.4	28	55	28
HF	18.9	12	22	30
Ear	16.7	13	20	30
GLS	31.6	28.2	34.5	14

Saccostomus mearnsi Heller, 1910

Diagnosis

This species differs from *S. campestris* by its grey ventral pelage, relatively longer tail, longer upper molar row (in *S. campestris* upper molar row < 5 mm, Hubert, 1978b), different karyotype (see Corti et al., 2004a, 2005), and non-overlapping distribution.

Taxonomic Issues

There is some karyotypic variation in East Africa with the following diploid numbers having been reported: 2n = 40-42, Somali (Hubert, 1978b) and 2n = 32, Tanzania (Fadda et al., 2001a). The taxonomic implications of this variability are not yet known.

Distribution

This species has a restricted distribution in the savannas of East Africa.

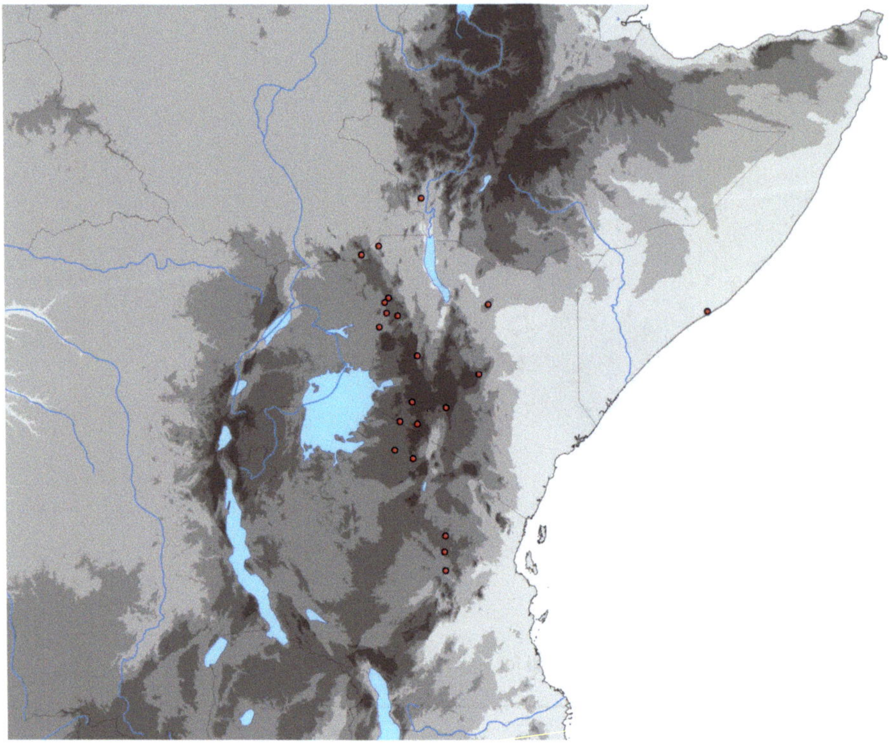

Figure 2.221 Distribution map

Measurements

External and cranial measurements (mm) and mass (g) for *Saccostomus mearnsi*, sexes combined

	Mean	Min	Max	N
Mass	57.3	43	69	3
HB	129.7	101	150	15
Tail	51.5	38	74	15
HF	21.0	18	25	18
Ear	18.7	14	22	14
GLS	32.0	27.3	34.9	9

Subfamily: Delanymyinae

Genus Account

Delanymys Hayman, 1962 (Delany's Swamp Mouse)

Diagnosis

This very small rodent (ca. HB 55-60 mm) is characterized by a very long prehensile tail (180% of HB) and long hindfeet. The pelage is long, soft and dense and brown dorsally, paler ventrally. The skull is characterized by a very short rostrum, a narrow interorbital constriction, and a rounded braincase without crests. The anterior palatal foramina are large and long reaching the first root of M^1. The foramen magnum is wide and situated ventrally. The dentition is specialized and has a unique pattern. *Delanymys* is restricted to swamps, bamboo and montane forests of south-western Uganda, Rwanda and Burundi.

Figure 2.222 Photograph of *Delanymys brooksi* museum specimen (FHNH 157491) (Julian Peterhans Kerbis).

Cranial Characters

The skull is small in size (GLS ca. 17–19 mm) and very compact with a short rostrum and frontal region. The interorbital constriction is narrow. The braincase is rounded and high especially in the parietal region. The foramen magnum is situated ventrally. The incisors are orthodont or slightly opisthodont and ungrooved. The anterior palatal foramina are long and wide and end at the first root of M^1. The toothrows are parallel. The bullae are small (3.5-3.7 mm). The mandible is short and has a low coronoid process.

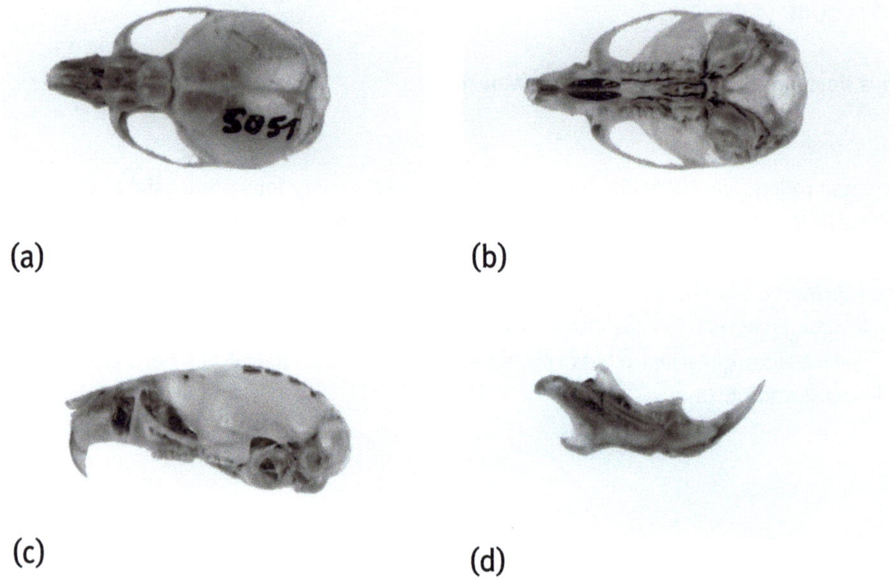

Figure 2.223 a-d Cranium of *Delanymys brooksi* (RMCA RG 5051) (from top left to bottom right: dorsal view, ventral view, lateral view and lower jaw).

Dentition

The cusps on the upper molars are high and round, situated in an alternating arrangement with a trace of a longitudinal crest connecting them. Molar M^3 is the smallest in the toothrow. Molar M^1 has three lobes. The first lobe has a single cusp connected to the second lobe by a small medium longitudinal crest. The second lobe has two large well separated cusps and a t4. The third lobe has two large cusps and is connected to the second lobe by a longitudinal median crest. Molar M^2 has two lobes with cusp t3, but without t1 and t4. Molar M^3 has two rows of cusps connected by a lingual crest and a tiny t3; t1 is absent. Molar M_1 is the longest tooth in the lower toothrow with two lobes. The first lobe has two oblique cusps with a trace of a longitudinal connection to the second lobe. The second and third lobes are also oblique. Molar M_2 has an antero-external cingulum, two rows of cusps and a posterior cingulum. The reduced M_3 has a tiny antero-external cingulum and shows two rows of oblique cusps connected by a longitudinal median crest.

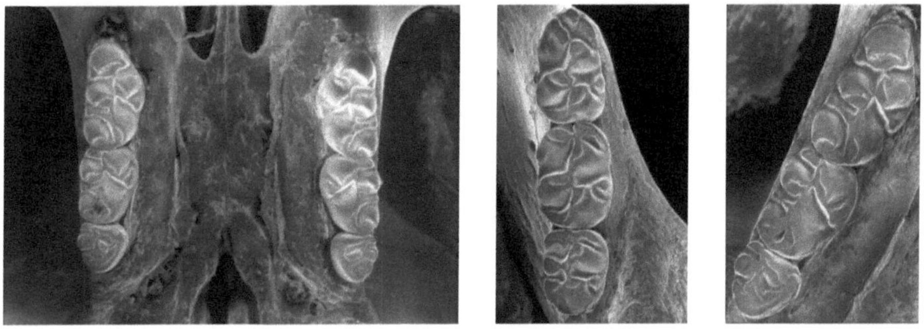

(a) (b) (c)

Figure 2.224 Cheek teeth of *Delanymys brooksi* (RMCA RG 5051) (A: upper cheek teeth; B: lower right toothrow; C: upper right toothrow).

Habitat & Biogeography

Delanymys inhabits marshes and swamps within montane and bamboo forests or grasslands at 1700-3100 m in the Lake Kivu region of eastern DRC, western Rwanda and south-western Uganda. It is generally found in swampy habitat in high altitude grasslands.

Fossil Record

None.

Biology & Ecology

This nocturnal and arboreal rodent is a good climber and eats grass and seeds (Dieterlen, 1969). This species is rare and no studies on its social and reproductive behaviour have been published. Owls and other small carnivores may be predators.

Systematic Notes

This genus was initially classified as a Dendromurinae by Hayman (1962), but was subsequently moved to Petromyscinae in Cricetidae based on dental features (Petter, 1967). A revision of the Dendromurinae highlighted the morphological differences of *Delanymys* (Denys et al., 1995). The description of an extinct fossil genus from South Africa *Stenodontomys* shows similarities with *Delanymys*, confirming differences with *Petromyscus* and *Mystromys* (Denys, 1994), resulting in the erection of a new subfamily Delanymyinae for these two genera (Musser & Carleton, 2005). Schenk et al. (2013) recently demonstrated on molecular grounds that *Delanymys* is placed closest to a clade grouping *Petromyscus* and *Mystromys* at the base of the Nesomyidae radiation.

Species List

This genus is monotypic.
Delanymys brooksi Hayman, 1962

Species Account

Delanymys brooksi Hayman, 1962

Diagnosis
See genus account.

Taxonomic Issues
See genus account.

Distribution
This species is only known from the Lake Kivu region of eastern DRC, western Rwanda and south-western Uganda where it has been collected from swamps in montane forests and grasslands (Delany, 1975; Van der Straeten & Verheyen, 1983).

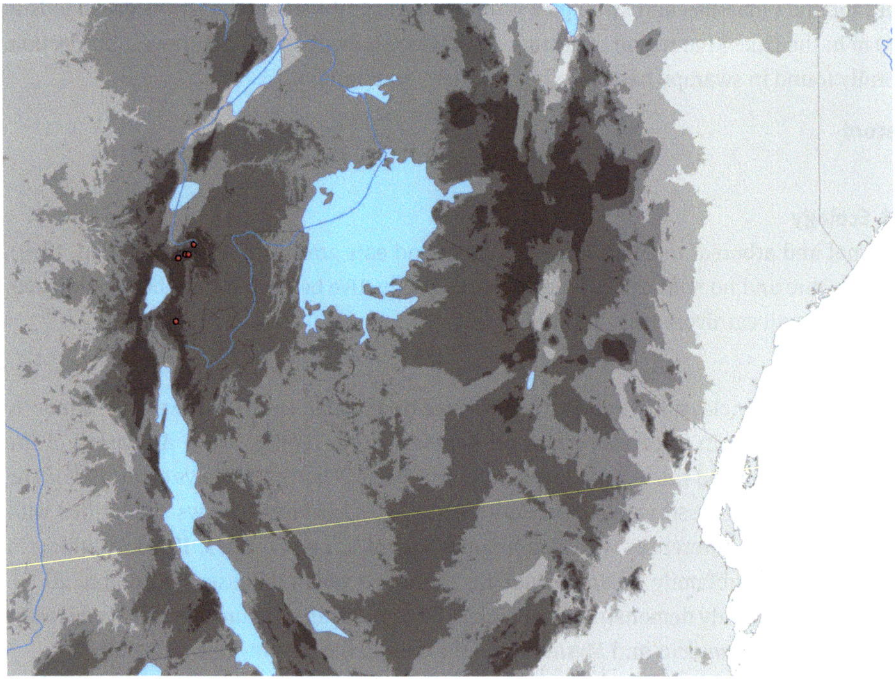

Figure 2.225 Distribution map

Measurements

External and cranial measurements (mm) and mass (g) for *Delanymys brooksi*, sexes combined

	Mean	Min	Max	N
Mass*	5.8	5.2	6.5	8
HB*	56.8	50	63	20
Tail*	103.3	87	111	20
HF*	18.4	17.0	19.5	20
Ear*	11.0	9	13	13
GLS**	18.1	17.4	19.0	4

*Delany (1975); **Van der Straeten & Verheyen (1983)

Subfamily: Dendromurinae

Genus Account

Dendromus Smith, 1829 (African Climbing Mice)

Diagnosis

The distinguishing characters of the genus are its very small to small size (HB ca. 65-85 mm), relatively long (100-160% of HB), thin and prehensile tail, and feet modified for climbing. Some species display a mid-dorsal black stripe (three stripes in *D. lovati*). There are five digits on the hindfoot: Digit 1 is reduced, and Digit 5 is long and semi-opposable with either a nail or a claw. The forefoot has five digits with greatly reduced Digits 1 and 5. The molars exhibit a typical Dendromurinae dental pattern. This genus occurs widely across Sub-Saharan Africa but avoids desert regions.

Figure 2.226 Photograph of *Dendromus melanotis* (Bill Branch).

Cranial Characters

The skull is small (ca. 19-24 mm in greatest length), rectangular in shape in dorsal view, with elongated nasals, opisthodont upper incisors each with a single groove. The upper palatal foramina extend beyond the anterior lobe of M^1. The palate between the molar tooth rows is wide anteriorly, narrowing posteriorly (divergent molar rows). The tympanic bullae are relatively small (3.2-5.2 mm).

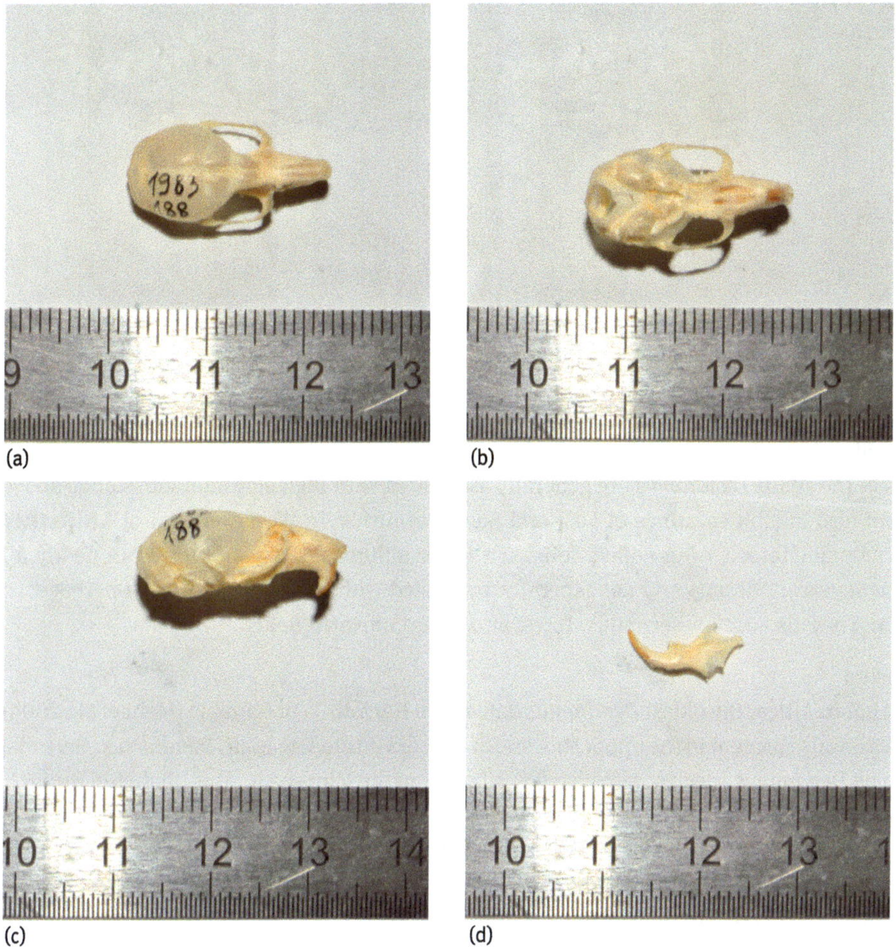

Figure 2.227 a-d Cranium of *Dendromus mystacalis* (MNHN-ZM-MO 1983-188) (from top left to bottom right: dorsal view, ventral view, lateral view and lower jaw).

Dentition

The upper and lower molar rows have a typical Dendromurinae cusp disposition (t4 is relatively low but the remaining cusps are high and project posteriorly). There is a deep longitudinal valley separating the median (t2, t5, t8) and labial (t3, t6, t9) cusps. Molar M^1 has a trace of a posterior cingulum connected to the t8. Molar M^2 has a t3, but t1 is absent. Molars M^3 and M_3 are very small. Molar M_1 has two lobes, with a single median cusp present on the first lobe connected to the second lobe by a longitudinal crest. The four other cusps of these two lobes are alternated and there may be supplementary cusps present. The posterior cingulum is conspicuous on M_1 and M_2.

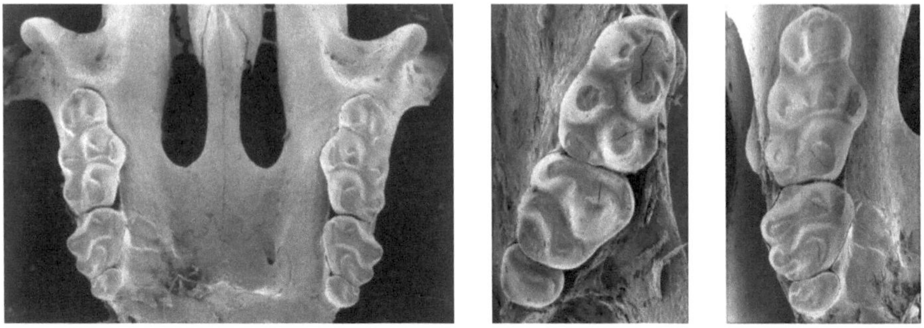

(a) (b) (c)

Figure 2.228 Cheek teeth of *Dendromus melanotis* (MNHN-ZM-MO 1962-1170) (A: upper cheek teeth; B: lower right toothrow; C: upper right toothrow).

Habitat & Biogeography

Members of the genus *Dendromus* are generally associated with high elevation grasslands above 2000m but also inhabit savannas of East and Southern Africa. In West and Central Africa they occur in montane forests or in gallery forests at lower altitudes. They are present in nearly all environments except deserts and are generally associated with wetlands, and moist herbaceous vegetation. They also occupy secondary forest as well as cultivated fields.

Fossil Record

In Sub-Saharan Africa, the oldest *Dendromus* dates from Harasib 3a in Namibia (Mein et al., 2004) and fossils are also present in the upper Miocene of Chorora (Ethiopia) as aff. *Dendromus* (Geraads, 2001). Some *Dendromus* species were also described from the Miocene in Abu-Dhabi (de Bruijn & Whybrow, 1984) and in Spain (Aguilar et al., 1984).

Dendromus denysae Harasib 3a (8.5-10 Ma), Namibia (Mein et al., 2004)

Dendromus sp. Chororora (8-10 Ma), Ethiopia (Geraads, 2001)

Dendromys averyi and *D. darti* Langebaanweg (4.5 Ma) (Denys, 1994).

Dendromus sp. Lukeino (6.1-5.8 Ma) Kenya Lukeino (Mein & Pickford, 2006; Winkler, 1990)

Dendromus sp. Aramis (4.4 Ma) Ethiopia (Louchart et al., 2009)

Dendromus sp. East and South African Plio-Pleistocene sites (Olduvai Bed I, Laetoli, Peninj) (Denys, 1987a,b)

Dendromus sp. Plio-Pleistocene sites of Namibia, Botswana, South Africa (Bolt's farm Waypoint 160) (Pickford & Mein, 1988; Pickford et al., 1994; Senegas, 2000)

Dendromus cf *melanotis* Plio-Pleistocene South Africa (Makapansgat, Sterkfontein, Kromdraai, Gladysvale, Haasgat, Gondolin, Drimolen, Swartkrans) (Pocock, 1987; Senegas, 2000)

Biology & Ecology

These nocturnal rodents are either granivorous or omnivorous and some species include insects in variable proportions (Dieterlen, 1971, 1976a; Hanney, 1965; Monadjem, 1997; Rowe Rowe, 1986). They are arboreal and construct grass nests attached about 1m above the ground or are able to dig burrows in the soil (Smithers, 1971). Their social and reproductive behaviours not well known. There are some indications that they are predominantly solitary and territorial (Smithers, 1971; Kingdon, 1974). Females have 2-8 embryos with a gestation period of 23-27 days, and reproduction is usually seasonal (Dieterlen, 1971; Hanney, 1965; Monadjem & Perrin, 2003; Taylor, 1998). Many occurrences of predation by owls have been reported in the literature; small mammalian carnivores and snakes are also potential predators of this genus.

Systematic Notes

This genus displays a wide variability of external characters and still needs a comprehensive taxonomic revision. There are about 44 names available in the literature that have been synthesised into 12 species by Musser & Carleton (2005). Thomas (1916) proposed three sub genera: *Poemys, Chortomys* and *Dendromus* based upon the existence of a nail or a claw on Digit 5 of the hindfoot. These subgeneric distinctions were followed by many authors including Bohmann (1942), Roberts (1951), Rosevear (1969), Dieterlen (1971), and Ansell (1974) but still need to be verified on molecular grounds. Bohmann (1942) drastically reduced the number of species of *Dendromus* to three based upon external characters only; a classification that is no longer valid. Musser & Carleton (2005) elevated some taxa to full species based on morphological grounds. Dieterlen (2009) revised the genus in Ethiopia and Sudan and described a new species, *D. ruppi*. Denys & Aniskine (2012) described a new species from West Africa, *D. lachaisei*, and suggested the occurrence of further cryptic diversity within the genus. Solano et al. (2014) demonstrated that *D. melanotis* comprised at least four lineages in South Africa based on molecular evidence.

Species List

14 species currently recognised.
Dendromus insignis (Thomas, 1903)
Dendromus kahuziensis Dieterlen, 1969
Dendromus lachaisei Denys & Aniskine, 2012
Dendromus leucostomus Monard, 1933
Dendromus lovati (de Winton, 1900)
Dendromus melanotis Smith, 1834
Dendromus mesomelas (Brants, 1827)
Dendromus messorius (Thomas, 1903)
Dendromus mystacalis Heuglin, 1863
Dendromus nyasae Thomas, 1916
Dendromus nyikae Wroughton, 1909
Dendromus oreas Osgood, 1936
Dendromus ruppi Dieterlen, 2009
Dendromus vernayi Hill & Carter, 1941

Key to Species

1	Three dorsal stripes down the back; endemic to Ethiopia	*D. lovati*
	Single dorsal stripe present or absent	2
2	Digit 5 of hindfoot with nail	3
	Digit 5 of hindfoot with claw	5
3	Size larger (GLS > 21 mm); mid-dorsal stripe irregular and sometimes inconspicuous; ventral pelage pure white or white with grey base	*D. nyikae*
	Size smaller (GLS < 21 mm); mid-dorsal stripe present or absent; ventral pelage greyish with grey base	4
4	Mid-dorsal stripe present and conspicuous	*D. melanotis*
	Mid-dorsal stripe absent	*D. leucostomus*
5	Ventral pelage pure white (with white base)	6
	Ventral greyish (sometimes white), but always with grey base	9
6	Mid-dorsal stripe conspicuous and always present; dorsal pelage bright rufous-brown; occurs in Southern and East Africa	7
	Mid-dorsal stripe absent or inconspicuous; dorsal pelage ginger-brown; occurs in West Africa and marginally in East Africa	8
7	Size larger (GLS > 22 mm); occurs in montane habitats (> 1800 m)	*D. ruppi*
	Size smaller (GLS < 22 mm); typically a lowland species (< 2000 m, except in Ethiopia)	*D. mystacalis*
8	Occurs from Ghana eastwards	*D. messorius*
	Occurs from Ivory Coast westwards	*D. lachaisei*
9	Restricted to the highlands of Cameroon	*D. oreas*
	Occurs elsewhere in Africa	10
10	Restricted to the highlands of central Angola	*D. vernayi*
	Occurs elsewhere in Africa	11
11	Restricted to Southern Africa	*D. mesomelas*
	Restricted to East Africa (including eastern DRC)	12
12	Ventral pelage white with grey base; tail relatively very long (> 150% of HB); mid-dorsal stripe very broad; endemic to Mt Kahuzi (eastern DRC)	*D. kahuziensis*
	Ventral pelage greyish with grey base; tail relatively shorter (< 130% of HB); mid-dorsal not particularly broad	13
13	Size larger (typically GLS > 23 mm, HB > 80 mm); ventral pelage darker (greyish)	*D. insignis*
	Size smaller (typically GLS < 23 mm, HB < 80 mm); ventral pelage off-white with grey base	*D. nyasae*

Species Accounts

Dendromus insignis (Thomas, 1903)

Diagnosis

This is one of three large *Dendromus* species occurring in the highlands of East Africa (*D. mesomelas* is restricted to Southern Africa). It has a rich brown dorsal pelage and grey or brown-grey ventral pelage (in *D. kahuziensis* the ventral pelage is white with grey base). There is a mid-dorsal stripe extending from the back of the head to the rump (in *D. kahuziensis* this stripe extends anteriorly to the forehead and is at least twice as thick; absent in *D. messorius*). *Dendromus nyasae* is very similar in size and appearance, but is generally smaller (GLS < 23 mm) and has greyish-white ventral pelage with grey base. *Dendromus ruppi* is immediately distinguishable by its pure white ventral pelage. Digit 5 of the hindfoot has a claw (in *D. nyikae* it is a nail).

Taxonomic Issues

This species was previously listed as a synonym of *D. mesomelas*, but the latter species is restricted to Southern Africa.

Distribution

This species occurs widely in East Africa typically at higher elevations.

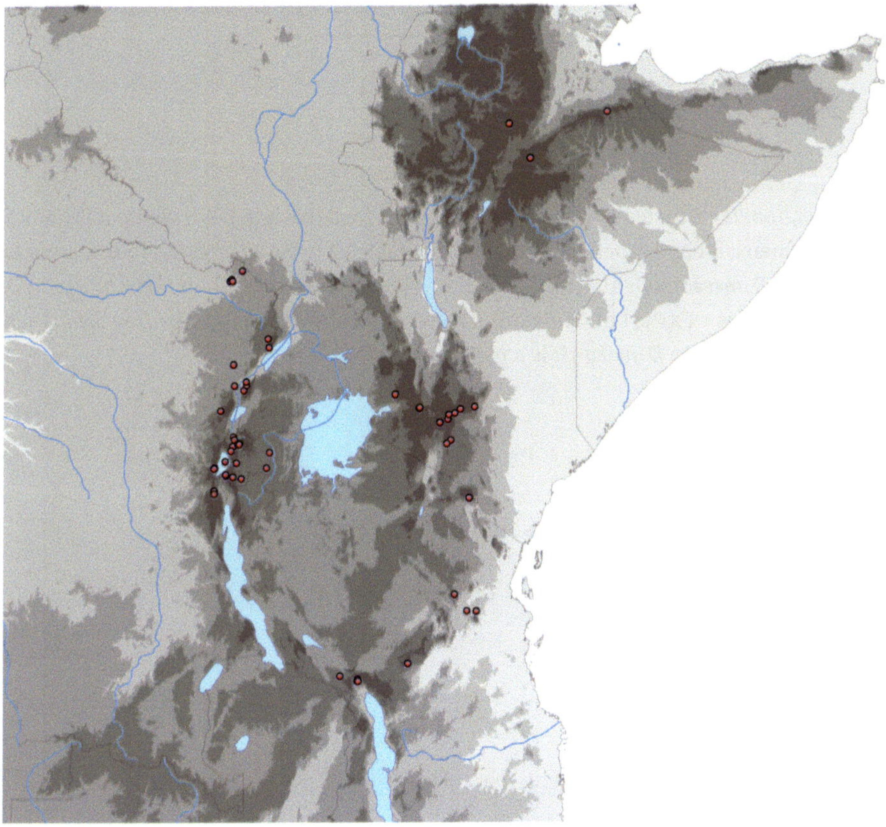

Figure 2.229 Distribution map

Measurements

External and cranial measurements (mm) and mass (g) for *Dendromus insignis*, sexes combined

	Mean	Min	Max	N
Mass	14.5	12	17	2
HB	85.8	78	92	10
Tail	99.5	82	111	10
HF	20.8	20	23	10
Ear	15.5	12	18	9
GLS	23.7	21.8	25.0	4

Dendromus kahuziensis Dieterlen, 1969

Diagnosis

This is a large *Dendromus* restricted to Mt Kahuzi, near Lake Kivu (eastern DRC) where it is known from just two specimens. It has a broad mid-dorsal stripe (at least twice as broad as in *D. nyasae* or *D. insignis*; absent in *D. messorius*). Also its tail is relatively very long; more than 150% of HB (in *D. nyasae* and *D. insignis* it is < 130%, tail < 115 mm). The ventral pelage is white with grey base. Digit 5 of the hindfoot has a claw (in *D. nyikae* it is a nail).

Taxonomic Issues

None.

Distribution

This species is known only from the type locality at Mt Kahuzi, DRC.

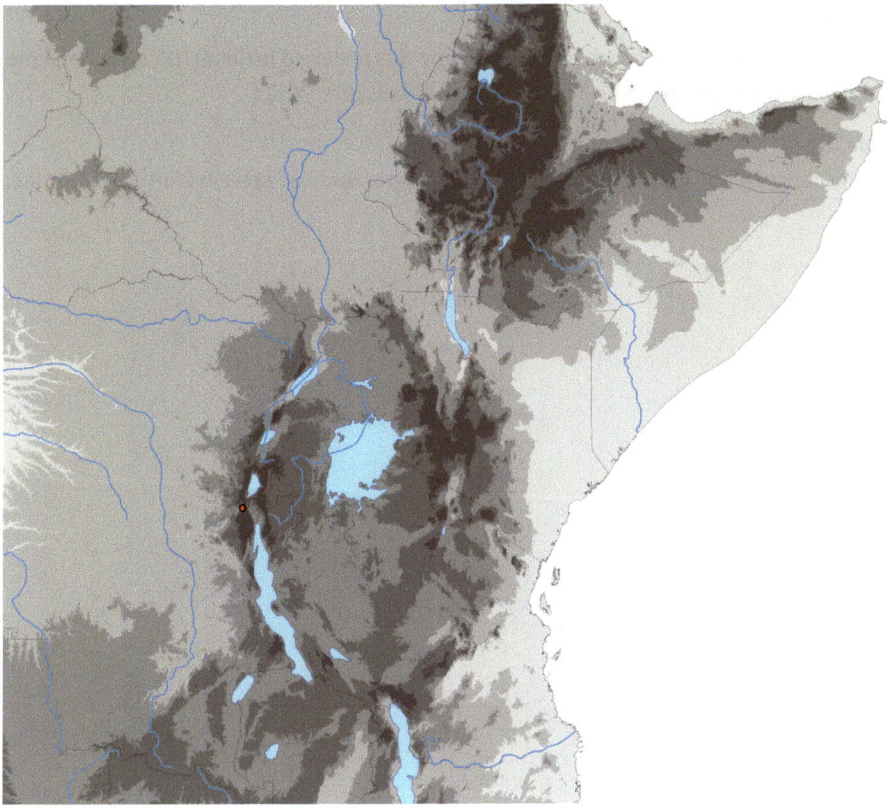

Figure 2.230 Distribution map

Measurements

External and cranial measurements (mm) and mass (g) for *Dendromus kahuziensis*, sexes combined

	Mean	Min	Max	N
Mass	11	10	12	2
HB	79.5	77	82	2
Tail	126.0	120	132	2
HF	21.5	21	22	2
Ear	14.5	14	15	2
GLS	23.6	23.4	23.7	2

Dendromus lachaisei Denys & Aniskine, 2012

Diagnosis

This is the only species of *Dendromus* within its range. It is relatively small, usually lacks a mid-dorsal stripe (which may occasionally be present but inconspicuous). The dorsal pelage is rufous-brown and the ventral pelage is pure white with white base (Denys & Aniskine, 2012).

Taxonomic Issues

This species was recently described where its taxonomy was reviewed in detail (Denys & Aniskine, 2012). Chromosome number 2n = 30, aFN = 56 (Denys & Aniskine, 2012).

Distribution

This species occurs patchily in the western parts of the Upper Guinea Forest in Ivory Coast, Guinea and Liberia.

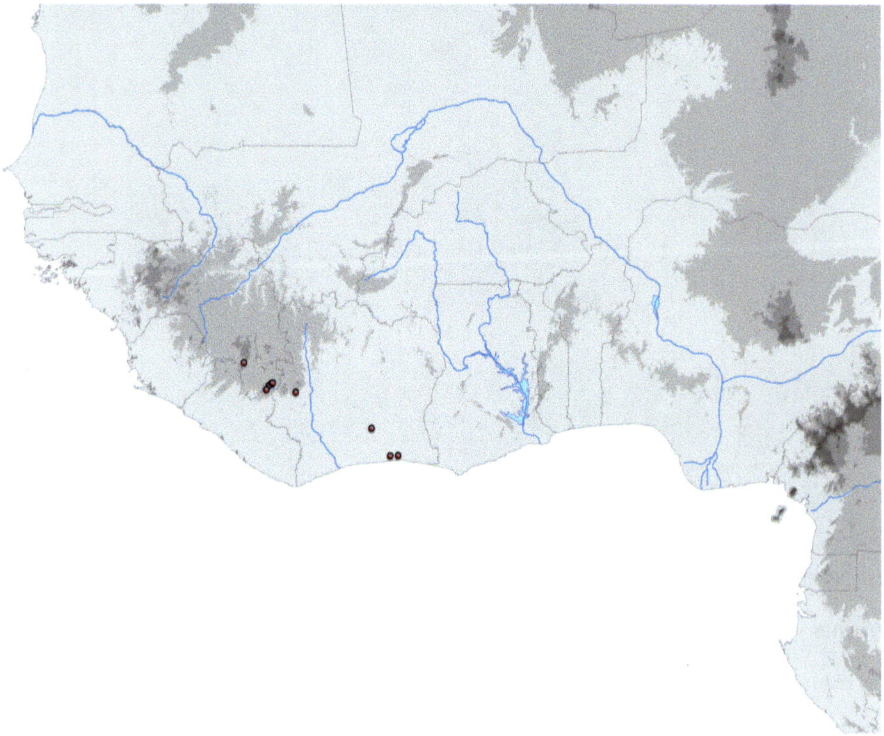

Figure 2.231 Distribution map

Measurements

External and cranial measurements (mm) and mass (g) for *Dendromus lachaisei*, sexes combined

	Mean	Min	Max	N
Mass	11	-	-	1
HB*	67.4	55	78	19
Tail*	90.4	79	98	19
HF*	17.2	15	19	19
Ear*	11.8	10	13	19
GLS*	20.6	18.6	22.1	20

*Denys & Aniskine (2012)

Dendromus leucostomus Monard, 1933

Diagnosis

This is a small *Dendromus* known from the type locality in the highlands of Angola. It differs from *D. mystacalis* by lacking a mid-dorsal stripe, and having grey ventral pelage with grey base (in *D. mystacalis*, the mid-dorsal stripe is present and the ventral pelage is pure white with white base). *Dendromus vernayi* and *D. nyikae* are larger (typically GLS > 21 mm) and possess mid-dorsal stripes. *Dendromus melanotis* also possesses a mid-dorsal stripe and differs in having a greyer dorsal pelage (richer brown in *D. leucostomus*).

Taxonomic Issues

It was previously listed as a subspecies of *D. melanotis* (Happold, 2013), from which it differs by the absence of a mid-dorsal stripe. The taxonomic affinities of this species are currently unclear and we retain this taxon as a valid species (as did Musser & Carleton, 2005) until the group is revised.

Distribution

This species is only known from the type locality Caluquembe, Angola.

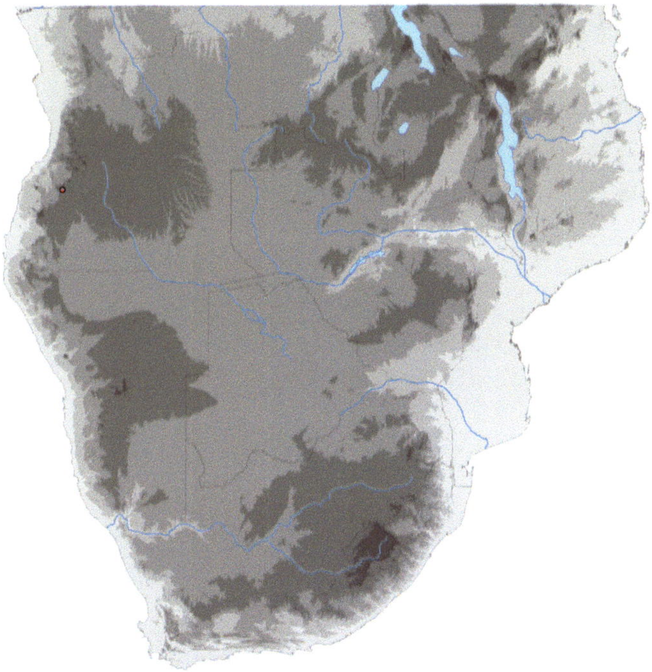

Figure 2.232 Distribution map

Measurements

External and cranial measurements (mm) and mass (g) for *Dendromus leucostomus*, sexes combined

	Mean	Min	Max	N
Mass	-	-	-	-
HB*	-	47	69	10
Tail*	-	57	80	10
HF*	-	16.5	18	10
Ear*	-	14	15	10
GLS*	19.3	-	-	1

*Hill & Carter (1941)

Dendromus lovati (de Winton, 1900)

Diagnosis

This is the only species of *Dendromus* with three dorsal stripes (in all other species there is either a single stripe or the stripe is absent).

Taxonomic Issues

Chromosome number 2n = 44 (Lavrenchenko et al., 1997).

Distribution

This species is endemic to Ethiopia, occurring widely in the highlands (Sillero-Zubiri et al., 1995a).

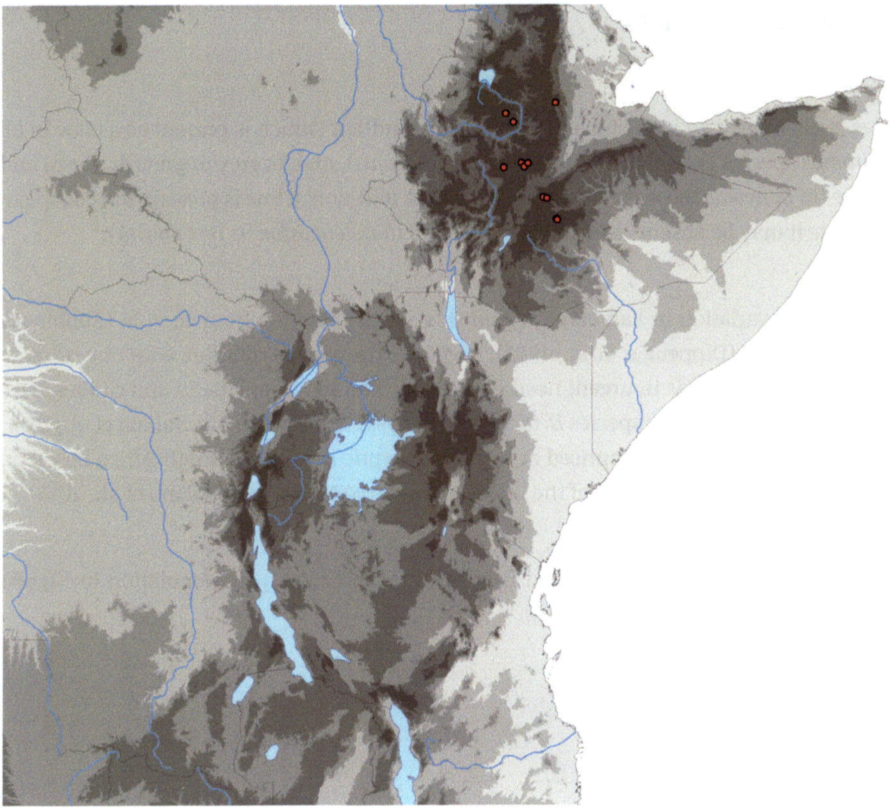

Figure 2.233 Distribution map

Measurements

External and cranial measurements (mm) and mass (g) for *Dendromus lovati*, sexes combined

	Mean	Min	Max	N
Mass*	16.4	11	23	11
HB*	74	57	95	21
Tail*	73	57	87	21
HF*	18	17	20	17
Ear*	16	15	18	16
GLS*	19.4	18.0	20.2	6

*Happold (2013)

Dendromus melanotis Smith, 1834

Diagnosis

This species is distinguished by a nail on Digit 5 of the hindfoot (which is present as a claw in all other *Dendromus* species except *D. nyikae* and *D. leucostomus*), and its grey or greyish dorsal and ventral pelage (in *D. nyikae* the ventral pelage is white). A mid-dorsal line is present and conspicuous (in *D. nyikae* it may be absent or inconspicuous, and in *D. leucostomus* it is absent).

Taxonomic Issues

There is significant variation in karyotype within this species ranging from 2n = 36 (Robbins & Baker, 1978) to 2n = 52 (Dippenaar et al., 1983), suggesting a species complex with one or more cryptic species. This species is in urgent need of revision. Records from Liberia and Guinea have been recently described as a new species *D. lachaisei* (Denys & Aniskine, 2012). Solano et al. (2014) demonstrated that *D. melanotis* comprised at least four cryptic lineages in South Africa based on molecular evidence. Karyotype of one of the lineages was 2n=42, FNa = 60 (Solano et al., 2014).

Distribution

This species occurs widely in southern and East Africa with an outlying population in Nigeria (which may refer to an undescribed species).

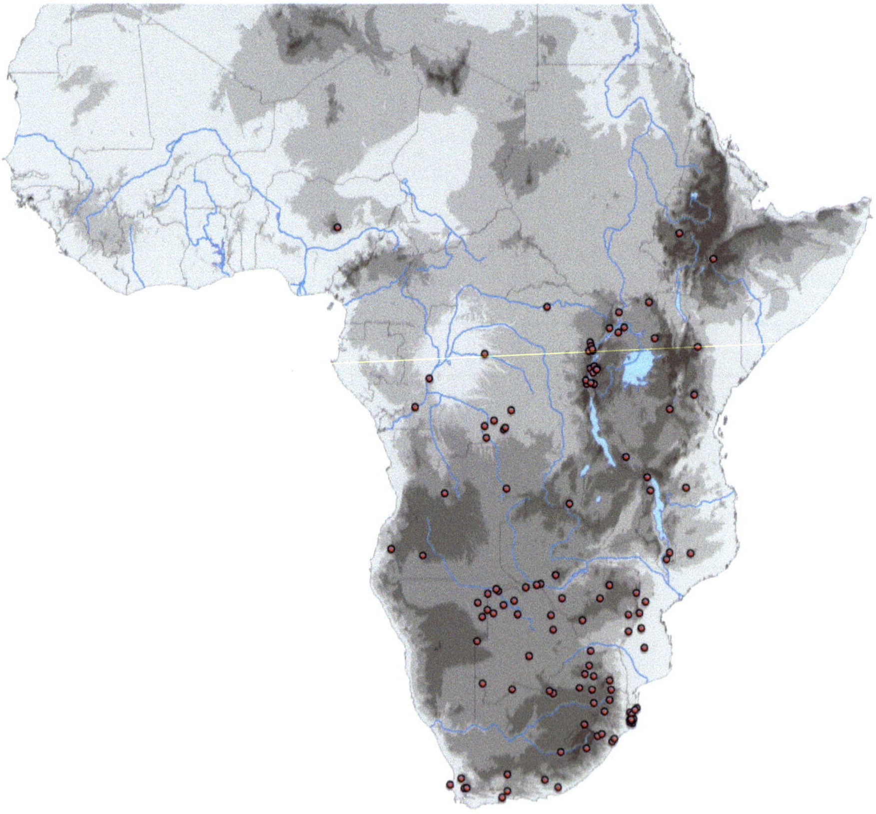

Figure 2.234 Distribution map

Measurements

External and cranial measurements (mm) and mass (g) for *Dendromus melanotis*, sexes combined

	Mean	Min	Max	N
Mass	9.0	6	15	15
HB	61.8	50	77	30
Tail	74.6	55	96	33
HF	16.9	12	20	25
Ear	13.8	8	18	30
GLS	20.1	18.2	21.3	14

Dendromus mesomelas (Brants, 1827)

Diagnosis

This is the large *Dendromus* of Southern Africa (*D. insignis, D. kahuziensis* and *D. nyasae* occur in East Africa). It has a rich brown dorsal pelage (in *D. melanotis* dorsal pelage is grey or greyish-brown) and white ventral pelage with grey base (in *D. nyikae* and *D. mystacalis* the ventral pelage is pure white with white base). There is a mid-dorsal stripe extending from the back of the head to the rump, which is variable in thickness and may occasionally be inconspicuous (De Graaff, 1981). Digit 5 of the hindfoot has a claw (in *D. nyikae* it is a nail).

Taxonomic Issues

This species was previously considered to occur widely across African highlands and included several taxa now listed as valid species including: *D. insignis, D. nyasae, D. oreas* and *D. vernayi.* The populations in northern Botswana, Zambia and Mozambique are widely isolated from the South African populations may refer to different species. Karyotype: 2n = 48, FNa = 60 (Solano et al., 2014).

Distribution

This species occurs widely in moist habitats within montane and temperate grasslands of southern Africa.

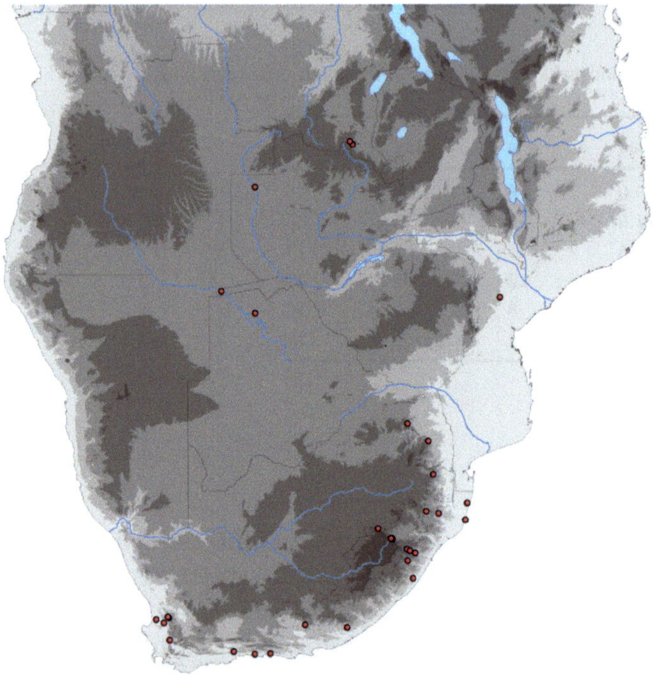

Figure 2.235 Distribution map

Measurements

External and cranial measurements (mm) and mass (g) for *Dendromus mesomelas*, sexes combined

	Mean	Min	Max	N
Mass	11.8	6	15	13
HB	75.4	67	88	20
Tail	102.3	77	121	19
HF	20.6	16	23	20
Ear	13.9	11	17	19
GLS	22.4	20.5	23.7	10

Dendromus messorius **(Thomas, 1903)**

Diagnosis

This small *Dendromus* is immediately distinguishable from all other species by its pure white ventral pelage with white base, and the lack of a mid-dorsal stripe; *D. mystacalis* and *D. lachaisei* are the only other *Dendromus* with a pure white ventral pelage but the former has a conspicuous mid-dorsal stripe, whereas the latter does not overlap in distribution. Digit 5 of the hindfoot has a claw (in *D. nyikae* and *D. melanotis* it is a nail).

Taxonomic Issues

It is not clear whether the eastern populations in Uganda and Kenya refer to the same species as those in Central and West Africa. This group is in urgent need of revision.

Distribution

This species occurs widely in tropical African from Ghana in the west to Kenya in the east.

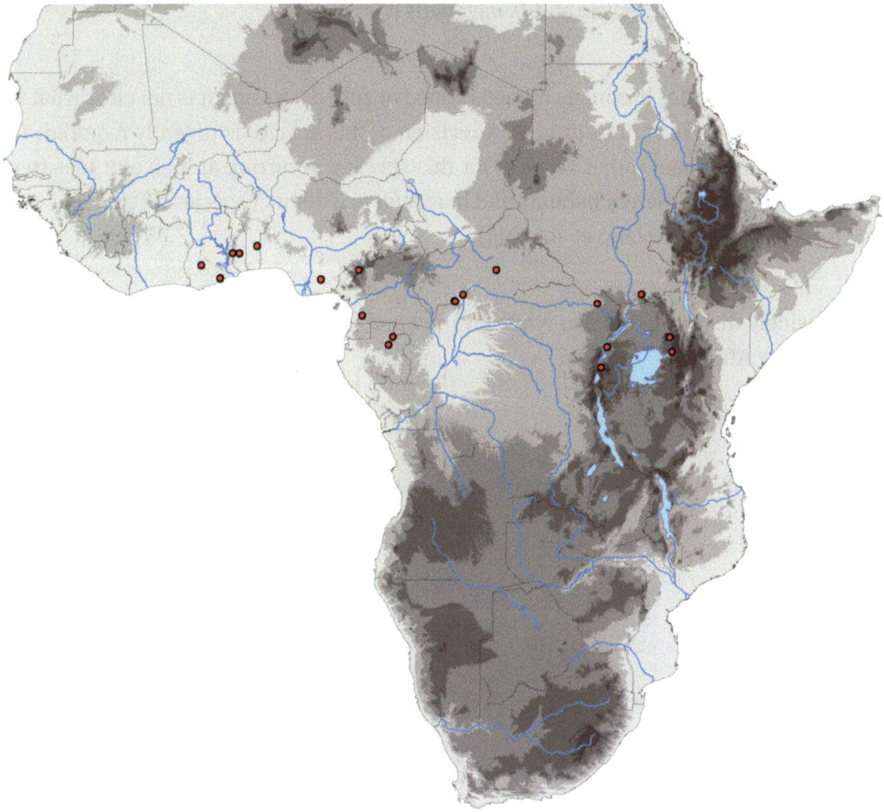

Figure 2.236 Distribution map

Measurements

External and cranial measurements (mm) and mass (g) for *Dendromus messorius*, sexes combined

	Mean	Min	Max	N
Mass	8	-	-	1
HB	63.7	57	70	6
Tail	86.6	78	95	5
HF	17.8	16	20	6
Ear	11.8	11	12	6
GLS	20.7	19.4	22.2	5

Dendromus mystacalis Heuglin, 1863

Diagnosis

This species is immediately recognisable by its pure white ventral pelage with white base and the presence of a mid-dorsal stripe. The mid-dorsal stripe is absent in *D. messorius* and *D. lachaisei* (although both have pure white ventral pelage). The dorsal pelage is bright orange-brown. Digit 5 of the hindfoot has a claw (in *D. nyikae* and *D. melanotis* it is a nail).

Taxonomic Issues

This widely distributed species has many named formed or subspecies and it is not clear what the relationship is between these various groups. A revision of this species is urgently required. The Ethiopian records of an animal rather atypical for this species, occurring at high altitudes (typically above 2000 m) and in different habitats and may refer to a distinct species (Dieterlen, 2009). Karyotype: 2n = 52, FNa = 60 (Solano et al., 2014).

Distribution

This species is widely distributed in low-lying savanna and grasslands (typically below 1000 m) of Southern and East Africa.

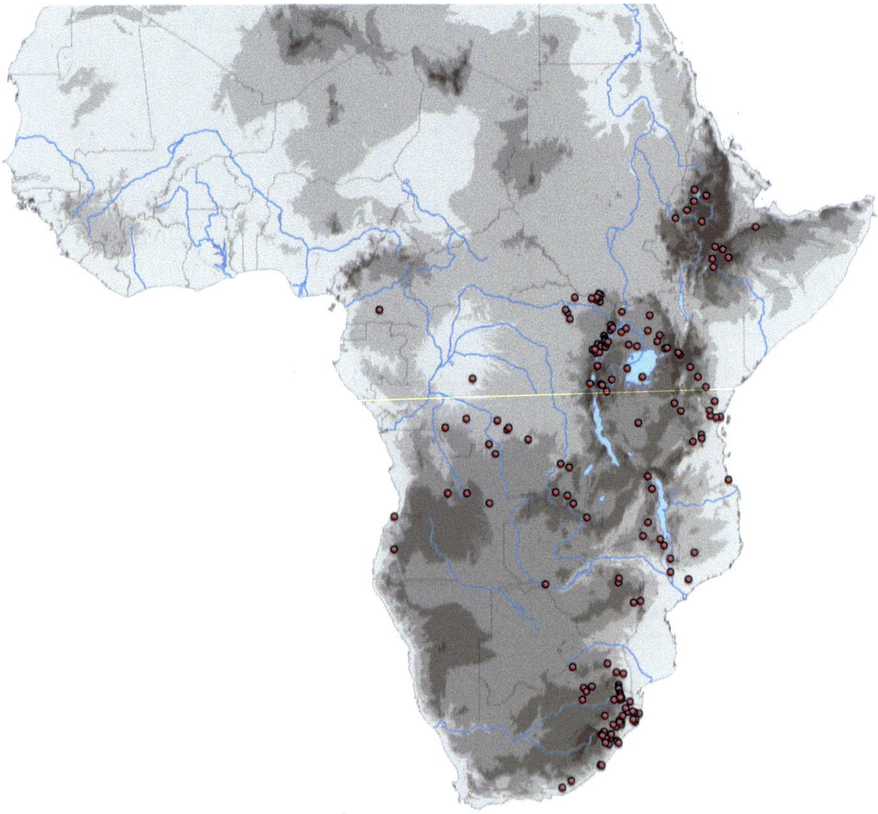

Figure 2.237 Distribution map

Measurements

External and cranial measurements (mm) and mass (g) for *Dendromus mystacalis*, sexes combined

	Mean	Min	Max	N
Mass	7.9	4	11	30
HB	61.8	47	80	60
Tail	82.4	64	101	63
HF	17.4	14	20	52
Ear	12.4	7	18	52
GLS	20.2	18.8	21.8	17

Dendromus nyasae Thomas, 1916

Diagnosis

This is a large *Dendromus* restricted to East Africa (*D. mesomelas* is restricted to Southern Africa). It has a rich brown dorsal pelage and grey-white ventral pelage with grey base (in *D. nyikae* and *D. mystacalis* the ventral pelage is pure white with white base). There is a mid-dorsal stripe extending from the back of the head to the rump (absent in *D. messorius*). *Dendromus insignis* is similar in size and appearance but is slightly larger (typically GLS > 23 mm). Digit 5 of the hindfoot has a claw (in *D. nyikae* it is a nail).

Taxonomic Issues

The taxonomy of this species is confused and there may be more than one species involved. The taxon *kivu* (from the eastern DRC) is retained in *D. nyasae* (from the Nyika Plateau, Malawi) but may represent a separate species (see Musser & Carleton, 2005; Happold, 2013). The relationship between *D. nyasae* and *D. insignis* is also not clear. This group is urgently in need of revision and a molecular phylogeny.

Distribution

This species occurs widely but patchily in highland grasslands and forests of East Africa.

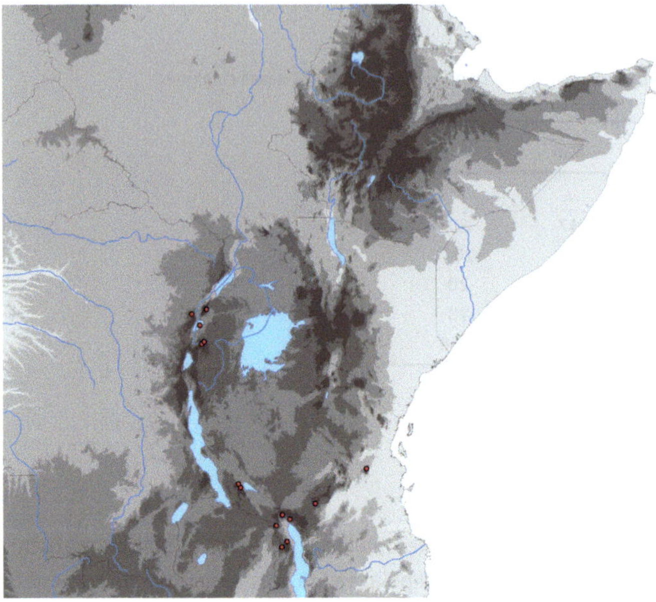

Figure 2.238 Distribution map

Measurements

External and cranial measurements (mm) and mass (g) for *Dendromus nyasae*, sexes combined

	Mean	Min	Max	N
Mass*	10.5	6	20	40
HB*	73.8	64	80	40
Tail*	91.8	84	105	38
HF*	18.9	17.0	21.5	40
Ear*	13.6	11.0	15.5	40
GLS*	22.2	20.7	23.2	32

*Happold (2013)

Dendromus nyikae Wroughton, 1909

Diagnosis

This species is distinguished by a nail on Digit 5 of the hindfoot (which is present as a claw in all other *Dendromus* species except *D. melanotis* and *D. leucostomus*), and its white ventral pelage (in *D. melanotis* the ventral pelage is greyish-white with grey base). A mid-dorsal line is absent or inconspicuous (in *D. melanotis* it is always present and conspicuous). This species may possibly co-occur with *D. leucostomus* in Angola from which it can be distinguished by its larger size (in *D. leucostomus* GLS < 21 mm, HB < 70 mm, Tail < 80 mm) and pure white ventral pelage (off-white with grey base in *D. leucostomus*).

Taxonomic Issues

This species was previously listed as a synonym of *D. melanotis* but differs sufficiently from it to warrant specific recognition (Musser & Carleton, 2005). However, the relationships between these two species and between populations within each species are still not clear and require revision.

Distribution

This species occurs widely but patchily throughout Southern Africa, extending into the Eastern Arc mountains of Tanzania in the north.

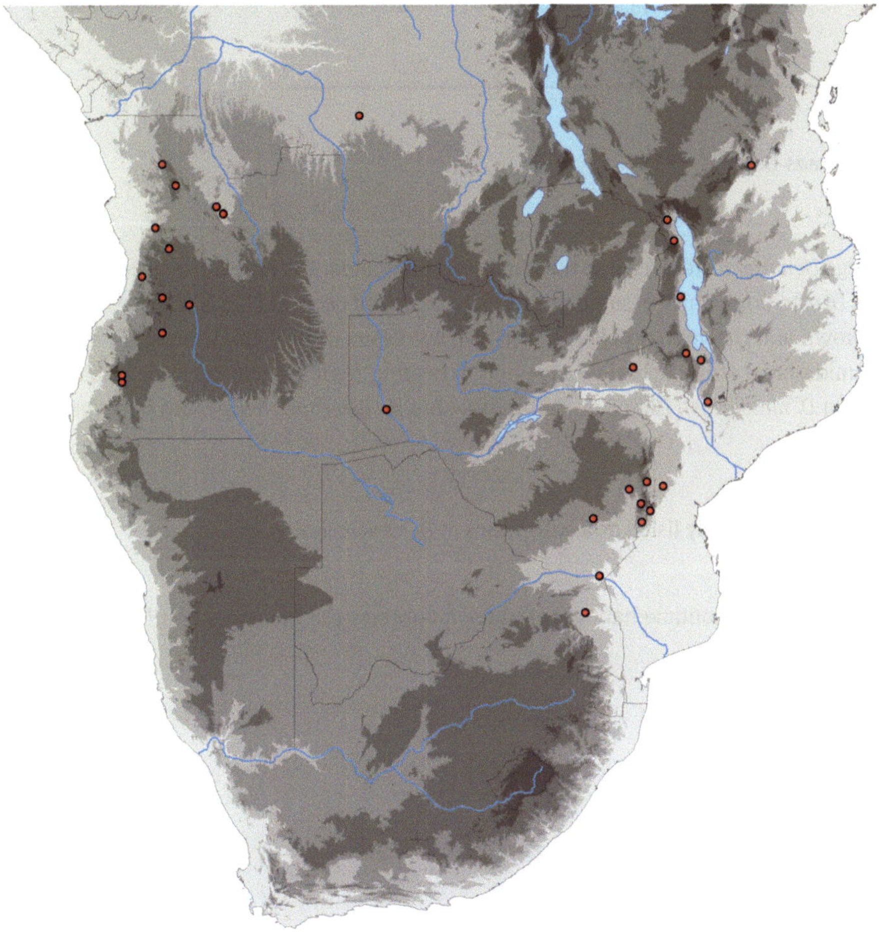

Figure 2.239 Distribution map

Measurements

External and cranial measurements (mm) and mass (g) for *Dendromus nyikae*, sexes combined

	Mean	Min	Max	N
Mass	12.4	10	14	6
HB	71.3	61	79	9
Tail	93.7	85	97	9
HF	18.4	17	20	7
Ear	14.8	12	16	9
GLS	22.0	21.5	22.9	6

Dendromus oreas Osgood, 1936

Diagnosis

This is a large *Dendromus* restricted to the Cameroon highlands. The only other *Dendromus* occurring here is *D. messorius* which is smaller (GLS < 21.5 mm) and occurs at lower altitudes. The dorsal pelage is brown and the ventral pelage is variable but never white (pure white in *D. messorius*). There is a mid-dorsal stripe (usually indistinct) extending from the mid-back to the base of the tail (absent in *D. messorius*). Digit 5 of the hindfoot has a claw (in *D. nyikae* and *D. melanotis* it is a nail).

Taxonomic Issues

This species was previously listed as a subspecies of *D. mesomelas*.

Distribution

This species occurs in montane habitats on three mountains (Mt Cameroon, Mt Manenguba and Mt Kupe) in Cameroon.

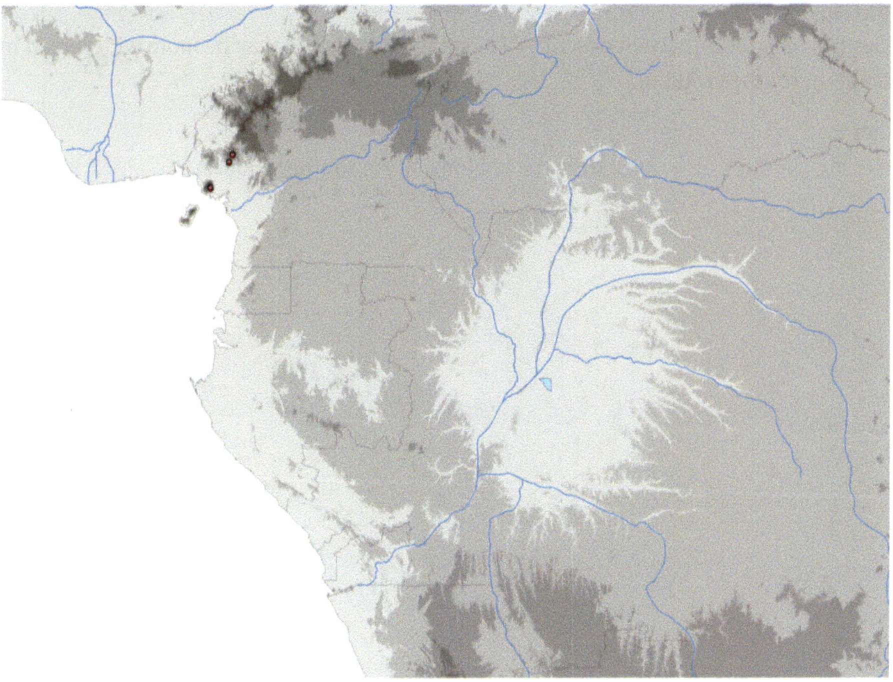

Figure 2.240 Distribution map

Measurements

External and cranial measurements (mm) and mass (g) for *Dendromus oreas*, sexes combined

	Mean	Min	Max	N
Mass*	11.1	9	13	5
HB*	69.3	60	74	6
Tail*	95.3	89	104	7
HF*	19.0	18.0	20.5	7
Ear*	15.3	13	19	6
GLS*	22.0	21.6	22.5	4

*Happold (2013)

Dendromus ruppi Dieterlen, 2009

Diagnosis

A large *Dendromus* restricted to a few localities in the Imatong Mts of southern Sudan (> 1800 m). It is distinguishable from other *Dendromus* species occurring in the region by: the presence of a mid-dorsal stripe (absent in *D. messorius*); pure white ventral pelage (grey or brown with grey base in *D. insignis*); and larger size (GLS < 22 mm in *D. melanotis, D. messorius* and *D. mystacalis*). Digit 5 has a claw (in *D. melanotis* it is a nail).

Taxonomic Issues

This species has only recently been described (Dieterlen, 2009), and may be more widespread in the mountains of southern Sudan.

Distribution

This species is currently only known from the Imatong Mts of southern Sudan.

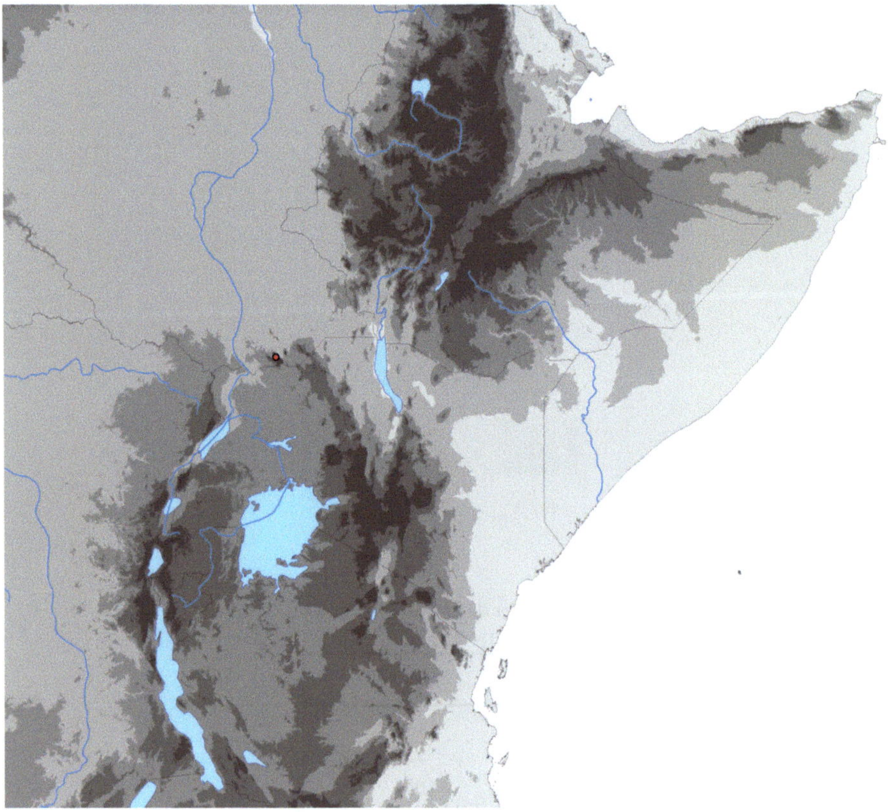

Figure 2.241 Distribution map

Measurements

External and cranial measurements (mm) and mass (g) for *Dendromus ruppi*, sexes combined

	Mean	Min	Max	N
Mass*	11.2	8	15	10
HB*	73.6	63	82	16
Tail*	103.4	91	115	16
HF*	19.8	18	20	16
Ear*	14.9	13	18	16
GLS*	22.7	22.1	23.9	8

*Dieterlen (2009)

Dendromus vernayi Hill & Carter, 1941

Diagnosis

This is a large *Dendromus* restricted to highlands of Angola, where it is only known from the type locality. The only other *Dendromus* species occurring here are: *D. melanotis, D. mystacalis* and *D. nyikae*. There is a mid-dorsal stripe (usually indistinct) extending from the back to the base of the tail (absent in *D. messorius* and *D. leucostomus*; the latter is known from further south in Angola). Digit 5 of the hindfoot has a claw (in *D. melanotis* and *D. nyikae* it is a nail). The ventral pelage is grey or grey-brown with grey base (in *D. mystacalis* and *D. nyikae* it is pure white with white base).

Taxonomic Issues

This species is currently only known from the type locality and its relationship to other large *Dendromus* of the *mesomelas* complex is not known.

Distribution

This species is only known from a series of specimens from near Chitau in the central Angolan highlands.

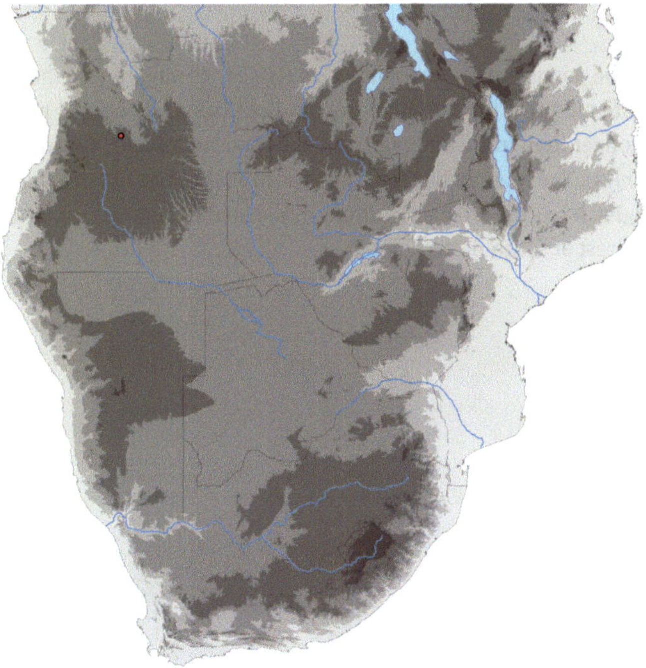

Figure 2.242 Distribution map

Measurements

External and cranial measurements (mm) and mass (g) for *Dendromus vernayi*, sexes combined

	Mean	Min	Max	N
Mass	-	-	-	-
HB*	67.8	63	77	4
Tail*	83.0	80	86	4
HF*	20.0	20	20	4
Ear	-	-	-	-
GLS*	21.8	21.4	22.1	2

*Hill & Carter (1941)

Genus Account

Dendroprionomys Petter, 1966 (Velvet Climbing Mouse)

Diagnosis

The distinguishing characters of the genus are its very small size (HB ca. 77 mm), and relatively long tail (140% of HB). The feet are modified for climbing: the hindfoot has an elongated Digit 1 (shorter than in *Prionomys*, but longer than in *Dendromus*), and an opposable Digit 5 (as in *Dendromus*, but in contrast to *Prionomys*). The forefoot has four digits as in *Prionomys*, but in contrast to *Dendromus* which has three digits. It has the typical Dendromurinae dental pattern. This monotypic genus is found only in rainforest of south-eastern Congo.

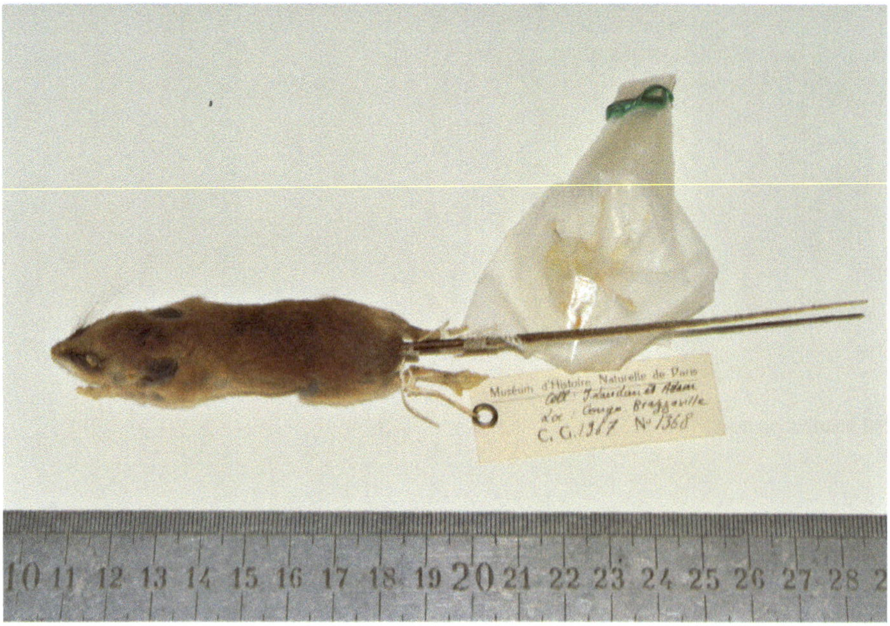

Figure 2.243 Photograph of *Dendroprionomys rousselotti* museum specimen (MNHN-ZM-MO 1967-1368) (Ara Monadjem).

Cranial Characters

The skull is small (GLS ca. 22-24 mm) with a globular-shaped braincase, orthodont upper incisors, each with a single groove, and palatal foramina which reach M^1. The rostrum is long and narrow. The zygomatic arch is thin and round. The tympanic bullae are small (4.0-4.2 mm).

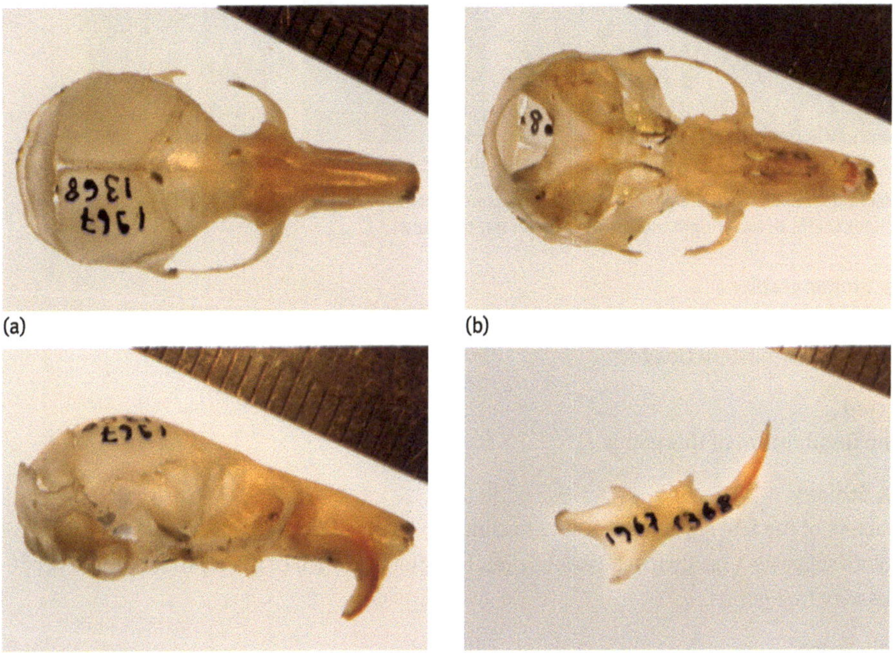

(a) (b) (c) (d)

Figure 2.244 a-d Cranium of *Dendroprionomys rousselotti* (MNHN-ZM-MO 1967-1368) (from top left to bottom right: dorsal view, ventral view, lateral view and lower jaw).

Dentition

The upper incisors are orthodont (as opposed to pro-odont in *Prionomys*), each with a single groove. The molars are tall with cusps separated by deep longitudinal valleys. The upper and lower molar rows have a typical Dendromurinae cusp disposition (only t4 on lingual side) with small M^3 and M_3. In contrast to *Prionomys* the cusps have a round aspect. Molars M_1 and M_2 are narrow and elongated with a continuous longitudinal valley from the anterior part M_1 to the posterior of M_3. A conspicuous external cingular crest is present on M_2 and M_3.

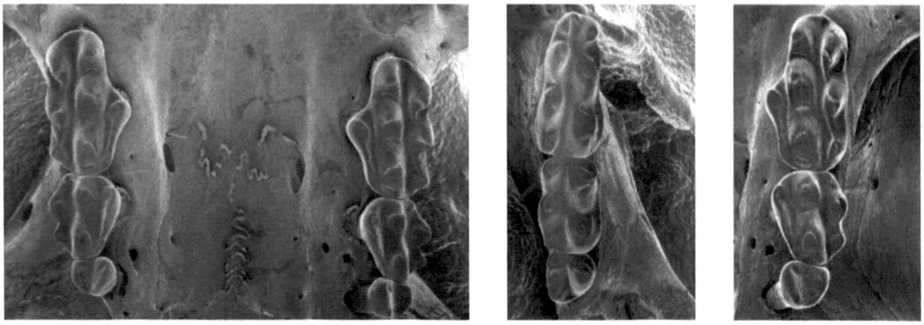

(a) (b) (c)

Figure 2.245 Upper cheek teeth of *Dendroprionomys rousselotti* (MNHN-ZM-MO 1948-18) (A: upper cheek teeth; B: lower right toothrow; C: upper left toothrow).

Habitat & Biogeography

Dendroprionomys is a very rare rodent known only by eight specimens from the same sand forest in Brazaville, south-eastern Congo.

Fossil Record

There is no fossil record of this genus.

Biology & Ecology

The characters of the feet suggest an ability to climb on stems of grasses and herbs. The structure of the molars suggests a partly insectivorous diet. Nothing is known of the reproduction or predators of this very rare genus.

Systematic Notes

The single species in the genus was described by Petter (1966b) on the basis of three specimens. It was initially placed in the Dendromuridae (Chaline et al., 1977). Phylogenetic analysis of its dental patterns suggested that *Dendroprionomys* is closely related to *Prionomys* (Denys et al., 1995). The species is now placed in the Nesomyidae (subfamily Dendromurinae) (Musser & Caleton, 2005). No cytogenetic or molecular analyses have yet been conducted.

Species List

Only one species is recognised.
Dendroprionomys rousselotti Petter, 1966

Species Account

Dendroprionomys rousselotti Petter, 1966

Diagnosis

See genus accounts.

Taxonomic Issues

See genus accounts.

Distribution

This species is known only from the type locality at Brazaville, Congo.

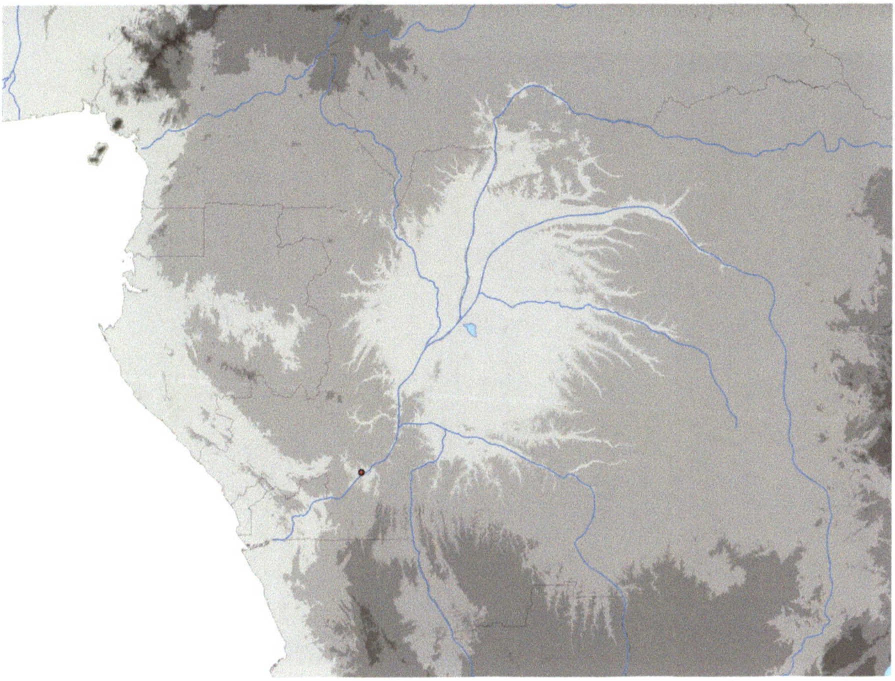

Figure 2.246 Distribution map

Measurements

External and cranial measurements (mm) and mass (g) for *Dendroprionomys rousseloti*, sexes combined

	Mean	Min	Max	N
Mass	-	-	-	-
HB*	77	-	-	1
Tail*	108	-	-	1
HF*	18	-	-	1
Ear*	16	-	-	1
GLS*	-	22.6	24.6	3

*Petter (1966b)

Genus Account

Malacothrix Wagner, 1843 (Long-eared Mouse)

Diagnosis

The distinguishing characters of the genus are its very small size (HB ca. 75 mm), relatively short tail (45% of HB), and large ears (ca. 19 mm). The dorsal pelage is very soft with three black stripes (one mid-dorsal and two lateral ones). The forefoot has four digits as in *Prionomys* but in contrast to *Dendromus* which has three digits. It has a typical Dendromurinae dental pattern. This monotypic genus is found only in arid regions of south-western Africa.

Figure 2.247 Photograph of *Malacothrix typica* (Beryl Wilson).

Cranial Characters

The skull is small (GLS ca. 24 mm) with a pronounced interorbital constriction, a narrow and straight edged rostrum, a large braincase, and a prominent masseter knob. The anterior palatal foramina are long and end just anterior to the second lobe of M^1. The upper molar rows are broadly parallel and separated by a broad palate. The tympanic bullae are relatively small (4-5 mm in greatest diameter).

(a) (b) (c) (d)

Figure 2.248 a-d Cranium of *Malacothrix typica* (TM 5498) (from top left to bottom right: dorsal view, ventral view, lateral view and lower jaw).

Dentition

The upper incisors are opisthodont, narrow, and with a single groove. The upper and lower molars have a Dendromurinae cusp disposition (only t4 is present) and broadly parallel. The molars display round bunodont cusps that are well separated. Molars M^1 and M_1 are very long and occupy half of the length of the molar toothrows. There is a strong antero-median cusp anteriorly on M^1. The posterior cingulum is present on M^1 and M_1 as well as on M^2 and M_2. Molar M_1 is narrow and displays a peculiar pattern with highly alternated cusps separated by a longitudinal crest. Molar M_2 has a conspicuous antero-labial cusp. Molar M^3 is greatly reduced and has a conspicuous t3 and a unique oblique row of cusps. Molar M_3 is tiny and has a single cusp.

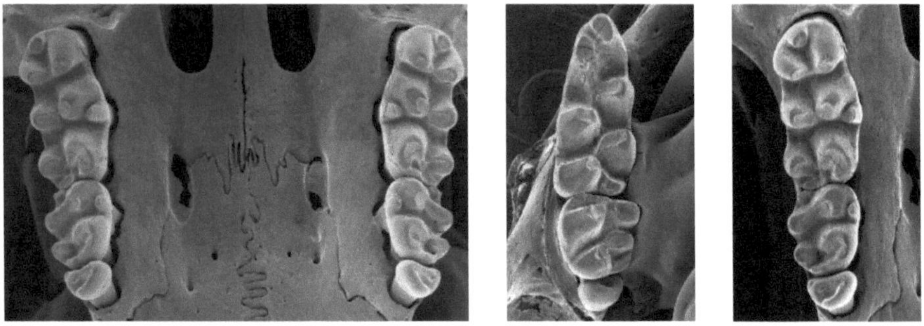

Figure 2.249 Upper cheek teeth of *Malacothrix typica* (MNHN-ZM-MO 1964-292) (A: upper cheek teeth; B: lower left toothrow; C: upper right toothrow).

Habitat & Biogeography

This genus inhabits semi-arid savannas and deserts in South Africa, Namibia, and Botswana, extending marginally into south-western Angola. Its main habitat is short grass on hard ground or pans with karroid vegetation.

Fossil Record

The oldest record of this genus is 3.3 Ma (Makapansgat) (Pocock, 1987). *Malacothrix* sp. has been recorded in Plio-Pleistocene caves of South Africa (Kromdraai, Sterkfontein, Taung, Border Cave, Gondolin, Swartkrans) (Denys, 1990c; Senegas, 2000) as well as in more recent caves (Avery, 1982).

Biology & Ecology

The diet, behaviour and reproductive biology of this genus has been mostly studied in captivity Smithers (1971). This species eats mainly green vegetable matter, grass seeds and some insects (Kerley, 1992). Knight & Skinner (1981) record a mean litter size of 4 (range: 2-8) and a gestation period of 22-26 days. Many predators have been recorded feeding on this genus including small mammalian carnivores and the barn owl *Tyto alba* (Nel & Nolte, 1965).

Systematic Notes

The single species has been placed in the Dendromurinae (Chaline et al., 1977) and a phylogenetic analysis of its morphology has confirmed this position (Denys et al., 1995). A recent revision of this genus has yet to be conducted.

Species List

This genus is monotypic.
Malacothrix typica (A. Smith, 1834)

Species Account

Malacothrix typica (A. Smith, 1834)

Diagnosis
See genus account.

Taxonomic Issues
See genus account.

Distribution
This species occurs widely in the arid parts of Southern Africa.

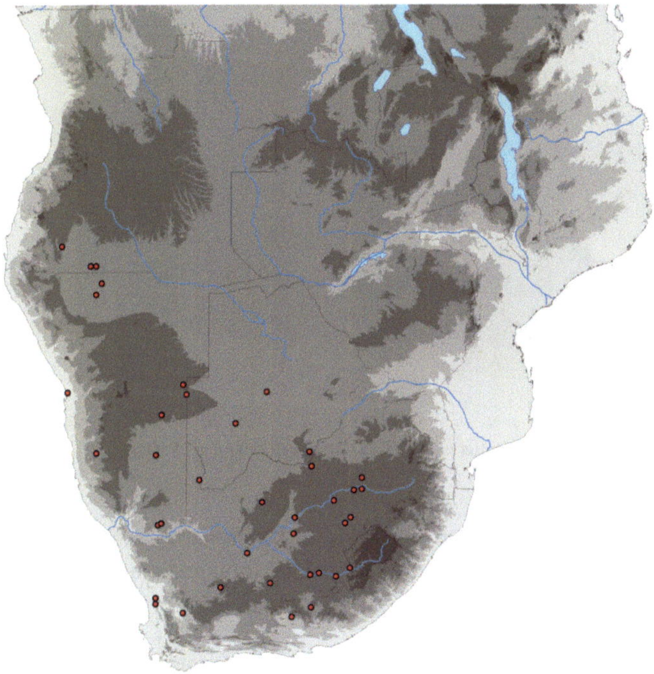

Figure 2.250 Distribution map

Measurements

External and cranial measurements (mm) and mass (g) for *Malacothrix typica*, sexes combined

	Mean	Min	Max	N
Mass	19.4	15	23	4
HB	75.5	60	87	13
Tail	33.4	29	41	12
HF	18.0	15	20	13
Ear	19.0	14	21	13
GLS	24.0	22.0	25.3	10

Genus Account

Megadendromus Dieterlen & Rupp 1978 (Giant Climbing Mouse)

Diagnosis

The distinguishing characters of the genus are its small size (HB ca. 120 mm), relatively long tail (80% of HB), and the presence of a well-marked black mid-dorsal stripe. It is also the largest representative of Dendromurinae. The feet are modified for climbing. The forefoot has four digits as in *Dendromus*, with Digit 5 being short and Digit 1 absent altogether. The hindfoot has five digits. Its molars exhibit typical Dendromurinae morphology. This monotypic genus is found only in the southern Ethiopian highlands.

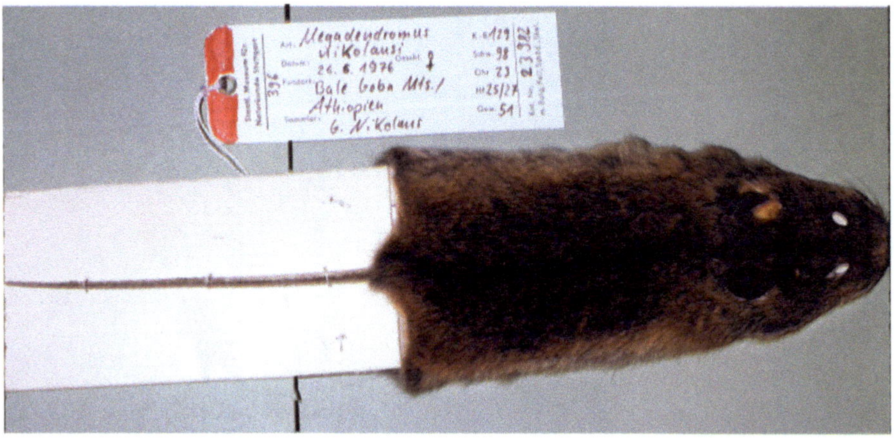

Figure 2.251 Photograph of the holotype of *Megadendromus nikolausi* (SMNS 23982) (Francis Petter).

Cranial Characters

The skull is medium-sized (GLS ca. 30 mm) with a long rostrum. The braincase is large and square, the upper incisors are opisthodont and grooved. The anterior palatal foramina are long and end at second lobe of M^1. The tympanic bullae are medium-sized (6.6 mm in length). The molar rows are parallel.

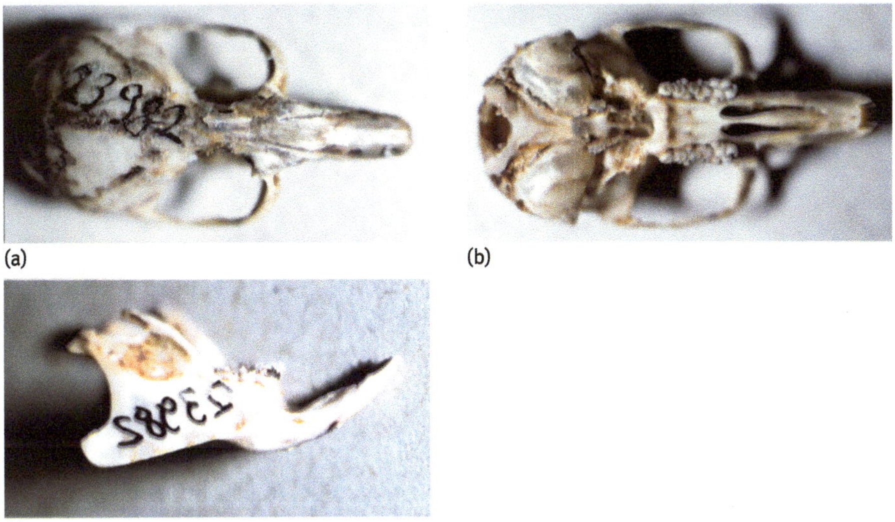

(a) (b) (c)

Figure 2.252 a-c Cranium of the holotype of *Megadendromus nikolausi* (SMNS 23982) (from top left to bottom left: dorsal view, ventral view and lower jaw).

Dentition

The molars show a typical Dendromurinae cusp disposition with the presence of cusp t4 on M^1 and M^2, and highly reduced M^3 and M_3. However, M^3 and M_3 are comparatively less reduced than in *Dendromus* (16-21% in *Megadendromus* compared with 11-14% in *Dendromus*). Molar M^1 has a t7 cusp with a well-developed posterior cingulum connected to the t8, which delimits a wide fovea. There are two stephanodont crests on the t4 and t7 cusps. Molars M_1 and M_2 are characterized by a large posterior cingulum; the first lobe of M_1 has two cusps and M_2 has a conspicuous antero-external cingulum. The labial cingular margin is well developed and accessory cusps are visible.

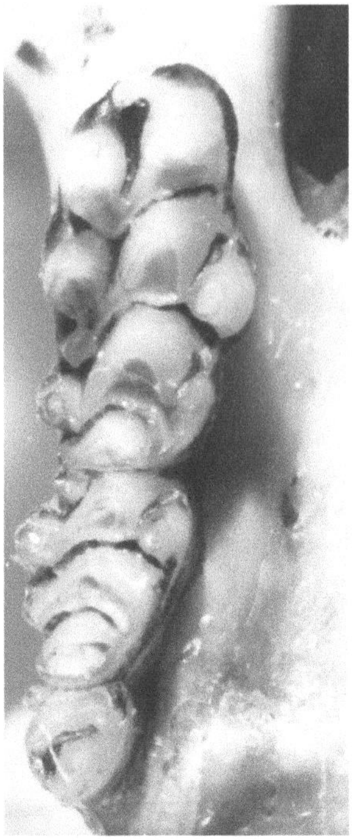

Figure 2.253 Upper cheek teeth of the holotype of *Megadendromus nikolausi* (SMNS 23982) (Francis Petter).

Habitat & Biogeography

Megadendromus is restricted to the high plateau of south-eastern Ethiopia at 3000-3900 m in the Afromontane and Afroalpine belts. It inhabits *Hagenia-Erica arborea* forest and *Erica arborea* scrub (Dieterlen & Rupp, 1978).

Fossil Record

There is no fossil record of this genus.

Biology & Ecology

This genus has rarely been collected and its biology and ecology are not known.

Systematic Notes

Phylogenetic analysis of its morphological and dental characters suggest that *Megadendromus* is closely related to *Dendromus, Malacothrix* and *Steatomys* (Denys et al., 1995), but is distinguished by the presence of cusp t7 on M^1. No cytogenetic or molecular work has been conducted on this genus.

Species List

This genus is monotypic.
Megadendromus nikolausi Dieterlen & Rupp, 1978

Species Account

Megadendromus nikolausi Dieterlen & Rupp, 1978

Diagnosis
See genus account.

Taxonomic Issues
See genus account.

Distribution
This species is only known from two localities (Mt Chilalo and the Bale Mts) in the highlands of Ethiopia, east of the Rift Valley.

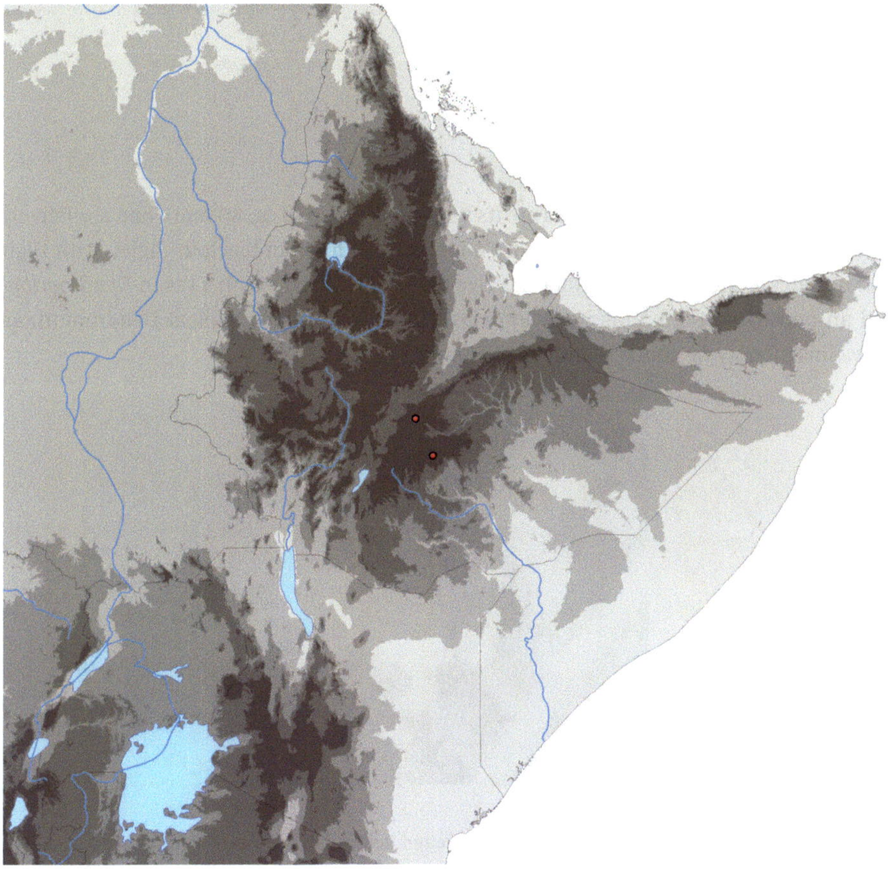

Figure 2.254 Distribution map

Measurements

External and cranial measurements (mm) and mass (g) for *Megadendromus nikolausi*, sexes combined

	Mean	Min	Max	N
Mass*	55.3	49	66	3
HB*	123.0	117	129	3
Tail*	98.7	92	106	3
HF*	26.7	26	27	3
Ear*	24.0	23	26	3
GLS*	30.3	-	-	1

*Dieterlen & Rupp (1978)

Genus Account

Prionomys Dollman 1910 (Bate's Climbing Mouse)

Diagnosis

The distinguishing characters of this genus are its very small size (HB ca. 80 mm), and a relatively long tail (122% of HB). The feet are modified for climbing; the forefoot has four digits, with Digit 1 absent and Digit 5 reduced. The hindfoot has five digits with a long Digit 1 (as in *Dendroprionomys*), and an opposable Digit 5 (as in *Dendromus*). Its molars display typical Dendromurinae morphology. This monotypic genus is endemic to Central Africa.

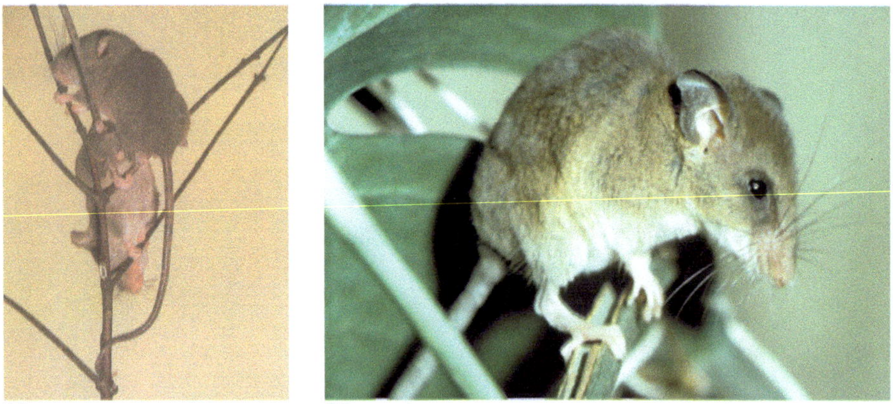

(a) (b)

Figure 2.255 a and b Photograph of *Prionomys batesi* (Francis Petter) (left and right).

Cranial Characters

The skull is small (GLS ca. 25 mm), very wide and with a short rostrum. The zygomatic arches are low and the braincase is large and square, the incisors are pro-odont and ungrooved. The anterior palatal foramina are short and end well before M^1. The tympanic bullae are small (4.3-5.2 mm length). The mandible is low and the angular articulation is poorly developed. The molar rows are nearly parallel.

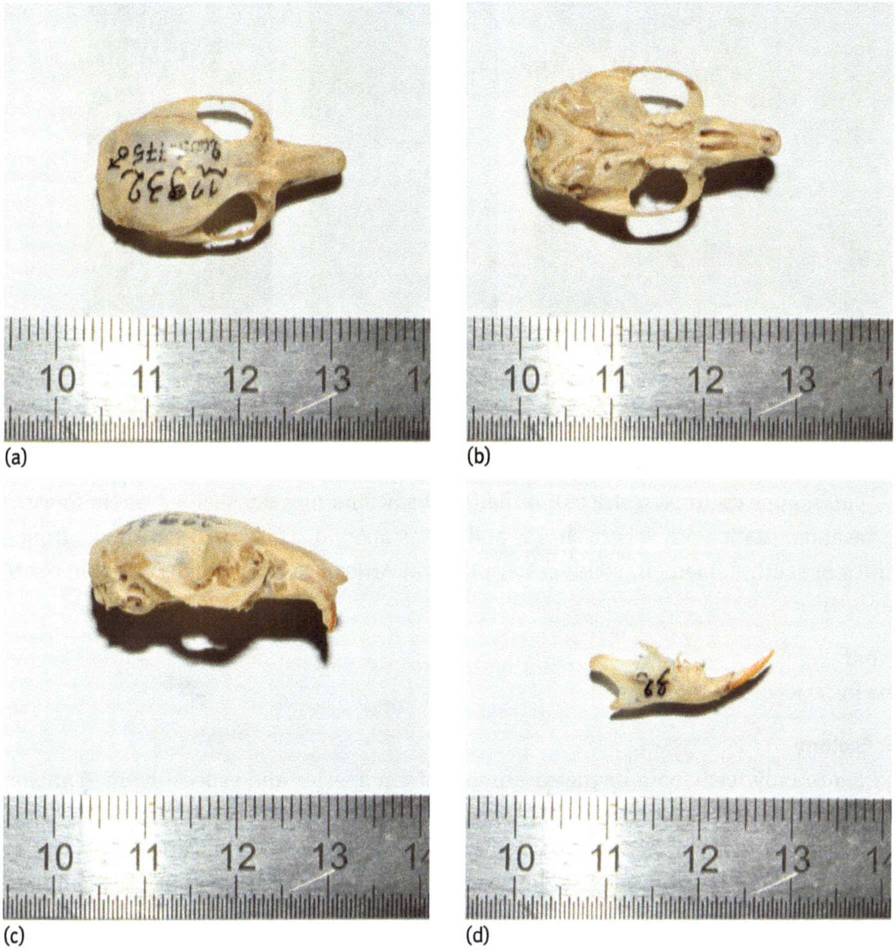

Figure 2.256 a-d Cranium of *Prionomys batesi* (MNHN-ZM-MO 2005-775) (from top left to bottom right: dorsal view, ventral view, lateral view and lower jaw).

Dentition

The molars have tall, well separated and pointed cusps that are aligned in longitudinal rows and separated by a conspicuous median longitudinal valley running along the upper molar toothrow. The upper and lower molars have a cusp disposition reminiscent of *Dendromus* (with only a t4 on the lingual side of M^1) and *Dendroprionomys* but with much narrower cusps. Molars M^3 and M_3 are highly reduced. Molar M_1 also displays a medium longitudinal groove separating the rows of cusps. There is a well-developed antero-external cingulum on M_2 and M_3.

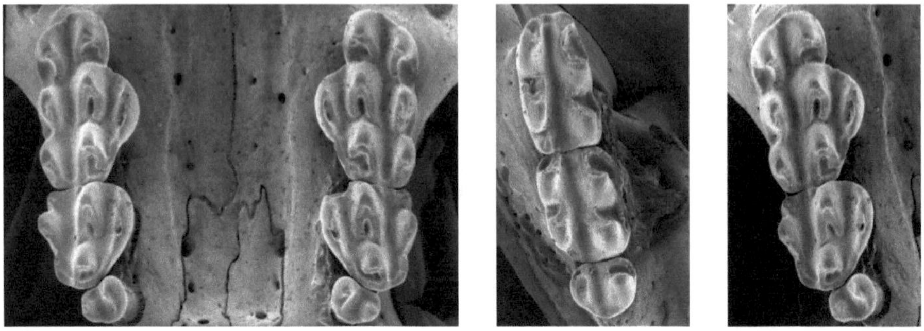

(a) (b) (c)

Figure 2.257 Upper cheek teeth of *Prionomys batesi* (MNHN-ZM-MO 1979-136) (A: upper cheek teeth; B: lower right toothrow; C: upper right toothrow).

Habitat & Biogeography

Prionomys batesi appears to be restricted to the forest-savanna mosaic, such as where forest is replacing savanna, particularly where shrubs and lianas abound. This species is known from a few localities in south Cameroon, south-eastern Central African Republic and Congo (Denys et al., 2006).

Fossil Record

There is no fossil record of this genus.

Biology & Ecology

Prionomys batesi only feeds on a restricted number of ant species and especially on *Tetramorium aculeatum* Mayr 1866 (= *Macromischoides aculeatus;* Genest-Villard, 1980). It is a nocturnal rodent, resting in burrows between tree roots during the day. Burrows are constructed by digging in forest soil using the forward-projecting incisors; each burrow is indicated by a little hillock of soil, ca. 30 cm in diameter, without a visible entrance hole. The burrow contains a small nest chamber lined with dry leaves. It is capable of climbing on thin branches, twigs and lianas using the prehensile tail, long digits, callosity on the forefoot and opposable Digit 1 of the hindfoot. One female was found with a single embryo. Two pairs of nipples are present.

Systematic Notes

This genus was initially placed in the Dendromuridae (Chaline et al., 1977). Phylogenetic analysis of its dental patterns suggested that *Prionomys* is closely related to *Dendroprionomys* (Denys et al., 1995), but this work also demonstrated that the Dendromuridae are not monophyletic. The genus is now placed in the Nesomyidae (subfamily Dendromurinae) (Musser & Carleton, 2005). No cytogenetic or molecular studies have yet been conducted on this genus.

Species List

This genus is monotypic.
Prionomys batesi Dollman, 1910

Species Account

Prionomys batesi Dollman, 1910

Diagnosis

See genus account.

Taxonomic Issues

See genus account.

Distribution

This species is only known from a handful of forest localities in Central Africa.

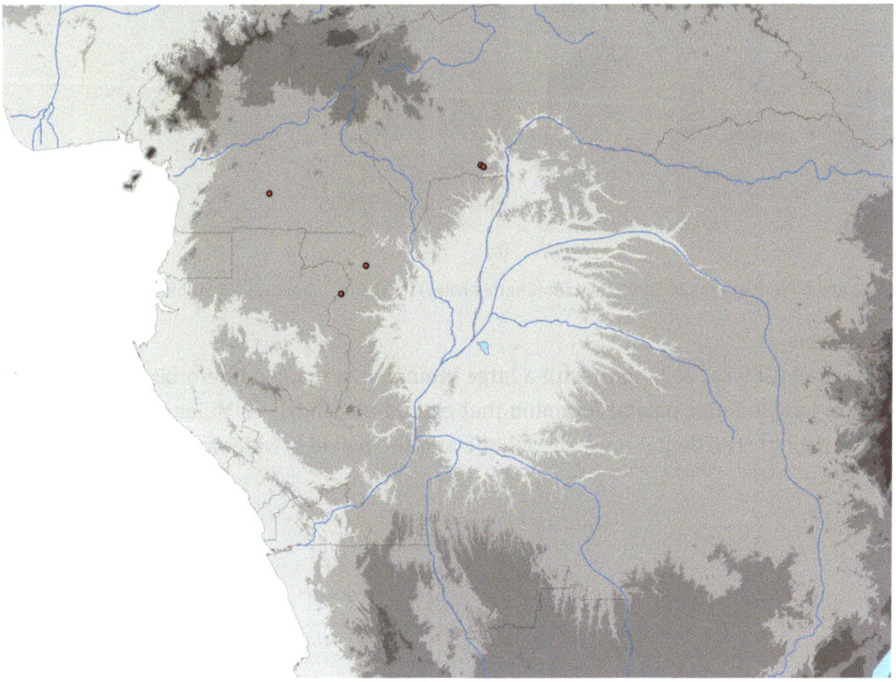

Figure 2.258 Distribution map

Measurements

External and cranial measurements (mm) and mass (g) for *Prionomys batesi*, sexes combined

	Mean	Min	Max	N
Mass	18.0	15	21	2
HB	82.0	79	85	4
Tail	107.3	103	112	4
HF	16.2	15	17	4
Ear	12.4	11	14	4
GLS	25.8	24.8	27.2	5

Genus Account

Steatomys Peters 1846 (Fat Mice)

Diagnosis

The distinguishing characters of the genus are its small to medium size (HB ca. 70-120 mm), relatively short tail (35-60% of HB), and short hindfeet. It has the typical Dendromurinae dental pattern. The small skull displays a wide palate, well developed tympanic bullae and highly reduced M^3 and M_3. This genus typically occurs in savanna habitats.

(a) (b)

Figure 2.259 a and b Photographs of *Steatomys* sp. (Charles Foley) (left) and *S. parvus* (Bill Stanley) (right).

Cranial Characters

The skull is small (GLS ca. 22-28 mm) with a large braincase, a marked interorbital constriction, short rostrum, long anterior palatal foramina that extend posteriorly to M^1, and large tympanic bullae (5-6 mm long). The dental rows are oblique and the palate is wider anteriorly than posteriorly.

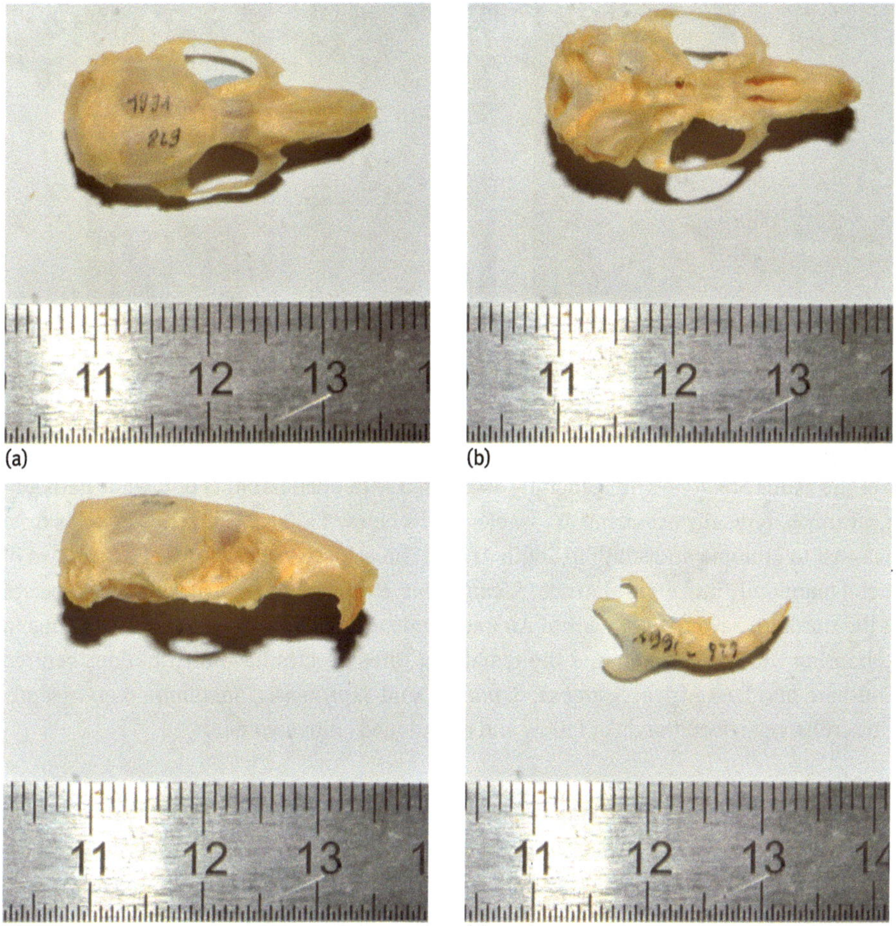

(a) (b) (c) (d)

Figure 2.260 a-d Cranium of *Steatomys caurinus* (MNHN-ZM-MO 1991-829) (from top left to bottom right: dorsal view, ventral view, lateral view and lower jaw).

Dentition

The upper incisors are opisthodont and grooved. The molars have a typical Dendromurinae cusp disposition (only t4 is present on the lingual side). The cusps are round, high and connected transversely. Molar M^1 has three lobes, the first with two connected cusps that project posteriorly. Molar M^2 has two lobes and an antero-external cusplet. Molar M^3 is reduced to a small cusp. Molar M_1 has a single oblique anterior cusp connected by a longitudinal crest to the second row of cusps which are alternated and make an oblique row. There is a posterior median cingulum. Molar M_2 has two rows of cusps and M_3 is reduced to a single cusp.

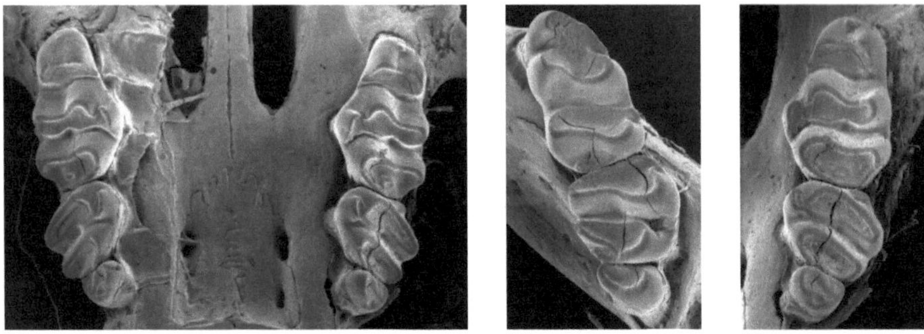

(a) (b) (c)

Figure 2.261 Upper cheek teeth of *Steatomys pratensis* (MNHN-ZM-MO 1969-58) (A: upper cheek teeth; B: lower right toothrow; C: upper left toothrow).

Habitat & Biogeography

Members of the genus *Steatomys* are generally associated with open habitats of the northern and southern savannas, typically occurring at low elevations below 1500 m. The genus extends from West Africa east to Ethiopia and south to South Africa. Three species occur in West Africa, two of which extend marginally into Central Africa: *S. cuppedius*, *S. caurinus* and *S. jacksoni*; one species occurs in the northern savannas of Central Africa: *S. opimus*; one species is restricted to Angola and neighbouring DRC: *S. bocagei*; and the remaining three species have distributions centred around Southern and East Africa: *S. krebsii*, *S. parvus*, and *S. pratensis*. *Steatomys* prefers sandy grounds in scrubs, open woodlands or fallow and abandoned cultivated fields.

Fossil Record

The oldest *Steatomys* is from the late Miocene in Harasib (9.5 - 10.5 MYA) in Namibia (Mein et al., 2004; Senut et al., 1992). Fossils also occur in South and East Africa during the Plio-Pleistocene. Mein et al. (2004) described *S. harasibensis* and *S. jaegeri* from Harasib deposits. The East African sites of Chorora and Lukeino yielded *S. minus* and *S. intermedius* (Geraads, 2001; Mein et al., 2006). More modern representatives have been described from Makapansgat, Swartkrans, Sterkfontein, Kromdraai, Gondolin, Drimolen, Border Cave, Laetoli, Olduvai and are either bearing modern species names or remain unidentified (Pocock, 1987; Denys, 1987a; Denys, 1990; Senegas, 2000).

Biology & Ecology

Fat mice live exclusively on the ground and prefer soft sandy soils where they construct deep burrows (De Graaff & Nel, 1992) allowing them to hibernate (aestivate) during the coolest seasons (Petter, 1966c). The physiology of several *Steatomys* species has been studied (Richardson & Perrin, 1992; Ellison, 1995) and they are able to store fat under their skin. *Steatomys* are nocturnal and solitary animals. They are graminivores but some may also eat vegetable matter and insects. Females have 4-6 pairs of nipples and they may have 2-6 embryos. Species of *Steatomys* are generally not abundant (Monadjem, 1999) and the biology and ecology of many species is unknown. They constitute prey for nocturnal avian predators such as owls (e.g. Vernon, 1972; Mendelsohn, 1982).

Systematic Notes

Steatomys has been recognized as a member of the Dendromurinae based on morphological and molecular grounds (Denys et al., 1995; Dubois et al., 1996; Jansa & Weksler, 2004), however the taxonomy of the genus is confused. Coetzee (1977) recognised only three species. A revision of the West and South Africa species has somewhat clarified the situation (Rosevear, 1969; Swanepoel & Schlitter, 1978; Taylor, 1998) but for other regions of tropical Africa there remain difficulties in delimiting and identifying species (Musser & Carleton, 2005).

Species List

Eight species are currently recognised.
Steatomys bocagei Thomas, 1892
Steatomys caurinus Thomas, 1912
Steatomys cuppedius Thomas & Hinton, 1920
Steatomys jacksoni Hayman, 1936
Steatomys krebsii Peters, 1852
Steatomys opimus Poursagues, 1894
Steatomys parvus Rhoads, 1896
Steatomys pratensis Peters, 1846

Key to Species

1	Restricted to West Africa (eastern limits northern Nigeria and Niger)	2
	Does not occur in West Africa	4
2	Size larger (typically GLS > 27 mm); nipples: 12	*S. jacksoni*
	Size smaller (typically GLS < 27 mm); nipples: 8 or variable (> 8)	3
3	Size larger (GLS > 24 mm); tail relatively shorter (typically about 40% of HB); dorsal pelage dark brown	*S. caurinus*
	Size smaller (GLS < 24 mm); tail relatively longer (always > 50% of HB); dorsal pelage pale sandy-brown	*S. cuppedius*
4	Size larger (GLS > 27 mm, HB > 105 mm)	5
	Size smaller (GLS < 27 mm, HB < 105 mm)	6
5	Restricted to Angola and neighbouring DRC (south of the Congo Basin); nipples: 8	*S. bocagei*
	Restricted to Central African Republic and neighbouring parts of DRC, Sudan and Cameroon (north of the Congo Basin); nipples: 10	*S. opimus*
6	More nipples: 10-14; size tends to be larger but with considerable overlap	*S. pratensis*
	Fewer nipples: 8; tends to be smaller	7
7	Tail bicoloured (brown dorsally and paler ventrally); tends to be larger but with considerable overlap	*S. krebsii*
	Tail unicoloured (white dorsally and ventrally)	*S. parvus*

Species Accounts

Steatomys bocagei Thomas, 1892

Diagnosis

This is the only species of *Steatomys* across most of its range. The larger size of *S. bocagei* distinguishes it from *S. pratensis, S. krebsii* (in both species, GLS < 27 mm, HB < 105 mm) and *S. parvus* (GLS < 24 mm, HB < 85 mm), and it has fewer nipples (8) than *S. pratensis* (10 – 14). *Steatomys opimus* is similar in size but differs by having grey or brown dorsal pelage (in *S. bocagei* dorsal pelage is a brighter reddish-brown), 10 nipples, and their distributions do not overlap. *Steatomys jacksoni* has 12 nipples.

Taxonomic Issues

This species was previously considered a synonym of *S. pratensis* (Musser & Carleton, 1993).

Distribution

This species is restricted to northern Angola and neighbouring regions of the DRC.

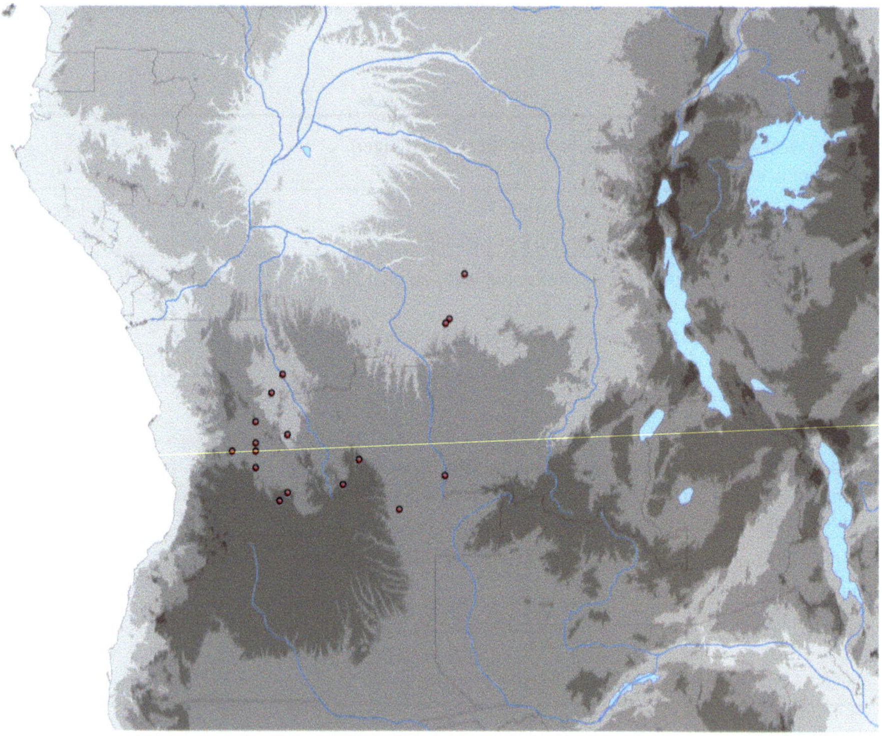

Figure 2.262 Distribution map

Measurements

External and cranial measurements (mm) and mass (g) for *Steatomys bocagei*, sexes combined

	Mean	Min	Max	N
Mass	-	-	-	-
HB	117	-	-	1
Tail	67	-	-	1
HF	18	-	-	1
Ear	17	-	-	1
GLS*	28.8	27.5	29.8	8

*Hill & Carter (1941)

Steatomys caurinus Thomas, 1912

Diagnosis

This is one of three *Steatomys* species occurring in West Africa. *Steatomys caurinus* can be distinguished from *S. cuppedius* by its larger size (in *S. cuppedius*, typically GLS < 24 mm, HB < 95 mm), shorter tail which is about 40% of HB (in *S. cuppedius* it is > 50%), darker pelage (in *S. cuppedius* it is pale sandy to grey-brown), and greater number of nipples (> 8) (in *S. cuppedius* it is always 8). *Steatomys jacksoni* is usually larger (GLS > 27 mm) and has 12 nipples. *Steatomys opimus* is marginally larger and does not overlap in distribution.

Taxonomic Issues

This species was previously listed as a subspecies of *S. pratensis* (Coetzee, 1977) and elevated to species rank by Swanepoel & Schlitter (1978).

Distribution

This species occurs in West Africa, extending east to northern parts of Central Africa.

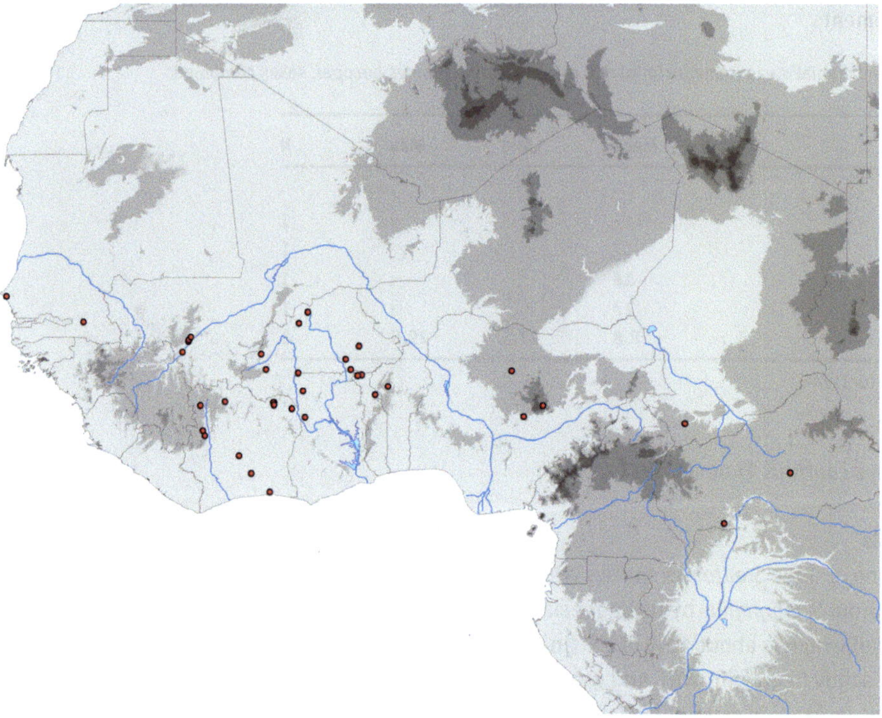

Figure 2.263 Distribution map

Measurements

External and cranial measurements (mm) and mass (g) for *Steatomys caurinus*, sexes combined

	Mean	Min	Max	N
Mass	46.3	30	66	3
HB	110.1	96	122	8
Tail	42.3	35	50	7
HF	19.3	18	20	7
Ear	16.4	15	19	8
GLS	26.3	24.3	27.8	8

Steatomys cuppedius Thomas & Hinton, 1920

Diagnosis

This is one of three *Steatomys* species occurring in West Africa. *Steatomys cuppedius* can be distinguished from *S. caurinus* by its smaller size (in *S. caurinus*, typically GLS > 24 mm, HB > 95 mm), longer tail which is always more than 50% of HB (in *S. caurinus* it is usually about 40%), paler, sandy-brown pelage (in *S. caurinus* it is dark reddish-brown), and fewer number of nipples (8) (in *S. caurinus* it is variable but always >8). *Steatomys jacksoni* is much larger (GLS > 27 mm) and has 12 nipples. *Steatomys opimus* is also much larger (HB > 110) and does not overlap in distribution.

Taxonomic Issues

It was previously listed as a subspecies of *S. parvus* (Coetzee, 1977).

Distribution

This species is patchily distributed in the northern parts of West and Central Africa.

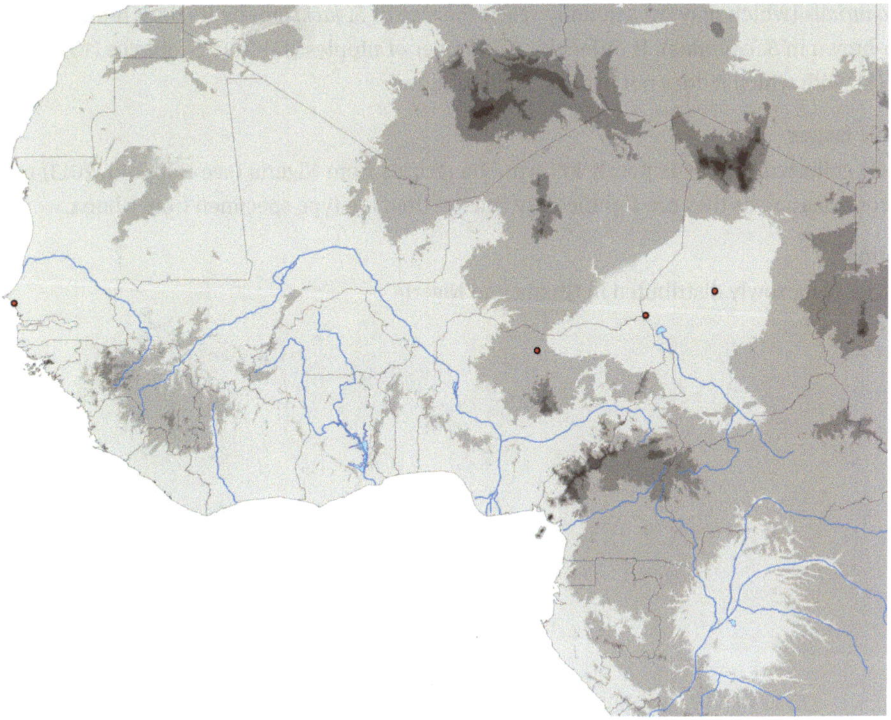

Figure 2.264 Distribution map

Measurements

External and cranial measurements (mm) and mass (g) for *Steatomys cuppedius*, sexes combined

	Mean	Min	Max	N
Mass	17.7	14	24	3
HB	83.8	78	93	4
Tail	45.8	41	50	4
HF	15.3	15	16	4
Ear	12.5	11	13	4
GLS	21.6	21.0	22.4	3

Steatomys jacksoni Hayman, 1936

Diagnosis

This species is distinguishable by its large size from the only other two *Steatomys* occurring in West Africa (in *S. caurinus*, typically GLS < 27 mm, *S. cuppedius*, GLS < 24 mm). It further differs from *S. caurinus* (which may occasionally reach the size of *S. jacksoni*) by its dark brown pelage (reddish-brown in *S. caurinus*). It differs by its number of nipples (12) from *S. opimus* (10) and *S. bocagei* (8), with which it does not overlap in distribution.

Taxonomic Issues

This rarely collected species is poorly known. The records from Nigeria (see Happold, 2013) may require confirmation as they are significantly smaller than the type specimen from Ghana.

Distribution

This species is narrowly distributed in Ghana and Nigeria.

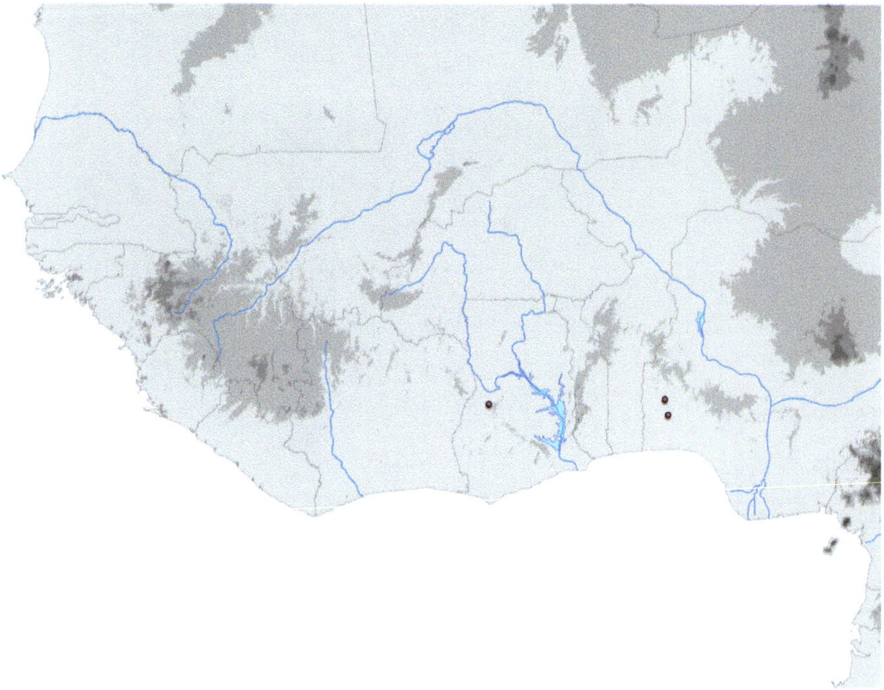

Figure 2.265 Distribution map

Measurements

External and cranial measurements (mm) and mass (g) for *Steatomys jacksoni*, sexes combined

	Mean	Min	Max	N
Mass	-	-	-	-
HB*	120	-	-	1
Tail*	50	-	-	1
HF*	18	-	-	1
Ear*	18	-	-	1
GLS*	28.9	-	-	1

*Rosevear (1969), measurements of the holotype

Steatomys krebsii Peters, 1852

Diagnosis

This species can be distinguished from *S. bocagei* by its smaller size (in *S. bocagei*, GLS > 27 mm, HB > 105 mm). Across most of its range it overlaps either *S. parvus* or *S. pratensis* from which it is not easily distinguished. *Steatomys krebsii* differs from *S. pratensis* in the number of nipples (8 in *S. krebsii*, 10-14 in *S. pratensis*); and from *S. parvus* by its bicoloured tail (in *S. parvus* it is white dorsally and ventrally), white ventral pelage (in *S. parvus* the ventral pelage tends to be buffy-brown in the area of overlap), and larger size (in *S. parvus*, typically GLS < 23 mm, HB < 80 mm, but there is considerable overlap in size of these two species).

Taxonomic Issues

The phylogenetic relationships of the various *Steatomys* species are not known, and this genus urgently needs revision.

Distribution

This species occurs widely in southern Africa, but avoiding the subtropical eastern regions.

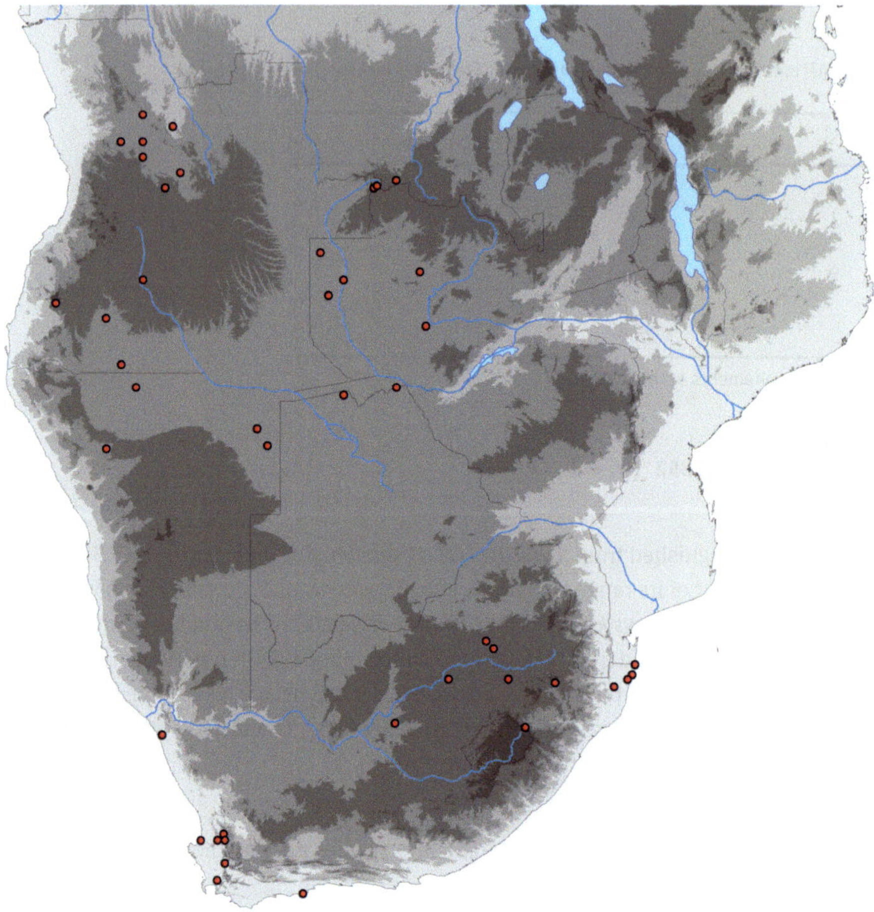

Figure 2.266 Distribution map

Measurements

External and cranial measurements (mm) and mass (g) for *Steatomys krebsii*, sexes combined

	Mean	Min	Max	N
Mass	27.1	15	42	5
HB	83.3	73	100	15
Tail	46.1	32	56	17
HF	16.4	14	21	17
Ear	15.2	13	18	17
GLS	23.9	22.5	25.4	10

Steatomys opimus Poursagues, 1894

Diagnosis

This is the only species of *Steatomys* occurring within its range. It is further distinguished by its large size (typically HB > 110 mm), which only overlaps that of *S. jacksoni* and *S. bocagei*. These three species can be readily distinguished on the number of nipples: *S. jacksoni* (12), *S. opimus* (10), and *S. bocagei* (8), and non-overlapping distributions.

Taxonomic Issues

This species was previously considered a subspecies of *S. pratensis* (Musser & Carleton, 1993).

Distribution

This species occurs in the savannas north of the Congo Basin in the DRC, Central African Republic and Sudan.

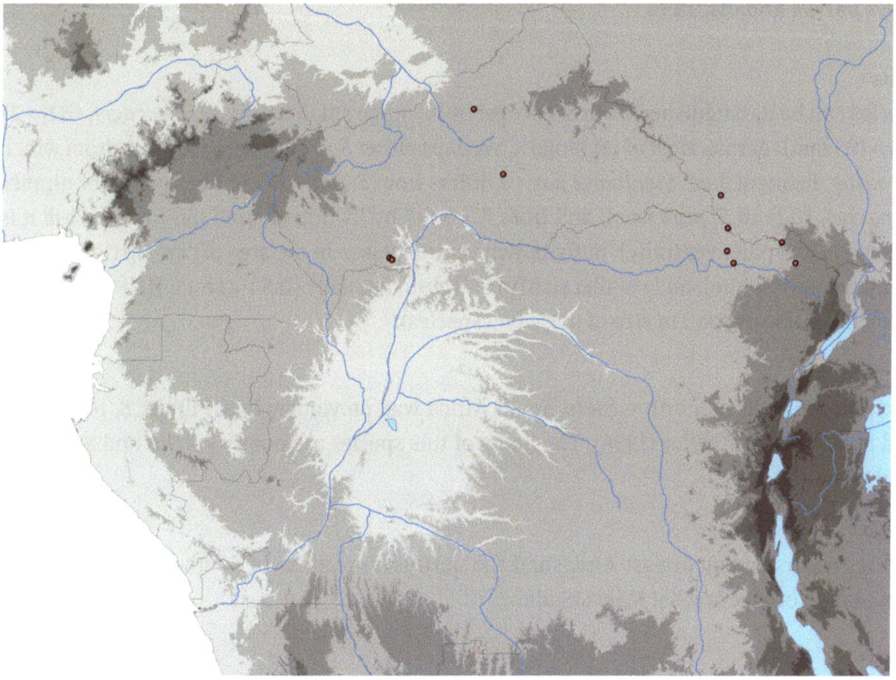

Figure 2.267 Distribution map

Measurements

External and cranial measurements (mm) and mass (g) for *Steatomys opimus*, sexes combined

	Mean	Min	Max	N
Mass*	-	30	50	-
HB*	120	115	125	5
Tail*	60	52	65	5
HF*	19	18	20	5
Ear*	18	17	20	5
GLS*	26.9	24.8	30.2	8

*Happold (2013)

Steatomys parvus Rhoads, 1896

Diagnosis

This species can be distinguished from *S. bocagei* by its much smaller size (in *S. bocagei*, GLS > 27 mm, HB > 105 mm). Across most of its range it overlaps either *S. krebsii* or *S. pratensis* from which it is not easily distinguished. *Steatomys parvus* differs from *S. pratensis* in the number of nipples (8 in *S. parvus*, 10-14 in *S. pratensis*); and from *S. krebsii* by its unicoloured tail (in *S. krebsii* it is brown dorsally and paler ventrally), buffy-brown ventral pelage in the area of overlap (in *S. krebsii* the ventral pelage is white), and smaller size (in *S. krebsii*, typically GLS > 23 mm, HB > 80 mm, but there is considerable overlap in size of these two species).

Taxonomic Issues

The taxon *tongensis*, from north-eastern South Africa was previously listed under *S. parvus* but assigned to *S. krebsii* by Taylor (1998). The limits of this species are poorly known and a revision is urgently required.

Distribution

This species occurs widely in East Africa and the northern savannas of Southern Africa. An isolated population occurs in Jebel Marra (Sudan).

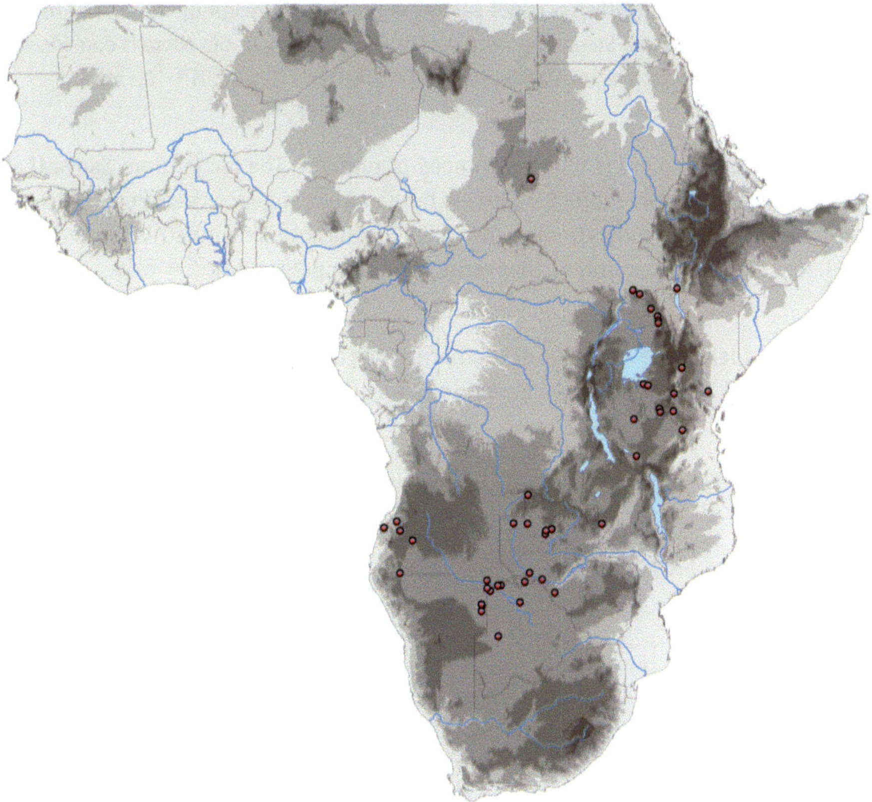

Figure 2.268 Distribution map

Measurements

External and cranial measurements (mm) and mass (g) for *Steatomys parvus*, sexes combined

	Mean	Min	Max	N
Mass	11.5	8	15	2
HB	69.9	55	83	8
Tail	42.1	34	51	9
HF	13.9	12	16	8
Ear	12.9	11	15	9
GLS*	22.6	22.0	23.0	4

*Roberts (1951) together with one extra skull measured at 22.5 mm

Steatomys pratensis Peters, 1846

Diagnosis

This species can be distinguished from *S. bocagei* by its smaller size (in *S. bocagei*, GLS > 27 mm, HB > 105 mm) a greater number of nipples (10-14) (8 in *S. bocagei*). Across most of its range it overlaps either *S. krebsii* or *S. parvus* from which it is not easily distinguished, except by number of nipples (8 in both *S. krebsii* and *S. parvus*). *Steatomys pratensis* tends to be larger, but there is considerable overlap in size, especially with *S. krebsii* (in *S. parvus*, typically GLS < 23 mm, HB < 80 mm).

Taxonomic Issues

This species (or species complex) is urgently in need of revision. The record from Kenya (Lake Naivasha) may represent a misidentified *S. parvus*, but is rather large for that species (HB = 89 mm). The record from Ethiopia comes from an owl pellet that has been identified as *S. pratensis* by Musser & Carleton (2005), but requires further verification (Happold, 2013).

Distribution

This species occurs widely in the savannas of Southern Africa and marginally into East Africa. The records from Kenya and Ethiopia may be based on mis-identifications and may refer to *S. parvus* (see above).

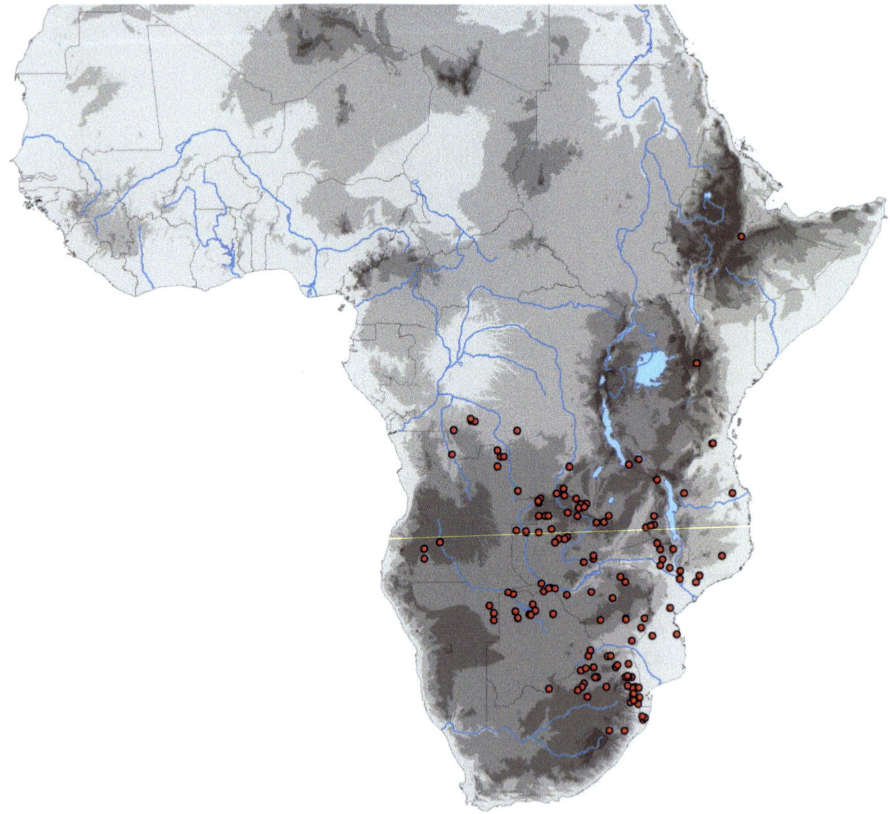

Figure 2.269 Distribution map

Measurements

External and cranial measurements (mm) and mass (g) for *Steatomys pratensis*, sexes combined

	Mean	Min	Max	N
Mass	23.6	9	40	23
HB	88.3	70	102	36
Tail	42.6	34	56	37
HF	15.6	13	19	32
Ear	14.2	12	17	23
GLS	24.5	22.9	26.1	12

Subfamily: Mystromyinae

Genus Account

Mystromys Wagner, 1841 (African White-tailed Rat)

Diagnosis

This medium sized to large rat (HB ca. 150 mm) is characterized by a short white tail (35-45% of HB), pale greyish-brown woolly dorsal pelage, white ventral pelage, and short feet with four digits on the forefoot. The skull is characterized by a narrow interorbital constriction and wide zygomatic breadth. The infraorbital foramina are long and enter deeply in the palate between the molar rows. The dentition of *Mystromys* is reminiscent of a Cricetinae dental pattern. *Mystromys* is endemic to the grasslands of South Africa.

Figure 2.270 Photograph of *Mystromys albicaudatus* museum specimen (MNHN-ZM-MO 1964-407) (Ara Monadjem).

Cranial Characters

The skull is relatively short in size (GLS ca. 35 mm), and broad at the level of the zygomatic arches; the rostrum is short and as wide as the interorbital constriction which is very narrow. The square braincase is large and high. The zygomatic plate has an anterior vertical margin. The incisors are orthodont. The anterior palatal foramina are long and extend posteriorly past M^1. The bullae are large (6.0-7.5 mm in diameter).

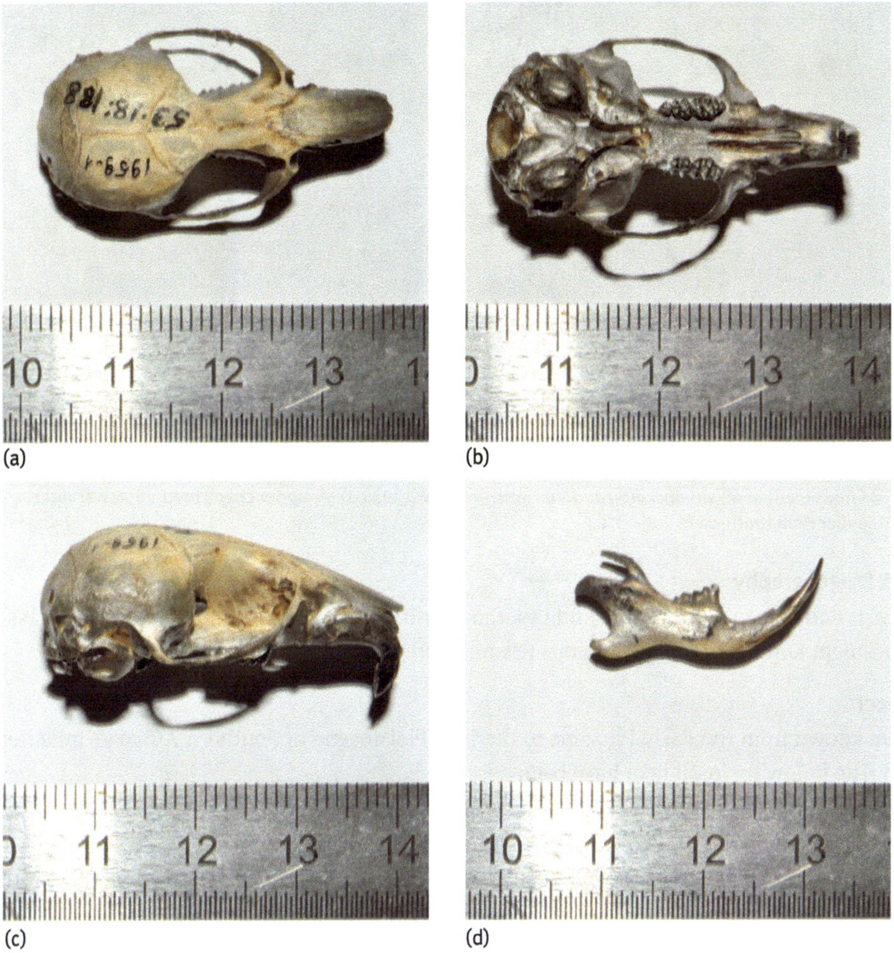

Figure 2.271 a-d Cranium of *Mystromys albicaudatus* (MNHN-ZM-MO 1959-1) (from top left to bottom right: dorsal view, ventral view, lateral view and lower jaw).

Dentition

The molar cusps are well separated and connected by a longitudinal crest that zigzags between the cusps from anterior to posterior (in an alternate arrangement). Molar M^1 displays three lobes of cusps: the first has one or two well fused cusps, the second and third have two cusps each, with a trace of a posterior cingulum on the latter. Molar M^2 has two lobes each with two cusps, and a tiny posterior cingulum. Molar M^3 is the smallest in the toothrow and has a tiny cusp and three main cusps connected on the lingual side. Molar M_1 is the longest tooth in the lower molar toothrow and has three lobes. The first lobe has a single cusp with a tiny antero-external crestiform cingulum and is connected to the second lobe by a longitudinal crest. The second and third lobes are also connected by a large median longitudinal crest. Molar M_2 has an antero-external cingulum, two rows of alternated cusps and a large posterior cingulum. Molar M_3, which is highly reduced, has a tiny antero-external cingulum and displays three cusps.

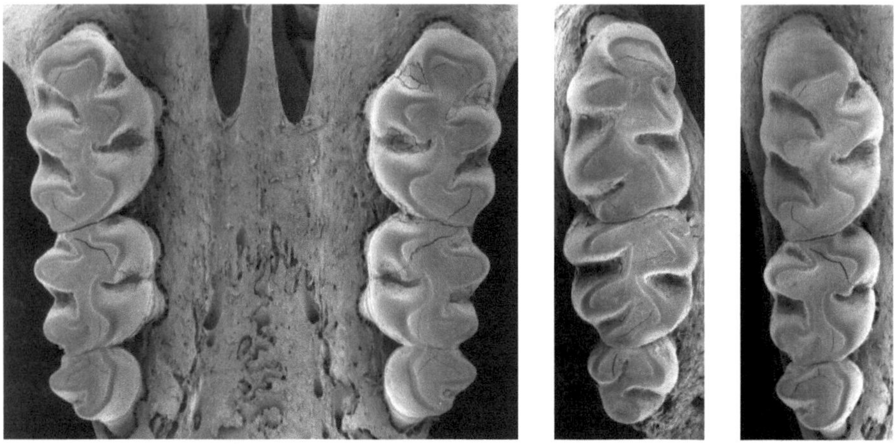

Figure 2.272 Cheek teeth of *Mystromys albicaudatus* (MNHN-ZM-MO 1959-1) (A: upper cheek teeth; B: lower right toothrow; C: upper right toothrow).

Habitat & Biogeography

Mystromys is endemic to South Africa and Lesotho, south of 25°S where it inhabits savanna grasslands, succulent Karoo and fynbos habitats (Dean, 1978).

Fossil Record

Fossils are known from the Early Pliocene to the Late Pleistocene in Southern Africa (Winkler et al., 2010). The following fossil taxa have been described:
Mystromys pocokei Langebaanweg, 5 MYA (Denys, 1991).
M. cf *pocockei* Waypoint 160, 4 MYA (Senegas, 2000).
M. cf *hausleitneri* Waypoint 160, 4 MYA (Senegas, 2000).
M. hausleitneri Broom, 1937 (3.3-1.0) (De Graaf, 1960; Pocock, 1987; Senegas, 2000).
M. antiquus Taung, Sterkfontein, Kromdraai, Makapansgat, Bolt's Farm, Gladysvale (De Graaff, 1960).
Proodontomys cookei Sterkfontein, Kromdraai, Makapansgat, Bolt's Farm, Gladysvale, Swartkrans, Taung, Drimolen, Gondolin, 3.3-1.0 MYA (Pocock, 1987).
Many more recent Upper Pleistocene fossil *Mystromys* attributed to the modern species have been recovered across South Africa and listed in Avery (1998).

Biology & Ecology

This terrestrial, nocturnal rat is herbivorous, exhibiting specialised features associated with a typical ruminant digestive system (Maddock & Perrin, 1981; Downs & Perrin 1995). It has a high metabolic rate well adapted to the temperate conditions prevalent on the South African highveld especially during winter. The female has two pairs of mammae and gives birth to 1-6 offspring after 37 days of gestation, and may be reproductively active throughout the year (Hallett & Meester, 1971; Rautenbach, 1982; Lynch, 1983). Several species of owl are known predators of this genus (Dean, 1978).

Systematic Notes

This genus was initially classified as a Cricetidae, the genus was then placed in its own subfamily Mystromyinae by Vorsontsov (1966). Lavocat (1973) suggested its probable relationships with the extinct subfamily of Myocricetodontinae rodents and its origins to be of ancestral Cricetinae stock. Molecular studies have confirmed that *Mystromys* belongs in the Nesomyidae family and represents a separate ancient lineage (the Mystromyinae) divergent from the Cricetomyinae, Dendromurinae, Petromyscinae and Nesomyinae (Jansa et al., 1999). Based on these analyses the closest relative of *Mystromys* is *Petromyscus*. Denys (1994) also stressed the dental differences between *Mystromys, Delanymys* and *Petromyscus* with regards to the available South African fossils *Stenodontomys, Proodontomys* and *Mystromys*. Recently, Schenk et al. (2013) confirmed that *Mystromys* belongs to a basal clade of Nesomyidae and is taxonomically closer to *Petromyscus* than *Delanymys*.

Species List

This genus is monotypic and only one species is presently recognized
Mystromys albicaudatus (A. Smith, 1834)

Species Account

Mystromys albicaudatus (A. Smith, 1834)

Diagnosis

See genus account.

Taxonomic Issues

See genus account.

Distribution

This species is endemic to the grasslands, Karoo, and fynbos habitats of South Africa and Lesotho.

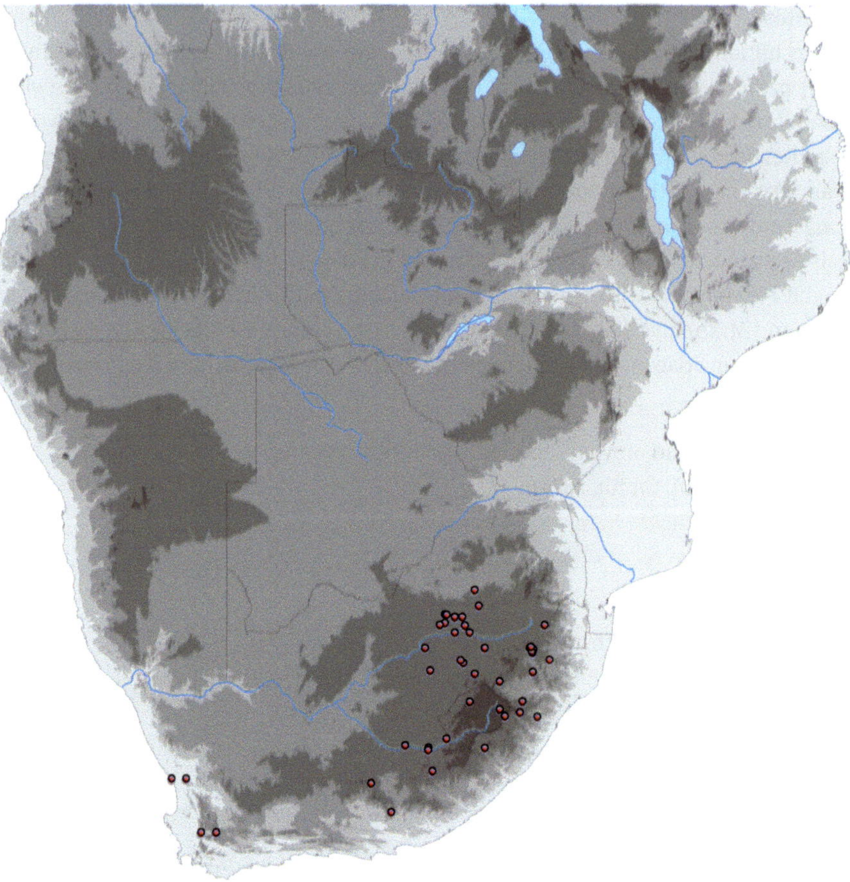

Figure 2.273 Distribution map

Measurements

External and cranial measurements (mm) and mass (g) for *Mystromys albicaudatus*, sexes combined

	Mean	Min	Max	N
Mass	80.8	49	111	6
HB	150.2	116	199	23
Tail	65.3	46	87	23
HF	26.8	23	32	20
Ear	23.5	18	28	22
GLS	35.0	32.2	37.7	11

Subfamily: Petromyscinae

Genus Account

Petromyscus Thomas, 1926 (Pygmy Rock Mice)

Diagnosis

These very small to small rodents (HB ca. 75-85 mm) are characterized by a relatively long tail (98 -113% of HB), fine, soft and silky brown pelage, short and broad feet. The skull is characterized by a long rostrum and the braincase is wide and flat. The anterior palatal foramina are short and end at the M^1, and the teeth are relatively small. The dentition is highly specialized and has a unique pattern with a t4 and a longitudinal crest connecting the rows of cusps. *Petromyscus* is adapted to arid regions of south-western Africa.

Figure 2.274 Photograph of *Petromyscus* sp. (Galen Rathbun).

Cranial Characters

The skull is small in size (GLS ca. 22-26 mm), with a short rostrum, a wide interorbital constriction and a rounded braincase. The zygomatic arches are thin and parallel-sided. The incisors are opisthodont. The anterior palatal foramina reach the anterior root of M^1 or in some species extend between them. The molar rows are parallel. The bullae are small and non-inflated (4.7-5.0 mm in diameter).

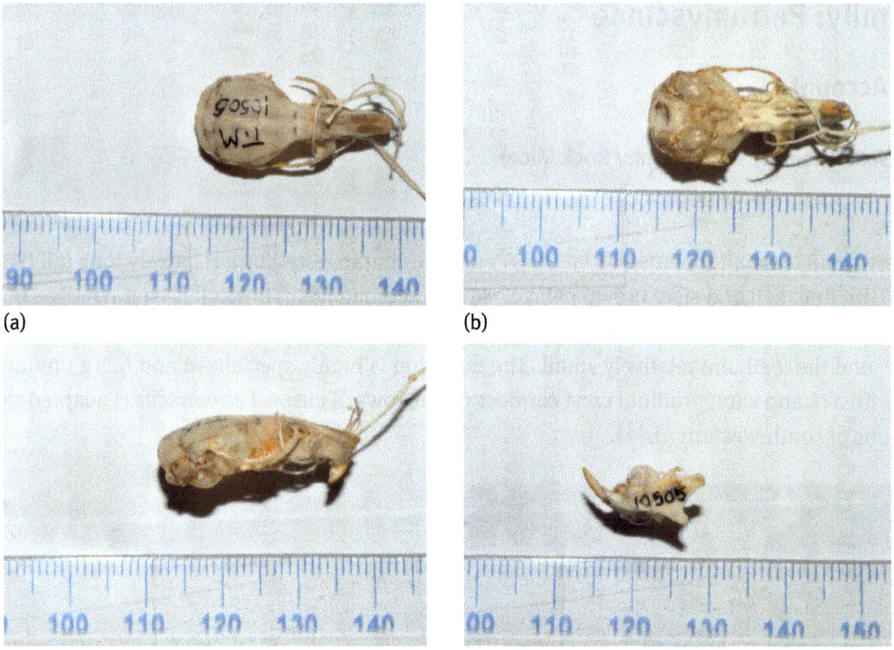

(a) (b) (c) (d)

Figure 2.275 a-d Cranium of *Petromyscus collinus* (TM 10505) (from top left to bottom right: dorsal view, ventral view, lateral view and lower jaw).

Dentition

The upper molars have round cusps that are connected by a median longitudinal crest running the length of the tooth; the cusps are also connected by transverse crests. Molar M^1 has three lobes, the first has one or two fused cusps, the second has three cusps (the most lingual being the t4) and the third has two cusps. Molar M^2 has two lobes connected longitudinally and an antero-external cusp (t3) connected to the t5; the t1 and t4 are absent. Molar M^3 molar is the smallest in the toothrow and has a tiny t3, no t1 and two rows of cusps connected on the lingual side of the tooth. Molar M_1 is the longest tooth in the lower toothrow with three lobes that have high and well separated cusps. The first and second lobes have their three cusps fused in their middle by a longitudinal crest; the third lobe has alternated cusps and a small posterior cingulum. Molar M_2 has an antero-external crestiform cingulum, two rows of longitudinaly alternated cusps and a posterior cingulum. Molar M_3 is highly reduced and has a tiny antero-external cingulum and two cusps that are connected lingually.

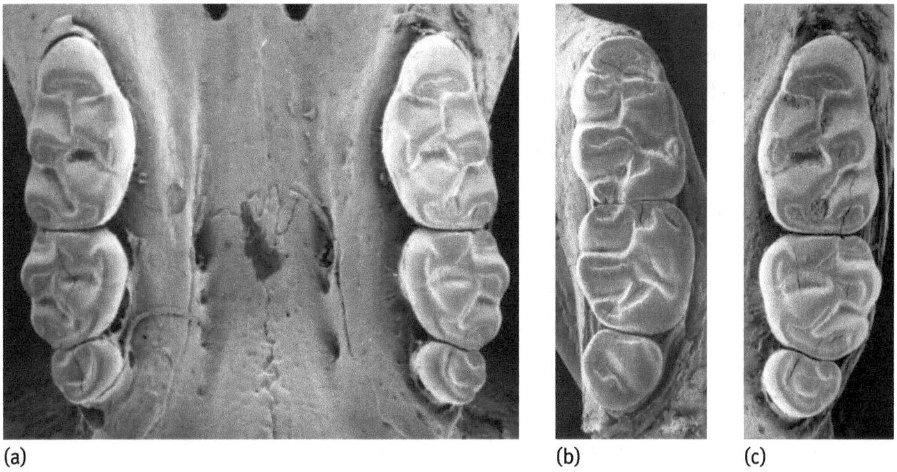

Figure 2.276 Cheek teeth of *Petromyscus collinus* (MNHN-ZM-MO 1959-200) (A: upper cheek teeth; B: lower right toothrow; C: upper left toothrow).

Habitat & Biogeography

Petromyscus is a rupicolous rodent restricted to rocky habitats in arid regions of south-western Africa. It inhabits the Namib Desert in Namibia and south-western Angola, and the Karoo of north-western South Africa occurring up to 1500 m.

Fossil Record

Harimyscus from the late Miocene of Namibia (11 MYA) is described as an early Petromyscinae (Mein et al., 2004).

Biology & Ecology

These nocturnal rodents shelter during the day in fissures and crevices on rocky outcrops (Withers, 1979). Their diet consists mostly of seeds but also includes stems and insects (Withers, 1979). They are generally solitary; males and females occupy separate home ranges, with little overlap between those of males (Withers, 1979). Reproduction is seasonal. Females have 2-6 pairs of nipples and give birth to 2-8 offspring per litter (Dempster & Perrin, 1989; Withers, 1983). Small carnivores, owls and snakes are the main predators (Happold, 2013).

Systematic Notes

Roberts (1951) recognised the Petromyscinae as a separate family based upon morphological characters. Hayman (1962) then placed *Petromyscus* and *Delanymys* into Dendromuridae while Petter (1967) suggested their relationships within Petromyscinae sensu Roberts due to their dental characters. Carleton & Musser (1984) and Denys (1994) suggested that *Petromyscus* is the single representative of Petromyscinae. Molecular analyses confirmed that it is only distantly related to the Dendromurinae (Jansa et al., 1999). Schenk et al. (2013) confirmed that it belongs to the family Nesomyidae where it is closer to *Mystromys* than to *Delanymys*. This genus is urgently in need of revision.

Species List

Four species are presently recognized.

Petromyscus barbouri Shortridge & Carter, 1938
Petromyscus collinus (Thomas & Hinton, 1925)
Petromyscus monticularis (Thomas & Hinton, 1925)
Petromyscus shortridgei Thomas, 1926

Key to Species
This key has been adapted from Skinner & Chimimba (2005).

1	Relative tail length shorter (tail typically < 100% of HB)	*P. monticolaris*
	Tail relatively longer (ca. 105% or more of HB)	2
2	3 pairs of nipples (1 + 2 = 6)	*P. collinus*
	2 pairs of nipples (0 + 2 = 4)	3
3	Size on average larger (GLS typically ca. 25-27 mm); occurring in northern Namibia	*P. shortridgei*
	Size on average smaller (GLS typically ca. 22-23 mm); occurring south of the Gariep River in South Africa	*P. barbouri*

Species Accounts

Petromyscus barbouri Shortridge & Carter, 1938

Diagnosis
This species co-occurs only with *P. collinus* from which it can be distinguished by relatively shorter, bicoloured tail (tail ca. 110% of HB), and 2 pairs of nipples (the axillary pair being absent). In *P. collinus* the tail is not bicoloured, is relatively longer (ca. 115% of HB), and it has 3 pairs of nipples.

Taxonomic Issues
This species was previously listed as a subspecies of *P. collinus* (De Graaff, 1981) but recognised as a valid species by Musser & Carleton (2005).

Distribution
This species is restricted to the arid parts of south-western South Africa.

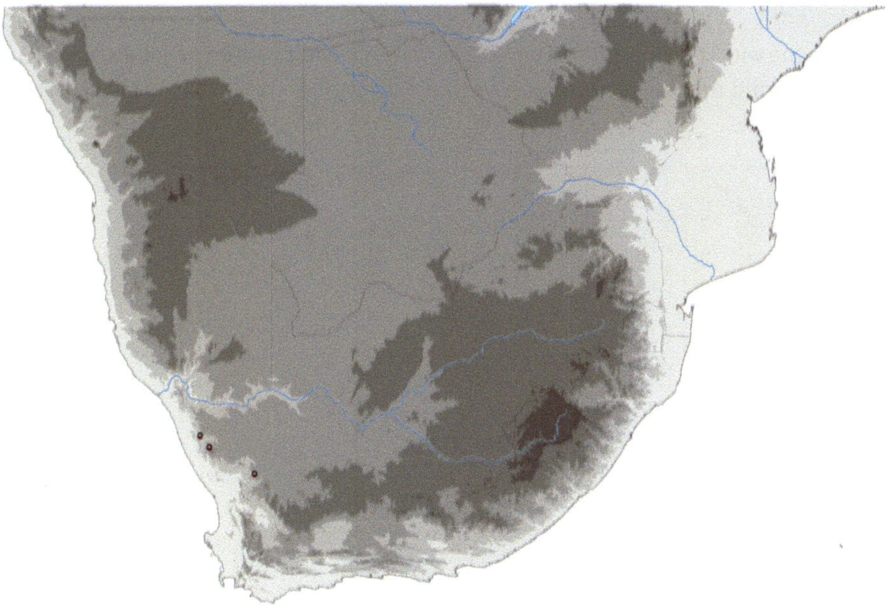

Figure 2.277 Distribution map

Measurements

External and cranial measurements (mm) and mass (g) for *Petromyscus barbouri*, sexes combined

	Mean	Min	Max	N
Mass	15.3	14	16	3
HB	73.7	71	75	3
Tail	81.7	76	89	3
HF	18.0	17	19	3
Ear	14.3	14	15	3
GLS*	22.7	22.4	22.9	2

*Happold (2013) presented a mean of 23.0 (range: 21.5-26.3, n = 21)

Petromyscus collinus (Thomas & Hinton, 1925)

Diagnosis

This species co-occurs with all the other three species of *Petromyscus* including: *P. barbouri* (tail relatively shorter, ca. 110% of HB; tail is bicoloured; 2 pairs of nipples); *P. monticolaris* (tail relatively shorter, ca. 100% or less of HB; ears shorter, typically < 13 mm); and *P. shortridgei* (tail relatively shorter, ca. 105% of HB; tail is bicoloured; 2 pairs of nipples). In *P. collinus* the uniformly coloured tail is relatively longer (ca. 115% of HB) and there are 3 pairs of nipples.

Taxonomic Issues

The taxa *barbouri* and *shortridgei* were previously listed as subspecies of *P. collinus*, but are considered valid species here until a phylogenetic analysis has been conducted for the species.

Distribution

This species has the widest distribution of any *Petromyscus* extending from the southern Karoo of South Africa northward to south-western Angola where it occupies rocky outcrops in desert and semi-arid environments.

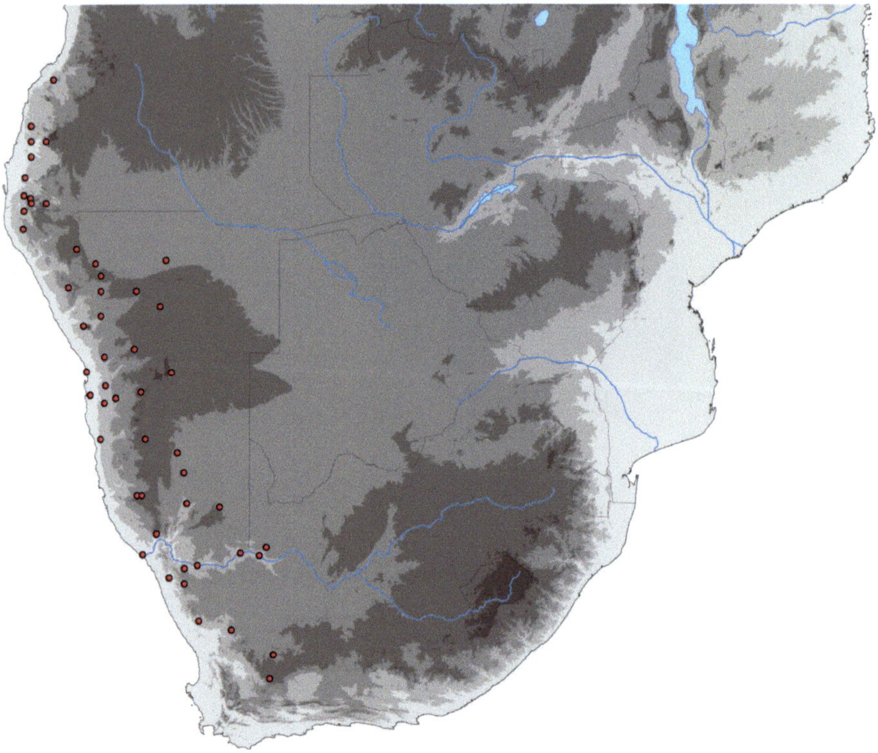

Figure 2.278 Distribution map

Measurements

External and cranial measurements (mm) and mass (g) for *Petromyscus collinus*, sexes combined

	Mean	Min	Max	N
Mass	19.9	14	24	17
HB	82.4	72	93	20
Tail	95.7	78	122	20
HF	16.5	15	21	18
Ear	15.2	12	18	19
GLS	25.3	24.4	26.0	11

Petromyscus monticularis (Thomas & Hinton, 1925)

Diagnosis

This species co-occurs only with *P. collinus* from which it can be distinguished by its shorter tail (typically < 100% of HB) and shorter ears (in *P. collinus* relative tail length is ca. 115% of HB and ears typically > 13 mm). This species has 3 pairs of nipples.

Taxonomic Issues

The phylogenetic relationships of the various species of *Petromysus* are currently unknown and a molecular analysis is urgently needed.

Distribution

This species is restricted to arid habitats in southern Namibia and neighbouring South Africa where it occurs in rocky outcrops.

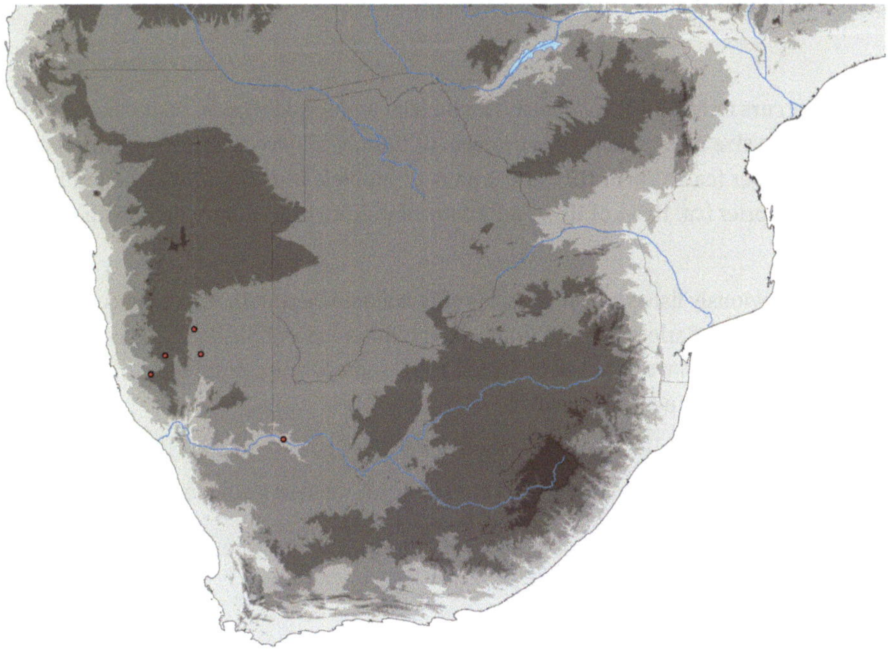

Figure 2.279 Distribution map

Measurements

External and cranial measurements (mm) and mass (g) for *Petromyscus monticularis*, sexes combined

	Mean	Min	Max	N
Mass	-	-	-	-
HB*	75.3	63	93	27
Tail*	73.5	66	84	26
HF*	15.6	13	17	28
Ear*	11.6	10	14	28
GLS*	22.6	19.0	26.2	28

*Happold (2013)

Petromyscus shortridgei Thomas, 1926

Diagnosis

This species co-occurs only with *P. collinus* from which it cannot easily be distinguished. The two species are similar in size but *P. collinus* has a longer, uniformly coloured tail (ca. 95 mm in length) that is relatively longer (ca. 115% of HB), and 3 pairs of nipples. In *P. shortridgei* the tail is bicoloured, relatively shorter (ca. 105% of HB), and 2 pairs of nipples (the axillary pair being absent).

Taxonomic Issues

This species was previously listed as a subspecies of *P. collinus* (De Graaff, 1981), but considered a valid species by Musser & Carleton (2005) from which it differs morphologically. The two species appear to be sympatric but have not been collected at the same locality (Happold, 2013). A molecular analysis of this genus is urgently needed.

Distribution

This species is known from a few scattered localities in north-western Namibia and south-western Angola.

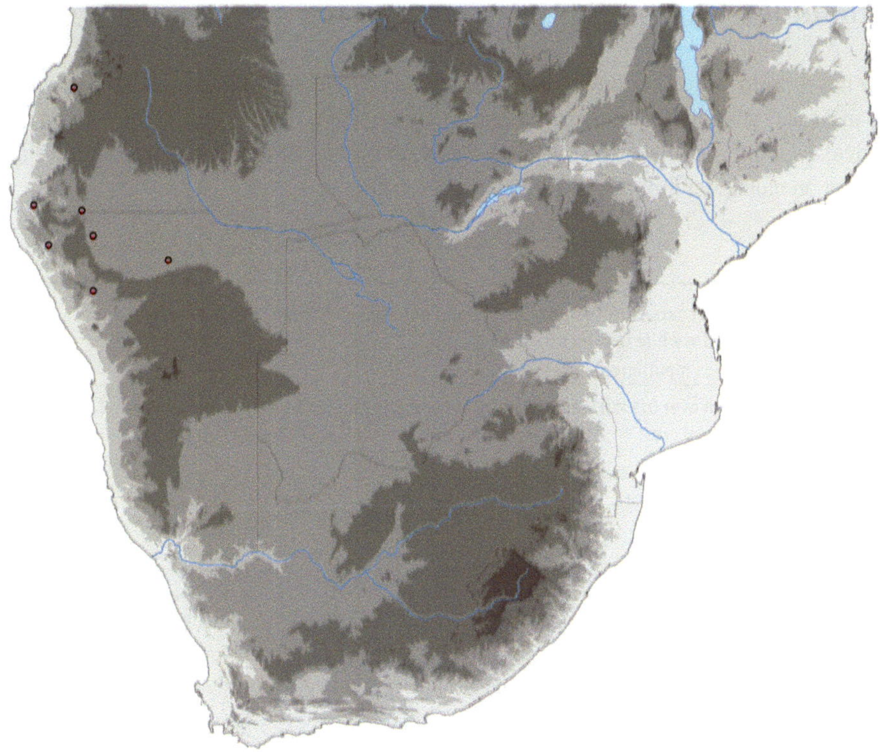

Figure 2.280 Distribution map

Measurements

External and cranial measurements (mm) and mass (g) for *Petromyscus shortridgei*, sexes combined

	Mean	Min	Max	N
Mass*	18.9	12	31	36
HB*	83.4	65	98	85
Tail*	87.4	74	104	81
HF*	16.5	15	20	82
Ear*	15.0	13	18	82
GLS*	25.7	24.0	28.8	58

*Happold (2013)

Family: Muridae

This family has a world-wide distribution. It is represented by five subfamilies and 309 species in Sub-Saharan Africa, making it the largest rodent family in Africa. There are currently no known morphological features that distinguish it from the Nesomyidae, although the two families are clearly distinct on molecular grounds (Jansa & Weksler, 2004).

Key to Subfamilies

1	Dorsal pelage with spiny or coarse fur; OR muzzle long and pointed, ears large and oval, ventral pelage pure white (*Deomys*)	Deomyinae
	Not possessing the features mentioned above	2
2	Large rodent (typically well over 0.5 kg) with distinctive and boldly marked black and white pattern with dorsal crest	Lophiomyinae
	Not as above	3
3	M^3 the largest molar; molars laminate	Murinae (*Otomys* and *Parotomys*)
	M^1 the largest molar; molars not laminate	4
4	First lobe of M^1 with three cusps (t1 present)	Murinae (except *Otomys* and *Parotomys*)
	First lobe of M^1 with two cusps (t1 absent)	5
5	Upper molars cuspidate or prismatic; tail typically long; hindfoot relatively long; widespread in Africa	Gerbillinae
	Upper molars with simple laminae; tail short (30% of HB); hindfoot not relatively long; known from a single locality in Togo	Leimacomyinae

Subfamily: Deomyinae

Key to Genera

1	First lobe of M^1 with two cusps (t1 absent); muzzle long and pointed	*Deomys*
	First lobe of M^1 with three cusps (t1 present)	2
2	Dorsal pelage spiny	*Acomys*
	Dorsal pelage not spiny (but may be coarse and brush-like)	3
3	Ventral pelage bright orange, red or yellowish-orange; nipples 2 or 3 pairs; size larger (HF > 18 mm); upper incisors opisthodont	*Lophuromys*
	Ventral pelage white or off-white; nipples 6 pairs; size smaller (HF 16-18 mm); upper incisors pro-odont	*Uranomys*

Genus Account

Acomys I. Geoffroy, 1838 (Spiny Mice)

Diagnosis

These small to medium sized rodents (HB ca. 85-120 mm), display a long-tail equal or slightly smaller than HB (76-100% of HB), and have characteristic spiny dorsal pelage. They also possess an extended palate which closes the mesopterygoid fossa and is constituted by an elongation of the palatine bone and partial fusion of the pterygoids, a character shared with *Uranomys*. Members of the genus can be distinguished from *Uranomys* by the longer tail, and the spiny hairs on the back. They are distinct from *Deomys* by their smaller tail which does not end in a tuft of hair. They have white ventral pelage and a longer tail compared to *Lophuromys* species which often have a red or dark brown pelage and a very short tail. Most *Acomys* species are inhabitants of savanna and desert zones often associated with rocky situations, extending from the Sahara to the Cape fynbos and are widespread across Africa except in forests and high altitude montane habitats. Outside Africa, *Acomys* species are found in Turkey, Arabia, Middle East, India and Afghanistan.

Figure 2.281 Photograph of *Acomys spinosissimus* (Ara Monadjem).

Cranial Characters

The skull is of small to medium size (GLS ca. 23–31 mm), displaying ridges in old individuals, especially in the case of the supraorbital crest. The zygomatic arches are robust and not parallel, converging anteriorly. The brain case is inflated and wide in the middle in lateral view and the skull is arched in this view. The nasal bone is long, more or less narrow depending on the species, extending anterior of the incisors. The incisors are opisthodont and ungrooved. The anterior palatal foramina are long and reach the t4 of the upper first molar. The palate itself is extended to cover the pterygoid fossae and reaches the tympanic bone. The anterior portion of the zygomatic plate is vertical. The bullae are relatively small (3.5-6.5 mm in greatest diameter).

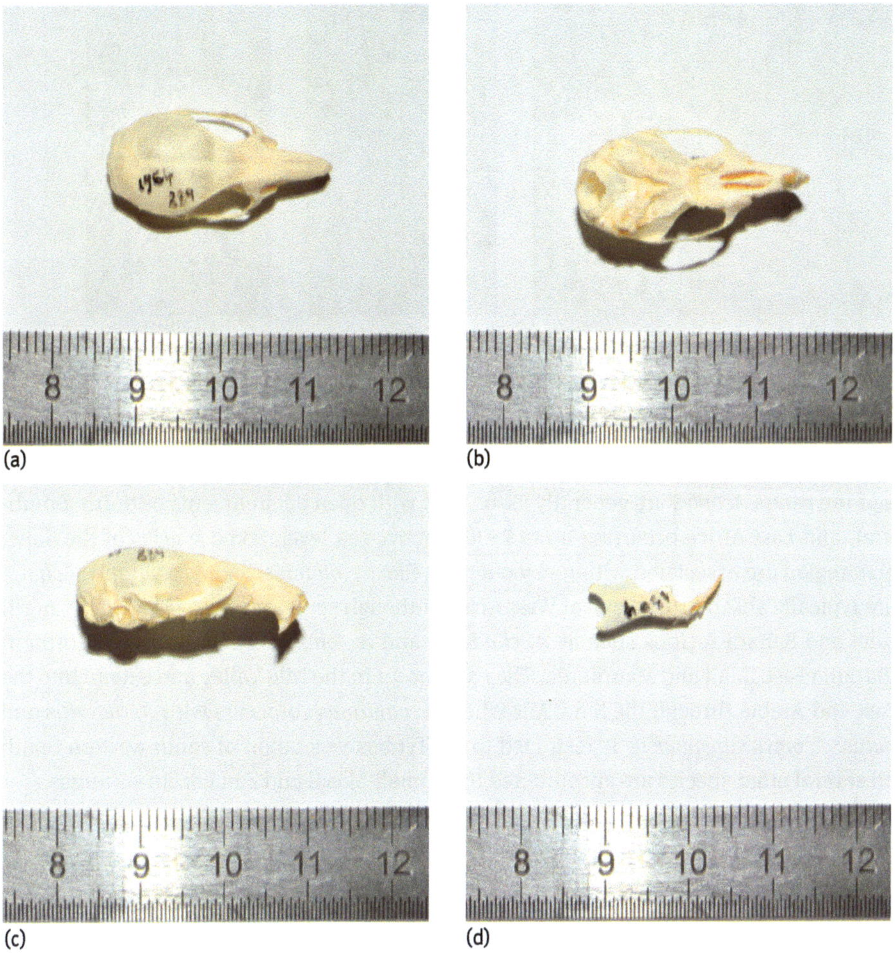

Figure 2.282 a-d Cranium of *Acomys ngurui* (MNHN-ZM-MO 1964-889) (from top left to bottom right: dorsal view, ventral view, lateral view and lower jaw).

Dentition

The individual cusps of the molars are fused into a number of convex chevrons that superficially resemble the Murinae. The first molar is the longest of the toothrow and has two anterior lobes with three cusps, including a t1 and a t4 on the internal side, the third lobe has two cusps plus a posterior cingulum connected to the latter. Some species show a supplementary cusp on the third lobe of the M^1 (t7). The M^3 is the smallest in the toothrow (18% of the dental row), and displays an antero-external cusp or cingulum (t3) and no internal one (t1) in contrast to the true Murinae (such as *Mus*, Praomyini or Arvicanthini) (Denys & Michaux, 1992). The M_1 is the longest in the lower toothrow, having two anterior laminae and in some species a median anterior cusplet connected to the first row of cusps. An antero-external cingulum is present on the M^2 and M^3. The M_2 and M_3 always have two rows of cusps. There is a posterior cingulum on M_1 and M_2. The two upper toothrows are nearly parallel.

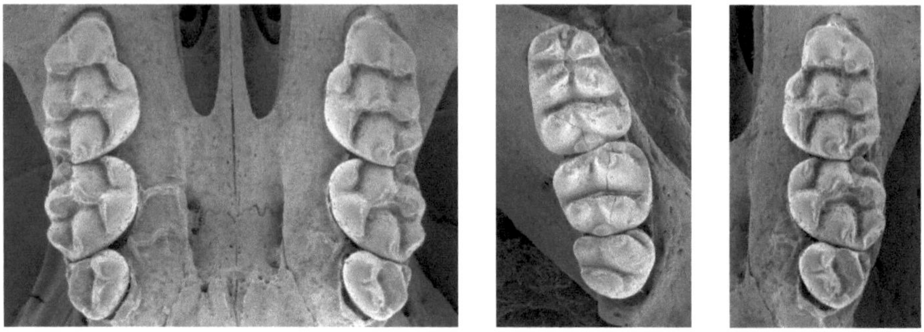

(a) (b) (c)

Figure 2.283 Cheek teeth of *Acomys chudeaui* (MNHN-ZM-MO 1975-424) (A: upper cheek teeth; B: lower right toothrow; C: upper left toothrow).

Habitat & Biogeography

Members of the genus *Acomys* are generally associated with open environments of North, Southern, Central, and East Africa occurring up to 2500m above sea level. Some species of the Sahelo-Sudanian region are associated with rocky outcrops like *A. johannis,* and *A. cahirinus. Acomys* species are typically absent from tropical West Africa although several species occur to the north in the Sahel and Sahara regions such as *A. chudeaui* and *A. johannis* in Chad, Niger, northern Nigeria, Burkina Faso, Mali and Mauritania. They also occur in the Nile Valley and extend into the Middle East and Arabia through the Sinaï Mts where *A. cahirinus* co-occurs with *A. russatus* and *A. dimidiatus. Acomys subspinosus* is restricted to the fynbos vegetation of south-western South Africa and several other species are encountered the Somali-Masai and Zambezian savannas.

Fossil Record

Different extinct genera have been described whose affinities with the Deomyinae are probable but not fully clear like *Preacomys* found in Chorora (Ethiopia), Harasib (Namibia), and Lukeino (Kenya) between 11-6 MYA; or *Tectonomys* at around 5 MYA in Tanzania. Among the different species of these genera, *P. griffini* from Harasib may be the direct ancestor of *A. mabele* in that they share very similar dental patterns but differ in size.

The oldest fossil attributed to the modern genus is recorded from Lemudong'o (Kenya) at 6 MYA (Manthi, 2007) and at around 6-4.5 MYA in South Africa (Pocock, 1987; Denys, 1990c) and at around 4.5-3 MYA in Ethiopia and Tanzania (Denys, 1987a; Louchart et al., 2009). The genus is also recorded in Pliocene sites of South and East Africa, and in the Holocene of Egypt (Pocock, 1987; Senegas, 2000).

Biology & Ecology

Spiny mice are diurnal or semi-diurnal rodents and their diet comprised mainly seeds, insects and sometimes grasses. Females of *A. cahirinus* give birth after 38 days of gestation usually to two to five offspring which are nidifugous and are mature at the age of 2-3 months (Dieterlen, 1961). In the same species in captivity, Dieterlen (1962, 1963) also observed females that assisted with labour, birth and adoption of offspring of other females. They have a number of predators including felids, jackals, mongooses, genets, snakes and different owl species but once taken by a predator they can lose parts of their fragile skin or their tail which can regenerate faster than in rats or mice (Seifert et al., 2012). In Egypt, *A. cahirinus* is a frequent inhabitant of houses.

Systematic Notes

Initially considered as a sister taxon of *Mus* by Ellerman (1941), however the earliest molecular studies showed that *Acomys* is more closely related to gerbils than to the Murinae (Chevret et al., 1993a). These authors proposed that *Acomys, Lophuromys,* and *Uranomys* belong to a monophyletic clade, which was also evident morphologically (Denys et al., 1992; Denys & Michaux, 1992). Later, with the inclusion of *Deomys*, Michaux & Catzeflis (2000) proposed the name Acomyinae for this clade. Musser & Carleton (2005) changed the subfamily name to Deomyinae for nomenclatural reasons. However, no morphological synapomorphy has yet been found that is shared by these four taxa.

A morphological and cytogenetical review of all *Acomys* species was made by Denys et al. (1994) who provided the external, skull and dental characteristics of all the type specimens available and validated *A. cahirinus* as distinct from *A. dimidiatus*. These authors also pointed that *A. subspinosus* is distinct from other *Acomys* species and the subgenus *Subacomys* was proposed which may in fact represent a distinct genus. Petter & Roche (1981), and Petter (1983) created the subgenus *Peracomys* for the Ethiopian species displaying a t7 on the upper M1 (*A. louisae*), a distinction supported by Denys et al. (1994), but not yet on molecular grounds.

A Cytochrome-b phylogeny of most of the African species has allowed Barome et al. (2000) to confirm the basal position of *A. subspinosus*. In the *cahirinus-dimidiatus* species complex some cytogenetic characters allowed the delimitation of these two species (Volobouev et al., 2002a, 2007a) and to exclude the presence of *A. cahirinus* from Sinai. A revision of the *A. spinosissimus* complex of East and Southern Africa by Barome et al (2001) suggested that *A. selousi* may be a distinct species. A recent study of some Sahelian and Saharan taxa based on molecular, cytogenetic and morphological analyses has shown that *A. airensis* is conspecific with *A. chudeaui* and is its junior synonym (Nicolas et al., 2009). Verheyen et al. (2011) described two new species (*A. muzei* and *A. ngurui*) from Tanzania and elevated *A. selousi* as a valid species. The taxon *A. johannis* may represent a species complex with two putative species listed by Barome et al. (2000) from Burkina Faso, named *Acomys* sp1 and sp2, that differ in karyotype; whether this karyotypic variability represents specific distinction has yet to be determined.

Species List

Fifteen species are known to occur in the region.

Acomys cahirinus (Desmarest, 1819)
Acomys chudeaui (Kollman, 1911)
Acomys cineraceus Heuglin, 1877
Acomys ignitus Dollman, 1910
Acomys johannis Thomas, 1912
Acomys kempi Dollman, 1911
Acomys louisae Thomas, 1896
Acomys mullah Thomas, 1904
Acomys muzei Verheyen et al., 2011
Acomys ngurui Verheyen et al., 2011
Acomys percivali Dollman, 1911
Acomys selousi De Winton, 1896
Acomys spinosissimus Peters, 1852

Acomys subspinosus (Waterhouse, 1838)
Acomys wilsoni Thomas, 1892

Key to Species

The species within this genus have yet to be fully resolved and the taxonomy presented here may well be out of date soon after publication of this book. Species limits are still poorly known and some species are only distinguishable on morphological grounds and distribution. This key relies on the assumption that species distributions are not overlapping. It should also be kept in mind that there is much overlap in size between the different species, so a single individual may be problematic to identify, and therefore this key is more likely to work when a series of (e.g. 5 or 6) individuals are available from a locality.

1	Restricted to West Africa (Chad west to Mauritania)	2
	Restricted to East Africa (Sudan, Ethiopia, Somalia, Uganda, Kenya and Tanzania)	3
	Restricted to Southern Africa (Zambia, southern DRC, Mozambique, Malawi and southwards to South Africa)	12
2	Dorsal pelage paler; occurs on the left bank of the Niger River; 2n = 40-46	*A. chudeaui*
	Dorsal pelage darker; occurs on the right bank of the Niger River; 2n = 66-68	*A. johannis*
3	Size larger (typically GLS > 27 mm)	4
	Size smaller (typically GLS < 27 mm)	8
4	Tail relatively very long (> 100% of HB); t7 cusp present on first upper molar	*A. louisae*
	Tail relatively shorter (ca. 100% or less of HB)	5
5	No orange or russet in dorsal pelage; restricted to north-eastern Ethiopia, Djibouti and north-western Somalia	*A. mullah*
	Some orange or russet present in dorsal pelage or flanks; may occur in northern Ethiopia	6
6	Flanks bright orange; tail relatively short (< 80% of HB); 2n = 50	*A. ignitus*
	Flanks not bright orange; tail usually ca. 90% or more of HB	7
7	Tail shorter (< 90 mm); occurs mostly in the southern parts of Sudan	*A. cineraceus*
	Tail longer (> 100 mm); occurs in central and northern parts of Sudan and northwards into Egypt and east into Ethiopia; 2n = 36-38	*A. cahirinus*
8	Dorsal pelage grey or greyish-brown without any orange or russet	*A. percivali*
	Dorsal pelage or flanks with some bright orange or russet	9
9	Tail relatively short (< 80% of HB); 2n = 60	*A. wilsoni*
	Tail relatively longer (typically ca. 90% or more of HB)	10
10	Occurs in rocky outcrops in semi-deserts of northern Kenya, southern Somalia and extreme southern Ethiopian	*A. kempi*
	Occurs exclusively in Tanzania	11
11	Tail relatively longer (ca. 100% of HB); occurs at higher altitudes west of the Eastern Arc mountains	A. muzei
	Tail relatively shorter (ca. 90% of HB); occurs at lower altitudes east of the Eastern Arc mountains	*A. ngurui*
12	Restricted to the fynbos in the southern parts of South Africa; palatal foramina are short; molars are bunodont; 2n=64	*A. subspinosus*
	Occurring in savanna in the northern parts of South Africa and northward	13
13	Occurring in southern Zimbabwe, eastern Botswana and northern South Africa; X chromosome very large	*A. selousi*
	Occurring further north	*A. spinosissimus*

Species Accounts

Acomys cahirinus (Desmarest, 1819)

Diagnosis
This is the only species of *Acomys* across most of its distribution, possibly overlapping with *A. cineraceus* (mean tail length shorter < 90 mm) to the south and *A. louisae* (tail relatively long > 100% of HB) and *A. mullah* (dorsal pelage without russet or cinnamon) in the east. Its dorsal pelage is variable but usually with russet or cinnamon.

Taxonomic Issues
A record from central Ethiopia has a radically different karyotype (2n = 68, FN = 68) from typical *A. cahirinus* (2n = 36, FN = 68) (Sokolov et al., 1993) and may refer to a separate species. At present only four specimens from the Rift Valley of Ethiopia have been karyotyped, three referring to the typical *A. cahirinus* karyotype and one to the 2n = 68 karyotype. Therefore the specific status of records from Ethiopia is currently unclear and numerous cryptic species are suspected (Lavrenchenko et al., 2011). Elsewhere in its range, 2n = 36-38 (Sokolov et al., 1993).

Distribution
This species occurs widely in the Sahara, the Nile Valley and extending into the Middle East. Within the region it occurs in Sudan and Ethiopia, although the status of the central and southern Ethiopian records (as plotted by Yalden et al., 1976) is currently not known and may refer to different species (Sokolov et al., 1993; Lavrenchenko et al., 2011).

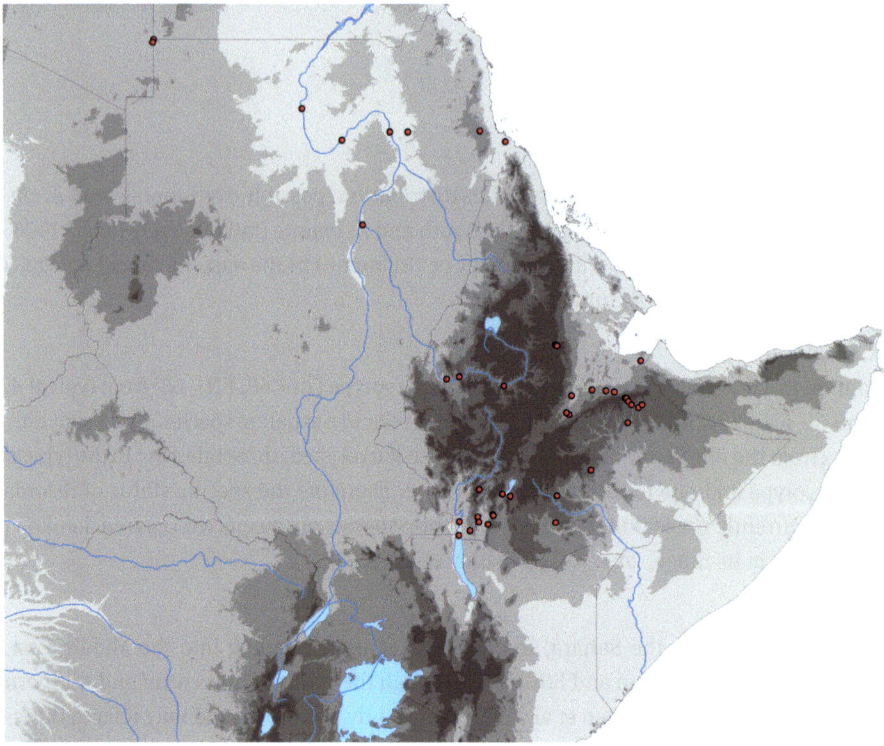

Figure 2.284 Distribution map

Measurements

External and cranial measurements (mm) and mass (g) for *Acomys cahirinus*, sexes combined

	Mean	Min	Max	N
Mass*	40.8	29.9	51.5	20
HB*	111	100	120	34
Tail*	102	92	137	30
HF*	19.9	18	21	27
Ear*	20.8	18	24	35
GLS*	29.9	27.6	31.4	34

*Happold (2013)

Acomys chudeaui (Kollman, 1911)

Diagnosis

This is the only species of *Acomys* within its Sub-Saharan range. *Acomys johannis* occurs on the right bank of the Niger River, its dorsal pelage is darker, and it differs in karyotype: 2n = 66-68 (Granjon & Duplantier, 2009).

Taxonomic Issues

This taxon was previously listed as a synonym of *A. cahirinus*, but recent work shows that it is distinct from that species and conspecific with *A. airensis*, the latter becoming the junior synonym. 2n = 40-46 (Granjon & Duplantier, 2009).

Distribution

This species is widely distributed in the western Sahel and north to the edge of the Sahara, and south to the Niger River (occurring only on the left bank).

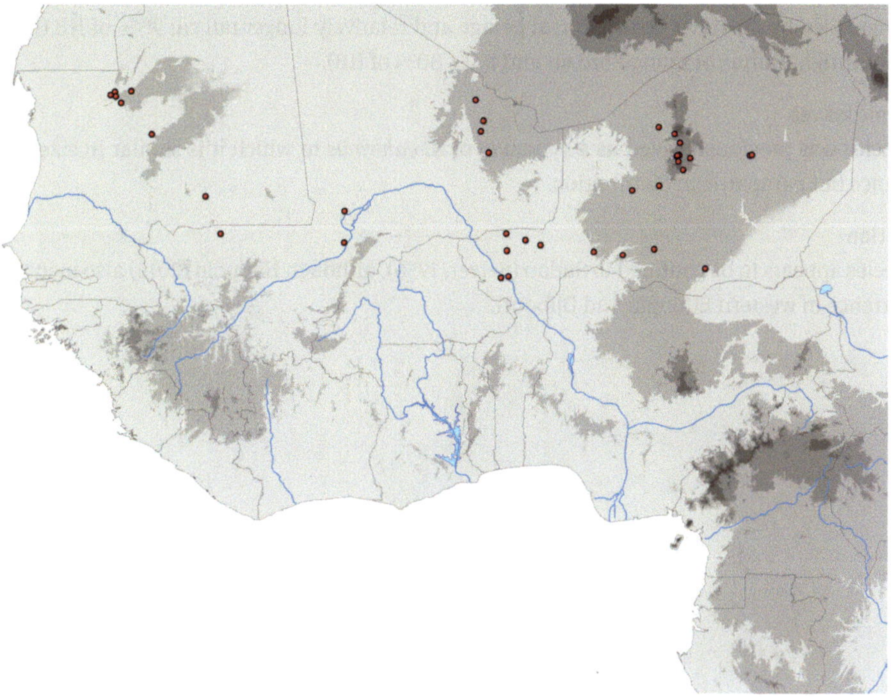

Figure 2.285 Distribution map

Measurements

External and cranial measurements (mm) and mass (g) for *Acomys chudeaui*, sexes combined

	Mean	Min	Max	N
Mass	40.0	27	53	3
HB	96.0	81	118	7
Tail	91.2	80	110	6
HF	17.1	16	18	7
Ear	14.8	12	19	7
GLS	28.0	26.0	30.1	11

Acomys cineraceus Heuglin, 1877

Diagnosis

This is the only species of *Acomys* throughout most of its range, but overlaps with *A. percivali* in southern Sudan from which it can be distinguished by its cinnamon or russet flanks and rump (in *A. percivali* there is no russet anywhere in the pelage). It may also overlap with *A. cahirinus* in the north of its range from which it can be distinguished by its shorter tail (in *A. cahirinus* mean tail length > 100 mm, in *A. cineraceus* < 90 mm); and with *A. wilsoni* in the south from which it can be distinguished by its duller brown dorsal pelage and relatively longer tail ca. 90% of HB (in *A. wilsoni* it is bright rufous or orange-brown and tail < 80% of HB).

Taxonomic Issues

This species was previously listed as a synonym of *A. cahirinus* to which it is similar in size and appearance but parapatric in distribution.

Distribution

This species appears to be confined to Sudan (Setzer, 1956), although Happold (2013) also mention its occurrence in western Ethiopia and Djibouti.

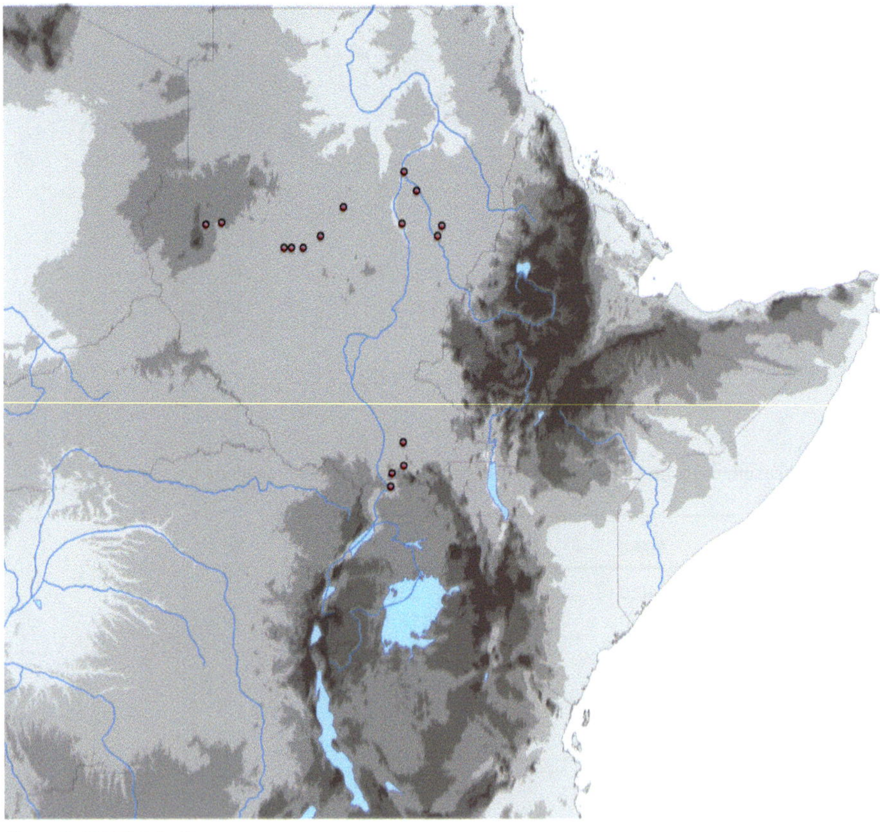

Figure 2.286 Distribution map

Measurements

External and cranial measurements (mm) and mass (g) for *Acomys cineraceus*, sexes combined

	Mean	Min	Max	N
Mass*	34.6	26	61	30
HB*	97.2	89	116	30
Tail*	87.6	72	121	23
HF*	17.2	13	19	27
Ear*	15.3	13	18	29
GLS*	27.6	26.3	31.3	21

*Happold (2013)

Acomys ignitus Dollman, 1910

Diagnosis

This is a brightly coloured species with orange-brown dorsal pelage, reddish-orange flanks and relatively short tail (< 80% of HB). It overlaps with several other *Acomys* species in Kenya and north-east Tanzania including: *A. percivali* (dorsal pelage grey or greyish brown); *A. kempi* (tail longer > 90 mm, and relatively longer ca. 95% of HB); *A. wilsoni* (darker chestnut brown without bright reddish-orange flanks).

Taxonomic Issues

This species was previously considered conspecific with *A. kempi*. 2n = 50, FN = 66-68 (Matthey, 1965).

Distribution

This species is restricted to East Africa where it has mostly been recorded in and around Tsavo National Park, restricted to rocky situations in dry savanna.

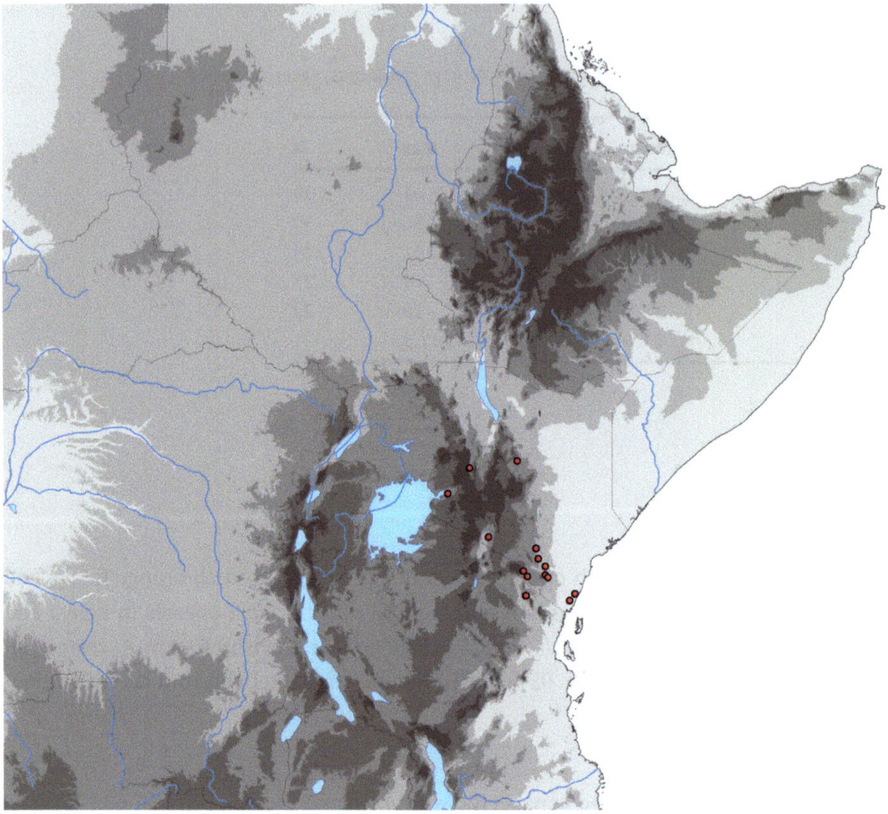

Figure 2.287 Distribution map

Measurements

External and cranial measurements (mm) and mass (g) for *Acomys ignitus*, sexes combined

	Mean	Min	Max	N
Mass	33.5	24	43	4
HB	94.8	85	112	6
Tail	76.5	74	81	4
HF	16.0	15	17	6
Ear	16.2	14	19	7
GLS	28.2	27.1	30.0	3

Acomys johannis Thomas, 1912

Diagnosis

This is the only species of *Acomys* within its Sub-Saharan range (*A. chudeaui* occurs on the left bank of the Niger River, its dorsal pelage is paler, and it differs in karyotype: 2n = 40-46) (Granjon & Duplantier, 2009).

Taxonomic Issues

This taxon may represent a species complex, with two putative species listed by Barome et al. (2000) from Burkina Faso, named *Acomys* sp1 and sp2, which differ in karyotype. Whether this karyotypic variability represents specific distinction has yet to be determined. The karyotype of this species is 2n = 66-68, FN = 66-72 (Granjon & Duplantier, 2009).

Distribution

This species occurs widely in West African Sahelo-Sudanian savannas on the right bank of the Niger River, northern Nigeria and an isolated recorded from Zakouma National Park, Chad.

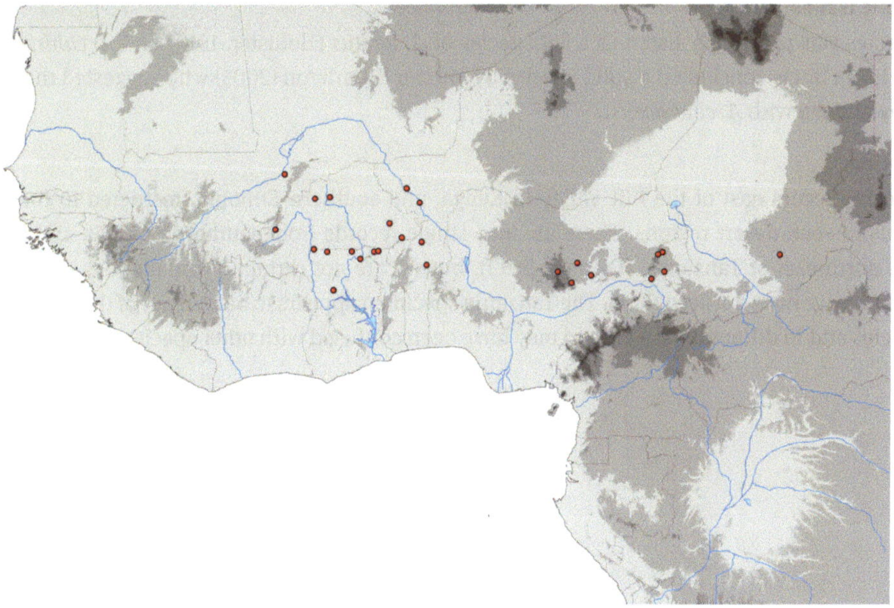

Figure 2.288 Distribution map

Measurements

External and cranial measurements (mm) and mass (g) for *Acomys johannis*, sexes combined

	Mean	Min	Max	N
Mass	43.8	38	49	4
HB	103.6	90	119	8
Tail	93.5	73	107	4
HF	18.0	17	19	8
Ear	16.4	11	18	8
GLS	29.8	29.0	30.7	8

Acomys kempi Dollman, 1911

Diagnosis

This is a relatively brightly coloured species with orange-brown dorsal pelage, dull orange flanks and relatively long tail (> 90% of HB). It overlaps with several other *Acomys* species in Kenya including: *A. percivali* (dorsal pelage grey or greyish brown); *A. ignitus* (tail shorter < 90 mm, and relatively shorter < 80% of HB); *A. wilsoni* (tail relatively shorter < 80% of HB; darker chestnut brown without orange flanks).

Taxonomic Issues

This species was previously listed as a subspecies of *A. ignitus* (Hollister, 1919) and *A. cahirinus* (Setzer, 1975), but is considered a valid species by Musser & Carleton (2005) who suggested that it is closely aligned with *A. cineraceus.*

Distribution

This species occurs east of the Rift Valley in Kenya, and southern Ethiopia, restricted to rocky situations in semi-desert regions. There are also single records from southern Somalia and the extreme north-east of Tanzania (Happold, 2013), but we are not certain of the identity of these specimens as we have not seen them. The Tanzania specimen appears to be well out of the range of this species and in different habitat (and may have been confused with other species of *Acomys*).

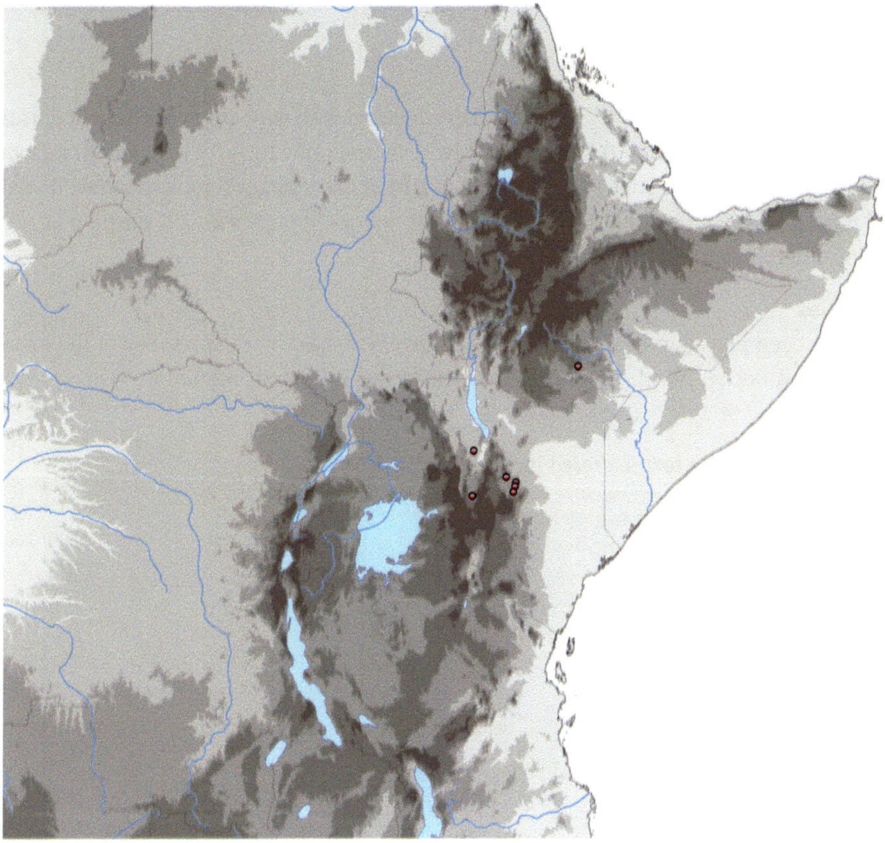

Figure 2.289 Distribution map

Measurements

External and cranial measurements (mm) and mass (g) for *Acomys kempi*, sexes combined

	Mean	Min	Max	N
Mass	-	-	-	-
HB*	98.8	90	110	12
Tail*	91.9	82	100	12
HF*	15.5	15	16	12
Ear	-	-	-	-
GLS*	25.4	23.8	27.1	10

*Hollister (1919)

Acomys louisae Thomas, 1896

Diagnosis
This species may co-occur with *A. mullah* (larger, HB > 100 mm with grey or greyish-brown dorsal pelage); *A. cahirinus* (larger, HB > 100 mm, but relatively shorter tail < 100% of HB); and *A. wilsoni* (smaller, GLS < 26 mm; tail relatively shorter < 80% of HB). It has a bright rufous or brown dorsal pelage and relatively very long tail (> 100% of HB). Furthermore, it possesses a t7 cusp on M[1] not present in any other *Acomys* species (Denys et al., 1994).

Taxonomic Issues
This species has been placed in its own subgenus *Peracomys* based on the presence of the extra cusp on M[1] (Petter & Roche, 1981). A full revision of this genus is urgently required.

Distribution
This species is known from a few scattered localities in the Horn of Africa.

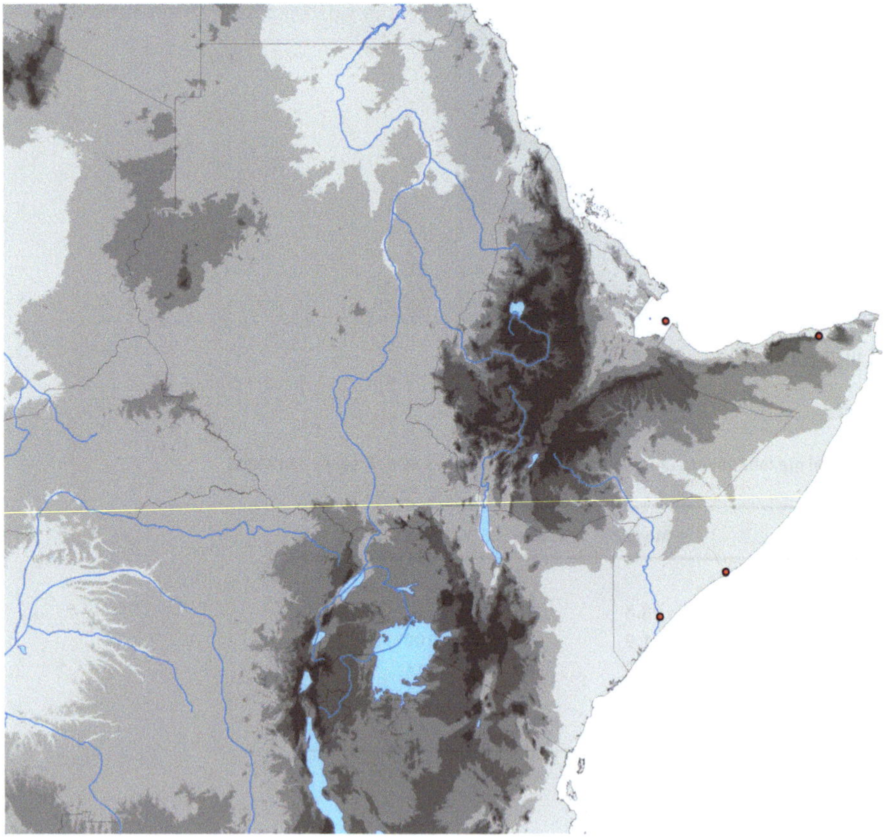

Figure 2.290 Distribution map

Measurements

External and cranial measurements (mm) and mass (g) for *Acomys louisae*, sexes combined

	Mean	Min	Max	N
Mass	-	-	-	-
HB	103.0	90	110	3
Tail	96.3	85	117	3
HF	18.3	18	19	3
Ear	16.7	16	17	3
GLS	29.9	28.8	30.7	3

Acomys mullah Thomas, 1904

Diagnosis

This species may co-occur with *A. louisae* (smaller, HB < 100 mm with brighter dorsal pelage which may show some rufous) and *A. cahirinus* (dorsal pelage russet). It has a grey or greyish-brown dorsal pelage and relatively long tail (> 90% of HB).

Taxonomic Issues

Previously, it has been listed as a synonym of *A. cahirinus* (Musser & Carleton, 1993).

Distribution

This species has been recorded from a few scattered records in Eritrea, Ethiopia and Somalia.

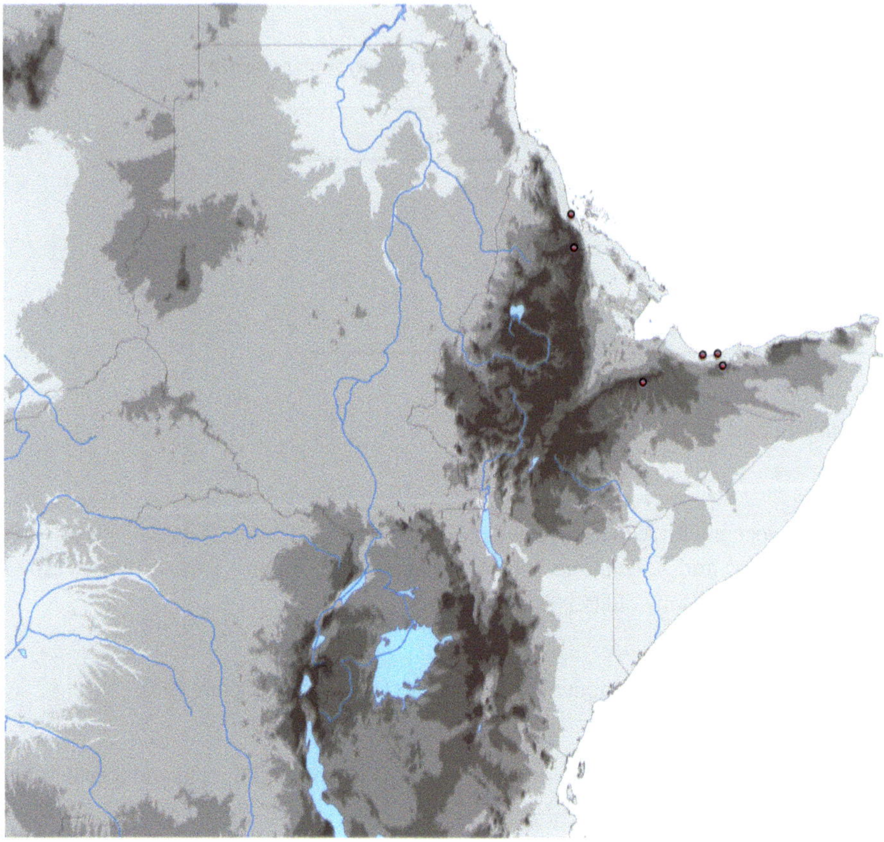

Figure 2.291 Distribution map

Measurements

External and cranial measurements (mm) and mass (g) for *Acomys mullah*, sexes combined

	Mean	Min	Max	N
Mass	-	-	-	-
HB*	118	106	134	7
Tail*	108	95	117	7
HF*	16.9	15	18	7
Ear*	17	14	18	7
GLS*	28.8	28.3	29.3	3

*Happold (2013)

Acomys muzei Verheyen, Hulselmans, Wendelen, Leirs, Corti, Backeljau & Verheyen, 2011

Diagnosis

This species is very similar to *A. ngurui* from which it differs in its relatively longer tail (ca. 100% of HB; in *A. ngurui* < 90% of HB) and non-overlapping distribution. *Acomys wilsoni* may co-occur with this species from which it can be distinguished by its smaller size (typically GLS < 24 mm) and its absolutely and relatively shorter tail (tail < 60 mm, < 80% of HB; in *A. muzei* it is ca. 100% of HB).

Taxonomic Issues

The species has only recently been described as a split from *A. spinosissimus* based on significant divergence of the cytochrome *b* gene (Verheyen et al., 2011). Its closest relative is *A. ngurui* with which it occurs parapatrically. 2n = 58-62, FN = 68-76 (Verheyen et al., 2011).

Distribution

This species is currently only known from Tanzania where it occurs at higher elevations than *M. ngurui*, to the west of the Eastern Arc Mts.

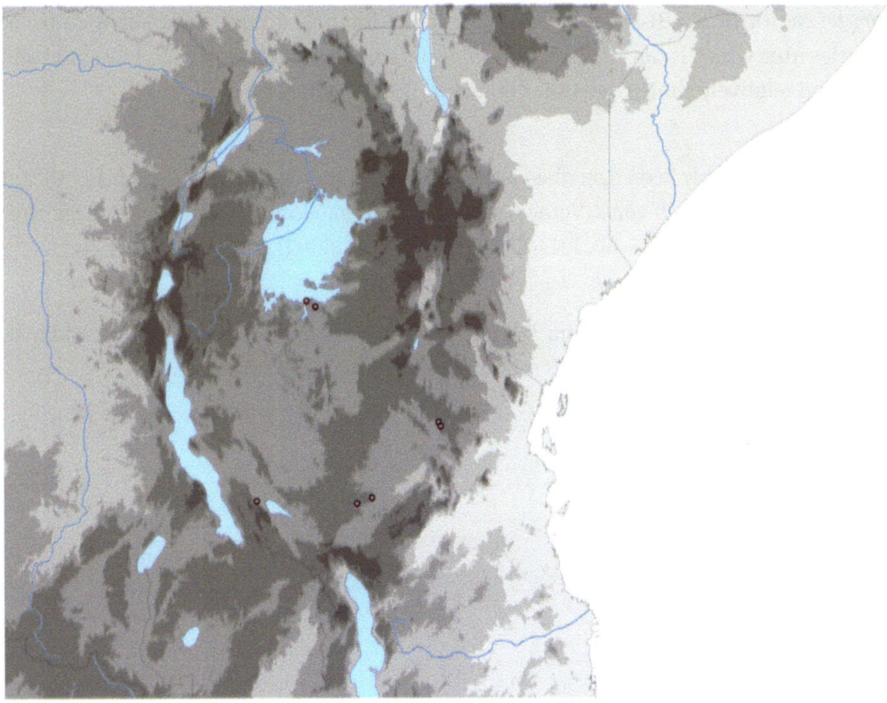

Figure 2.292 Distribution map

Measurements

External and cranial measurements (mm) and mass (g) for *Acomys muzei*, sexes combined

	Mean	Min	Max	N
Mass*	19.1	14	30	39
HB*	86.0	75	97	38
Tail*	87.0	78	103	35
HF*	16.5	14	20	37
Ear*	13.6	9	17	35
GLS*	25.1	23.8	26.8	47

*Verheyen et al. (2011)

Acomys ngurui Verheyen, Hulselmans, Wendelen, Leirs, Corti, Backeljau & Verheyen, 2011

Diagnosis

This species is very similar to *A. muzei* from which it differs in its relatively shorter tail (ca. 84% of HB; in *A. muzei* ca. 100% of HB) and non-overlapping distribution. *Acomys wilsoni* may co-occur with this species from which it can be distinguished by its smaller size (typically GLS < 24 mm) and its absolutely and relatively shorter tail (tail < 60 mm, < 80% of HB).

Taxonomic Issues

The species has only recently been described as a split from *A. spinosissimus* based on significant divergence of the cytochrome *b* gene (Verheyen et al., 2011). Its closest relative is *A. muzei* with which it occurs parapatrically. 2n = 60, FN = 68 (Verheyen et al., 2011).

Distribution

This species is currently only known from Tanzania where it occurs at lower elevations than *M. ngurui*, to the east of the Eastern Arc Mts. It may also occur in northern Mozambique.

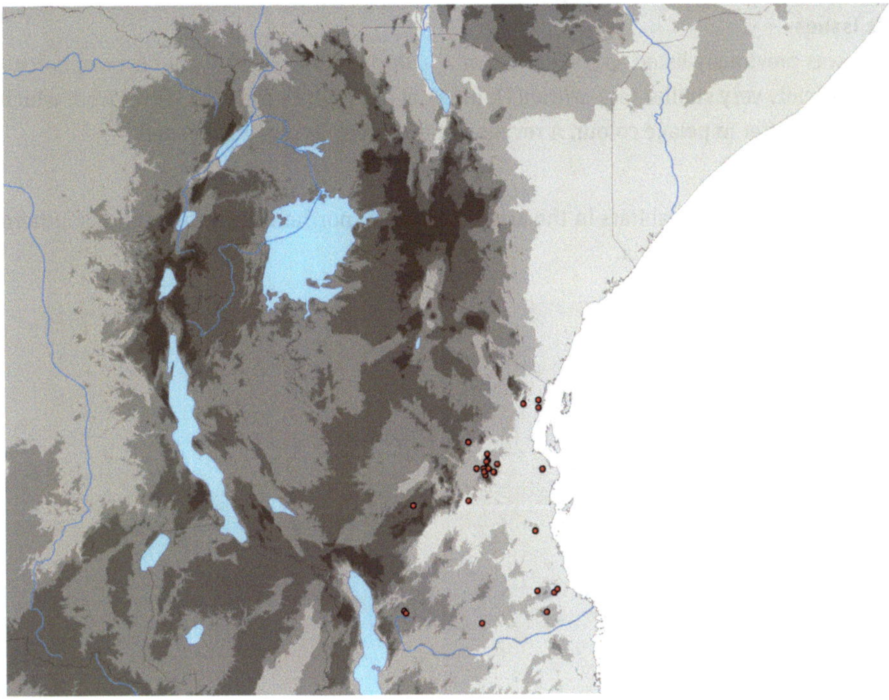

Figure 2.293 Distribution map

Measurements

External and cranial measurements (mm) and mass (g) for *Acomys ngurui*, sexes combined

	Mean	Min	Max	N
Mass*	23.2	11	38	172
HB*	91.4	69	110	166
Tail*	76.5	62	90	112
HF*	15.3	12	19	165
Ear*	14.0	11	19	133
GLS*	25.5	23.5	27.3	228

*Verheyen et al. (2011)

Acomys percivali Dollman, 1911

Diagnosis

This is a relatively dull coloured species with grey dorsal pelage lacking any russet or orange. The tail is relatively short (< 80% of HB). It overlaps with several other *Acomys* species in Kenya including: *A. kempi* (dorsal pelage with dull orange flanks; tail relatively longer > 90% of HB); *A. ignitus* (dorsal pelage brighter with bright orange flanks); *A. wilsoni* (dorsal pelage chestnut brown).

Taxonomic Issues

This species was previously listed as a synonym of *A. cahirinus* (Setzer, 1975), but is morphologically and genetically very similar to *A. wilsoni* (Janecek et al., 1991; Denys et al., 1994) from which it appears to only differ in pelage colour. A revision of this group is urgently needed.

Distribution

This species occurs in rocky habitats in the arid savannas of northern Kenya, Uganda, southern Sudan, and south-western Ethiopia.

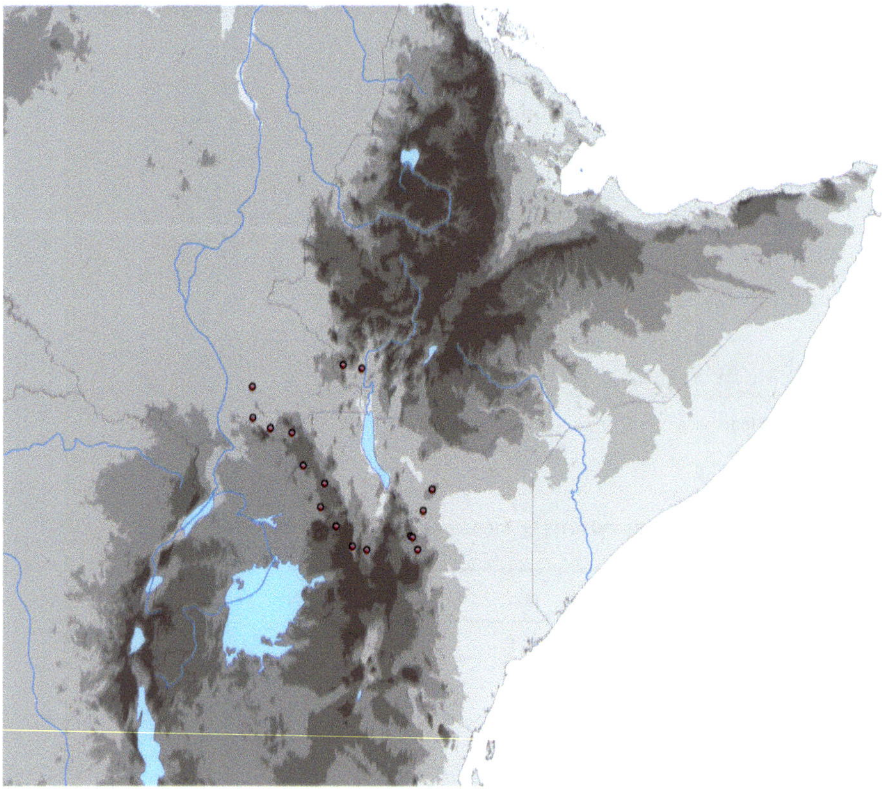

Figure 2.294 Distribution map

Measurements

External and cranial measurements (mm) and mass (g) for *Acomys percivali*, sexes combined

	Mean	Min	Max	N
Mass	31.0	27	38	3
HB	89.7	82	100	7
Tail	77.3	67	100	7
HF	15.3	14	17	7
Ear	14.9	12	19	7
GLS*	24.9	24.7	25.0	2

*A sample of 11 skulls measured by Hollister (1919) had mean GLS = 24.9 mm; in contrast Happold (2013) reported mean GLS = 27.4 mm for a sample of six skulls, which to us appears rather large for this species.

Acomys selousi De Winton, 1896

Diagnosis

This is the only species of *Acomys* across its range, although it possibly could co-occur with *A. spinosissimus* in the northern parts of its range, from which it can currently only be distinguished on genetic grounds (and by its non-overlapping distribution). This species is unique in having a giant X chromosome (20% of the length of the haploid chromosome set) (Matthey, 1965; Barome et al., 2001; Denys et al., 2011b).

Taxonomic Issues

This species has been split from the *A. spinosissimus* species complex (Barome et al., 2001; Verheyen et al., 2011), and the limits of its distribution are not yet fully resolved. 2n = 60 (Dippenaar & Rautenbach, 1986).

Distribution

This species is restricted to the savannas of northern South Africa, eastern Botswana, and southern Zimbabwe. Due to confusion with *A. spinosissimus*, the northern limits of this species are currently not known and records from south-eastern Zimbabwe and southern Mozambique may refer to *A. spinosissimus*.

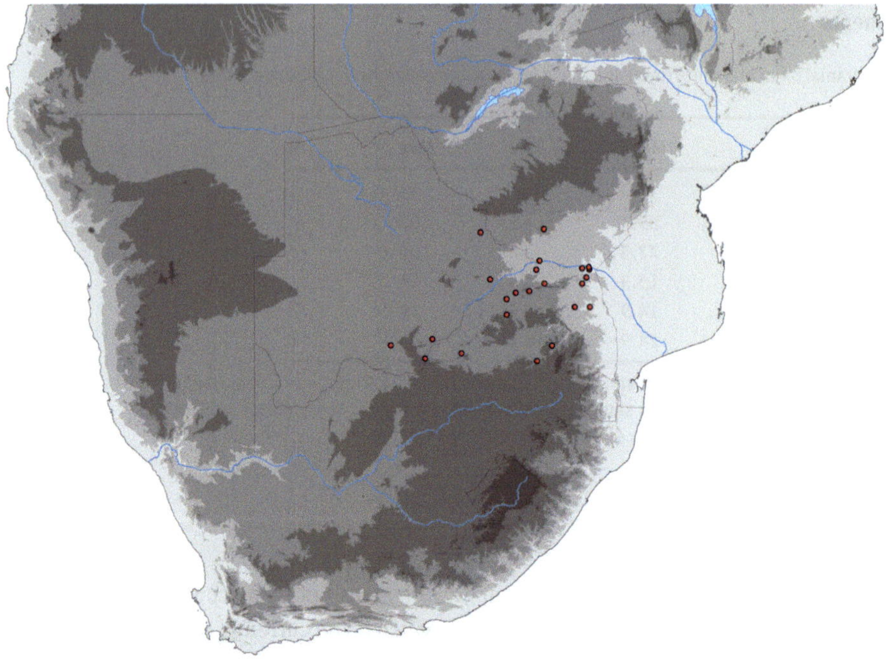

Figure 2.295 Distribution map

Measurements

External and cranial measurements (mm) and mass (g) for *Acomys selousi*, sexes combined

	Mean	Min	Max	N
Mass	22.2	19	27	10
HB	90.0	74	102	11
Tail	76.4	57	86	10
HF	16.0	15	17	11
Ear	13.0	11	14	11
GLS	24.8	23.2	25.6	12

Acomys spinosissimus Peters, 1852

Diagnosis

This is the only species of *Acomys* across its range, although it possibly could co-occur with *A. selousi* in the southern parts of its range from which it can currently only be distinguished on molecular grounds (and by its non-overlapping distribution).

Taxonomic Issues

This species has recently been shown to comprise a species complex including at least three distinct cryptic species in addition to *A. spinosissimus*: *A. muzei*, *A. ngurui* and *A. selousi* (Verheyen et al., 2011).

Distribution

This species has a wide distribution in South-Central Africa from northern Malawi and southern DRC south to Zimbabwe and Mozambique. The southern limits of its range are currently unknown and records from south-eastern Zimbabwe and southern Mozambique may refer to *A. selousi*.

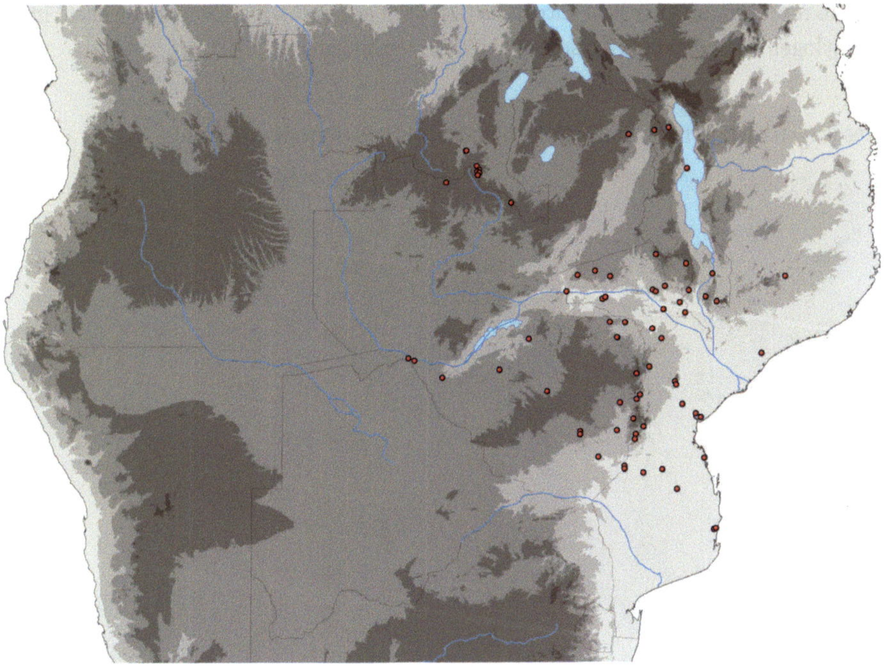

Figure 2.296 Distribution map

Measurements

External and cranial measurements (mm) and mass (g) for *Acomys spinosissimus*, sexes combined

	Mean	Min	Max	N
Mass	21.0	19	24	3
HB	87.0	72	105	12
Tail	77.0	55	98	9
HF	17.3	16	18	10
Ear	13.8	12	17	12
GLS	25.3	22.7	27.9	16

Acomys subspinosus (Waterhouse, 1838)

Diagnosis

This is the only species of *Acomys* within its range and therefore should not be confused with any other species. It has dark russet-brown dorsal pelage which is less spiny than in some other *Acomys* species, and a relatively long tail (> 90% of HB). Characteristic features include its bunodont molars, the round and isolated t1 and t4 which are not fused to t5-6 or t2-3 (Denys et al., 1994). The palatal foramina are short and the maxilla-palatal suture distinct (Petter, 1983).

Taxonomic Issues

Based on its unique morphology, this species has been placed in its own subgenus *Subacomys* and was claimed to represent an ancestral karyotype (Denys et al., 1994). It appears to be basal in the molecular phylogenies of *Acomys* (Barome et al., 2001; Verheyen et al., 2011).

Distribution

This species is restricted to rocky habitats in fynbos, south-western South Africa.

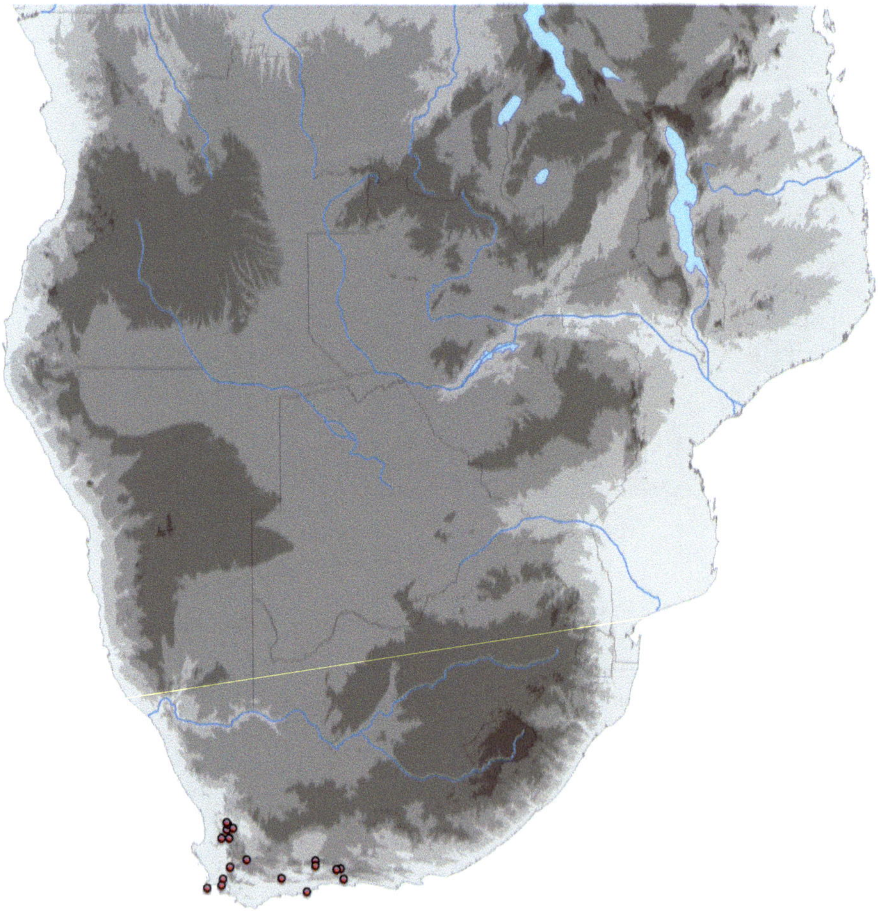

Figure 2.297 Distribution map

Measurements

External and cranial measurements (mm) and mass (g) for *Acomys subspinosus*, sexes combined

	Mean	Min	Max	N
Mass	20.4	16	27	16
HB	89.9	75	98	14
Tail	81.8	54	98	14
HF	18.0	16	20	15
Ear	12.0	9	15	15
GLS	25.4	24.2	26.6	10

Acomys wilsoni Thomas, 1892

Diagnosis

This is a relatively brightly coloured species with cinnamon or chestnut-brown dorsal pelage, and relatively short tail (< 80% of HB). It overlaps with several other *Acomys* species in Kenya including: *A. percivali* (dorsal pelage grey or greyish brown); *A. ignitus* (bright orange flanks; larger, typically GLS > 27 mm); *A. kempi* (dull orange flanks). It may also overlap *A. muzei/A. ngurui* in northern Tanzania from which it can be distinguished by its smaller size and relatively short tail (in *A. muzei/A. ngurui tail* is ca. 90% or more of HB) and different dorsal pelage colour.

Taxonomic Issues

This species has been included in *A. subspinosus* (Setzer, 1975), but it is closest in morphology and genetics to *A. percivali* (Janecek et al., 1991; Denys et al., 1994), from which it only seems to differ in pelage colour. Three divergent clades were recovered for members of this species by Verheyen et al. (2011), suggesting the presence of cryptic species. This group is urgently in need of revision. 2n = 50 (Matthey, 1968).

Distribution

This species occurs widely in rocky habitats throughout East Africa.

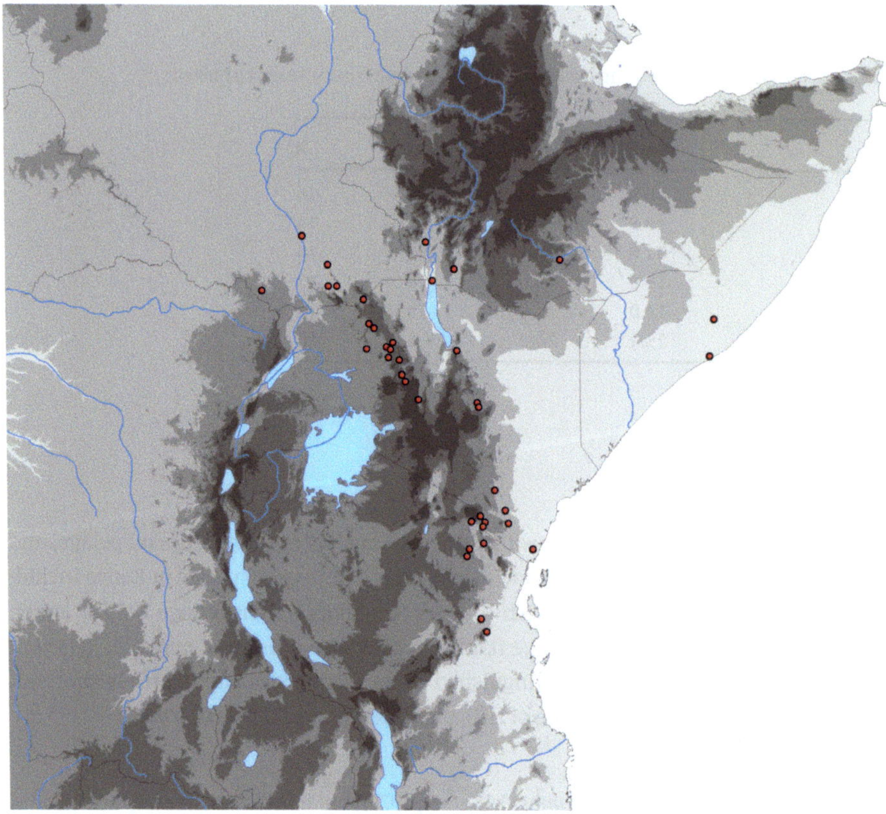

Figure 2.298 Distribution map

Measurements

External and cranial measurements (mm) and mass (g) for *Acomys wilsoni*, sexes combined

	Mean	Min	Max	N
Mass	32.5	28	37	2
HB	82.4	75	98	11
Tail	49.2	35	56	10
HF	13.2	12	15	11
Ear	13.4	11	15	11
GLS*	23.4	21.8	25.4	8

*A sample of 12 skulls measured by Hollister (1919) had mean GLS = 22.6 mm; in contrast Happold (2013) reported mean GLS = 27.9 mm for a sample of 10 skulls, which to us appears rather large for this species. The holotype of *A. wilsoni* has a damaged skull; the skulls of *A. w. argillaceus* and *A. w. enid* are small: GLS: 22.4 and 23.8 mm

Genus Account

Deomys Thomas, 1888 (Link Rats)

Diagnosis

This elegant medium sized rat (HB 100-150 mm) is characterized by a long tail (140% of HB) ending in a small tuft of hair, long feet and relatively large ears. The narrow skull is characterized by a very long rostrum and a wide palate, with teeth displaying high cusps. The upper incisors are weakly grooved. The dentition is specialized and is reminiscent of the typical Dendromurinae pattern. The genus is distinguished from *Colomys* by its short Digit 5, longer ears and the tuft at the end of the tail and from *Malacomys* by its pure white ventral pelage, its smaller hindfoot and the tuft at the end of the tail. It is distinguished from *Grammomys* species with white ventral pelage by its longer hindfoot. *Deomys* occurs in rainforests of Central Africa.

Figure 2.299 Photograph of *Deomys ferrugineus* (Jan Kennis).

Cranial Characters

The skull is long in size (34–40mm in greatest length), relatively narrow with a very long rostrum. The supraorbital is narrow with a sharp crest. The zygomatic arches are thin and parallel-sided. The nasal bone is prominent and expanded anteriorly. The incisors are orthodont; the upper incisors have two weak grooves, while the lower incisors lack grooving. The anterior palatal foramen is short and stops well anterior of the first upper molar. The palate is depressed between inflated maxillaries and two parallel grooves run longitudinally adjacent to the molar rows, each ending in a tiny palatine foramen. The palate extends beyond the molar row but, in contrast to *Acomys* and *Uranomys*, the pterygoids are not fused. The anterior surface of the zygomatic plate is oblique. The bullae are small (5-6 mm in greatest diameter).

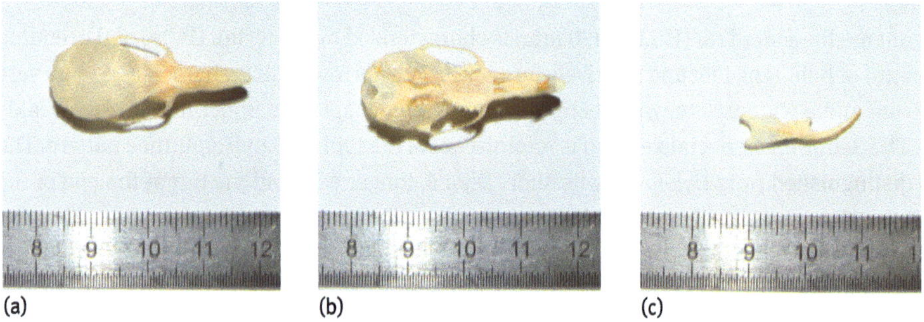

Figure 2.300 a-c Cranium of *Deomys ferrugineus* (MNHN-ZM-MO 1991-72) (from left to right: dorsal view, ventral view and lower jaw).

Dentition

Cusps are high and separated. The M^3 molar is the smallest in the toothrow. The M^1 displays three lobes with different numbers of cusps. The first lobe has two large cusps separated by a deep longitudinal valley, the second one displays three cusps (the most lingual being the t4) and the third one has two cusps and a posterior cingulum. The longitudinal valley separates external from central cusps and is nearly rectilinear. The M^2 displays two lobes and an antero-external cusp (t3), a t4, but no t1. There is also a posterior cingulum on the M^2. The M^3 has a tiny t3, no t1, and two rows of cusps. The M_1 is the longest tooth in the lower toothrow and is very narrow. The first and second lobes have their 4 cusps fused in the middle by a longitudinal crest and the third lobe has two cusps with a small posterior cingulum. The M_2 has an antero-external cingulum ending in a crest, and two rows of cusps with a posterior cingulum are visible. The highly reduced M_3 has a tiny antero-external cingulum and shows two cusps.

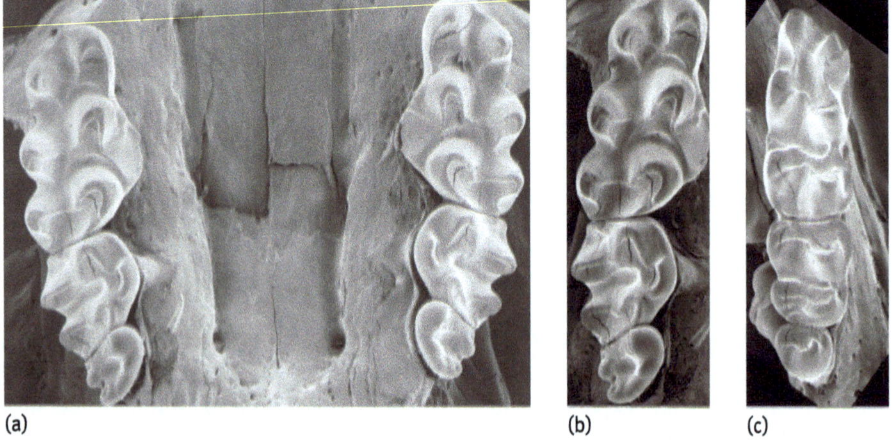

Figure 2.301 Cheek teeth of *Deomys ferrugineus* (MNHN-ZM-MO 1983-232) (A: upper cheek teeth; B: upper right toothrow; C: lower right toothrow).

Habitat & Biogeography

Deomys inhabits tropical forests from Cameroon to Uganda and south-western Congo, on the right bank of the Congo River. It is generally found in low-lying forests (up to 1500 m), including swamp forests and prefers undisturbed situations. This is confirmed by the fact that survival is higher in undisturbed forest compared with adjacent fallow fields (Kennis et al., 2012).

Fossil Record

None.

Biology & Ecology

It constructs nests out of leaves in holes or crevasses at the base of trees (Rosevear, 1969). It is terrestrial, crepuscular and nocturnal. Its diet is comprised of insects, small invertebrates, small vertebrates and some vegetable matter (Rahm, 1972; Genest-Villard, 1980). Females give birth to 1-4 young and appear to be reproductively active during the whole year. Predators include small mammalian carnivores.

Systematic Notes

This genus was initially classified in the Dendromurinae with which it does not share morphological features (Rosevear, 1969; Denys et al., 1995). Later, based on molecular analyses, it was placed in the Deomyinae along with *Acomys, Uranomys* and *Lophuromys* (see *Acomys* account for details and references). However, morphologically *Deomys* is very different from these taxa (Denys et al., 1995).

Species List

This genus is monotypic and only one species is presently recognized
Deomys ferrugineus Thomas, 1888

Species Account

Deomys ferrugineus Thomas, 1888

Diagnosis

See genus account.

Taxonomic Issues

It is not clear whether the eastern and western populations are connected or represent two separate evolutionary lineages.

Distribution

The species occurs widely in the Lower Guinea Forest of Central Africa and in the forests of north-eastern DRC, Uganda and Rwanda, but occurring only on the right bank of the Congo River (Katuala et al., 2008).

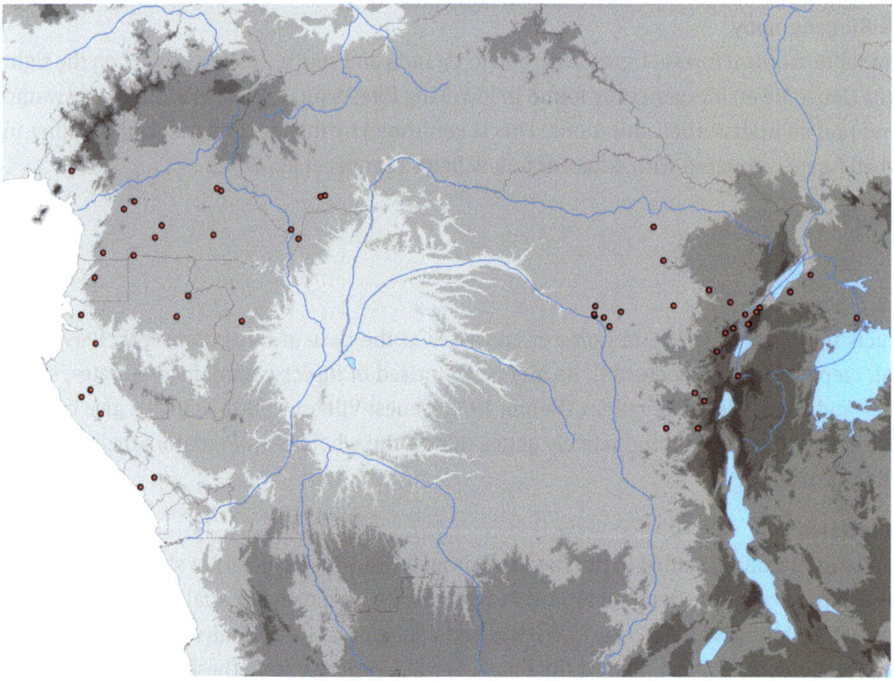

Figure 2.302 Distribution map

Measurements

External and cranial measurements (mm) and mass (g) for *Deomys ferrugineus*, sexes combined

	Mean	Min	Max	N
Mass	44.3	34	54	4
HB	121.7	105	146	7
Tail	180.4	165	191	8
HF	33.6	32	36	8
Ear	23.6	22	26	8
GLS	35.2	33.4	36.7	8

Genus Account

Lophuromys Peters, 1874 (Brush-furred Rats)

Diagnosis

The distinguishing characters of the genus are its small to large size (HB ca. 90-180 mm), with a relatively short tail (50-85% of HB) except in two species (*L. luteogaster* and *L. woosnami*) that have longer tails (100-105% of HB). The pelage is highly variable between and within species, and can be bright reddish-brown to dark brown. There are some brush-like hairs in the dorsal pelage like in *Uranomys* but not as stiff as in *Acomys*. It has an *Acomys*-like dental pattern.

(a)

(b)

Figure 2.303 a and b Photographs of *Lophuromys sikapusi* (Ara Monadjem) (left) and *L.* cf *brevicaudus* (Marian Polak) (right).

Cranial Characters

The skull is medium sized (24-34 mm in greatest length), and characterized by a wide interorbital constriction as a result of oblique fronto-parietal crests delimitating a relatively large braincase. The nasals may be long or short depending on the species, but are relatively enlarged anteriorly. In ventral view, the long palatal foramen reaches the level of the upper M1 and there is no rearward extension of the palate beyond the toothrow as in *Acomys*. The upper molar rows are parallel and the molars are relatively large. The tympanic bullae are short and not inflated (5-6mm in length).

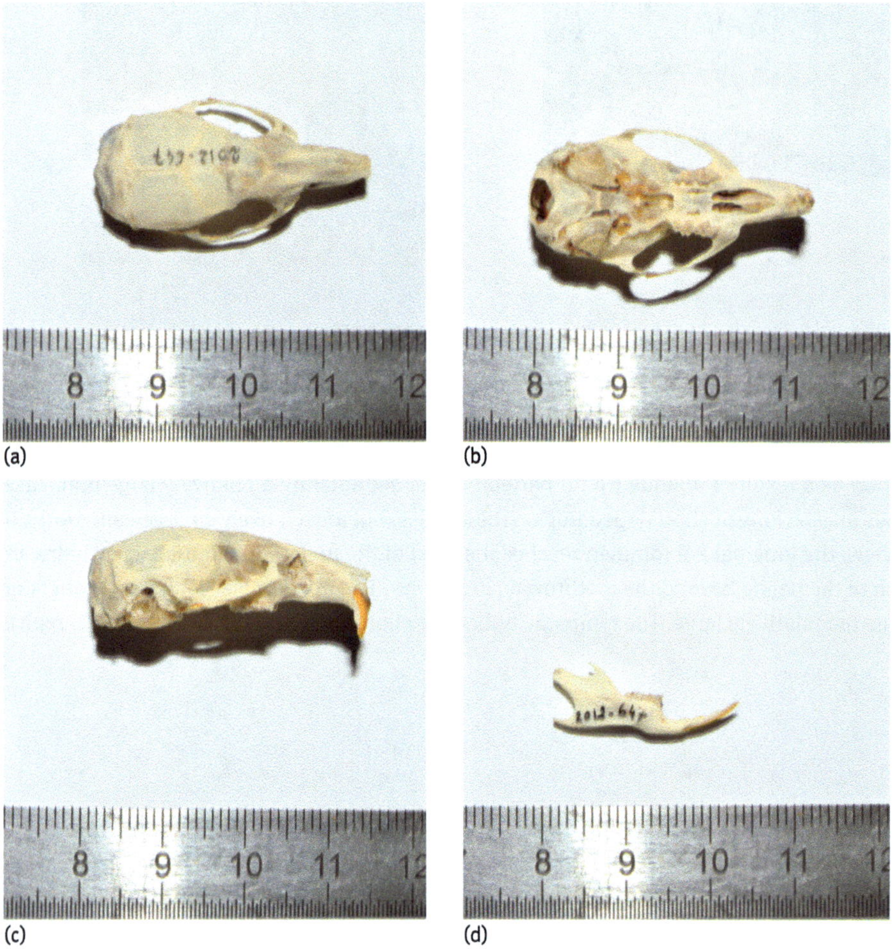

Figure 2.304 a-d Cranium of *Lophuromys sikapusi* (MNHN-ZM-MO 2012-647) (from top left to bottom right: dorsal view, ventral view, lateral view and lower jaw).

Dentition

The upper incisors are pro-odont or orthodont, the long palatal foramen reaches the level of the M^1. The molars display the typical Acomyinae dental pattern. The t1 and t4 typically have small stephanodont crests and are situated posterior to the t2-t3 and t5-t6 cusps, respectively. The t3 and the t6 may also bear small crests and there may be a small posterior cingulum on the M^1. The M^3 and M_3 are reduced in size and are variable in pattern (Denys & Michaux, 1992). The cusps of the first lobe of the lower M_1 are fused longitudinally with those of the second lobe. There is a posterior cingulum on the M_1. There is an antero-labial cingulum on the M_2.

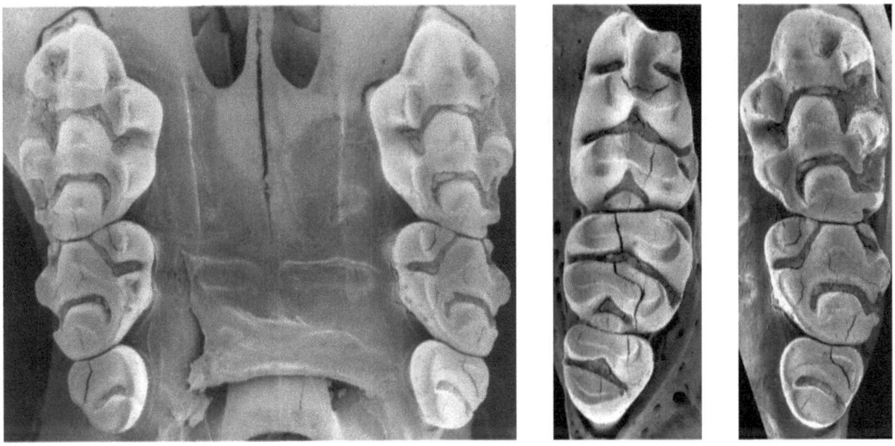

(a) (b) (c)

Figure 2.305 Cheek teeth of *Lophuromys flavopunctatus* (MNHN-ZM-MO 1972-286) (A: upper cheek teeth; B: lower right toothrow; C: upper left toothrow).

Habitat & Biogeography

Members of the genus *Lophuromys* are generally associated with moist environments of tropical West, Central and East Africa, from sea level up to 4500 m. They are associated with secondary forests, montane forests, marshes, cultivated fields, moist grasslands and swamps, as well as with dense herbaceous cover in the alpine zone. This genus extends from Guinea in West Africa to Kenya and Ethiopia in the east, and then south to Malawi and northern Mozambique. Some species are endemic to a single mountain such as: *L. brevicaudus* (Bale Mts, Ethiopia), *L. cinereus* (Kahuzi-Biega, Kivu Mts, eastern DRC), *L. dieterleni* (Mt Oku, Cameroon), *L. eisentrauti* (Mt Lefo, Cameroon), *L. medicaudatus* (Kivu, DRC), *L. melanonyx* (Bale Mts, Ethiopia), *L. roseveari* (Mt Cameroon, Cameroon), and *L. woosnami* (Rwenzori Mts, DRC and Uganda).

Fossil Record

There is no fossil record of this genus.

Biology & Ecology

These rodents are terrestrial and nocturnal. They do not dig burrows or climb off the ground, but can create "tunnels" by pushing their way through the vegetation. From the few species that have been studied, it would appear that *Lophuromys* species typically have an omnivorous diet which includes insects, molluscs, vegetable matter and seeds (Happold, 2013). After 30 to 32 days of gestation, 2-4 young are born which are raised communally by several females, as in *Acomys* spp. (Dieterlen, 1976b).

Systematic Notes

This genus was previously placed within the Murinae, however recent molecular analyses have placed *Lophuromys* with *Acomys, Uranomys* and *Deomys* in the Deomyinae subfamily (Steppan et al., 2005). The taxonomy of the genus has been the object of several revisions by Dieterlen (1976b, 1979, 1987), who divided the genus into two subgenera: *Kivumys (woosnami, luteogaster, medicaudatus)* which have long tails and long hindfeet; and *Lophuromys (*the remaining species) characterized by short tails, short hindfeet, and a special stomach morphology. This subgeneric distinction was retained by Carleton & Musser (2005). The *Lophuromys* subgenus is further divided into the *sikapusi* group (unspeckled) and the *flavopunctatus* group (speckled). Numerous new species have recently been described within both these groups in Central and East Africa (Verheyen et al., 1997, 2000, 2002, 2007; Lavrenchenko et al., 1998a, 2007). As an indication of this explosion of new species, Happold (2013) recognised 15 species, whereas 34 species are listed here. Whether all these taxa will continue to be recognised as valid species is hard to say, although the molecular divergence between many of them seems rather large. Therefore, we retain these 34 species with the knowledge that future work may result in synonymy of some of these taxa.

Species List

34 species are currently recognised.

Lophuromys angolensis W. Verheyen, Dierckx & Hulselmans, 2000
Lophuromys ansorgei De Winton, 1896
Lophuromys aquilus True, 1892
Lophuromys brevicaudus Osgood, 1936
Lophuromys brunneus Thomas, 1906
Lophuromys chercherensis Lavrenchenko, W. Verheyen, E. Verheyen, Hulselmans & Leirs, 2007
Lophuromys chrysopus Osgood, 1936
Lophuromys cinereus Dieterlen and Gelmroth, 1974
Lophuromys dieterleni W. Verheyen, Hulselmans, Colyn & Hutterer, 1997
Lophuromys dudui W. Verheyen, Hulselmans Dierckx & E. Verheyen, 2002
Lophuromys eisentrauti Dieterlen, 1978
Lophuromys flavopunctatus Thomas, 1888
Lophuromys huttereri W. Verheyen, Colyn & Hulselmans, 1996
Lophuromys kilonzoi W. Verheyen, Hulselmans, Dierckx, Mulungu, Leirs, Corti & E. Verheyen, 2007
Lophuromys laticeps Thomas et al., 1907
Lophuromys luteogaster Hatt, 1934
Lophuromys machangui W. Verheyen, Hulselmans, Dierckx, Mulungu, Leirs, Corti & E. Verheyen, 2007
Lophuromys makundii W. Verheyen, Hulselmans, Dierckx, Mulungu, Leirs, Corti & E. Verheyen, 2007
Lophuromys margarettae Heller, 1912
Lophuromys medicaudatus Dieterlen, 1975
Lophuromys melanonyx F. Petter, 1972
Lophuromys menagashae Lavrenchenko, W. Verheyen, E. Verheyen, Hulselmans & Leirs, 2007
Lophuromys nudicaudus Heller, 1911
Lophuromys pseudosikapusi Lavrenchenko, W. Verheyen, E. Verheyen, Hulselmans & Leirs, 2007
Lophuromys rahmi Verheyen, 1964
Lophuromys rita Dollman, 1910
Lophuromys roseveari W. Verheyen, Hulselmans, Colyn & Hutterer, 1997
Lophuromys sabunii W. Verheyen, Hulselmans, Dierckx, Mulungu, Leirs, Corti & E. Verheyen, 2007
Lophuromys sikapusi (Temminck, 1853)
Lophuromys simiensis Osgood, 1936

Lophuromys stanleyi W. Verheyen, Hulselmans, Dierckx, Mulungu, Leirs, Corti & E. Verheyen, 2007
Lophuromys verhageni W. Verheyen, Hulselmans, Dierckx, Mulungu, Leirs, Corti & E. Verheyen, 2007
Lophuromys woosnami Thomas, 1906
Lophuromys zena Dollman, 1909

Key to Species

At this stage it is not possible to present a key to the *Lophuromys* species, mostly due to the fact that the number of recognised species has more than doubled in the past decade. However, very few species overlap in distribution and therefore most can be identified (at least tentatively) based on location. Hence we suggest that the distribution maps are studied carefully to see which possible species may co-occur at a particular site, and then use the diagnostic features (mentioned under "Diagnosis") to distinguish them. This genus is in urgent need of review. We provide a key to the three "groups" within this genus below.

1	Tail longer (ca. 85% of HB and longer)	Subgenus *Kivumys*
	Tail shorter (ca. 60% of HB)	Subgenus *Lophuromys*
2	Dorsal pelage speckled (either conspicuously or faintly)	*flavopunctatus* group
	Dorsal pelage not speckled	*sikapusi* group

Species Accounts

Lophuromys angolensis **W. Verheyen, Dierckx & Hulselmans, 2000**

Diagnosis

This is the only species of *Lophuromys* currently known within its range, although it may possibly co-occur with *L. ansorgei* from which it can be differentiated by its smaller size (in *L. ansorgei* from the Congo basin, GLS > 33 mm).

Taxonomic Issues

This species is in the *sikapusi* group, and has only recently been described (Verheyen et al., 2000).

Distribution

This species is restricted to northern Angola and neighbouring regions of the DRC, on the left bank of the Congo River.

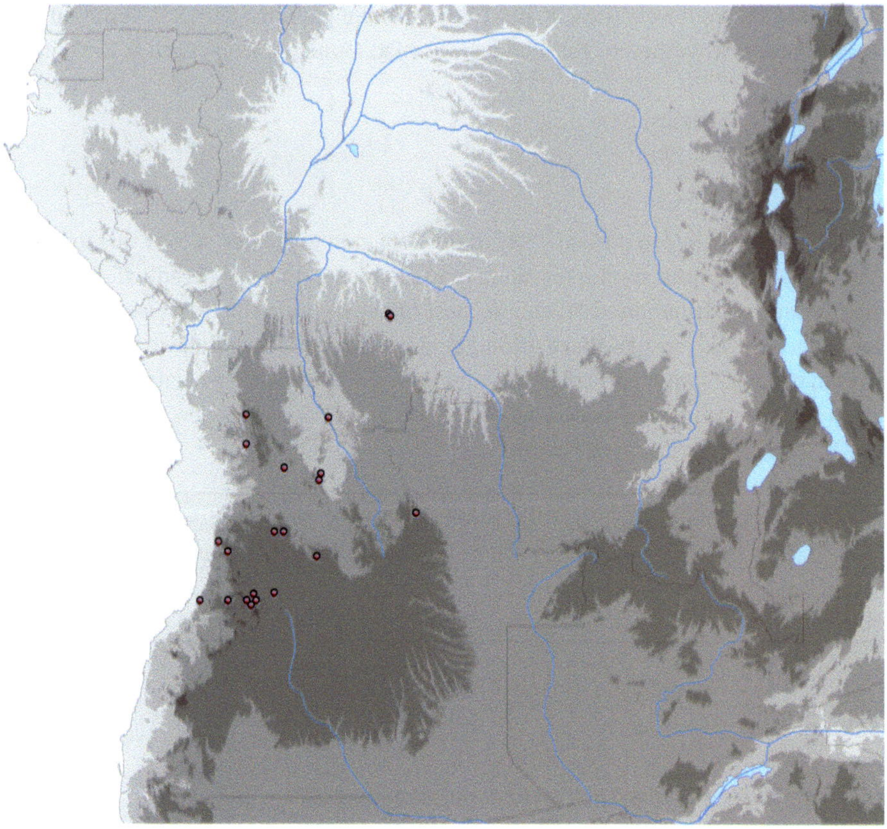

Figure 2.306 Distribution map

Measurements

External and cranial measurements (mm) and mass (g) for *Lophuromys angolensis*, sexes combined

	Mean	Min	Max	N
Mass*	54.9	43	71	44
HB*	134.8	115	154	43
Tail*	75.5	51	88	39
HF*	22.8	20	24	45
Ear*	16.6	15	18	46
GLS*	30.5	28.8	32.5	41

*Verheyen et al. (2000)

Lophuromys ansorgei De Winton, 1896

Diagnosis

This species co-occurs with several other *Lophuromys* species across its range including: *Lophuromys dudui* (smaller, GLS < 30 mm; dorsal pelage speckled); *L. luteogaster* and *L. woosnami* (in both these species tail is relatively longer ca. 85% or more of HB, in *L. angorsei* it is < 60% of HB); *L. nudicaudus* (smaller, GLS < 30 mm); It can be distinguished from members of the *flavopunctatus* group including *L. rita* (left bank of the Congo River), *L. laticeps* (higher elevations in the Albertine Rift), *L. stanleyi* (higher elevations in the Ruwenzori Mts), *L. dudui* (right bank of the Congo River), and *L. margarettae* (southern Sudan, eastern Uganda and Kenya) by the lack of speckling of its dorsal pelage. The dorsal pelage is reddish-brown and ventral pelage is bright reddish-brown.

Taxonomic Issues

This taxon is in the *sikapusi* group, and has only recently been elevated to species status (Verheyen et al., 2000).

Distribution

This species occurs widely in the Congo Basin extending north to the Sanaga River (Cameroon), and east into Uganda and the extreme western regions of Kenya. The limits of its distribution are still poorly known.

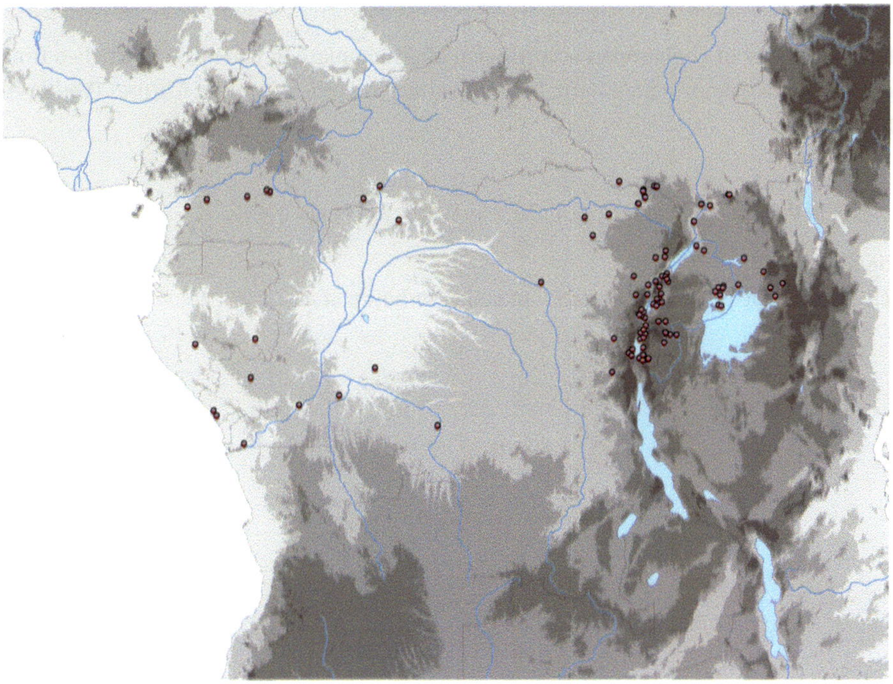

Figure 2.307 Distribution map

Measurements

External and cranial measurements (mm) and mass (g) for *Lophuromys ansorgei*, sexes combined

	Mean	Min	Max	N
Mass*	76.7	45	111	50
HB*	134.3	100	165	50
Tail*	71.1	49	91	50
HF*	22.7	20	26	50
Ear*	14.8	10	20	67
GLS**	33.2	30.5	35.3	46

*Delany (1975) as *L. sikapusi ansorgei*; **Verheyen et al. (2000)

Lophuromys aquilus True, 1892

Diagnosis
This is the only species of *Lophuromys* occurring on Mt Kilimanjaro. It can be distinguished from *L. verhageni*, from neighbouring Mt Meru, by multivariate analysis of cranial features (Verheyen et al., 2007).

Taxonomic Issues
This species is in the *flavopunctatus* group; 2n = 68 (Verheyen et al., 2007).

Distribution
This species is endemic to Mt Kilimanjaro, Tanzania (Verheyen et al., 2007).

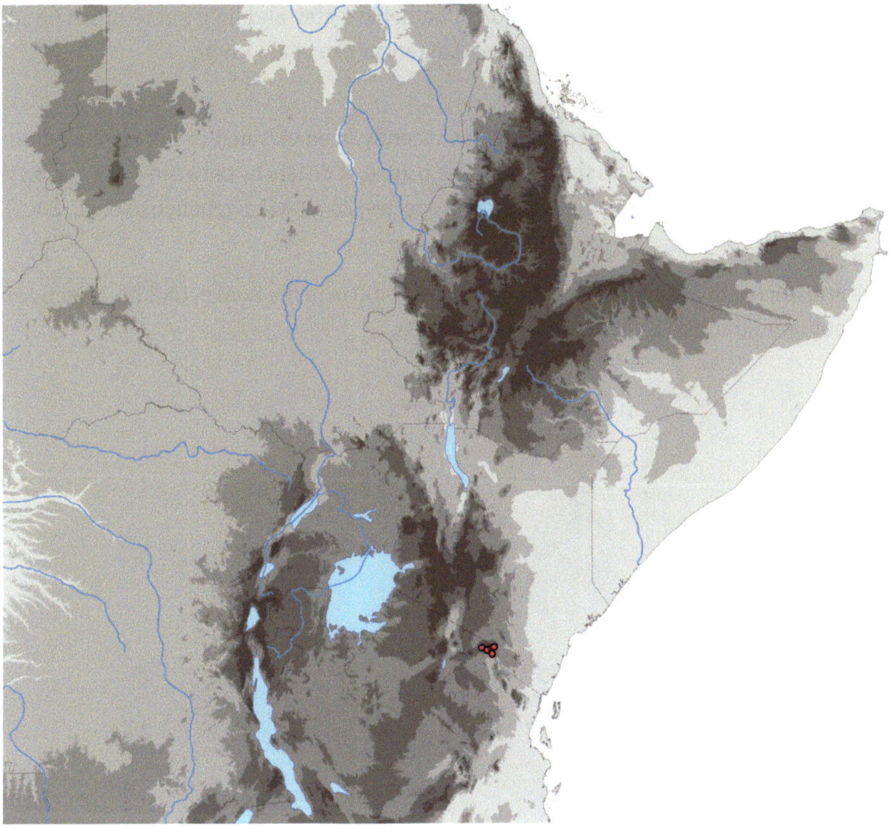

Figure 2.308 Distribution map

Measurements

External and cranial measurements (mm) and mass (g) for *Lophuromys aquilus*, sexes combined

	Mean	Min	Max	N
Mass	53.9	36	80	69
HB	128.1	117	145	69
Tail	75.3	66	84	50
HF	22.1	20	24	69
Ear	17.6	14.5	19.7	67
GLS	30.5	29.1	32.2	64

Lophuromys brevicaudus Osgood, 1936

Diagnosis

This is one of nine species endemic to Ethiopia. Only two other *Lophuromys* species co-occur with it: *L. chrysopus* (tail longer, > 70 mm; dorsal hairs with dark and light subterminal bands giving a speckled appearance; in forests at 1600-2700m; 2n = 54); and *L. melanonyx* (larger, GLS > 31 mm, mass > 60 g; 2n = 60). In *L. brevicaudus*, the dorsal hairs have grey and light subterminal bands giving a paler speckled appearance to this animal (Lavrenchenko et al., 1998).

Taxonomic Issues

This species is in the *flavopunctatus* group and has recently been elevated to full species status (Lavrenchenko et al., 1998). Based on the cytochrome b gene, it differs by between 6.5% and 9.1% in genetic distance from other Ethiopian *Lophuromys* species; 2n = 68 (Lavrenchenko et al., 2007).

Distribution

This species is restricted to Afroalpine healthlands above 2400 m in the Bonke, Chilalo, Gedeb and Bale Mts of southern Ethiopia (Lavrenchenko et al., 1998).

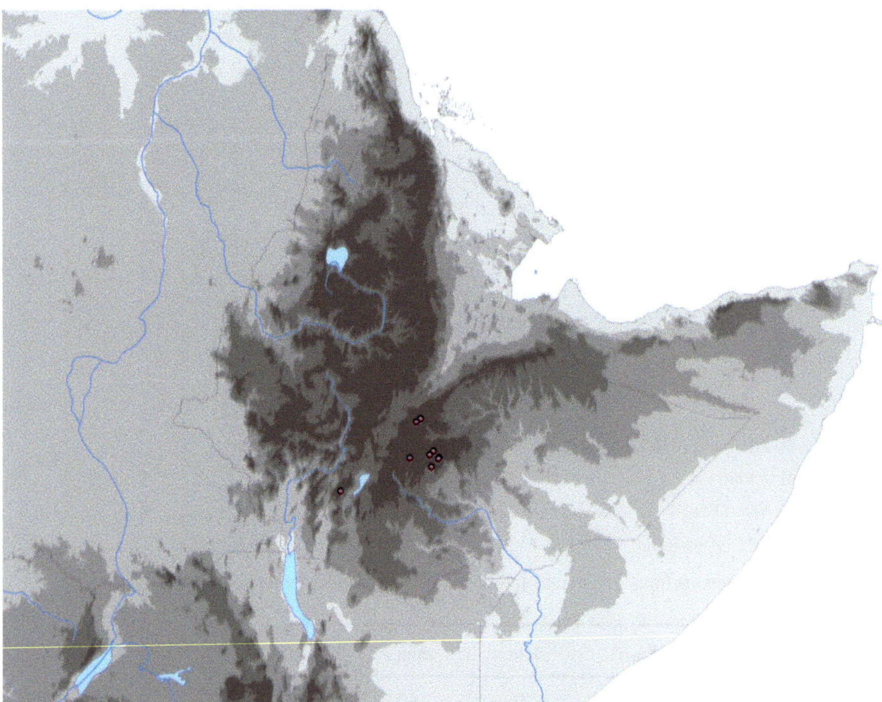

Figure 2.309 Distribution map

Measurements

External and cranial measurements (mm) and mass (g) for *Lophuromys brevicaudus*, sexes combined

	Mean	Min	Max	N
Mass*	43.3	29	60	86
HB*	114.2	96	134	87
Tail*	59.3	50	66	60
HF*	23.5	21.3	25.0	87
Ear*	17.1	15	19	87
GLS*	29.4	28.0	31.0	33

*Lavrenchenko et al. (1998)

Lophuromys brunneus Thomas, 1906

Diagnosis

This is one of nine species endemic to Ethiopia. The only other *Lophuromys* species occurring within its range is *L. chrysopus* (on average small, GLS < 31 mm; 2n = 54). In *L. chrysopus*, the dorsal hairs have dark and light subterminal bands giving a dark speckled appearance similar to *L. brunneus* (Lavrenchenko et al., 2007).

Taxonomic Issues

This species is in the *flavopunctatus* group and only recently elevated to full species status (Lavrenchenko et al., 1998a). Based on the cytochrome b gene, it differs by between 2.2% and 9.0% in genetic distance from other Ethiopian *Lophuromys* species; 2n = 68 (Lavrenchenko et al., 2007). The genetic difference between *L. brunneus* and *L. flavopunctatus* is rather small and these two taxa may possibly be conspecific.

Distribution

This species occurs in highland forests west of the Rift Valley generally above 1800 m (Lavrenchenko et al., 2007). It is syntopic with *L. chrysopus* in the Beletta Forest.

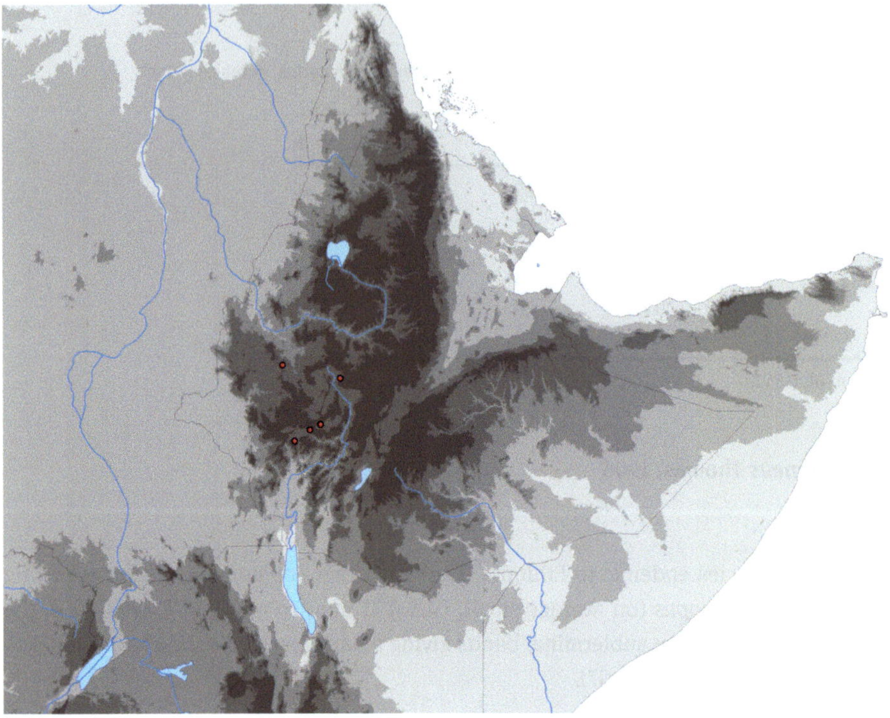

Figure 2.310 Distribution map

Measurements

External and cranial measurements (mm) and mass (g) for *Lophuromys brunneus*, sexes combined

	Mean	Min	Max	N
Mass*	61.9	38	76	31
HB*	124.4	101	138	31
Tail*	70.2	55	84	17
HF*	24.6	22.8	26.7	31
Ear*	18.6	17	20	31
GLS*	31.1	28.8	33.8	29

*Lavrenchenko et al. (2007)

Lophuromys chercherensis Lavrenchenko, W. Verheyen, E. Verheyen, Hulselmans & Leirs, 2007

Diagnosis

This is one of nine species endemic to Ethiopia, and is the only species occurring in the Chercher Mts east of the Rift Valley. The dorsal hairs have dark and light subterminal bands giving a dark speckled appearance similar to *L. chryopus* which occurs further south (Levrenchenko et al., 2007). Its closest relatives seem to be *L. simiensis* and *L. menageshae*, which are morphologically similar but differ by having an obviously bicoloured tail (in *L. chercherensis* the tail is not conspicuously bicoloured).

Taxonomic Issues

This newly described species is in the *flavopunctatus* group (Lavrenchenko et al., 2007). Based on the cytochrome b gene, it differs by between 3.8% and 10.4% in genetic distance from other Ethiopian *Lophuromys* species; 2n = 70 (Lavrenchenko et al., 2007).

Distribution

This species is currently only known from the Chercher Mts where it has been collected in montane forest between 2000 - 2700 m (Lavrenchenko et al., 2007).

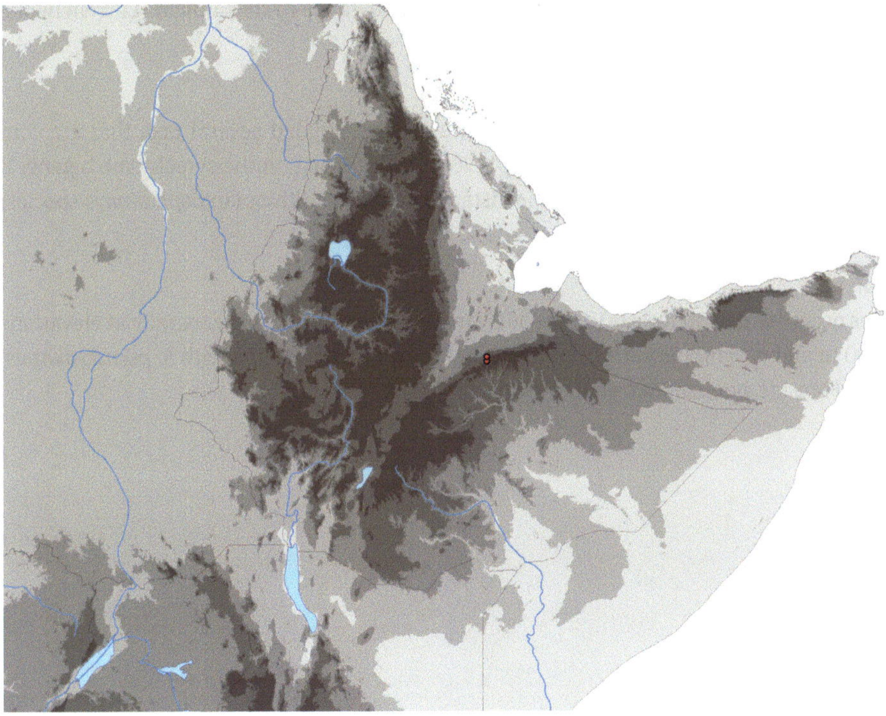

Figure 2.311 Distribution map

Measurements

External and cranial measurements (mm) and mass (g) for *Lophuromys chercherensis*, sexes combined

	Mean	Min	Max	N
Mass*	60.3	36	78	9
HB*	128.3	114	145	9
Tail*	62.5	60	65	4
HF*	22.9	22.0	24.0	9
Ear*	17.8	17.0	18.0	5
GLS*	29.8	27.8	31.9	9

*Lavrenchenko et al. (2007)

Lophuromys chrysopus Osgood, 1936

Diagnosis

This is one of nine species endemic to Ethiopia. The other *Lophuromys* species that co-occur with it include: *L. brevicaudus* (tail shorter, < 70 mm; dorsal hairs with grey and light subterminal bands giving a paler speckled appearance; in Afroalpine heathlands above 2400 m; 2n = 68); *L. melanonyx* (larger, GLS > 31 mm, mass > 60 g; dorsal hairs with grey and light subterminal bands giving a paler speckled appearance; occurring above 3000 m, 2n = 60); *L. pseudosikapusi* (larger, GLS > 31 mm, HB > 130 mm; unspeckled due to lack of white or light subterminal band; 2n = 70); and *L. brunneus* (on average larger, GLS > 30 mm; 2n = 68). In *L. chrysopus*, the dorsal hairs have dark and light subterminal bands giving a darker speckled appearance to this animal (Lavrenchenko et al., 1998a).

Taxonomic Issues

This species is in the *flavopunctatus* group and previously included several taxa that are considered full species here (Lavrenchenko et al., 1998a, 2007). Based on the cytochrome b gene, it differs by between 8.0% and 10.4% in genetic distance from other Ethiopian *Lophuromys* species; 2n = 54 (Lavrenchenko et al., 2007).

Distribution

This species occurs widely in forests in the southern highlands of Ethiopia generally at elevations of 1600-2700 m and is syntopic with *L. brunneus* in the Beletta Forest and with *S. pseudosikapusi* in the Sheko Forest (Lavrenchenko et al., 1998a, 2007).

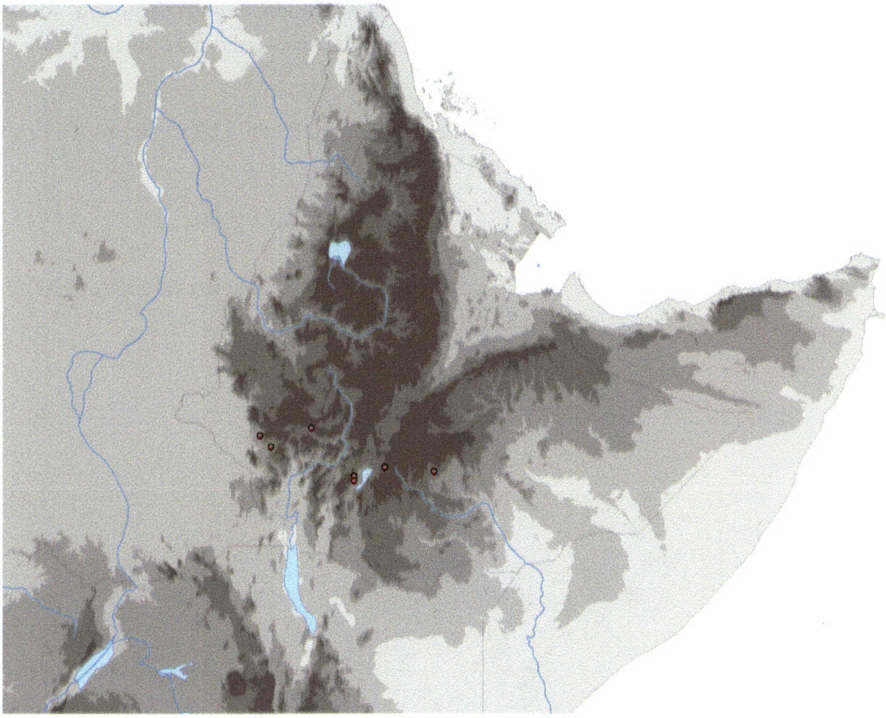

Figure 2.312 Distribution map

Measurements

External and cranial measurements (mm) and mass (g) for *Lophuromys chrysopus*, sexes combined

	Mean	Min	Max	N
Mass*	41.0	32	59	47
HB*	113.2	100	130	47
Tail*	78.8	72	87	22
HF*	21.3	20.0	22.3	42
Ear*	17.5	16.0	19.0	47
GLS*	29.5	27.8	31.1	66

*Verheyen et al. (2007)

Lophuromys cinereus Dieterlen & Gelmroth, 1974

Diagnosis

This species is distinguished from other members of the *flavopunctatus* group (e.g. *L. laticeps* and *L. dudui*) by its brown-grey dorsal and paler grey ventral pelage. Other Albertine Rift endemics include: *L. rahmi* (smaller GLS < 27 mm; dorsal and ventral pelage reddish or orange); *L. medicaudatus* (relatively tail length longer ca. 85% of HB; in *L. cinereus* it is ca. 60% or less of HB);

L. woosnami (relatively tail length much longer ca. 100% of HB); and *L. stanleyi* (endemic to the Ruwenzori Mts). It is distinguished from the *sikapusi* group by its speckled dorsal pelage.

Taxonomic Issues

This species is in the *flavopunctatus* group, and the possibility that it might be an aberrant morph of that species has been raised (Dieterlen, 1987). It is retained as a full species until further information comes to light.

Distribution

This species is only known from the type locality Kahuzi-Biega NP, DRC (Musser & Carleton, 2005), where it was collected in montane swamps at about 2300 m (Dieterlen, 1987).

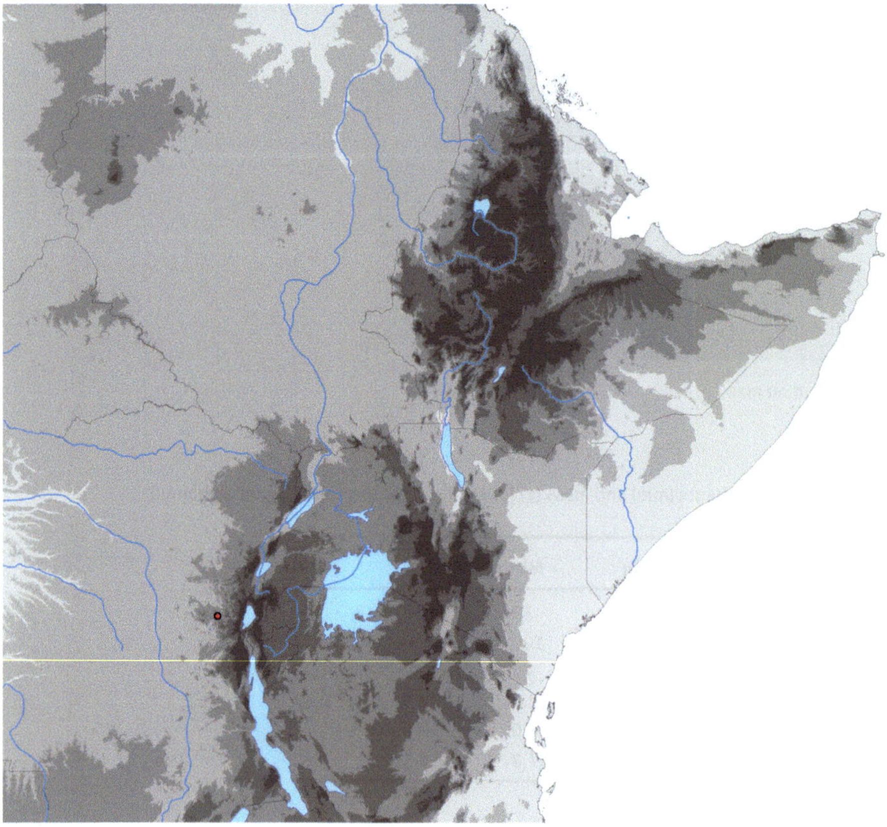

Figure 2.313 Distribution map

Measurements

External and cranial measurements (mm) and mass (g) for *Lophuromys cinereus*, sexes combined

	Mean	Min	Max	N
Mass*	52	45	56	3
HB*	122	110	129	3
Tail*	69.5	65	74	3
HF*	23.5	20	25	3
Ear*	19.1	17	21	3
GLS*	29.8	28.0	31.0	3

*Happold (2013)

Lophuromys dieterleni W. Verheyen, Hulselmans, Colyn & Hutterer, 1997

Diagnosis

This is the only species of *Lophuromys* occurring in montane forest on Mt Oku. *Lophuromys sikapusi* is very similar in size and appearance but occurs at lower elevations, and the two species show differences in the shape of the skull (Verheyen et al., 1997).

Taxonomic Issues

This species has recently been described and is tentatively included in the *flavopunctatus* group based on cranio-dental features, despite its unspeckled dorsal pelage (Verheyen et al., 1997). Its closest relative appears to be the diminutive *L. eisentrauti* from Mt Lefo approximately 50 km to the south-west (Verheyen et al., 1997).

Distribution

This species is restricted to Mt Oku, occurring in montane forest above 2000 m (Verheyen et al., 1997).

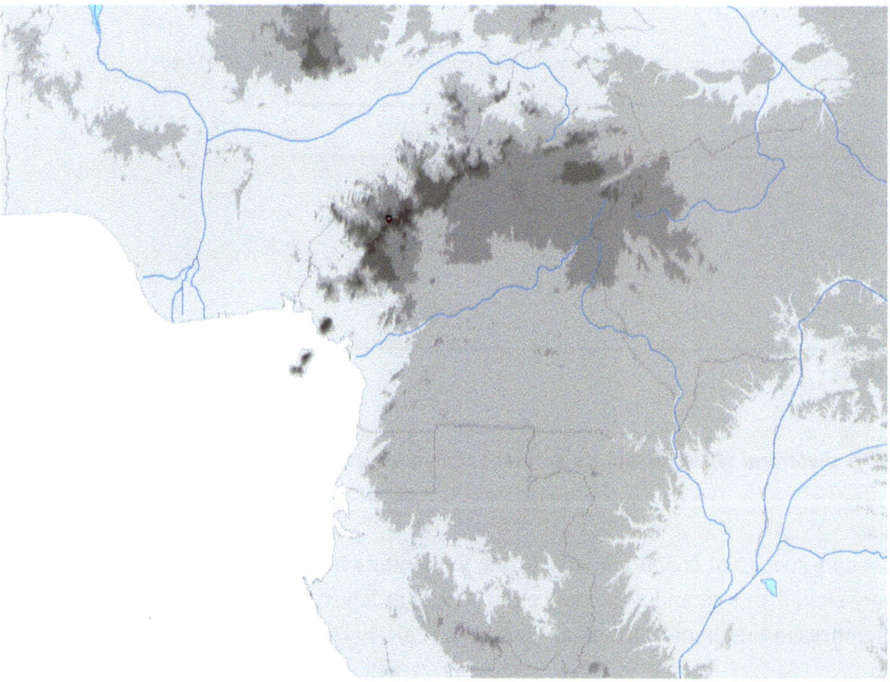

Figure 2.314 Distribution map

Measurements

External and cranial measurements (mm) and mass (g) for *Lophuromys dieterleni*, sexes combined

	Mean	Min	Max	N
Mass*	58.8	49	62	5
HB*	120.8	113	128	5
Tail*	75.3	74	77	3
HF*	21.4	20	24	5
Ear*	17.4	17	18	5
GLS*	30.9	30.9	31.0	2

*Verheyen et al. (2007)

Lophuromys dudui W. Verheyen, Hulselmans Dierckx & E. Verheyen, 2002

Diagnosis

This low-lying forest species co-occurs with several *Lophuromys* species including: *Lophuromys ansorgei* (larger, GLS > 30 mm; dorsal pelage not speckled); *L. luteogaster* (tail relatively longer ca. 85% of HB, in *L. dudui* it is < 60% of HB); and *L. nudicaudus* (dorsal pelage not speckled). It can be distinguished from other members of the *flavopunctatus* group including *L. rita* (left bank of Congo River), *L. laticeps* (higher elevations in the Albertine Rift), and *L. stanleyi* (higher elevations in the Ruwenzori Mts) by multivariate analysis of cranial features (Verheyen et al., 2007).

Taxonomic Issues

This recently described species belongs to the *flavopunctatus* group (Verheyen et al., 2002), but its relationship to *L. rita* (which occurs on the left bank of the Congo River) and to *L. laticeps* (which occurs at higher elevations in the Albertine Rift) is not fully resolved (Verheyen et al., 2007).

Distribution

This species occurs widely in the Congo Basin on the right bank of the Congo River where it typically occurs in low-lying forest below 1000 m. The eastern limits of its distribution are not currently known but it is thought that it does not occur in the higher elevations of the Albertine Rift where it is replaced by *L. laticeps*.

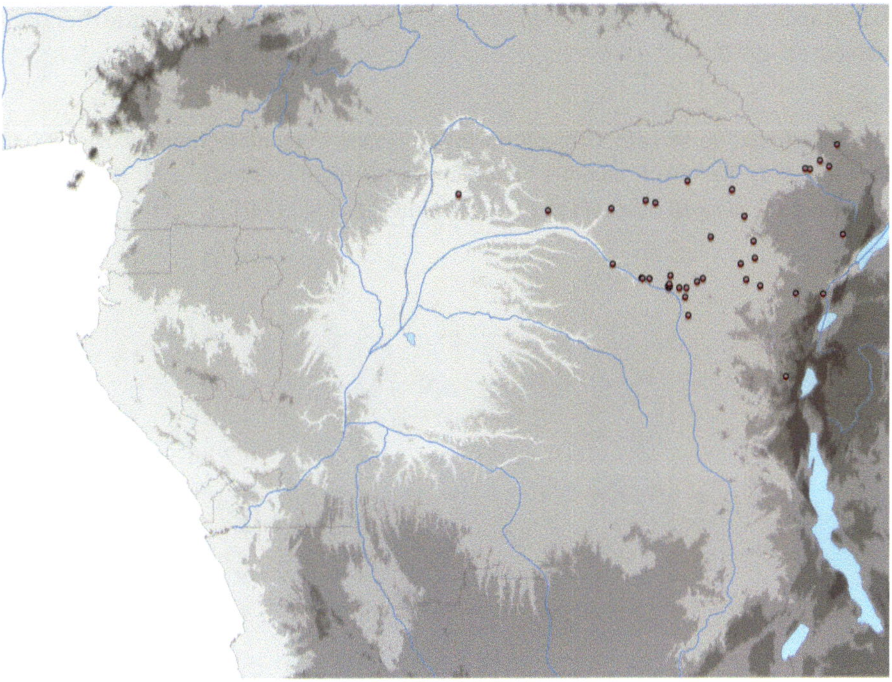

Figure 2.315 Distribution map

Measurements

External and cranial measurements (mm) and mass (g) for *Lophuromys dudui*, sexes combined

	Mean	Min	Max	N
Mass	-	-	-	-
HB*	108.9	89	124	92
Tail*	61.0	51	72	52
HF*	18.9	17.2	20.9	85
Ear*	13.5	11.0	16.2	61
GLS*	28.0	26.3	29.8	87

*Verheyen et al. (2007)

Lophuromys eisentrauti Dieterlen, 1978

Diagnosis

This is the only species of *Lophuromys* occurring in montane forest on Mt Lefo. *Lophuromys sikapusi* is much larger (GLS > 30 mm, HF > 22 mm), and occurs at lower elevations (Verheyen et al., 1997). *Lophuromys dieterleni* is slightly larger (GLS > 30 mm, HF > 20 mm) and occurs only on Mt Oku.

Taxonomic Issues

This species has recently been described and is tentatively included in the *flavopunctatus* group based on cranio-dental features, despite its unspeckled dorsal pelage (Verheyen et al., 1997). Its closest relative appears to be the larger *L. dieterleni* from Mt Oku approximately 50 km to the north-east (Hutterer et al., 1992).

Distribution

This species is restricted to Mt Lefo, Cameroon where it occurs in montane forest above 2000 m (Verheyen et al., 1997). Records from other mountains in the highlands such as Mt Kupe and Mt Manenguba refer to *L. sikapusi* (Hutterer et al., 1992; Verheyen et al., 1997).

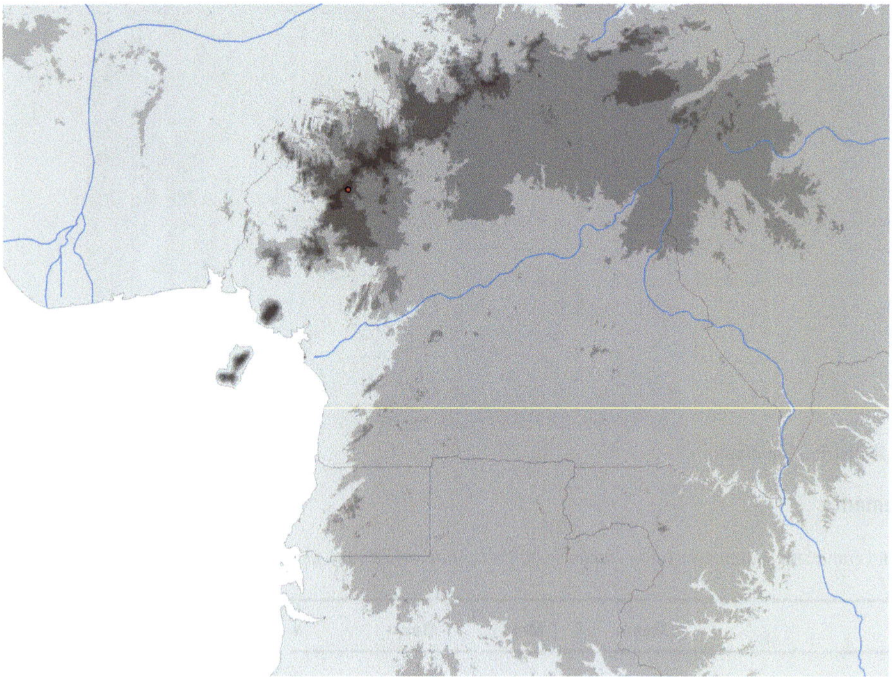

Figure 2.316 Distribution map

Measurements

External and cranial measurements (mm) and mass (g) for *Lophuromys eisentrauti*, sexes combined

	Mean	Min	Max	N
Mass	-	-	-	-
HB*	94	-	-	1
Tail*	53	-	-	1
HF*	19	-	-	1
Ear*	14	-	-	1
GLS*	26.8	-	-	1

*Dieterlen (1979), measurements of the holotype

Lophuromys flavopunctatus Thomas, 1888

Diagnosis

This is one of nine species endemic to Ethiopia. Only one other *Lophuromys* species occurs within its range: *L. menageshae* (larger, typically GLS > 31 mm; 2n = 70). Its closest relative seems to be *L. brunneus* from which it cannot easily be distinguished (see below).

Taxonomic Issues

This newly described species is in the *flavopunctatus* group (Lavrenchenko et al., 2007). Based on the cytochrome b gene, it differs by between 2.2% and 9.3% in genetic distance from other Ethiopian *Lophuromys* species; 2n = 68 (Lavrenchenko et al., 2007). The genetic difference between *L. flavopunctatus* and *L. brunneus* is rather small and these two taxa may possibly be conspecific, although their karyotypes are distinct.

Distribution

This species occurs at several localities west of the Rift Valley where it has been collected in various habitats between 1900 - 3000 m (Lavrenchenko et al., 2007).

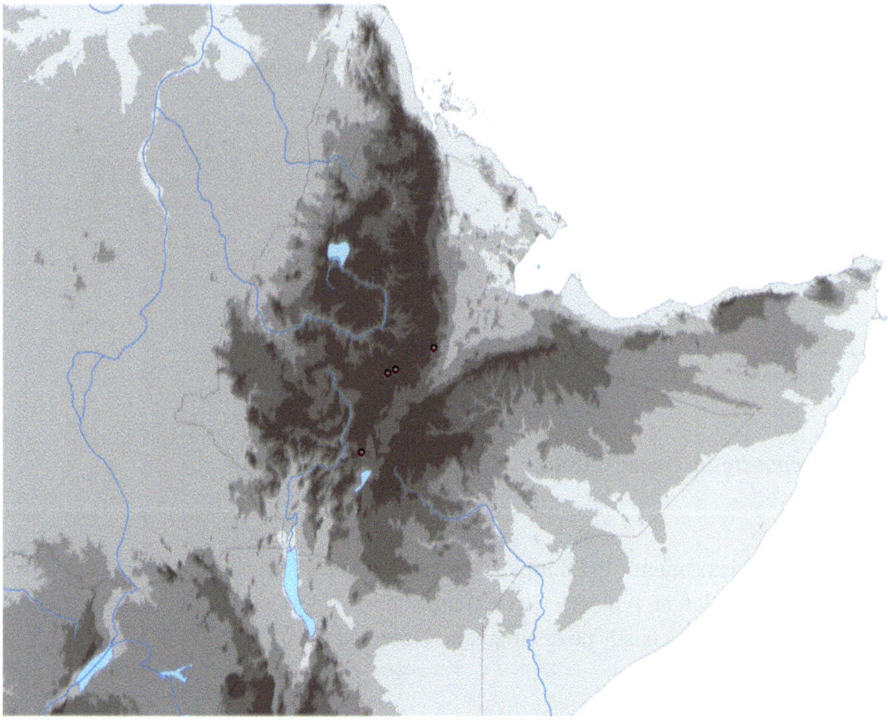

Figure 2.317 Distribution map

Measurements

External and cranial measurements (mm) and mass (g) for *Lophuromys flavopunctatus*, sexes combined

	Mean	Min	Max	N
Mass*	69.0	69	69	2
HB*	127.0	125	129	3
Tail*	62.3	58	68	3
HF*	24.2	23.0	25.3	3
Ear*	18.0	17.0	19.0	3
GLS*	29.8	28.6	31.4	5

*Lavrenchenko et al. (2007)

Lophuromys huttereri W. Verheyen, Colyn & Hulselmans, 1996

Diagnosis

This species co-occurs with *L. ansorgei* (larger, GLS > 30 mm), and *L. rita* (dorsal pelage speckled). *Lophuromys nudicaudus* is very similar in size and appearance but occurs exclusively on the right bank of the Congo River (Verheyen et al., 1996b).

Taxonomic Issues

This species was recently described (Verheyen et al., 1996b) and is in the *sikapusi* group, but its phylogenetic relationship with *L. nudicaudus* remains unresolved.

Distribution

This species is known from a few localities in the Congo Basin on the left bank of the Congo River (Verheyen et al., 1996b).

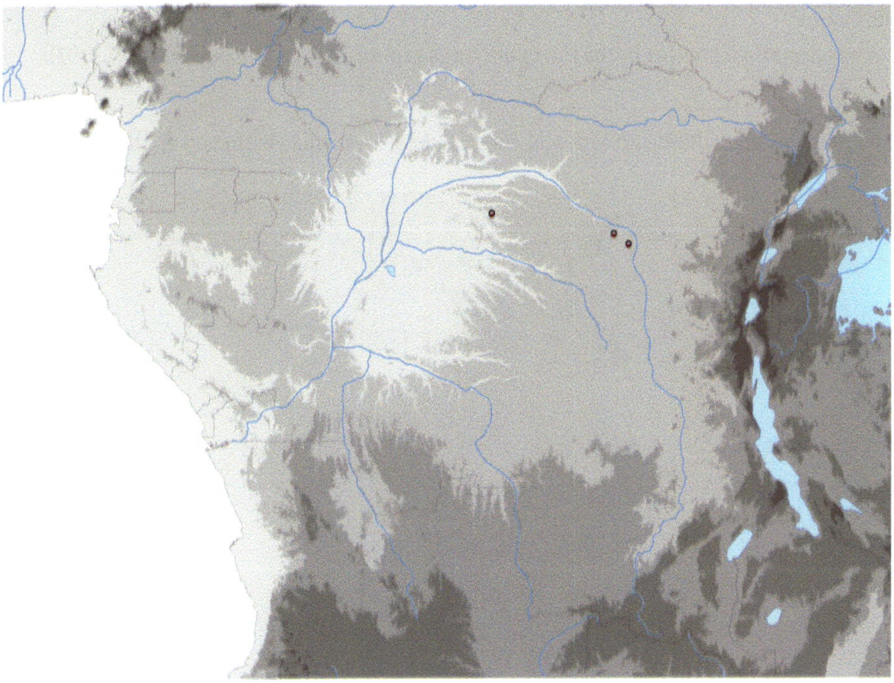

Figure 2.318 Distribution map

Measurements

External and cranial measurements (mm) and mass (g) for *Lophuromys huttereri*, sexes combined

	Mean	Min	Max	N
Mass	-	-	-	-
HB*	-	93	114	8-11
Tail*	-	59	61	8-11
HF*	-	18	20	8-10
Ear*	-	-	-	-
GLS*	28.8	28.0	29.9	7

*Verheyen et al. (1996b)

Lophuromys kilonzoi W. Verheyen, Hulselmans, Dierckx, Mulungu, Leirs, Corti & E. Verheyen, 2007

Diagnosis

This is the only species of *Lophuromys* occurring on the Usambara and Uluguru Mts. It can be distinguished from *L. aquilus* and *L. verhageni* (from Mts Kilimanjaro and Meru, respectively), by multivariate analysis of cranial features (Verheyen et al., 2007).

Taxonomic Issues

This is a newly described species in the *flavopunctatus* group; 2n = 68 (Verheyen et al., 2007).

Distribution

This species is endemic to the Usambara and Uluguru Mts, Tanzania (Verheyen et al., 2007).

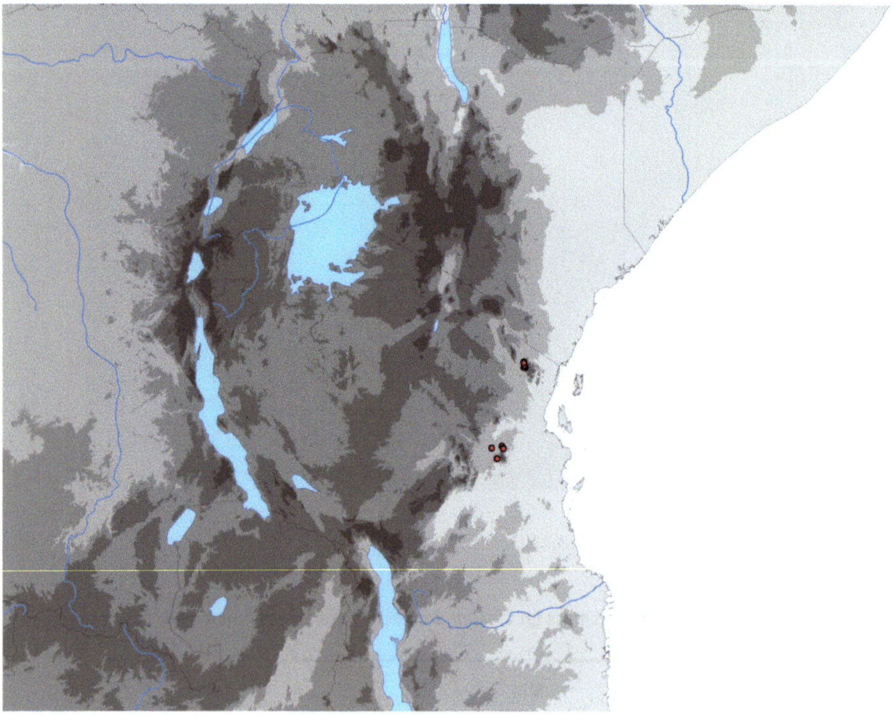

Figure 2.319 Distribution map

Measurements

External and cranial measurements (mm) and mass (g) for *Lophuromys kilonzoi*, sexes combined

	Mean	Min	Max	N
Mass*	54.5	40	74	51
HB*	126.3	115	139	52
Tail*	75.3	50	94	42
HF*	21.7	19.1	23.9	50
Ear*	19.1	15.5	21.7	51
GLS*	30.1	28.4	31.3	45

*Verheyen et al. (2007)

Lophuromys laticeps Thomas et al., 1907

Diagnosis

This Albertine Rift endemic co-occurs with a number of species including: *L. woosnami* (tail absolutely and relatively longer ca. 100% of HB; in *L. laticeps* it is < 60% of HB); *L. medicaudatus* (relative tail length longer ca. 85% of HB; dorsal pelage olive-brown). *Lophuromys ansorgei* is similar in size but does not have a speckled dorsal pelage. It can be distinguished from other members of the *flavopunctatus* group (including *L. dudui* which generally occurs at lower elevations in the Congo Basin and *L. stanleyi* which occurs at higher elevations in the Ruwenzori Mts), by multivariate analysis of cranial features (Verheyen et al., 2007).

Taxonomic Issues

This taxon, in the *flavopunctatus* group, has recently been elevated to full species status (Verheyen et al., 2007).

Distribution

This species is endemic to the Albertine Rift, where it is relatively widely distributed, occurring over a range of habitats and altitudes.

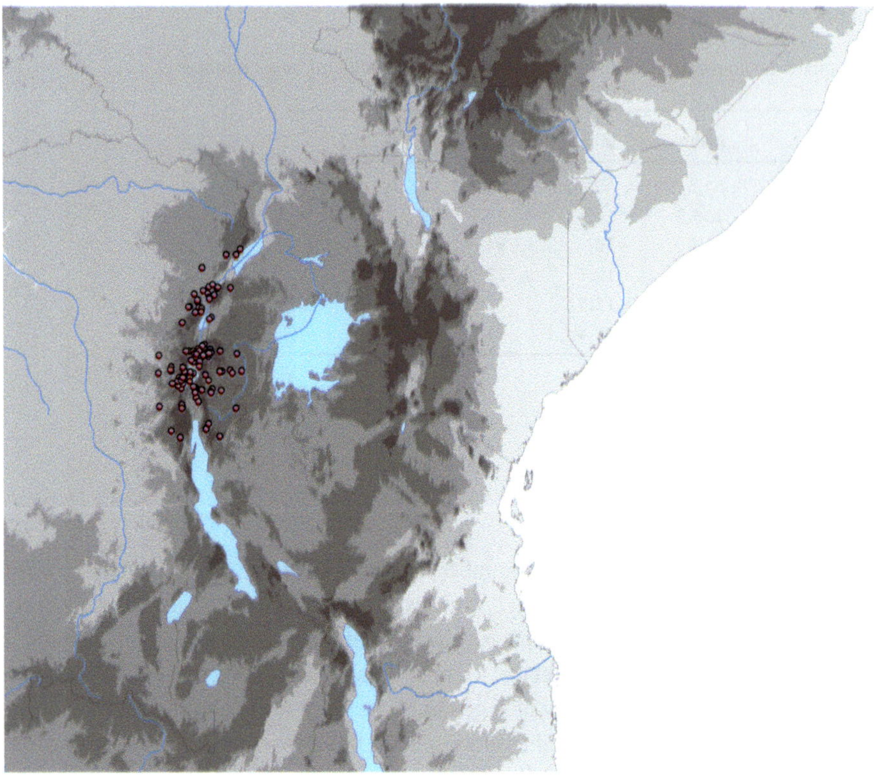

Figure 2.320 Distribution map

Measurements

External and cranial measurements (mm) and mass (g) for *Lophuromys laticeps*, sexes combined

	Mean	Min	Max	N
Mass*	59.2	43	73	96
HB*	115.0	99	136	96
Tail*	65.9	54	76	73
HF*	21.1	19.1	23.3	95
Ear*	17.0	15.1	18.6	93
GLS*	30.1	27.7	31.2	103

*Verheyen et al. (2007)

Lophuromys luteogaster Hatt, 1934

Diagnosis

The *Kivumys* subgenus is distinguished from members of the *flavopunctatus* and *sikapusi* groups by its longer tail (ca. 85% or more of HB; ca. 60% of HB in other species). Two other species in the subgenus *Kivumys* could be confused with this species: *L. woosnami* (larger, GLS > 30 mm;

occurring in montane habitats above 1800 m) and *L. medicaudatus* (tail is ca. 85% of HB; in *L. luteogaster* it is ca. 100% of HB). Other Albertine Rift endemics include: *L. rahmi* (smaller GLS < 27 mm); *L. cinereus* and *L. stanleyi* (relatively tail length shorter ca. 60% or less of HB). The dorsal pelage is uniform medium to dark brown with reddish or yellowish base and the ventral pelage is dark cinnamon (Dieterlen, 1975).

Taxonomic Issues

This species belongs in the *Kivumys* subgenus (Dieterlen, 1987). It originally included records of *L. medicaudatus* (Verheyen, 1964) which was subsequently described as a valid species based on differences in morphology and ecology (Dieterlen, 1975).

Distribution

This species is endemic to the north-eastern DRC where it is known from four localities in low-lying forest (below 1200 m) (Dieterlen, 1987).

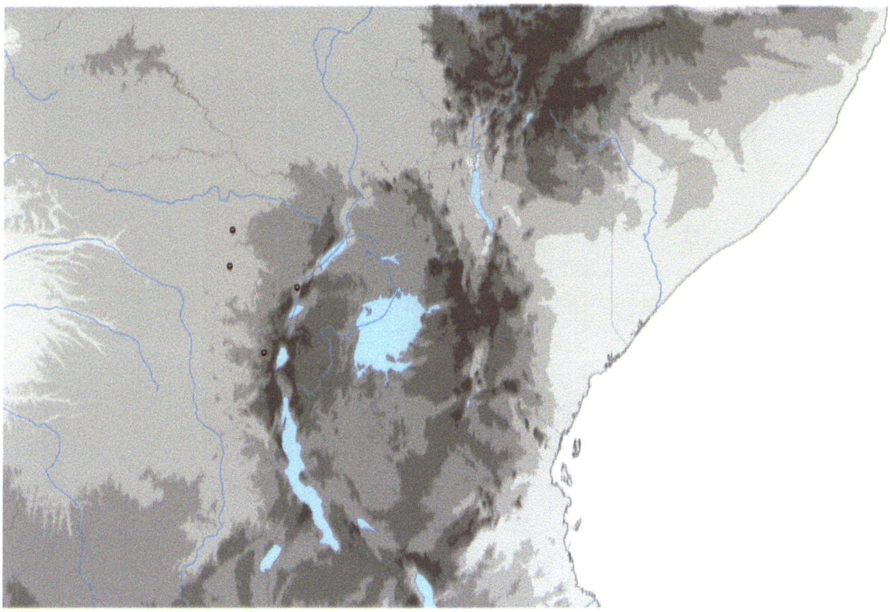

Figure 2.321 Distribution map

Measurements

External and cranial measurements (mm) and mass (g) for *Lophuromys luteogaster*, sexes combined

	Mean	Min	Max	N
Mass*	34.0	28	41	25
HB*	102	90	113	25
Tail*	106	90	117	25
HF*	21	19	23	25
Ear*	17.5	16	19	25
GLS*	28.6	27.2	29.6	12

*Dieterlen (1975, 1976b)

Lophuromys machangui W. Verheyen, Hulselmans, Dierckx, Mulungu, Leirs, Corti & E. Verheyen, 2007

Diagnosis

This is the only species of *Lophuromys* occurring across its range in southern Tanzania, northern Malawi and Mozambique. It can be distinguished from other members of the *flavopunctatus* group in Tanzania (including *L. sabunii* from the neighbouring Ufipa Plateau), by multivariate analysis of cranial features (Verheyen et al., 2007).

Taxonomic Issues

This is a newly described species in the *flavopunctatus* group (Verheyen et al., 2007).

Distribution

This species is known from Mt Rungwe in southern Tanzania where it was collected in a variety of forested and scrubby habitats above 2000 m (Verheyen et al., 2007). Records from neighbouring mountains are thought to represent this species.

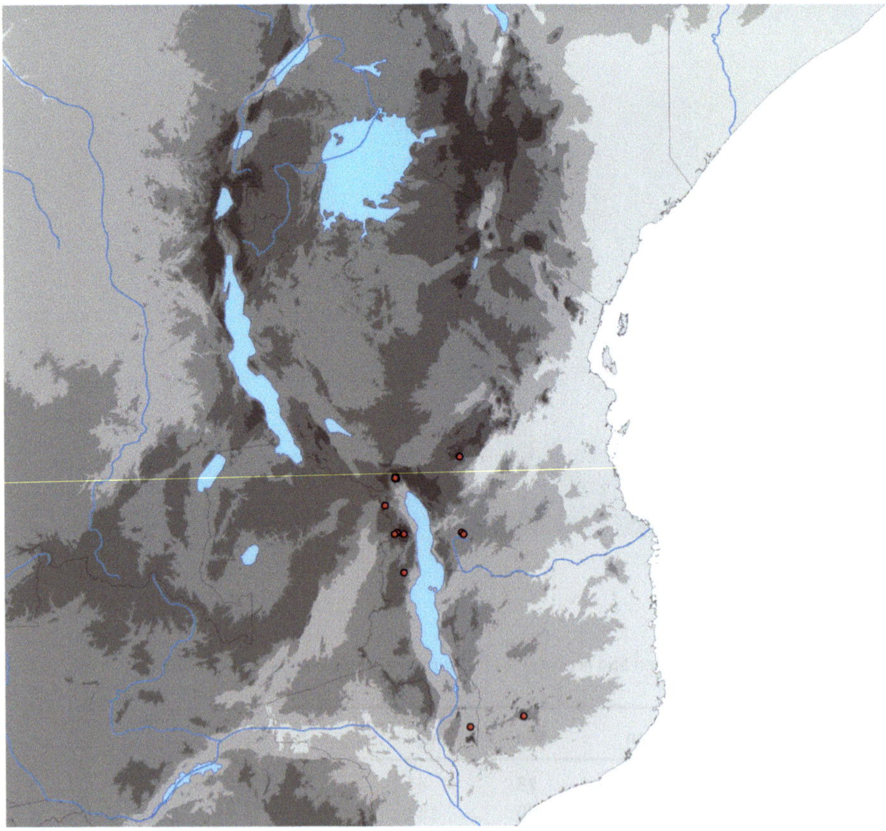

Figure 2.322 Distribution map

Measurements

External and cranial measurements (mm) and mass (g) for *Lophuromys machangui*, sexes combined

	Mean	Min	Max	N
Mass*	52.3	38	84	63
HB*	123.4	105	137	63
Tail*	64.3	24	81	58
HF*	20.9	17.6	22.7	61
Ear*	16.9	13.4	19.1	58
GLS*	29.7	28.1	31.3	53

*Verheyen et al. (2007)

Lophuromys makundii W. Verheyen, Hulselmans, Dierckx, Mulungu, Leirs, Corti & E. Verheyen, 2007

Diagnosis

This is the only species of *Lophuromys* occurring at Mt Hanang, north-central Tanzania. It can be distinguished from other members of the *flavopunctatus* group in Tanzania (including *L. verhageni* from Mt Meru and *L. aquilus* from Mt kilimanjaro), by multivariate analysis of cranial features (Verheyen et al., 2007).

Taxonomic Issues

This is a newly described species in the *flavopunctatus* group (Verheyen et al., 2007).

Distribution

This species is endemic to Mt Hanang, north-central Tanzania where it has been recorded from a variety of forested and shrubby habitats above 2000 m (Verheyen et al., 2007).

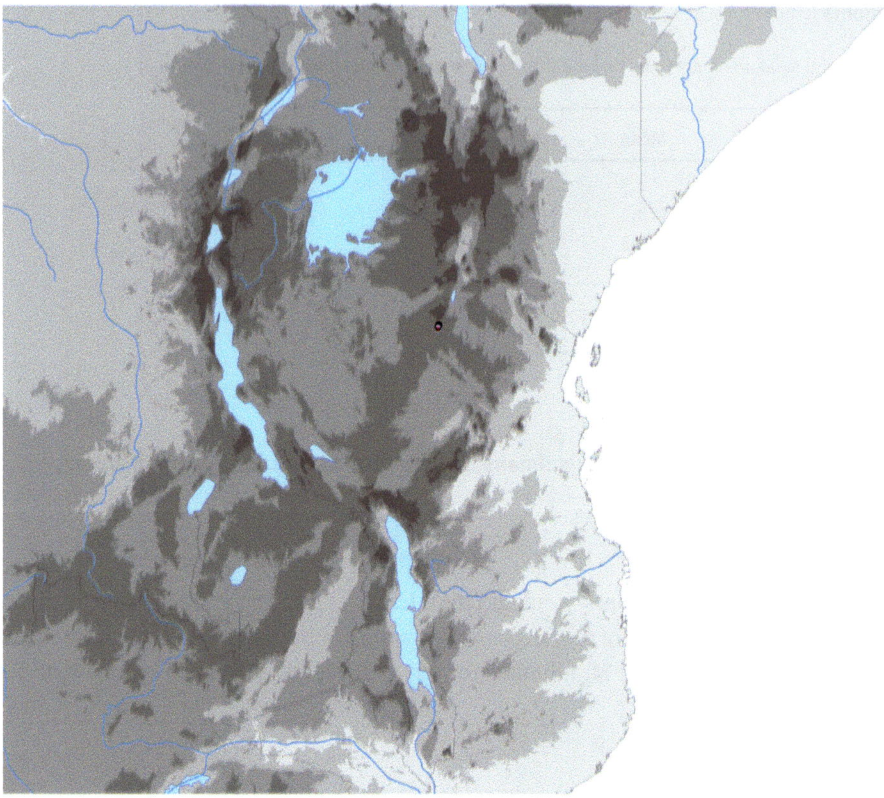

Figure 2.323 Distribution map

Measurements

External and cranial measurements (mm) and mass (g) for *Lophuromys makundii*, sexes combined

	Mean	Min	Max	N
Mass*	48.8	33	58	19
HB*	124.9	118	134	19
Tail*	76.3	59	85	17
HF*	21.9	21.0	23.1	19
Ear*	18.8	16.5	20.7	19
GLS*	29.5	27.6	30.9	15

*Verheyen et al. (2007)

Lophuromys margarettae Heller, 1912

Diagnosis

This is the only species of *Lophuromys* across most of its range, but co-occurs with *L. zena* at Mt Kenya and in the Aberdare Range (Kenya) and with *L. ansorgei* in the border region between Uganda and Kenya, and in southern Sudan. These two species can be distinguished from *L. marga-*

rettae as follows: *L. ansorgei* is larger (typically GLS > 31 mm), and dorsal pelage is not speckled; *L. zena* is similar in size but can be distinguished on molecular grounds and by multivariate analysis of cranial characters, and occurs at high elevations (Verheyen et al., 2007).

Taxonomic Issues

This taxon is in the *flavopunctatus* group and has recently been elevated to full species status (Verheyen et al., 2007).

Distribution

This species is endemic to the highlands of Kenya and Western Uganda, with an outlying population in the mountains of southern Sudan (which may refer to a different species).

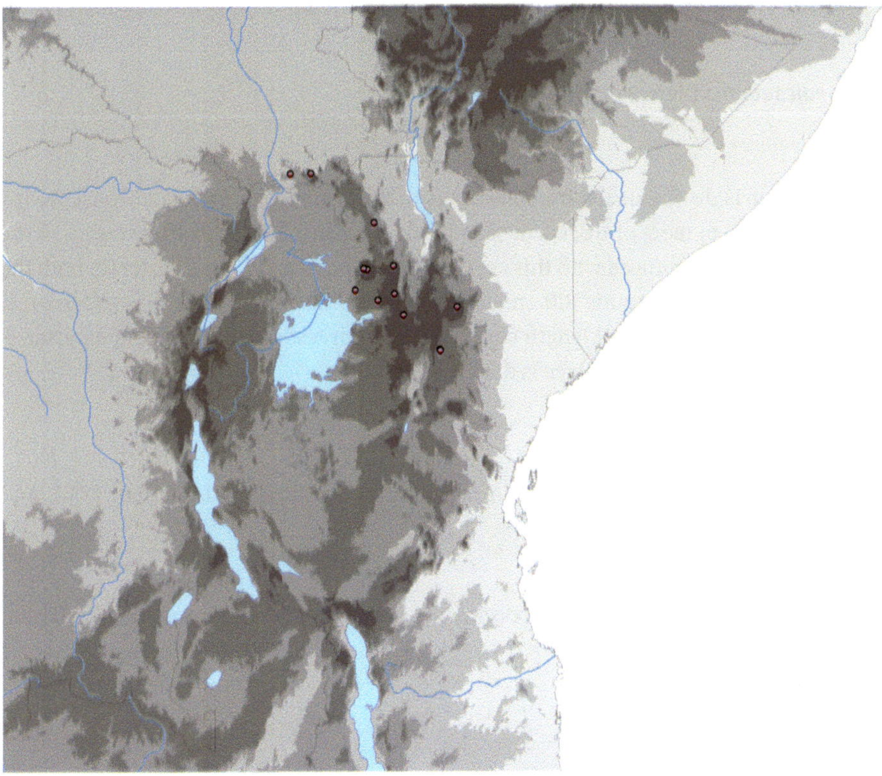

Figure 2.324 Distribution map

Measurements

External and cranial measurements (mm) and mass (g) for *Lophuromys margarettae*, sexes combined

	Mean	Min	Max	N
Mass	-	-	-	-
HB*	120	-	-	1
Tail*	80	-	-	1
HF*	21	-	-	1
Ear*	18	-	-	1
GLS**	29.9	28.3	31.4	42

*Heller (1913), measurements of the holotype; **Verheyen et al. (2007)

Lophuromys medicaudatus Dieterlen, 1975

Diagnosis
The *Kivumys* subgenus is distinguished from members of the *flavopunctatus* and *sikapusi* groups by its longer tail (ca. 85% or more of HB; ca. 60% of HB in other species). Two other species in the subgenus *Kivumys* could be confused with this species: *L. woosnami* and *L. luteogaster* (in both the tail is ca. 100% of HB). Other Albertine Rift endemics include: *L. rahmi* (smaller GLS < 27 mm); *L. cinereus* and *L. stanleyi* (relatively tail length shorter ca. 60% or less of HB). The dorsal pelage is uniform dark brown and the ventral pelage is dark orange.

Taxonomic Issues
This species belongs in the *Kivumys* subgenus (Dieterlen, 1987). It was originally listed as *L. luteogaster* (Verheyen, 1964) from which it differs in morphology and ecology (Dieterlen, 1975), and their distributions do not overlap.

Distribution
This species is endemic to the Albertine Rift, occurring in montane swamps around Lake Kivu at 1800-2500 m (Dieterlen, 1975).

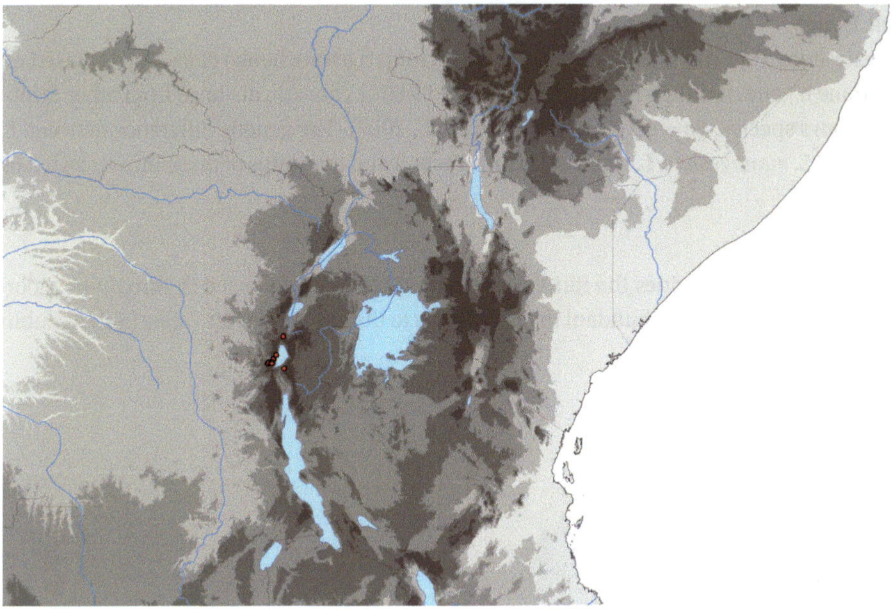

Figure 2.325 Distribution map

Measurements

External and cranial measurements (mm) and mass (g) for *Lophuromys medicaudatus*, sexes combined

	Mean	Min	Max	N
Mass*	35.4	29	43	27
HB*	103.2	92	112	27
Tail*	87.2	73	95	20
HF*	20.8	18	23	27
Ear*	16.5	15	19	26
GLS*	29.1	27.7	30.2	19

*Dieterlen (1975)

Lophuromys melanonyx F. Petter, 1972

Diagnosis

This is one of nine species endemic to Ethiopia. Three other *Lophuromys* species occur within its range: *L. flavopunctatus* (smaller, typically GLS < 31 mm; dorsal hairs with black and light subterminal bands giving a darker speckled appearance; 2n = 68); *L. chrysopus* (smaller, typically GLS < 31 mm; dorsal hairs with dark and light subterminal bands giving a darker speckled appearance; 2n = 54); *L. brevicaudus* (smaller, GLS < 31 mm; 2n = 68). Its closest relatives seem to be *L. simiensis* (smaller, typically GLS < 31 mm; dorsal hairs with dark and light subterminal bands giving a darker speckled appearance; 2n = 70). In *L. melanonyx*, the dorsal hairs with grey and light subterminal bands giving a paler speckled appearance similar to *L. brevicaudus* (Lavrenchenko et al., 2007).

Taxonomic Issues

This newly described species is in the *flavopunctatus* group (Lavrenchenko et al., 2007). Based on the cytochrome b gene, it differs by between 1.8% and 9.8% in genetic distance from other Ethiopian *Lophuromys* species; 2n = 60 (Lavrenchenko et al., 2007). The genetic difference between *L. melanonyx*, and *L. manageshae*/*L. simiensis* is rather small and the status of these three taxa needs to be re-assessed.

Distribution

This species occurs on both sides the Rift Valley where it has been collected in Afroalpine moorland above 3200 m, but is most abundant in the Bale Mts to the east of the Rift Valley (Sillero-Zubiri et al., 1995a).

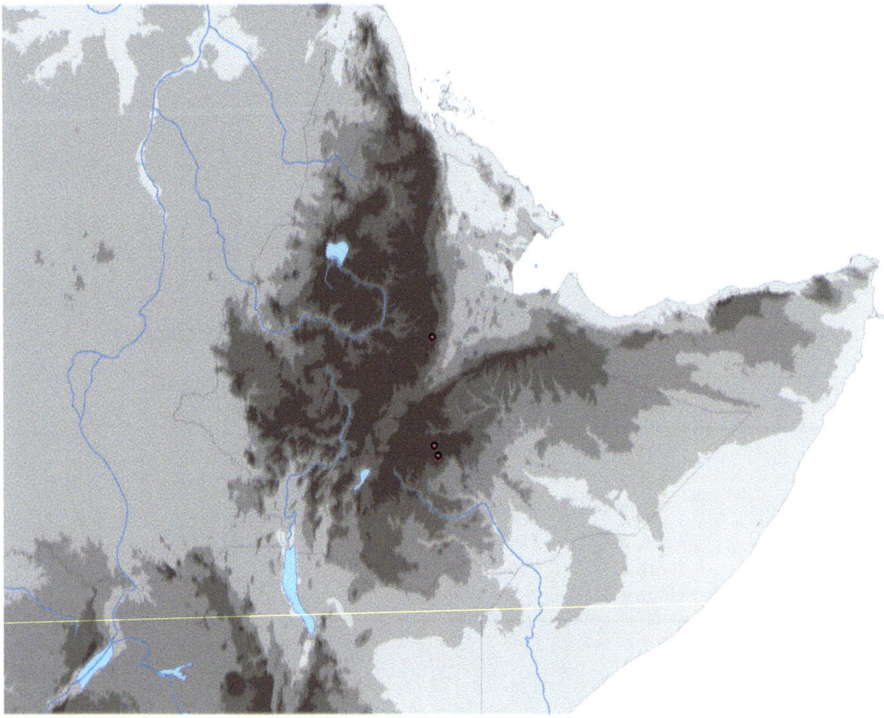

Figure 2.326 Distribution map

Measurements

External and cranial measurements (mm) and mass (g) for *Lophuromys melanonyx*, sexes combined

	Mean	Min	Max	N
Mass*	94.6	60	142	717
HB*	145.7	120	180	700
Tail*	63.7	30	99	671
HF*	22.8	21	25	234
Ear*	22.1	19	26	235
GLS**	33.0	31.6	34.4	14

*Sillero-Zubiri et al. (1995); **Dieterlen (1976b)

Lophuromys menagashae Lavrenchenko, W. Verheyen, E. Verheyen, Hulselmans & Leirs, 2007

Diagnosis

This is one of nine species endemic to Ethiopia. Only one other *Lophuromys* species occurs within its range: *L. flavopunctatus* (smaller, typically GLS < 31 mm; 2n = 68). Its closest relative seems to be *L. simiensis* from which it can be distinguished by its larger size (in *L. simiensis*, typically GLS < 31 mm).

Taxonomic Issues

This newly described species is in the *flavopunctatus* group (Lavrenchenko et al., 2007). Based on the cytochrome b gene, it differs by between 1.5% and 10.4% in genetic distance from other Ethiopian *Lophuromys* species; 2n = 70 (Lavrenchenko et al., 2007). The genetic difference between *L. manageshae* and *L. simiensis* is rather small and these two taxa may possibly be conspecific.

Distribution

This species occurs at several northern localities west of the Rift Valley where it has been collected in montane forest between 2100 - 2600 m (Lavrenchenko et al., 2007).

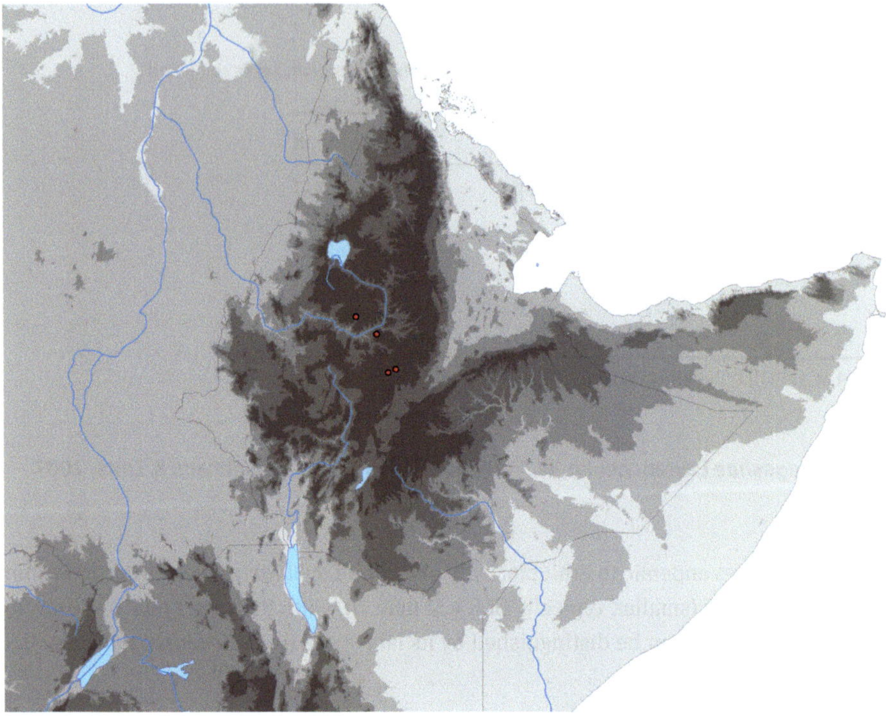

Figure 2.327 Distribution map

Measurements

External and cranial measurements (mm) and mass (g) for *Lophuromys menagashae*, sexes combined

	Mean	Min	Max	N
Mass*	74.5	72	77	2
HB*	134.3	129	143	6
Tail*	67.8	60	75	6
HF*	22.6	22.0	24.0	8
Ear*	17.4	15.0	20.0	9
GLS*	31.6	31.0	33.2	10

*Lavrenchenko et al. (2007)

Lophuromys nudicaudus Heller, 1911

Diagnosis

This species co-occurs with *L. ansorgei* (larger, GLS > 30 mm), marginally with *L. sikapusi* in Cameroon (larger, GLS > 30 mm), and possibly with *L. dudui* (dorsal pelage speckled). *Lophuromys huttereri* is very similar in size and appearance but occurs exclusively on the left bank of the Congo River (Verheyen et al., 1996b).

Taxonomic Issues

This species was previously listed as a subspecies of *L. sikapusi*, but was elevated to a full species by Rosevear (1969) and reviewed by Verheyen et al. (1996b). 2n = 56 (Verheyen & Van der Straeten, 1980).

Distribution

This species is widely distributed in the Lower Guinea Forest of Central Africa, with an isolated record from the right bank of the Congo Basin (Verheyen et al., 1996b).

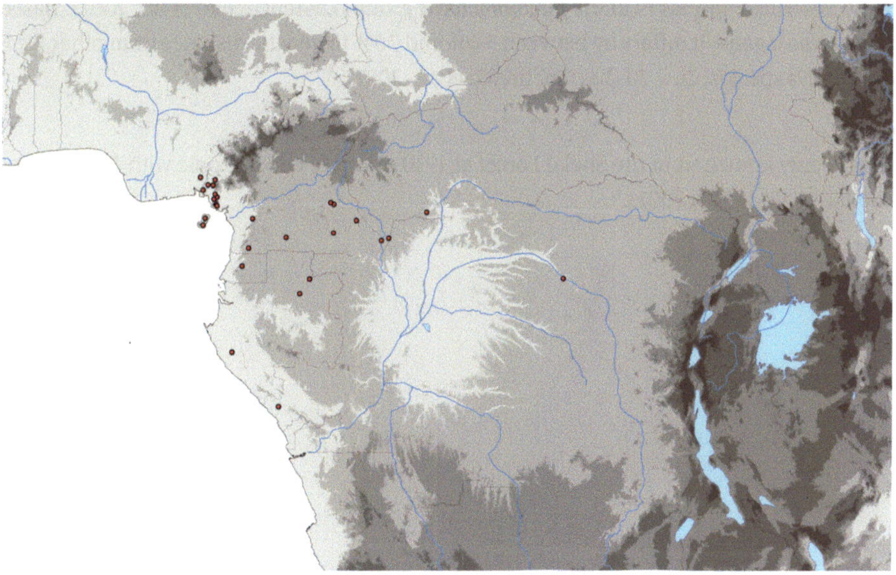

Figure 2.328 Distribution map

Measurements

External and cranial measurements (mm) and mass (g) for *Lophuromys nudicaudus*, sexes combined

	Mean	Min	Max	N
Mass*	39.5	29	52	23
HB*	106.2	89	119	41
Tail*	61.8	47	74	33
HF*	18.6	16.3	21.0	44
Ear*	14.6	10	18	40
GLS*	27.9	26.5	29.3	34

*Verheyen et al. (1996b)

Lophuromys pseudosikapusi Lavrenchenko, W. Verheyen, E. Verheyen, Hulselmans & Leirs, 2007

Diagnosis

This is one of nine species endemic to Ethiopia. The only other *Lophuromys* species that occurs within its range is *L. chrysopus* (smaller, GLS < 31 mm, HB < 130 mm; dorsal pelage speckled; 2n = 54).

Taxonomic Issues

This recently described species is in the *flavopunctatus* group (Lavrenchenko et al., 2007). Based on the cytochrome b gene, it differs by between 3.6% and 7.0% in genetic distance from other Ethiopian *Lophuromys* species; 2n = 70 (Lavrenchenko et al., 2007).

Distribution

This species is only known from the Sheko Forest at 1930 m where it is syntopic with *L. chrysopus* (Lavrenchenko et al., 1998a, 2007).

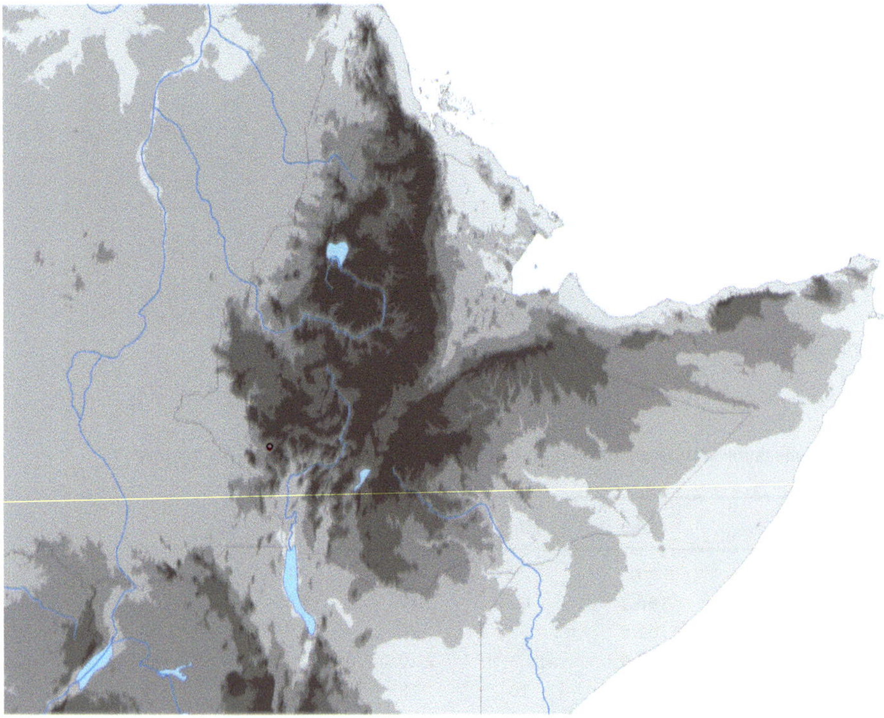

Figure 2.329 Distribution map

Measurements

External and cranial measurements (mm) and mass (g) for *Lophuromys pseudosikapusi*, sexes combined

	Mean	Min	Max	N
Mass*	60.0	53	67	2
HB*	134	130	138	2
Tail*	82.0	78	86	2
HF*	24.3	23.5	25.0	2
Ear*	19.8	19.5	20.0	2
GLS*	31.5	31.3	31.8	2

*Lavrenchenko et al. (2007)

Lophuromys rahmi Verheyen, 1964

Diagnosis

This Albertine Rift endemic co-occurs with a number of species including: *L. woosnami* (larger, GLS > 30 mm, with a longer absolute and relative tail ca. 100% of HB; in *L. rahmi* it is < 60% of HB); *L. medicaudatus* (relative tail length longer ca. 85% of HB; dorsal pelage olive-brown). It differs from other members of the *flavopunctatus* group (e.g. *L. dudui* and *L. laticeps*) by not having speckled pelage. *Lophuromys ansorgei* is larger (GLS > 30 mm), and *L. stanleyi* is restricted to the Ruwenzori Mts. The dorsal pelage in *L. rahmi* is reddish-brown and ventral pelage is bright reddish-orange.

Taxonomic Issues

This species is in the *flavopunctatus* group.

Distribution

This species is endemic to the Albertine Rift around Lake Kivu in the eastern DRC and Rwanda in montane forest above 1800 m (Dieterlen, 1987).

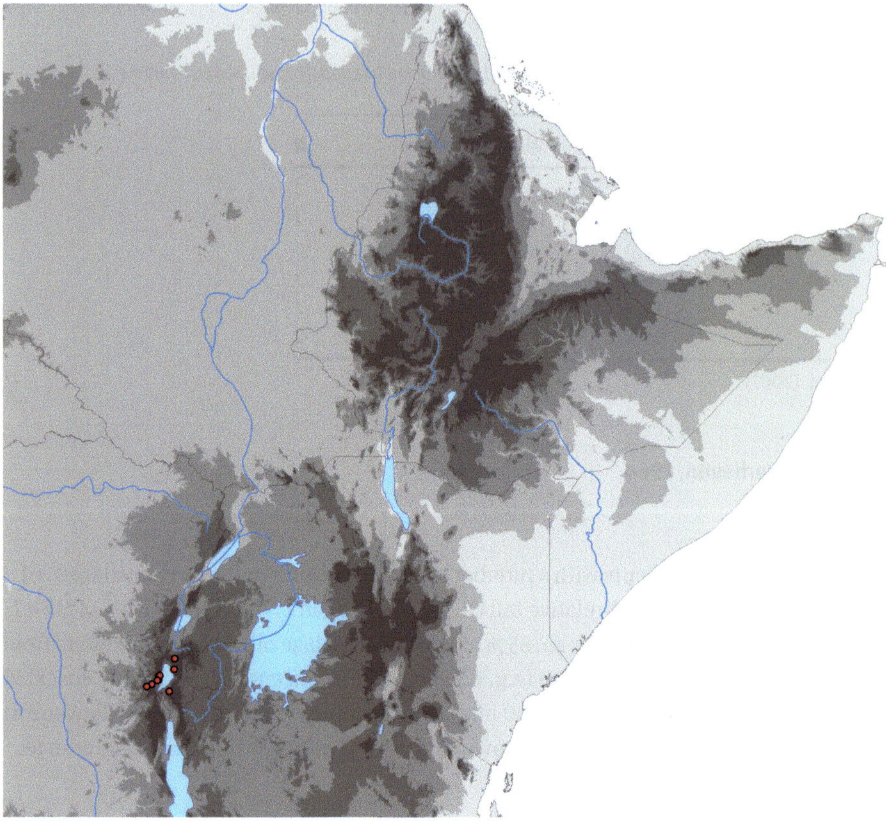

Figure 2.330 Distribution map

Measurements

External and cranial measurements (mm) and mass (g) for *Lophuromys rahmi*, sexes combined

	Mean	Min	Max	N
Mass*	32.5	30	45	11-16
HB*	102	95	116	11-16
Tail*	52.6	48	56	11-16
HF*	16.3	13	18	11-16
Ear*	12.5	10	15	11-16
GLS*	25.3	24.7	25.9	21

*Dieterlen (1976b)

Lophuromys rita Dollman, 1910

Diagnosis

This is the only member of the *flavopunctatus* group occurring on the left bank of the Congo River. It co-occurs with several species of *Lophuromys* across its range including: *L. huttereri*, *L. ansorgei*, and *L. angolensis* all of which are in the *sikapusi* group and do not have a speckled dorsal pelage.

Taxonomic Issues

This taxon, in the *flavopunctatus* group, was recently elevated to full species status (Verheyen et al., 2007) based mostly on molecular grounds.

Distribution

This species is widely distributed in the Congo Basin, on the left bank of the Congo River where it has been recorded from rainforest (Verheyen et al., 2007). The taxonomic affinities of the Zambian records are not clear, but these specimens were recorded as *L. flavopunctatus rita* by Ansell (1957).

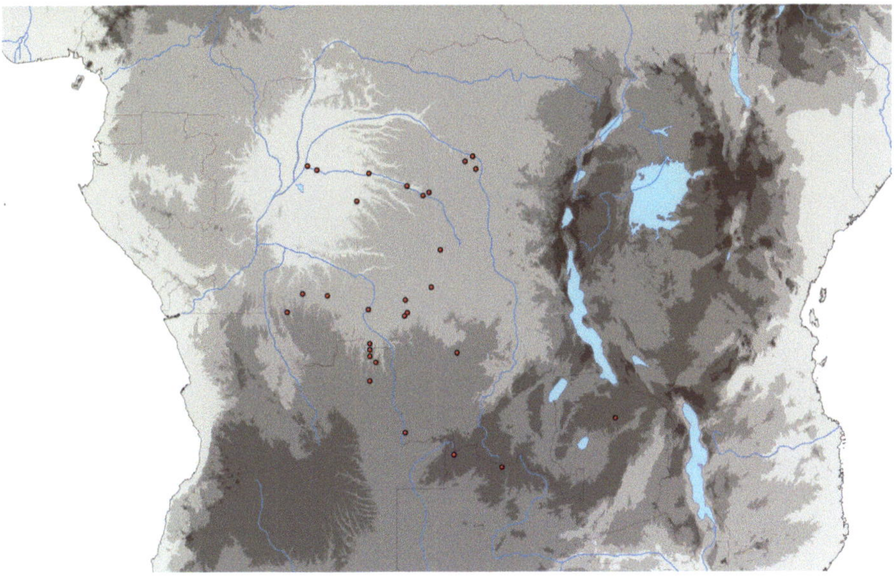

Figure 2.331 Distribution map

Measurements

External and cranial measurements (mm) and mass (g) for *Lophuromys rita*, sexes combined

	Mean	Min	Max	N
Mass	-	-	-	-
HB	-	-	-	-
Tail	-	-	-	-
HF	-	-	-	-
Ear	-	-	-	-
GLS*	29.8	28.8	31.1	22

*Verheyen et al. (2007)

Lophuromys roseveari W. Verheyen, Hulselmans, Colyn & Hutterer, 1997

Diagnosis

This is the only species of *Lophuromys* occurring in montane habitats on Mt Cameroon. *Lophuromys sikapusi* is very similar in size and appearance, but occurs at the base of mountain in low-lying habitats, and the two species show subtle differences in the shape of the cranium (Verheyen et al., 1997).

Taxonomic Issues

This species has recently been described and is included in the *sikapusi* group based on cranio-dental features (Verheyen et al., 1997).

Distribution

This species is only known from Mt Cameroon where it occurs in montane habitats above 1200 m (Verheyen et al., 1997).

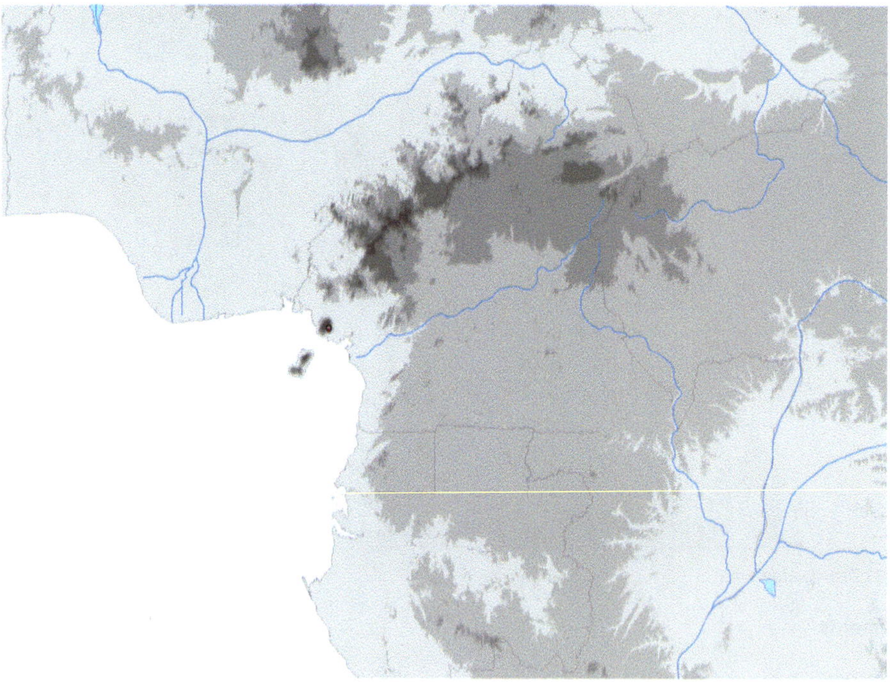

Figure 2.332 Distribution map

Measurements

External and cranial measurements (mm) and mass (g) for *Lophuromys roseveari*, sexes combined

	Mean	Min	Max	N
Mass*	63.5	49	88	24
HB*	127.5	104	141	32
Tail*	66.0	50	78	29
HF*	22.5	20.0	25.1	32
Ear*	18.5	16	21	28
GLS*	31.3	30.3	32.6	15

*Verheyen et al. (1997)

Lophuromys sabunii W. Verheyen, Hulselmans, Dierckx, Mulungu, Leirs, Corti & E. Verheyen, 2007

Diagnosis

This is the only species of *Lophuromys* occurring in the Ufipa Plateau, south-western Tanzania. It can be distinguished from other members of the *flavopunctatus* group in Tanzania (including *L. machangui* from neighbouring Mt Rungwe), by multivariate analysis of cranial features (Verheyen et al., 2007).

Taxonomic Issues

This is a newly described species in the *flavopunctatus* group (Verheyen et al., 2007).

DISTRIBUTION

This species is endemic to the Ufipa Plateau, south-western Tanzania where it has been recorded in various forest-edge and agricultural habitats at 1750 m (Verheyen, 2007).

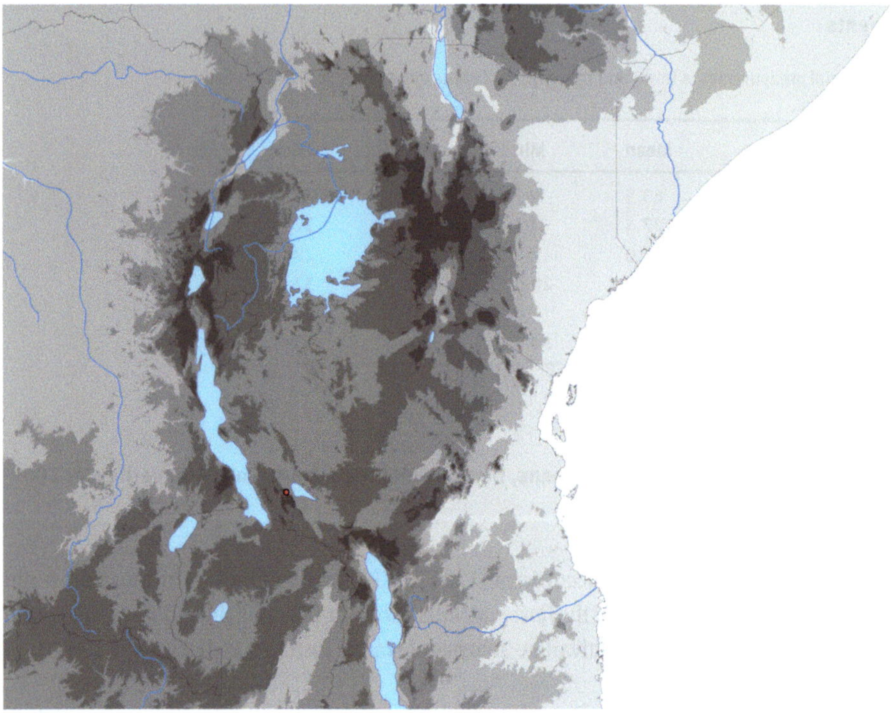

Figure 2.333 Distribution map

Measurements

External and cranial measurements (mm) and mass (g) for *Lophuromys sabunii*, sexes combined

	Mean	Min	Max	N
Mass*	52.0	38	87	29
HB*	127.7	112	145	29
Tail*	72.7	30	80	25
HF*	21.7	20.0	22.9	28
Ear*	16.9	12.2	19.5	28
GLS*	30.2	28.9	31.7	25

*Verheyen et al. (2007)

Lophuromys sikapusi (Temminck, 1853)

Diagnosis

This is the only species of *Lophuromys* occurring in West Africa, west of the Sanaga River (Cameroon) except for the species endemic to the Cameroon highlands: *L. roseveari*, *L. dieterleni* and *L. eisentrauti* all of which occur at higher elevations (typically above 2000 m).

Taxonomic Issues

Considering the dramatic diversity recently unearthed in the *flavopunctatus* group (see Verheyen et al., 2002, 2007; Lavrenchenko, 1998a, 2007), it would not be surprising if *L. sikapusi* also consisted of two or more cryptic species. This species is in need of revision.

Distribution

This species inhabits various moist habitats (such as swamps, fallow fields, and grasslands) within low-lying rainforest (typically below 1600 m) from coastal Guinea (Denys et al., 2009a) in the west, to the Sanaga River, Cameroon in the east.

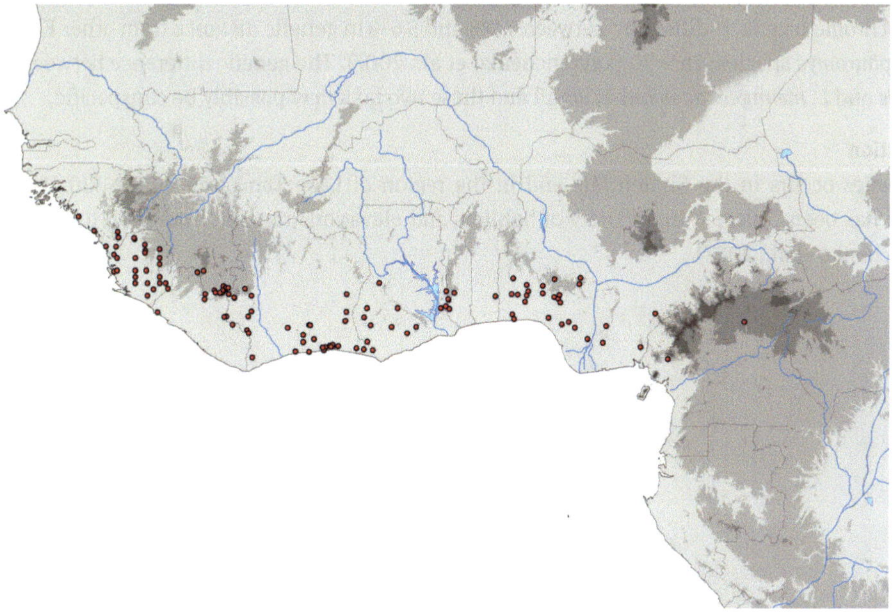

Figure 2.334 Distribution map

Measurements

External and cranial measurements (mm) and mass (g) for *Lophuromys sikapusi*, sexes combined

	Mean	Min	Max	N
Mass	63.7	34	87	18
HB	129.9	110	145	32
Tail	69.0	58	75	30
HF	23.0	20	25	31
Ear	16.2	14	18	26
GLS	31.1	27.6	33.5	33

Lophuromys simiensis Osgood, 1936

Diagnosis

This is one of nine species endemic to Ethiopia. Only one other *Lophuromys* species occurs within its range: *L. flavopunctatus* (smaller, typically GLS < 31 mm; 2n = 68). Its closest relative seems to be *L. menageshae* from which it can be distinguished by its smaller size (in *L. menageshae*, GLS > 31 mm).

Taxonomic Issues

This newly described species is in the *flavopunctatus* group (Lavrenchenko et al., 2007). Based on the cytochrome b gene, it differs by between 1.5% and 9.6% in genetic distance from other Ethiopian *Lophuromys* species; 2n = 70 (Lavrenchenko et al., 2007). The genetic difference between *L. simiensis* and *L. menageshae* is rather small and these two taxa may possibly be conspecific.

Distribution

This species occurs in the Simien Mts and in the region of Lake Tana west of the Rift Valley where it has been collected in a variety of habitats and elevations ranging from 1600 to 3800 m (Lavrenchenko et al., 2007).

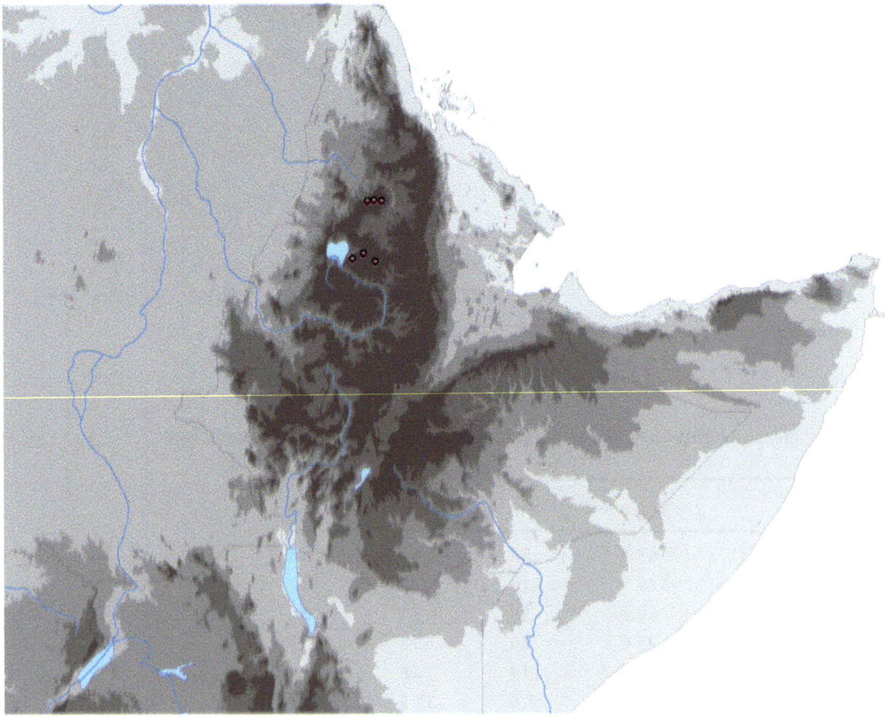

Figure 2.335 Distribution map

Measurements

External and cranial measurements (mm) and mass (g) for *Lophuromys simiensis*, sexes combined

	Mean	Min	Max	N
Mass*	57.1	37	74	60
HB*	133.4	114	145	60
Tail*	71.4	58	85	42
HF*	23.1	21.0	25.0	60
Ear*	17.5	15.5	20.0	60
GLS*	30.1	28.4	31.7	58

*Lavrenchenko et al. (2007)

Lophuromys stanleyi W. Verheyen, Hulselmans, Dierckx, Mulungu, Leirs, Corti & E. Verheyen, 2007

Diagnosis

This is one of four *Lophuromys* species that occurs in the Ruwenzori Mts, the other two species being: *L. woosnami* (larger, GLS > 30 mm; tail relatively longer ca. 100% of HB; dorsal pelage not speckled); *L. ansorgei* (larger, GLS > 30 mm; dorsal pelage not speckled); and *L. laticeps* (can only be distinguished on multivariate analysis of cranial features, but occurs at lower elevations) (Verheyen et al., 2007). *Lophuromys dudui* is smaller (HF < 21 mm) and occurs in low-lying forest habitats below 1000 m.

Taxonomic Issues

This is a newly described species in the *flavopunctatus* group (Verheyen et al., 2007).

Distribution

This species is endemic to the Ruwenzori Mts, eastern DRC and western Uganda (Verheyen et al., 2007).

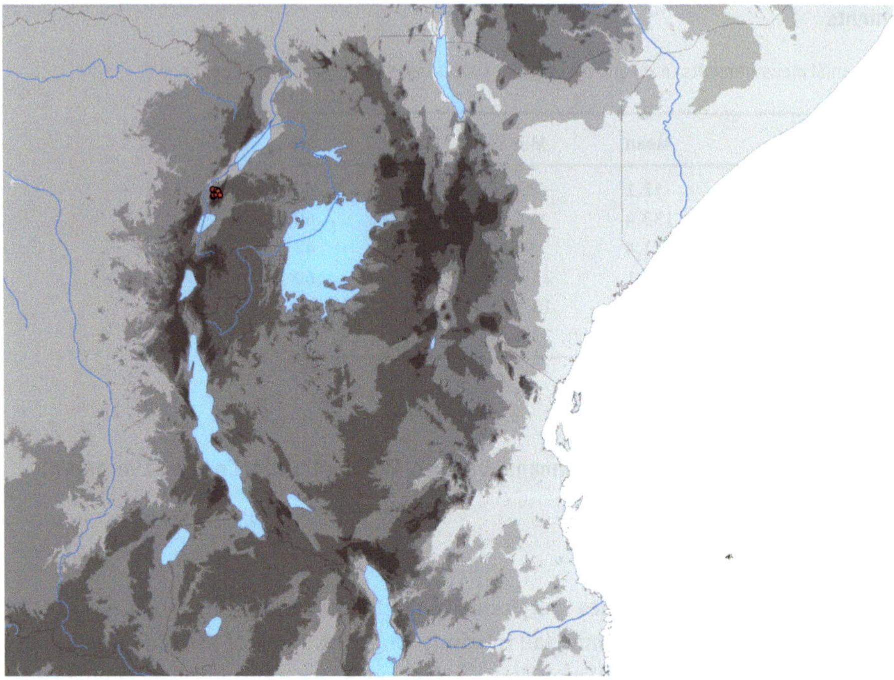

Figure 2.336 Distribution map

Measurements

External and cranial measurements (mm) and mass (g) for *Lophuromys stanleyi*, sexes combined

	Mean	Min	Max	N
Mass*	45.9	36	55	18
HB*	118.5	113	126	18
Tail*	68.3	40	80	18
HF*	22.9	22.0	24.0	19
Ear*	18.2	16.0	19.0	18
GLS*	28.9	26.9	30.2	31

*Verheyen et al. (2007)

Lophuromys verhageni W. Verheyen, Hulselmans, Dierckx, Mulungu, Leirs, Corti & E. Verheyen, 2007

Diagnosis

This is the only species of *Lophuromys* occurring on Mt Meru. It can be distinguished from *L. aquilus*, from neighbouring Mt Kilimanjaro, by multivariate analysis of cranial features (Verheyen et al., 2007).

Taxonomic Issues

This species is in the *flavopunctatus* group. 2n = 68 (Verheyen et al., 2007).

Distribution

This species is endemic to Mt Meru, Tanzania (Verheyen et al., 2007).

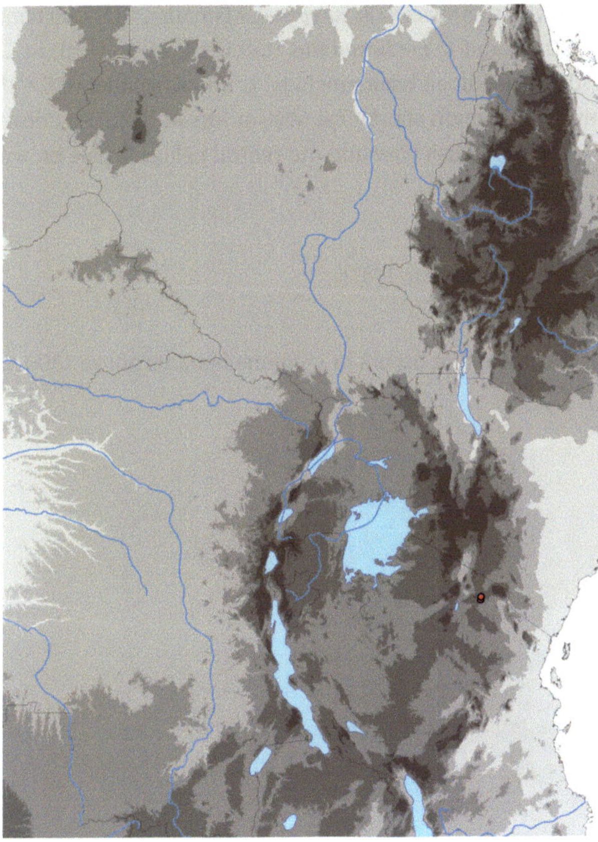

Figure 2.337 Distribution map

Measurements

External and cranial measurements (mm) and mass (g) for *Lophuromys verhageni*, sexes combined

	Mean	Min	Max	N
Mass*	53.2	40	64	37
HB*	126.2	112	140	37
Tail*	70.1	43	81	33
HF*	22.5	21.0	23.5	35
Ear*	18.7	14.3	22.0	36
GLS*	30.4	29.0	31.6	37

*Verheyen et al. (2007)

Lophuromys woosnami Thomas, 1906

Diagnosis

The *Kivumys* subgenus is distinguished from members of the *flavopunctatus* and *sikapusi* groups by its longer tail (ca. 85% or more of HB; ca. 60% of HB in other species). Two other species in the subgenus *Kivumys* could be confused with this species: *L. luteogaster* (smaller, GLS < 30 mm; occurring in low-lying forests below 1200 m) and *L. medicaudatus* (tail is ca. 85% of HB; in *L. woosnami* it is ca. 100% of HB). Other Albertine Rift endemics include: *L. rahmi* (smaller GLS < 27 mm); *L. cinereus* and *L. stanleyi* (relatively tail length shorter ca. 60% or less of HB). The dorsal pelage is uniform medium to dark brown with reddish base and the ventral pelage is pale brown with reddish hue (Dieterlen, 1976b).

Taxonomic Issues

This species belongs in the *Kivumys* subgenus (Dieterlen, 1987).

Distribution

This species is endemic to the Albertine Rift where it occurs in montane habitats above 1800 m (Dieterlen, 1987).

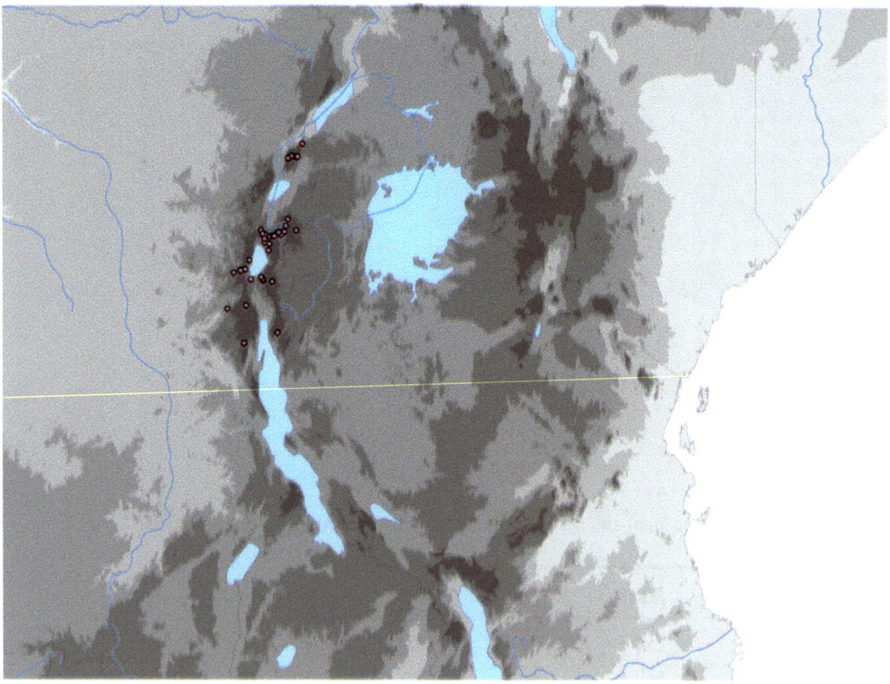

Figure 2.338 Distribution map

Measurements

External and cranial measurements (mm) and mass (g) for *Lophuromys woosnami*, sexes combined

	Mean	Min	Max	N
Mass*	47.0	38	64	20
HB*	117.5	106	135	40
Tail*	118.0	104	133	40
HF*	25.8	24.0	27.5	40
Ear*	22.4	20	25	40
GLS*	31.3	30.2	33.4	40

*Dieterlen (1976b)

Lophuromys zena Dollman, 1909

Diagnosis

The only other species of *Lophuromys* occurring within the range of this species is *L. margarettae* (similar in size but can be distinguished on molecular grounds and by multivariate analysis of cranial characters), which occurs at lower elevations (Verheyen et al., 2007).

Taxonomic Issues

This taxon is in the *flavopunctatus* group and has recently been elevated to full species status (Verheyen et al., 2007).

Distribution

This species is restricted to high altitudes on Mt Kenya and in the Aberdare Range, Kenya.

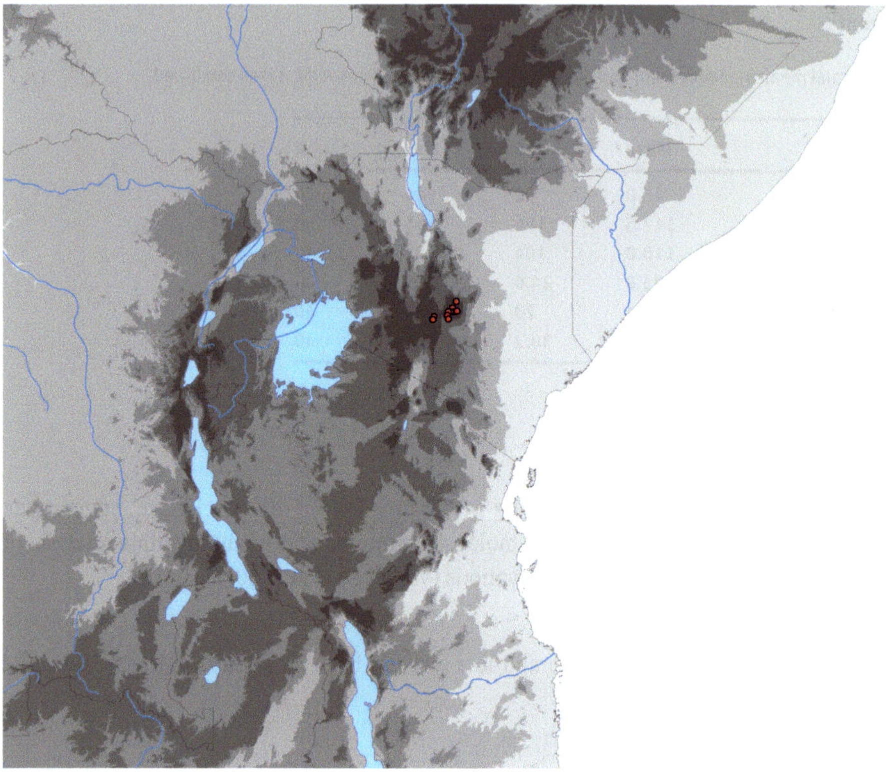

Figure 2.339 Distribution map

Measurements

External and cranial measurements (mm) and mass (g) for *Lophuromys zena*, sexes combined

	Mean	Min	Max	N
Mass*	51.7	34	70	52
HB*	123.4	108	139	52
Tail*	70.7	42	88	51
HF*	21.5	19.0	23.2	52
Ear*	18.2	16.3	20.5	51
GLS*	30.1	28.1	32.1	55

*Verheyen et al. (2007)

Genus Account

Uranomys Dollman, 1909 (Rudd's Brush-furred Mouse)

Diagnosis

The distinguishing characters of the genus are its small size (HB ca. 105 mm) and short tail relative to HB (54-58 %). The dorsal pelage has some brush-like hairs. The palate extends beyond M^3 and covers the mesopterygoid fossa as in *Acomys*. It has a typical Deomyinae dental pattern. Ventral pelage is white or off-white.

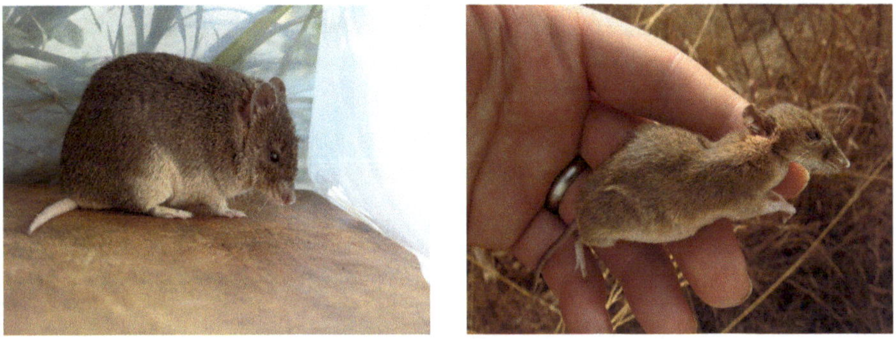

(a) (b)

Figure 2.340 a and b Photographs of *Uranomys ruddi* (Leonid Lavrenchenko, left and Jan Decher, right).

Cranial Characters

It has a small sized skull (24-26 mm) with a relatively short rostrum, pro-odont incisors, and a fronto-parietal crest running from the interorbital constriction to the braincase. In ventral view the palatal foramen is long and reaches the t4 of the upper first molar. The palate itself is extended to partly cover the pterygoid fossa. The bullae are relatively small (4-5 mm in greatest diameter).

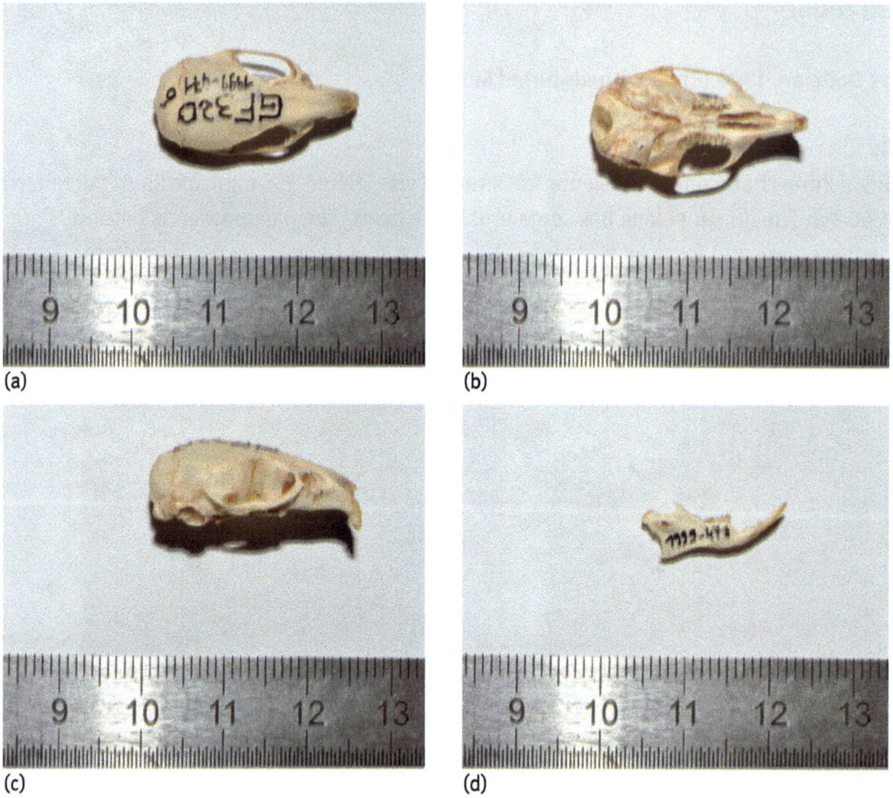

Figure 2.341 a-d Cranium of *Uranomys ruddi* (MNHN-ZM-MO 1999-47) (from left top to bottom right: dorsal view, ventral view, lateral view and lower jaw).

Dentition

The upper and lower incisors are ungrooved. The upper and lower molar rows have an *Acomys*-like cusp disposition (presence of a t1 and t4 on the lingual side of the M^1). The M^3 and M_3 are the smallest teeth of the toothrow. The M^3 has an antero-external cusp (t3) but no antero-internal one (t1). The M_1 has a single anterior cusp compared with two in *Acomys, Deomys* and *Lophuromys*; these two cusps are fused at an early stage of wear and connected to an antero-external cingulum. The posterior rows of cusps are transversally fused and aligned. The posterior cingulum is very small on the M_1 and is absent on the M_2. The M_2 and M_3 have an antero-external cingulum.

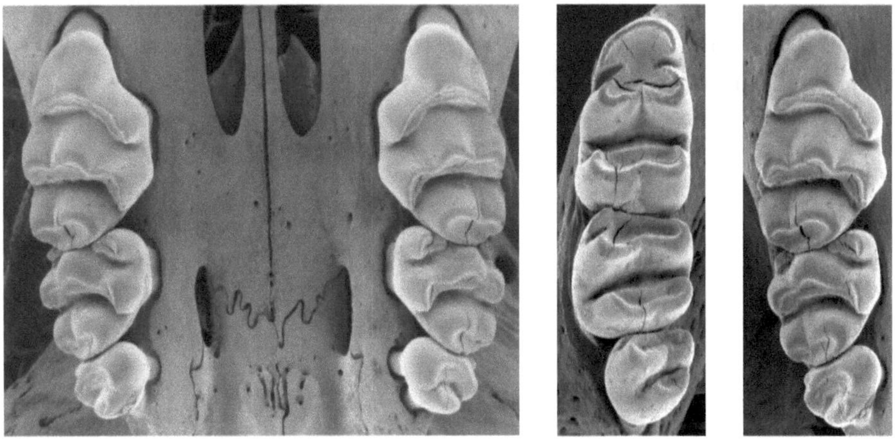

(a) (b) (c)

Figure 2.342 Cheek teeth of *Uranomys ruddi* (MNHN-ZM-MO 1971-601) (A: upper cheek teeth; B: lower left toothrow; C: upper right toothrow).

Habitat & Biogeography

This genus occurs in Guinean savannas, the ecotone between forest and savanna, Somali-Masai savannas, and Zambezian savannas. The distribution is disjuncted extending from Senegal in West Africa east to Nigeria, and patchily to Uganda, and south through Tanzania and Malawi to Zimbabwe and Mozambique.

Fossil Record

There are few fossil records of this genus. The oldest *Uranomys* is from Aramis (4.3 Ma) in Ethiopia (Louchart et al., 2009, Wesselman, personal communication). In Southern Africa *Uranomys* occurs later in Plio-Pleistocene caves of Angola (Winkler et al., 2010).

Biology & Ecology

These terrestrial and nocturnal animals live in deep burrows and may be partly fossorial (Bellier, 1968). Little is known of its biology but it is partly insectivorous but may also eat cultivated crops. It prefers savannas with some bushes such as riverine forests (Denys et al., 2009a) and at the ecotone between regenerating forest and grassland (Bellier, 1968). Females have 6 pairs of mammae and between 2 to 6 embryos. They are caught by owls and probably small carnivores but there are very few such records of predation.

Systematic Notes

This genus was previously placed in the Murinae (Musser & Carleton, 1993). It has since been moved to the Acomyinae based on molecular grounds (Chevret et al., 1993a, 2001; Dubois et al., 1999; Steppan et al., 2005), as a sister taxon of *Acomys* (Denys & Michaux, 1992). The Acomyinae has been synonymized with the Deomyinae by Carleton & Musser (2005). *Uranomys* displays significant variability in morphology and chromosome number across its geographical range (Verheyen, 1964; Granjon et al., 1992), suggesting the probable existence of cryptic species. Multiple species have been described within this genus in the past (Verheyen, 1964). 2n = 50-58 (Viegas-Péquignot et al., 1983).

Species List

Only one species is currently recognised.
Uranomys ruddi Dollman, 1909

Species Account

Uranomys ruddi Dollman, 1909

Diagnosis

See genus account.

Taxonomic Issues

See genus account.

Distribution

This species is widely but patchily distributed across the savannas of West and East Africa, extending south into Zimbabwe and Mozambique.

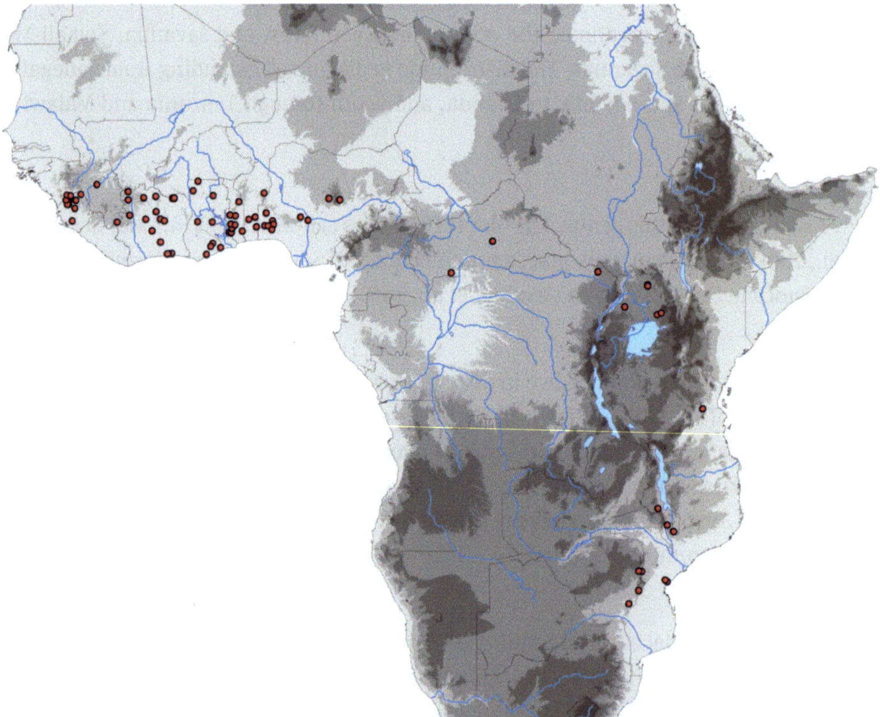

Figure 2.343 Distribution map

Measurements

External and cranial measurements (mm) and mass (g) for *Uranomys ruddi*, sexes combined

	Mean	Min	Max	N
Mass	33.1	26	43	7
HB	106.4	95	119	12
Tail	68.6	60	73	7
HF	17.1	16	19	9
Ear	13.4	10	15	5
GLS	26.4	25	27.7	9

Subfamily: Gerbillinae

Key to Genera

This key has been modified from Petter (1975).

1	Mandible lacking a coronoid process	*Ammodillus*
	Mandible with coronoid process	2
2	Tail short (ca. 55% of HB) and thickened	*Pachyuromys*
	Tail not short (ca. 70% of HB or longer) and not thickened	3
3	Size very small (HB < 60 mm); lower jaw with 2 molars	*Desmodilliscus*
	Size larger (HB > 60 mm); lower jaw with 3 molars	4
4	Molars prismatic (pattern consisting of deep, equal-sized re-entrant folds separating transverse rows)	5
	Molars not as above	7
5	Size smaller (HB < 90 mm)	*Gerbillus (Dipodillus)*
	Size larger (HB > 110)	6
6	Upper incisors ungrooved	*Psammomys*
	Upper incisors grooved	*Meriones*
7	Bullae not inflated	8
	Bullae inflated	9
8	Posterior palatal foramina shorter (<3 mm)	*Gerbilliscus* (part)
	Posterior palatal foramina longer (> 3 mm)	*Taterillus*
9	Tail longer (>100% of HB)	10
	Tail shorter (<80% of HB)	11
10	Cusps not connected by longitudinal crests; restricted to arid parts of Southern Africa (Namibia and South Africa)	*Gerbilliscus (Gerbillurus)*
	Cusps connected by longitudinal crests; not occurring in Southern Africa	*Gerbillus (Gerbillus)*
11	M^3 with 2 cusps; restricted to Southern Africa	*Desmodillus*
	M^3 with 3-4 cusps; restricted to the Horn of Africa	*Microdillus*

Genus Account

Ammodillus Thomas, 1904 (Ammodille, Walo or Somali Gerbil)

Diagnosis

These are small-sized gerbils (HB 85-110 mm; Roche & Petter, 1968; Happold, 2013) with a soft, pale chestnut dorsal pelage, white ventral pelage extending onto the flanks and cheeks, small ears with white pre and postaural spots and a long (ca. 135-150 %) black tail with the terminal third of the tail marked with a tuft ("pencil") of longer hairs. The soles of the hindfeet are naked. These gerbils are highly unique in craniodental characters which easily distinguish them from all other gerbils, lacking a coronoid process on the mandible, possessing two primitive cusplets on the M_1 and M_2, and having a highly modified zygomatic and temporal (masticatory) region and auditory bullae. There are three pairs of nipples (1 + 2 = 6).

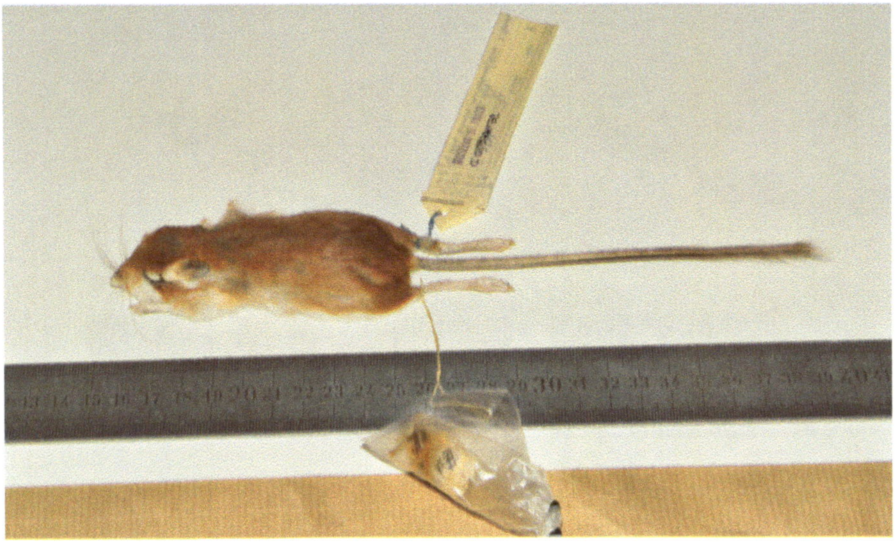

Figure 2.344 Photograph of *Ammodillus imbellis* museum specimen (MNHN-ZM-MO 1977-32) (Ara Monadjem).

Cranial Characters

The skull is medium-sized (ca. 30 mm) and triangular in shape with narrow, elongated nasals and large tympanic and inflated mastoid chambers of the bullae (30% of GSL) which are just visible when viewing the skull dorsally. The braincase is compressed laterally (as in Taterillini). The zygomatic arches are slender over most of their length but expanded abruptly anteriorly into an "orbital shield". The zygomatic plate is wide and convex in shape and cut-away above, leaving a narrow infraorbital foramen as typical for gerbils. The frontals are marked by raised supraorbital ridges which follow the outline of the orbits down to the zygomatic arches, apparently providing the attachment for the origin of the masseter muscle (Pavlinov et al., 1990; Pavlinov, 2008). The coronoid process of the mandible is completely lacking, a unique feature among muroids. The auditory bulla is slightly inflated and large, with the mastoid and tympanic portions being equal-sized. The anterior palatal foramina just reach the alveoli margin of the M^1. The posterior palatal foramina are well developed and extend anteriorly to the point of contact of M^1 and M^2. The hamular process ("mastoid strut") is present, with a very narrow ventral margin defining a large subsquamosal (=stapedial) foramen.

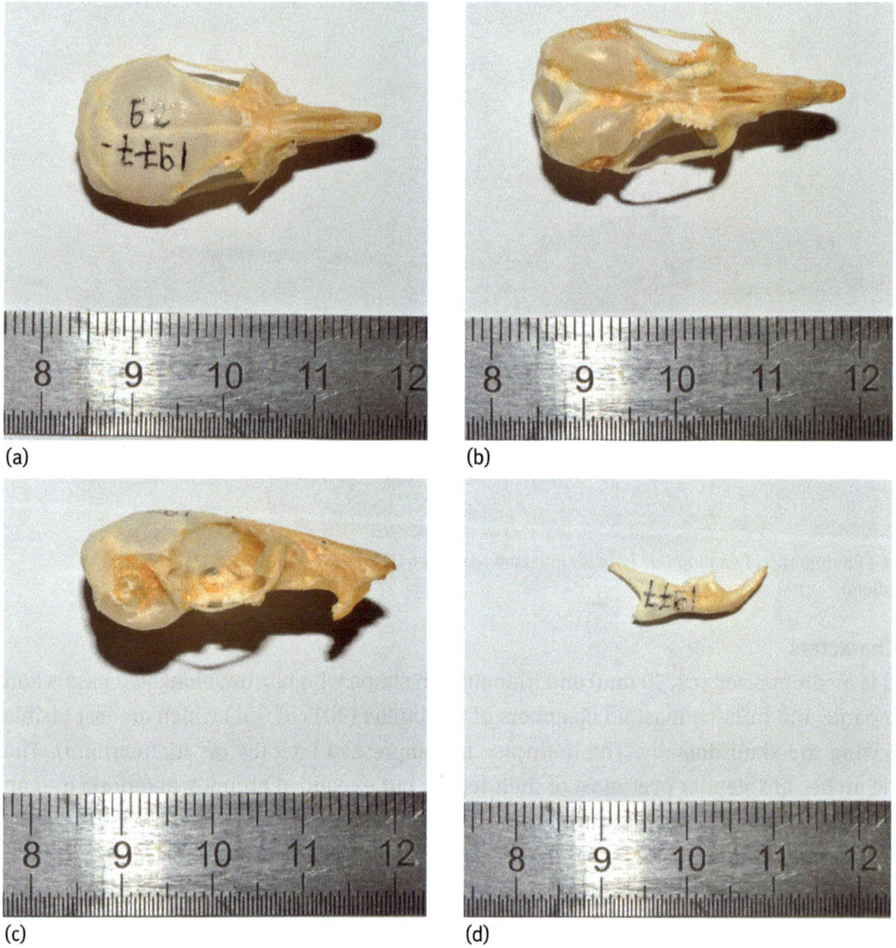

Figure 2.345 a-d Cranium of *Ammodillus imbellis* (MNHN-ZM-MO 1977-29) (from top left to bottom right: dorsal view, ventral view, lateral view and lower jaw).

Dentition

Upper incisors are distinctly opisthodont and grooved. The molar rows converge strongly posteriorly, more so than in other gerbils. The upper molars are relatively high and generally laminate with a faint longitudinal crest at the back of the first lobe of M^1. The lower molars show an asymmetric pattern (termed "teniodont" by Pavlinov, 2008). The anteroconid of M^1 is composed of two small tubercles which are connected to the second (oblique) row of cusps by a wide longitudinal crest. The M^1 (three laminae) is considerably larger than M^2 (two laminae) which is turn is considerably larger than M^3 (one lamina plus a tiny cusp). There are accessory cusplets on the lower molars, termed protoconulid (on M_1) and hypoconulid (on the lower M_1 and M_2) respectively by Pavlinov (2008). The anterior lamina (termed "anteroconid" by Pavlinov, 2008) on M_1 is large but separate from the anterior cusplet (protoconulid). The hypoconulid is absent in all other gerbillines except for *Desmodillus*.

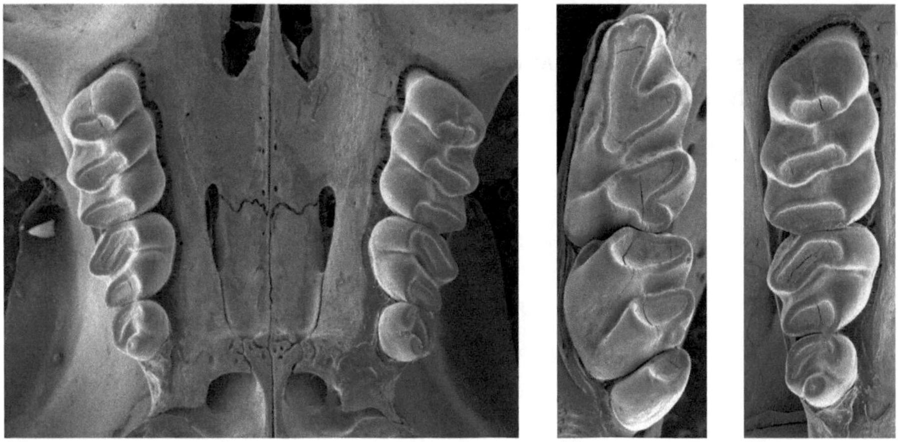

(a) (b) (c)

Figure 2.346 Upper cheek teeth of *Ammodillus imbellis* (MNHN-ZM-MO 1977-28) (A: upper cheek teeth; B: lower left toothrow; C: upper right toothrow).

Habitat & Biogeography

The single species occurs in open dry grasslands and shrublands in Somalia and eastern Ethiopia (Yalden et al., 1976; Granjon & Schlitter, 2008).

Fossil Record

No fossils are known.

Biology & Ecology

Very little is known about the ecology of this species.

Systematic Notes

Pavlinov (1987) considered this genus to be distinct enough to warrant its own tribe, Ammodillini, a viewpoint endorsed by Musser & Carleton (2005). Cladistic analysis of craniodental characters indicates that this genus is a basal lineage of Gerbillinae and its relationship with other tribes and genera of gerbils is uncertain (Pavlinov et al., 1990; Pavlinov, 2008).

Annotated Species List

One species is currently recognised:
Ammodillus imbellis (de Winton, 1898)

Species Account

Ammodillus imbellis (de Winton, 1898)

Diagnosis

See genus account.

Taxonomic Issues

The karyotype is 2n = 18 (Capanna & Merani, 1981).

Distribution

This species is restricted to the Horn of Africa where it is known from just a few localities in Somalia and eastern Ethiopia.

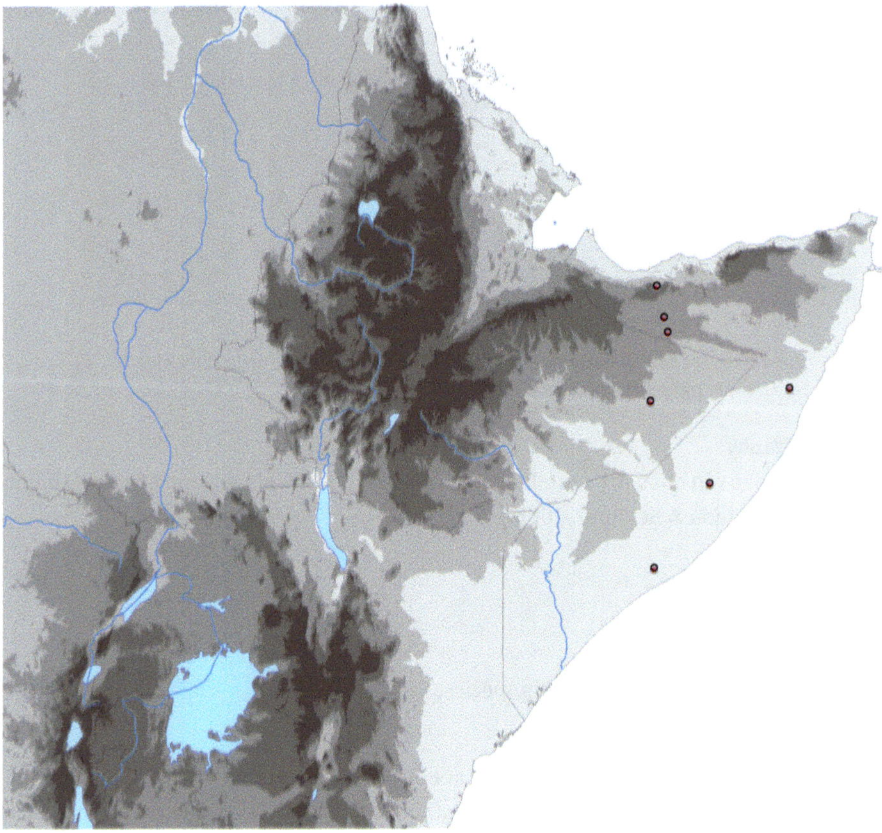

Figure 2.347 Distribution map

Measurements

External and cranial measurements (mm) and mass (g) for *Ammodillus imbellis*, sexes combined

	Mean	Min	Max	N
Mass	31.0	-	-	1
HB	97.7	84	105	7
Tail	140.0	129	150	6
HF	27.7	26	29	7
Ear	15.5	14	16	6
GLS	30.9	29.5	32.7	7

Genus Account

Desmodilliscus Wettstein, 1916 (Dwarf Gerbils or Pouched Gerbils)

Diagnosis

They are unmistakable gerbils in their very small body size (HB 40-75 mm) and tail which is shorter (60-70 %) than the head and body, white and finely-furred, lacking the terminal tuft of long hairs. The dorsal pelage is soft and brownish to sandy grey in colour with white markings above the eyes and anterior (preaural) and posterior (postaural) to the ears. The forearms, forefeet and hindfeet are white, the hindfoot soles are naked but the forefeet show a fine covering of short white hairs on the plantar pads. The type specimen apparently possessed clear cheek pouches. On craniodental characters, the genus can immediately be distinguished by the absence of the M_3 molar and the greatly enlarged posterior palatal foramina. They have four pairs of nipples (2 + 2 = 8).

Figure 2.348 Photograph of *Desmodilliscus braueri* museum specimen (MNHN-ZM-MO 1986-563) (Ara Monadjem).

Cranial Characters

The skull is small (ca. 22 mm) and triangular in shape, with greatly inflated bullae including the mastoid compartment which can be seen projecting beyond the foramen magnum when viewed dorsally, and a short and widened rostrum compared with other gerbils. The zygomatic arches are slender and not expanded anteriorly into an orbital shield. The zygomatic plate is rather narrow and the anterior edge is slightly convex, ending before the premaxilla-maxilla suture (it is usually expanded more anteriorly in other gerbils beyond this suture). The anterior palatal foramina end well before M^1 and the posterior palatal foramina are very wide (with round margins) and extend from M^2 till the anterior edge of M^1, there being a separation of only 1 mm between the anterior and posterior palatal foramina.

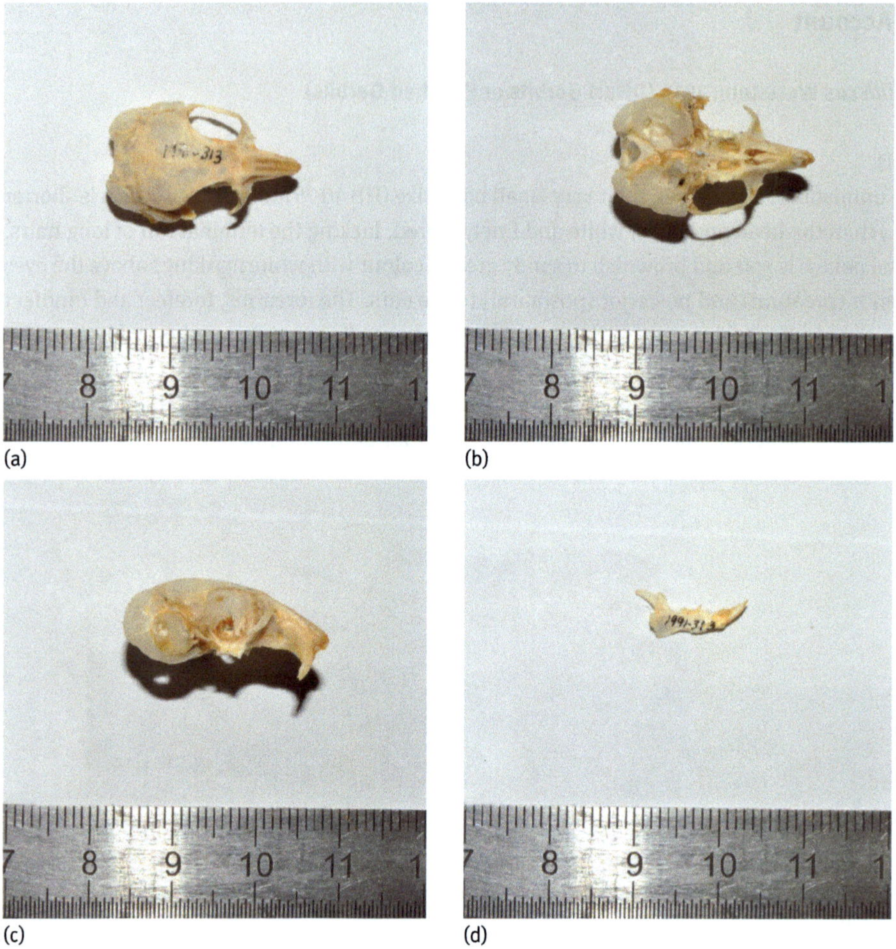

Figure 2.349 a-d Cranium of *Desmodilliscus braueri* (MNHN-ZM-MO 1991-313) (from top left to bottom right: dorsal view, ventral view, lateral view and lower jaw).

Dentition

The upper incisors are opisthodont, grooved and narrow. The groves appear on the lateral face of the incisors. The molars represent the advanced lophodont pattern described by Pavlinov (2008), with asymmetry of cusps in M^1 and especially M_1. The absence of the M_3 makes the dentition unique amongst muroids.

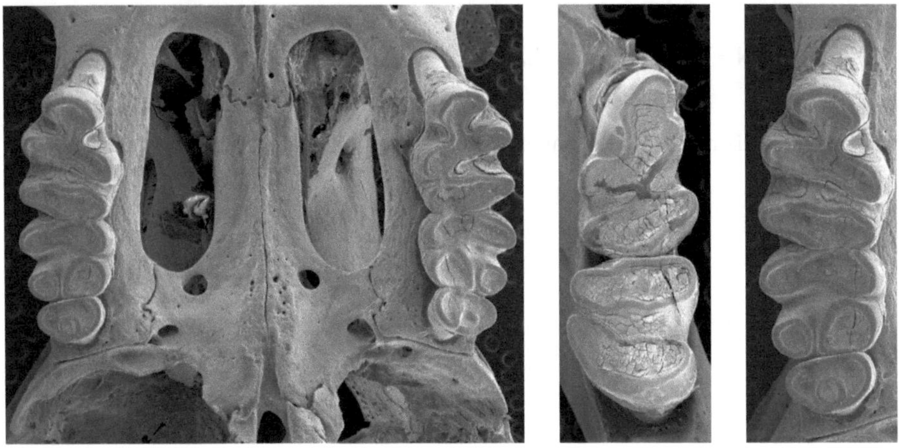

(a) (b) (c)

Figure 2.350 Upper cheek teeth of *Desmodilliscus braueri* (MNHN-ZM-MO 2001-1582) (A: upper cheek teeth; B: lower right toothrow; C: upper right toothrow).

Habitat & Biogeography

They are distributed across Sahelian savannas (200-500 mm of rainfall) from Mauritania to Sudan.

Fossil Record

No fossils are known.

Biology & Ecology

They are nocturnal, terrestrial and live in burrows. Reproduction is seasonal in captivity and litter size is two or three (Poulet, 1984).

Systematic Notes

This genus has been assigned to the tribe Gerbillini and Subtribe Desmodilliscina (Musser & Carleton, 2005). Morphological (Pavlinov et al., 1990; Tong, 1989), chromosomal (Qumsiyeh et al., 1987), allozyme (Bennazou et al., 1984) and immunological (Chevret, 1994) data suggest a basal position for *Desmodilliscus*. Based on mitochondrial DNA sequences, Chevret & Dobigny (2005) suggested *Pachyuromys* to be the closest relative of *Desmodilliscus*.

Annotated Species List

One species is recognised:
Desmodilliscus braueri Wettstein, 1916

Species Account

Desmodilliscus braueri Wettstein, 1916

Diagnosis

See genus account.

Taxonomic Issues

Karyotype is 2n = 78, FN = 104 (Granjon et al., 1992).

Distribution

This species is widely distributed in the Sahel savanna zone from Mauritania to Sudan.

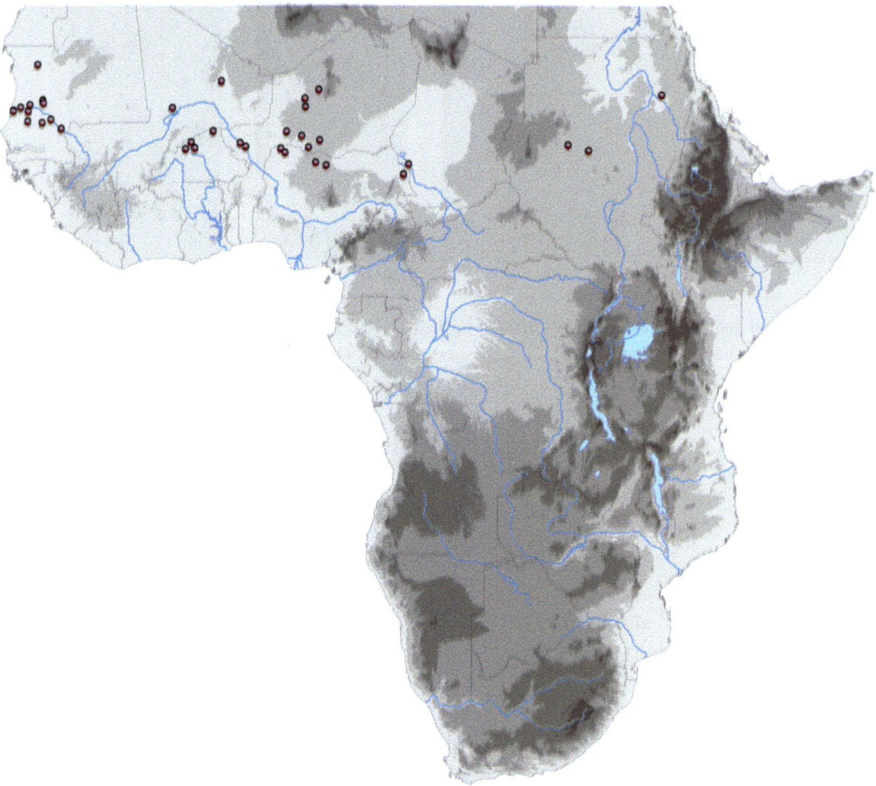

Figure 2.351 Distribution map

Measurements

External and cranial measurements (mm) and mass (g) for *Desmodilliscus braueri*, sexes combined

	Mean	Min	Max	N
Mass*	5.4	4	8	4
HB	59.7	50	64	11
Tail	40.9	35	46	10
HF	14.6	13	16	11
Ear	7.6	6	9	11
GLS	21.7	20.2	22.8	10

*Happold (2013) reports a mean mass of 9.6 g (n = 19; range: 6-14g)

Genus Account

Desmodillus Thomas & Schwann, 1904 (Cape Short-tailed Gerbils)

Diagnosis

These small gerbils (Mass ca. 30-80 g; HB ca. 85-130 mm) have a glossy chocolate brown to reddish brown dorsal pelage and white ventral pelage extending to the cheeks and throat, with white markings above and below the eye and before and after the ear. The ear is small and oval. The tail is sparsely haired with a tuft and is shorter than the head and body (ca. 70-80 % of HB). The hind-feet have hairy soles. They have four pairs of nipples (2 + 2 = 8). The bullae are greatly enlarged proportionate to the skull size.

Figure 2.352 Photograph of *Desmodillus auricularis* (Galen Rathbun).

Cranial Characters

Medium-sized (GSL 35-38 mm), triangular-shaped skull with elongated and narrow rostrum and large bullae (41% of GSL) with inflated ("fully pneumatised") mastoid chamber extending beyond the occipital condyles and easily visible when the skull is viewed from above (a derived feature not present in other Taterillini). The supraoccipital ridges are slightly developed extending backwards to the posterior margin of the parietal bones. The zygomatic arch is slender. The zygomatic plate is relatively narrow compared to other members of the tribe Taterillini, with a convex anterior edge. The anterior palatal foramina end short of the M^1, while the posterior palatal foramina reach from the second lamina of $M^1$to the M^3. The hamular process ("mastoid strut") is present, with a very narrow ventral margin defining a large subsquamosal (=stapedial) foramen.

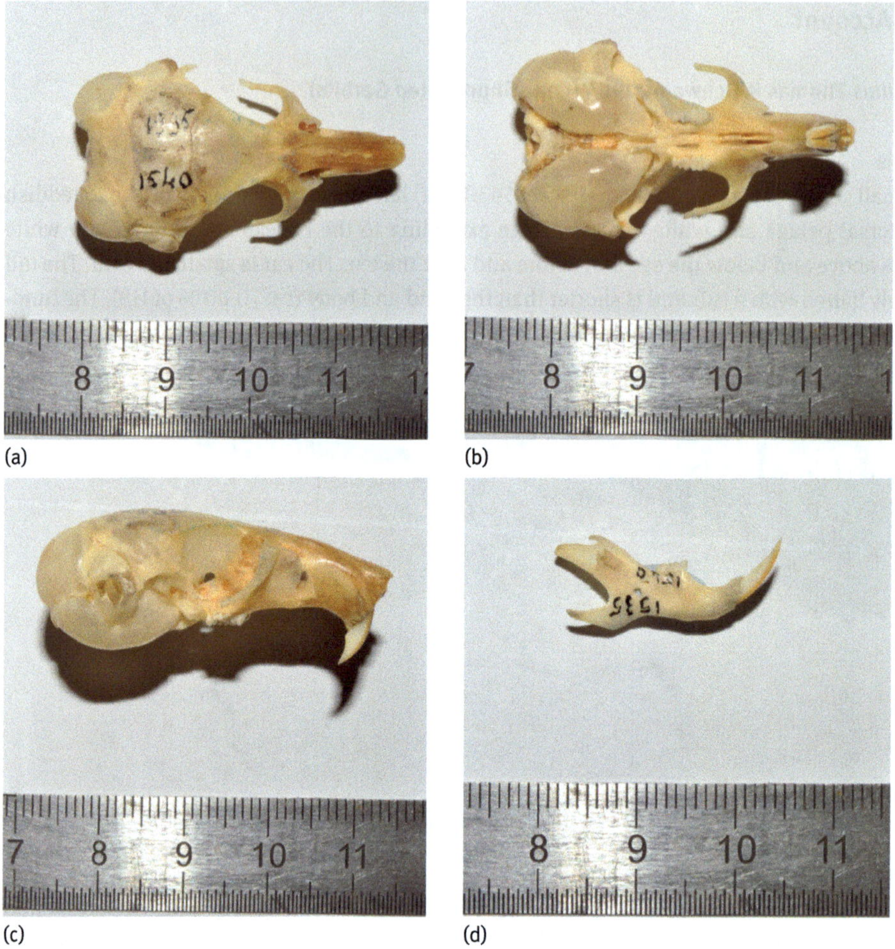

Figure 2.353 a-d Cranium of *Desmodillus auricularis* (MNHN-ZM-MO 1935-1540) (from top left to bottom right: dorsal view, ventral view, lateral view and lower jaw).

Dentition

The upper incisors are opisthodont and ungrooved. The molar rows are more or less parallel and molars show a "semi-lophodont" (laminate) pattern of dentition, without longitudinal ridges between laminae. The cusp arrangement is symmetrical unlike the case in genera with "advanced lophodont" dentition where asymmetry occurs (*Ammodillus* and *Desmodilliscus*). In young animals, there are two cusps in the first lobe of the M_1 (the anterior "protoconulid" of M_1 is separate from the anteroconid) but in older animals they become fused to form the typical horseshoe shape of the molar wear surface of the anteroconid, which is typical for Taterillini. The M_3 is very small and reduced to a tiny cusp.

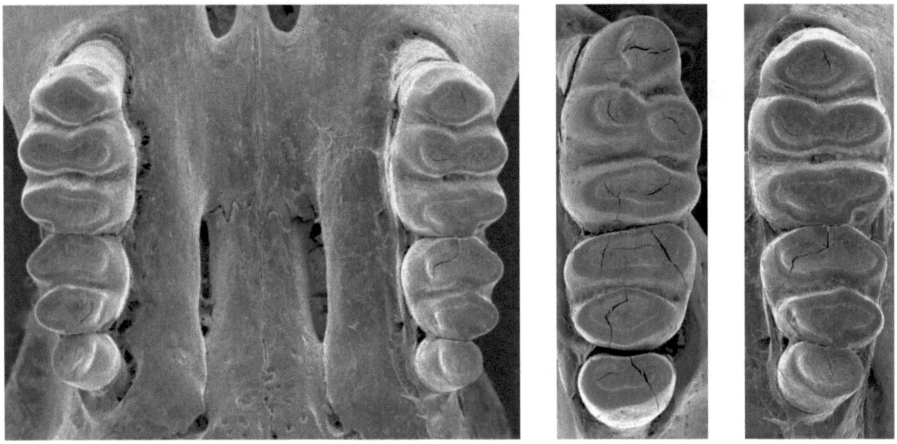
(a) (b) (c)

Figure 2.354 Cheek teeth of *Desmodillus auricularis* (MNHN-ZM-MO 1983-814) (A: upper cheek teeth; B: lower right toothrow; C: upper left toothrow).

Habitat & Biogeography

They occur in areas with calcareous ground or fine soils and sparse grass or shrub cover, extending throughout the semi-arid Karoo, Namaqualand and Kalahari areas from just north of the Cape Fold Mountains in the south through western South Africa, much of Botswana and Namibia and extending narrowly into south-western Angola.

Fossil Record

Fossils are known from the Early Pliocene to Middle Pleistocene of Southern Africa. The earliest fossils are known from Langebaanweg in South Africa (Pocock, 1976).

Biology & Ecology

They are nocturnal and terrestrial, do not use saltatorial locomotion, and excavate complex burrows. They are omnivores eating mostly seeds in winter and insects and green leaves in summer (Happold, 2013). Breeding may be seasonal when conditions are good, but usually restricted to the hot wet season. Mean litter size varies geographically ca. 2-4 (Happold, 2013).

Systematic Notes

The genus has been placed in tribe Taterillini and (together with *Gerbillurus*) in Subtribe Gerbillurina (Pavlinov et al., 1990; Musser & Carleton, 2005; Pavlinov, 2008). Phylogenetic analysis of morphological, chromosomal and molecular data have not confidently resolved the phylogenetic position of *Desmodillus* but recent molecular studies show it to be a basal lineage in the Taterillini, a conclusion supported by its retention of the primitive protoconulid cusp found also in the Miocene genus *Protatera* (Qumsiyeh et al., 1987; Pavlinov et al., 1990; Chevret et al., 2005, Colangelo et al., 2007; Pavlinov, 2008; Granjon et al., 2012).

Annotated Species List

One species is recognised:
Desmodillus auricularis (Smith, 1834)

Species Account

Desmodillus auricularis (Smith, 1834)

Diagnosis
See genus account.

Taxonomic Issues
Karyotype is 2n = 52 (Qumsiyeh et al., 1987).

Distribution
This species is widely distributed in the arid savanna and desert zone of south-western Africa.

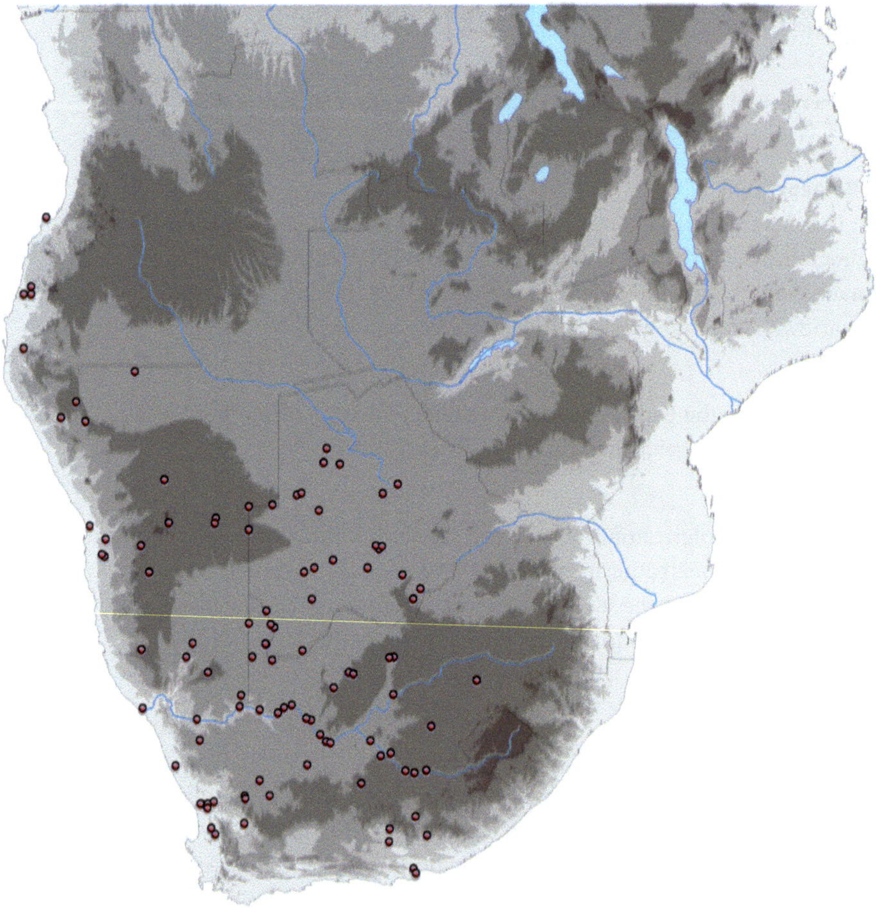

Figure 2.355 Distribution map

Measurements

External and cranial measurements (mm) and mass (g) for *Desmodillus auricularis*, sexes combined

	Mean	Min	Max	N
Mass	46.1	18	71	23
HB	109.2	96	126	29
Tail	88.7	63	106	30
HF	24.9	23	27	29
Ear	11.7	10	15	32
GLS	34.9	30.6	38.3	86

Genus Account

Gerbilliscus Thomas, 1897 (Gerbils & Hairy-footed Gerbils)

Diagnosis

Multiple recent molecular and chromosomal studies show *Gerbillurus* to be a clade nested within *Gerbilliscus* and therefore synonymous with it (see under Systematic notes, below). However, given the clear morphological differences between these two groups they are described separately in this account. Individuals of *Gerbilliscus* (excluding *Gerbillurus*) are large (Mass ca. 90-130 g; HB ca. 130-160 mm) and heavily built compared to most gerbils and also have a darker colouration although the colour varies considerably. The tail varies from shorter to longer than the head and body (80-140% of HB) and has a distinctive white tip in *G. boehmi*. The hindfoot soles are naked. The posterior mastoid chamber of the bulla is not inflated (and not visible from above). In contrast to the above description, species formerly assigned to *Gerbillurus* are smaller in size (Mass ca. 30-40 g; HB ca. 95-100 mm), have tails that are always longer than the head and body (120-140% of HB), hindfoot soles that are partly furred (having a naked central line) and the bulla with an expanded mastoid chamber (visibly projecting beyond the posterior margin of the skull when viewed from above). The dorsal colour of *Gerbillurus* is usually a pale blonde to reddish brown colour while the ventral colour is white extending to the chin and throat. White markings are evident before and after the ear and above the eye in *Gerbillurus* which are lacking in *Gerbilliscus*. There may be three (1 + 2 = 6) or four (2 + 2 = 8) pairs of nipples in *Gerbilliscus*.

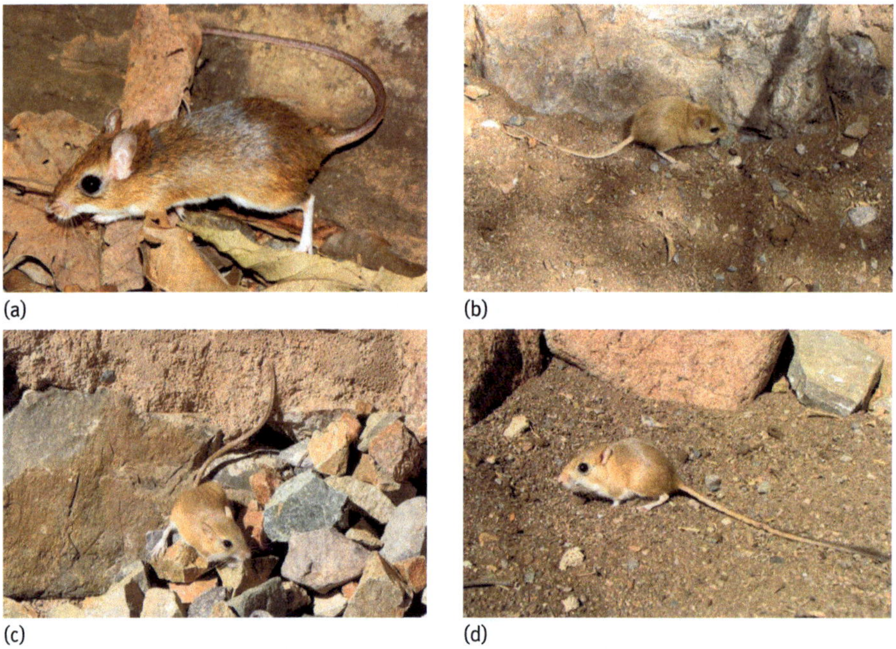

Figure 2.356 a-d Photograph of *Gerbilliscus* species, from top left to bottom right: *G.* cf *guineae* (James Harvey), *G. leucogaster*, *G. (Gerbillurus) paeba*, *G. (Gerbillurus) vallinus* (Galen Rathbun, three images).

Cranial Characters

The skull sizes of the two groups do not overlap, with the *Gerbillurus* group being smaller (GSL ca. 27-33 mm) compared to *Gerbilliscus* excluding *Gerbillurus* (GSL ca. 35-45 mm). Supraorbital ridges are well developed and extend backwards to the posterior margin of the parietals in *Gerbilliscus*, but are very faint and restricted to the interorbital region in *Gerbillurus*. The bullae are somewhat primitive in *Gerbilliscus* with the mastoid completely uninflated, but inflated in *Gerbillurus*. The anterior palatal foramina barely reach the alveolus of M^1 in *Gerbilliscus* but extend to the back on the first laminae of M^1 in *Gerbillurus*, and the posterior palatal foramina are more developed in the latter where they extend from the back of M^2 to the middle of M^1. The zygomatic plate is very broad in *Gerbilliscus* and has a convex anterior edge and evenly curved dorsal angle but in *Gerbillurus* it is narrower, forward sloping and cut away above. The hamular process and associated subsquamosal foramen are absent in *Gerbilliscus* but present in *Gerbillurus*.

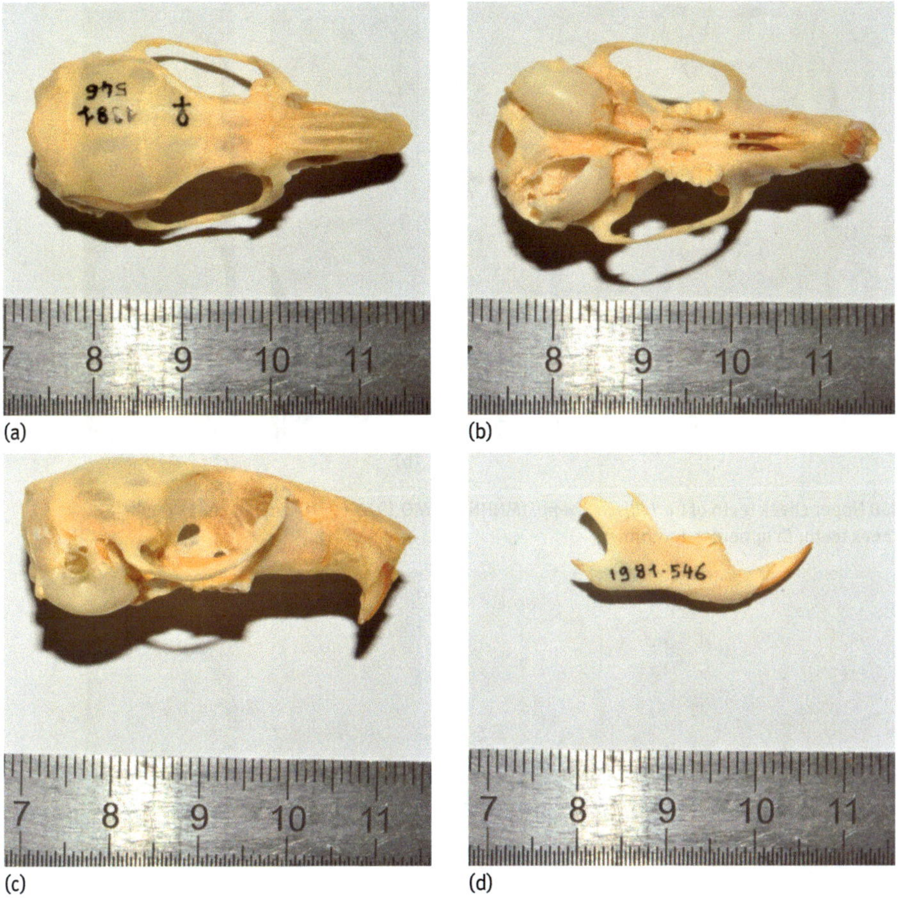

Figure 2.357 a-d Cranium of *Gerbilliscus kempi* (MNHN-ZM-MO 1981-546) (from top left to bottom right: dorsal view, ventral view, lateral view and lower jaw).

Dentition

The upper incisors are narrow, grooved (double-grooved in *G. boehmi*) and slightly (*Gerbilliscus*) to very (*Gerbillurus*) opisthodont. The molars have an "advanced lophodont" pattern in *Gerbilliscus*, lacking longitudinal connections between laminae. The horseshoe-shaped appearance of the anteroconid of M_1 is present from an early age but is highly variable. A "semi-lophodont" pattern has been described for the molar crown of *Gerbillurus*, also having the horseshoe-shaped anteroconid on M_1 which becomes more circular ("rhomboid") with age (as in other *Gerbilliscus*). Cusps are round and high and do not fuse together in transverse laminae in *Gerbillurus* whereas they do fuse more easily and earlier in other *Gerbilliscus* species.

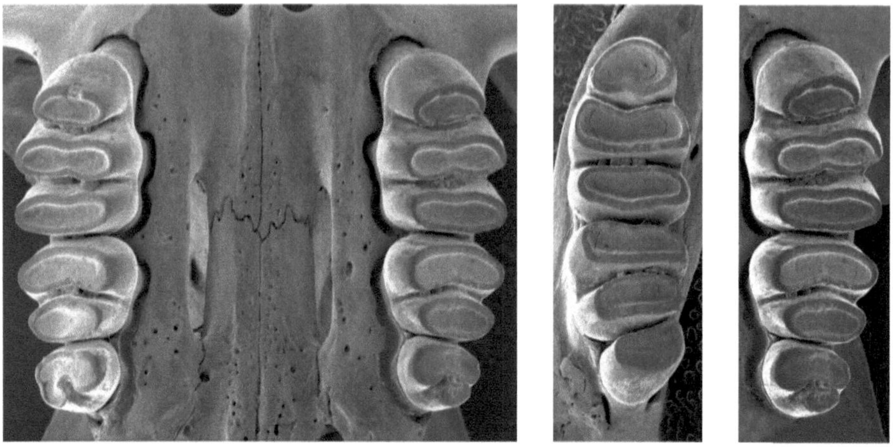

(a) (b) (c)

Figure 2.358 Upper cheek teeth of *Gerbilliscus kempi* (MNHN-ZM-MO 1973-695) (A: upper cheek teeth; B: lower cheek teeth; C: upper left toothrow).

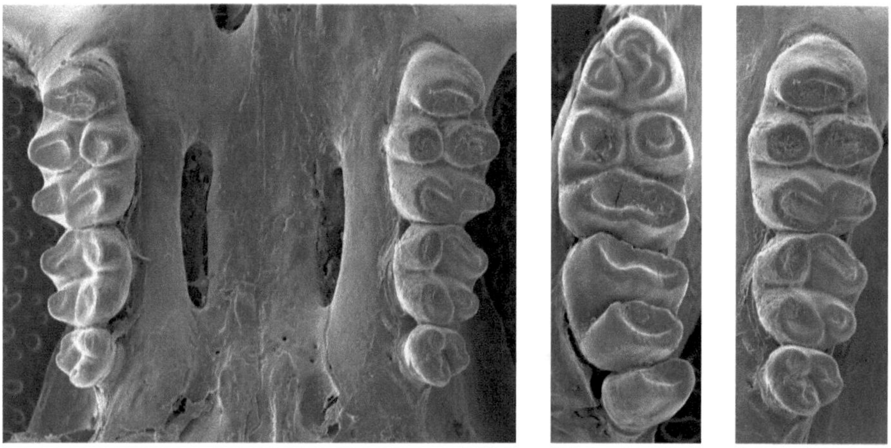

(a) (b) (c)

Figure 2.359 Cheek teeth of *Gerbillurus paeba* (MNHN-ZM-MO 1969-66) (A: upper cheek teeth; B: lower left toothrow; C: upper left toothrow).

Habitat & Biogeography

Gerbilliscus are found in a broad range of northern and southern savannas and semi-arid to arid habitats throughout Sub-Saharan Africa from Cape Town in the south to Sudan and Senegal in the north, while *Gerbillurus* are restricted to sparsely vegetated arid and semi-arid regions of western Southern Africa.

Fossil Record

Fossils are recorded from the Late Miocene to Middle Pleistocene of East Africa and Early Pliocene to Late Pleistocene in Southern and Central Africa (Winkler et al., 2010). The earliest known fossils are from 4 MYA deposits at Kanapoi in East Africa (Denys, 1999).

Biology & Ecology

They are nocturnal and terrestrial (*Gerbilliscus*) to quadripedally saltatorial (*Gerbillurus*) and occupy simple (*Gerbillurus*) to complex (*Gerbilliscus*) burrows. The diet is omnivorous and opportunistic but mostly granivorous in *Gerbilliscus* and mostly herbivorous in *Gerbillurus* although it varies seasonally in both groups and insects can form an important component during certain parts of the year (Happold, 2013). In *Gerbilliscus* sensu lato breeding may be strictly seasonal (e.g. *G. afra*, *G. gambianus*, *G. nigricauda*, *G. robusta*, *G. validus*), or continuous with a peak at certain times of the year (e.g. cool dry season in *G. brantsi*, and summer in *G. leucogaster*). Similarly in *Gerbillurus* sensu lato breeding may be seasonal (e.g. *G. paeba* or *G. tytonis* in the Namib Desert) or continuous (*G. setzeri*). Although up to eight or nine embryos have been observed, litter size is usually 1-6 (Happold, 2013).

Systematic Notes

Gerbillliscus was described by Shortridge (1934) as a subgenus of *Gerbillus* but was later promoted to its own genus (Musser & Carleton, 1993). *Gerbilliscus* was also previously included in *Tatera*, but a considerable body of evidence suggests that the genus *Tatera* is restricted to a single Asian species and that all African species formerly grouped in *Tatera* form a distinct clade that corresponds with *Gerbilliscus* (Pavlinov et al., 1990; Pavlinov, 2001; Musser & Carleton, 2005; Colangelo et al., 2007). Up until recently *Gerbillurus* and *Gerbilliscus* were treated as separate genera (Musser & Carleton, 2005; Pavlinov, 2008). However, chromosomal (Qumsiyeh et al., 1987; Volobouev et al., 2007b) and mitochondrial and nuclear molecular data (Chevret & Dobigny, 2005; Colangelo et al., 2007; Granjon et al., 2012) suggest that they should be combined into one genus. Musser & Carleton (2005) do not recognise *G. gambianus* (Thomas, 1910) which they treat as a synonym of *G. kempi*; but it is recognised by Happold (2013) based on chromosomal (Matthey & Petter, 1970) and morphological grounds (shorter ear than in *G. kempi*) grounds. Many taxonomic problems remain to be solved in this genus but Granjon et al. (2012) provided a useful molecular analysis of several species which revealed clear western, eastern and southern African geographical groups.

Annotated Species List

Eighteen species are recognised from Sub-Saharan Africa including four which were previously regarded as *Gerbillurus* (marked with asterisk):

Gerbilliscus afra (Gray, 1830)
Gerbilliscus boehmi (Noack, 1887)
Gerbilliscus brantsii (Smith, 1836)
Gerbilliscus gambianus (Thomas, 1910)
Gerbilliscus giffardi (Wroughton, 1906)
Gerbilliscus guineae (Thomas, 1910)
Gerbilliscus inclusus Thomas & Wroughton, 1908
Gerbilliscus kempi (Wroughton, 1906)
Gerbilliscus leucogaster (Peters, 1852)
Gerbilliscus nigricaudus (Peters, 1878)
**Gerbilliscus paeba* (A. Smith, 1836)
Gerbilliscus phillipsi (de Winton, 1898)
Gerbilliscus robustus (Cretzschmar, 1826)
**Gerbilliscus setzeri* (Schlitter, 1973)

**Gerbilliscus tytonis* (Bauer & Niethammer, 1960)
Gerbilliscus validus (Bocage, 1890)
**Gerbilliscus vallinus* (Thomas, 1918)
Gerbilliscus vicinus (Peters, 1878)

Key to Species

Our understanding of the species limits within this genus has recently been significantly improved through a number of studies that have focussed on chromosomal and molecular analyses. As a result, many of the species cannot currently be diagnosed on morphological grounds. For this reason, the following key must be used with caution.

1	Size smaller (GLS < 34 mm); hindfeet with hairy soles; tail relatively longer (ca. 120% or more of HB); restricted to the arid zones of western Southern Africa. "*Gerbillurus*" group	2
	Size larger (GLS > 34 mm); hindfeet without hairy soles; tail relatively shorter (ca. 115% or less of HB, but with the exception of *G. boehmi*); occurring widely across Sub-Saharan Africa. "*Gerbilliscus*" group	5
2	Pencil of hair in terminal part of tail present (pencil may be grey, red or black); 2n = 60	3
	Pencil of hair in terminal part of tail absent; 2n = 36	4
3	Dorsal pelage pale sandy-brown (with no reddish or orange colour); pencil on tail grey; occurs on the coast of northern Namibia	*G. setzeri*
	Dorsal pelage darker reddish-brown; pencil on tail reddish or black; occurs in southern Namibia and north-western South Africa	*G. vallinus*
4	Dorsal pelage lighter, typically pale reddish-brown; hindfeet shorter, typically < 30 mm without distinctive fringe of hairs on sides of digits	*G. paeba*
	Dorsal pelage darker, typically reddish-brown; hindfeet longer, typically > 30 mm with distinctive fringe of hairs on sides of digits	*G. tytonis*
5	Tail white in terminal half	6
	Tail not white in terminal half (but may occasionally have a white tip)	7
6	Size larger (GSL > 42 mm); tail relatively very long (ca. 130% of HB)	*G. boehmi*
	Size smaller (GLS < 42 mm); tail relatively shorter (ca. 105% of HB	*G. brantsii*
7	Tail entirely black or partially black on terminal half	8
	Tail typically brown above and paler below	9
8	Tail entirely black along its whole length, dorsally and ventrally; size larger (GSL typically > 40 mm; HF > 37 mm); width of M^1 2.4-2.5 mm	*G. nigricaudus*
	Tail not usually black along its whole length, and paler ventrally; size smaller (GSL typically < 41 mm; HF < 36 mm); width of M^1 2.0-2.4 mm	*G. robustus*
9	Tail with black dorsal line running along its length	*G. leucogaster*
	Tail without black dorsal line running along its length	10
10	Restricted to the extreme south-western parts of South Africa	*G. afra*
	Not occurring in the south-western parts of South Africa	11
11	Tail without a pencil and relatively short (ca. 95% of HB)	*G. validus*
	Tail with a pencil and typically longer (ca. 100% or more of HB)	12
12	Occurring in East Africa	13
	Occurring in West Africa (with one species extending east to Sudan)	15
13	Restricted to east and south of Lake Malawi in southern Tanzania, Mozambique and marginally into Zimbabwe	*G. inclusus*
	Occurring in northern Tanzania, Kenya and Ethiopia	14

14	Size smaller (GLS < 39 mm, HF < 35 mm)	*G. phillipsi*
	Size larger (GLS > 39 mm, HF > 35 mm)	*G. vicinus*
15	Tail relatively longer (ca. 110-120% of HB)	16
	Tail relatively shorter (ca. 90-100% of HB)	17
16	Ear slightly shorter (ca. 18 mm); 2n = 52	*G. gambianus*
	Ear slightly longer (ca. 20 mm); 2n = 36	*G. kempi*
17	Size larger (GLS ca. 40 mm); 2n = 46	*G. giffardi*
	Size smaller (GLS ca. 35 mm); 2n = 50	*G. guinea*

Species Accounts

Gerbilliscus afra (Gray, 1830)

Diagnosis
This is the only species of *Gerbilliscus* within its range. Usually four pairs of nipples (2 + 2 = 8) are present.

Taxonomic Issues
Chromosome number 2n = 44, FN = 66 (Qumsiyeh, 1986).

Distribution
This species is endemic to the Western and Northern Cape provinces of South Africa where it occupies sandy soils and has adapted well to agricultural fields.

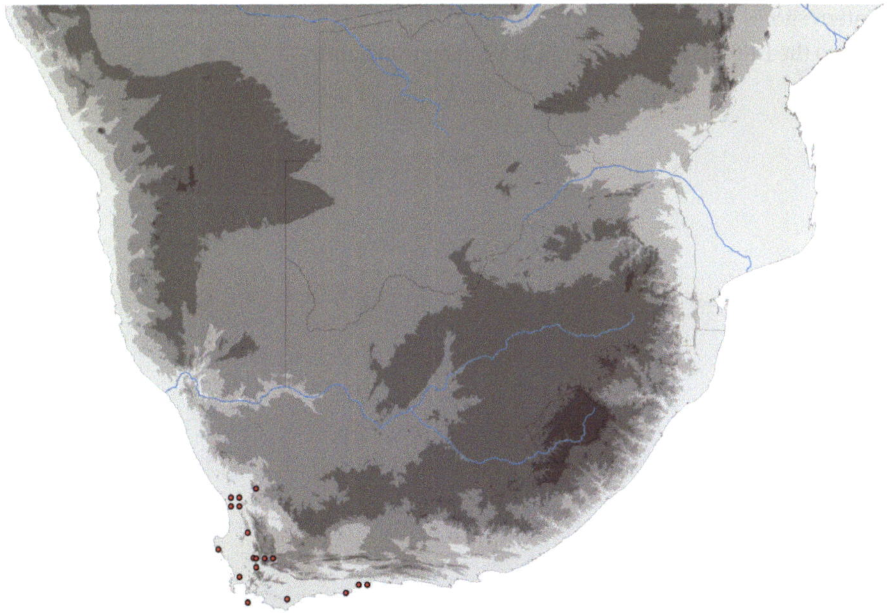

Figure 2.360 Distribution map

Measurements

External and cranial measurements (mm) and mass (g) for *Gerbilliscus afra*, sexes combined

	Mean	Min	Max	N
Mass	89.0	74	104	2
HB	136.7	122	149	12
Tail	160.1	150	168	12
HF	39.8	38	43	12
Ear	24.8	23	28	12
GLS	39.3	37.8	40.2	11

Gerbilliscus boehmi (Noack, 1887)

Diagnosis

This species is readily distinguishable by its long (ca. 130% of HB), white-tipped tail (no other species of *Gerbilliscus* within its range has a white-tipped tail). Three or four pairs of nipples (1 = 2 = 6 or 2 + 2 = 8) are present.

Taxonomic Issues

See genus account.

Distribution

This species occurs in a relatively narrow range from Zambia and Malawi in the south northwards (on the western shores of Lake Malawi) to Rwanda, Uganda and Tanzania in the north, extending marginally into the DRC. It typically occurs in Miombo woodland.

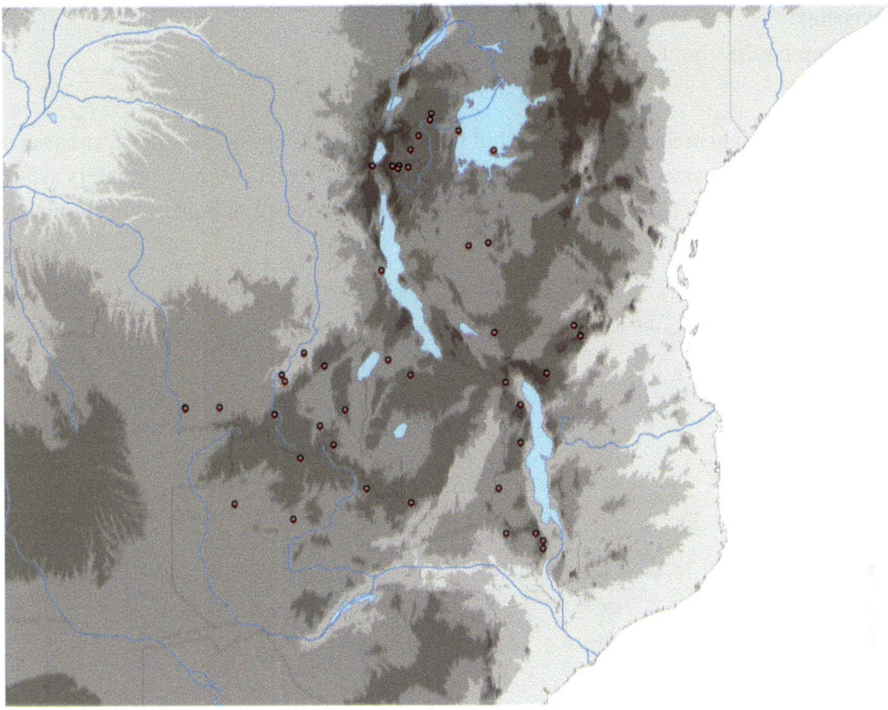

Figure 2.361 Distribution map

Measurements

External and cranial measurements (mm) and mass (g) for *Gerbilliscus boehmi*, sexes combined

	Mean	Min	Max	N
Mass	-	-	-	-
HB	157.3	143	175	7
Tail	207.0	190	229	7
HF	42.7	42	44	6
Ear	24.2	22	26	6
GLS	40.5	39.5	42.1	4

Gerbilliscus brantsii (Smith, 1836)

Diagnosis

This species can be distinguished from other *Gerbilliscus* species within its range (*G. afra* and *G. leucogaster*) by its tail which is white in the terminal half of its length and is marginally longer than its body (ca. 105% of HB); no other *Gerbilliscus* within its range has white in the tail. Some individuals lack white in the tail but can be distinguished from the sympatric *G. leucogaster* by a lack of a black line running along the dorsal length of the tail. Three or four pairs of nipples (1 + 2 = 6 or 2 + 2 = 8) are present.

Taxonomic Issues

Chromosome number 2n = 44, FN = 66 (Qumsiyeh, 1986).

Distribution

This species occurs widely in Southern Africa (but is absent from the mesic eastern savannas and the arid western regions), extending marginally into southern Angola, where it inhabits sandy soils where it is typically associated with open grasslands and agricultural fields where it may be a considerable pest.

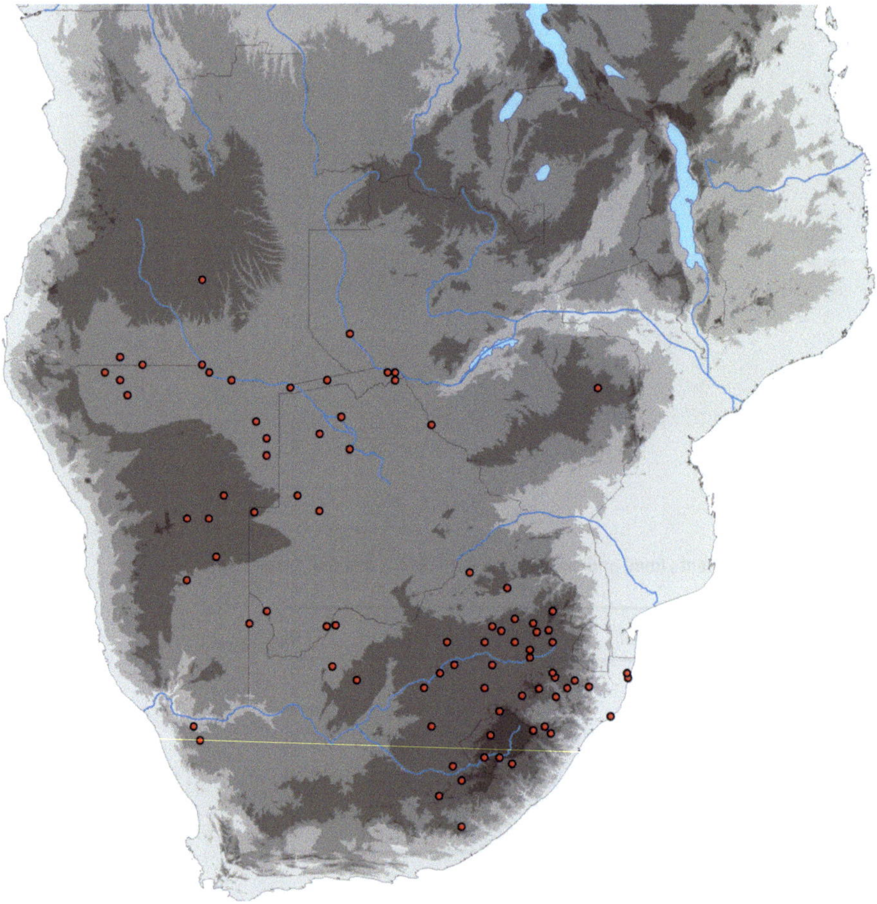

Figure 2.362 Distribution map

Measurements

External and cranial measurements (mm) and mass (g) for *Gerbilliscus brantsii*, sexes combined

	Mean	Min	Max	N
Mass	87.3	48	170	11
HB	138.8	114	172	21
Tail	147.5	103	171	20
HF	37.1	34	41	18
Ear	21.4	18	26	21
GLS	38.3	35.4	41.5	14

Gerbilliscus gambianus (Thomas, 1910)

Diagnosis

This is one of four *Gerbilliscus* species occurring in West Africa that can only be reliably identified on chromosomal and molecular grounds (Colangelo et al., 2007; Volobouev et al., 2007b), the other three species include: *G. guineae* (2n = 50; tail relatively longer, ca. 115-120% of HB); *G. giffardi* (2n = 46; tail relatively longer, ca. 120% of HB); *G. kempi* (2n = 48; ears relatively longer with limited overlap, ca. 21 mm). In this species the tail is relatively short (ca. 90-95% of HB). Four pairs of nipples (2 + 2 = 8) are present.

Taxonomic Issues

Chromosome number 2n = 52, FN = 64 (Matthey & Petter, 1970; Volobouev et al., 2007b).

Distribution

This species occurs in a narrow belt within Sudano-Sahelian savanna in southern Senegal and Mali, with an isolated population around Lake Chad and south-eastern Niger (Dobigny et al., 2002a; Colangelo et al., 2007; Volobouev et al., 2007b; Granjon & Duplantier, 2009).

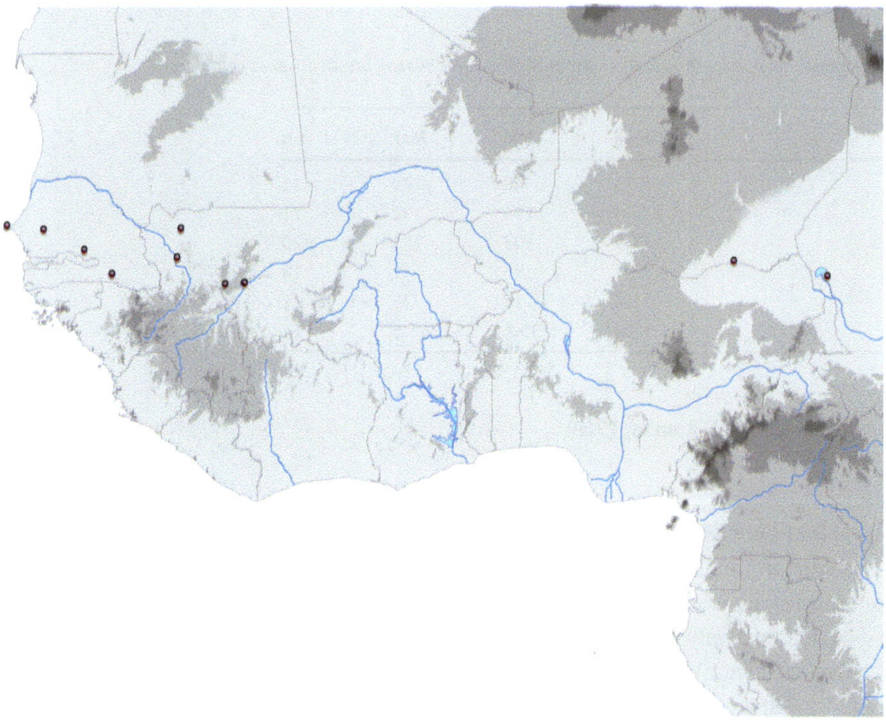

Figure 2.363 Distribution map

Measurements

External and cranial measurements (mm) and mass (g) for *Gerbilliscus gambianus*, sexes combined

	Mean	Min	Max	N
Mass*	94.5	61	193	283
HB*	157.1	125	196	252
Tail*	146.3	115	175	227
HF*	30.8	27	34	241
Ear*	18.9	16	23	250
GLS**	37.2	34.0	40.7	24

*Granjon & Duplantier (2009); **Happold (2013)

Gerbilliscus giffardi (Wroughton, 1906)

Diagnosis

This is one of four *Gerbilliscus* species occurring in West Africa that can only be reliably identified on chromosomal and molecular grounds (Granjon et al., 2012), the other three species include: *G. guineae* (2n = 50; size smaller, GLS ca. 35 mm, but probably with overlap); *G. gambianus* (2n = 52; tail relatively shorter, ca. 90-95% of HB); *G. kempi* (2n = 48; tail relatively shorter, ca. 95-100% of HB). In this species the tail is relatively long (ca. 120% of HB).

Taxonomic Issues

Chromosome number 2n = 46, FN = 64 (Volobouev et al., 2007b; Granjon et al., 2012).

Distribution

This West African endemic is known from a few widely scattered localities in Guinea woodland, extending north into Sahelo-Sudanian savanna. It is probably more widespread than the map suggests, but has only recently been diagnosed on molecular grounds (Granjon et al., 2012).

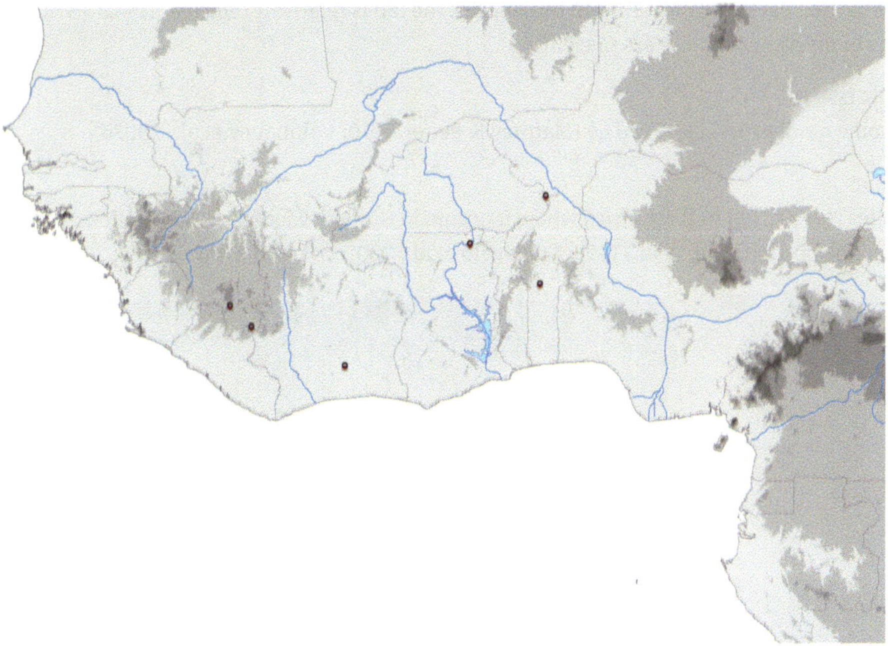

Figure 2.364 Distribution map

Measurements

External and cranial measurements (mm) and mass (g) for *Gerbilliscus giffardi*, sexes combined

	Mean	Min	Max	N
Mass*	111.3	-	-	6
HB*	144.7	-	-	6
Tail*	172.3	-	-	4
HF*	32.8	-	-	6
Ear	-	-	-	-
GLS*	40.9	-	-	6

*Granjon et al. (2012)

Gerbilliscus guineae (Thomas, 1910)

Diagnosis

This is one of four *Gerbilliscus* species occurring in West Africa that can only be reliably identified on chromosomal and molecular grounds (Colangelo et al., 2007; Volobouev et al., 2007b), the other three species include: *G. gambianus* (2n = 52; tail relatively shorter, ca. 90-95% of HB); *G. giffardi* (2n = 46; size larger, GLS ca. 40 mm, but probably with overlap); *G. kempi* (2n = 48; tail relatively shorter, ca. 95-100% of HB). In this species the tail is relatively long (ca. 115-120% of HB). Four pairs of nipples (2 + 2 = 8) are present.

Taxonomic Issues

Chromosome number 2n = 50, FN = 64 (Matthey & Petter, 1970; Volobouev et al., 2007b).

Distribution

This species is restricted to West Africa where it occurs in Guinea woodland and Sahelo-Sudanian savanna to the north (Grubb et al., 1998; Colangelo et al., 2007; Granjon & Duplantier, 2009).

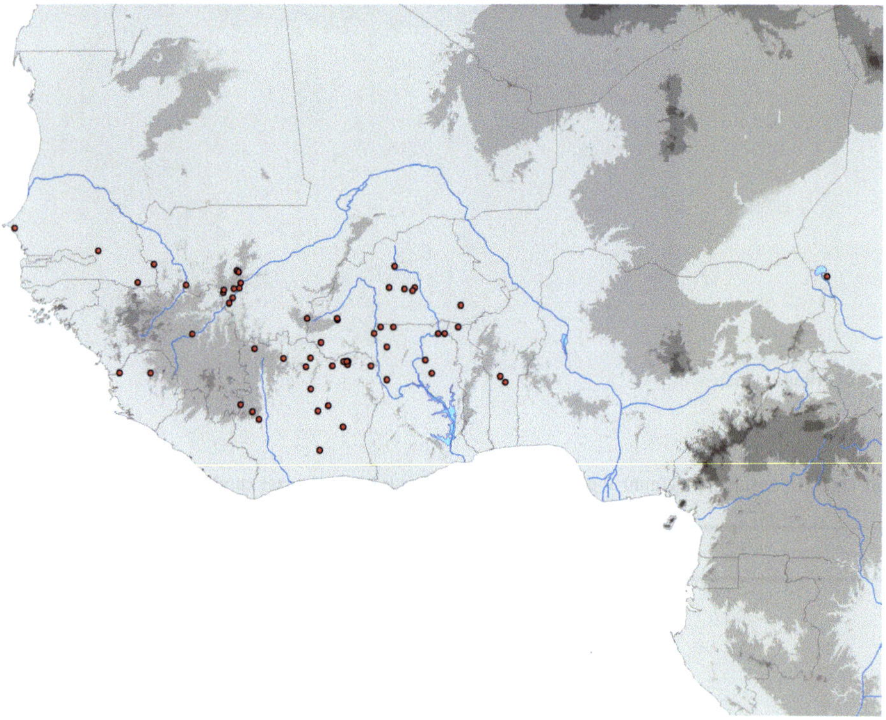

Figure 2.365 Distribution map

Measurements

External and cranial measurements (mm) and mass (g) for *Gerbilliscus guineae*, sexes combined

	Mean	Min	Max	N
Mass*	72.8	50	136	159
HB*	140.9	102	178	125
Tail*	173.5	151	198	96
HF*	33.0	28	38	108
Ear*	20.7	18	24	119
GLS	35.0	33.0	37.5	5

*Granjon & Duplantier (2009)

Gerbilliscus inclusus Thomas & Wroughton, 1908

Diagnosis
This is a large, dark *Gerbilliscus* species that only seems to co-occur with *G. leucogaster* (size smaller, GLS typically < 39 mm; HF typically < 39 mm; dorsal black line present on tail). Its tail is relative long (ca. 105-110% of HB) and lacks a black dorsal line on it (although it is darker above than below). *Gerbilliscus boehmi* occurs on the other side of Lake Malawi (the two species are not known to co-occur) and has the terminal length of its tail white (occasional specimens of *G. inclusus* have white-tipped tails, but not as extensive as in *G. boehmi*).

Taxonomic Issues
This species has not yet been sequenced and its phylogenetic affinities are currently unknown (see Granjon et al., 2012).

Distribution
This species occurs widely but patchily in the coastal savannas of South-central Africa from eastern Zimbabwe in the south to southern Tanzania in the north, with an isolated record from Muheza in north-eastern Tanzania (Hubbard, 1970). Its distribution is probably not as patchy as suggested by the map.

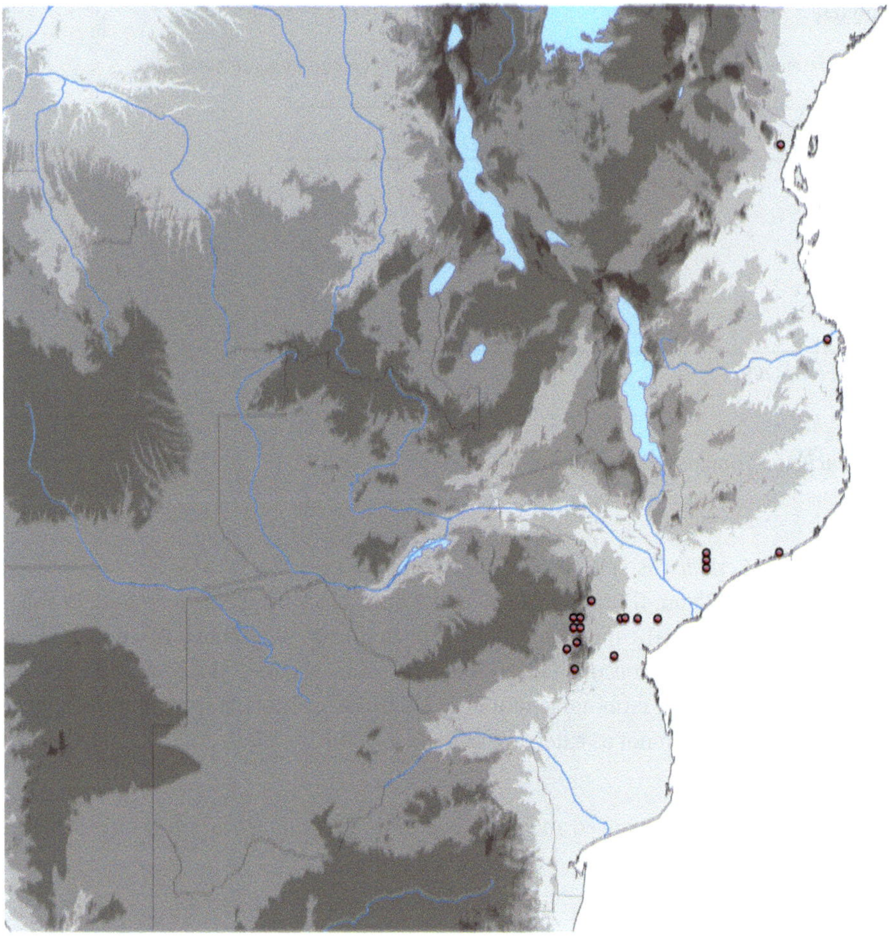

Figure 2.366 Distribution map

Measurements

External and cranial measurements (mm) and mass (g) for *Gerbilliscus inclusus*, sexes combined

	Mean	Min	Max	N
Mass*	121.6	99	154	13
HB*	156.0	-	-	13
Tail*	164.1	135	191	13
HF*	41.2	39	44	13
Ear*	25.0	21	29	13
GLS	42.5	39.7	44.6	3

*Smithers & Wilson (1979); HB estimated from Total length and Tail length

Gerbilliscus kempi (Wroughton, 1906)

Diagnosis

This is one of four *Gerbilliscus* species occurring in West Africa that can only be reliably identified on chromosomal and molecular grounds (Colangelo et al., 2007; Volobouev et al., 2007b), the other three species include: *G. gambianus* (2n = 52; tail ear relatively shorter with limited overlap, ca. 18 mm); *G. giffardi* (2n = 46; tail relatively longer, ca. 120% of HB); *G. guineae* (2n = 50; tail relatively longer, ca. 115-120% of HB). In this species the tail is relatively long (ca. 95-100% of HB).

Taxonomic Issues

Chromosome number 2n = 36, FN = 64 (Volobouev et al., 2007b).

Distribution

This species is widespread in the northern savannas from southern Senegal in the west to Sudan in the east where it predominantly occupies Guinea woodland, extending into Sahelo-Sudanian savannas to the north (Grubb et al., 1998; Volobouev et al., 2007b; Granjon & Duplantier, 2009).

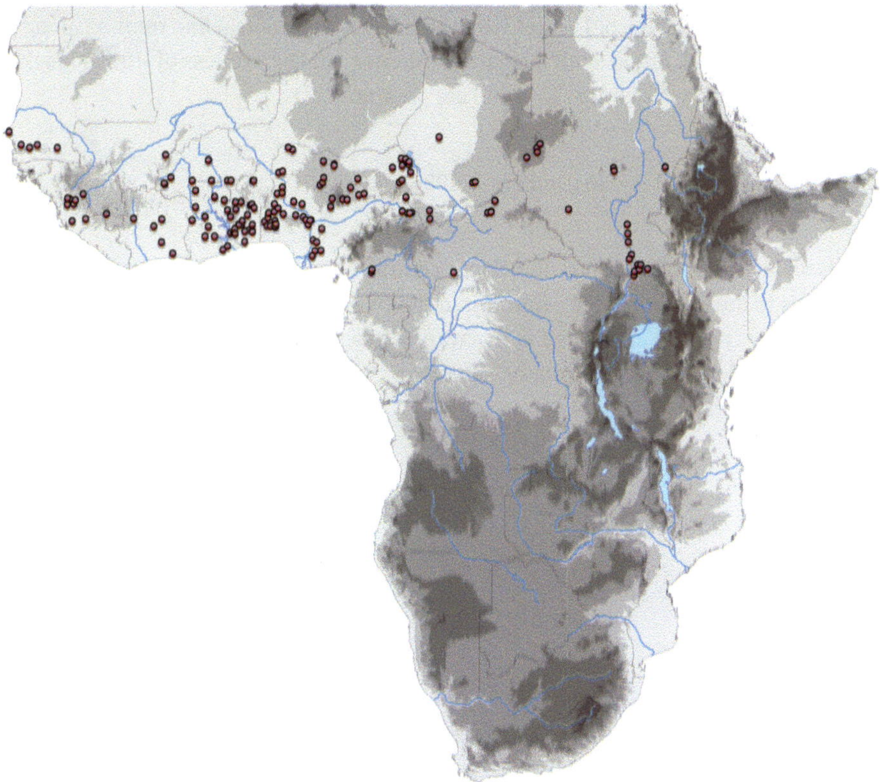

Figure 2.367 Distribution map

Measurements

External and cranial measurements (mm) and mass (g) for *Gerbilliscus kempi*, sexes combined

	Mean	Min	Max	N
Mass	97.3	74	128	7
HB	154.1	125	184	27
Tail	152.7	115	188	24
HF	35.5	32	41	28
Ear	19.6	17	23	28
GLS	40.1	36.3	43.3	29

Gerbilliscus leucogaster (Peters, 1852)

Diagnosis
This species can be distinguished from other *Gerbilliscus* species within its range (*G. afra* and *G. brantsii*) by its relatively long tail (ca. 115% of HB) which has a black dorsal line running along its length (in both *G. afra* and *G. brantsii* tail is relatively shorter, ca. 110% or less of HB, and lacks a dorsal black line). Four pairs of nipples (2 + 2 = 8) are present.

Taxonomic Issues
Chromosome number 2n = 40, FN = 66 (Qumsiyeh, 1986).

Distribution
This species is widespread in Southern and South-central Africa from the Gariep River, South Africa, northwards to the borders of the rainforest in the DRC and to Morogoro, Tanzania (Colangelo et al., 2005). It is typically associated with savanna habitats, but avoids semi-arid and arid environments (except along rivers and drainage lines).

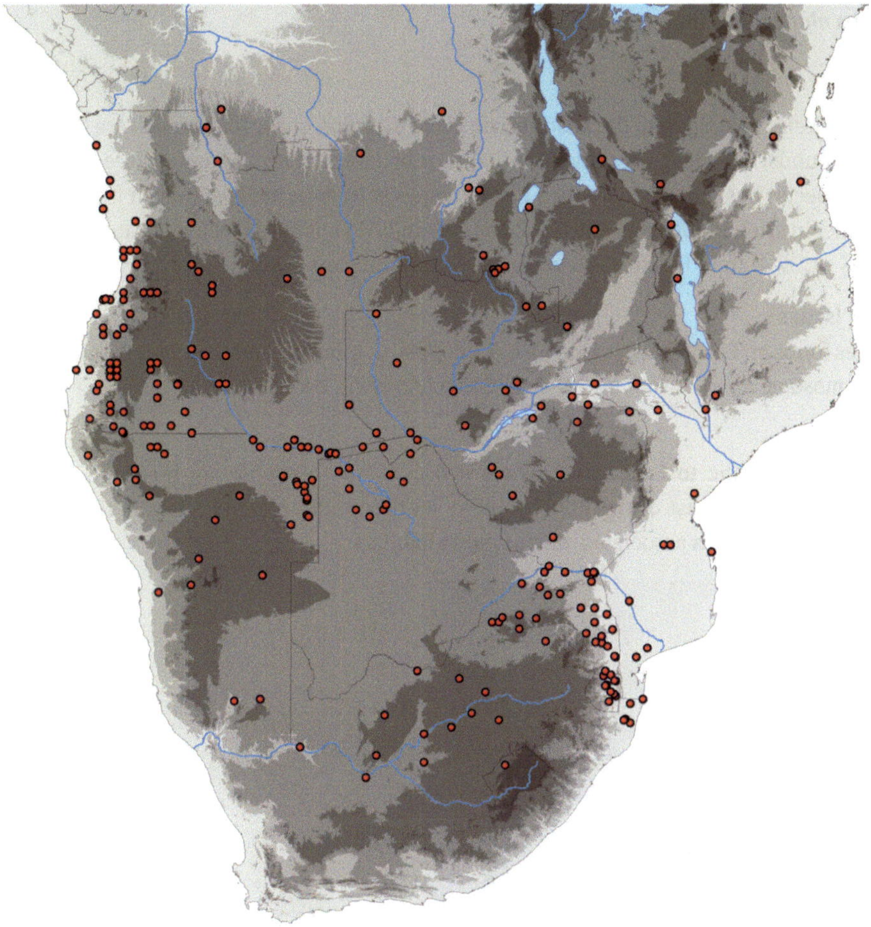

Figure 2.368 Distribution map

Measurements

External and cranial measurements (mm) and mass (g) for *Gerbilliscus leucogaster*, sexes combined

	Mean	Min	Max	N
Mass	72.2	48	100	39
HB	130.9	97	171	62
Tail	152.4	124	175	60
HF	34.7	30	40	46
Ear	20.5	16	24	47
GLS	37.7	34.7	40.2	22

Gerbilliscus nigricaudus (Peters, 1878)

Diagnosis

One of four species belonging to the East African clade (Granjon et al., 2012) which are distinct on molecular grounds, but which are difficult to diagnose on morphological grounds (see Bates, 1985, 1988). This species has a black tail which distinguishes is from *G. phillipsi* (which is also significantly smaller in size); *G. robustus* and *G. vicinus* generally do not have black tails, and are smaller in size (GLS ca. 39-41 mm; HF ca. 35-36 mm), but with significant overlap.

Taxonomic Issues

This species forms an East African clade with *G. phillipsi*, *G. robustus* and *G. vicinus* (Granjon et al., 2012). Chromosome number 2n = 36, FN = 68 (Colangelo et al., 2005).

Distribution

This species is endemic to East Africa where it occurs from extreme north-eastern Tanzania through eastern Kenya (east of the Rift Valley) to southern Ethiopia and Somalia (Bates, 1988; Colangelo et al, 2005). It is associated with savanna habitats.

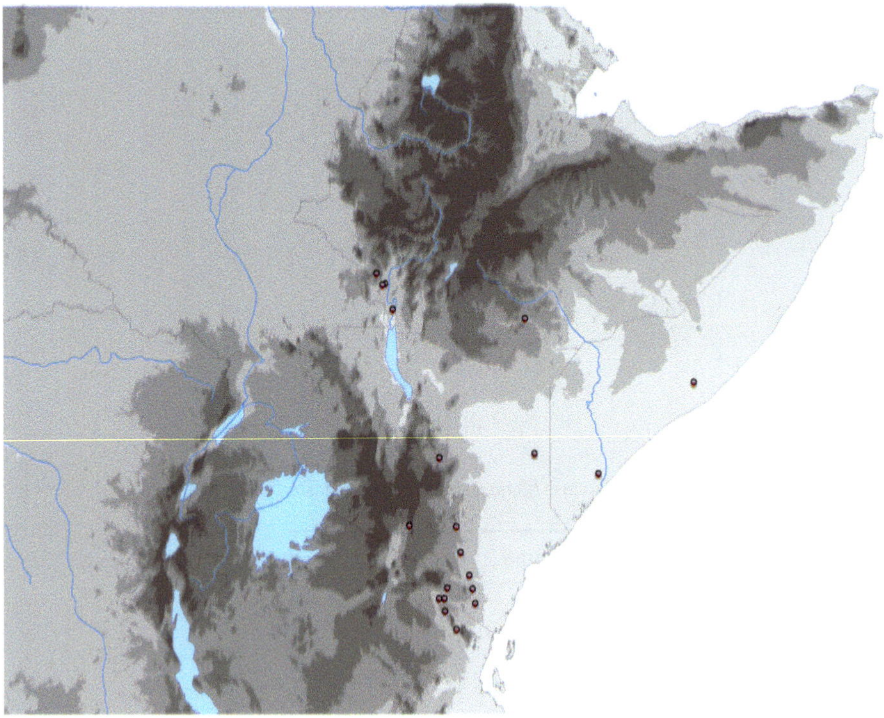

Figure 2.369 Distribution map

Measurements

External and cranial measurements (mm) and mass (g) for *Gerbilliscus nigricaudus*, sexes combined

	Mean	Min	Max	N
Mass*	123	80	195	-
HB	149.8	137	171	8
Tail	188.3	170	213	8
HF	39.1	37	41	8
Ear	22.9	22	25	8
GLS	43.8	41.5	47.0	8

*Neal (1982)

Gerbilliscus paeba (A. Smith, 1836)

Diagnosis

This is one of four species belonging to the "*Gerbillurus*" group that can readily be distinguished from all other *Gerbilliscus* species by their small size (GLS < 35 mm), relatively long tails (typically 120% or more of HB) and hairy soles in the hindfeet. This species can be distinguished from other members of the *Gerbillurus* group as follows: *G. setzeri* (pencil of grey hairs on terminal end of tail; 2n = 60); *G. tytonis* (dorsal pelage darker, typically reddish-brown; hindfeet longer, typically > 30 mm with distinctive fringe of hairs on sides of digits); *G. vallinus* (dorsal pelage darker, typically reddish-brown; hindfeet longer, > 30 mm; pencil of red or black hairs on terminal end of tail; 2n = 60). In this species the dorsal pelage varies geographically but is usually light reddish-brown to reddish-grey. Three pairs of nipples (1 + 2 = 6) are present.

Taxonomic Issues

This species was previously listed in the genus *Gerbillurus* (see genus account). Chromosomal and molecular data both indicate that *G. tytonis* is the closest relative of this species (Qumsiyeh et al., 1991; Granjon et al., 2012). Chromosome number 2n = 36, FN = 68 (Qumsiyeh, 1986).

Distribution

This species is widespread in the arid regions of Southern Africa, extending east along the Limpopo River and into Mozambique (Perrin et al., 1999a).

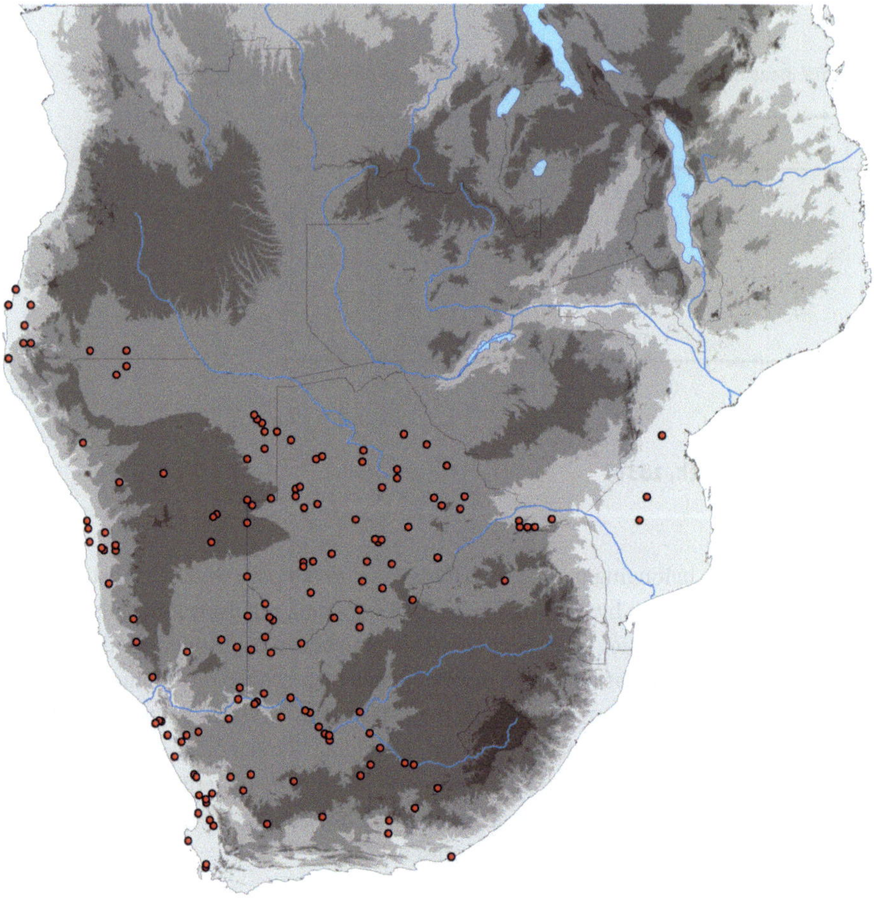

Figure 2.370 Distribution map

Measurements

External and cranial measurements (mm) and mass (g) for *Gerbilliscus paeba*, sexes combined

	Mean	Min	Max	N
Mass	28.3	21	36	11
HB	96.6	75	110	14
Tail	111.7	99	120	14
HF	27.0	25	30	13
Ear	16.1	14	18	14
GLS	28.7	26.8	30.9	14

Gerbilliscus phillipsi (de Winton, 1898)

Diagnosis

One of four species belonging to the East African clade (Granjon et al., 2012) which are distinct on molecular grounds, but which are difficult to diagnose on morphological grounds (see Bates, 1985, 1988). This species can be distinguished from *G. nigricaudus* by its bicoloured tail which is brown above and paler below and smaller size (in *G. nigricaudus*, GLS > 40 mm); *G. robustus* and *G. vicinus* are typically larger (GLS ca. 39-41 mm; HF ca. 35-36 mm), but with some overlap.

Taxonomic Issues

This species was previously listed as a subspecies of *G. robustus* but elevated to species status on morphological grounds by Bates (1985, 1988), from which it also differs on molecular grounds (Granjon et al., 2012).

Distribution

This species is patchily recorded from Ethiopia and Somalia with a single record from Kenya (Bates, 1988) where it occurs in semi-arid savannas.

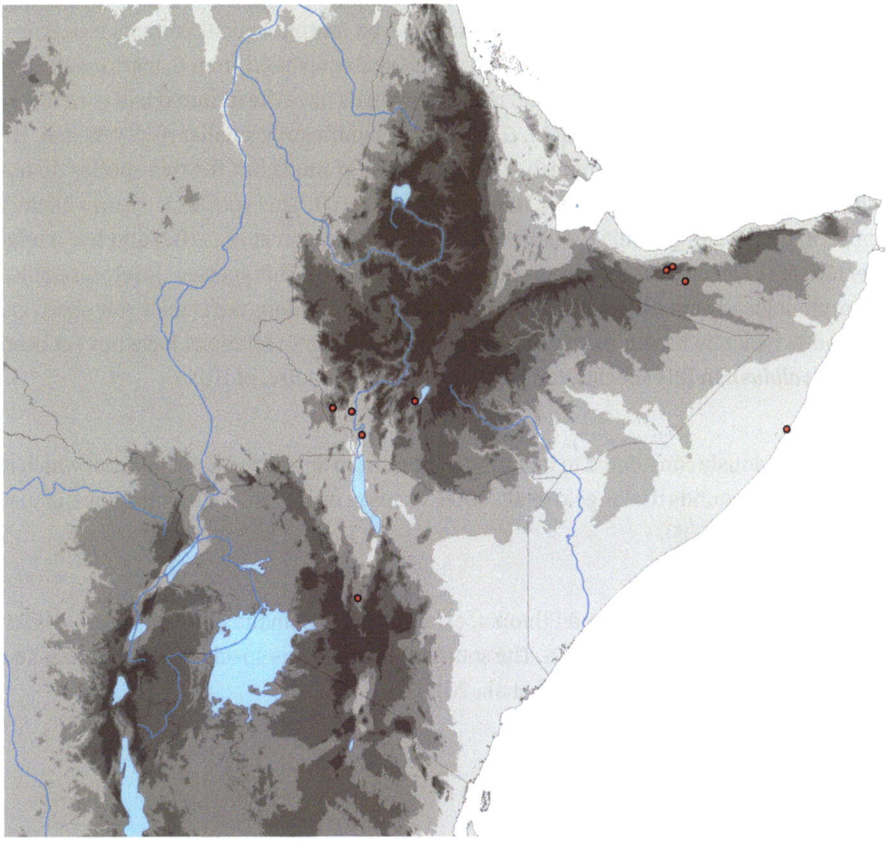

Figure 2.371 Distribution map

Measurements

External and cranial measurements (mm) and mass (g) for *Gerbilliscus phillipsi*, sexes combined

	Mean	Min	Max	N
Mass	-	-	-	-
HB*	132.4	120	148	11
Tail*	167.0	138	178	11
HF*	32.9	32	35	11
Ear*	19.7	17	22	11
GLS*	37.7	36.0	39.4	9

*Bates (1985)

Gerbilliscus robustus (Cretzschmar, 1826)

Diagnosis

One of four species belonging to the East African clade (Granjon et al., 2012) which are distinct on molecular grounds, but which are difficult to diagnose on morphological grounds (see Bates, 1985, 1988). This species typically has a brown tail which distinguishes it from *G. nigricaudus* (but tail colour is highly variable in *G. robustus* and most individuals have the terminal half with black, and some may occasionally be totally black); *G. phillipsi* is significantly smaller (GLS < 39 mm, HF < 35 mm); *G. vicinus* can only be distinguished on molecular grounds but the two species do not appear to have overlapping distributions (Colangelo et al., 2007). This species is clearly distinct from *G. kempi* on molecular grounds (Colangelo et al., 2007; Granjon et al., 2012) and has a relatively longer tail (in *G. kempi* tail is relatively short, ca. 100% of HB). This species closely resembles *G. validus* morphologically but differs in tail characteristics (in *G. validus* tail is relatively short, ca. 95% of HB, and is typically brown in colour), but their phylogenetic relations have not yet been resolved (see *G. validus*). In this species, tail is relatively long (ca. 120% of HB).

Taxonomic Issues

This species was previously considered to include *G. vicinus* from Kenya and Tanzania, but which is distinct on molecular grounds (Colangelo et al., 2005; Granjon et al., 2012). Chromosome number 2n = 36 (Colangelo et al., 2005).

Distribution

This species occurs widely in Sudan and Ethiopia, extending into Somalia and Chad (Bates, 1988; Colangelo et al., 2005; Granjon et al., 2012). The southern limits of this species are not known and it may extend into northern Kenya (Granjon et al., 2012).

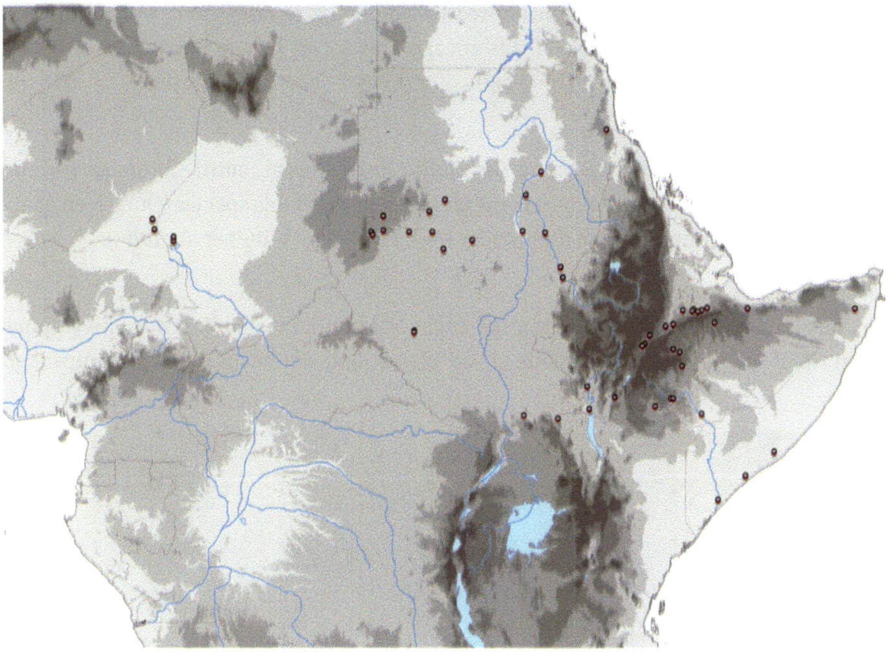

Figure 2.372 Distribution map

Measurements

External and cranial measurements (mm) and mass (g) for *Gerbilliscus robustus*, sexes combined

	Mean	Min	Max	N
Mass	-	-	-	-
HB*	148.6	111	180	24
Tail*	178.1	150	203	24
HF*	35.4	25	40	24
Ear*	21.1	15	25	24
GLS*	41.7	39.0	44.7	68

*Bates (1985); this sample may have included some specimens of *G. vicinus* (which was considered a synonym of *G. robustus*).

Gerbilliscus setzeri (Schlitter, 1973)

Diagnosis

This is one of four species belonging to the "*Gerbillurus*" group that can readily be distinguished from all other *Gerbilliscus* species by their small size (GLS < 35 mm), relatively long tails (typically 120% or more of HB) and hairy soles in the hindfeet. This species can be distinguished from other members of the *Gerbillurus* group as follows: *G. paeba* (pencil of grey hairs on terminal end of tail absent; 2n = 36); *G. tytonis* (pencil of grey hairs on terminal end of tail absent; dorsal pelage darker, typically reddish-brown; hindfeet with distinctive fringe of hairs on sides of digits; 2n =

36); *G. vallinus* (dorsal pelage darker, typically reddish-brown; size larger, but with significant overlap; non-overlapping distributions). In this species the dorsal pelage is pale sandy-brown (without any reddish or orange colour). Three pairs of nipples (1 + 2 = 6) are present.

Taxonomic Issues

This species was previously listed in the genus *Gerbillurus* (see genus account). Chromosomal and molecular data both indicate that *G. vallinus* is the closest relative of this species (Qumsiyeh et al., 1991; Granjon et al., 2012). Chromosome number 2n = 60, FN = 76 (Schlitter et al., 1984).

Distribution

This species has a narrow distribution along the coast of Namibia, extending into the extreme south-western tip of Angola where it inhabits gravel plains in desert environments (Dempster et al., 1998).

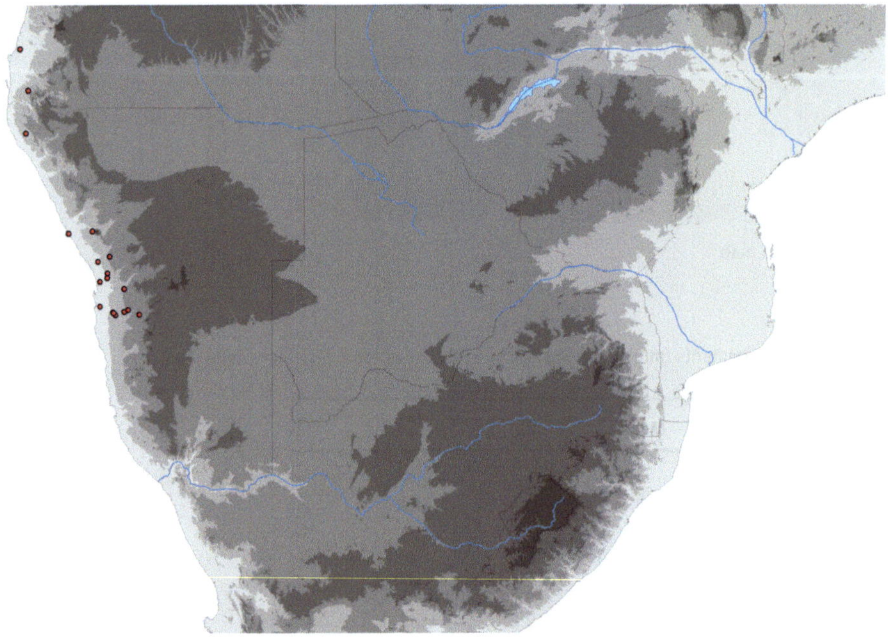

Figure 2.373 Distribution map

Measurements

External and cranial measurements (mm) and mass (g) for *Gerbilliscus setzeri*, sexes combined

	Mean	Min	Max	N
Mass	36.5	23	54	6
HB	100.9	86	114	10
Tail	125.8	110	141	10
HF	32.6	30.5	35.0	10
Ear	14.4	12	18	10
GLS	31.7	29.9	32.9	9

Gerbilliscus tytonis (Bauer & Niethammer, 1960)

Diagnosis

This is one of four species belonging to the "*Gerbillurus*" group that can readily be distinguished from all other *Gerbilliscus* species by their small size (GLS < 35 mm), relatively long tails (typically 120% or more of HB) and hairy soles in the hindfeet. This species can be distinguished from other members of the *Gerbillurus* group as follows: *G. setzeri* (pencil of grey hairs on terminal end of tail; 2n = 60); *G. paeba* (dorsal pelage lighter, typically light reddish-brown; hindfeet shorter, typically < 30 mm without distinctive fringe of hairs on sides of digits); *G. vallinus* (hindfeet without distinctive fringe of hairs on sides of digits; pencil of red or black hairs on terminal end of tail; 2n = 60). In this species the dorsal pelage is usually dark reddish-brown. Three pairs of nipples (1 + 2 = 6) are present.

Taxonomic Issues

This species was previously listed in the genus *Gerbillurus* (see genus account). Chromosomal and molecular data both indicate that *G. paeba* is the closest relative of this species (Qumsiyeh et al., 1991; Granjon et al., 2012). Chromosome number 2n = 36, FN = 68 (Schlitter et al., 1984).

Distribution

This species is endemic to Namibia where it occurs in the southern Namib desert typically in sand dunes (Perrin et al., 1999b).

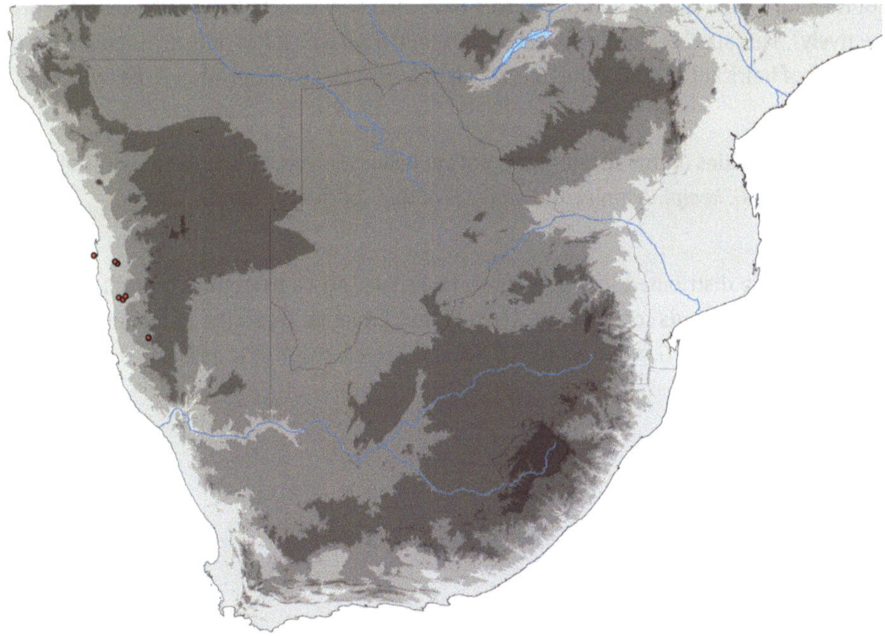

Figure 2.374 Distribution map

Measurements

External and cranial measurements (mm) and mass (g) for *Gerbilliscus tytonis*, sexes combined

	Mean	Min	Max	N
Mass	28.7	24	33	6
HB	99.3	90	109	7
Tail	121.4	118	125	5
HF	32.9	32	35	7
Ear	13.9	12	15	7
GLS	28.9	27.8	30.0	6

Gerbilliscus validus (Bocage, 1890)

Diagnosis

This species can be distinguished from *G. robustus* by the length and colour of its tail (in *G. robustus* tail is relatively long, ca. 120% of HB and usually has at least the terminal half with black). In this species tail is relatively short (ca. 95% of HB) without any black. It is also distinguished from the other members of the East African clade (*G. phillipsi*, *G. vicinus*, and *G. nigricaudus*) by its shorter tail (in East African clade, tail ca. 120% of HB). *Gerbilliscus boehmi* has white in the terminal half of the tail; *G. leucogaster* and *G. inclusus* have relatively long tails (ca. 115% and 105% of HB, respectively). This species closely resembles *G. kempi* from which it is only distinguishable by its lack of a tuft of hairs at the end of the tail (which may also be absent in adult *G. kempi*).

Taxonomic Issues

The phylogenetic affinities of this species have not yet been resolved, although on morphological grounds it resembles *G. kempi*. A molecular analysis of this species is urgently needed.

Distribution

This species is widely distributed in South-central and East Africa extending from the Zambezi River in the south northwards to Tanzania, Uganda and Ethiopia. Its status in Sudan is unclear where it might overlap with *G. robustus* and *G. kempi* (Setzer, 1956). It is associated with sandy soils in grasslands.

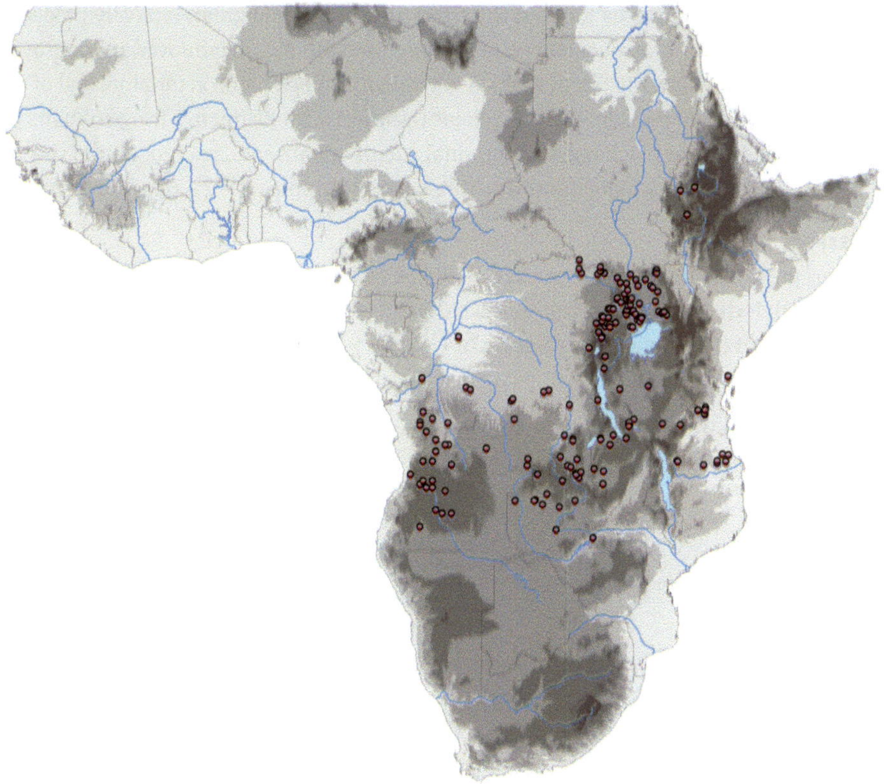

Figure 2.375 Distribution map

Measurements

External and cranial measurements (mm) and mass (g) for *Gerbilliscus validus*, sexes combined

	Mean	Min	Max	N
Mass*	129	90	160	-
HB	154.9	131	175	14
Tail	150.8	128	185	14
HF	37.4	35	43	14
Ear	22.3	20	25	13
GLS	42.1	39.2	45.2	12

*Neal (1982)

Gerbilliscus vallinus (Thomas, 1918)

Diagnosis

This is one of four species belonging to the "*Gerbillurus*" group that can readily be distinguished from all other *Gerbilliscus* species by their small size (GLS < 35 mm), relatively long tails (typically 120% or more of HB) and hairy soles in the hindfeet. This species can be distinguished from other members of the *Gerbillurus* group as follows: *G. setzeri* (dorsal pelage lighter, typically pale sandy-brown without any red or orange colour; size smaller, but with significant overlap; non-overlapping distributions); *G. paeba* (distinctive pencil of reddish or black hairs on terminal end of tail absent; hindfeet shorter, < 30 mm; 2n = 36); *G. tytonis* (hindfeet with distinctive fringe of hairs on sides of digits; pencil of red or black hairs on terminal end of tail absent; 2n = 36). In this species the dorsal pelage is usually dark reddish-brown. Three pairs of nipples (1 + 2 = 6) are present.

Taxonomic Issues

This species was previously listed in the genus *Gerbillurus* (see genus account). Chromosomal and molecular data both indicate that *G. setzeri* is the closest relative of this species (Qumsiyeh et al., 1991; Granjon et al., 2012). Chromosome number 2n = 60, FN = 70-74 (Qumsiyeh, 1986).

Distribution

This species occurs in southern Namibia and north-western South Africa where it inhabits gravel plains or shallow sandy soils (Dempster et al., 1999).

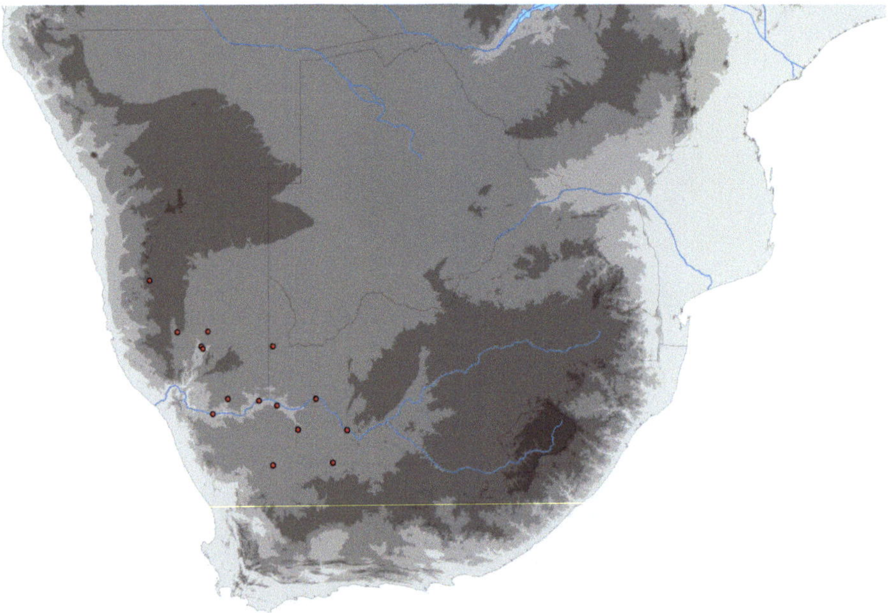

Figure 2.376 Distribution map

Measurements

External and cranial measurements (mm) and mass (g) for *Gerbilliscus vallinus*, sexes combined

	Mean	Min	Max	N
Mass	30.8	18	36	11
HB	98.2	90	106	13
Tail	139.2	129	151	13
HF	30.4	29	32	12
Ear	14.2	12	15	13
GLS	30.4	28.9	31.5	13

Gerbilliscus vicinus (Peters, 1878)

Diagnosis

One of four species belonging to the East African clade (Granjon et al., 2012) which are distinct on molecular grounds, but which are difficult to diagnose on morphological grounds (see Bates, 1985, 1988). This species typically has a brown tail which distinguishes is from *G. nigricaudus*; *G. phillipsi* is significantly smaller (GLS < 39 mm, HF < 35 mm); *G. robustus* can only be distinguished on molecular grounds but the two species do not appear to have overlapping distributions (Colangelo et al., 2007).

Taxonomic Issues

This species was previously considered a synonym of *G. robustus* but is clearly distinct on molecular grounds (Colangelo et al., 2005; Granjon et al., 2012). Chromosome number 2n = 36, FN = 68.

Distribution

This species is widespread in Tanzania and Kenya (Colangelo et al., 2005, 2007), but its northern limits are not currently known and it may extend further north than suggested by the map. It may be a pest of crops in parts of its range (Massawe et al., 2011).

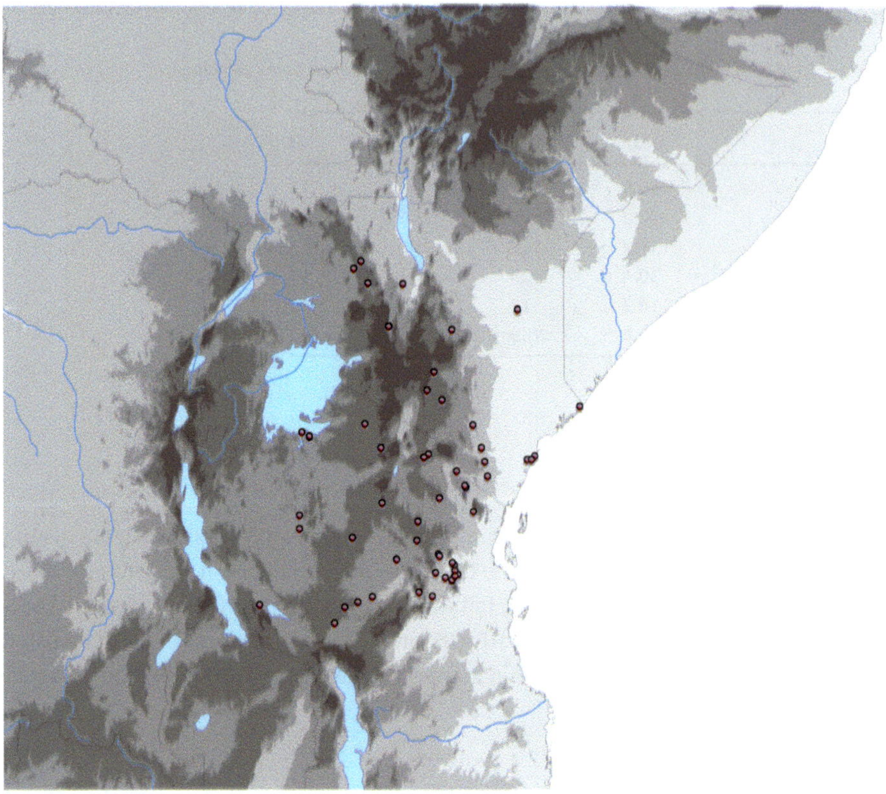

Figure 2.377 Distribution map

Measurements

External and cranial measurements (mm) and mass (g) for *Gerbilliscus vicinus*, sexes combined

	Mean	Min	Max	N
Mass*	74.3	54	104	6
HB	138.9	109	157	20
Tail	168.1	139	195	20
HF	36.1	32	43	20
Ear	19.9	12	24	19
GLS	39.7	35.8	44.1	21

*Neal (1982) reported a mean mass of 103 g for "*Tatera robusta*" from Kenya, which presumably refers to *G. vicinus*.

Genus Account

Gerbillus Desmarest, 1804 (Gerbils)

Diagnosis

They are small to very small gerbils (Mass ca. 10-45 g; HB ca. 65-105 mm) with long to very long tails (115-160% of HB) which often possess a terminal tuft ("pencil"). The dorsal pelage is a smooth pale sandy colour while the ventral pelage is pure white. There may be white markings around the eye and chin. The elongated hindfeet may have naked, partly furred or fully furred soles. They have four pairs of nipples (2 + 2 = 8).

(a) (b)

Figure 2.378 a and b Photographs of *Gerbillus* sp. (Ara Monadjem, left and David Hegner, right).

Cranial Characters

The skull is usually small to medium in size (ca. 20-33 mm) but may be slightly larger, e.g. in *G. pyramidum* (ca. 35 mm) which is the largest species in the genus. It is roughly triangular in shape with rounded braincase, moderately inflated bullae with fully (*Gerbillus* sensu lato) to partially (*Dipodillus* sensu lato) inflated ('pneumatised") mastoid chamber and narrow rectilinear nasals which are rounded at the end. The supraorbital ridges are well developed but do not reach as far as the posterior margin on the parietals. The zygomatic arch is slender and the zygomatic plate is moderately wide, does not cover, or barely covers the premaxillary-maxillary suture (not as expanded laterally as in *Gerbilliscus*) with a convex anterior edge. The palatal foramina barely reach the alveoli of M^1 and the posterior palatal foramina extend from the middle or back of M^2 to the front of M^1 being hardly separated from the anterior foramina (as opposed to a much greater separation in *Gerbilliscus*).

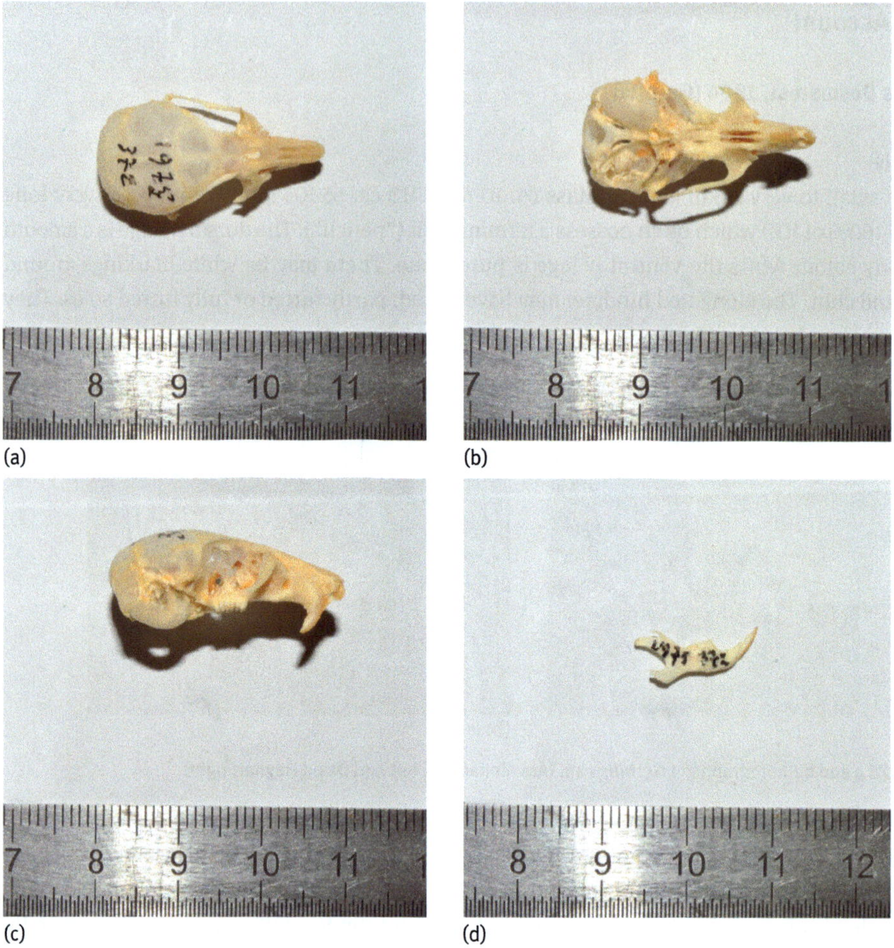

Figure 2.379 a-d Cranium of *Gerbillus nigeriae* (MNHN-ZM-MO-1975-372) (from top left to bottom right: dorsal view, ventral view, lateral view and lower jaw).

Dentition

The upper incisors are opisthodont and well grooved. The upper molar rows are parallel and the molars are generally arranged in laminae with M^1 (3 laminae) > M^2 (2 laminae) > M^1 (1 lamina). The dentition is rather primitive for gerbils, with the symmetrical laminate pattern not fully formed and cusps still separate and showing asymmetry. The pattern was described by Pavlinov et al. (1990) as bunodont, retaining several well separated cusps which are usually fully merged transversely into laminae in more advanced forms. However, in members of *Dipodillus* sensu lato, the crown pattern is described as being a more advanced "semi-prismatic" form whereby adjacent cusps are alternated, with longitudinal crests formed between them, still giving an asymmetrical appearance to the molar crown (Fig. 4 of Pavlinov et al., 1990).

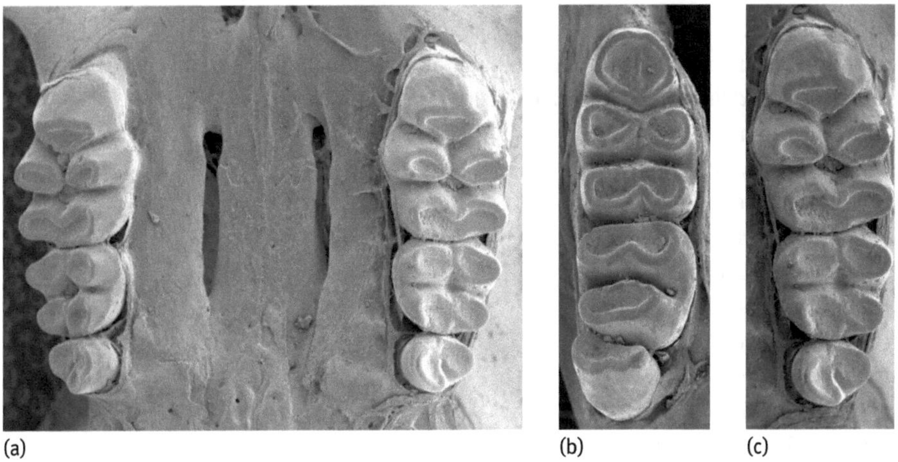

Figure 2.380 Cheek teeth of *Gerbillus dunni* (MNHN-ZM-MO-1884-1762) (A: upper cheek teeth; B: lower right toothrow; C: upper left toothrow).

Habitat & Biogeography

Most of the species are typically associated with desert and semi-desert habitats across the Sahara Desert and the Sahelian region of North Africa with many species occurring extralimitally to Africa, extending throughout the Middle East to India. In Sub-Saharan Africa the genus occurs as far south as northern Tanzania in the east, and Senegal and Burkina Faso in the west. One species, *G. rupicola*, uniquely occurs in rocky habitats (Granjon et al., 2002).

Fossil Record

Fossils are known from the Late Pliocene to Middle Pleistocene in North Africa and the Early Pliocene to Early Pleistocene in East Africa. The earliest known fossils are from 2.5-3 MYA deposits at Al Oughlam, North Africa (Geraads, 1995) and from 2.08 MYA in Omo Shungura member F, Ethiopia (Wesselman, 1984).

Biology & Ecology

They are nocturnal and terrestrial and dig extensive burrows. Although predominantly granivorous, some species also feed on green vegetation and insects. Some may become pests in agriculture, e.g., *G. nigeriae*. Reproduction is usually seasonal and litter size typically 2-6 (Happold, 2013).

Systematic Notes

The genus has been assigned to tribe Gerbillini and subtribe Gerbillina, together with *Microdillus* (Musser & Carleton, 2005; based on Pavlinov et al., 1990). The genus *Dipodillus* was formerly regarded as a distinct genus but clearly belongs under *Gerbillus* on morphological (Lay, 1983) and molecular (Abiadh et al., 2010) grounds. The genus *Gerbillus* is highly speciose and very problematic taxonomically with many names available, and uncertainties about the validity of many forms, some of which have been described only from restricted localities and few specimens. Morphological differences between species are slight (Lay, 1983) and chromosomal characters are very important for species diagnosis and identification (Aniskin et al., 2005; Ba et al., 2006; Hima et al., 2011). The genus probably contains many cryptic species and new species have been described in recent years (e.g., see Granjon et al., 1999; 2002a).

Annotated Species List

Twenty-three species are recognised from Sub-Saharan Africa including seven which were previously regarded as *Dipodillus* (marked with asterisk):

Gerbillus acticola Thomas, 1918
Gerbillus agag Thomas, 1903
**Gerbillus bottai* Lataste, 1882
Gerbillus brockmani (Thomas, 1910)
**Gerbillus campestris* (Loche, 1867)
Gerbillus dunni Thomas, 1904
Gerbillus gerbillus (Oliver, 1801)
**Gerbillus harwoodi* Thomas, 1901
Gerbillus henleyi (de Winton, 1903)
**Gerbillus lowei* (Thomas & Hinton, 1923)
**Gerbillus mackilligini* (Thomas, 1904)
Gerbillus nancillus Thomas & Hinton, 1923
Gerbillus nanus Blanford, 1875
Gerbillus nigeriae Thomas & Hinton, 1920
Gerbillus pulvinatus Rhoads, 1896
Gerbillus pusillus Peters, 1878
Gerbillus pyramidum Geoffroy, 1803
Gerbillus rosalinda St Leger, 1929
**Gerbillus rupicola* Granjon, Aniskin, Volobouev & Sicard, 2002
**Gerbillus somalicus* (Thomas, 1910)
**Gerbillus stigmonyx* (Heuglin, 1877)
Gerbillus tarabuli Thomas, 1902
Gerbillus watersi de Winton, 1901

[Note: *cosensi* (we include in *agag*), *percivali* (we include in *pusillus*) and *juliana* (we include in *watersi*) were recognised by Happold (2013) who treated agag as a *incertae sedis*]

Key to Species

1	Distribution limited to north-eastern Africa (excluding Sudan)	2
	Distribution limited to the Sudan region including the Nile Valley	8
	Distribution limited to West Africa	11
2	Soles of hindfeet partially or fully hairy	3
	Soles of hindfeet naked	4
3	Soles of hindfeet partially hairy; toes with fringe of stiff hairs; HF < 28 mm; 2n = 62	*G.pulvinatus*
	Soles of hindfeet fully hairy; HF > 29 mm; 2n = 74	*G. dunni*
4	Size larger, GLS > 28 mm, HB typically >90 mm	*G. acticola*
	Size smaller, GLS < 27 mm, HB typically < 90 mm	5
5	White ventral pelage clearly and sharply delineated from the darker flanks	6
	White ventral pelage not sharply delineated from the darker flanks, instead merging with it	7
6	Size slightly smaller, GLS ca. 24 mm, but with significant overlap; tail relatively longer, ca. 155% of HB; occurring widely in East Africa but absent from northern Sudan	*G. pusillus*

	Size slightly larger, GLS ca. 25 mm, but with significant overlap; tail relatively shorter, ca. 135% of HB; endemic to northern Sudan	*G. watersi*
	Tail relatively shorter, ca. 125% of HB; endemic to southern Kenya and northern Tanzania	*G. harwoodi*
7	Upper molar toothrow shorter, 3.1-3.4 mm	*G. brockmani*
	Upper molar toothrow longer, 3.5-4.1 mm	*G. somalicus*
8	Soles of hindfeet naked	9
	Soles of hindfeet hairy	10
9	Size larger, GLS > 28 mm; tail relatively longer (ca. 140% of HB)	*G. lowei*
	Size smaller, GLS < 27 mm; tail relatively shorter (ca. 115% of HB)	*G. stigmonyx*
	Size smaller, GLS < 28 mm; tail relatively longer (ca. 155% of HB)	*G. mackilligini*
10	Size larger, GLS > 27 mm	*G. rosalinda* / *G. agag*
	Size smaller, GLS < 27 mm	*G. bottai*
11	Soles of hindfeet naked or partly naked	12
	Soles of hindfeet hairy	16
12	Size larger, GLS > 29mm	13
	Size smaller, GLS < 26 mm	14
13	Occurring on rocky substrates; 2n = 52	*G. rupicola*
	Occurring on sandy substrates; 2n = 56-58	*G. campestris*
14	Chromosome number 2n = 52; completely naked sole	15
	Chromosome number 2n = 56; partly naked sole	*G. nancillus*
15	Hindfoot larger (HF > 20 mm), conspicuous terminal pencil on very long tail (ca. 140-150% of HB)	*G. nanus*
	Hindfoot shorter (HF < 20 mm), lacking conspicuous terminal pencil on a shorter tail (ca. 130% of HB)	*G. henleyi*
16	Chromosome number 2n = 42-43	*G. gerbillus*
	Chromosome number 2n = 60-74	*G. nigeriae*
	Chromosome number 2n = 40	*G. tarabuli*
	Chromosome number 2n = 38	*G. pyramidum*

Species Accounts

Gerbillus acticola Thomas, 1918

Diagnosis

This is one of eight *Gerbillus* species endemic to north-eastern Africa, which can be distinguished from the other six as follows: *G. brockmani* (size smaller, GLS < 26 mm, HB < 85 mm); *G. dunni* (soles of hindfeet hairy); *G. harwoodi* (size smaller, GLS < 26 mm, HB < 90 mm); *G. pulvinatus* (soles of hindfeet partially hairy; toes with fringe of hairs); *G. pusillus* (size smaller, GLS < 25 mm, HB < 80 mm); *G. somalicus* (size smaller, GLS < 27 mm, HB < 90 mm); *G. watersi* (size smaller, GLS < 26 mm, HB typically < 90 mm). This species has hindfeet with naked soles, and the white ventral pelage is clearly delineated from that of the flanks.

Taxonomic Issues

This taxon has been considered a subspecies of *G. pyramidum*, but considering the > 1300 km of separation between the two (and the large difference in size), it is likely that this is a valid species (Lay, 1983).

Distribution

This species is endemic to northern Somalia.

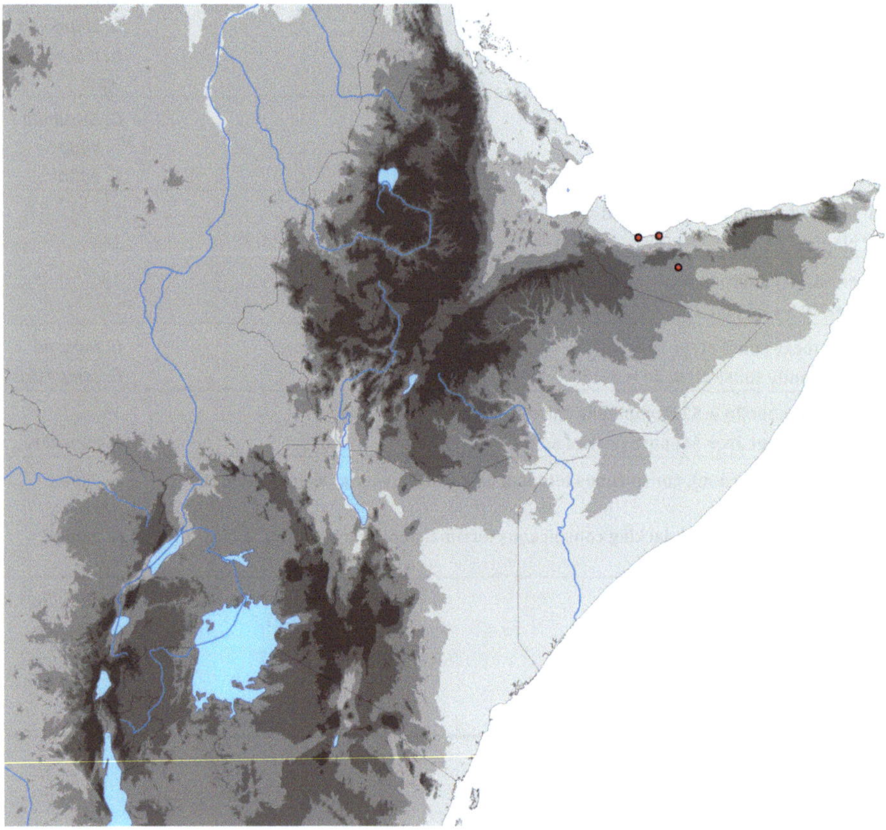

Figure 2.381 Distribution map

Measurements

External and cranial measurements (mm) and mass (g) for *Gerbillus acticola*, sexes combined

	Mean	Min	Max	N
Mass	-	-	-	-
HB*	94	90	104	5
Tail*	129	124	142	5
HF*	27	26	29	5
Ear*	15	15	15	5
GLS*	30.2	28.3	31.6	5

*Happold (2013)

Gerbillus agag Thomas, 1903

Diagnosis

This is one of six species of the genus endemic or near-endemic to the Sudan region, none of which have been karyotyped. *Gerbillus agag* is a relatively large gerbil having a hairy hindfoot sole, a relatively long tail (ca. 125% of HB) and a considerable area of white fur extending onto the chin, cheeks, throat, limbs and feet. It is larger than the hairy-soled *G. bottae* (GLS < 27 mm), and the naked-soled *G. stigmonyx* (GLS < 27 mm) and *G. makilligini* (GLS < 28 mm). It can be distinguished from similar-sized *G. lowei* which has a naked sole and a much more conspicuous terminal pencil on a longer (ca. 140% of HB) tail, but cannot easily be distinguished from similar-sized *G. rosalinda* which has similar coloration and tail length.

Taxonomic Issues

Although Happold (2013) treats *G. agag* as *incertae cedis*, we follow Lay (1983), Pavlinov (1990) and Musser & Carleton (2005) in retaining it as a species pending revision. We follow Musser & Carleton (2005) in including *cosensi* (from Kenya) in this species. *Gerbillus rosalinda* has been previously considered to be a subspecies of *G. agag* (Petter, 1975).

Distribution

This species occurs in Sudan (Setzer, 1956), northern Kenya (where it is represented by *G. cosensi* treated here as a synonym of *G. agag*) (Petter 1975) and northern Chad (represented by another synonym *dalloni*).

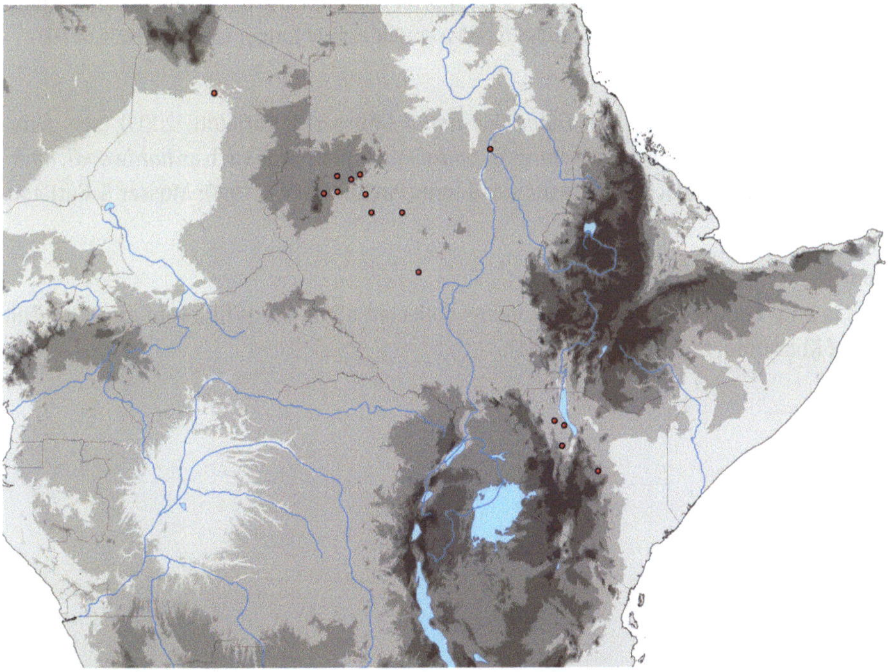

Figure 2.382 Distribution map

Measurements

External and cranial measurements (mm) and mass (g) for *Gerbillus agag*, sexes combined

	Mean	Min	Max	N
Mass*	32	30	34	2
HB*	100	90	135	7
Tail*	125	100	143	7
HF*	26	25	30	7
Ear*	15	13	20	7
GLS*	29.7	29.6	29.8	3

* Happold (2013), as *G. cosensi*

Gerbillus bottai Lataste, 1882

Diagnosis

This is one of six species of the genus endemic or near-endemic to the Sudan region, none of which have been karyotyped. *Gerbillus agag* is a relatively small gerbil (GLS ca. 25-27 mm) having a hairy or partly hairy hindfoot sole, a relatively long tail (ca. 125% of HB) and a white chin, throat and limbs. It is smaller than the naked-soled *G. lowei* (GLS > 28 mm), as well as being smaller than the hairy-soled *G. agag* (GLS > 29 mm) and *G. rosalinda* (GLS > 27 mm). It can be distinguished from similar-sized *G. stigmonyx* which has a naked sole, shorter tail (ca. 115% of HB) and a conspicuous terminal pencil of the tail. *Gerbillus mackilligini* has a considerably longer tail (ca. 155% of HB).

Taxonomic Issues

This species was previously listed as *Dipodillus bottai* (Musser & Carleton, 2005) (see genus account). There is uncertainty about the status of *bottai* and other forms such as *hardwoodi*, *luteolus*, *stigmonyx* and *watersi* which may be included in it (Pavlinov et al., 1990; Musser & Carleton, 2005).

Distribution

This species is only known from a few specimens collected along the White Nile and Blue Nile rivers south of Khartoum, Sudan (Setzer, 1956).

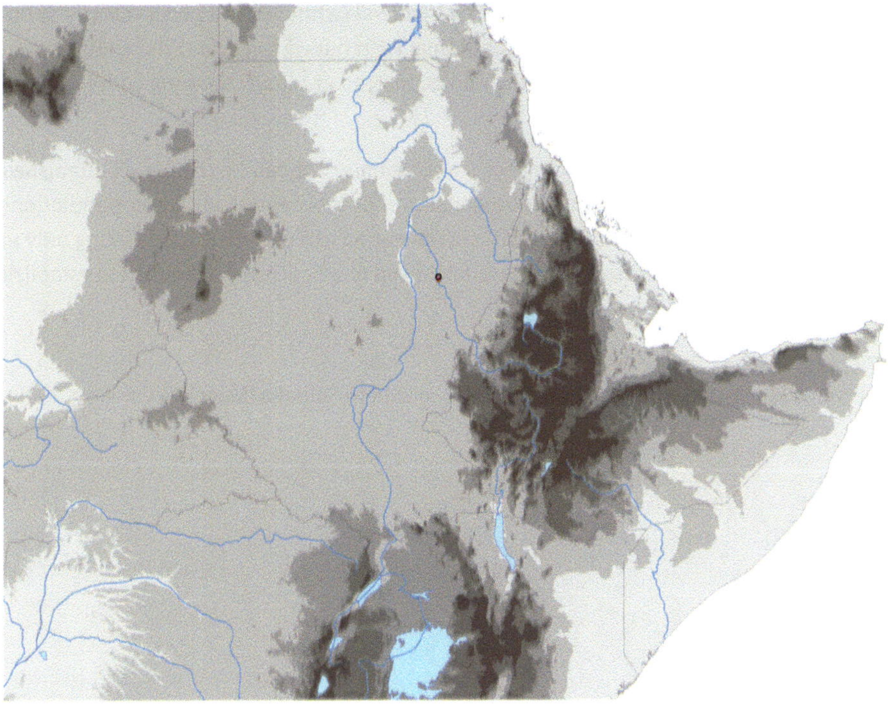

Figure 2.383 Distribution map

Measurements

External and cranial measurements (mm) and mass (g) for *Gerbillus bottai*, sexes combined

	Mean	Min	Max	N
Mass*	23	19	26	?
HB*	86	83	92	4
Tail*	106	101	110	4
HF*	20	19	21	4
Ear*	10.7	10	11	4
GLS*	26.2	25.4	27	4

*Happold (2013)

Gerbillus brockmani (Thomas, 1910)

Diagnosis

This is one of eight *Gerbillus* species endemic to north-eastern Africa, which can be distinguished from the other six as follows: *G. acticola* (size larger, GLS > 28 mm, HB > 90 mm); *G. dunni* (soles of hindfeet hairy); *G. pulvinatus* (soles of hindfeet partially hairy; toes with fringe of hairs); *G. harwoodi* (white ventral pelage clearly delineated from that of the flanks; upper molar toothrow longer, 3.6-3.9 mm); *G. pusillus* (size smaller, GLS < 25 mm, HB < 80 mm); *G. somalicus* (size similar, but with longer upper molar toothrow, 3.5-4.1 mm); *G. watersi* (white ventral pelage clearly delineated from that of the flanks). This species has hindfeet with naked soles, and the white ventral pelage is not sharply delineated from that of the flanks, instead gradually merging with it.

Taxonomic Issues

This species was previously listed as a subspecies of *G. nanus*, but can be readily distinguished from that species on morphological grounds (larger size of *G. brockmani*), and their populations are separated by at least 3000 km. It is very similar in appearance to *G. somalicus*, differing only in the length of the upper molar toothrow; these two taxa need to be critically examined to establish whether they are separate species or not.

Distribution

This species is only known from the type locality, Burao, northern Somalia (Musser & Carleton, 2005).

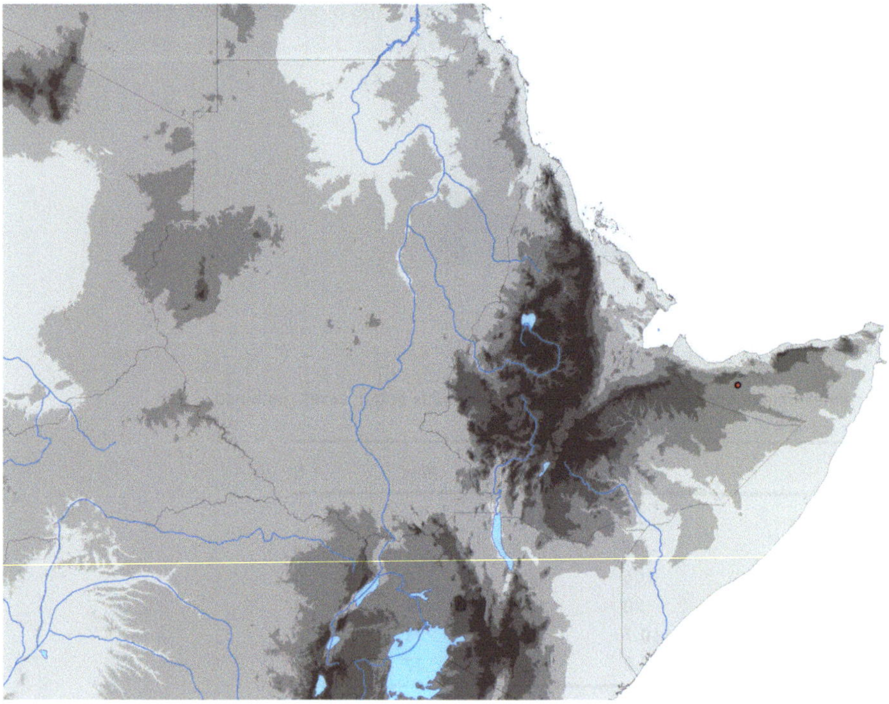

Figure 2.384 Distribution map

Measurements

External and cranial measurements (mm) and mass (g) for *Gerbillus brockmani*, sexes combined

	Mean	Min	Max	N
Mass	-	-	-	-
HB*	78	71	84	4
Tail*	110	106	117	4
HF*	21	21	22	4
Ear*	12	11	13	4
GLS*	25.4	25.0	25.7	4

*Happold (2013)

Gerbillus campestris (Loche, 1867)

Diagnosis

This is one of nine species of the genus endemic to Sub-Saharan West Africa (Granjon & Duplantier, 2009). This species is a relatively large-sized gerbil having naked hindfoot soles. It is distinctly larger than other naked-soled or partly-naked-soled gerbils (*G. henleyi*, *G. nanus* and *G. nancillus*) within its range. It is very similar to the North African *G. rupicola* which is found in rocky substrate habitats (as opposed to sandy habitats) has a different karyotype (2n = 52).

Taxonomic Issues

This species was previously listed as *Dipodillus campestris* (Musser & Carleton, 2005; Granjon & Duplantier, 2009) (see genus account). Chromosome number 2n = 56/58 (Dobigny et al., 2002a; Granjon & Duplantier, 2009). A recent molecular phylogeny has demonstrated its monophyly (Nicolas et al., 2014).

Distribution

This species occurs widely but patchily in the Sahara, marginally extending into Sub-Saharan Africa in northern Mauritania, Mali, Niger, Chad and Sudan.

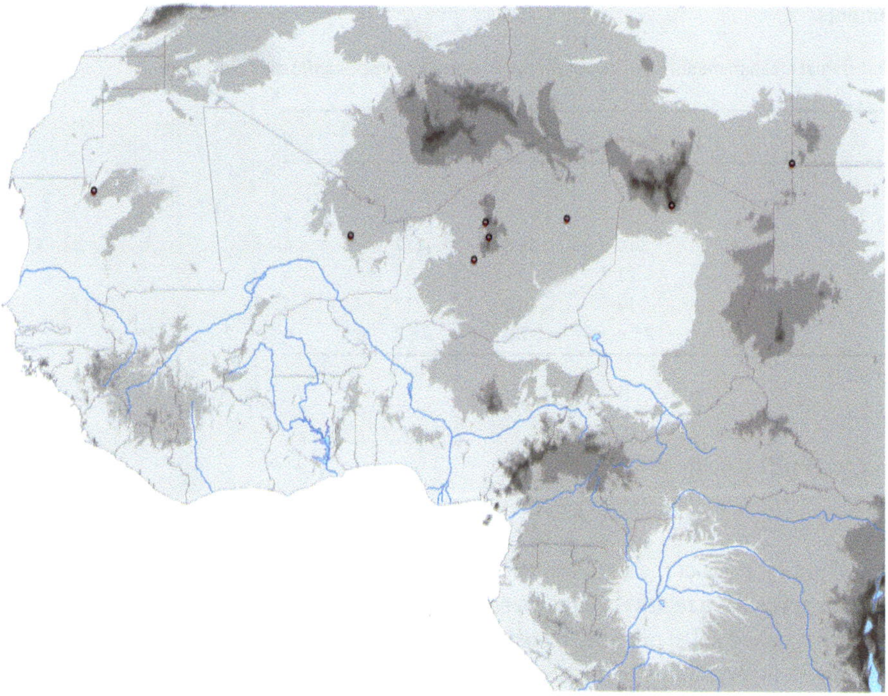

Figure 2.385 Distribution map

Measurements

External and cranial measurements (mm) and mass (g) for *Gerbillus campestris*, sexes combined

	Mean	Min	Max	N
Mass*	29	21	38	39
HB*	103	99	112	59
Tail*	133	118	153	51
HF*	27	25	29	58
Ear*	17	15	20	59
GLS*	29.6	27.9	31.9	51

* Happold (2013)

Gerbillus dunni Thomas, 1904

Diagnosis

This is one of eight *Gerbillus* species endemic to north-eastern Africa, which can be distinguished from the other six as follows: *G. acticola* (soles of hindfeet naked); *G. bockmani* (size smaller, GLS < 26 mm, HB < 85 mm; soles of hindfeet naked); *G. harwoodi* (soles of hindfeet naked); *G. pulvinatus* (size similar, but HF < 28 mm; soles of hindfeet only partially hairy; toes with fringe of hairs; 2n = 62); *G. pusillus* (size smaller, GLS < 25 mm, HB < 80 mm; soles of hindfeet naked); *G. somalicus* (size smaller, GLS < 26 mm, HB < 85 mm; soles of hindfeet naked); *G. watersi* (size smaller, GLS < 27 mm;

soles of hindfeet naked). This species has hindfeet with hairy soles, and the white ventral pelage is clearly delineated from that of the flanks.

Taxonomic Issues

This species has been considered a subspecies of *G. pyramidum* (Yalden et al., 1976), but can be distinguished from it on morphological (*G. dunni* is significantly smaller) and chromosomal grounds (Capanna & Merani, 1981). Chromosome number 2n = 74 (Capanna & Merani, 1981).

Distribution

This species is known from a few scattered records in Ethiopia, Eritrea and Somalia, with most records from the low-lying parts of the Rift Valley (Yalden et al., 1976).

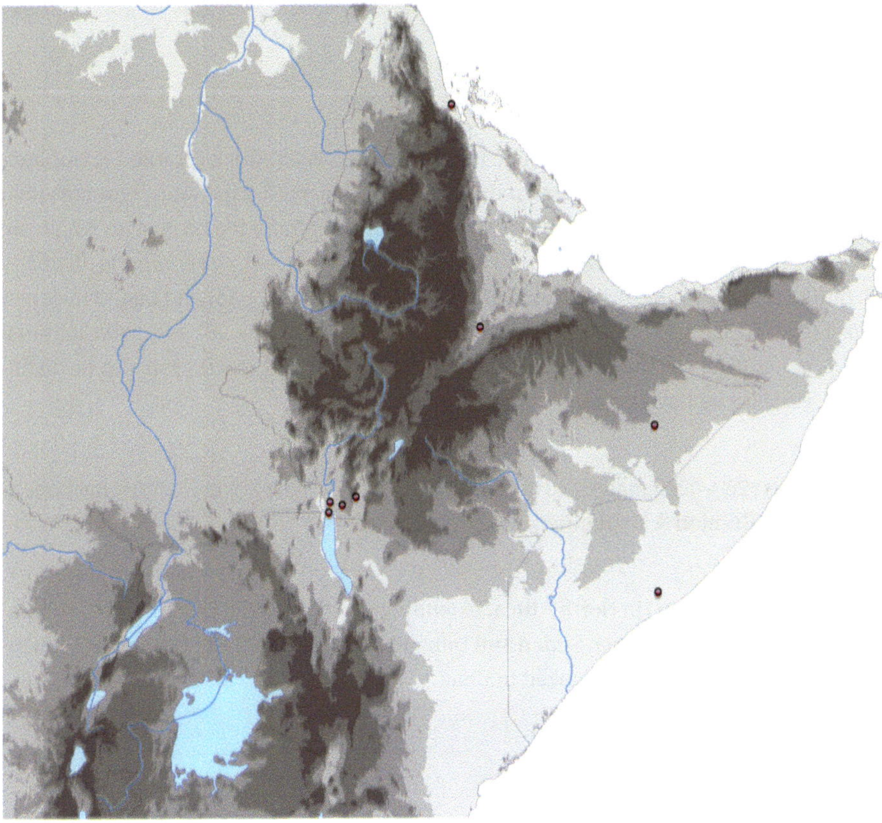

Figure 2.386 Distribution map

Measurements

External and cranial measurements (mm) and mass (g) for *Gerbillus dunni*, sexes combined

	Mean	Min	Max	N
Mass	38.7	34	41	3
HB	97.8	93	102	4
Tail	134.5	130	143	4
HF	30.0	29	31	4
Ear	14.3	14	15	4
GLS	30.6	30.3	30.8	4

Gerbillus gerbillus (Oliver, 1801)

Diagnosis

This is one of nine species of the genus endemic to Sub-Saharan West Africa (Granjon & Duplantier, 2009). It is a relatively large gerbil having hairy hindfoot soles. Within its range, chromosomal diploid numbers are required to distinguish this species from other hairy-footed species which are similar in size, including *G. nigeriae* (2n = 60-74) and *G. tarabuli* (2n = 40). *G. gerbillus* is distinctly smaller and different in karyotype from *G. pyramidum* (GLS > 30 mm; 2n = 38) and distinctly larger than *G. henleyi* (GSL < 23mm; 2n = 52), *G. nancillus* (GLS < 22.0 mm; 2n = 56) and, with minimal overlap, than *G. nanus* (GLS < 25 mm; 2n = 52). The three last-mentioned species can also be distinguished from *G. gerbillus* by having naked or partly naked soles.

Taxonomic Issues

Chromosome number: 2n = 42 (females), 43 (males); aFN = 72 – 74. There is an X-autosome Robertsonian translocation (Granjon & Duplantier, 2009).

Distribution

This species is widely distributed in North Africa, extending into the Middle East. It enters Sub-Saharan Africa in Mauritania, Mali, Niger, Sudan and Djibouti (Yalden et al., 1976; Granjon & Duplantier, 2009) and has probably been overlooked in Chad.

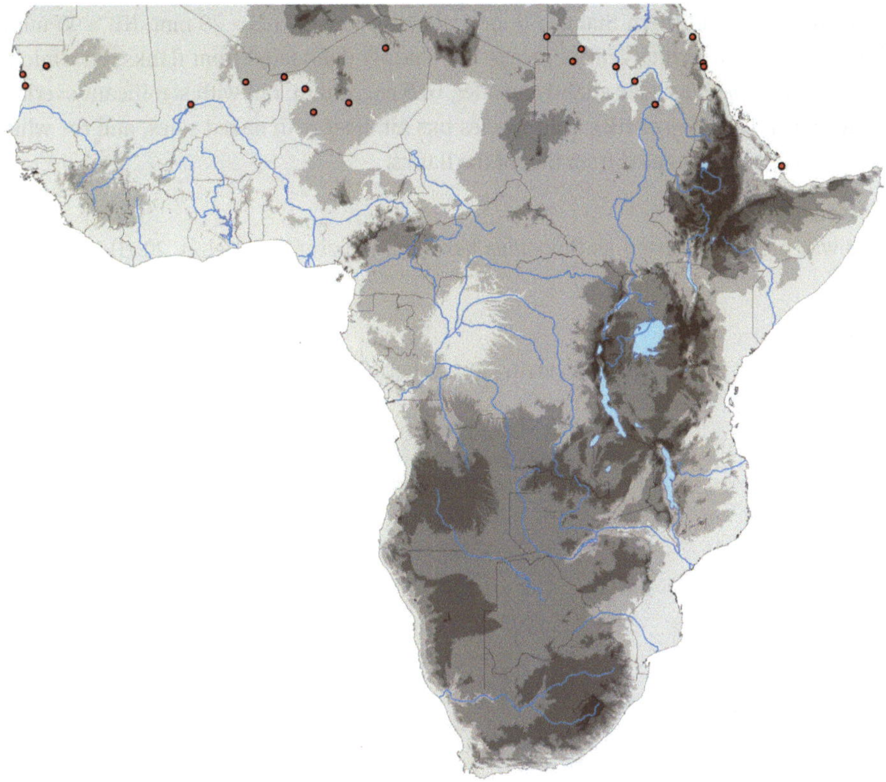

Figure 2.387 Distribution map

Measurements

External and cranial measurements (mm) and mass (g) for *Gerbillus gerbillus*, sexes combined

	Mean	Min	Max	N
Mass	21.7	14.5	29	5
HB	84.5	73	104	10
Tail	123.6	107	135	7
HF	29.5	28	32	10
Ear	12.0	11	13	9
GLS	27.2	24.8	28.9	8

Gerbillus harwoodi Thomas, 1901

Diagnosis

This is one of eight *Gerbillus* species endemic to north-eastern Africa, which can be distinguished from the other six as follows: *G. brockmani* (size similar; white ventral pelage not sharply delineated from flanks; endemic to northern Somalia); *G. dunni* (soles of hindfeet hairy); *G. pulvinatus* (soles of hindfeet partially hairy; toes with fringe of hairs); *G. watersi* (tail relatively longer, ca.

135% of HB; endemic to northern Sudan); *G. acticola* (size larger, GLS > 28 mm, HB > 90 mm); *G. somalicus* (size similar; white ventral pelage not sharply delineated from flanks; endemic to northern Somalia); *G. pusillus* (size slightly smaller, GLS ca. 24 mm, but with significant overlap; tail relatively longer, ca. 155% of HB). This species has hindfeet with naked soles, and the white ventral pelage is clearly delineated from that of the flanks.

Taxonomic Issues

This species was previously listed as *Dipodillus harwoodi* (Musser & Carleton, 2005) (see genus account).

Distribution

This species is endemic to East Africa occurring in northern Tanzania and southern Kenya.

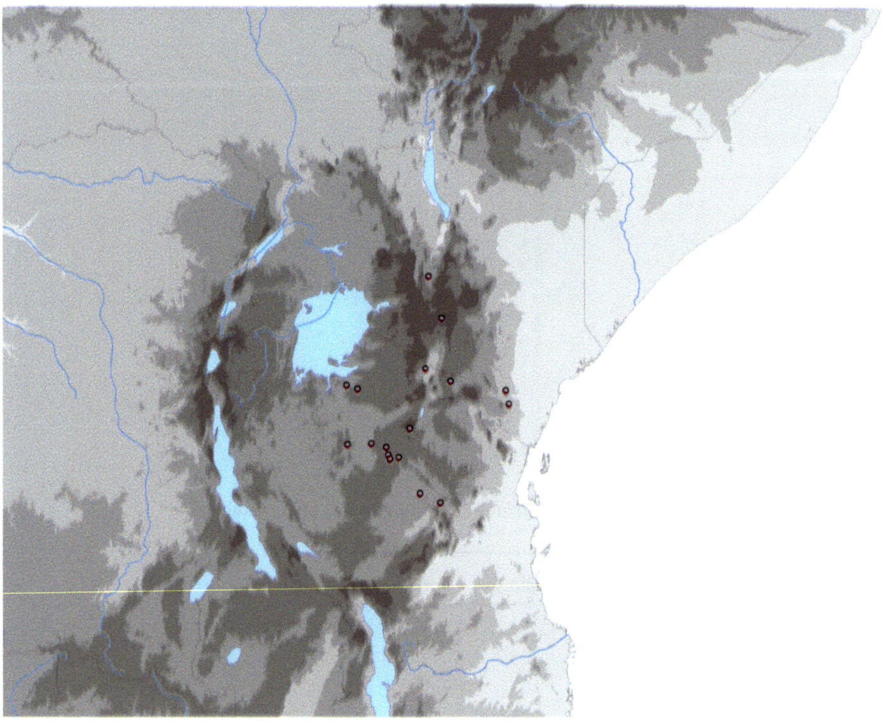

Figure 2.388 Distribution map

Measurements

External and cranial measurements (mm) and mass (g) for *Gerbillus harwoodi*, sexes combined

	Mean	Min	Max	N
Mass	14.0	14	14	2
HB	75	60	89	9
Tail	95	85	112	9
HF	20.5	19	25	9
Ear	9.4	9	10	9
GLS	24.0	22.9	25.3	7

Gerbillus henleyi (de Winton, 1903)

Diagnosis

This is one of nine species of the genus endemic to Sub-Saharan West Africa (Granjon & Duplantier, 2009). It is a relatively very small gerbil with delicate and naked hindfoot soles. Within its range, chromosomal diploid number is required to distinguish this species from similar-sized, partly naked-soled *G. nancillus* (2n = 56). It is on average smaller than *G. nanus*, which has a distinctly larger hindfoot (HF > 20 mm) and a slightly longer tail (ca. 150% versus 130% of HB in *G. henleyi*), but the two species share the same karyotype. It differs karyologically and is distinctly smaller-sized compared to the naked-soled *G. rupicola* (GLS > 29 mm; 2n = 52; aFN = 68) and *G. campestris* (GLS > 27 mm; 2n = 56/58).

Taxonomic Issues

Chromosomal number: 2n = 52; aFN = 58-62 mm (Volobouev et al., 1995).

Distribution

Within Sub-Saharan Africa, this species occurs in a narrow band on the southern edge of the Sahara from Mauritania in the east to Sudan in the west with an isolated record from Djibouti (Pearch et al., 2001; Granjon & Duplantier, 2009).

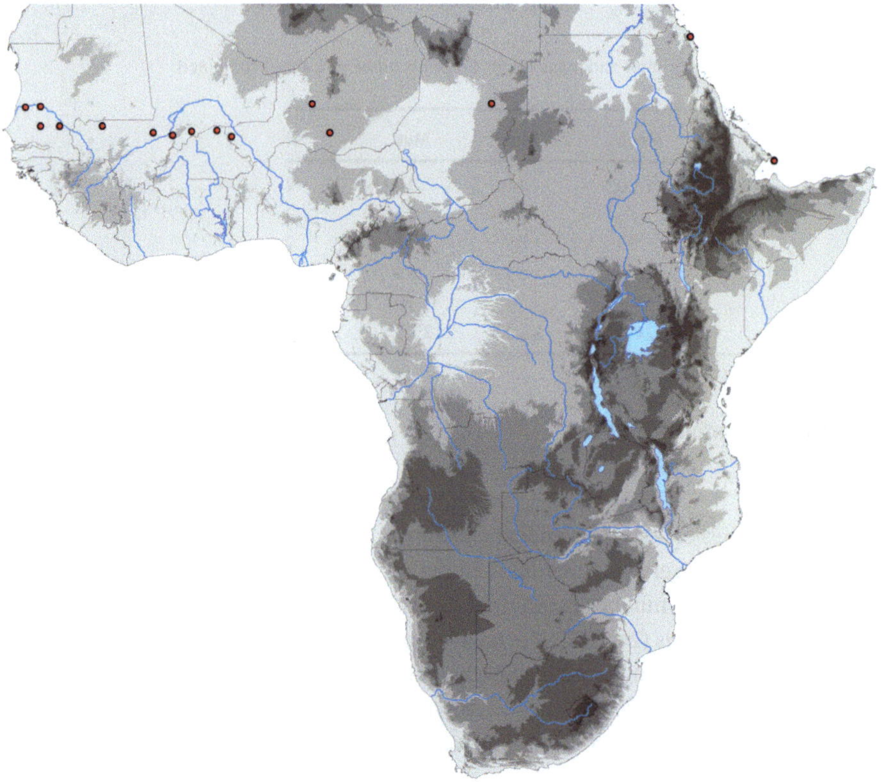

Figure 2.389 Distribution map

Measurements

External and cranial measurements (mm) and mass (g) for *Gerbillus henleyi*, sexes combined

	Mean	Min	Max	N
Mass	-	-	-	-
HB	62.9	60	65	3
Tail	84.7	77	92	3
HF	17.2	17	17.5	3
Ear	8.3	8	8.5	2
GLS	20.7	19.7	21.7	3

Gerbillus lowei (Thomas & Hinton, 1923)

Diagnosis

This is one of six species of the genus endemic or near-endemic to the Sudan region, none of which have been karyotyped. *Gerbillus lowei* is a relatively large gerbil (GLS ca. 28-31 mm) having a naked hindfoot sole, a very long tail (ca. 140% of HB) with a conspicuous terminal pencil and a white chin and throat. It can be distinguished from three of the other gerbils from the same region having hairy or partly hairy soles (*G. agag*, *G. bottai*, *G. rosalinda*). It is distinctly larger in size from the naked-soled *G. stigmonyx* (GLS < 27 mm) and *G. mackilligini* (GLS < 28 mm), and the former also has a much shorter tail (ca. 115% of HB).

Taxonomic Issues

This species was included in *G. campestris* by Petter (1975), but treated as a valid species by Lay (1983). Previously listed as *Dipodillus lowei* (Musser & Carleton, 2005) (see genus account).

Distribution

This species is only known from Jebel Marra where it has been recorded at high elevations above 2300 m (Setzer, 1956).

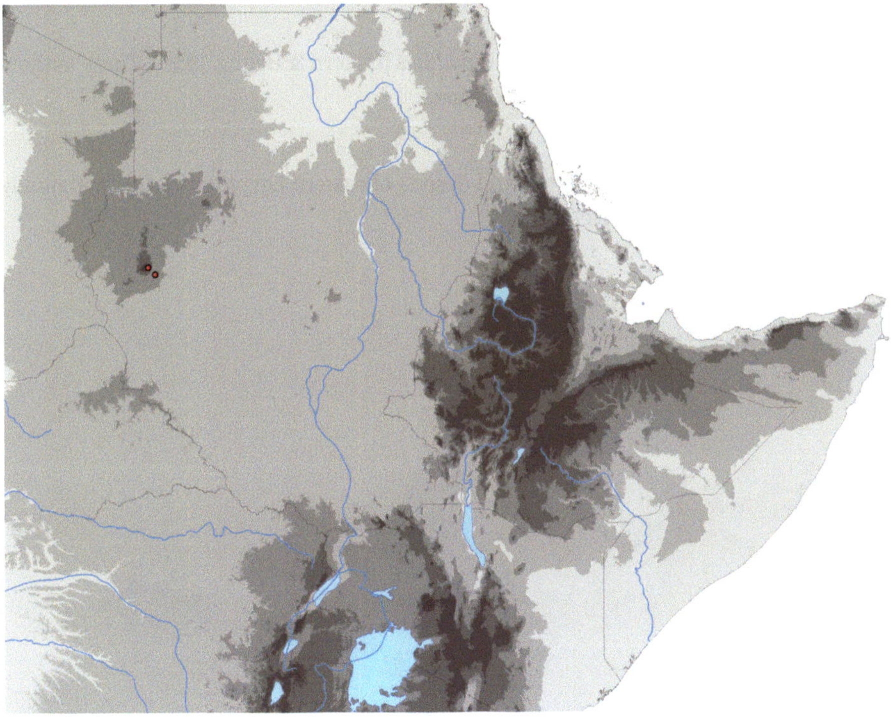

Figure 2.390 Distribution map

Measurements

External and cranial measurements (mm) and mass (g) for *Gerbillus lowei*, sexes combined

	Mean	Min	Max	N
Mass*	36	32	40	3
HB*	104	89	113	10
Tail*	148	141	158	10
HF*	26	25	27	10
Ear*	15	14	16	10
GLS*	30.2	28.5	31.1	10

* Happold (2013)

Gerbillus mackilligini (Thomas, 1904)

Diagnosis

This is one of six species of the genus endemic or near-endemic to the Sudan region, none of which have been karyotyped. *Gerbillus mackilligini* is a relatively small gerbil (GLS < 28 mm) with naked hindfeet, very long tail (ca. 155% of HB) with a conspicuous terminal pencil and a white chin and throat. Its naked soles readily distinguish it from three of the other gerbils from the same region having hairy or partly hairy soles (*G. agag*, *G. bottai*, *G. rosalinda*). It is distinguishable from the other two naked soled species as follows: *S. stigmonyx* (tail relatively shorter, ca 115% of HB); *G. lowei* (size larger, GLS > 28 mm; tail relatively shorter, ca. 140% of HB).

Taxonomic Issues

This taxon was previously listed as a species of *Dipodillus* (Osborn & Helmy, 1980). These authors also demonstrated its specific distinction from *G. nanus*.

Distribution

This species has a narrow distribution in the arid deserts of eastern Egypt, extending marginally into north-eastern Sudan (Osborn & Helmy, 1980).

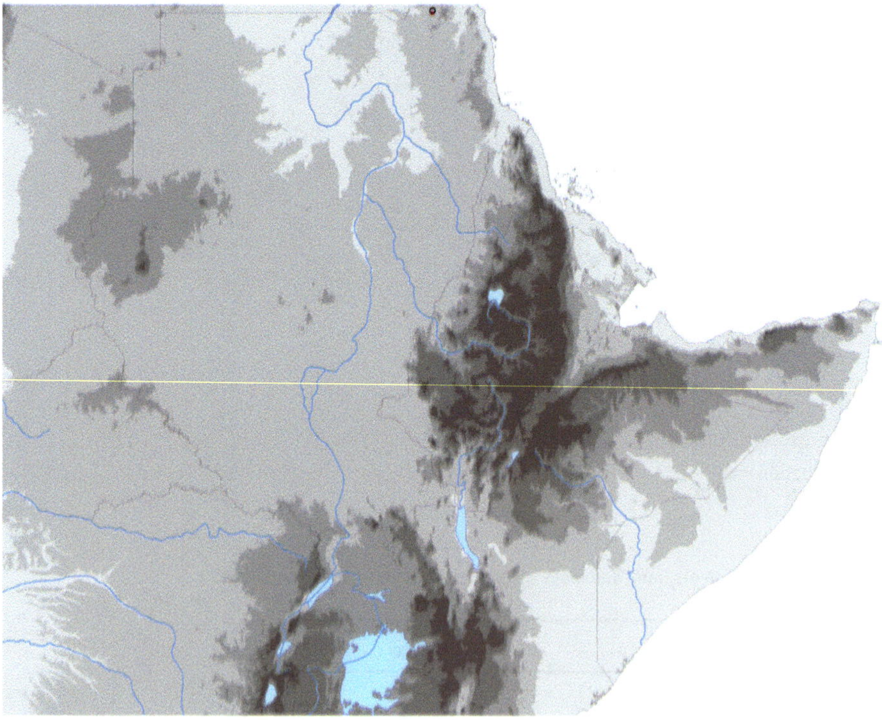

Figure 2.391 Distribution map

Measurements

External and cranial measurements (mm) and mass (g) for *Gerbillus mackilligini*, sexes combined

	Mean	Min	Max	N
Mass	-	-	-	-
HB*	78	72	86	6
Tail*	120	99	138	6
HF*	24.2	22	26	6
Ear*	12.9	12	14	6
GLS*	26.9	26.2	27.7	3

*Osborn & Helmy (1980)

Gerbillus nancillus Thomas & Hinton, 1923

Diagnosis

This is one of nine species of the genus endemic to Sub-Saharan West Africa (Granjon & Duplantier, 2009). It is a relatively small gerbil with partly-furred soles (posterior portion naked). Within its range, chromosomal diploid number is required to distinguish this species from similar-sized naked-soled *G. henleyi* (2n = 52) and *G. nanus* (2n = 52); the latter species also has a distinctly larger hindfoot (HF > 20 mm) and an almost naked hindfoot sole (it is only partially haired in *G. nancillus*). *Gerbillus nancillus* differs karyologically and is distinctly smaller-sized compared to naked soled *G. rupicola* (GLS > 29 mm; 2n = 52; aFN = 68) and *G. campestris* (GLS > 27 mm; 2n = 56-58).

Taxonomic Issues

Chromosomal number: 2n = 56; aFN = 54 (Dobigny et al., 2002a).

Distribution

This species is only known from a few scattered localities on the southern margin of the Sahara from Mali, Niger and Sudan (Setzer, 1956; Dobigny et al., 2002a; Granjon & Duplantier, 2009).

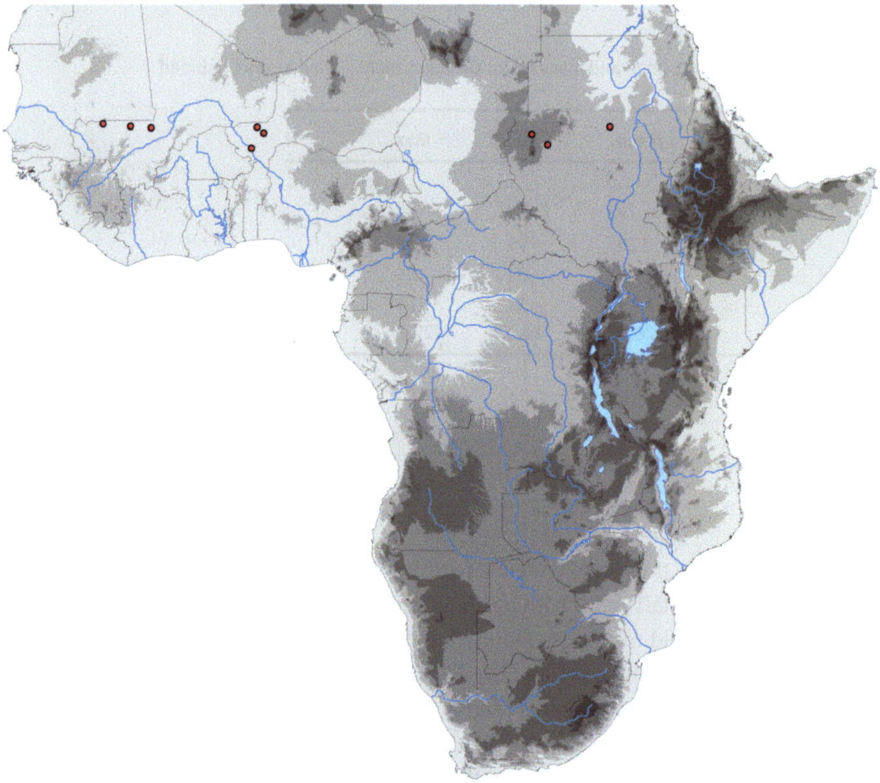

Figure 2.392 Distribution map

Measurements

External and cranial measurements (mm) and mass (g) for *Gerbillus nancillus*, sexes combined

	Mean	Min	Max	N
Mass	8.8	8	10	3
HB	65.3	61	69	3
Tail	91	-	-	1
HF	17.8	17	19	3
Ear	10.5	10	11	3
GLS	21.3	20.7	21.8	3

Gerbillus nanus Blanford, 1875

Diagnosis

This is one of nine species of the genus endemic to Sub-Saharan West Africa (Granjon & Duplantier, 2009). It is a relatively small gerbil with naked soles. Within its range, chromosomal diploid number is required to distinguish *G. nanus* from similar-sized *G. nancillus* (2n = 56) which has partly-furred soles. *Gerbillus henleyi* is on average smaller and shares the same karyotype as *G.*

nanus but differs in having a more slender hindfoot (< 20 mm) and a shorter tail (ca. 130% versus 150% of HB in *G. henleyi*). *Gerbillus nanus* differs karyologically and is distinctly smaller-sized compared to the naked-soled *G. rupicola* (GLS > 29 mm; 2n = 52; aFN = 68) and *G. campestris* (GLS > 27 mm; 2n = 56-58).

Taxonomic Issues

Chromosome number: 2n = 52; aFN = 58-62 (Volobouev et al., 1995).

Distribution

This species has a wide distribution in the arid deserts of West Africa extending from Mauritania in the west to Chad in the east.

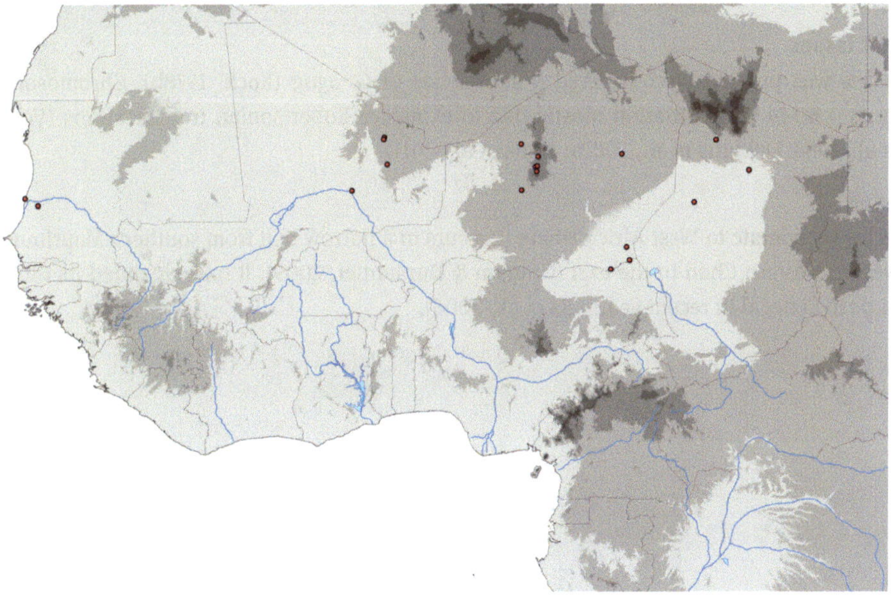

Figure 2.393 Distribution map

Measurements

External and cranial measurements (mm) and mass (g) for *Gerbillus nanus*, sexes combined

	Mean	Min	Max	N
Mass	17.5	12	23	2
HB	74	63	87	7
Tail	113.3	100	125	6
HF	21.9	20.5	24	7
Ear	11.3	10.5	13	7
GLS	24.4	22.7	25.4	7

Gerbillus nigeriae Thomas & Hinton, 1920

Diagnosis

This is one of nine species of the genus endemic to Sub-Saharan West Africa (Granjon & Duplantier, 2009). It is a relatively large gerbil having hairy hindfoot soles. Within its range, chromosomal diploid numbers are required to distinguish this species from other hairy-footed species which are similar in size, including *G. gerbillus* (2n = 42-43), *G. pyramidum* (2n = 38) and *G. tarabuli* (2n = 40). *G. nigeriae* is distinctly larger than *G. henleyi* (GSL < 23 mm; 2n = 52), *G. nancillus* (GLS < 22.0 mm; 2n = 56) and, with minimal overlap, with *G. nanus* (GLS < 26 mm; 2n = 52). These last three-mentioned species also have naked or semi-naked soles of the hindfeet, which are completely haired over along their length in *G. nigeriae*.

Taxonomic Issues

This species was previously considered a subspecies of *G. agag* (Kock, 1978b). Chromosome number: 2n = 60-74 with variation mostly due to extensive Robertsonian translocations (Volobouev et al., 1988; Dobigny et al., 2002a; Hima et al., 2011).

Distribution

This species is endemic to West Africa where it occurs in a narrow belt from southern Mauritania in the west to western Chad in the east (Granjon & Duplantier, 2009). It has expanded its range southwards in Senegal in recent years (Ba et al., 2006).

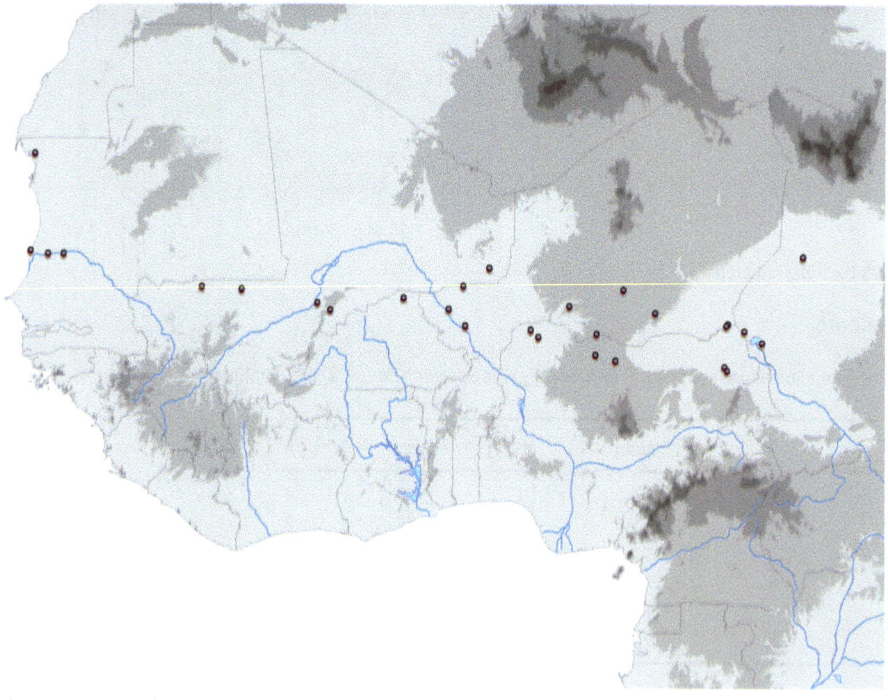

Figure 2.394 Distribution map

Measurements

External and cranial measurements (mm) and mass (g) for *Gerbillus nigeriae*, sexes combined

	Mean	Min	Max	N
Mass	24	16	35	8
HB	93.3	82	105	11
Tail	114.3	104	137	10
HF	25.4	25	27.5	11
Ear	14.1	14	15	9
GLS	28.7	26.6	31.3	14

Gerbillus pulvinatus Rhoads, 1896

Diagnosis

This is one of eight *Gerbillus* species endemic to north-eastern Africa, which can be distinguished from the other six as follows: *G. acticola* (soles of hindfeet naked); *G. brockmani* (size smaller, GLS < 26 mm, HB < 85 mm; soles of hindfeet naked); *G. dunni* (size similar, but HF > 29 mm; soles of hindfeet fully hairy; 2n = 74); *G. harwoodi* (soles of hindfeet naked); *G. pusillus* (size smaller, GLS < 25 mm, HB < 80 mm; soles of hindfeet naked); *G. somalicus* (size smaller, GLS < 26 mm, HB < 85 mm; soles of hindfeet naked); *G. watersi* (size smaller, GLS < 27 mm; soles of hindfeet naked). This species has hindfeet with partially hairy soles (toes with fringe of stiff hairs), and the white ventral pelage is not sharply delineated from that of the flanks, instead merging with it.

Taxonomic Issues

This species was previously included as a subspecies of *G. pyramidum* (Yalden et al., 1976), but considered a valid species by Musser & Carleton (2005). Chromosome number 2n = 62, FN = 84 (Hubert, 1978c).

Distribution

This species is known from a few scattered localities in northern Kenya, southern Ethiopia and Djibouti (Pearch et al., 2001).

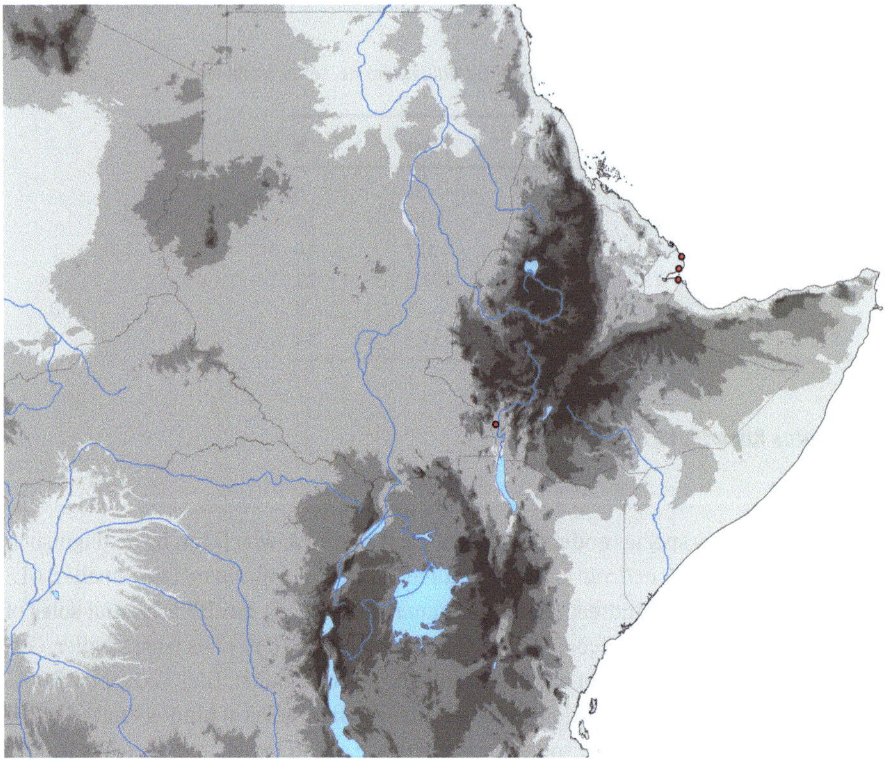

Figure 2.395 Distribution map

Measurements

External and cranial measurements (mm) and mass (g) for *Gerbillus pulvinatus*, sexes combined

	Mean	Min	Max	N
Mass	-	-	-	-
HB*	94.6	86	101	7
Tail*	133.0	118	142	7
HF*	27.6	27	29	7
Ear*	15.1	15	16	7
GLS*	30.4	30.2	30.6	7

*Happold (2013)

Gerbillus pusillus Peters, 1878

Diagnosis

This is one of eight *Gerbillus* species endemic to north-eastern Africa, which can be distinguished from the other six as follows: *G. brockmani* (size slightly larger, GLS > 25 mm; white ventral pelage not sharply delineated from flanks); *G. dunni* (soles of hindfeet hairy); *G. pulvinatus* (soles of hind-feet partially hairy; toes with fringe of hairs); *G. harwoodi* (tail relatively shorter, ca. 125% of HB);

G. acticola (size larger, GLS > 28 mm, HB > 90 mm); *G. somalicus* (size slightly larger, GLS > 24.5 mm, HB > 80 mm; white ventral pelage not sharply delineated from flanks); *G. watersi* (size lightly larger, GLS ca. 25 mm, but with significant overlap; tail relatively shorter, ca. 135% of HB; endemic to northern Sudan). This species has hindfeet with naked soles, and the white ventral pelage is clearly delineated from that of the flanks , and a long tail (ca. 155% of HB).

Taxonomic Issues

Chromosome number 2n = 34 (Capanna & Merani, 1981).

Distribution

This species has a patchy distribution in East Africa extending from northern Tanzania to Ethiopia and Somalia.

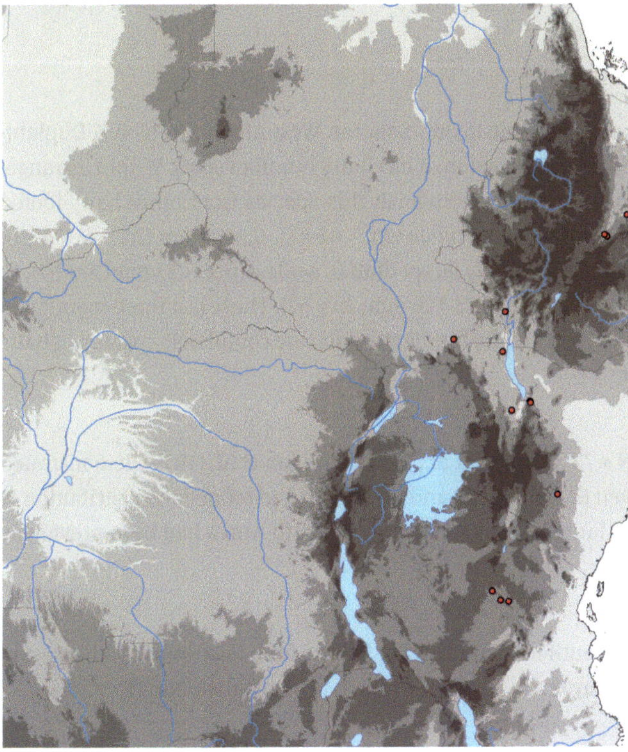

Figure 2.396 Distribution map

Measurements

External and cranial measurements (mm) and mass (g) for *Gerbillus pusillus*, sexes combined

	Mean	Min	Max	N
Mass	15	-	-	1
HB	65.8	54	77	9
Tail	99.3	82	115	7
HF	19.8	18	21	9
Ear	9.5	9	11	9
GLS	22.9	20.5	24.3	9

Gerbillus pyramidum Geoffroy, 1803

Diagnosis

This is one of nine species of the genus endemic to Sub-Saharan West Africa (Granjon & Duplantier, 2009). It is the largest-sized gerbil from the region. It has hairy hindfoot soles. Within its range, chromosomal diploid numbers are required to distinguish this species from other hairy-footed species which are similar in size, including *G. gerbillus* (2n = 42-43), *G. nigeriae* (2n = 60-74), *G. tarabuli* (2n = 40). *Gerbillus pyramidum* is distinctly larger than *G. henleyi* (GSL < 23mm; 2n = 52), *G. nancillus* (GLS < 22.0 mm; 2n = 56) and *G. nanus* (GLS < 26 mm; 2n = 52). These last three-mentioned species also have naked or semi-naked soles of the hindfeet, which are completely covered in hairs along their length in *G. pyramidum.*

Taxonomic Issues

Chromosome number: 2n = 38; aFN = 72 (Granjon et al., 1999). Granjon et al. (1999) demonstrated the chromosomal distinction between *G. pyramidum* and *G. tarabuli* and refined the distribution of the two species; specimens from the latter species from much of North Africa had been previously misclassified as *G. pyramidum.*

Distribution

This species occurs widely along the southern margin of the Sahara from Mauritania in the west to Sudan in the east (Setzer, 1956; Dobigny et al., 2001a,b; Granjon & Duplantier, 2009)

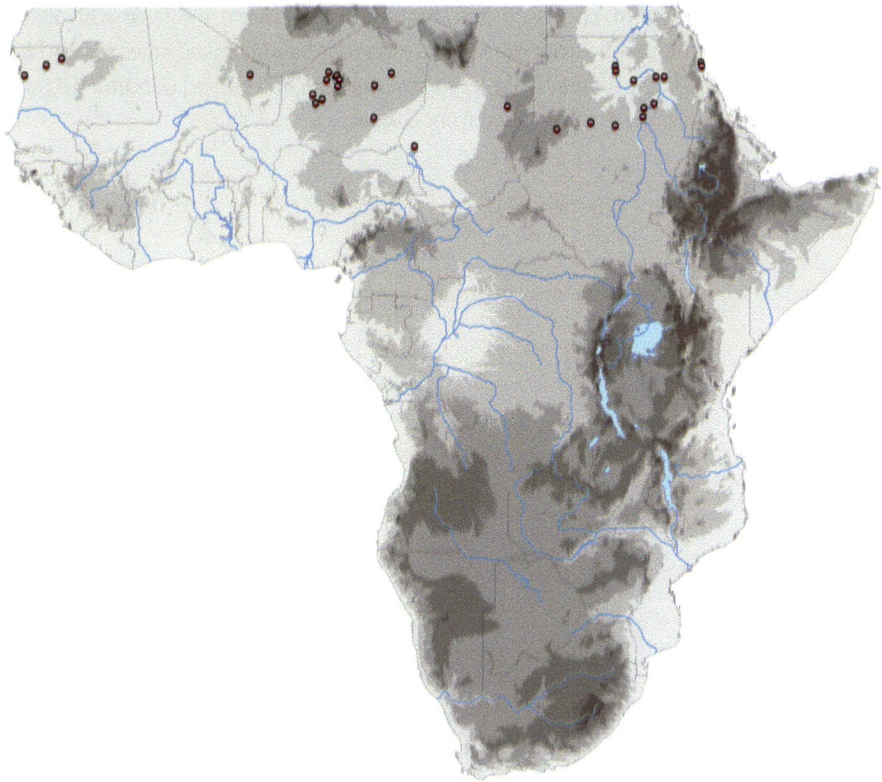

Figure 2.397 Distribution map

Measurements

External and cranial measurements (mm) and mass (g) for *Gerbillus pyramidum*, sexes combined

	Mean	Min	Max	N
Mass*	43.3	27	64	80
HB	106.2	90	122	9
Tail	148.8	140	160	7
HF	31.4	28	34	9
Ear	14.7	12	16	9
GLS	33.3	29.8	35.8	9

* Granjon & Duplantier (2009)

Gerbillus rosalinda St Leger, 1929

Diagnosis

This is one of six species of the genus endemic or near-endemic to the Sudan region, none of which have been karyotyped. *Gerbillus rosalinda* is a relatively large gerbil (GLS ca. 27-32 mm) having a hairy hindfoot sole, a long tail (ca. 120% of HB) without a conspicuous terminal pencil and an extensive white area including chin, cheeks, throat, limbs and feet. Its hairy soles readily distinguish it from three of the other gerbils from the same region that having naked soles (*G. lowei, G. mackilligini* and *G. stigmonyx*). It is distinctly larger in size than the hairy-soled *G. bottai* (GLS < 27 mm). It cannot be morphologically distinguished from *G. agag*.

Taxonomic Issues

This taxon may be a subspecies or synonym of *G. agag* (Petter, 1975).

Distribution

This species is known from just a few records in Sudan (Setzer, 1956).

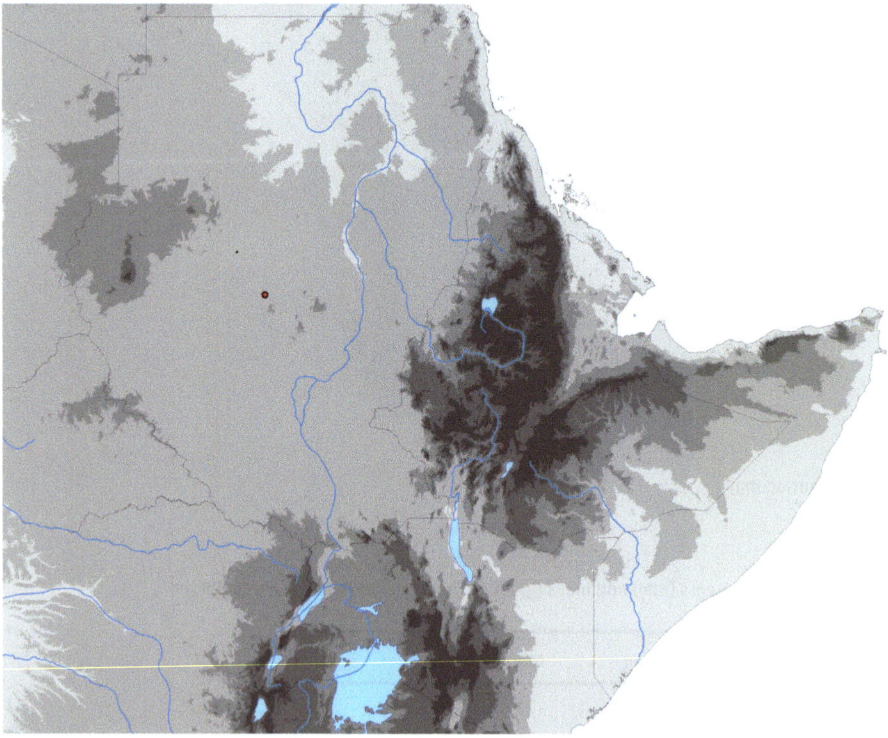

Figure 2.398 Distribution map

Measurements

External and cranial measurements (mm) and mass (g) for *Gerbillus rosalinda*, sexes combined

	Mean	Min	Max	N
Mass	28.5	20	35	11
HB	97	87	106	11
Tail	117	105	130	11
HF	25.3	23	27	11
Ear	15	13	16	11
GLS	29.1	27.5	31.6	20

Gerbillus rupicola Granjon, Aniskin, Volobouev & Sicard, 2002

Diagnosis

This is one of nine species of the genus endemic to Sub-Saharan West Africa (Granjon & Duplantier, 2009). This distinctive species is a relatively large-sized gerbil having naked hindfoot soles and being unusual amongst members of *Gerbillus* in its habitat of rocky rather than sandy substrates and in having a distribution limited to two closely-spaced localities in Mali. It is distinctly larger than other naked-soled or partly-naked-soled gerbils (*G. henleyi*, *G. nanus* and *G. nancillus*) within its range. It is very similar to the North African *G. campestris* which has a different karyotype (2n = 56-58).

Taxonomic Issues

Chromosome number: 2n = 52; FN = 72 (Granjon et al., 2002a). This species was previously listed as *Dipodillus rupicola* (Musser & Carleton, 2005; Granjon & Duplantier, 2009) (see genus account).

Distribution

This species is currently known from just two neighbouring localities in Mali where it exclusively inhabits rocky outcrops (Granjon et al., 2002a).

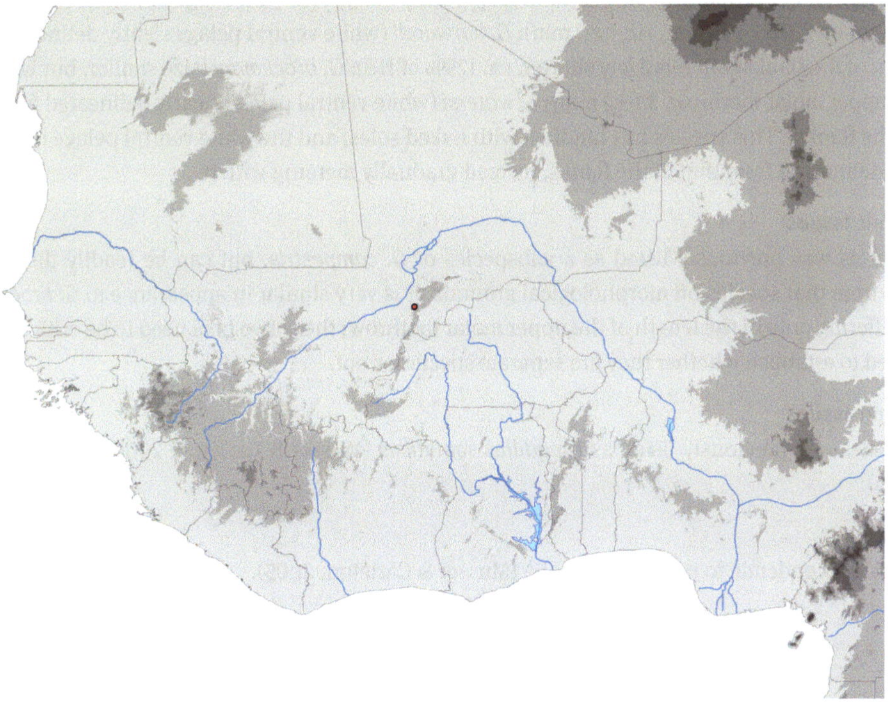

Figure 2.399 Distribution map

Measurements

External and cranial measurements (mm) and mass (g) for *Gerbillus rupicola*, sexes combined

	Mean	Min	Max	N
Mass	37	33	39	3
HB	102.7	100	106	3
Tail	138.5	135	142	2
HF	27.7	27	28	3
Ear	15.3	14.5	16.5	3
GLS	30	29.5	30.6	3

Gerbillus somalicus (Thomas, 1910)

Diagnosis

This is one of eight *Gerbillus* species endemic to the Horn of Africa, which can be distinguished from the other six as follows: *G. acticola* (size larger, GLS > 28 mm, HB > 90 mm); *G. dunni* (soles of hindfeet hairy); *G. pulvinatus* (soles of hindfeet partially hairy; toes with fringe of hairs); *G. pusillus* (size smaller, GLS < 25 mm, HB < 80 mm); *G. harwoodi* (white ventral pelage clearly delineated from that of the flanks; tail relatively shorter, ca. 125% of HB); *G. brockmani* (size similar, but with shorter upper molar toothrow, 3.1-3.4 mm); *G. watersi* (white ventral pelage clearly delineated from that of the flanks). This species has hindfeet with naked soles, and the white ventral pelage is not sharply delineated from that of the flanks, instead gradually merging with it.

Taxonomic Issues

This species was previously listed as a subspecies of *G. campestris*, but can be readily distinguished from that species on morphological grounds. It is very similar in appearance to *G. brockmani*, differing only in the length of the upper molar toothrow; these two taxa need to be critically examined to establish whether they are separate species or not.

Taxonomic Issues

This species was previously listed as *Dipodillus somalicus* (Musser & Carleton, 2005) (see genus account).

Distribution

This species is endemic to northern Somalia (Musser & Carleton, 2005).

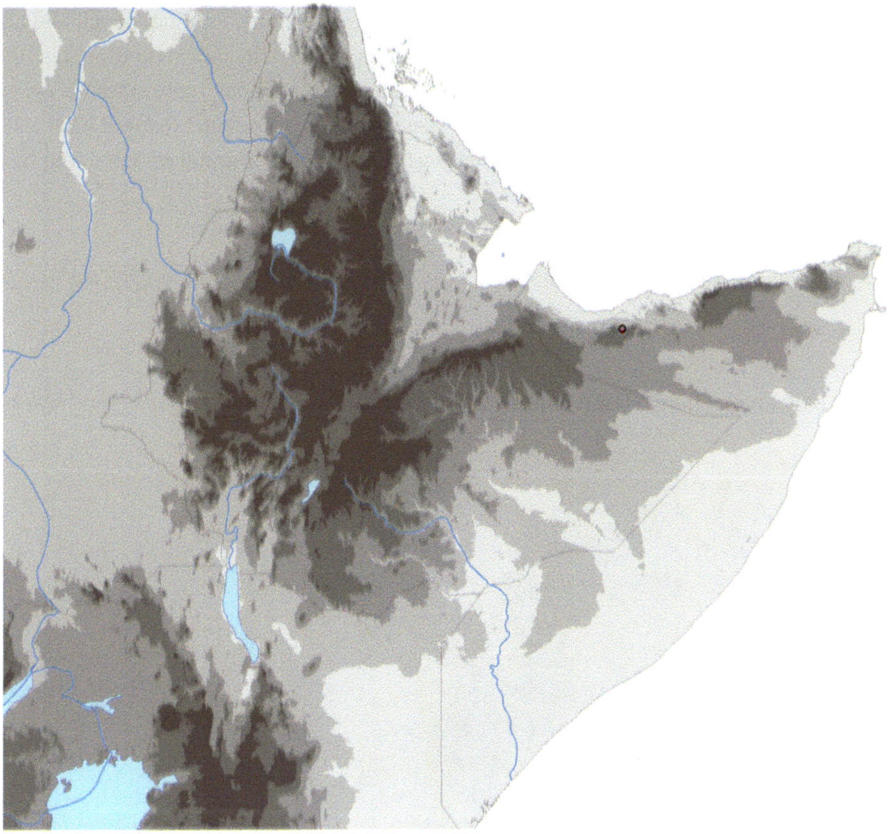

Figure 2.400 Distribution map

Measurements

External and cranial measurements (mm) and mass (g) for *Gerbillus somalicus*, sexes combined

	Mean	Min	Max	N
Mass	-	-	-	-
HB*	86	82	90	5
Tail*	117	112	124	5
HF*	24.6	24	25	5
Ear*	11.8	11	12	5
GLS*	25.6	24.6	26.5	5

*Happold (2013)

Gerbillus stigmonyx (Heuglin, 1877)

Diagnosis

This is one of six species of the genus endemic or near-endemic to the Sudan region, none of which have been karyotyped. *Gerbillus stigmonyx* is a relatively small gerbil (GLS ca. 26-27 mm)

having a naked hindfoot sole, a long tail (115% of HB) with a conspicuous terminal pencil and an extensive white area including chin, cheeks and throat, limbs and feet. Its naked hindfoot readily distinguishes it from three of the other gerbils from the same region having hairy soles (*G. agag*, *G. bottai* and *G. rosalinda*). It is distinctly smaller in size than the naked-soled *G. lowei* (GLS > 28 mm) which also has a longer tail (140% of HB). *Gerbillus mackilligini* is also naked-soled, but has a relatively much longer tail (ca. 155% of HB).

Taxonomic Issues

This species was previously listed as *Dipodillus stigmonyx* (Musser & Carleton, 2005) (see genus account).

Distribution

This species is only known from a few localities along the White Nile from Khartoum southwards (Setzer, 1956).

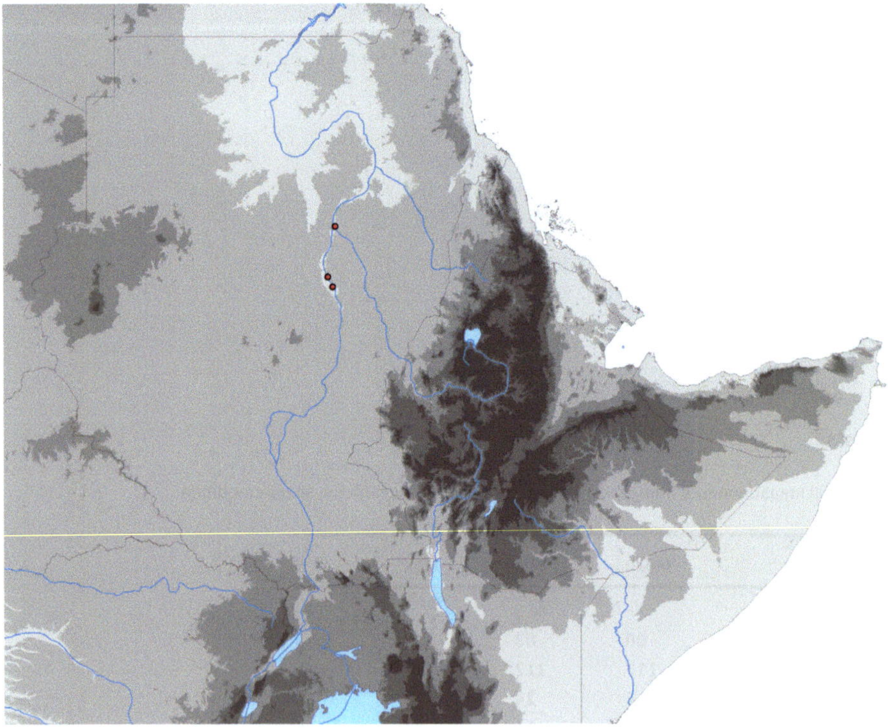

Figure 2.401 Distribution map

Measurements

External and cranial measurements (mm) and mass (g) for *Gerbillus stigmonyx*, sexes combined

	Mean	Min	Max	N
Mass	-	-	-	-
HB*	90	86	95	5
Tail*	102	97	106	5
HF*	21	20	23	5
Ear*	11	8	13	5
GLS*	26.5	25.9	26.9	6

* Happold (2013)

Gerbillus tarabuli Thomas, 1902

Diagnosis

This is one of nine species of the genus endemic to Sub-Saharan West Africa (Granjon & Duplantier, 2009). It is a relatively large gerbil with hairy hindfoot soles. Within its range, chromosomal diploid numbers are required to distinguish this species from other hairy-footed species which are similar in size, including *G. gerbillus* (2n = 42-43), *G. nigeriae* (2n = 60-74) and *G. pyramidum* (2n = 38). *Gerbillus tarabuli* is distinctly larger than *G. henleyi* (GSL < 23mm; 2n = 52), *G. nancillus* (GLS < 22.0 mm; 2n = 56) and *G. nanus* (GLS < 26 mm; 2n = 52). These last three-mentioned species also have naked or semi-naked soles of the hindfeet, which are completely hair-covered along their length in *G. tarabuli*.

Taxonomic Issues

Chromosome number 2n = 40; aFN = 74 (Granjon et al., 1999). *Gerbillus tarabuli* was previously considered to be a subspecies of *G. pyramidum*. Granjon et al. (1999) demonstrated the chromosomal distinction between *G. pyramidum* and *G. tarabuli* and refined the distribution of the two species; specimens of *G. tarabuli* species from much of North Africa and Senegal had been previously misclassified as *G. pyramidum*.

Distribution

This species is widespread in the western parts of the Sahara, occurring marginally into Sub-Saharan Africa in Mauritania, Mali and Niger (Granjon & Duplantier, 2009).

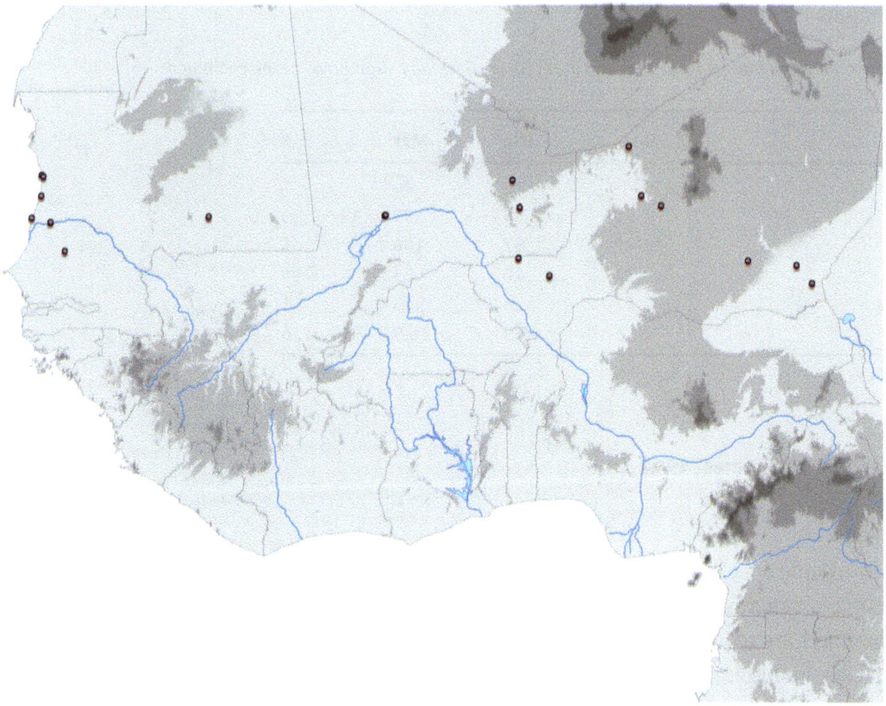

Figure 2.402 Distribution map

Measurements

External and cranial measurements (mm) and mass (g) for *Gerbillus tarabuli*, sexes combined

	Mean	Min	Max	N
Mass	32.6	26	39	11
HB	100.5	93	116	13
Tail	129.2	114	159	12
HF	29.2	27	31	13
Ear	13.5	12.5	15	13
GLS	29.6	27.7	31.8	11

Gerbillus watersi de Winton, 1901

Diagnosis

This is one of eight *Gerbillus* species endemic to north-eastern Africa, which can be distinguished from the other six as follows: *G. brockmani* (size similar; white ventral pelage not sharply delineated from flanks; endemic to northern Somalia); *G. dunni* (soles of hindfeet hairy); *G. pulvinatus* (soles of hindfeet partially hairy; toes with fringe of hairs); *G. harwoodi* (tail relatively shorter, ca. 125% of HB; endemic to southern Kenya and northern Tanzania); *G. acticola* (size larger, GLS > 28 mm, HB > 90 mm); *G. somalicus* (size similar; white ventral pelage not sharply delineated from flanks; endemic to northern Somalia); *G. pusillus* (size slightly smaller, GLS ca. 24 mm, but with significant overlap; tail relatively longer, ca. 155% of HB; occurring widely in East Africa but absent from northern Sudan). This species has hindfeet with naked soles, and the white ventral pelage is clearly delineated from that of the flanks.

Taxonomic Issues

The taxon *juliani* is considered a valid species by Happold (2013), but we follow Musser & Carleton (2005) in including it as a synonym of *G. watersi* pending a taxonomic revision of the group.

Distribution

This species occurs patchily in northern Sudan, Djibouti and Somalia.

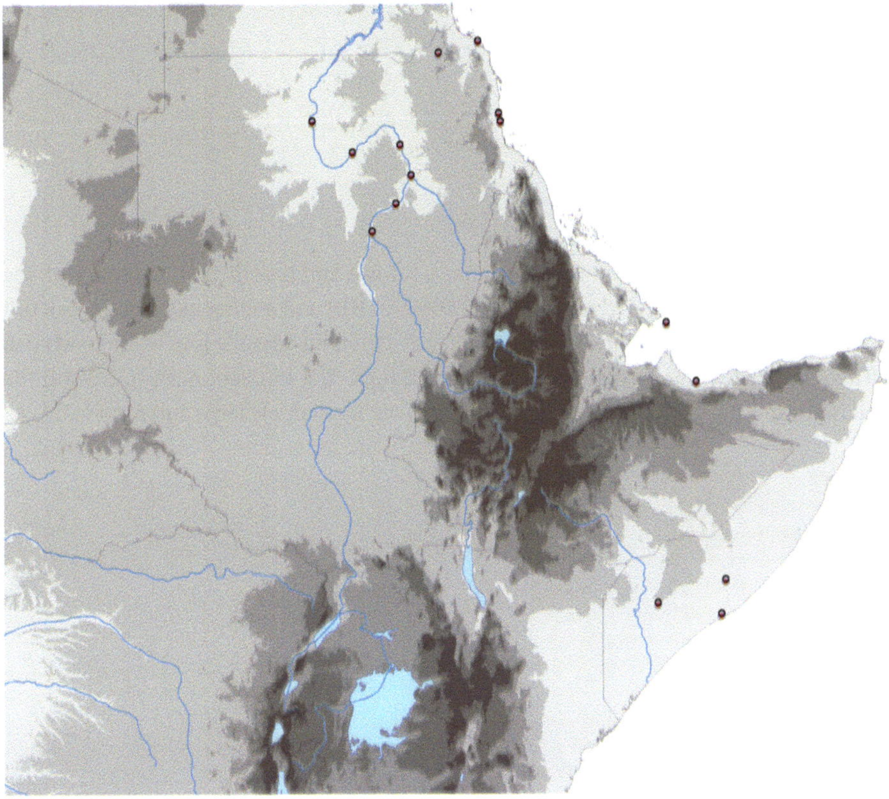

Figure 2.403 Distribution map

Measurements

External and cranial measurements (mm) and mass (g) for *Gerbillus watersi*, sexes combined

	Mean	Min	Max	N
Mass*	17	14	20	5
HB*	81.5	70	96	11
Tail*	108	90	125	11
HF*	20.6	18	23	11
Ear*	11.4	10	13	11
GLS*	25.1	24.2	25.9	9

*Happold (2013)

Genus Account

Meriones Illeger, 1811 (Jirds)

Diagnosis

Jirds are medium-sized (Mass 65-110 g; HB 115-155 mm), rat-like and heavily built rodents with the tail about equal in length to the head and body (ca. 100% of HB) and ending in a small black tuft ("pencil"). The pelage is noticeably long and silky, sandy-coloured dorsally and white to whitish ventrally. There are distinctive postauricular white markings. The hindfeet have soles with the anterior half hairy and the posterior half naked. The ears are moderately large and rounded.

Figure 2.404 Photograph of *Meriones crassus* (Ara Monadjem).

Cranial Characters

The skull is medium to large in size (GSL 36-42 mm), with a broad braincase tapering to a long narrow rostrum. The bullae are greatly inflated with fully inflated ("pneumatised") mastoid chamber which projects beyond the posterior margin of the skull and can be viewed from above. In the west African form, the zygomatic arch is slender and the zygomatic plate not so wide, with the anterior edge straight or slightly sinuous and not reaching the maxilla-premaxilla suture. The anterior palatal foramina are long and reach to the alveolus of M^1, while the posterior palatal foramina reach from the middle of M^2 to the middle of M^1. The skull possesses a roughly triangular area of bone immediately above the auditory opening, termed the suprameatal triangle. This is well developed in this genus but appears also to be present in other gerbil genera such as *Desmodilliscus*, *Microdillus* and *Pachyuromys* and greatly reduced or absent in *Desmodillus*, *Gerbilliscus* and *Gerbillus*.

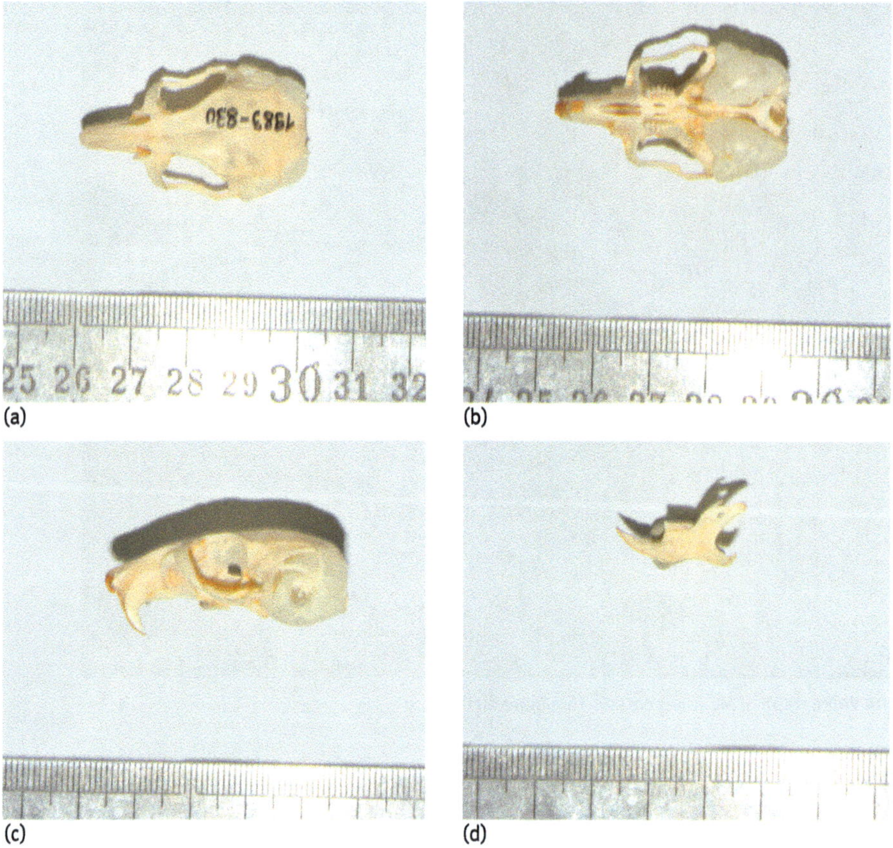

Figure 2.405 a-d Cranium of *Meriones crassus* (MNHN-ZM-MO 1983-830) (from top left to bottom right: dorsal view, ventral view, lateral view and lower jaw).

Dentition

The upper incisors are opisthodont and grooved. The molars have a well-developed, advanced high-crowned "prismatic" shape (Fig 4 of Pavlinov, 2008), termed "lozenge-shaped" by Rosevear (1969), whereby cusps are completely merged transversely and also connected by longitudinal bridges between laminae. The M^3 is greatly reduced relative to M^1 and M^2.

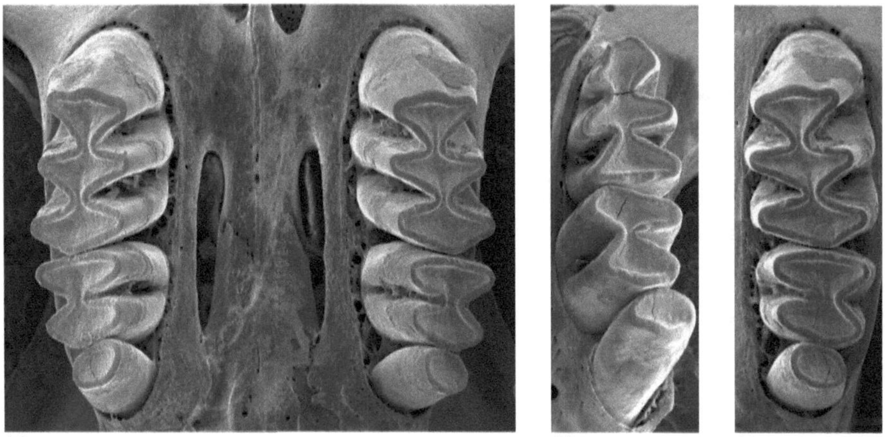
(a) (b) (c)

Figure 2.406 Cheek teeth of *Meriones crassus* (MNHN-ZM-MO 1987-326) (A: upper cheek teeth; B: left toothrow; C: upper left toothrow).

Habitat & Biogeography

Although 16 species are known from North Africa, the Middle East, the Indian subcontinent, central Asia and China, only one species, *M. crassus* occurs in the Sub-Saharan region where it has been recorded from sparsely vegetated semi-arid and desert habitats in Mauritania and Sudan (Setzer 1956).

Fossil Record

Fossils are known from the Early to Late Pleistocene of North Africa (Tong, 1989; Wessels, 1999; Winkler et al., 2010). Pavlinov (2008) discusses the extinct genera, *Pseudomeriones* and *Epimeriones*, indicating that the similarity with *Meriones* may be just superficial.

Biology & Ecology

They are nocturnal and terrestrial, excavating large and complex burrows (Petter, 1961).

Systematic Notes

The genus is placed in tribe Gerbillini and subtribe Rhombomyina, along with *Psammomys*, *Brachiones*, *Rhombomys* and *Sekeetamys* (Pavlinov et al., 1990; Musser & Carleton, 2005; Happold, 2013).

Annotated Species List

Only one species occurs in the Sub-Saharan region; 16 species are known in total of which 15 are extralimital to our region:
Meriones crassus Sundevall, 1842

Species Account

Meriones crassus Sundevall, 1842

Diagnosis

This species co-occurs with two other *Meriones* species in North Africa (beyond the scope of this book): *M. libycus* (claws black; 2n = 44); and *M. shawii* (smaller, GLS < 37 mm; 2n = 44).

Taxonomic Issues

Karyotype is 2n = 60, FN = 70-74 (Matthey, 1957).

Distribution

This species occurs marginally in the very northern parts of Sub-Saharan Africa typically in desert and arid habitats.

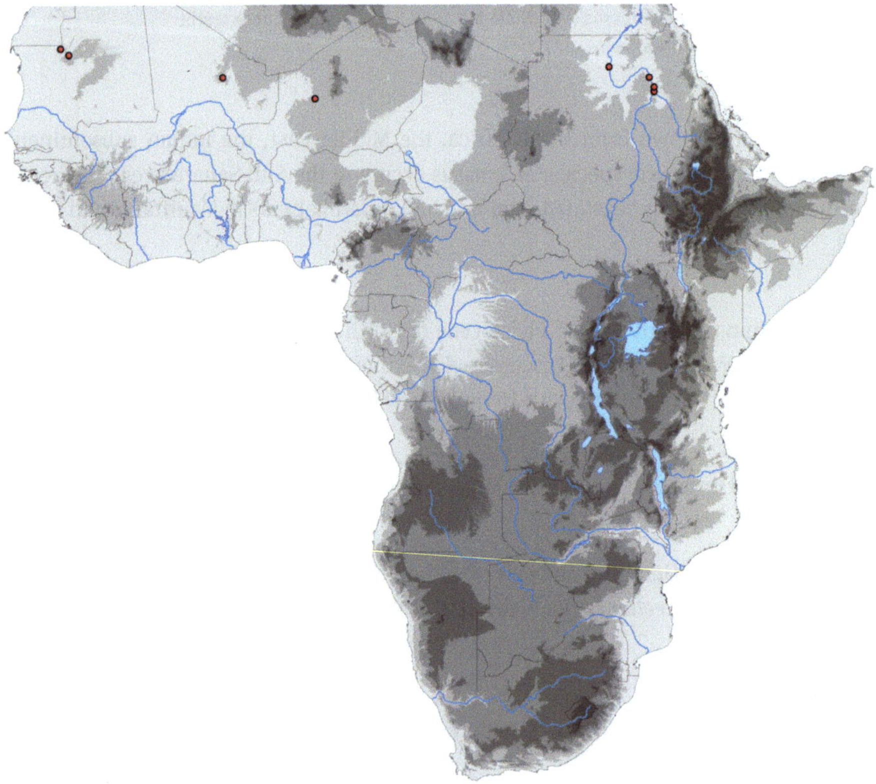

Figure 2.407 Distribution map

Measurements

External and cranial measurements (mm) and mass (g) for *Meriones crassus*, sexes combined

	Mean	Min	Max	N
Mass*	84.0	65	109	32
HB*	136.0	114	153	47
Tail*	137.0	105	158	40
HF*	34.0	31	37	48
Ear*	19.0	14	22	46
GLS*	39.1	36.6	42.2	26

*Happold (2013)

Genus Account

Microdillus Thomas, 1910 (Somali Pygmy Gerbil)

Diagnosis

This is a very small gerbil (HB 66-80 mm) with a long shaggy pelage, dull orange-brown above, pure white below, extending to the chin, throat, cheeks and nose and with conspicuous white supraorbital and postauricular markings. The tail is shorter (ca. 80% of HB) than the head and body, lacking a terminal tuft ("pencil") and apparently quite swollen and fatty in life, similar in some respects to *Pachyuromys* (Roche & Petter, 1968). The soles of the hindfeet are naked. This gerbil can easily be distinguished in its geographic range (endemic to Somalia) in having a much shorter tail than similar-sized gerbils (except for *Desmodilliscus braueri*), as well as in cranial and dental features (see below).

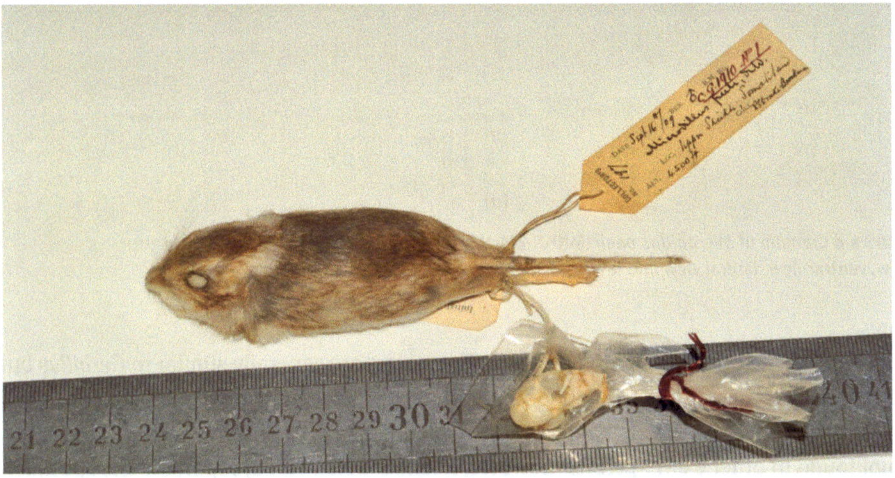

Figure 2.408 Photograph of *Microdillus peeli* museum specimen (MNHN-ZM-MO 1910-1) (Ara Monadjem).

Cranial Characters

The skull is small (ca. 24-25 mm), diagnostically strongly bowed in lateral profile, with a narrow interorbital region, narrow and short nasals and highly enlarged bullae including the mastoid chamber which extends beyond the occiput and is visible from above (with exception of a small individual from central Somalia examined by Roche & Petter (1968) where the mastoid chamber was apparently not visible dorsally). The zygomatic plate is relatively narrow compared to other gerbils and has a vertical anterior edge. Both anterior and palatal foramina are very well developed (the latter is almost as long as the upper toothrow) and separated by only a narrow bony bridge.

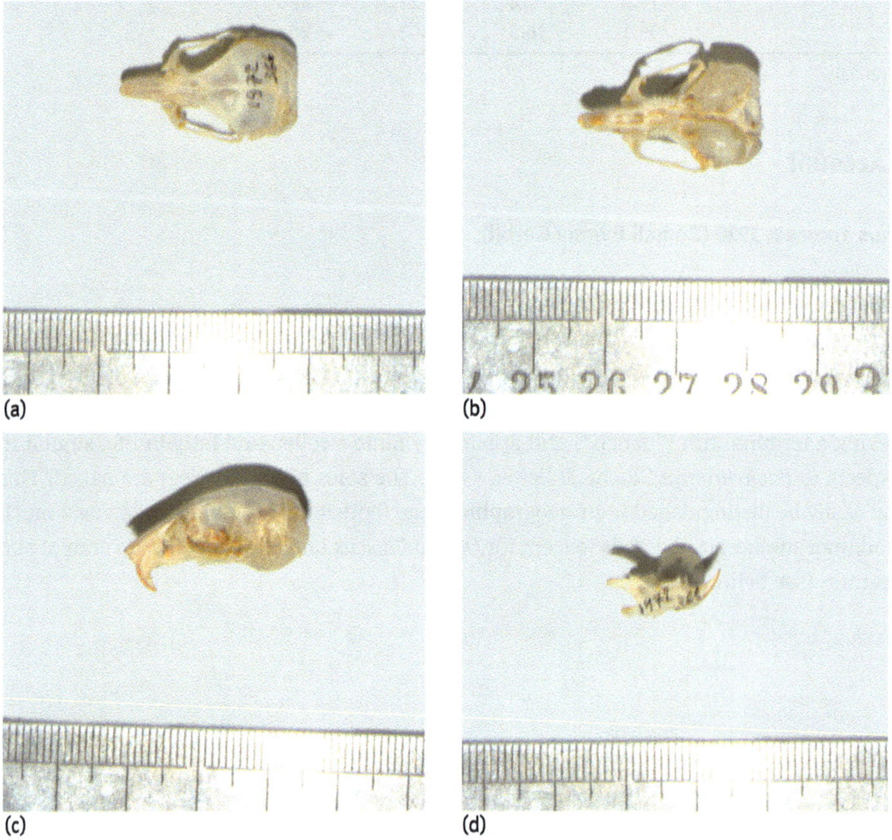

Figure 2.409 a-d Cranium of *Microdillus peeli* (MNHN-ZM-MO 1972-366) (from top left to bottom right: dorsal view, ventral view, lateral view and lower jaw).

Dentition

The upper incisors are opisthodont and grooved. The molars are generally similar to *Gerbillus* but more flattened. They are more or less lophodont, although M^1 may have longitudinal connections between laminae (Roche & Petter, 1968). The M^3 typically has four cusps, which is a primitive feature not found in other extant gerbils (however, Roche & Petter (1968) report on one specimen that had three rather than four cusps). The lower molars display a strong anteroconid connected to the first row of cusps by a median longitudinal crest. The M_2 and M_3 display oblique well fused rows of cusps.

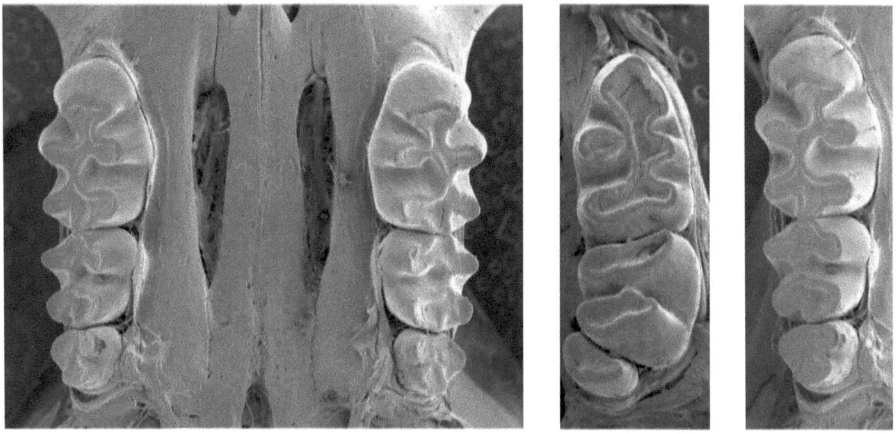

(a) (b) (c)

Figure 2.410 Cheek teeth of *Microdillus peeli* (MNHN-ZM-MO 1977-366) (A: upper cheek teeth; B: lower right toothrow; C: upper right toothrow).

Habitat & Biogeography

They are endemic to the Horn of Africa where they have been recorded from northern and central Somalia, and the extreme western parts of Ethiopia. The species is known from four sites in both sandy and hilly habitats.

Fossil Record

No fossils are known.

Biology & Ecology

Almost nothing is known of these rare gerbils.

Systematic Notes

The genus has been placed in subtribe Gerbillina of tribe Gerbillini and is closely related to *Gerbillus* (including *Dipodillus*) (Pavlinov et al., 1990; Musser & Carleton, 2005).

Annotated Species List

Only one species is known:
Microdillus peeli (de Winton, 1898)

Species Account

Microdillus peeli (de Winton, 1898)

Diagnosis

See genus account.

Taxonomic Issues

See genus account.

Distribution

This species is restricted to the Horn of Africa where it is known from Somalia and northern Ethiopia.

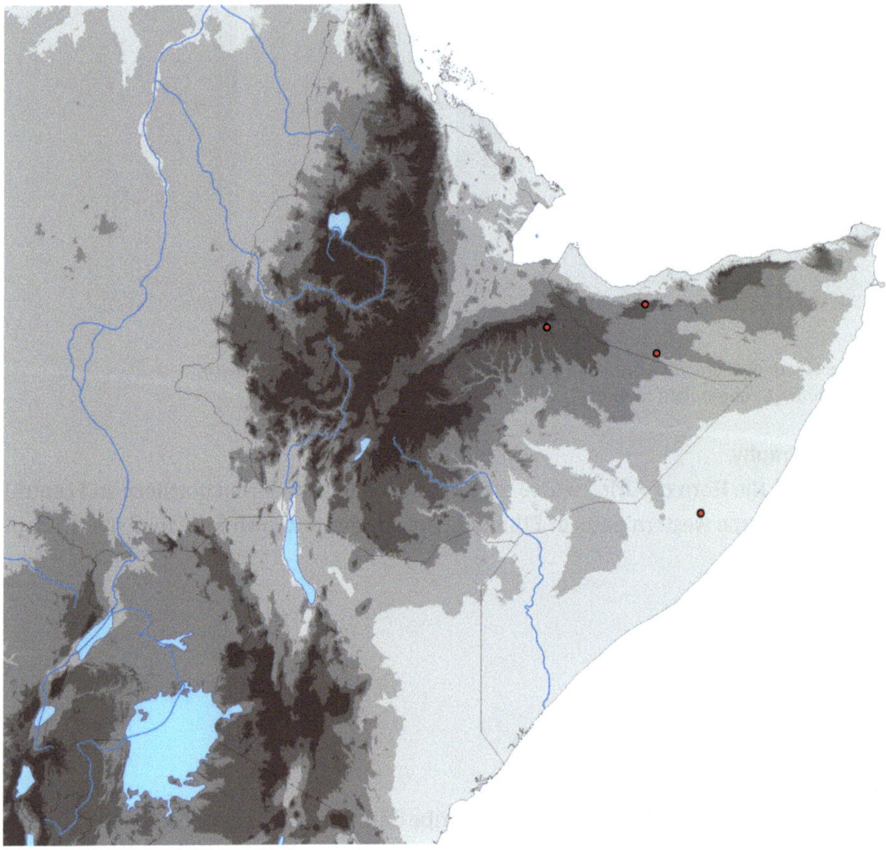

Figure 2.411 Distribution map

Measurements

External and cranial measurements (mm) and mass (g) for *Microdillus peeli*, sexes combined

	Mean	Min	Max	N
Mass	13.5	13	14	2
HB	72.0	71	73	3
Tail	59.3	57	61	3
HF	18.0	-	-	1
Ear	10.7	10	11	3
GLS	24.8	24.1	25.3	3

Genus Account

Pachyuromys Lataste, 1880 (Fat-tailed Jird)

Diagnosis

With its conspicuous short fat, flesh-coloured tail and shortened ears and limbs, and very inflated auditory bullae, this genus is one of the gerbils most morphologically specialised to a desert environment. These small rodents (Mass ca. 20-45 g; HB ca. 90-120 mm) have a long shaggy pelage, beige or sandy coloured above, pure white below extending to the chin, throat, muzzle and lips. The tail is diagnostically much shorter (ca. 55% of HB) than the head and body, and is thick, naked and club-shaped. The soles of the hindfeet are naked.

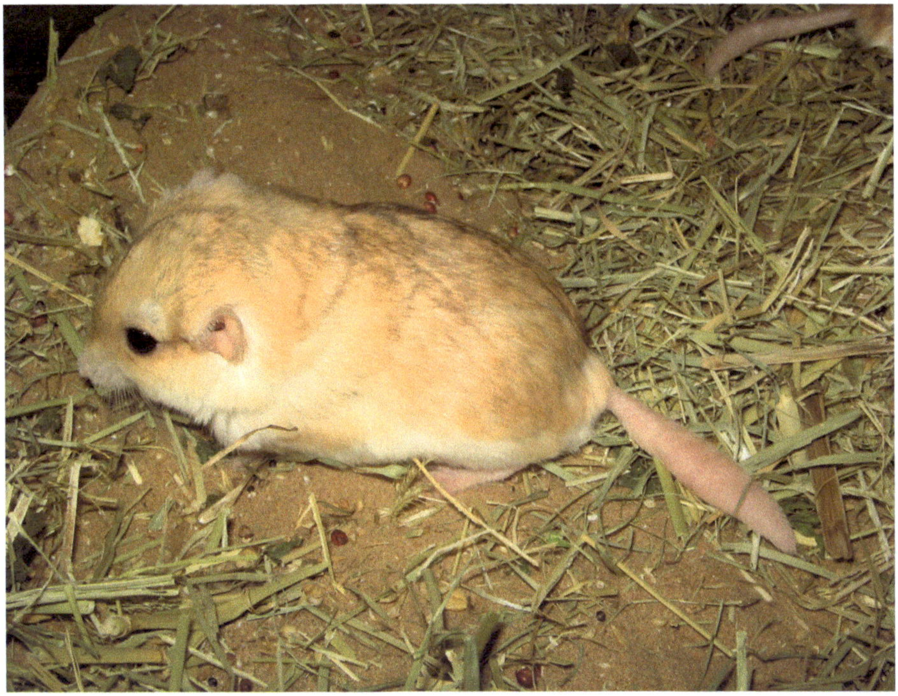

Figure 2.412 Photograph of *Pachyuromys duprasi* (Ara Monadjem).

Cranial Characters

The skull is medium-sized (ca. 32-37 mm), with a broad braincase and very inflated bullae (comprising 47% of GSL, larger than in any other gerbils) including the mastoid chamber which extends beyond the occiput and is visible from above. The zygomatic arch is quite slender and the zygomatic plate quite wide with a convex anterior edge. The anterior palatal foramina reach to just short of the alveolus of M^1 while the posterior palatal foramina extend from about the posterior margin of M^2 to the middle of M^1. The suprameatal triangle is well developed, as in *Meriones*.

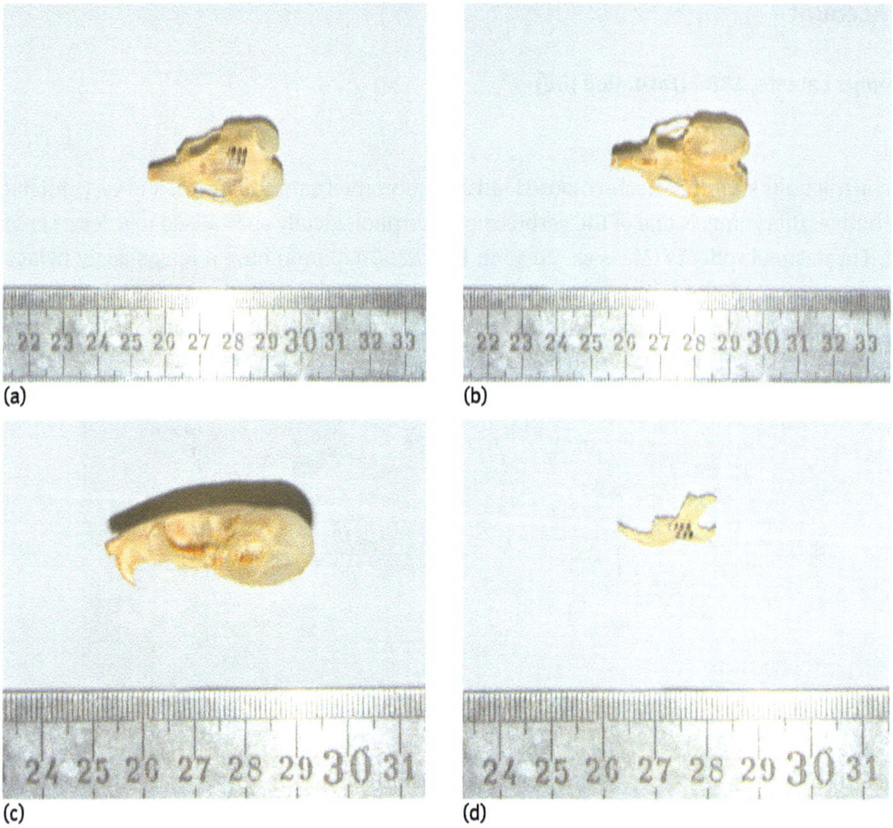

Figure 2.413 a-d Cranium of *Pachyuromys duprasi* (MNHN-ZM-MO 1958-239) (from top left to bottom right: dorsal view, ventral view, lateral view and lower jaw).

Dentition

The upper incisors are faintly grooved. The molar crown has been described as both "advanced bunodont" and "semi-lophodont" (as found also in *Gerbillurus* and *Desmodillus*), with transverse laminae forming early in life and no accessory cusplets but they retain traces of longitudinal crests. On the M_1 the anteroconid is connected to the first lobe.

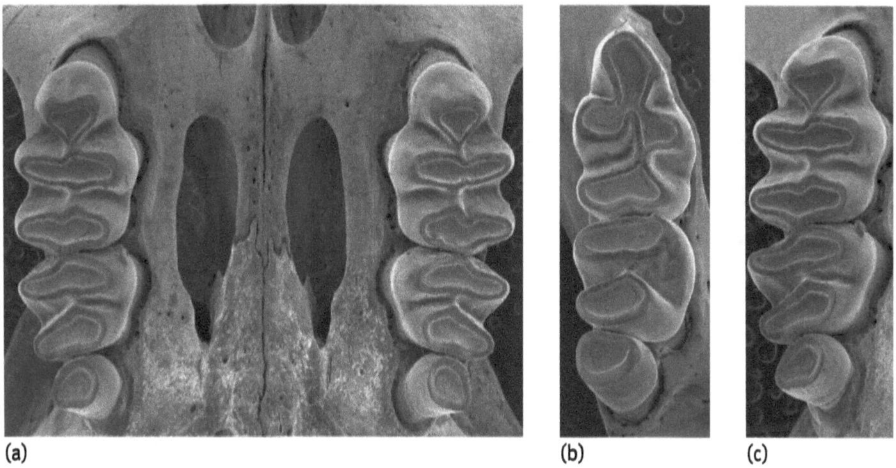

Figure 2.414 Cheek teeth of *Pachyuromys duprasi* (MNHN-ZM-MO 1990-113) (A: upper cheek teeth; B: lower right toothrow; C: upper right toothrow).

Habitat & Biogeography
They occur in “hamadas” (hard plains with pebbles and boulders and sparse vegetation) and dry river beds in the Sahara desert, but only isolated records from Mauritania and Mali are known from the Sub-Saharan region as defined by this book.

Fossil Record
No fossils are known.

Biology & Ecology
They are mainly herbivorous but may also to be omnivorous. The tail is used to store fat and they may enter into torpor below 32-35°C in captivity (Petter, 1961). Litter size in captivity is 3-9. Seasonality of breeding is unknown in the wild (Happold et al., 2013).

Systematic Notes
The genus is placed in tribe Gerbillini and subtribe Pachyuromyina (Pavlinov et al., 1990; Musser & Carleton, 2005). According to mitochondrial DNA sequences (Chevret & Dobigny, 2005), *Pachyuromys* and *Desmodilliscus* form a well supported clade at the base of the gerbil tree.

Annotated Species List
Only one species is known:
Pachyuromys duprasi Lataste, 1880

Species Account

Pachyuromys duprasi Lataste, 1880

Diagnosis
See genus account.

Taxonomic Issues

Karyotype is 2n = 54 (Qumsiyeh & Schlitter, 1991).

Distribution

Within Sub-Saharan Africa, this species has a restricted and patchy distribution in Mauritania and Mali, but is more widespread in other parts of the Saharan.

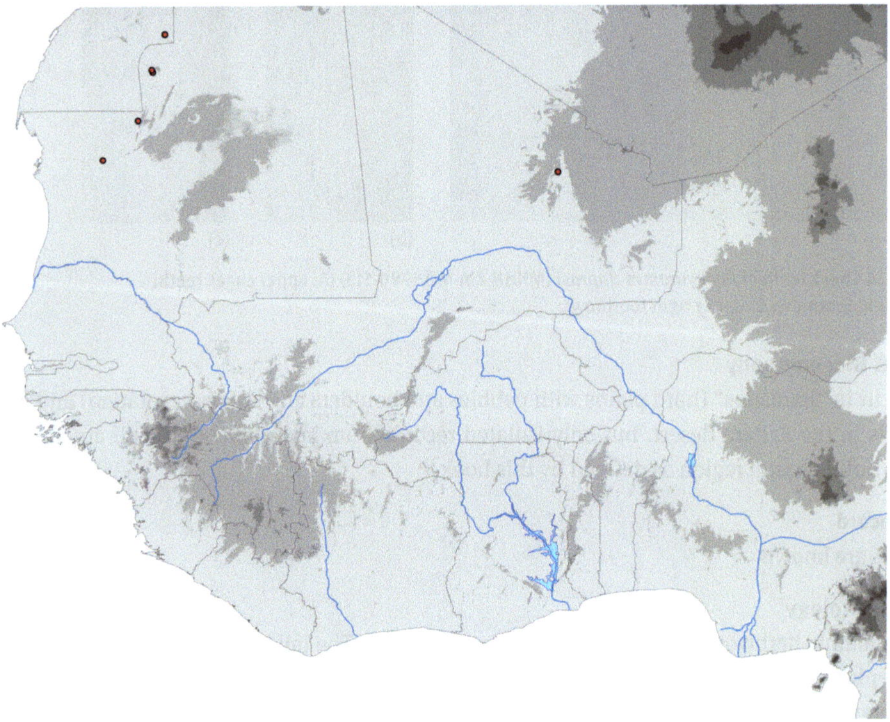

Figure 2.415 Distribution map

Measurements

External and cranial measurements (mm) and mass (g) for *Pachyuromys duprasi*, sexes combined

	Mean	Min	Max	N
Mass*	36.5	22	45	3
HB*	108.3	93	121	4
Tail*	58.2	55	62	4
HF*	23.3	22	24	4
Ear*	14.0	12	16	4
GLS*	24.9	32.4	36.5	4

*Happold (2013)

Genus Account

Psammomys Cretzschmar, 1828 (Sand Rats)

Diagnosis

They are heavily-built, rat-like gerbils (Mass ca. 80-240 g; HB ca. 115-185 mm) with a long pelage that is sandy-coloured dorsally and pale ochre ventrally and on the flanks. There are white postauricular markings. The tail is thick, hairy, shorter (ca. 70% of HB) than the head and body and terminates in a black tuft ("pencil"). The hindfoot soles are partly furred. They have four pairs of nipples (2 + 2 = 8).

Figure 2.416 Photograph of *Psammomys obesus* (Ara Monadjem).

Cranial Characters

The skull is medium-sized (ca. 37-43 mm) and robust in shape, with a wide and rather flat braincase and highly inflated bullae including a fully inflated mastoid chamber which extends beyond the occiput and is visible when viewed from above. The braincase and interorbital region bears well developed supraorbital and supraoccipital ridges. The anterior and posterior palatal foramina are reduced in length and slit-like, the anterior ending far short of the M^1 and the posterior extending from about the first lamina of M^2 to the second lamina of M^1.

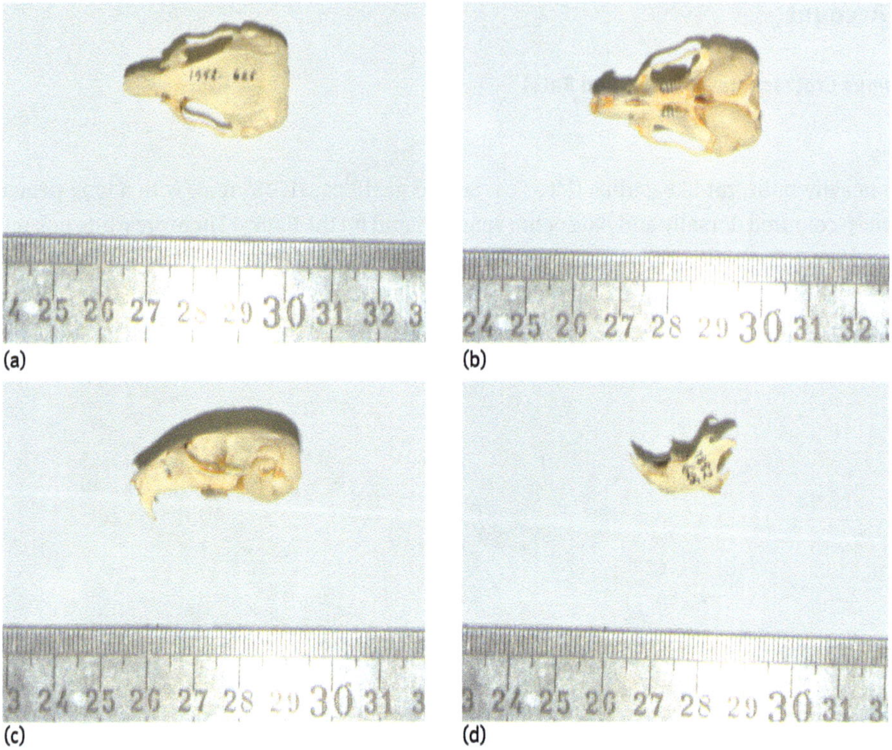

Figure 2.417 a-d Cranium of *Psammomys obesus* (MNHN-ZM-MO 1952-625) (from top left to bottom right: dorsal view, ventral view, lateral view and lower jaw).

Dentition

The upper incisors are ungrooved or very faintly grooved. The molars are high crowned (hypsodont) and arranged in diamond-shaped laminae in a symmetrical "prismatic" or "lozenge-shaped" formation (as in *Meriones*), with longitudinal connections between laminae.

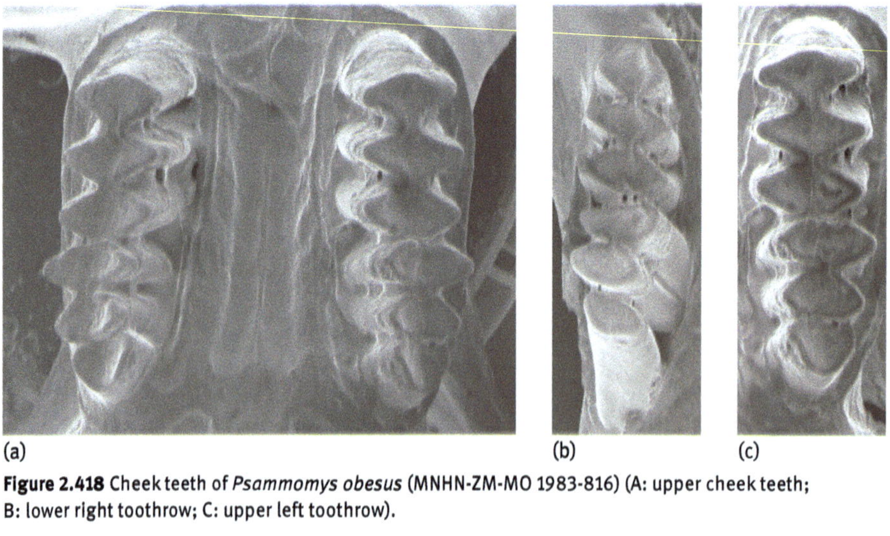

Figure 2.418 Cheek teeth of *Psammomys obesus* (MNHN-ZM-MO 1983-816) (A: upper cheek teeth; B: lower right toothrow; C: upper left toothrow).

Habitat & Biogeography

They are closely associated with wadis, salt-flats and coastal lowlands within the Sahara Desert having damp and saline soils with presence of salt bushes (Chenopodiaceae). They occur mostly in the northern Sahara and only marginally extending into Sub-Saharan Africa in Mauritania and Sudan.

Fossil Record

The extinct genus *Parameriones* described from the Pleistocene of the Levant as being close to *Meriones* is included by Pavlinov (2008) in *Psammomys*.

Biology & Ecology

They are terrestrial and both diurnal and nocturnal. Sand rats are herbivorous, feeding predominantly on succulent saltbushes, for which their kidneys are highly adapted (Daly & Daly, 1973, 1974). Breeding is seasonal, with female reproductive condition being dependent on rainfall and the number of rain days in the preceding month, and litter size is 2-8 (Fichet-Calvet et al., 1999). These rodents have great importance to humans. Since they develop diabetes on a standard laboratory diet, they have become a model for studying diabetes and obesity in humans (e.g., see Kalderon et al., 1986). They are also the main reservoir for leishmaniasis (Fichet-Calvet et al., 2000).

Systematic Notes

The genus has been assigned to tribe Gerbillini and subtribe Rhombomyina, together with *Rhombomys*, *Meriones* and *Sekeetamys* (Musser & Carleton, 2005, based on Pavlinov et al., 1990). Morphological (Pavlinov et al., 1990; Tong, 1989), chromosomal (Bennazou et al., 1984), and mitochondrial DNA sequences, (Chevret & Dobigny, 2005) suggested a close relationship with *Rhombomys* and/or *Meriones*.

Annotated Species List

Although two species are recognised from North Africa, only one species is recorded from Sub-Saharan Africa:
Psammomys obesus Cretzschmar, 1828

Species Account

Psammomys obesus Cretzschmar, 1828

Diagnosis

This is the only species of *Psammomys* occurring in the Sub-Saharan region. Extralimitally, it co-occurs with *P. vexillaris* (smaller, GLS < 37 mm) in North Africa.

Taxonomic Issues

Karyotype is 2n = 48, FN = 74-78 (Qumsiyeh & Schlitter, 1991).

Distribution

This species occurs in deserts extending marginally into Sub-Saharan Africa in the east (Sudan) and the west (Mauritania).

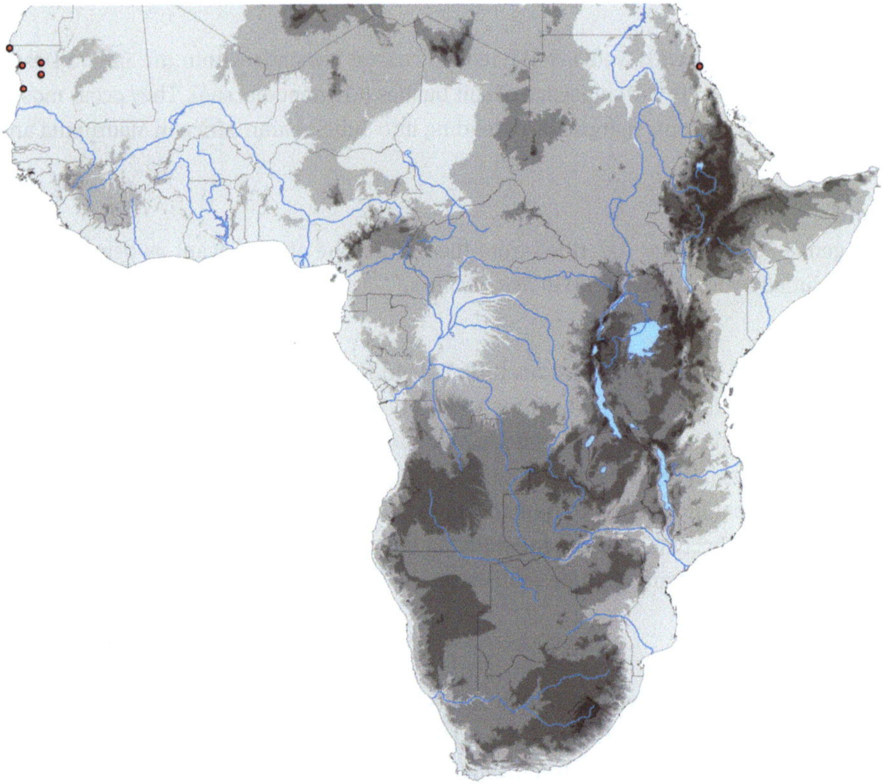

Figure 2.419 Distribution map

Measurements

External and cranial measurements (mm) and mass (g) for *Psammomys obesus*, sexes combined

	Mean	Min	Max	N
Mass*	150.2	82	237	429
HB*	159.3	116	185	428
Tail*	116.2	88	140	406
HF*	34.3	32	36	103
Ear*	15.2	13	17	83
GLS*	40.6	36.9	43.4	99

*Happold (2013)

Genus Account

Taterillus Thomas, 1910 (Taterils or Gerbils)

Diagnosis

They are small to medium-sized gerbils (Mass ca. 20-80 g; HB ca. 90-170 mm) with a pointed muzzle, large eyes, elongated ears and a pale yellow to reddish-brown dorsal pelage, white ventral pelage, cheeks, hands and feet, and white supraorbital, postorbital and occasionally postauricular markings. The tail is much longer (ca. 125-145 % of HB) than the head and body, well-haired throughout its length with a long terminal tuft ("pencil") of hairs. The soles of the elongated hindfeet are usually naked (although some may have a transverse narrow band of hairs in the ankle region). Females have four pairs of nipples (2 + 2 = 8). These rodents are very similar in appearance to *Gerbilliscus* but have a slighter build, longer tail and longer posterior palatal foramina plus a triangular anteroconid on the M_1 (see below).

Figure 2.420 Photograph of *Taterillus congicus* (Christiane Denys).

Cranial Characters

The skull is medium-sized (ca. 28-35 mm) with a rounded braincase, very long and narrow, tapering nasals and inflated bullae but with the mastoid chamber not extending beyond the occiput and hence not visible when viewed from above. The interorbital region bears well developed supraorbital ridges. The zygomatic arch is gracile and the zygomatic plate is very wide, extending beyond the maxillary-premaxillary suture and having a convex anterior edge. The anterior palatal foramina end far short of the M^1 but the narrow, slit-like posterior foramina are long, extending from the end of M^2 to the front of M^1, i.e. almost as long as the upper molar row.

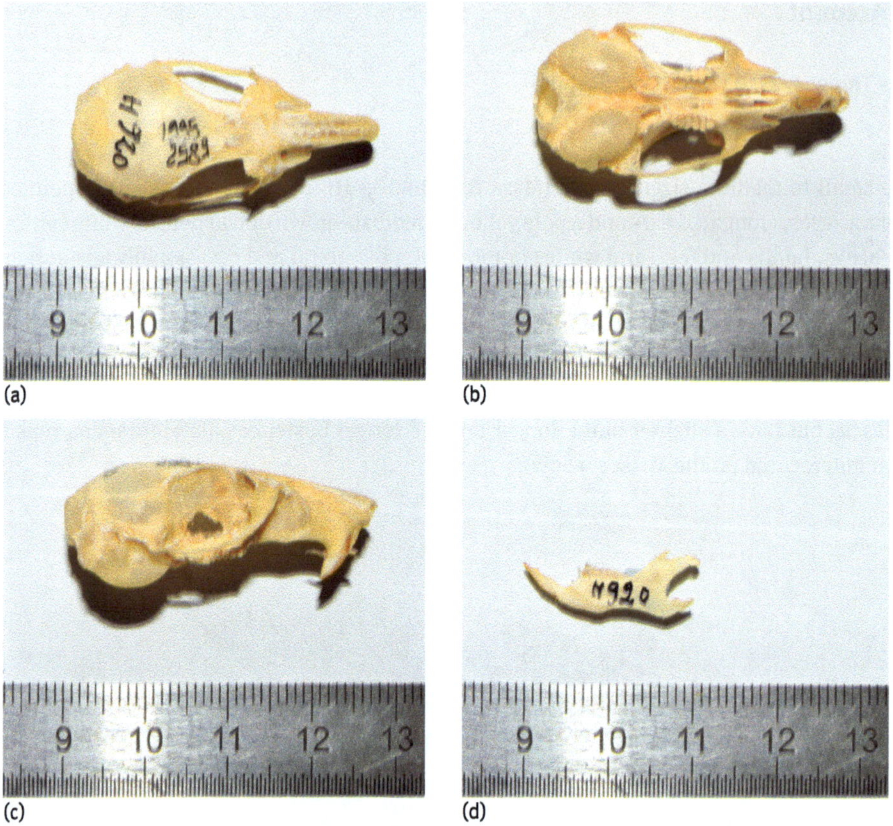

Figure 2.421 a-d Cranium of *Taterillus gracilis* (MNHN-ZM-MO 1995-2589) (from top left to bottom right: dorsal view, ventral view, lateral view and lower jaw).

Dentition

The upper incisors are opisthodont and strongly grooved. The molars are arranged in laminae in a symmetrical lophodont formation (as in *Gerbilliscus*). The only difference with *Gerbilliscus* is the shape of the anteroconid of the M_1 which is triangular in unworn or slightly worn molars, becoming rhomboid in worn teeth, whereas it is rounded anteriorly in *Gerbilliscus*.

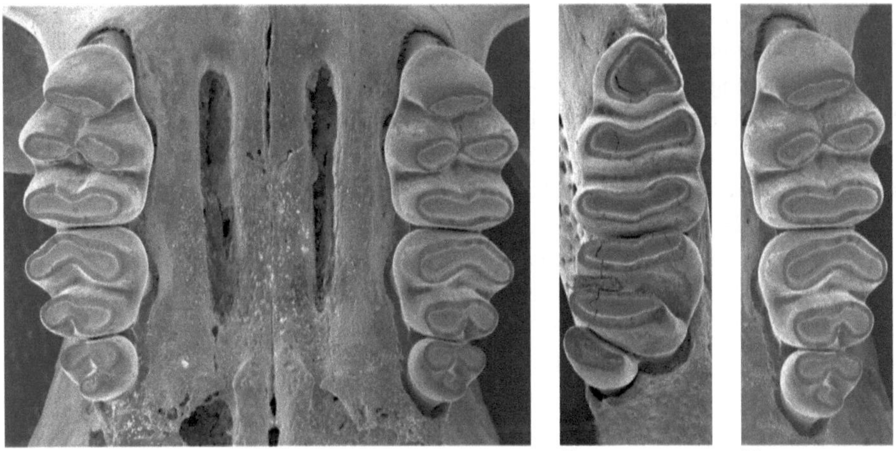

(a) (b) (c)

Figure 2.422 Cheek teeth of *Taterillus lacustris* (MNHN-ZM-MO 1973-786) (A: upper cheek teeth; B: lower right toothrow; C: upper left toothrow).

Habitat & Biogeography

The distribution range of this genus is closely associated with the northern Sudan, Guinea and Sahel savannas, extending from Mauritania in the west to Somalia in the east and south to central Kenya. They occur on sandy to sandy clay soils with the vegetation ranging from thorn scrub to wooded savannas and they are also often found on fallow lands and the edges of crop fields, even occurring in houses (Happold, 2013). Breeding is seasonal but the length of the breeding season is very variable depending on rainfall. Litter sizes in captivity are ca. 2-6 but the range of embryo numbers is broader (1-8) (Happold, 2013 and references therein).

Fossil Record

Taterillus has been reported from Makapansgat in northern South Africa (Pocock, 1987) and north-western Botswana (Pickford and Mein, 1988). It seems more probable that these records are misidentified *Gerbillurus* or *Desmodillus* (Winkler et al., 2010).

Biology & Ecology

They are nocturnal, terrestrial, excavate burrows and feed on seeds, leaves, stems and insects (see Granjon & Duplantier, 2009 for a recent review).

Systematic Notes

The genus has been assigned to tribe Taterillini and subtribe Gerbillurina, together with *Tatera*, *Gerbilliscus* and *Desmodillus* (Musser & Carleton, 2005, based on Pavlinov et al., 1990). Whilst morphological (Pavlinov et al., 1990; Tong, 1989) and chromosomal data (Bennazou et al., 1984) support this association, allozyme (Bennazou et al., 1984), DNA hybridisation (Chevret, 1994) and mitochondrial DNA sequences (Chevret & Dobigny, 2005) suggest an association with members of the tribe Gerbillini and in particular the subtribes, Gerbillina (*Gerbillus*) and Rhombomyina (*Rhombomys*, *Sekeetamys*, *Meriones* and *Psammomys*).

Species of *Taterillus* cannot be distinguished on morphological or molecular grounds and have been described and diagnosed purely on chromosomal grounds (Tranier et al., 1973; Robbins,

1974; Sicard et al., 1988; Dobigny et al., 2002a,b,c, 2003, 2005, 2011). Species of *Taterillus* in west Africa (*T. arenarius*, *T. gracilis*, *T. lacustris*, *T. petteri*, *T. pygargus*, and *T. tranieri*) comprise a monophyletic complex of sibling species defined by a unique autosome-gonosome translocation, which has not been recorded in species from Central and East Africa (*T. congicus* and *T. emini*; Dobigny et al., 2005; 2011). The East African species *T. harringtoni* is recognised by Musser & Carleton (2005) but included as a synonym of *T. emini* by Happold (2013) since the two have the same karyotype. Dobigny et al. (2002c) discovered two additional unnamed cryptic species from west Africa (Lake Chad area) on chromosomal grounds, but later (Dobigny et al., 2011) assigned one of these to *T. lacustris*, at the same time correcting Tranier et al.'s (1973) karyotypic interpretation of *T. lacustris*. A reinterpretation of the karyotype of *T. lacustris* showed that it in fact possesses the same gonosome-autosome translocation found in all other West African *Taterillus* (Dobigny et al., 2011). Dobigny et al. (2003) recently named a new species from Mali and Mauritania, *T. tranieri*.

Annotated Species List

Eight species are recognised:
Taterillus arenarius Robbins, 1974
Taterillus congicus Thomas, 1915
Taterillus emini (Thomas, 1892)
Taterillus gracilis (Thomas, 1892)
Taterillus lacustris (Thomas & Wroughton, 1907)
Taterillus petteri Sicard, Tranier & Gautun, 1988
Taterillus pygargus (F. Cuvier, 1838)
Taterillus tranieri Dobigny, Granjon, Aniskin, Ba & Volobouev, 2003

Key to Species

Due to considerable overlap in external and cranial characteristics, the eight species of the genus *Taterillus* can only be unambiguously distinguished by their karyotype (see Dobigny et al., 2003), and therefore a key to the genus is not presented here. Instead, we tabulate the number of chromosomes for each of the species (taken from Matthey & Jotterand, 1972; Tranier et al., 1973; Dobigny et al., 2005) which is currently the only reliable way to differentiate between these eight species.

Species	Chromosome number
Taterillus arenarius	2n = 30-31
Taterillus congicus	2n = 54
Taterillus emini	2n = 44
Taterillus gracilis	2n = 37-39
Taterillus lacustris	2n = 28
Taterillus petteri	2n = 18-19
Taterillus pygargus	2n = 22-23
Taterillus tranieri	2n = 14-15

Species Accounts

Taterillus arenarius Robbins, 1974

Diagnosis

This species can only be differentiated by its chromosome number. At present it is the only *Taterillus* species recorded from south-western Mauritania. Two species, *T. gracilis* and *T. pygargus* occur close by to the south in Senegal, and *T. tranieri* is known from eastern Mauritania.

Taxonomic Issues

There are putative records from Mali and Niger based on morphological similarities but these require confirmation from karyological analysis (Granjon & Duplantier, 2009).

Distribution

Based on chromosome number, this species is only known from Mauritania where it occurs in arid sandy plains; the putative records from Mali and Niger are not shown on the map.

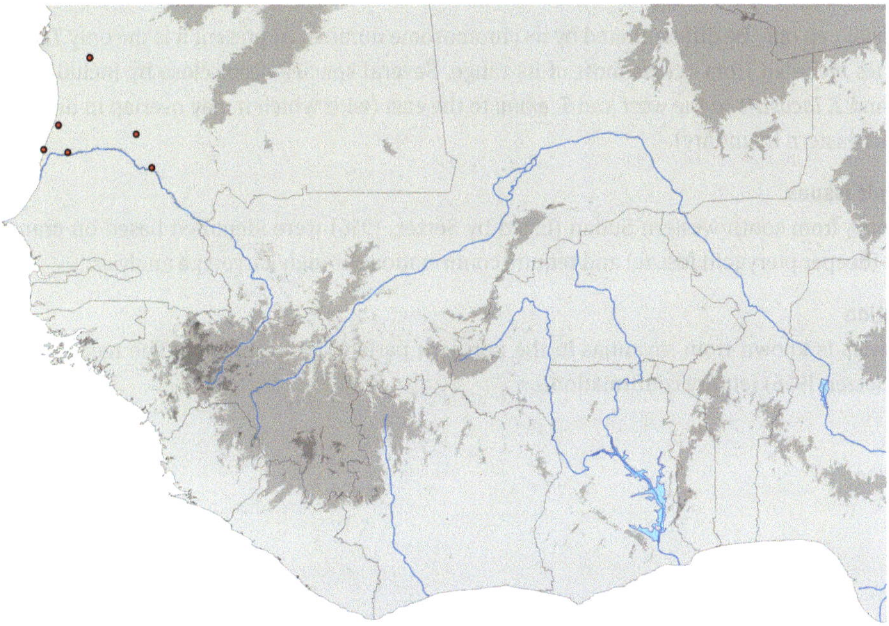

Figure 2.423 Distribution map

Measurements

External and cranial measurements (mm) and mass (g) for *Taterillus arenarius*, sexes combined

	Mean	Min	Max	N
Mass*	42.4	28	66	11
HB*	121.0	110	142	12
Tail*	149.7	141	155	10
HF*	30.6	29	32	11
Ear*	20.8	20	23	13
GLS**	33.7	33.6	33.8	2

*Granjon & Duplantier (2009); **Happold (2013)

Taterillus congicus Thomas, 1915

Diagnosis

This species can only be differentiated by its chromosome number. At present it is the only *Taterillus* species recorded from across most of its range. Several species occur close by including: *T. gracilis* and *T. lacustris* to the west and *T. emini* to the east (with which it may overlap in distribution on its eastern boundary).

Taxonomic Issues

The records from south-western Sudan (listed by Setzer, 1956) were identified based on cranial features (deeper pterygoid fossae) and require confirmation through karyotype analysis.

Distribution

This species is known from savannas in the northern parts of Central Africa. The records from Sudan (Setzer, 1956) require confirmation.

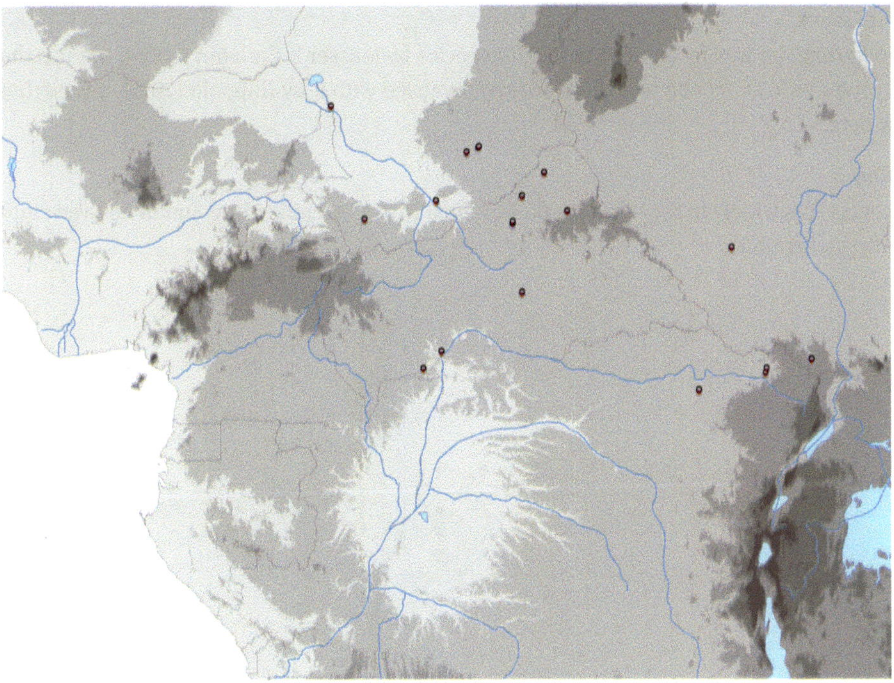

Figure 2.424 Distribution map

Measurements

External and cranial measurements (mm) and mass (g) for *Taterillus congicus*, sexes combined

	Mean	Min	Max	N
Mass*	48.6	32	77	60
HB*	119.9	95	142	87
Tail*	158.7	131	188	56
HF*	30.0	28	32	60
Ear*	19.8	18	23	87
GLS**	33.1	30.2	34.7	18

*Granjon & Duplantier (2009); **Happold (2013)

Taterillus emini (Thomas, 1892)

Diagnosis

This species can only be differentiated by its chromosome number. At present it is the only *Taterillus* species recorded from across most of its range. Its distributional limits to the west are currently unknown, although the record from the Central African Republic is based on a karyotyped specimen (Genest & Petter, 1973). There may be some overlap in its distribution with *T. congicus* from which it can only safely be distinguished by chromosome number.

Taxonomic Issues

The taxon *harringtoni* was recognised as a valid species by Musser & Carleton (2005), but has the same karyotype as *T. emini* and was therefore synonymised with it by Happold (2013); a view that we follow here.

Distribution

This species occurs widely in the northern parts of East Africa extending into Sudan and west into Central African Republic.

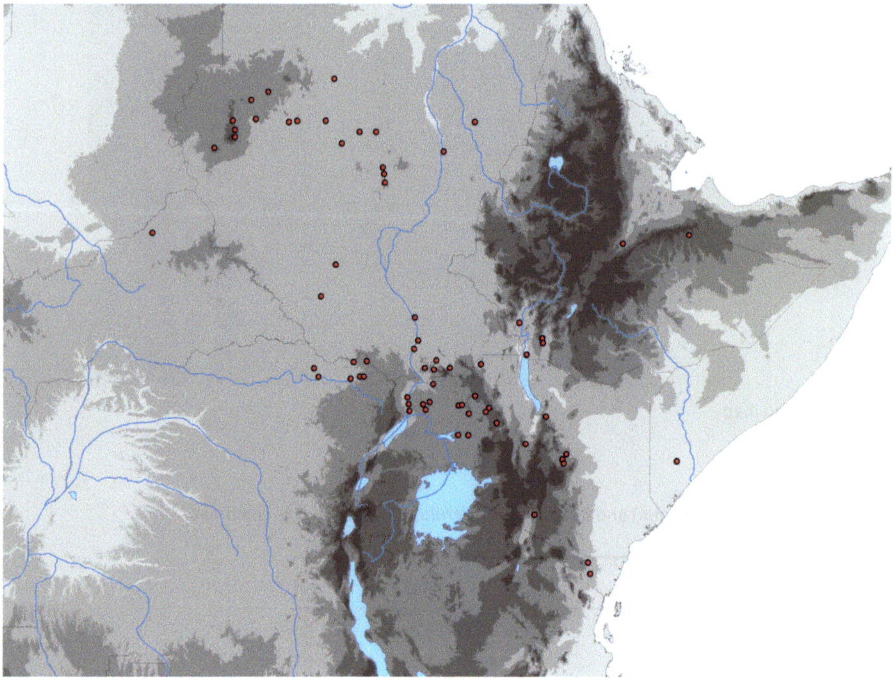

Figure 2.425 Distribution map

Measurements

External and cranial measurements (mm) and mass (g) for *Taterillus emini*, sexes combined

	Mean	Min	Max	N
Mass	-	-	-	-
HB	108.4	78	130	8
Tail	134.3	108	165	8
HF	30.3	27	34	8
Ear	17.8	15	20	7
GLS	34.4	32.6	36.9	4

Taterillus gracilis (Thomas, 1892)

Diagnosis

This species can only be differentiated by its chromosome number. It co-occurs with several other *Taterillus* from which it cannot be distinguished on any external or cranial features (Dobigny et al., 2003, 2005). Other species within its range include: *T. lacustris* (currently only known from the Lake Chad region), *T. petteri* and *T. pygargus*. Immediately to the north is *T. arenarius* and to its east *T. congicus*.

Taxonomic Issues

Most of the specimens of *T. gracilis* examined have not been karyotyped and therefore may represent misidentifications that refer to one of the other cryptic species of *Taterillus*.

Distribution

This species is widely distributed in West African savannas, but due to morphological confusion with other *Taterillus* species, many of these records may refer to another species.

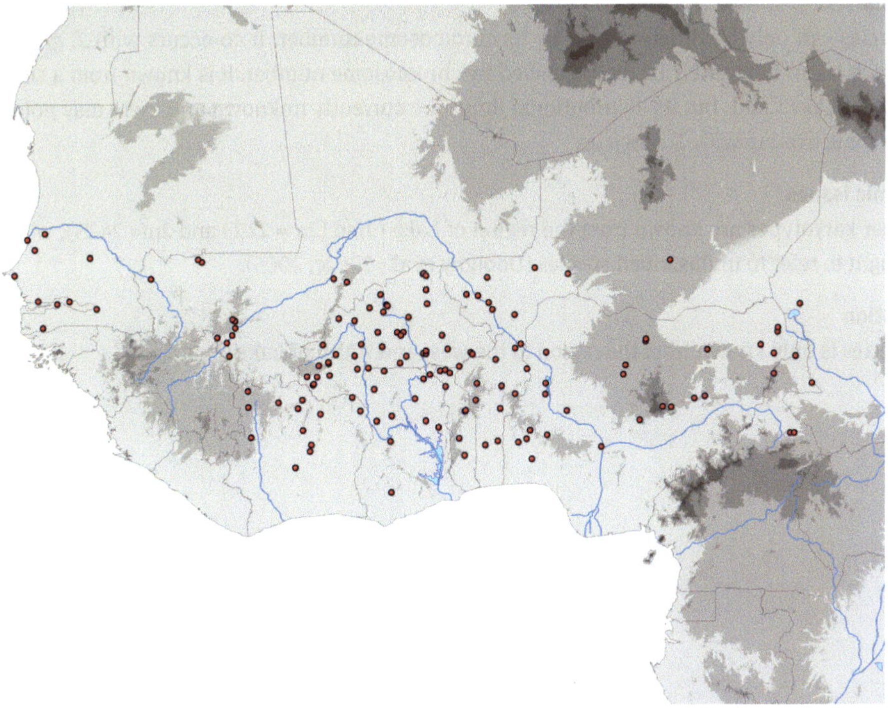

Figure 2.426 Distribution map

Measurements

External and cranial measurements (mm) and mass (g) for *Taterillus gracilis*, sexes combined

	Mean	Min	Max	N
Mass	54.8	26	74	13
HB	118.6	95	134	20
Tail	149.2	132	164	17
HF	31.0	28	34	20
Ear	20.0	16	24	20
GLS*	34.5	32.0	36.9	18

*Happold (2013) presents a mean of 31.9 mm (n = 32; range: 29.7-33.8)

Taterillus lacustris (Thomas & Wroughton, 1907)

Diagnosis
This species can only be differentiated by its chromosome number. It co-occurs with *T. gracilis* from which it can only safely be distinguished by chromosome number. It is known from a small area around Lake Chad, but its distributional limits are currently unknown and there may potentially be some overlap with *T. congicus*.

Taxonomic Issues
Two other karyotypes are known from the region of Lake Chad (2n = 22-23 and 2n = 24-25), which are thought to refer to undescribed species (Dobigny et al., 2002c, 2005).

Distribution
This species is only known from the region to the west and south of Lake Chad.

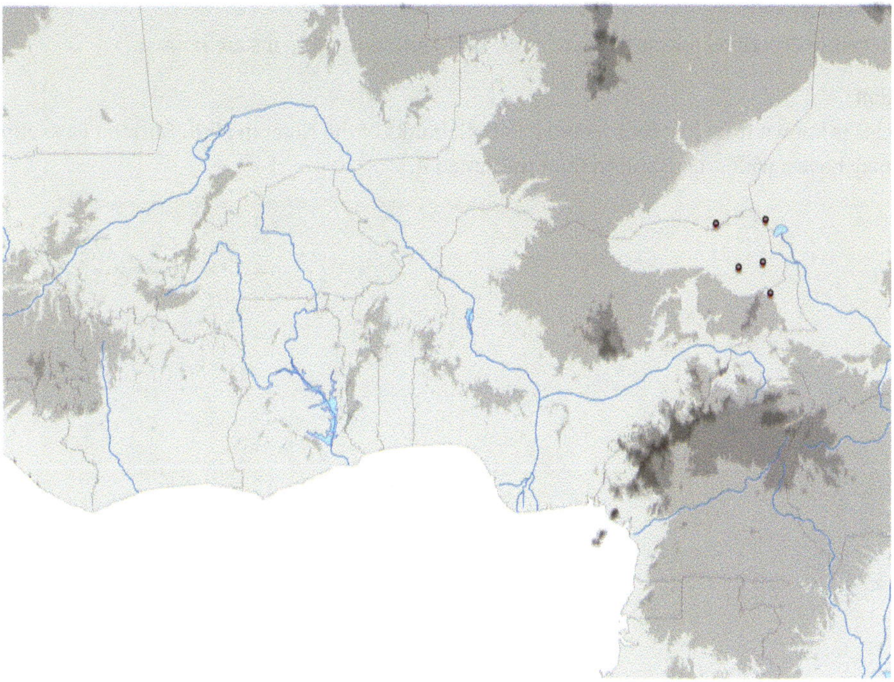

Figure 2.427 Distribution map

Measurements

External and cranial measurements (mm) and mass (g) for *Taterillus lacustris*, sexes combined

	Mean	Min	Max	N
Mass*	45.7	43	48	3
HB*	116.0	105	122	5
Tail*	151.0	145	162	5
HF*	31.2	30	32	4
Ear*	20.0	16	22	5
GLS**	33.8	29.7	38.2	4

*Happold (2013); **Happold (2013) plus two specimens measured by the authors

Taterillus petteri Sicard, Tranier & Gautun, 1988

Diagnosis

This species can only be differentiated by its chromosome number. It co-occurs with *T. gracilis* and possibly with *T. tranieri* from which it can only safely be distinguished by chromosome number. It is known from both banks of the Niger River, but its distributional limits are currently unknown and there may potentially be some overlap with other *Taterillus* species.

Taxonomic Issues

This is a recently described species (Sicard et al, 1988) and no synonyms are known.

Distribution

This species is known from just four records on both banks of the Niger in Mali, Burkina Faso and Niger. It may have a wider distribution than presented it.

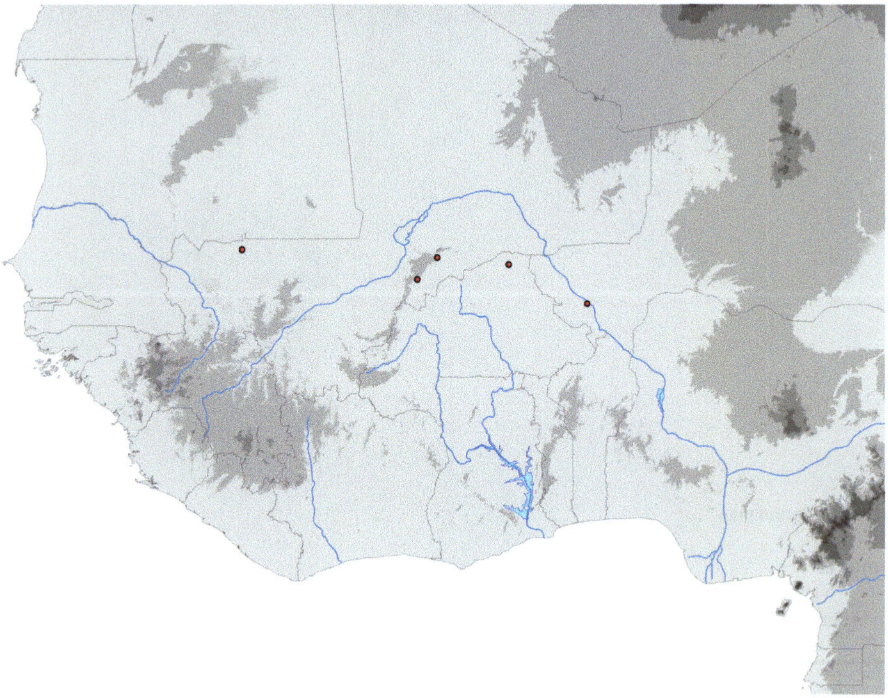

Figure 2.428 Distribution map

Measurements

External and cranial measurements (mm) and mass (g) for *Taterillus petteri*, sexes combined

	Mean	Min	Max	N
Mass	47.7	30	61	7
HB	114.6	101	135	17
Tail	144.3	125	155	12
HF	30.4	28	33	16
Ear	18.9	17	21	17
GLS*	33.8	30.7	36.0	17

*Happold (2013) presents a mean of 31.1 mm (n = 11; range: 28.2-33.3)

Taterillus pygargus (F. Cuvier, 1838)

Diagnosis

This species can only be differentiated by its chromosome number. It co-occurs with *T. gracilis* from which it can only safely be distinguished by chromosome number. Its distributional limits are currently unknown and there may potentially be some overlap with other *Taterillus* species.

Taxonomic Issues

None.

Distribution

This species is known from a few scattered localities in Senegal and Niger, but may well be more widely distributed on the left bank of the Niger River.

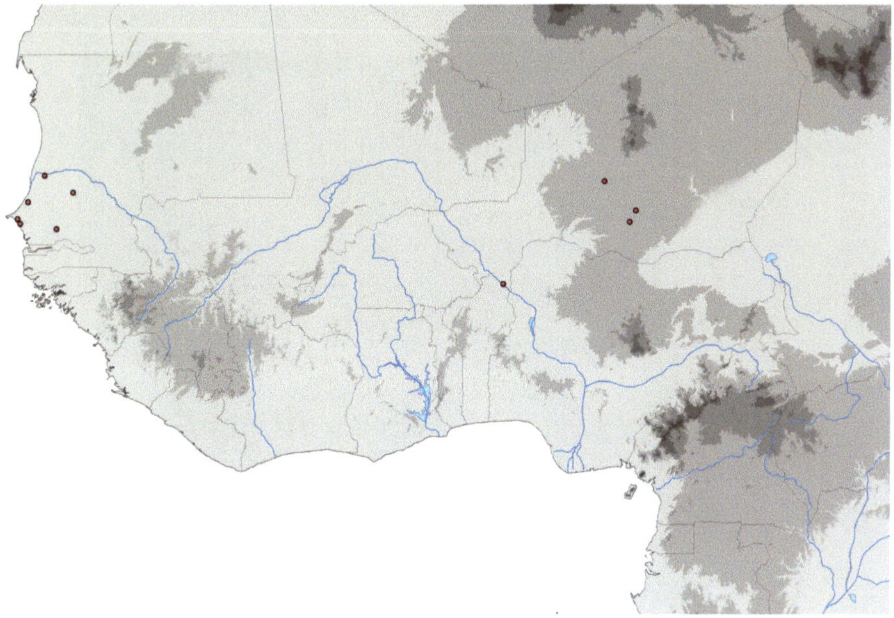

Figure 2.429 Distribution map

Measurements

External and cranial measurements (mm) and mass (g) for *Taterillus pygargus*, sexes combined

	Mean	Min	Max	N
Mass*	48.1	31	68	49
HB*	116.7	102	134	83
Tail*	147.4	128	165	41
HF*	29.7	28	32	48
Ear*	19.4	17	22	83
GLS**	31.7	29.9	34.2	36

*Granjon & Duplantier (2009); **Happold (2013)

Taterillus tranieri Dobigny, Granjon, Aniskin, Ba & Volobouev, 2003

Diagnosis

This species can only be differentiated by its chromosome number. It co-occurs with *T. gracilis* and possibly *T. petteri* from which it can only safely be distinguished by chromosome number. Its distributional limits are currently unknown and there may potentially be some overlap with other *Taterillus* species.

Taxonomic Issues

This is a recently described species (Dobigny et al., 2003).

Distribution

This species is currently only known from two localities in Mauritania and Mali (Dobigny et al., 2005).

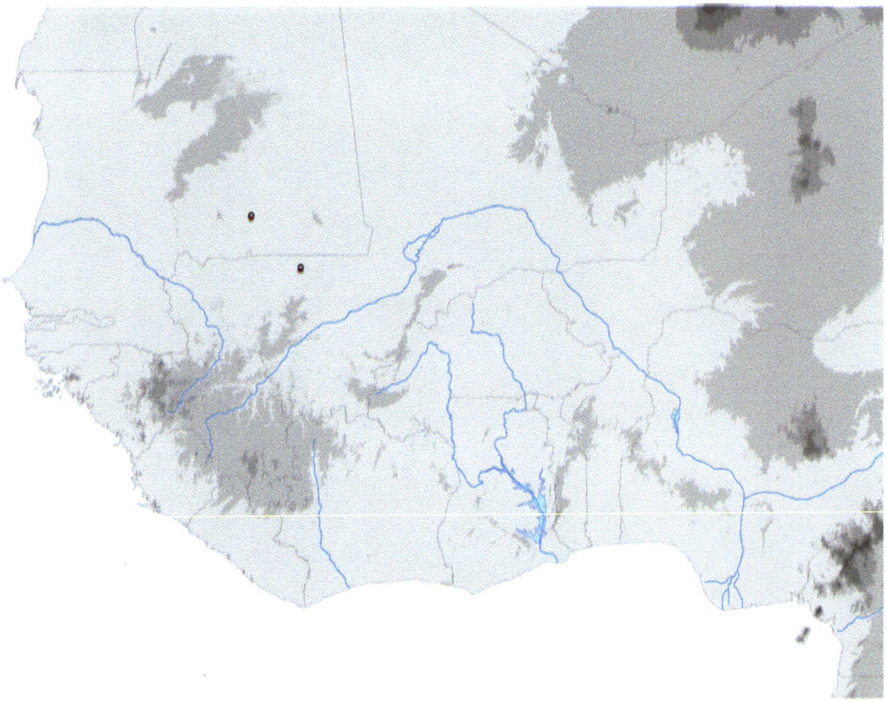

Figure 2.430 Distribution map

Measurements

External and cranial measurements (mm) and mass (g) for *Taterillus tranieri*, sexes combined

	Mean	Min	Max	N
Mass*	46.8	40	54	4
HB*	123.5	111	134	4
Tail*	160.5	145	180	4
HF*	30.6	30	31	3
Ear*	20.3	19.5	20.5	4
GLS*	31.2	30.2	32.5	3

*Happold (2013)

Subfamily: Leimacomyinae

Genus Account

Leimacomys Matschie, 1893 (Büttner's Forest Mouse or Büttner's Togo Mouse)

Diagnosis

This rare (possibly extinct), small sized rodent (Mass ca.30 g; HB ca. 118 mm) is characterised by a very short, sparsely-haired tail (30% of HB) and a dense, "bristly" pelage lacking guard hairs, golden brown dorsally and grey-brown ventrally. The claws on the forefeet and hindfeet are long (> 10% of hindfoot length) and straight, except for the 5th digit of the forefeet which is reduced with a flat nail. The upper incisors are shallowly-grooved and markedly pro-odont, with the cutting edges forming a semi-circle. The species is known from only two specimens obtained in Togo in 1890.

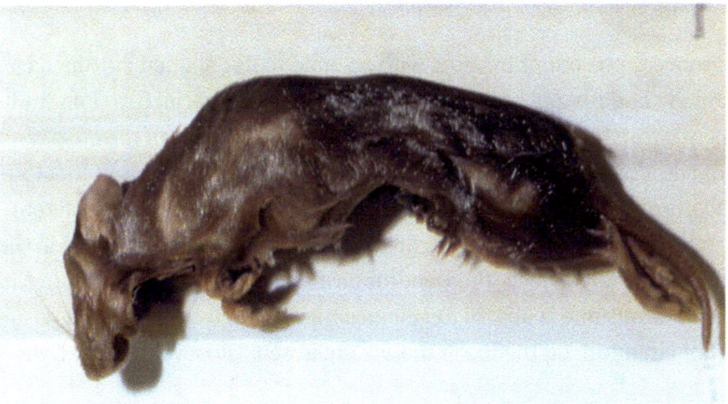

Figure 2.431 Photograph of *Leimacomys buettneri* holotype (ZMB 6856) (Jan Decher).

Cranial Characters

The skull is robust and small-sized (GLS 30 mm), with a moderately long and wide rostrum which tapers in lateral view. The slight supraorbital ridges extend backwards across the braincase in an amphoral shape. The zygomatic arch is robust but not bowed outwards and the zygomatic plate is well developed and broad with a curved anterior edge (distinguishing it from Dendromurinae in which it is weakly developed). The anterior palatal foramina are narrow and very long, extending to the middle of M^1 and ending just before the posterior palatal foramina. The posterior margin of the palate extends far beyond the posterior margins of the upper toothrow. The bullae are small.

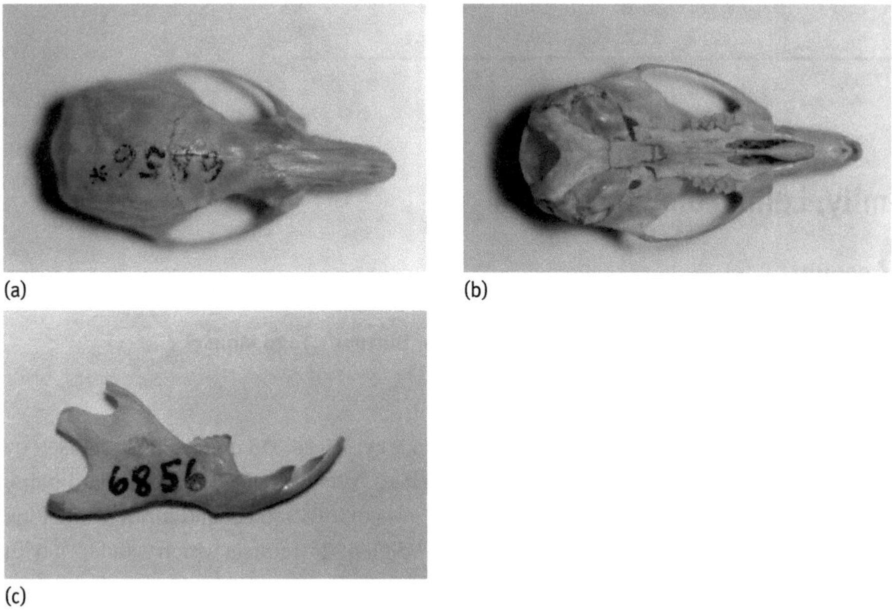

Figure 2.432 a-c Cranium of the holotype of *Leimacomys buettneri* (ZMB 6856) (from top left to bottom left: dorsal view, ventral view and lower jaw).

Dentition

Leimacomys has shallowly grooved, pro-odont incisors with a semicircular-shaped cutting edge. The molar cusp pattern is strongly reminiscent of the Dendromurinae in its biserial (two longitudinal rows of cusps) arrangement (which is triserial in murines) and in the absence of the cusp t1 on M^1, as well as the presence of additional lingual cusps on the second lobe of M^1 (t4) and first lobe of M^2 (t1). The upper molar toothrows diverge slightly anteriorly and cusps are not well differentiated, forming simple transverse laminae. The M^3 and M_3 are moderately reduced compared to the first and second molars, but not as reduced as in the Dendromurinae. The longitudinal groove in the upper molars found in Dendromurinae is absent in *Leimacomys*. The antero-labial and postero-labial cusplets and posterior cingulum on the lower molars are absent (Rosevear, 1969; Denys, 1993; Denys et al., 1995).

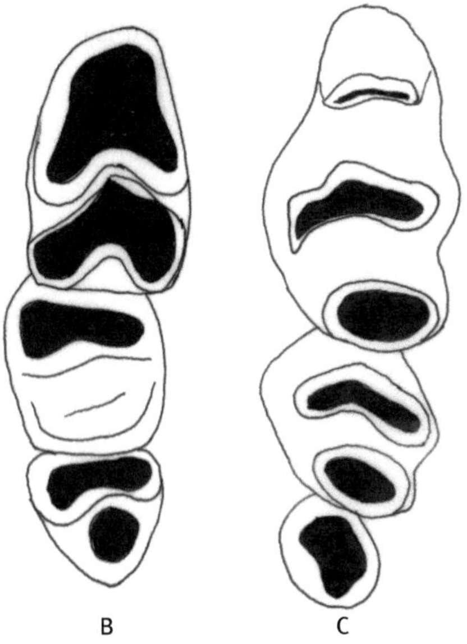

Figure 2.433 Cheek teeth of *Leimacomys buettneri* drawn from the Holotype (ZMB 6856), reprinted from Denys (1993) (B: lower left toothrow; C: upper right toothrow).

Habitat & Biogeography

This genus is only known from one locality, Bismarckburg in Togo in the Rainforest-Savanna Mosaic biotic zone (Happold, 2013). According to the 50-year definition of extinction adopted by IUCN and CITES, this species could be regarded as extinct (Schlitter, 1989) but as pointed out by Musser & Carleton (2005), insectivorous murids are often very difficult to trap in conventional traps, and intensive surveys of the surrounds of the type locality should be conducted using pitfall traps and drift fences. Two expeditions in the 1990s by the Oxford University and National Geographic Society using interrogations of indigenous knowledge and various trapping methods failed to locate this species (Decher et al., 2001).

Fossil Record

No fossils are known (Winkler et al., 2010).

Biology & Ecology

Nothing is known about the biology of this rare rodent. The contents of a single stomach comprised termites, suggesting it is insectivorous (Dieterlen, 1976c).

Systematic Notes

Because of its rarity, and the absence of molecular data, the relationships of this enigmatic genus remain obscure. It was originally assigned to Dendromurinae by Matschie (1893), a position maintained until recently, albeit with reservations (Rosevear, 1969). An affiliation with Dendromurinae is suggested by dental characters such as the possession of an extra lingual cusp on the second lamina of M^1 and anterior lamina of M^2 (Denys, 1993). However, in terms of cranial and external characters the genus is unlike Dendromurinae (see above), more closely resembling a deomyine or murine rodent. Some of its dental patterns are reminiscent of the Gerbillinae. The skull shape is superficially very similar to *Lophuromys* but this may be a result of convergence to a similar insectivorous diet. Cladistic analysis of craniodental and external characters by Denys et al. (1995)

revealed that *Leimacomys* is not a true Dendromurinae (lingual cusps on M^1 and M^2 are convergent), probably representing an ancient lineage of forest-endemic rodents. Hence it is currently recognised in its own subfamily (Musser & Carleton, 2005).

Species List

A single species is known.
Leimacomys buettneri Matschie, 1893

Species Accounts

Leimacomys buettneri Matschie, 1893

Diagnosis

See genus account.

Taxonomic Issues

See genus account.

Distribution

This species is known from two specimens both collected at the type locality in the "high forest belt" of western Togo, but possibly in the forest-savanna ecotone (see Happold, 2013).

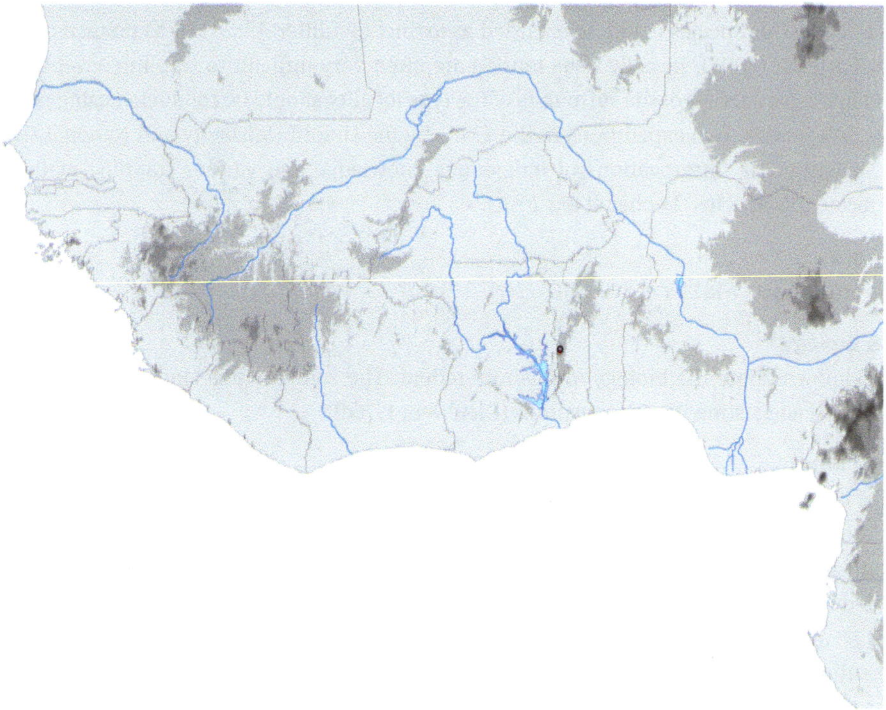

Figure 2.434 Distribution map

Measurements

External and cranial measurements (mm) and mass (g) for *Leimacomys buettneri*, sexes combined

	Mean	Min	Max	N
Mass	-	-	-	-
HB*	118	-	-	1
Tail*	37	-	-	1
HF*	21	-	-	1
Ear*	14	-	-	1
GLS*	30.6	-	-	1

*Rosevear (1969), measurements of the type specimen

Subfamily: Lophiomyinae

Genus Account

Lophiomys Milne-Edwards, 1867 (African Maned Rats)

Diagnosis

This unmistakable monotypic genus is medium to large in size (Mass ca. 600-800 g). As a result of its shaggy coat, *Lophiomys* gives the impression of having a squat body, with short legs and tail. The pelage is long and shaggy, with individual hairs being banded (from base to tip) white, dark grey and white giving the animal an overall grey colour. The lateral band, tapering from the head to the rump, is cinnamon-brown centrally, flanked on both sides by white and black. The dorsal fur generally obscures the brown band which is only visible when the fur is erected. This lateral band corresponds to a glandular strip of skin (Kingdon, 1974), but is not responsible for toxin production (Kingdon et al., 2012; see below under "Biology & Ecology"). The ventral hairs are paler grey. The ears are relatively small, rounded and fringed with short white hairs. The face is black with white eyebrow and moustachial patches. There are five digits on the hindfoot and four digits on the forefoot, all with sharp and curved claws. The first digit on the forefoot is greatly reduced but present and without a claw. The tail is shorter than body length (ca. 75% of HB), with a similar pattern to the dorsal fur.

Figure 2.435 Photograph of *Lophiomys imhausi* (M.F. Kinnaird).

Cranial Characters

The large skull (mostly 61-65 mm in greatest length) is robust and unique among rodents by having the dorsal part covered with minute rounded projections giving it a granulated appearance which is rough to the touch. It is relatively flat and broad for its length. The palate ends just beyond the posterior margin of the third molars. The zygomatic arch is long but robust, deepening posteriorly due to the attachment of a lateral projection of the parietal. The infraorbital foramen is oval shaped. The anterior palatal foramina are relatively long, almost reaching the anterior margin of the first molar. The paraoccipital process is well developed. The bullae are moderate in size and not inflated.

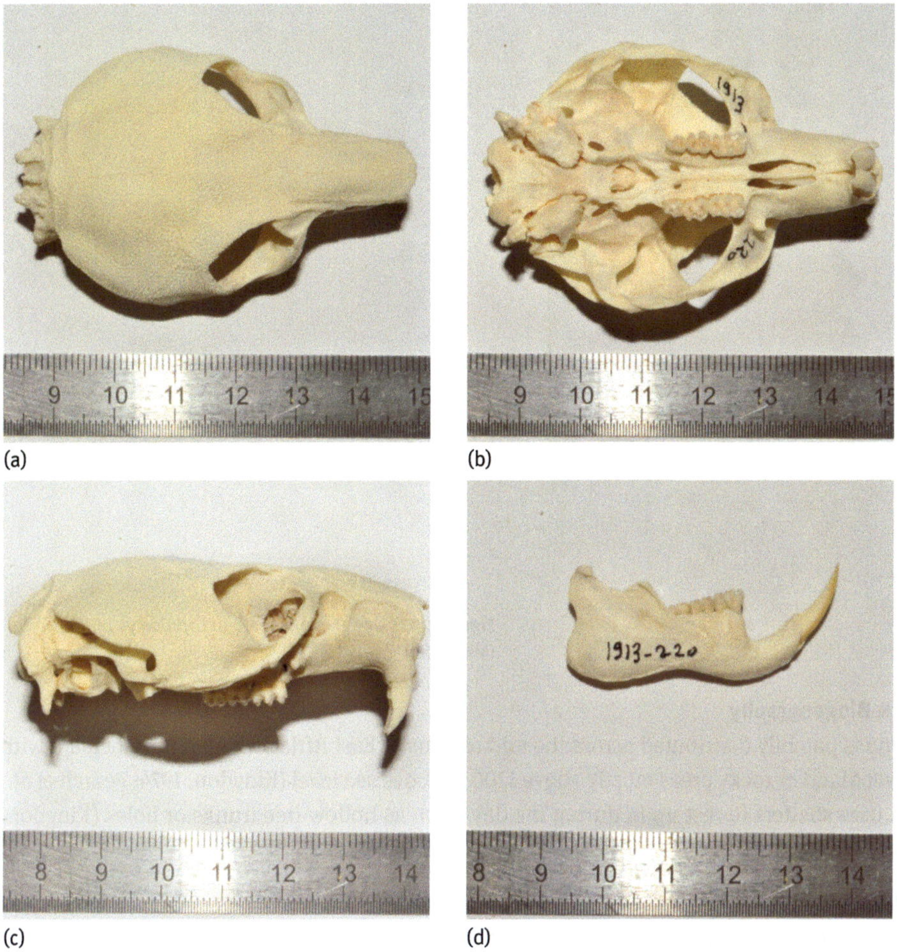

Figure 2.436 a-d Cranium of *Lophiomys imhausi* (MNHN-ZM-MO 1913-220) (from top left to bottom right: dorsal view, ventral view, lateral view and lower jaw).

Dentition

Lophiomys has ungrooved, orthodont incisors. The dental formula is: 1003/1003. The second and third molars are smaller than the first molar. The first upper molar has six cusps arranged in pairs with a ridge between each pair forming a transverse loph and deep troughs between lophs. A longitudinal crest runs across the troughs medially joining the lophs. A full description of the dentition of *Lophiomys* is provided by Wahlert (1984).

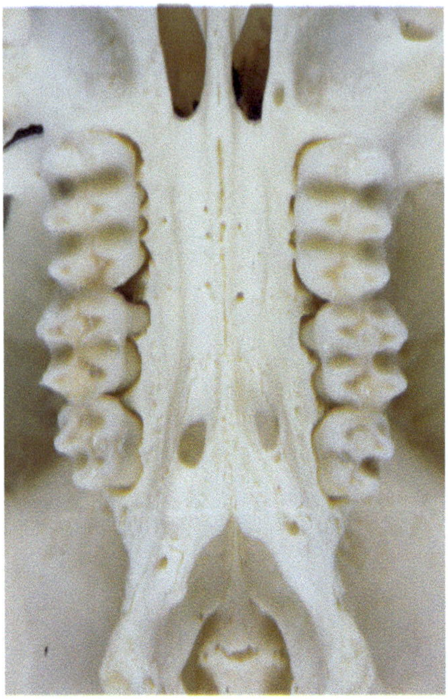

Figure 2.437 Upper cheek teeth of *Lophiomys imhausi* (MNHN-ZM-MO 1913-220).

Habitat & Biogeography

Lophiomys is patchily distributed across the arid regions of East Africa where it is associated with forest, woodland or rocky areas mostly above 1200 m above sea level (Kingdon, 1974; Pearch et al., 2001). It uses shelters to rest up in during the day, such as hollow tree trunks or holes (Kingdon, 1974).

Fossil Record

The oldest *Lophiomys* fossil is from 4.8-5.7 MYA, Ethiopia (Wesselman et al., 2009). In North Africa, it has been reported from the Middle Pliocene; in Eastern Africa from the Late Miocene (Winkler et al, 2010).

Biology & Ecology

The biology and ecology of *Lophiomys* is poorly known. It has several morphological features that serve an anti-predation function including loose, shaggy fur, a heavily armoured cranium and toxin secretions (Kingdon et al., 2012), which may explain its slow movement (Kingdon, 1974). It is unusual among rodents in having a four-chambered stomach (Scopin et al., 2011). It feeds on leaves and seeds (Hollister, 1919), but may also take fruits and tubers (Kingdon, 1974). Based on a single captive female, litter size is 1 (Delany, 1975).

Systematic Notes

The taxonomic position of *Lophiomys* has long been debated (see Happold, 2013, for a recent review) and remained unresolved until recently. Previously, it was been placed in its own family (Lophiomyidae), or in the subfamily Lophiomyinae within the Cricetidae (e.g. Happold, 2013). However, a recent molecular phylogeny for the muroids clearly shows *Lophiomys* embedded within the Muridae clade, possibly closest to the Gerbillinae (Jansa & Weksler, 2004), and is divergent enough to be considered in its own subfamily. Schenk et al. (2013) found *Lophiomys* to be the basal taxon of a clade constituting the Deomyinae and Gerbillinae, and this entire clade was the sister taxon of the Murinae. So here, we recognise *Lophiomys* in its own subfamily Lophiomyinae within the Muridae. Numerous forms have been described within *Lophiomys* but it is not clear how these related to each other and all recent authorities only recognise a single species (Musser & Carleton, 2005). The diploid number of chromosomes is not known in *Lophiomys*.

Species List

One species is described within the genus *Lophiomys*, restricted to Africa.
Lophiomys imhausi Milne-Edwards, 1867

Species Accounts

Lophiomys imhausi Milne-Edwards, 1867

Diagnosis

See genus account.

Taxonomic Issues

See genus account.

Distribution

This species is patchily distributed in East Africa, with an isolated record from the eastern shores of Lake Tanganyika (Tanzania).

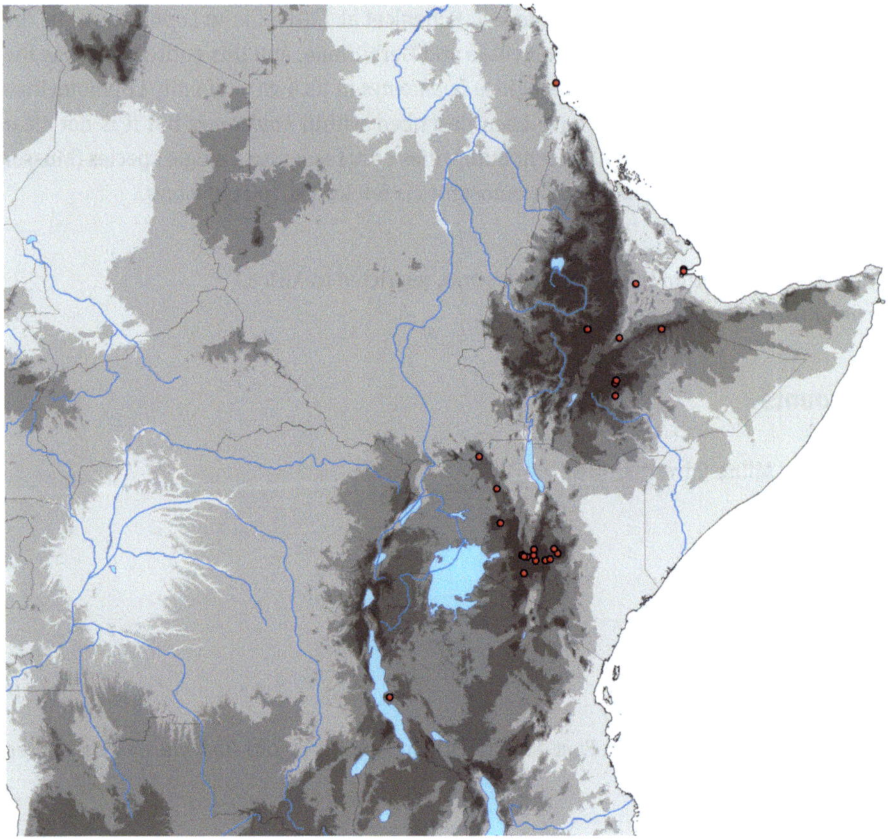

Figure 2.438 Distribution map

Measurements

External and cranial measurements (mm) and mass (g) for *Lophiomys imhausi*, sexes combined

	Mean	Min	Max	N
Mass*	-	590	920	-
HB	277.3	252	301	4
Tail	180.8	145	205	4
HF	47.5	34	56	4
Ear	26.8	30	39	4
GLS	63.1	55.9	67.0	8

*Kingdon (1974)

Subfamily: Murinae

To assist with the identification of the genera within the Murinae, we present novel information pertaining to the presence or absence of macrodonty and stephanodonty within each genus. This information is summarised in Table 10 (below) and is discussed within each genus account in this subfamily. Macrodonty refers to the relative size of the molars. Genera with relatively large teeth (in relation to the size of the skull) are referred to as macrodont whereas genera with relatively small teeth are microdont; those in between are normal. Stephanodonty refers to the presence of ridges connecting the various cusps in longitudinal rows (in other words: cusps t1 and t4; t4 and t7; or t3 and t6). Genera with such connections are termed stephanodont; all others are normal. Some genera such as *Thamnomys* and *Thallomys* have connecting ridges between some of the transverse cusps; these are considered stephanodont here.

Table 10. The presence or absence of macrodonty or stephanodonty in genera within the Murinae.

	Stephanodont	Not Stephanodont (normal)
Microdont	*Grammomys; Thallomys*	*Heimyscus; Malacomys*
Normal	*Thamnomys; Zelotomys*	*Aethomys; Colomys; Dephomys; Hybomys; Hylomyscus; Lemniscomys; Mastomys; Micaelamys; Mus; Myomyscus; Nilopegamys; Pelomys; Praomys, Rattus; Rhabdomys; Stochomys*
Macrodont	*Desmomys; Lammotemys; Oenomys; Stenocephalemys*	*Arvicanthis; Dasymys; Mylomys*

In addition to macrodonty and stephanodonty, it is possible to group these genera by the structure of the 3rd upper molar (Denys & Michaux, 1992). These groups are presented in Figure 2.439.

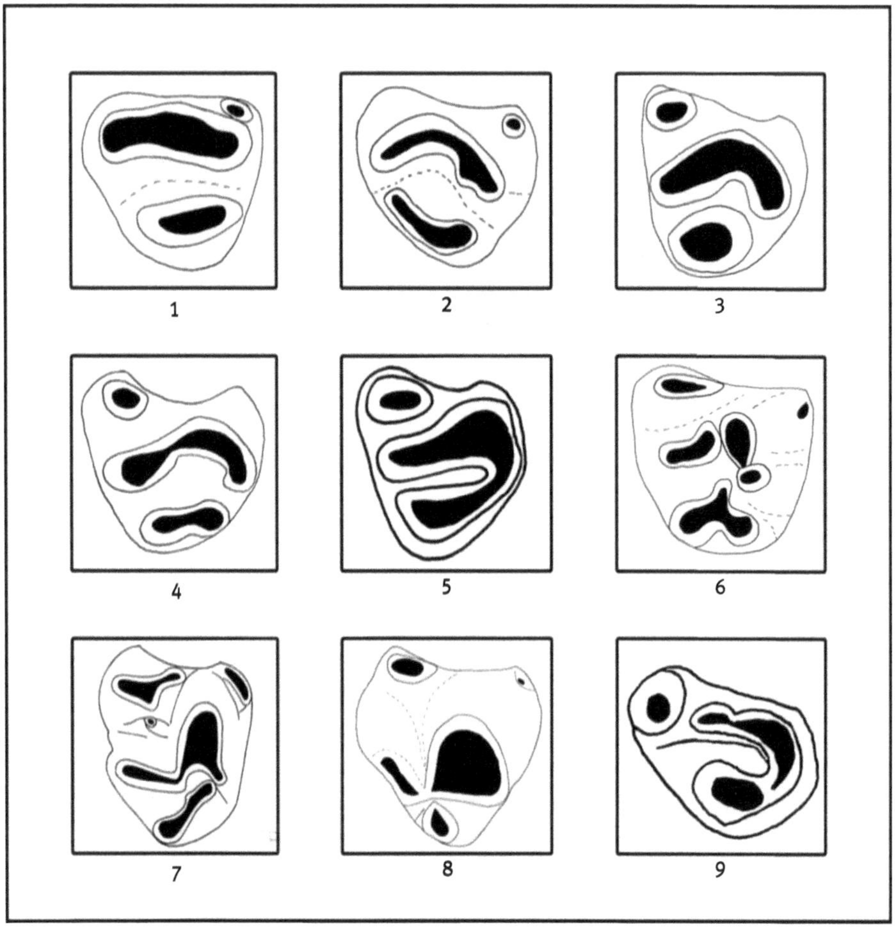

Figure 2.439 The shape and structure of the upper left M³ of the families Muridae and Nesomyidae illustrating morphological categories (after Denys & Michaux 1992) (not drawn to scale).

1: t3 present, t1 absent, t3 connected to the first row of cusps: Cricetomyinae (*Saccostomus, Cricetomys, Beamys*), Dendromurinae (*Steatomys, Malacothrix, Dendroprionomys, Prionomys*), Petromyscinae (*Petromyscus*), Delanymyinae (*Delanymys*), and Deomyinae (*Deomys*)

2: t3 present, t1 absent, t3 isolated: Deomyinae (*Acomys, Uranomys, Lophuromys*)

3: t1 present, a distal cusp, t3 absent: Murinae (*Rhabdomys, Aethomys, Mus (Muriculus), Malacomys, Mastomys, Praomys, some Lemniscomys, Stenocephalemys*)

4: t1 present, t3 absent, second lobe with two distal cusps: Murinae (*Arvicanthis, Dasymys, Stochomys, some Lemniscomys*)

5: rather large t1, labial link between first lobe and second lobe: Murinae (*Colomys, Dephomys, Hybomys, Hylomyscus, Zelotomys*)

6: presence of t1 and tiny t3, trace of median longitudinal link: Murinae (*Grammomys, Thallomys*)

7: presence of a t1 and a t3, longitudinal link between first and second lobe, stephanodont upper first molars: Murinae (*Oenomys, Lamottemys, Thamnomys, Desmomys*)

8: presence of a large t1, only 3 main high round cusps: Murinae (*Mylomys, Pelomys*)

9: small molars with t1 and link between first and second lobe, cusps poorly differentiated: Murinae: *Mus (Nannomys)*

We hope that these provide additional morphological features for identifying genera within this traditionally difficult and diverse subfamily.

Key to Genera

This key has been adapted from Misonne (1974) and Meester et al. (1986).

1	M^3 the largest molar in the upper toothrow; molars laminate	2
	M^1 the largest molar in the upper toothrow; molars not laminate	3
2	Bullae slightly or not at all enlarged, maximum length 9.5 mm	*Otomys*
	Bullae greatly enlarged, minimum length 10.8 mm in adults	*Parotomys*
3	Upper incisors grooved	4
	Upper incisors ungrooved	5
4	Forefoot with 3 digits	*Mylomys*
	Forefoot with 4 digits (although Digit 5 may be reduced)	*Pelomys*
5	Dorsal pelage with lines or stripes	6
	Dorsal pelage without lines or stripes	12
6	Ventral pelage white	7
	Ventral pelage not white	10
7	Dorsal pelage with single dark line	8
	Dorsal pelage with more than one dark line	9
8	Tail longer (> 90% of HB)	*Lemniscomys* (part)
	Tail shorter (< 85% of HB)	*Arvicanthis* (part)
9	Dorsal pelage with 4 dark lines	*Rhabdomys*
	Dorsal pelage with many stripes (sometimes broken into dots)	*Lemniscomys* (part)
10	Size smaller (HB < 80 mm)	*Mus (Muriculus)*
	Size larger (HB > 100 mm)	11
11	Dorsal pelage dark brown (chocolate) without speckling	*Hybomys*
	Dorsal pelage greyish-brown (not chocolate) with specking	*Arvicanthis* (part)
12	Hindfoot very long (> 26% of HB)	13
	Hindfoot not very long (< 26% of HB)	14
13	Ventral pelage greyish-white	*Malacomys*
	Ventral pelage pure white	*Colomys/Nilopegamys*
14	Mouth and nose red	*Oenomys*
	Mouth and nose not red	15
15	Tail white or whitish, with very little hair	*Zelotomys*
	Tail not white or whitish	16
16	Strong interorbital constriction	17
	No strong interorbital constriction	18
17	Anterior part of zygomatic plate concave	*Dasymys*
	Anterior part of zygomatic plate straight	*Stenocephalemys*
18	M^1 with t7 cusp (or at least replaced by a strong ridge)	19
	M^1 without t7 cusp or ridge	20
19	t7 well developed; restricted to the Albertine Rift	*Thamnomys*
	t7 reduced to a ridge; widespread in Africa	*Grammomys*

20	Rows of cusps on M^1 markedly distorted	21
	Rows of cusps on M^1 not markedly distorted	22
21	Size smaller (typically HB < 70 mm; except commensal *M. musculus* < 90 mm)	*Mus*
	Size larger (typically HB > 80 mm)	*Praomys* group*
22	Dark markings ("spectacles") around eyes	*Thallomys*
	No dark markings around eyes	23
23	Dorsal pelage dark brown (chocolate)	*Hybomys*
	Dorsal pelage greyish or reddish (not chocolate)	24
24	Anterior palatal foramina penetrating between molars	*Aethomys*
	Anterior palatal foramina not penetrating between molars (but may just reach molars)	25
25	Tail shorter (< 85% of HB)	*Arvicanthis* (part)
	Tail longer (> 85% of HB)	26
26	Dorsal pelage with long bristles; occupying forest habitat	*Stochomys*
	Dorsal pelage without long bristles; commensal with humans	*Rattus*

**Praomys* group includes the genera: *Praomys, Mastomys, Hylomyscus,* and *Myomyscus* which are best separated on molecular or chromosomal grounds (but see Diagnosis section of each genus for additional features).

Genus Account

Aethomys Thomas, 1915 (Veld Rats)

Diagnosis

In their rat-like appearance, reddish to brownish-grey dorsal pelage with distinctively coarse black guard hairs, pure white or grey ventral pelage, and sparsely-haired, coarse-scaled and relatively long tails (ca. 80-120 % of HB), these medium to large-sized rodents (Mass ca. 70-120 g; HB ca. 110-160 mm) lack diagnostic external characters and can easily be confused with members of *Rattus* and other genera. They can be distinguished from *Micaelamys* (which was until recently included in the same genus) as the latter have smaller body size (HB < 120 mm for *Micaelamys*, > 135 mm for *Aethomys*). The fleshy, larger hindfoot of *Rattus* (HF ca. 40 mm) is a useful character to distinguish members of this genus from *Aethomys* (HF < 30 mm). Furthermore, *Rattus* is completely commensal whereas members of *Aethomys* are very rarely found inside human habitations. Members of *Aethomys* are terrestrial; the arboreal habit of *Grammomys* and *Thallomys* helps to distinguish individuals of these genera from *Aethomys*.

Figure 2.440 Photograph of *Aethomys chrysophilus* (Ara Monadjem).

Cranial Characters

The skull is medium-large in size (GSL ca. 35–41 mm) and ovoid in shape, with poor to well-developed supraorbital crests and moderately robust zygomatic arches. According to Kingdon (1974) skulls vary between species from narrow (e.g. *A. chrysophilus*) to broad (e.g. *A. hindei*) with others (such as *kaiseri*) being intermediate. The nasal bone is of average length and width, tapering at the tip. Incisors are opisthodont to orthodont and ungrooved. The anterior palatal foramen reaches at least to the level of the first upper molars. The anterior surface of the zygomatic plate is vertical or very slightly tilted forwards and bluntly rounded dorsally. The bullae are relatively small (6-7 mm in greatest diameter).

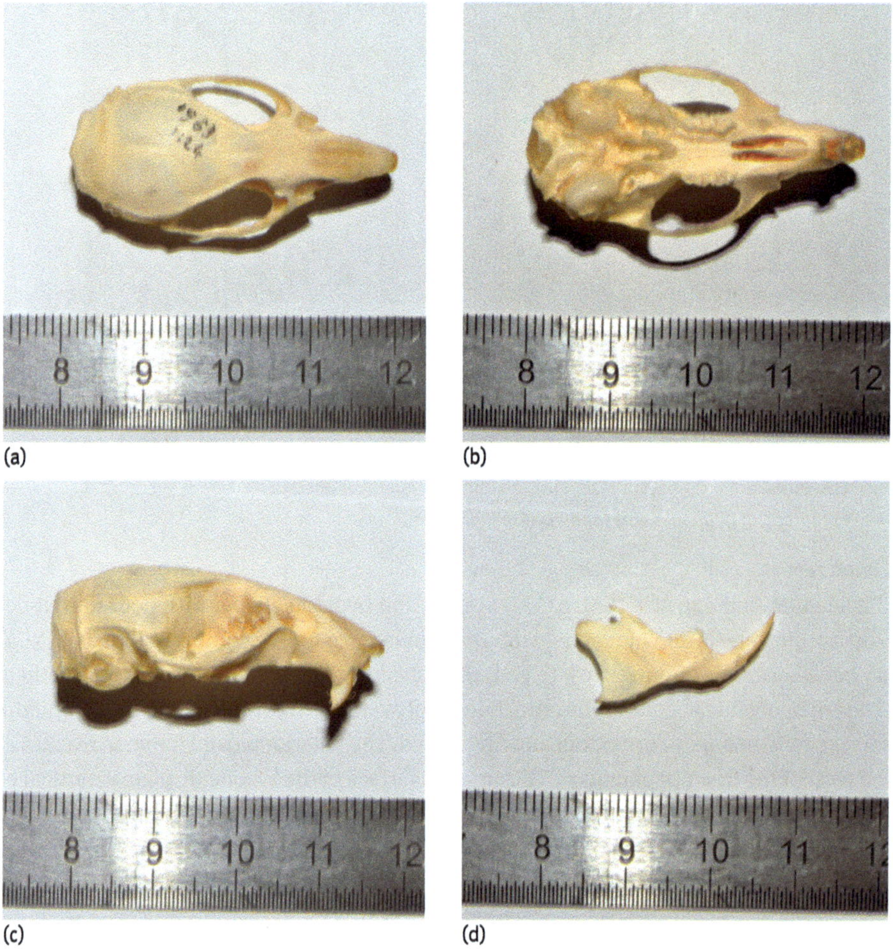

Figure 2.441 a-d Cranium of *Aethomys hindei medicatus* (MNHN-ZM-MO 1963-1124) (from top left to bottom right: dorsal view, ventral view, lateral view and lower jaw).

Dentition

Molars are moderately sized showing neither microdonty nor macrodonty (MTR/GSL ± 0.16-0.18, Table 10). Individual cusps are rounded and well-separated but not as distinctively large as in *Pelomys* and *Mylomys*. Some species (e.g. *A. chrysophilus*) display stephanodont crests (Denys, 1994). Rows of cusps on M^1 are only slightly distorted, with t1 situated behind t2 and t3 which form a high, round lobe. The first upper molar is the longest molar in the toothrow. Molar M^3 lacks the t3 cusp (Group 3 in Figure 2.439), a character found in all murids except for *Acomys*, *Uranomys*, *Lophuromys* and genera (*Desmomys*, *Lamottemys*, *Oenomys*, *Thamnomys*) showing stephanodonty (formation of longitudinal crests between transverse rows of cusps leading to larger and higher cusps and a more abrasive occlusal surface associated with a more herbivorous diet). The first lower molar M_1 has two cusps in the anterior row (as opposed to three in *Micaelamys*). Other subtle differences in molars between *Aethomys* and *Micaelamys* are elaborated by Misonne (1969).

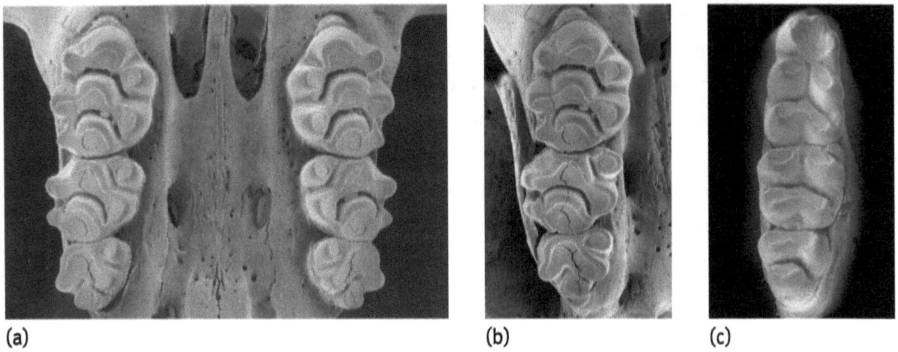

Figure 2.442 Cheek teeth of *Aethomys hindei medicatus* (MNHN-ZM-MO 1963-1124) (A: upper cheek teeth; B: upper right toothrow; C: lower right toothrow).

Habitat & Biogeography

This genus is associated with grassland to woodland savannas in southern, eastern and central Africa, with one species (*A. stannarius*) restricted to West Africa. Whilst *A. chrysophilus* has a very broad distribution in Africa, other species are much more localised to particular regions, e.g. *A. bocagei* and *A. thomasi* to Angola, *A. stannarius* to northern Nigeria and Cameroon and *A. silindensis* to eastern Zimbabwe.

Fossil Record

Fossils have been recorded as early as the late Miocene. In Southern Africa (Namibia), 10.5-9.5 MYA *Aethomys* fossil represents the earliest murine presence in the region (Winkler et al., 2010; Mathews & Stynder, 2011).
A. deheinzelini Omo (Wesselman, 1984)
A. lavocati Olduvai bed I, Natron (Denys, 1987b)
A. adamanticola & *A. modernis* Langebaanweg (Denys, 1990b)
A. sp. East Turkana (Black & Krishtalka, 1986)
A. sp. Harasib 3a (Mein et al., 2004)
A. sp. Lemudong'o (Manthi, 2007)
Aethomys cf *lavocati* from Hadar (Reed, 2011)

Biology & Ecology

Although they also feed on insects (Hanney, 1965), Kingdon (1974) classified them as predominantly unspecialised herbivores feeding on fallen fruits, seeds, grass seeds, leaves, buds and roots. In *A. chrysophilus*, breeding patterns vary geographically from seasonal to non-seasonal while in *A. ineptus* they are seasonal (Muteka et al., 2006a,b ; Chimimba & Linzey, 2008; Linzey & Chimimba, 2008 and references therein). In captive *A. chrysophilus* between one and five (mean 3.1) young comprised a litter (Brooks, 1972) whereas embryo numbers varied from one to three in *A. ineptus* (mean 2.7; Taylor, 1998).

Systematic Notes

Although Musser & Carleton (2005) placed *Aethomys* in its own division, molecular data (Ducroz et al, 2001; Lecompte et al., 2008) support the inclusion of *Aethomys* as a basal member of tribe Arvicanthini (together with *Arvicanthis* and several other genera). An association between *Aetho-*

mys and *Arvicanthis* based on molar occlusal patterns was already suggested by Misonne (1969). Although *Micaelamys* (represented by two species, *A. namaquensis* and *A. granti*) was previously included in *Aethomys*, both molecular (Russo, 2003) and morphological (Chimimba, 2005) data attest to the paraphyly of these two genera. On molecular grounds, members of *Aethomys* are phylogenetically closer to *Dasymys* than to *Micaelamys* (Russo, 2003). Therefore, we treat *Aethomys* and *Micaelamys* as separate genera. We also support the recognition of *A. ineptus* as a cryptic species distinct from *A. chrysophilus*, based on chromosomal, reproductive, biochemical (Visser & Robinson, 1986, 1987), morphological (Chimimba et al., 1999), molecular (Russo et al., 2006) and biogeographical (Linzey et al., 2003) grounds.

Annotated Species List

Nine species are recognised, all restricted to Sub-Saharan Africa:
Aethomys bocagei (Thomas, 1904)
Aethomys chrysophilus (de Winton, 1897)
Aethomys hindei (Thomas, 1902)
Aethomys ineptus (Thomas and Wroughton, 1908)
Aethomys kaiseri (Noack, 1887)
Aethomys nyikae (Thomas, 1897)
Aethomys silindensis Roberts, 1938
Aethomys stannarius (Thomas, 1913)
Aethomys thomasi (de Winton, 1897)

Key to Species

This key has been adapted from Davis (1975).

1	Tail relatively shorter (< 110% of HB)	2
	Tail relatively longer (> 110% of HB)	4
2	Incisors orthodont; M^1 wider (> 2.3 mm); skull broader (zygomatic width > 55% of GSL)	*A. kaiseri*
	Incisors opisthodont; M^1 narrower (< 2.2 mm); skull narrower (zygomatic width < 55% of GSL)	3
3	Five plantar pads on foot	*A. thomasi*
	Six plantar pads on foot	*A. hindei*/*A. stannarius*
4	Size larger (GSL > 39 mm)	5
	Size smaller (GSL < 39 mm)	6
5	Ears longer (ca. 25 mm)	*A. bocagei*
	Ears shorter (ca. 21 mm)	*A. silindensis*
6	Incisors orthodont; tail relatively shorter (ca. 115% of HB)	*A. nyikae*
	Incisors strongly opisthodont; tail relatively longer (ca. 120% of HB)	7
7	2n = 50; occurring north of the Limpopo River	*A. chrysophilus*
	2n = 44; restricted to to eastern parts of South Africa and Swaziland	*A. ineptus*

Species Accounts

Aethomys bocagei (Thomas, 1904)

Diagnosis

This is one of five species of *Aethomys* occurring in Angola, the others include: *A. thomasi* (shorter tail ca. 83% of HB; size smaller, GLS < 38 mm; ventral pelage greyish-white); *A. chrysophilus* (size smaller, GLS < 38 mm); *A. kaiseri* (shorter tail ca. 83% of HB; size smaller, GLS < 38 mm); *A. nyikae* (size smaller, GLS < 38 mm). This species has pale brown dorsal pelage with white ventral pelage and a long tail (ca. 106-110% of HB).

Taxonomic Issues

2n = 50 (Matthey, 1963).

Distribution

This species is restricted to north-eastern Angola and neighbouring DRC where it occurs in savanna habitats, often in a forest-savanna mosaic.

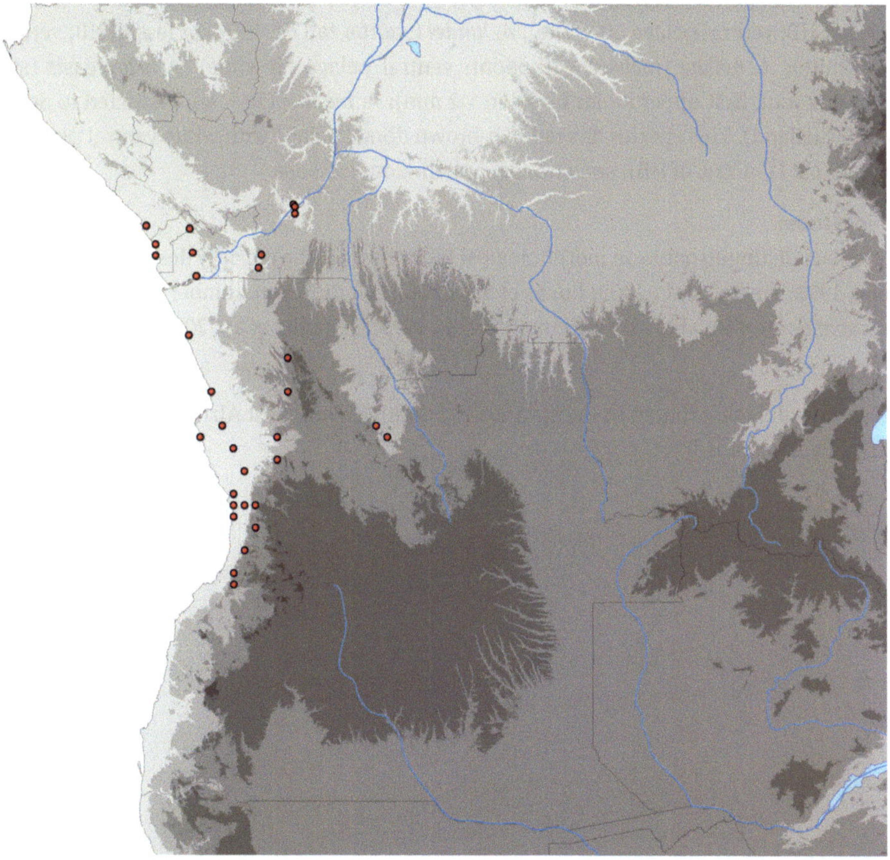

Figure 2.443 Distribution map

Measurements

External and cranial measurements (mm) and mass (g) for *Aethomys bocagei*, sexes combined

	Mean	Min	Max	N
Mass	-	-	-	-
HB	159.2	145	198	6
Tail	168.7	159	180	6
HF	35.3	34	37	6
Ear	24.0	21	32	6
GLS	39.9	39.0	41.3	5

Aethomys chrysophilus (de Winton, 1897)

Diagnosis

This species may co-occur with the following *Aethomys* species: *A. thomasi* (shorter tail ca. 83% of HB; restricted to the plateau of Angola); *A. bocagei* (size larger, GLS > 39 mm); *A. kaiseri* (shorter tail ca. 83% of HB; ventral pelage off-white); *A. hindei* (shorter tail ca. 100% or less of HB; ventral pelage off-white); *A. nyikae* (incisors orthodont; ventral pelage off-white); *A. silindensis* (size larger, GLS > 39 mm; first upper molar broader, > 2 mm); *A. ineptus* (2n = 44; restricted to South Africa and Swaziland). This species has reddish-brown dorsal pelage with white ventral pelage, a very long tail (ca. 115-120% of HB), and strongly opisthodont incisors.

Taxonomic Issues

This species is indistinguishable on morphological grounds from the recently described *A. ineptus* (Chimimba, 1998; Linzey et al., 2003; Russo et al., 2006), but differs in number of chromosomes. In *A. chrysophilus* 2n = 50, aFN = 60 (Chimimba et al., 1999; Castliglia et al., 2003).

Distribution

This species is widely distributed in the savannas of Southern and East Africa, but is replaced in South Africa by *A. ineptus*.

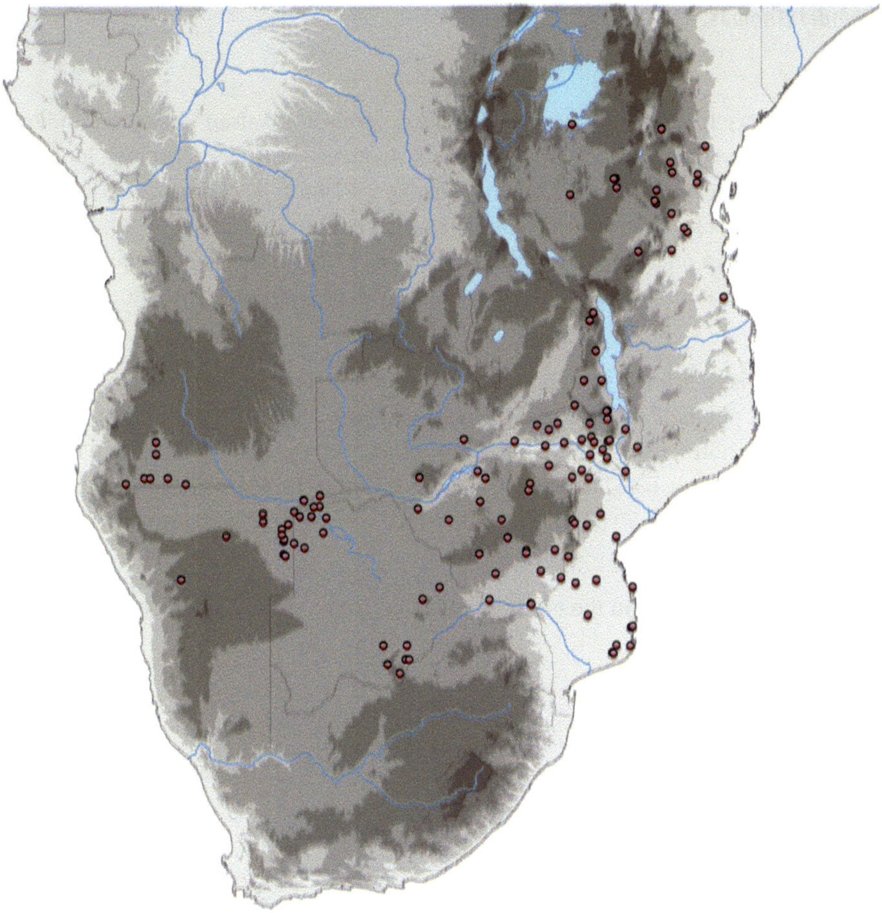

Figure 2.444 Distribution map

Measurements

External and cranial measurements (mm) and mass (g) for *Aethomys chrysophilus*, sexes combined

	Mean	Min	Max	N
Mass*	86.7	63	112	100
HB**	143.2	120	169	128
Tail**	163.3	135	190	128
HF**	29.2	28	35	128
Ear**	20.6	15	27	128
GLS**	36.4	32.1	40.6	365

*Smithers & Wilson (1979), **Chimimba et al. (1999)

Aethomys hindei (Thomas, 1902)

Diagnosis

This species may co-occur with the following *Aethomys* species: *A. kaiseri* (incisors orthodont; skull relatively wider, zygomatic width > 55% of GLS); *A. stannarius* (tail on average longer ca. 103% of HB; ventral pelage white; auditory bulla smaller, < 7mm); *A. chrysophilus* (longer tail ca. 115% of HB; ventral pelage white). This species has medium brown dorsal pelage with greyish-white ventral pelage, a relatively short tail (ca. 100% or less of HB), relatively large auditory bulla (> 7mm), relatively narrow skull (zygomatic width < 55% of GLS) and opisthodont incisors.

Taxonomic Issues

This taxon may comprise a species complex and a revision is urgently needed. Tail length varies considerably between populations, with the main break being the Rift Valley of East Africa; to the east, tail length is shorter (ca. 85%) than to the west (ca. 100%) (Denys & Tranier, 1992). The taxon *A. hindei* was previously listed as a subspecies of *A. kaiseri* (Hollister, 1919). The number of chromosomes in *A. hindei* is 2n = 50 (Matthey, 1963).

Distribution

This species is widely distributed in the northern savannas from central Cameroon east to Kenya and northern Tanzania.

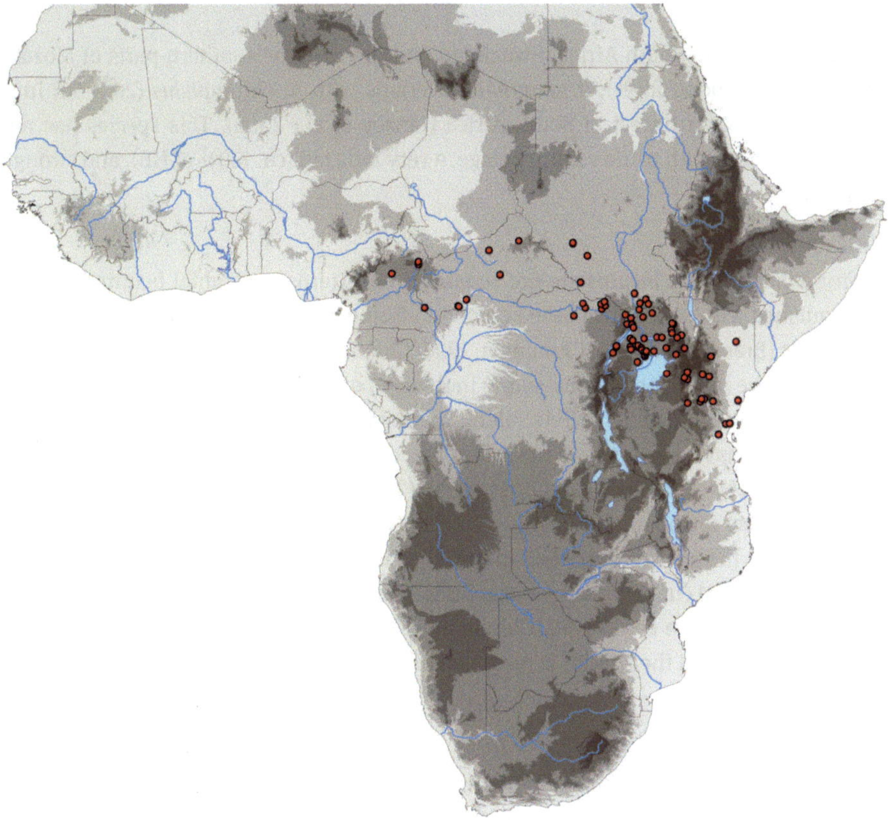

Figure 2.445 Distribution map

Measurements

External and cranial measurements (mm) and mass (g) for *Aethomys hindei*, sexes combined

	Mean	Min	Max	N
Mass	117.9	87	158	8
HB	151.4	120	175	25
Tail	151.7	120	188	25
HF	29.9	27	35	25
Ear	20.7	18	25	23
GLS	37.3	33.4	40.5	15

*Aethomys ineptus** (Thomas and Wroughton, 1908)

*This name is probably a synonym of *A. chrysophilus* and a new name will most likely need to be sought for this species (see Taxonomic Issues, below).

Diagnosis

This species is restricted to South Africa, Swaziland and the extreme southern parts of Mozambique (south of the Limpopo River); it marginally overlaps with *A. chrysophilus* (2n = 50; In *A. ineptus* 2n = 44) in the north-west of South Africa (Linzey et al., 2003). This species has reddish-brown dorsal pelage with white ventral pelage, a very long tail (ca. 120% of HB), and strongly opisthodont incisors.

Taxonomic Issues

Aethomys chrysophilus and *A. ineptus* are not separable on external or cranial features and can only be distinguished with confidence based on chromosome number, sperm morphology, or on molecular grounds (Breed et al., 1988; Chimimba, 1998; Chimimba et al., 1999; Linzey et al., 2003; Russo et al., 2006). The number of chromosomes in *A. ineptus* is 2n = 44 (Linzey et al., 2003).

The type locality for *A. ineptus* is Tete, along the Zambezi River in northern Mozambique. The species described as *A. ineptus* here has yet to be positively reported north of the Limpopo River [see maps in Linzey et al. (2003) and Russo et al. (2006)], and therefore is unlikely to occur further north in the Tete region of Mozambique. Hence, the name *ineptus* is probably a synonym of *A. chrysophilus* and the species described in this account will need a new name. It should be noted that based on cranial characteristics *A. ineptus* was originally thought to extend north of the Limpopo River (see Chimimba et al., 1999; Chimimba, 2001), but more recent chromosomal and molecular work suggests that the species is restricted to south of the Limpopo River (Linzey et al., 2003; Russo et al., 2006).

Distribution

This species is restricted to the savannas of South Africa, Swaziland and neighbouring Mozambique south of the Limpopo River. Records from the Gariep River (e.g. Lynch, 1983) have not been karyotyped and are assumed to belong to this species rather than *A. chrysophilus* which occurs further to the north.

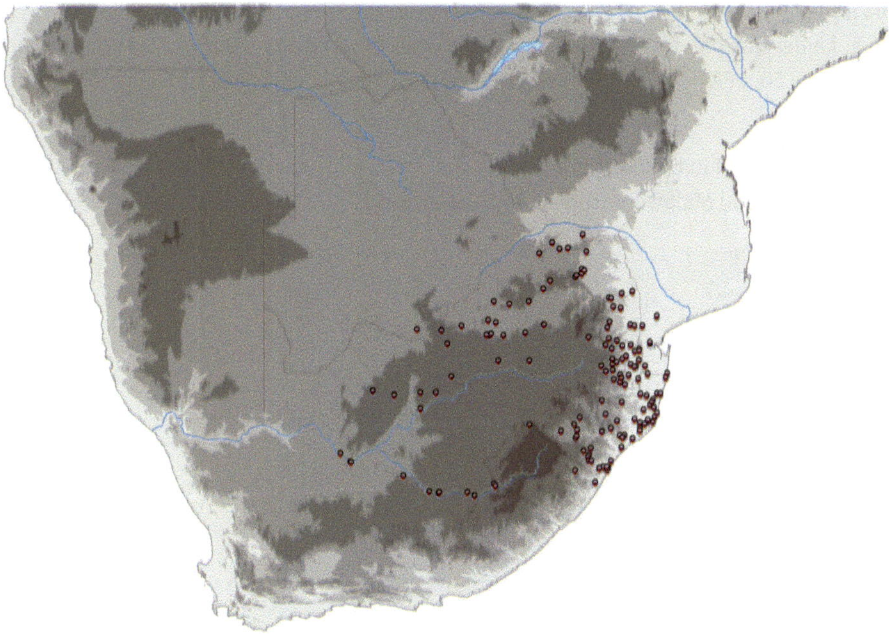

Figure 2.446 Distribution map

Measurements

External and cranial measurements (mm) and mass (g) for *Aethomys ineptus*, sexes combined

	Mean	Min	Max	N
Mass	77.7	50	110	55
HB	131.8	109	161	78
Tail	156.9	115	190	77
HF	29.9	25	33	53
Ear	20.2	16	24	47
GLS*	36.0	32.5	40.5	612

*Chimimba et al. (1999); it is important to note that this sample included specimens from north of the Limpopo River which probably refer to *A. chrysophilus* (see Linzey et al., 2003; Russo et al., 2006).

Aethomys kaiseri (Noack, 1887)

Diagnosis

This species may co-occur with the following *Aethomys* species: *A. hindei* (incisors opisthodont; skull relatively narrower, zygomatic width < 55% of GLS); *A. nyikae* (longer tail ca. 115% of HB); *A. chrysophilus* (longer tail ca. 115% of HB; ventral pelage white; incisors opisthodont); *A. bocagei* (size larger, GLS > 39 mm); *A. thomasi* (incisors opisthodont; skull relatively narrower, zygomatic width < 55% of GLS; upper molars narrower, M1 < 2.5 mm). This species has medium brown dorsal pelage with greyish-white ventral pelage, a relatively short tail (ca. 83% of HB), relatively wide skull (zygomatic width > 55% of GLS) and orthodont incisors.

Taxonomic Issues

The taxon *A. hindei* was previously listed as a subspecies of *A. kaiseri* (Hollister, 1919). Chromosome number 2n = 50, aFN = 60 (Castiglia et al., 2003).

Distribution

This species is distributed in the northern parts of the Zambezian woodlands from central Angola west through Zambia and Malawi northward to Lake Victoria.

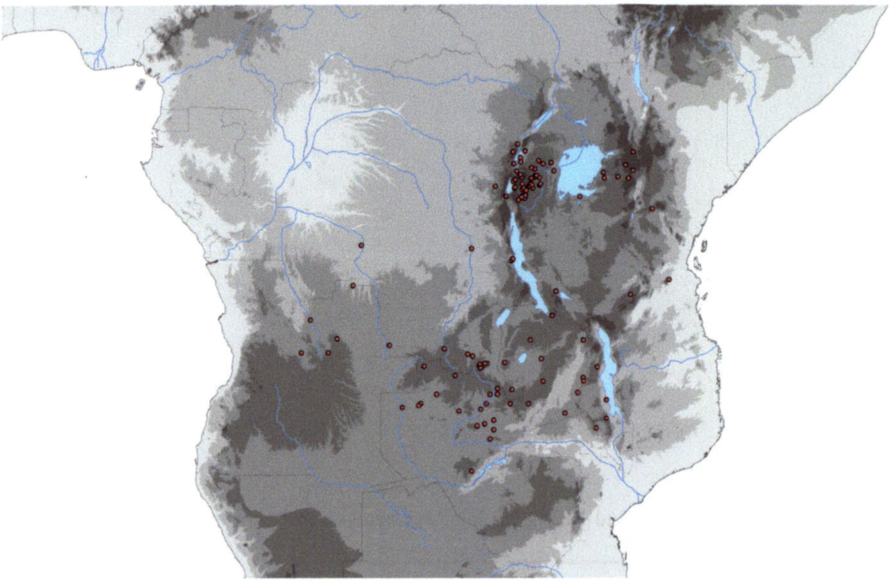

Figure 2.447 Distribution map

Measurements

External and cranial measurements (mm) and mass (g) for *Aethomys kaiseri*, sexes combined

	Mean	Min	Max	N
Mass	113.8	78	144	14
HB	162.5	139	196	23
Tail	134.9	124	167	23
HF	28.9	27	39	23
Ear	21.3	19	24	20
GLS*	36.4	34.0	38.0	9

*Ansell & Ansell (1973)

Aethomys nyikae (Thomas, 1897)

Diagnosis

This species may co-occur with the following *Aethomys* species: *A. thomasi* (shorter tail ca. 83% of HB; restricted to the plateau of Angola); *A. bocagei* (size larger, GLS > 39 mm); *A. kaiseri* (shorter tail ca. 83% of HB); *A. chrysophilus* (incisors opisthodont; ventral pelage white). This species has dark brown dorsal pelage tinged with red, off-white ventral pelage, a long tail (ca. 115% of HB), and orthodont incisors.

Taxonomic Issues

This species has previously been listed as a subspecies of *A. chrysophilus* and *A. kaiseri* (Hanney, 1965; Delany, 1975).

Distribution

This species occurs in the southern parts of the Zambezian woodland region, from northern Angola east to Lake Malawi. The record from eastern Zimbabwe (e.g. De Graaff, 1981) is not included here as it is based on a single juvenile specimen that may have been misidentified (Chimimba et al., 1999).

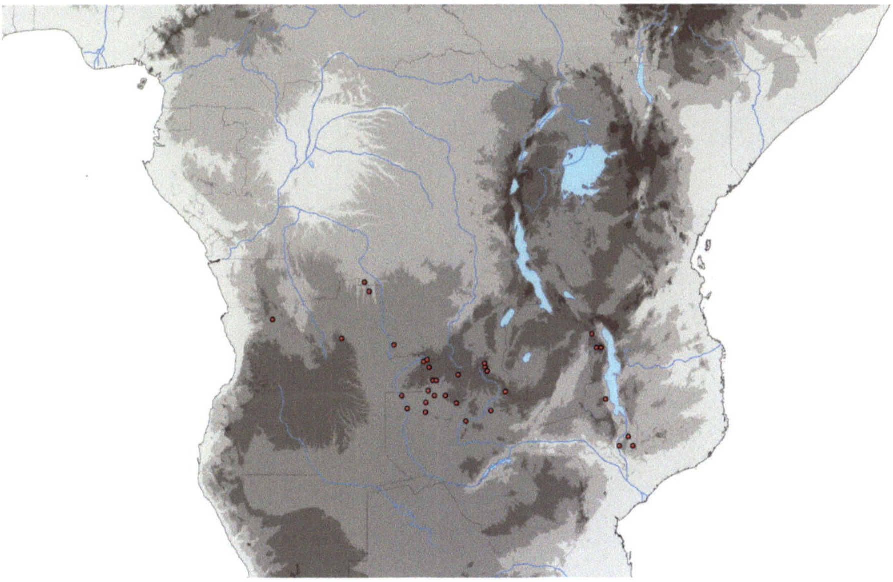

Figure 2.448 Distribution map

Measurements

External and cranial measurements (mm) and mass (g) for *Aethomys nyikae*, sexes combined

	Mean	Min	Max	N
Mass*	90.2	56	106	13
HB*	143.4	120	156	12
Tail*	165.8	145	178	12
HF*	28.6	28	29	14
Ear*	20.6	20	21	14
GLS*	35.1	34	38	8

*Happold (2013)

Aethomys silindensis, Roberts, 1938

Diagnosis
This species is restricted to the eastern highlands of Zimbabwe where it co-occurs only with *Aethomys chrysophilus* (size smaller, GLS < 39 mm; first upper molar narrower, < 2 mm). This species has buffy-brown dorsal pelage with greyish-white ventral pelage, a long tail (ca. 103-110% of HB), and slightly opisthodont incisors.

Taxonomic Issues
This taxon was previously listed as a subspecies of *A. chrysophilus* (Ellerman et al., 1953).

Distribution
This species is known from four localities in the eastern highlands of Zimbabwe and neighbouring Mozambique where it occurs in thick vegetation and rocky areas.

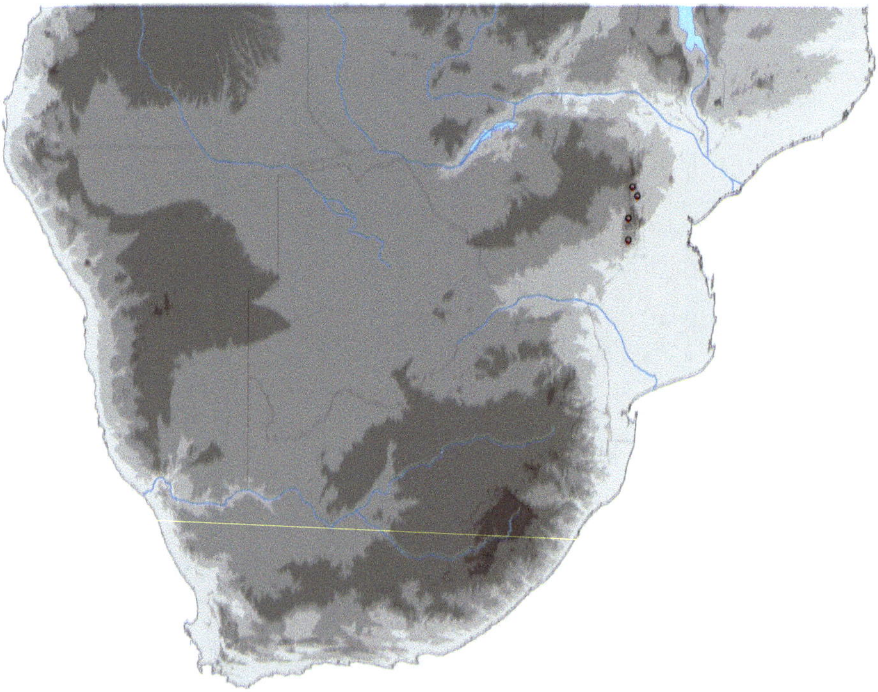

Figure 2.449 Distribution map

Measurements

External and cranial measurements (mm) and mass (g) for *Aethomys silindensis*, sexes combined

	Mean	Min	Max	N
Mass	-	-	-	-
HB*	173.6	155	200	7
Tail*	177.7	166	194	7
HF*	33.6	32	35	8
Ear*	22.6	21	25	8
GLS*	40.4	39.0	41.8	6

*Chimimba et al. (1999)

Aethomys stannarius (Thomas, 1913)

Diagnosis

This species is restricted to northern Nigeria and neighbouring Cameroon where it may co-occur with *A. hindei* (tail on average shorter ca. 100% of HB in Cameroon; ventral pelage greyish-white; auditory bulla larger, > 7mm). This species has yellowish-brown dorsal pelage with white ventral pelage, a relatively long tail (ca. 103-110% of HB) and relatively small auditory bulla (< 7mm).

Taxonomic Issues

This species was previously listed as a subspecies of *A. hindei* (Happold, 1987).

Distribution

This species is restricted to northern Nigeria, extending marginally into northern Cameroon where it occurs in savanna habitats.

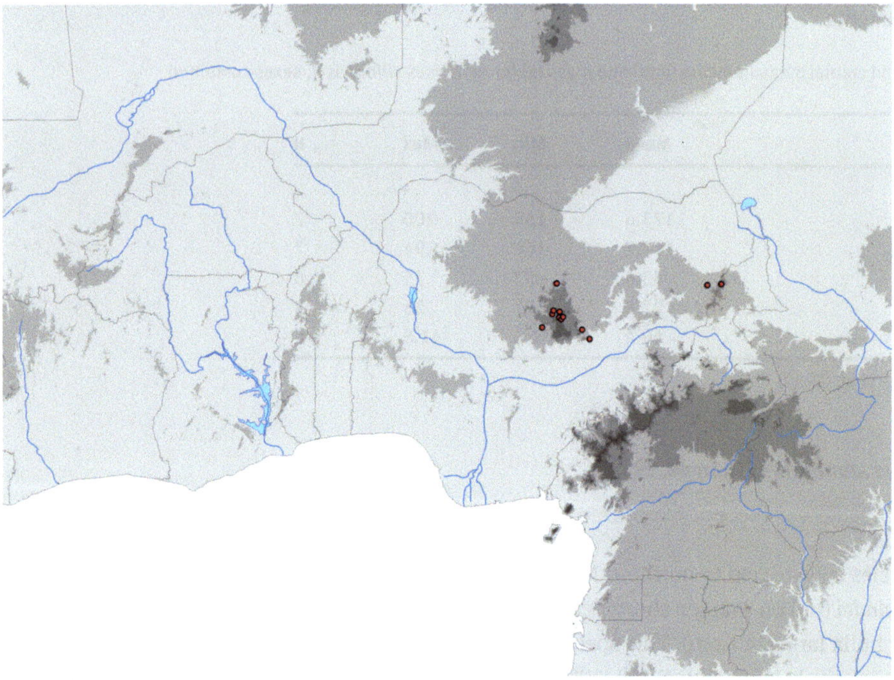

Figure 2.450 Distribution map

Measurements

External and cranial measurements (mm) and mass (g) for *Aethomys stannarius*, sexes combined

	Mean	Min	Max	N
Mass*	92	60	120	3
HB*	148	126	169	4
Tail*	152	124	203	4
HF*	31	28	33	4
Ear*	22	20	24	4
GLS*	36.4	34.4	39.8	6

*Happold (2013)

Aethomys thomasi (de Winton, 1897)

Diagnosis

This is one of five species of *Aethomys* occurring in Angola, the others include: *A. bocagei* (longer tail > 100% HB; size larger, GLS > 39 mm; ventral pelage white); *A. chrysophilus* (longer tail, ca. 115% or more of HB; ventral pelage white); *A. kaiseri* (incisors opisthodont; skull relatively broader, zygomatic width > 55% of GLS; upper molars broader, M^1 > 3 mm); *A. nyikae* (longer tail, ca. 115% of HB). This species has grizzled rufous-brown dorsal pelage with greyish-white ventral pelage and a long tail (ca. 106% of HB).

Taxonomic Issues

This species was previously listed as a subspecies of *A. kaiseri* (Ellerman et al., 1953), and is very similar to the form *vernayi* listed as *A. kaiseri* (Crawford-Cabral, 1999).

Distribution

This species is endemic to Angola where it occurs on the central plateau.

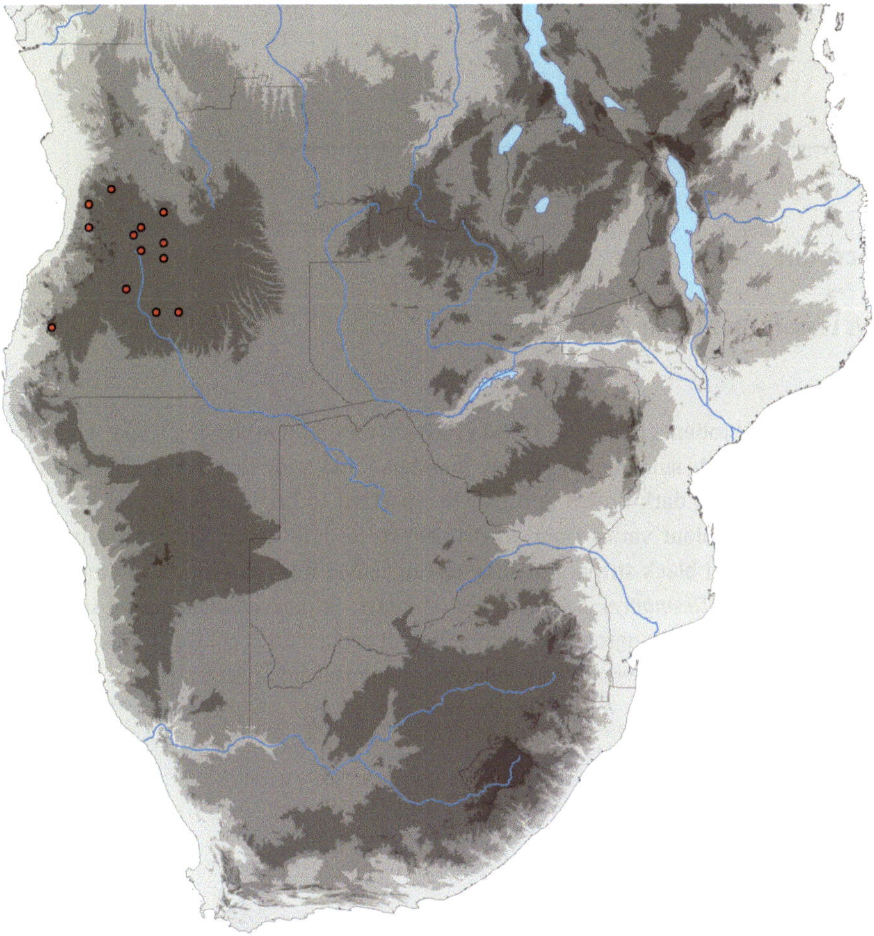

Figure 2.451 Distribution map

Measurements

External and cranial measurements (mm) and mass (g) for *Aethomys thomasi*, sexes combined

	Mean	Min	Max	N
Mass	-	-	-	-
HB*	149.2	140	162	10
Tail*	124.5	112	135	11
HF*	28.6	25	31	10
Ear*	22.0	20	25	11
GLS*	36.0	-	-	1

*Happold (2013)

Genus Account

Arvicanthis Lesson, 1842 (Grass Rats)

Diagnosis

Medium to large-sized rodents (Mass ca. 35-130 g, HB ca. 100-160 mm) having a sparsely-haired, coarse-scaled tail which is subequal or shorter than head and body length (ca. 65-95% of HB). Dorsal colour varies from dark brown and heavily speckled to brighter buffy and less darkly speckled, and ventral colour varies from greyish brown to pure white. There is sometimes a barely visible mid-dorsal black stripe. It can be distinguished from closely related members of the *Arvicanthis* division (*Desmomys, Lemniscomys, Mylomys, Pelomys, Rhabdomys* by its reduced mammary formula (1 + 2 = 6), ungrooved incisors, lack of conspicuous pelage stripes and highly reduced fifth finger. Some members of the genus are commensal crop pests.

Figure 2.452 Photograph of *Arvicanthis niloticus* (Ara Monadjem).

Cranial Characters

The skull is medium-large in size (GLS ca. 28-38 mm), robust and ovoid in shape, with well-developed supraorbital crests and robust zygomatic arches which are thickened in the middle (as in Otomyinae). The nasal bone is rather short and narrow, tapering at the tip. The anterior palatal foramina reach just to the first upper molars. The anterior surface of the zygomatic plate is vertical and bluntly rounded dorsally. The bullae are relatively small (greatest diameter ± 20% of GSL).

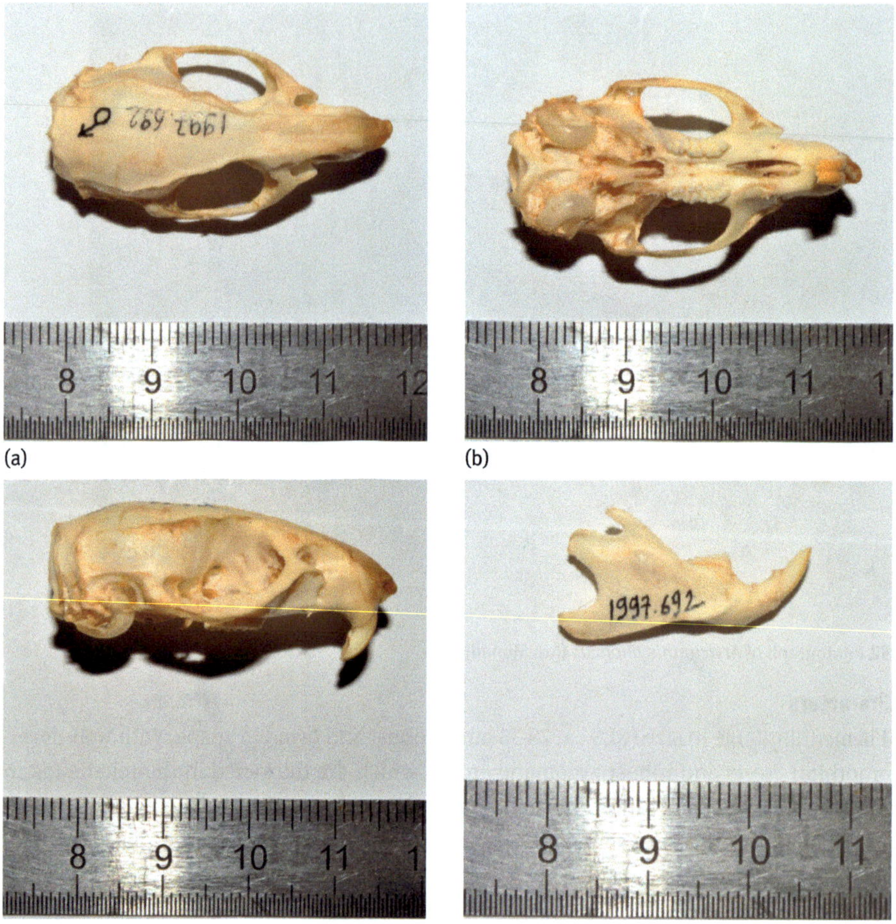

(a) (b) (c) (d)

Figure 2.453 a-d Cranium of *Arvicanthis niloticus* (MNHN-ZM-MO 1997-692) (from left top to bottom right: dorsal view, ventral view, lateral view and lower jaw).

Dentition

The upper incisors are orthodont to slightly pro-odont and ungrooved. The molars show macrodonty (MTR/GSL ± 0.21, Table 10); this varies between species being most developed in *A. blicki* and least developed in *A. neumanni*. Individual cusps are large and well-separated but not as distinctively large as in *Pelomys* (although *A. blicki* approaches this) and *Mylomys*. Rows of cusps on M^1 are barely distorted, with t1 almost in line with t2 and t3. Molar M^1 is the longest molar in the toothrow. Molar M^3 lacks the t3 cusp (Group 4 in Figure 2.439), a character found in all murids except for *Acomys*, *Uranomys*, *Lophuromys* and genera (*Desmomys*, *Lamottemys*, *Oenomys*, *Thamnomys*) showing stephanodonty (formation of longitudinal crests between transverse rows of cusps leading to larger and higher cusps and a more abrasive occlusal surface associated with a more herbivorous diet).

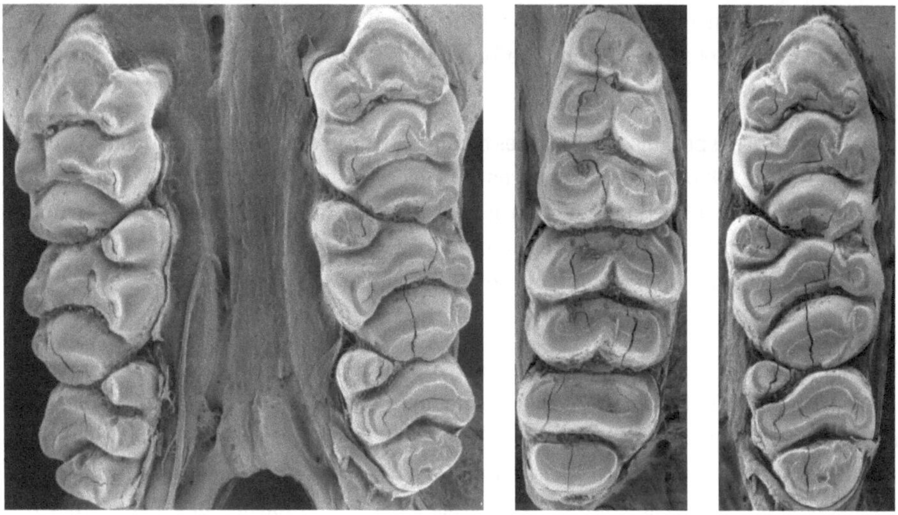

(a) (b) (c)

Figure 2.454 Cheek teeth of *Arvicanthis niloticus* (MNHN-ZM-MO 1978-12) (A: upper cheek teeth; B: lower right toothrow; C: upper left toothrow).

Habitat & Biogeography

They are typically associated with grassland and bush within the northern savannas, often associated with anthropogenic habitats. Habitat associations include semi-deserts along the Nile River and the Sahelian zone (*A. niloticus*), Sudanian and Guinean savannas of West Africa (*A. rufinus* and *A. ansorgei*), mesic to arid savannas in East Africa (*A. nairobae* and *A. neumanni*) and montane grasslands and alpine moorlands on the Ethiopian Plateau (*A. abyssinicus* and *A. blicki*). The genus distribution ranges from Egypt (along the Nile Valley) and south-western Arabia in the north to eastern Zambia in the south and from Senegal in the west to Ethiopia in the east.

Fossil Record

Fossils have been recovered from the Middle to Late Pleistocene of North Africa, the Late Miocene to Middle Pleistocene of East Africa and the Early to Late Pleistocene of Southern Africa (northern South Africa and Zambia) (Winkler et al., 2010). Thus, whilst the present distribution is confined to Sub-Saharan Africa and the south-western portion of the Arabian Peninsula, the genus had a wider distribution in northern and southern Africa during the Plio-Pleistocene and it was also known from the Middle to Late Pleistocene of Israel, Afghanistan (Sen et al., 1979), North Africa (Jaeger, 1975).

A. musisii Kazinga Channel, Uganda (Mein, 1994)
A. primaevus Olduvai Bed I (Jaeger, 1976)
A. cf *niloticus* Olduvai Bed II, Lake Natron (Denys 1987b)
A. niloticus chroudiae (Jaeger 1975)
A. sp. Garba IV (Sabatier, 1982)
A. sp. Omo Member B (Wesselman, 1984)
A. ectos Israel – Middle to Late Pleistocene (Tchernov, 1968)

Arvicanthis has also been reported, but not yet described, from Lemudong'o (Manthi, 2007), Aramis (4.4 Ma, Louchart et al., 2009), and the Tugen Hills (Winkler, 2002).

Biology & Ecology

These diurnal rodents are specialised herbivores feeding on the leaves, seeds and shoots of grasses (Kingdon, 1974; Delany, 1986). Breeding patterns vary geographically from seasonal (further from the Equator) to aseasonal (close to the Equator); photoperiod change stimulated reproduction in populations from 13-14° N but not from those close to the Equator (Sicard et al., 1992; Nunes et al., 2002). Mean litter size varies from around four to six (Delany, 1986).

Systematic Notes

Molecular data (Ducroz et al., 2001; Lecompte et al., 2008) support the existence of the tribe Arvicanthini comprising 13 African genera. This tribe includes a closely associated clade (*Arvicanthis* division) of six genera (*Arvicanthis, Desmomys, Mylomys, Pelomys, Lemniscomys, Rhabdomys*) and an additional eight genera (*Aethomys, Micaelamys, Dasymys, Hybomys, Stochomys, Grammomys, Oenomys* and *Golunda*). Musser and Carleton (2005) suggest the addition of four genera to this tribe (*Dephomys, Lamottemys, Thallomys, Thamnomys*) and one fossil genus (*Malpaisomys*). Ducroz et al. (2001) mapped several morphological characters onto the molecular clade for Arvicanthini but found none to be diagnostic for either the tribe Arvicathini or the Division *Arvicanthis*. However, within the tribe, the reduction in the number of mammae was diagnostic for the genus *Arvicanthis*.

Previous treatments recognised as few as one species of *Arvicanthis* in Africa, *A. niloticus* (Misonne, 1974; Honacki et al., 1982). Subsequent studies have shown that this species complex contains at least seven species, and additional taxa may still be undescribed within *A. niloticus* sensu stricto as currently defined (Musser & Carleton, 2005). Indeed, *A. niloticus* from the Nile Valley is distinguished by a cytochrome-b genetic distance of approximately 9% (representing a 1.8 MYA divergence) from West African populations (Ducroz et al., 2001; Abdel Rahman Ahmed et al., 2008). Detailed chromosomal, molecular, and interbreeding studies by Volobouev et al. (1987, 2002a,b), Ducroz (1998), and Ducroz et al. (1998) revealed the existence of at least three additional species in West Africa based on chromosomal grounds; two of these have been described as *A. rufinus* and *A. ansorgei*, whilst a third chromosomal form, "ANI-2" from central Africa remains to be described. Although a number of authors considered two species of *Arvicanthis* to occur in the Nile Valley (*A. niloticus* and *A. testicularis*: e.g., Delany, 1975; Bekele et al., 1993; Fadda & Corti, 1998), the taxon *testicularis* has been shown conclusively to be a synonym of *niloticus* and only one species occurs in the Nile Valley (Abdel Rahman Ahmed, 2005; Abdel Rahman Ahmed et al., 2008; 2009).

Annotated Species List

Seven species described:

Arvicanthis abyssinicus (Rüppell, 1842)
Arvicanthis ansorgei Thomas, 1910
Arvicanthis blicki Frick, 1914
Arvicanthis nairobae J. A. Allen, 1909
Arvicanthis neumanni (Matschie, 1894)
Arvicanthis niloticus (Desmarest, 1822)
Arvicanthis rufinus (Temminck, 1853)

Key to Species

The species belonging to this genus are still poorly diagnosed morphologically and for this reason a key cannot be presented here. The seven species can be distinguished on chromosomal and molecular grounds. However, most species appear to be allopatric or parapatric, and can therefore be tentatively identified on geographic location. See the species accounts for additional information that can be used to distinguish the species.

Species Accounts

Arvicanthis abyssinicus (Rüppell, 1842)

Diagnosis

This is one of four *Arvicanthis* species occurring in Ethiopia from which it can be distinguished as follows: *A. blicki* (size larger, GLS > 35 mm; HF > 34 mm; typically occurs at higher elevations above 3500 m in moorland; 2n = 48); *A. neumanni* (size smaller, GLS typically < 31 mm, HF < 27 mm; occurs at lower elevations below 1500 m; 2n = 54); *A. niloticus* (typically occurs below 2000 m; 2n = 62).

Taxonomic Issues

Chromosome number 2n = 62, FN = 64 (Ducroz et al., 1998).

Distribution

This species is endemic to Ethiopia where it occurs on both sides of the Rift Valley (but more widespread on the west), inhabiting montane grasslands above 2000 m (Yalden et al., 1976).

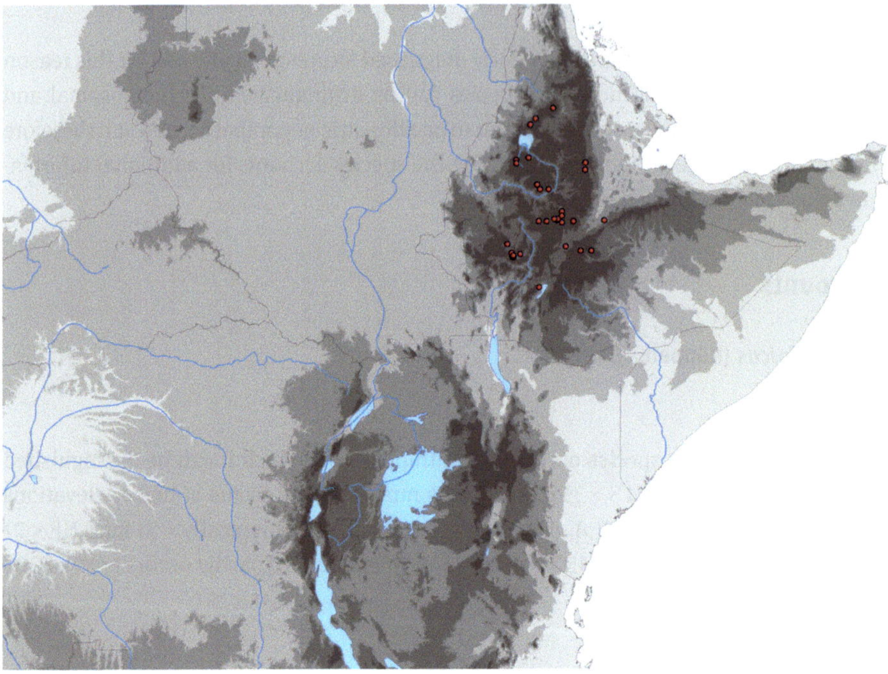

Figure 2.455 Distribution map

Measurements

External and cranial measurements (mm) and mass (g) for *Arvicanthis abyssinicus*, sexes combined

	Mean	Min	Max	N
Mass	99.5	72	125	6
HB	140.8	125	150	6
Tail	102.2	98	109	6
HF	30.7	28	32	6
Ear	15.8	13	17	6
GLS	32.4	30.5	34.1	6

Arvicanthis ansorgei Thomas, 1910

Diagnosis

This is one of three species of *Arvicanthis* currently described from West Africa from which it differs only on molecular and chromosomal grounds (Ducroz et al., 1998; Volobouev et al., 2002b). The three species appear to be parapatric with *A. ansorgei* sandwiched between the other two species (*A. rufinus* in Guinea woodland to the south and *A. niloticus* in drier savanna to the north). This species cannot be distinguished from *A. niloticus* on morphological grounds, but differs in karyotype and on molecular grounds (Ducroz et al., 1998; Dobigny et al., 2013).

Taxonomic Issues

This taxon was previously listed as a synonym of *A. niloticus*. Chromosome number 2n = 62, FN = 74-76 (Volobouev et al., 2002b).

Distribution

This species is known from only a few scattered localities in Sudanian savanna from Senegal and Guinea-Bissau in the west through southern Mali to Burkina Faso (Granjon & Duplantier, 2009). It is sympatric with *A. niloticus* in southern Mali.

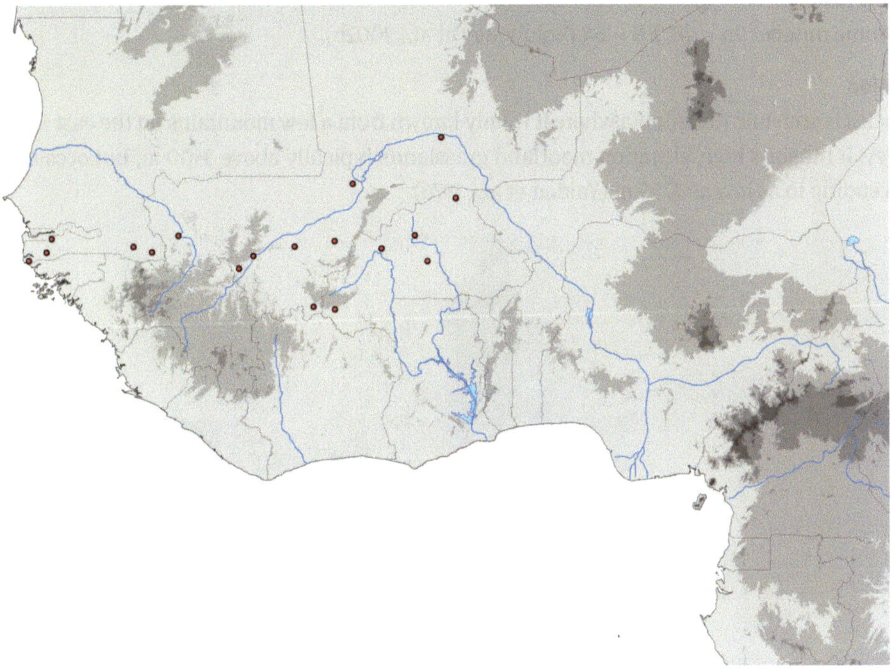

Figure 2.456 Distribution map

Measurements

External and cranial measurements (mm) and mass (g) for *Arvicanthis ansorgei*, sexes combined

	Mean	Min	Max	N
Mass*	103.6	68	180	88
HB*	156.4	129	177	83
Tail*	133.0	110	159	59
HF*	30.3	27	34	84
Ear*	20.0	16	23	82
GLS	37.1	34.4	39.7	2

*Granjon & Duplantier (2009)

Arvicanthis blicki Frick, 1914

Diagnosis
This is one of four *Arvicanthis* species occurring in Ethiopia from which it can be distinguished as follows: *A. abyssinicus* (size smaller, GLS < 35 mm; HF < 34 mm; typically occurs at lower elevations below 3500 m in montane grassland; 2n = 62); *A. neumanni* (size smaller, GLS < 34 mm; occurs at lower elevations below 1500 m; 2n = 54); *A. niloticus* (typically occurs below 2000 m; 2n = 62).

Taxonomic Issues
Chromosome number 2n = 48, FN = 64 (Volobouev et al., 2002b).

Distribution
This species is endemic to Ethiopia where it is only known from a few mountains on the east of the Rift Valley. It inhabits high-elevation moorland grasslands typically above 3500 m, but occasionally descending to as low as 2750 m (Yalden et al., 1976).

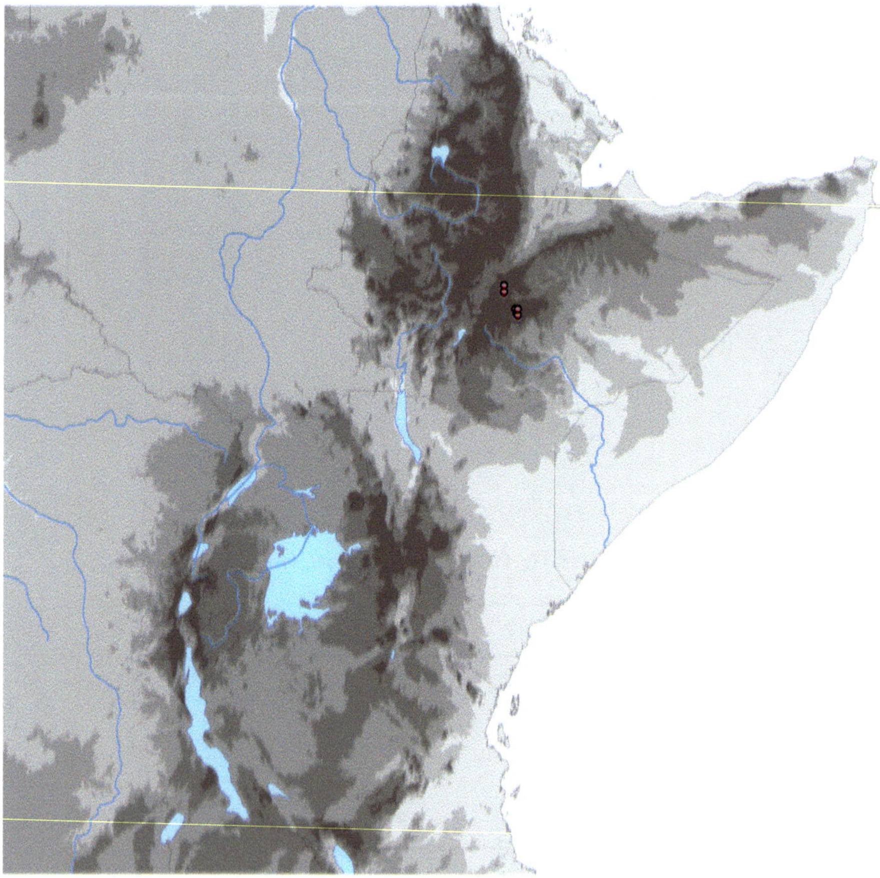

Figure 2.457 Distribution map

Measurements

External and cranial measurements (mm) and mass (g) for *Arvicanthis blicki*, sexes combined

	Mean	Min	Max	N
Mass	143.3	140	150	3
HB	169.0	160	175	5
Tail	107.2	102	112	5
HF*	35.4	34	37	5
Ear	19.6	19	20	5
GLS	35.8	35.4	36.4	5

*Happold (2013) reports a mean HF of 30.0 and 30.8 mm (range 29-32 mm) for male and female *A. blicki* which in our opinion is far too small for this species

Arvicanthis nairobae J. A. Allen, 1909

Diagnosis

This species co-occurs in East Africa with *A. niloticus* (size larger, GLS typically > 32 mm; occurs west of the Rift Valley) and *A. neumanni* (differs in karyotype; 2n = 53-54 in area of overlap).

Taxonomic Issues

On molecular and chromosomal grounds, this species is more closely related to West African *Arvicanthis* (*A. rufinus* and *A. ansorgei*) than to East African species (*A. niloticus, A. neumanni, A. abyssinicus* and *A. blicki*). Chromosome number 2n = 62, FN = 78 (Castiglia et al., 2006).

Distribution

This species is endemic to East Africa where it occurs in a narrow band from north-eastern Tanzania to central and western Kenya. Due to possible confusion with *A. niloticus* and *A. neumanni*, the distributional limits of this species are unresolved. The specimens from northern Tanzania have been confirmed by karyological data (Fadda et al., 2001a). The taxonomic identity of the Kenyan specimens is uncertain, and may they may refer to an as yet undescribed species (Castiglia et al., 2006). Musser & Carleton (2005) mention a record from southern Ethiopia which is not mapped here.

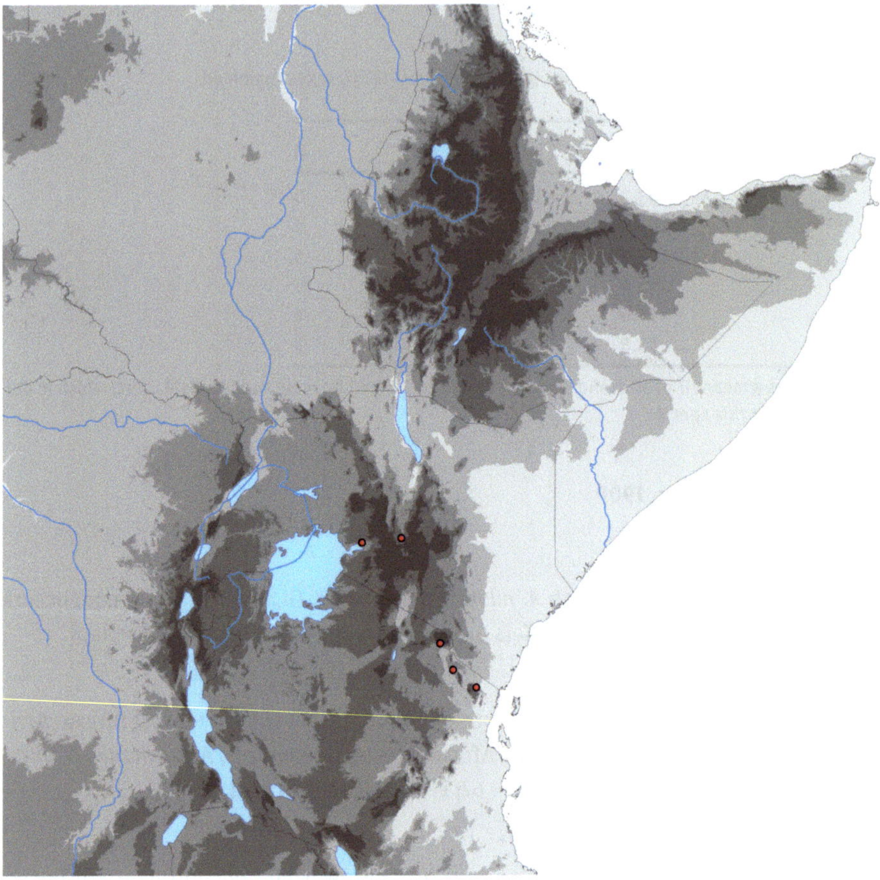

Figure 2.458 Distribution map

Measurements

External and cranial measurements (mm) and mass (g) for *Arvicanthis nairobae*, sexes combined

	Mean	Min	Max	N
Mass	-	-	-	-
HB*	144	-	-	4
Tail*	116	105	120	4
HF*	26	-	-	1
Ear	-	-	-	-
GLS*	34	-	-	1

*Allen (1909); HB calculated by subtracting Tail from Total length

Arvicanthis neumanni (Matschie, 1894)

Diagnosis

This is one of four *Arvicanthis* species occurring in Ethiopia from which it can be distinguished as follows: *A. abyssinicus* (size slightly larger, HF > 25 mm but with considerable overlap; occurs at higher elevations above 2000 m in montane grassland; 2n = 62); *A. blicki* (size much larger, GLS > 35 mm, HF > 34 mm; occurs at higher elevations above 2750 m in moorland; 2n = 48); *A. niloticus* (2n = 62). Outside of Ethiopia this species also co-occurs with *A. niloticus* (see above) and *A. nairobae* (2n = 62).

Taxonomic Issues

Chromosome number 2n = 62, FN = 66-67 in Ethiopia (Baskevich & Lavrenchenko, 2000), and 2n = 53-54, FN = 62 in Tanzania (Fadda et al., 2001a), suggesting a species complex (Castiglia et al., 2006). The name *A. somalicus* is available for the northern taxon (occurring in Ethiopia and Somalia). We retain these two taxa in *A. neumanni* pending further systematic studies.

Distribution

This species has been recorded widely in East Africa, extending from northern Tanzania in the south to Ethiopia and Somalia in the north. This is a low-lying species typically occurring in arid savanna below 1500 m. Due to possible confusion with *A. nairobae* and *A. niloticus* (Musser & Carleton, 2005), the distributional limits of this species are currently unresolved. The specimens from Tanzania have been karyotyped and refer to *A. neumanni* (Fadda et al., 2001a; Castiglia et al., 2006). The specimens from Ethiopia and Somalia may refer to a different species (for which the name *A. somalicus* is available). The taxonomic status of the Kenyan specimens is uncertain and may represent either of these two species or some other species.

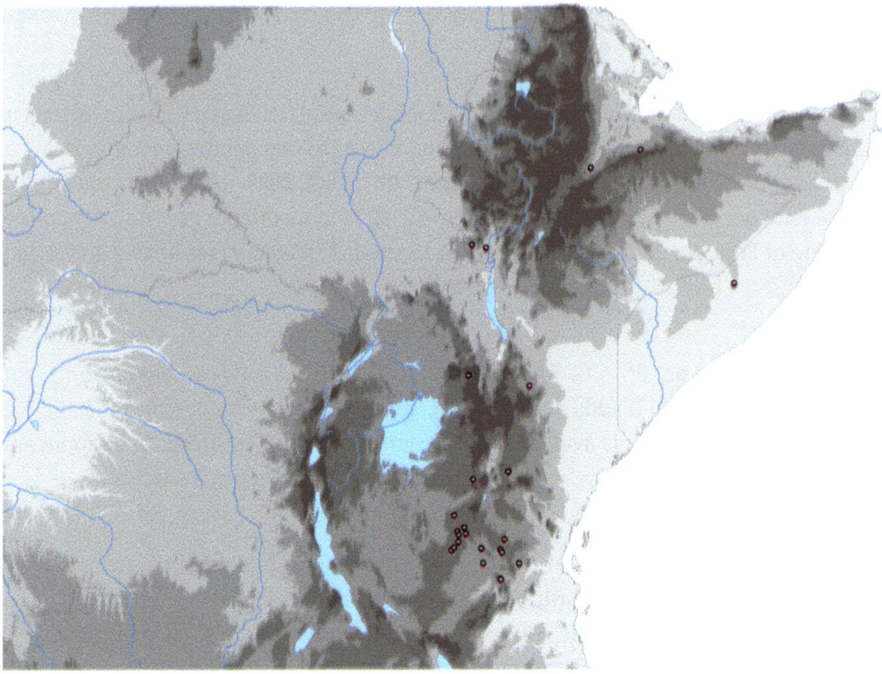

Figure 2.459 Distribution map

Measurements

External and cranial measurements (mm) and mass (g) for *Arvicanthis neumanni**, sexes combined

	Mean	Min	Max	N
Mass	62.7	48	72	3
HB	129.6	115	160	7
Tail	105.3	88	118	7
HF	25.3	24	27	6
Ear	16.0	14	18	6
GLS	30.4	28.8	31.8	8

*The values presented here are larger than those in Happold (2013), potentially reflecting differences between specimens from the southern and northern parts of this species range.

Arvicanthis niloticus (Desmarest, 1822)

Diagnosis

This species is widely distributed throughout tropical Africa where it overlaps, albeit marginally, with perhaps all other *Arvicanthis* species. There are currently no morphological features that can be used to confidently distinguish this species, and it may in fact represent a species complex (see Taxonomic Issues). See the other *Arvicanthis* species accounts for further information.

Taxonomic Issues

This species was previously thought to be the sole representative of the genus *Arvicanthis* (see genus account). The Ethiopian populations were previously referred to as *A. deembensis* (Yalden et al., 1976), but thought to be conspecific with *A. niloticus* by Musser & Carleton (2005). There are at least three different karyotypes within what is referred to as *A. niloticus* present in Ethiopia (2n = 62, 2n = 60, 2n = 56), suggesting the presence of two cryptic species, named ANI-6 and ANI-7 by Castiglia et al. (2006). One of these, ANI-6, is in fact more closely related to *A. ansorgei* (from which it differs by 6-8% on the cytochrome b gene) than to *A. niloticus*, further supporting its specific status. A molecular study suggested that the species is monophyletic with some geographic structure in three clades (Dobigny et al., 2013), but did not explore the East African diversity in Tanzania. This species complex is in urgent need of complete integrative taxonomic review.

Distribution

This species occurs widely in tropical Africa, extending from Senegal and Mauritania in the west to Ethiopia in the east and south to Tanzania, with an isolated population in western Zambia. It is associated with savanna habitats.

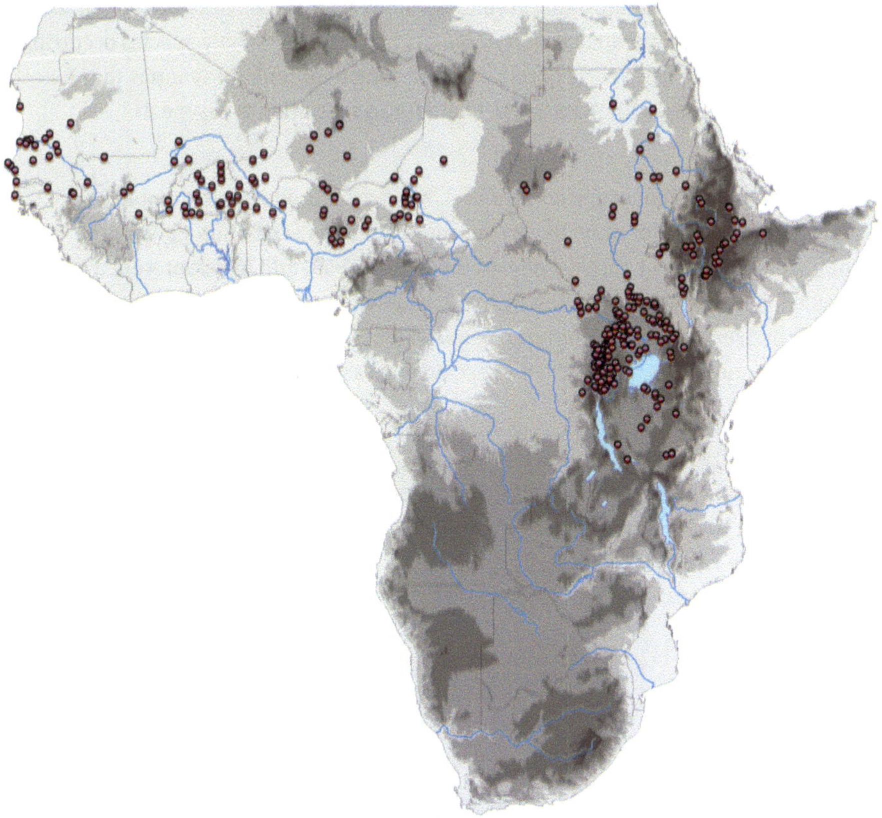

Figure 2.460 Distribution map

Measurements

External and cranial measurements (mm) and mass (g) for *Arvicanthis niloticus*, sexes combined

	Mean	Min	Max	N
Mass	114.8	89	160	17
HB	156.6	127	188	33
Tail	129.6	92	155	30
HF	33.0	30	36	32
Ear	18.3	13	22	31
GLS	34.9	29.8	40.2	43

Arvicanthis rufinus (Temminck, 1853)

Diagnosis

This is one of three species of *Arvicanthis* currently described from West Africa from which it differs only on molecular and chromosomal grounds (Ducroz et al., 1998; Volobouev et al., 2002b). The three species appear to be parapatric and *A. rufinus* is typically the only species occurring in Guinea woodland (the other two species occur in drier Sahelo-Sudanian savannas to the north, with perhaps marginal overlap with *A. ansorgei* where these vegetation types meet).

Taxonomic Issues

This taxon was previously listed as a synomym of *A. niloticus*. Chromosome number 2n = 62, FN = 76 (Volobouev et al., 2002b) for animals west of the Niger River. The animals from the Central African Republic have a different chromosome number 2n = 56-58, and may represent a different, as yet undescribed, species (corresponding to ANI-2 in Dobigny et al., 2013). However, these two populations differ by 5.4% on the cytochrome b gene, which is rather small for inter-species differentiation. The taxonomic status of the Nigerian specimens east of the Niger River and those from Cameroon are at present not known.

Distribution

This species occurs widely in West Africa where it is associated with Guinea woodland. The specimens west of the Niger River (south-western Nigeria to Sierra Leone) refer to *A. rufinus*; the specimens to the east of the river are temporarily placed here, but probably refer to an undescribed species (see Taxonomic Issues).

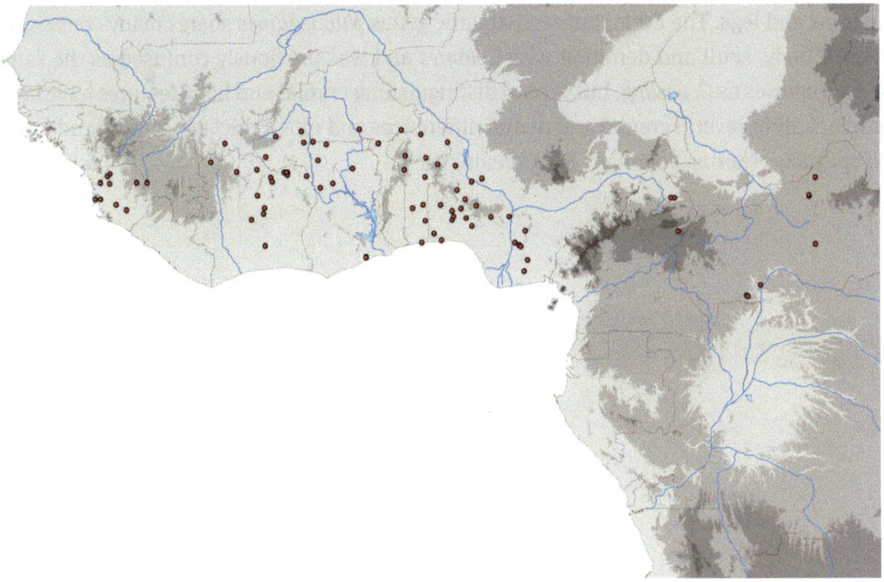

Figure 2.461 Distribution map

Measurements

External and cranial measurements (mm) and mass (g) for *Arvicanthis rufinus*, sexes combined

	Mean	Min	Max	N
Mass*	103.3	59	168	13
HB*	143.8	115	177	13
Tail*	138.2	127	146	4
HF*	32.7	29	38	14
Ear*	19.8	16	24	11
GLS	37.9	34.7	40.5	8

*Happold (2013)

Genus Account

Colomys Thomas & Wroughton, 1907 (African Wading Rats)

Diagnosis

This is a distinctive rodent with long hindfeet (HF > 32 mm or 26% of HB), partly webbed toes, enlarged foramen magnum, a proportionately large brain and dense mystacial vibrissae on a thickened muzzle, adapted for foraging on aquatic insects by wading in shallow streams (Dieterlen & Statzner, 1981; Stephan & Dieterlen, 1982; Dieterlen, 1983; Kerbis Peterhans & Patterson, 1995). The tail is longer (130% of HB) than head and body length. The dorsal pelage is dense and velvety in texture (making it buoyant and/or waterproof) and buffy-brown in colour with a darker brown cap on the head. The ventral pelage is pure white extending laterally onto the sides of the body and arms and legs. The Ethiopian semi-aquatic genus *Nilopegamys* shares many convergent similarities of body, skull and dentition with *Colomys* and was previously confused as the same genus or even species as *C. gosling*, but several distinguishing cranial and body features have been described, including several cranial and dental differences and much thicker hindfeet and larger size in *Nilopegamys* (Kerbis Peterhans & Patterson, 1995).

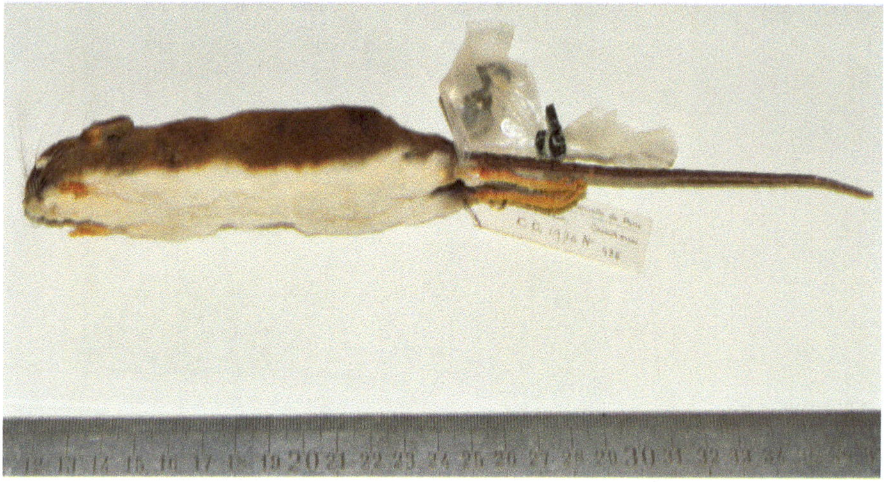

Figure 2.462 Photograph of *Colomys goslingi* museum specimen (MHNH-ZM-MO 1970-436) (Ara Monadjem).

Cranial Characters

The skull is medium in size (GSL ca. 34 mm), long, thin, smooth (without crests) and flattish in shape, with very gracile zygomatic arches. The foramen magnum is enlarged. The nasal bone is long, thin and tapering at the tip. The anterior palatal foramen just reaches the first upper molars. The zygomatic plate is narrow with the basal portion barely expanded and the anterior surface vertical. The bullae are relatively small (greatest diameter ± 13% of GSL). A mastoid strut is present.

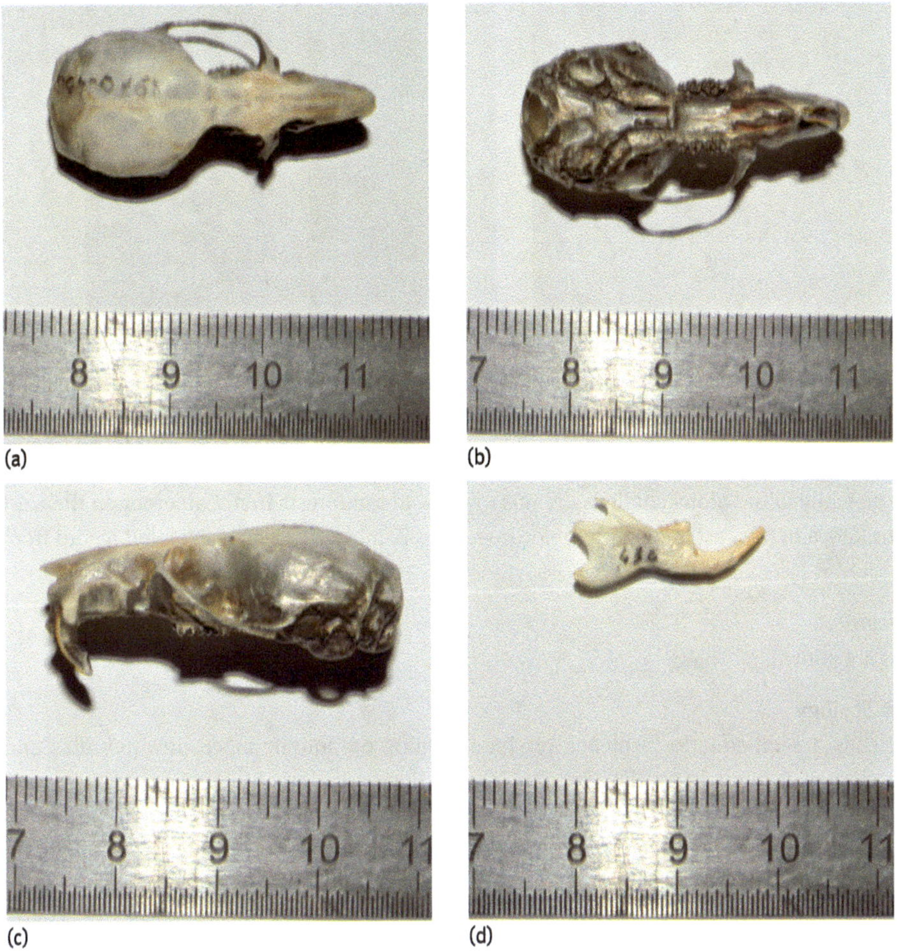

Figure 2.463 a-d Cranium of *Colomys goslingi* (MNHN-ZM-MO 1970-436) (from top left to bottom right: dorsal view, ventral view, lateral view and lower jaw).

Dentition

Incisors are opisthodont, narrow, very long and ungrooved. Molars are average-sized (MTR/GSL ± 0.15, Table 10). Individual cusps are well-separated. Rows of cusps on M^1 are arranged in transverse laminae, with t1 and t3 at the same level and slightly posterior to t2. Molar M^1 is the longest in the toothrow, and much broader than M^2 (150-190%). M^3 is highly reduced with only a single distal cusp in the position of t8 and t9. Molar M^3 lacks the t3 cusp (Group 5 in Figure 2.439), a character found in all murids except for *Acomys*, *Uranomys*, *Lophuromys* and genera (*Desmomys*, *Lamottemys*, *Oenomys*, *Thamnomys*) showing stephanodonty (formation of longitudinal crests between transverse rows of cusps leading to larger and higher cusps and a more abrasive occlusal surface associated with a more herbivorous diet).

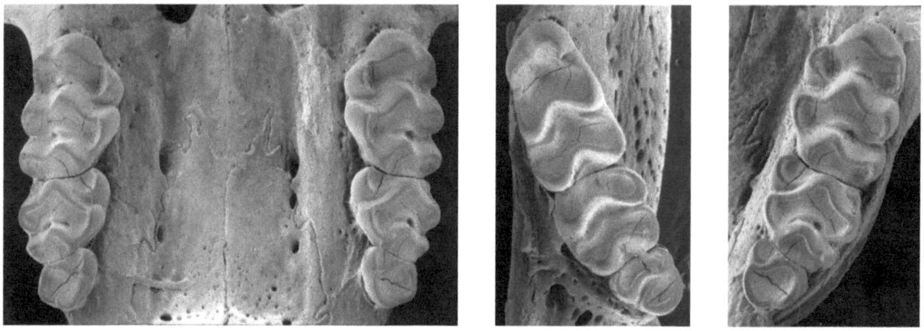

(a) (b) (c)

Figure 2.464 Cheek teeth of *Colomys goslingi* (MNHN-ZM-MO 1970-436) (A: upper cheek teeth; B: lower left toothrow; C: upper left toothrow).

Habitat & Biogeography

They are typically found along shallow streams in tropical rainforests from Cameroon in the west to western Kenya in the east and south to north-western Zambia. There is an isolated record from Liberia.

Fossil Record

No fossils are known.

Biology & Ecology

The species is a semi-aquatic carnivous predator feeding on aquatic macro-invertebrates and small vertebrates. It hunts in shallow streams by standing on its long hindfeet in shallow water with its vibrissae spread out over the water like a sensory net to detect prey and using its fingers to sift through mud and detritus (Dieterlen, 1983).

Systematic Notes

Morphological (Misonne, 1969; Musser & Carleton, 2005) and molecular (Lecompte et al., 2002a,b; 2008; Schenk et al., 2013) data suggest that *Zelotomys* and *Colomys* are sister species. Lecompte et al. (2008) place these two genera into the *Colomys* Division within tribe Praomyini. Ethiopian *Nilopegamys* was previously included in *Colomys* but this has been clearly refuted by Kerbis Peterhans & Patterson (1995).

Annotated Species List

One species is known, although Musser & Carleton (2005) consider that cryptic species may be present:
Colomys goslingi Thomas & Wroughton, 1907

Species Account

Colomys goslingi Thomas & Wroughton, 1907

Diagnosis

See genus account.

Taxonomic Issues

See genus account.

Distribution

This species is patchily distributed in tropical rainforest from Cameroon to Kenya and south to north-western Zambia, with an isolated record from Liberia.

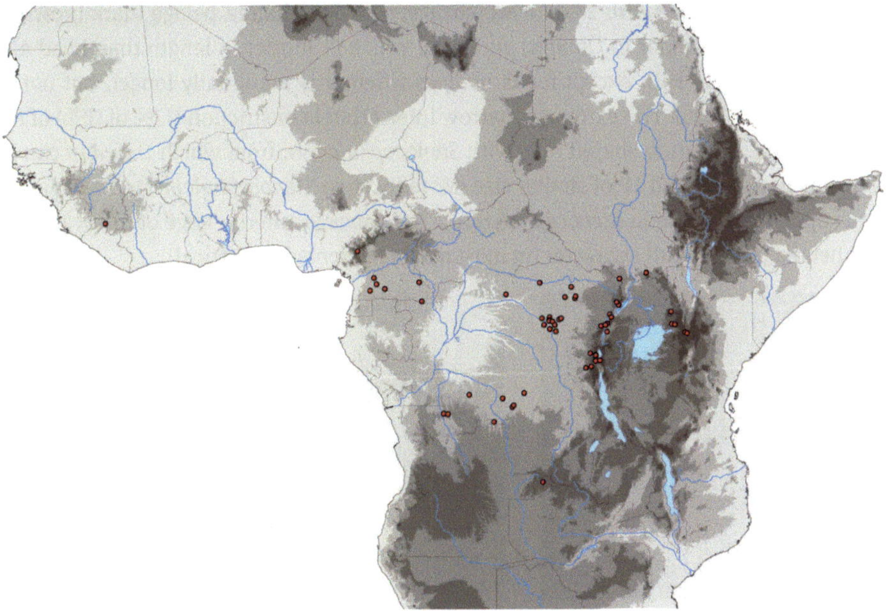

Figure 2.465 Distribution map

Measurements

External and cranial measurements (mm) and mass (g) for *Colomys goslingi*, sexes combined

	Mean	Min	Max	N
Mass	54.6	45	64	4
HB	127.5	120	132	6
Tail	163.8	155	180	6
HF	35.7	32	38	7
Ear	19.4	18	20	5
GLS	33.8	31.7	34.9	6

Genus Account

Dasymys Peters, 1875 (Shaggy Rats)

Diagnosis

These distinctive medium-large sized (ca. Mass 80-120 g; HB ca. 130-170 mm) rodents are characterised by a flat, disc-like face with small eyes, shaggy, soft, long-haired pelage (dark brown to reddish brown in colour, similar to *Otomys*) and a tail which is shorter in length than head and body length (ca. 73-95% of HB), except for *D. nudipes* where it is marginally longer, but not as short as in *Otomys* (45 - 73%). The extremely narrow inter-orbital constriction (12% of GSL) of the cranium is diagnostic, a feature shared only with *Stenocephalemys*, from which *Dasymys* can be distinguished by having M^3 and M^2 of similar size (M^3 reduced in *Stenocephalemys*) and molars of large size (as opposed to those in *Stenocephalemys*). The concave anterior edge of the zygomatic plate is another diagnostic skull character (in most murine rodents it is vertical).

(a) (b)

Figure 2.466 a and b Photographs of *Dasymys* sp. (Enviro-Insight, left and Bill Stanley, right).

Cranial Characters

The skull is medium sized (GSL. 32-40 mm), ovoid in shape, with robust zygomatic arches and a very marked interorbital constriction (12% of GSL). The nasal bone is quite wide and short. The anterior palatal foramina end just before the roots of the first upper molars. The zygomatic plate has a concave anterior surface. The bullae are relatively small (greatest diameter ca. 13% of GSL).

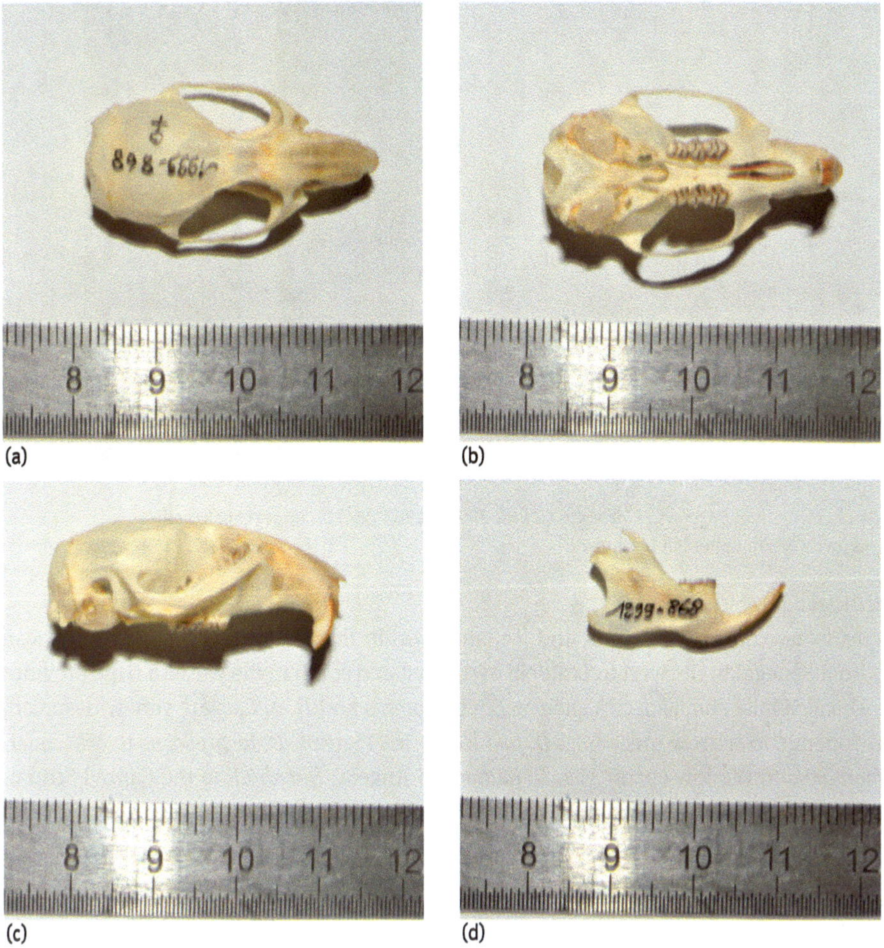

Figure 2.467 a-d Cranium of *Dasymys rufulus* (MNHN-ZM-MO 1999-868) (from top left to bottom right: dorsal view, ventral view, lateral view and lower jaw).

Dentition

Incisors are slightly opisthodont and ungrooved. Molars show macrodonty (MTR/GSL ± 0.21, Table 10). Individual cusps are well-separated in the upper toothrow but in the lower toothrow, cusps of the posterior lobe of M_1, M_2 and M_3 are poorly differentiated.Rows of cusps are arranged in transverse rows, with t1 in line with t2 and t3. Molar M^1 is the longest molar in the toothrow. Molar M^3 lacks the t3 cusp (Group 4 in Figure 2.439), a character found in all murids except for *Acomys*, *Uranomys*, *Lophuromys* and genera (*Desmomys*, *Lamottemys*, *Oenomys*, *Thamnomys*) showing stephanodonty (formation of longitudinal crests between transverse rows of cusps leading to larger and higher cusps and a more abrasive occlusal surface associated with a more herbivorous diet).

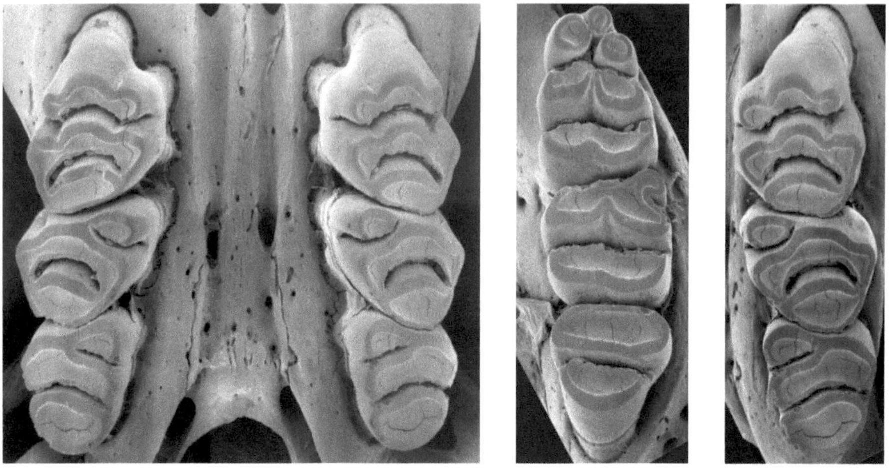

(a) (b) (c)

Figure 2.468 Cheek teeth of *Dasymys* cf *incomtus* (MNHN-ZM-MO 1983-884) (A: upper cheek teeth; B: lower right toothrow; C: upper left toothrow).

Habitat & Biogeography

This species is associated with rivers and wetlands within the northern and southern African savannas from Senegal in the west to Ethiopia in the east and south to the Western Cape province of South Africa. Whilst some species such as *D.* cf *incomtus* and *D. rufulus* are very widespread, many are endemic to narrow areas (e.g. *D. foxi* to the Jos Plateau, *D. longipilosus* to Mt Cameroon, *D. montanus* to the Ruwenzori Mts, *D. nudipes* to Angola, *D. cabrali* to the Caprivi Strip of north-eastern Namibia and neighbouring Okavango Swamps of Botswana, *D. griseifrons* to Ethiopia, and *D. sua* to Tanzania).

Fossil Record

Dasymys is known from the early Pliocene to the Late Pleistocene in Southern African deposits (from South Africa, Namibia and Angola).

Dasymys bolti Plio-Pleistocene, South Africa (Pocock, 1987; Denys, 1990; Senegas, 2000)

Biology & Ecology

These rodents are herbivores and good swimmers, adapted to living in very marshy habitats where they build runways and nests in dense ground cover (Kingdon, 1974; Happold, 2013). Based on a captive study, the gestation period is 29 days and litter size varies from two to five (mean 2.9). Maximum reproductive output during one artificial breeding season was 18 young. Young are altricial and nipple-cling. Reproductive output is low compared to other African rodents (Pillay, 2003).

Systematic Notes

Although Musser & Carleton (2005) recognised only five species, consideration of recent morphometric (Verheyen et al., 2003; Mullin et al., 2002, 2004a,b), allozyme and chromosomal (Mullin et al., 2002) and molecular (Verheyen et al., 2003) studies resulted in the recognition of at least 14 morphological species. At least until more detailed molecular and chromosomal data can be brought to bear, this classification represents a convenient hypothesis for further testing. Mullin et al. (2005) provided a biogeographical framework for 11 of these morphological species, showing that many of the range-restricted endemics were associated with endemic hotspots for other species of mammals. The increase in number of species has been as a result of the splitting of *D. incomtus* (*D. alleni, D. cabrali, D. capensis, D. griseifrons, D. medius, D. nudipes, D. robsertsii, D. rwandae, D. sua*) (Verheyen et al., 2003; Mullin et al., 2004a) and *D. rufulus* (*D. longipilosus*) (Mullin et al., 2004b). Most of these new species were recognised based on morphological grounds. Where molecular analyses were included (Verheyen et al., 2003), the genetic distances between some species pairs was very low, casting doubt on the validity of those species. In particular, the group including: *D. alleni, D. medius, D. rwandae* and *D. sua* (the *medius* group) typically differ 1-3% in the cytochrome b gene, and therefore may refer to a single species (which is distinct from other *D. incomtus* populations to the west and south). *Dasymys incomtus*, as understood here, refers to the population restricted to eastern South Africa (the type locality for this species), leaving the populations outside of this region without a name. We refer to them as *D.* cf *incomtus* here, but note that a number of names are available (see species account). This genus is in urgent need of a continent-wide review.

Annotated Species List

Fifteen species are currently recognised:
Dasymys alleni Lawrence & Loveridge, 1953
Dasymys cabrali Verheyen et al., 2003 [= *shortridgei* Mullin et al., 2004a)
Dasymys capensis Roberts, 1936
Dasymys foxi Thomas, 1912
Dasymys griseifrons Osgood, 1936
Dasymys incomtus (Sundevall, 1847)
Dasymys cf *incomtus* (possibly *Dasymys bentleyae*)
Dasymys longipilosus Eisentraut, 1963
Dasymys medius Thomas, 1905
Dasymys montanus Thomas, 1906
Dasymys nudipes (Peters, 1870)
Dasymys robertsii Mullin et al., 2004a
Dasymys rufulus Miller, 1900
Dasymys rwandae Verheyen et al., 2003
Dasymys sua Verheyen et al., 2003

Key to Species

At this stage it is not possible to provide a key for *Dasymys*. The genus is in urgent need of a synthetic revision employing karyotypic, molecular and morphological data. Recent reviews (e.g. Verheyen et al., 2003; Mullin et al., 2002, 2004a,b) have depended heavily on morphometric variation and have been complicated by different geographical biases and interpretations, leading to some confusion about the status and limits of described species. Apart from relative tail length, which serves to distinguish *D. nudipes* and *D. montanus* from the rest (see Key below), overlap between morphometric characters does not allow us to unambiguously distinguish the remaining species. Nevertheless, with a few exceptions, most species occur allopatrically and geographical distribution together with comments below under Diagnosis in the following species accounts will help to reach at least a tentative identification.

1	Tail longer than head and body length, six plantar pads, restricted to Angola	*D. nudipes*
	Tail shorter than head and body length, five plantar pads, not restricted to Angola	2
2	Tail very short, < 75%, fur dark and long, restricted to Mt Ruwenzori	*D. montanus*
	Tail not very short, > 75%, fur not particularly long and dark (except for *D. longipilosus* which is moderately dark and long-haired), not restricted to Mt Ruwenzori	Other *Dasymys*

Species Accounts

Dasymys alleni Lawrence & Loveridge, 1953

Diagnosis

According to Verheyen et al. (2003) this species can be distinguished craniometrically (without overlap) from *incomtus, sua* and *rwandae* occurring in central-east Africa.

Taxonomic Issues

This taxon was recently elevated to full species status (Verheyen et al., 2003) based on molecular and craniometric features. However, the species has not yet been properly diagnosed and its distributional limits are currently unknown. Furthermore, it is only 1.0-1.6% divergent from *D. medius* and *D. rwandae*, and 2.1-2.5% from *D. sua* in its cytochrome b gene (Verheyen et al., 2003) which is rather low for species recognition. This species group is urgently in need of revision.

Distribution

As defined by Verheyen et al. (2003), the species occurs on Mt Rungwe and nearby mountains of the southern highlands in south-western Tanzania as well as along the western rim of the Albertine Rift (Tshibati, Bukavu, Albertville) (not plotted).

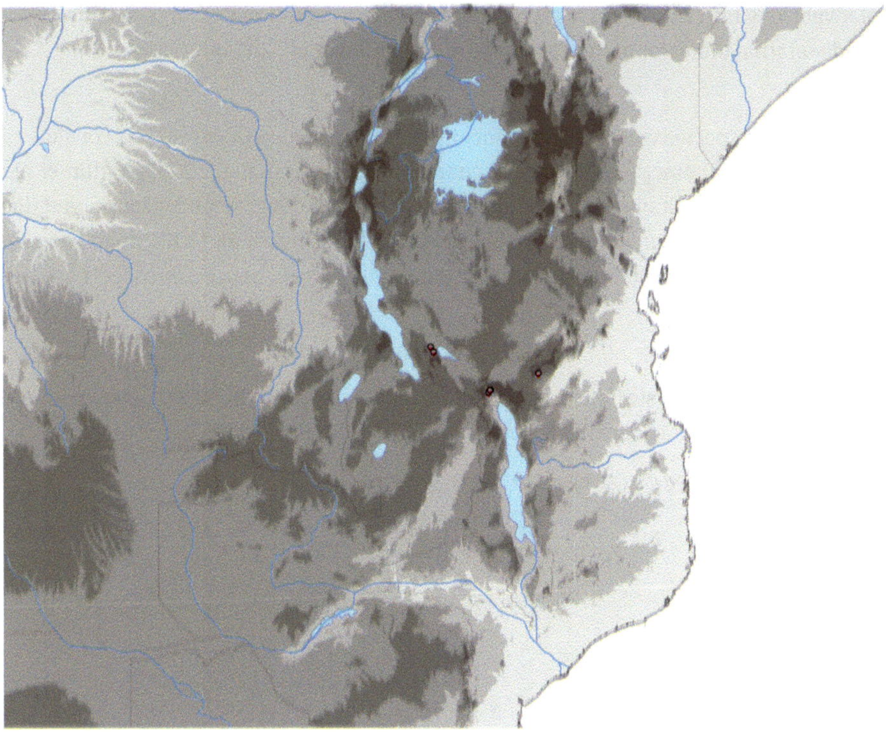

Figure 2.469 Distribution map

Measurements

External and cranial measurements (mm) and mass (g) for *Dasymys alleni*, sexes combined

	Mean	Min	Max	N
Mass	95.3	48	135	34
HB	140.1	112	179	34
Tail	124.5	103	140	34
HF	30.2	25.0	34.0	35
Ear	19.5	17.0	21.0	33
GLS*	34.3	29.5	37.2	33

*Verheyen et al. (2003)

Dasymys cabrali Verheyen et al., 2003 [= *shortridgei* Mullin et al., 2004a)

Diagnosis

This species is intermediate in body and cranial size between *incomtus* and *nudipes* and craniometrically distinct (without overlap) from both (Verheyen et al., 2003). Of particular diagnostic importance for *cabrali* is the combination of its very narrow interorbital region relative to well developed zygomatic plate (Verheyen et al., 2003). Furthermore, this species is characterised as having a relatively long tail (ca. 95% of head and body length) in addition to a large maxillary toothrow and wide internal diameter of the zygomatic arch (Mullin et al., 2004a).

Taxonomic Issues

This recently described species was previously included in *D. incomtus*. However, it is considered a valid species based on distinct morphological characters (Verheyen et al., 2003; Mullin et al., 2004a). This species was simultaneously described by two independent research groups, and due to the rules of taxonomic nomenclature the name given by Verheyen et al. (2003) takes priority over the one given by Mullin et al. (2004a; who named it *D. shortridgei*); hence *shortridgei* becomes a synonym of *D. cabrali*.

Distribution

This species has a narrow distribution in northern Botswana, north-eastern Namibia and south-western Angola where it occurs in swampy habitats (Mullin et al., 2005).

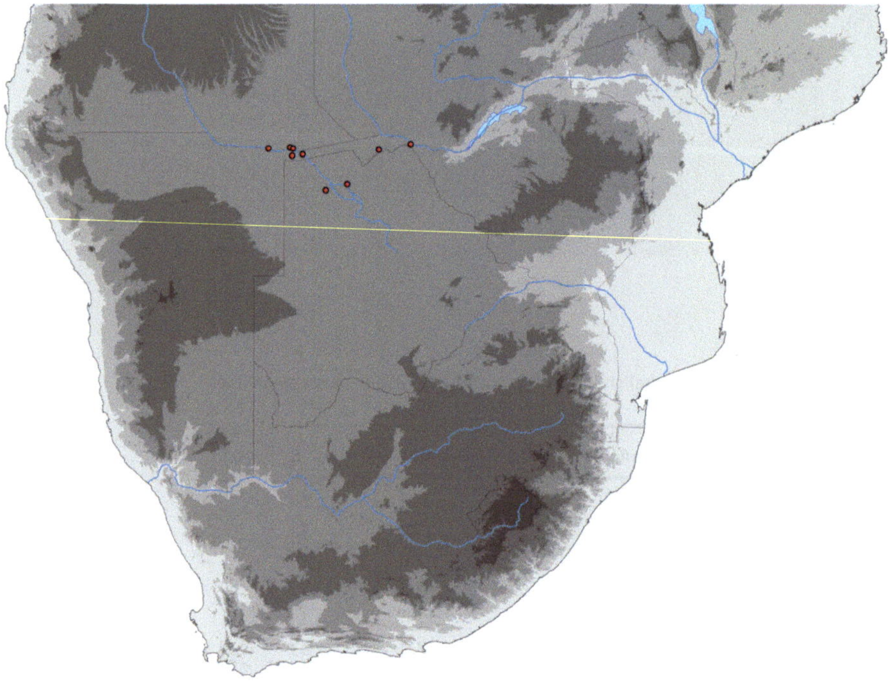

Figure 2.470 Distribution map

Measurements

External and cranial measurements (mm) and mass (g) for *Dasymys cabrali*, sexes combined

	Mean	Standard deviation	N
Mass*	119.3	20.4	5
HB*	163.5	15.0	110
Tail*	153.4	15.9	109
HF*	35.6	2.3	110
Ear*	19.2	1.6	110
GLS*	37.3	1.5	56

*Mullin et al. (2004a)

Dasymys capensis Roberts, 1936

Diagnosis

After *D. nudipes* and *D. cabrali*, in terms of body and cranial size this is the third-largest species of *Dasymys*. Compared to *D. incomtus*, it has a significantly wider zygomatic arch and braincase and significantly longer maxillary toothrow (Mullin et al., 2004a).

Taxonomic Issues

This taxon has recently been elevated to full species status based on differences in cranial morphology and its isolated distribution (Mullin et al., 2004a).

Distribution

This species is known from just a few localities in the Western Cape province of South Africa where it occurs in marshy habitats (Mullin et al., 2005).

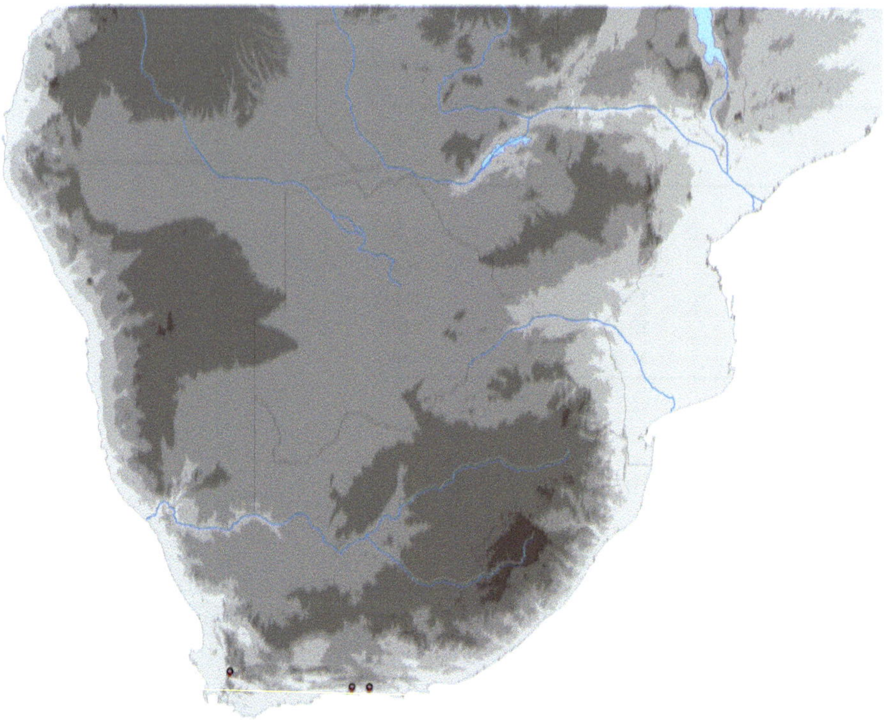

Figure 2.471 Distribution map

Measurements

External and cranial measurements (mm) and mass (g) for *Dasymys capensis*, sexes combined

	Mean	Standard deviation	N
Mass*	111.3	18.3	4
HB*	152.9	18.3	8
Tail*	132.0	11.5	8
HF*	34.6	0.9	8
Ear*	19.6	0.8	7
GLS*	38.0	1.9	7

*Mullin et al. (2004a)

Dasymys foxi Thomas, 1912

Diagnosis

This species is morphologically (slightly larger) and biogeographically distinct from *D. rufulus* in West Africa (Carleton & Martinez, 1991).

Taxonomic Issues

This taxon was previously listed as a synonym of *D. incomtus* (Rosevear, 1969; Happold, 1987) and was only elevated to full species status by Carleton & Martinez (1991).

Distribution

This species is restricted to the Jos Plateau of north-central Nigeria where it occurs in marshy habitats (Happold, 1987).

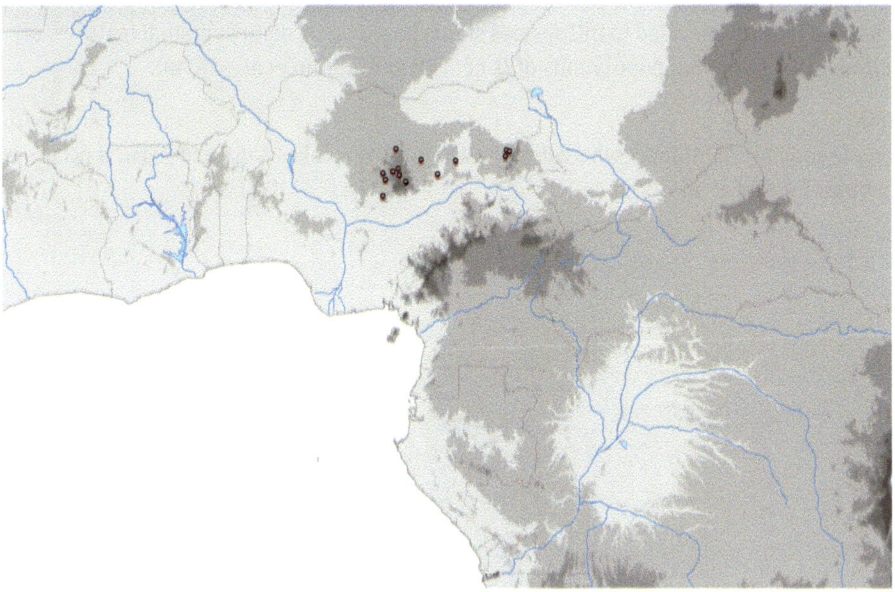

Figure 2.472 Distribution map

Measurements

External and cranial measurements (mm) and mass (g) for *Dasymys foxi*, sexes combined

	Mean	Standard deviation	N
Mass*	116.6	25.2	56
HB*	151.8	17.4	66
Tail*	133.8	11.2	66
HF*	33.0	1.7	66
Ear*	20.7	1.1	67
GLS*	36.1	1.7	13

*Mullin et al. (2004a)

Dasymys griseifrons Osgood, 1936

Diagnosis

This species has the widest interorbital constriction of all *Dasymys* and an overall large size in terms of both external and cranial measurements, although not as large as *nudipes*, *cabrali* and *capensis* (Mullin et al., 2004a).

Taxonomic Issues

This species was recently elevated to full species status based on its distinct cranial features, separating it from other Ethiopian populations of *D.* cf *incomtus* (Mullin et al., 2004a).

Distribution

This species is endemic to Ethiopia where it is known only from Lakes Tana and Jigga; it occurs in marshy habitats (Mullin et al., 2005).

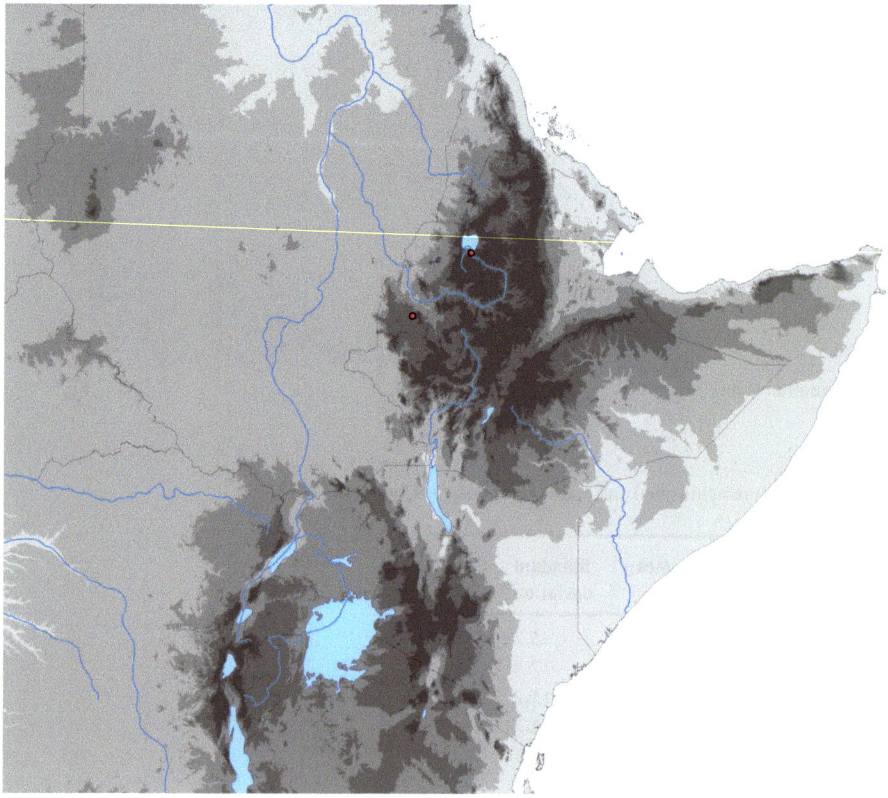

Figure 2.473 Distribution map

Measurements

External and cranial measurements (mm) and mass (g) for *Dasymys griseifrons*, sexes combined

	Mean	Standard deviation	N
Mass	-	-	-
HB*	165.7	11.7	10
Tail*	135.0	14.1	10
HF*	34.4	1.8	10
Ear	-	-	-
GLS*	36.5	1.8	5

*Mullin et al. (2004a)

Dasymys incomtus (Sundevall, 1847)

Diagnosis

Dasymys incomtus sensu stricto from eastern South Africa can be distinguished on chromosomal (2n = 38, FN = 46), allozymic (fixed allelic differences at three loci) and morphometric grounds from *D. robertsii* occurring in northern South Africa, and on morphometric grounds from *D. capensis* occurring in the Western Cape (Mullin et al., 2002; 2004a)

Taxonomic Issues

This species was previously thought to range widely across Africa, but recent genetic and morphometric analyses show that it is endemic to eastern South Africa (Mullin et al., 2004a). Chromosomal number 2n = 38, FN = 46 (Mullin et al., 2002).

Distribution

This species is restricted to the eastern regions of South Africa and Swaziland where it occurs in marshy habitats (Mullin et al., 2005). Based on morphological similarity, the species also occurs in southern Zimbabwe (Mullin et al., 2005), but this has not been confirmed by chromosomal or molecular analysis.

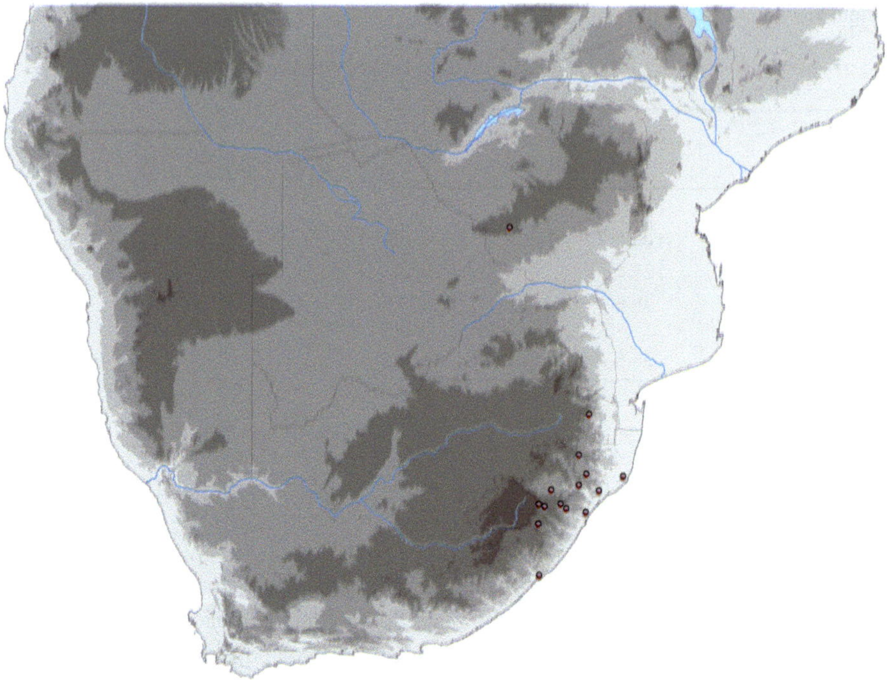

Figure 2.474 Distribution map

Measurements

External and cranial measurements (mm) and mass (g) for *Dasymys incomtus*, sexes combined

	Mean	Standard deviation	N
Mass*	130.5	33.5	22
HB*	156.9	19.8	47
Tail*	139.2	13.9	45
HF*	31.9	3.3	43
Ear*	20.7	2.6	30
GLS*	37.0	2.8	36

*Mullin et al. (2004a)

Dasymys cf *incomtus*

Diagnosis

There are no defining characters for this group.

Taxonomic Issues

This species (or species complex) represents the remaining populations in Central and East Africa previously named *D. incomtus*. The chromosome number is mostly unknown but populations from Ethiopia are variable with 2n = 38-40, FN = 44 (Volobouev et al., 2000). The populations from the

Congo Basin may refer to the taxon *bentleyae* Thomas 1892 which is distinct from the *D. medius* group (East Africa), true *D. incomtus* (eastern South Africa), and *D. rufulus* (West Africa) on molecular grounds (Verheyen et al., 2003). Hence, the name *D. bentleyae* is available for this species, at least for the population in the Congo Basin.

Distribution

The limits of this species are currently poorly defined and its southern and eastern limits are not known. It may also occur in West Africa where two separate karyotypes are known (Matthey, 1958; Tranier & Gautun, 1979), one of which may refer to *D.* cf *incomtus*. It inhabits swampy habitats in forest and woodland (Mullin et al., 2005).

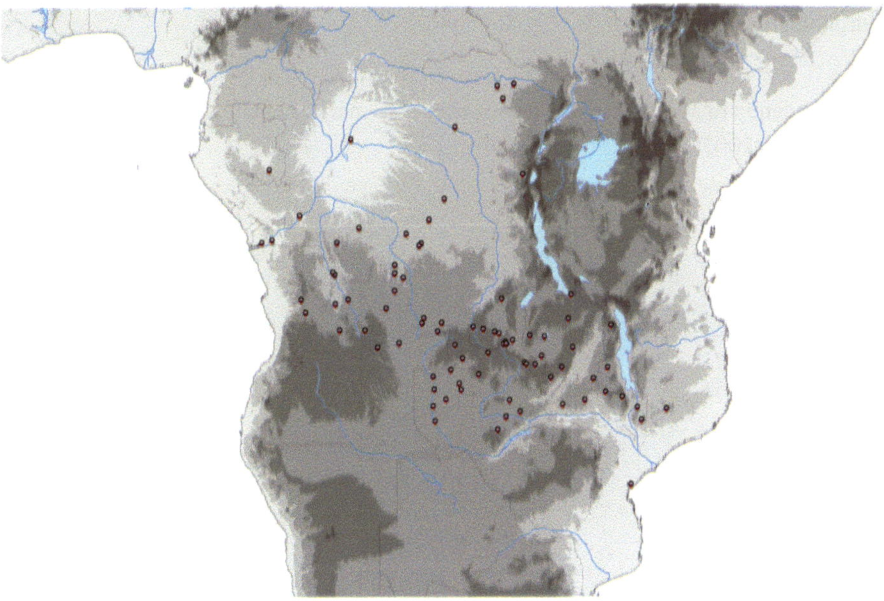

Figure 2.475 Distribution map

Measurements

External and cranial measurements (mm) and mass (g) for *Dasymys* cf *incomtus*, sexes combined

	Mean	Min	Max	N
Mass	-	-	-	-
HB	-	-	-	-
Tail	-	-	-	-
HF	-	-	-	-
Ear	-	-	-	-
GLS*	33.1	30.4	35.5	80

*Verheyen et al. (2003), based on measurements of *Dasymys incomtus bentleyae*

Dasymys longipilosus Eisentraut, 1963

Diagnosis

This species has dark pelage and a short tail (ca. 80% of HB), although not as pronounced in either aspect as seen in *D. montanus*. Distinguishing features include a small hindfoot and total body length in addition to a small condylobasal length, upper toothrow length, internal diameter of the zygomatic arch, greatest mandible length and mandibular toothrow length (Mullin et al., 2004a).

Taxonomic Issues

This taxon was previously considered a synonym of *D. incomtus* (Verheyen et al., 2003) or *D. rufulus* (Musser & Carleton, 2005), but elevated to full species status by Mullin et al. (2004a) based on clear morphological differences.

Distribution

This species is endemic to Mt Cameroon where it occurs in marshy habitats at high elevations > 2000 m (Mullin et al., 2005).

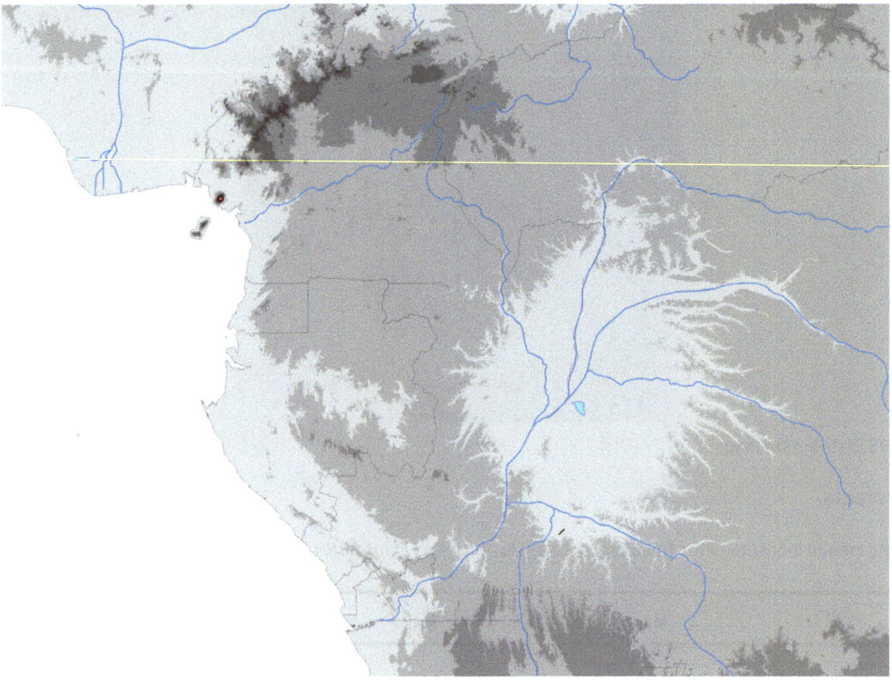

Figure 2.476 Distribution map

Measurements

External and cranial measurements (mm) and mass (g) for *Dasymys longipilosus*, sexes combined

	Mean	Standard deviation	N
Mass	-	-	-
HB*	158.0	2.8	3
Tail*	130.0	0	3
HF*	22.0	4.2	3
Ear	22.5	7.8	3
GLS*	30.6	0.4	3

*Mullin et al. (2004a)

Dasymys medius Thomas, 1906

Diagnosis

According to Mullin et al. (2004a), this species is distinct in terms of its small upper toothrow length and small upper zygomatic arch width.

Taxonomic Issues

This taxon has recently been elevated to full species status based on molecular (Verheyen et al., 2003) and craniometric (Mullin et al., 2004a) data. However, it is only 1.3-1.9% divergent from *D. rwandae* and *D. alleni* in its cytochrome b gene (Verheyen et al., 2003) which is rather low for species recognition. Mullin et al. (2004a; 2005) used this name to describe all East African *Dasymys* in their sample from the DRC, Burundi, Rwanda, Uganda and Kenya (which however excluded critical series referable to *sua* and *rwandae*). This species group is urgently in need of revision.

Distribution

This species occurs patchily but widely in high altitude marshes of East Africa (Mullin et al., 2005).

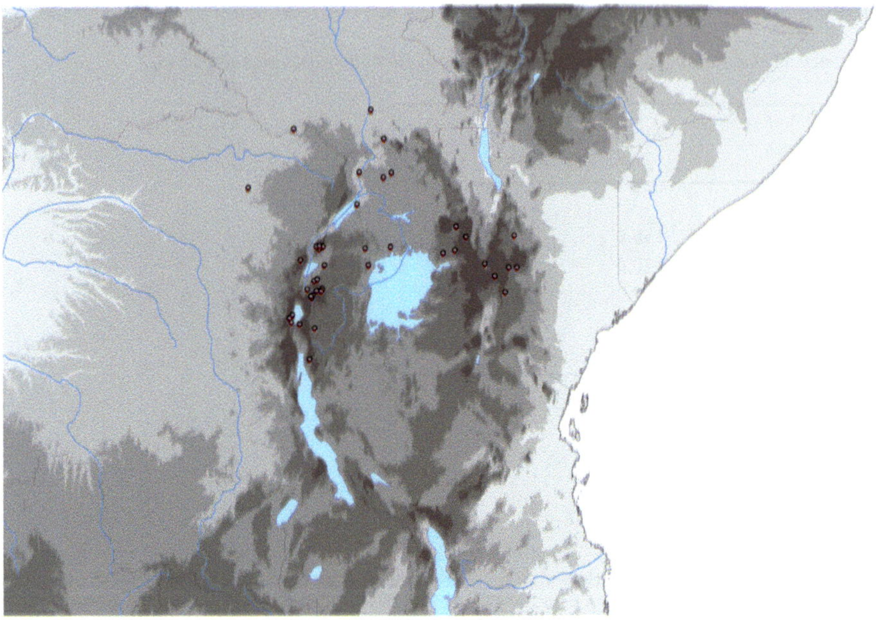

Figure 2.477 Distribution map

Measurements

External and cranial measurements (mm) and mass (g) for *Dasymys medius*, sexes combined

	Mean	Standard deviation	N
Mass	106.6	30.3	64
HB	150.0	16.0	255
Tail	133.6	14.6	252
HF	30.5	2.8	244
Ear	20.5	2.5	198
GLS	35.9	1.6	112

Dasymys montanus Thomas, 1906

Diagnosis

This distinctive, dark-coloured and long-haired species can also be distinguished by its small size overall (both in terms of cranial and external measurements), very short tail (ca. 75% of HB, cf > 80% in other species), short rostrum, low, squat cranium and wide zygomatic breadth (Mullin et al., 2004a; Musser & Carleton, 2005).

Taxonomic Issues

This species was previously listed as a subspecies of *D. incomtus* (Delany, 1975), but elevated to full species status based on morphological grounds (Musser & Carleton, 2005); it also occurs at higher elevations but the two species co-occur at 2600 m (Kerbis Peterhans et al., 1998) which is the highest record for *D. incomtus* and the lowest for *D. montanus*. Based on craniometric similarity of type specimens, Verheyen et al. (2003) included *montanus* within their *incomtus bentleyae*.

Distribution

This species is endemic to the Ruwenzori Mts where it occurs in high elevation marshes above 2600 m (Musser & Carleton, 2005; Mullin et al., 2005).

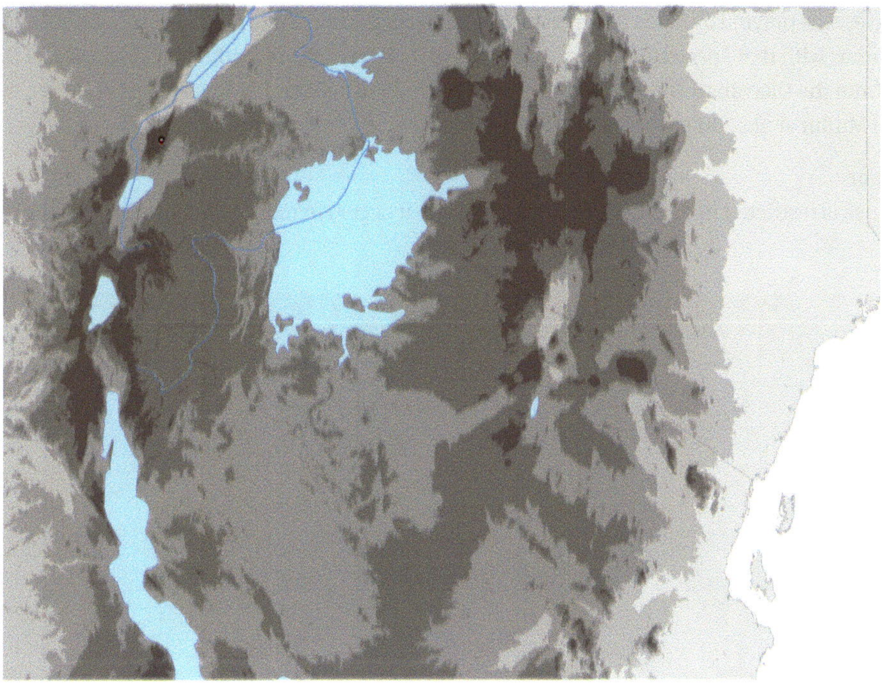

Figure 2.478 Distribution map

Measurements

External and cranial measurements (mm) and mass (g) for *Dasymys montanus*, sexes combined

	Mean	Standard deviation	N
Mass*	82.0	18.4	3
HB*	141.8	8.3	6
Tail*	103.5	6.8	6
HF*	28.0	1.7	6
Ear*	18.3	1.0	6
GLS*	32.5	1.5	3

*Mullin et al. (2004a)

Dasymys nudipes (Peters, 1870)

Diagnosis
This species can be clearly diagnosed by having five plantar pads as opposed to six in other *Dasymys* species, in addition to being the largest-sized member of *Dasymys* (GLS ca. 40 mm). It is also unique amongst *Dasymys* in having a tail that is longer than the head and body length

Taxonomic Issues
This species was previously listed as a synonym of *D. incomtus* but is morphologically distinct and may co-occur with that species in the northern part of its range. The recently described species *D. cabrali* (from the Okavango and Chobe rivers) was previously included in *D. nudipes* (Verheyen et al., 2003; Mullin et al., 2004a).

Distribution
This species is restricted to the plateau of Angola where it occurs in marshy habitats (Mullin et al., 2005).

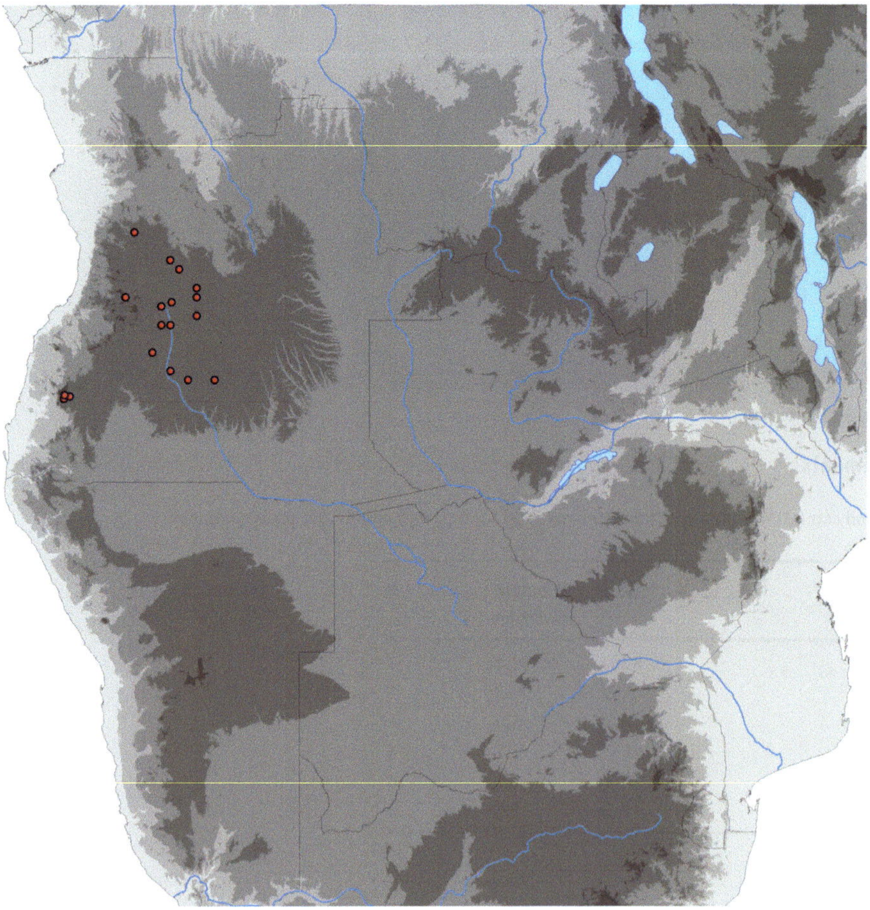

Figure 2.479 Distribution map

Measurements

External and cranial measurements (mm) and mass (g) for *Dasymys nudipes*, sexes combined

	Mean	Standard deviation	N
Mass	-	-	-
HB*	167.3	12.7	9
Tail*	175.1	9.7	10
HF*	41.8	1.8	10
Ear*	24.0	2.9	10
GLS*	39.8	1.1	10

*Mullin et al. (2004a)

Dasymys robertsii Mullin et al., 2004a

Diagnosis

The species can be distinguished from *D. incomtus* sensu stricto on chromosomal (in *D. incomtus* 2n = 38), allozymic (fixed allelic differences at three genetic loci) and craniometric grounds.

Taxonomic Issues

This recently described species was previously known as *D. incomtus* but is chromosomally and morphologically distinct from it (Mullin et al., 2002; 2004a). Chromosomal number 2n = 46 (Mullin et al., 2002).

Distribution

This species is patchily distributed in northern South Africa and Zimbabwe where it occurs in marshy habitats (Mullin et al., 2005).

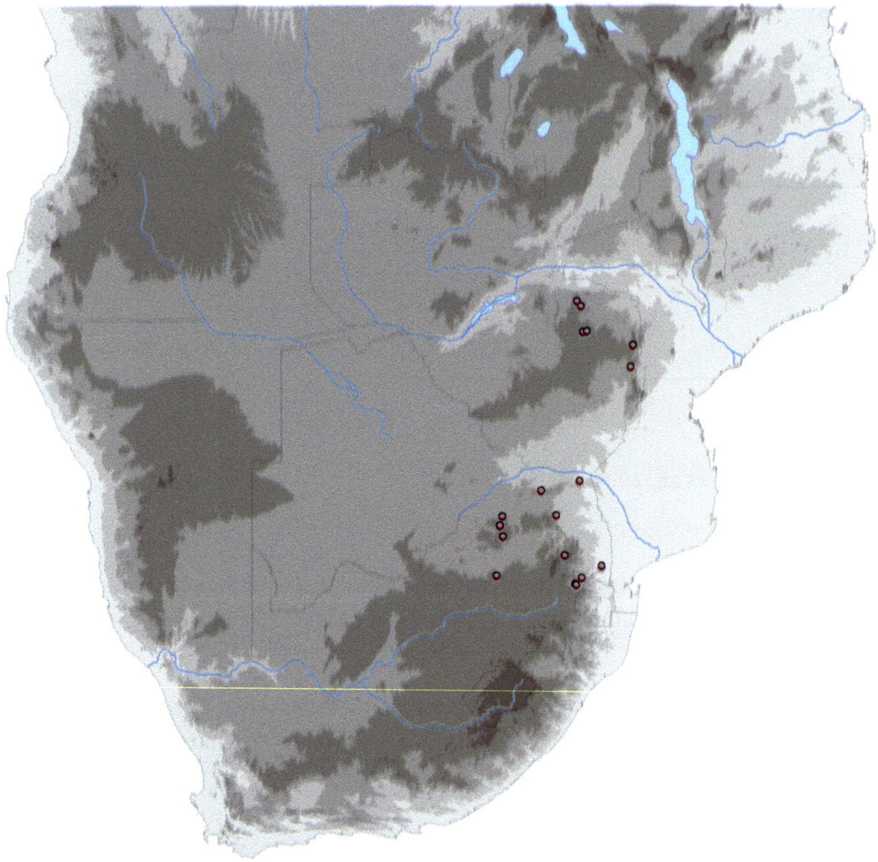

Figure 2.480 Distribution map

Measurements

External and cranial measurements (mm) and mass (g) for *Dasymys robertsii*, sexes combined

	Mean	Standard deviation	N
Mass*	90.2	26.7	27
HB*	148.6	17.1	49
Tail*	127.8	13.3	49
HF*	32.0	4.9	49
Ear*	20.5	2.0	48
GLS*	36.1	1.8	50

*Mullin et al. (2004a)

Dasymys rufulus Miller, 1900

Diagnosis
This is the only species of *Dasymys* known from West Africa, although *D.* cf *incomtus* may also be present in which case the two species can only be distinguished on molecular or chromosomal grounds.

Taxonomic Issues
This taxon may comprise a species complex. Chromosome number is variable 2n = 36-39, FN = 44-50, resulting from the presence of supernumerary "B" chromosomes (Volobouev et al., 2000); without these Bs, the karyotype is 2n = 34, FN = 44.

Distribution
This species is widely but patchily distributed in West Africa from western Nigeria to Senegal, where it occurs in swampy habitats (Mullin et al., 2005).

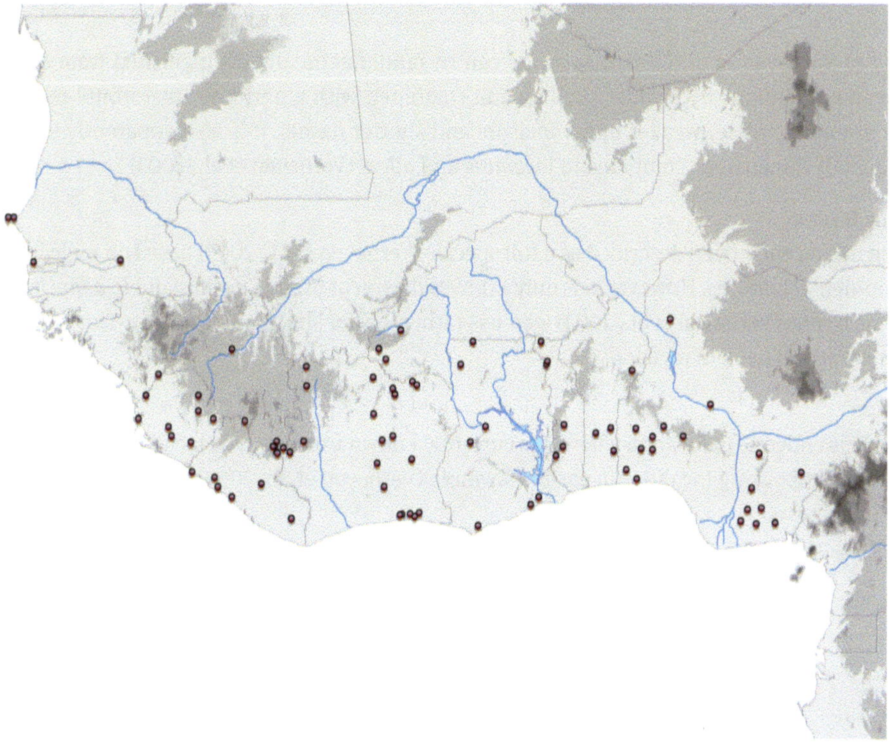

Figure 2.481 Distribution map

Measurements

External and cranial measurements (mm) and mass (g) for *Dasymys rufulus*, sexes combined

	Mean	Min	Max	N
Mass	-	-	-	-
HB*	141	122	159	19
Tail*	139	131	152	19
HF*	28.5	25	31	19
Ear*	19	15	23	19
GLS*	34.4	32.5	37.2	19

*Rosevear (1969)

Dasymys rwandae Verheyen et al., 2003

Diagnosis

According to Verheyen et al. (2003), *rwandae* can be craniometrically distinguished from other East African populations by a wider skull overall combined with a narrower interorbital region, a weaker zygomatic plate, weaker upper incisors and shorter nasals. It is also apparently smaller-sized in body dimensions compared to *incomtus* and *alleni* (Verheyen et al., 2003).

Taxonomic Issues

This taxon has recently been described as a full species (Verheyen et al., 2003) based on molecular and cranio-dental features. However, it is only 1.0-1.9% divergent from *D. medius* and *D. alleni* in its cytochrome b gene (Verheyen et al., 2003) which is rather low for species recognition. This species group is urgently in need of revision.

Distribution

This species is endemic to Rwanda where it is currently known from the type locality of Kinigi and a few other neighbouring localities in eastern Rwanda (Verheyen et al., 2003). Its habitat associations are not known.

Figure 2.482 Distribution map

Measurements

External and cranial measurements (mm) and mass (g) for *Dasymys rwandae*, sexes combined

	Mean	Min	Max	N
Mass*	77.1	27	140	42
HB*	121.5	87	151	42
Tail*	111.4	86	140	41
HF*	27.4	23.6	30.0	42
Ear*	17.6	13.7	20.5	42
GLS*	32.5	27.0	35.7	38

*Verheyen et al. (2003)

Dasymys sua Verheyen et al., 2003

Diagnosis

According to Veheyen et al. (2003), *sua* can be distinguished craniometrically from *incomtus* by its heavier rostrum, larger zygomatic plate and wider zygomatic width, narrower interorbital breadth, narrower palate, lower ramus height, smaller bullae, finer and more pro-odont upper incisors and shorter upper molar tooth row.

Taxonomic Issues

This taxon has recently been described as a full species (Verheyen et al., 2003) based on molecular and cranio-dental features. However, it is only 2.0-2.5% divergent from *D. rwandae* and *D. alleni* in its cytochrome b gene (Verheyen et al., 2003) which is rather low for species recognition. This species group is urgently in need of revision.

Distribution

This species is currently only known from a few localities on the flanks of the Uluguru Mts at mid-elevations (600-1600 m) (Verheyen et al., 2003). Its habitat associations are not known.

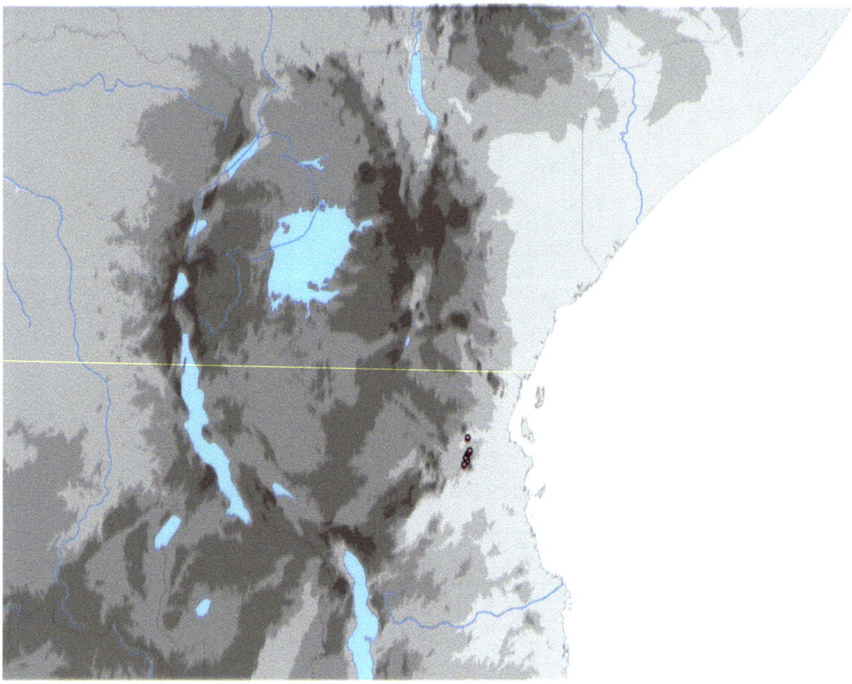

Figure 2.483 Distribution map

Measurements

External and cranial measurements (mm) and mass (g) for *Dasymys sua*, sexes combined

	Mean	Min	Max	N
Mass*	78.0	38	115	25
HB*	141.1	108	160	25
Tail*	127.7	109	140	25
HF*	27.5	25.0	29.1	25
Ear*	18.3	15.0	22.0	25
GLS*	33.4	30.1	36.0	23

*Verheyen et al. (2003)

Genus Account

Dephomys Thomas, 1926 (Defua Rats)

Diagnosis

This is a medium-sized (Mass ca. 50-60 g; HB ca. 125-130 mm) West African endemic rodent with reddish-brown dorsal pelage comprising long and sleek fur with sparse black-tipped guard hairs (mostly confined to the rump which is often a brighter orange colour). Ventral hairs are white-tipped with grey base. The tail is much longer (160% of HB) than head and body length, black in colour with rings of small scales and bristle. Members of this genus could be confused with *Grammomys, Thamnomys* or *Stochomys*. However, *Grammomys* and *Thamnomys* have pure white bellies (deep grey in *Dephomys*) and *Stochomys* has a denser covering of black guard hairs dorsally (these are paler-tipped and mostly restricted to the posterior dorsal surface in *Dephomys*) as well as a chocolate-brown ventral pelage (deep grey in *Dephomys*). *Stochomys* also has a much longer hindfoot (about 30 mm, cf 25 mm in *Dephomys*) (Rosevear, 1969).

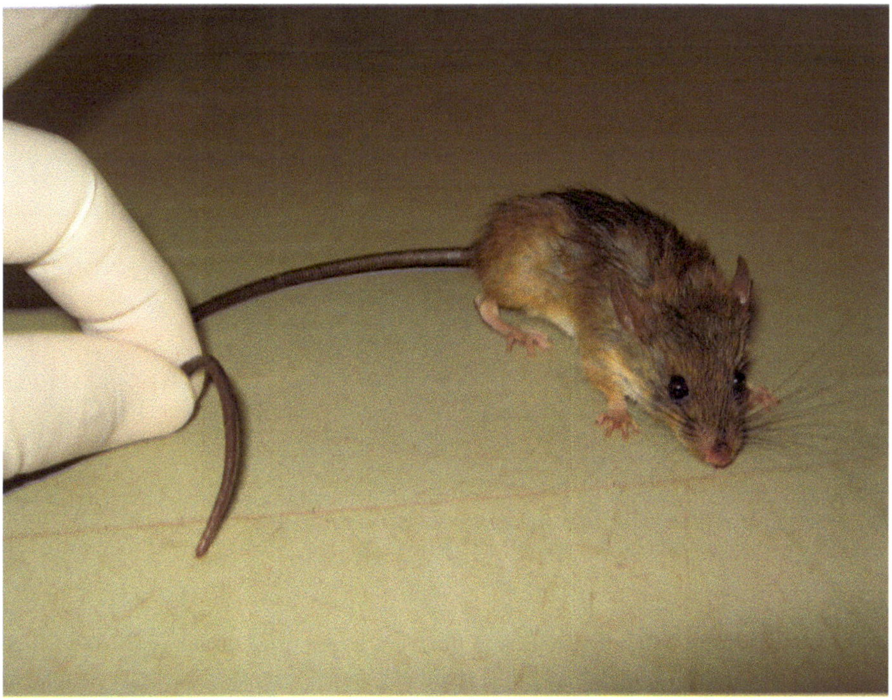

Figure 2.484 Photograph of *Dephomys defua* (Ara Monadjem).

Cranial Characters

The skull is medium sized (GSL ca. 33-34 mm) and flattish but oval in shape, with broad braincase, mostly smooth (without crests), but having a distinct supraorbital ridge which extends onto the braincase. The zygomatic arches are gracile. The nasal bone is quite deep and long, not obviously tapering. The anterior palatal foramina just reach the level of the first upper molars. The zygomatic plate is expanded below and has a vertical anterior surface. The bullae are relatively small (greatest diameter ± 14% of GSL).

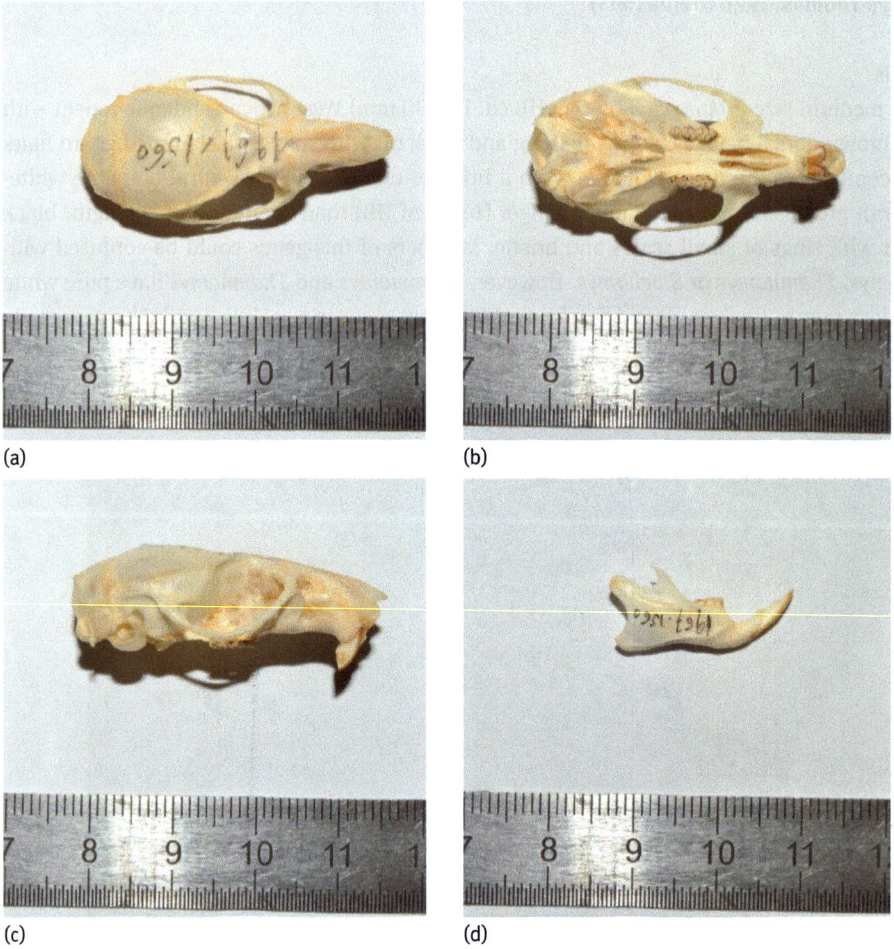

Figure 2.485 a-d Cranium of *Dephomys defua* (MNHN-ZM-MO 1967-1560) (from top left to bottom right: dorsal view, ventral view, lateral view and lower jaw).

Dentition

The incisors are opisthodont and ungrooved, and the molars are normal-sized (MTR/GSL ± 0.15-0.16, Table 10). Rows of cusps are distorted, with t1 behind t2 and t3. The M^1 is the longest molar in the toothrow. The cusps t2 and t3 on M^1 are poorly separated; t1 and t4 are distinctly larger in size than t2 and t5. There are traces of stephanodont crests on the labial side of the upper molars and the t9 is small on the M^1 and nearly absent on the M^2. The t3 cusp may (*D. defua*) or may not (*D. eburneae*) be present on M^3. The configuration of M^3 corresponds to Group 5 in Figure 2.439). The M_1 and M_2 display many suplementary cusplets and a wide posterior cingulum.

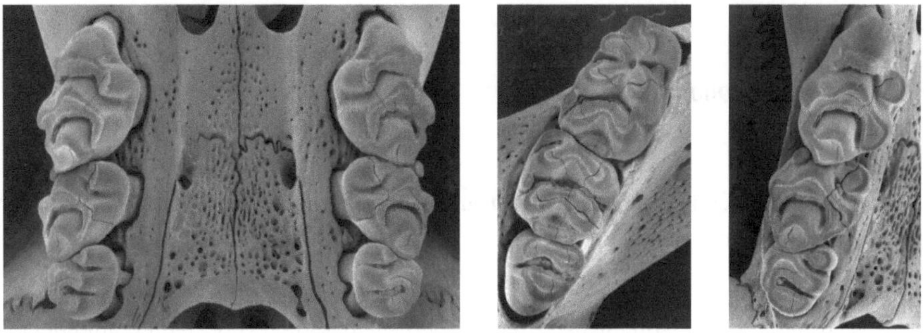

(a) (b) (c)

Figure 2.486 Cheek teeth of *Dephomys eburneae* (MNHN-ZM-MO 1980-117) (A: upper cheek teeth; B: lower left toothrow; C: upper right toothrow).

Habitat & Biogeography

It is known from the high forest zone of West Africa from Sierra Leone to Ghana.

Fossil Record

No fossil record.

Biology & Ecology

This is a rare genus whose ecology and life history is very poorly known (Happold, 2013).

Systematic Notes

Although *Dephomys* has been previously included under *Stochomys*, Rosevear (1969) and Misonne (1969) have pointed out differences which justify its recognition as a distinct genus. Although no molecular studies have been completed, it has been assigned to the *Hybomys* division of the Arvicanthini tribe (Musser & Carleton, 2005; Lecompte et al., 2008). According to Van der Straeten (1984), *Dephomys* is close to *Stochomys* on external characters but close to *Hybomys* on cranial characters. *Dephomys eburneae* was originally included as a subspecies of *D. defua*, but the two species can be separated on the grounds of rostrum height and the presence (in *defua*) or absence (in *eburneae*) of the t3 on M^1 (Van der Straeten, 1984) and on chromosomal grounds (Tranier & Dosso, 1979).

Annotated Species List

Two species are currently recognised.

Dephomys defua (Miller, 1900)

Dephomys eburneae (Heim de Balsac et Bellier, 1967)

Key to Species

1	The cusp t3 present on the first upper molar; 2n = 54	*D. defua*
	The cusp t3 absent on the first upper molar; 2n = 42-44	*D. eburneae*

Species Accounts

Dephomys defua (Miller, 1900)

Diagnosis

This species can be distinguished from *D. eburneae* by the presence of the t3 on M^1 (Van der Straeten, 1984) and on chromosomal grounds.

Taxonomic Issues

This species is currently only distinguished from *D. eburneae* based on very slight cranio-dental differences (Van der Straeten, 1984), but molecular analyses have yet to be conducted. Since they differ in karyotype, we retain both species until this genus has been thoroughly revised. Chromosome number: 2n = 54, FN = 102 (Tranier & Dosso, 1979).

Distribution

This species occurs in the Upper Guinea Forest from Sierra Leone in the west to Togo in the east, where it inhabits rainforest.

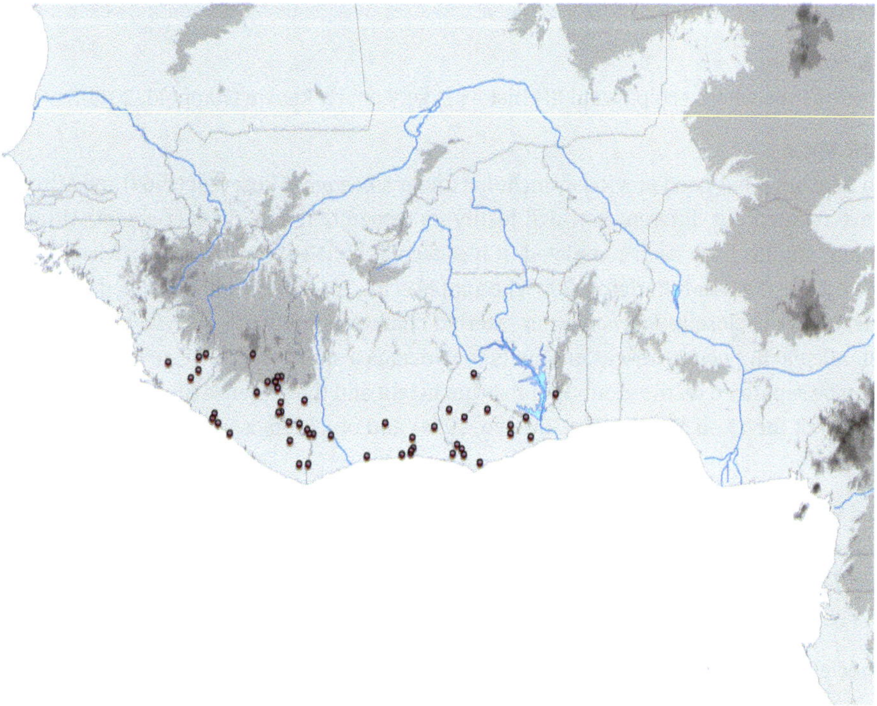

Figure 2.487 Distribution map

Measurements

External and cranial measurements (mm) and mass (g) for *Dephomys defua*, sexes combined

	Mean	Min	Max	N
Mass*	62.5	57	68	2
HB	124.8	112	136	4
Tail	191.0	184	195	4
HF	26.3	25	27	4
Ear	17.9	16	20	4
GLS	32.5	28.4	35.0	8

*Happold (2013) presents a mean mass of 47 g (range: 41-53, n = 8)

Dephomys eburneae (Heim de Balsac & Bellier, 1967)

Diagnosis

This species can be distinguished from *D. defua* by the absence of the t3 on M^1 (Van der Straeten, 1984) and on chromosomal grounds.

Taxonomic Issues

This species is currently only distinguished from *D. defua* based on very slight cranio-dental differences (Van der Straeten, 1984) and on chromosomal grounds, but molecular analyses have yet to be conducted. Since they differ in karyotype, we retain both species until this genus has been thoroughly revised. Chromosome number: 2n = 42-44, FN = 56 (Tranier & Dosso, 1979).

Distribution

This species is known from just a few records in southern Ivory Coast and western Liberia, and may be sympatric with *D. defua*.

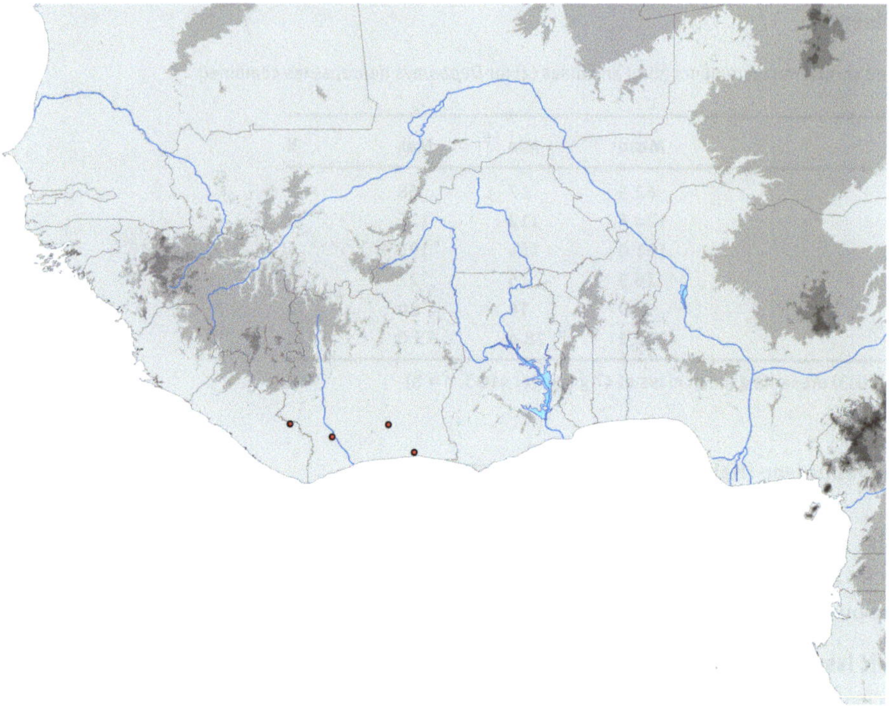

Figure 2.488 Distribution map

Measurements

External and cranial measurements (mm) and mass (g) for *Dephomys eburneae*, sexes combined

	Mean	Min	Max	N
Mass	53.0	47	59	2
HB	128.3	120	142	4
Tail	194.3	182	203	4
HF	27.5	26	28	4
Ear	18.7	18	20	3
GLS	34.1	32.7	35.7	4

Genus Account

Desmomys Thomas, 1910 (Scrub Rats)

Diagnosis

These high-altitude Ethiopian endemics are medium-sized rodents (mass 45-100g; HB 110-155mm; Lavrenchenko, 2003) with tails which are either equal (*D. harringtoni*) or slightly longer (115% of HB; *D. yaldeni*) than head and body length. They were previously regarded as members of *Pelomys* (Ellerman, 1941; Misonne, 1974) but their semi-arboreal habit as well as certain karyological, molecular and dental characters (see below) set them apart (Lavrenchenko, 2003; Lavrenchenko & Verheyen, 2005).

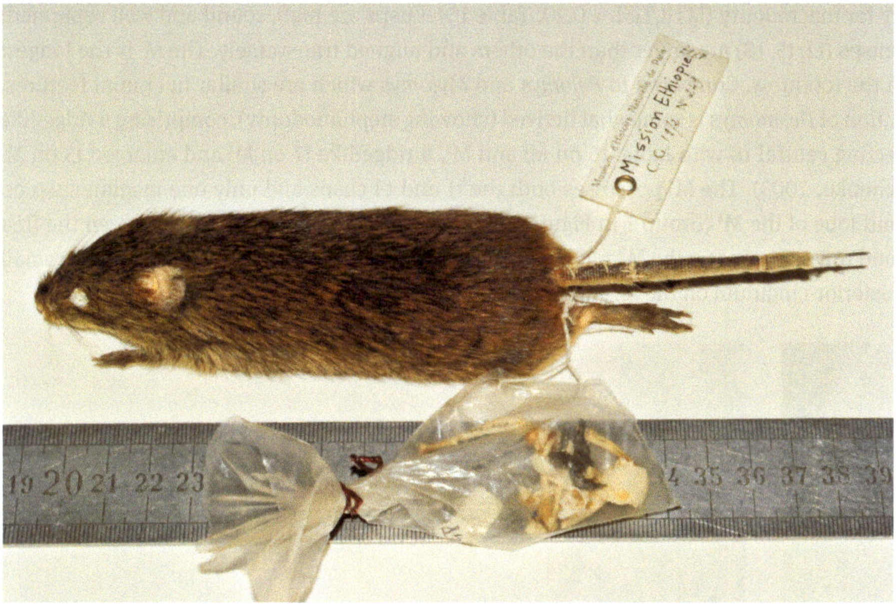

Figure 2.489 Photograph of *Desmomys harringtoni* (MNHN-ZM-MO 1972-249) (Ara Monadjem).

Cranial Characters

The skull closely resembles that of *Pelomys* (see that account) and is medium-sized (condylo-basal skull length 28-30 mm; Lavrenchenko, 2003).

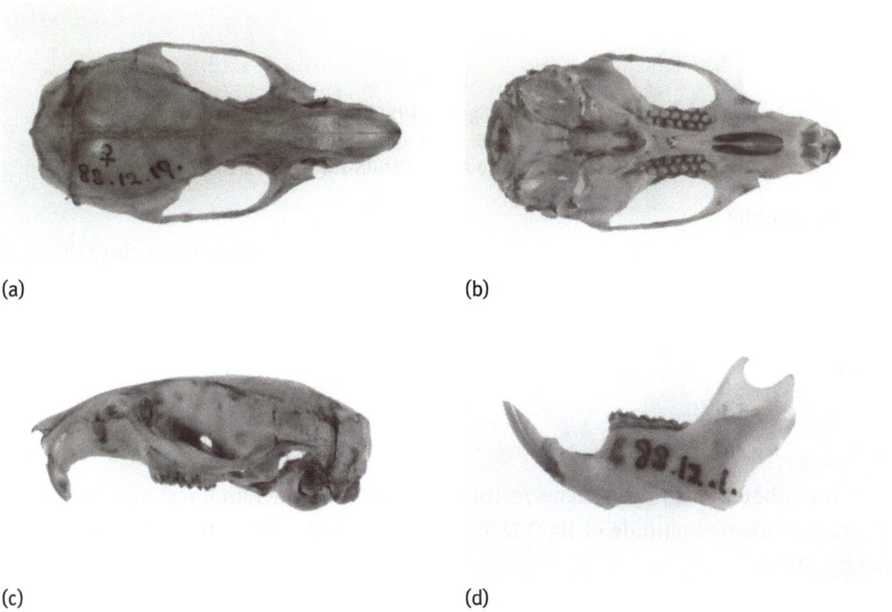

Figure 2.490 a-d Cranium of *Desmomys harringtoni* (BMNH 88.1.19) (from top left to bottom right: dorsal view, ventral view, lateral view and lower jaw).

Dentition

The upper incisors show faint ridges (not grooves as indicated for *Pelomys*), and the molars show tendency for macrodonty (MTR/GSL ± 0.20, Table 10). Cusps are high, round and well separated; central cusps (t2, t5, t8) are larger than the others and aligned transversely. The M^1 is the longest molar in the toothrow. Compared to *Pelomys* and *Mylomys* which are similar in cranial features, the dentition of *Desmomys* is somewhat derived (showing stephanodonty), comprising a ridge-like t9 connecting central t8 with labial t6 on M^1 and M^2, a ridge-like t7 on M^2 and enlarged t5 on M^3 (Lavrenchenko, 2003). The M^3 possesses both the t1 and t3 cusps and only one median cusp on the second lobe of the M^3 (Group 7 in Figure 2.439). There is a longitudinal link between the first and second rows of cusps on the M_1 and M_2. There is an antero-median cusp on the M_1, and a small round posterior-cingulum on the M_1 and M_2. The M_3 is similar in size to the M_2.

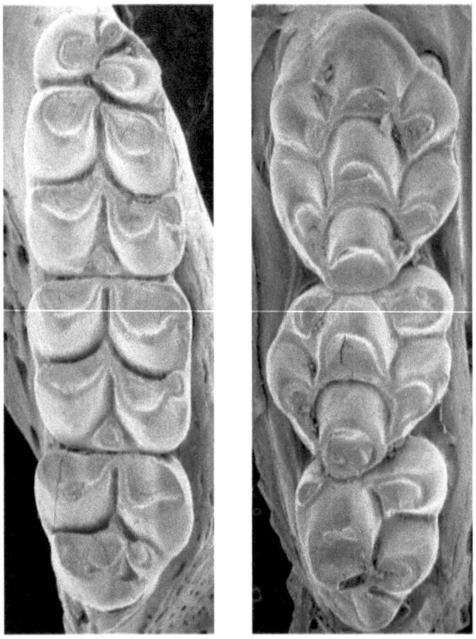

B C

Figure 2.491 Cheek teeth of *Desmomys harringtoni* (B: lower right toothrow (BMNH 8.6.88); C: upper right toothrow (MNHN-ZM-MO 1972-249).

Habitat & Biogeography

While *D. harringtoni* occurs more widely in forest and bush habitats throughout the Ethiopian plateau, *D. yaldeni* is known only from a degraded humid rainforest at Sheko Forest in south-western Ethiopia.

Fossil Record

None.

Biology & Ecology

Apparently semi-arboreal with two species restricted to Afromontane rain forest and grasslands of the Ethiopian Plateau at altitude of 1800-2800 m (Yalden et al., 1976; Yalden & Largen, 1992; Lavrenchenko, 2003).

Systematic Notes

Molecular data confirm the placement of *Desmomys* in the *Arvicanthis* division as sister species to *Rhabdomys* (Ducroz et al., 2001; Lavrenchenko & Verheyen, 2005), as well as the validity of the two recognised species (Lavrenchenko, 2003; Lavrenchenko & Verheyen, 2005). *Desmomys yaldeni* is a smaller animal with more gracile molars, shorter dorsal fur. Moreover the two species are genetically very distinct, having diverged from each other during the late Pliocene (Lavrenchenko & Verheyen, 2005). A possible third species, *D. rex* was shown to be probably a mistaken *Mylomys* (Lavrenchenko, 2003).

Annotated Species List

Two species are currently recognised.
Desmomys harringtoni (Thomas, 1902)
Desmomys yaldeni Lavrenchenko, 2003

Key to Species

1	Tail longer (ca. 115% of HB), narrower M1 (< 1.9 mm)	*D. yaldeni*
	Tail shorter (ca. 95% of HB), broader M1 (> 1.9 mm)	*D. harringtoni*

Species Accounts

Desmomys harringtoni (Thomas, 1902)

Diagnosis

This species is distinct on morphological and molecular grounds from *D. yaldeni* (tail longer ca. 115% of HB; M^1 narrower, < 1.9 mm). In this species relative tail length is ca. 95% of HB and M^1 is broader, > 1.9 mm in width.

Taxonomic Issues

See genus account.

Distribution

This species is restricted to the Ethiopian highlands occurring in forested and bushy habitats above 1800 m.

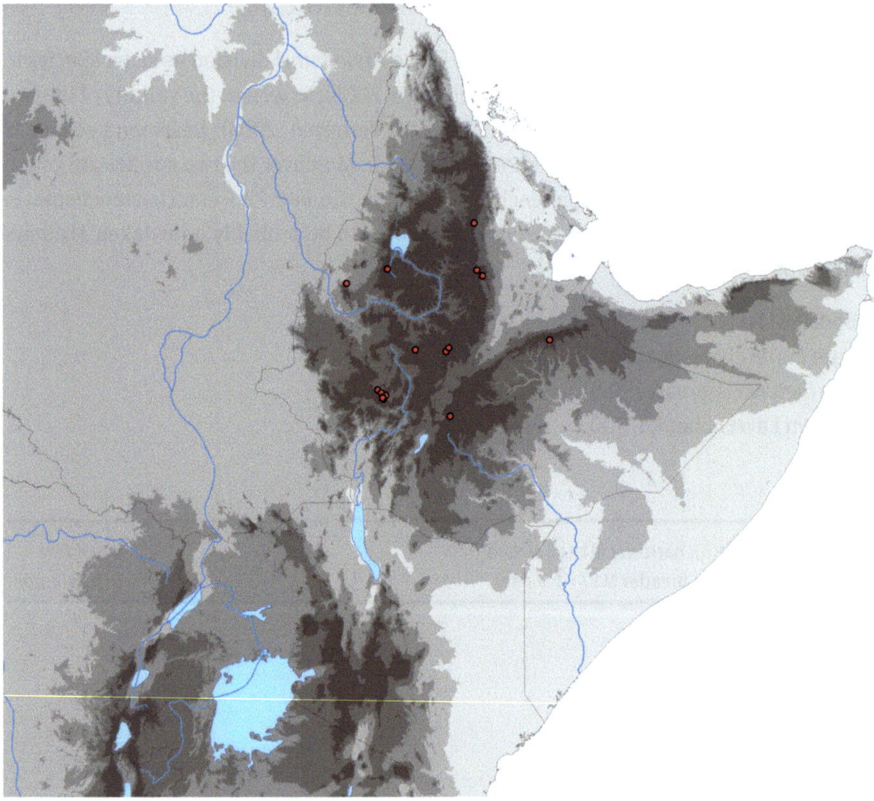

Figure 2.492 Distribution map

Measurements

External and cranial measurements (mm) and mass (g) for *Desmomys harringtoni*, sexes combined

	Mean	Min	Max	N
Mass*	64.9	53	77	21
HB*	136.3	128	144	21
Tail*	128.3	122	135	21
HF*	27.6	27	29	21
Ear*	17.4	17	18	21
GLS*	32.4	30.8	34.9	21

*Happold (2013)

Desmomys yaldeni Lavrenchenko, 2003

Diagnosis

This species is distinct on morphological and molecular grounds from *D. harringtoni* (tail shorter ca. 95% of HB; M^1 broader, > 1.9 mm). In this species relative tail length is ca. 115% of HB and M^1 is narrower, < 1.9 mm in width.

Taxonomic Issues

See genus account.

Distribution

This species is only known from the Sheko Forest in south-western Ethiopia where it occupies montane forest above 1800 m.

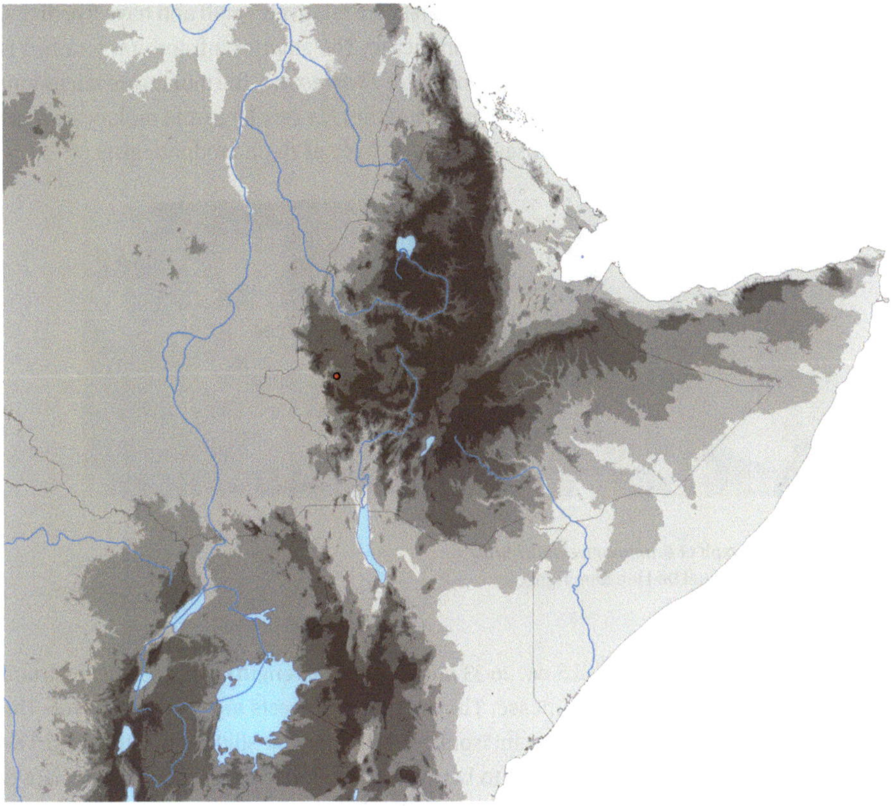

Figure 2.493 Distribution map

Measurements

External and cranial measurements (mm) and mass (g) for *Desmomys yaldeni*, sexes combined

	Mean	Min	Max	N
Mass*	47.0	45	49	2
HB*	124.5	117	132	2
Tail*	143.0	141	145	2
HF*	28.0	27	29	2
Ear*	17.5	17	18	2
GLS	-	-	-	-

*Lavrenchenko (2003)

Genus Account

Grammomys Thomas, 1915 (Thicket Rats)

Diagnosis

This small to medium-sized rat (HB ca. 85-140 mm) is characterized by a very long tail (120-163% of HB) ending with an inconspicuous but visible tuft. The feet are relatively small and have five digits. The skull is small with a short rostrum and broad braincase. The anterior infraorbital foramen is small. The dentition is typical of the Murinae. In contrast to *Thamnomys*, the molars are microdont and a small crestiform t7 is usually present (in *Thamnomys* the t7 is very large and molars are macrodont; whereas in most Murinae the t7 is absent). There is trace of stephanodont crests.

(a)

(b)

Figure 2.494 a and b Photograph of *Grammomys dolichurus* (Andrew Carter, left and Emmanuel Do Linh San, right).

Cranial Characters

The skull is small to medium in size (GLS ca. 26-33 mm in greatest length), with a short rostrum, large interorbital region and a broad braincase. The supraorbital crests are weak or absent. The zygomatic arches are thin and oblique. The incisors are orthodont or slightly opisthodont. The anterior palatal foramina are long and extend to between the anterior edges of M^1. The palate is wide and the molar rows short (microdonty) and parallel. The bullae are small (4.5-6.4 mm in diameter).

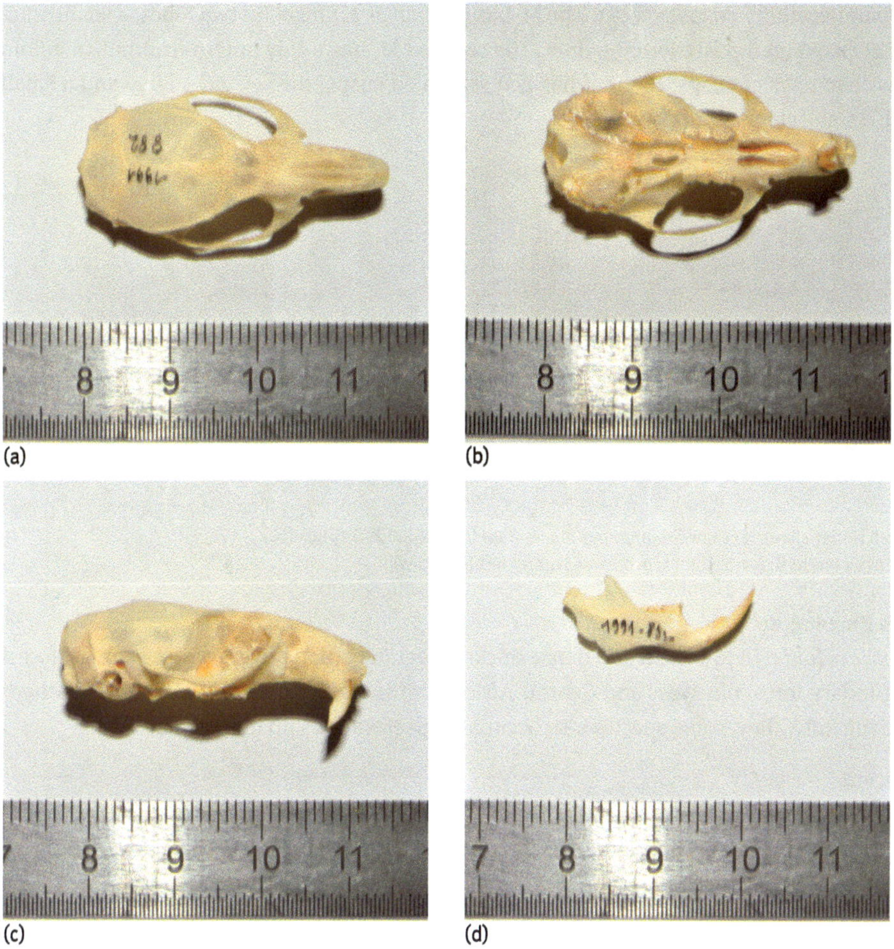

Figure 2.495 a-d Cranium of *Grammomys poensis* (MNHN-ZM-MO 1991-882) (from top left to bottom right: dorsal view, ventral view, lateral view and lower jaw).

Dentition

Molars M^1 are the largest and M^3 the smallest in the upper toothrow. Molar M^1 displays three lobes, with three cusps each. The first lobe has a t1, a large t2 and an oblique t3 with a small stephanodont crest; the second has cusps t4, t5, and t6; and the third has a small crestiform t7, a t8, and a t9 plus a small posterior cingulum. There is a stephanodont crest linking the t6 to the t9; the t4 and t7 are very close. The second molar displays two lobes of three cusps (including t4 and a t7) and the t1 and t3 are present. There is also a tiny posterior cingulum on the M^2. The M^3 is as long as the M^2 but narrower, and has a tiny t3 and a large t1 and two lobes connected on the external side with a crest uniting the t5, t6 and t9. The M_1 is the longest tooth in the lower toothrow. It is composed of a first lobe with two well separated cusps and a small antero-median cusplet. There is a stephanodont crest uniting the first lobe with the second lobe and the cusps are high and well separated. The third lobe is composed of two cusps united by a stephanodont crest and a small round posterior cingulum is visible. The external cingular margin of the tooth is wide and there are some suplementary cusps present. The M_2 has four cusps arranged in two lobes, a small antero-external cusp and a posterior cingulum. The reduced M_3 has a tiny antero-external cingulum and shows two rows of cusps; the first has two separated cusps, the second a large and a small cusp with a small stephanodont crest.

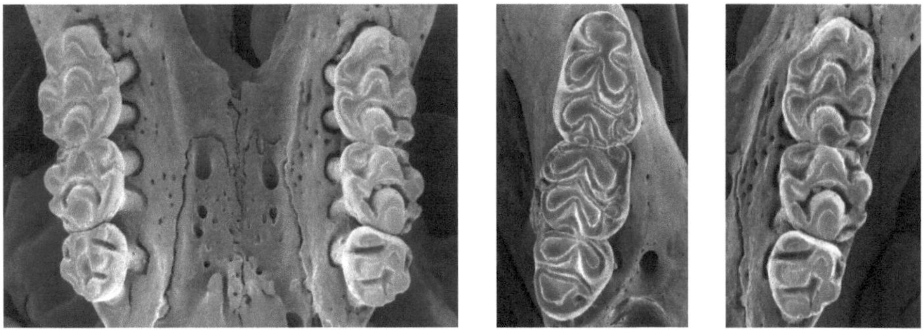

Figure 2.496 Upper cheek teeth of *Grammomys macmillani* (MNHN-ZM-MO 1980-410) (A: upper cheek teeth; B: lower right toothrow; C: upper left toothrow).

Habitat & Biogeography

Grammomys is found in low bush and dense thicket vegetation along rivers or in savannas, forest edge, secondary forests in West and Central Africa, and in coastal forest in East and Southern Africa. In the Rift Valley, some species also occur at high elevations up to 3000 m.

Fossil Record

Grammomys sp. Berg Aukas 54, ca. 1.5 MYA, Namibia (Conroy et al., 1992); Makapansgat, ca. 3.0-3.3 MYA, Swartkrans (SK1), Sterkfontein Extension, Kromdraai B, (Denys, 1990c); Olduvai Bed I, 1.7 MYA (Jaeger, 1976).

Grammomys cf *dolichurus* , Isenya , Kenya, 0.6 MYA (Brugal & Denys, 1989).

Biology & Ecology

This genus is arboreal and nocturnal. It is an agile climber with a semi-prehensile tail. It constructs spherical nests of shredded grass or lianas at 2-4 m above the ground (Rahm, 1967; Roche et al., 1984; Malcolm & Ray, 2000), or in abandoned weaver (*Ploceus*) nests. Its diet is frugivorous or vegetarian (leaves, stems, bark), with some insects (Hanney, 1965; Genest-Villard, 1980; Wirminghaus & Perrin, 1992). Females give birth to 1-6 offspring per litter (Hanney, 1965; Delany, 1975; Rahm 1967) after a gestation period of 24 days. Breeding seems to vary seasonally locally according to the climate. Known predators include small and medium sized carnivores (Taylor, 1998).

Systematic Notes

Initially considered as subgenus of *Thamnomys* by Thomas in 1915, this opinion was followed by subsequent authors including Allen (1939), Hatt (1940), Hollister (1919), and Petter & Tranier (1975). However, other authors have elevated it to genus rank based on dental and skull morphology (Ellerman, 1941; Misonne, 1969; Rosevear, 1969), an opinion which was followed by Musser & Carleton (2005) and Happold (2013). Recent phylogenetic analyses of African Murinae by Lecompte et al. (2008) and Schenk et al. (2013) have placed *Grammomys* in the Arvicanthini tribe close to

Micaelamys. Most species of the genus are rarely collected and require further information on their ecology. The genus is also in urgent need of taxonomic revision. In a preliminary revision, Hutterer & Dieterlen (1984) and Musser & Carleton (2005) have elevated some taxa to full species status in Central and East Africa. Recently two new species have been described by Krystufek (2008) and Denys et al. (2011b). Furthermore, Krystufek et al. (2008) provide a phylogeny for four of the 14 species (*G. dolichurus, G. cometes, G. macmillani,* and *G. ibeanus*), which shows that, in contrast to what was previously thought, *G. cometes* and *G. ibeanus* are not each other's closest relatives.

Species List

Fourteen species are currently recognized.
Grammomys aridulus Thomas & Hinton, 1923
Grammomys brevirostris Krystufek, 2008
Grammomys buntingi (Thomas, 1911)
Grammomys caniceps Hutterer & Dieterlen, 1984
Grammomys cometes Thomas & Wroughton, 1908
Grammomys dolichurus (Smuts, 1832)
Grammomys dryas (Thomas, 1907)
Grammomys gigas Dollman, 1911
Grammomys ibeanus Osgood, 1910
Grammomys kuru (Thomas & Wroughton, 1907)
Gramommys macmillani (Wroughton, 1907)
Grammomys minnae Hutterer & Dieterlen, 1984
Grammomys poensis (Eisentraut, 1965)
Grammomys selousi Denys et al., 2011

Key to Species

At this stage it is not possible to provide a key for *Grammomys*. The genus is in urgent need of a synthetic revision employing karyotypic, molecular and morphological data. Nevertheless, with a few exceptions, most species occur allopatrically and geographical distribution together with comments below under Diagnosis in the following species accounts will help to reach at least a tentative identification.

Species Accounts

Grammomys aridulus Thomas & Hinton, 1923

Diagnosis

This is the only species of *Grammomys* occurring in the Jebel Marra region. It has an olive-brown dorsal pelage with bright orange-brown flanks and creamy-white ventral pelage.

Taxonomic Issues

This species was previously listed as a subspecies of *G. macmillani* (Happold, 1966) but considered a valid species based on morphological grounds by Hutterer & Dieterlen (1984).

Distribution

This species is restricted to Jebel Marra, western Sudan where it occurs in bushy savanna above 1000 m.

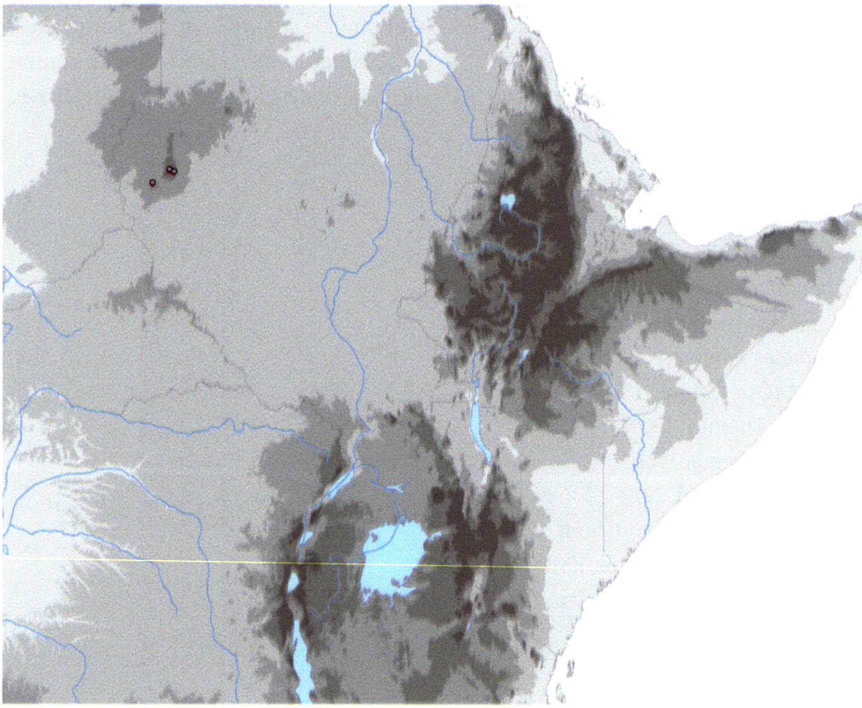

Figure 2.497 Distribution map

Measurements

External and cranial measurements (mm) and mass (g) for *Grammomys aridulus*, sexes combined

	Mean	Min	Max	N
Mass*	43	-	-	1
HB*	114	-	-	1
Tail*	185	-	-	1
HF*	23.3	-	-	1
Ear	-	-	-	-
GLS*	30.1	-	-	1

*Hutterer & Dieterlen (1984), measurements of the type specimen

Grammomys brevirostris Krystufek, 2008

Diagnosis

This species is currently only known from the type specimen which differs from other *Grammomys* by having a short rostrum, a long masseteric plate (3.3 mm in length) with convex anterior margin, t1 cusp of first upper molar separated from t2, and t9 of first upper molar greatly reduced (Krystufek, 2008). This species may co-occur with the following species: *G. ibeanus* (on average narrower skull with zygomatic width 53.5% of condylo-basal length; shorter masseteric plate 2.3-3.1 mm; anterior margin of plate straight or concave; t1 and t2 on M^1 connected); *G. dolichurus* (shorter masseteric plate 2.3-2.7 mm; anterior margin of plate straight or concave; t1 and t2 on M^1 connected). This species has a reddish-brown dorsal pelage and pure white ventral pelage with sharp demarcation of grey flanks from white ventral pelage (without orange lateral line). White subauricular tufts are absent and the tail is dark without a terminal tuft. The zygomatic width is 55.7% of the condylo-basal length.

Taxonomic Issues

This species was recently described based on a single specimen discovered in the collections of the Natural History Museum, London (Krystufek, 2008). The species was characterised on slight differences in cranio-dental morphology without supporting molecular data. We retain this species until further evidence comes to light.

Distribution

This species is only known from a single specimen from the type locality in the Loita Plains of south-western Kenya. Its habitat associations and distributional limits are currently unknown (Krystufek, 2008).

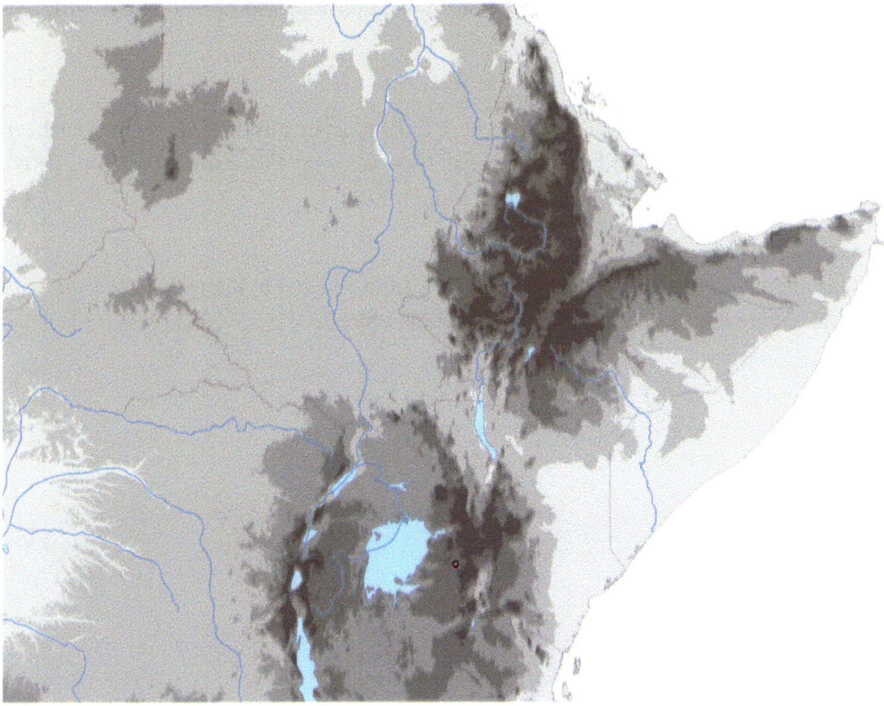

Figure 2.498 Distribution map

Measurements

External and cranial measurements (mm) and mass (g) for *Grammomys brevirostris*, sexes combined

	Mean	Min	Max	N
Mass	-	-	-	-
HB	-	-	-	-
Tail	-	-	-	-
HF	-	-	-	-
Ear	-	-	-	-
GLS*	28.9	-	-	1

*This is the condylo-basal length of the type specimen (Krystufek, 2008). No external measurements were presented of this specimen.

Grammomys buntingi (Thomas, 1911)

Diagnosis

The only other species of *Grammomys* occurring within the range of *G. buntingi* is *G. macmillani* which has a straight anterior margin to the masseteric plate. *Grammomys poensis* may be parapatric with *G. buntingi* in Ivory Coast but differs by being larger (GLS > 32 mm) and having a different number of chromosomes (2n = 36). This species has an olive-brown dorsal pelage with reddish-brown rump and pure white ventral pelage.

Taxonomic Issues

This species was previously listed as a synonym of *G. dolichurus* by Misonne (1974) but has been recognised as a separate species for several decades (Rosevear, 1969, Musser & Carleton, 2005). Chromosome number is 2n = 52, FN = 56 (Petter & Tranier, 1975).

Distribution

This species is endemic to the Upper Guinea Rainforest from Senegal in the west to central Ivory Coast in the east, where it occupies primary and secondary forest.

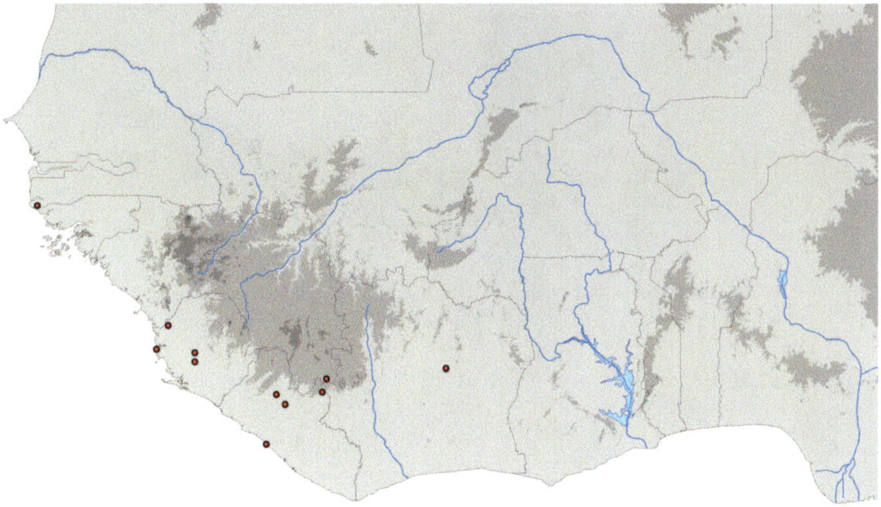

Figure 2.499 Distribution map

Measurements

External and cranial measurements (mm) and mass (g) for *Grammomys buntingi*, sexes combined

	Mean	Min	Max	N
Mass	-	-	-	-
HB	114.4	100	125	5
Tail	170.0	165	180	4
HF	25.0	24	26	5
Ear	18.4	16	20	5
GLS	30.0	28.6	31.4	6

Grammomys caniceps Hutterer & Dieterlen, 1984

Diagnosis

This is the only species of *Grammomys* occurring within its range. *Grammomys dolichurus* occurs in coastal habitats further to the south and can be distinguished by its larger size (GLS > 28 mm) and different number of chromosomes (2n = 52). In this species the dorsal pelage is greyish-brown with scattered black hairs, a bright cinnamon rump and pure white ventral pelage.

Taxonomic Issues

This species was previously listed as *G. dolichurus* (Roche et al., 1984) but described as a new species based on morphological and chromosomal grounds by Hutterer & Dieterlen (1984). Chromosome number is 2n = 54-61, FN = 78 (Hutterer & Dieterlen, 1984; Roche et al., 1984).

Distribution

This species is known from just three localities along the coast of East Africa from Somalia and Kenya where its habitat associations are not currently know (Hutterer & Dieterlen, 1984).

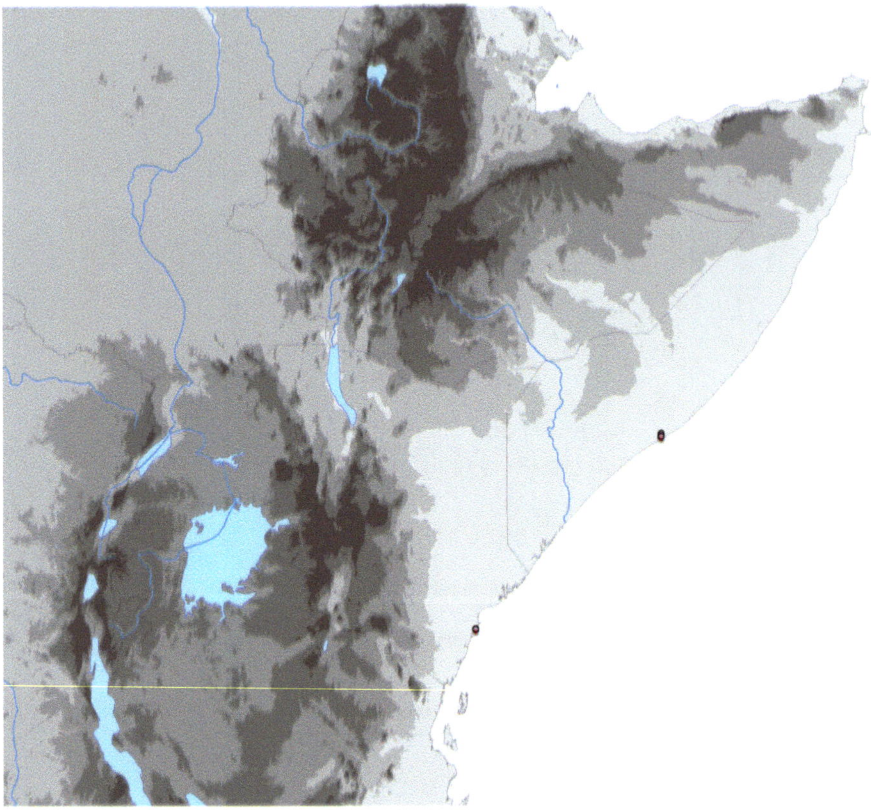

Figure 2.500 Distribution map

Measurements

External and cranial measurements (mm) and mass (g) for *Grammomys caniceps*, sexes combined

	Mean	Min	Max	N
Mass*	28.4	18	40	9
HB*	97.4	86	105	9
Tail*	150.7	147	155	9
HF*	21.1	19	22	9
Ear*	16.0	14	17	9
GLS*	26.3	25.8	27.8	9

*Hutterer & Dieterlen (1984)

Grammomys cometes Thomas & Wroughton, 1908

Diagnosis

The only other *Grammomys* species to co-occur within its range is *G. dolichurus* (size smaller, GLS typically < 31 mm; white postauricular patch always absent). However, these two species can only confidently be separated on molecular grounds (Krystufek et al., 2008). This species has a reddish-brown dorsal pelage with pure white ventral pelage.

Taxonomic Issues

This species was previously thought to extend into East Africa as the taxon *ibeanus* which is now considered a valid species. *Grammomys cometes* has been restricted to south of the Zambezi River and *G. ibeanus* north of it (Musser & Carleton, 2005). However, a recent phylogeny for the genus suggests that its closest relative is actually *G. dolichurus* from which it differs by 3.4% divergence on the cytochrome b gene (Krystufek et al., 2004), and not *G. ibeanus*. Chromosome number is 2n = 52, FN = 62 (Krystufek et al. (2008).

Distribution

This species occurs in the eastern parts of Southern Africa, south of the Zambezi River where it occupies coastal, riverine and montane forests.

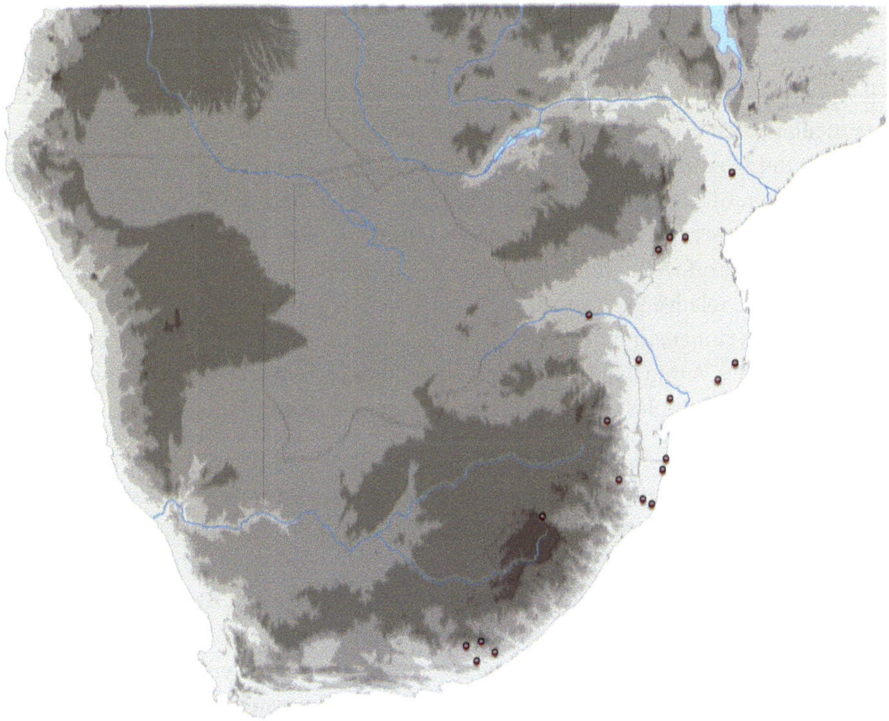

Figure 2.501 Distribution map

Measurements

External and cranial measurements (mm) and mass (g) for *Grammomys cometes*, sexes combined

	Mean	Min	Max	N
Mass*	52	-	-	1
HB	124.8	110	145	4
Tail	179.7	165	192	3
HF	23.5	23	24	2
Ear	18.5	17	20	2
GLS	31.6	31.2	31.9	4

*Taylor (1998)

Grammomys dolichurus (Smuts, 1832)

Diagnosis

This species co-occurs with the following species: *G. cometes* (size larger, GLS typically > 31 mm; white postauricular patch usually, but not always, present); *G. dryas* (two pairs of nipples (0 + 2 = 4); restricted to the Albertine Rift where it occupies montane forests above 1000 m); *G. ibeanus* (size on average larger, GLS typically > 31 mm and molar toothrow longer > 4.9 mm; white subauricular patch usually present); *G. kuru* (tail relatively longer ca. 160% of HB; two pairs of nipples (0 + 2 = 4); ventral pelage off-white with orange tinge); *G. macmillani* (size on average smaller, GLS typically ca. 29 mm; tail relatively longer ca. 160% of HB; reddish-brown subauricular patch; 2n = 68-76). This species has cinnamon-brown dorsal pelage, pure white ventral pelage and lacks white postauricular ear patches. A thin band of orange is sometimes present on the flanks. The tail is ca. 150% of HB. Three pairs of nipples (1 + 2 = 6) present. The molar toothrow is < 4.7 mm.

Taxonomic Issues

This species cannot be confidently distinguished from *G. cometes* on morphological grounds but is genetically, and perhaps ecologically, distinct (Krystufek et al., 2008); 2n = 52, FN = 62 (Krystufek et al., 2008). A second karyotype has also been reported from South Africa: 2n = 44 from Woodbush in the north-east of the country (Dippenaar et al., 1983); and two new species have recently been described from Kenya and Tanzania which may be from the *G. dolichurus* group. This suggests the presence of cryptic species within this group and casts doubt on the distribution map presented here. This species is in urgent need of re-appraisal and revision.

Distribution

This species occurs widely in Southern and East Africa where it occupies a variety of forest and woodland habitats, particularly in thick vegetation, but is absent from the rainforest zone.

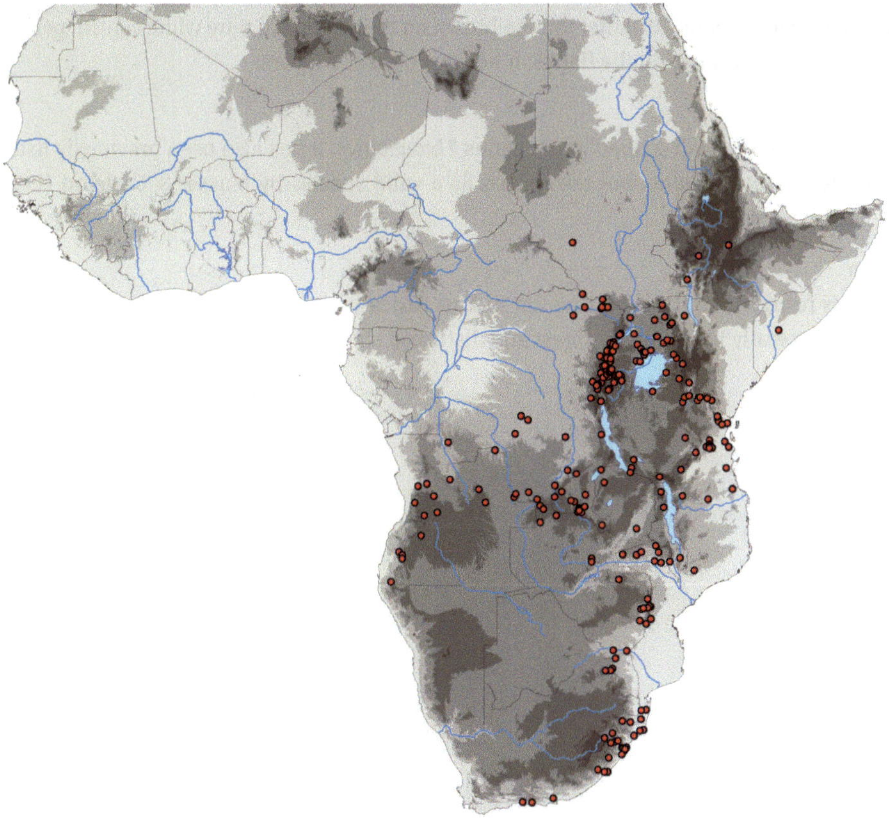

Figure 2.502 Distribution map

Measurements

External and cranial measurements (mm) and mass (g) for *Grammomys dolichurus*, sexes combined

	Mean	Min	Max	N
Mass	36.1	20	54	17
HB	107.0	90	136	35
Tail	163.2	129	185	31
HF	23.7	21	26	29
Ear	17.5	15	23	31
GLS	29.7	27.8	32.1	25

Grammomys dryas (Thomas, 1907)

Diagnosis

This is the only species of *Grammomys* occurring in montane habitats in the Albertine Rift typically above 1500 m. *Grammomys kuru* is a rainforest species occurring at lower altitudes, and *G. dolichurus* also occurs at lower altitudes and has three pairs of nipples. This species has rufous-brown dorsal pelage becoming brighter orange-brown on the rump, and pure white ventral pelage. There is a thin band of orange on the flanks. It has two pairs of nipples.

Taxonomic Issues

This species was originally described in the genus *Thamnomys*. It has also been previously listed as a subspecies of *G. dolichurus* but recognised as a valid species by Musser & Carleton (1993, 2005).

Distribution

This species is restricted to the Albertine Rift of East Africa where it occurs in montane habitats above 1000 m, typically in secondary forest and disturbed shrubby habitats.

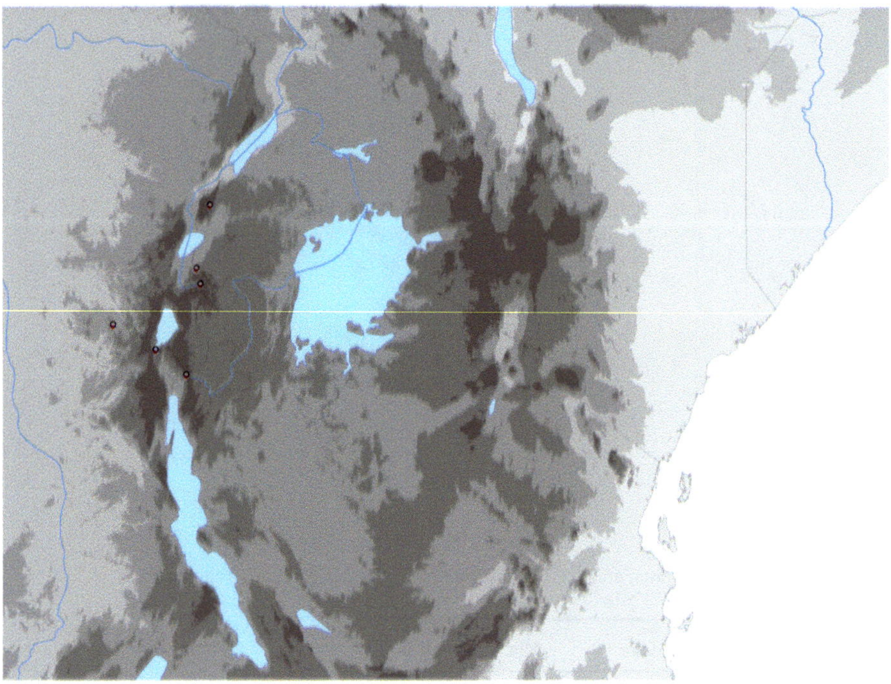

Figure 2.503 Distribution map

Measurements

External and cranial measurements (mm) and mass (g) for *Grammomys dryas*, sexes combined

	Mean	Min	Max	N
Mass*	40.8	30	59	65
HB*	112	100	130	36
Tail*	160	142	177	36
HF*	24.3	22	28	36
Ear*	17.3	15	21	36
GLS*	29.2	27.2	31.8	33

*Happold (2013)

Grammomys gigas Dollman, 1911

Diagnosis

This species can be distinguished from all other *Grammomys* species by its large skull (GLS of the holotype is 35.5 mm). This species has rufous-brown dorsal pelage becoming brighter reddish-brown on rump, and white ventral pelage. There is a thin band of orange on the flanks.

Taxonomic Issues

The validity of this species is uncertain. It is only known from a single specimen which is typically large for a *Grammomys*. However most of its measurements fall within that of *G. ibeanus*, but its skull is distinctly larger. We retain this species pending the collection of further specimens from Mt Kenya.

Distribution

This species is only known from the type specimen collected from Solai on the slopes of Mt Kenya.

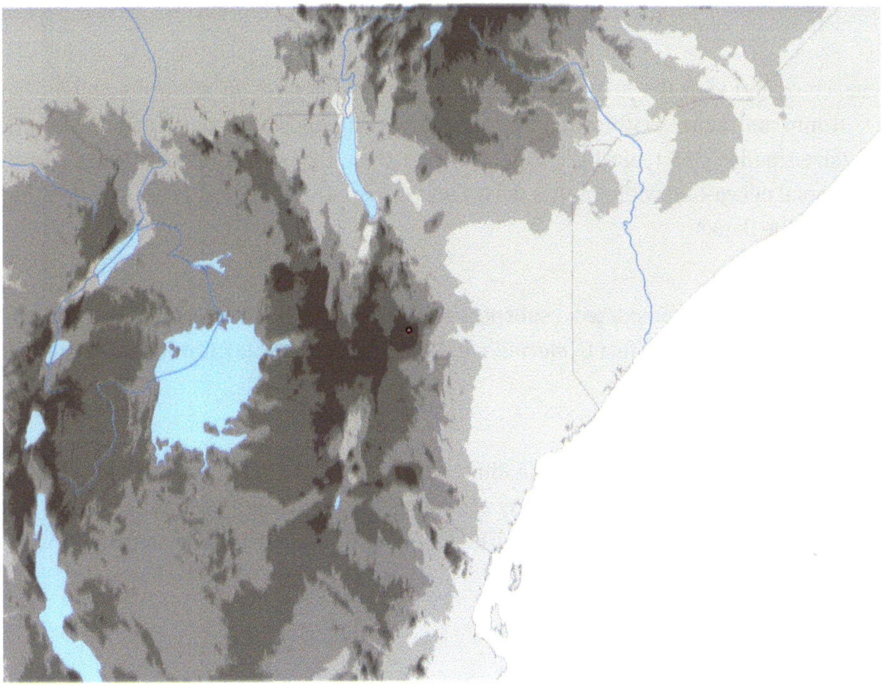

Figure 2.504 Distribution map

Measurements

External and cranial measurements (mm) and mass (g) for *Grammomys gigas*, sexes combined

	Mean	Min	Max	N
Mass	-	-	-	-
HB*	132	-	-	1
Tail*	201	-	-	1
HF*	26.5	-	-	1
Ear*	19	-	-	1
GLS*	35.5	-	-	1

*Happold (2013), measurements of the holotype.

Grammomys ibeanus Osgood, 1910

Diagnosis
This species co-occurs with the following species: *G. dolichurus* (size on average smaller, GLS typically < 31 mm and molar toothrow shorter < 4.7 mm; white postauricular patch absent); *G. macmillani* (size smaller, GLS typically < 30 mm, HB < 110 mm; 2n = 68-76). This species has an olive-brown dorsal pelage and white ventral pelage and three pairs of nipples. There may be a thin line of yellow on the flanks.

Taxonomic Issues
This species was previously considered a subspecies of *G. cometes*, but a molecular analysis suggests that its closest relative is in fact *G. macmillani* (Krystufek et al., 2008). Chromosome number is 2n = 44-48 (Krystufek et al., 2008).

Distribution
This species occurs widely in East Africa extending from southern Sudan in the north to southern Malawi in the south where it occupies a variety of forested habitats including montane forest and thickets. It appears to occupy relatively undisturbed forest in the Usambara Mts compared with *G. macmillani* (Stanley & Goodman, 2011).

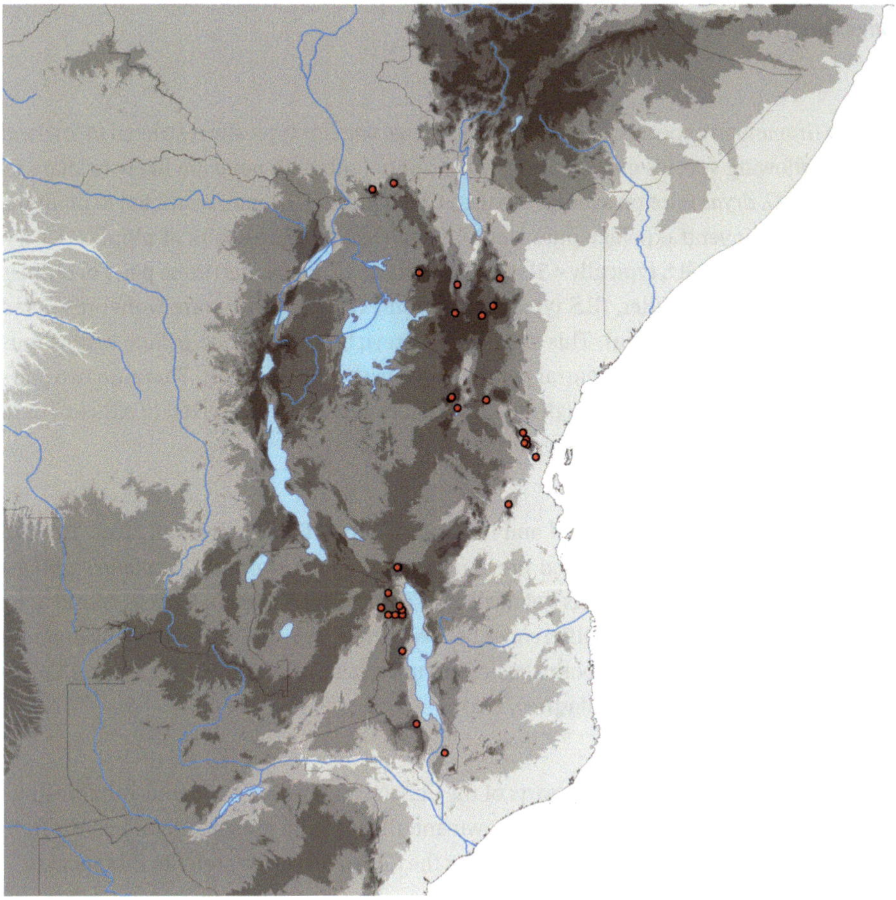

Figure 2.505 Distribution map

Measurements

External and cranial measurements (mm) and mass (g) for *Grammomys ibeanus*, sexes combined

	Mean	Min	Max	N
Mass*	44.8	27	60	24
HB*	121.8	112	130	6
Tail*	174.6	129	220	25
HF*	24.3	21	26	24
Ear*	18.9	17	21	24
GLS**	32.0	30.1	33.0	6

*Stanley & Goodman (2011); **Happold (2013)

Grammomys kuru (Thomas & Wroughton, 1907)

Diagnosis

This is the only *Grammomys* species across most of its range and it is typically restricted to lowland rainforest. The following species may co-occur with *G. kuru* (although none are likely to share its rainforest habitat): *G. dryas* (size on average smaller, GLS typically < 31 mm; restricted to montane habitats above 1000 m; ventral pelage pure white); *G. macmillani* (three pairs of nipples (1 + 2 = 6); size on average smaller, GLS typically < 31 mm; 2n = 68-76); *G. dolichurus* (three pairs of nipples (1 + 2 = 6); size on average smaller, GLS typically < 31 mm; ventral pelage pure white; restricted to forested habitats within savanna). This species has greyish-brown dorsal pelage with rufous towards the rump and creamy-white ventral pelage tinged with orange or buff. There are two pairs of nipples (0 + 2 = 4). The t7 cusp, although small, is visible and is not connected to either t4 or t8 (a character unique to this species and *G. poensis*)

Taxonomic Issues

This species is currently poorly diagnosed and its relationship to *G. poensis* is not clear. The two species may be conspecific and have been previously listed as the species *Thamnomys rutilans* (Rosevear, 1969); but the name *rutilans* is not available. The closely related *G. poensis* is said to be larger (Musser & Carleton, 2005), but differences between these two species have not been quantified and the distributional limits of both species are uncertain, especially in the central Congo Basin. Chromosome number 2n = 50 (Matthey, 1963).

Distribution

This species occurs widely in the eastern parts of the Congo Basin where it inhabits rainforest. Its southern and western limits are currently unknown and the material assigned to *G. poensis* from the southern regions of the Congo Basin (left bank of the Congo River) may refer to this species.

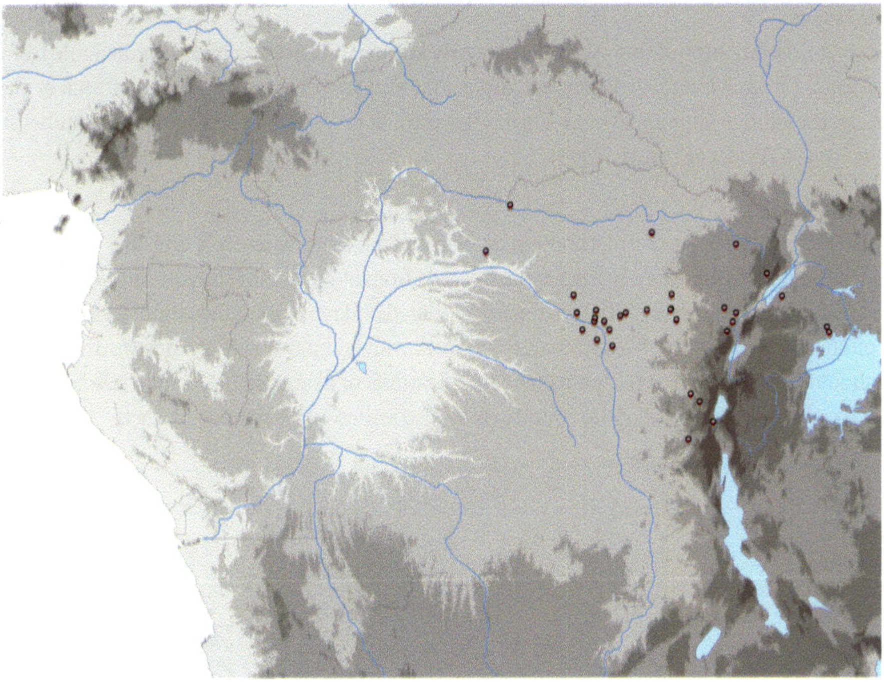

Figure 2.506 Distribution map

Measurements

External and cranial measurements (mm) and mass (g) for *Grammomys kuru*, sexes combined

	Mean	Min	Max	N
Mass*	47.1	31	60	16
HB*	122.2	109	145	16
Tail*	159.9	140	172	16
HF*	23.7	22	26	16
Ear*	14.5	12	17	16
GLS	-	-	-	-

*Delany (1975)

Gramommys macmillani (Wroughton, 1907)

Diagnosis

This species co-occurs with the following species: *G. dolichurus* (size on average larger, GLS typically ca. 30 mm; tail relatively shorter ca. 150% of HB; subauricular patch absent; 2n = 52); *G. ibeanus* (size on average larger, GLS typically > 31 mm, HB > 110 mm; 2n = 44-48); *G. kuru* (two pairs of nipples (0 + 2 = 6); size on average larger, GLS typically > 30 mm; 2n = 50; ventral pelage white with orange or rufous tinge; occupies rainforest habitats). This species has an olive-brown dorsal pelage and white ventral pelage and three pairs of nipples. A rufous subauricular patch is present.

Taxonomic Issues

The taxonomic relationships of this species are poorly known. It is not easily distinguishable from *G. dolichurus*, but from which it is distinct on molecular and karyotypic grounds (Krystufek et al., 2008). Its closest relative appears to be *G. ibeanus* (Krystufek et al., 2008). Chromosome number is variable 2n = 68-76 (Petter & Tranier, 1975).

Distribution

This species has a wide but patchy distribution across tropical Africa where it is typically associated with forested habitats within the savanna zone. Records from Malawi, Mozambique and Zimbabwe mentioned by Musser & Carleton (2005) are unsubstantiated and require confirmation; as a result we have not plotted them here. The West African records are widely separated from those further to the east and should be re-examined (either as misidentifications or as a possible cryptic species). The Tanzanian records (all in the east of the country) may also represent a separate species.

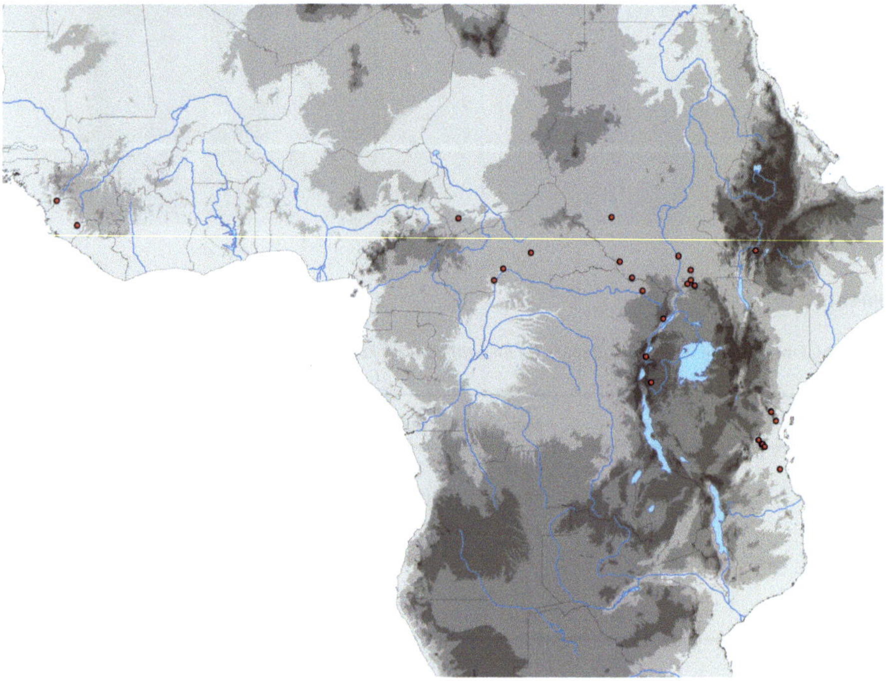

Figure 2.507 Distribution map

Measurements

External and cranial measurements (mm) and mass (g) for *Grammomys macmillani*, sexes combined

	Mean	Min	Max	N
Mass				
HB	105.1	98	110	8
Tail	164.3	149	186	8
HF	24.1	22	27	8
Ear	17.3	15	21	8
GLS*	29.0	26.5	30.6	10

*Happold (2013) reported a mean of 28.1 (range: 26.0-29.0, n = unknown) from southern Sudan

Grammomys minnae Hutterer & Dieterlen, 1984

Diagnosis
This is the only *Grammomys* species currently known to occur within its range and only one of three species recorded from Ethiopia. The other species of *Grammomys* known from Ethiopia are: *G. dolichurus* (2n = 52); *G. macmillani* (2n = 68-76). This species has greyish-brown dorsal pelage with creamy-white ventral pelage.

Taxonomic Issues
Attention was originally drawn to an undescribed species of *Grammomys* with an unusual karyotype by Olert et al., 1978). The species was subsequently described and named by Hutterer & Dieterlen (1984). Chromosome number is 2n = 32, FN = 64 (Olert et al., 1978).

Distribution
This species is only known from two montane localities in southern Ethiopia: Bulcha Forest and Arba Minch (Hutterer & Dieterlen, 1984). It appears to occupy gallery forest within savanna.

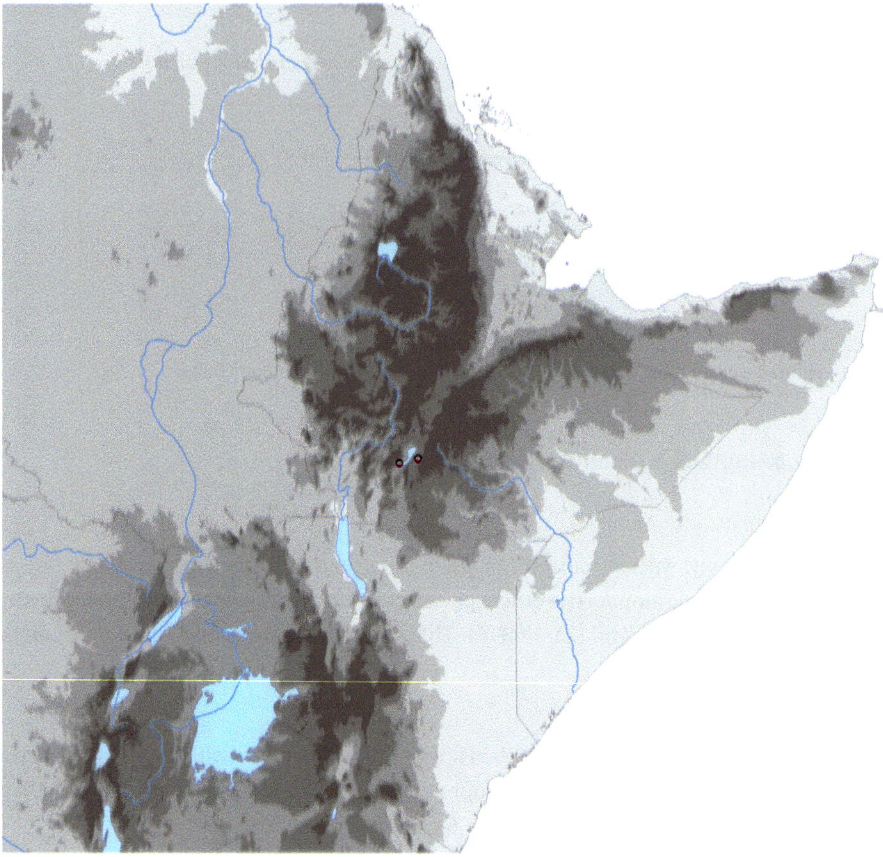

Figure 2.508 Distribution map

Measurements

External and cranial measurements (mm) and mass (g) for *Grammomys minnae*, sexes combined

	Mean	Min	Max	N
Mass*	36	-	-	1
HB*	110	-	-	1
Tail*	173	-	-	1
HF*	23	-	-	1
Ear*	18	-	-	1
GLS*	29.4	29.2	29.5	2

*Hutterer & Dieterlen (1984), measurements of the holotype and paratype.

Grammomys poensis (Eisentraut, 1965)

Diagnosis

This is the only species of *Grammomys* that occurs across most of its range. In West Africa the following species may co-occur with it: *G. buntingi* (size smaller GLS < 30 mm; 2n = 52); *G. macmillani* (size smaller GLS < 31 mm; 2n = 68-76). In northern Angola and neighbouring DRC, this species may co-occur with *G. dolichurus* which differs in karyotype (2n = 52) and in having pure white ventral pelage. This species has greyish-brown dorsal pelage with rufous towards the rump and creamy-white ventral pelage tinged with orange or buff. There are two pairs of nipples (0 + 2 = 4). The t7 cusp, although small, is visible and is not connected to either t4 or t8 (a character unique to this species and *G. kuru*).

Taxonomic Issues

This species is currently poorly diagnosed and its relationship to *G. kuru* is not clear. The two species may be conspecific and have been previously listed as the species *Thamnomys rutilans* (Rosevear, 1969). The name *rutilans* was abandoned because it is a junior synonym of a South American rodent and was replaced with *poensis* by Musser & Carleton (2005). The closely related *G. kuru* is said to be smaller (Musser & Carleton, 2005), but differences between these two species have not been quantified and the distributional limits of both species are uncertain, especially in the central Congo Basin. Chromosome number is 2n = 36 (Tranier & Dosso, 1979).

Distribution

This species occurs widely in the Upper and Lower Guinea Forest zones and is thought to extend into the Congo Basin (left bank of the Congo River), although these records may refer to *G. kuru*, and into northern Angola. There is a break in its distribution in Nigeria. It inhabits rainforest.

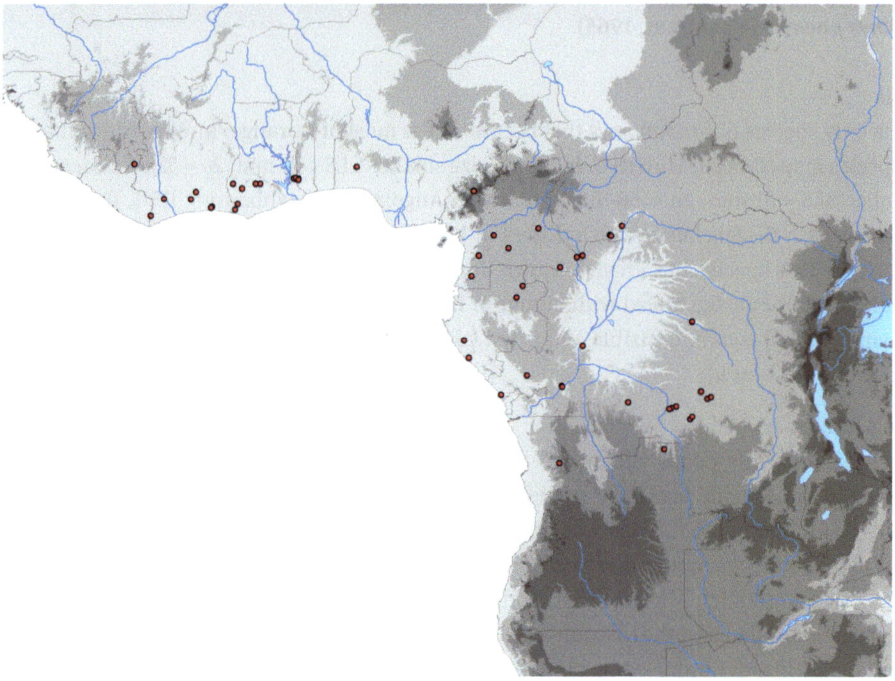

Figure 2.509 Distribution map

Measurements

External and cranial measurements (mm) and mass (g) for *Grammomys poensis*, sexes combined

	Mean	Min	Max	N
Mass	70.2	55	92	6
HB	127.3	112	147	24
Tail	184.2	163	200	24
HF	25.7	22	29	24
Ear	16.3	15	18	19
GLS	33.9	32.0	36.8	24

Grammomys selousi Denys et al., 2011

Diagnosis

This species may co-occur with several other *Grammomys* in Tanzania, but can be differentiated from all of them by its chromosome number. This species has a light yellow-brown dorsal pelage, a creamy-white ventral pelage and lacks a postauricular white ear patch (Denys et al., 2011b). The t7 cusp is not visible on the M^1, not even as a crest connected to the t4.

Taxonomic Issues

Although this species is chromosomally distinct from other members of the genus, it only differs slightly in morphology and has not yet been examined on molecular grounds. This species is probably closely related to *G. dolichurus*; the entire *G. dolichurus* group is in urgent need of revision. Chromosome number is 2n = 49-50, FN = 56 (Denys et al., 2011b).

Distribution

This species is only known from two specimens collected from the type locality, Kichi Forest of north-eastern Tanzania. The specimens were collected in Miombo woodland (Denys et al., 2011b).

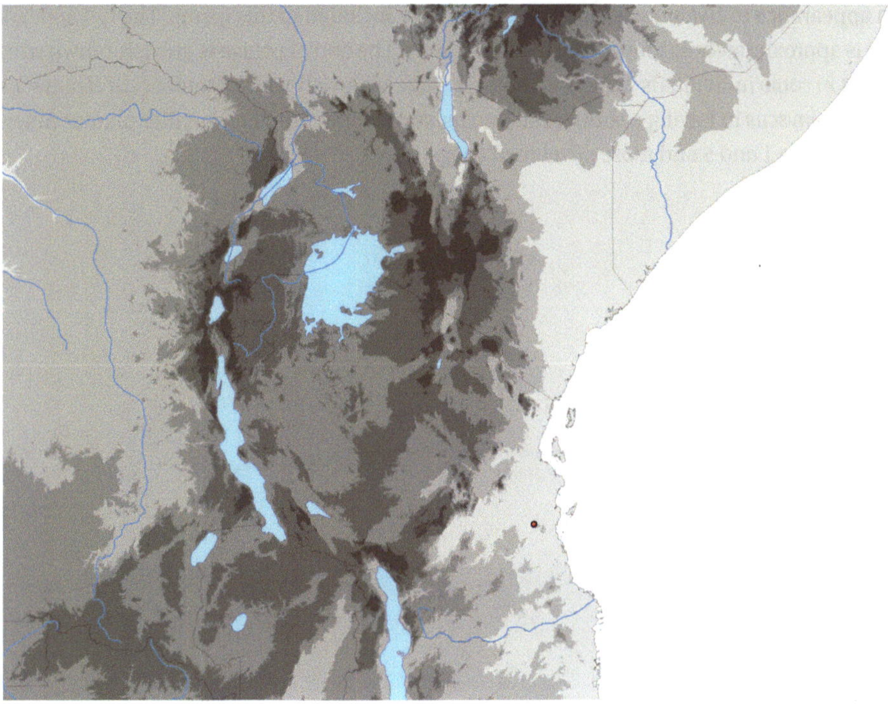

Figure 2.510 Distribution map

Measurements

External and cranial measurements (mm) and mass (g) for *Grammomys selousi*, sexes combined

	Mean	Min	Max	N
Mass*				
HB*	107	-	-	1
Tail*	167	-	-	1
HF*	23.5	-	-	1
Ear*	17.5	-	-	1
GLS*	29.7	-	-	1

*Denys et al. (2011b), measurements of the holotype.

Genus Account

Heimyscus Misonne, 1969 (African Smoky Mouse)

Diagnosis

This monotypic genus comprises small delicate mice (Mass ca. 23 g; HB ca. 90 mm) which are similar in appearance to *Hylomyscus* and were previously included in this genus. The tail appears naked and is approximately as long as the head and body. The dorsal pelage is greyish-brown with a reddish tint in some individuals while the ventral pelage is greyish-white. Members of *Heimyscus* differ from *Hylomyscus* in having a shorter tail, longer and thinner hindfeet, smaller plantar tubercles, reduced Digits 1 and 5 and more developed cusps.

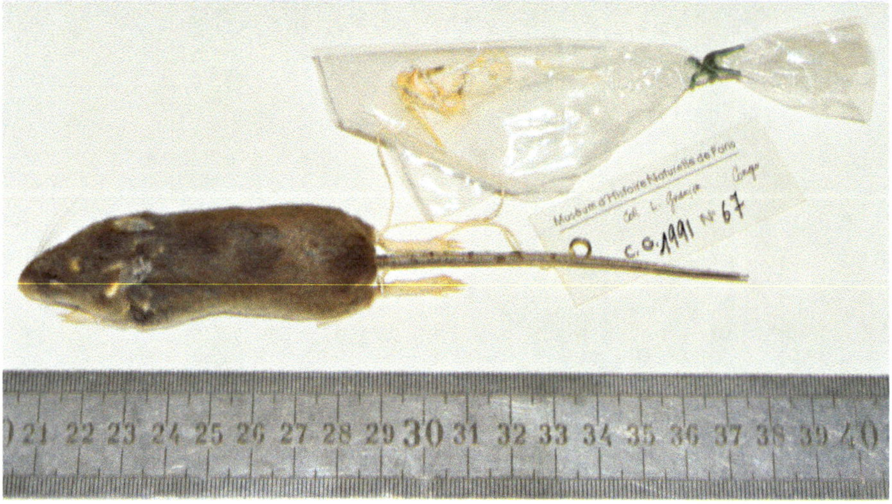

Figure 2.511 Photograph of *Heimyscus fumosus* museum specimen (MNHN-ZM-MO 1991-67) (Ara Monadjem).

Cranial Characters

The skull is very similar to that of *Hylomyscus*, being small in size (GSL ca. 26 mm) and long and narrow in shape with a long, narrow tapering rostrum. Apart from a trace of the supraorbital ridges, it is smooth and without crests. The zygomatic arches are slender, the palatal foramina are short and end well before the M^1, and the anterior edge of the zygomatic plate slopes forward and is cut away above. A small mastoid strut is present.

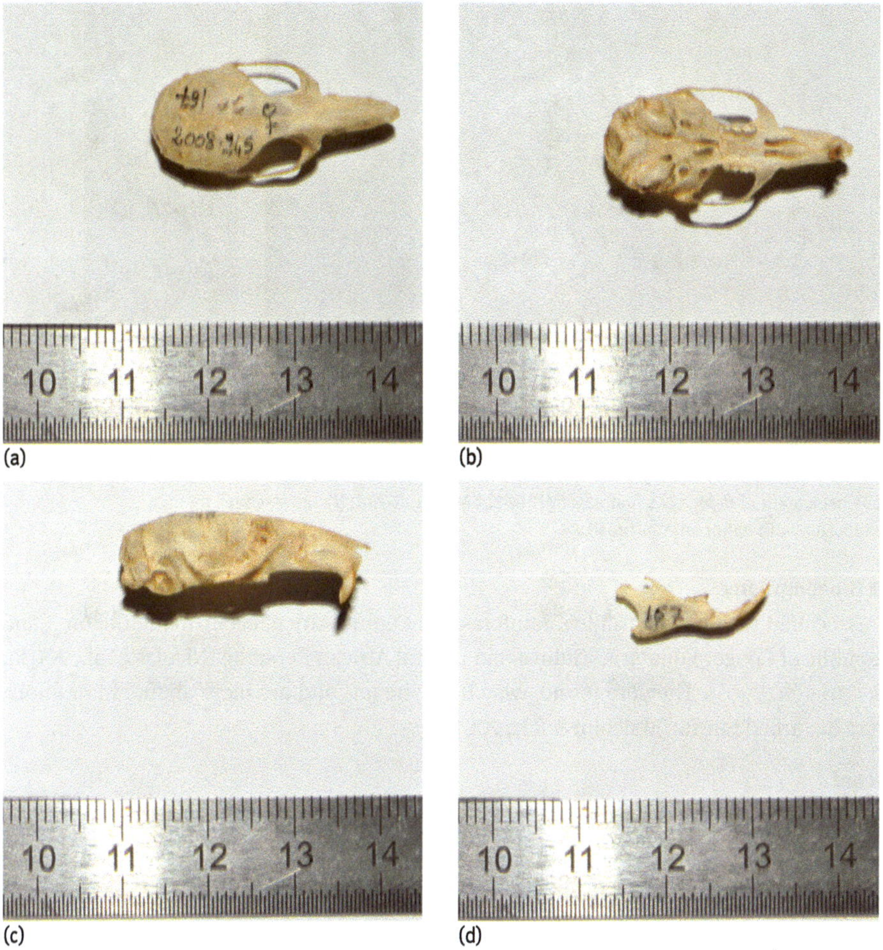

Figure 2.512 a-d Cranium of *Heimyscus fumosus* (MNHN-ZM-MO 2008-245) (from top left to bottom right: dorsal view, ventral view, lateral view and lower jaw).

Dentition

Incisors are opisthodont and ungrooved. Cusps are well developed (more so than in *Hylomyscus*) and the rows are distorted. In other respects the dentition is as described for *Hylomyscus* with molar row of average length (MTR/GSL ± 0.13, Table 10) and M^3 configuration as illustrated in Group 5 (Figure 2.439).

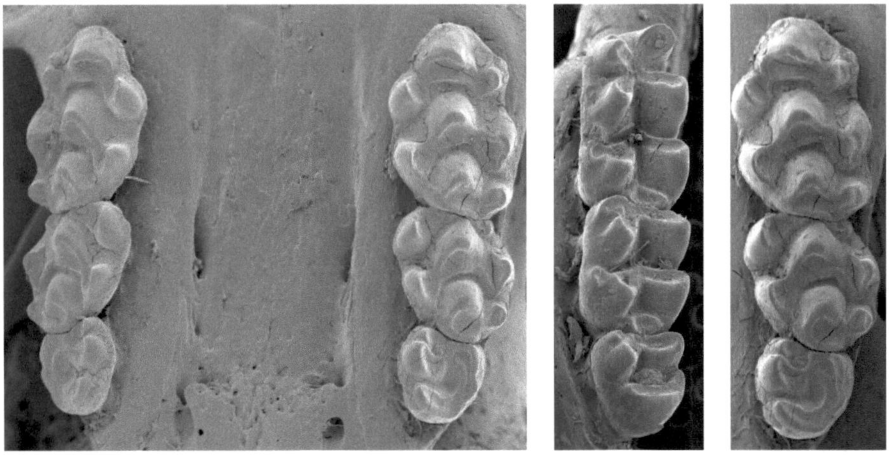

(a) (b) (c)

Figure 2.513 Cheek teeth of *Heimyscus fumosus* (MNHN-ZM-MO 2008-286) (A: upper cheek teeth; B: lower left toothrow; C: upper left toothrow).

Habitat & Biogeography

They are associated with climax tropical rainforests in Central Africa, occurring in Gabon, Cameroon, Republic of Congo, Equatorial Guinea and Central African Republic (Nicolas et al., 2003a; Musser & Carleton, 2005). They are found away from streams and are more abundant in undisturbed than disturbed forests (Malcolm & Ray, 2000).

Fossil Record

None.

Biology & Ecology

They are known to feed on insects, fruit and seeds (Nicolas et al., 2003b). In Gabon, breeding may be continuous (Duplantier, 1989) or seasonal (August to March: Nicolas & Colyn, 2003; Nicolas et al., 2003b). Litter sizes vary from one to four with the mean usually between two and three (Nicolas et al., 2003b). Males reach sexual maturity before females (Nicolas & Colyn, 2003; Nicolas et al., 2003b).

Systematic Notes

Heimyscus can be distinguished from *Hylomyscus* on DNA (Lecompte et al., 2002a), chromosomal, morphological (Misonne, 1969; Robbins et al., 1980) and ecological (Duplantier, 1989) grounds. Chromosome number 2n = 40, aFN = 48 (Lecompte et al., 2005c).

Annotated Species List

Only one species recognised:

Heimyscus fumosus (Brosset, Dubost & Heim de Balsac, 1965)

Species Account

Heimyscus fumosus (Brosset, Dubost & Heim de Balsac, 1965)

Diagnosis
See genus account.

Taxonomic Issues
See genus account.

Distribution
This species has a narrow distribution in Central Africa where it occupies rainforest, being rather more abundant in primary than disturbed forest.

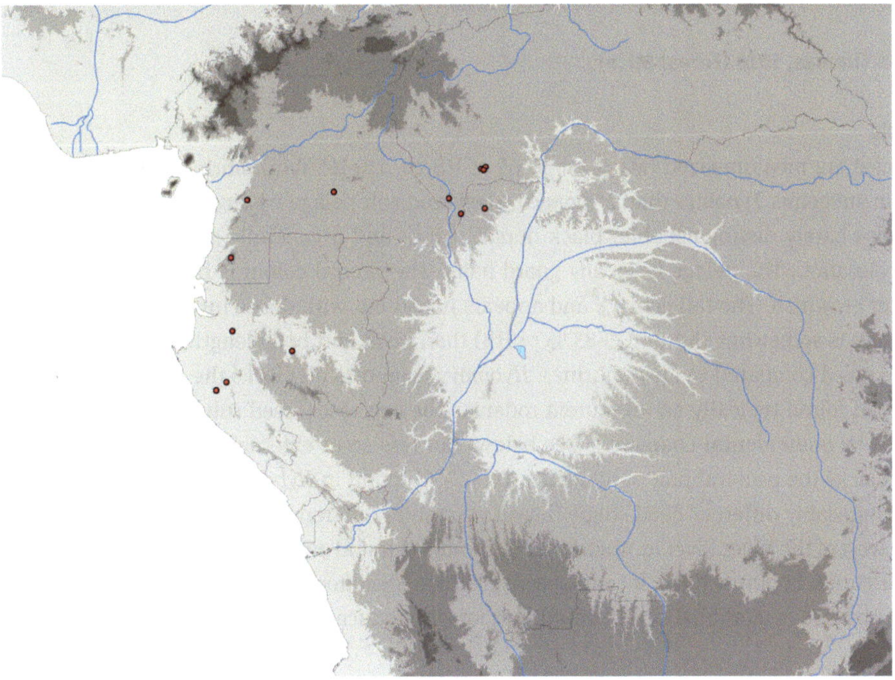

Figure 2.514 Distribution map

Measurements

External and cranial measurements (mm) and mass (g) for *Heimyscus fumosus*, sexes combined

	Mean	Min	Max	N
Mass	23.0	16	31	11
HB	89.2	80	98	12
Tail	98.2	86	111	12
HF	20.7	19	22	12
Ear	17.0	14	18	12
GLS	25.9	23.8	27.6	13

Genus Account

Hybomys Thomas, 1910 (Forest Mice)

Diagnosis

Forest mice are medium-sized (Mass ca. 50-65 g; HB ca. 110-165 mm) murines with fairly short (longer in subgenus *Typomys* than *Hybomys*), sleek and soft pelage having one or three variable (sometimes barely distinguishable) black dorsal stripes and a generally speckled yellowish to rufous coloration. The pelage lacks any guard hairs. The ventral colour is highly variable, from greyish to brownish. The tail is black and appears naked but with dense inconspicuous rings of bristles and is somewhat shorter (75-85 % of HB) than head and body length. Digits 1 and 5 are highly reduced compared to most murines. *Hybomys* diagnostically lacks the sixth (hypothenar) plantar pad found typically on all muroid rodents. The two recognised subgenera can be easily separated by craniodental characters (see below) but less so in external appearance. However, *Typomys* lacks the pectoral pair of mammae (0 + 2 = 4) found in *Hybomys* (1 + 2 = 6), has longer fur, a recognisably different head shape (based on cranial differences listed below) and it is distributed west of the Niger River in West Africa (*Hybomys* is found east of the Niger River) (Rosevear, 1969; Carleton & Robbins, 1985). It can be easily distinguished from similar *Dephomys* by its dorsal striping, absence of guard hairs, narrower hindfoot with reduced 5th digit and shorter tail.

Figure 2.515 Photograph of *Hybomys trivirgatus* (Ara Monadjem).

Cranial Characters

Rosevear (1969) commented that the cranial differences between the two subgenera were so profound as to render impossible a general description of the skull of *Hybomys*. However, the skull is medium sized (GSL ca. 32-36 mm) with a smoothly contoured and rounded braincase lacking obvious ridges (but with some interorbital "beading"), a long broad rostrum and gracile zygomatic arches (Carleton & Robbins, 1985). Cranial differences between the two subgenera are comprehensively documented by Rosevear (1969) and include *inter alia* the following: cranial profile arched continuously from front to back (*Hybomys*) to flat anteriorly (*Typomys*), rostrum deep, broad and blunt (*Hybomys*) versus shallow, narrow and tapering (*Typomys*), supraorbital ridges parallel (*Hybomys*) versus curved (*Typomys*), posterior margin of the palatal foramina just reaching M^1 (*Hybomys*) versus stopping well anterior of M^1 (*Typomys*), and the anterior edge of the zygomatic plate vertical (*Hybomys*) versus curving backwards from the base (*Typomys*).

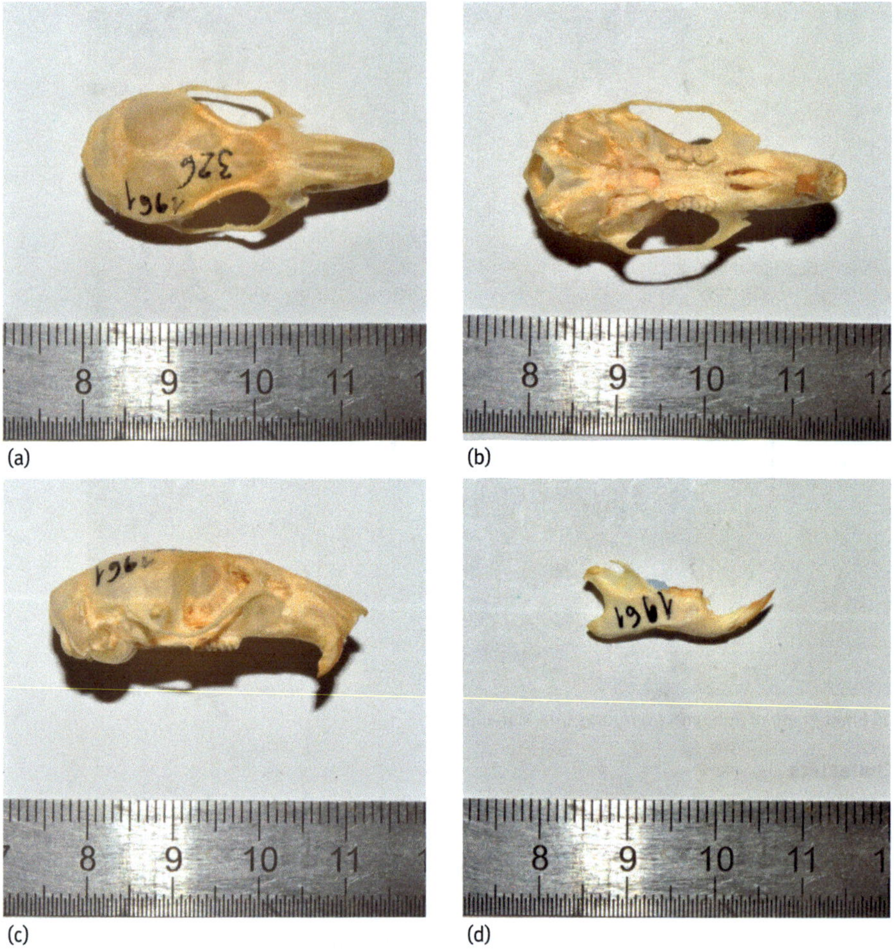

Figure 2.516 Cranium of *Hybomys univittatus* (MNHN-ZM-MO 1991-326) (from top left to bottom right: dorsal view, ventral view, lateral view and lower jaw).

Dentition

The incisors are slightly opisthodont and un-grooved. Cusps are well developed and the transverse rows are only slightly distorted. Molar tooth row is of average size (MTR/GSL ± 0.15, Table 10). Dentition is apparently very similar to *Dephomys* but the cusp t3 of M^1 is larger and separate from cusp t2 (Carleton & Musser, 2005). The last molar M^3 is half the size of M^2 and cusp arrangement of M^3 is as illustrated (Group 5 of Figure 2.439). Dental differences between the subgenera include the following (from Rosevear, 1969): lower incisors projecting upwards (*Hybomys*) versus forwards, i.e. pro-odont (*Typomys*), the presence of t9 on M^1 and M^2 (*Hybomys*) versus its absence or extreme reduction on M^1 and especially M^2 (*Typomys*).

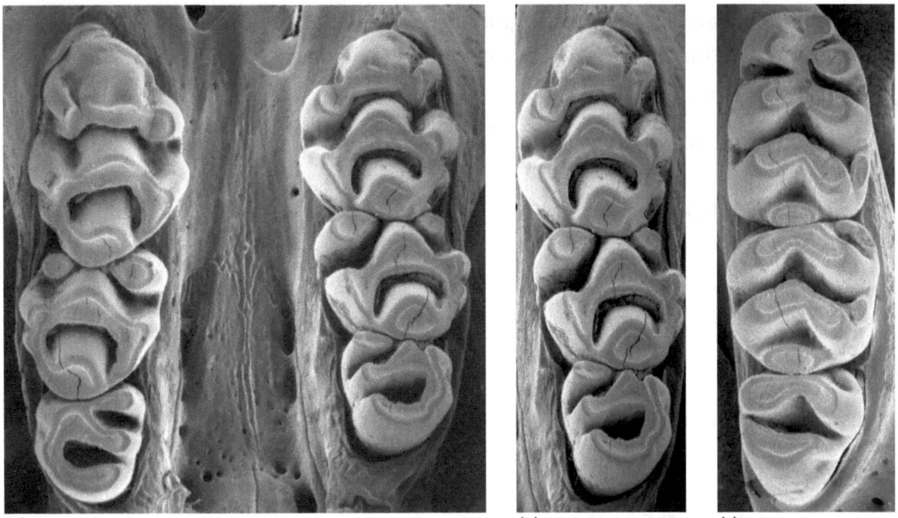

(a) (b) (c)

Figure 2.517 Upper cheek teeth of *Hybomys univittatus* (MNHN-ZM-MO 1966-67) (A: upper cheek teeth; B: upper right toothrow; C: lower right toothrow).

Habitat & Biogeography

They are associated with primary and secondary tropical rainforests and montane forest, particularly where ground cover is abundant, extending from Sierra Leone and Guinea in the west to Uganda in the east and south to north-western Zambia.

Fossil Record

None.

Biology & Ecology

They are terrestrial (although apparently capable of swimming), primarily nocturnal but sometimes diurnal, and live in tunnels constructed under roots, stumps and rotting logs. Their diet is predominantly insectivorous although fruits and seeds are eaten particularly in the wet season. (Happold, 2013). In West Africa, reproduction in *H. trivirgatus* occurs during the wet season whereas in central Africa *H. univittatus* reproduces throughout the year with a peak in the wet season (Happold, 2013). Litter size varies from one to six, with a mean usually of around 2.5 (Happold, 2013).

Systematic Notes

Molecular data show *Hybomys* to belong to the Arvicanthini tribe, with *Stochomys* as closest relative (Lecompte et al., 2008). Although it was previously considered to be only subspecifically different from *H. univittatus*, *H. lunarius* has been shown to be specifically distinct from it on chromosomal (Verheyen & Van der Straeten, 1985) and morphometric (Van der Straeten et al., 1986) grounds. As pointed out by Musser & Carleton (2005), the Uinka, DRC population assigned to "*lunarius*" by Van der Straeten and colleagues is actually distinct from true *lunarius* occurring in the Albertine Rift and represents an undescribed species. *Hybomys planifrons* has been placed as a subspecies of either *H. univittatus* or *H. trivirgatus* but Carleton & Robbins (1985) showed it to be distinct on chromosomal and morphometric grounds. A molecular phylogeny for Central African specimens demonstrated that *H. univittatus* is a complex of cryptic species (Missoup, 2010).

Annotated Species List

Six species are recognised in two subgenera:
Subgenus *Hybomys:*
Hybomys badius Osgood, 1936
Hybomys basilii Eisentraut, 1965
Hybomys lunaris (Thomas, 1906)
Hybomys univittatus (Peters, 1876)
Subgenus *Typomys*:
Hybomys planifrons (Miller, 1900)
Hybomys trivirgatus (Temminck, 1853)

Key to Species

1	Rostrum narrow and tapering; two pairs of nipples (0 + 2 = 4); lower incisors pro-odont; cranial profile flat anteriorly (subgenus *Typomys*)	2
	Rostrum broad and blunt; three pairs of nipples (1 + 2 = 4); lower incisors not pro-odont; cranial profile arched along its entire length (subgenus *Hybomys*)	3
2	A single mid-dorsal stripe extending from the head to the base of the tail	*H. planifrons*
	Three mid-dorsal stripes present (the two lateral stripes usually shorter and sometimes inconspicuous)	*H. trivirgatus*
3	Typically montane species with distributions restricted to Bioko Island, Cameroon highlands or Ruwenzori Mts (Uganda)	4
	Not occurring in the montane regions mentioned above	*H. univittatus*
4	Restricted to Bioko Island	*H. basilii*
	Restricted to the Cameroon highlands	*H. badius*
	Restricted to the Ruwenzori Mts	*H. lunaris*

Species Accounts

Hybomys badius Osgood, 1936

Diagnosis

The only other *Hybomys* occurring within the range of this species is *H. univittatus* which differs slightly in cranio-dental characters that can be only be illustrated in multivariate analyses (Van der Straeten & Hutterer, 1986). In fact, the distinction between this species and *H. univittatus* has not been clearly quantified. We follow Musser & Carleton (2005) in recognising the species occurring in the Cameroon highlands above 1500 m as *H. badius* where it is the only species of *Hybomys*. Three pairs of nipples present (1 + 2 = 6).

Taxonomic Issues

This taxon was described from Mt Cameroon but originally listed as a subspecies of *H. univittatus*. The taxon *eisentrauti* Van der Straeten & Hutterer (1986) was described from Mt Lefo and Mt Oku

(Cameroon) but considered a synonym of *H. badius* by Musser & Carleton (2005) and Happold (2013). We follow this arrangement pending further study. The *H. univittatus* group is in urgent need of revision.

Distribution

This species is endemic to Cameroon where it occurs in forested habitats generally at higher elevations.

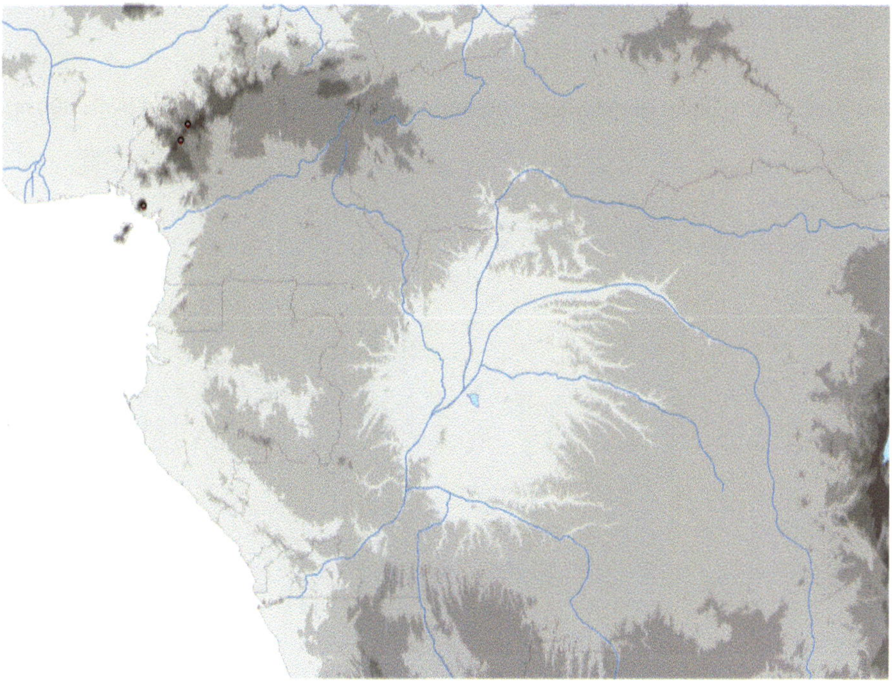

Figure 2.518 Distribution map

Measurements

External and cranial measurements (mm) and mass (g) for *Hybomys badius*, sexes combined

	Mean	Min	Max	N
Mass*	63.7	55	75	16
HB*	122.4	105	137	16
Tail*	114.3	108	121	12
HF*	29.3	27	31	16
Ear*	16.7	15	18	16
GLS*	34.7	33.3	35.9	12

*Happold (2013)

Hybomys basilii Eisentraut, 1965

Diagnosis
This is the only species of *Hybomys* occurring on Bioko Island. The mid-dorsal stripe is indistinct and may be absent; when present, it is relatively short, starting from the mid-back to base of tail.

Taxonomic Issues
This taxon was originally listed as a subspecies of *H. univittatus* but elevated to full species status by Van der Straeten (1985).

Distribution
This species is endemic to Bioko Island where it occurs in lowland and montane forest (450-2000 m).

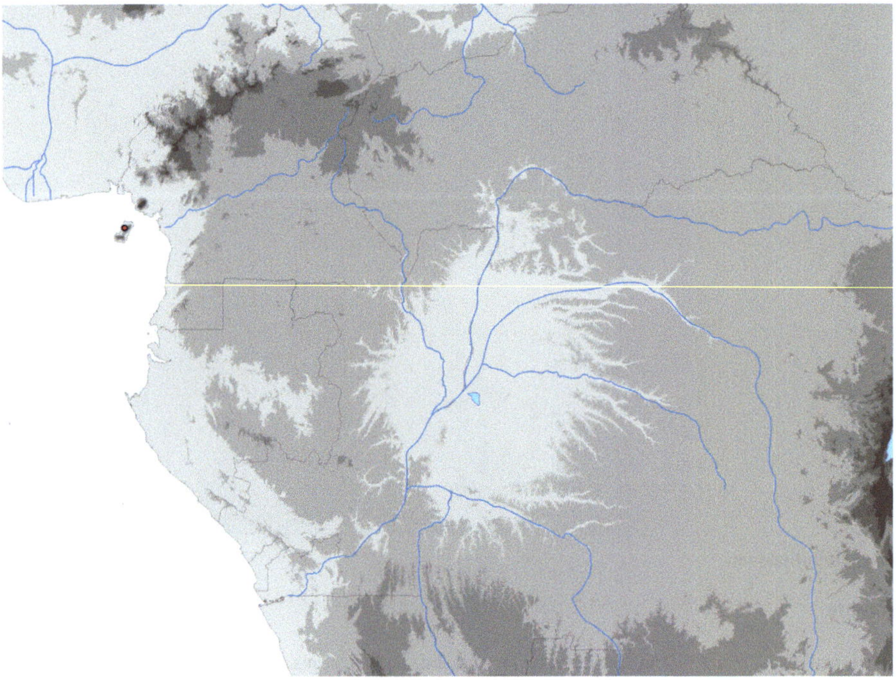

Figure 2.519 Distribution map

Measurements

External and cranial measurements (mm) and mass (g) for *Hybomys basilii*, sexes combined

	Mean	Min	Max	N
Mass*	86.2	70	95	6
HB**	137.0	118	160	34
Tail**	114.3	90	132	30
HF**	31.2	29	33	34
Ear**	17.2	15	20	32
GLS**	35.9	33.4	37.8	33

*Happold (2013); **Van der Straeten (1985)

Hybomys lunaris (Thomas, 1906)

Diagnosis

This species is very similar to *H. univittatus* in appearance, differing only slightly in cranio-dental features (Van der Straeten et al., 1986). However, this species is currently thought to be endemic to the Ruwenzori Mts where it is the only *Hybomys* present.

Taxonomic Issues

This species was originally described as a subspecies of *H. univittatus*, but is distinct from it on morphological and chromosomal grounds (Verheyen & Van der Straeten, 1985; Van der Straeten et al., 1986). Musser & Carleton (2005) emphasis its smaller size in relation to *H. univittatus*, but Delany (1975) argues that variation within Uganda is large and encompasses both taxa. We retain this species until further taxonomic (particularly molecular) analyses have been conducted. Chromosome number 2n = 48 (Verheyen & Van der Straeten, 1985)

Distribution

This species is currently only known from the Ruwenzori Mts in montane forest above 1800m. The larger distributional range portrayed by Van der Straeten et al. (1986) appears to be in error (Musser & Carleton, 2005), although the latter authors mention that two specimens from Kanyawara (Kibale Forest, Uganda about 60km from the Ruwenzori Mts but at lower altitude) fit the description of *H. lunaris*.

Figure 2.520 Distribution map

Measurements

External and cranial measurements (mm) and mass (g) for *Hybomys lunaris*, sexes combined

	Mean	Min	Max	N
Mass*	-	-	-	-
HB*	108	-	-	1
Tail*	115	-	-	1
HF*	25	-	-	1
Ear*	17	-	-	1
GLS*	32.0	-	-	1

*Happold (2013), measurements of the holotype

Hybomys planifrons (Miller, 1900)

Diagnosis

This species co-occurs with *H. trivirgatus* from which it can be distinguished by the presence of a single mid-dorsal stripe and chromosome number (in *H. trivirgatus* there are three mid-dorsal stripes and 2n = 40-43). Two pairs of nipples are present (0 + 2 = 4).

Taxonomic Issues

This species is morphologically similar to *H. trivirgatus* but can be distinguished from it on both morphological and chromosomal grounds (Carleton & Robbins, 1985). Chromosome number is 2n = 35-39, FN = 40-42 (Carleton & Robbins, 1985).

Distribution

This species is restricted to the Upper Guinea Forest in West Africa where it occurs in forested habitats west of the Sassandra River, Ivory Coast.

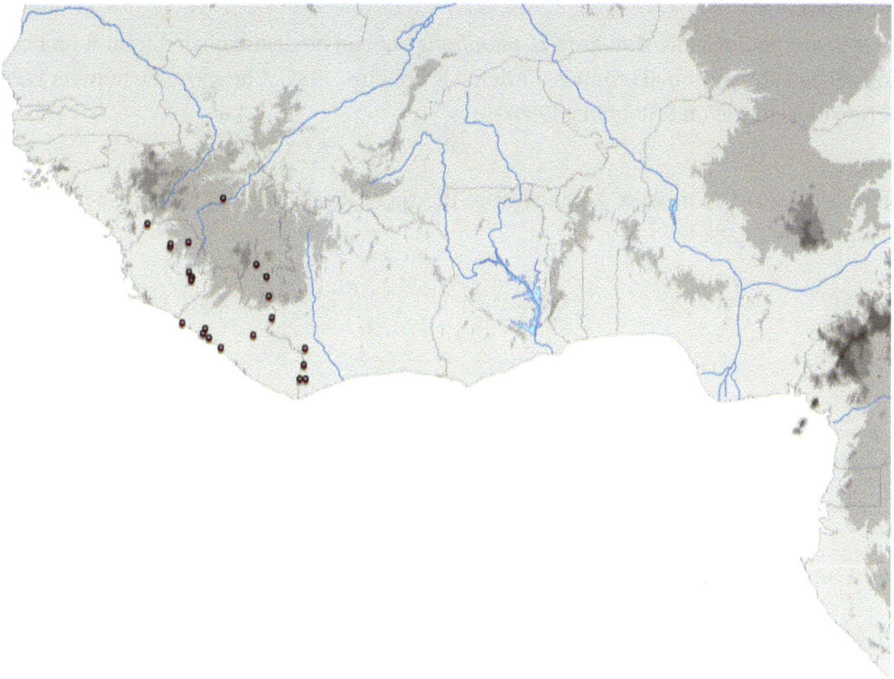

Figure 2.521 Distribution map

Measurements

External and cranial measurements (mm) and mass (g) for *Hybomys planifrons*, sexes combined

	Mean	Min	Max	N
Mass*	51	46	60	8
HB*	124	120	130	8
Tail*	97	91	104	8
HF*	30.6	28	32	8
Ear*	17.4	16	19	8
GLS*	32.8	32.0	33.8	8

*Happold (2013)

Hybomys trivirgatus (Temminck, 1853)

Diagnosis

This species co-occurs with *H. planifrons* from which it can be distinguished by the presence of three mid-dorsal stripes and chromosome number (in *H. planifrons* a single mid-dorsal stripe and 2n = 35-39). Two pairs of nipples are present (0 + 2 = 4).

Taxonomic Issues

This species is morphologically similar to *H. planifrons* but can be distinguished from it on both morphological and chromosomal grounds (Carleton & Robbins, 1985). Chromosome number is 2n = 40-43, FN = 40-43 (Carleton & Robbins, 1985).

Distribution

This species occurs widely in West Africa west of the Niger River (Nigeria) where it occupies a variety of forested habitats.

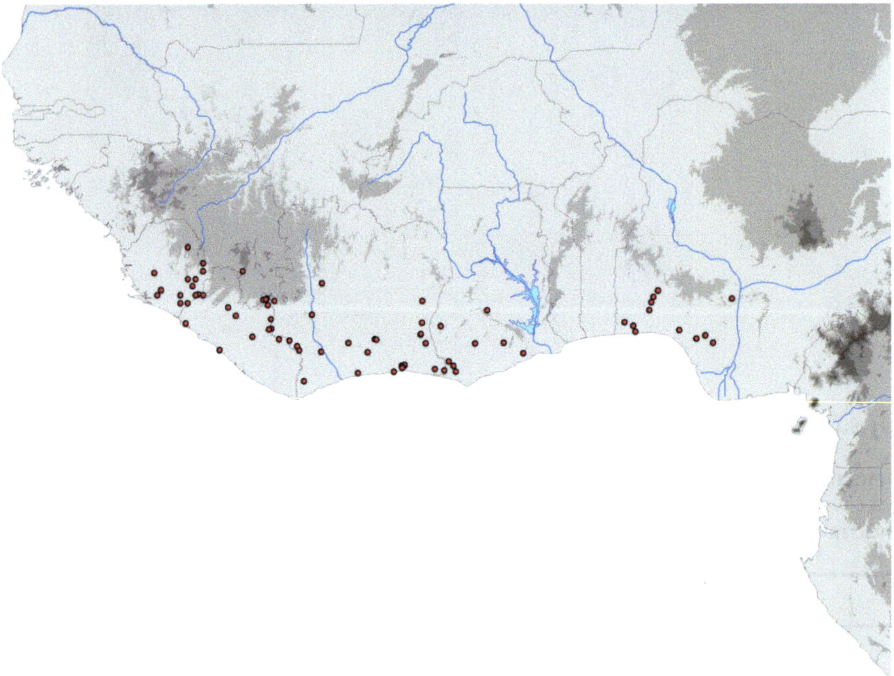

Figure 2.522 Distribution map

Measurements

External and cranial measurements (mm) and mass (g) for *Hybomys trivirgatus*, sexes combined

	Mean	Min	Max	N
Mass	49.6	29	63	7
HB	118.7	105	130	13
Tail	97.0	85	110	13
HF	30.6	27	34	13
Ear	16.3	15	18	10
GLS	33.4	28.1	36.1	13

Hybomys univittatus (Peters, 1876)

Diagnosis

This is the only species of *Hybomys* across most of its range. It co-occurs with *H. badius* in Cameroon and with *H. lunaris* in the Albertine Rift, however it has not yet been recorded syntopically (ie at the same location) with them. *Hybomys univittatus* is typically a lowland species, but may occur at higher elevations in parts of its range. Diagnostic features for separating these species have yet to be identified for this group and identification may have to be based on distribution until such time as these have been quantified.

Taxonomic Issues

Hybomys univittatus as currently recognised is likely to comprise a species complex (Musser & Carleton, 2005; Missoup, 2010). There is some chromosomal variability within this species with specimens from Rwanda (previously named *H. lunaris*) and north-western DRC being 2n = 48, whereas specimens from Cameroon 2n = 44-46, and Gabon 2n = 48 (Carleton & Robbins, 1985) or 2n = 44 (Lecompte et al., 2005b). This species group is in urgent need of revision (particularly on molecular grounds) and preliminary molecular analyses indicate the existence of at least two cryptic species in Central Africa (Missoup, 2010).

Distribution

This species has a wide range in the forests of the Congo Basin, extending from the Niger River (Nigeria) in the west to the shores of Lake Victoria in the east, with an isolated population in north-western Zambia and neighbouring DRC.

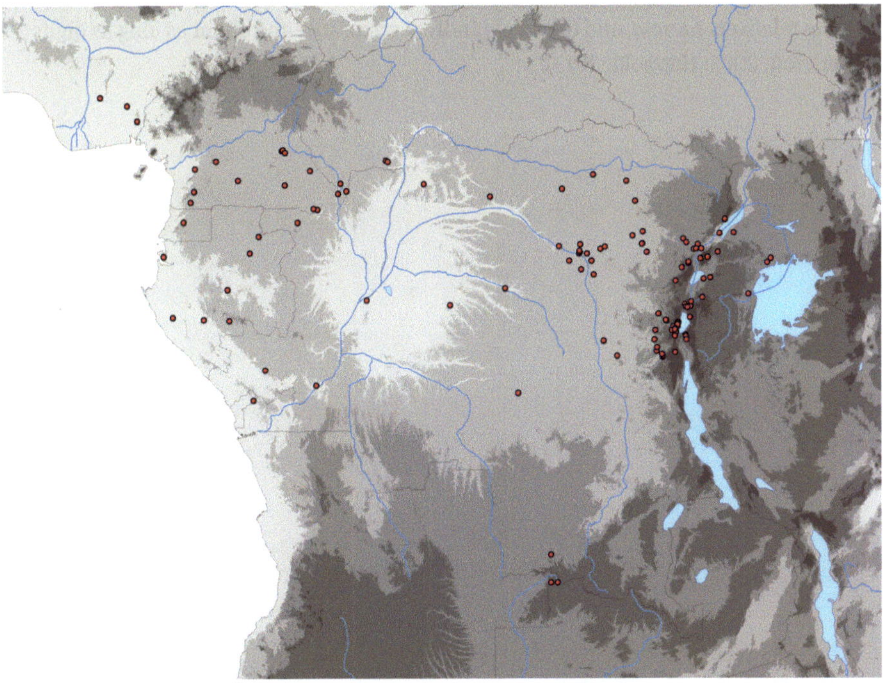

Figure 2.523 Distribution map

Measurements

External and cranial measurements (mm) and mass (g) for *Hybomys univittatus*, sexes combined

	Mean	Min	Max	N
Mass	62.8	46	70	4
HB	127.7	117	140	11
Tail	113.6	105	121	10
HF	29.9	28	31	10
Ear	16.2	15	18	9
GLS	34.1	31.0	35.7	7

Genus Account

Hylomyscus Thomas, 1926 (Wood Mice)

Diagnosis

These small to very small (Mass ca. 10-35 g; HB ca. 70-120 mm) soft-furred arboreal mice lack guard hairs (except in *H. aeta*). The smallest member of the genus is *H. parvus* whilst the largest is *H. endorobae*. Dorsal colour is reddish-brown to grey-brown while the ventral pelage is white tipped and sharply divided from the flanks, often by a pinky-buff line (Rosevear, 1969). The tail is usually very long (120-165% of HB) and appears naked having very fine scales and bristles. Characters considered diagnostic for this genus include small body and skull size, reduced cusps t3 and t9, zygomatic plate relatively narrow, and digit 5 of hindfoot opposable and about as long as digits 2-4 (Musser & Carleton, 2005; Happold, 2013).

Figure 2.524 Photograph of *Hylomyscus simus* (Ara Monadjem).

Cranial Characters

The skull is small (GSL 22-27 mm) with rounded and smoothly contoured braincase (only the *H. aeta* group has supraorbital "beading") with small bullae. A distinctive feature is the narrow zygomatic plate and the backward sloping anterior edge which is cut away above and not evenly curved, giving rise to a narrow zygomatic notch. A "mastoid strut" (hamular process) is present and varies from a shorter process with small, partly occluded subsquamosal foramen (e.g. *H. denniae*) to a long process with pronounced foramen (e.g. *H. arcimontensis*) (Carleton & Stanley, 2005).

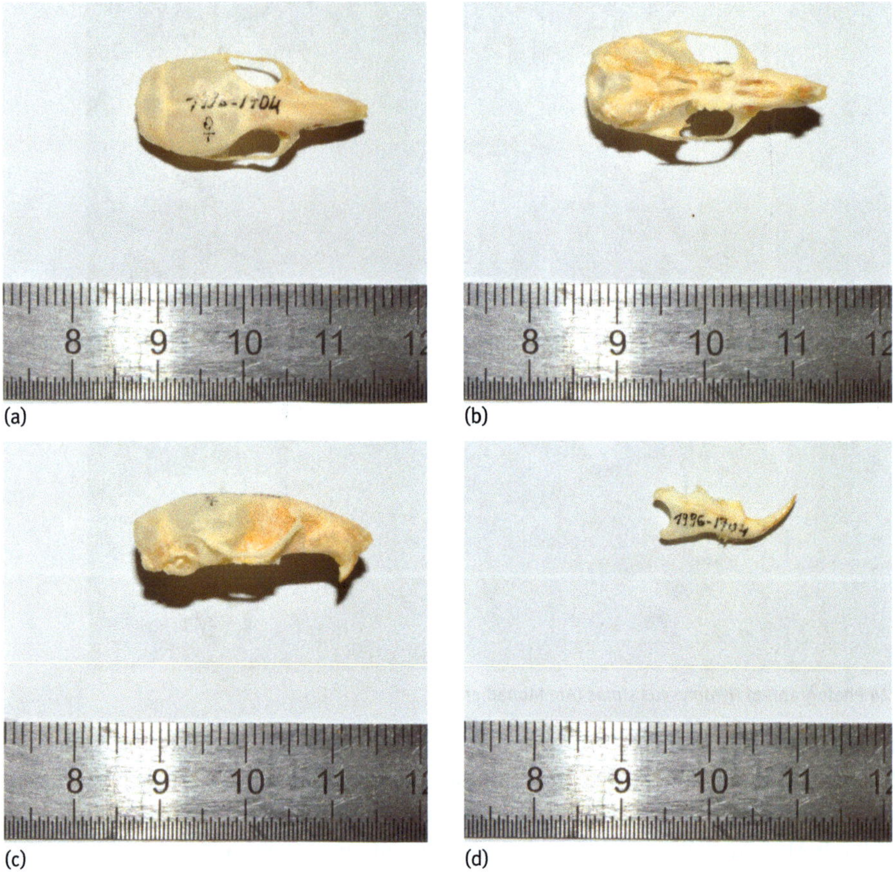

Figure 2.525 a-d Cranium of *Hylomyscus simus* MNHN-ZM-MO 1996-1704) (from top left to bottom right: dorsal view, ventral view, lateral view and lower jaw).

Dentition

Incisors vary between species from orthodont to opisthodont to pro-odont (Rosevear, 1969) but are ungrooved. The molars are of average size (MTR/GSL ± 0.16, Table 10). Cusp rows are distorted with t1 and t4 situated well posterior to t3 and t5 respectively. The cusp t3 is poorly developed on M^1 and M^2 and not well differentiated from t2, although this varies between species (Carleton et al., 2006); cusp t9 is also very poorly developed on M^1 and M^3. The last molar M^3 is half the size of M^2 and shows elements of three laminae and three distinct lingual cusps; its cusp arrangement is as illustrated in Group 5 (Figure 2.439).

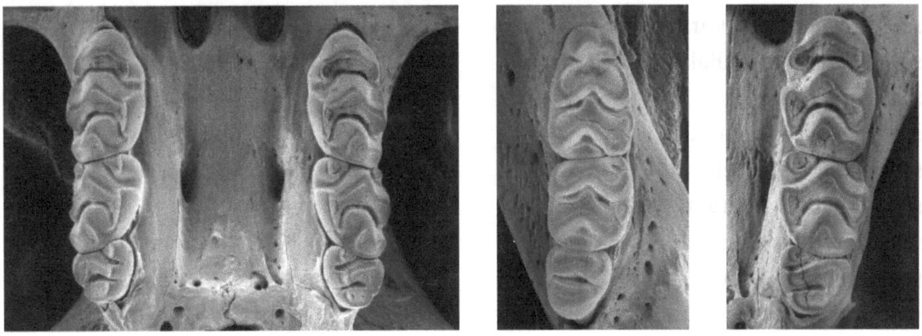

(a) (b) (c)

Figure 2.526 Cheek teeth of *Hylomyscus pamfi* (MNHN-ZM-MO 1965-658) (A: upper cheek teeth; B: lower right toothrow; C: upper left toothrow).

Habitat & Biogeography

They are widespread in lowland and montane tropical rainforest from Guinea in the west to Kenya and Tanzania in the east and south to Angola and north-west Zambia. In East Africa they are restricted to Afromontane habitats.

Fossil Record

None.

Biology & Ecology

They are arboreal, nocturnal and omnivorous. Breeding is usually continuous with a peak in the wet season and litter size varies from one to four (Genest Villard, 1980; Happold, 2013).

Systematic Notes

Morphological (Lecompte et al., 2002a) and molecular (Lecompte et al., 2002b; 2008b) data support the placement of *Hylomyscus* in the Praomyini tribe, with *Heimycus* and *Mastomys pernanus* as its nearest relatives (Lecompte et al., 2008; Schenk et al., 2013). From just eight species recognised by Musser & Carleton (2005), recent morphological, morphometric and molecular studies have increased the number of recognised species to 16 (Carleton & Stanley, 2005; Carleton et al., 2006; Nicolas et al., 2006, 2008c, 2010c). Molecular data indicate that additional cryptic species still remain to be described (Nicolas et al., 2006). Recognised species were arranged into six relatively homogenous species-groups by Carleton et al. (2006) which can be distinguished based on number of mammae, condition of incisors (pro-odont, opisthodont or orthodont), development of cusps t3 and t9 on M^1, condition of the interorbital region (development of a supraorbital crest), rostral length, length of incisive foramen and condition of the subsquamosal foramen. The six recognised species groups are *aeta, alleni, anselli, baeri, denniae* and *parvus*. Species added after Carleton et al. (2006) have all been from the *H. alleni* group (Nicolas et al., 2006, 2008c, 2010c).

Annotated Species List

Sixteen species are recognised:
Hylomyscus aeta (Thomas, 1911)
Hylomyscus alleni (Waterhouse, 1838)
Hylomyscus anselli (Bishop, 1979)

Hylomyscus arcimontensis Carleton & Stanley, 2005
Hylomyscus baeri Heim de Balsac & Aellen, 1965
Hylomyscus carillus (Thomas, 1904)
Hylomyscus denniae (Thomas, 1906)
Hylomyscus endorobae (Heller, 1910)
Hylomyscus grandis Eisentraut, 1969
Hylomyscus kaimosae (Heller, 1912)
Hylomyscus pamfi Nicolas et al., 2010
Hylomyscus parvus Brosset, Dubost & Heim de Balsac, 1965
Hylomyscus simus (Allen & Coolidge, 1930)
Hylomyscus stella (Thomas, 1911)
Hylomyscus vulcanorum Lönnberg & Gyldenstolpe, 1925
Hylomyscus walterverheyeni Nicolas et al., 2008

Key to Species

A full key to all the species is currently not possible, since many species can only be confidently identified based on molecular grounds. Instead we present a key to the six species groups. See "Diagnosis" under each account for further details on how to distinguish species within each group (many of which do not occur sympatrically and therefore can be distinguished based on geographic location).

1	Supraorbital ridges present distinctly present; t9 on M^1 large and distinct	2
	Supraorbital ridges absent or weakly developed; t9 on M^1 absent or indistinct	3
2	Ventral pelage pure white (base of hairs with white); tip of tail without obvious hairs; occurs west of the Dahomey Gap	*H. baeri* group
	Ventral pelage not pure white (base of hairs with grey); tip of tail with obvious but short hairs; occurs east of the Dahomey Gap	*H. aeta* group
3	Incisors orthodont (at 90°) or pro-odont (projecting forwards)	4
	Incisors opisthodont (projecting backwards)	5
4	Size smaller (GLS < 22 mm; molar toothrow 3.0 – 3.4 mm); incisors pro-odont	*H. parvus* group
	Size larger (GLS typically > 22 mm; molar toothrow > 3.3 mm); incisors orthodont or slightly pro-odont	*H. alleni* group
5	Three pairs of nipples (1 + 2 = 6); restricted to southern Tanzania, Zambia and possibly Angola	*H. anselli* group
	Four pairs of nipples (2 + 2 = 8); restricted to the Albertine Rift, Kenya and northern Tanzania	*H. denniae* group

The groups are composed of the following species:
Hylomyscus baeri group: *H. baeri*
Hylomyscus aeta group: *H. aeta*; *H. grandis*
Hylomyscus parvus group: *H. parvus*; *H. kaimosae*
Hylomyscus alleni group: *H. alleni; H. stella; H. walterverheyeni; H. simus; H. pamfi; H. carillus*
Hylomyscus denniae group: *H. denniae; H. volcanorum; H. endorobe*
Hylomyscus anselli group: *H. anselli; H. arcimontensis*

Species Accounts

Hylomyscus aeta (Thomas, 1911)

Diagnosis

At present, this species can only be identified with certainty on molecular grounds, although a multivariate analysis of its cranio-dental features can separate it from other syntopic species of *Hylomyscus* (Nicolas et al., 2008c). These species include: *H. walterverheyeni* and *H. alleni* (size on average smaller, GLS < 26 mm, molar toothrow 3.3-4.2 mm; supraorbital ridges absent on cranium; hairs not present on end of tail); *H. parvus* (size smaller, GLS < 22 mm, molar toothrow 3.0-3.4 mm). This species has hairs on the terminal end of the tail, well developed supraorbital ridges and molar toothrow is 4.2-4.6 mm in length (Nicolas et al., 2008c). Incisors are opisthodont. Three pairs of nipples (1 + 2 = 6) are present.

Taxonomic Issues

At least one cryptic species has been reported, but not yet described, within *H. aeta* (Nicolas et al., 2006).

Distribution

This species is distributed widely in the Congo Basin from Gabon and southern Cameroon in the east to the Albertine Rift in the west where it occupies forested habitats; typically arboreal.

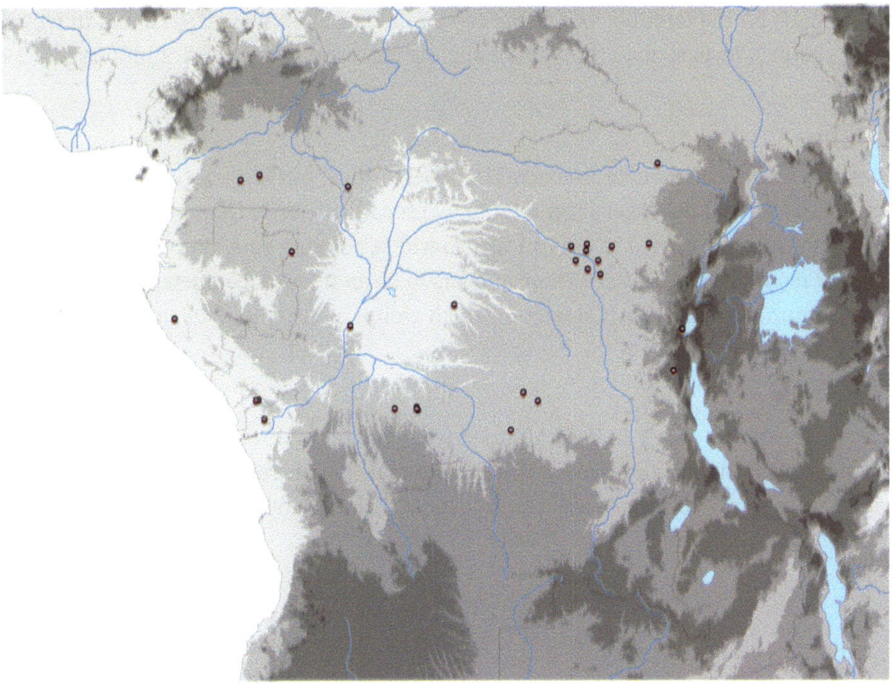

Figure 2.527 Distribution map

Measurements

External and cranial measurements (mm) and mass (g) for *Hylomyscus aeta*, sexes combined

	Mean	Min	Max	N
Mass	-	-	-	-
HB*	96.0	-	-	7
Tail*	127.5	115	142	7
HF*	20.3	19	22	7
Ear*	14.9	14	16	7
GLS*	26.4	25.1	28.1	10

*Robbins et al. (1980); HB calculated by substracting Tail from Total Length

Hylomyscus alleni (Waterhouse, 1838)

Diagnosis

At present, this species can only be identified with certainty on molecular grounds, although a multivariate analysis of its cranio-dental features can separate it from other syntopic species of *Hylomyscus* (Nicolas et al., 2008c). These species include: *H. aeta* (size on average larger, GLS > 26 mm; supraorbital ridges present on cranium; hairs present on end of tail); *H. walterverheyeni* (incisors orthodont); *H. parvus* (size smaller, GLS < 22 mm, molar toothrow 3.0-3.4 mm). This species does not have hairs on the terminal end of the tail and molar toothrow is 3.3-4.2 mm in length (Nicolas et al., 2008c). Incisors are orthodont or slightly pro-odont. Four pairs of nipples (2 + 2 = 8) are present.

Taxonomic Issues

This species was previously thought to extend into West Africa, but recent studies have shown that the specimens west of the Niger River refer to two separate species (*H. pamfi* and *H. simus*). This species is therefore restricted to Central Africa, with a possible outlying population on the Jos Plateau that requires investigation. There appears to be an undescribed new species (named Taxon 2) from Korup National Park (Cameroon) which groups as sister to *H. alleni* (Nicolas et al., 2008c).

Distribution

This species is distributed widely in Central Africa where it occupies rainforest. It has been collected at a wider range of localities than indicated here (only sequenced specimens are included in this map) (Nicolas et al., 2008c). Records from eastern Nigeria (Happold, 1987) may refer to this species or to *H. walterverheyeni*.

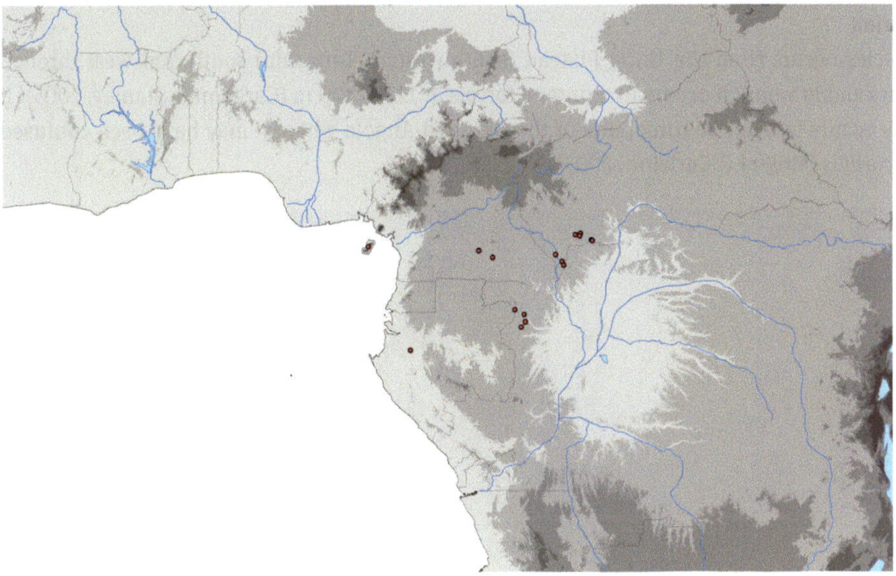

Figure 2.528 Distribution map

Measurements

External and cranial measurements (mm) and mass (g) for *Hylomyscus alleni*, sexes combined

	Mean	Min	Max	N
Mass*	19.0	10	30	115
HB*	82.0	67	96	122
Tail*	128.0	101	152	120
HF*	17.8	15.1	19.9	119
Ear*	14.2	10.3	17.3	112
GLS*	23.7	21.0	26.1	188

*Nicolas et al. (2008c)

Hylomyscus anselli (Bishop, 1979)

Diagnosis

This is the only *Hylomyscus* species occurring within its range. It is on average larger than the very similar *H. arcimontensis* (GLS typically ca. 24-26 mm), but can readily be distinguished from it by its non-overlapping distribution (they occupy separate mountain ranges). This species has 3 pairs of nipples (1 + 2 = 6), and the incisors are orthodont (Carleton & Stanley, 2005).

Taxonomic Issues

This recently described species is closely related to *H. arcimontensis* which occurs in the Eastern Arc Mts of Tanzania (Carleton & Stanley, 2005).

Distribution

This species occurs from the Ufipa Plateau (south-western Tanzania) south and west through northern Zambia where it occurs in montane forests at 1220-2300 m (Carleton & Stanley, 2005). It probably occurs in neighbouring parts of the DRC and Angola where it may have been confused with *H. carillus* (Musser & Carleton, 2005).

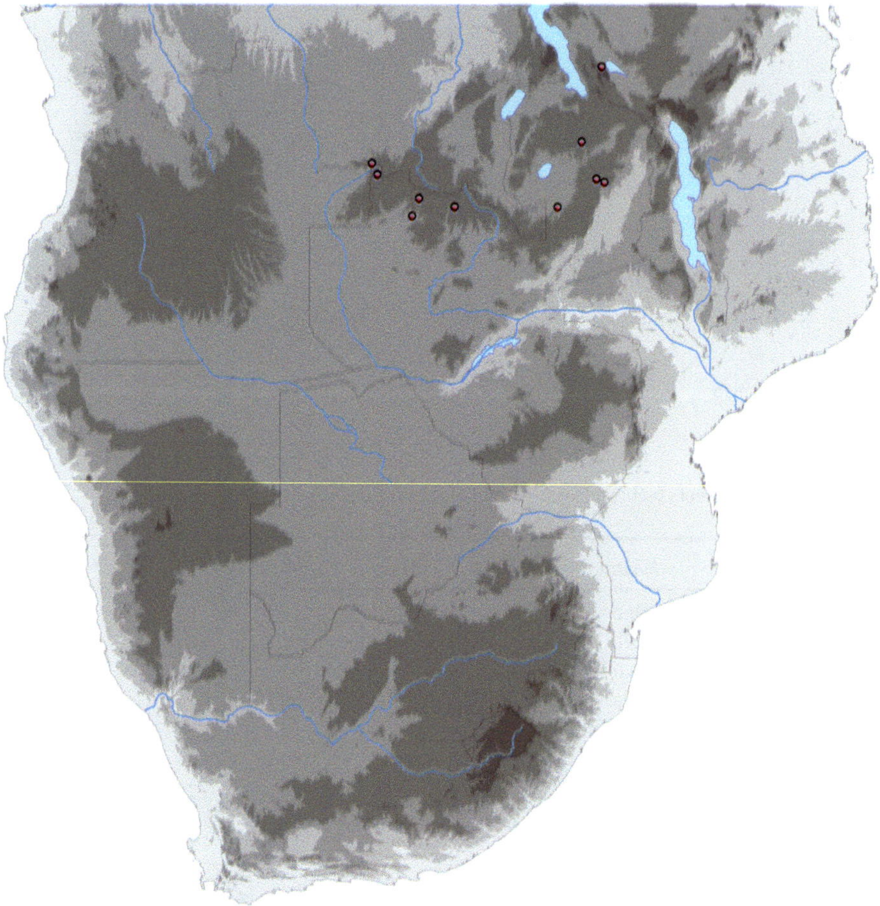

Figure 2.529 Distribution map

Measurements

External and cranial measurements (mm) and mass (g) for *Hylomyscus anselli*, sexes combined

	Mean	Min	Max	N
Mass*	28.5	22	35	12
HB*	101.0	95	109	12
Tail*	146.0	141	159	12
HF*	21.0	20	22	12
Ear*	20.0	19	21	12
GLS*	27.4	26.3	28.6	12

*Carleton & Stanley (2005)

Hylomyscus arcimontensis Carleton & Stanley, 2005

Diagnosis

This is the only *Hylomyscus* species occurring within its range. It is on average smaller than the very similar *H. anselli* (GLS typically ca. 27-28 mm), but can readily be distinguished from it by its non-overlapping distribution (they occupy separate mountain ranges). *Hylomyscus endorobae* occurs further north in Kenya and has 4 pairs of nipples. This species has orthodont incisors and three pairs of nipples (1 + 2 = 6).

Taxonomic Issues

This recently described species is in the *H. anselli* group (Carleton & Stanley, 2005).

Distribution

This species is endemic to the Eastern Arc Mts of Tanzania occurring from the South Pare Mts in the north to Rungwe Mt in the south, and possibly extending to the extreme north of Malawi into the Misuku Mts. The record from Malawi is based on a juvenile specimen that was identified as *Praomys delectorum* (Carleton & Stanley, 2005). Further material from northern Malawi is needed to throw light on the status of *Hylomyscus* in that country.

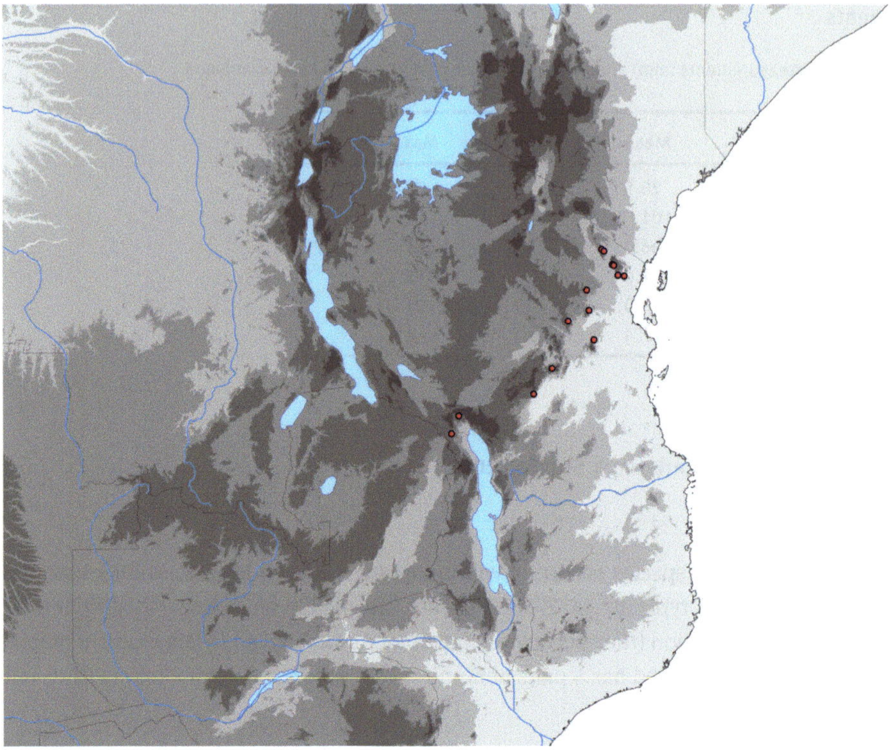

Figure 2.530 Distribution map

Measurements

External and cranial measurements (mm) and mass (g) for *Hylomyscus arcimontensis*, sexes combined

	Mean	Min	Max	N
Mass*	20.5	13.5	28.5	105
HB*	91.3	75	105	103
Tail*	132.7	127	137	102
HF*	20.2	18	22	105
Ear*	18.1	16	21	104
GLS*	25.4	23.9	27.0	109

*Carleton & Stanley (2005)

Hylomyscus baeri Heim de Balsac & Aellen, 1965

Diagnosis

This is the only species of *Hylomyscus* with a pure white ventral pelage with the base of the hairs white; in all other *Hylomyscus* species the base of the ventral hairs are grey. The supraorbital ridges are moderately developed and obvious. Incisors are opisthodont. Four pairs of nipples (2 + 2 = 8) are present.

Taxonomic Issues

This species was described half a century ago, but is still known by just a few specimens which is unusual considering its relatively wide distribution.

Distribution

This species is endemic to the Upper Guinea Forest region of West Africa where it is known from a few scattered localities in secondary forest, forest edge and plantations (Gautun & Bellier, 1970).

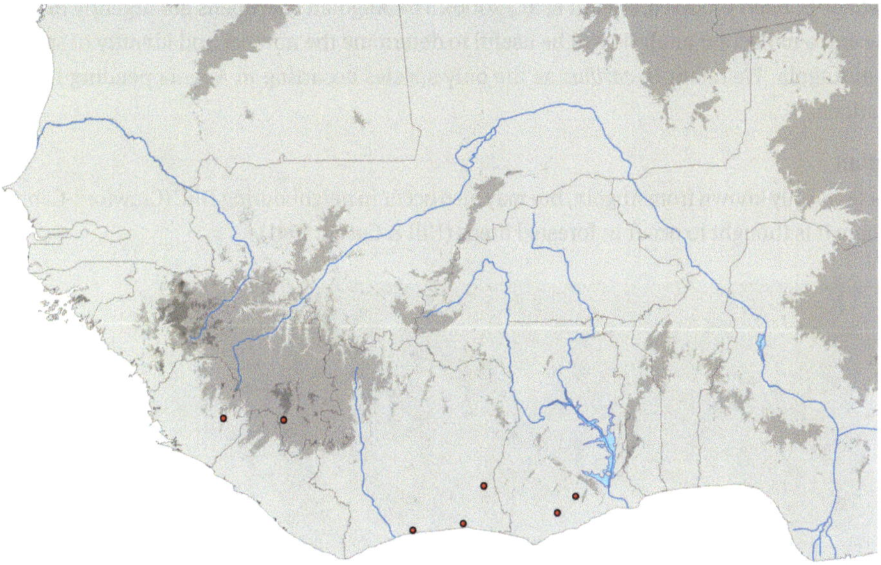

Figure 2.531 Distribution map

Measurements

External and cranial measurements (mm) and mass (g) for *Hylomyscus baeri*, sexes combined

	Mean	Min	Max	N
Mass*	21.8	21	23	10
HB*	105.9	98	115	10
Tail*	128.1	111	141	10
HF*	21.8	21	23	10
Ear*	16.3	15	19	10
GLS*	25.8	24.2	26.9	10

*Gautun & Bellier (1970)

Hylomyscus carillus (Thomas, 1904)

Diagnosis

It is not clear what the diagnostic features of this species are. It appears to be the only *Hylomyscus* species occurring in Angola, although Musser & Carleton (2005) also consider "*H. denniae*" (= *H.*

anselli) and "*H. stella*" to be present. Four pairs of nipples (2 + 2 = 8) are present, and incisors are orthodont (Hill & Carter, 1941).

Taxonomic Issues

The taxonomy of this species is confused and its diagnostic characters have not been quantified. This species is in the *H. alleni* group (Carleton et al., 2006), although Happold (2013) mentions the "skull with strong supraorbital ridges", which is not a feature of the *H. alleni* group, but rather of the *H. aeta* or *H. baeri* groups (Carleton et al., 2006). The Angolan specimens are urgently in need of review and a molecular analysis will be useful to determine the number and identity of species present in Angola. We retain *H. carillus* as the only species occurring in Angola pending further investigations.

Distribution

This species is only known from Angola, but may also occur in neighbouring DRC (Crawford-Cabral, 1998) where it is thought to occur in forested areas (Hill & Carter, 1941).

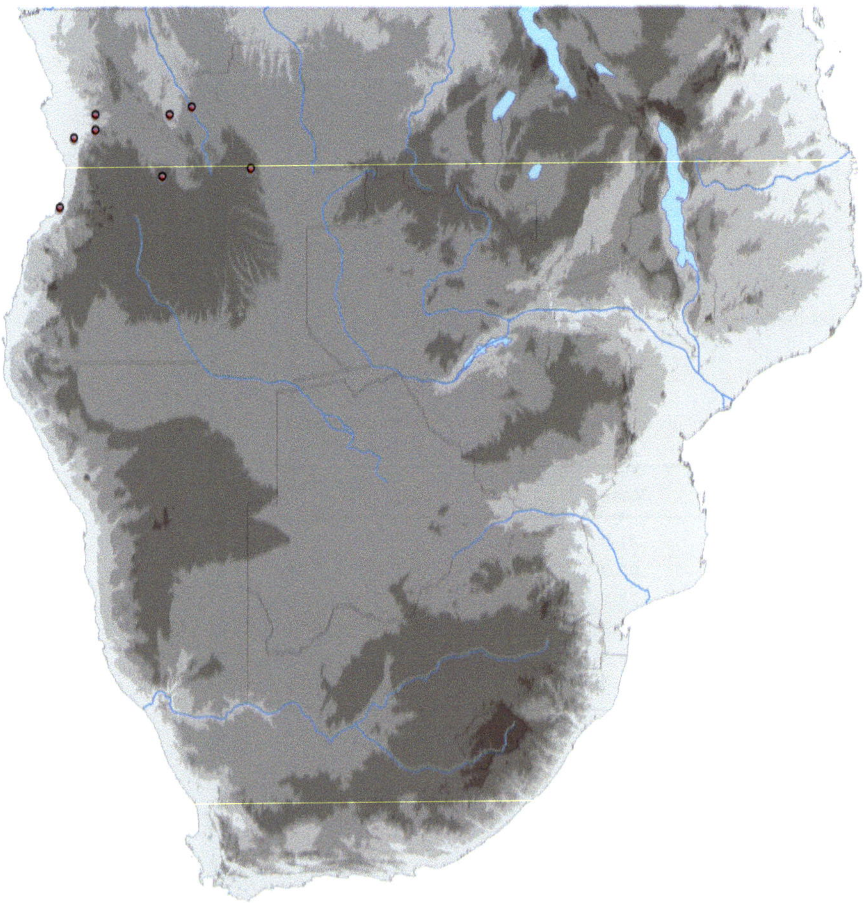

Figure 2.532 Distribution map

Measurements

External and cranial measurements (mm) and mass (g) for *Hylomyscus carillus*, sexes combined

	Mean	Min	Max	N
Mass	-	-	-	-
HB*	92.0	87	97	4
Tail*	133.3	118	146	4
HF*	19.1	19.0	19.5	4
Ear*	18.3	17	19	3
GLS*	25.9	25.4	26.4	4

*Hill & Carter (1941)

Hylomyscus denniae (Thomas, 1906)

Diagnosis

This is the only species of the *H. denniae* group occurring within its range, but may co-occur with: *H. stella* (incisors orthodont; occurs at lower altitudes); and *H. aeta* (tail with hairy tip; supraorbital ridges well developed and obvious; three pairs of nipples; occurs at lower altitudes). Four pairs of nipples (2 + 2 = 8) are present, and incisors are opisthodont (Carleton et al., 2006).

Taxonomic Issues

The species *H. volcanorum* and *H. endorobae* were previously listed as synonyms of this species (Carleton et al., 2006).

Distribution

This species is endemic to the upper slopes of the Ruwenzori Mts (both in Uganda and DRC) where it has been recorded above 1900 m in montane forests and other alpine habitats (Carleton et al., 2006). The identification of specimens reported from Mt Elgon (Clausnitzer & Kityo, 2001) are uncertain and may refer to another species (Carleton et al., 2006), and hence have not been plotted.

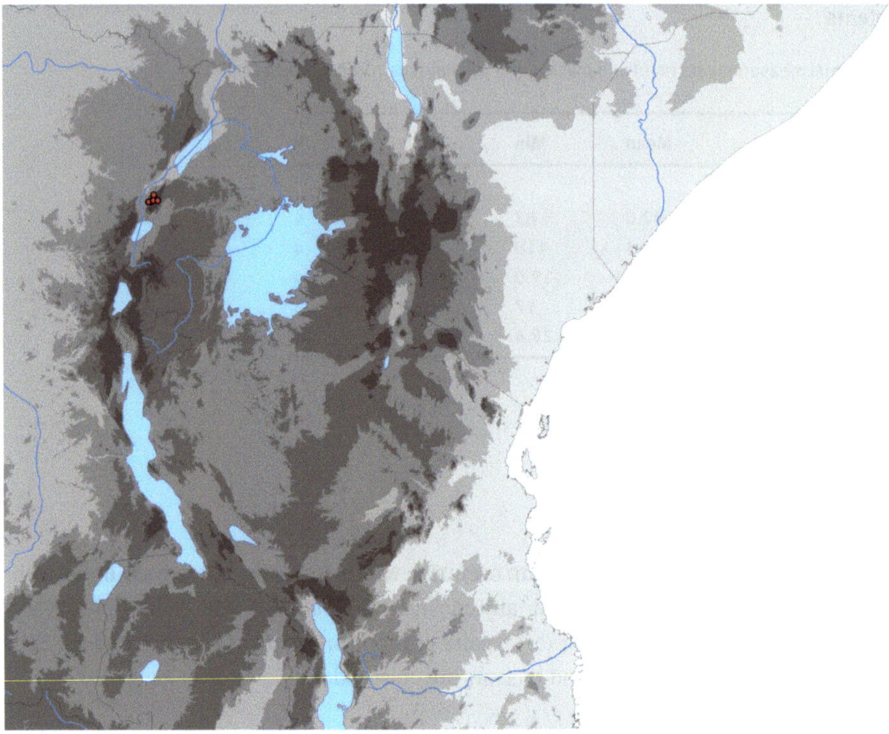

Figure 2.533 Distribution map

Measurements

External and cranial measurements (mm) and mass (g) for *Hylomyscus denniae*, sexes combined

	Mean	Min	Max	N
Mass*	31.5	17.5	65.0	56
HB*	101.7	90	115	56
Tail*	135.7	122	157	56
HF*	21.4	19	23	56
Ear*	19.9	18	23	56
GLS*	26.7	24.3	28.7	56

*Carleton & Stanley (2006)

Hylomyscus endorobae (Heller, 1910)

Diagnosis

This is the only species of the *Hylomyscus* occurring within its range. Four pairs of nipples (2 + 2 = 8) are present, and incisors are opisthodont (Carleton et al., 2006).

Taxonomic Issues

This species is in the *H. denniae* group and was previously listed as a synonym of that species.

Distribution

This species is currently only known from highlands of Kenya including Mt Kenya, the Aberdare Range and the Mau Escarpment above 2100 m where it occurs in montane forest. The status of *Hylomyscus* specimens from Mt Elgon are currently uncertain (Carleton et al., 2006), and may possibly refer to this species.

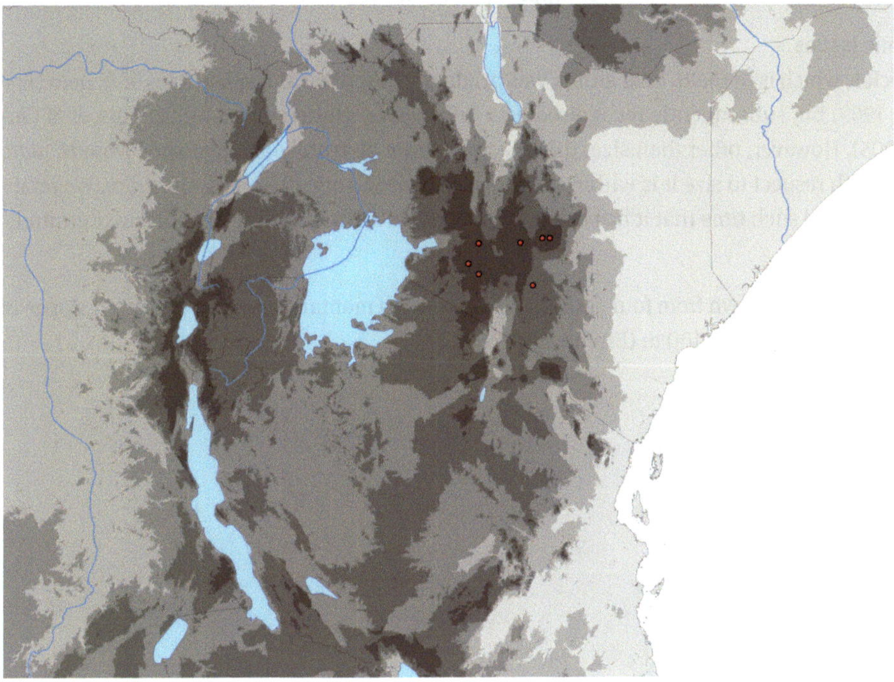

Figure 2.534 Distribution map

Measurements

External and cranial measurements (mm) and mass (g) for *Hylomyscus endorobae*, sexes combined

	Mean	Min	Max	N
Mass*	-	-	-	-
HB*	103.8	92	120	45
Tail*	149.4	123	172	45
HF*	22.8	18	26	45
Ear	-	-	-	-
GLS*	27.7	25.6	29.4	45

*Carleton & Stanley (2006)

Hylomyscus grandis Eisentraut, 1969

Diagnosis
This species can be distinguished from *H. aeta* which also occurs in Cameroon by its larger size (upper molar toothrow is 4.0-4.3 mm in *H. aeta* from Cameroon and 4.4-4.7 mm in *H. grandis*) (Hutterer et al., 1992). There are two pairs of nipples (1 + 1 = 4, Happold, 2013).

Taxonomic Issues
This species, which is in the *H. aeta* group, was originally described as a subspecies of *H. aeta* (Eisentraut, 1969) but subsequently recognised as a full species (Hutterer et al., 1992; Musser & Carleton, 2005). However, other than size, this species cannot be readily distinguished from *H. aeta*, and even with respect to size it is within the range of *H. aeta* from other parts of Africa. We retain this species until such time that it has been properly diagnosed, including on molecular grounds.

Distribution
This species is only known from four specimens collected in montane forest on the upper slopes of Mt Oku (Cameroon) at ca. 2100 m (Eisentraut, 1969).

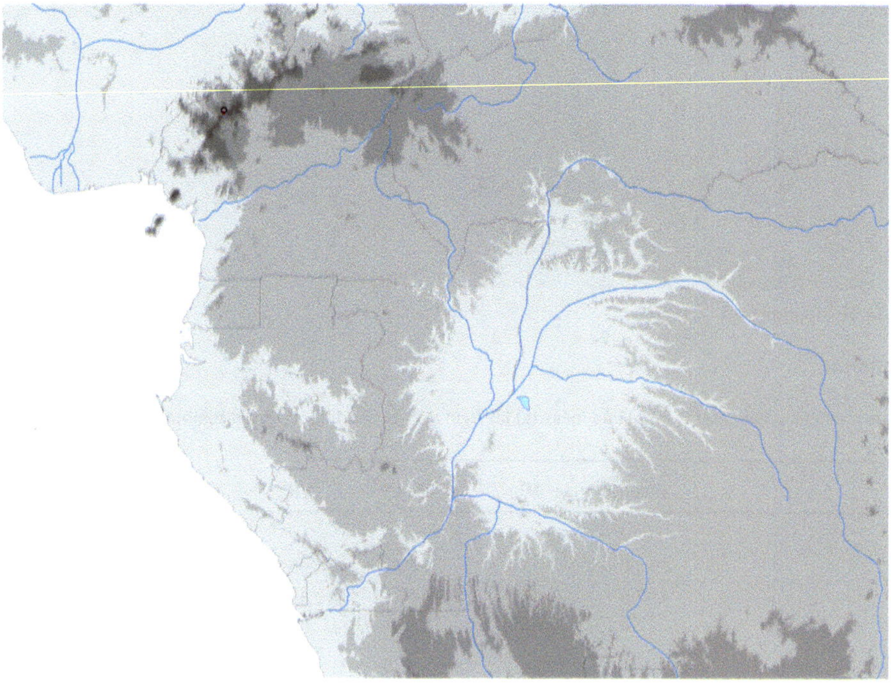

Figure 2.535 Distribution map

Measurements

External and cranial measurements (mm) and mass (g) for *Hylomyscus grandis*, sexes combined

	Mean	Min	Max	N
Mass*	33.0	25	41	4
HB*	96.8	89	110	4
Tail*	140.0	130	150	4
HF*	20.0	18	22	4
Ear*	16.5	15	20	4
GLS*	27.6	27.3	27.9	4

*Eisentraut (1969)

Hylomyscus kaimosae (Heller, 1912)

Diagnosis
This species has not yet been clearly diagnosed and therefore it is not possible to distinguish it with certainty from other *Hylomyscus* species within its range. In fact, even its range is not yet known (see Taxonomic Issues, below). At present, this species is only distinguishable on molecular grounds.

Taxonomic Issues
This taxon was traditionally listed as a synonym of *H. stella* (Musser & Carleton, 2005; Happold, 2013), however it is distinct on molecular grounds and its closest relative is actually *H. parvus* (Nicolas et al., 2006; Nicolas et al., 2010c). Hence, this species is in the *H. parvus* group. The molecular characterisation of this species was based on samples from Tanzania at Mufindi and Mbete (Nicolas et al., 2006), and not the type locality of *kaimosae* which is Kakamega Forest, Kenya. Therefore, at this stage, it would appear that the Tanzanian records certainly refer to a distinct species (named *H. kaimosae* here) that may or may not be conspecific with the true *H. kaimosae* from Kakamega. The records from South Sudan are tentatively included here, but may refer to *H. stella* or an undescribed species. Both the Kakamega and South Sudan populations require urgent molecular characterisation.

Distribution
This species is thought to occur patchily in East Africa (see Taxonomic Issues).

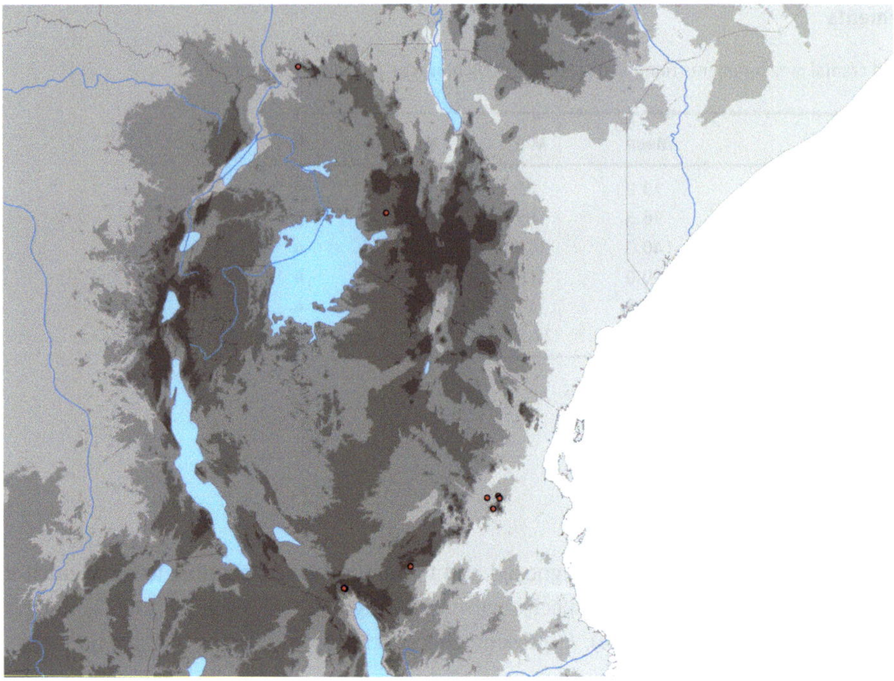

Figure 2.536 Distribution map

Measurements

External and cranial measurements (mm) and mass (g) for *Hylomyscus kaimosae*, sexes combined

	Mean	Min	Max	N
Mass	-	-	-	-
HB*	86	-	-	1
Tail*	135	-	-	1
HF*	17.5	-	-	1
Ear*	16	-	-	1
GLS*	25	-	-	1

*Heller (1912), measurements of the holotype

Hylomyscus pamfi Nicolas et al., 2010

Diagnosis

At present, this species can only be identified with certainty on molecular grounds, although it is not known to co-occur with any other *Hylomyscus* species (Nicolas et al., 2010c). Incisors are orthodont. Four pairs of nipples (2 + 2 = 8) are present.

Taxonomic Issues

This recently described species is in the *H. alleni* group (Nicolas et al., 2010c).

Distribution

The occurrence of this species has been confirmed by molecular analysis from the Dahomey Gap in Benin and western Nigeria, and is assumed to also occur in Togo and the extreme eastern regions of Ghana (east of the Volta River); it has been collected from dense forest (Nicolas et al., 2010c).

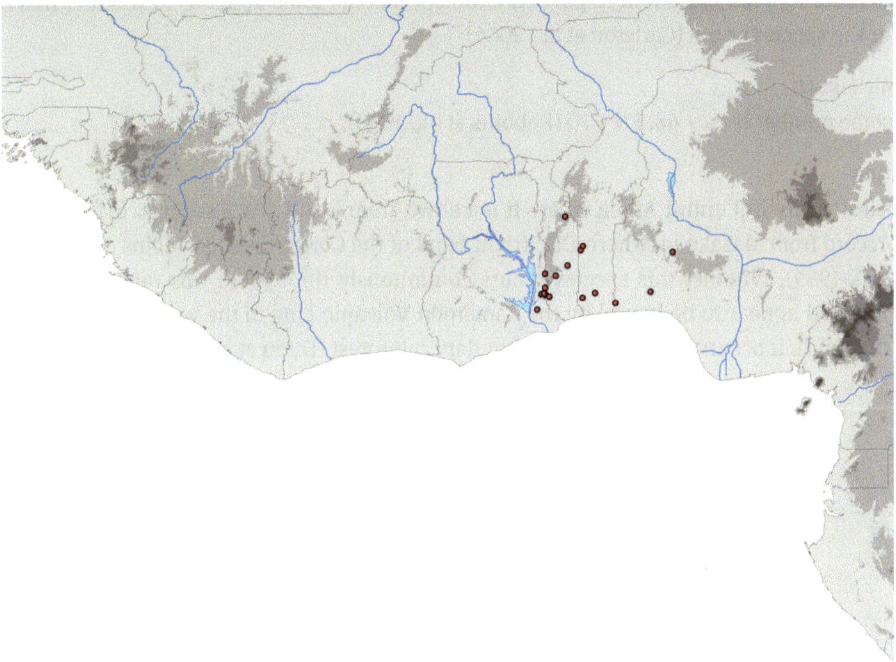

Figure 2.537 Distribution map

Measurements

External and cranial measurements (mm) and mass (g) for *Hylomyscus pamfi*, sexes combined

	Mean	Min	Max	N
Mass	-	-	-	-
HB*	82.0	56	105	19
Tail*	104.0	59	121	17
HF*	18.0	16	19	19
Ear*	12.0	11	14	18
GLS*	24.2	22.8	25.8	36

*Nicolas et al. (2010c)

Hylomyscus parvus Brosset, Dubost & Heim de Balsac, 1965

Diagnosis

This is the smallest species of *Hylomyscus* and can generally be distinguished by its small size (GLS < 22 mm) and pro-odont incisors. It is also distinct on molecular grounds (Nicolas et al., 2008c).

It co-occurs with three other species across its range: *H. aeta* (size larger, GLS > 26 mm; supraorbital ridges present on cranium; hairs present on end of tail); *H. walterverheyeni* (size larger, GLS typically > 22 mm; molar toothrow 3.3-4.1 mm; incisors orthodont; four pairs of nipples); *H. alleni* (size larger, GLS typically > 22 mm, molar toothrow 3.3-4.2 mm; four pairs of nipples). Three pairs of nipples (1 + 2 = 6) are present in *H. parvus*; although there is some doubt as to whether this is typical for the species or not (Carleton et al., 2006).

Taxonomic Issues

Chromosome number is 2n = 46, FN = 70 (Robbins et al., 1980).

Distribution

This species occurs in Central Africa where it is known from just a few localities, including an isolated record from Masako, northern DRC (right bank of the Congo River) (Robbins et al., 1980; Dudu et al., 1989). Whether this species occurs continuously throughout this range is not yet known. It would appear to be bound by the Cameroon Volcanic Line in the west and the Congo River in the south. It occurs in primary and secondary rainforest (Dudu et al., 1989).

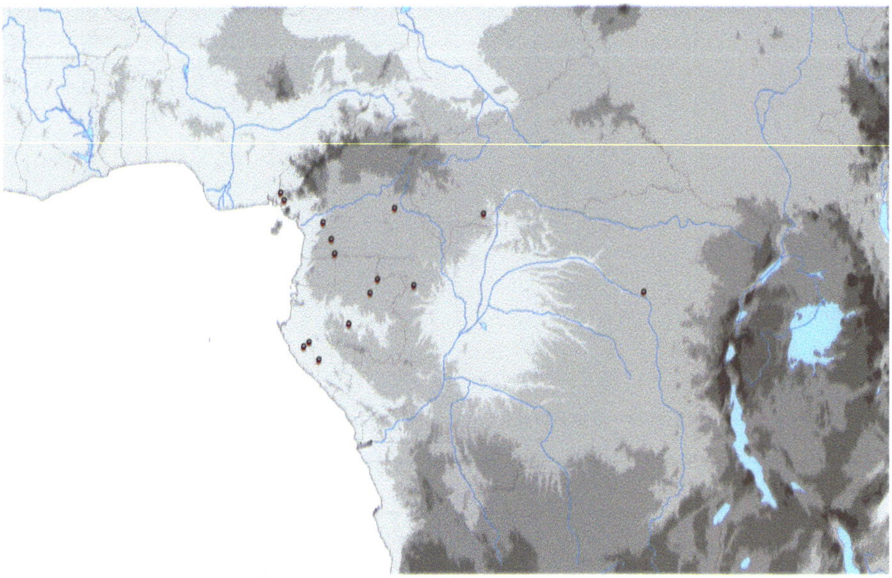

Figure 2.538 Distribution map

Measurements

External and cranial measurements (mm) and mass (g) for *Hylomyscus parvus*, sexes combined

	Mean	Min	Max	N
Mass*	12.7	10	16	16
HB**	67.4	56	77	21
Tail**	105.7	87	124	29
HF**	16.2	14	19	31
Ear**	14.1	13	15	28
GLS**	20.7	19.4	21.9	40

*Happold (2013); **Robbins et al. (1980) and Dudu et al. (1989) combined

Hylomyscus simus (Allen & Coolidge, 1930)

Diagnosis

At present, this species can only be identified with certainty on molecular grounds, although it is not known to co-occur with any other *Hylomyscus* species (Nicolas et al., 2010c). Incisors are slightly pro-odont. Four pairs of nipples (2 + 2 = 8) are present.

Taxonomic Issues

This species has recently been elevated to full species status and is in the *H. alleni* group (Nicolas et al., 2010c).

Distribution

This species is restricted to the Upper Guinea Forest, occurring from Sierra Leone and Guinea in the west to Ghana in the east (the Volta River is assumed to be the eastern boundary), where it occupies rainforest.

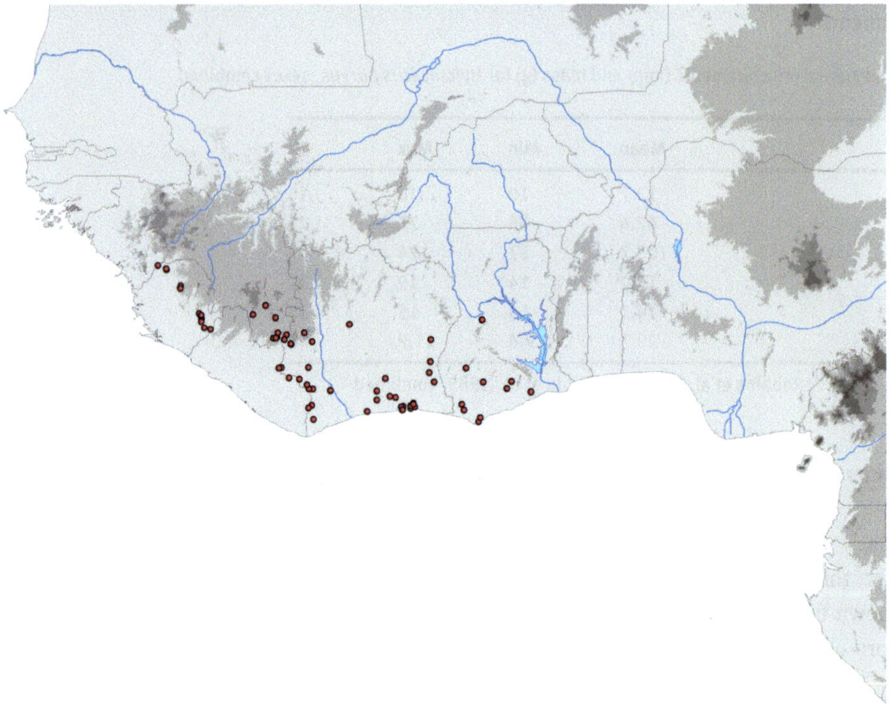

Figure 2.539 Distribution map

Measurements

External and cranial measurements (mm) and mass (g) for *Hylomyscus simus*, sexes combined

	Mean	Min	Max	N
Mass	18.5	12	22	5
HB*	89.0	66	128	35
Tail*	123.0	52	143	32
HF*	19.0	16	20	35
Ear*	16.0	13	18	35
GLS*	23.8	20.7	25.3	37

*Nicolas et al. (2010c)

Hylomyscus stella (Thomas, 1911)

Diagnosis

At present, this species can only be identified with certainty on molecular grounds, although a multivariate analysis of its cranio-dental features can separate it from other syntopic species of *Hylomyscus* (Nicolas et al., 2008c). These species include: *H. aeta* (size on average larger, GLS > 26 mm; supraorbital ridges present on cranium; hairs present on end of tail); *H. parvus* (size smaller, GLS < 22 mm, molar toothrow 3.0-3.4 mm; three pairs of nipples). Members of the *H. denniae* group typically occur at higher altitudes without a zone of overlap (Carleton & Stanley, 2005). This species does not have hairs on the terminal end of the tail and molar toothrow is 3.3-4.1 mm in length (Nicolas et al., 2008c). Incisors are orthodont. Four pairs of nipples (2 + 2 = 8) are present.

Taxonomic Issues

This species is in the *H. alleni* group.

Distribution

This species is currently known from the western parts of the Congo Basin, extending eastwards to the shores of Lake Victoria (Nicolas et al., 2008c).

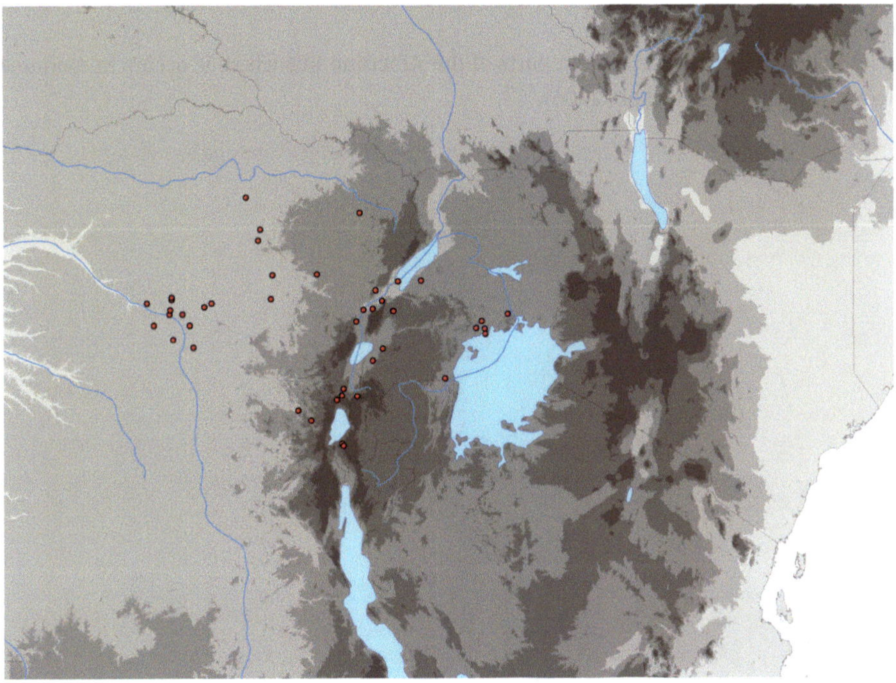

Figure 2.540 Distribution map

Measurements

External and cranial measurements (mm) and mass (g) for *Hylomyscus stella*, sexes combined

	Mean	Min	Max	N
Mass*	18.0	14	24	17
HB*	90.0	72	104	18
Tail*	129.0	111	140	17
HF*	17.7	17	20	17
Ear*	14.3	11	17	16
GLS*	24.4	21.9	26.6	90

*Nicolas et al. (2008c)

Hylomyscus vulcanorum Lönnberg & Gyldenstolpe, 1925

Diagnosis
This is the only species of the *H. denniae* group occurring within its range, but may co-occur with: *H. stella* (incisors orthodont; occurs at lower altitudes); and *H. aeta* (tail with hairy tip; supraorbital ridges well developed and obvious; three pairs of nipples; occurs at lower altitudes). Four pairs of nipples (2 + 2 = 8) are present, and incisors are opisthodont (Carleton et al., 2006).

Taxonomic Issues
This species is in the *H. denniae* group (Carleton et al., 2006).

Distribution
This species is restricted to the southern parts of the Albertine Rift where it occurs in montane forests above 1670 m (Carleton et al., 2006).

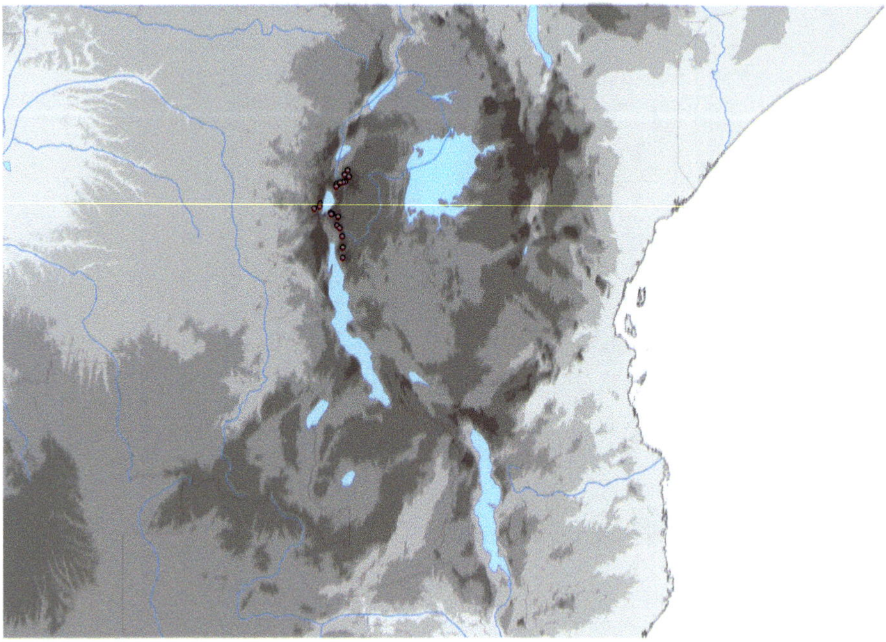

Figure 2.541 Distribution map

Measurements

External and cranial measurements (mm) and mass (g) for *Hylomyscus vulcanorum*, sexes combined

	Mean	Min	Max	N
Mass*	21.5	16	30	36
HB*	94.7	81	107	36
Tail*	132.8	117	155	36
HF*	20.5	19	22	36
Ear*	17.6	16	22	36
GLS*	25.0	23.1	26.8	36

*Carleton & Stanley (2006)

Hylomyscus walterverheyeni Nicolas et al., 2008

Diagnosis

At present, this species can only be identified with certainty on molecular grounds, although a multivariate analysis of its cranio-dental features can separate it from other syntopic species of *Hylomyscus* (Nicolas et al., 2008c). These species include: *H. aeta* (size on average larger, GLS > 26 mm; supraorbital ridges present on cranium; hairs present on end of tail); *H. alleni* (very similar in appearance and only distinguishable based on multivariate analysis of cranio-dental features); *H. parvus* (size smaller, GLS < 22 mm, molar toothrow 3.0-3.4 mm). This species does not have hairs on the terminal end of the tail and molar toothrow is 3.3-4.1 mm in length (Nicolas et al., 2008c). Incisors are orthodont (or slightly opisthodont). Four pairs of nipples (2 + 2 = 8) are present.

Taxonomic Issues

This recently described species is in the *H. stella* group (Nicolas et al., 2008c).

Distribution

This species occurs widely in Central Africa from the Cameroon Volcanic Line in the west to the Oubangui River in the east (Nicolas et al., 2008c; Missoup et al., 2009). Records from eastern Nigeria (Happold, 1987) may refer to this species or to *H. alleni*.

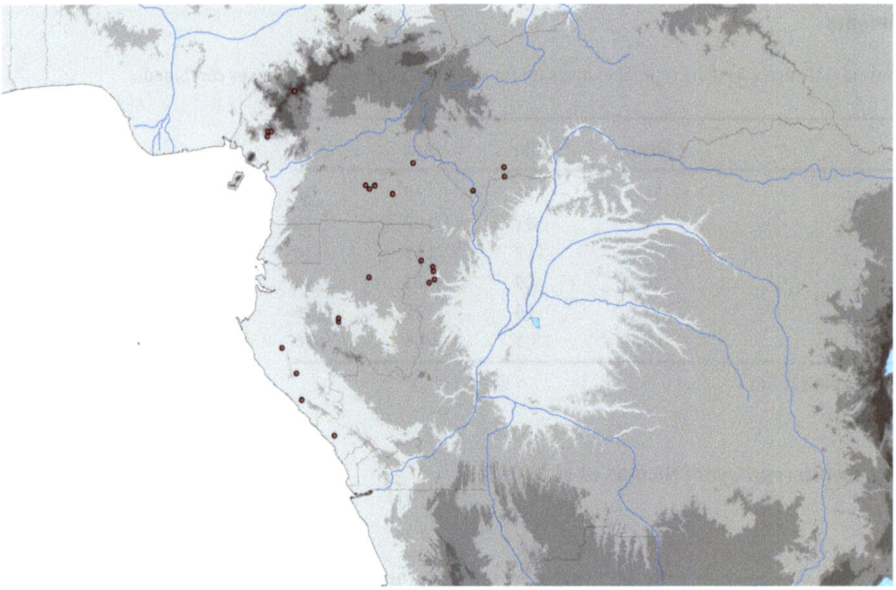

Figure 2.542 Distribution map

Measurements

External and cranial measurements (mm) and mass (g) for *Hylomyscus walterverheyeni*, sexes combined

	Mean	Min	Max	N
Mass*	18.0	11	29	249
HB*	86.0	67	99	263
Tail*	129.0	103	153	259
HF*	17.6	15.5	19.9	263
Ear*	14.8	11.4	19.7	250
GLS*	24.6	21.5	27.2	260

*Nicolas et al. (2008c)

Genus Account

Lamottemys F. Petter, 1986 (Mount Oku Rat)

Diagnosis

This small-sized rodent (Mass ca. 75 g; HB ca. 140 mm) has uniformly brown dorsal pelage without any striping, very similar in colour and texture to *Hybomys*. The ventral pelage is a paler yellowish grey. The tail is about equal in length to the head and body. The ear is dark with short hairs. The mammary formula is 0+2 = 4 as opposed to 1+2 = 6 in *Hybomys badius* (Denys et al., 2014). The 5th digit of the forefeet is reduced in size with a nail, not claw (claw present in *Hybomys*).

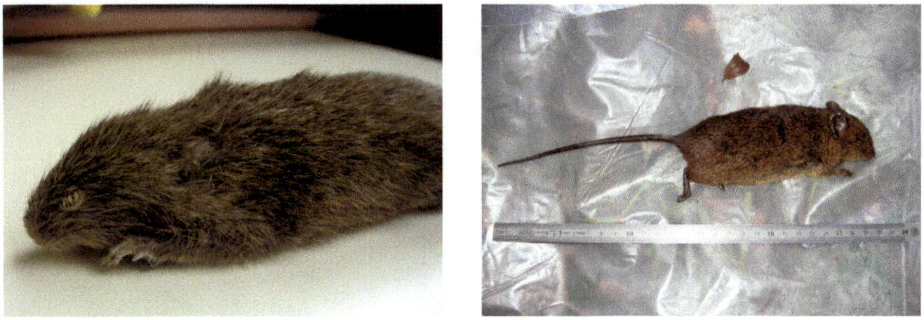

(a) (b)

Figure 2.543 a and b Photograph of *Lamottemys okuensis* holotype (left) and specimen (right) (Christiane Denys).

Cranial Characters

The skull is large (ca. 33 mm) and massively built with a broad interobital region lacking any constriction but having a weakly developed supraorbital ridge extending on the parietals, long broad and blunt-ending nasals, and fairly robust zygomatic arches which are expanded dorso-ventrally in the middle. The zygomatic plate is quite narrow with a vertical anterior edge and smooth rounded dorsal angle. The posterior margin of the palatal foramina ends well anterior of the M^1 alveoli. The bullae are not inflated (17% GSL). A detailed account and illustrations of the skull are presented in Denys et al. (2014)

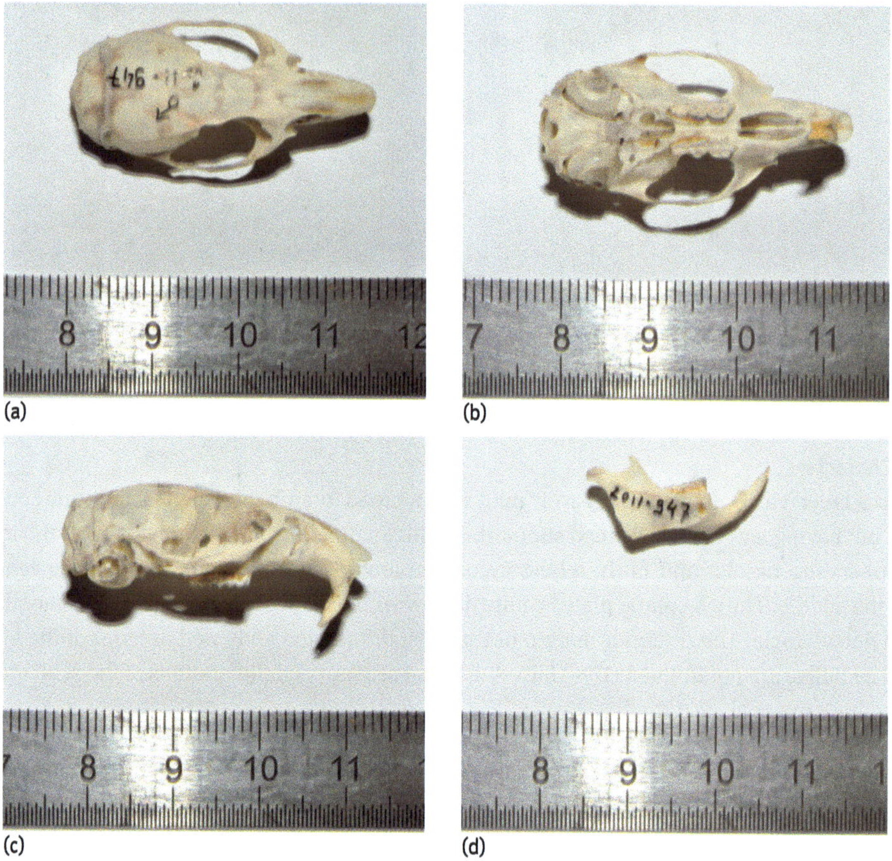

Figure 2.544 a-d Cranium of *Lamottemys okuensis* (MNHN-ZM-MO 2011-947) (from top left to bottom right: dorsal view, ventral view, lateral view and lower jaw).

Dentition

The upper incisors are opisthodont and may have a very faint groove, with a faint groove on the lower pair. The species is characterised by its massive molar dentition (macrodonty: MTR/GSL ± 0.21; Table 10) which also displays stephanodonty (extreme development of longitudinal crests) leading to the central row of cusps being much larger than the inner and outer rows and arranged in a strict linear formation. This is well developed on M^1 and M^2 but not on M^3 which does not show any crests connecting the first and second lobes. The M^3 is slightly smaller than M^2 (not much reduced) and retains cusps t1 and t3 (Group 7 on Figure 2.439). Stephanodonty is a feature shared with *Oenomys* which is very similar in molar cusp pattern to *Lamottemys* but does not have such developed macrodonty (MTR/GSL = 0.19). On the other hand *Desmomys* does not have a stephanodont dental pattern but displays strong macrodonty (MTR/GSL = 0.22). Detailed descriptions and illustrations of the dentition of *Lamottemys* and related genera are provided in Petter (1986) and Denys et al. (2014)

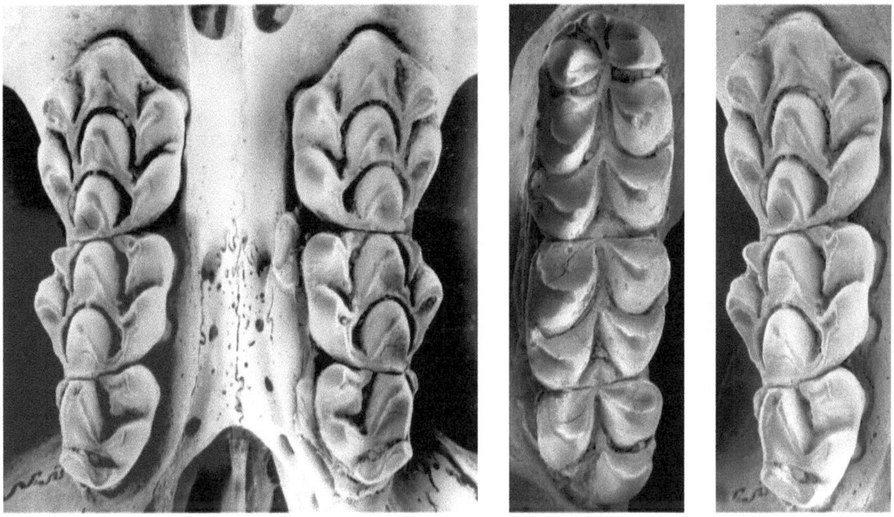

(a) (b) (c)

Figure 2.545 Cheek teeth of *Lamottemys okuensis* (MNHN-ZM-MO 1984-494) (A: upper cheek teeth; B: lower left toothrow; C: upper right toothrow).

Habitat & Biogeography

These rodents are confined to dense montane forests at high elevations (2100-2900 m) on Mt Oku, Cameroon.

Fossil Record

None.

Biology & Ecology

Very little is known of the ecology of these rodents. Denys et al. (2014) recorded contents of one stomach comprising only green plant matter suggesting an entirely herbivorous diet.

Systematic Notes

Although no molecular data have yet been analysed for this species (Lecompte et al., 2008) it was placed in the *Oenomys* division by Musser & Carleton (2005), which falls in the tribe Arvicanthini as defined by Lecompte et al. (2008).

Annotated Species List

One species is recognised:
Lamottemys okuensis F. Petter, 1986

Species Account

Lamottemys okuensis F. Petter, 1986

Diagnosis

See genus account.

Taxonomic Issues

See genus account.

Distribution

This species is only known from Mt Oku, Cameroon, where it occurs in montane forest above 2100 m (Dieterlen & Van der Straeten, 1988).

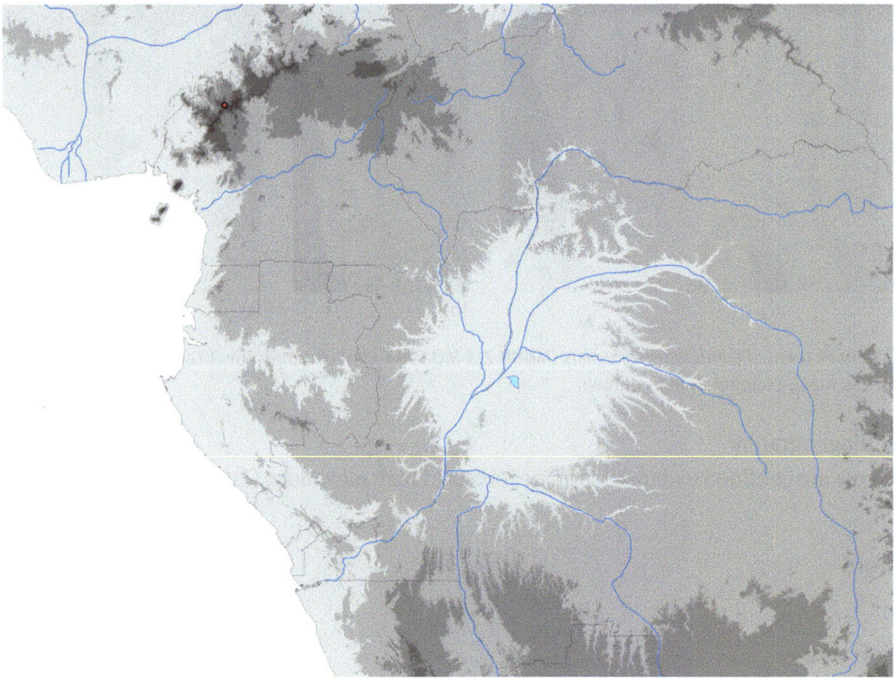

Figure 2.546 Distribution map

Measurements

External and cranial measurements (mm) and mass (g) for *Lamottemys okuensis*, sexes combined

	Mean	Min	Max	N
Mass	71.8	62	85	5
HB	139.2	130	149	5
Tail	129.2	123	140	5
HF	30.5	29.5	32.0	5
Ear	17.4	16.0	18.5	5
GLS	33.0	31.7	33.9	5

Genus Account

Lemniscomys Trouessart, 1881 (Grass Mice)

Diagnosis

They are medium-sized rodents (Mass 40-80 g; HB 90-140 mm) with short, tawny-furred ears, and a short-haired, sleek pelage which has three characteristic dorsal colour patterns: 1) multiple longitudinal rows of yellowish spots situated close together or just coalescing (*striatus* group); 2) multiple longitudinal rows of yellowish stripes (*barbarus* group); 3) single mid-dorsal (spinal) black stripe (*griselda* group). In all cases, the striping extends from neck to rump and contrasts with a yellowish dark brown background coloration. The ventral fur is pure white contrasting with the darker flanks and back. The tail is whitish ventrally, marked with spare black hairs and small concentric scales dorsally and may be sub-equal or longer (100-125% of HB) than head and body length. There is typically a white ring of hair around the eyes. Females have four pairs of nipples (2 + 2 = 8). The forefeet appear three-fingered with a vestigial 1st digit and highly reduced 5th digit possessing a nail rather than a claw. The 1st and 5th digits of the hindfoot are also reduced but not as much, reaching the base of the other digits.

(a)

(b)

(c)

Figure 2.547 a-c Photographs of *Lemniscomys zebra* (James Harvey) (top left) and *L. rosalia* (Bill Stanley) (bottom left) and *Lemiscomys* specimens (Ara Monadjem) (right): from top to bottom: *L. bellieri*, *L. macculus*, *L. striatus*, *L. zebra*, *L. griselda* and *L. linulus*.

Cranial Characters

Skull is medium-sized (ca 25-35 mm) and rather long and narrow, with supraorbital ridges from indistinct to developed and occipital ridges sometimes present. The rostrum is fairly short, narrow and tapering and the zygomatic arches are parallel-sided and vary from quite gracile to quite robust. The zygomatic plate is characterised by an acute antero-dorsal angle and leading edge tilted slighted backgrounds (in *L. hoogstraali* the antero-dorsal surface is markedly indented in its lower part resulting in a distinctly concave angle). The palatal foramina may reach just to the alveolus of M^1 or as far as the second row of cusps on M^1. The bullae are not inflated (16-20 % of GSL). The membrane (periosteum) surrounding the braincase is deeply pigmented (black), possibly an adaptation to reduced ultraviolet radiation in these diurnal rodents (Rosevear, 1969); this adaptation apparently also occurs in related diurnal genera, *Rhabdomys*, *Arvicanthis* and *Pelomys* (Happold, 2013).

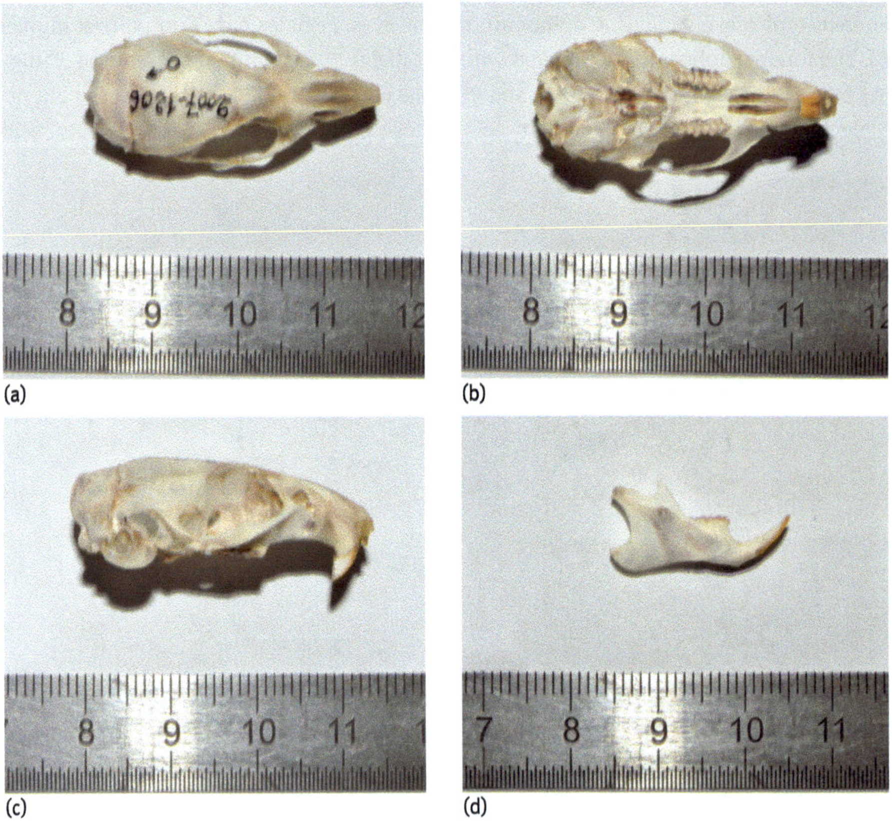

Figure 2.548 a-d Cranium of *Lemniscomys rosalia* (MNHN-ZM-MO 2007-1206) (from top left to bottom right: dorsal view, ventral view, lateral view and lower jaw).

Dentition

The incisors are opisthodont to orthodont and ungrooved. The molars are not quite as broad as in *Arvicanthis* (MTR/GSL ± 0.17-0.18; Table 10) and rows of cusps are distorted with t1 and t4 situated posterior to their corresponding rows, much more so than in *Arvicanthis*. Molar M^3 is approximately 66% of the length of M^2 and cusp configuration is as shown for Group 4 (Figure 2.439). Anteromedian and labial cusplets on M^1 and M^2 small and variably developed; posterior cingulum on M^1 and M^2 reduced or absent.

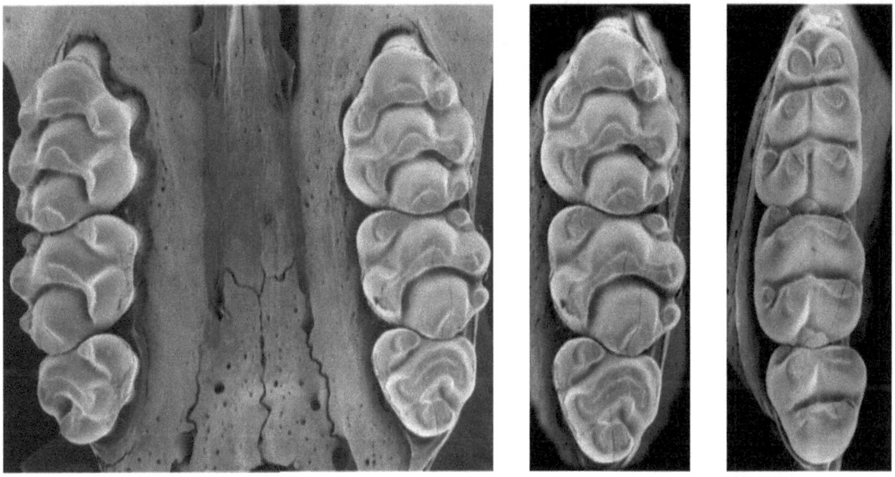

(a) (b) (c)

Figure 2.549 Cheek teeth of *Lemniscomys barbarus* (MNHN-ZM-MO 1981-631) (A: upper cheek teeth; B: upper left toothrow; C: lower left toothrow).

Habitat & Biogeography

They occur typically in savannas and forest clearings over much of Sub-Saharan Africa from Senegal in the west to Ethiopia, Kenya and Tanzania in the east and south to the KwaZulu-Natal province of South Africa. They can also be found in disturbed rainforest as well as in alpine habitats > 3000 m in Ethiopia. They are dependent on dense grass cover (Monadjem & Perrin, 2003).

Fossil Record

Fossils are recorded from the Late Miocene of Ethiopia and Kenya to the Late Pliocene of Ethiopia (Winkler, et al., 2010). The earliest fossils are known from 6 MYA deposits at Lemudon'go in East Africa (Manthi, 2007).

Lemniscomys sp. is known from Plio-Pleistocene caves of South Africa (Pocock, 1987; Senegas, 2000).

Lemniscomys aff. *striatus* was described in the Omo member F (Wesselman, 1984).

Lemniscomys from Hadar, 2.4 MYA (Reed, 2011)

Biology & Ecology

They are terrestrial and crepuscular. Reproduction is seasonal with breeding occurring in the wet season (Neal, 1977; Happold, 2013). Mean litter size is five (Delany, 1986) but may vary from 1 to 11 (Skinner & Chimimba, 2005; Happold, 2013). Diet varies seasonally including fruit, seeds and insects in the breeding season and grass, stems and leaves in the non-breeding season.

Systematic Notes

Morphological (Musser & Carleton, 2005) and molecular (Ducroz et al., 2001; Lecompte et al., 2008; Schenk et al., 2013) data support the inclusion of *Lemniscomys* as a member of the *Arvi-*

canthis division which also includes *Arvicanthis*, *Pelomys*, *Mylomys*, *Desmomys* and *Rhabdomys*. Its sister genus seems to be *Arvicanthis* (Lecompte et al., 2008; Schenk et al., 2013). Some species have been separated based on only craniometric grounds, often with overlap, and in the absence biogeographical foundation or qualitative or molecular or chromosomal characters. These we believe should be considered as conspecific until evidence can be provided to the contrary. Examples include *L. bellieri* Van der Straeten, 1976 (which may be conspecific with *L. macculus* (Thomas & Wroughton, 1910)), *L. hoogstraali* Dieterlen, 1991 (which may be a large geographic variant of *L. barbarus*), and *L. roseveari* Van der Straeten, 1980a and *L. rosalia* (Thomas, 1904) (which may be conspecific with *L. griselda*) (also see Castiglia et al., 2002).

Annotated Species List

Ten species are currently recognised from Sub-Saharan Africa (an additional species from North Africa, *L. barbarus*, is not here considered), although we are doubtful about the validity of four of these (marked with asterisks, below):

**Lemniscomys bellieri* Van der Straeten, 1975
Lemniscomys griselda (Thomas, 1904)
**Lemniscomys hoogstraali* Dieterlen, 1991
Lemniscomys linulus (Thomas, 1910)
Lemniscomys macculus (Thomas & Wroughton, 1910)
Lemniscomys mittendorfi Eisentraut, 1968
**Lemniscomys rosalia* (Thomas, 1904)
**Lemniscomys roseveari* Van der Straeten, 1980
Lemniscomys striatus (Linnaeus, 1758)
Lemniscomys zebra (Heuglin, 1864)

Key to Species

1	Dorsal pelage with multiple lines that are broken into spots (*L. striatus* group)	2
	Dorsal pelage with single or multiple lines which are not broken into spots	4
2	Mid-dorsal line flanked by eight rows of spots arranged in neat lines; size on average larger (GLS, typically > 29 mm)	*L. striatus*
	Mid-dorsal line flanked by eight rows of spots not arranged into neat lines and/or spots in lower rows tending to connect to form lines; size on average smaller (GLS, typically < 29 mm)	3
3	Restricted to East Africa	*L. macculus*
	Restricted to Mt Oku, Cameroon	*L. mittendorfi*
	Restricted to West Africa, west of the Dahomey Gap	*L. bellieri*
4	Dorsal pelage with multiple rows of lines (*L. barbarus* group)	5
	Dorsal pelage with a single mid-dorsal stripe (*L. griselda* group)	6
5	Dorsal pelage with five lateral stripes (with spots in between)	*L. hoogstraali*
	Dorsal pelage with six or seven lateral stripes	*L. zeba*
6	Restricted to West Africa	*L. linulus*
	Occurring in Southern or East Africa	7
7	Occurring mostly in Angola and neighbouring DRC	*L. griselda*
	Only known from north-western Zambia	*L. roseveari*
	Widespread in Southern and East Africa, but absent from Angola and north-western Zambia	*L. rosalia*

Species Accounts

Lemniscomys bellieri Van der Straeten, 1975

Diagnosis

This species is in the *L. striatus* group and can be distinguished from other members of this group as follows: *L. striatus* (size on average larger, GLS ca. 30-31 mm; differs on karyological (Van der Straeten & Verheyen, 1978a) and molecular grounds (Ducroz et al., 2001); all the lateral rows of spots (including the lower ones) are arranged in lines); *L. macculus* (restricted to East Africa); *L. mittendorfi* (endemic to Mt Oku, Cameroon). A dark mid-dorsal stripe is flanked by eight lateral stripes that are broken into pale buffy spots; the lower lines may be inconspicuous and seen as scattered spots rather than lines.

Taxonomic Issues

The original description of this species is inadequate and insufficient to allow a proper diagnosis of it. Van der Straeten (1975) mentioned in his description of this species that "this note has to be considered as a short preliminary description and the complete discussion of the exact systematic status of this new species will follow in a later publication". However, no subsequent publication has been forthcoming. Based on this paper, it is not possible to identify this species with certainty. Furthermore, Ducroz (2001) noted that the genetic distance (based on the cytochrome b gene) between *L. bellieri* and *L. macculus* was very low at 3.3%, suggesting that these two species might be conspecific; a view repeated by Musser & Carleton (2005). These two species also share the same number of chromosomes and are morphologically similar (Van der Straeten & Verheyen, 1980). Chromosome number is 2n = 56 (Van der Straeten, 1975). We cautiously retain this taxon as a full species pending further investigations.

Distribution

This species is restricted to West Africa, west of the Dahomey Gap where it occurs in grasslands, particularly in moist savanna, avoiding forest and forest edge habitats (Van der Straeten, 1975).

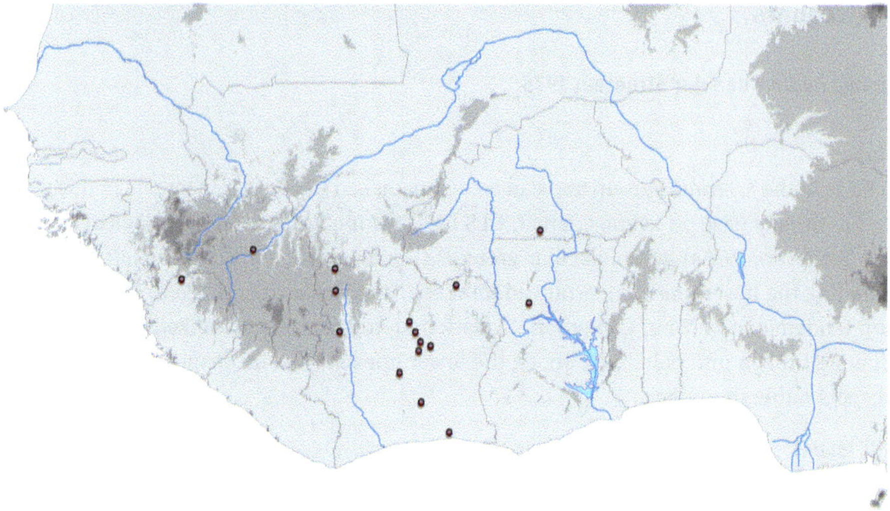

Figure 2.550 Distribution map

Measurements

External and cranial measurements (mm) and mass (g) for *Lemniscomys bellieri*, sexes combined

	Mean	Min	Max	N
Mass	-	-	-	-
HB*	109	91	127	251
Tail*	112	94	134	251
HF*	25.4	23	27	251
Ear*	15.9	13	19	251
GLS*	28.2	24.4	31.3	251

*Van der Straeten (1975)

Lemniscomys griselda (Thomas, 1904)

Diagnosis

This species can currently only be distinguished from other members of the *L. griselda* group by its distribution which is centred on Angola; it is the only member of this species group occurring within its range, although the eastern limits of its distribution have not been resolved (see below). In this species, the dorsal pelage has a single mid-dorsal stripe which may be flanked by faint lateral stripes.

Taxonomic Issues

This species is poorly diagnosed and currently cannot be distinguished from *L. rosalia* or *L. roseveari* with any confidence. Van der Straeten (1980a) used multivariate analysis of cranio-dental measurements to separate these three species, but measurements overlap significantly and no diagnostic features were identified. Molecular and karyological analyses have not yet been conducted. We strongly question the validity of recognising these three taxa as separate species, but have chosen to retain them pending further investigation.

Distribution

This species is restricted to Angola and neighbouring DRC. The eastern limits of this species are not known and the records from the DRC may refer to *R. rosalia*.

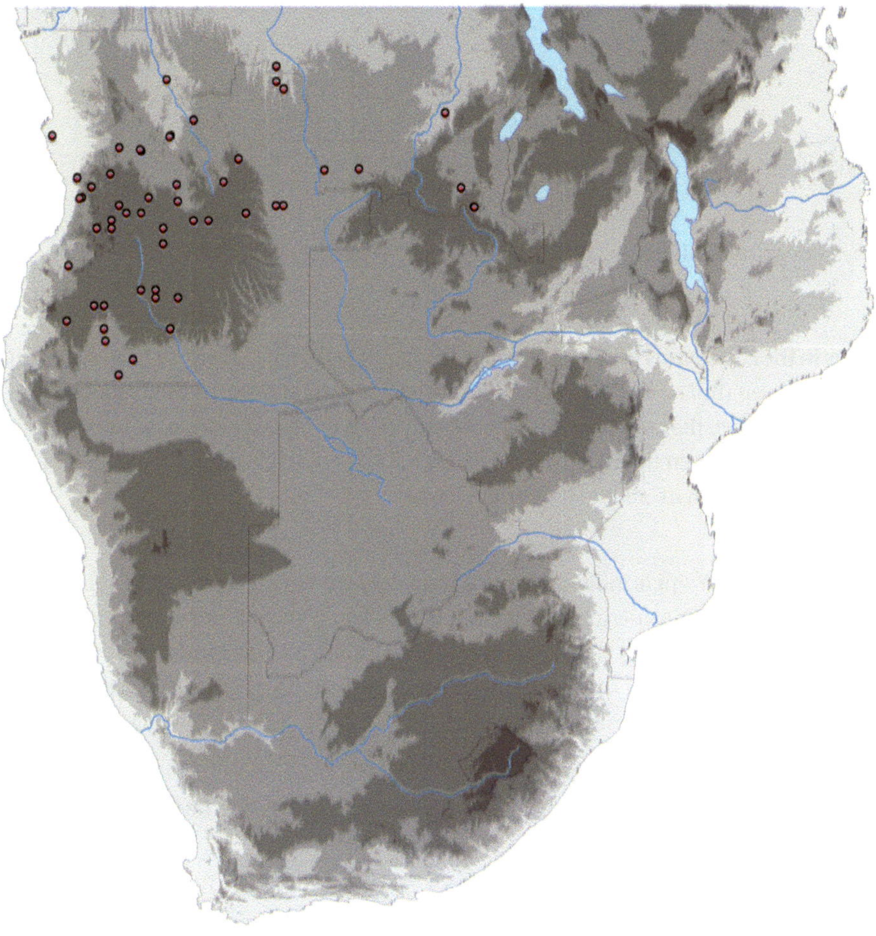

Figure 2.551 Distribution map

Measurements

External and cranial measurements (mm) and mass (g) for *Lemniscomys griselda*, sexes combined

	Mean	Min	Max	N
Mass	-	-	-	-
HB*	127.4	116	138	5
Tail*	134.0	118	149	5
HF*	28.8	28	30	5
Ear	-	-	-	-
GLS*	32.6	31.0	34.7	5

*Hill & Carter (1941)

Lemniscomys hoogstraali Dieterlen, 1991

Diagnosis

This species is in the *L. barbarus* group, and can be distinguished from *L. zebra* by its unique stripe pattern and larger size (in *L. zebra*, the dorsal pelage has 6-7 lateral lines on the flanks, and GLS < 30 mm). In this species, there are five dark lateral stripes on either side of the mid-dorsal stripe with white stripes in between them. Furthermore, each dark stripe has a thin white line (broken into spots) running through its middle (Dieterlen, 1991).

Taxonomic Issues

This species is known by a single individual and it is unclear whether it represents a valid species or an aberrant individual. Considering the fact that it is still only known by the single type specimen, and the fact that *L. zebra* has been collected widely around the type locality of *L. hoogstraali* (Carleton & Van der Straeten, 1997), we strongly question the validity of this species. However, we retain this species pending further investigations.

Distribution

This species is known from a single specimen collected along the Nile River in South Sudan (Musser & Carleton, 2005).

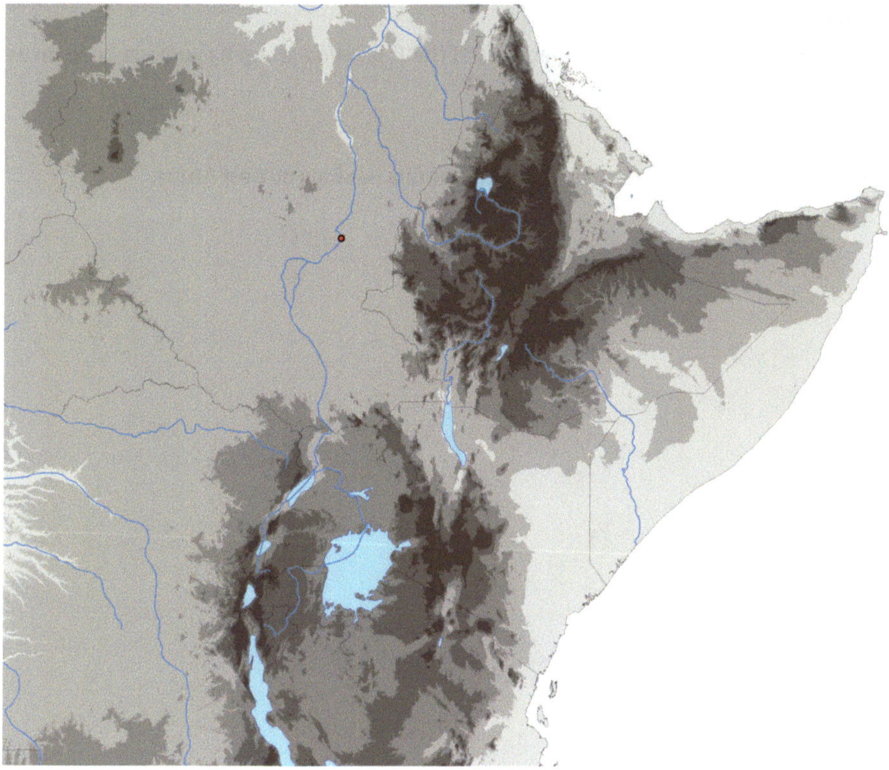

Figure 2.552 Distribution map

Measurements

External and cranial measurements (mm) and mass (g) for *Lemniscomys hoogstraali*, sexes combined

	Mean	Min	Max	N
Mass	-	-	-	-
HB*	127	-	-	1
Tail*	128	-	-	1
HF*	28	-	-	1
Ear*	16	-	-	1
GLS*	31.9	-	-	1

*Dieterlen (1991), measurements of the holotype

Lemniscomys linulus (Thomas, 1910)

Diagnosis

This species is the only member of the *L. griselda* group occurring in West Africa. The dorsal pelage has a single mid-dorsal stripe.

Taxonomic Issues

The relationship of this species to other members of the *L. griselda* group has not yet been investigated.

Distribution

This species has a narrow distribution in the drier savanna habitats of West Africa.

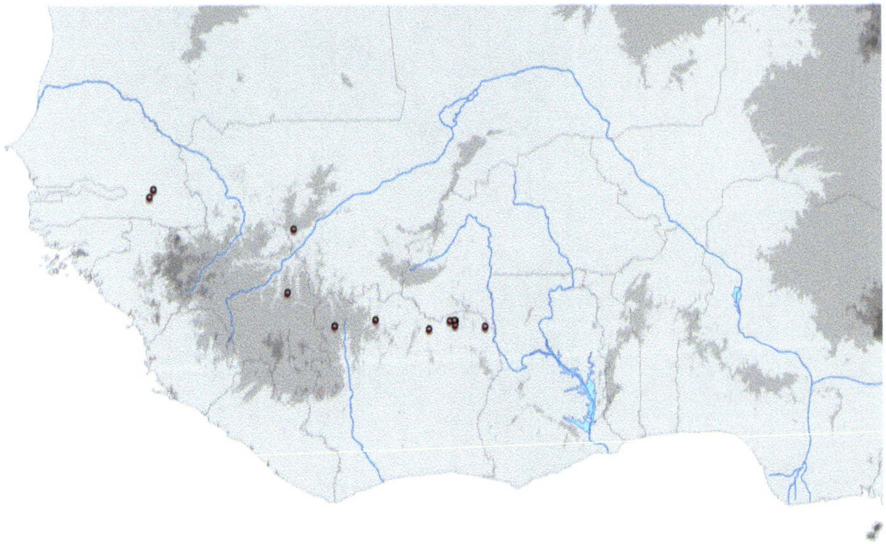

Figure 2.553 Distribution map

Measurements

External and cranial measurements (mm) and mass (g) for *Lemniscomys linulus*, sexes combined

	Mean	Min	Max	N
Mass	38.7	37	41	3
HB*	105.4	93	126	117
Tail*	111.4	95	135	109
HF*	26.0	23	28	116
Ear*	16.6	15	18	115
GLS*	27.3	25.2	29.5	98

*Van der Straeten (1980b)

Lemniscomys macculus (Thomas & Wroughton, 1910)

Diagnosis

This species is in the *L. striatus* group and can be distinguished from other members of this group as follows: *L. striatus* (size on average larger, GLS typically > 29 mm; differs on molecular grounds (Ducroz et al., 2001)); *L. bellieri* (restricted to West Africa); *L. mittendorfi* (endemic to Mt Oku, Cameroon). A dark mid-dorsal stripe is flanked by eight lateral stripes that are broken into pale buffy spots; the lower four rows tend not to be broken into spots and appear as lines.

Taxonomic Issues

This species may be conspecific with *L. bellieri*; the two species are only 3.3% divergent on the cytochrome b gene, share the same chromosome number, and are similar in cranio-dental and external morphology (Ducroz et al., 2001) (see *L. bellieri* account for further details). Chromosome number is 2n = 56, FN = 60 (Castiglia et al., 2002; Bulatova et al., 2002).

Distribution

This species is restricted to East Africa, extending from northeastern DRC, through Uganda and Sudan to Ethiopia and south to Kenya and the extreme north-west of Tanzania. It occurs in dry Acacia grasslands, avoiding cultivated areas (Verheyen & Verschuren, 1966).

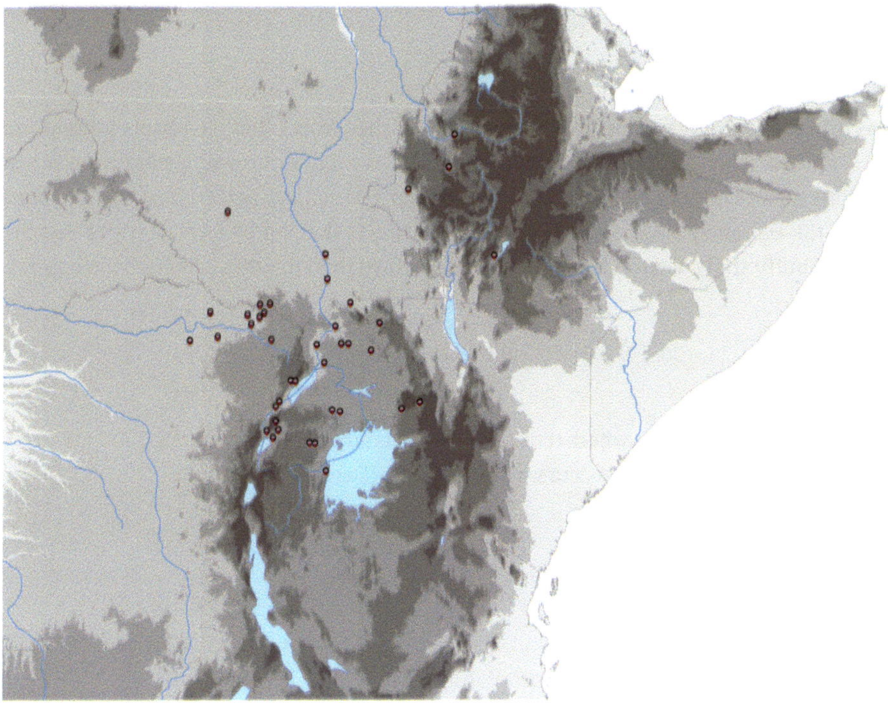

Figure 2.554 Distribution map

Measurements

External and cranial measurements (mm) and mass (g) for *Lemniscomys macculus*, sexes combined

	Mean	Min	Max	N
Mass	37.3	31	41	3
HB*	99.2	81	118	45
Tail*	109.0	70	150	41
HF*	22.3	19.8	25.5	55
Ear*	14.6	10	20	45
GLS*	27.1	25.0	29.0	55

*Van der Straeten & Verheyen (1979b)

Lemniscomys mittendorfi Eisentraut, 1968

Diagnosis

This species is in the *L. striatus* group and can be distinguished from other members of this group as follows: *L. striatus* (size larger, GLS > 28 mm; dorsal pelage lighter brown; generally occurs at lower elevations on Mt Oku); *L. macculus* (restricted to East Africa); *L. bellieri* (restricted to West Africa, west of the Dahomey Gap). A dark mid-dorsal stripe is flanked by lateral stripes that are broken into pale buffy spots. Its dorsal pelage is darker brown than in other *Lemniscomys* species.

Taxonomic Issues

Chromosome number is 2n = 56, FN = 66-72 (Fülling, 1992).

Distribution

This species is endemic to Mt Oku where it occurs in grassland at high elevations (2300 m) above the montane forest zone, where it co-occurs with *L. striatus* (Denys et al., 2014).

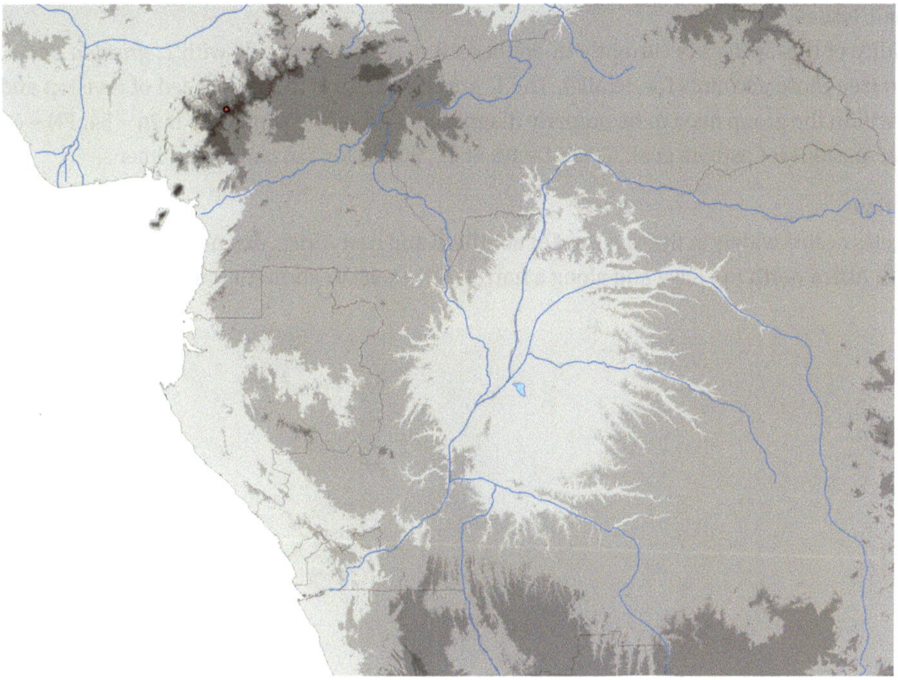

Figure 2.555 Distribution map

Measurements

External and cranial measurements (mm) and mass (g) for *Lemniscomys mittendorfi*, sexes combined

	Mean	Min	Max	N
Mass*	28	22	34	4
HB*	92	85	98	5
Tail*	79	75	84	5
HF*	20.6	19	22	5
Ear*	14.2	13	15	5
GLS*	25.9	24.9	26.9	4

*Happold (2013)

Lemniscomys rosalia (Thomas, 1904)

Diagnosis

This species can currently only be distinguished from other members of the *L. griselda* group by its distribution which includes Southern and East Africa; it is the only member of this species group occurring within most of its range, although the south-western limits of its distribution (e.g. Angola, DRC and Zambia) have not been resolved (see below). In this species, the dorsal pelage has a single mid-dorsal stripe.

Taxonomic Issues

The validity of this species is currently in doubt as it may be conspecific with *L. griselda* and *L. roseveari* (see those accounts for details). The *L. griselda* group is in urgent need of revision and species within the group need to be properly diagnosed. Chromosome number is 2n = 54, FN = 64 (Fadda et al., 2001a; Castiglia et al., 2002; Denys et al., 2011b) for Tanzanian specimens.

Distribution

This species occurs widely in the savannas of Southern and East Africa, extending from north-eastern South Africa north to Kenya and along a narrow strip west to northern Namibia.

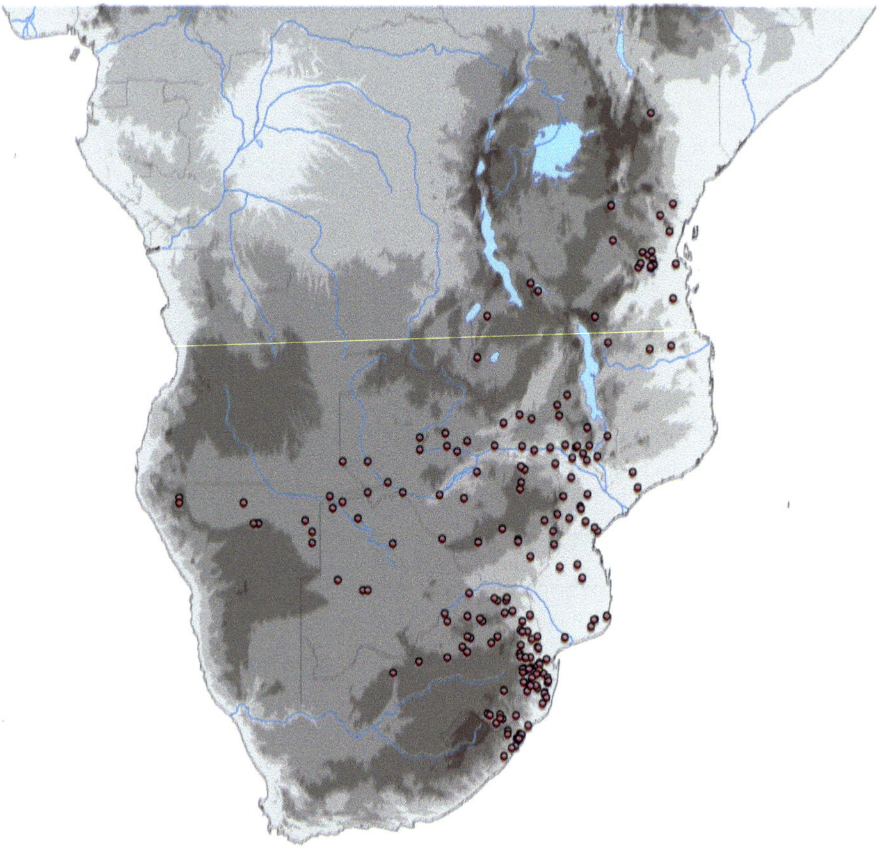

Figure 2.556 Distribution map

Measurements

External and cranial measurements (mm) and mass (g) for *Lemniscomys rosalia*, sexes combined

	Mean	Min	Max	N
Mass	57.8	33	82	57
HB	122.8	96	161	66
Tail	130.9	88	160	66
HF	28.0	24	33	52
Ear	16.1	7	20	45
GLS	31.4	29.6	34.1	13

Lemniscomys roseveari Van der Straeten, 1980

Diagnosis

This species can currently only be distinguished from other members of the *L. griselda* group by its distribution which is restricted to north-western Zambia; it is the only member of this species group occurring within its range, although the limits of its distribution are unknown (see below). In this species, the dorsal pelage has a single mid-dorsal stripe which may be flanked by faint lateral stripes.

Taxonomic Issues

This species is poorly diagnosed and currently cannot be distinguished from *L. rosalia* or *L. roseveari* with any confidence. In the original description of the species Van der Straeten (1980a) used multivariate analysis of cranio-dental measurements to separate these three species, but measurements overlap significantly and no diagnostic features were identified. Molecular and karyological analyses have not yet been conducted. We strongly question the validity of recognising these three taxa as separate species, but have chosen to retain them pending further investigation.

Distribution

This species is only known from north-western Zambia (Van der Straeten, 1980a).

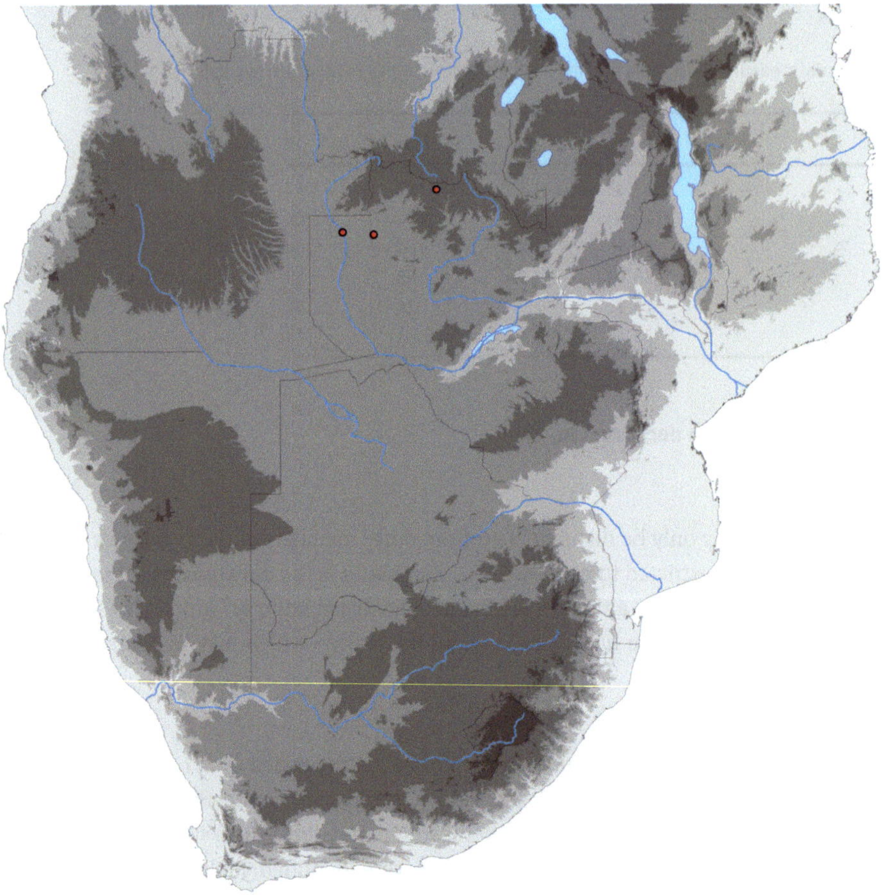

Figure 2.557 Distribution map

Measurements

External and cranial measurements (mm) and mass (g) for *Lemniscomys roseveari*, sexes combined

	Mean	Min	Max	N
Mass	-	-	-	-
HB*	124.5	113	137	26
Tail*	137.9	123	153	25
HF*	27.8	26.0	29.5	26
Ear*	16.3	13	19	26
GLS*	32.5	30.7	33.9	28

*Van der Straeten (1980a)

Lemniscomys striatus (Linnaeus, 1758)

Diagnosis

This species is in the *L. striatus* group and can be distinguished from other members of this group as follows: *L. mittendorfi* (size smaller, GLS < 28 mm; dorsal pelage darker brown; restricted to high elevations on Mt Oku); *L. macculus* (size on average smaller, GLS typically < 29 mm; differs on molecular grounds (Ducroz et al., 2001)); *L. bellieri* (size on average smaller, GLS ca. 28-29 mm; differs on karyological (Van der Straeten & Verheyen, 1978a) and molecular grounds (Ducroz et al., 2001); the lower lateral rows of spots are not arranged in neat lines). A dark mid-dorsal stripe is flanked by eight lateral stripes that are broken into pale buffy spots.

Taxonomic Issues

This species appears to consist of at least two distinct species; two individuals sequenced from Ivory Coast and Congo differed by 8% on the cytochrome b gene (Ducroz et al., 2001), which is very high for intra-specific variation. A more recent study demonstrated clear genetic structure within *L. striatus* with the existence of four geographically separated clades: West Africa; Benin-Nigeria; Central Africa; and East Africa (Nicolas et al., 2008a). This provides further evidence for *L. striatus* representing a species complex. Chromosome number in specimens from West Africa is 2n = 44, FN = 68-74 (Van der Straeten & Verheyen, 1978a; Capanna et al., 1997; Castiglia et al., 2002); chromosome number has not been reported from Central or East Africa. This species requires urgent revision.

Distribution

This species is widely distributed in tropical Africa from Guinea and Sierra Leone in the west to Ethiopia in the east and south to northern Angola, southern DRC and Tanzania. It occurs in a variety of grassland habitats, often at the edge of forest or clearings within it.

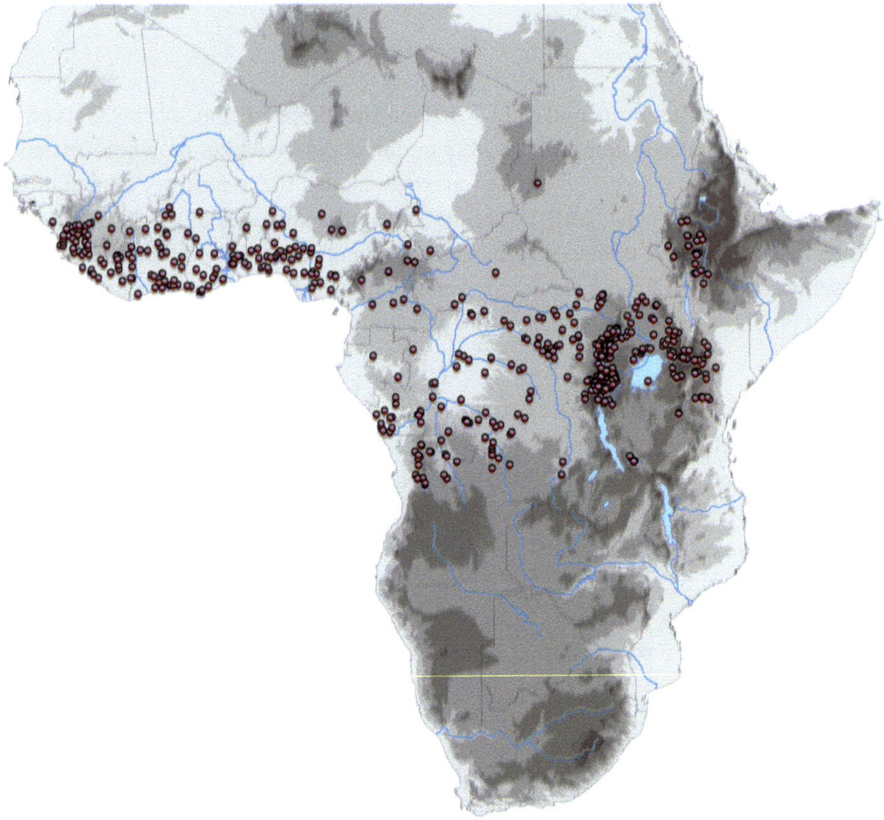

Figure 2.558 Distribution map

Measurements

External and cranial measurements (mm) and mass (g) for *Lemniscomys striatus*, sexes combined

	Mean	Min	Max	N
Mass	41.1	28	67	12
HB	115.6	97	139	39
Tail	124.4	96	153	34
HF	26.7	22	29	39
Ear	15.9	12	21	37
GLS	30.7	27.6	33.1	20

Lemniscomys zebra (Heuglin, 1864)

Diagnosis

This species is in the *L. barbarus* group, and can be distinguished from *L. hoogstraali* by its unique stripe pattern and larger size (in *L. hoogstraali*, the dorsal pelage has five lateral lines on the flanks with spots in between, and GLS of the holotype is 31.2 mm). In this species, there are 6-7 dark lateral stripes on either side of the mid-dorsal stripe, with light stripes in between.

Taxonomic Issues

This species may be conspecific with *L. hoogstraali* (see the latter species account for more details). Chromosome number is 2n = 54, FN = 58 (Fadda et al., 2001; Castiglia et al., 2002; Dobigny et al., 2002a).

Distribution

This species is widely distributed in the drier savannas of the Sahel, extending into East Africa.

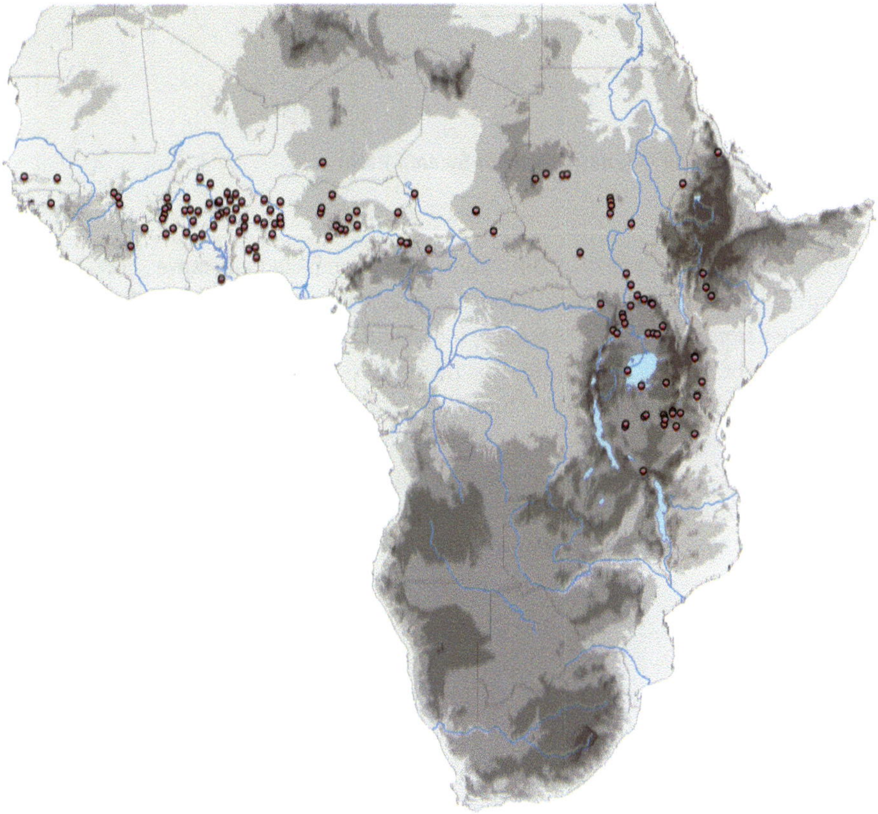

Figure 2.559 Distribution map

Measurements

External and cranial measurements (mm) and mass (g) for *Lemniscomys zebra*, sexes combined

	Mean	Min	Max	N
Mass	33.4	24	60	7
HB	106.1	96	119	9
Tail	120.7	110	133	7
HF	25.0	22	27	8
Ear	14.5	13	16	8
GLS	27.6	25.1	29.2	8

Genus Account

Malacomys Milne-Edwards, 1977 (Long-footed or Swamp Rats)

Diagnosis

These large rodents (Mass 60-80 g; HB 130-160 mm) with large naked ears have short, soft, fine fur lacking guard hairs and a bright reddish brown to chocolate brown dorsal colouration. The ventral pelage varies from white (in *M. cansdalei*) to grey (in other species). The tail is bicoloured (dark above and paler below), sparsely haired and relatively very long (130-136% of HB). The most distinctive features of this genus are its very long and slender, pale-coloured hindfoot (> 31 mm; about 25% of HB), large ears (> 22 mm), a long rostrum and relatively small molars. Among similar-sized African rodents, only *Rattus* and *Gerbilliscus* have such long hindfeet but these are much thicker in the case of *Rattus* and the tail ends in a terminal tuft in *Gerbilliscus*. The 1st digit on the forefoot is rudimentary with a flattened claw, while the 5th digit is slightly larger, reaching the first joint of the 4th digit.

Figure 2.560 Photograph of *Malacomys edwardsi* (Ara Monadjem).

Cranial Characters

The skull is large (35-38 mm), flattish in profile, smoothly contoured (without crests) and very long and thin in shape due to the long, thin tapering rostrum. The palatal foramina are noticeably short and end far anterior of M^1 and the zygomatic arches are fairly robust. The anterior edge of the zygomatic plate is convex in shape with a rounded anterior-dorsal angle. Subsquamosal foramen (and "mastoid strut") is usually present.

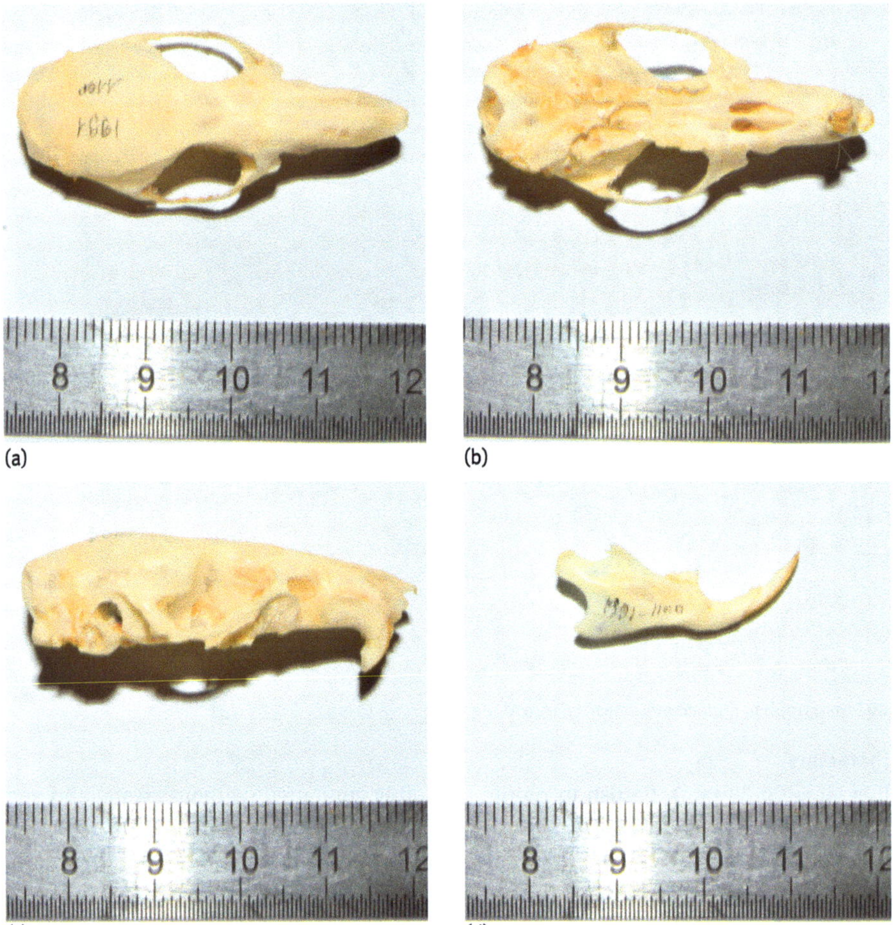

(a) (b) (c) (d)

Figure 2.561 a-d Cranium of *Malacomys longipes* (MNHN-ZM-MO 1991-1100) (from top left to bottom right: dorsal view, ventral view, lateral view and lower jaw).

Dentition

According to Misonne (1969) the unusual dentition of this genus distinguishes it from all other African genera. The upper incisors are orthodont, short and ungrooved. The molars are remarkably small and narrow for the size of the cranium, tending towards microdonty (MTR/GSL ± 0.13-0.14, Table 10), and rows of cusps are slightly distorted. Apart from cusp t1 which is separate in all upper molars, cusps t2 and t3 are fused and t9 appears to be absent or reduced to a trace. The cusp t3 on M^2 is absent or very reduced (cf. present in *Praomys*). Molar M^3 has cusp configuration shown for Group 3 (Figure 2.439). Molars, M_1 and M_2 have a posterior cingulum but no supplementary cingular cusplets.

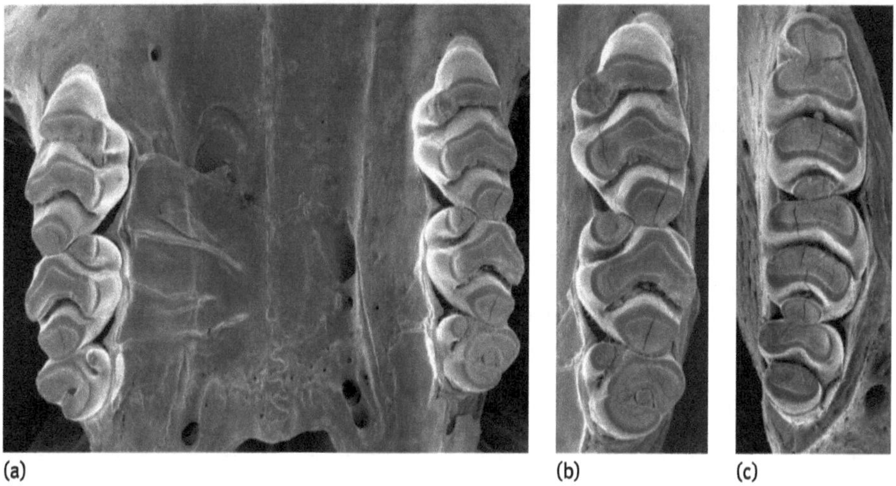

Figure 2.562 Cheek teeth of *Malacomys edwardsi* (MNHN-ZM-MO 1979-151) (A: upper cheek teeth; B: upper left toothrow; C: lower right toothrow).

Habitat & Biogeography

They are associated with lowland tropical rainforests, typically along shallow streams or where moist soils are present, from Sierra Leone and Guinea (Mt Nimba) in the west to the extreme west of Uganda and Rwanda in the east and south through the Congo Basin to north-western Zambia and north-eastern Angola.

Fossil Record

Fossils have been recorded from Plio-Pleistocene deposits in Angola (Winkler et al., 2010), but need confirmation.

Biology & Ecology

They are terrestrial, nocturnal and feed on slugs, earthworms and plant parts (Grubb et al., 1998; Happold, 1987; 2013). They are waders according to Kerbis Peterhans & Patterson (1995). Breeding is known to occur during the dry season and early wet season while litter size may vary from 1 to 5 (Happold, 1977, 2013).

Systematic Notes

Based on its phylogenetic distinctness on morphological (Misonne, 1969), chromosomal (Viegas-Péquignot et al., 1983) and molecular (Lecompte et al, 2002a; 2008) grounds, the genus has been allocated its own tribe, Malacomyini (Lecompte et al., 2008); although Schenk et al. (2013) found it to be part of a clade that includes *Apodemus* and *Tokkudaia*. The species *Praomys lukolelae* and *P. vershureni* possess characters (such as a long hindfoot) which link them with *Malacomys* and in fact these species were previously included in this genus (e.g. see Dieterlen & Van der Straeten, 1984); however, molecular analyses place these two species in *Praomys* (Lecompte et al., 2008). *Malacomys cansdalei* was previously considered to be a subspecies of *M. longipes* (Rautenbach & Schlitter, 1978) but shown to be morphometrically distinct by Van der Straeten & Verheyen (1979a).

Annotated Species List

Three species are recognised:
Malacomys cansdalei Ansell, 1958
Malacomys edwardsi Rochebrune, 1885
Malacomys longipes Milne-Edwards, 1877

Key to Genus

1	Size larger (HF > 35 mm), 3 pairs of nipples (1 + 2 = 6)	2
	Size smaller (HF < 35 mm), 2 pairs of nipples (0 + 2 = 4)	*M. edwardsi*
2	Tail relatively longer (ca. 136% of HB); ears longer (typically > 27 mm); hindfoot longer (> 42 mm); restricted to West Africa (west of the Volta River)	*M. cansdalei*
	Tail relatively shorter (ca. 117% of HB); ears shorter (typically < 27 mm); hindfoot shorter (< 42 mm); occurs in Central Africa (east of the Niger River)	*M. longipes*

Species Accounts

Malacomys cansdalei Ansell, 1958

Diagnosis

This species co-occurs with *M. edwardsi* throughout most of the Upper Guinea Forest zone, where it can be distinguished by its larger size and greater number of nipples (in *M. edwardsi*, HF < 35 mm and only two pairs of nipples are present). Three pairs of nipples (1 + 2 = 6) are present in *M. cansdalei*.

Taxonomic Issues

This taxon was previously listed as a synonym of *M. longipes* (Rautenbach & Schlitter, 1978), but elevated to full species rank by Van der Straeten & Verheyen (1979a).

Distribution

This species is endemic to the Upper Guinea Forest west of the Volta River where it occurs in forested habitats.

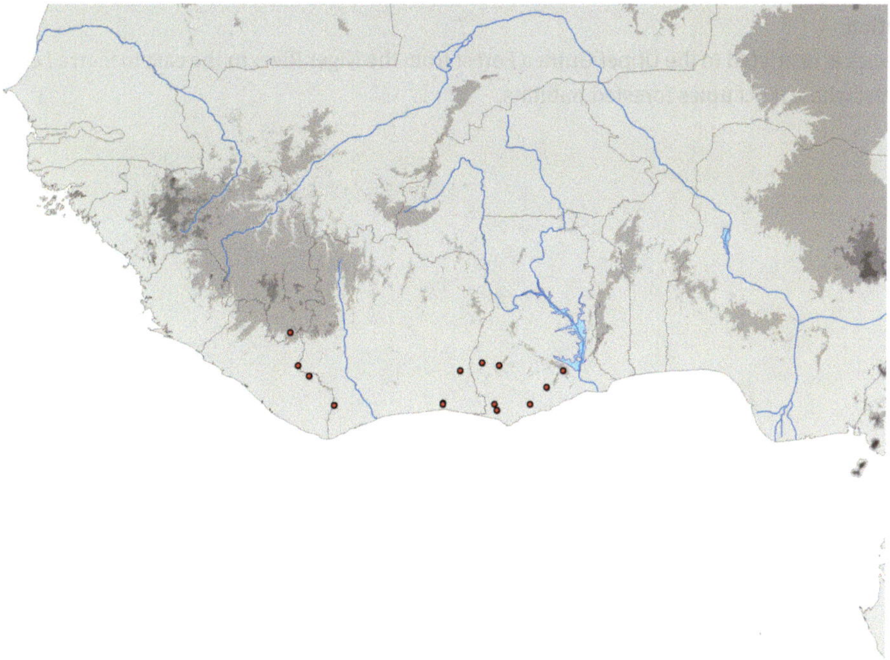

Figure 2.563 Distribution map

Measurements

External and cranial measurements (mm) and mass (g) for *Malacomys cansdalei*, sexes combined

	Mean	Min	Max	N
Mass*	81.7	63	101	14
HB*	144	133	162	17
Tail*	196	178	208	16
HF*	42.7	40	44	17
Ear*	28.2	27	31	15
GLS*	40.2	39.1	42.2	11

*Cole (1972)

Malacomys edwardsi Rochebrune, 1885

Diagnosis

This species co-occurs with *M. cansdalei* throughout most of the Upper Guinea Forest zone, where it can be distinguished by its smaller size and lower number of nipples (in *M. cansdalei*, HF > 38 mm and three pairs of nipples are present). This species does not co-occur with *M. longipes* (which occurs east of the Niger River). Two pairs of nipples (0 + 2 = 6) are present in *M. edwardsi*.

Taxonomic Issues

Chromosome number is 2n = 48, FN = 52 (Matthey, 1958; Van der Straeten & Verheyen, 1979a).

Distribution

This species is restricted to the Upper Guinea Forest from the Niger River in the east to Sierra Leone in the west where it occupies forested habitats.

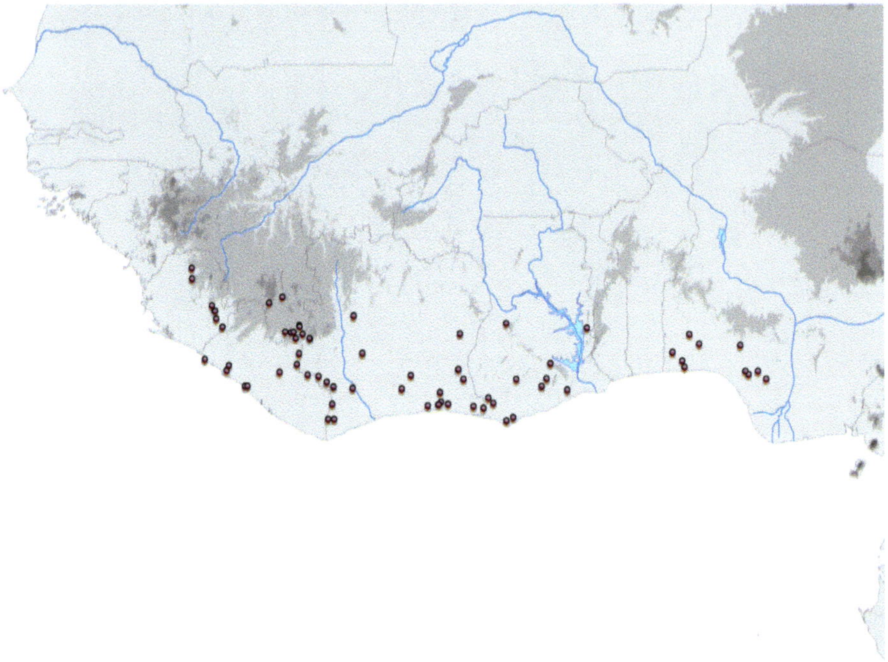

Figure 2.564 Distribution map

Measurements

External and cranial measurements (mm) and mass (g) for *Malacomys edwardsi*, sexes combined

	Mean	Min	Max	N
Mass	64.5	46	80	6
HB	140.1	126	160	9
Tail	168.8	155	180	9
HF	34.2	31	37	9
Ear	25.7	24	28	7
GLS	37.9	34.9	39.6	10

Malacomys longipes Milne-Edwards, 1877

Diagnosis

This is the only species of *Malacomys* occurring within its range. It can be distinguished from *M. edwardsi* by its larger size and greater number of nipples (in *M. edwardsi* GLS < 40 mm, and there are two pairs of nipples), and from *M. cansdalei* by its relatively shorter tail, shorter HF and shorter ears (see key). In this species, there are three pairs of nipples (1 + 2 = 6).

Taxonomic Issues

This species is distinct from *M. edwardsi* on molecular grounds (Lecompte et al., 2008), but a phylogeny for the genus has not yet been developed. Chromosome number is 2n = 48 (Primus et al., 2006). Chromosome number 2n = 48, aFN = 48 (Lecompte et al., 2005b).

Distribution

This species extends from the Niger River in the west to the shores of Lake Victoria in the east and south through north-eastern Angola to north-western Zambia.

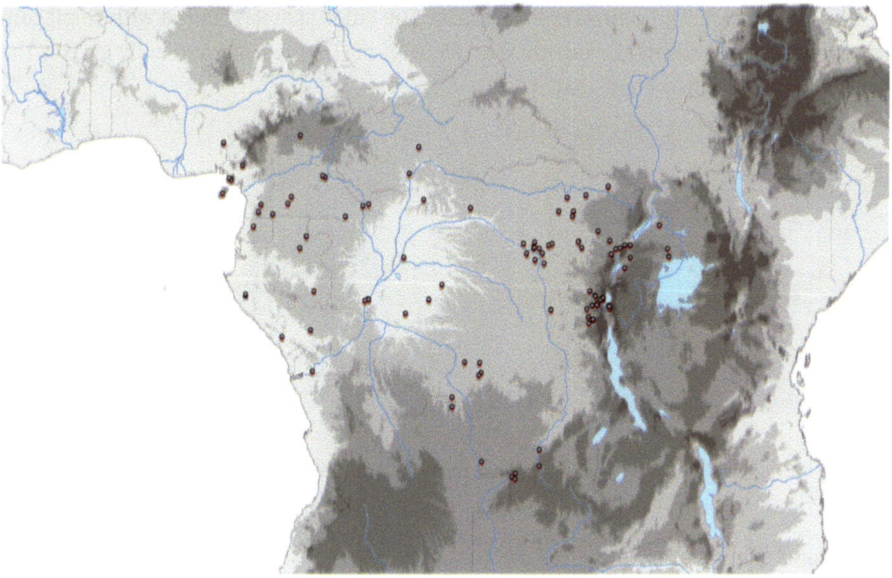

Figure 2.565 Distribution map

Measurements

External and cranial measurements (mm) and mass (g) for *Malacomys longipes*, sexes combined

	Mean	Min	Max	N
Mass	107.9	94	120	7
HB	157.1	135	175	9
Tail	183.9	170	200	9
HF*	36.2	34	39	8
Ear	24.2	21	26	6
GLS	42.0	40.4	43.1	7

*Happold (2013) reported a mean value of 39.4 mm (range: 36-41 mm; n = 14)

Genus Account

Mastomys Thomas, 1915 (Multimammate Mice)

Diagnosis

Females of these medium-sized rodents (Mass ca. 20-100 g; HB ca. 70-140 mm) are immediately recognisable by having 16 to 24 pairs of mammae, much more than any other rodent or probably, mammal. The dorsal pelage is soft-furred with sparse guard hairs and variable colour, typically dark greyish to grey-brown. The ventral colour is grey. The tail is sparsely haired with small concentric scales and typically shorter or subequal to (85-103% of HB) the head and body length. The ears are quite large and slightly elongated.

(a)

(b)

Figure 2.566 a and b Photographs of *Mastomys* sp. (Charles Foley) (left) and *M. pernanus* (Bill Stanley) (right).

Cranial Characters

The skull is medium-sized and rather long and narrow in shape, with poorly to moderately developed supraorbital ridges. According to Rosevear (1969) the skull of *Mastomys* can be distinguished from similar-sized West African genera by the combination of having a palatine bone which extends no further than the anterior cusps of M^2 and palatal foramina which extend as far back as the second or sometimes the third row of cusps on M^1. In Southern Africa, the skull of this species can be immediately distinguished by having a "mastoid strut" associated with the subsquamosal foramen (Meester at al., 1986). The zygomatic arches are gracile in this species and the anterior edge of the zygomatic plate slightly convex with rounded dorsal edge. The bullae are not inflated. The mesopterygoid fossa is narrow (usually < 1 mm but slightly larger in *M. kollmanspergeri*).

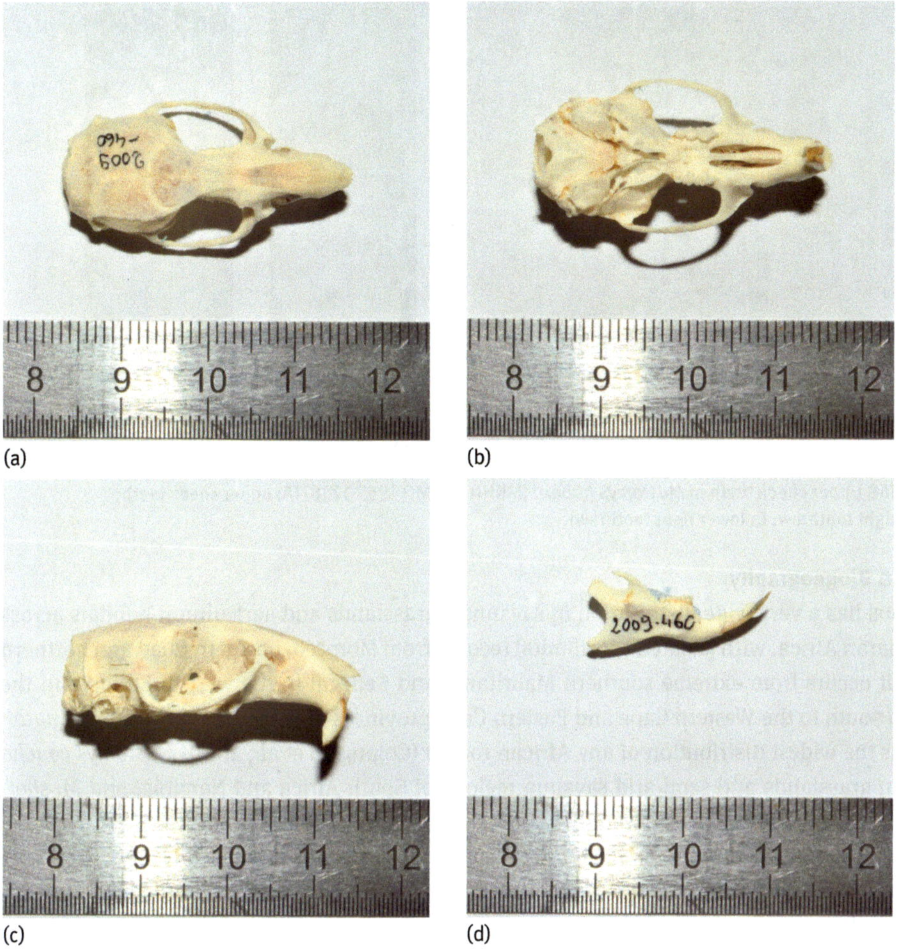

Figure 2.567 a-d Cranium of *Mastomys natalensis* (MNHN-ZM-MO 2009-460) (from top left to bottom right: dorsal view, ventral view, lateral view and lower jaw).

Dentition

The upper incisors are orthodont to slightly opisthodont and ungrooved. The molars are of small to average length (MTR/GSL ± 0.13-0.17, Table 10). Rows of cusps on M^1 are clearly distorted with t1 situated far behind t2 and t3 and just in front of t4 which may be reduced. In M^2, t3 is reduced or absent. Molar M^3 is much reduced and cusp configuration is as shown in Group 3 (Figure 2.439). There is a posterior cingulum on M_1 and an antero-external cusp on the lower molars.

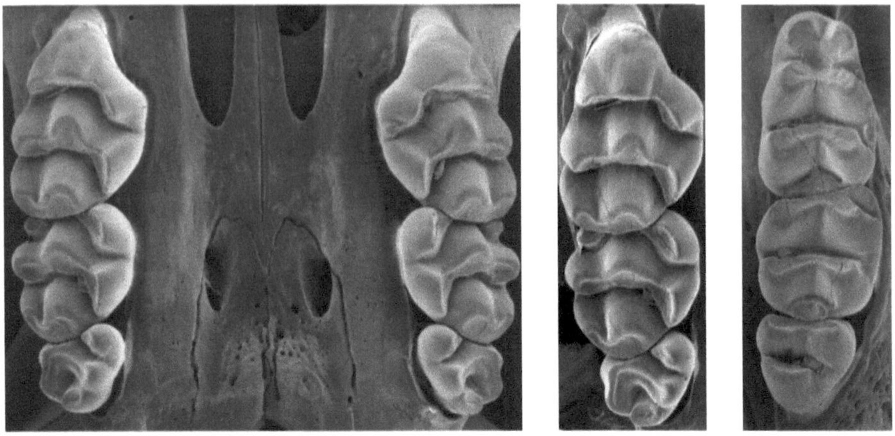

Figure 2.568 Upper cheek teeth of *Mastomys huberti* (MNHN-ZM-MO 1963-1248) (A: upper cheek teeth; B: upper right toothrow; C: lower right toothrow).

Habitat & Biogeography

The genus has a very wide distribution in savannas, grasslands and agricultural habitats across Sub-Saharan Africa, with isolated extralimital records from Morocco, northern Chad and northern Sudan. It occurs from extreme southern Mauritania and Senegal in the west to Ethiopia in the east and south to the Western Cape and Eastern Cape provinces of South Africa. *Mastomys natalensis* has the widest distribution of any African rodent (Colangelo et al., 2013). *Mastomys coucha* occurs in grasslands and semi-arid savanna regions of South Africa and Namibia, and *M. shortridgei* is restricted to localised swampy habitats in northern Botswana, north-western Namibia and south-eastern Angola, whilst most other members of the genus are typically associated with anthropogenic habitats and are often major crop pests as well as vectors of a range of diseases including plague.

Fossil Record

This genus has been recorded from Early Pliocene to Late Pleistocene fossil deposits in Southern and Central Africa (Winkler et al., 2010). The earliest fossils are from 6 MYA at Lemudon'go in East Africa (Manthi, 2007).

Mastomys cineraceus Laetoli (Denys, 1987a, 2011)
Mastomys cf *minor* Omo B (Wesselman, 1984)
Mastomys minor Olduvai Bed I (Jaeger, 1976)
Mastomys sp. Hadar (Reed, 2011)
Mastomys sp. Plio-Pleistocene caves of RSA (Senegas, 2000; Avery, 1995, 1998)

Biology & Ecology

They are terrestrial and nocturnal. Diet varies from granivorous to omnivorous, sometimes including arthropods and carrion. Multimammate mice are significant crop pests and frequently dig up newly planted maize seeds or climb maize stalks to feed on the seeds, as well as feeding on stored grains inside village dwellings. Breeding is typically seasonal and strongly correlated with rainfall, with litter sizes varying from 1 to 27. The breeding and population biology of *M. natalensis* has been intensively studied in Tanzania over multiple years (see e.g., Leirs, 1999; Leirs et al., 1994, 1996) while that of *M. erythroleucus* (Crespin et al., 2008) and *M. huberti* (Granjon et al., 2005a) have similarly been studied over multiple years in response to the flood regime of the Niger River in Mali.

Systematic Notes

A good review of the systematics of *Mastomys* is provided by Granjon et al. (1997). *Mastomys* has been assigned on molecular grounds to the *Stenocephalemys* division of the tribe Praomyini (Musser & Carleton, 2005; Lecompte et al., 2008; Schenk et al., 2013). Species of this genus are cryptic and difficult to distinguish morphologically but clearly separable by molecular (Lecompte et al., 2002a,b, 2005a) and chromosomal (Britton-Davidian, 1995; Volobouev et al., 2002a; Dobigny et al., 2008) markers. *Mastomys* was previously included in *Praomys* but multiple lines of evidence show it to be a monophyletic clade (except for *M. pernanus*, see below) distinct from the latter genus (see Musser & Carleton, 2005 and references therein). A phylogeny for six of the eight species is presented by Colangelo et al. (2010b); the two remaining species (*M. pernanus* and *M. shortridgei*) have also been sequenced and appear in separate phylogenies (Lecompte et al., 2002a, 2005a; Eiseb 2014). The taxon *pernanus* has been assigned to *Mastomys* on morphological grounds (Van der Straeten, 1999; Granjon et al., 1997) but it groups with *Heimyscus* and *Hylomyscus* based on mitochondrial DNA sequences (Lecompte et al., 2002a, 2005a), and therefore should not be included in this genus. However, we have retained it in *Mastomys* pending further investigation and the sequencing of additional specimens of *M. pernanus*. *Mastomys shortridgei* appears to group with *M. coucha* based on diploid number but the fundamental numbers are different (Granjon et al., 1997). This species has recently been sequenced and cytochrome b data show paraphyly between *coucha* and *shortridgei*; however the species can be diagnosed on chromosomes and skull shape (Eiseb, 2014).

Annotated Species List

Eight species are recognised:
Mastomys awashensis Lavrenchenko, Likhnova & Baskevich (in Lavrenchenko et al., 1998a)
Mastomys coucha (Smith, 1834)
Mastomys erythroleucus (Temminck, 1853)
Mastomys huberti (Wroughton, 1909)
Mastomys kollmannspergeri (Petter, 1957)
Mastomys natalensis (Smith, 1834)
Mastomys pernanus (Kershaw, 1921)
Mastomys shortridgei (St. Leger, 1933)

Key to Species

Members of the genus *Mastomys* cannot be distinguished based on external or cranio-dental characters and require molecular and/or karyological analysis. Instead of a key, we present the chromosome number, mammary formula, and other species of *Mastomys* co-occurring with each species. For further details on identification see the species accounts.

Mastomys species	Chromosome number	Mammary formula	Co-occurring species
M. awashensis	2n = 32	Not known	*M. erythroleucus; M. natalensis*
M. coucha	2n = 36, FN = 52-56	12 pairs	*M. natalensis*
M. erythroleucus	2n = 38, FN = 50-60	12 pairs	*M. awashensis; M. huberti; M. kollmanspergeri; M. natalensis*
M. huberti	2n = 32, FN = 44-46	12 pairs	*M. erythroleucus; M. natalensis*
M. kollmannspergeri	2n = 38, FN = 40-41	9-12 pairs	*M. erythroleucus; M. natalensis*
M. natalensis	2n = 32, FN = 52-54	12 pairs	All
M. pernanus	Not known	Not known	*M. erythroleucus*(?); *M. natalensis*
M. shortridgei	2n = 36, FN = 51/52*	Not known	*M. natalensis*

* Based on Eiseb (2014); 2n was reported as 2n = 50 by Granjon (1997), from a conference abstract presentation (Gordon, 1985).

Species Accounts

Mastomys awashensis Lavrenchenko, Likhnova & Baskevich (in Lavrenchenko et al., 1998)

Diagnosis

This species is endemic to Ethiopia where it can be distinguished from *M. natalensis* and *M. erythroleucus* on molecular (Colangelo et al., 2010b) and slight morphological (Lavrenchenko et al., 1998b) grounds.

Taxonomic Issues

Although morphologically similar, this species differs by > 8% based on the cytochrome b gene from other *Mastomys* species (Colangelo et al., 2010b). Chromosome number is 2n = 32.

Distribution

This species is endemic to Ethiopia where it is only known from a few localities. It may be more widespread than presented here, but currently cannot be easily differentiated from *M. natalensis* or *M. erythroleucus* (both of which also occur in Ethiopia) on morphological grounds.

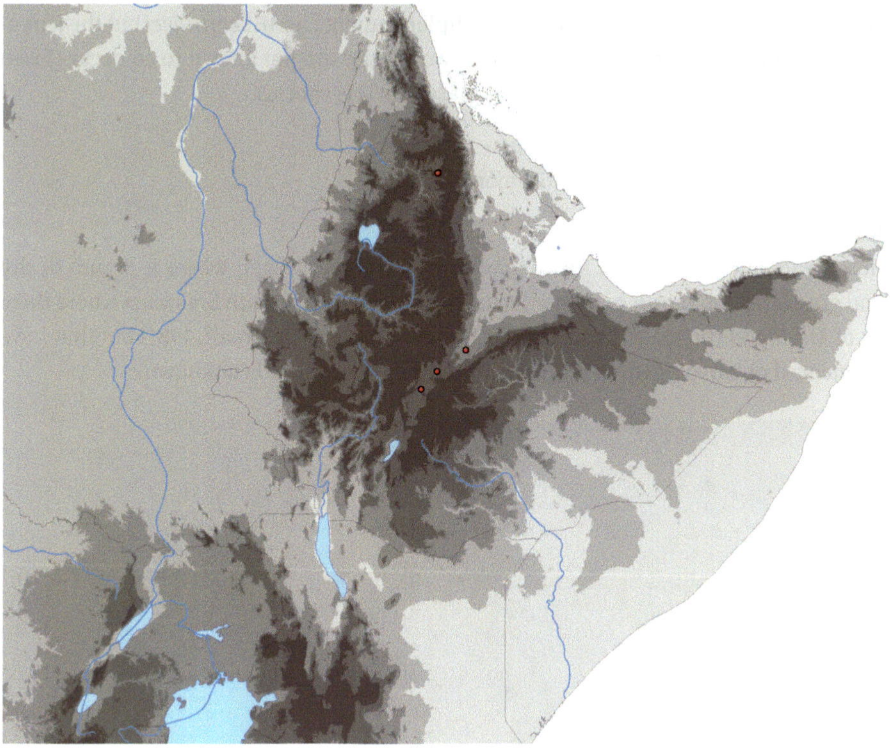

Figure 2.569 Distribution map

Measurements

External and cranial measurements (mm) and mass (g) for *Mastomys awashensis*, sexes combined

	Mean	Min	Max	N
Mass	-	-	-	-
HB*	122	116	127	6
Tail*	118	111	123	6
HF*	23.8	23.5	24.3	6
Ear*	19.0	18	20	6
GLS*	29.8	28.8	30.7	6

*Lavrenchenko et al. (1998b)

Mastomys coucha (Smith, 1834)

Diagnosis

This species co-occurs only marginally with *M. natalensis* in South Africa (Venturi et al., 2004), but overlaps more extensively in southern Zimbabwe (Gordon, 1978) and northern Namibia (S. Eiseb, personal communication). The two species can be distinguished on karyological grounds (in *M. natalensis* 2n = 32) but not on cranio-dental features, not even in a multivariate analysis

(Dippenaar et al., 1993). This species has the same diploid number as *M. shortridgei* but the karyotypes are distinct in fundamental number and the two species are not known to co-occur (Eiseb 2014). Twelve equally spaced pairs of nipples are present in this species.

Taxonomic Issues

Chromosome number 2n = 36, FN = 52-56 (Granjon et al., 1997).

Distribution

This species is endemic to Southern Africa south of the Zambezi River, where it occurs in dry savanna and grassland habitats. It probably occurs in eastern and southern Botswana where there are records of *Mastomys* (previously assigned to *M. natalensis* e.g. De Graaff, 1981) but that have not been sequenced or karyotyped. Its status in Mozambique is currently unknown.

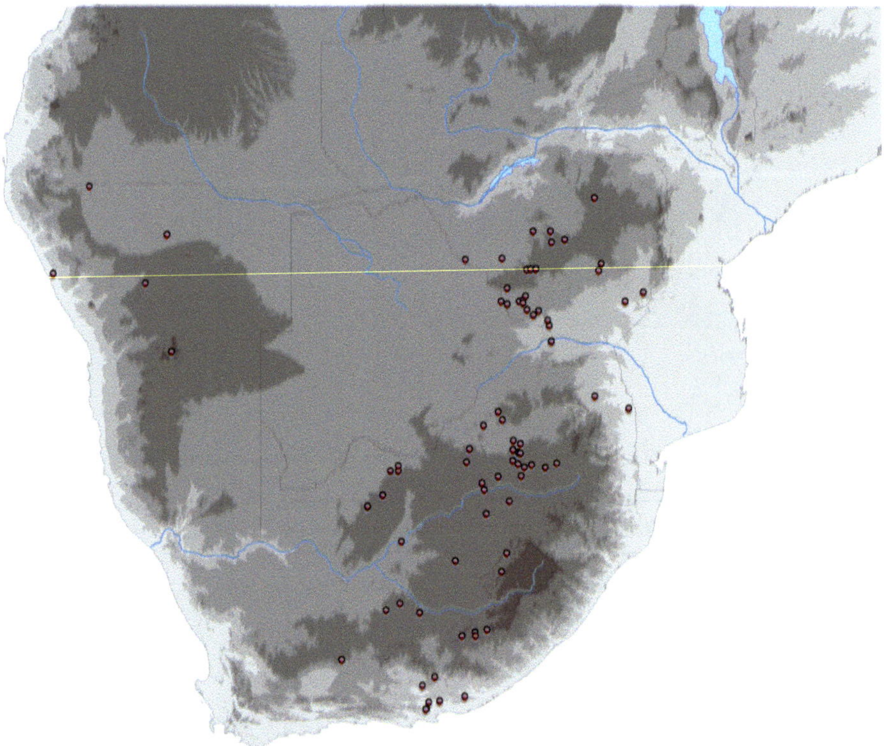

Figure 2.570 Distribution map

Measurements

External and cranial measurements (mm) and mass (g) for *Mastomys coucha*, sexes combined

	Mean	Min	Max	N
Mass	31.8	22	40	6
HB	100.3	82	119	7
Tail	88.6	70	107	8
HF	21.4	19	24	8
Ear	17.1	13	20	8
GLS*	26.9	-	-	21

*Njobe (1997)

Mastomys erythroleucus (Temminck, 1853)

Diagnosis

This species can only be distinguished with certainty from other species of *Mastomys*, including syntopic *M. huberti* and *M. natalensis* (2n = 32 in both species), on chromosomal or molecular grounds (Lalis et al., 2009; Denys et al., 2011a). Twelve equally spaced pairs of nipples are present in this species.

Taxonomic Issues

Four "phylogroups" were identified in this species which differed from each other by 3.0-3.6% based on the cytochrome b gene (Brouat et al., 2009), although it is not clear what taxonomic implications this result may have. Chromosome number is 2n = 38, FN = 50-60 (Volobouev et al., 2002a).

Distribution

This species is widely distributed in the northern savannas extending from Senegal in the west to Ethiopia in the east and south the shores of Lake Victoria. This map is based entirely on sequenced specimens (Brouat et al., 2009), and hence the gap in records from Chad to Ethiopia and Uganda is probably not real and most likely due to a lack of sequenced specimens (for example large series of *Mastomys* have been collected from southern Sudan (Setzer, 1956) but these could represent either *M. erythroleucus* or *M. natalensis* and therefore are not mapped here).

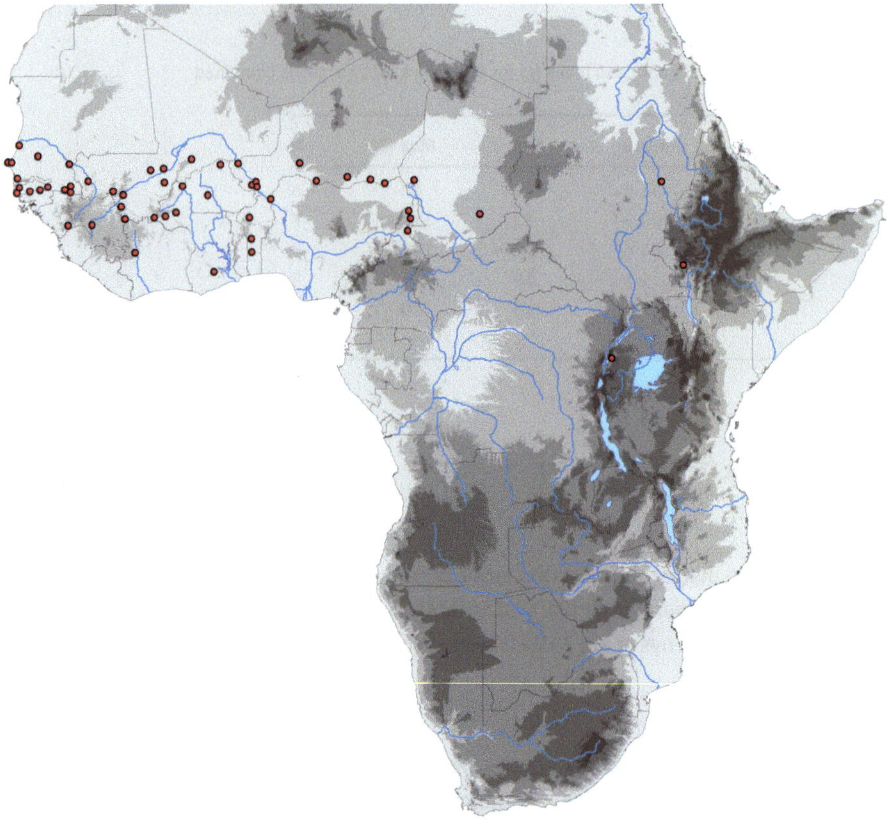

Figure 2.571 Distribution map

Measurements

External and cranial measurements (mm) and mass (g) for *Mastomys erythroleucus*, sexes combined

	Mean	Min	Max	N
Mass*	51.3	32	107	2798
HB*	129.9	92	178	2119
Tail*	115.8	81	151	1965
HF*	23.4	20	27	1918
Ear*	18.8	15	23	2012
GLS**	28.8	23.7	34.3	41

*Granjon & Duplantier (2009); ** Denys et al. (2011a)

Mastomys huberti (Wroughton, 1909)

Diagnosis

This species co-occurs with two to three other *Mastomys* species from which it can only be identified with certainty based on chromosomal or molecular grounds. There are slight differences in cranio-dental characters which may be used in a multivariate analysis to separate these species but this is not 100% effective (Lalis et al., 2009; Denys et al., 2011a). However, it does differ somewhat in its habitat, preferring wetter situations such as flooded ricefields and (in contrast to *M. erythroleucus* and *M. natalensis*) avoids human habitation (Granjon et al., 2005a; Granjon & Duplantier, 2009; Denys et al., 2011a). Twelve equally spaced pairs of nipples are present in this species.

Taxonomic Issues

The type specimen of this species is from northern Nigeria, but it has not been karyotyped and therefore it is not possible to confirm whether it corresponds to the specimens referred to as *M. huberti* here, with chromosome number 2n = 32; FN = 44-46 (Granjon et al., 1997; Denys et al., 2009a). The possibility therefore exists that *M. huberti* is actually a synonym of *M. natalensis*, in which case the species currently referred to as *M. huberti* requires a new name. Collection of *Mastomys* specimens from the type locality of *M. huberti* is therefore critical to addressing this issue.

Distribution

This species is largely confined to the Senegal and Niger River deltas, but also occurs in fragmented populations elsewhere in Senegal, the extreme southern Mauritania, and western Guinea (Mouline et al., 2008; Denys et al., 2011a). The type locality is in northern Nigeria, but this specimen has not been karyotyped (not plotted here). The species is thought to be present in Burkina Faso and Niger (Granjon & Duplantier, 2009), but this requires confirmation.

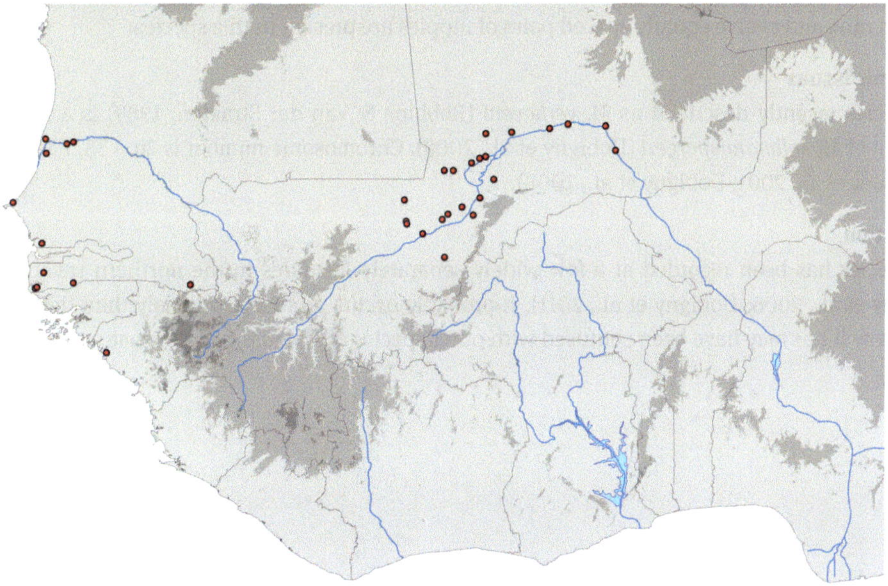

Figure 2.572 Distribution map

Measurements

External and cranial measurements (mm) and mass (g) for *Mastomys huberti*, sexes combined

	Mean	Min	Max	N
Mass*	41.6	10	91	68
HB*	113.2	74	146	68
Tail*	91.6	62	120	68
HF*	22.9	20	25	68
Ear*	17.0	14	20	68
GLS*	27.0	21.9	31.4	73

*Denys et al. (2011a)

Mastomys kollmannspergeri (Petter, 1957)

Diagnosis

This species is currently known to co-occur with *M. erythroleucus* and *M. natalensis* from which it can only be distinguished with certainty on chromosomal and molecular grounds (in *M. erythroleucus* and *M. natalensis*, 2n = 38 and 32, respectively. Furthermore the ventral pelage of these two species is typically grey or creamy whereas it is grey with white tips in *M. kollmannspergeri*). Between nine and twelve equally spaced pairs of nipples are present in this species.

Taxonomic Issues

The species recently described as *M. verheyeni* (Robbins & Van der Straeten, 1989) is a junior synonym of *M. kollmannspergeri* (Dobigny et al., 2008). Chromosome number is 2n = 38, FN = 40 (Volobouev et al., 2001; Dobigny et al., 2008).

Distribution

This species has been recorded at a few widely separately localities in the northern savannas (Dobigny et al., 2008; Dobigny et al., 2011). It probably occurs more continuously than the map suggests as it has may have been confused with other species of *Mastomys* in the past.

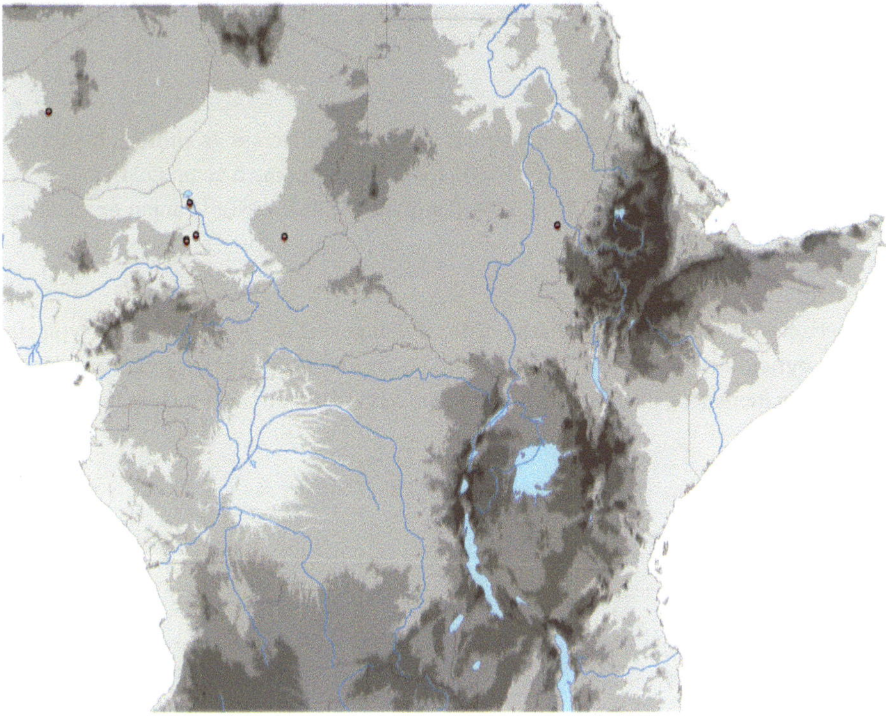

Figure 2.573 Distribution map

Measurements

External and cranial measurements (mm) and mass (g) for *Mastomys kollmannspergeri*, sexes combined

	Mean	Min	Max	N
Mass*	93.9	54	160	47
HB*	136.7	118	166	47
Tail*	115.3	96	137	44
HF*	26.6	24	29	47
Ear*	17.7	15	20	47
GLS*	33.8	30.4	38.9	50

*Happold (2013)

Mastomys natalensis (Smith, 1834)

Diagnosis

This species can only be distinguished with certainty from other *Mastomys* species on chromosomal and molecular grounds (Granjon et al., 1997; Colangelo et al., 2013); whereas on morphological grounds it is not possible to get a complete separation between co-occurring *Mastomys* species (Dippenaar et al., 1993; Lalis et al., 2009). Even sperm morphology is very similar between

M. natalensis and *M. coucha* (Breed, 1995). Nine to twelve equally spaced pairs of nipples are present in this species.

Taxonomic Issues

In a recent study, six "phylogroups" of *M. natalensis* were reported across its entire Sub-Saharan African range, with sequence divergence of 2.1-3.8% for the cytochrome b gene (Colangelo et al., 2013). Chromosome number is 2n =32, FN = 52-54 (Duplantier et al., 1990; Granjon et al., 1996; Dobigny et al., 2008; Denys et al., 2011b).

Distribution

This species occurs widely across Sub-Saharan Africa (Colangelo et al., 2013) where it is typically associated with agricultural fields and homes, but also occurs in natural savannas and grasslands, often sharply increasing in abundance after some form of disturbance such as fire.

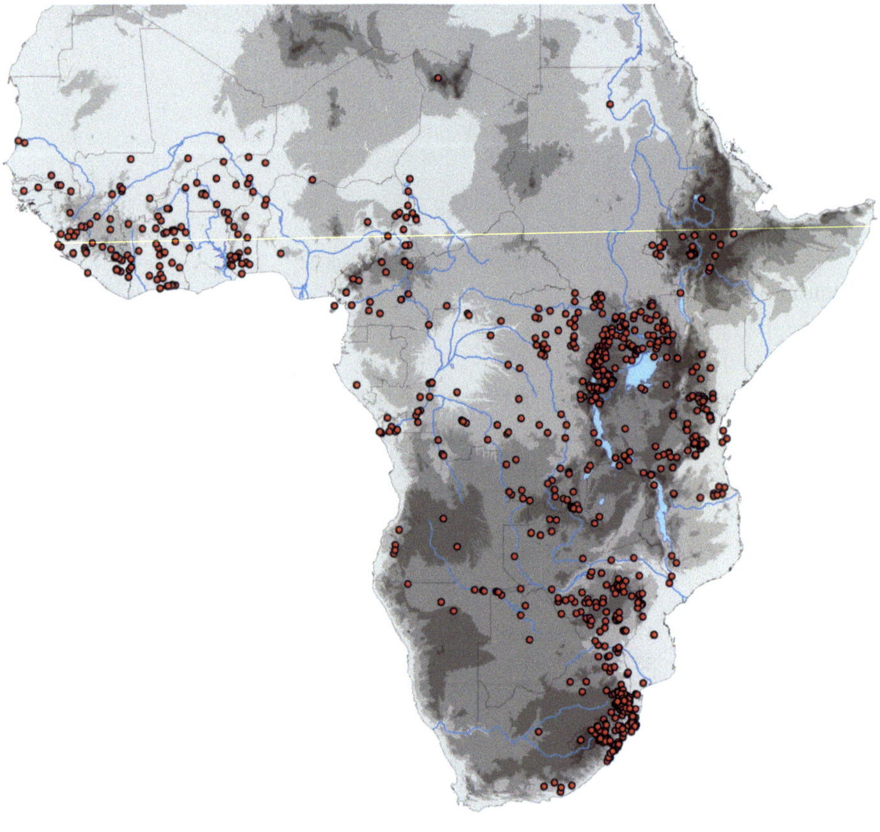

Figure 2.574 Distribution map

Measurements

External and cranial measurements (mm) and mass (g) for *Mastomys natalensis*, sexes combined

	Mean	Min	Max	N
Mass	40.8	15	98	109
HB	108.1	76	155	160
Tail	102.1	70	174	162
HF	22.9	18	26	109
Ear	17.3	12	22	114
GLS*	28.3	22.3	34.8	1453

*Happold (2013)

Mastomys pernanus (Kershaw, 1921)

Diagnosis

This species co-occurs with *M. natalensis* from which is can be distinguished on molecular grounds (Lecompte et al., 2002a). The two species also differ in size (adult *M. natalensis* typically have GLS > 25 mm and HB > 90 mm) and there is a prominent postauricular white patch present in *M. pernanus* (absent in other *Mastomys* species).The ventral pelage is grey with white tips. The number of nipples is not known.

Taxonomic Issues

The taxonomic status of this species is still not clear. It has been listed as a species of *Mastomys* based on morphological similarities (e.g. see Van der Straeten, 1999), but a single sequenced specimen grouped with *Hylomyscus* (and not with *Mastomys*) (Lecompte et al., 2002a, 2005a). More specimens of this species need to be collected and examined, including karyological and molecular analyses, before its status will be resolved (for example, the number of chromosomes in this species is not known).

Distribution

This species is known from just a handful of specimens most of which were collected in the Serengeti ecosystem in the border region between Tanzania and Kenya (Van der Straeten, 1999).

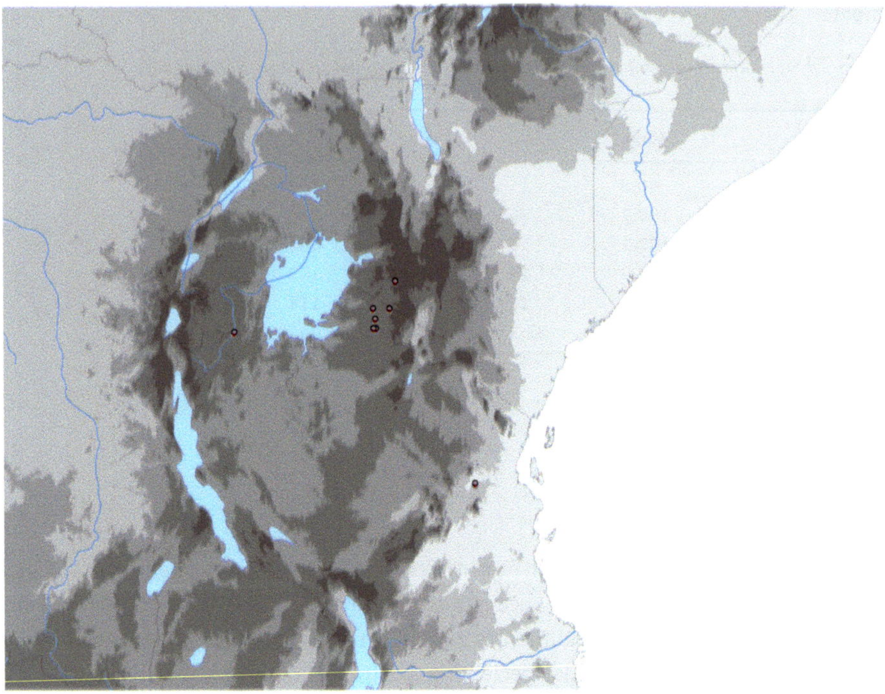

Figure 2.575 Distribution map

Measurements

External and cranial measurements (mm) and mass (g) for *Mastomys pernanus*, sexes combined

	Mean	Min	Max	N
Mass	-	-	-	-
HB*	78.6	73	88	7
Tail*	66.8	60	78	6
HF*	16.1	15.0	18.4	7
Ear*	14.9	14	16	7
GLS*	24.1	23.2	25.1	3

*Van der Straeten (1999)

Mastomys shortridgei (St. Leger, 1933)

Diagnosis

This species can only be distinguished from the co-occurring *M. natalensis* and allopatric *M. coucha* on chromosomal and subtle cranial shape grounds (Eiseb, 2014). It is clearly distinct from *M. natalensis*, but only partially distinct from *M. coucha* on the grounds of cytochrome b sequences, indicating a very recent split between *coucha* and *shortridgei*, probably around 0.7 million years ago linked to drainage evolution in the region currently occupied by the Okavango Delta and Caprivi (Eiseb, 2014). With respect to its karyotype, it is very similar to *M. coucha* (2n =

36, FN = 52-56) but clearly different from *M. natalensis* (2n = 32) (Granjon et al., 1997). Either four or five equally spaced pairs of nipples are present in this species (Granjon et al., 1997; Happold, 2013).

Taxonomic Issues

Chromosome number is 2n = 36, FN = 50 (Granjon et al., 1997).

Distribution

This species is known from just a few scattered localities in Angola, and along the Kavango River in the Caprivi Strip (Namibia) and the extreme north-western region of Botswana. It appears to be associated with swampy or marshy habitats, and has been recently trapped on small islands within the Kavango River (Eiseb, personal communication).

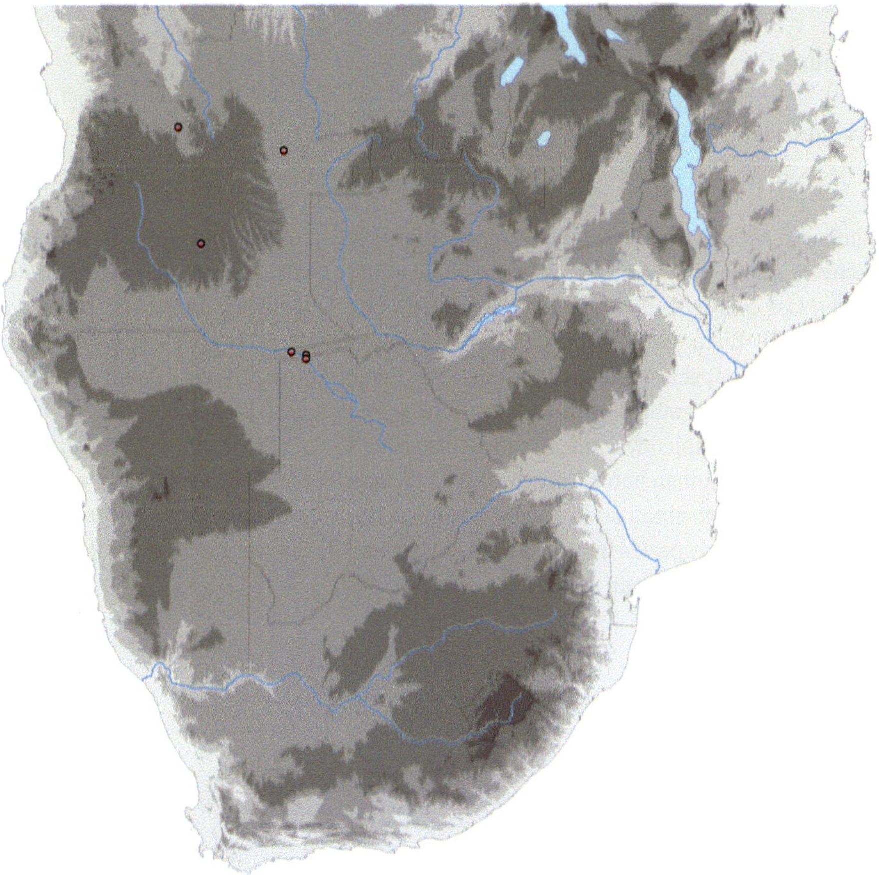

Figure 2.576 Distribution map

Measurements

External and cranial measurements (mm) and mass (g) for *Mastomys shortridgei*, sexes combined

	Mean	Min	Max	N
Mass	-	-	-	-
HB*	110	-	-	1
Tail*	96	-	-	1
HF*	24	-	-	1
Ear*	17.5	-	-	1
GLS*	-	-	-	-

*Roberts (1951), measurements of the holotype

Genus Account

Micaelamys Ellerman, 1941 (Lesser Veld Rats)

Diagnosis

They are similar to *Aethomys* in their rat-like appearance, reddish to brownish-grey dorsal pelage with distinctively coarse black guard hairs, pure white or grey ventral pelage, but are somewhat smaller-sized (Mass ca. 40-50 g; HB ca. 110-115 mm) and have a relatively longer tail in *M. namaquensis* (135% of HB) but not in *M. granti* (105% of HB); however, the latter species does not co-occur with any *Aethomys* species. The fleshy, larger hindfoot of *Rattus* (HF ca. 40mm) is a useful character to distinguish members of this genus from *Aethomys* and *Micaelamys* (HF < 30mm). Furthermore, *Rattus* is completely commensal whereas members of *Aethomys* and *Micaelamys* are very rarely found inside human habitations. Members of *Aethomys* and *Micaelamys* are terrestrial; the arboreal habit of *Grammomys* and *Thallomys* helps to distinguish individuals of these genera from *Micaelamys*. The two species in *Micaelamys* are exclusively Southern African in their distribution while *Aethomys* occurs more broadly.

Figure 2.577 Photograph of *Micaelamys namaquensis* (Enviro-Insight).

Cranial Characters

The skull is medium-large in size (27-35 mm in greatest length) and very similar to *Aethomys* and *Pelomys*, ovoid in shape, with poor to well-developed supraorbital crests and moderately robust zygomatic arches. The nasal bone is of average length and width, tapering at the tip. The anterior palatal foramina reach at least to M^1. The anterior surface of the zygomatic plate is vertical or very slightly tilted forwards and bluntly rounded dorsally. The bullae are relatively small (6-7 mm in greatest diameter).

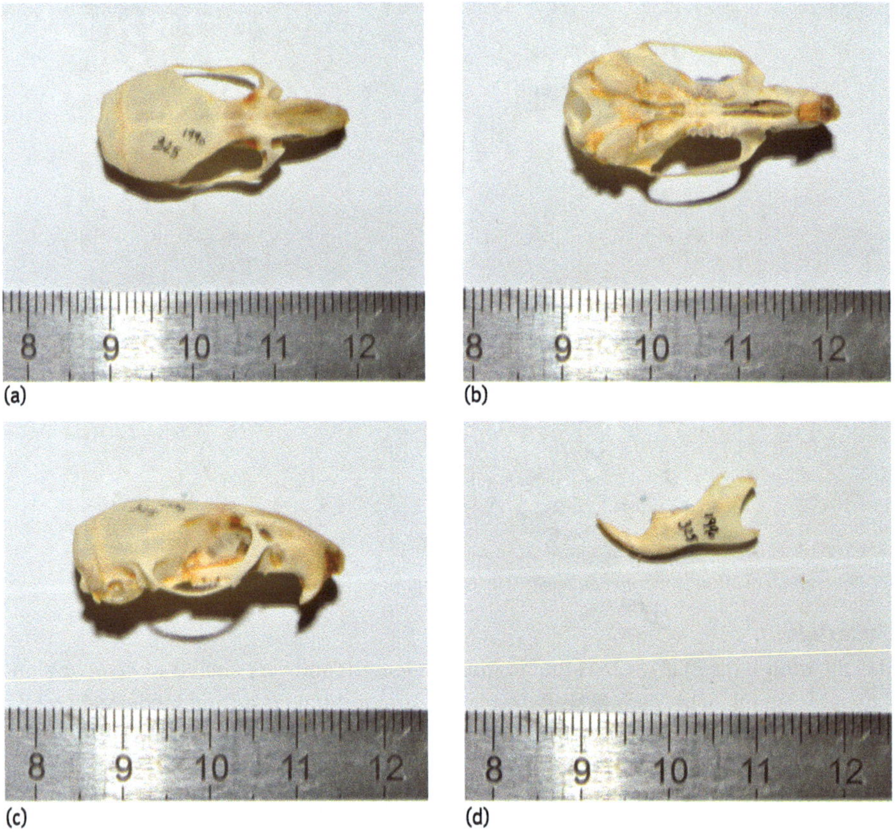

Figure 2.578 a-d Cranium of *Micaelamys namaquensis* (MNHN-ZM-MO 1990-325) (from top left to bottom right: dorsal view, ventral view, lateral view and lower jaw).

Dentition

The upper incisors are opisthodont to orthodont and ungrooved. The molars are moderately sized showing neither microdonty nor macrodonty (MTR/GSL ± 0.16-0.18, Table 10). Individual cusps are rounded and well-separated but not as distinctively large as in *Pelomys* and *Mylomys*. Rows of cusps on M^1 are only slightly distorted, with t1 situated behind t2 and t3 which form a high, round first lobe. The first upper molar is the longest molar in the toothrow. Molar M^3 lacks the t3 cusp (Group 3 in Figure 2.439). The first lower molar M_1 has three cusplets in the anterior row (as opposed to two in *Aethomys*). Other subtle differences in molar morphology between *Aethomys* and *Micaelamys* are pointed out by Misonne (1969).

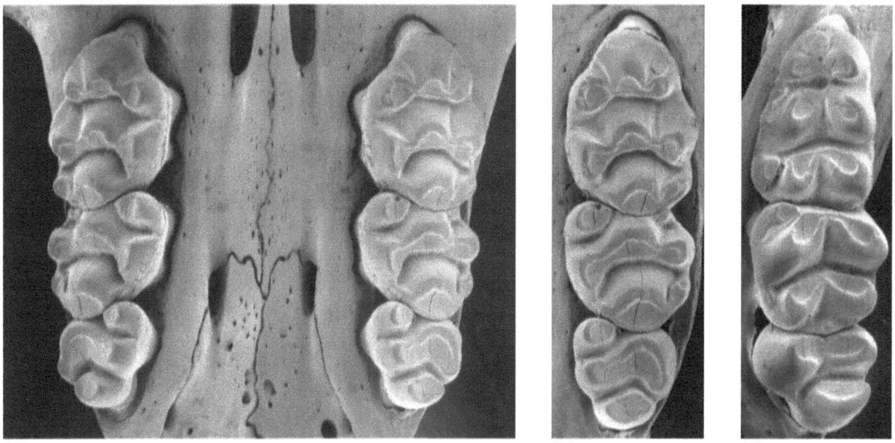

(a) (b) (c)

Figure 2.579 Upper cheek teeth of *Micaelamys namaquensis* (MNHN-ZM-MO 1964-49) (A: upper cheek teeth; B: upper left toothrow; C: lower left toothrow).

Habitat & Biogeography

The genus is associated with rocky outcrops in semi-arid to mesic savannas of Southern Africa, with *M. granti* restricted to South Africa south of the Gariep River.

Fossil Record

Pocock (1987) described *A.* cf *namaquensis* from Makapansgat , Swartkrans and Kromdraai, South Africa.

Biology & Ecology

They are nocturnal, terrestrial and the diet is omnivorous and granivorous (Kerley et al., 1990; Monadjem, 1997). In *A. namaquensis*, breeding is seasonal, occurring during the wet summer months (Muteka et al., 2006c).

Systematic Notes

Musser & Carleton (2005) placed *Aethomys* (including *Micaelamys*) in its own division. Chromosomal (Baker et al., 1988), morphological (Chimimba, 2005) and molecular (Ducroz et al., 2001; Russo, 2003; Lecompte et al., 2008) data attest to the paraphyly of *Aethomys* sensu lato and support recognition of *Micaelamys* as a distinct genus. Whilst *Micaelamys* clearly belongs to the Arvicanthini clade relationships with other genera in this clade were unresolved by the combined mitochondrial and nuclear sequence analysis of Lecompte et al. (2008). Schenk et al. (2013) confirmed its position in the Arvicanthini and found it close to *Grammomys*.

Annotated Species List

Two species are recognised:
Micaelamys granti (Wroughton, 1908)
Micaelamys namaquensis (A. Smith, 1834)

Key to Species

1	Tail relatively longer (ca. 135% of HB); 2n = 24; 3 pairs of nipples; ventral pelage varies considerably but is often pure white (hairs with white base)	*M. namaquensis*
	Tail relatively shorter (ca. 105% of HB); 2n = 32; 5 pairs of nipples; ventral pelage never pure white (hairs always with grey base)	*M. granti*

Species Accounts

Micaelamys granti (Wroughton, 1908)

Diagnosis

This species can be can be distinguished from *M. namaquensis* on morphological and chromosomal grounds (Visser & Robinson, 1986, 1987; Chimimba et al., 1999). In *M. namaquensis* the tail is relatively longer (ca. 135% of HB), the ventral pelage is often pure white, and there are only three pairs of nipples. In *M. granti* the relative tail length is ca. 105% (but can vary greatly from less than 100% to almost 120%), the ventral pelage is never pure white (usually grey or greyish), and there are five pairs of nipples.

Taxonomic Issues

Chromosome number is 2n =32 (Visser & Robinson, 1986).

Distribution

This species is endemic to South Africa, south of the Gariep River, where it occupies rocky outcrops in arid or semi-arid landscapes.

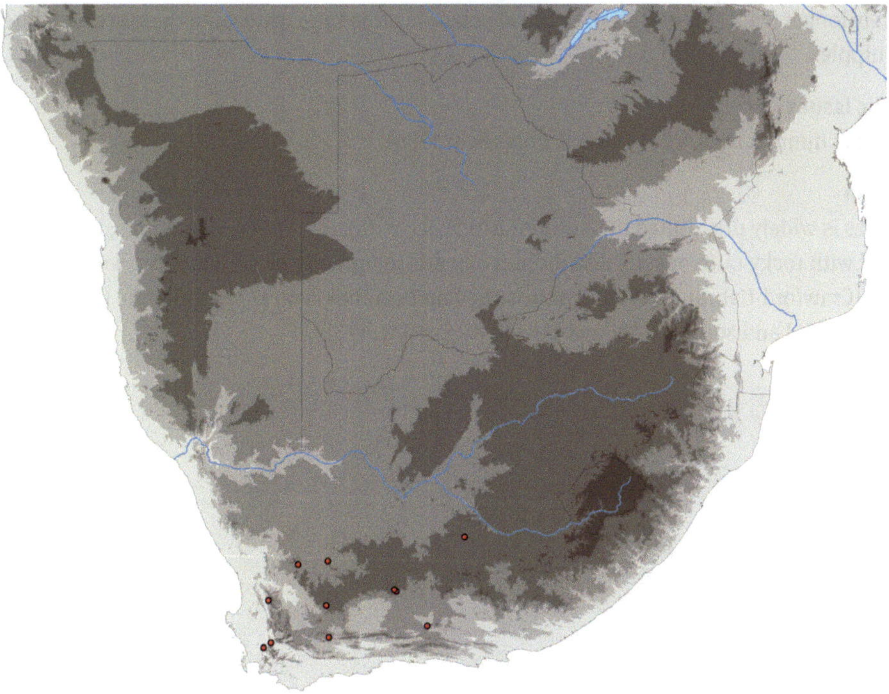

Figure 2.580 Distribution map

Measurements

External and cranial measurements (mm) and mass (g) for *Micaelamys granti*, sexes combined

	Mean	Min	Max	N
Mass	41.4	19	65	5
HB*	111.8	93	125	55
Tail*	116.8	96	138	55
HF*	24.1	22	26	55
Ear*	17.7	16	20	55
GLS*	30.0	28.0	31.9	86

*Chimimba et al. (1999)

Micaelamys namaquensis (A. Smith, 1834)

Diagnosis

This species can be can be distinguished from *M. granti* on morphological and chromosomal grounds (Visser & Robinson, 1986, 1987; Chimimba et al., 1999). In *M. granti* the relative tail length is ca. 105% (but can vary greatly from less than 100% to almost 120%), the ventral pelage is never pure white (usually grey or greyish), and there are five pairs of nipples. In *M. namaquensis* the tail is relatively longer (ca. 135% of HB), the ventral pelage is often pure white, and there are only three pairs of nipples.

Taxonomic Issues

Chromosome number is 2n = 24 (Visser & Robinson, 1986).

Distribution

This species is widely distributed in Southern Africa occurring in a variety of habitats, but usually associated with rocky outcrops. Its distribution extends marginally beyond into the western parts of Angola (Crawford-Cabral, 1998) and across the Zambezi into central Mozambique (Smithers & Labao Tello, 1976) and southern Malawi (Ansell & Dowsett, 1988).

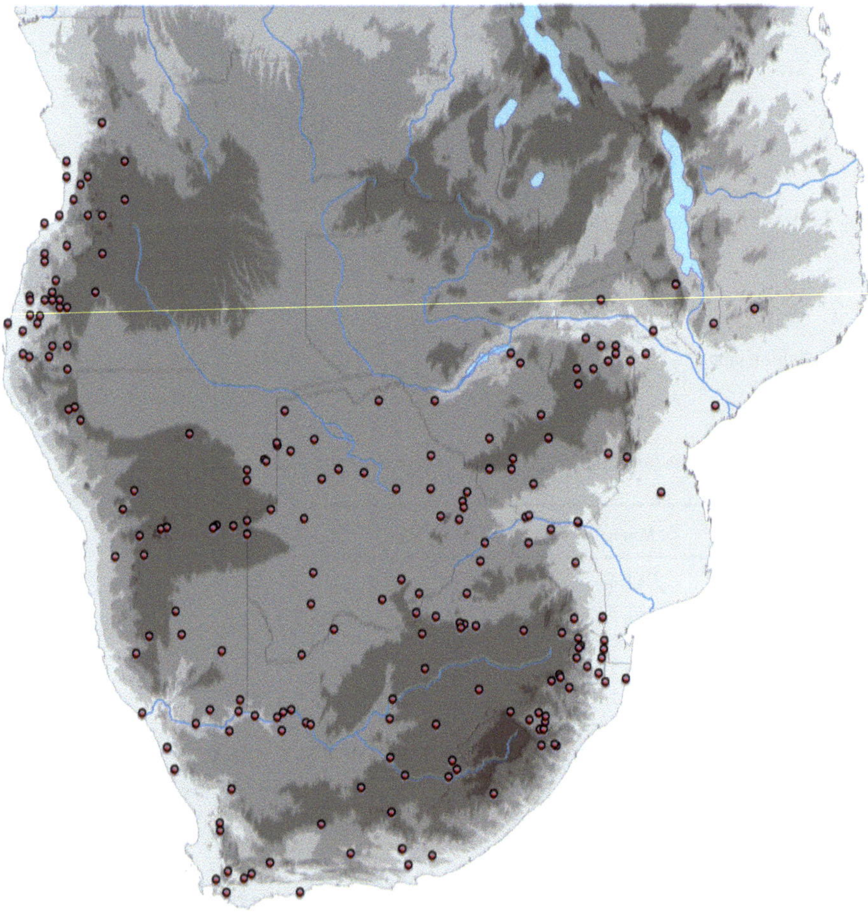

Figure 2.581 Distribution map

Measurements

External and cranial measurements (mm) and mass (g) for *Micaelamys namaquensis*, sexes combined

	Mean	Min	Max	N
Mass	47.8	33	89	20
HB*	113.4	80	147	2012
Tail*	153.7	107	197	2012
HF*	25.8	16	32	2012
Ear*	17.9	11	24	2007
GLS*	31.3	27.1	34.9	2730

*Chimimba et al. (1999)

Genus Account

Mus Linnaeus, 1758 (Old World Mice and Pygmy Mice)

Diagnosis

Externally, these rodents are most easily diagnosed on their very small to small body size (Mass 2-18 g; HB 37-97 mm in the subgenus *Nannomys* (Britton-Davidian et al., 2012); Mass 13-30 g; HB 72-92 mm in the subgenus *Mus* (Rosevear, 1969)) with naked tails subequal (in the subgenus *Mus*) or just shorter than head and body length (43-97 % in the subgenus *Nannomys*; Britton-Davidian et al., 2012). The dorsal pelage comprises short sleek fur which varies in colour from dark brownish-grey to more sandy in species from semi-arid habitats. The ventral colour is greyish to white. In *M. imberbis* there is a dark mid-dorsal stripe. From cranio-dental characters, the genus is distinguished by its small cranium (GSL < 25 mm), incisors which are notched on the posterior surface (except in *M. imberbis*), a tiny, much reduced M^3 and a prominent masseteric knob located just anterior to the zygomatic plate. A mastoid strut (and subsquamosal or stapedial foramen) is present.

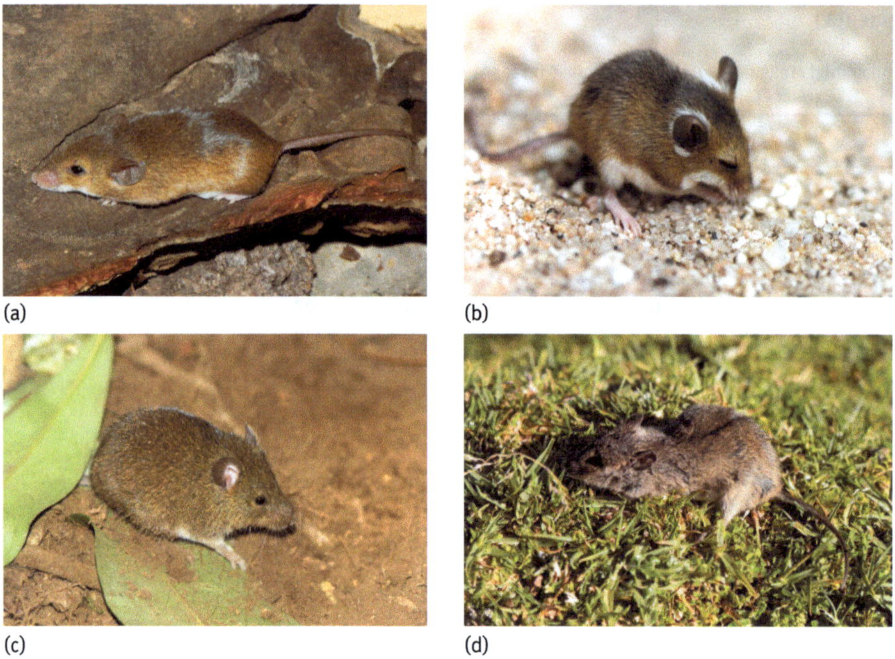

Figure 2.582 a-d Photographs of *Mus*, from top left to bottom right: *M. (Nannomys)* cf *musculoides* (James Harvey), *M. (Nannomys) tenellus* (Bill Stanley), *M. (Nannomys) triton* (Bill Stanley) and *M. (Muriculus) imberbis* (Marian Polak) (bottom).

Cranial Characters

The cranium is small-sized (GSL 16-25 mm) and delicate, narrow and flattish in shape, with narrow tapering nasals, gracile zygomatic arches and a slightly convex to vertical anterior edge of the zygomatic plate with bluntly rounded dorsal edge. This can be variable, e.g. in *M. setulosus* the anterior face is backward sloping and has a sharp dorsal angle. The braincase is smoothly contoured without crests or with faint supraorbital ridges. The palatal foramina reach the second row of cusps in M^1. In the case of *M. imberbis*, the coronoid process of the mandible is conspicuously low.

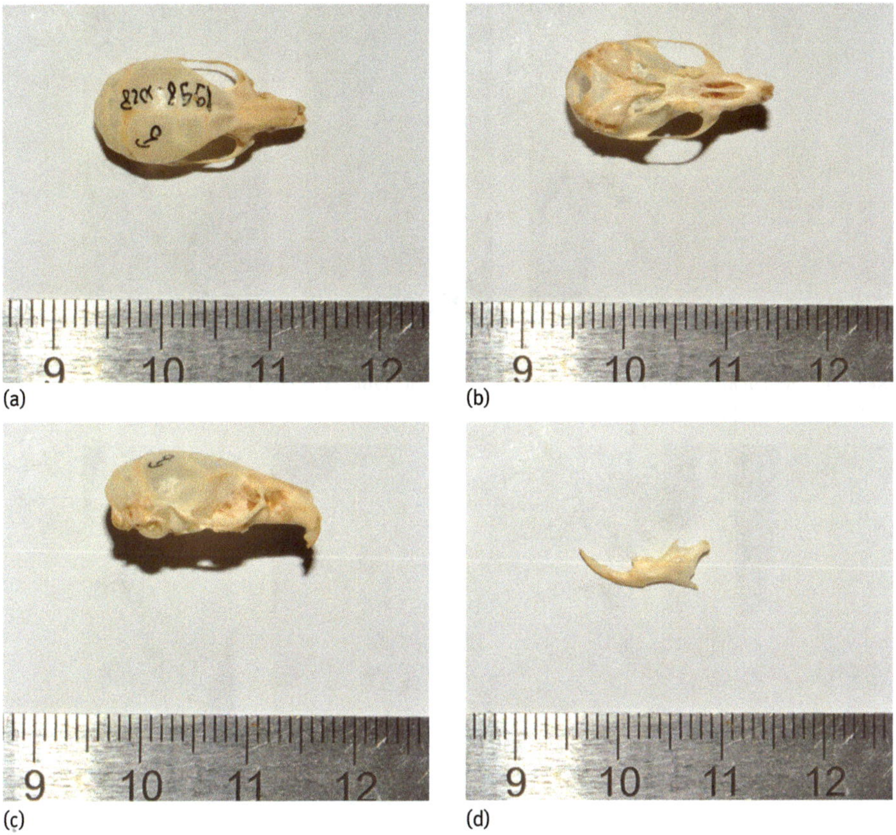

Figure 2.583 a-d Cranium of *Mus setulosus* (MNHN-ZM-MO 1998-1028) (from top left to bottom right: dorsal view, ventral view, lateral view and lower jaw).

Dentition

The upper incisors vary from opisthodont to highly pro-odont (e.g., in *M. imberbis*) and are ungrooved and characteristically notched on the posterior surface (except for *M. imberbis*). The molar row is of average size (MTR/GSL ± 0.18, Table 10). The cusps on M^1 and M^2 are extremely distorted, with t1 positioned far posterior of the anterior lobe (which is more or less elongated and tilted posteriorly), in line with t5 and t6, and t4 is similarly displaced posteriorly. Molars $M^3$and M_3 are highly reduced, M^3 having a particularly reduced cusp configuration (Group 9 in Figure 2.439) although *M. imberbis* has a different cusp configuration (Group 3 of Figure 2.439) more similar to *Praomys*. The antero-lingual cusplet on M_1 and M_2 is smaller than the labial.

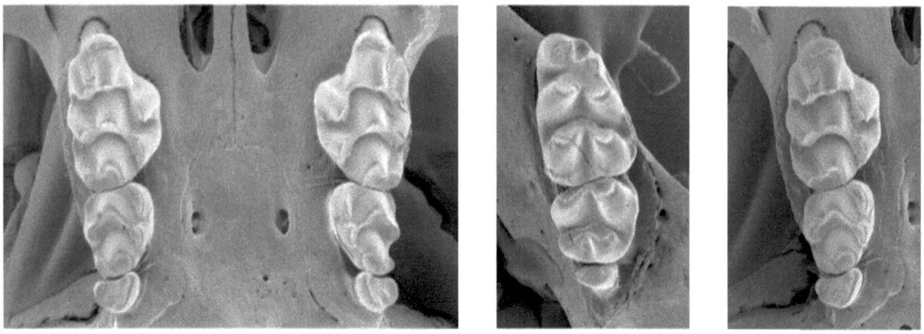

(a) (b) (c)

Figure 2.584 Cheek teeth of *Mus (Nannomys) indutus* (MNHN-ZM-MO 1964-23) (A: upper cheek teeth; B: lower right toothrow; C: upper right toothrow).

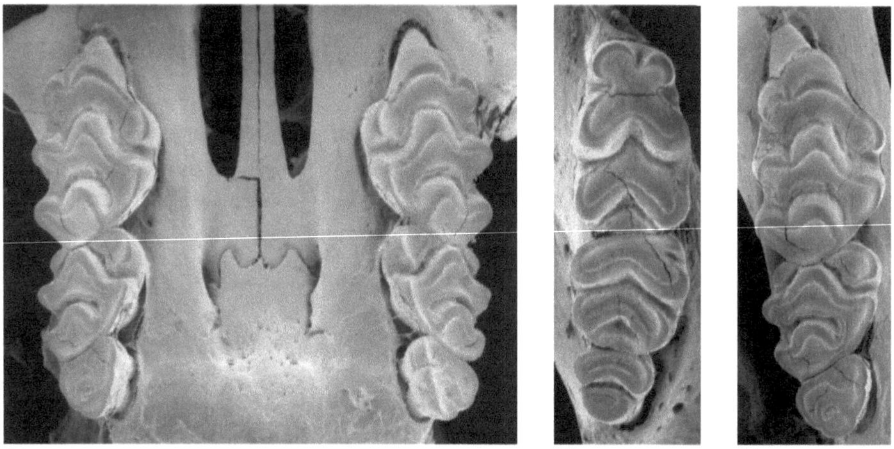

(a) (b) (c)

Figure 2.585 Upper cheek teeth of *Mus (Muriculus) imberbis* (MNHN-ZM-MO 1972-229) (A: upper cheek teeth; B: lower right toothrow; C: upper right toothrow).

Habitat & Biogeography

They are known to be catholic in their habitat preferences, occupying a range of savanna, forest semi-arid, Afromontane, Fynbos, cultivated habitats and houses, and are very widespread from Senegal and Mauritiania in the west to Eritrea, Ethiopia and Somalia in the east and southwards to Cape Town, South Africa. The house mouse (*M. musculus*) is commensal and is a major pest causing devastating losses to stored grains.

Fossil Record

Fossil records are known from the Early Pliocene to Middle Pleistocene of North Africa and the Early Pliocene to Late Pleistocene of Southern and Central Africa (Winkler et al., 2010). The oldest *Mus* remains in Africa come from the Lissafsa deposits of North Africa (Geraads, 1998b). The earliest *Mus* in Sub-Saharan Africa is found in Aramis, Ethiopia at 4.4 MYA (Louchart et al., 2009).
Mus sp. Omo Shungura member B (Ethiopia) (Wesselman, 1984).
Mus petteri Olduvai Bed I (Jaeger, 1976)
Mus sp. Botswana (Pickford & Mein, 1988)

Mus sp. East Turkana (Black & Krishtalka, 1986)
Mus sp. Hadar (Ethiopia) (Sabatier, 1982; Reed, 2011),
Mus sp. Plio-Pleistocene caves of RSA (Pocock, 1987; Sénégas, 2000)

Biology & Ecology

They are nocturnal and terrestrial and the diet is granivorous, herbivorous and omnivorous. To take two examples, a study of stomach contents of *M. minutoides* in Swaziland showed 44% vegetable matter, 40% seeds and 16% arthropods (Monadjem, 1997) while in a similar study of stomach contents of *M. triton* in Malawi, 81% of stomachs included insects (including beetles and ants), 28% contained white vegetable matter (seeds) and 8% contained earthworms (Hanney, 1965). *Mus musculus* is highly omnivorous and will eat anything edible in human environments. Breeding is closely tied to rainfall and is usually seasonal, ceasing in the late dry season, but in some cases it is continuous (Happold, 2013). Mean litter size (or embryo number) is usually around 4 with a range of 1-8 (Happold, 2013). In *M. musculus*, litter size can vary from 4 to 7, with 5 to 10 litters a year, a gestation of 18-20 days and age of sexual maturity of 3 months, resulting in population explosions (or plagues, e.g. in Australia) when conditions are suitable (Rosevear, 1969).

Systematic Notes

Recent molecular studies have resolved the monophyly and relationships among the four subgenera of *Mus* which are thought to have diverged from each other from an Asian ancestor at similar times (6.7-7.8 MYA) (Chevret et al., 2005; Veyrunes et al., 2006; Lecompte et al., 2008). According to molecular data, Lecompte et al. (2008) placed *Mus* into its own tribe Murini together with *Muriculus* (which we here include in *Mus* based on Meheretu et al, 2014). There is still much taxonomic confusion about the validity and limits of some of the species of the *Nannomys* subgenus (see Musser & Carleton, 2005; Veyrunes et al., 2004; 2005; Kan Kouassi, et al., 2008; Mboumba et al., 2011; Britton-Davidian et al., 2012; Lamb et al., 2014). We follow Britton-Davidian et al. (2012) in not recognising *M. orangiae* (listed in Musser & Carleton, 2005; Happold, 2013) based on the fact that it cannot be diagnosed on either molecular or morphological grounds; specimens identified as *M. orangiae* in the past probably belong to *M. indutus* and/or *M. minutoides* (see Britton-Davidian et al., 2012 for further details).

Annotated Species List

Nineteen Sub-Saharan African species are recognised in two subgenera, comprising one species formerly included in *Muriculus*, 18 species of *Nannomys* and one species of *Mus*.
Formerly included in *Muriculus* (now in subgenus *Mus*):
Mus imberbis (Rüppell, 1842)
Nannomys subgenus:
Mus baoulei (Vermeiren & Verheyen, 1980)
Mus bufo (Thomas, 1906)
Mus callewaerti (Thomas, 1925)
Mus goundae F. Petter & Genest, 1970
Mus haussa (Thomas & Hinton, 1920)
Mus indutus (Thomas, 1910)
Mus mahomet Rhoads, 1896
Mus mattheyi F. Petter, 1969

Mus minutoides Smith, 1834
*Mus musculoides*Temminck, 1853
Mus neavei (Thomas, 1910)
Mus oubanguii F. Petter & Genest, 1970
Mus setulosus Peters, 1876
Mus setzeri F. Petter, 1978
Mus sorella (Thomas, 1909)
Mus tenellus (Thomas, 1903)
Mus triton (Thomas, 1909)
Mus subgenus:
Mus musculus Linnaeus, 1758

Key to Species

It is not possible at this stage to provide a complete diagnostic key to all the species in the genus; instead we provide a key to the two described African subgenera, and certain clearly defined species groups and species.

1	Dorsal pelage with dark mid-dorsal stripe, upper incisors lacking notch on posterior edge; mandible with reduced coronoid process	*Mus (Mus) imberbis*
	Dorsal pelage without stripes, upper incisors with posterior notch; mandible with normal coronoid process	2
2	Size larger, anterior surface of zygomatic plate vertical with masseteric knob at the lower anterior corner of the plate, and M^3 with two laminae visible	*Mus (Mus) musculus*
	Size smaller, convex anterior surface of zygomatic plate with masseteric knob at the lower centre of the plate, and M^3 reduced, without clear laminae	3 (*Nannomys*)
3	Upper incisors pro-odont	4 (*M. sorella* group)
	Upper incisors orthodont or opisthodont	6
4	Dorsal pelage dark greyish-brown colour	*M. sorella*
	Dorsal pelage brighter rich tawny to reddish-brown colour	5
5	Restricted to Gounda River, Central African Republic	*M. goundae*
	Restricted to Oubangui River, Central African Republic	*M. oubanguii*
	Occurring in Tanzania, DRC, Zimbabwe, Zambia, Mozambique, South Africa	*M. neavei*
6	Size larger, greatest skull length > 20 mm	7
	Size smaller, greatest skull length < 20 mm	10
7	Size larger, skull length > 23 mm, restricted to Angola	*M. callewaerti*
	Size small, skull length < 23 mm, not restricted to Angola	8
8	Belly white	*M. setulosus*
	Belly grey	9
9	Very long tail, ca 95% of head and body length	*M. bufo*
	Shorter tail, ca. 70% of head and body length	*M. triton*
10	Pure white coloration of ventrum extended dorsally to include part of the rump posteriorly as well as much of the muzzle posterior to the ear	*M. setzeri*
	White coloration not as extensive*	*M. baoulei, M. haussa, M. indutus, M. mahomet, M. mattheyi, M. minutoides, M. musculoides, M. tenellus*

*Some of these small-sized species can be distinguished on molecular, karyotypic or distributional grounds (see species accounts). Note that although *M. mahomet* is here classified as "small" (skull length < 20 mm), it is somewhat intermediate (18 – 21 mm); the species can be distinguished from other *Nannomys* on a combination of its buff-coloured belly and its range which is restricted to higher elevations > 1500m in Ethiopia.

Species Accounts

Mus baoulei (Vermeiren & Verheyen, 1980)

Diagnosis

This species can be readily distinguished on molecular (Kan Kouassi et al, 2008) and chromosomal grounds (Happold, 2013), and differs from other West African *Mus* species by having very long anterior palatal foramina (3.65-4.70 mm) in addition to other morphological features which include red markings close to the ear openings, a short tail (50-65% of HB), orthodont upper incisors, a narrow rostrum and an isolated distal t1 on M^1.

Taxonomic Issues

Chromosome number 2n = 20/21, FN = 32 (Kan Koussi et al., 2008; Happold, 2013)

Distribution

This species is known from a few scattered localities in West Africa (Kan Kouassi et al., 2008) where it has been collected from savanna habitats.

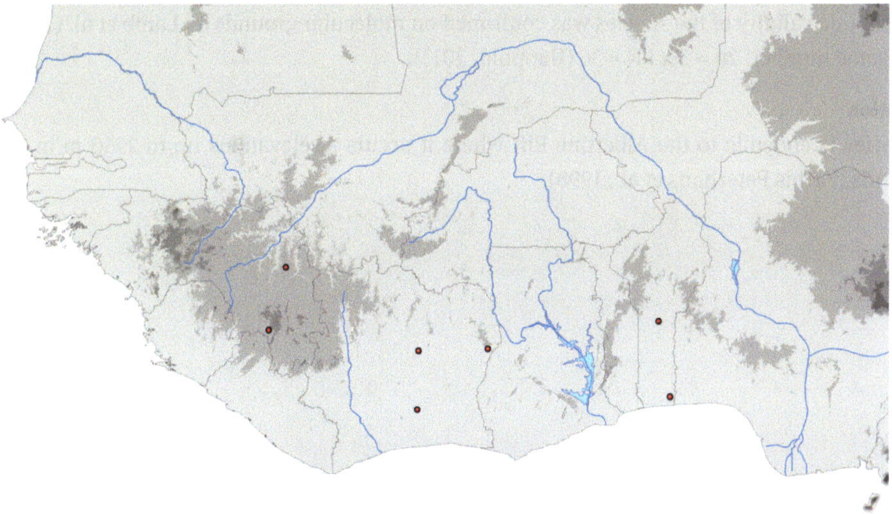

Figure 2.586 Distribution map

Measurements

External and cranial measurements (mm) and mass (g) for *Mus baoulei*, sexes combined

	Mean	Min	Max	N
Mass	6.8	6	8	4
HB	65.9	56	70	7
Tail	42.6	38	45	7
HF	13.1	12	15	7
Ear	10.6	10	12	4
GLS	18.7	18.0	20.3	9

Mus bufo (Thomas, 1906)

Diagnosis

The species can be distinguished by its very long tail (ca. 97% of HB) which easily distinguishes it from co-occurring *M. triton* with a shorter tail (ca. 70% of HB). Although Britton-Davidian et al. (2012) considered that *M. bufo* and *M. triton* could be separated on size ("small" and "large", respectively), our data (below) and those of Happold (2013) show considerable overlap in both body and cranial size between these two species.

Taxonomic Issues

The taxonomic validity of the species was confirmed on molecular grounds by Lamb et al. (2014). Chromosome number: 2n = 36; FN = 36 (Happold, 2013).

Distribution

This species is endemic to the Albertine Rift where it occurs at elevations up to 2960 m in the Ruwenzoris (Kerbis Peterhans et al., 1998).

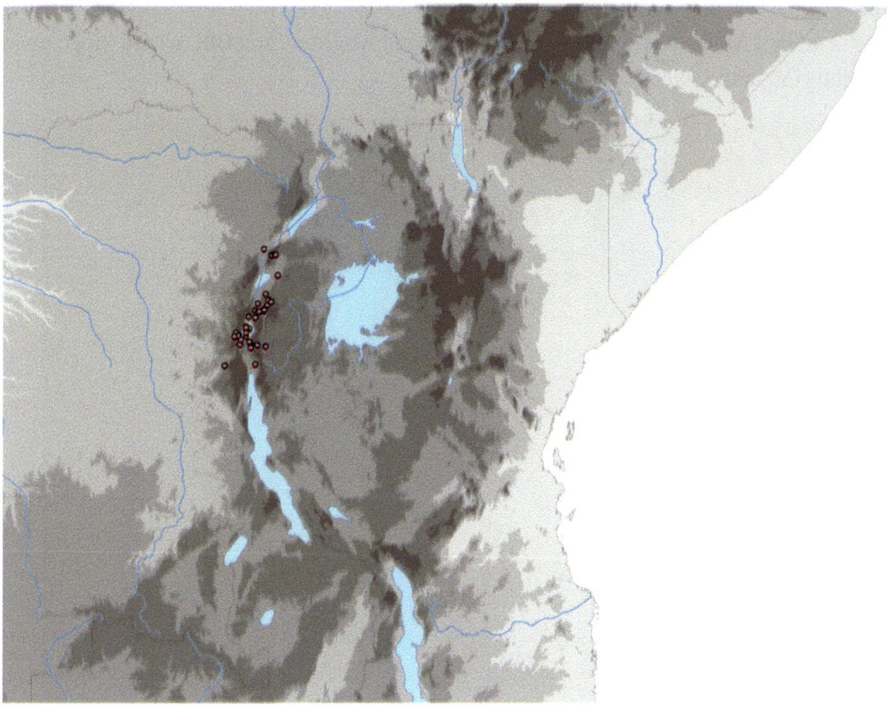

Figure 2.587 Distribution map

Measurements

External and cranial measurements (mm) and mass (g) for *Mus bufo*, sexes combined

	Mean	Min	Max	N
Mass	-	-	-	-
HB	71.2	69	77	5
Tail	63.4	60	66	5
HF	17.1	16	18	5
Ear	11.6	10	12	5
GLS	21.1	20.3	21.6	6

Mus callewaerti (Thomas, 1925)

Diagnosis

This is the largest species of the genus in Africa (HB > 84 mm; GLS > 23 mm), only narrowly overlapping with co-occurring smaller *M. triton*. The tail is extremely short (ca. 60% of head and body length, cf. 70% in *M. triton*).

Taxonomic Issues

Misonne (1965) details additional cranial and dental differences between *M. triton* and *M. callewaerti*.

Distribution

This species is known from a few records from Angola and neighbouring DRC where it occurs in savanna habitats (Misonne, 1965).

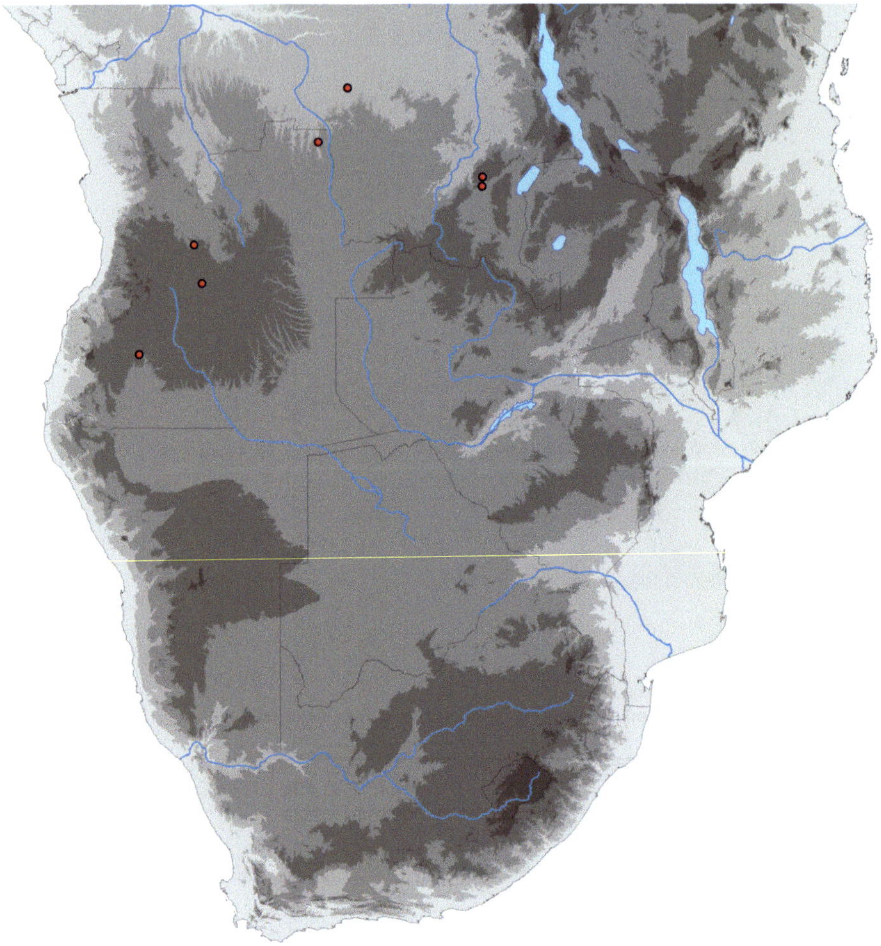

Figure 2.588 Distribution map

Measurements

External and cranial measurements (mm) and mass (g) for *Mus callewaerti*, sexes combined

	Mean	Min	Max	N
Mass*	16.2	14	18	4
HB*	84.2	80	89	6
Tail*	50.4	43	57	7
HF*	16.7	16.5	17	3
Ear*	12.1	11	14	6
GLS*	24.3	23.4	25.0	3

*Hill & Carter (1941) and Lamb et al. (2014) combined

Mus goundae F. Petter & Genest, 1970

Diagnosis

This species has been defined on chromosomal grounds (2n = 16-19; FN = 30; Petter, 1981). Known only from the type series which is small-sized (HB ca. 55-62 mm; GLS < 20 mm) and has a rich ochraceous-brown dorsal colour with orange-rufous flanks and pure white belly and relatively short tail (55% of HB).

Taxonomic Issues

It is considered part of the *M. sorella* group characterised by a long snout, greatly reduced molars and pro-odont incisors (Petter, 1981).

Distribution

This species is only known from the type locality on the banks of the Gounda River, Central African Republic (Petter & Genest, 1970; Petter, 1981).

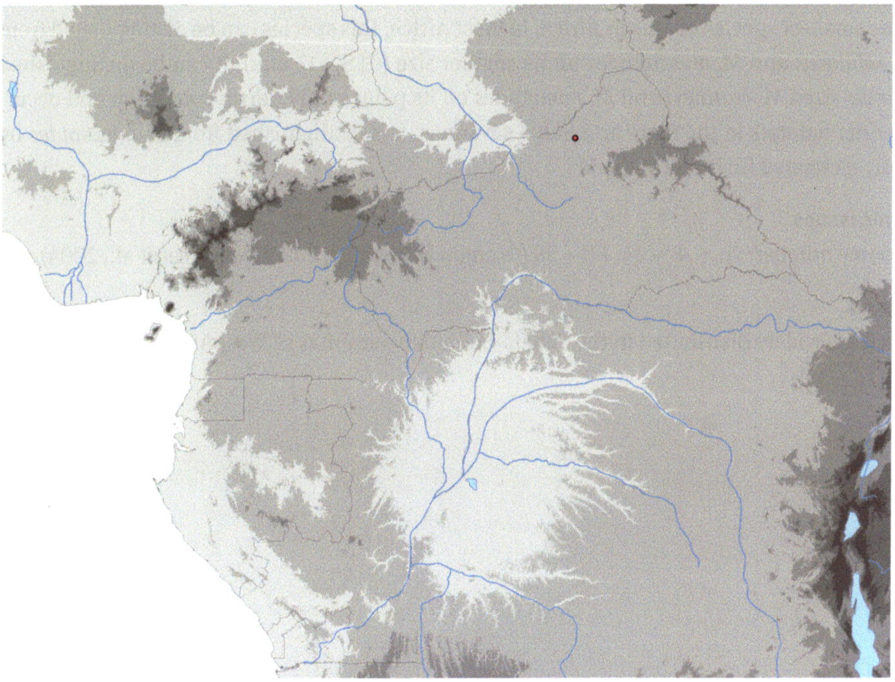

Figure 2.589 Distribution map

Measurements

External and cranial measurements (mm) and mass (g) for *Mus goundae*, sexes combined

	Mean	Min	Max	N
Mass	-	-	-	-
HB	58.3	55	62	4
Tail	30.8	30	32	4
HF	13.1	12	14	4
Ear	11.3	10	13	4
GLS	18.6	17.8	20.0	4

Mus haussa (Thomas & Hinton, 1920)

Diagnosis

This is the smallest species of *Mus* in Africa. In West Africa, this species can be distinguished from both *M. setulosus* and *M. musculoides* on its smaller size (GLS < 17 mm). It can be distinguished from similar-sized *M. mattheyi* and *M. minutoides* on its paler sandy dorsal coloration and occurrence in drier habitats in the more arid Sahel savannas. It is distinguished from *M. minutoides* by the greatly elongated first lobe of the M^1.

Taxonomic Issues

Chromosome number: 2n = 28 – 34; FN = 38 (Granjon & Dobigny, 2003; Veyrunes et al., 2004).

Distribution

This species is widely distributed in the Sahelo-Sudanian savannas of West Africa.

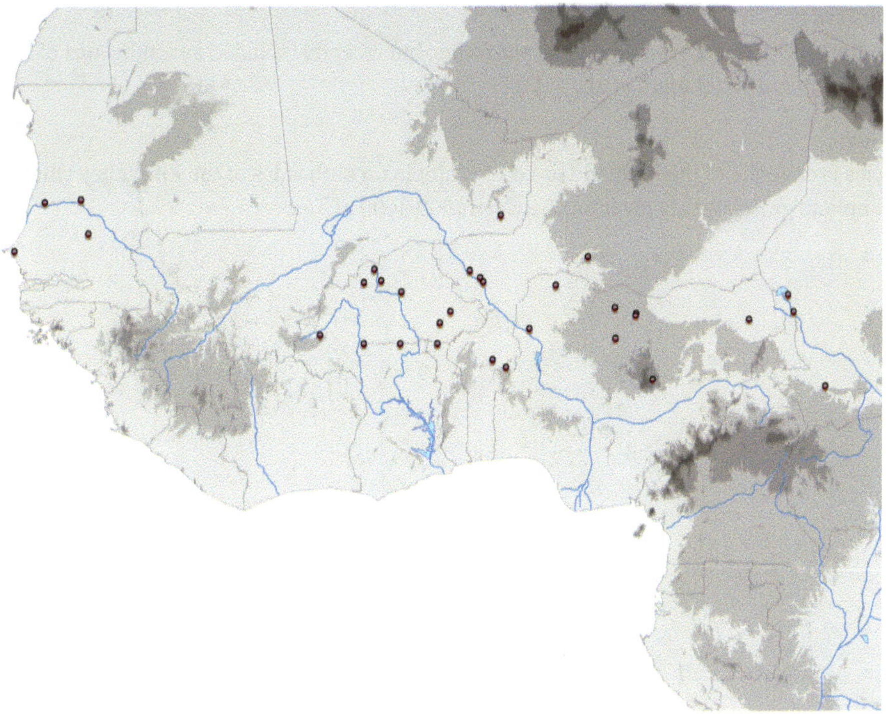

Figure 2.590 Distribution map

Measurements

External and cranial measurements (mm) and mass (g) for *Mus haussa*, sexes combined

	Mean	Min	Max	N
Mass	-	-	-	-
HB	49.9	41	58	8
Tail	37.6	31	44	8
HF	12.0	11	13	8
Ear	8.9	8	10	8
GLS	16.5	15.3	17.2	8

Mus imberbis (Rüppell, 1842)

Diagnosis

This Ethiopian endemic is easily distinguished from other *Mus* by the mid-dorsal black stripe, as well the pro-odont incisors which lack the posterior notch found in other *Mus* species, and a distinctive mandible which has a very low coronoid process at the same height as the articular condyle.

Taxonomic Issues

Until recently included in a separate genus *Muriculus*, but recently included as a subgenus of *Mus* on molecular grounds (Meheretu et al., 2014).

Distribution

This species is endemic to the highland areas of Ethiopia on both sides of the Rift Valley where it occurs in upland grassland at elevations between 1900-3400 m.

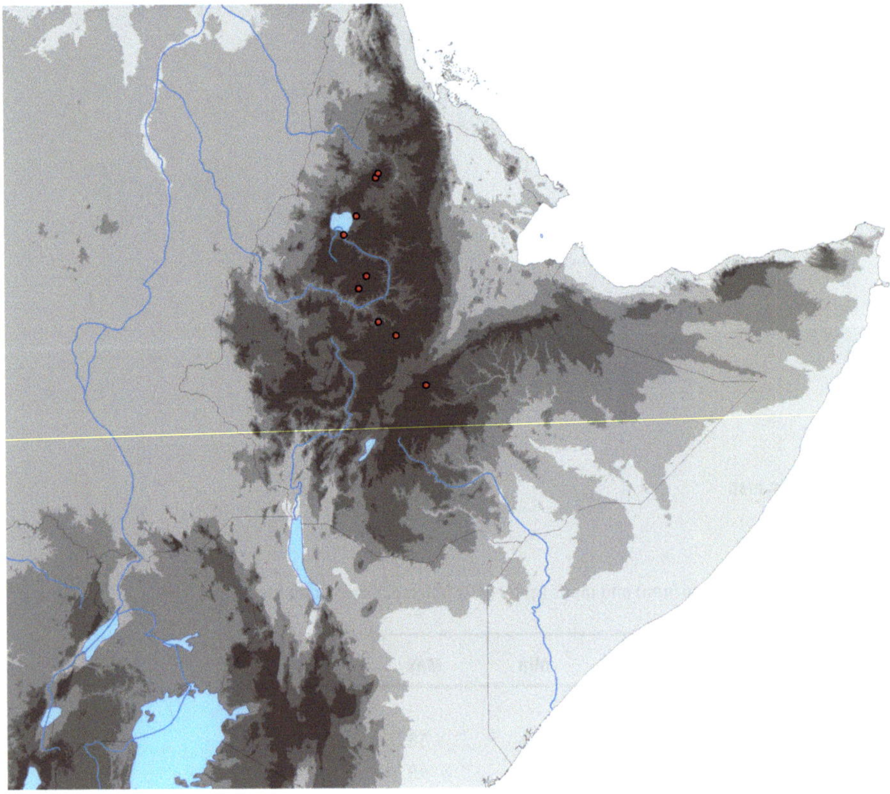

Figure 2.591 Distribution map

Measurements

External and cranial measurements (mm) and mass (g) for *Mus imberbis*, sexes combined

	Mean	Min	Max	N
Mass	16	-	-	1
HB	80	-	-	1
Tail	46	-	-	1
HF	26	-	-	1
Ear	13	-	-	1
GLS	21.9	-	-	1

Mus indutus (Thomas, 1910)

Diagnosis

They can be distinguished chromosomally from *M. minutoides* by the absence of the diagnostic ex-autosome (X.1) transolocation, and genetically by a large cytochrome b genetic distance (12-16%; Lamb et al., 2014), but morphologically they are not as clearly defined. Generally, *M. indutus* is identified morphologically from *M. minutoides* by having a pinkish-buff dorsal pelage speckled with black and white markings at the base of the ears, but specimens from KwaZulu-Natal and Namibia having these characteristics were shown by Lamb et al. (2014) to represent undescribed species within the *M. minutoides* clade.

Taxonomic Issues

Chromosome number: 2n = 36; FN = 36-37 (Veyrunes et al., 2004).

Distribution

Due to confusion with *M. minutoides* in the past (see Lamb et al., 2014), the distribution of this species has not been resolved. It has been confirmed, based on molecular grounds, from three widely separated localities in Southern Africa (Botswana and South Africa), suggesting that it has a wide distribution in the region. In spite of having samples from north-eastern Namibia and southern Angola within the previously suggested range of *M. indutus* (Happold, 2013), Lamb et al. (2014) did not record *M. indutus* from these regions based on molecular sequences.

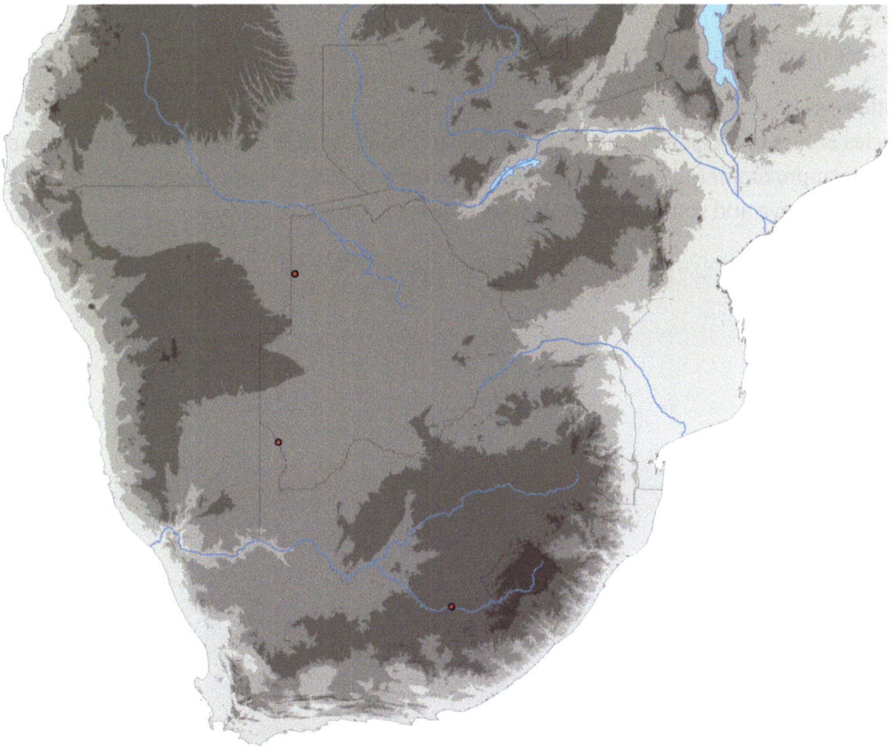

Figure 2.592 Distribution map

Measurements

External and cranial measurements (mm) and mass (g) for *Mus indutus*, sexes combined

	Mean	Min	Max	N
Mass*	5.4	3	8	85
HB*	53.2	45	65	12
Tail*	42	30	52	90
HF*	14	13	16	90
Ear*	11	8	12	90
GLS**	17.7	16.5	19.5	11

*Smithers (1971); **Roberts (1951)

Mus mahomet Rhoads, 1896

Diagnosis

The species is not very clearly defined morphologically, being slightly larger in size than *M. minutoides* (GSL ca. 18-21 mm cf. 17-20 mm) with a distinct dorsal pelage colour (dark greyish-brown) and a greyish-white (as opposed to pure white) belly (Yalden & Largen, 1992; Musser & Carleton, 2005).

Taxonomic Issues

Chromosome number: 2n = 36; FN = 36 (Musser & Carleton, 2005).

Distribution

This species occurs in forest-edge, scrub and grassland habitats in the highlands of Ethiopia and Eritrea from approximately 1500-3200 m (Yalden & Largen, 1992). It apparently also occurs in south-western Kenya and south-western Uganda (Musser & Carleton, 2005).

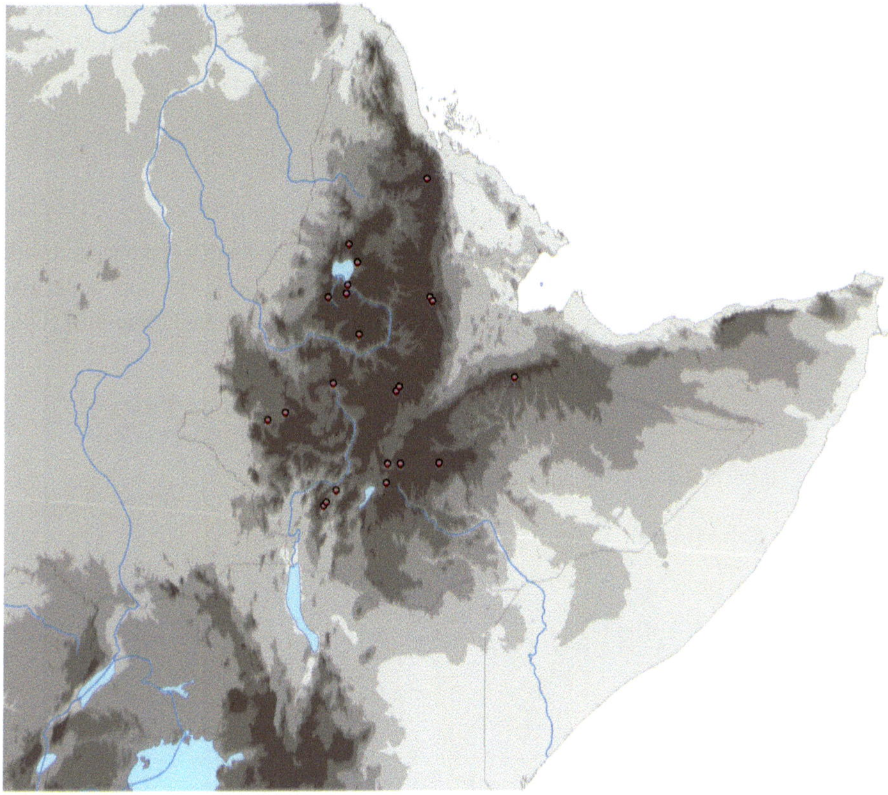

Figure 2.593 Distribution map

Measurements

External and cranial measurements (mm) and mass (g) for *Mus mahomet*, sexes combined

	Mean	Min	Max	N
Mass*	10.3	6	13	17
HB*	67	63	73	21
Tail*	53.5	46	60	21
HF*	15	14	15	21
Ear*	11.8	9	13	21
CBL*	19.9	18.4	21.0	21

*Happold (2013)

Mus mattheyi F. Petter, 1969

Diagnosis

Musser & Carleton (2005) considered that the diagnostic traits used in the description of *M. mattheyi* by Petter (1969) overlapped with those of *M. haussa*, making it difficult to distinguish the two species on morphological grounds. In general, *M. haussa* has a paler sandy dorsal coloration, is slightly smaller in size and occupies more arid habitats. However, the two species can clearly be distinguished on chromosomal and molecular grounds (Veyrunes et al., 2004; Britton-Davidian et al., 2012; Lamb et al., 2014).

Taxonomic Issues

Chromosomal number: 2n = 36; FN = 36 (Veyrunes et al., 2004).

Distribution

This species is widely distribution at scattered locations in West Africa where it is associated with savanna habitats. The distribution below was based only on voucher specimens from molecular and chromosomal studies where the species could be unequivocably identified (Veyrunes et al., 2004; Kan Kouassi et al., 2008; Britton-Davidian, 2012) and may be wider.

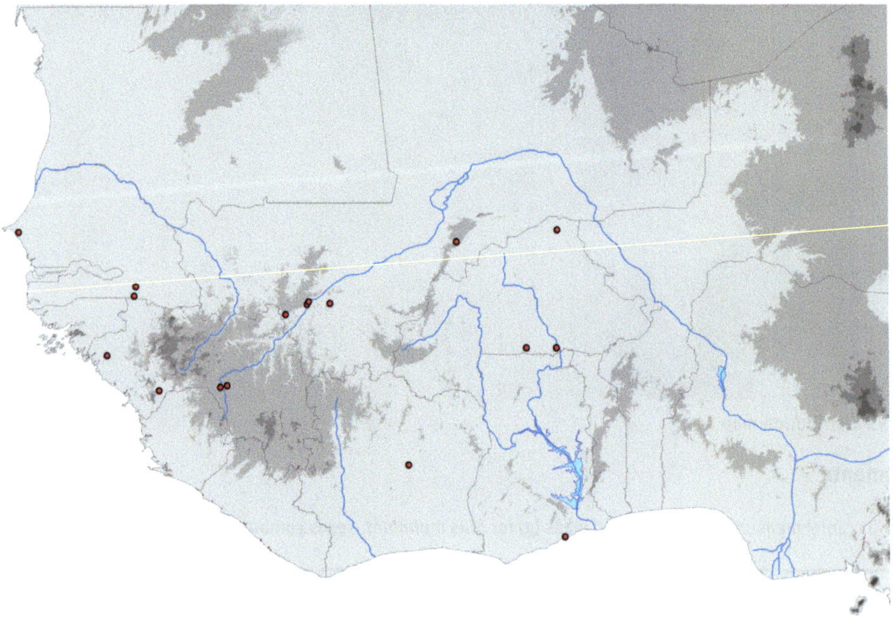

Figure 2.594 Distribution map

Measurements

External and cranial measurements (mm) and mass (g) for *Mus mattheyi*, sexes combined

	Mean	Min	Max	N
Mass	6.5	6	7	2
HB	52.9	45	61	16
Tail	37.6	32	41	14
HF	12.1	11	14	16
Ear	8.8	7	10	16
GLS	17.4	15.3	19.7	14

Mus minutoides Smith, 1834

Diagnosis

While it can be diagnosed on chromosomal and molecular grounds it remains poorly resolved morphologically and can easily be confused with other small *Mus* (*Nannomys*) species such as *M. musculoides, M. haussa, M. mattheyi, M. indutus* and *M. mahomet*. In general it has a short tail (75% of HB), a pure white belly, brownish-buff to brownish-orange dorsal pelage and upper incisors than are orthodont or opisthodont but never pro-odont (distinguishing it from the *M. sorella* complex).

Taxonomic Issues

Chromosome number: 2n = 18-36, FN = 34. There is a diagnostic Robertsonian translocation between the X chromosome and autosomal pair 1 (Veyrunes et al., 2004). Based on molecular and chromosomal evidence, it is clearly a species complex comprising three distinct geographical clades (from West, West-central and Southern and East Africa) within which several undescribed cryptic species probably occur (Lamb et al., 2014). Populations from the type locality of Cape Town having 2n = 18 comprise a molecular clade which could be described as *M. minutoides* sensu stricto, probably specifically distinct from another distinct clade from the north of South Africa having 2n = 34 (Veyrunes et al., 2010). The status of East African populations requires further study as only three chromosomally and genetically vouchered specimens (from Kenya, Tanzania and Ethiopia) can be unambiguously assigned to the *M. minutoides* complex (from the South and East African clade). Corti et al. (2005) described four novel karyotypes of *Mus* from East Africa, two of which (from Nairobi, Kenya, and several Zambian localities) were assigned to the *M. minutoides* complex based on inferred autosome-sex chromosome translocations. A specimen from Berega, Tanzania tentatively assigned to *M. minutoides* by Lamb et al. (2014) was shown to differ considerably in cytochrome b genetic distance (9-18%) from both the *M. minutoides* and *M. musculoides* complexes.

Distribution

This species is widely distributed across Sub-Saharan Africa where it occurs in a wide variety of savanna and grassland habitats. Based on chromosomal and molecular evidence (Veyrunes et al., 2004, 2005; Kan Kouassi et al., 2008; Mboumba et al., 2011; Brittan-Davidian 2012; Lamb et al., 2014; see vouchered map below), its distribution is more extensive than previously thought (e.g. compare the two distributions below). For example, it occurs in Central and West Africa (where it is sympatric with *M. musculoides*), as well as in the arid parts of Southern Africa (in sympatry with *M. indutus*). As mentioned above, its distribution in East Africa remains poorly resolved with only very few vouchered specimens currently available. The red dots on the map are based on molecular identification; the white dots are based on morphological identification only.

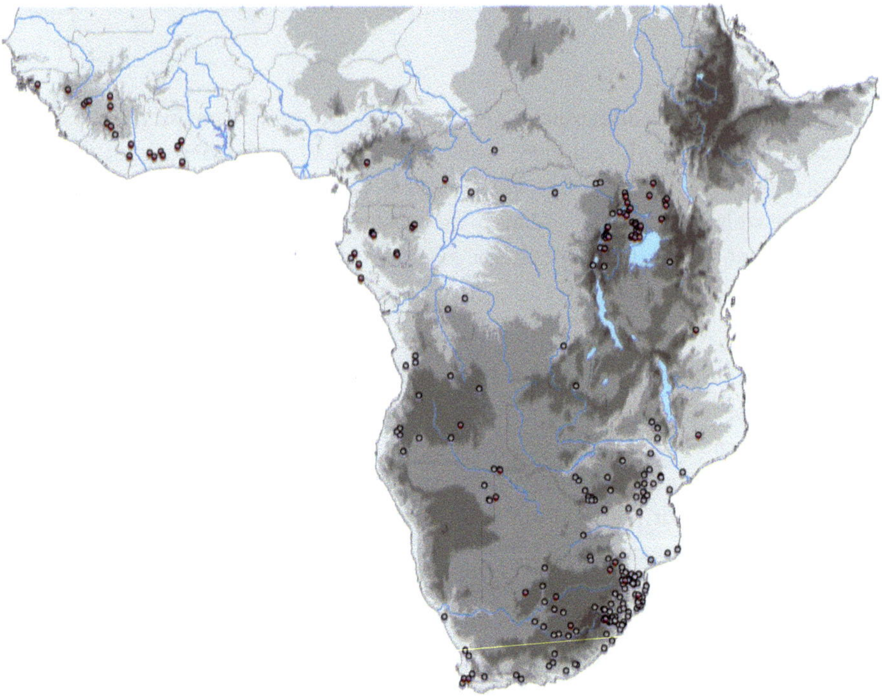

Figure 2.595 Distribution map

Measurements

External and cranial measurements (mm) and mass (g) for *Mus minutoides*, sexes combined

	Mean	Min	Max	N
Mass	6.8	3	12	42
HB	56.5	40	72	73
Tail	42.7	27	57	72
HF	13.1	9	16	47
Ear	9.3	6	12	55
GLS	18.2	17.0	19.2	5

Mus musculoides Temminck, 1853

Diagnosis

Although it can be defined unambiguously from chromosomal and molecular markers, it cannot be diagnosed morphologically from similar-sized *Mus* (*Nannomys*) such as *M. minutoides*, with which it has been shown to be sympatric in West Africa (Kan Kouassi et al., 2008; Brittan-Davidian 2012; Lamb et al., 2014).

Taxonomic Issues

Musser & Carleton (2005) considered *M. musculoides* to be a synonym within the *M. minutoides* complex whilst others (e.g. Happold 2013) considered all West African populations of the *M. minutoides* complex to be *M. musculoides*. However, it has been shown to be chromosomally and genetically distinct from *M. minutoides* (Veyrunes et al., 2004, 2005; Kan Kouassi et al., 2008). Chromosome number: 2n = 18-19; FN = 36. There is a diagnostic Robertsonian translocation between the X chromosome and autosomal pair 7.

Distribution

Due to confusion with *M. minutoides* in the past (see above) the distributional limits of this species have not been fully resolved. However, based on molecular and chromosomal evidence (Veyrunes et al., 2004, 2005; Kan Kouassi et al., 2008; Mboumba et al., 2011; Brittan-Davidian 2012; Lamb et al., 2014; map below), it is restricted to West Africa where it probably has an extensive range (occurring sympatrically with the similar *M. minutoides*).

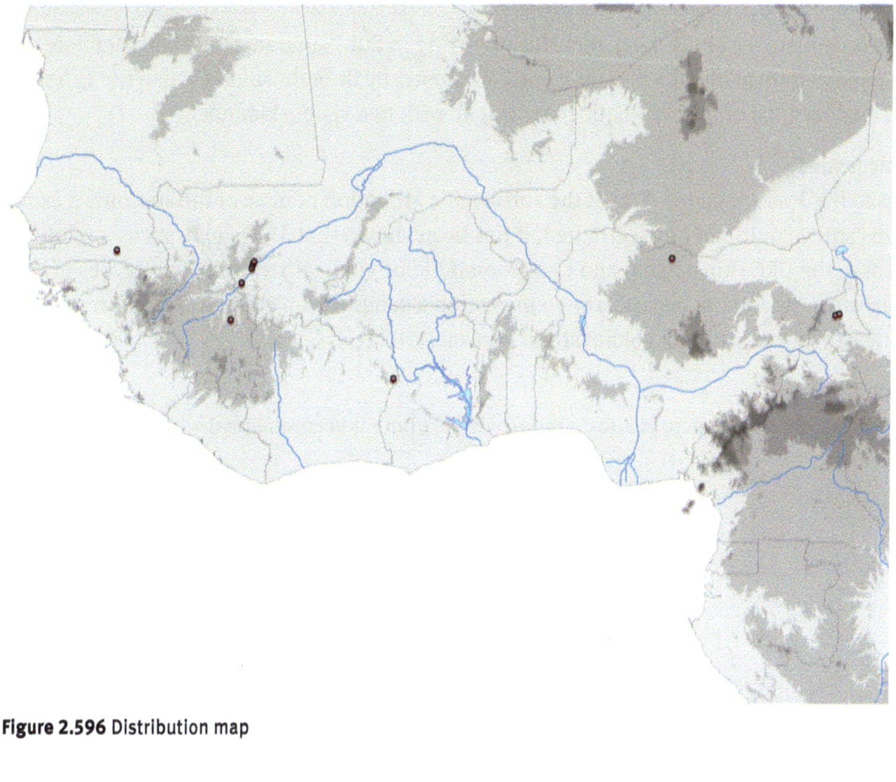

Figure 2.596 Distribution map

Measurements

External and cranial measurements (mm) and mass (g) for *Mus musculoides*, sexes combined

	Mean	Min	Max	N
Mass	6.8	6	9	6
HB	63.8	58	75	6
Tail	49.4	47	53	6
HF	13.0	12	14	5
Ear	8.5	7	10	4
GLS	17.3	16.7	18.1	4

Mus musculus Linnaeus, 1758

Diagnosis

This is the only member of the subgenus *Mus* occurring in Sub-Saharan Africa and therefore can be distinguished from all members of subgenus *Nannomys* by the subgenus key above (e.g. having a vertical leading edge of the zygomatic plate and M^3 with two visible laminae).

Taxonomic Issues

It belongs to the *Mus* subgenus. Whilst the subspecific affiliation of most of the specimens examined in this study could not be ascertained, it has been shown that *M. m. domesticus* haplotypes are present in the DRC (Kinshasha) and Cameroon (Kaleme et al., 2011) whilst haplotypes of both *M. m. domesticus* and *M. m. musculus* were present in a sample from Durban, South Africa, indicating at least two independent colonisations (Downs, 2012).

Distribution

This species occurs widely across Sub-Saharan Africa where it is commensal with humans.

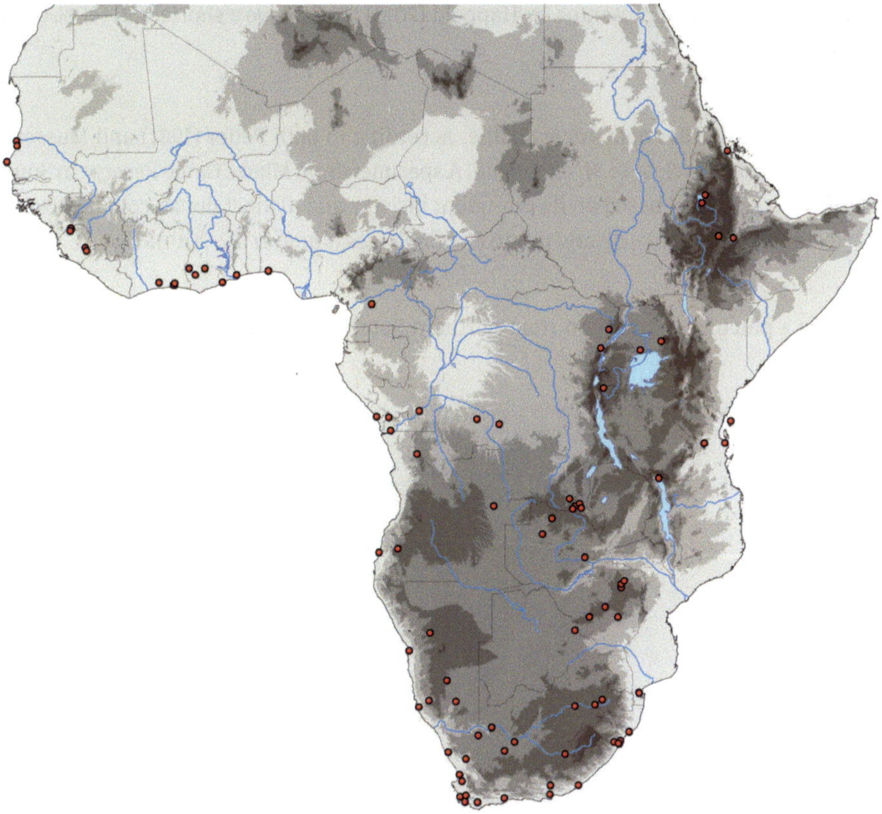

Figure 2.597 Distribution map

Measurements

External and cranial measurements (mm) and mass (g) for *Mus musculus*, sexes combined

	Mean	Min	Max	N
Mass	14.5	10	20	6
HB	77.4	61	109	14
Tail	78.6	62	88	14
HF	17.3	15	20	12
Ear	12.7	12	15	14
GLS	20.5	18.2	21.8	5

Mus neavei (Thomas, 1910)

Diagnosis

This species can be distinguished morphologically from *M. sorella* from its brighter "rich tawny" dorsal pelage compared with a dark grey pelage in *M. sorella*, its smaller size and shorter maxillary toothrow (3.0-3.2 mm in *M. neavei*, n = 7, cf. 3.2-3.7 mm, n = 9, in *M. sorella*) (Petter, 1981; Musser &

Carleton, 2005). Based on data presented by Happold (2013) it may have a shorter tail (40% of HB cf. 65% of HB in *M. sorella*).

Taxonomic Issues

No chromosomal or molecular data are available. According to both Petter (1981) and Musser & Carleton (2005), *M. neavei* is close to *M. oubanguii*. A specimen from Mkuze Game Reserve in South Africa, identified as *M. neavei* based on its pro-odont incisors and distinct dorsal colour was very similar in cytochrome b sequence divergence to *M. minutoides* (Lamb et al., 2014).

Distribution

This species has a patchily distribution from northern South Africa northward to southern and eastern Tanzania.

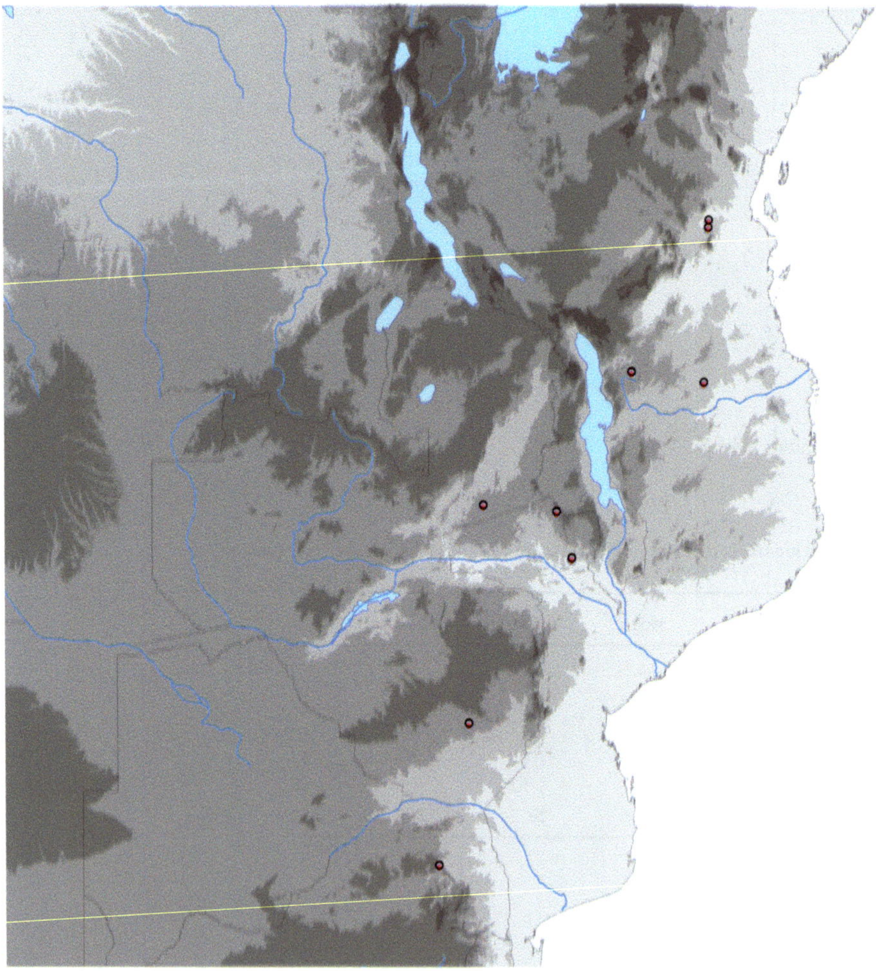

Figure 2.598 Distribution map

Measurements

External and cranial measurements (mm) and mass (g) for *Mus neavei*, sexes combined

	Mean	Min	Max	N
Mass	-	-	-	-
HB*	88.8	58	106	5
Tail*	38.4	33	48	5
HF*	13	12	14	6
Ear*	11	10	12	6
GLS*	18.5	18	18.9	6

*Happold (2013)

Mus oubanguii F. Petter & Genest, 1970

Diagnosis

Apparently diagnosable from other members of the *M. sorella* group on its reddish-brown dorsal pelage coloration and the large, blackish-coloured and slightly pointed ears (Petter, 1981).

Taxonomic Issues

It is a member of the *M. sorella* group having pro-odont incisors. Chromosome number: 2n = 28; FN variable (Petter, 1981).

Distribution

This species is known from just three localities in the Central African Republic, where it has been recorded from savanna habitats (Genest-Villard, 1973; Petter, 1981).

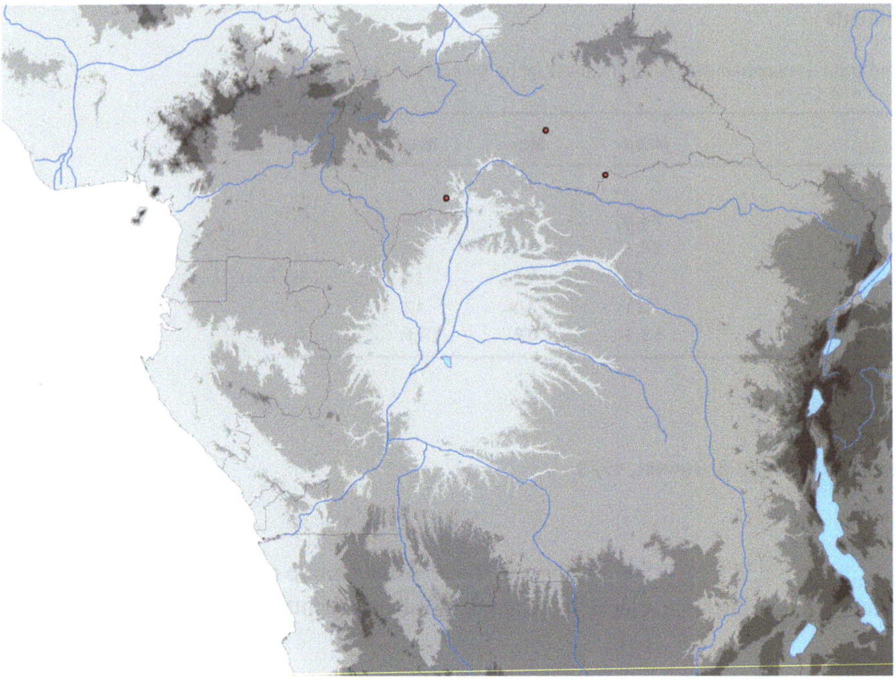

Figure 2.599 Distribution map

Measurements

External and cranial measurements (mm) and mass (g) for *Mus oubanguii*, sexes combined

	Mean	Min	Max	N
Mass	-	-	-	-
HB	62.9	50	69	8
Tail	33.8	26	41	6
HF	13.8	13	14	8
Ear	11.6	9	13	8
GLS	19.5	17.1	20.6	8

Mus setulosus Peters, 1876

Diagnosis

This is the largest West and Central African species of the subgenus *Nannomys*, where it can be distinguished by having GSL > 20 mm, a maxillary toothrow length > 3.5 mm and an additional small anterio-external cusp in M^2 (Rosevear, 1969). The dorsal pelage colour is a "dull" blackish-brown, not as "bright" as in *M. minutoides* and *M. musculoides*.

Taxonomic Issues

Lamb et al. (2014) suggested that *M. setulosus* is a species complex.

Distribution

This species is widely but patchily recorded from West and Central Africa (Kan Kouassi et al., 2008; Mboumba et al., 2011; Britton-Davidian et al., 2012) where it occurs in forested habitats. Only records based on sequenced specimens have been included, except for the Ethiopian records which are based on morphological identification (Yalden et al., 1976).

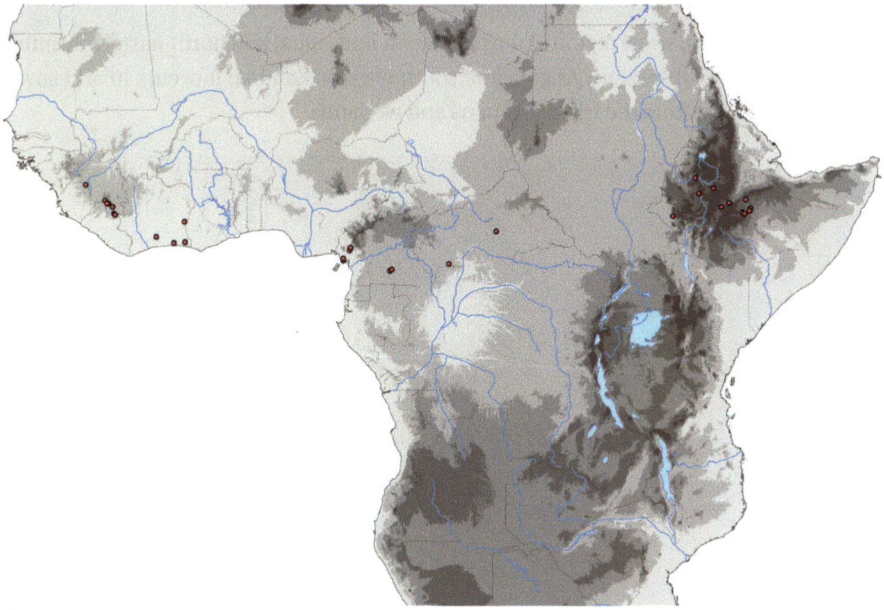

Figure 2.600 Distribution map

Measurements

External and cranial measurements (mm) and mass (g) for *Mus setulosus*, sexes combined

	Mean	Min	Max	N
Mass	13.7	8	23	16
HB	72.3	52	84	32
Tail	53.4	43	63	29
HF	15.4	13	17	31
Ear	11.0	8	14	31
GLS	20.9	19.9	22.4	27

Mus setzeri F. Petter, 1978

Diagnosis

Distinguishable from sympatrically-occurring *M. indutus* mostly on the basis of its large ears (Ear > 13 mm), "pinkish-clay" (pale buffy) and less grizzled dorsal pelage coloration and shorter tail (60% versus 80% of HB). In addition, the white underfur colour extends further up the flanks in

M. setzeri and continues on to the upper parts of the rump and across the muzzle, posterior to the nose (Skinner & Chimimba, 2005).

Taxonomic Issues

Chromosome number not known.

Distribution

This species is endemic to a narrow region in north-western Botswana, and north-eastern Namibia; possibly extending into south-eastern Angola and south-western Zambia. It occurs in arid savannas and has been collected from the fringes of pans and wetlands.

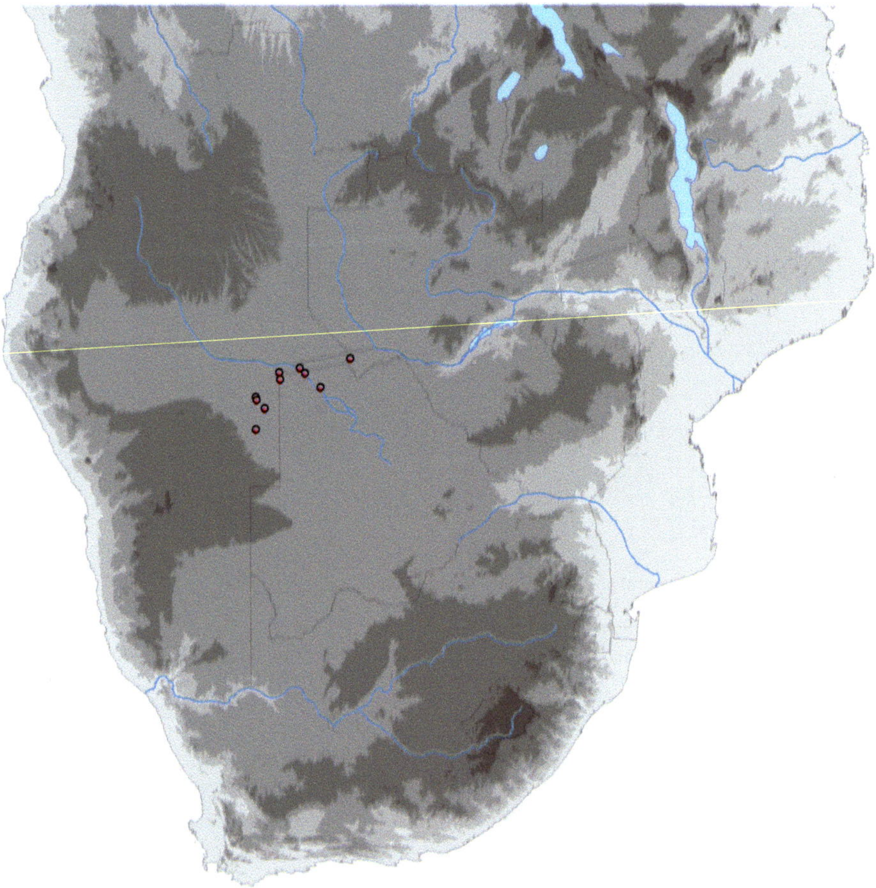

Figure 2.601 Distribution map

Measurements

External and cranial measurements (mm) and mass (g) for *Mus setzeri*, sexes combined

	Mean	Min	Max	N
Mass*	6.8	5	9	11
HB*	54			9
Tail*	36	31	48	9
HF*	14	13	15	11
Ear*	13.6	13	14	11
GLS	18.6	17.9	19.3	2

*Smithers (1971); HB estimated from total length and tail length

Mus sorella (Thomas, 1909)

Diagnosis

Thomas described this species with the following characteristics: "Skull long, unusually slender, especially in the muzzle; incisors more thrown forward than usual; braincase narrow, smooth and rounded. Palatal foramina long, reaching past the antero-internal cusp of M^1". The pro-odont incisors distinguish members of the *sorella* group from others of the subgenus *Nannomys*. However, *M. sorella* can only be distinguished from other members of the *sorella* group (*M. goundae*, *M. neavei* and *M. oubanguii*) on its duller dark greyish-brown dorsal pelage (dorsal coloration of other members is brighter).

Taxonomic Issues

The taxonomy of this species was reviewed by Petter (1981). The relationship between members of the *M. sorella* group is unresolved and requires a comprehensive systematic revision (Musser & Carleton, 2005). Chromosome number not known.

Distribution

This species has been recorded patchily from tropical savannas from Angola and Zambia in the south, northward to East Africa (Petter, 1981; Crawford-Cabral, 1998)

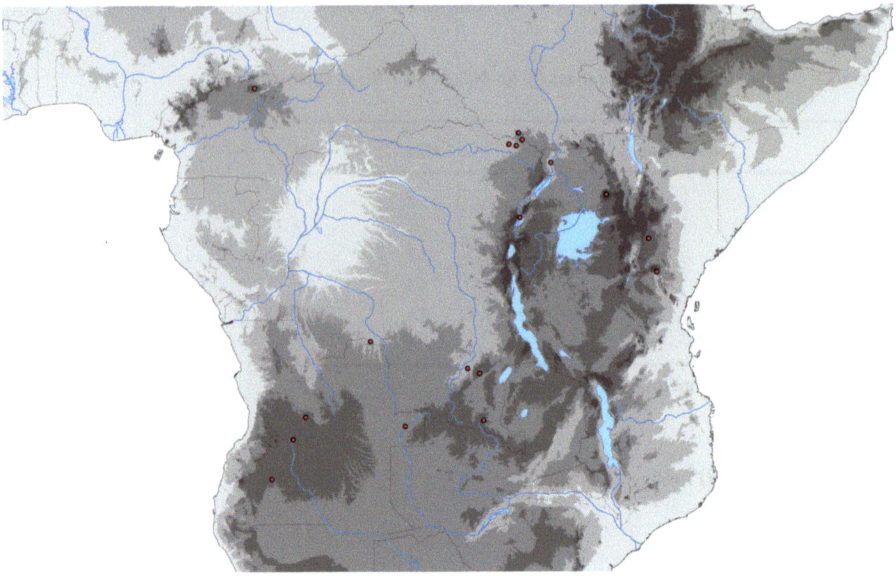

Figure 2.602 Distribution map

Measurements

External and cranial measurements (mm) and mass (g) for *Mus sorella*, sexes combined

	Mean	Min	Max	N
Mass	-	-	-	-
HB*	59.9	51	73	22
Tail*	39.8	34	46	21
HF*	13.1	12	14	22
Ear*	10.8	10	12	22
GLS*	19.7	17.8	21.0	26

*Happold (2013)

Mus tenellus (Thomas, 1903)

Diagnosis

Distinguished by its restricted distribution, grassy steppe habitat, small size, pale sandy-coloured dorsal pelage and extensive patch of white postauricular hair (more than any other *Mus*), sometimes forming a white ring around the base of the ear.

Taxonomic Issues

Chromosome number not known.

Distribution

This species has been recorded widely in Ethiopia, extending patchily south into Kenya, Tanzania, South Sudan and Somalia where it inhabits savannas at elevations 600-2000 m (Petter,

1972; Yalden et al., 1976; Musser & Carleton, 2005). Records from Ethiopian should be treated with caution as Musser & Carleton (2005) have indicated, based on critical examination of holotypes and museum series, that most published Ethiopian records of *M. tenellus* are in fact *M. minutoides*.

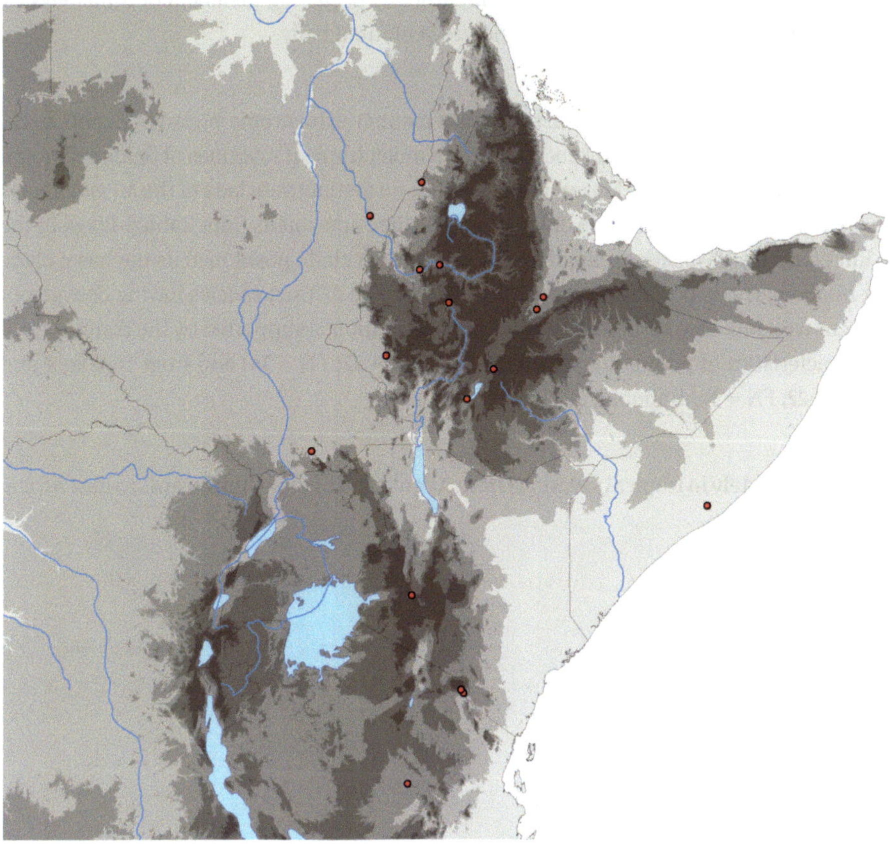

Figure 2.603 Distribution map

Measurements

External and cranial measurements (mm) and mass (g) for *Mus tenellus*, sexes combined

	Mean	Min	Max	N
Mass	-	-	-	-
HB*	53.8	44	62	8
Tail*	38.4	34	43	8
HF*	12.6	12	14	8
Ear*	8.6	8	10	8
GLS*	16.1	15.5	16.5	6

*Happold (2013)

Mus triton (Thomas, 1909)

Diagnosis

The species can be easily distinguished from sympatrically-occurring members of *Nannomys* by the combination of larger size (GSL > 20 mm) and grey-coloured belly.

Taxonomic Issues

It is a species complex comprising at least two very distinct cytochrome b clades (Lamb et al., 2014). One clade of large, grey-bellied specimens from moist tropical savannas at Selous National Park in Tanzania and Mt Namuli in Mozambique comprise a distinct subclade of the *M. minutoides* sensu lato clade whereas another clade of large, grey-bellied specimens from Kahuzi-Biega in the Albertine Rift comprise the sister-group of *M. callewaerti* in a clade positioned as the basal clade to all other *Nannomys*. It is possible that two phylogenetically distinct species having convergent similarity occur west and east of the Rift Valley. This supposition is supported by the distinct chromosomal characteristics of populations from the DRC (2n = 32; FN = 34) and from Tanzania and Malawi (2n = 20-22; FN = 34; Happold, 2013).

Distribution

This species occurs widely in tropical savannas from Angola and Zambia northwards to East Africa.

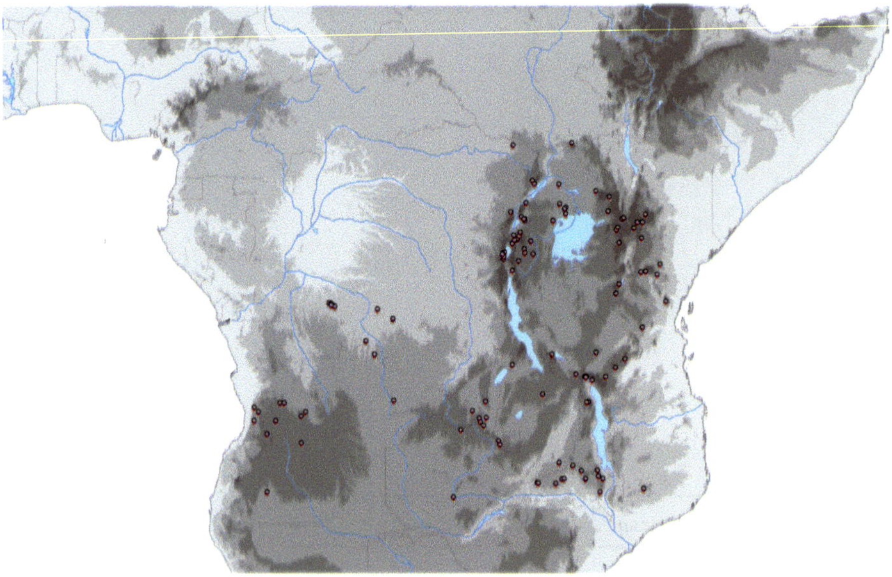

Figure 2.604 Distribution map

Measurements

External and cranial measurements (mm) and mass (g) for *Mus triton*, sexes combined

	Mean	Min	Max	N
Mass	11.1	8	16	4
HB	72.3	56	84	23
Tail	52.7	42	65	23
HF	15.9	14	17	22
Ear	11.8	8	14	23
GLS	21.4	20.2	23.5	16

Genus Account

Mylomys Thomas, 1906 (Three-toed Grass Rats)

Diagnosis

These rodents are large-sized (Mass ca. 115 g; HB ca. 160-210 mm) and have a coarse buffy-brown dorsal colour similar to *Arvicanthis* and a pure white ventral colour. There are striking similarities between members of *Mylomys* and those of *Arvicanthis* and *Pelomys*. The tail is bicoloured (darker above) and coarsely scaled. Diagnostically, the forefeet have only three functional digits (hence the vernacular name).

(a)

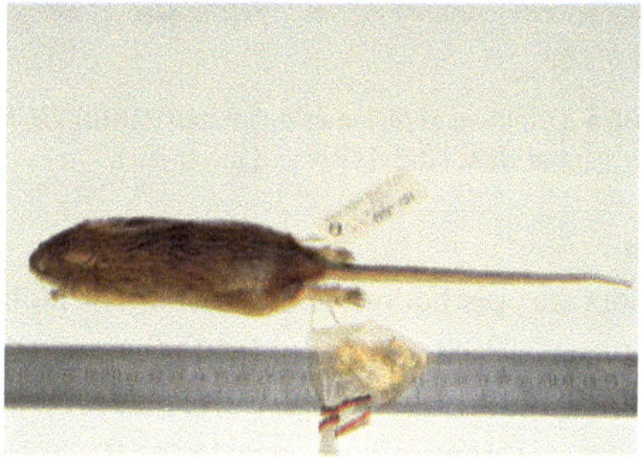

(b)

Figure 2.605 a and b Photographs of *Mylomys dybowskii* (Jan Decher) (left) and museum specimen (MNHN-ZM-MO 1979-101) (Ara Monadjem) (right).

Cranial Characters

The skull is large-sized (GSL 35-40 mm) and ovoid in shape with short and wide nasals which taper sharply at the tip. The braincase is mostly smoothly contoured but with supraorbital and occipital crests present. The zygomatic arches are robust, the zygomatic plate is narrow and barely expanded at the base, and the palatal foramina are short, barely reaching the alveolus of the M^1. The bullae are not expanded (16% of GSL).

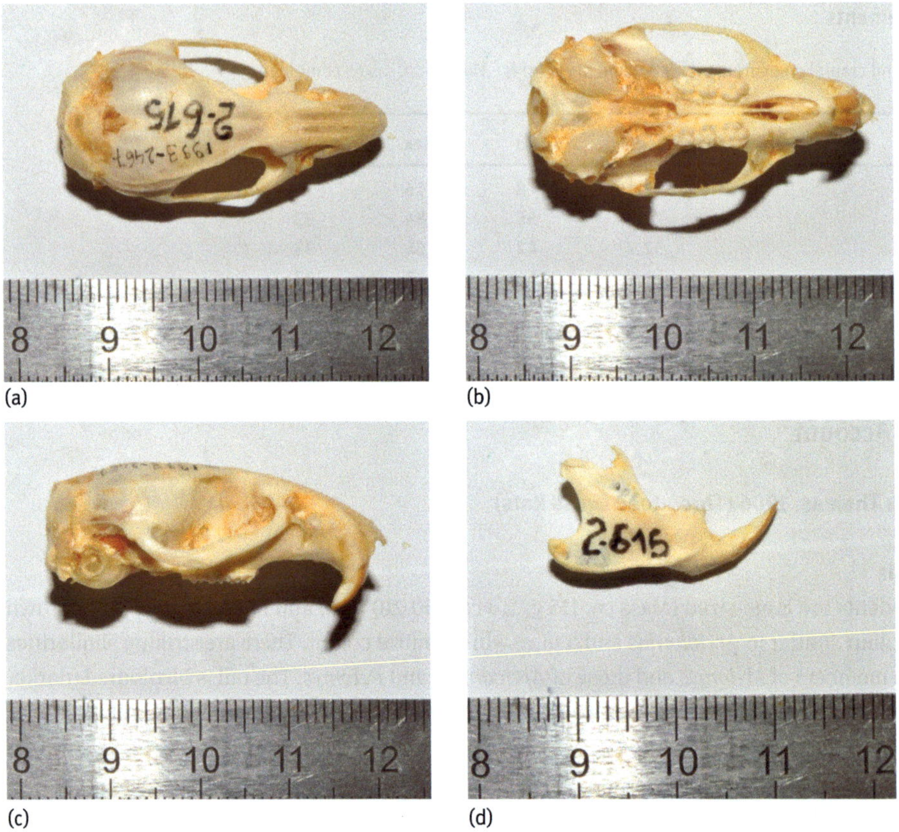

Figure 2.606 a-d Cranium of *Mylomys dybowskii* (MNHN-ZM-MO 1993-2467) (from top left to bottom right: dorsal view, ventral view, lateral view and lower jaw).

Dentition

The incisors are opisthodont and grooved. The molars are very robust and high-crowned, showing macrodonty (MTR/GSL ± 0.24, Table 10). Cusps are high and separate and arranged in longitudinal rows, with the three medial cusps (t2, t5 and t8 on M^1 and M^2) being enlarged and facing backwards in the upper toothrow and forwards in the lower toothrow. The cusp t9 is small but present on M^1 and M^2, while t3 is present on M^3 (Group 8 of Figure 2.439). The M^3 is small compared with M^2. No posterior cingulum is present on M^1 or M^2 but these (and an antero-external cusplet on M_2) are present on M_1 and M_2.

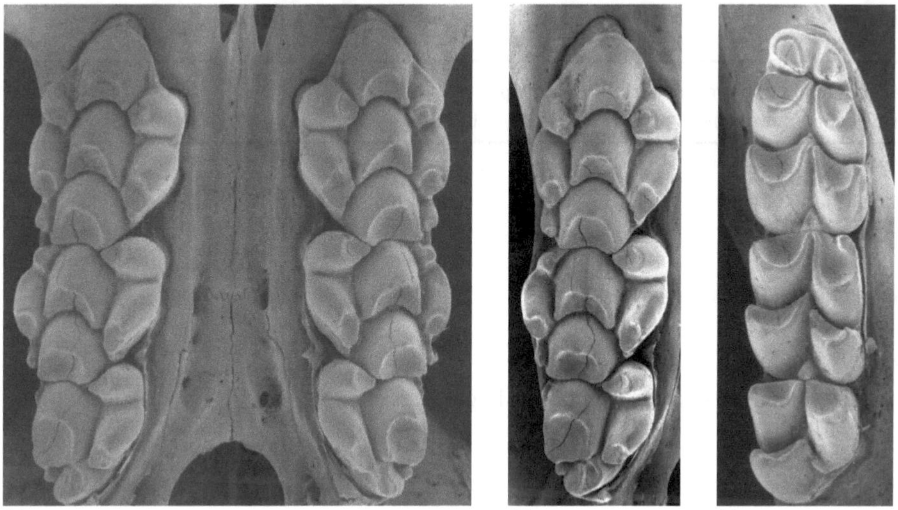
(a) (b) (c)

Figure 2.607 Cheek teeth of *Mylomys dybowskii* (MNHN-ZM-MO 1971-525) (A: upper cheek teeth; B: upper right toothrow; C: lower right toothrow).

Habitat & Biogeography

Mylomys dybowskii has a scattered distribution in grassy clearings in forests and is distributed patchily from Mt Nimba in the west through Ghana, Cameroon, north-western and eastern DRC, southern Sudan, Uganda, Kenya, Rwanda and Tanzania. *Mylomys rex* is only known from an isolated forest (Charada Forest) at 1800 m in Ethiopia.

Fossil Record

There are no known fossil records.

Biology & Ecology

The diet is essentially herbivorous, thus explaining the high crowned and powerfully built grinding molars (the genus name comes from the Greek word *myle* for mill). In Uganda, reproduction appears to occur throughout most months of the year (Happold, 2013).

Systematic Notes

Based on morphological and molecular (Ducroz et al., 2001; Lecompte et al., 2008) grounds, *Mylomys* has been associated with *Pelomys, Arvicanthis, Lemniscomys, Rhabdomys* and *Desmomys* of the *Arvicanthis* division of tribe Arvicanthini. *Mylomys rex* is known only from the type specimen and was previously included in *Desmomys* but has been shown to be a *Mylomys* (Lavrenchenko 2003).

Annotated Species List

Mylomys dybowskii (Pousargues, 1893)
Mylomys rex (Thomas, 1906)

Key to Species

1	Size smaller (HB < 200 mm); not occurring in Ethiopia	*M. dybowskii*
	Size larger (HB > 200 mm); restricted to Charada Forest, Ethiopia	*M. rex*

Species Accounts

Mylomys dybowskii (Pousargues, 1893)

Diagnosis

This species is not known to co-occur with the larger *M. rex* (HB > 200 mm).

Taxonomic Issues

Chromosome number is 2n = 42 (Matthey, 1970; Gautun et al., 1986).

Distribution

This species is widely but patchily distributed in tropical Africa occurring from Mt Nimba to western Kenya, and south to northern Malawi. It occupies grasslands in the forest-savanna mosaic and may also occur in montane situations.

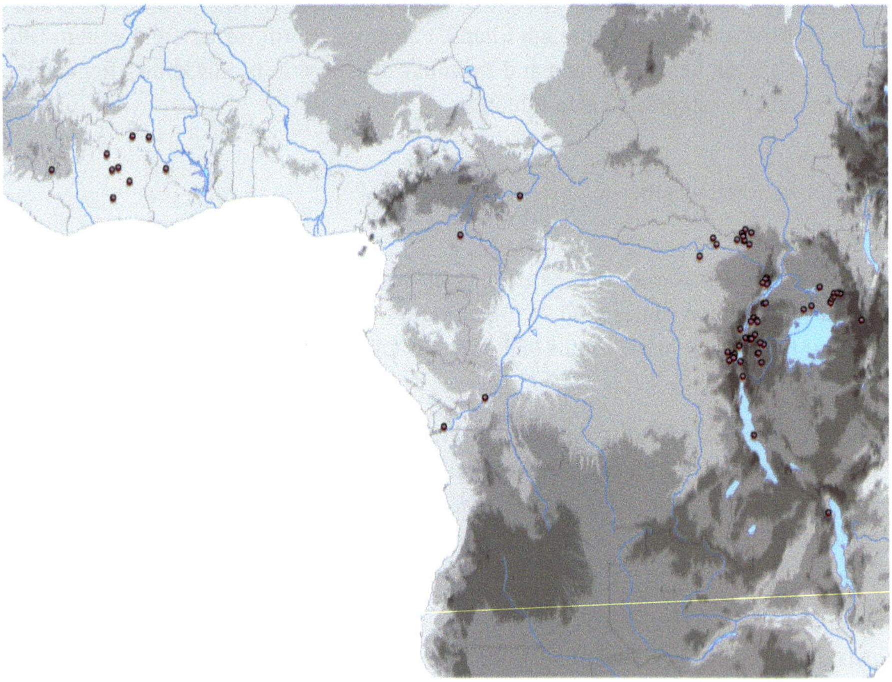

Figure 2.608 Distribution map

Measurements

External and cranial measurements (mm) and mass (g) for *Mylomys dybowskii*, sexes combined

	Mean	Min	Max	N
Mass*	113.5	46	190	84
HB*	159.7	122	194	87
Tail*	146.7	104	180	85
HF*	32.7	29	37	87
Ear*	20.0	14	25	83
GLS	37.4	35.0	40.5	6

*Delany (1975)

Mylomys rex (Thomas, 1906)

Diagnosis
This species is not known to co-occur with the smaller *M. dybowskii* (HB < 200 mm).

Taxonomic Issues
This species, which is currently only known by the holotype, was originally described in the genus *Arvicanthis*, moved to *Pelomys* and finally to *Mylomys* (see Happold, 2013 for a taxonomic history of this species).

Distribution
This species is only known from the Charada Forest in south-western Ethiopia (1800 m) where it presumably occupies montane forest.

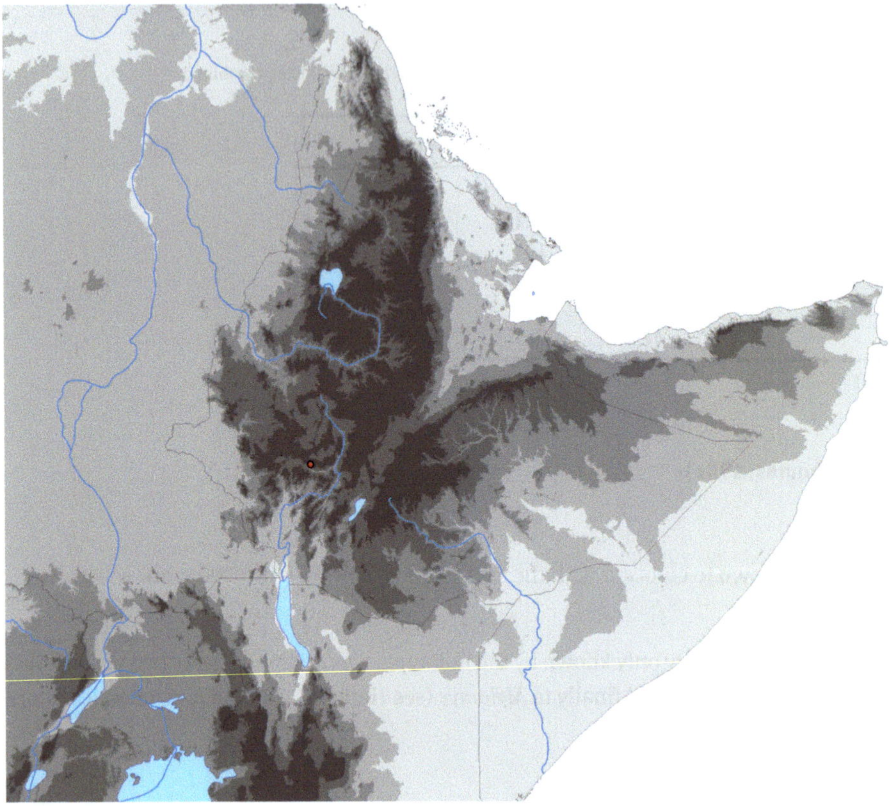

Figure 2.609 Distribution map

Measurements

External and cranial measurements (mm) and mass (g) for *Mylomys rex*, sexes combined

	Mean	Min	Max	N
Mass	-	-	-	-
HB*	212	-	-	1
Tail*	175	-	-	1
HF*	35	-	-	1
Ear*	24	-	-	1
GLS	-	-	-	-

*Thomas (1906)

Genus Account

Myomyscus Shortridge, 1942 (Meadow Mice)

Diagnosis

These small-sized (Mass: ca. 30-40 g; HB ca. 100-120 mm) mice are similar to *Praomys* and *Mastomys* in having short soft fur with few guard hairs and a generally buffy brown or grey dorsal colouration. They can be distinguished from *Mastomys* in having a tail longer than head and body (100-140% of HB), only 4 or 5 pairs of nipples (not "multimammate") and the ventral colour whiter. Since the genus as currently comprised (three species) is polyphyletic, opinion has been expressed (e.g. Musser & Carleton, 2005) that the genus should be confined to the type species, *M. verreauxii* which is endemic to the Fynbos biome of South Africa. If this is true it would make the genus readily diagnosable by its restricted distribution.

(a) (b)

Figure 2.610 a and b Photographs of *Myomyscus brockmani* (Lauren E. Helgen) (left) and museum specimen (MHNH-ZM-MO 1964-287) *M. verreauxii* (MHNH-ZM-MO 1964-287) (Ara Monadjem) (right).

Cranial Characters

The skull is very similar to that of *Mastomys*, although the mesopterygoid fossa is broader (> 1 mm) (Hill & Carter, 1941; Kingdon, 1974).

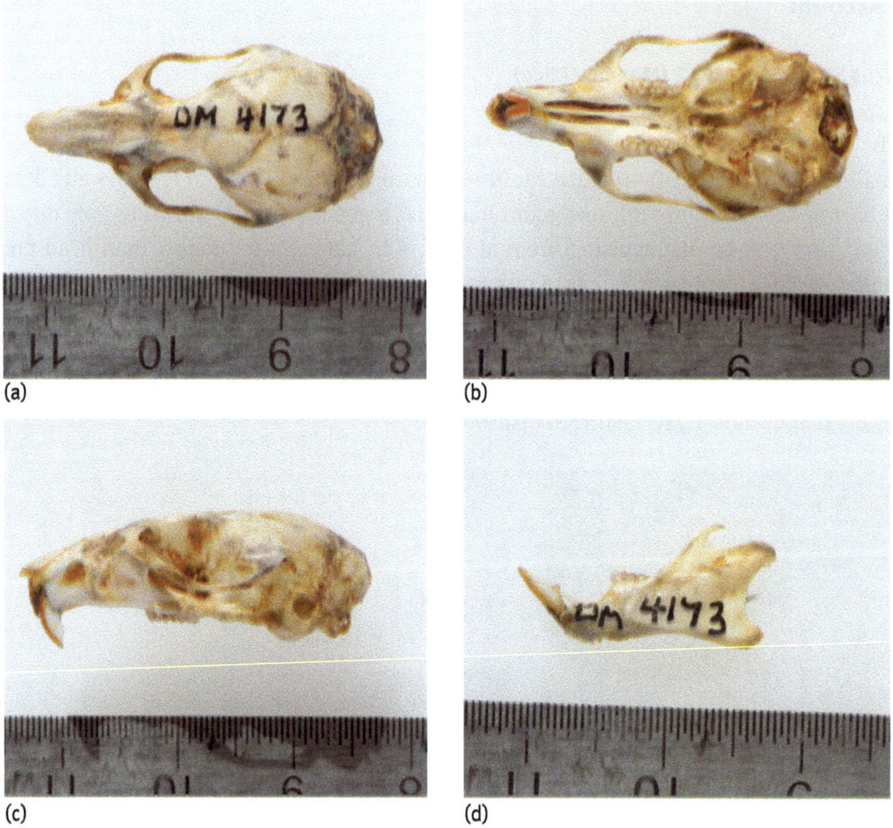

Figure 2.611 a-d Cranium of *Myomyscus verreauxii* (DM 4173) (from top left to bottom right: dorsal view, ventral view, lateral view and lower jaw).

Dentition

The rows of cusps are slightly distorted, with t1 apparently less displaced posterior to t2 and t3 than is the case for *Mastomys*. The cusps t2 and t3 are well separated with the latter situated slightly behind the former. The posterior cingulum is present on M_1 and M_2 and the anterior cusplet is present on M_2.

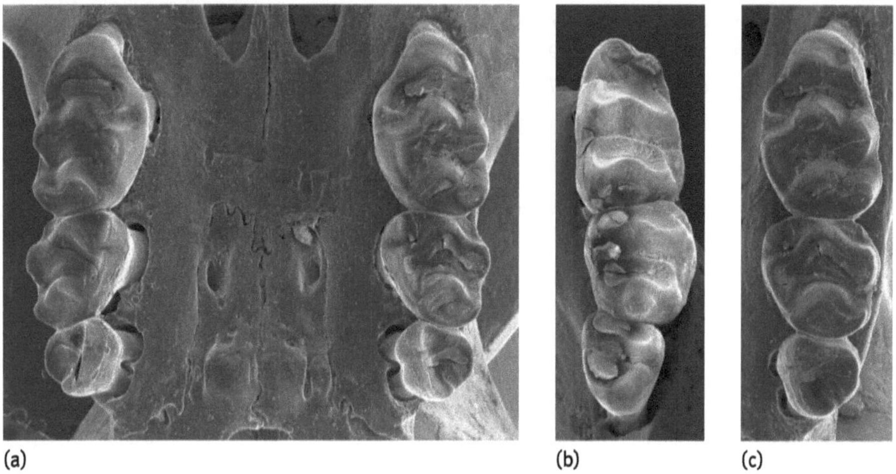

Figure 2.612 Upper cheek teeth of *Myomyscus brockmani* (RMCA 2026) (A: upper cheek teeth; B: lower right toothrow; C: upper left toothrow).

Habitat & Biogeography

The range of the genus is discontinuous in East, Central and Southern Africa. The three species are typically associated with grasslands, heathlands and drier savannas in rocky and mountainous habitats, in the Cape Fold Mts (Fynbos) of South Africa (*M. verreauxii*), the central highlands of Angola (*M. angolensis*), the Imatong Mts of South Sudan, Ethiopian lowland savannas and highland plateaux from 600-1700 m (Yalden, 1976), Uganda and the Rift Valley of Kenya and Tanzania (*M. brockmani*).

Fossil Record

Fossils are known from Late Pliocene to Late Pleistocene deposits of South Africa (Winkler et al., 2010).

Biology & Ecology

They are nocturnal and terrestrial. Very little is known about their diet or reproduction but *M. verreauxii* eats insects and *Protea* seeds. Births seem to occur in the late wet season in *M. brockmani* and two pregnant specimens had 6 and 7 embryos (Happold, 2013).

Systematic Notes

Opinions have differed widely on the relationships and species composition of this genus (see Musser & Carleton, 2005 and references therein for a recent discussion). The genus was previously known as *Myomys* but this name was shown to be a synonym of *Mastomys*. The *Myomys* species *daltoni* and *derooi* recognised by Musser & Carleton (1993) were transferred to the genus *Praomys* by Musser & Carleton (2005) while the *Myomys* species *ruppi* and *albipes* recognised by Musser & Carleton (1993) were transferred to the genus *Stenocephalemys* by Musser & Carleton (2005). Recently molecular data agree that *Myomys* belongs to the *Stenocephalemys* division of tribe Praomyini (Lecompte et al., 2002a,b, 2005b, 2008). The genus *Myomyscus* as currently defined (with three African species as well as the Arabian species, *M. yemeni*) is clearly polyphyletic (Lecompte et al., 2008) with *M. verreauxi* allied with *Colomys* and *Zelotomys*, and *M. brockmani* and *M. yemeni* forming a sister pair. The name *M. fumatus* was long used for *M. brockmani* but was shown to be a *Mastomys* (Van der Straeten & Robbins, 1997). We retain all three species in *Myomyscus* until a new genus has been described for the taxa *brockmani* and the extralimital *yemeni* (the taxon *angolensis* has not yet been sequenced and therefore its phylogenetic position is currently unclear).

Annotated Species List

Three species are recognised in Sub-Saharan Africa:
Myomyscus angolensis (Bocage, 1890)
Myomyscus brockmani (Thomas, 1908)
Myomyscus verreauxii (Smith, 1834)

Key to Species

1	Tail shorter (ca. 100% of HB); Endemic to Angola	*M. angolensis*
	Tail longer (> 120% of HB); occurring elsewhere	2
2	Ventral pelage greyish-white; restricted to the Fynbos biome of South Africa	*M. verreauxii*
	Ventral pelage pure white; widely distributed in East Africa	*M. brockmani*

Species Accounts

Myomyscus angolensis (Bocage, 1890)

Diagnosis

This species can readily be distinguished from other *Myomyscus* species by its distribution (*M. verreauxii* is limited to the extreme south of South Africa; *M. brockmani* occurs in East Africa). This species is very similar to *Mastomys shortridgei* with which it shares many cranio-dental features and has the same number of nipples (five pairs), but the tail is distinctly shorter in the latter species (ca. 87% of HB in *M. shortridgei* versus ca. 100% of HB in *M. angolensis*). The only other species of *Mastomys* to co-occur with *M. angolensis* is *M. natalensis* which differs in number of nipples (usually 12 pairs). The ventral pelage is grey at the base with white tips.

Taxonomic Issues

The taxonomic affinities of this taxon are not yet resolved. We have retained it in *Myomyscus* pending further molecular investigations (see genus account for further details).

Distribution

This species is endemic to Angola. Although the limits of its distribution are poorly known it appears to occur throughout the western highlands of the country (above 1000 m) where it occurs in various woodland habitats.

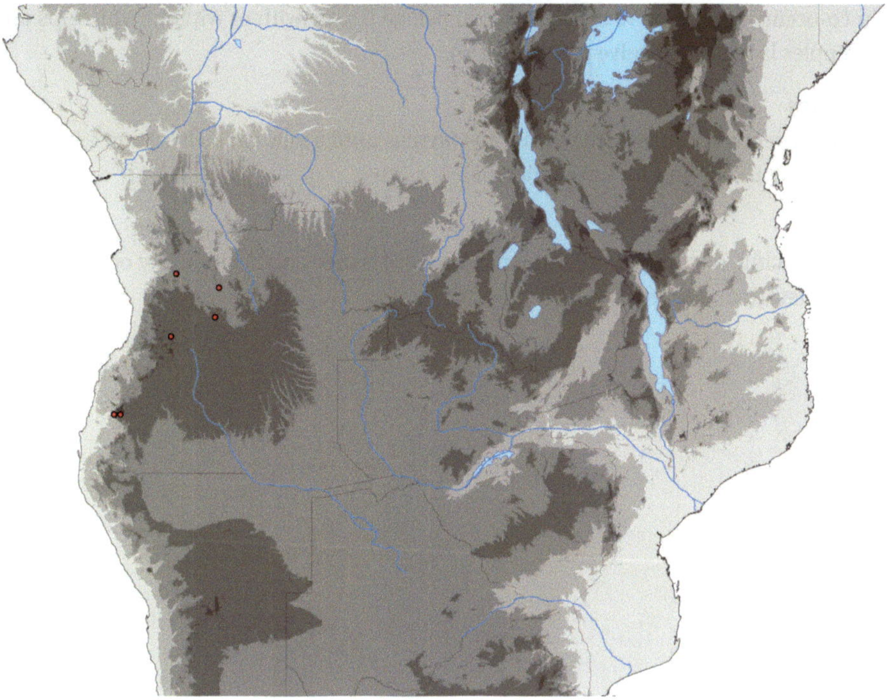

Figure 2.613 Distribution map

Measurements

External and cranial measurements (mm) and mass (g) for *Myomyscus angolensis*, sexes combined

	Mean	Min	Max	N
Mass	-	-	-	-
HB*	120.5	109	127	6
Tail*	121.5	103	128	6
HF*	24.2	24	25	6
Ear*	-	-	-	-
GLS*	31.1	30.5	32.0	5

*Hill & Carter (1941)

Myomyscus brockmani (Thomas, 1908)

Diagnosis

This species can readily be distinguished from other *Myomyscus* species by its distribution (*M. verreauxii* is limited to the extreme south of South Africa; *M. angolensis* is endemic to Angola). This species is very similar to *Mastomys* spp. from which it can be distinguished on molecular (Lecompte et al., 2005b) and morphological (ventral pelage and number of nipples) grounds. In this species, the ventral pelage is pure white (hairs with white base) and it has four or five pairs of nipples (in co-occurring *Mastomys* species ventral pelage is grey with or without white tips and number of nipples is typically twelve pairs or more).

Taxonomic Issues

This species is currently retained in the genus *Myomyscus* until a new genus has been erected for this species (see genus account for further details).

Distribution

This species is patchily distributed in East Africa where it is associated with rocky outcrops typically at higher elevations above 1000 m.

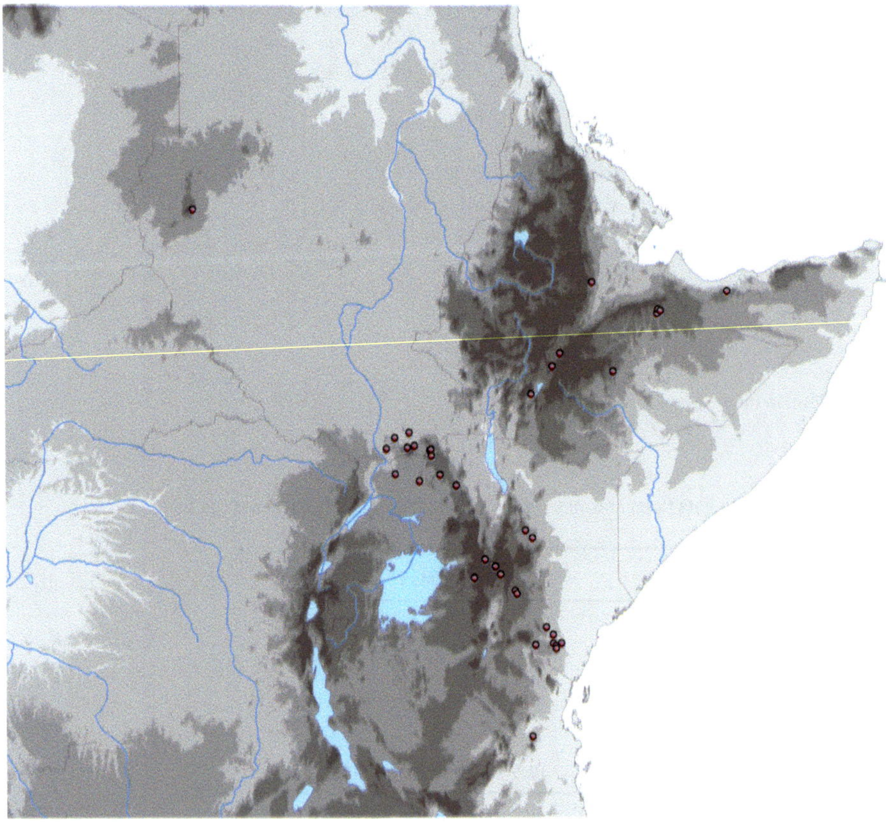

Figure 2.614 Distribution map

Measurements

External and cranial measurements (mm) and mass (g) for *Myomyscus brockmani*, sexes combined

	Mean	Min	Max	N
Mass*	29.7	22	39	11
HB*	104	93	125	15
Tail*	146	115	175	15
HF*	20.7	18	22	15
Ear*	17.0	14	20	15
GLS*	27.7	25.5	29.9	14

*Happold (2013)

Myomyscus verreauxii (Smith, 1834)

Diagnosis

This species can readily be distinguished from other *Myomyscus* species by its distribution (*M. brockmani* occurs in East Africa; *M. angolensis* is endemic to Angola). This species is very similar to *Mastomys* spp. from which it can be distinguished on molecular (Lecompte et al., 2005b) and morphological (relative tail length and number of nipples) grounds. In this species, the tail is relatively long (ca. 130-140% of HB) and it has five pairs of nipples (in co-occurring *Mastomys* species relative tail length is ca. 100% or less of HB and number of nipples is typically twelve pairs or more).

Taxonomic Issues

This taxon is probably the sole representative of the genus *Myomyscus* (see genus account for further details).

Distribution

This species is endemic to the southern parts of South Africa where it is associated with a variety of habitats including grasslands, heaths and forest.

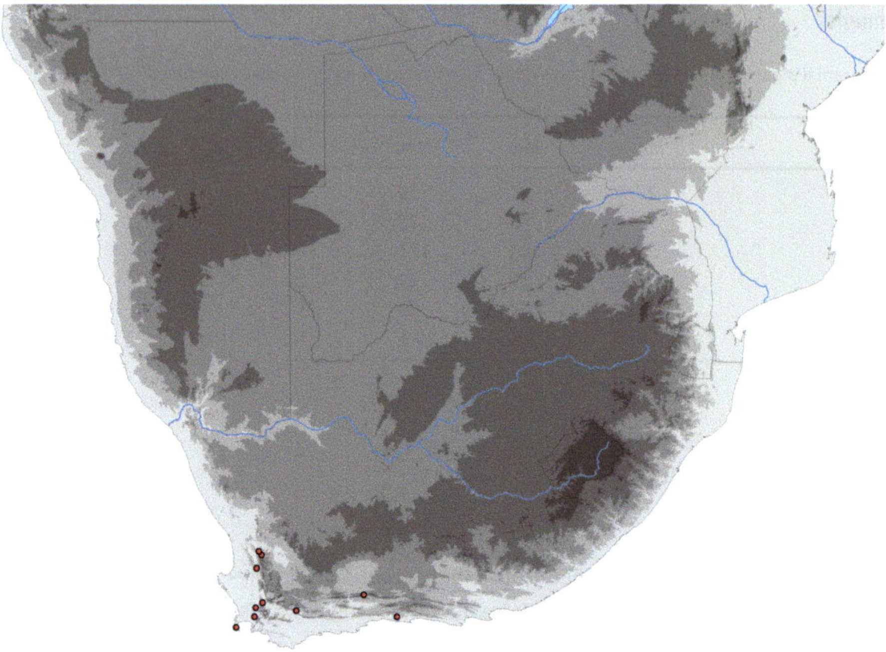

Figure 2.615 Distribution map

Measurements

External and cranial measurements (mm) and mass (g) for *Myomyscus verreauxii*, sexes combined

	Mean	Min	Max	N
Mass	36.3	22	52	5
HB	102.2	90	113	6
Tail	140.8	130	148	6
HF	25.8	25	27	6
Ear	17.5	17	19	6
GLS	29.2	28.4	30.5	5

Genus Account

Nilopegamys Osgood, 1928 (Ethiopian Water Rat)

Diagnosis

This medium sized rodent (Mass ca. 80 g; HB 148 mm) is characterised by a long (120% of HB), bicoloured tail having a mid-ventral line of black hairs, thick, soft pelage (buffy above and pure white below), small, densely-haired ears with white edging on its postero-ventral edge and broad and long (40 mm) hindfeet faintly edged with hairs. This genus is known only from the type specimen of its only species, *N. plumbeus*, collected from a tributary of the Blue Nile in north-western Ethiopia (Kerbis Peterhans & Patterson, 1995). The forelimbs and hindlimbs are densely haired and there are distinctive carpal vibrissae. The hindfeet have six plantar pads, including the hypothenar (which is lacking in *Colomys*). The morphology of *Nilopegamys* is adapted to its unique aquatic carnivorous life history, and many of its characteristics are shared with *Colomys* which has a similar life history but narrower hindfeet lacking hairs, larger ears and also lacks the distinctive carpal bristles of *Nilopegamys* (Kerbis Peterhans & Paterson, 1995).

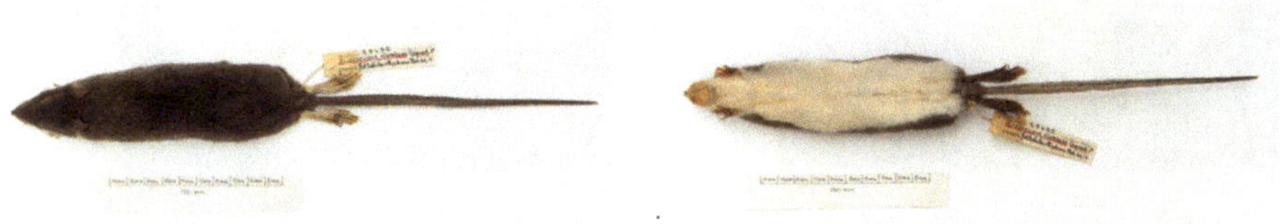

(a) (b)

Figure 2.616 a and b Photographs of *Nilopegamys plumbeus* museum specimen (FHNH 28633), dorsal (left) and ventral (right) (Julian Peterhans Kerbis).

Cranial Characters

The skull is large (GSL of holotype 35.1 mm; Kerbis Peterhans & Patterson, 1995) and slightly flattened in shape with high-set orbits, a long and broad braincase, long, slightly tapering nasals, well developed supraorbital ridges extending onto the rostrum (less developed in *Colomys*), fairly robust zygomatic arches, broad zygomatic plate (much narrower in *Colomys*) and palatal foramina that extend to the first laminae of M^1. There is a small subsquamosal foramen present, bordering a thick hamular process (in *Colomys* the subsquamosal foramen is huge, and the hamular process (= "mastoid strut") long and thin). The foramen magnum is proportionately large, as is the brain size, these characters being characteristic of semi-aquatic rodents.

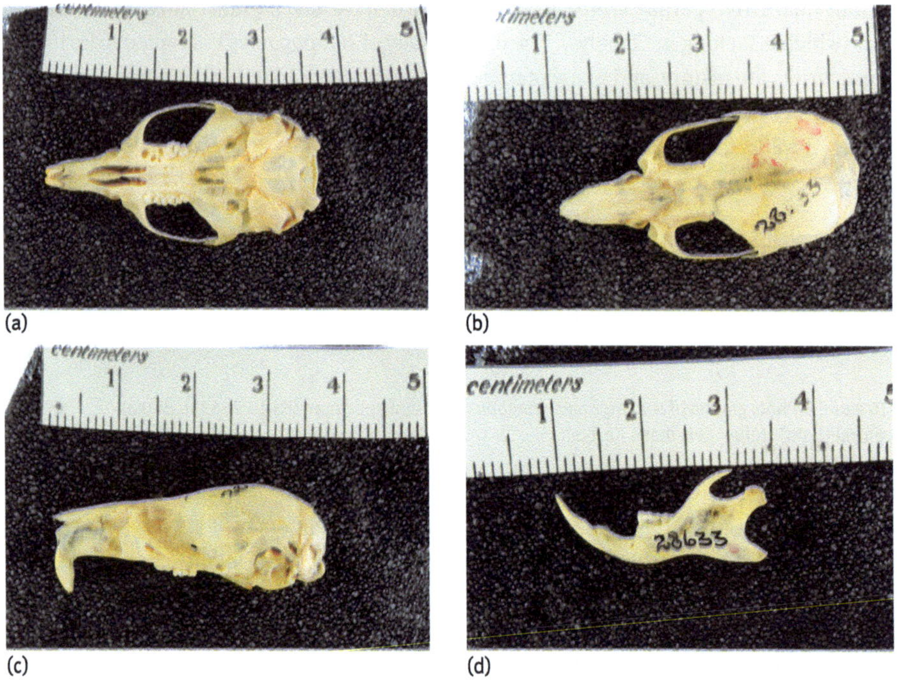

Figure 2.617 a-d Cranium of *Nilopegamys plumbeus* (FMNH 28633) (from top left to bottom right: dorsal view, ventral view, lateral view and lower jaw) (Photographs courtesy of Bruce Patterson, Field Museum of Natural History, Chicago).

Dentition

The upper incisors are orthodont, robust and ungrooved. The molars are moderately developed (MTR/GSL 0.16). In the holotype at least, cusps are not well separated but joined transversely to form laminae. There is only slight distortion of the transverse rows. The cusp t3 on M^2 is very well developed.

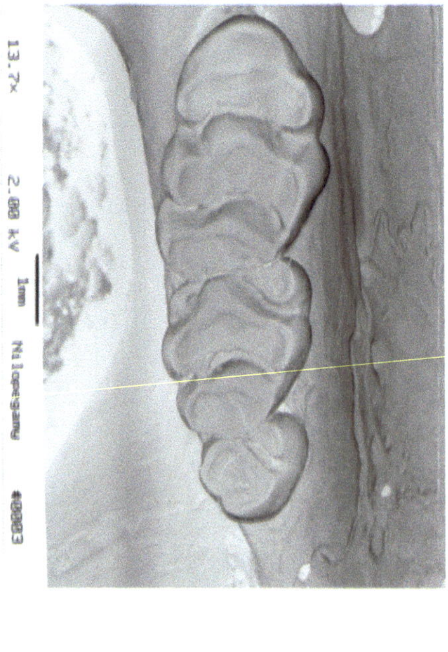

Figure 2.618 Upper left toothrow of *Nilopegamys plumbeus* (FMNH 28633).

Habitat & Biogeography
This genus is only known from one stream in north-western Ethiopia.

Fossil Record
None.

Biology & Ecology
The sole species is supremely adapted for wading in shallow streams and feeding on a carnivous diet but otherwise nothing is known of its life history. The stomach contents of the holotype had a “fishy smell” suggesting a diet of fish (Kerbis Peterhans & Patterson, 1995).

Systematic Notes
Nilopegamys was until 1995 treated as a subspecies of *Colomys goslingi* (Hayman, 1966; Misonne, 1974; Musser & Carleton, 1993). However, Kerbis Peterhans & Patterson (1995) showed that *Nilopegamys* can be distinguished from *Colomys* on external, cranial, dental and postcranial evidence and reflects a more extreme adaptation to its amphibious lifestyle than *Colomys*.

Annotated Species List
One species is recognised:
Nilopegamys plumbeus Osgood, 1928

Species Account

Nilopegamys plumbeus Osgood, 1928

Diagnosis
See genus account.

Taxonomic Issues
See genus account.

Distribution
This species is known only by the holotype which was collected from a tributary of the Blue Nile in north-western Ethiopia (Kerbis Peterhans & Patterson, 1995).

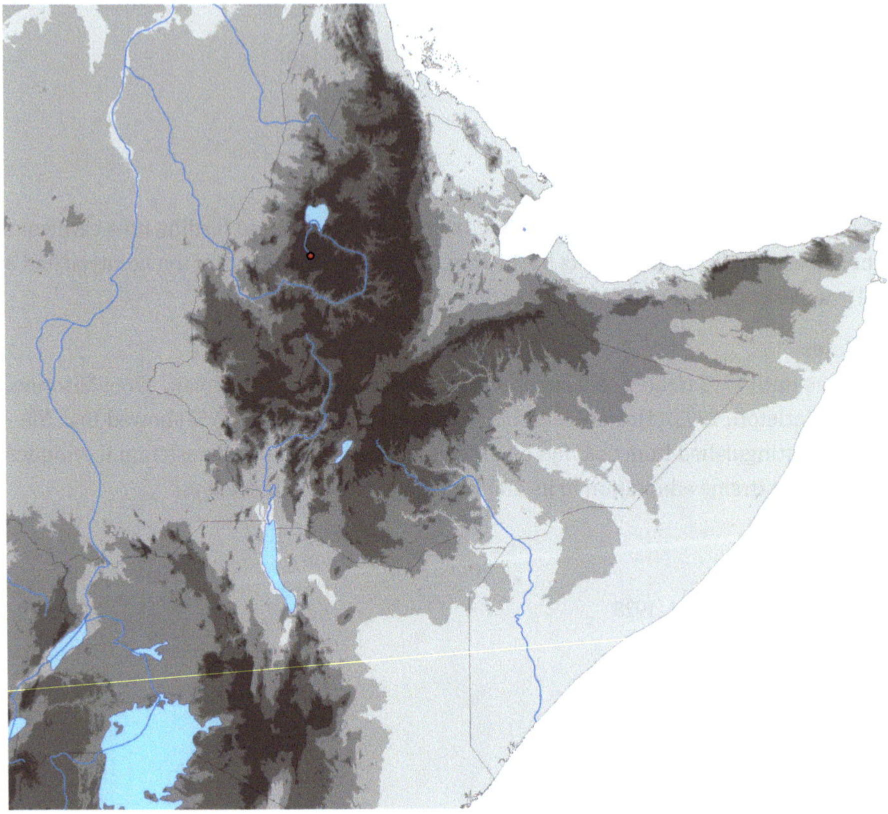

Figure 2.619 Distribution map

Measurements

External and cranial measurements (mm) and mass (g) for *Nilopegamys plumbeus*, sexes combined

	Mean	Min	Max	N
Mass*	71-81	-	-	1
HB*	148	-	-	1
Tail*	180	-	-	1
HF*	40	-	-	1
Ear*	14.4	-	-	1
GLS*	35.1	-	-	1

*Kerbis Peterhans & Patterson (1995), measurements of the holotype: mass and ear length are based on estimates by the authors

Genus Account

Oenomys Thomas, 1904 (Rufous-nosed Rats)

Diagnosis

These medium to large-sized rats (Mass ca. 90-110 g; HB ca. 140-160) have a tail which is sparsely haired with fine concentric scales, bicoloured and relatively long (110% of HB). The ears are small and densely furred. The pelage is soft with long guard hairs and typically greyish brown to bright reddish dorsally with prominently darker orange around the nose, rump and anal area, and pure white to pinkish coloured ventrally.

(a) (b)

Figure 2.620 a and b Photographs of *Oenomys hypoxanthus* (Bill Stanley) (left) and museum specimen (MNHN 1991-84) (Ara Monadjem) (right)."

Cranial Characters

The cranium is large (GSL 34-40 mm), long and narrow in shape, with well-developed supraorbital and supraoccipital ridges and rectilinear nasals which widen slightly anteriorly. The profile is flattish with the rostrum deflected downwards. The zygomatic arches are fairly robust and the zygomatic plate has a vertical to slightly concave anterior edge. The palatal foramina extend just to the alveolus of M^1.

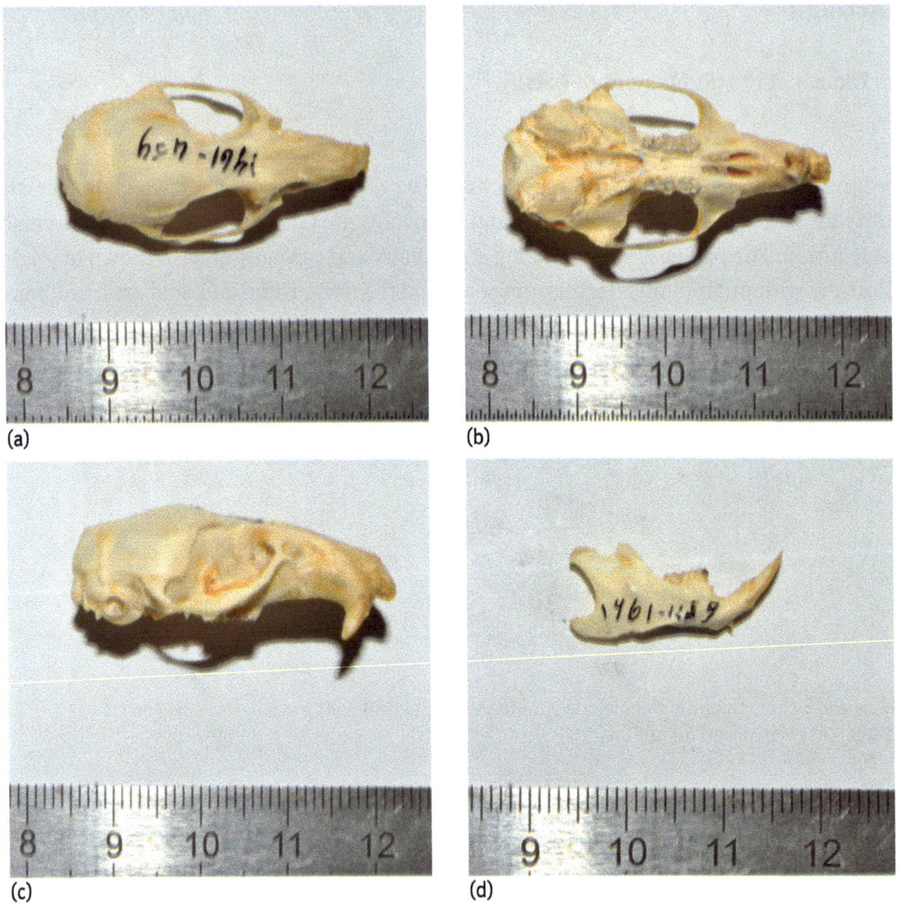

Figure 2.621 a-d Cranium of *Oenomys ornatus* (MNHN-ZM-MO 1964-459) (from top left to bottom right: dorsal view, ventral view, lateral view and lower jaw).

Dentition

The upper incisors are opisthodont and ungrooved. The molars show macrodonty (MTR/GSL ± 0.19, Table 10) as well as extremely developed stephanodonty (development of longitudinal crests); crests linking the second and third laminae (cusps t6 and t9 and t4 and t8). The medial row of cusps is enlarged and arranged in a strict longitudinal row. The cusp t9 is present on M^1, but absent or highly reduced on M^2 and M^3. Molar M^3 is distinguished in having clear traces of all three laminae, with two cusps in each row, including t3 in the first row (Group 7 in Figure 2.439). A number of accessory labial cusplets are present in the lower molars; the posterior cingulum is present on M_1 and M_2 while an anterior cusplet is present on M_2 and M_3.

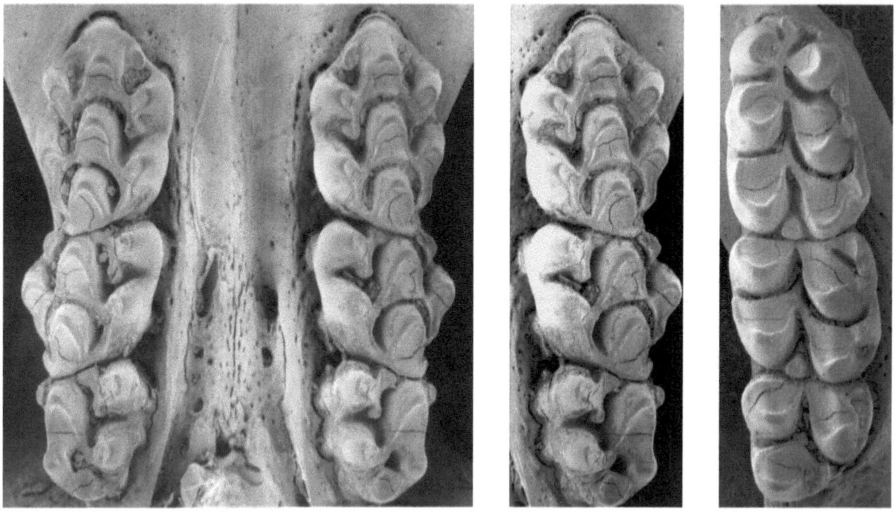

(a) (b) (c)

Figure 2.622 Cheek teeth of *Oenomys hypoxanthus* (MNHN-ZM-MO 1981-538) (A: upper cheek teeth; B: upper left toothrow; C: lower right toothrow).

Habitat & Biogeography

They occur in grassy clearings within or at the edge of rainforests, from Sierra Leone and Guinea in the west across the Congo Basin to Ethiopia, Kenya and western Tanzania in the east and to Angola in the south.

Fossil Record

Fossils are known from the Middle Pliocene to Middle Pleistocene of East Africa. The earliest fossil is from 3.3 MYA at Hadar (Sabatier, 1982).
Oenomys tiercelini Hadar (Sabatier, 1982)
Oenomys olduvaiensis Olduvai Bed I (Jaeger, 1976)
Oenomys sp. Omo Shungura (Wesselman, 1984)

Biology & Ecology

They are terrestrial but capable of climbing, and seem to be both diurnal and nocturnal. Their diet is herbivorous, comprising predominantly grass, herbs and forest and cultivated fruits. In the DRC, breeding was seasonal at lower altitudes (< 2000 m) and aseasonal at higher altitudes (> 2000 m) (Happold, 2013).

Systematic Notes

On molecular grounds, *Oenomys* was found to belong to a basal clade of the tribe Arvicanthini (Lecompte et al., 2008). Within this tribe it has been assigned to the *Oenomys* division which also includes *Grammomys* (on molecular grounds) as well as (based on morphological characters) *Thamnomys*, *Thallomys*, *Lamottemys* and the extinct *Malpaisomys* (Musser & Carleton, 2005). Schenk et al. (2013) found it basal to the Arvicanthini tribe. Renaud (1999) showed that the two currently recognised species can be distinguished clearly on the basis of molar size and shape, and further, that two cryptic species can be recognised corresponding to Central and East African populations of *O. hypoxanthus*. We retain *O. hypoxanthus* as a single species pending further investigation.

Annotated Species List

Two species are recognised:
Oenomys hypoxanthus (Pucheran, 1855)
Oenomys ornatus Thomas, 1911

Key to Species

1	Tail relatively shorter (ca. 115% of HB); 2n = 32; occurs east of Dahomey Gap	*O. hypoxanthus*
	Tail relatively longer (ca. 130% of HB); 2n = 46; occurs west of Dahomey Gap	*O. ornatus*

Species Accounts

Oenomys hypoxanthus (Pucheran, 1855)

Diagnosis

This species is easily distinguished from *O. ornatus* based on their non-overlapping distributions, the relatively longer tail of *O. ornatus* (ca. 130% of HB), and different karyotype (see key).

Taxonomic Issues

Based on morphological grounds, this species may comprise two cryptic species: one occurring in East Africa and extending westwards into the Albertine Rift; the other occurring in Central Africa (Renaud, 1999). Chromosome number is 2n = 32 (Maddalena et al., 1989).

Distribution

This species has a wide distribution in tropical Africa occurring from the Dahomey Gap eastwards to Ethiopia and south to northern Angola and southern DRC. It occupies dense, tall grasslands or tangled vegetation typically on the edge of forests such as along swamps or clearings.

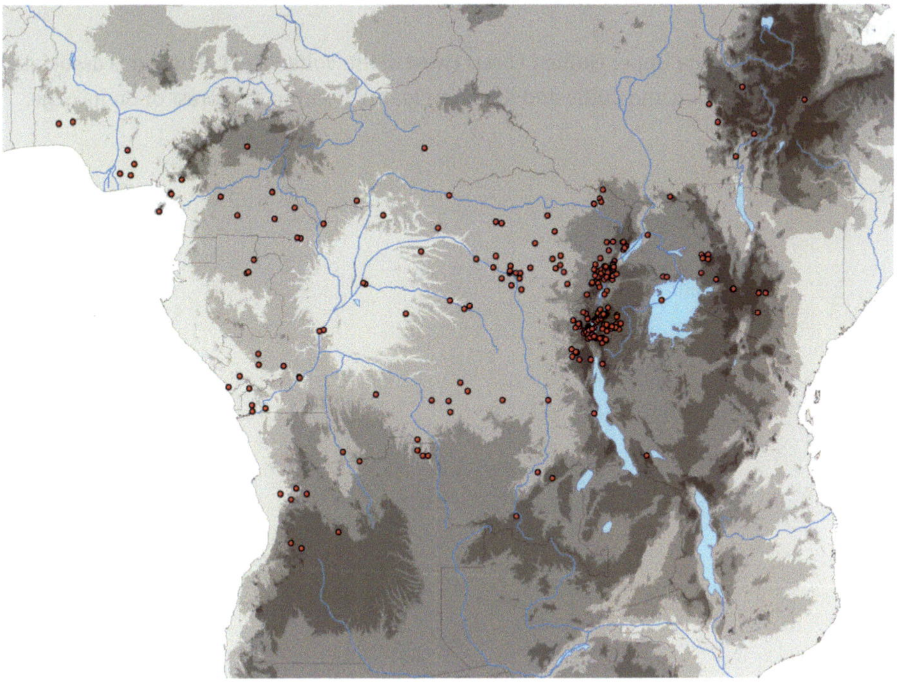

Figure 2.623 Distribution map

Measurements

External and cranial measurements (mm) and mass (g) for *Oenomys hypoxanthus*, sexes combined

	Mean	Min	Max	N
Mass	110.8	74	138	6
HB	154.1	136	185	16
Tail	179.3	152	199	16
HF	31.6	29.0	35.0	16
Ear	19.6	13	25	16
GLS	37.4	34.0	40.8	12

Oenomys ornatus Thomas, 1911

Diagnosis

This species is easily distinguished from *O. hypoxanthus* based on their non-overlapping distributions, the relatively shorter tail of *O. hypoxanthus* (ca. 115% of HB), and different karyotype (see key).

Taxonomic Issues

This taxon was previously listed as a subspecies of *O. hypoxanthus* (Rosevear, 1969) but differs on chromosomal and morphological grounds (Renaud, 1999). Number of chromosomes is 2n = 46 (Tranier & Gautun, 1979).

Distribution

This species is endemic to the Upper Guinea Forest (west of the Volta River), where it is associated with dense tangled vegetation and cultivated areas in clearings within the forest or on its edge.

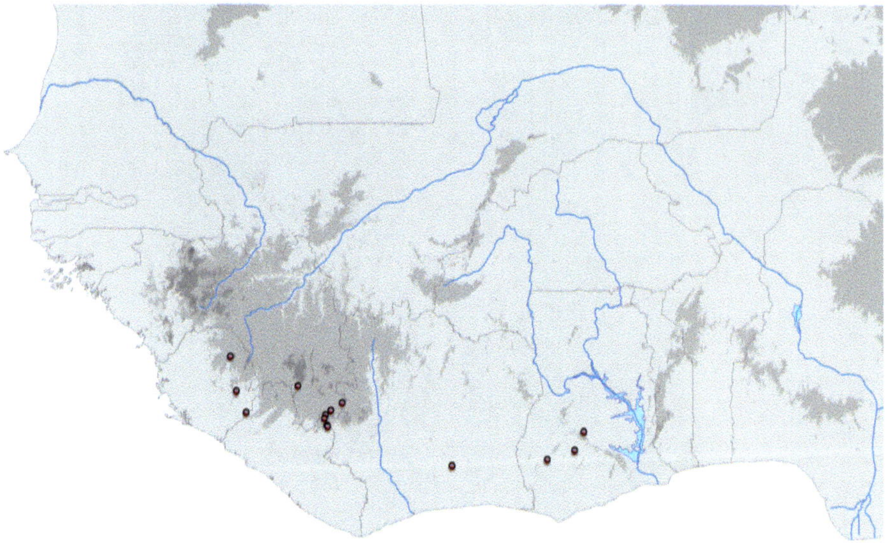

Figure 2.624 Distribution map

Measurements

External and cranial measurements (mm) and mass (g) for *Oenomys ornatus*, sexes combined

	Mean	Min	Max	N
Mass*	95	80	103	4
HB	140.7	130	157	6
Tail	183.8	175	200	6
HF	32.7	30	35	6
Ear	16.3	12	18	6
GLS	35.1	33.4	36.5	6

*Happold (2013)

Genus Account

Otomys Cuvier, 1824 (Vlei Rats or Laminate-toothed Rats)

Diagnosis

Their large size (HB 120-210 mm), stocky, short-tailed (45-73% of HB), vole-like appearance and shaggy coarse brownish-grey pelage make members of this genus unmistakeable. The robust skull and laminate dentition are also very characteristic of this genus. Members of the genus can be distinguished from members of *Parotomys* by the more pallid coloration of the latter (although

O. unisculcatus also has a similar pallid pelage) as well as by the marked expansion of the bullae in *Parotomys* (bullae diameter is <9 mm in *Otomys* and > 10 mm in *Parotomys*). *Parotomys* species are adapted to arid environments in the western regions of southern Africa whilst most *Otomys* species are adapted to mesic and/or montane habitats, or semi-arid Karroo habitats in the case of *O. unisculcatus*.

Figure 2.625 a-e Photographs of *Otomys*, from top left to bottom right: *O. unisulcatus* (Enviro-Insight), *O.* cf *helleri* (Marian Polak), *O. denti* (Bill Stanley), *O. lacustris* (Bill Stanley) and *O. tropicalis* (Bill Stanley)."

Cranial Characters

The skull is large in size (GSL 30-50 mm), robust, ovoid to approximately square-shaped and often heavily ridged, especially in the case of the supraorbital and supraoccipital crests. The zygomatic arches are robust and parallel-sided ("boxy"). The nasal bone is prominent and expanded anteriorly to a varying degree in different species (from barely expanded in species such as *O. unisulcatus* to right-angled in species such as *O. irroratus*). The anterior palatal foramina are relatively short and do not reach M^1. The palate itself is folded. The anterior surface of the zygomatic plate is vertical and bluntly rounded dorsally. The bullae are relatively small (6-9 mm in greatest diameter)

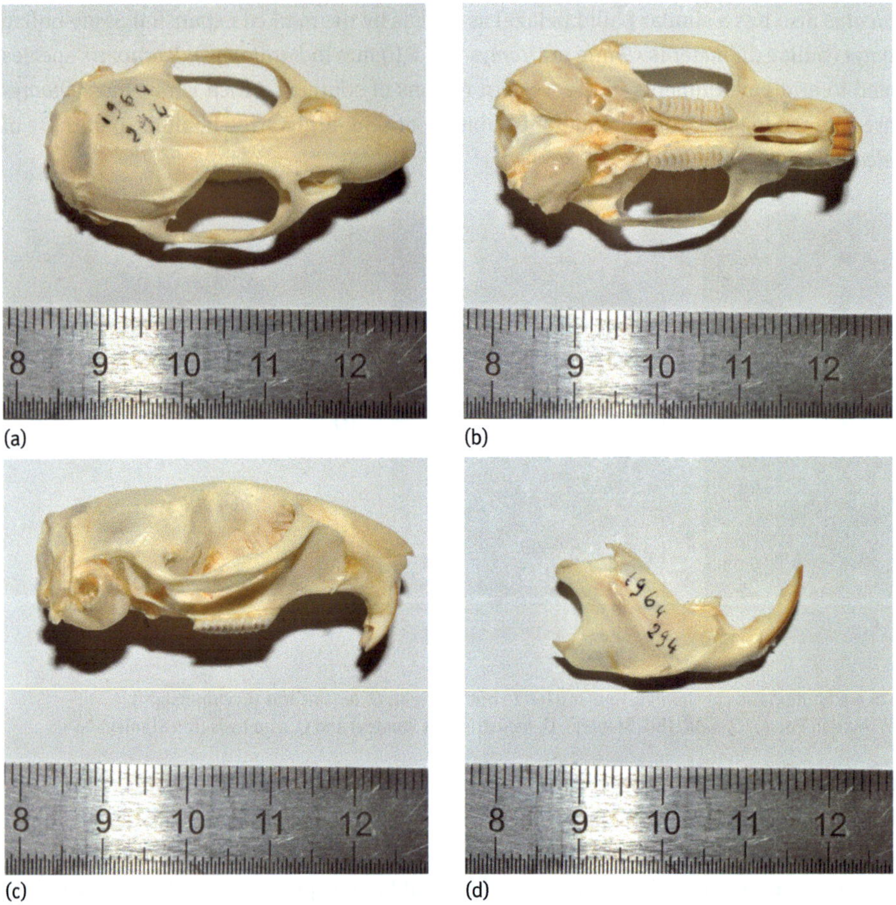

Figure 2.626 a-d Cranium of *Otomys irroratus* (MNHN-ZM-MO 1964-294) (from top left to bottom right: dorsal view, ventral view, lateral view and lower jaw).

Dentition

The incisors are opisthodont; upper incisors are always grooved, while lower incisors lack grooving in *O. unisulcatus* but have one to two grooves in the remaining species. Individual cusps are fused into a number of transverse laminae. The M^3 is the longest molar in the toothrow, having between four and ten laminae in different species. The M^1 has three laminae and the M^2 has two. The M_1 is the longest tooth in the lower toothrow, having between three and seven laminae in different species. The M_2 and M_3 always have two laminae. The two upper toothrows are barely diverging posteriorly.

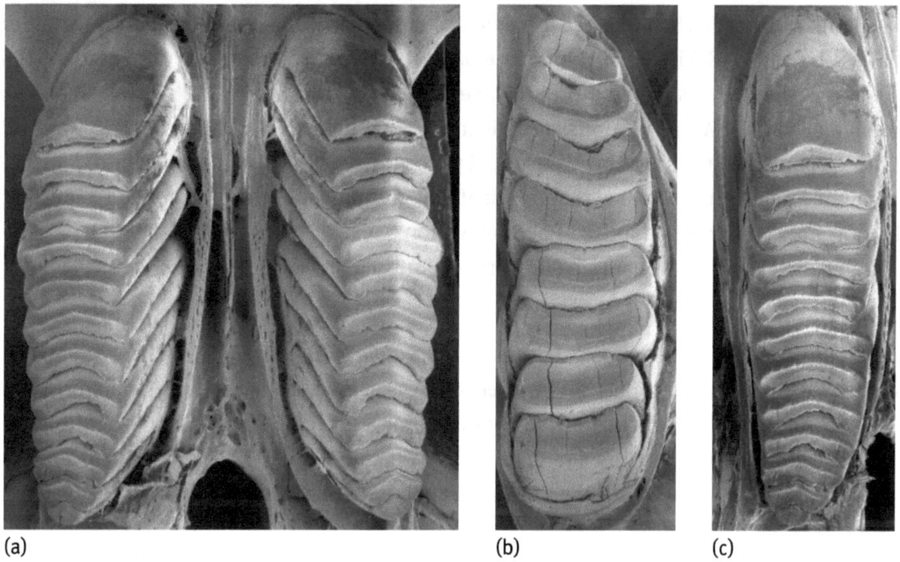

Figure 2.627 Cheek teeth of *Otomys tropicalis* sensu lato (MNHN-ZM-MO 1961-24) (A: upper cheek teeth; B: lower right toothrow; C: upper right toothrow).

Habitat & Biogeography

With the exception of *O. unisulcatus* which occurs in semi-arid Karoo habitats, members of the genus *Otomys* are generally associated with mesic grasslands and moorlands within alpine, montane and sub-montane regions of Southern, Central, East and West Africa. *Otomys* species occur more or less continuously throughout much of South Africa but, further north, are restricted to isolated mountain ranges associated with the African Rift from Mozambique to Ethiopia, as well as the highlands of Angola and the Cameroon Volcanic Line in Central Africa (Mt Oku and Mt Cameroon in Cameroon and the Gotel Mts in Nigeria). With one exception (*O. angoniensis* from mesic savanna and submontane and montane habitats across Southern, Central and East Africa), the otomyine fauna of southern, eastern, north-eastern (Ethiopia) and western Africa are completely distinct from each other, i.e. each region has its own endemic fauna. Because of their narrow ecological niches, many species are limited to one or two adjacent mountain ranges, e.g. *O. burtoni* from Mt Cameroon, *O. occidentalis* from Mt Oku and the Gotel Mts of West Africa, *O. barbouri* and *O. jacksoni* from Mt Elgon of Kenya and Uganda, *O. simiensis* from the Simien Mts of Ethiopia, *O. helleri* and *O. yaldeni* from the Bale Mts of Ethiopia, *O. dartmouthi* from the Ruwenzori Mts of eastern DRC, *O. dollmani* from Mt Gargues, *O. zinki* from Mt Kilimanjaro, and *O. lacustris* and *O. uzungwensis* from the Eastern Arc Mts of Tanzania and associated mountains of southern Tanzania and the Nyika Plateau of Malawi. Within African mountain ranges where *Otomys* occurs, there are typically multiple species occurring locally within different elevational zones. Thus, in the South African Drakensberg Range, *O. angoniensis* occurs on the lower slopes in savanna habitats, *O. auratus* and *O. laminatus* at mid-elevation in grasslands and *O. sloggetti* at the highest elevations in alpine heath habitats. In the Eastern Arc Mts, *O. lacustris* is found at lower elevations and *O. uzungwensis* at higher elevations. In the Ruwenzori Mts, *O. denti* is found at lower altitudes while *O. dartmouthi* occurs at higher elevations. On Mt Elgon, *O. tropicalis* is found at lower elevations while *O. barbouri* and *O. jacksoni* occur sympatrically but in different vegetation associations (Clausnitzer & Kityo, 2001) in alpine habitats above the treeline. In the Aberdare Mts and Mt Kenya, *O. angoniensis* is found on the lower slopes, *O. tropicalis* at mid-elevation and *O. orestes* at high elevation. While previous classifications (Bohmann, 1952; Petter, 1982) lumped all alpine-adapted populations under one widespread polytypic species (*O. typus* sensu lato) molecular phylogenetic analysis has revealed that this artificial grouping is highly polyphyletic (Taylor et al., 2011). Morphological traits adaptive to high-elevation (such as pallid coloration and a particular skull shape) are clearly convergent, having arisen multiple times.

Fossil Record

The oldest South African Miocene fossil species of *Euryotomys* (5-4 MYA) listed below display a dentition (with residual cusps) which can be interpreted as a link between an ancestral murid and a modern otomyine rodent, providing strong evidence for a murid, South African origin for the subfamily (Pocock, 1976) . The earliest *Otomys* fossils (*O.* cf *sloggetti*, 3.5-3.0 MYA) attest to a South African origin too for this genus, with the first fossil appearance of *Otomys* in East Africa (eastern DRC) at 2.3 MYA.

Euryotomys pelomyoides, 6-4.5 MYA, Langebaanweg "E"Quarry, South Africa (Pocock, 1976; Denys et al., 1987)
E. bolti, 4.0-5.0 MYA, Bolt's Farm, South Africa (Sénégas & Avery, 1998; Sénégas, 2001)
O. cf *gracilis*, 3.5-3.0 MYA, Makapansgat, South Africa (Denys, 1989, 1990, 2003)
O. cf *sloggetti*, 3.5-3.0 MYA, Makapansgat, South Africa (Denys, 1989, 1990, 2003)
O. sp, 2.3 MYA, eastern DRC (Denys, 2003)
O. petteri, 2.0-1.5 MYA, Olduvai Bed I (Denys, 1989, 1990, 2003)
O. cf *angoniensis*, 0-8-0.3 MYA, Olduvai Bed I (Denys, 1989c)

Biology & Ecology

Vlei rats are exclusively herbivorous, with a diet mainly comprised of grasses. They are generally K-selected, giving birth usually to one or two offspring (maximum five) which are precocial and born with erupted incisors enabling them to nipple-cling to their mother immediately after birth. They have a number of predators including felids, jackals, mongooses, genets, snakes and different owl species. *Otomys* skulls typically comprise the bulk of owl pellets.

Systematic Notes

The murid origin of Otomyinae suggested from fossil evidence is supported by DNA evidence which places it within a murine clade as a sister group to *Arvicanthis* and its relatives (Chevret et al., 1993b; Ducroz et al., 2001; Lecompte et al., 2008). Senegas (2001) and Taylor et al. (2004) suggested that *Otomys* and *Parotomys* should be placed within a murid tribe (Otomyini) rather than a separate murid subfamily but Happold (2013) suggested that a decision on the taxonomic rank should await broader sampling of Asian and Australian genera. There is also debate about the delimitation of genera, with three (*Otomys*, *Myotomys* and *Parotomys*) recognised by Musser & Carleton (2005). Due to the demonstrated paraphyly of *Myotomys* species (*unisulcatus* and *sloggetti*: Maree, 2002; Taylor et al., 2004), only two genera (*Otomys* and *Parotomys*) are currently recognised pending a final phylogenetic analysis of species relationships (Happold, 2013). Recent systematic revisions involving molecular, chromosomal, morphometric and morphological characters have revealed multiple cryptic evolutionary species within several previously recognised biological species and species complexes (defined under the polytypic concept of the Biological Species Concept), including *O. tropicalis* sensu lato (Carleton & Byrne, 2006; Taylor et al., 2011), *O. irroratus* sensu lato (Taylor et al., 2009a; Engelbrecht *et al.*, 2011), *O. denti* and *O. lacustris* (Taylor et al., 2009b) and *O. typus* sensu lato (Taylor et al., 2011). As a result, whereas only 15 species were formally recognised by Happold (2013), we recognise 29 species here, with an additional two undescribed taxa which we recommend for formal description.

Species List

Thirty-one species are recognised here of which two are undescribed taxa (*O.* cf *lacustris* and *O.* cf *sungae*) which warrant formal description.

Otomys anchietae (Bocage, 1882)
Otomys angoniensis Wroughton, 1906
Otomys auratus Wroughton, 1906
Otomys barbouri Lawrence & Loveridge, 1953
Otomys burtoni Thomas, 1918
Otomys cf *lacustris* Allen & Loveridge, 1933
Otomys cf *sungae* Bohmann, 1943
Otomys cheesmani Taylor et al., 2011
Otomys cuanzensis Hill & Carter, 1937
Otomys dartmouthi Thomas, 1906
Otomys denti Thomas, 1906
Otomys dollmani Heller, 1912
Otomys fortior Thomas, 1906
Otomys helleri Frick, 1914
Otomys irroratus (Brants, 1827)
Otomys jacksoni Thomas, 1891
Otomys karoensis Roberts, 1931
Otomys lacustris Allen & Loveridge, 1933
Otomys laminatus Thomas & Schwann, 1905
Otomys occidentalis Dieterlen & Van der Straaten, 1992
Otomys orestes Thomas, 1900
Otomys simiensis Taylor et al., 2011
Otomys sloggetti Thomas, 1902
Otomys sungae Bohmann, 1943
Otoyms thomasi Osgood, 1910
Otomys tropicalis Thomas, 1902
Otomys typus (Heuglin, 1877)
Otomys unisulcatus F. Cuvier, 1829
Otomys uzungwensis Lawrence & Loveridge, 1953
Otomys yaldeni Taylor et al., 2011
Otomys zinki Bohmann, 1943

Key to Species

1	Five or more laminae on M_1 (*anchietae* complex)	2
	Four laminae on M_1	4
2	Larger sized, GSL > 45 mm, restricted to Angola	*O. anchietae*
	Smaller-sized, GSL < 40mm	3
3	Restricted to Mt Elgon (Uganda/Kenya)	*O. barbouri*
	Restricted to Cameroon Volcanic Line in West Africa	*O. occidentalis*
	Restricted to southern Eastern Arc and southern Rift Mts of Tanzania and bordering Malawi and Zambia	*O. lacustris*
4	Two deep grooves on lower incisors (*typus* complex)	17
	Less than two deep grooves	5
5	One deep outer and one shallow inner groove on lower incisors	6
	No such grooves on lower incisors	14
6	Nine to ten laminae on M^3	*O. laminatus*
	Five to eight laminae on M^3	7
7	Petrotympanic foramen slit-like in shape	8
	Petrotympanic foramen is a round hole	9
8	Confined to Mt Cameroon	*O. burtoni*
	Widespread in East and Southern Africa	*O. angoniensis*
9	Dorsal pelage pallid, buffy (Munsell Hue 10YR), size smaller, MTR typically < 9.0 mm, restricted to Fynbos biome apart from marginal intrusion into grasslands in the south-western Free State Province of South Africa	*O. karoensis*
	Dorsal pelage darker, less buffy (Munsell Hue 7.5YR), size larger, MTR typically > 9.0 mm, not restricted to Fynbos biome	10
10	Restricted to Southern Africa	11
	Restricted to East Africa	13
11	Restricted to Angola	*O. cuanzensis*
	Restricted to South Africa and Zimbabwe/Mozambique	12
12	Distributed in the Fynbos and Thicket biomes in southern South Africa	*O. irroratus*
	Distributed in the Grassland biome of South Africa and eastern Zimbabwe and extreme western Mozambique	*O. auratus*
13	Restricted to Mt Gargues, Kenya	*O. dollmani*
	Restricted to higher elevation alpine meadows on Mau Escarpment and Uasin Gishu Plateau,Kenya	*O. thomasi*
	Restricted to higher elevation, alpine meadows in Mts Kenya and Aberdares	*O. orestes*
	Restricted to montane and alpine zones of Mt Kilimanjaro	*O. zinki*
	More widespread in mid-elevation zones of Albertine Rift Mountains, Mt Elgon and the Kenya Rift Mountains	*O. tropicalis*
14	One deep groove on lower incisors (*denti* complex)	15
	One faint groove or no groove on lower incisors	16
15	Restricted to the Albertine Rift Mts	*O. denti*
	Restricted to the Northern Eastern Arc Mts	*O. sungae*
	Restricted to the Nyika Plateau of Malawi and Zambia	*O.* cf *sungae*
16	Tail short, usually < 50% of HB, petrotympanic foramen slit-like in shape	*O. sloggetti*
	Tail longer, usually >60% of HB, petrotympanic foramen a round hole	*O. unisulcatus*

17	Larger in size, GSL > 43 mm	*O. cheesmani*
	Smaller in size, GSL < 43 mm	18
18	Restricted alpine habitats > 3300m in the Rwenzori Mountains, 6 laminae in M^3; GSL > 35 mm	*O. dartmouthi*
	Restricted to montane evergreen forests of south-western Ethiopia; 8-9 laminae in M^3; GSL > 35 mm	*O. fortior*
	Restricted to alpine habitats in the Bale Mts and Arussi Plateau of eastern Ethiopia; 8-9 laminae in M^3; GSL > 35 mm	*O. helleri*
	Restricted to the alpine zone of Mt Elgon, 7 laminae in M^3; small size GSL < 35 mm	*O. jacksoni*
	Restricted to mid-elevation slopes in the Simien Mts of northern Ethiopia, 6-7 laminae in M^3, GSL > 35 mm	*O. simiensis*
	Restricted to the Mau Escarpment and Uasin Gishu Plateau on the western edge of the Kenyan Rift, at elevations of approximately 2450-2750 m, GSL > 35 mm	*O. thomasi*
	Distributed from Simien Mts in the north (where they occur at higher altitudes than *simiensis*) to the mountains of the western rim of the Ethiopian Rift; 8-9 laminae in M^3; GSL > 35 mm	*O. typus*
	Restricted to Udzungwa Mts of west-central Tanzania, extending to the Nyika Plateau of Malawi and bordering Zambia; 7 laminae in M^3; GSL > 35 mm	*O. uzungwensis*
	Restricted to the Bale Massif in south-eastern Ethiopia; 6-7 laminae in M^3; small size GSL < 35 mm	*O. yaldeni*

Species Accounts

Otomys anchietae (Bocage, 1882)

Diagnosis

Conspicuously larger than all other *Otomys* species (HB > 200 mm; skull length > 45 mm). The M_1 has five or six laminae.

Taxonomic Issues

Previously grouped with other species having 5 laminae in M_1 (*O. barbouri, O. lacustris, O. occidentalis*) but considered as a distinct species by modern reviews (Taylor et al., 2014).

Distribution

Confined to the highlands of central Angola, associated with rivers within both the Congo and Okavango catchments.

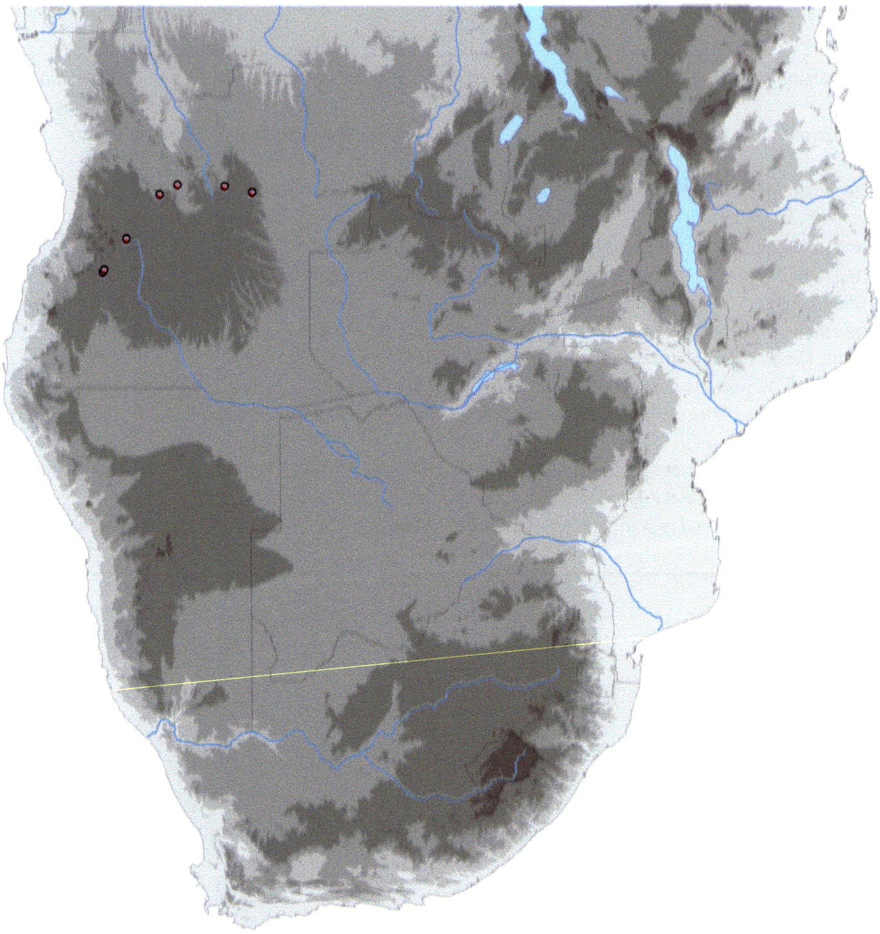

Figure 2.628 Distribution map

Measurements

External and cranial measurements (mm) and mass (g) for *Otomys anchietae*, sexes combined

	Mean	Min	Max	N
Mass	-	-	-	-
HB*	205	197	217	11
Tail*	115	87	127	11
HF*	39	36	41	11
Ear	-	-	-	-
GLS**	48.7	45.2	52.8	7

*Happold (2013); **Taylor et al. (2014)

Otomys angoniensis Wroughton, 1906

Diagnosis

The slit-like petrotympanic foramen distinguishes this species from all other *Otomys* except for the two West African species (*O. occidentalis* and *O. burtoni*) which can be distinguished from *O. angoniensis* on biogeographic grounds as well as the possession of five laminae in M_1 of *O. occidentalis* (cf four in *O. angoniensis*).

Taxonomic Issues

Unpublished molecular data (Taylor & Maree, unpublished) show that East and Southern Africa *O. angoniensis* may be distinct species. *O. a. maximus* has been regarded as a separate species (Musser & Carleton, 2005) but is provisionally included in *O. angoniensis* pending a revision of this species.

Distribution

This is the most widespread of all *Otomys* species, occurring in eastern mesic savannas in Southern and Central Africa and foothills and lower reaches of major Rift mountain ranges in Malawi, the western slopes of the Albertine Rift, Uganda, Kenya and northern Tanzania.

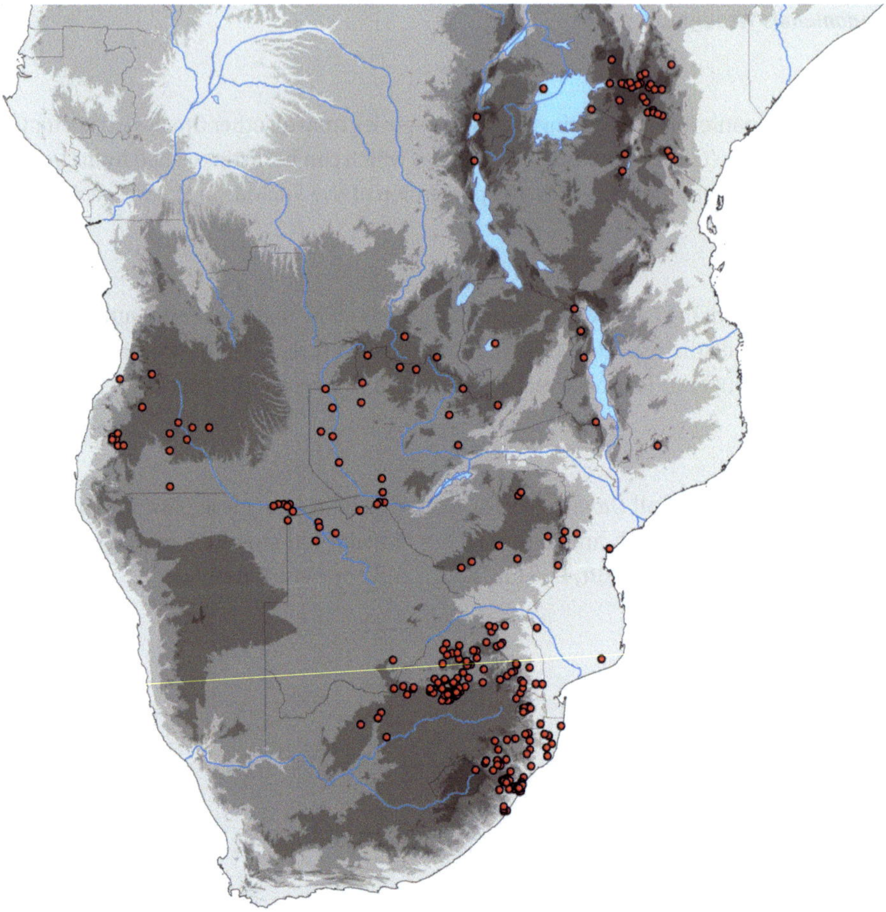

Figure 2.629 Distribution map

Measurements

External and cranial measurements (mm) and mass (g) for *Otomys angoniensis*, sexes combined

	Mean	Min	Max	N
Mass	114.3	60	182	17
HB	154.9	124	192	37
Tail	78.0	40	100	33
HF	27.9	21	32	30
Ear	20.3	15	25	38
GLS	39.4	33.2	49.6	63

Otomys auratus Wroughton, 1906

Diagnosis

This species was until recently included in *O. irroratus* and the two cannot be distinguished on morphological grounds; however *O. auratus* is closely associated with the Grassland biome in South Africa and the Eastern Highlands of Zimbabwe whilst *O. irroratus* is closely associated with the Fynbos and Thicket biomes of the Western Cape and Eastern Cape provinces of South Africa. *Otomys auratus* can be distinguished from other co-occurring *Otomys* by having a round petrotympanic foramen (as opposed to slit-like in *O. angoniensis*), sharply expanded nasal bones (as opposed to more moderately expanded nasal bones in *O. angoniensis*) and almost always six laminae in upper M^3 (as opposed to seven in *O. angoniensis* and 8-10 in *O. laminatus*). In the Drakensberg, *O. auratus* co-occurs with *O. sloggetti* but the latter occurs at higher elevations (> 2000 m) and can be distinguished by its pallid coloration, barely expanded nasal bones, slit-like petrotympanic foramen and possession of 4-5 laminae in M^3.

Taxonomic Issues

Otomys tropicalis was formerly included in *O. irroratus* sensu lato but the two can be distinguished on morphological and molecular grounds (Taylor & Kumirai, 2001; Taylor et al., 2011). *Otomys auratus* was shown to differ from *O. irroratus* on molecular, chromosomal and ecological grounds (Taylor et al., 2009a; Engelbrecht et al., 2011).

Distribution

The species is widely distributed throughout the Highveld grasslands of the central plateau and Drakensberg Escarpment of South Africa, with isolated populations in the Soutpansberg Mts of northern South Africa and the Eastern Highlands of Zimbabwe and neighbouring Mozambique.

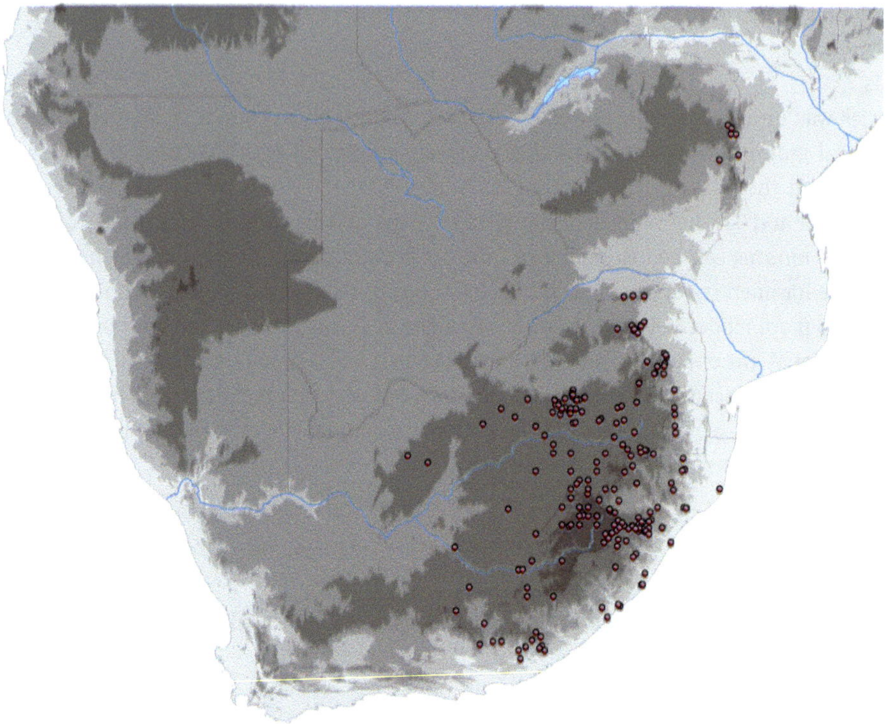

Figure 2.630 Distribution map

Measurements

External and cranial measurements (mm) and mass (g) for *Otomys auratus*, sexes combined

	Mean	Min	Max	N
Mass	127.3	63	211	32
HB	168.0	133	202	43
Tail	84.7	58	120	48
HF	29.3	21	38	41
Ear	21.4	17.5	26	49
GLS	40.9	32.6	45.9	113

Otomys barbouri Lawrence & Loveridge, 1953

Diagnosis

This species is distinguished from most other *Otomys* (including sympatric *O. tropicalis* and *O. jacksoni* from Mt Elgon) by having 5 laminae in M_1. Among other species having a 5-laminated M_1, *O. anchietae* is much larger (GLS > 45 mm) and restricted to Angola, *O. occidentalis* has a slit-like petrotympanic foramen (cf. round in *O. barbouri*) and is restricted to the Cameroon Highlands, while *O. lacustris* has a relatively longer tail (65% of HB, cf. 45% of HB in *O. barbouri*).

Taxonomic Issues

O. barbouri was previously included as a synonym of *O. anchietae* (Musser & Carleton, 1993) but modern molecular studies have shown that this *O. barbouri* is unrelated to other members of the *anchietae* group and is placed in a basal position in the Otomyini phylogeny (Taylor et al., 2011, 2014).

Distribution

This species has only been recorded from Mt Elgon in Kenya and Uganda.

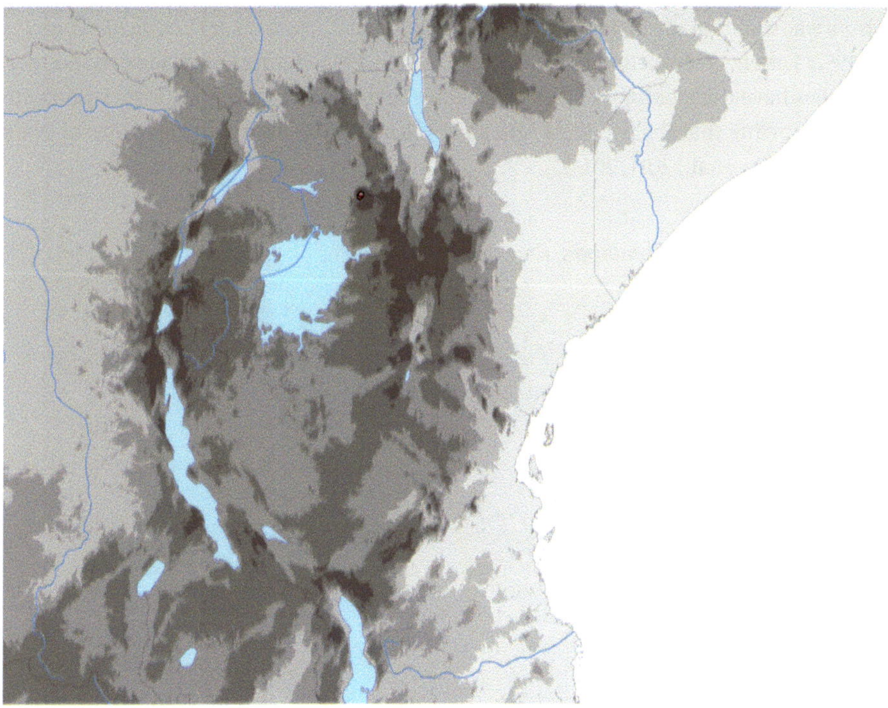

Figure 2.631 Distribution map

Measurements

External and cranial measurements (mm) and mass (g) for *Otomys barbouri*, sexes combined

	Mean	Min	Max	N
Mass*	112	78	161	92
HB*	166	125	210	24
Tail*	74.5	57	96	24
HF*	27	23	30	24
Ear*	24	21	28	24
GLS	37.0	33.6	39.0	15

*Clausnitzer (2003)

Otomys burtoni Thomas, 1918

Diagnosis

The slit-shaped petrotympanic foramen is an unusual character shared with *O. occidentalis* (which can be distinguished by having 5 laminae in M_1, cf 4 in *O. burtoni*), *O. sloggetti* (which is found only in South Africa and has only 4-5 M^3 laminae cf 6-7 in *O. burtoni*) and *O. angoniensis* (which is very distinct on molecular grounds and does not occur in West Africa; Taylor et al., 2004, 2014).

Taxonomic Issues

Otomys burtoni was traditionally included as part of a polytypic *O. irroratus* or *O. tropicalis* complex (Bohmann, 1952; Petter, 1982; Dieterlen & Van der Straeten, 1992) but recent molecular and morphometric studies have demonstrated its specific status and close phylogenetic association with *O. occidentalis*, implying that the two West African species resulted from a single dispersal event from East Africa (Taylor et al., 2011, 2014).

Distribution

This species only occurs on Mt Cameroon, Cameroon.

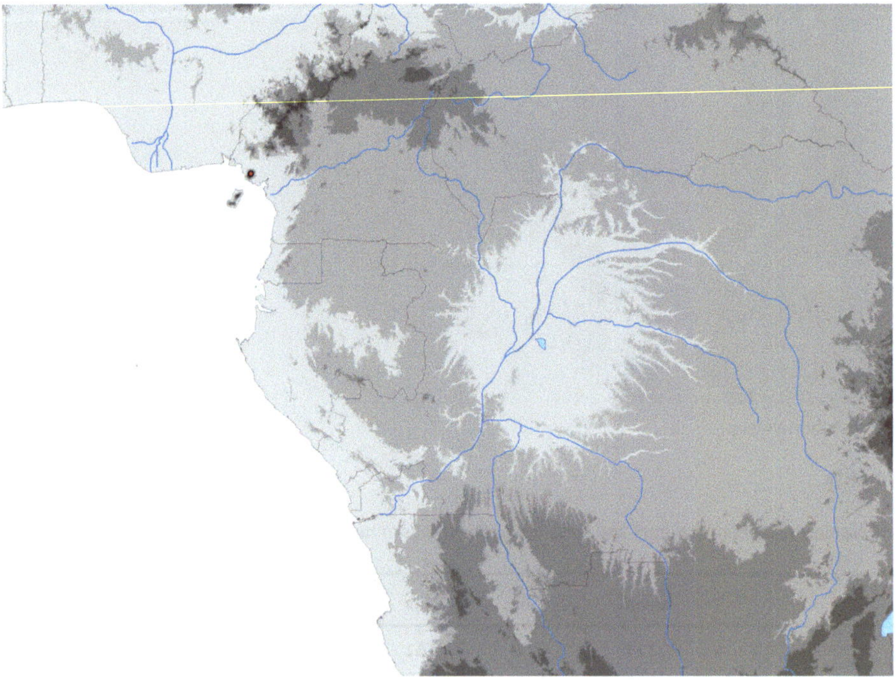

Figure 2.632 Distribution map

Measurements

External and cranial measurements (mm) and mass (g) for *Otomys burtoni*, sexes combined

	Mean	Min	Max	N
Mass*	90	71	105	3
HB*	156	145	167	6
Tail*	78	75	84	6
HFcu*	30	29	32	4
Ear*	24	22	26	6
GLS**	38.4	37.2	39.0	6

*Dieterlen & Van der Straaten (1992); **Taylor et al. (2014)

Otomys cf *lacustris*

Diagnosis

This species is morphologically indistinguishable from *O. lacustris* sensu stricto, having 5 M^3 laminae, but separated from this taxon by the Makambako Gap which separates the Eastern Arc (*O.* cf *lacustris*) from the Southern Rift Highlands (*O. lacustris* sensu stricto). It is distinguished from the co-occurring *O. uzungwensis* by the presence of two deep grooves in the lower incisors and 4 M^3 laminae in the latter species, and from co-occurring *O. sungae* by the blackish coloration and presence of 4 M^3 laminae in the latter species.

Taxonomic Issues

Separated from *O. lacustris* sensu stricto on molecular and biogeographical grounds (Taylor et al., 2009b).

Distribution

Known only from the Ukagura, Uluguru and Udzungwa Mts of the Eastern Arc.

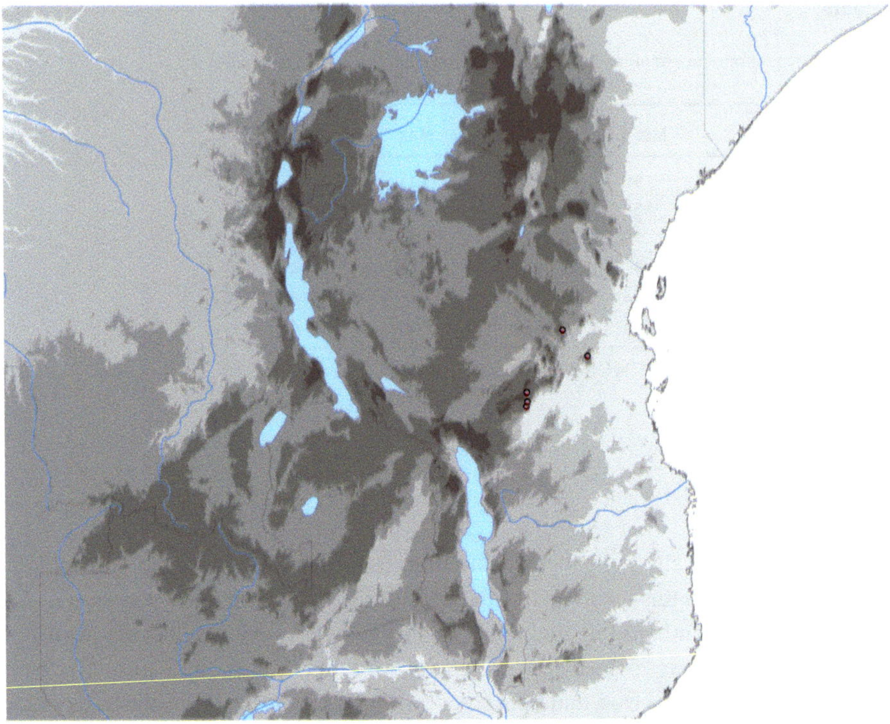

Figure 2.633 Distribution map

Measurements

External and cranial measurements (mm) and mass (g) for *Otomys* cf *lacustris*, sexes combined

	Mean	Min	Max	N
Mass*	132.0	106	180	3
HB*	166.5	160	180	4
Tail*	100.0	93	108	4
HF*	32.0	30	34	4
Ear*	22.2	20	24	4
GLS	38.2	36.3	40.6	7

*Taylor et al. (2009b)

Otomys cf *sungae*

Diagnosis

Otomys cf *sungae* from Nyika Plateau is morphometrically distinguishable from *O. sungae* sensu stricto from the northern Eastern Arc Mts on complex ventral cranial shape characters, including a narrower zygomatic arch, proportionately longer rostrum and toothrow and palatal foramina which are relatively short. Additionally, *O. sungae* has a modal number of 7 M^3 laminae, compared to 6 in *O.* cf *sungae* (Taylor et al., 2009b).

Taxonomic Issues

Taylor et al. (2011) showed clearly that topotypic *O. sungae* from Usambara Mts was phylogenetically distantly related to the Nyika Plateau population formerly classified as *O. sungae*. No prior name exists for this un-described species which awaits formal description.

Distribution

Known only from the Nyika Plateau of Malawi and neighbouring Zambia.

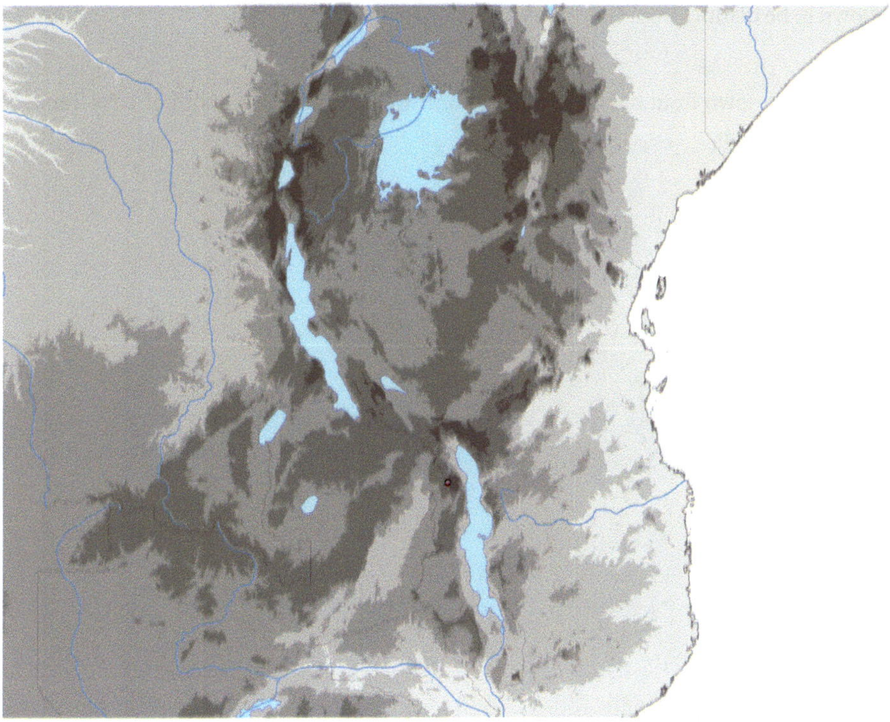

Figure 2.634 Distribution map

Measurements

External and cranial measurements (mm) and mass (g) for *Otomys* cf *sungae*, sexes combined

	Mean	Min	Max	N
Mass	-	-	-	-
HB	-	-	-	-
Tail	-	-	-	-
HF	-	-	-	-
Ear	-	-	-	-
GLS*	39.0	37.4	40.3	8

*Taylor et al. (2014)

Otomys cheesmani Taylor et al., 2011

Diagnosis
This species is distinguished by the presence of two deep grooves in the lower incisors, together with its large size (GLS > 43 mm).

Taxonomic Issues
Recently shown to be distinct from *O. typus* sensu lato (Taylor et al., 2011).

Distribution
This species is only known from two localities in north-western Ethiopia, south of Lake Tana, at elevations of 2100-2500 m.

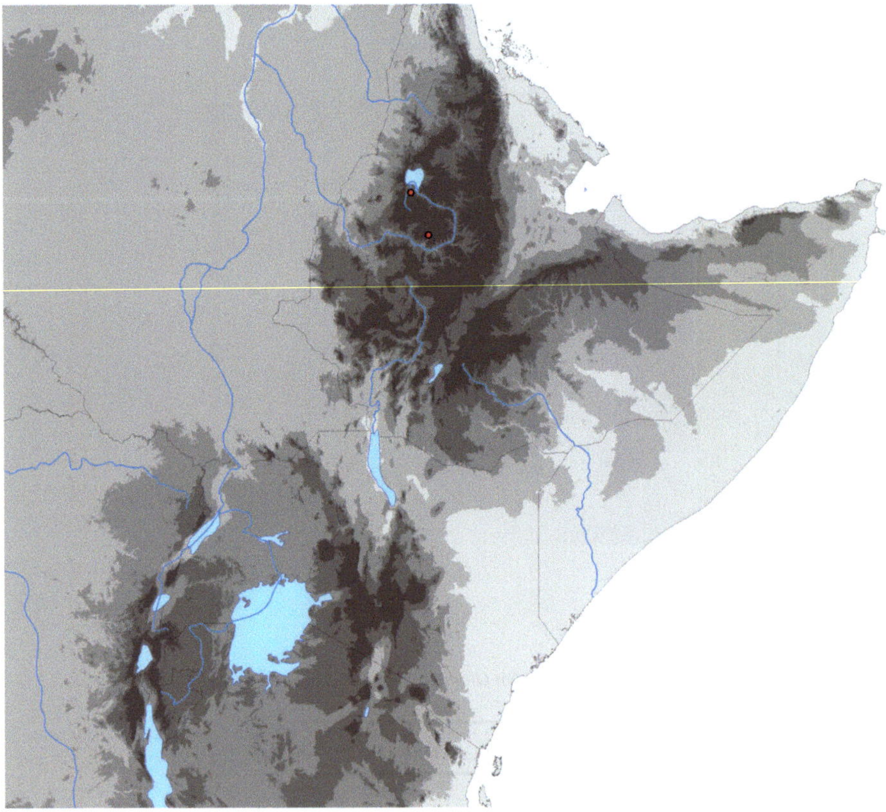

Figure 2.635 Distribution map

Measurements

External and cranial measurements (mm) and mass (g) for *Otomys cheesmani*, sexes combined

	Mean	Min	Max	N
Mass	-	-	-	-
HB*	188.8	165	210	6
Tail*	93.0	77	106	6
HF*	30.2	28	31	6
Ear*	23.2	22	24	6
GLS*	45.3	43.3	48.2	6

*Taylor et al. (2009b)

Otomys cuanzensis Hill & Carter, 1937

Diagnosis

It co-occurs narrowly with *O. anchietae*, which is much larger in size (GLS > 45 mm), as well as with *O. angoniensis maximus*, which has a slit-like petrotympanic foramen (round in *O. cuanzensis*). It can be distinguished from *O. irroratus* by having nasal bones which are not sharply expanded, and from *O. tropicalis* by having a modal number of 6 laminae in upper M^3 as opposed to 7 in *O. tropicalis*.

Taxonomic Issues

It was mistakenly included as a synonym of *O. angoniensis* by Musser & Carleton (1993) but is currently treated as a distinct species pending a detailed taxonomic review (Musser & Carleton, 2005).

Distribution

This species occurs throughout the drainage basin of the Cuanza River and other rivers in north-western Angola flowing into the Atlantic Ocean.

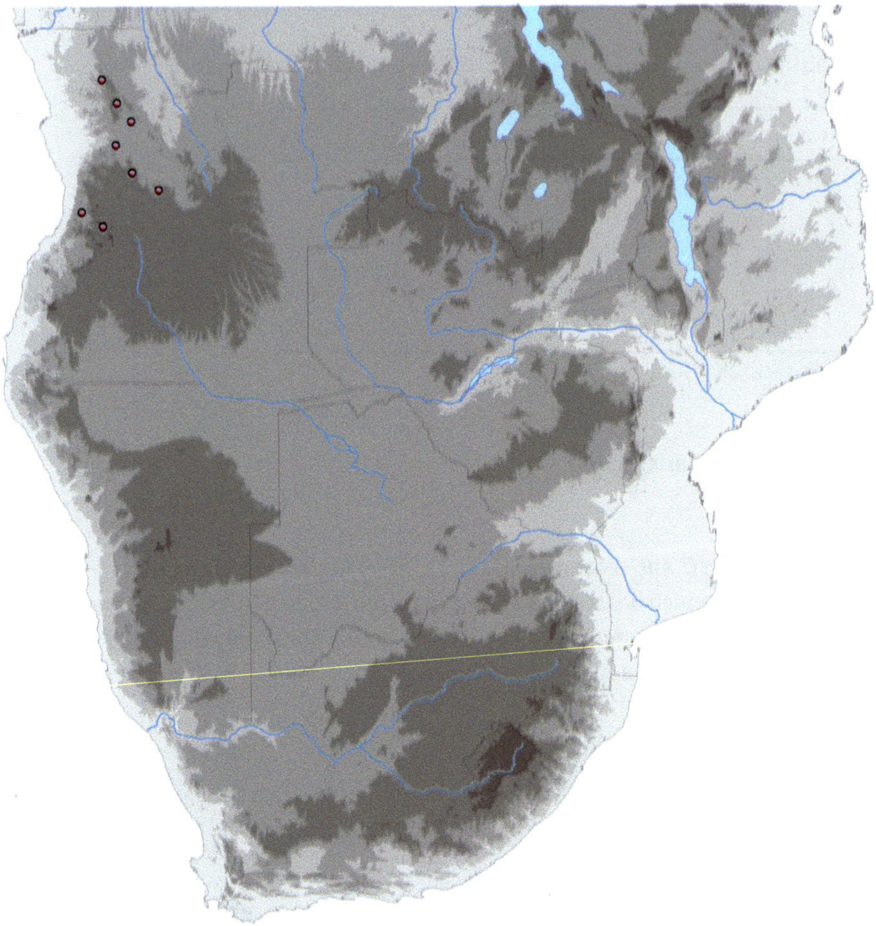

Figure 2.636 Distribution map

Measurements

External and cranial measurements (mm) and mass (g) for *Otomys cuanzensis*, sexes combined

	Mean	Min	Max	N
Mass	-	-	-	-
HB*	180.3	178	184	3
Tail*	104.7	100	109	3
HF*	32	31	33	3
Ear*	22	22	22	3
GLS	39.8	37.7	43.3	7

*Happold (2013)

Otomys dartmouthi Thomas, 1906

Diagnosis
Defined on molecular and biogeographic grounds, being the only *Otomys* species confined to the alpine zone (> 3300 m) in the Ruwenzori Mts. Morphologically it can be distinguished by having two deep grooves on the lower incisors (former *typus* complex) as well as consistently only 6 laminae in M^3. This combination of characters is found elsewhere only in a few individuals of *O. yaldeni* which usually has 7 laminae in M^3 and is also distinctly smaller in size (GLS < 36 mm).

Taxonomic Issues
Previously included within polytypic *O. typus* (Musser & Carleton, 2005) but shown to be specifically distinct on molecular grounds, most closely related to *O. tropicalis* from Kenya (Taylor et al., 2011).

Distribution
This species is confined to the alpine zone (3300-4400 m) of Ruwenzori Mts of south-western Uganda and contiguous DRC.

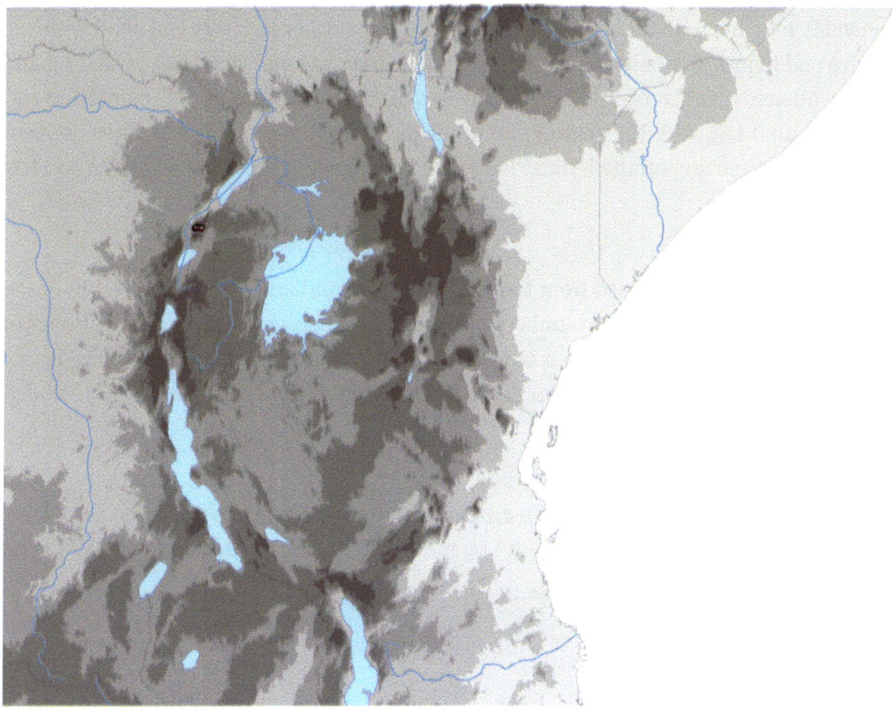

Figure 2.637 Distribution map

Measurements

External and cranial measurements (mm) and mass (g) for *Otomys dartmouthi*, sexes combined

	Mean	Min	Max	N
Mass	-	-	-	-
HB*	150	150	150	1
Tail*	93	93	93	1
HF*	26.5	26.5	26.5	1
Ear*	25	25	25	1
GLS*	37.8	36.6	39.1	9

*Taylor et al. (2011)

Otomys denti Thomas, 1906

Diagnosis

The species can be diagnosed by having a single groove on the lower incisors, 4 laminae in M_1 and 5-6 laminae in M^3, together with dark blackish-brown dorsal pelage, black feet and a tail which is blackish above and below (not paler ventrally). Although *O. sungae* and *O.* cf *sungae* share many of the above-mentioned characters, they are slightly less blackish in overall pelage coloration and also have a bicoloured tail, with the ventral surface slightly paler than the dorsal. They are geographically distinct, occurring on the Nyika Plateau (*O.* cf *sungae*) and northern Eastern Arc Mts (*O. sungae*).

Taxonomic Issues

Two subspecies have been recognised from the south (*O. d. kempi*) and north (*O. d. denti*) of the Albertine Rift based on having modal numbers of 6 and 5 laminae in M^3, respectively. However these populations cannot be distinguished morphologically or genetically and do not merit specific status based on current evidence (Taylor et al., 2009b). Until recently, *O. sungae* was included within *O. denti* (Taylor et al., 2009b).

Distribution

They occur throughout the Ruwenzori Mts (*O. d. denti*) and the mountains to the south associated with the Albertine Rift, extending from DRC and Uganda in the north, south through Rwanda and Burundi.

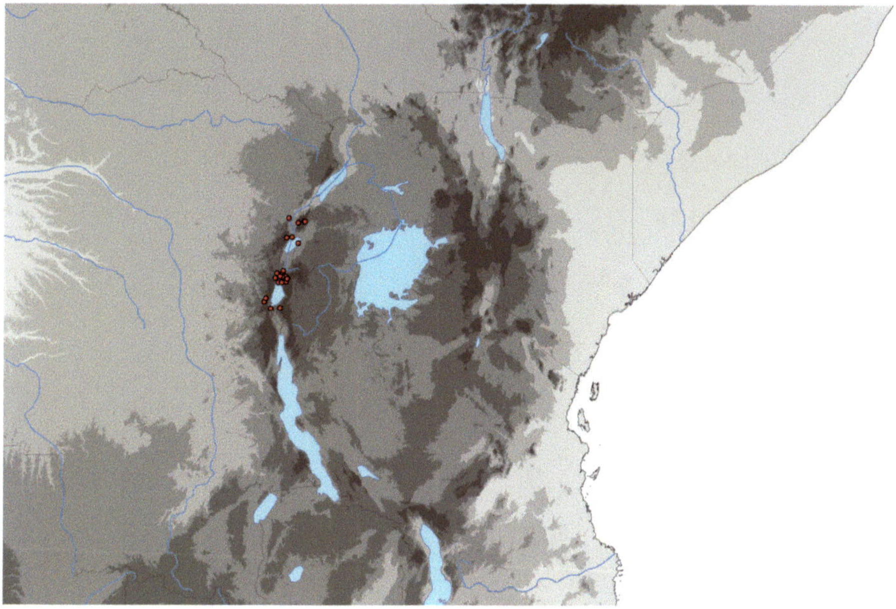

Figure 2.638 Distribution map

Measurements

External and cranial measurements (mm) and mass (g) for *Otomys denti*, sexes combined

	Mean	Min	Max	N
Mass*	92.4	67	106	6
HB*	152.2	139	165	11
Tail*	88.2	78	102	11
HF*	29.8	27	31.5	11
Ear*	21.6	20	23	11
GLS*	37.8	34.0	41.8	61

*Taylor et al. (2009b)

Otomys dollmani Heller, 1912

Diagnosis

A dark-coloured and diminutive *Otomys* with a relatively long tail (63% of HB) and 6 laminae in M^3.

Taxonomic Issues

The validity of this species, as distinct from *O. orestes*, was discussed by Carleton & Byrne (2006).

Distribution

This species is confined to Mt Gargues, Kenya.

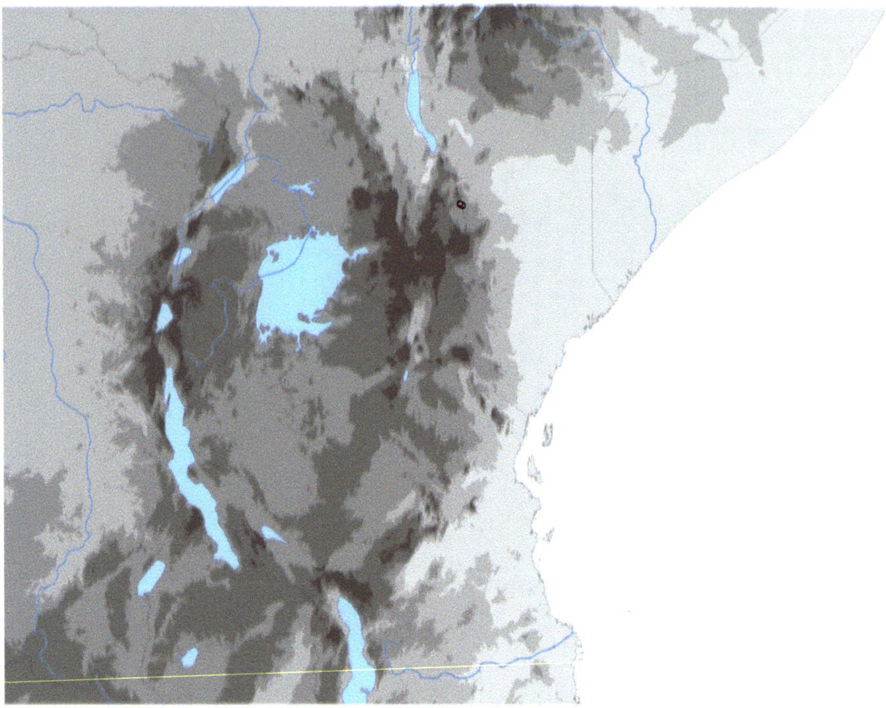

Figure 2.639 Distribution map

Measurements

External and cranial measurements (mm) and mass (g) for *Otomys dollmani*, sexes combined

	Mean	Min	Max	N
Mass	-	-	-	-
HB*	144.7	138	150	6
Tail*	92.0	85	105	6
HF*	25.1	23	27	6
Ear	-	-	-	-
GLS*	35.5	33.7	37.4	6

*Carleton & Byrne (2006)

Otomys fortior Thomas, 1906

Diagnosis

This species is defined on molecular, chromosomal and ecological grounds but cannot easily be diagnosed on morphological grounds. Like all members of the *O. typus* complex it has two deep grooves in the lower incisors. In terms of size, chromosomes and number of M^3 laminae (modal number 8), it is most like *O. typus* and *O. helleri*; however it is slightly larger in cranial size and has a narrower interorbital region and a shallower braincase compared to these species (Taylor et al.,

2011). It is the only forest-dwelling *Otomys* and the only *Otomys* occurring in the south-western montane forests of Ethiopia.

Taxonomic Issues

Previously included in *O. typus* but shown to be distinct on molecular, chromosomal, ecological and subtle cranial shape differences (Taylor et al., 2011).

Distribution

This species is restricted to montane evergreen forests in south-western Ethiopia.

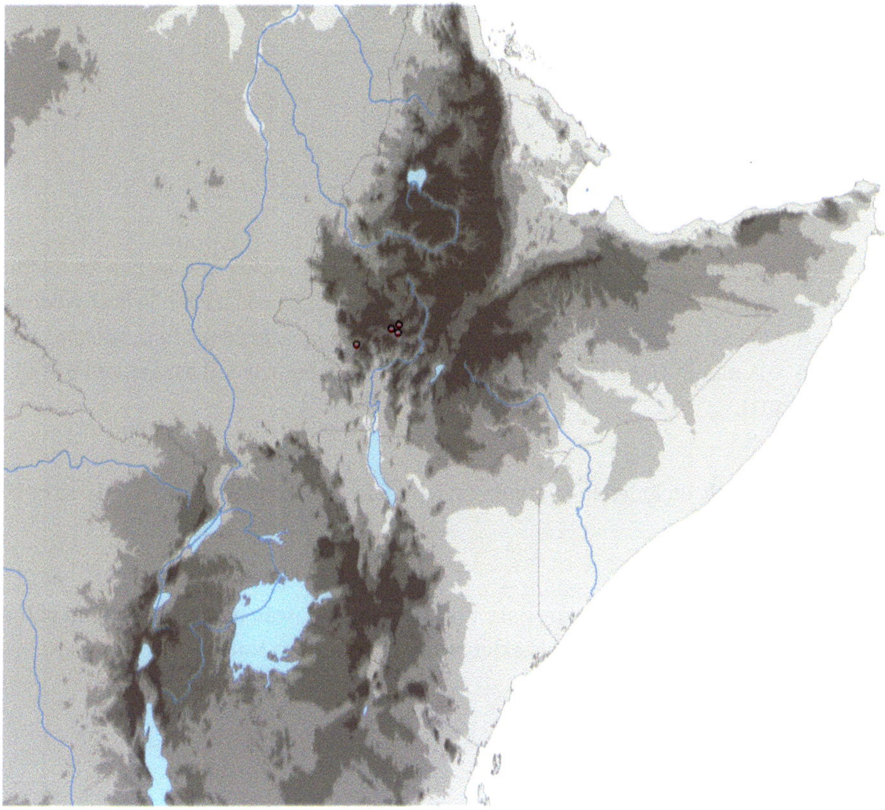

Figure 2.640 Distribution map

Measurements

External and cranial measurements (mm) and mass (g) for *Otomys fortior*, sexes combined

	Mean	Min	Max	N
Mass	-	-	-	-
HB*	160.2	145	182	8
Tail*	93.7	85	98	8
HF*	28.8	28	30	8
Ear*	21.3	20	26	8
GLS*	39.6	36.6	42.6	8

*Taylor et al. (2011)

Otomys helleri Frick, 1914

Diagnosis

It co-occurs with *O. yaldeni* in the region of the Bale Mts, where the two species can be distinguished on chromosomal characters (2n = 58; aFN = 58 in *O. helleri*; 2n = 56, aFN = 54 in *O. yaldeni*) as well as on cranio-dental (modal number of laminae in M^3 is 8 in *O. helleri*; 7 in *O. yaldeni*) and craniometric (GLS > 35 mm in *O. helleri*; < 35 mm in *O. yaldeni*) characters (Lavrenchenko et al., 1997; Taylor et al., 2011).

Taxonomic Issues

This species was previously included in *O. typus* (Taylor et al., 2011), and recorded as “Species B” by Lavrenchenko et al. (1997).

Distribution

This species occurs in alpine habitats in the Bale Mts and Arussi Plateau on the eastern rim of the Ethiopian Rift in southern Ethiopia, possibly extending southwards to Malka.

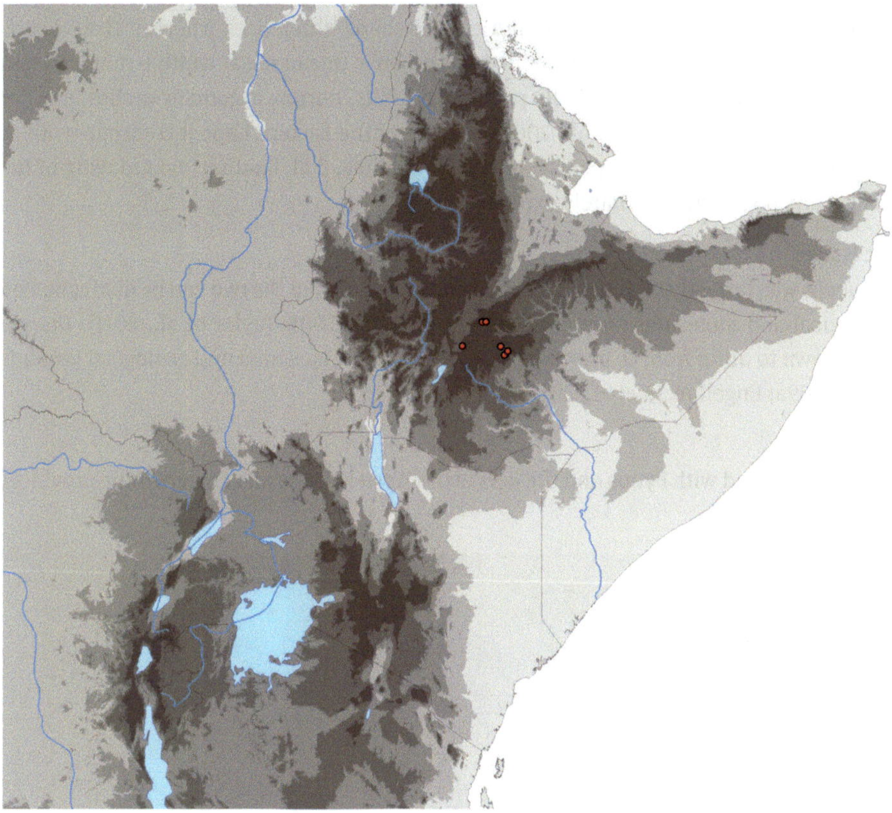

Figure 2.641 Distribution map

Measurements

External and cranial measurements (mm) and mass (g) for *Otomys helleri*, sexes combined

	Mean	Min	Max	N
Mass	-	-	-	-
HB*	151.6	123	185	12
Tail*	80.8	68	95	12
HF*	27.0	23	30	12
Ear*	21.8	20	24.5	12
GLS*	38.7	34.6	43.3	12

*Taylor et al. (2011)

Otomys irroratus (Brants, 1827)

Diagnosis

In the Fynbos biome of the Western Cape, it co-occurs narrowly with *O. laminatus*, which has 8-10 M^3 laminae as opposed to 6 in *O. irroratus*, as well as with *O. karoensis*, which is distinguished by its pallid buffy coloration, smaller size and cranial shape differences whereby *O. karoensis* has shorter and narrower, moderately expanded nasal bones (mean nasal width 6 mm, cf 7 mm in *O. irroratus*), more anterior and medially positioned orbits, narrow zygomatic arches, a wider braincase and larger bullae (Taylor et al., 2005). In parts of the Eastern Cape it co-occurs with *O. unisulcatus* which has a much paler pelage, only 4 M^3 laminae, 3 M_1 laminae, no grooving of the lower incisors, and nasals that are barely expanded.

Taxonomic Issues

Otomys tropicalis was formerly included in *O. irroratus* sensu lato but the two can be distinguished on morphological and molecular grounds (Taylor & Kumirai, 2001; Taylor et al., 2011). *Otomys auratus* was shown to differ from *O. irroratus* on molecular, chromosomal and ecological grounds (Taylor et al., 2009a; Engelbrecht et al., 2011).

Distribution

This species is associated with Fynos and thickets in the Western Cape and Eastern Cape provinces of South Africa.

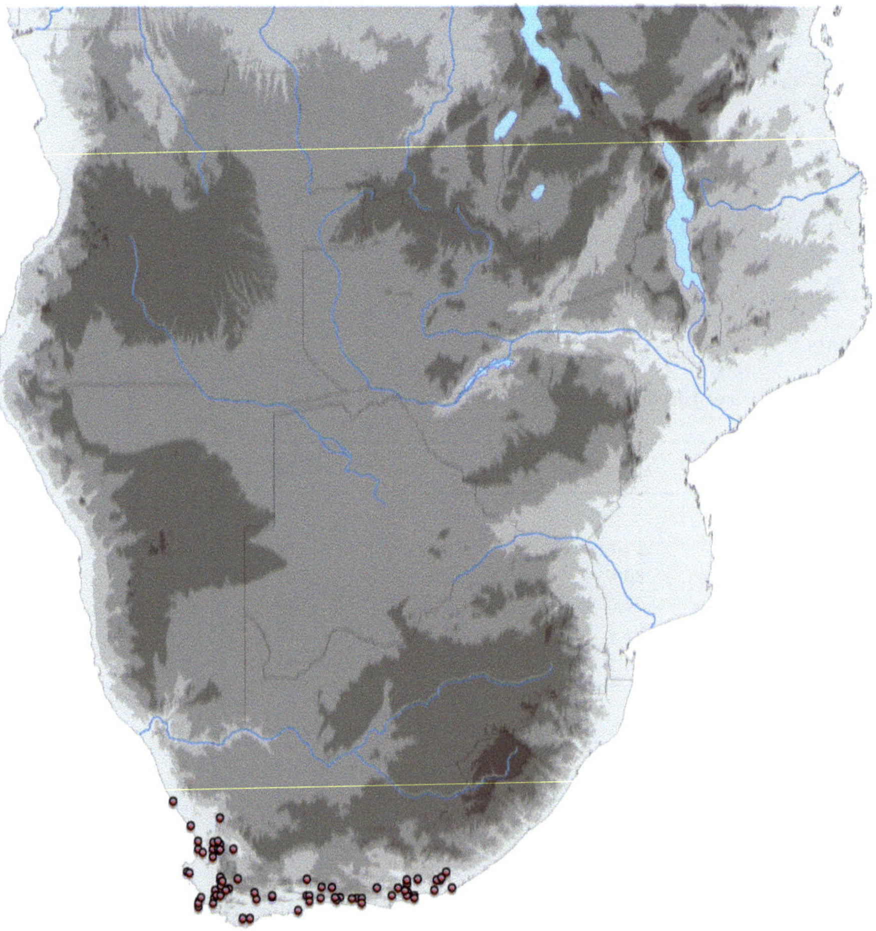

Figure 2.642 Distribution map

Measurements

External and cranial measurements (mm) and mass (g) for *Otomys irroratus*, sexes combined

	Mean	Min	Max	N
Mass	132.4	80	201	8
HB	164.0	130	190	12
Tail	101.7	85	114	11
HF	30.4	27	35	13
Ear	22.1	19	25	18
GLS	40.0	34.1	45.3	88

Otomys jacksoni Thomas 1891

Diagnosis

This species is the smallest-sized (GLS < 36 mm), pallid-coloured, very short-tailed (43% of HB) member of the *O. typus* "complex", having two deep grooves on the lower incisors. It can be distinguished from *O. barbouri* co-occurring in the alpine zone of Mt Elgon by the larger size, single lower incisor groove and possession of 5 M_1 laminae in the latter species. It can be distinguished from *O. tropicalis elgonis* occurring on the lower slopes of Mt Elgon (below the bamboo line) by the larger size and possession of one deep and one shallow groove on the lower incisor of the latter species.

Taxonomic Issues

Until recently considered to be a synonym or subspecies of *O. typus* (Taylor et al., 2011).

Distribution

This species occurs in the alpine zone (3300-4200 m) of Mt Elgon which straddles Kenya and Uganda.

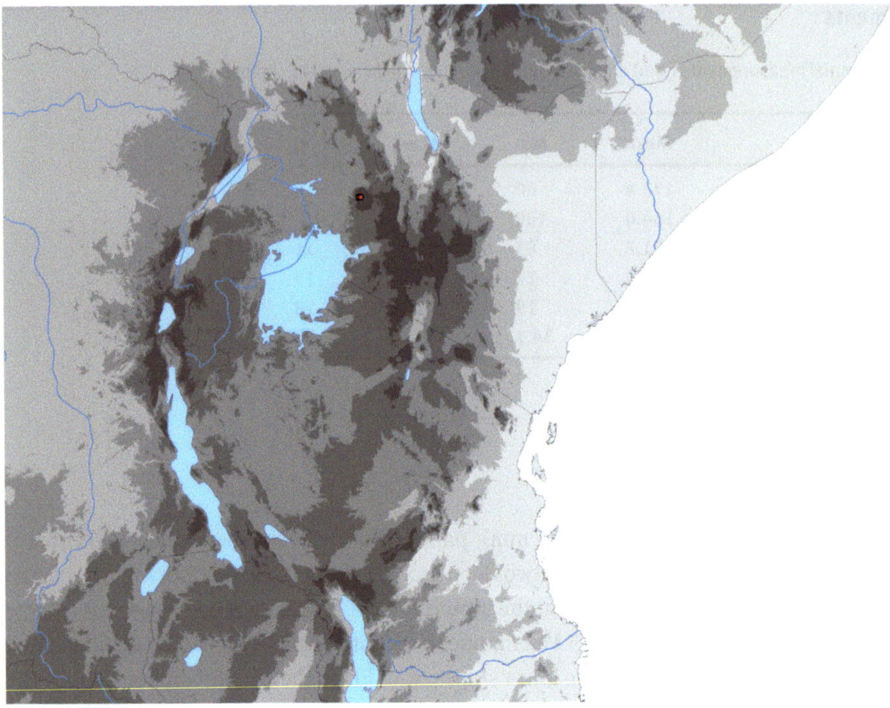

Figure 2.643 Distribution map

Measurements

External and cranial measurements (mm) and mass (g) for *Otomys jacksoni*, sexes combined

	Mean	Min	Max	N
Mass*	89.2	70	121	44
HB*	143	120	171	35
Tail*	67.5	57	82	36
HF*	23	19	26	36
Ear*	22.5	19	26	35
GLS**	35.1	34.3	35.5	3

*Clausnitzer (2000); **Taylor et al. (2011)

Otomys karoensis Roberts, 1931

Diagnosis

This species can be distinguished from both *O. irroratus*, co-occurring in the Fynbos of the Western Cape, as well as *O. auratus* co-occurring in grasslands of the southern Drakensberg, by having narrow and short nasal bones which are not sharply expanded, smaller size on average and a more faded pallid to pallid-buffy coloration. Taylor et al. (1993) provide a discriminant function equation employing 10 cranial variables, which completely separates *O. karoensis* and *O. irroratus*.

Otomys karoensis can be separated from *O. laminatus*, with which it co-occurs narrowly in the Western Cape, by the large size and high number of M^3 laminae in the latter species (8-10, cf 5-6 in *O. karoensis*).

Taxonomic Issues

Up until recently, *karoensis* was included as a synonym or subspecies of *O. saundersiae* (Meester et al., 1986; Taylor et al., 2005, but see Taylor et al., 1993, and Musser & Carleton, 2005). Taylor et al. (2009a) finally showed from chromosomal and molecular evidence that *O. saundersiae* from the Eastern Cape is a synonym of *O. irroratus*, leaving *O. karoensis* as the valid name for this taxon. The two disjunct populations of the species in Western Cape and northern Eastern Cape and Free State (see below) may be distinct species although molecular and chromosomal data are currently lacking. Individuals from the Free State are conspicuously smaller than other members of this species (Taylor et al., 2005).

Distribution

The species occurs in two disjunct regions, the Fynbos of the Cape Fold Belt Mts of the Western Cape as well as in the grasslands of southern Drakensberg in northern Eastern Cape and the central plateau grasslands of the southern Free State.

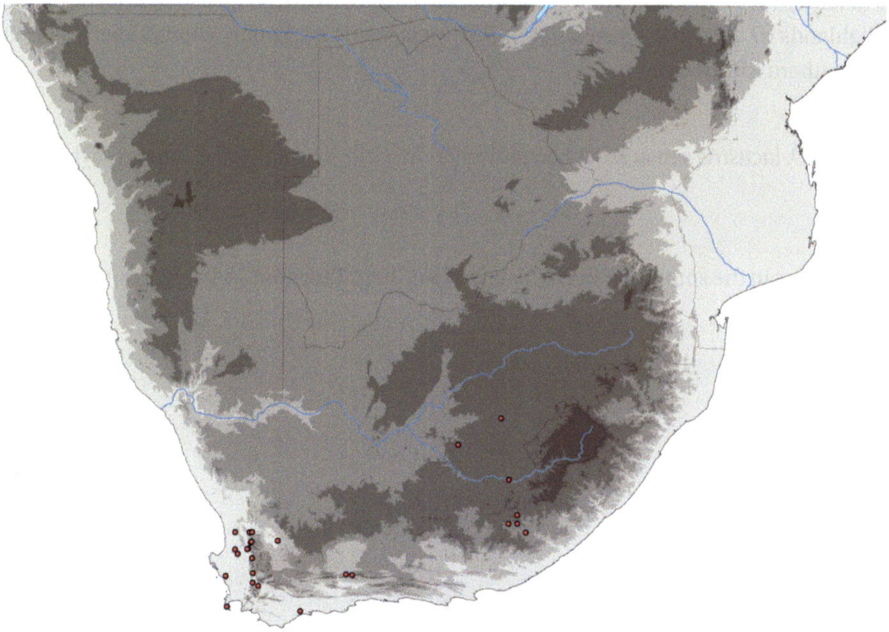

Figure 2.644 Distribution map

Measurements

External and cranial measurements (mm) and mass (g) for *Otomys karoensis*, sexes combined

	Mean	Min	Max	N
Mass*	108	78	139	4
HB*	134	110	154	11
Tail*	98	86	111	10
HF*	27	25	30	11
Ear*	21	18	23	11
GLS	36.2	31.3	39.6	38

*Happold (2013)

Otomys lacustris Allen & Loveridge, 1933

Diagnosis

This species is morphologically indistinguishable from *O.* cf *lacustris* sensu stricto but separated from this taxon by the Makambako Gap separating the Eastern Arc (*O.* cf *lacustris*) from the southern Rift Highlands (*O. lacustris* sensu stricto). *Otomys lacustris* is the only *Otomys* species occurring in the southern Rift Highlands.

Taxonomic Issues

Separated from *O. lacustris* sensu stricto on molecular and biogeographical grounds (Taylor et al., 2009b).

Distribution

This species occurs in the southern Rift Highlands of southern Tanzania and neighbouring Zambia and Malawi.

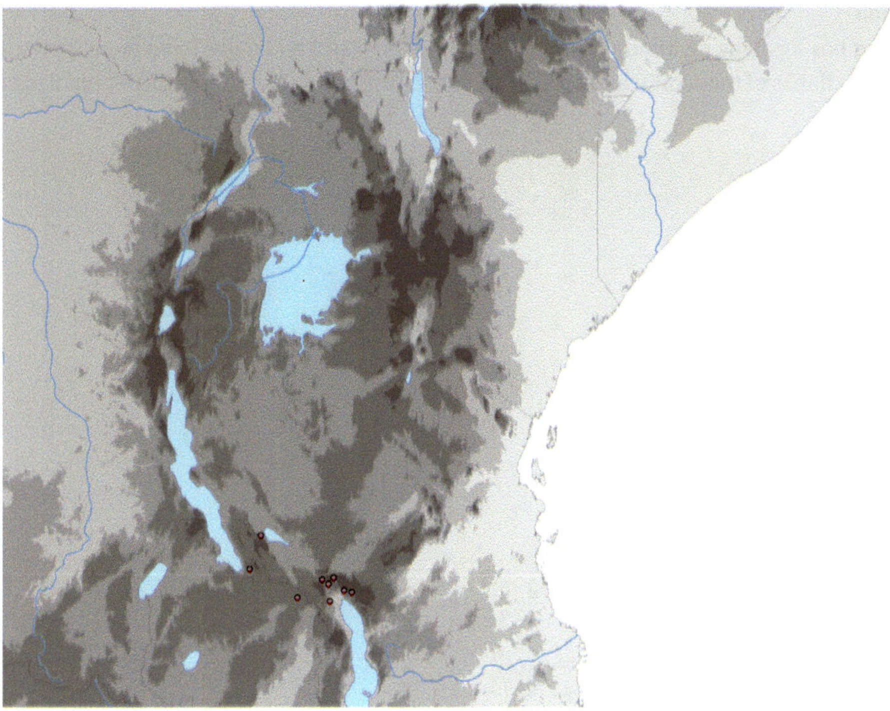

Figure 2.645 Distribution map

Measurements

External and cranial measurements (mm) and mass (g) for *Otomys lacustris*, sexes combined

	Mean	Min	Max	N
Mass	-	-	-	-
HB*	167.9	155	185	10
Tail*	100.0	93	108	4
HF*	30.2	30	32	10
Ear*	22.5	20	25	10
GLS*	38.5	35.2	41.4	14

*Taylor et al. (2009b)

Otomys laminatus Thomas & Schwann, 1905

Diagnosis

Although similar in external appearance to co-occurring *O. irroratus* and *O. auratus*, it can be distinguished by the high number of laminae in both the M^3 (8-10) and M_1 (7) molars.

Taxonomic Issues

It is possible that the two disjunct populations represent different species.

Distribution

This species has a patchy distribution in the Western Cape (Paarl and Cape Town areas) as well as the eastern flanks of the central and northern Drakensberg in KwaZulu-Natal and Mpumalanga provinces.

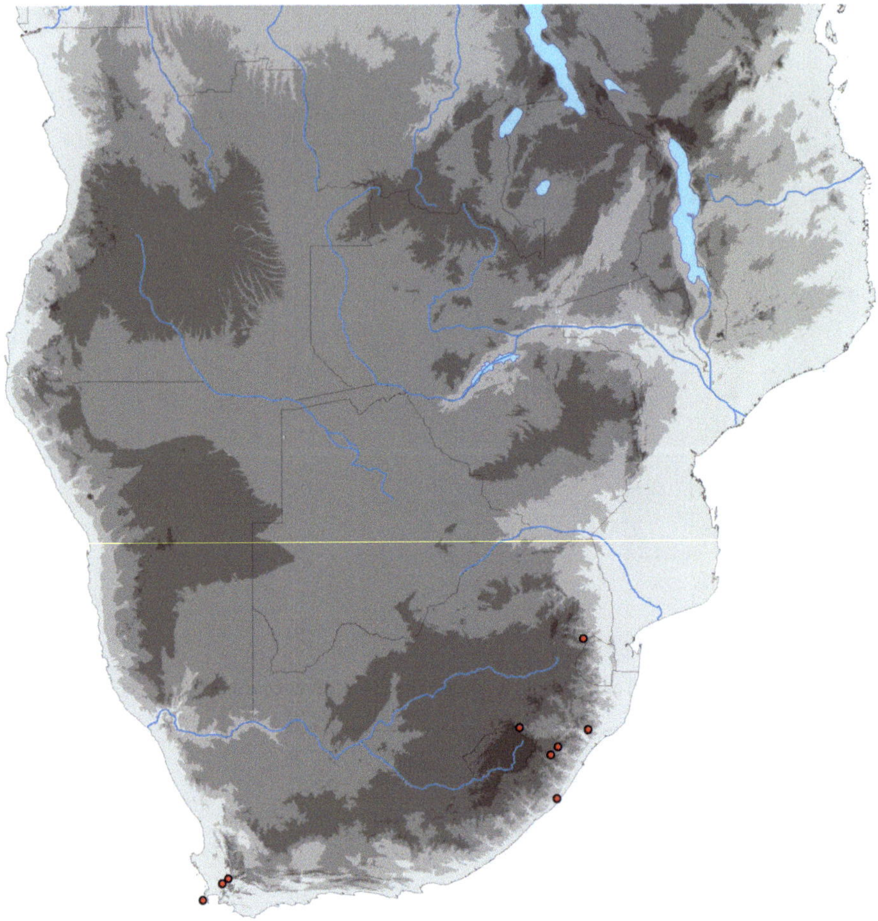

Figure 2.646 Distribution map

Measurements

External and cranial measurements (mm) and mass (g) for *Otomys laminatus*, sexes combined

	Mean	Min	Max	N
Mass*	165	140	264	5
HB*	188.5	158	213	8
Tail*	108	97	115	8
HF*	30	22	34	8
Ear*	22.5	17	26	8
GLS*	44.6	42.7	47.1	7

*Happold (2013)

Otomys occidentalis Dieterlen & Van der Straaten, 1992

Diagnosis

It is the only *Otomys* species found on Mt Oku and the Gotel Mts and can be easily distinguished from *O. burtoni* found on nearby Mt Cameroon by its smaller size and having 5 laminae in M^3.

Taxonomic Issues

This species was relatively recently described by Dieterlen & Van der Straeten (1992). Its closest relative appears to be *O. burtoni* (Taylor et al., 2014).

Distribution

This species is only found in Mt Oku and Gotel Mts of the Cameroon Volcanic Line in Cameroon and Nigeria.

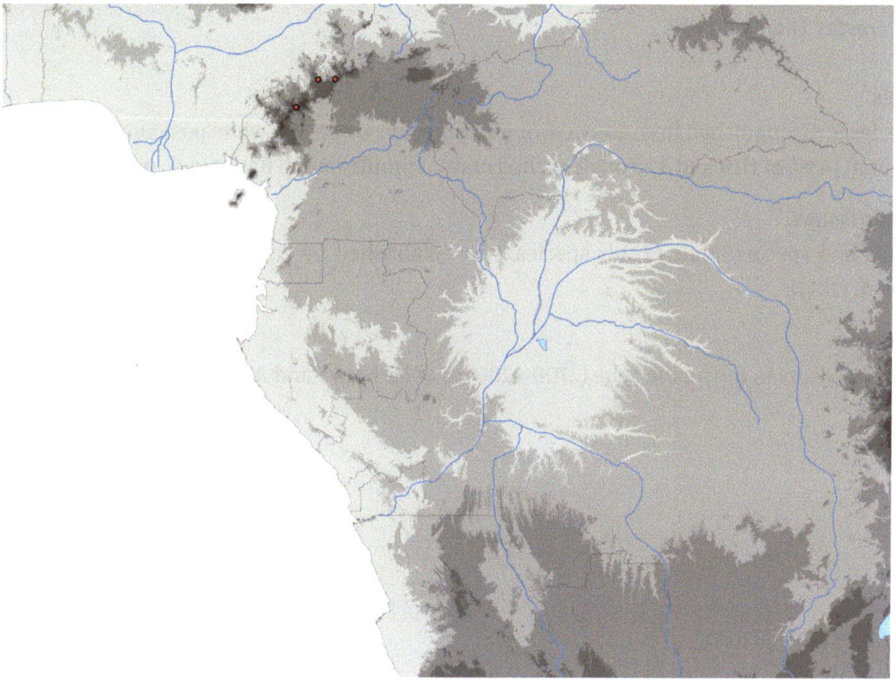

Figure 2.647 Distribution map

Measurements

External and cranial measurements (mm) and mass (g) for *Otomys occidentalis*, sexes combined

	Mean	Min	Max	N
Mass*	77	69	88	3
HB*	143	131	160	4
Tail*	78	68	88	4
HF*	28	24	29	4
Ear*	20	18	22	5
GLS**	36.6	34.5	38.7	24

*Happold (2013); **Taylor et al. (2014)

Otomys orestes Thomas, 1900

Diagnosis

This species is distinguished from co-occurring *O. tropicalis* by having white post-aural markings, a shorter tail (43% of HB) and a strongly arched cranial profile (much flatter in *O. tropicalis*).

Taxonomic Issues

The status and species limits of this species were established by Carleton & Byrne (2006) and Taylor et al. (2011).

Distribution

This species occurs in the alpine zone (2700-4200 m) of Mt Kenya and Aberdare Mts in Kenya.

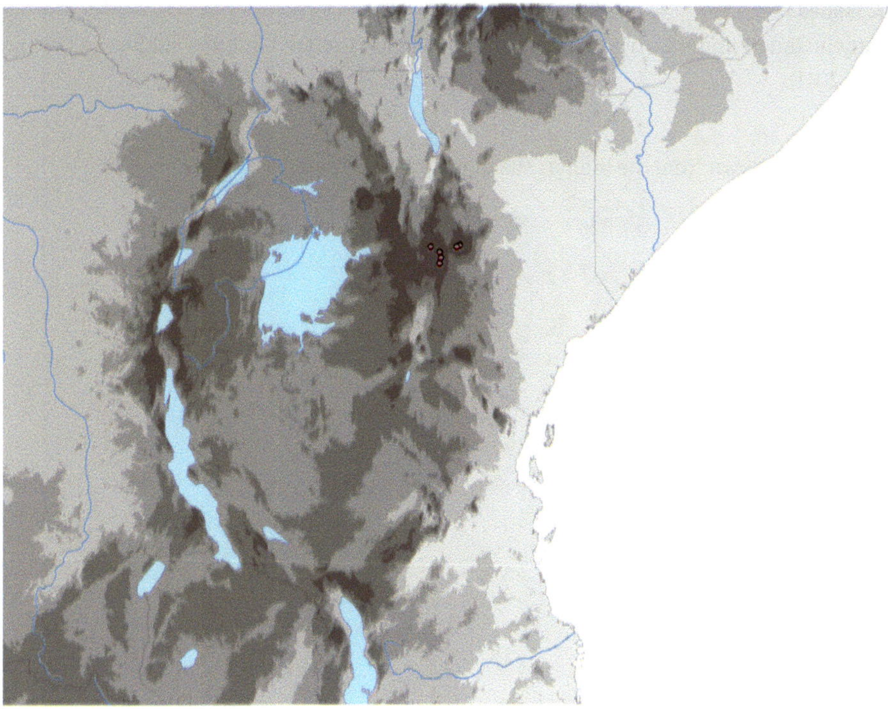

Figure 2.648 Distribution map

Measurements

External and cranial measurements (mm) and mass (g) for *Otomys orestes*, sexes combined

	Mean	Min	Max	N
Mass	-	-	-	-
HB*	159.6	135	175	10
Tail*	73.1	61	93	10
HF*	28.5	25	30	10
Ear**	22.0	21	25	5
GLS**	38.4	37.3	39.5	14

*Carleton & Byrne (2006); **Taylor et al. (2011)

Otomys simiensis Thomas, 1902

Diagnosis

This relatively small-sized, dark-brown coloured *Otomys* species has a unique karyotype (2n = 54; aFN = 54). It co-occurs in the Simien Mts of Ethiopia with *O. typus* from which it differs by having only 7 laminae in M^3 as opposed to 8-9 in *O. typus*. It also has a double-layered pelage due to the preponderance of longer black bristles among the guard hairs (Taylor et al., 2011).

Taxonomic Issues

Until recently included under *O. typus*, the species is distantly related to other Ethiopian *Otomys* (Taylor et al., 2011).

Distribution

This species is only known from the Simien Mts of northern Ethiopia.

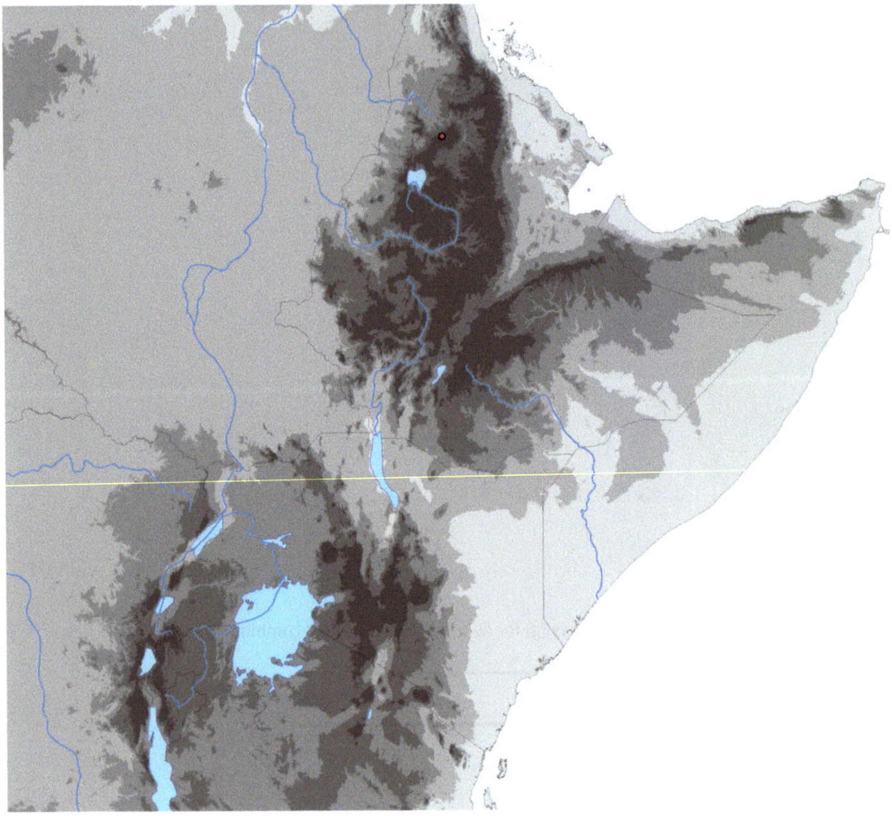

Figure 2.649 Distribution map

Measurements

External and cranial measurements (mm) and mass (g) for *Otomys simiensis*, sexes combined

	Mean	Min	Max	N
Mass	-	-	-	-
HB*	155.2	135	172	5
Tail*	82.8	77	88	5
HF*	25.0	24	26	5
Ear*	19.4	17.5	21.8	5
GLS*	37.0	35.2	39.1	5

*Taylor et al. (2011)

Otomys sloggetti Thomas, 1902

Diagnosis
This species is a medium-sized, pale greyish-buff-coloured *Otomys* with slit-like petrotympanic foramen, a single groove on lower incisors and only 4-5 laminae in M^3.

Taxonomic Issues
Some authors placed *O. sloggetti* and *O. unisulcatus* in the genus *Myotomys* (Musser & Carelton, 2005). Although *O. sloggetti* is considered to be defined by primitive dental characters, its position on the phylogeny of Otomyinae remains unresolved (Maree, 2002; Taylor et al., 2004).

Distribution
It is found at high elevations (> 2000 m) in the Drakensberg Mts of the Eastern Cape and KwaZulu Natal provinces of South Africa as well as Lesotho, with isolated populations from mountains in the Karoo.

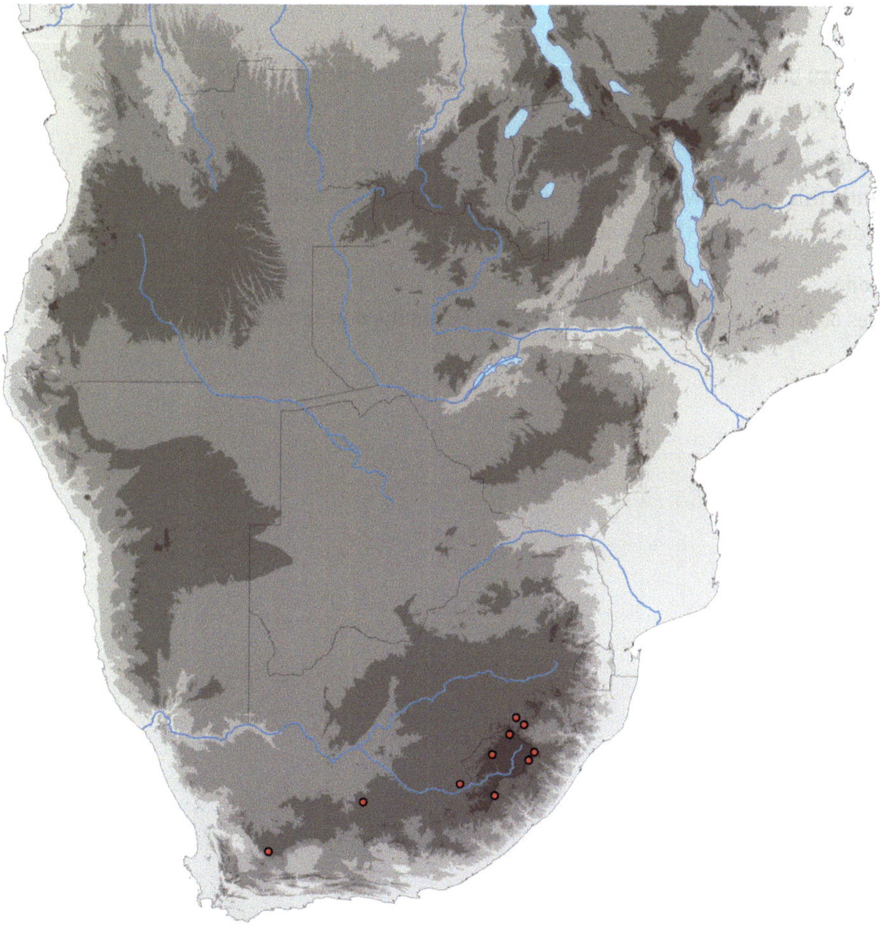

Figure 2.650 Distribution map

Measurements

External and cranial measurements (mm) and mass (g) for *Otomys sloggetti*, sexes combined

	Mean	Min	Max	N
Mass*	84	45	155	61
HB*	127	93	170	73
Tail*	62	47	84	73
HF*	28	21	30	73
Ear*	18	14	22	72
GLS*	37.6	35	40	11

*Happold (2013)

Otomys sungae Bohmann, 1943

Diagnosis
This species can be distinguished from co-occurring *O.* cf *lacustris* by the presence of 5 M^3 laminae in the latter species.

Taxonomic Issues
Until recently this species was included in *O. denti* but is diagnosable on molecular, morphological and ecological grounds (Taylor et al., 2009b).

Distribution
This species occurs on the Uluguru and Usambara Mts in the northern Eastern Arc Mts, Tanzania.

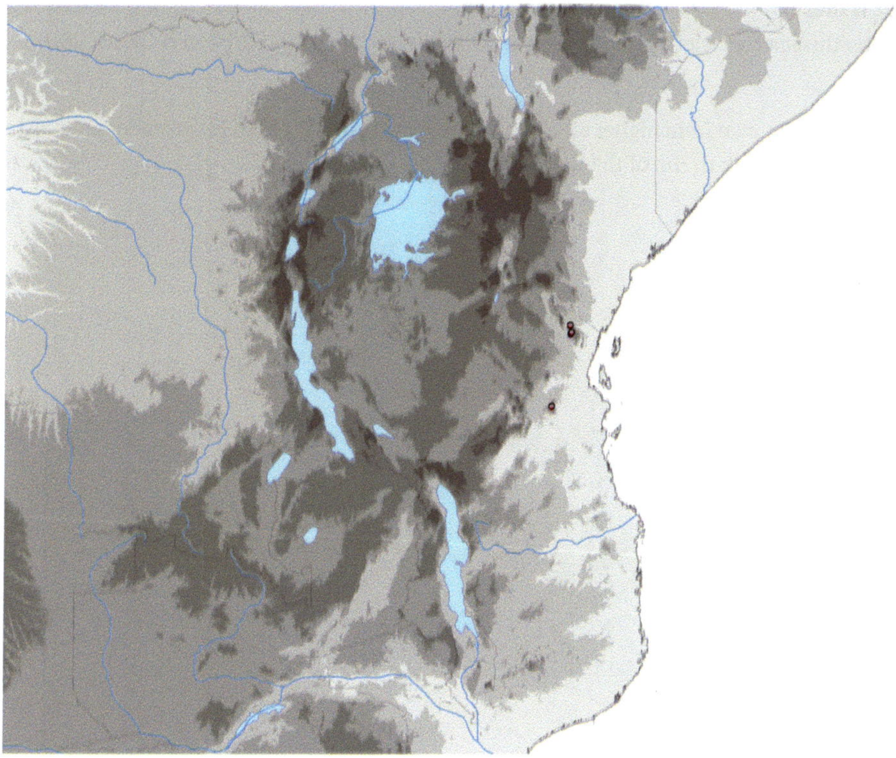

Figure 2.651 Distribution map

Measurements

External and cranial measurements (mm) and mass (g) for *Otomys sungae*, sexes combined

	Mean	Min	Max	N
Mass	-	-	-	-
HB*	164	150	180	5
Tail*	91	80	100	5
HF*	30.2	30	31	4
Ear*	23.6	20	28	5
GLS*	38.0	35.4	40.7	10

*Taylor et al. (2009b)

Otomys thomasi Osgood, 1910

Diagnosis

This large, buffy-coloured *Otomys* shares the same distinctively arched cranial profile as *O. orestes*, but differs from this species by its conspicuously larger size and absence of post-aural tufts (Taylor et al., 2011). It differs from co-occurring *O. angoniensis* by having a round petrotympanic foramen (slit-like in *O. angoniensis*).

Taxonomic Issues

Until recently this species was included in either *O. typus* or *O. orestes* (Carleton & Byrne, 2006).

Distribution

This species is only known from higher elevations (2450-2700 m) in the Mau Escarpment and Uasin Gishu Plateau of the western rim of the Kenyan Rift.

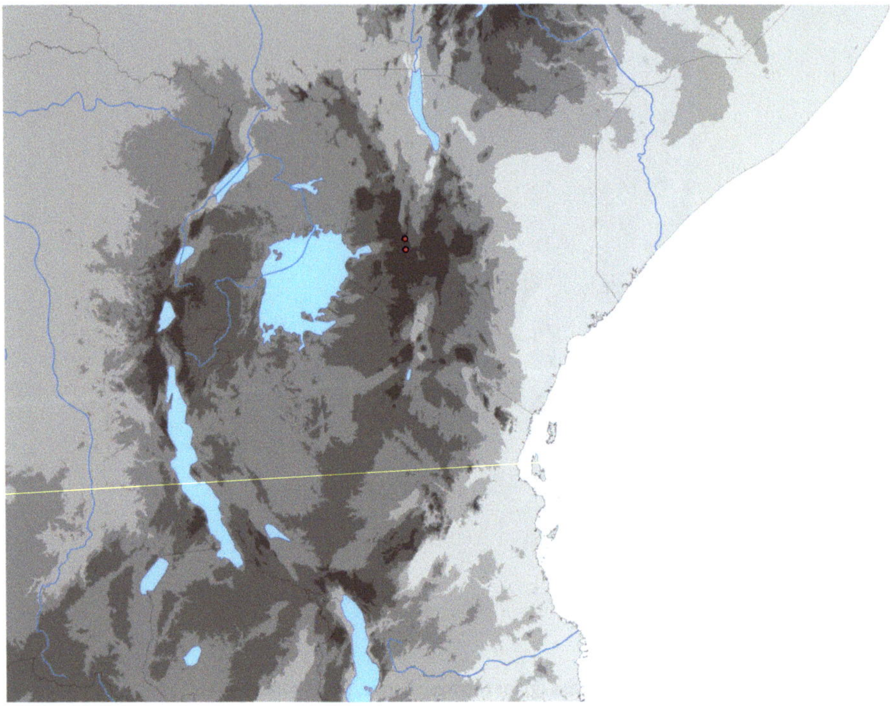

Figure 2.652 Distribution map

Measurements

External and cranial measurements (mm) and mass (g) for *Otomys thomasi*, sexes combined

	Mean	Min	Max	N
Mass	-	-	-	-
HB*	170.7	162	182	3
Tail*	88.3	84	93	3
HF*	30.3	30	31	3
Ear*	21.3	20	22	3
GLS*	40.8	40.2	41.4	5

* Taylor et al. (2011)

Otomys tropicalis Thomas, 1902

Diagnosis

This species is distinguished by having a round petrotympanic foramen, 7 M^3 laminae and one shallow and one deep groove on lower incisors. Throughout its East African distribution it can be distinguished from lower elevation populations of *O. angoniensis* (slit-shaped petrotympanic foramen) and higher elevation populations of *O. jacksoni*, *O. dartmouthi* (smaller size and possessing two lower incisor grooves), *O. orestes* (post-aural tufts and arched cranial profile), *O. thomasi* (arched cranial profile, buffy coloration and larger size, GLS > 40 mm) and *O. zinki* (more sharply angled nasals and lateral incisor groove very faint, restricted in distribution to Mt Kilimanjaro).

Taxonomic Issues

This taxon is clearly a composite of several species which still require revision (Taylor et al., 2011).

Distribution

As currently recognised, this species occurs throughout the Albertine Rift, Mt Elgon and the Kenyan Rift.

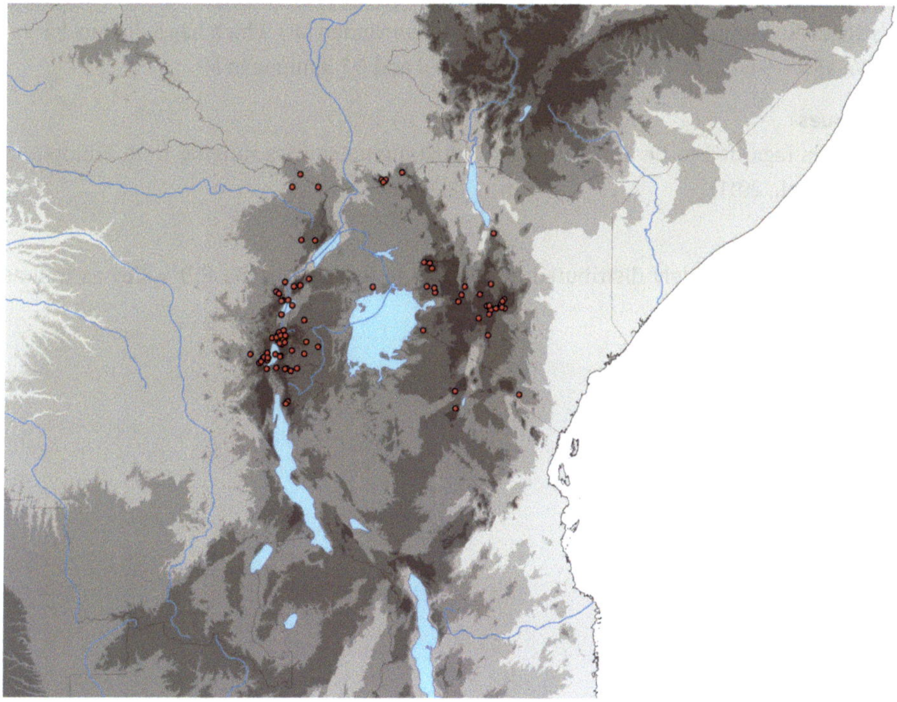

Figure 2.653 Distribution map

Measurements

External and cranial measurements (mm) and mass (g) for *Otomys tropicalis*, sexes combined

	Mean	Min	Max	N
Mass*	111.3	81	133	9
HB*	159.0	125	205	37
Tail*	86.7	52	102	37
HF*	29.9	26	34	36
Ear*	21.9	17	25	30
GLS*	39.6	35.8	44.0	124

* Taylor et al. (2011)

Otomys typus (Heuglin, 1877)

Diagnosis

This species is distinguished by having two deep grooves on the lower incisors and 8-9 laminae in M^3. It is distinguished from *O. simiensis* co-occurring in the Simien Mts which has a distinct karyotype (2n = 54; aFN = 54; cf. 2n = 58; aFN = 58 in *O. typus*) and 6-7 laminae in M^3.

Taxonomic Issues

Previously it was regarded to be a broadly distributed alpine species ranging from Ethiopia to Malawi (Taylor et al., 2011).

Distribution

This species is relatively widely distributed in Ethiopia to the west of the Rift valley as well as Hirna to the east of the Rift.

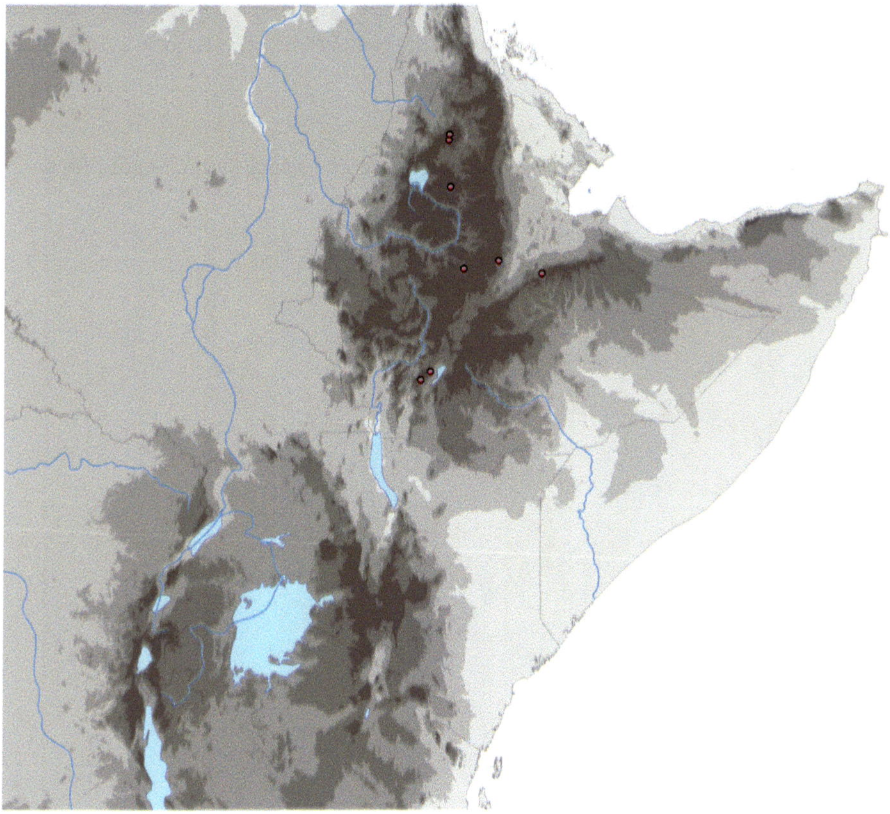

Figure 2.654 Distribution map

Measurements

External and cranial measurements (mm) and mass (g) for *Otomys typus*, sexes combined

	Mean	Min	Max	N
Mass	--	-	-	-
HB*	163.2	155	178	4
Tail*	83.7	77	92	4
HF*	26.0	25	27	4
Ear*	21.8	20	25	4
GLS*	38.3	37.1	40.2	4

* Taylor et al. (2011)

Otomys unisulcatus F. Cuvier, 1829

Diagnosis

This species is distinguished by its pallid buffy coloration and absence of grooving on lower incisors.

Taxonomic Issues

Some authors include *O. unisulcatus* and *O. sloggetti* in the genus *Myotomys* (Pocock, 1976; Musser & Carleton, 2005).

Distribution

This species occurs throughout the semi-arid Succulent Karoo and Nama-Karoo, South Africa.

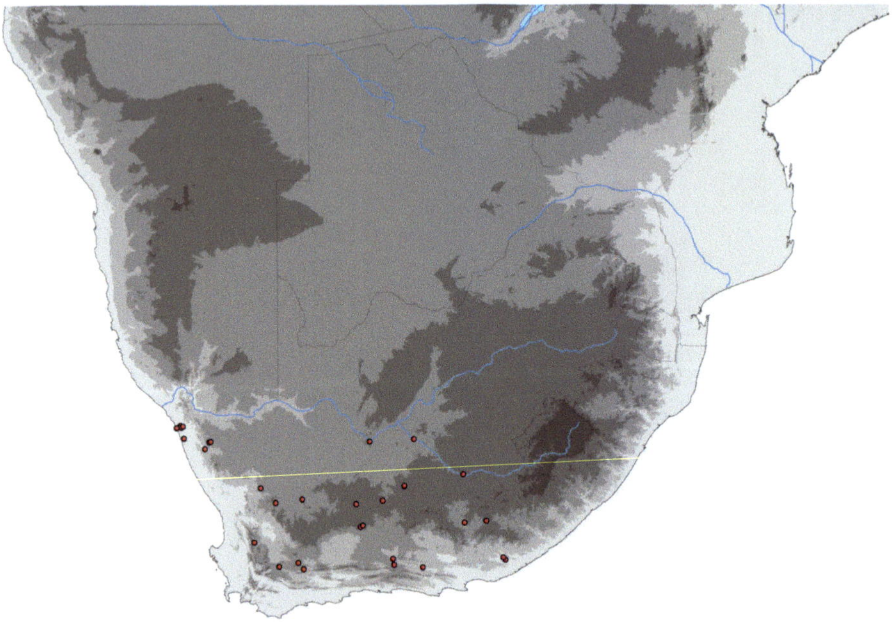

Figure 2.655 Distribution map

Measurements

External and cranial measurements (mm) and mass (g) for *Otomys unisulcatus*, sexes combined

	Mean	Min	Max	N
Mass	-	-	-	-
HB	152.5	110	205	40
Tail	92	63	106	40
HF	28.5	22	32	40
Ear	23.5	20	33	40
GLS	35.2	31.9	39.4	40

* Happold (2013)

Otomys uzungwensis Lawrence & Loveridge, 1953

Diagnosis

This species is distinguished by having two deep grooves on lower incisors, 4 M_1 laminae and 7 M^3 laminae, compared with co-occurring *O.* cf *lacustris* (one deep incisor groove and 5 M_1 laminae).

Taxonomic Issues

Until recently this taxon was included in *O. typus* but see Carleton & Byrne (2006) and Taylor et al. (2011).

Distribution

This species occurs on the Udzungwa Mts of Tanzania and the Nyika Plateau of Malawi and neighbouring Zambia.

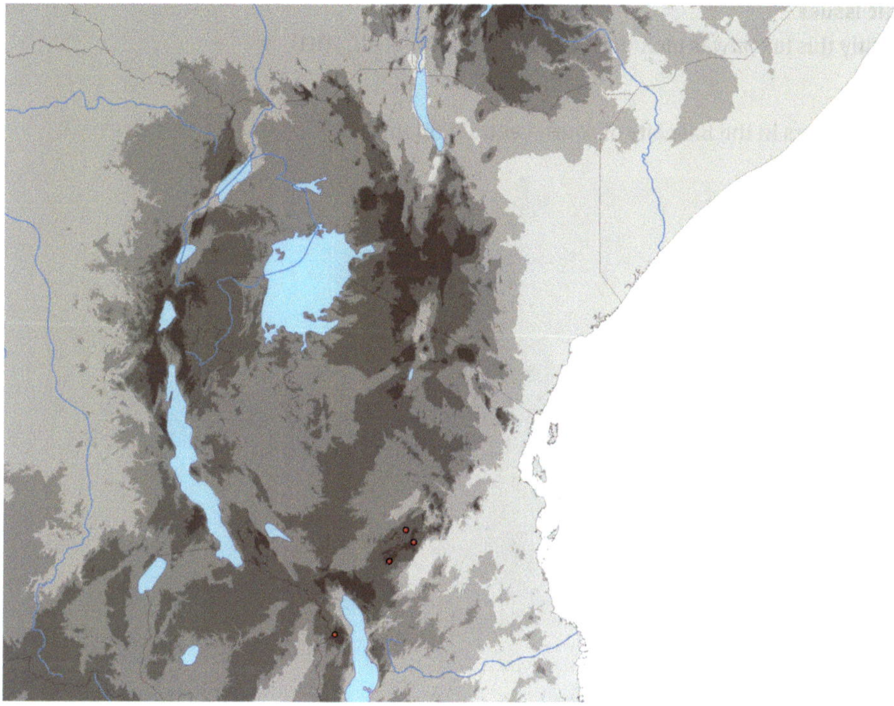

Figure 2.656 Distribution map

Measurements

External and cranial measurements (mm) and mass (g) for *Otomys uzungwensis*, sexes combined

	Mean	Min	Max	N
Mass	-	-	-	-
HB	-	-	-	-
Tail	-	-	-	-
HF	-	-	-	-
Ear	-	-	-	-
GLS*	37.3	36	38.9	10

* Taylor et al. (2011)

Otomys yaldeni Taylor et al. (2011)

Diagnosis

This species is distinguished by having two deep grooves on the lower incisors and a smaller size (GLS < 36 mm) than all other Ethiopian species. It also possesses a distinct karyotype (2n = 56; aFN = 54).

Taxonomic Issues

Until recently this taxon was included in *O. typus* (Taylor et al., 2011).

Distribution

This species occurs in the Bale Mts of south-central Ethiopia.

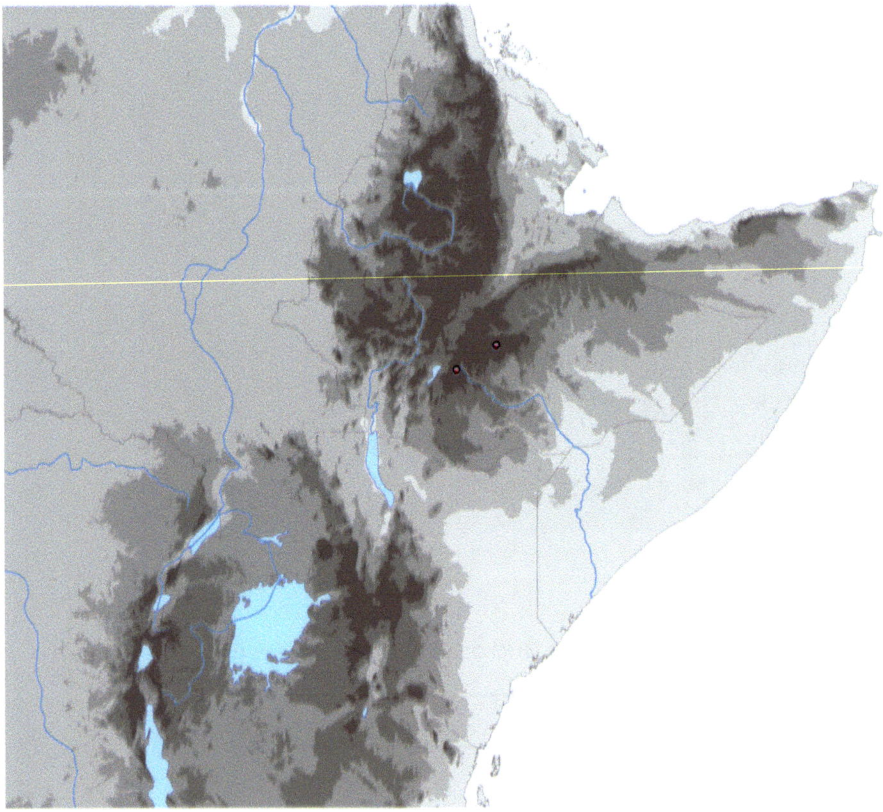

Figure 2.657 Distribution map

Measurements

External and cranial measurements (mm) and mass (g) for *Otomys yaldeni*, sexes combined

	Mean	Min	Max	N
Mass	-	-	-	-
HB*	143.3	125	155	3
Tail*	71.5	70	73	3
HF*	23.7	22	25	3
Ear*	19.2	18.5	20	3
GLS*	35.2	34.7	35.5	3

*Taylor et al. (2011)

Otomys zinki Bohmann, 1943

Diagnosis

They have one deep and one shallow lower incisor groove and are found only at higher elevations on Mt Kilimanjaro. *Otomys angoniensis* from lower elevations can be distinguished in having a slit-like petrotympanic foramen.

Taxonomic Issues

Until recently this taxon was included in *O. typus* (Taylor et al., 2011).

Distribution

This species occurs at higher elevations at Mt Kilimanjaro, Tanzania.

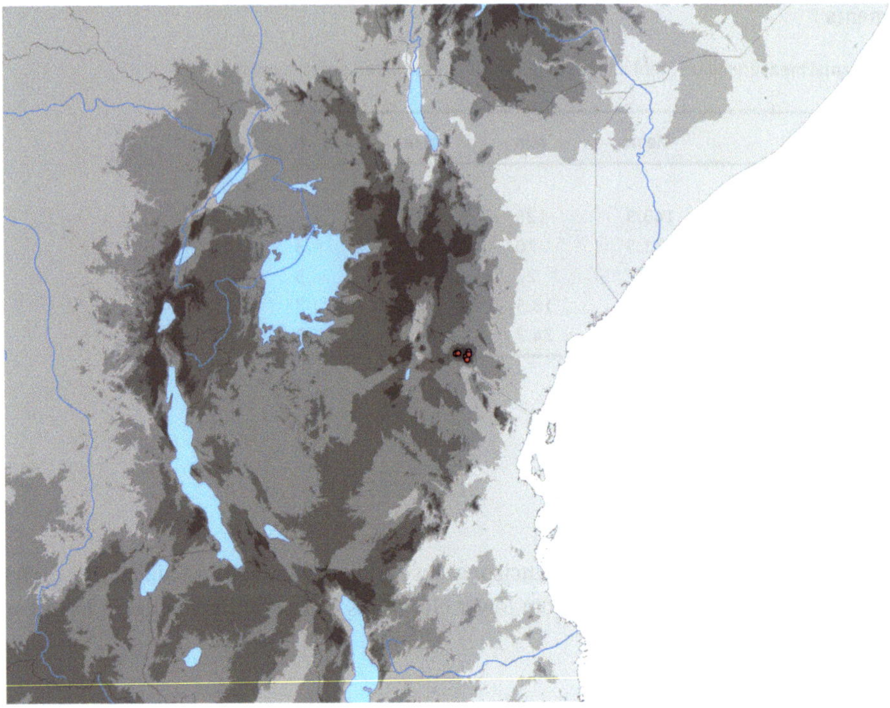

Figure 2.658 Distribution map

Measurements

External and cranial measurements (mm) and mass (g) for *Otomys zinki*, sexes combined

	Mean	Min	Max	N
Mass	-	-	-	-
HB*	161.6	135	186	16
Tail*	84.1	60	108	15
HF*	27.9	24	32	16
Ear*	-	-	-	-
GLS*	39.4	36.6	42.4	12

*Taylor et al. (2011)

Genus Account

Parotomys Thomas, 1918 (Whistling Rats)

Diagnosis

Along with members of the genus *Otomys*, the large size (HB 130-170 mm), stocky, short-tailed (60-80% of HB), vole-like appearance, shaggy coarse-grey pelage and the robust skull and laminate dentition make whistling rats unmistakeable. Their name "whistling rats" derives from the distinctive vocalization which comprises a single sharp high-pitched alarm call. *Parotomys* can be distinguished from *Otomys* by their more pallid pelage coloration (although *O. unisculcatus* also has a similar pallid pelage), the rectilinear shape of the nasal bones (more expanded anteriorly in *Otomys*) as well as by the considerable expansion of the bullae in *Parotomys* (bullae diameter is > 10 mm in *Parotomys* and < 9 mm in *Otomys*). *Parotomys* species occupy arid and semi-arid habitats in western Southern Africa whilst most *Otomys* species are adapted to mesic and/or montane habitats, or semi-arid Karoo habitats in the case of *O. unisculcatus* (Taylor et al., 2004).

(a)

(b)

Figure 2.659 a and b Photographs of *Parotomys brantsii* (Emmanuel Do Linh San).

Cranial Characters

The skull is large in size (GSL 30-40 mm), robust, ovoid to approximately square-shaped and often heavily ridged, especially in the case of the supraorbital and supraoccipital crests. The zygomatic arches are robust and parallel-sided ("boxy"). The nasal bone is prominent but rectilinear in shape and not expanded anteriorly as in *Otomys*. The anterior palatal foramen is relatively short and does not reach the first upper molar. The palate itself is folded. The anterior surface of the zygomatic plate is vertical and bluntly rounded dorsally. The bullae are large (> 10 mm in greatest diameter).

Figure 2.660 a-d Cranium of *Parotomys brantsii* (TM 22485) (from top left to bottom right: dorsal view, ventral view, lateral view and lower jaw).

Dentition

The incisors are opisthodont; upper incisors are grooved in *P. brantsii* but not *P. littledalei*, while lower incisors lack grooving in either species. Individual cusps are fused into a number of transverse laminae. The M^3 is the longest molar in the toothrow, having two (*P. brantsii*) or three (*P. littledalei*) laminae. The M^1 has three laminae and the M^2 has two. The M_1 is the longest tooth in the lower toothrow, having between three (*P. littledalei*) and four (*P. brantsii*) laminae. The M_2 and M_3 always have two laminae. The two upper toothrows are barely diverging posteriorly.

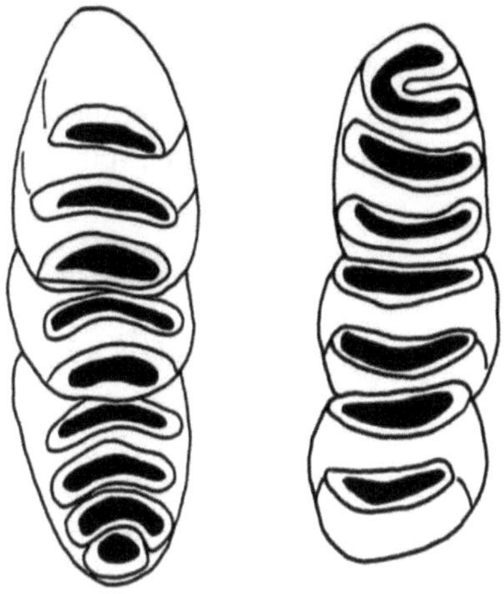

(a) (b)

Figure 2.661 Cheek teeth of *Parotomys brantsii* (TM 11176) (A: upper left toothrow; C: lower right toothrow) (drawn by C. Denys).

Habitat & Biogeography

Whistling rats are associated with the Namib and Kaokoveld deserts in Namibia, the semi-desert Succulent Karoo and Nama Karoo biomes of South Africa and southern Namibia and xeric Kalahari savannas of northern South Africa and the extreme south-west of Botswana.

Fossil Record

Parotomys fossils have been recovered from mid-Pleistocene deposits in South Africa (Winkler et al., 2010).

Biology & Ecology

Whistling rats are diurnal to crepuscular (Jackson, 1998a), generalist herbivores. *P. brantsii* has been shown to feed predominantly on a variety of annuals in spring, and mostly perennial succulents, grasses and shrubs in summer and winter (Nel & Rautenbach, 1974; Coetzee & Jackson, 1999; Jackson, 1998b; Jackson, 2001). They are central place foragers; smaller food items are eaten *in situ* whilst larger items are eaten at burrow entrances or stored underground (Jackson, 2001). Both species of *Parotomys* are burrow-dwellers that rely on the cooler temperatures of their complex burrows (with multiple nest chambers and numerous entrances) to escape high diurnal temperatures in their semi-desert environment (Du Plessis & Kerley, 1991; Du Plessis et al., 1992; Jackson, 2000; Jackson et al., 2002; Roper et al., 2002). *Parotomys brantsii* are gregarious but solitary, occupying separate burrows at the start of the breeding season, but burrow-sharing occurs with increasing densities in the post-breeding period. Males visit and solicit females during the breeding season indicating male-male competition but aggression appears limited (Jackson, 1999). *Parotomys brantsii* has a range of aerial and terrestrial predators and uses an "urgency-based" alarm call to alert conspecifics of predators depending on the immediate threat (Jackson, 2001).

Systematic Notes

The delimitation of genera with Otomyinae is unclear with three genera (*Otomys*, *Myotomys* and *Parotomys*) recognised by Musser & Carleton (2005). Due to the demonstrated paraphyly of *Myotomys* species (*unisulcatus* and *sloggetti*: Rambau et al., 1997; Maree, 2002; Taylor et al., 2004), only two genera (*Otomys* and *Parotomys*) are currently recognised pending a final phylogenetic analysis of species relationships (Happold, 2013). Although cranio-dental evidence suggests that *Parotomys* is primitive relative to *Otomys* (Pocock 1976; Taylor et al., 2004), fossil data do not support this and recent mtDNA-based molecular analyses fail to confirm the monophyly or presumed basal position of *Parotomys* or to resolve the deeper nodes of the Otomyinae (Taylor et al., 2009a, 2011, 2014).

Species List

Two species are recognised:
Parotomys brantsii (A. Smith, 1834)
Parotomys littledalei Thomas, 1918

Key to Species

1	Upper incisors grooved; M_1 with 4 laminae	*P. brantsii*
	Upper incisors not grooved; M_1 with 3 laminae	*P. littledalei*

Species Accounts

Parotomys brantsii (A. Smith, 1834)

Diagnosis

This species is similar in size and pelage to *P. littledalei* but distinguishable by having grooved upper incisors (ungrooved in *P. littledalei*) and four (not three) laminae in M_1. The alarm whistle is longer and higher pitched (10 kHz) compared with *P. littledalei* (7.7 kHz) (Le Roux et al., 2002).

Taxonomic Issues

Chromosome number 2n = 44; FN = 79/80 (Rambau et al., 1997).

Distribution

This species is restricted to the arid regions of Southern Africa occurring in western South Africa, southern Namibia and extreme south-western Botswana (Smithers, 1971). It is associated with open habitats and hard sandy substrates throughout the arid Karoo and southern Kalahari.

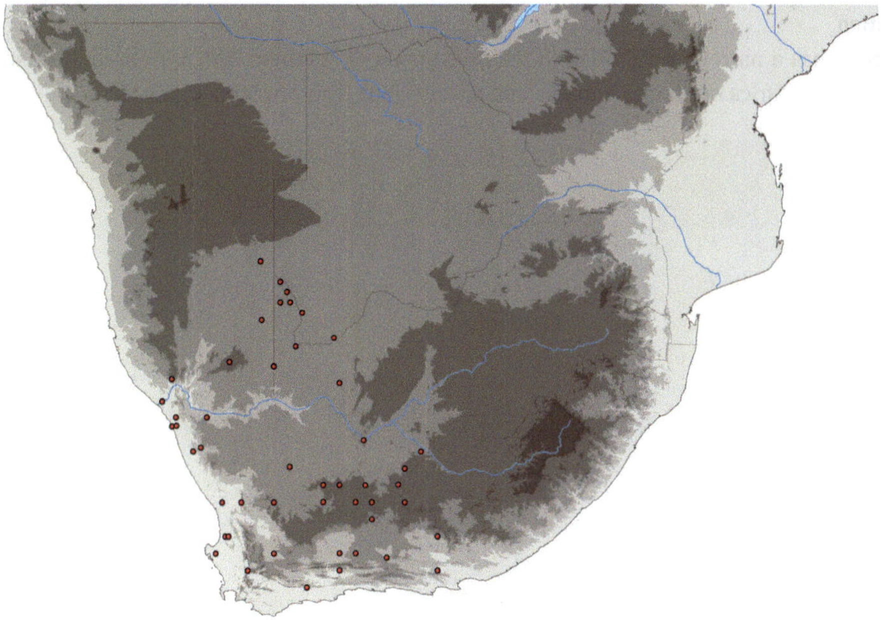

Figure 2.662 Distribution map

Measurements

External and cranial measurements (mm) and mass (g) for *Parotomys brantsii*, sexes combined

	Mean	Min	Max	N
Mass	94.6	75	111	5
HB	148.0	141	158	6
Tail	100.4	93	115	5
HF	31.6	28	33	4
Ear	17.3	15	22	6
GLS*	36.6	34.0	40.5	40

*Happold (2013)

Parotomys littledalei Thomas, 1918

Diagnosis

This species is similar in size and pelage to *P. brantsii* but distinguishable by having ungrooved upper incisors (grooved in *P. brantsii*) and three (not four) laminae in M_1. The alarm whistle is shorter and lower pitched (7.7 kHz) compared with *P. littledalei* (10 kHz) (Le Roux et al., 2002).

Taxonomic Issues

Chromosomal number 2n = 50 (Meester, 1988).

Distribution

This species has a narrow distribution in the driest parts of Southern Africa, from the western regions of South Africa north into Namibia and mostly along a narrow strip of desert.

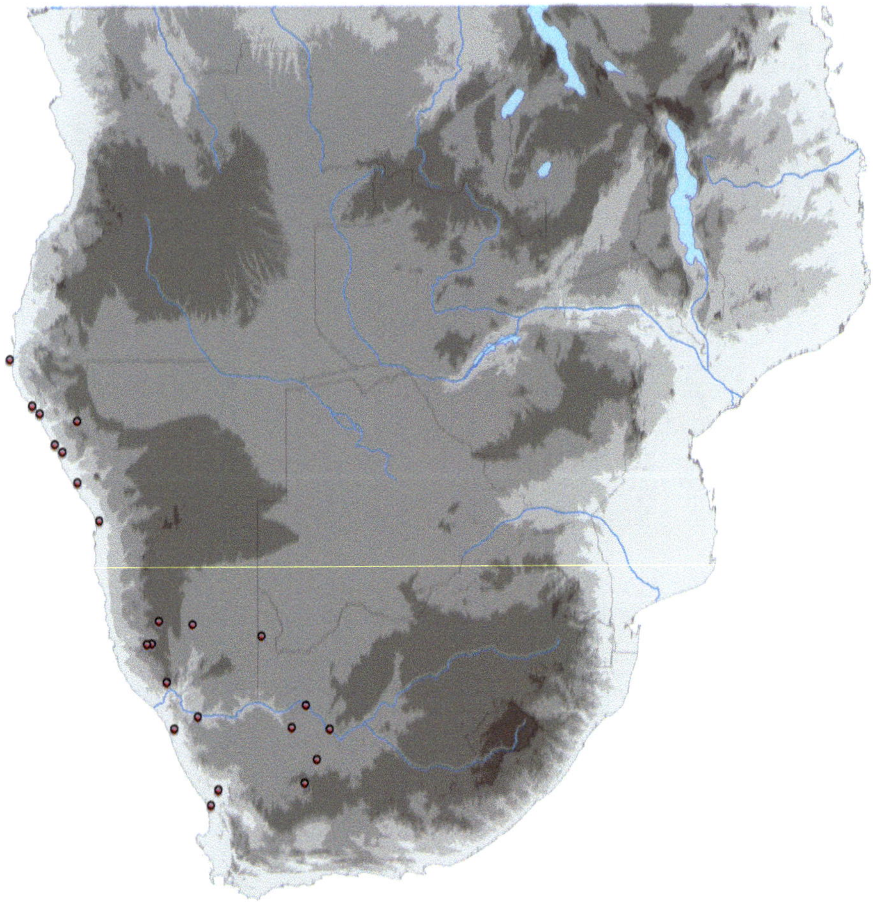

Figure 2.663 Distribution map

Measurements

External and cranial measurements (mm) and mass (g) for *Parotomys littledalei*, sexes combined

	Mean	Min	Max	N
Mass	105.4	97	110	5
HB	146.8	124	170	13
Tail	101.4	82	112	13
HF	29.1	26	34	12
Ear	19.6	17	25	13
GLS	35.7	32.5	38.8	10

Genus Account

Pelomys Peters, 1852 (Creek Rats)

Diagnosis

These medium to large sized rats (Mass ca. 50-140 g; HB ca. 120-150 mm) have a coarse buffy-brown to blackish dorsal pelage (similar in colour to *Arvicanthis*) with a black mid-dorsal stripe in some species, and a white, greyish or yellowish ventral pelage. This genus, however, is immediately recognisable and distinguishable from members of the genera *Arvicanthis*, *Lemiscomys*, *Desmomys* and *Rhabdomys* by grooved upper incisors. *Pelomys* differs from the similar sized *Mylomys* by having significantly shorter upper molar toothrow (> 7.7 mm in *Mylomys*, < 7.3 mm in *Pelomys*) and a mid-dorsal stripe (which may be faint or absent). The tail is coarsely scaled and usually just shorter or equal to head and body length (90-100% of HB) except for *P. isseli* (113%) and *P. hopkinsi* (135% of HB) where it is longer. Number of nipples, where known, is either 2 + 2 = 8 (*P. fallax* and *P. campanae*) or 1 + 2 = 6 (*P. hopkinsi*).

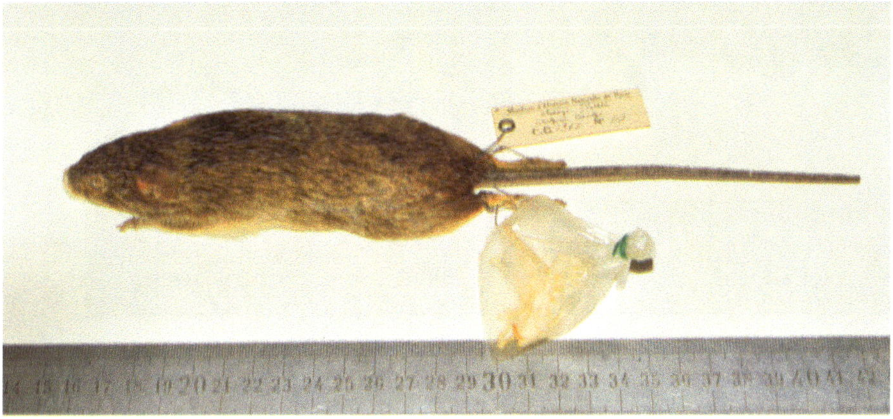

Figure 2.664 Photograph of *Pelomys* sp. museum specimen (MNHN-ZM-MO 1966-117) (Ara Monadjem).

Cranial Characters

The skull is medium to large in size (ca. 29-35 mm) and ovoid and well built in shape with distinct supraorbital and occipital ridges, and short and wide nasals which taper at the tip. The zygomatic arches are robust and the anterior edge of the zygomatic plate is concave. The palatal foramina stop just short of the M^1 alveolus.

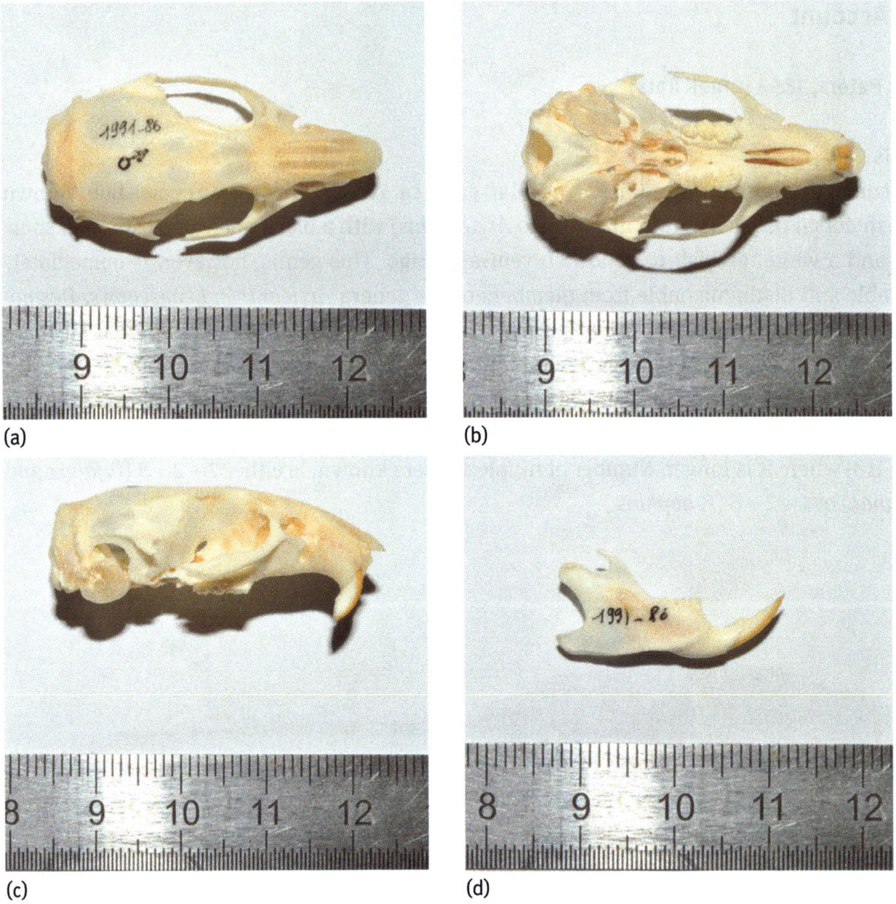

Figure 2.665 a-d Cranium of *Pelomys campanae* (MNHN-ZM-MO 1991-86) (from top left to bottom right: dorsal view, ventral view, lateral view and lower jaw).

Dentition

The upper incisors are orthodont to slightly opisthodont and grooved. The molars are quite heavy (MTR/GSL ± 0.16-0.17, Table 10) and the cusps are well separated with the transverse rows not distorted and the medial cusps (t2, t5, t8) on M^1 and M^2 enlarged and arranged in a straight line. The dentition is very similar to that of *Arvicanthis* but not as heavy in *Mylomys*. The cusp t3 is absent from M^3 (Group 8 of Figure 2.439). The M_1 displays an anterior median cusp before the first lobe and a posterior cingulum.

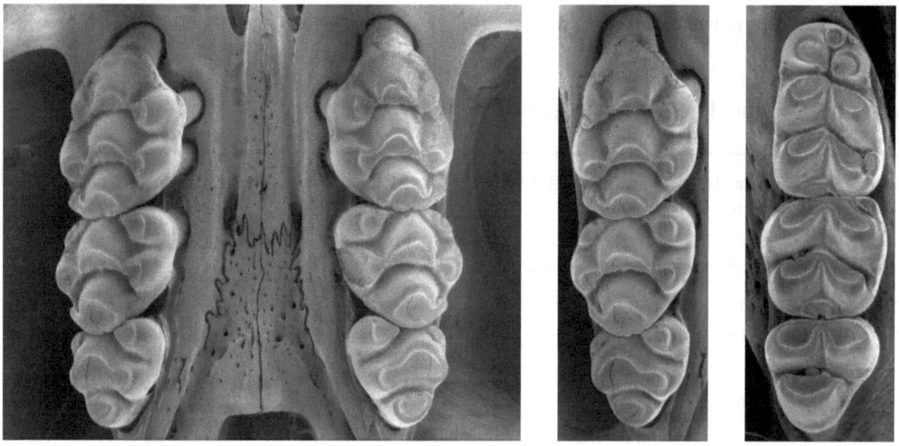

(a) (b) (c)

Figure 2.666 Cheek teeth of *Pelomys campanae* (MNHN-ZM-MO 1961-899) (A: upper cheek teeth; B: upper right toothrow; C: lower right toothrow).

Habitat & Biogeography

They are widespread in moist habitats within the southern savannas in East and Central Africa.

Fossil Record

Fossils are known from the Early Pliocene of North Africa, Early to Late Pliocene of East Africa, Early Pliocene to Late Pleistocene of Southern Africa (Winkler et al., 2010). The earliest fossils are known from 5.3 MYA at Amama3 in North Africa (Winkler et al., 2010). The earliest Sub-Saharan fossil is *Pelomys* sp. from Makapansgat (Denys, 1990c; Senegas, 2000).

P. dietrichi Olduvai bed I, Tanzania (Jaeger, 1976)

Pelomys sp. Omo Shungura (Wesselman, 1984)

Biology & Ecology

They are diurnal and feed mainly on the leaves, shoots and stems of grasses. In the eastern DRC, reproduction in *P. fallax* is continuous whereas it is seasonal in Southern Africa. Based on embryo counts as well as nests with intact litters, litter size varies from 2 to 7.

Systematic Notes

Based upon morphological grounds two subgenera have been described: subgenus *Komemys* (*P. isseli* and *P. hopkinsoni*), subgenus *Pelomys* (*P. campanae, P. fallax* and *P. minor*). The validity of these subgenera has not been tested and this distinction was not retained by Musser and Carleton (2005). Molecular data suggest that *Mylomys* is the sister species of *Pelomys* and that *Pelomys* belongs to the tribe Arvicanthini (Lecompte et al., 2008).

Annotated Species List

Five species are recognised:

Pelomys campanae (Huet, 1888)

Pelomys fallax (Peters, 1852)

Pelomys hopkinsi Hayman, 1955

Pelomys isseli (de Beaux, 1924)

Pelomys minor Cabrera & Ruxton, 1926

Key to Species

This key has been adapted from Misonne (1974).

1	Tail relatively longer (ca. 105% or more of HB); 5th digit of hindfoot longer (reaching the base of the 4th digit) and with claw (subgenus *Komemys*)	2
	Tail relatively shorter (ca. 102% or less of HB); 5th digit of hindfoot shorter (not reaching the base of the 4th digit) and with nail (subgenus *Pelomys*)	3
2	Feet light (or whitish) in colour; restricted to islands in Lake Victoria	*P. isseli*
	Feet dark in colour; not occurring on islands in Lake Victoria	*P. hopkinsi*
3	Size smaller (HF < 27 mm)	*P. minor*
	Size larger (HF > 30 mm)	4
4	Mid-dorsal line usually distinct and extending from mid-back to base of tail; tail relatively longer (ca. 102% of HB)	*P. campanae*
	Mid-dorsal line usually indistinct (especially in adults), extending from neck to base of tail (juveniles and subadults); tail relatively shorter (ca. 90-96% of HB)	*P. fallax*

Species Accounts

Pelomys campanae (Huet, 1888)

Diagnosis

This species is distinguishable from *P. minor* based on its larger size (in *P. minor*, HF < 27 mm). This species is not easily distinguished from *P. fallax* (tail relatively shorter, ca. 90-96% of HB; size on average larger, upper toothrow is typically > 6.5 mm in length but with significant overlap in measurements). This species has a tail which is ca. 102% of HB and the upper molar length is typically < 6.6 mm. The mid-dorsal line is also generally distinct in *P. campanae* but typically indistinct in adult *P. fallax*. *Pelomys campanae* also has a different karyotype from *P. fallax*.

Taxonomic Issues

The phylogenetic relationships of this species with other *Pelomys* (particularly within the subgenus *Pelomys*) are poorly known and a molecular study of this group is urgently needed. Chromosome number 2n = 48, aFN = 52 (Matthey 1963).

Distribution

This species occurs widely in western and northern Angola and western DRC where it inhabits grasslands.

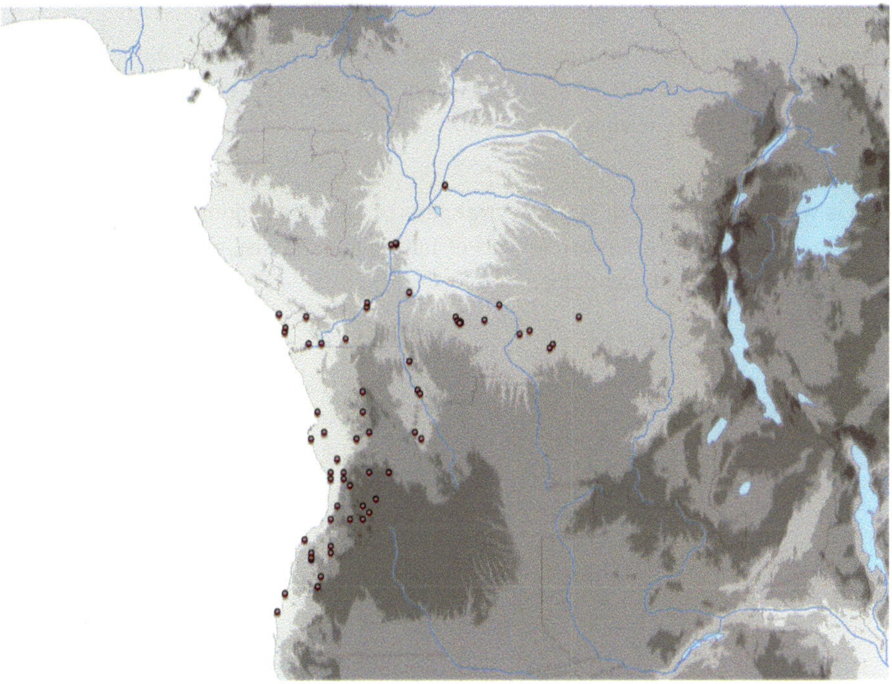

Figure 2.667 Distribution map

Measurements

External and cranial measurements (mm) and mass (g) for *Pelomys campanae*, sexes combined

	Mean	Min	Max	N
Mass	-	-	-	-
HB	137.0	120	170	5
Tail	141.0	138	146	4
HF	31.8	30	33	5
Ear	16.6	14	19	5
GLS	35.3	32.5	36.7	6

Pelomys fallax (Peters, 1852)

Diagnosis

This species is distinguishable from *P. minor* based on its larger size (in *P. minor*, HF < 27 mm). This species is not easily distinguished from *P. campanae* (tail relatively longer, ca. 102% of HB; size on average smaller, upper toothrow is typically < 6.6 mm in length but with significant overlap in measurements). This species has a tail which is ca. 90-96% of HB and the upper molar length is typically > 6.5 mm. The mid-dorsal line is also generally indistinct in adult *P. fallax* but typically distinct in *P. campanae*. Its relatively shorter tail readily distinguishes it from the *Komemys* subgenus (relative tail length ca. 105% or more of HB). *Pelomys fallax* has a different karyotype from *P. campanae*.

Taxonomic Issues

The phylogenetic relationships of this species with other *Pelomys* (particularly within the subgenus *Pelomys*) are poorly known and a molecular study of this group is urgently needed. Chromosome number 2n = 42, aFN = 56 (Denys et al., 2011b).

Distribution

This species is widely distributed in South-central parts of Africa, extending northwards to the Albertine Rift in the west and the Eastern Arc Mts of Tanzania in the east, and southwards to the Okavango Delta and southern Mozambique.

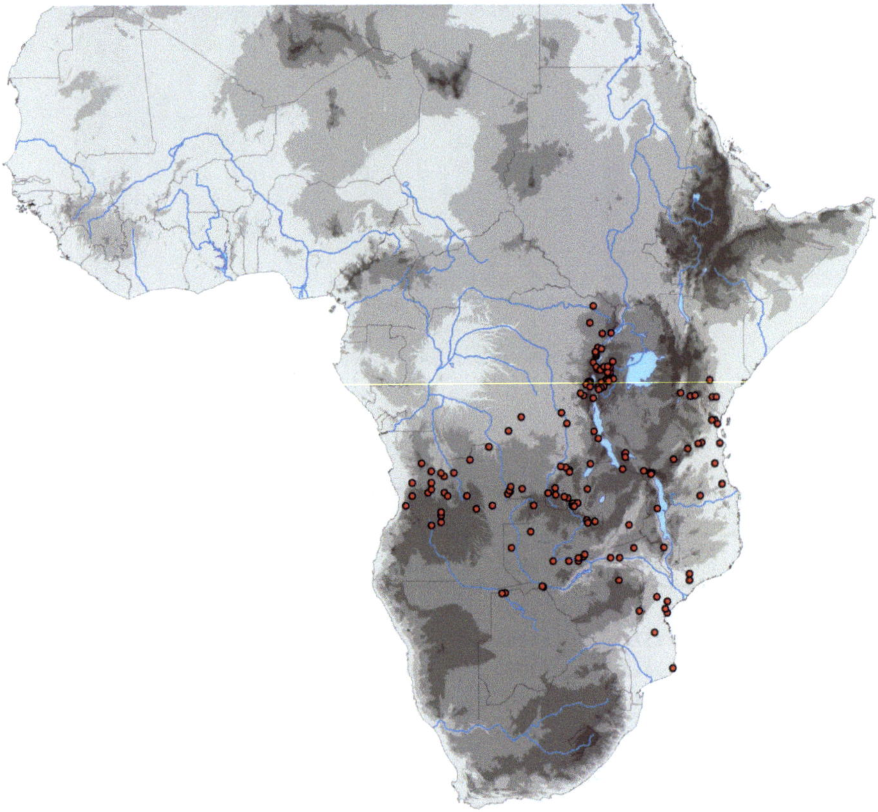

Figure 2.668 Distribution map

Measurements

External and cranial measurements (mm) and mass (g) for *Pelomys fallax*, sexes combined

	Mean	Min	Max	N
Mass*	131.4	100	170	17
HB	148.5	130	170	18
Tail	143.6	105	169	17
HF	31.5	27	36	18
Ear	18.0	15	21	18
GLS	34.4	30.2	36.8	10

*De Graaff (1981)

Pelomys hopkinsi Hayman, 1955

Diagnosis

This species is distinguishable from *P. minor* based on its larger size (in *P. minor*, HF < 27 mm), and from *P. fallax* by its relatively longer tail (in *P. fallax* relative tail length is ca. 90-96% of HB). This does not co-occur with *P. isseli*, from which it can be distinguished by its darker feet. This species has a tail which is ca. 125% of HB. The mid-dorsal line is distinct.

Taxonomic Issues

The species is closely related to *P. isseli*, both belonging to the subgenus *Komemys*.

Distribution

This species has a narrow range restricted to the Albertine Rift with a single possible locality in western Kenya (Bekele & Schlitter, 1989).

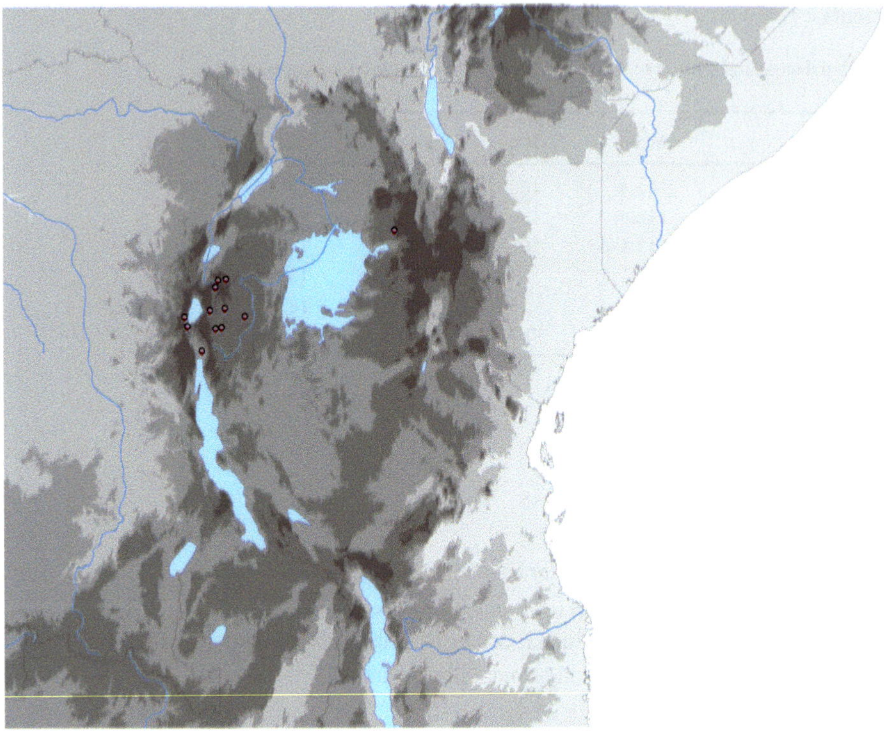

Figure 2.669 Distribution map

Measurements

External and cranial measurements (mm) and mass (g) for *Pelomys hopkinsi*, sexes combined

	Mean	Min	Max	N
Mass*	54.5	41	68	2
HB*	121.5	116	127	2
Tail**	154.0	135	166	3
HF**	31.2	30.2	32.0	3
Ear*	15.9	15.0	16.8	2
GLS*	33.8	30.7	36.2	4

*Bekele & Schlitter (1989); **also includes measurements from Delany (1975)

Pelomys isseli (de Beaux, 1924)

Diagnosis

This is the only species of *Pelomys* occurring on islands of Lake Victoria. It can be differentiated from *P. minor* based on its larger size (in *P. minor*, HF < 27 mm), and from *P. fallax* by its relatively longer tail (in *P. fallax* relative tail length is ca. 90-96% of HB). This does not co-occur with *P. hopkinsi*, from which it can be distinguished by its lighter-coloured feet. This species has a tail which is ca. 110% of HB. The mid-dorsal line is distinct.

Taxonomic Issues

This species is closely related to *P. hopkinsi*, both belonging to the subgenus *Komemys*.

Distribution

This species is restricted to a few islands in the northern parts of Lake Victoria where it has been collected from forest edge and scrub habitats (Delany, 1975).

Figure 2.670 Distribution map

Measurements

External and cranial measurements (mm) and mass (g) for *Pelomys isseli*, sexes combined

	Mean	Min	Max	N
Mass*	71.5	60	83	2
HB*	140.5	132	149	2
Tail*	152.0	148	156	2
HF*	31.5	31	32	2
Ear*	15.5	15	16	2
GLS*	33.7	32.7	34.7	2

*Bekele & Schlitter (1989)

***Pelomys minor* Cabrera & Ruxton, 1926**

Diagnosis

This species is distinguishable from all other *Pelomys* species by its diminutive size (HF > 30 mm for all other *Pelomys* species). This species has a tail which is ca. 95% of HB. The mid-dorsal line is distinct.

Taxonomic Issues

This species is currently only distinguishable based on its small size. A molecular analysis of this species is urgently needed to resolve its phylogenetic relationships.

Distribution

This species occurs in the southern parts of the DRC, marginally extending into neighbouring Zambia and Angola.

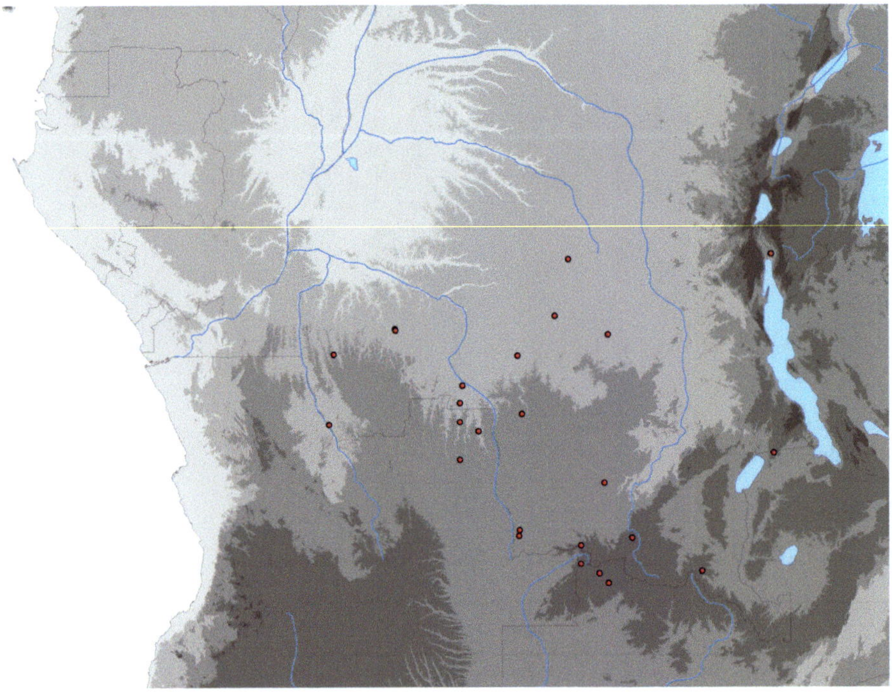

Figure 2.671 Distribution map

Measurements

External and cranial measurements (mm) and mass (g) for *Pelomys minor*, sexes combined

	Mean	Min	Max	N
Mass*	49.0	42	56	2
HB*	122.5	101	161	16
Tail*	113.3	100	131	19
HF*	24.2	21	27	13
Ear*	14.5	13	16	6
GLS*	29.2	27.4	32.3	40

*Happold (2013)

Genus Account

Praomys Thomas, 1915 (Soft-furred Mice)

Diagnosis

As the name suggests, these small to medium sized mice (Mass ca. 25-60 g; HB ca. 100-130 mm) possess a soft pelage which is greyish buffy to almost black (in *P. degraaffi*) to reddish brown dorsally and greyish to whitish ventrally (but always having grey bases to the hairs). The ears are large and rounded. The tail is finely scaled, appears naked and is longer than the head and body length (ca. 110-130% of HB). Members of the genus can be distinguished from the similar *Mastomys* by the reduced number of mammae in females (3-5 pairs as opposed to 8-12 pairs in *Mastomys*), longer tails and by having the palatine bone extending anteriorly to the middle of M^2, as opposed to only M^1 in *Mastomys*, and shorter palatal foramina which end before the M^1, as opposed to extending at least to the second lamina of M^1 in *Mastomys*. Whilst members of *Mastomys* are typically savanna species those of *Praomys* are typically forest and woodland species (with the exception of the *P. daltoni* group which occur in savanna).

Figure 2.672 Photograph of *Praomys rostratus* (Ara Monadjem).

Cranial Characters

There is considerable variation in cranial features which indicates the probable existence of multiple genera within *Praomys* as suggested by Musser & Carleton (2005). The skull is medium sized and long and narrow in shape due to the long and gradually tapering rostrum. The zygomatic arches are gracile and the anterior edge of the zygomatic plate is usually slightly concave. Supraorbital ridges are present but vary considerably in development depending on species and species-group. The subsquamosal foramen and hamular process ("mastoid strut") are usually present.

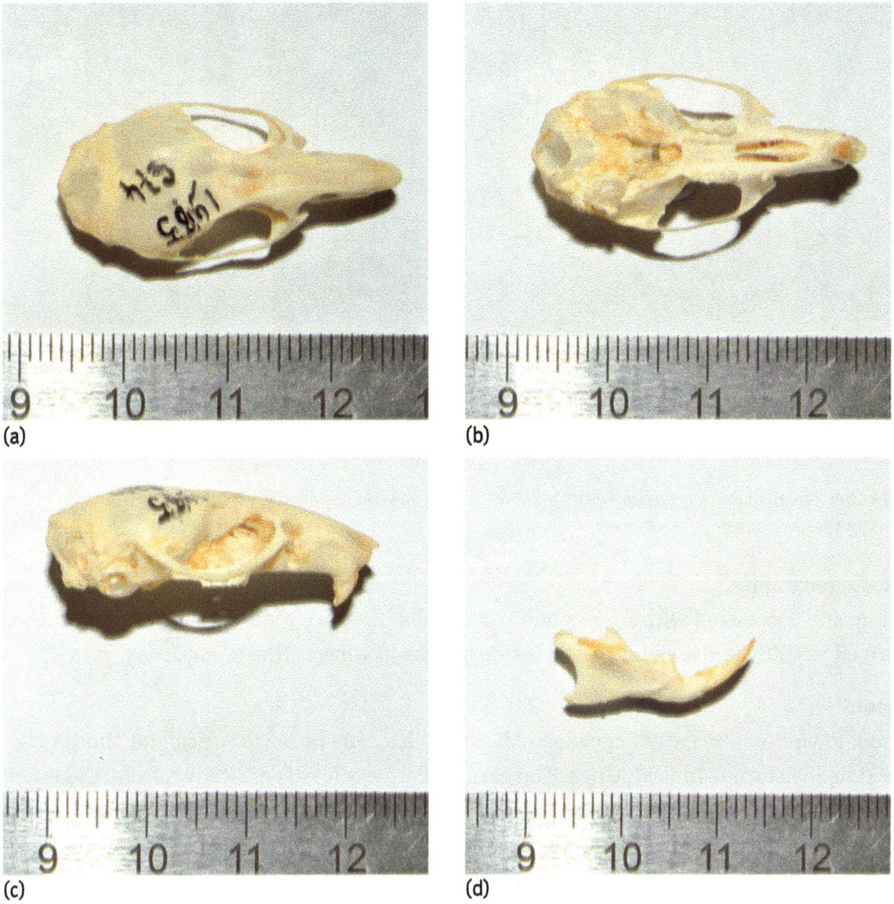

Figure 2.673 a-d Cranium of *Praomys tullbergi* (MNHN-ZM-MO 1965-674) (from top left to bottom right: dorsal view, ventral view, lateral view and lower jaw).

Dentition

The upper incisors are opisthodont and ungrooved. The molar row is quite gracile (MTR/GSL ± 0.14-0.15, Table 10) and rows of cusps are distorted. The cusp t3 on M^1 may be distinct or poorly distinguished from t2 depending on species. Molar M^3 is reduced but shows remnants of all three laminae (Group 3 of Figure 2.439). The lower molars have small cusplets and a posterior cingulum is present on M_1 and M_2.

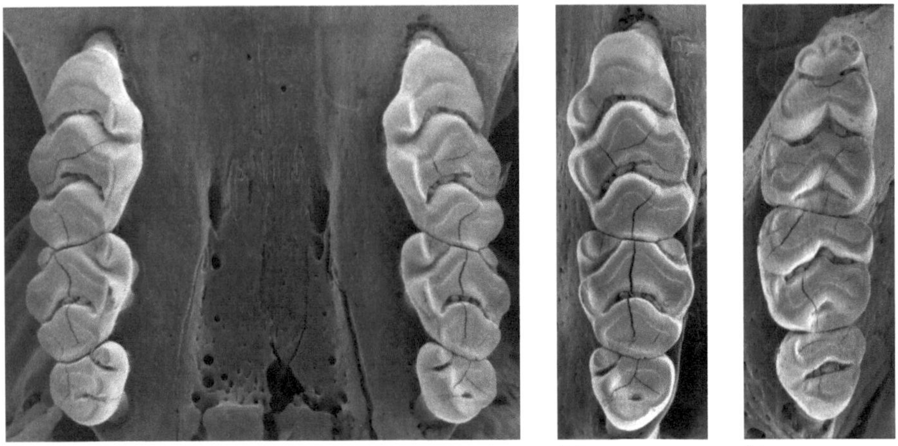

(a) (b) (c)

Figure 2.674 Cheek teeth of *Praomys petteri* (MNHN-ZM-MO 11952-445) (A: upper cheek teeth; B: upper left toothrow; C: lower left toothrow).

Habitat & Biogeography

Although mostly associated with forests and woodlands, they are more numerous in disturbed forests as well as cultivated areas and even inside houses in villages (Rosevear, 1969).

Fossil Record

Fossils are known from the Early Pliocene to Middle Pleistocene in North Africa and Middle Pliocene to Middle Pleistocene in East Africa. The earliest *Praomys* fossils are known from 3.3 MYA at Hadar, Ethiopia and Makapansgat, South Africa (Sabatier, 1982; Pocock, 1987; Sénégas, 2000).

Biology & Ecology

They are nocturnal terrestrial rodents. Various studies have shown that they forage on the ground but occasionally are good climbers and have a mixed diet including seeds, fruits, and insects (Hanney, 1965; Genest-Villard, 1980; Happold, 2013). They live either in old termite mounds or under fallen logs, roots or cracks in the soil and nests may be constituted of vegetable matter (Rahm, 1967). The reproductive biology of some species has been studied in detail (see Happold, 2013), and it appears that they are territorial (Happold 1978).

Systematic Notes

Based on cladistic analysis of molecular (Lecompte et al., 2002b, 2008) and morphological (Lecompte et al., 2002a) characters, *Praomys* is the core member of the Praomyini tribe and forms a close association with *Heimyscus*, *Hylomyscus*, *Stenocephalemys*, *Mastomys* and *Myomyscus* (*Stenocephalemys* division of Musser & Carleton, 2005). Based largely on preceding studies (e.g. Van der Straeten & Dudu, 1990; Lecompte et al., 1999; Van der Straeten & Kerbis Peterhans, 1999; Van der Straeten et al., 2003), Musser & Carleton (2005) identified five species groups, distinguished largely on size, the development of the supraorbital ridges, the length of palatal foramina, length of hindfeet and the development of cusp t3 on M^1, which they considered to be in some cases worthy of elevation to genus status (aspects of the *P. jacksoni* and *P. tullbergi* group were confirmed by the molecular study of Nicolas et al., 2005):

1) *P. jacksoni* group: *degraaffi, jacksoni, minor, montis**, *mutoni**, *peromyscus**, *sudanensis**, *viator**, and at least one undescribed species (* included as synomyns of *P. jacksoni* by Musser & Carleton, 2005, but *mutoni* was accepted as a valid species by Happold (2013) and retained as such here pending further investigation);
2) *P. tullbergi* group: *coetzeei, misonnei, morio, rostratus, tullbergi, petteri, obscurus*, and *hartwigi*.
3) *P. delectorum* group: *delectorum, melanotus**, *octomastis**, and *taitae** (*included as synonyms of *P. delectorum* by Musser & Carleton (2005); two of these were elevated to species status by Musser & Stanley (2012), but are weakly differentiated on molecular grounds (Bryja et al., 2014); see *P. delectorum* account for further details).
4) *P. lukolelae* group: *lukolelae* and *verschureni* (these two taxa were previously assigned to *Malacomys* due to their long hindfeet).
5) *P. daltoni* group: *daltoni*, and *P. derooi** (these two taxa were previously assigned to *Myomys* = *Myomyscus*; *included as a synonym of *P. daltoni* based on molecular analysis (Bryja et al., 2010)).

Annotated Species List

Sixteen species are recognised:
Praomys coetzeei Van der Straeten, 2008
Praomys daltoni (Thomas, 1892)
Praomys degraaffi Van der Straeten & Peterhans, 1999
Praomys delectorum (Thomas, 1910)
Praomys hartwigi Eisentraut, 1968
Praomys jacksoni (de Winton, 1897)
Praomys lukolelae (Hatt, 1934)
Praomys minor Hatt, 1934
Praomys misonnei Van der Straeten & Dieterlen, 1987
Praomys morio (Trouessart, 1881)
Praomys mutoni Van der Straeten & Dudu (1990)
Praomys obscurus Hutterer & Dieterlen, 1992
Praomys petteri Van der Straeten, Lecompte & Denys, 2003
Praomys rostratus (Miller, 1900)
Praomys tullbergi (Thomas, 1894)
Praomys verschureni (Verheyen & Van der Straeten, 1977)

Key to Species

The species within this genus cannot currently be identified with certainty based on external or cranio-dental characters and are therefore generally difficult to distinguish in the field or even from museum specimens. We provide a key to the five species groups here. For species identification within these groups, see the species accounts (in most cases this is only achievable on molecular grounds).

1	Palatal ridges: 2 + 7 = 9	*P. tullbergi* species group
	Palatal ridges: 2 + 5 = 7	2
2	Cusp t3 on 1st upper molar poorly-developed or absent; hindfoot longer (HF typically > 28 mm)	*P. lukolelae* species group
	Cusp t3 on 1st upper molar well-developed; hindfoot shorter (HF typically < 28 mm)	3

3	Supraoccipital ridges raised	*P. jacksoni* species group
	Supraoccipital ridges not raised	4
4	Occurs in montane forests of eastern Africa	*P. delectorum* species group
	Occurs in savannas and houses of western Africa and the Sahel	*P. daltoni* species group

Species Accounts

Praomys coetzeei Van der Straeten, 2008

Diagnosis

This is the only species of the *P. tullbergi* group occurring in Angola.

Taxonomic Issues

This recently described species was previously mentioned as *P. tullbergi* from northern Angola (Musser & Carleton, 2005), but was shown to be distinct in cranio-dental characters (Van der Straeten, 2008) and is also geographically separated from the rest of the species group (Nicolas et al., 2010a).

Distribution

This species is known from just a few specimens from northern Angola (Van der Straeten, 2008).

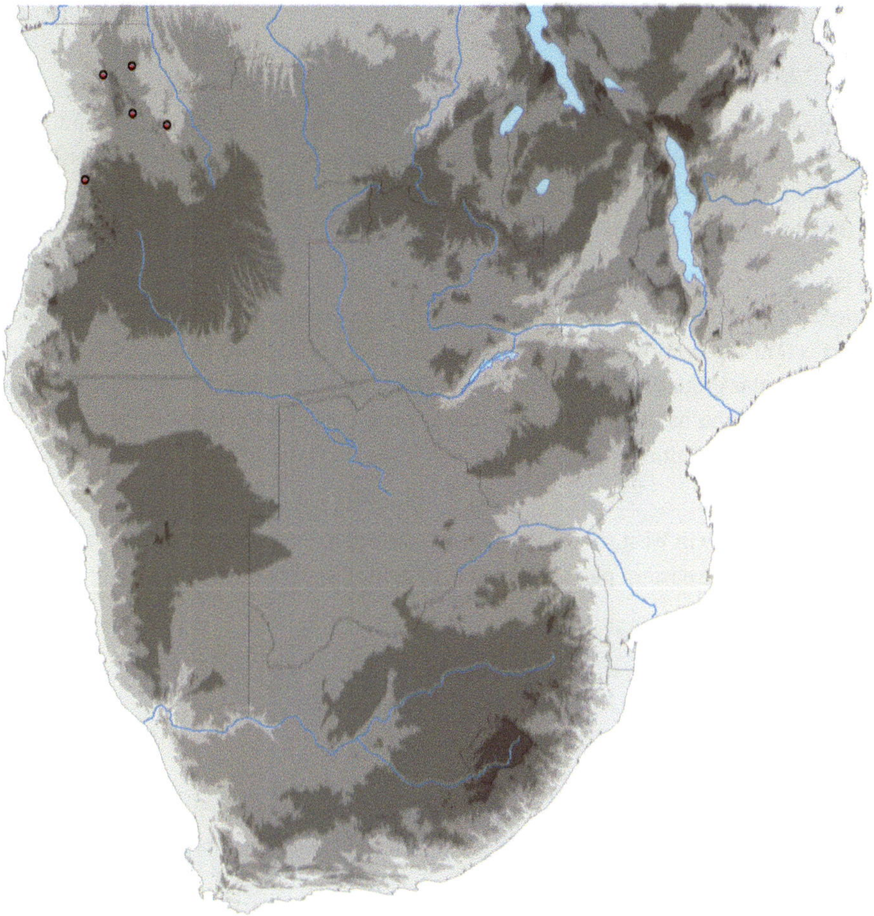

Figure 2.675 Distribution map

Measurements

External and cranial measurements (mm) and mass (g) for *Praomys coetzeei*, sexes combined

	Mean	Min	Max	N
Mass	-	-	-	-
HB*	117.8	90	140	18
Tail*	150.2	123	163	15
HF*	26.5	24	28	16
Ear*	18.6	14	21	17
GLS*	33.2	31.3	35.2	16

*Van der Straeten (2008)

***Praomys daltoni* (Thomas, 1892)**

Diagnosis

This is the only species within the *P. daltoni* group and can therefore be separated from other *Praomys* species groups based on the characters presented in the key. The form *daltoni* (see below) can easily be distinguished from all other *Praomys* species based on its ventral pelage which is pure white (hairs with white bases).

Taxonomic Issues

The taxon *derooi* described by Van der Straeten & Verheyen (1978b) was previously considered a separate species (Musser & Carleton, 2005), but has recently been shown to be conspecific with *P. daltoni* on molecular grounds (Bryja et al., 2010). The form *daltoni* has a pure white belly and occurs in natural savanna habitats whereas *derooi* has a grey ventral pelage and is commensal with humans. Chromosome number 2n = 36, aFN = 34 (Granjon et al., 1992; Dobigny et al., 2001b)

Distribution

This species occurs widely in West Africa, extending eastward to Sudan (Bryja et al., 2010). It may be more widespread in the eastern parts of its range than indicated in the map.

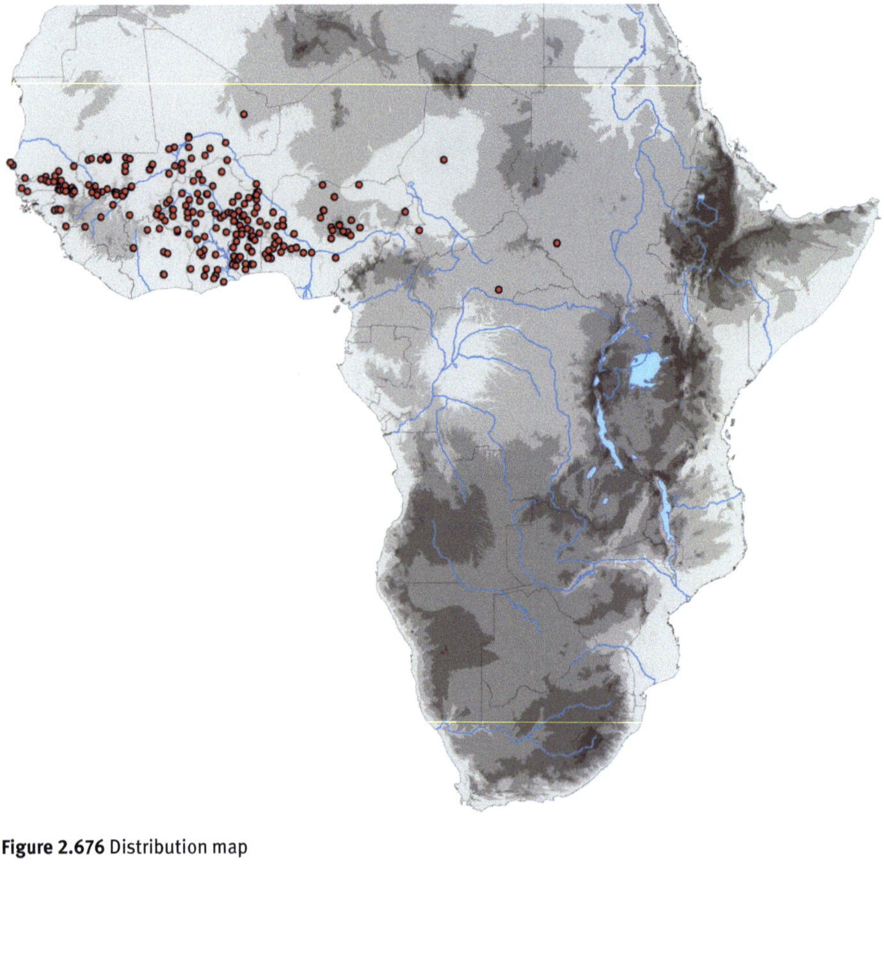

Figure 2.676 Distribution map

Measurements

External and cranial measurements (mm) and mass (g) for *Praomys daltoni*, sexes combined

	Mean	Min	Max	N
Mass	41.0	24	66	18
HB	112.6	98	133	18
Tail	125.1	107	142	17
HF	22.1	20	25	17
Ear	17.6	16	20	17
GLS	28.6	25.8	32.2	19

Praomys degraaffi Van der Straeten & Peterhans, 1999

Diagnosis

The only other member of the *P. jacksoni* group to co-occur with this species is *P. jacksoni* (Van der Straeten & Kerbis Peterhans, 1999) from which it can be readily distinguished on molecular grounds (Lecompte et al., 2002a,b, 2005b); morphological differences include: number of nipples and dorsal pelage colour (in *P. jacksoni* there are three pairs of nipples and the dorsal pelage of adults has a rufous hue). Four pairs of nipples are present in this species and the dorsal pelage is dark grey or blackish typically without a rufous hue in the adults (Van der Straeten & Kerbis Peterhans, 1999).

Taxonomic Issues

This recently described species is very similar to *P. jacksoni*, differing only subtly in external and cranio-dental features (Van der Straeten & Kerbis Peterhans, 1999), but the two species are distinct on molecular grounds (Lecompte et al., 2005b); it is in the *P. jacksoni* species group. Chromosome number 2n = 26, FN = 28 (Van der Straeten & Kerbis Peterhans, 1999).

Distribution

This species is endemic to the Albertine Rift from south-western Uganda to south-western Burundi, where it occurs in montane forests above 1800 m (Van der Straeten & Kerbis Peterhans, 1999).

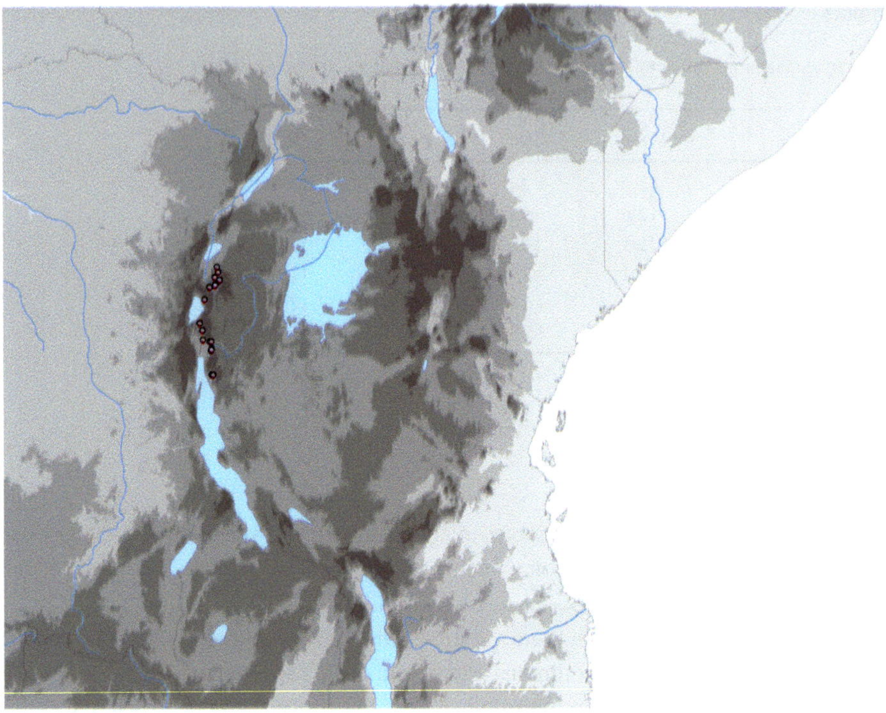

Figure 2.677 Distribution map

Measurements

External and cranial measurements (mm) and mass (g) for *Praomys degraaffi*, sexes combined

	Mean	Min	Max	N
Mass*	40.4	28	64	39
HB*	110.8	92	124	38
Tail*	136.3	115	151	40
HF*	26.4	24	29	48
Ear*	19.6	17	22	41
GLS*	31.3	29.1	32.9	48

*Van der Straeten & Kerbis Peterhans (1999)

Praomys delectorum (Thomas, 1910)

Diagnosis

This is the only species of *Praomys* in the *P. delectorum* group. Four pairs of nipples (2 + 2 = 8) are present.

Taxonomic Issues

The *P. delectorum* species group was recently revised based on extensive and careful morphological analysis by Carleton & Stanley (2012) who proposed the existence of three species with non-overlapping distributions: *P. delectorum* (southern Malawi and northern Mozambique), *P. melanotus* (south-western Tanzania and northern Malawi), and *P. taitae* (Eastern Arc Mts and northern highlands of Tanzania). However, a recent and thorough molecular analysis of this group showed that the genetic distance, based on the cytochrome b gene, between these three species was only 2.0-2.8% (Bryja et al., 2014), suggesting a single species rather than a species complex. Furthermore, the phylogroups did not fully match up to the morphogroups, with specimens from the southern mountains of the Eastern Arc range being genetically more closely related to those in the mountains of south-western Tanzania rather than to the northern mountains of the Eastern Arc range (Bryja et al., 2014); morphologically the opposite trend was observed (Carleton & Stanley, 2012). For this reason, we recognise a single species *P. delectorum* here, but note that further research may alter this conclusion. Chromosome number 2n = 48 (Matthey, 1965).

Distribution

This species occurs in the Eastern Arc Mts of Tanzania and Kenya, extending westward to north-central Tanzania, and southward to Malawi and northern Mozambique (Carleton & Stanley, 2012; Bryja et al., 2014). It occupies montane forests typically between 1000-2400 m but has been recorded from lower elevations such as on the foothills of the East Usambara Mts (Carleton & Stanley, 2012).

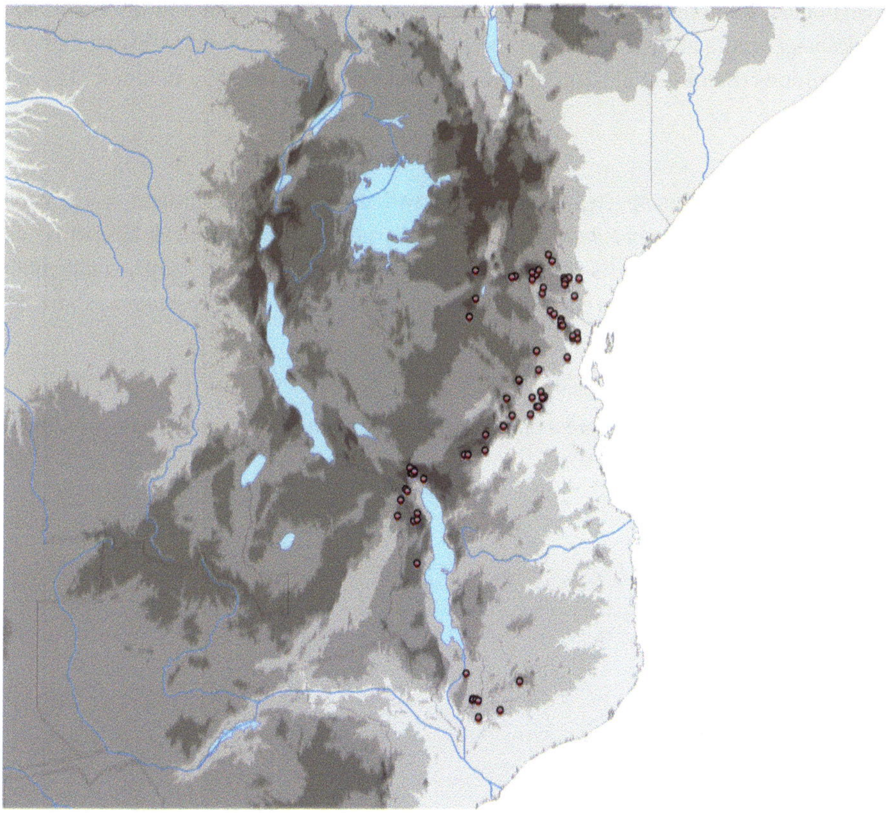

Figure 2.678 Distribution map

Measurements

External and cranial measurements (mm) and mass (g) for *Praomys delectorum*, sexes combined

	Mean	Min	Max	N
Mass*	28.1	20	38	51
HB*	102.6	90	120	86
Tail*	124.6	110	142	86
HF*	23.9	21	26	86
Ear*	20.1	18	25	85
GLS*	28.2	25.8	30.1	92

*Carleton & Stanley (2012)

Praomys hartwigi Eisentraut, 1968

Diagnosis

This is one of eight species in the *P. tullbergi* group. It can be distinguished from most of these species by its non-overlapping distribution (restricted to Mt Oku and Mt Manengouba in the Bamenda Highlands, Cameroon). The closely related *P. obscurus* only occurs in the Gotel Mts, Nigeria; *P. misonnei* occurs at lower altitudes (perhaps up to ca. 1000 m) and is distinguishable on molecular grounds (Nicolas et al., 2010a; Missoup et al., 2012). Three pairs of nipples (1 + 2 = 6) are present.

Taxonomic Issues

It has been suggested that this species is conspecific with *P. obscurus* by Hutterer et al. (1992) who described it as a subspecies of *P. hartwigi* based on morphological differences. Missoup et al. (2012) performed a molecular analysis of *P. hartwigi* from Mt Oku and of *P. obscurus*, and demonstrated their validity in the *P. tullbergi* complex. The closest relative of the *P. obscurus* - *P. hartwigi* clade is *P. petteri*.

Distribution

This species is restricted to montane forest on Mt Oku and Mt Manengouba in the Bamenda Highlands of Cameroon (Hutterer et al., 1992).

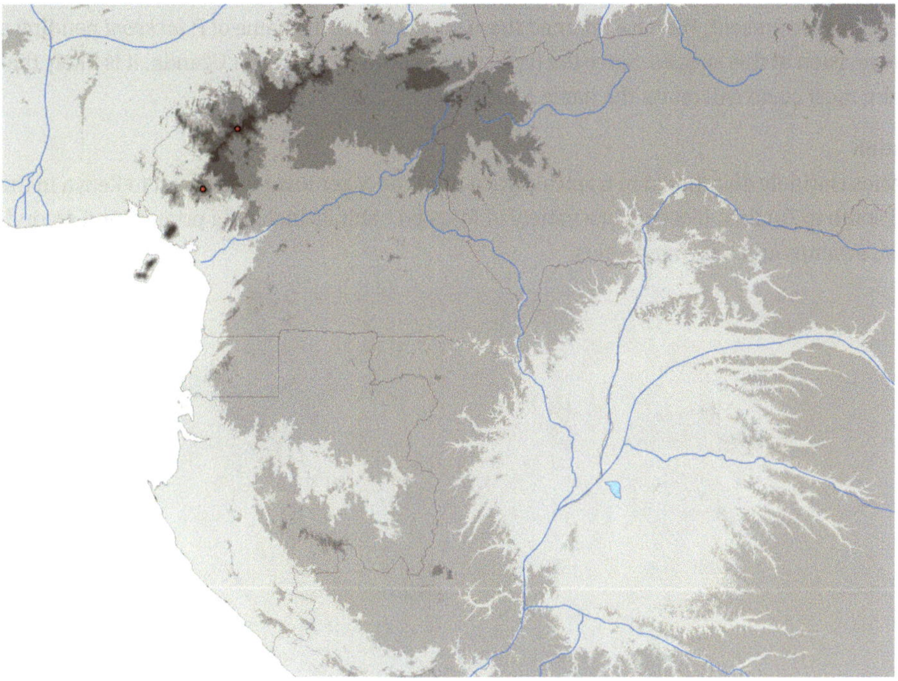

Figure 2.679 Distribution map

Measurements

External and cranial measurements (mm) and mass (g) for *Praomys hartwigi*, sexes combined

	Mean	Min	Max	N
Mass*	57.2	44	70	11
HB*	125.0	115	134	11
Tail*	162.1	152	173	11
HF*	26.5	24	28	11
Ear*	19.2	17	21	11
GLS*	32.6	31.6	33.6	8-9

*Hutterer et al. (1992)

Praomys jacksoni (de Winton, 1897)

Diagnosis

This species co-occurs with all other members of the *P. jacksoni* group (*P. degraaffi, P. minor* and *P. mutoni*) and can only be distinguished with certainty on molecular grounds (Kennis et al., 2011). Four pairs of nipples (2 + 2 = 8) are present.

Taxonomic Issues

Recent molecular studies have detected four deep clades within *P. jacksoni* with interclade divergence, based on the cytochrome b gene, of > 4% (Kennis et al., 2011), strongly suggesting that cryptic species are present. We have retained these clades under the name of *P. jacksoni* pending a complete revision of this species. Since the type specimen was collected in Uganda, it is likely that the eastern most clade will retain the name *P. jacksoni*.

Distribution

This species is widely distributed in tropical Africa from the Niger River in the west to Kenya in the east and south to Zambia. It occurs in a variety of forested habitats including primary and secondary forest, swamps and cultivated fields.

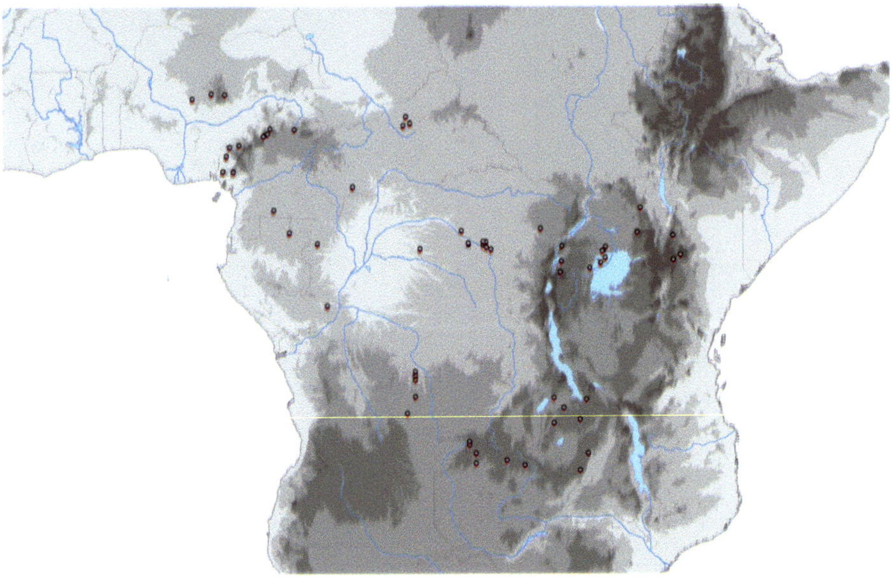

Figure 2.680 Distribution map

Measurements

External and cranial measurements (mm) and mass (g) for *Praomys jacksoni*, sexes combined

	Mean	Min	Max	N
Mass*	41.0	21	59	100
HB**	115.4	94	131	88
Tail**	146.9	125	168	82
HF**	25.0	22	28	88
Ear**	18.4	15	22	80
GLS**	31.0	28.1	34.2	89

*Delany (1975); **Van der Straeten & Dieterlen (1987)

Praomys lukolelae (Hatt, 1934)

Diagnosis

This species can be distinguished from the only other member of the *P. lukolelae* group (*P. verschureni*) by non-overlapping distribution (*P. verschureni* occurs on the right bank of the Congo River) and on molecular grounds (Kennis et al., 2011).

Taxonomic Issues

This is a poorly known species that has been rarely collected. A recent molecular analysis of specimens collected from the Kisangani area (left bank of Congo River) was assigned to this species but may be different from topotypical *P. lukolelae* (Kennis et al., 2011). The same study, based only on the cytochrome b gene, suggested that *P. verschureni* and *P. lukolelae* represent a separate clade (Kennis et al., 2011) which may warrant genus recognition.

Distribution

This species is known from a few scattered localities on the left bank of the Congo River where it predominantly occupies primary rainforest (Kennis et al., 2011).

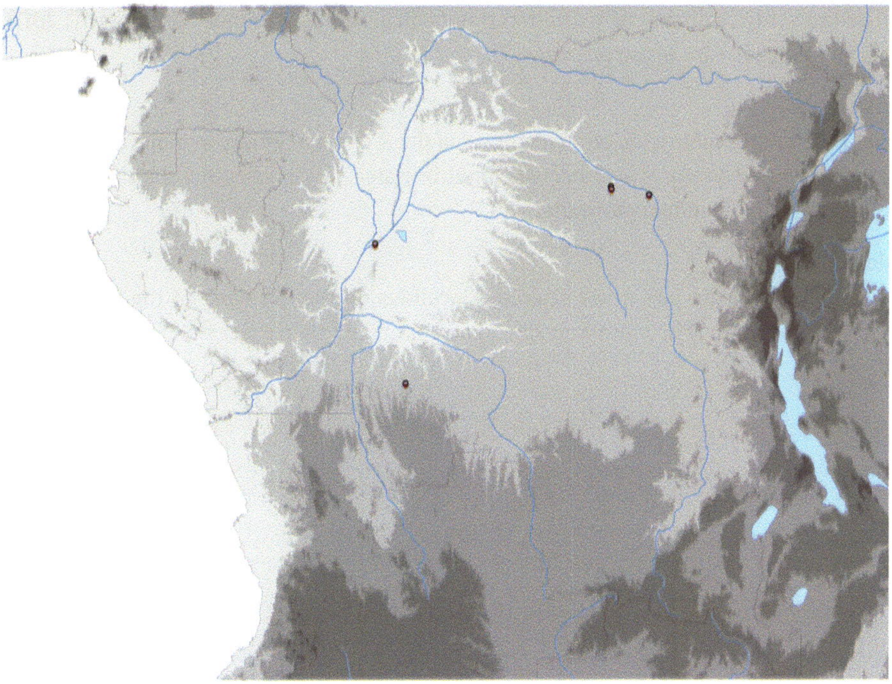

Figure 2.681 Distribution map

Measurements

External and cranial measurements (mm) and mass (g) for *Praomys lukolelae*, sexes combined

	Mean	Min	Max	N
Mass*	41.1	27	57	32
HB*	123.8	109	144	31
Tail*	141.2	116	159	32
HF*	30.2	28	32	34
Ear*	22.9	18	26	34
GLS*	31.5	28.5	33.7	32

*Happold (2013)

Praomys minor Hatt, 1934

Diagnosis

This species co-occurs with two of the other three members of the *P. jacksoni* group (*P. jacksoni* and *P. mutoni*) and can only be distinguished with certainty on molecular grounds (Kennis et al., 2011). Three pairs of nipples (1 + 2 = 6) are present.

Taxonomic Issues

This species is distinguishable from other members of the *P. jacksoni* group on molecular grounds and may be differentiated from *P. jacksoni* using multivariate analysis of cranio-dental characters (Kennis et al., 2011) and mammary formula.

Distribution

This species is known from just a few scattered localities on the left bank of the Congo River (Kennis et al., 2011), but may be more widespread and abundant than suggested from this map.

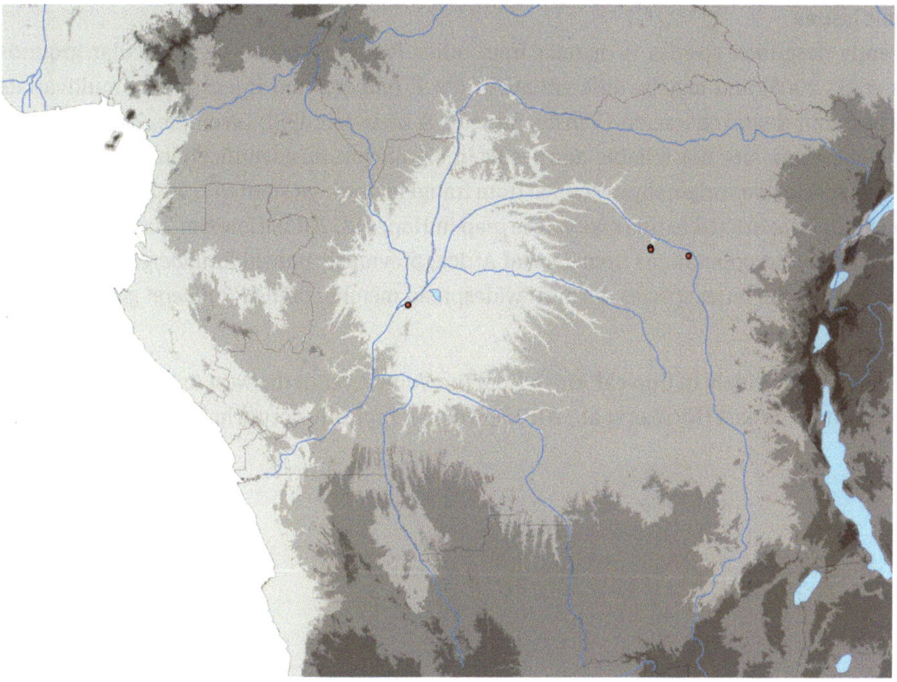

Figure 2.682 Distribution map

Measurements

External and cranial measurements (mm) and mass (g) for *Praomys minor*, sexes combined

	Mean	Min	Max	N
Mass*	26.5	18	-	26
HB*	100.0	80	118	26
Tail*	127.1	102	145	26
HF*	20.7	19	22	25
Ear*	16.3	13	18	24
GLS*	28.4	23.4	30.0	55

*Kennis et al. (2011); the maximum value given for mass is 22 g which is clearly wrong since the mean value presented is 26.5 g (unless the mean value is also incorrect) and has been omitted from this table.

Praomys misonnei Van der Straeten & Dieterlen, 1987

Diagnosis

This species belongs to the *P. tullbergi* group and co-occurs with *P. petteri* from which it can readily be distinguished on molecular grounds (Nicolas et al., 2005, 2012a); morphological identification is more difficult, requiring multivariate analysis of cranio-dental characters (see below). Three pairs of nipples (1 + 2 = 6) are present.

Taxonomic Issues

This recently described species is distinct from other *Praomys* species on molecular grounds (Nicolas et al., 2005) and may be differentiated from *P. tullbergi* and *P. jacksoni* by multivariate analysis of cranio-dental characters (Van der Straeten & Dieterlen, 1987; Lecompte, 2002a, 2005b), but these characters are not reliable and result in significant misidentification (Nicolas et al., 2010a). This species was originally described from Irangi in north-western DRC and its distribution was thought to be restricted to this region (see map in Happold, 2013). However, recent molecular work has shown that populations from Central Africa, previously named *P. tullbergi*, are in fact *P. misonnei*, making the latter species the most widespread member of the *P. tullbergi* group.

Distribution

This species occurs widely in tropical Africa from the Volta River in the west to Kakamega Forest, western Kenya in the east (Nicolas et al., 2010a), where it typically occurs in lowland rainforest.

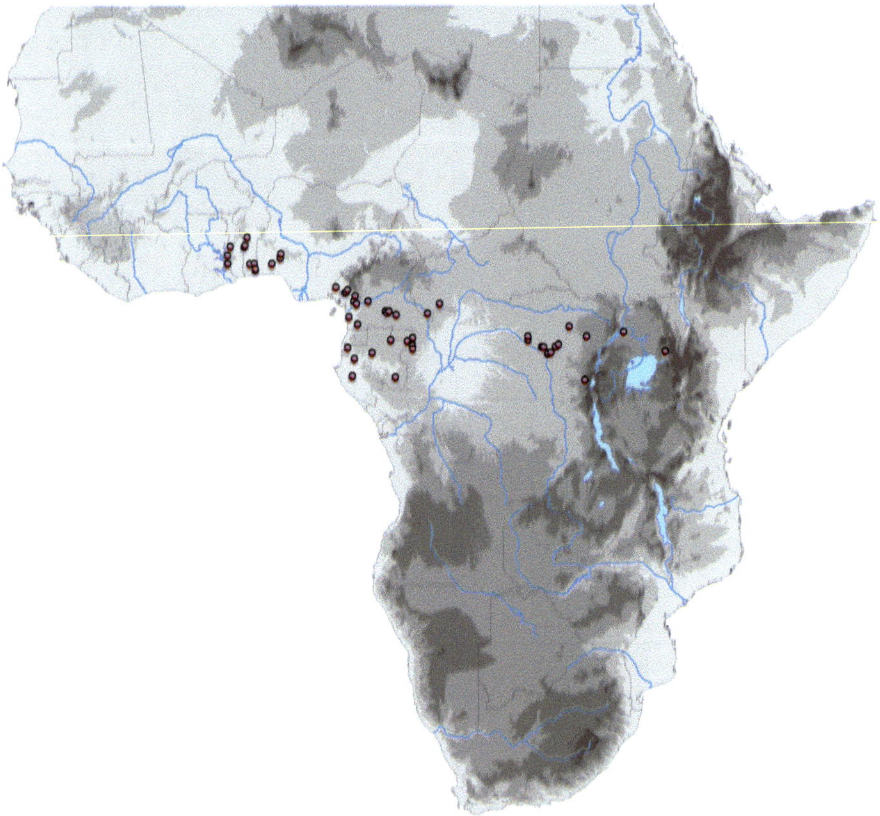

Figure 2.683 Distribution map

Measurements

External and cranial measurements (mm) and mass (g) for *Praomys misonnei*, sexes combined

	Mean	Min	Max	N
Mass	-	-	-	-
HB*	112.0	89	123	33
Tail*	141.3	113	163	32
HF*	24.4	21	26	31
Ear*	18.5	16	21	30
GLS*	31.0	29.2	32.4	33

*Van der Straeten & Dieterlen (1987)

Praomys morio (Trouessart, 1881)

Diagnosis
This is one of eight species in the *P. tullbergi* group. It can be distinguished from most of these species by its non-overlapping distribution (restricted to Mt Cameroon and Bioko Island). *Praomys misonnei* occurs at lower altitudes (below ca. 600 m) and is on average smaller and has a shorter upper molar toothrow (< 5.0 mm). Three pairs of nipples (1 + 2 = 6) are present and the upper molar toothrow is > 5.0 mm (Eisentraut, 1970).

Taxonomic Issues
A recent molecular analysis has shown that *P. morio* is a valid species within the *tullbergi* complex and its sister taxon is an unnamed *Praomys* species from lowland forests (Missoup et al., 2012). Chromosome number 2n = 34 (Happold, 2013).

Distribution
This species occurs in montane forests above 1000 m on Mt Cameroon and on Bioko Island.

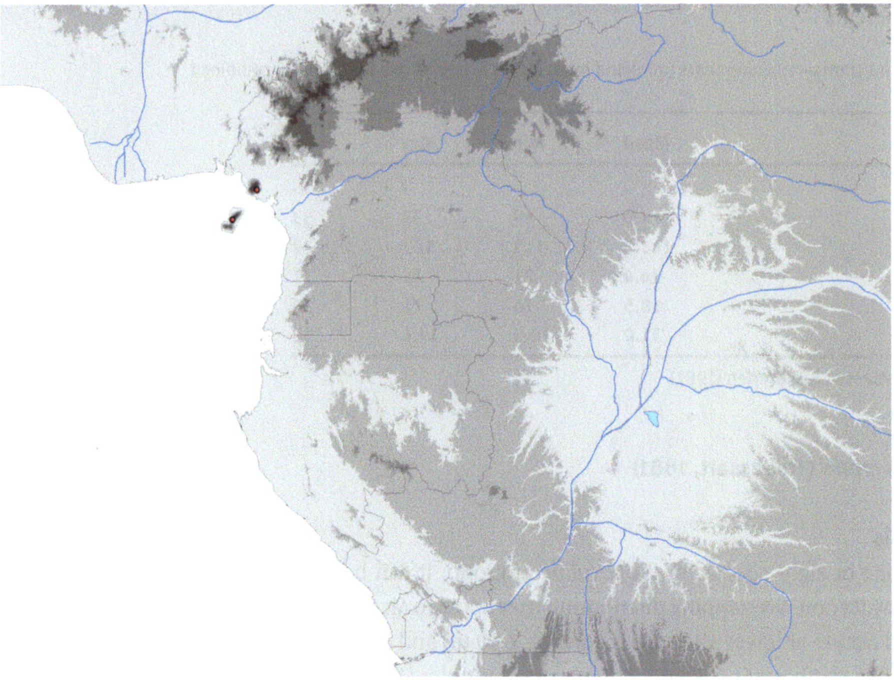

Figure 2.684 Distribution map

Measurements

External and cranial measurements (mm) and mass (g) for *Praomys morio*, sexes combined

	Mean	Min	Max	N
Mass*	43.0	27	62	83
HB*	113.5	100	130	83
Tail*	129.9	110	147	83
HF*	23.5	21	25	83
Ear*	18.3	16	20	83
GLS*	30.2	28.1	31.9	83

*Eisentraut (1970)

Praomys mutoni Van der Straeten & Dudu (1990)

Diagnosis

This species co-occurs with two of the other three members of the *P. jacksoni* group (*P. jacksoni* and *P. minor*) and can only be distinguished with certainty on molecular grounds (Kennis et al., 2011), although the dorsal pelage appears be slightly darker than that of *P. jacksoni* collected at the same locality (Van der Straeten & Dudu, 1990). Three pairs of nipples (1 + 2 = 6) are present.

Taxonomic Issues

This species is readily distinguishable on molecular grounds (Kennis et al., 2011), but external or cranio-dental features overlap extensively with *P. jacksoni* and the two species can only be distinguished by way of multivariate analysis (Van der Straeten & Dudu, 1990).

Distribution

This species occurs in a narrow distribution on both banks of the Congo River (Kennis et al., 2011), where it occupies wetter habitats, typically along river banks, than other *Praomys* species (Van der Straeten & Dudu, 1990).

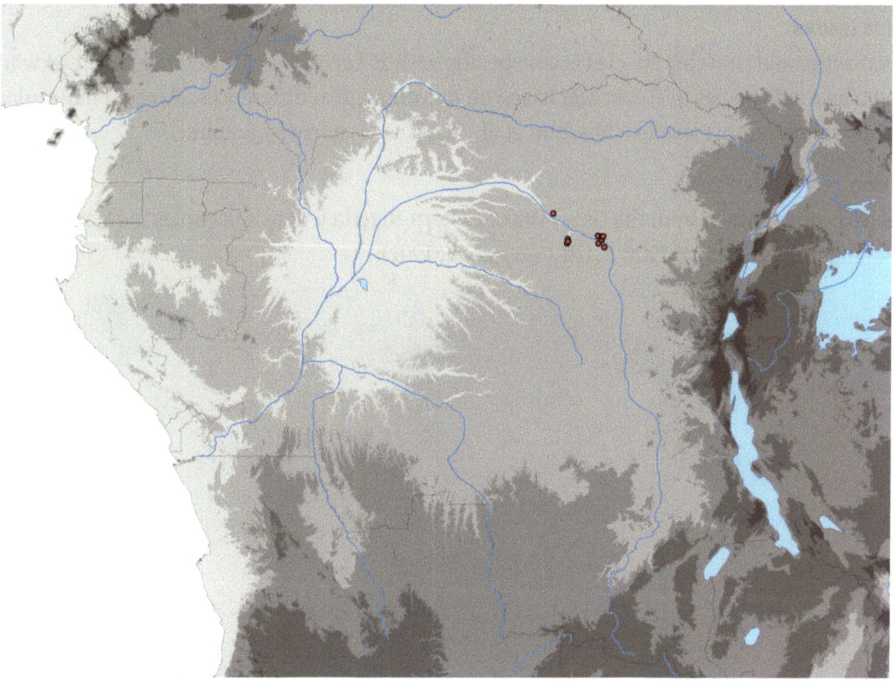

Figure 2.685 Distribution map

Measurements

External and cranial measurements (mm) and mass (g) for *Praomys mutoni*, sexes combined

	Mean	Min	Max	N
Mass*	45	31	57	29
HB**	119.9	110	139	11
Tail**	143.8	113	171	10
HF**	22.1	21	24	10
Ear**	18.2	17	19	8
GLS**	32.6	31.8	33.1	11

*Happold (2013); **Van der Straeten & Dudu (1990)

Praomys obscurus Hutterer & Dieterlen, 1992

Diagnosis
This is one of eight species in the *P. tullbergi* group. It can be distinguished from most of these species by its non-overlapping distribution (restricted to the Gotel Mts, Nigeria). The closely related *P. hartwigi* only occurs in the Bamenda Highlands, Cameroon; *P. misonnei* occurs at lower altitudes (perhaps up to ca. 1000 m), and has not yet been recorded from the Gotel Mts; the two species can be distinguished on molecular and morphometric grounds (Nicolas et al., 2010a; Missoup et al., 2012). Three pairs of nipples (1 + 2 = 6) are present.

Taxonomic Issues
It has been suggested that this species is conspecific with *P. hartwigi* by Hutterer et al. (1992) who described it as a subspecies of *P. hartwigi* based on morphological differences. A recent molecular analysis has validated this species (Missoup et al., 2012; see *P. hartwigi* account).

Distribution
This species is only known from the Gotel Mts of eastern Nigeria where it occupies montane forest and grassland habitats above 1900 m (Hutterer et al., 1992).

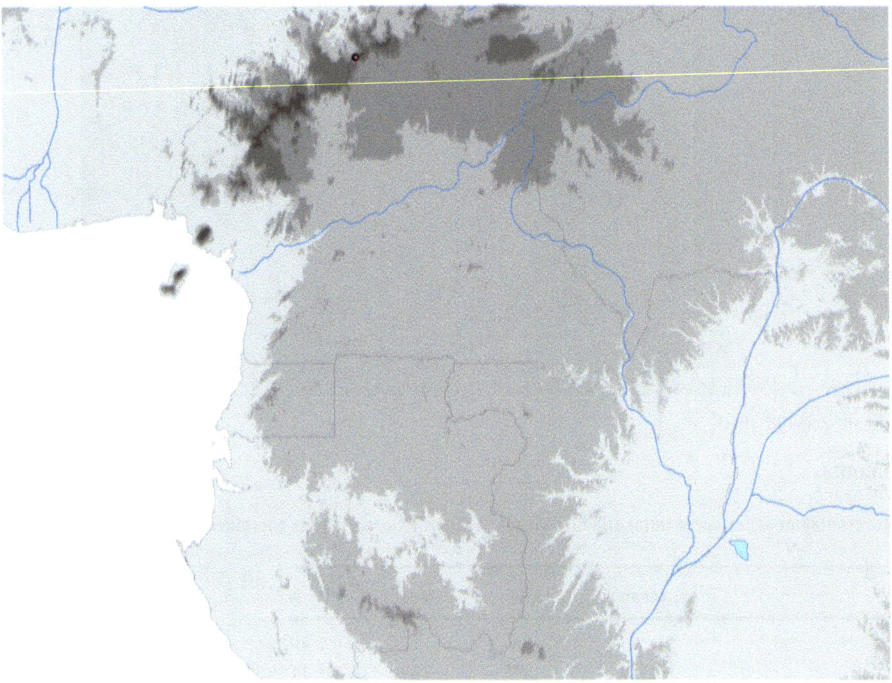

Figure 2.686 Distribution map

Measurements

External and cranial measurements (mm) and mass (g) for *Praomys obscurus*, sexes combined

	Mean	Min	Max	N
Mass*	47.2	34	60	22
HB*	121.0	108	137	22
Tail*	161.0	149	174	22
HF*	27.1	26	28	22
Ear*	20.1	19	21	22
GLS*	31.4	29.7	33.1	15-22

*Hutterer et al. (1992)

Praomys petteri Van der Straeten, Lecompte & Denys, 2003

Diagnosis

This species belongs to the *P. tullbergi* group and co-occurs with *P. misonnei* from which it can readily be distinguished on molecular grounds (Nicolas et al., 2005); morphological identification is more difficult, requiring multivariate analysis of cranio-dental characters (see below). Three pairs of nipples (1 + 2 = 6) are present.

Taxonomic Issues

This recently described species is distinct from other *Praomys* species on molecular grounds (Nicolas et al., 2005) and may be differentiated from other members of the *P. tullbergi* group by multivariate analysis of cranio-dental characters (Van der Straeten et al., 2003). Chromosome number 2n = 42 (Matthey, 1965).

Distribution

This species has a relatively restricted distribution in Central Africa between the Sanaga River in the west and Oubangui River in the east, where it occurs in primary and secondary rainforest (Van der Straeten et al., 2003; Nicolas et al., 2005).

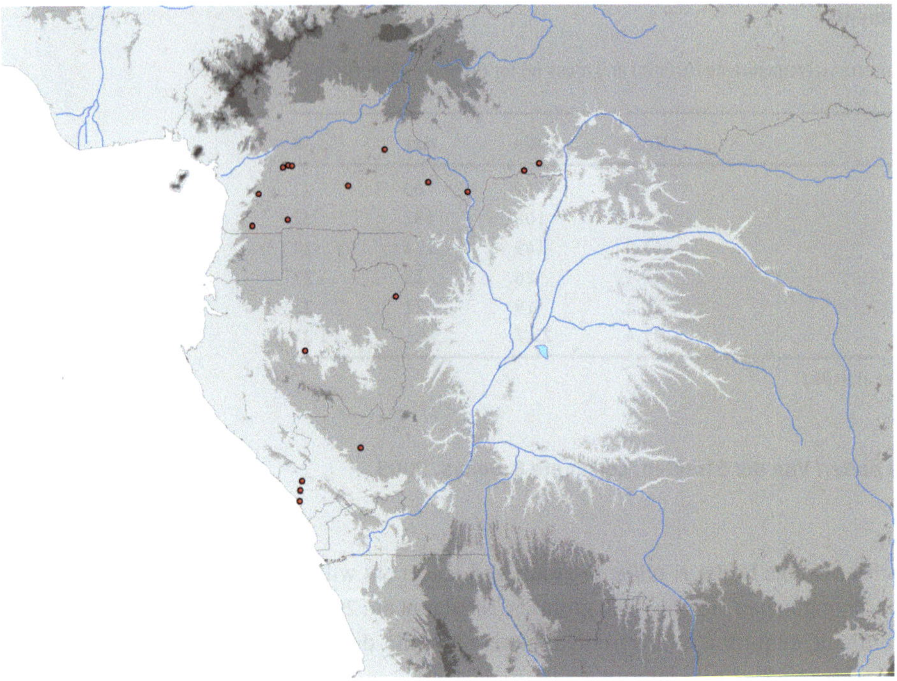

Figure 2.687 Distribution map

Measurements

External and cranial measurements (mm) and mass (g) for *Praomys petteri*, sexes combined

	Mean	Min	Max	N
Mass*	49.3	40	61	9
HB*	122.5	99	143	25
Tail*	143.3	100	165	24
HF*	26.3	24	28	27
Ear*	19.0	15	22	27
GLS*	34.6	30.5	36.7	34

*Van der Straeten et al. (2003)

Praomys rostratus (Miller, 1900)

Diagnosis

This is one of two species in the *P. tullbergi* group that occur in West Africa, west of the Volta River, Ghana. It can be distinguished from the similar *P. tullbergi* on molecular grounds (Nicolas et al., 2008b), from which it also differs by being slightly larger in external and cranio-dental measurements, but with significant overlap (Akpatou et al., 2007). Three pairs of nipples (1 + 2 = 6) are present.

Taxonomic Issues

The distinction between this species and *P. tullbergi* has recently been demonstrated on morphometric (Akpatou et al., 2007) and molecular (Nicolas et al., 2008b) grounds. Chromosome number 2n = 34, FN = 32 (Tranier & Dosso, 1979; Gautun et al., 1986).

Distribution

This species is widely distributed in the tropical parts of West Africa from Senegal and Guinea in the west to the Ivory Coast-Ghana boundary in the east where it occupies a variety of forests and woodlands (Nicolas et al., 2008b).

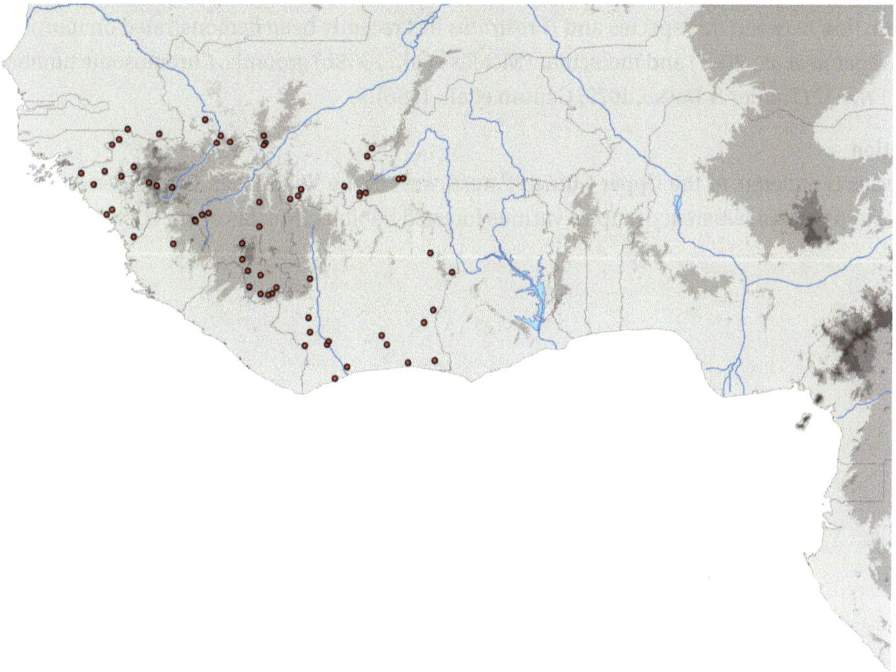

Figure 2.688 Distribution map

Measurements

External and cranial measurements (mm) and mass (g) for *Praomys rostratus*, sexes combined

	Mean	Min	Max	N
Mass*	50.5	31	73	51
HB*	129.5	109	147	51
Tail*	145.1	124	171	51
HF*	26.4	24	31	51
Ear*	18.9	17	21	51
GLS*	34.0	30.0	36.9	98

*Akpatou et al. (2007)

Praomys tullbergi (Thomas, 1894)

Diagnosis

This is one of two species in the *P. tullbergi* group that occur in West Africa, west of the Volta River, Ghana. It can be distinguished from the similar *P. rostratus* on molecular grounds (Nicolas et al., 2008b), from which it also differs by being slightly smaller in external and cranio-dental measurements, but with significant overlap (Akpatou et al., 2007). Three pairs of nipples (1 + 2 = 6) are present.

Taxonomic Issues

The distinction between this species and *P. rostratus* has recently been demonstrated on morphometric (Akpatou et al., 2007) and molecular (Nicolas et al., 2008b) grounds. Chromosome number 2n = 34, FN = 32 (Tranier & Dosso, 1979; Gautun et al., 1986).

Distribution

This species is endemic to the Upper Guinea Forest, west of the Volta River, Ghana, westward to south-eastern Guinea, where it occupies various forested habitats (Nicolas et al., 2008b).

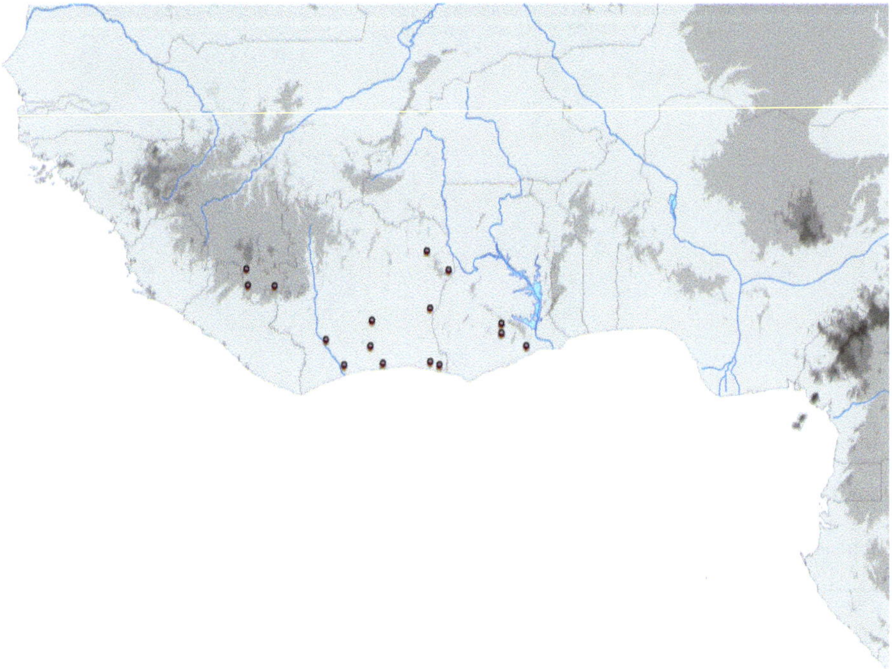

Figure 2.689 Distribution map

Measurements

External and cranial measurements (mm) and mass (g) for *Praomys tullbergi*, sexes combined

	Mean	Min	Max	N
Mass*	37.9	28	54	52
HB*	119.3	108	131	52
Tail*	139.5	115	153	52
HF*	24.1	22	26	52
Ear*	18.7	17	21	52
GLS*	31.5	29.1	34.3	78

*Akpatou et al. (2007)

Praomys verschureni (Verheyen & Van der Straeten, 1977)

Diagnosis

This species can be distinguished from the only other member of the *P. lukolelae* group (*P. lukolelae*) by non-overlapping distribution (*P. lukolelae* occurs on the left bank of the Congo River) and on molecular grounds (Kennis et al., 2011). Three pairs of nipples (1 + 2 = 6) are present.

Taxonomic Issues

This taxon was originally described in the genus *Malacomys* due to its long hindfoot and other characters shared with that genus (Verheyen & Van der Straeten, 1977), but placed in *Praomys* on molecular grounds (Lecompte et al., 2002b). A recent molecular study, based only on the cytochrome b gene, suggested that *P. verschureni* and *P. lukolelae* represent a separate clade (Kennis et al., 2011) which may warrant genus recognition. Chromosome number 2n = 58, FN = 58-60 (Robbins & Van der Straeten, 1982).

Distribution

This species is known from a few scattered localities on the right bank of the Congo River where it inhabits lowland and submontane rainforest up to 1800 m.

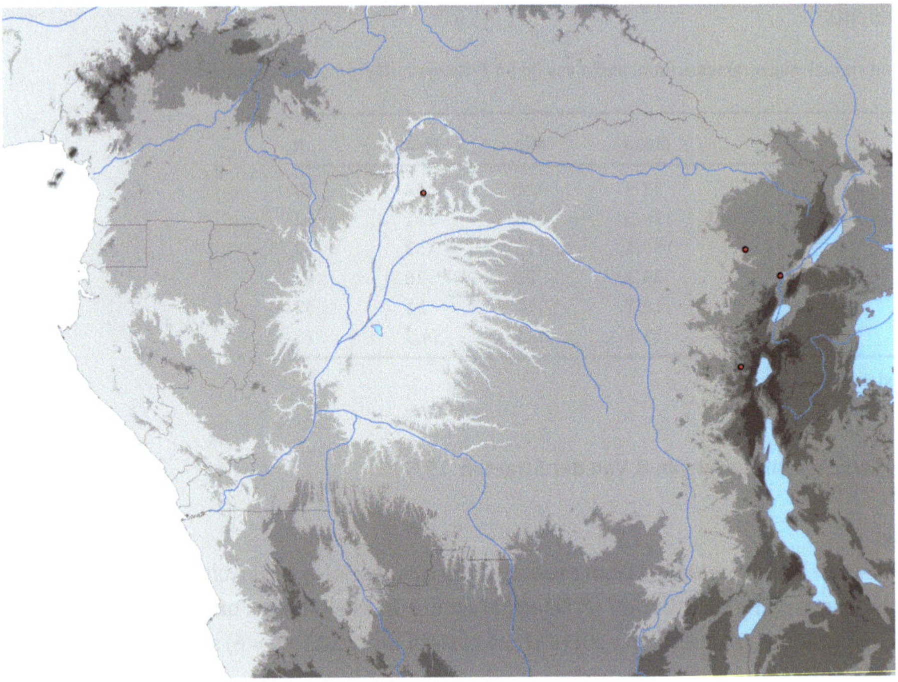

Figure 2.690 Distribution map

Measurements

External and cranial measurements (mm) and mass (g) for *Praomys verschureni*, sexes combined

	Mean	Min	Max	N
Mass*	50.8	40	65	28
HB*	121.3	109	135	29
Tail*	143.0	123	162	27
HF*	30.1	28	32	27
Ear*	20.3	19	23	24
GLS*	34.4	33.1	36.1	22

*Happold (2013)

Genus Account

Rattus Fischer, 1803 (Rats)

Diagnosis

These large to very large (Mass ca. 100-200 g or more in large adults; HB ca. 160-210 mm) invasive alien rodents have a characteristic coarse blackish-brown to brownish dorsal pelage, well supplied with guard hairs, and ventral pelage which is usually grey but white in the "frugivorous" colour phase of *R. rattus*. The tail is relatively shorter in *R. norvegicus* (ca. 90% of HB) and noticeably thickened, whereas it is longer and thinner in *R. rattus* (ca. 115-120 % of HB). The hindfeet are large and fleshy when compared with similar-sized indigenous rodents (HF typically much greater than 30 mm). The ears are larger and translucent in *R. rattus* (and capable of being pulled forward to cover the eyes) but much smaller in *R. norvegicus*. The face is more pointed in the case of *R. rattus*. There are usually five pairs of nipples (2 + 3 = 10) although there may occasionally be an extra thoracic pair. (3 + 3 = 12).

(a) (b)

Figure 2.691 a and b Photographs of *Rattus rattus* (Ara Monadjem, left and Bill Stanley, right).

Cranial Characters

The cranium is large (GSL ca. 35-50 mm; Table 10) and somewhat ovoid in shape (although the braincase appears rounder in *R. rattus*), and slightly curved in lateral profile (more so in *R. norvegicus*). Supraorbital, temporal and occipital ridges are well developed and differ in shape between species, with the supraorbital-temporal being "amphoral" in shape (curving outwards posteriorly from interorbital region and then curving inwards again on the braincase till it reaches the occipital ridge) in *R. rattus* versus two straight, almost parallel but slightly diverging lines in *R. norvegicus*. The nasals are long narrow and deep. The zygomatic arches are quite slender in *R. rattus* and but more robust and thickened anteriorly in *R. norvegicus*. The zygomatic plate has a sharp anteriodorsal angle in *R. rattus* but a blunter angle in *R. norvegicus*. The palatal foramina are noticeably short and stop at or just short of the root of M^1. The palate is broad (> twice the width of M^1) and extends well posterior to M^3.

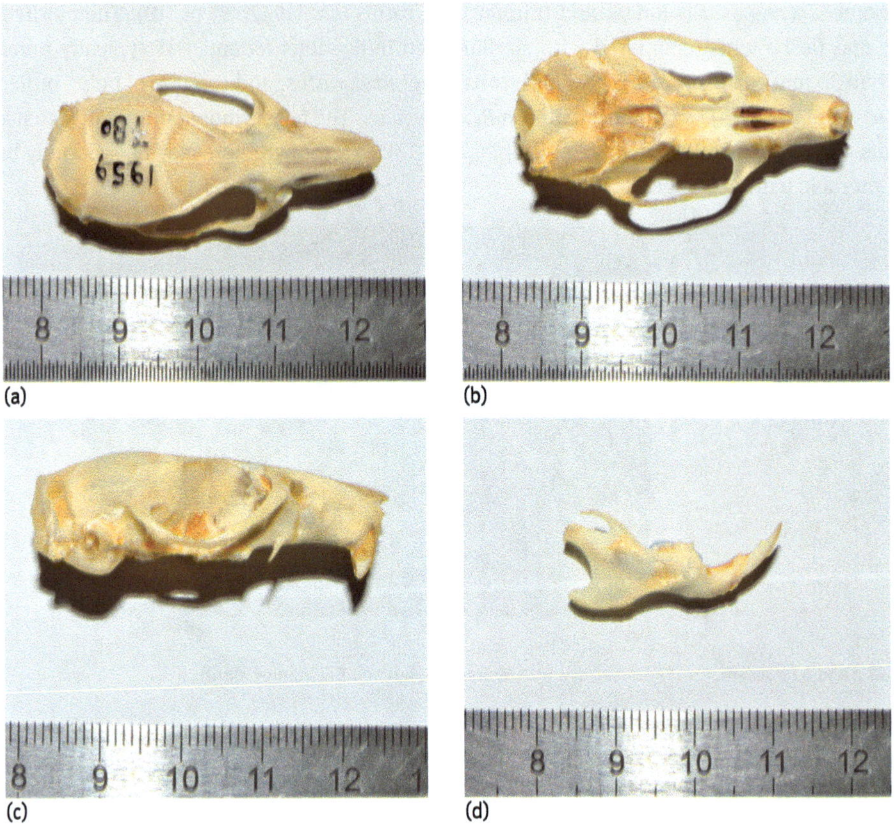

Figure 2.692 a-d Cranium of *Rattus rattus* (MNHN-ZM-MO1959-780) (from top left to bottom right: dorsal view, ventral view, lateral view and lower jaw).

Dentition

The upper incisors are orthodont and ungrooved. The molars are proportionately narrow and small (MTR/GSL 0.16) and comprise distinct and well separated cusps, with the medial row enlarged. Rows of cusps are strongly distorted transversely with cusp t1 located well behind t2 and joined to it by a long ridge. The cusp t3 on M^1 is reduced in *R. norvegicus* and an accessory cusplet may be present on t2 in *R. norvegicus*. The cusp t9 on M^1 and M^2 is small and closely associated with t8. Molar M^3 comprises two laminae with a large t1 but lacks a t3 cusp. No anterior cusplets are present but a small posterior cusplet is present on M_1 and M_2 and up to three supplementary labial cusplets may be present on these two molars (more developed in *R. rattus* than *R. norvegicus*).

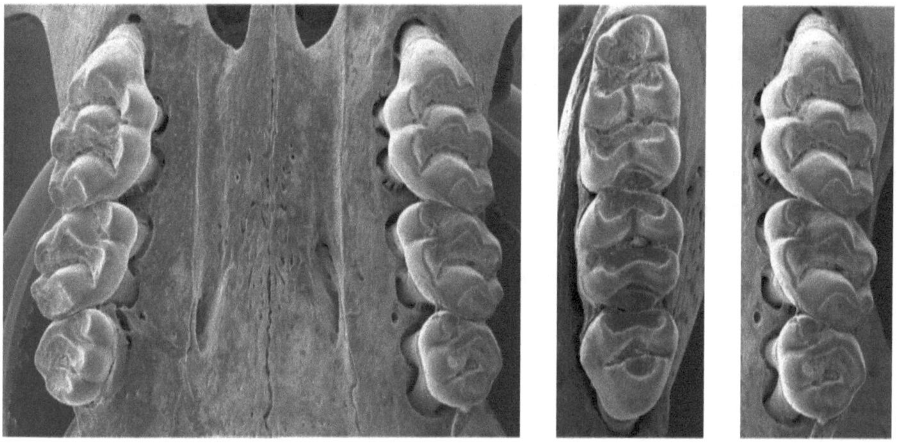

(a) (b) (c)

Figure 2.693 Cheek teeth of *Rattus rattus* (MNHN-ZM-MO 1994-751) (A: upper cheek teeth; B: lower left toothrow; C: upper left toothrow).

Habitat & Biogeography

Although usually strictly commensal in Madagascar and Africa, *R. rattus* has also been detected in forested habitats away from human settlements. Until recently it was considered that *R. norvegicus* was mostly confined to large port cities in Africa (e.g. de Graaff, 1981; Happold, 2013), but Bastos et al. (2011) showed that it was widespread in the interior Gauteng province of South Africa. Although previously considered not to occur in Africa, the Asian house rat (*R. tanezumi* or "Lineage II" of Aplin et al., 2011) was recently shown from molecular evidence (Bastos et al., 2011) to be present and widespread at least in South Africa and Swaziland, even though ecological niche models showed the climate of South Africa to be unsuitable for the species based on its current range in South-East Asia. Clearly the anthropogenic niche provided by transportation routes and cities was sufficient for the invasion of Africa by this species.

Fossil Record

No fossils are known in Africa as this genus evolved in Eurasia and invaded Africa during historical times.

Biology & Ecology

They are nocturnal and terrestrial but *Rattus rattus* is very capable of climbing and often found in roofs (hence the name "roof rat" which is sometimes used). These introduced commensal rodents are destructive pests with a highly omnivorous diet that benefits from human habitation, food and waste. They are prolific breeders capable of rapid population growth in favourable conditions. In Africa, females give birth to multiple litters throughout the year and litter size varies from about 5 to 12 (Happold, 2013). Predators are mammalian carnivores, raptors and snakes found in urban environments.

Systematic Notes

Until recently only two species of *Rattus* were known from Africa, *R. rattus* and *R. norvegicus*. However the presence of a third species (part of the *R. rattus* group) has been demonstrated on molecular grounds from Cape Town well as from the north-eastern parts of South Africa and Swaziland (Bastos et al., 2011; Aplin et al., 2011). This lineage was recognised by Bastos et al. (2011) as *R. tanezumi* but is currently classified as "Lineage II" of *R. rattus* (Aplin et al., 2011).

Annotated Species List

Three species are recognised in Sub-Saharan Africa:
Rattus norvegicus (Berkenhout, 1769)
Rattus rattus (Linnaeus, 1758)
Rattus tanezumi Temminck, 1844 (or "Lineage II" of the *R. rattus* group; Aplin et al., 2012).

Key to Species

1	Size larger (GLS > 45 mm); tail relatively shorter (< 100% of HB); ears shorter (ca. 21 mm)	*R. norvegicus*
	Size smaller (GLS < 45 mm); tail relatively longer (> 100% of HB); ears longer (ca. 22 mm)	2
2	*R. rattus* group (can only be distinguished on molecular grounds)	*R. rattus* *R. tanezumi*

Species Accounts

Rattus norvegicus (Berkenhout, 1769)

Diagnosis

This species can be distinguished from *R. rattus* and *R. tanezumi* by its larger size, shorter ears and relatively shorter tail (in *R. rattus* and *R. tanezumi*, GSL < 45 mm; HF typically ca. 32-35 mm in adults; tail typically > 100% of HB). Six pairs of nipples (3 + 3 = 12) are present.

Taxonomic Issues

Chromosome number 2n = 42.

Distribution

This species is widespread along the coastline of Africa, mostly associated with ports, but also occurs at interior sites, particularly in South Africa where it is commensal with humans (Bastos et al., 2011).

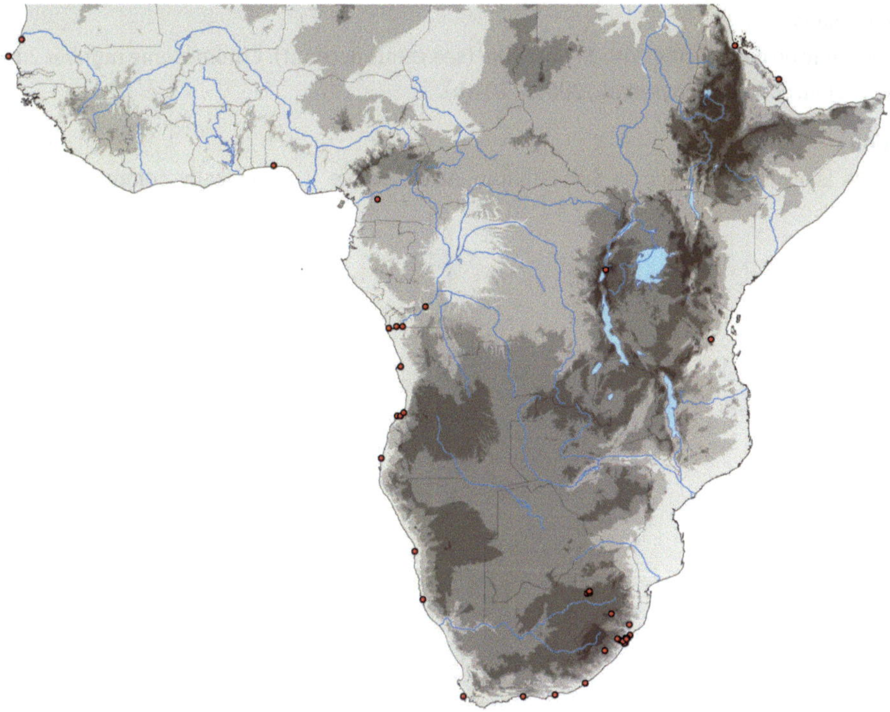

Figure 2.694 Distribution map

Measurements

External and cranial measurements (mm) and mass (g) for *Rattus norvegicus*, sexes combined

	Mean	Min	Max	N
Mass	195.7	65	462	19
HB	201.4	106	251	22
Tail	181.3	110	230	23
HF	39.1	31	45	20
Ear	20.0	15	24	22
GLS	48.5	46.5	51.2	5

Rattus rattus (Linnaeus, 1758)

Diagnosis

This species can be distinguished from *R. norvegicus* by its smaller size, longer ears and relatively shorter tail (in *R. norvegicus*, GSL > 45 mm; HF typically ca. 40 mm in adults; tail typically < 100% of HB). Five or six pairs of nipples (3 + 2 = 10 or 3 + 3 = 12) are present. This species can only be distinguished from *R. tanezumi* on molecular grounds (Bastos et al., 2011).

Taxonomic Issues

This species may be conspecific with *R. tanezumi* (see genus account). Chromosome number 2n = 38 (Dobigny et al., 2002a; Denys et al., 2009a, 2011b).

Distribution

This species is widespread across Sub-Saharan Africa where it is typically commensal with humans.

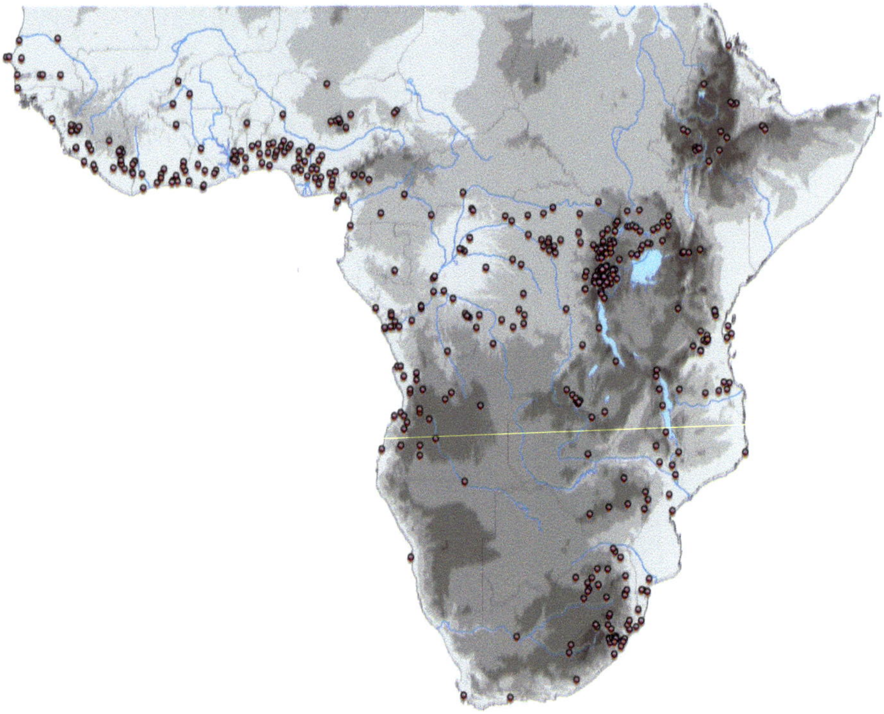

Figure 2.695 Distribution map

Measurements

External and cranial measurements (mm) and mass (g) for *Rattus rattus*, sexes combined

	Mean	Min	Max	N
Mass	115.2	56	257	28
HB	168.3	100	342	58
Tail	178.9	95	250	57
HF	33.8	23	42	53
Ear	21.2	16	31	45
GLS	39.9	34.9	44.7	23

Rattus tanezumi Temminck, 1844 (or "Lineage II" of the *R. rattus* group; Aplin et al., 2011).

Diagnosis

This species can be distinguished from *R. norvegicus* by its smaller size, longer ears and relatively shorter tail (in *R. norvegicus*, GSL > 45 mm; HF typically ca. 40 mm in adults; tail typically < 100% of HB). This species can only be distinguished from *R. rattus* on molecular grounds (Bastos et al., 2011).

Taxonomic Issues

This taxon is probably conspecific with *R. rattus* (Aplin et al., 2011; see genus account). However, we retain this taxon as a valid species pending further investigation into the taxonomy of the *R. rattus* group, especially considering its recent discovery in Southern Africa (Bastos et al, 2011).

Distribution

This species is currently only known from a few scattered localities in northern and eastern South Africa and Swaziland where it was only recently discovered and is commensal with humans (Bastos et al., 2011).

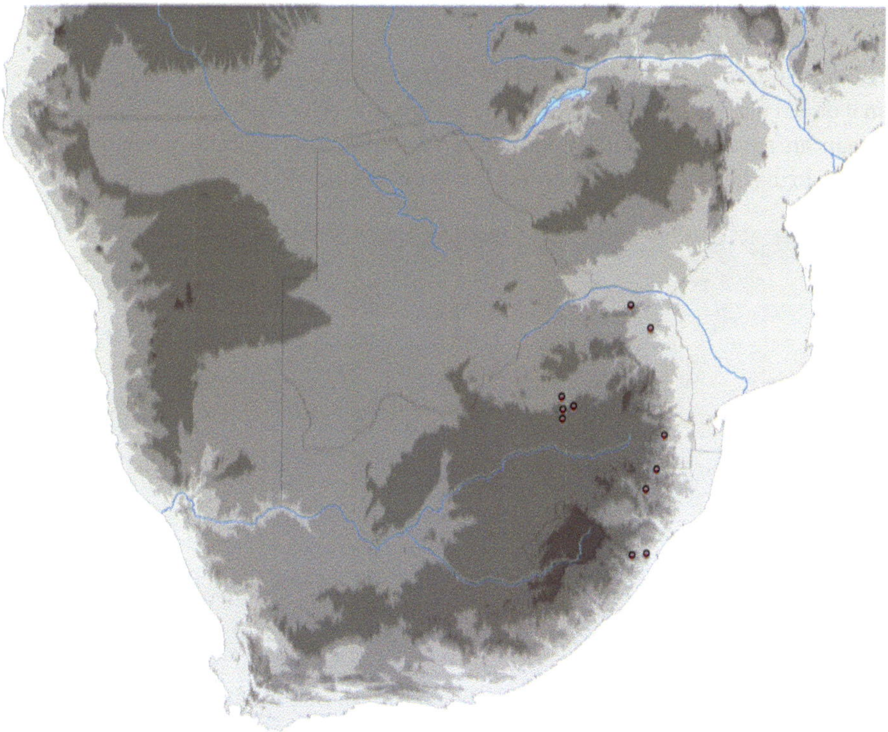

Figure 2.696 Distribution map

Measurements

External and cranial measurements (mm) and mass (g) for *Rattus tanezumi*, sexes combined*

	Mean	Min	Max	N
Mass	101.0	67.9	160.1	5
HB	149.7	108.1	189.6	5
Tail	169.5	142.1	203.3	5
HF	31.6	29	33	4
Ear	20.8	17.2	25.1	6
GLS	39.5	37.0	40.9	7

*These values are probably not representative of the species due to the small sample sizes and the inclusion of subadults. Adult measurements are typically similar to those of *R. rattus*.

Genus Account

Rhabdomys Thomas, 1916 (Four-striped Grass Mice)

Diagnosis

These small mice (Mass ca. 40-50 g; HB 105-115 mm) are immediately recognisable by the four longitudinal dorsal black stripes against a tawny background color. The ventral pelage is a dirty white. The ears are small and the tail is sparsely haired with concentric scales and approximately equal in length to the head and body. There are four pairs of nipples (2 + 2 = 8).

(a)

(b)

Figure 2.697 a and b Photograph of *Rhabdomys chakae* (Bill Branch) (left) and *R.* sp. (Emmanuel Do Linh San) (right).

Cranial Characters

The skull is small and fairly long and narrow in shape, with short and narrow nasals which taper at the end. Supraorbital ridges are well developed and extend onto the parietals. The zygomatic arches are relative robust and the zygomatic plate has an acute anterodorsal angle and slightly backwards sloping anterior edge. The palatal foramina reach just to the anterior root of the M^1. The bullae are moderately sized (6 mm or 20% of GSL) but apparently larger in *R. bechuanae* than in *R. dilectus* (Hill & Carter, 1941).

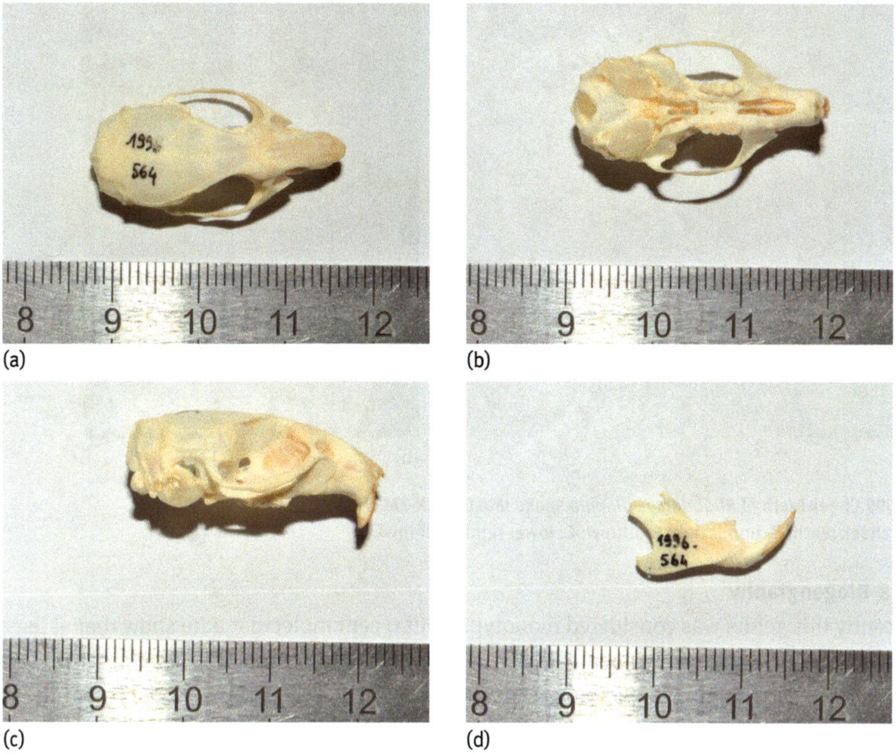

Figure 2.698 a-d Cranium of *Rhabdomys pumilio* sensu lato (MNHN-ZM-MO 1996-564) (from top left to bottom right: dorsal view, ventral view, lateral view and lower jaw).

Dentition

The upper incisors are pro-odont and ungrooved. Molars are of normal size (MTR/GSL ± 0.17, Table 10) with cusps on M^1 distorted, t1 situated behind t2, t2 and t3 well separated. The cusps t6 and t9 are joined on M^1 and M^2 with t9 being reduced on M^1 and reduced or absent on M^2. The cusps t1 and t3 are present on M^2. Labial cusplets are present on M_1 and M_2. Posterior cingulum is present on M_1 and M_2, usually smaller on the latter, and anterior plus cingular cusplets are present on M_2 and M_3, reduced in the latter. There are traces of three laminae on M_3, with only a single cusp apparent on the third lamina due to the fusion of the cusps t8 and t9 (Group 3, Figure 2.439).

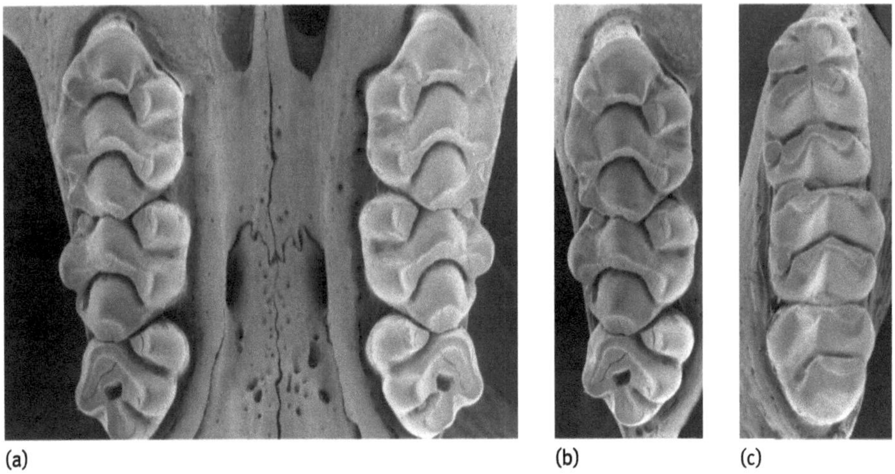

Figure 2.699 Cheek teeth of *Rhabdomys pumilio* sensu lato (MNHN-ZM-MO 1935-1539) (A: upper cheek teeth; B: upper right toothrow; C: lower right toothrow).

Habitat & Biogeography

Until recently this genus was considered monotypic, but recent molecular data show that at least four species can be recognised which correspond generally with different vegetation types (Du Toit et al., 2012). *Rhabdomys pumilio* sensu stricto is associated with Fynbos and Succulent Karoo in the winter rainfall climate region of South Africa whilst *R. intermedius* occurs in Nama Karoo of the central parts of South Africa, *R. bechuanae* is found in Nama Karoo and arid savanna of the Northern Cape province of South Africa, Botswana, Namibia and south-western Angola (Hill & Carter, 1941), and *R. dilectus* is associated with montane and submontane grassland distributed from the Great Escarpment of South Africa and similar montane regions of Zimbabwe, Angola northwards to Uganda. Hill & Carter (1941) pointed out morphological and ecological differences between *R. bechuanae* and *R. pumilio* (= *dilectus*) in Angola; the former occurred in more arid habitats and had a larger body size, paler colouration with indistinct stripes and larger bullae. According to Musser & Carleton (2005) no intergradation occurs between these species.

Fossil Record

Fossils are known from the Early Pliocene to Late Pleistocene in Southern Africa (Winkler et al., 2010). The earliest fossil remains are from 3.3 MYA at Makapansgat in South Africa (Pocock, 1976).

Biology & Ecology

Rhabdomys species are diurnal and crepuscular, terrestrial and scansorial. They are omnivores, eating seeds, fruits, leaves, stems, flowers, *Acacia* pods, *Protea* flower bracts, *Pinus* bark, insects, worms and snails. They can be pests to the forestry industry where they ring-bark young *Pinus* saplings. The diet varies considerably between seasons and insects are dominant during the period of reproduction (Perrin, 1980). Recent research by N. Pillay, C. Shradin and co-workers has documented interesting variation in life history between the mesic and arid species. For example, *R. dilectus* seems to be mostly solitary in behaviour and social structure whereas the more arid-adapted species seem to be communal (Schradin & Pillay, 2005). Reproduction is seasonal and litter size varies from 2 to 11, being lower in harsher environments (Happold, 2013).

Systematic Notes

Rhabdomys is most closely related to *Desmomys* on molecular grounds and both of these genera are included within the *Arvicanthis* division of tribe Arvicanthini (Musser & Carleton, 2005; Lecompte et al., 2008). Historically the genus has been considered to be monotypic, although Musser & Carleton (2005) pointed out the need for comprehensive and careful examination of museum collections in the light of molecular and chromosomal studies to detect probably cryptic species and to define their geographical limits. Based on recent molecular studies, we here recognise five species. Rambau et al. (2003) demonstrated specific separation between eastern mesic-adapted populations termed *R. dilectus* and more arid-adapted populations termed *R. pumilio* from the drier parts of South Africa, Botswana, Namibia and south-western Angola. Within *R. dilectus*, "northern" and "southern" subspecies (which we regard as species, since at least one case of sympatry is known) were defined on chromosomal and molecular grounds, termed *R. d. dilectus* and *R. d. chakae* respectively (Rambau et al., 2003; Castiglia et al., 2012). Du Toit et al. (2012) provided molecular evidence for the presence of three distinct taxa within *R. pumilio* which, with the removal *of R. dilectus* and *R. chakae*, becomes paraphyletic; *R. pumilio* sensu stricto is more closely related to the *R. dilectus* group than to the other two taxa. The following names are available for the three taxa above-mentioned taxa: *pumilio, bechuanae* and *intermedius* (Du Toit et al., 2012).

Annotated Species List

Five species are recognised:
Rhabdomys bechuanae (Thomas, 1893)
Rhabdomys chakae (Wroughton, 1905)
Rhabdomys dilectus (De Winton, 1897)
Rhabdomys intermedius (Wroughton, 1905)
Rhabdomys pumilio (Sparrman, 1784)

Key to Genus

At present these five species are only distinguishable on molecular grounds; morphologically there is great overlap in appearance and measurements, and most species share the same karyotype (Rambau et al., 2003; Castiglia et al., 2012; Du Toit et al., 2012). For this reason we are unable to provide a key, however, their distributions are typically allopatric or parapatric with minimal overlap between neighbouring species. The chromosome number and general distribution of each species is provided below (see species accounts for detailed maps).

***Rhabdomys* species**	**Chromosome number**	**Distribution**
R. bechuanae	2n = 38, 46	Northern Cape through Botswana to Namibia and extreme south-western Angola
R. chakae	2n = 48	Eastern South Africa, Lesotho and Swaziland
R. dilectus	2n = 48	Northern South Africa, Zimbabwe, East Africa and Angola (except extreme south-west).
R. intermedius	2n = 48	Inland regions of the Cape province, South Africa
R. pumilio	2n = 48	Coastal parts of the Cape province, South Africa

Species Accounts

Rhabdomys bechuanae (Thomas, 1893)

Diagnosis

This species can only be distinguished from other *Rhabdomys* species on molecular grounds (Du Toit et al., 2012).

Taxonomic Issues

This taxon has recently been elevated to species status following a molecular analysis of the genus (Du Toit et al., 2012). The closest relative of this species is *R. intermedius* (Du Toit et al., 2012). See genus account for further details.

Distribution

This species occurs widely in the central and western regions of South Africa mostly north of the Gariep River, northwards and westwards through Botswana to Namibia and extreme south-western Angola.

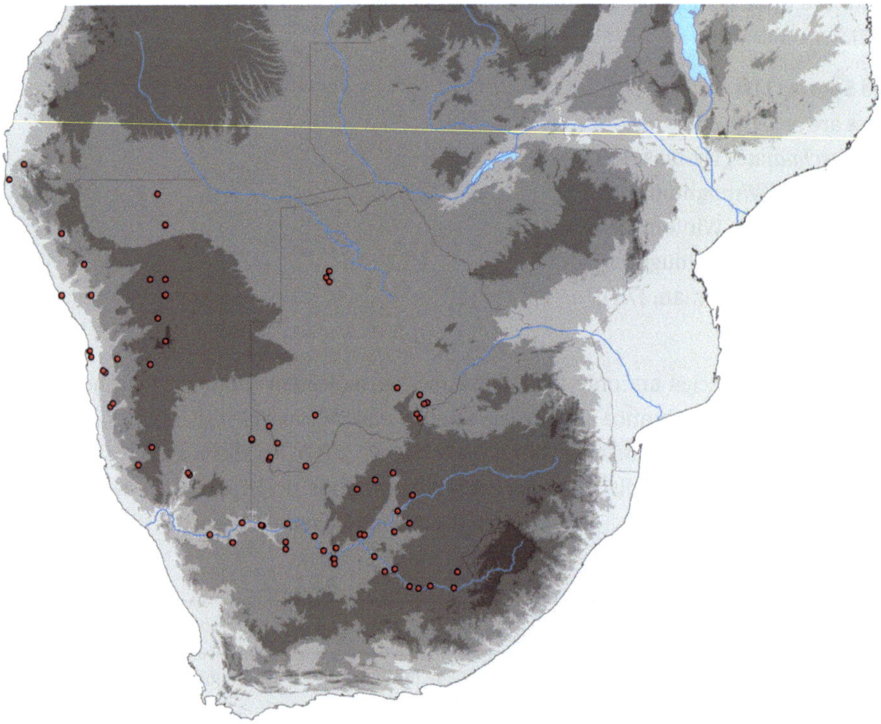

Figure 2.700 Distribution map

Measurements

External and cranial measurements (mm) and mass (g) for *Rhabdomys bechuanae*, sexes combined

	Mean	Min	Max	N
Mass*	44.5	32	55	14
HB*	106.9	-	-	33
Tail*	107.8	95	120	32
HF*	24.2	22	27	41
Ear*	12.4	11	15	41
GLS	25.8	25.7	25.9	2

*Smithers (1971); HB estimated from Total Length minus Tail Length

Rhabdomys chakae (Wroughton, 1905)

Diagnosis

This species can only be distinguished from other *Rhabdomys* species on molecular grounds (Du Toit et al., 2012).

Taxonomic Issues

This taxon has recently been elevated to species status following a molecular analysis of the genus (Rambau et al., 2003; Castiglia et al., 2012; Du Toit et al., 2012). The closest relative of this species is *R. dilectus* (Du Toit et al., 2012). See genus account for further details.

Distribution

This species occurs in the eastern regions of South Africa from the Soutpansberg Mts in the north to the Eastern Cape in the south, extending into Swaziland and Lesotho (Castiglia et al., 2012; Du Toit et al., 2012).

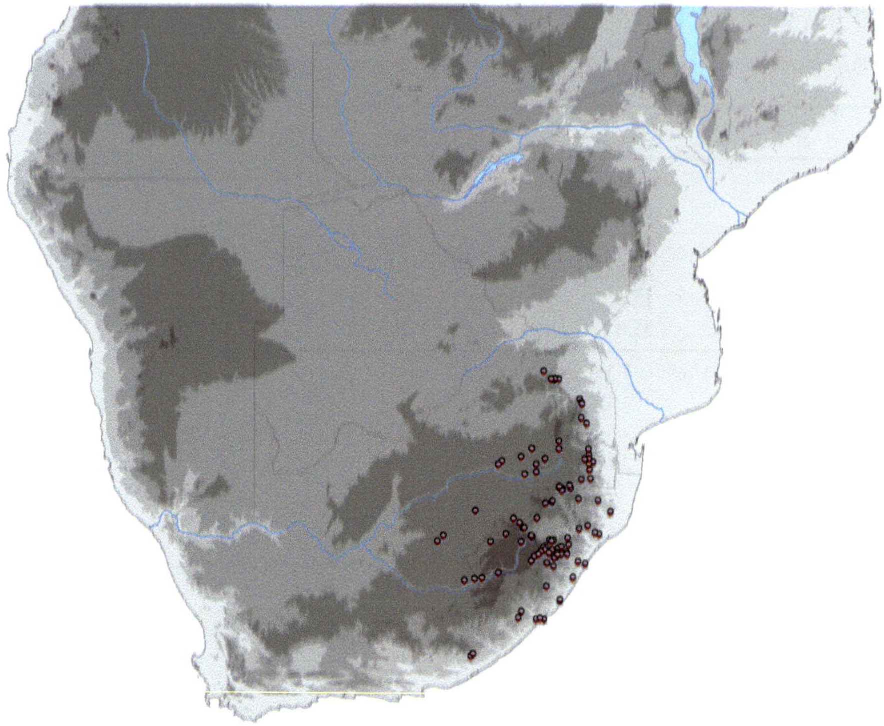

Figure 2.701 Distribution map

Measurements

External and cranial measurements (mm) and mass (g) for *Rhabdomys chakae*, sexes combined

	Mean	Min	Max	N
Mass	45.5	14	70	34
HB	109.2	75	140	54
Tail	85.4	67	102	48
HF	22.1	18	25	41
Ear	12.7	8	19	46
GLS	27.0	25.6	30.2	10

Rhabdomys dilectus (De Winton, 1897)

Diagnosis

This species can only be distinguished from other *Rhabdomys* species on molecular grounds (Du Toit et al., 2012).

Taxonomic Issues

This taxon has recently been elevated to species status following a molecular analysis of the genus (Rambau et al., 2003; Castiglia et al., 2012; Du Toit et al., 2012). The closest relative of this species is *R. chakae* (Du Toit et al., 2012). See genus account for further details.

Distribution

This species occurs from northern South Africa northwards through Zimbabwe, Malawi to East Africa with an apparently isolated population on the plateau of Angola (Castiglia et al., 2012).

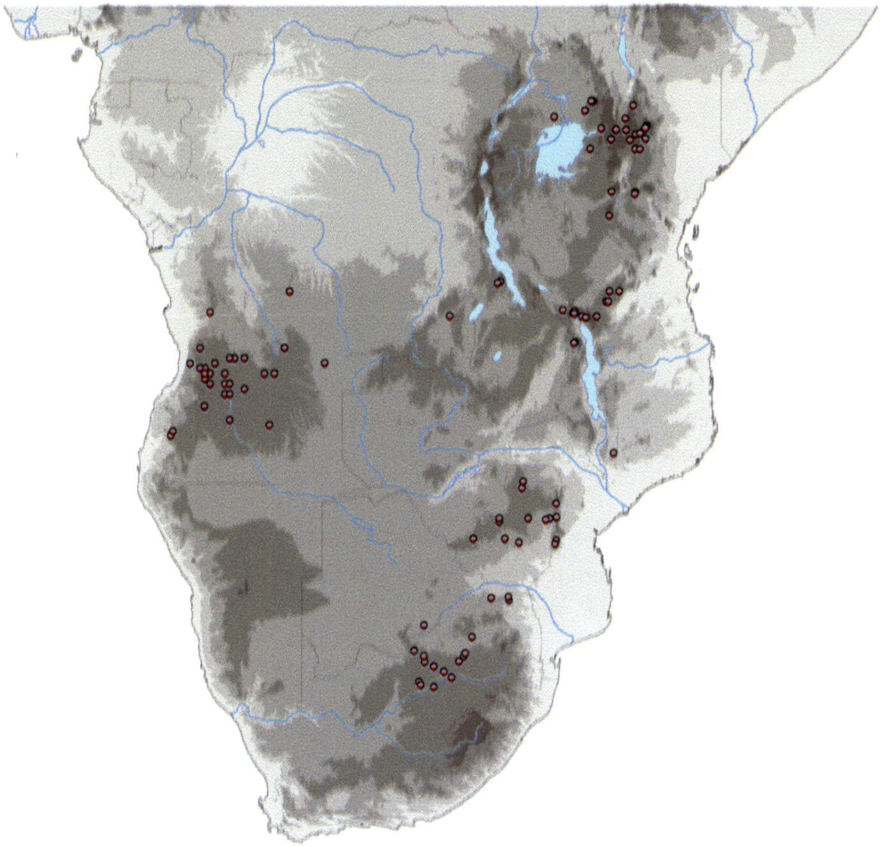

Figure 2.702 Distribution map

Measurements

External and cranial measurements (mm) and mass (g) for *Rhabdomys dilectus*, sexes combined

	Mean	Min	Max	N
Mass	39.0	30	47	4
HB	111.1	91	126	10
Tail	82.6	72	90	8
HF	20.9	18	24	8
Ear	13.1	10	20	11
GLS*	24.5	21.0	26.5	25

*Hollister (1919)

Rhabdomys intermedius (Wroughton, 1905)

Diagnosis

This species can only be distinguished from other *Rhabdomys* species on molecular grounds (Du Toit et al., 2012).

Taxonomic Issues

This taxon has recently been elevated to species status following a molecular analysis of the genus (Du Toit et al., 2012). The closest relative of this species is *R. bechuanae* (Du Toit et al., 2012). See genus account for further details.

Distribution

This species is endemic to the southern regions of South Africa where it occurs in a narrow band through the Karoo, wedged between *R. pumilio* on the coast and *R. bechuanae* further north (Du Toit et al., 2012).

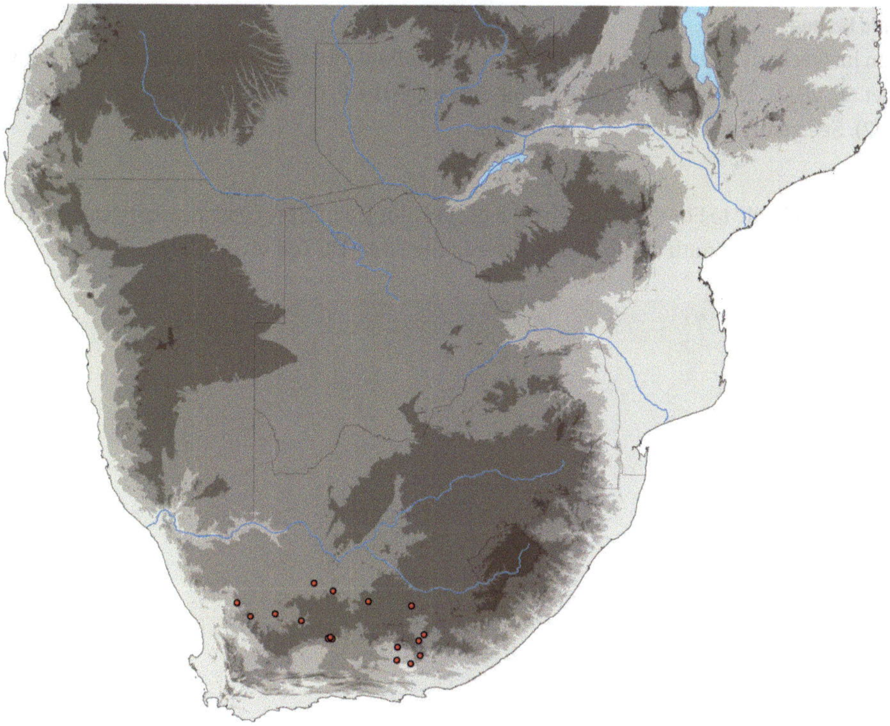

Figure 2.703 Distribution map

Measurements

External and cranial measurements (mm) and mass (g) for *Rhabdomys intermedius*, sexes combined

	Mean	Min	Max	N
Mass	44.0	28	64	8
HB	110.7	96	135	10
Tail	106.7	90	121	9
HF	25.0	22	27	7
Ear	13.8	12	15	10
GLS	28.0	26.6	29.3	3

Rhabdomys pumilio (Sparrman, 1784)

Diagnosis

This species can only be distinguished from other *Rhabdomys* species on molecular grounds (Du Toit et al., 2012).

Taxonomic Issues

This taxon has recently been elevated to species status following a molecular analysis of the genus (Du Toit et al., 2012). The closest relative of this species is the *R. dilectus-R. chakae* group (Du Toit et al., 2012). See genus account for further details.

Distribution

This species is endemic to South Africa where it occurs along a narrow coastal strip from Port Elisabeth in the east to the mouth of the Gariep River in the west (Du Toit et al, 2012).

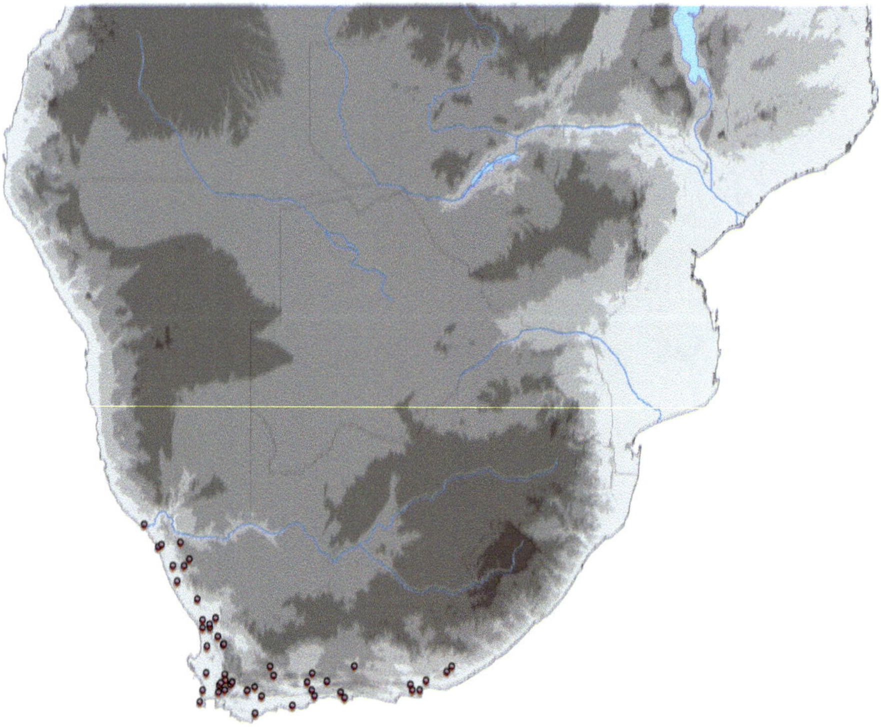

Figure 2.704 Distribution map

Measurements

External and cranial measurements (mm) and mass (g) for *Rhabdomys pumilio*, sexes combined

	Mean	Min	Max	N
Mass	41.8	24	54	13
HB	114.6	93	134	16
Tail	114.6	99	128	15
HF	25.4	22	28	16
Ear	14.5	13	16	16
GLS	29.3	28.0	31.6	4

Genus Account

Stenocephalemys Frick, 1914 (Ethiopian Rats)

Diagnosis

These Ethiopian endemics are medium to large in size (Mass ca. 60-130 g; HB ca. 130-170 mm) with a long and soft to woolly-haired pelage which is sandy to reddish-brown above and greyish white below. These mice are very similar in appearance to the other soft-furred rodents, *Praomys*, *Mastomys* and *Myomyscus*, but can be distinguished on molecular grounds (see Systematic Notes, below) and by their more dorsally positioned eyes (stenocephaly) resulting from a narrower interorbital constriction, more robust zygomatic arches and enlarged molar toothrow (see below). The tail varies in length from shorter than the head and body (ca. 90% of HB) in the moorland species to longer (ca. 120% of HB) in the forest species.

(a) (b)

Figure 2.705 a and b Photographs of *Stenocephalemys albipes* (Marian Polak, left and Ema Hrouzkova, right).

Cranial Characters

The skull is fairly large in size (GSL ca. 33-40 mm) and long and narrow in shape with a conspicuously narrow interorbital region. This latter aspect, termed "stenocephaly" (which gives its name to the genus) is most clearly developed in the two high-elevation moorland species (*S. albocaudata* and *S. griseicaudata*) and is an adaptation to having the orbits placed dorsally in the head to better detect potential avian predators in open habitats (Petter, 1972). The interorbital region (frontal bones) is more rounded in *S. albipes* and *S. ruppi* and more square in *S. albocaudata* and *S. griseicauda* (Yalden et al., 1976). The skull is smooth without obvious crests and with the braincase sutures are well defined even in adult individuals. The zygomatic arches are rather robust and the zygomatic plate has a vertical edge. The palatal foramina extend to between the M^1 molars, about midway down their length. The hamular process ("mastoid strut") and associated subsquamosal foramen is present.

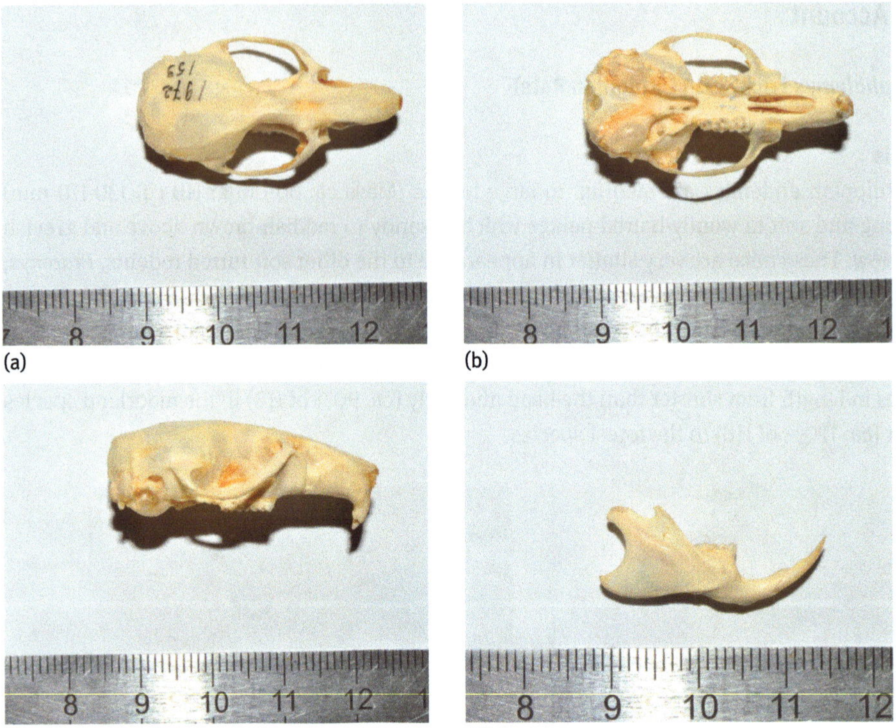

Figure 2.706 a-d Cranium of *Stenocephalemys griseicauda* (MNHN-ZM-MO 1972-135) (from top left to bottom right: dorsal view, ventral view, lateral view and lower jaw).

Dentition

The incisors are orthodont to slightly opisthodont and ungrooved. The molars are proportionally very large, i.e. showing macrodonty (MTR/GSL ± 0.19, Table 10) and the molars are well differentiated from each other with transverse rows only slightly distorted. The cusps t1 and t4 on M^1 and M^2 are large and well differentiated from each other, while t3 on M^1 is poorly differentiated from t2; t9 on M^1 is large and well differentiated from t6. Small traces of longitudinal crests are present on t1, t3 and t6 and between t4 and t8. The cusp t3 is reduced on M^2 and absent on M^3. Traces of three laminae are present on M^3 (Group 3; Figure 2.439). On the lower molars, the posterior cingulum is present on M_1 and M_2, although reduced in the latter, whilst a well-developed anterior cusplet and associated crest is present on the first lobe of M_2.

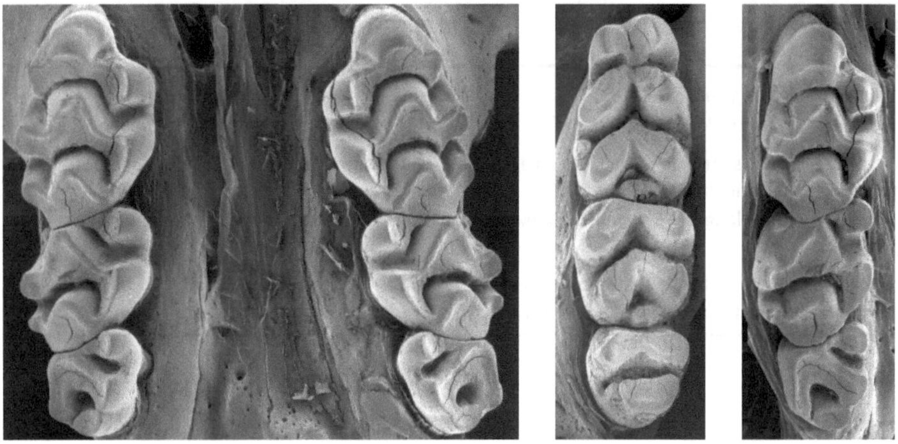
(a) (b) (c)

Figure 2.707 Cheek teeth of *Stenocephalemys albocaudata* (MNHN-ZM-MO 1972-211) (A: upper cheek teeth; B: lower left toothrow; C: upper right toothrow).

Habitat & Biogeography

All four species are Ethiopian endemics. Two of them (*S. griseicauda* and *S. albocaudata*) occur in moorlands and bushy habitats at higher elevations (2400-4400 m) in the eastern and western plateaux of Ethiopia, whilst the other (*S. albipes* and *S. ruppi*) two are forest-dwelling at lower altitudes (1500-3000m) in the eastern and western plateaux of Ethiopia.

Fossil Record

Fossils affiliated with *Stenocephalemys* have been recorded from the Late Miocene of Ethiopia; these are the earliest murine records for East Africa (Geraads, 2001).

Biology & Ecology

They are nocturnal with a diet which is granivorous and herbivorous. Reproduction is continuous in species from lower altitudes and seasonal in higher-elevation species (Happold, 2013).

Systematic Notes

Stenocephalemys belongs to tribe Praomyini defined on molecular grounds and within this tribe it was placed in the *Stenocephalemys* division along with *Heimyscus*, *Hylomyscus*, *Mastomys*, *Myomyscus* and *Praomys* (Musser & Carleton, 2005; Lecompte et al., 2008; Schenk et al., 2013). The forest-dwelling *S. albipes* and *S. ruppi* have been placed in both *Praomys* (e.g. Rupp, 1980; Yalden & Largen, 1992) and *Myomys* (= *Myomyscus*) (e.g. Musser & Carleton, 1993) but molecular, morphometric and morphological data suggest their inclusion in *Stenocephalemys* (e.g. Rupp, 1980; Van der Straeten & Dieterlen, 1983; Fadda & Corti, 2000; Fadda et al., 2001b; Musser & Carleton, 2005).

Annotated Species List

Four species are currently recognised:
Stenocephalemys albipes (Ruppell, 1842)
Stenocephalemys albocaudata Frick, 1914
Stenocephalemys griseicauda F. Petter, 1972
Stenocephalemys ruppi (Van der Straeten & Dieterlen, 1983)

Key to Species

1	Tail relatively long (ca. 120% of HB)	2
	Tail relatively short (< 100% of HB)	3
2	Upper molar toothrow shorter (< 5.9 mm); tail black above, white below; widespread in montane forests of Ethiopia	*S. albipes*
	Upper molar toothrow longer (> 6.3 mm); tail grey above, paler below; restricted to the Gughe Mts	*S. ruppi*
3	Tail white; typically most abundant in moorlands above 3500 m	*S. albocaudata*
	Tail grey; typically most abundant in moorlands ca. 3000 m	*S. griseicauda*

Species Accounts

Stenocephalemys albipes (Ruppell, 1842)

Diagnosis

This species is readily distinguished from the moorland species (*S. albocaudatus* and *S. griseicauda*) by its longer bicoloured tail and smaller size (in moorland species, tail < 100% of HB, GLS > 34 mm, HF > 25 mm, upper molar toothrow > 6.5 mm). It is distinguished from *S. ruppi*, with which it is syntopic in the Gughe Mts, by its smaller molars and greater contrasting bicoloured tail (in *S. ruppi*, upper molar toothrow > 6.3 mm; tail grey above and paler below). In this species, tail is relatively long (ca. 120-125% of HB) black above and white below, and upper toothrow is short (5.0-5.8 mm). Five pairs of nipples (3 + 2 = 10) are present.

Taxonomic Issues

This taxon was previously placed in the genus *Praomys* (Yalden et al., 1976), but is clearly groups with other members of the genus *Stenocephalemys* on molecular grounds (see genus account for details and references). Chromosome number 2n = 46, FN = 50-53 (Corti et al., 2005).

Distribution

This species is endemic to Ethiopia and Eritrea where it typically occurs in montane forests at 1500-3300 m, but down to 820 m at Godare Forest (Yalden et al., 1976; Bekele, 1996a,b).

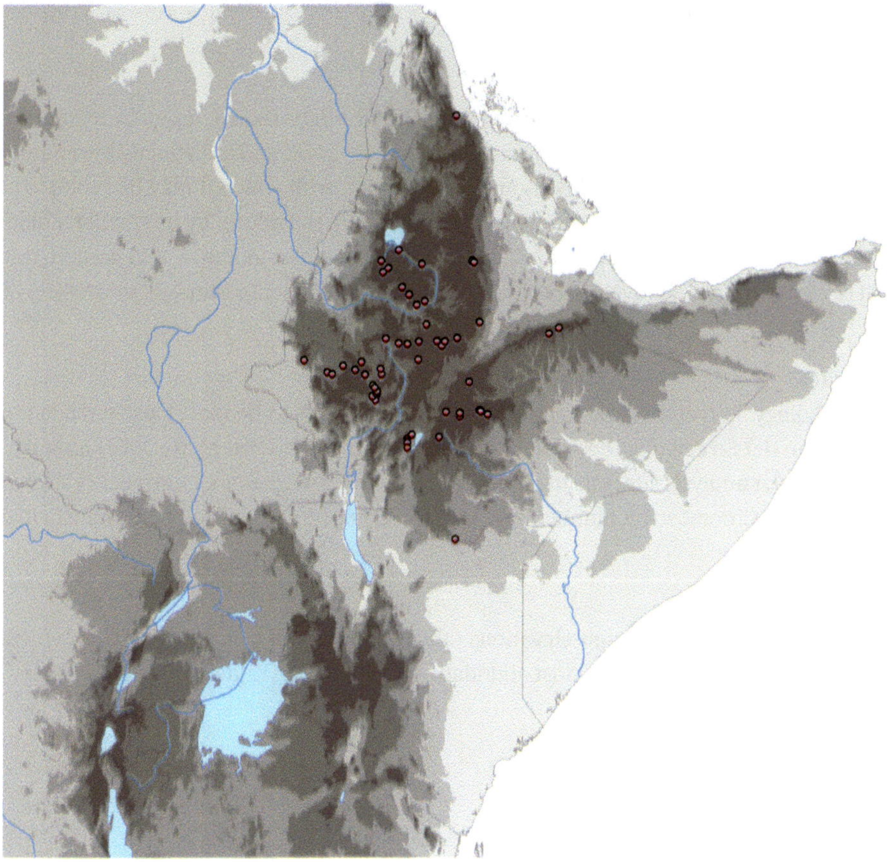

Figure 2.708 Distribution map

Measurements

External and cranial measurements (mm) and mass (g) for *Stenocephalemys albipes*, sexes combined

	Mean	Min	Max	N
Mass*	59.0	40	79	30
HB**	133.6	107	171	109
Tail**	161.3	134	195	105
HF**	28.9	25	32	108
Ear**	22.4	17	30	110
GLS**	33.3	29.6	36.7	119

*Happold (2013); **Van der Straeten & Dieterlen (1983)

Stenocephalemys albocaudata Frick, 1914

Diagnosis

This species is readily distinguished from the forest species (*S. albipes* and *S. ruppi*) by its shorter, unicoloured tail and smaller size (in forest species, tail ca. 120% of HB and is distinctly bicoloured, GLS typically < 34 mm). It is distinguished from *S. griseicauda*, with which it may be syntopic in the Bale Mts (although the two species are typically stratified by elevation), by its smaller molars and colour of the tail (in *S. albocaudata*, tail is white and upper molar toothrow > 7.5 mm). In this species, tail is grey in colour, relatively short (ca. 87% of HB), and upper toothrow is shorter (6.5-7.1 mm).

Taxonomic Issues

Based on morphology, karyotype, and habitat this species appears to be most closely related to *S. griseicauda* (Van der Straeten & Dieterlen, 1983; Corti et al., 1999; Corti et al., 2005) although preliminary molecular analysis suggests that *S. griseicauda* and *S. albipes* are more closely related (Fadda et al., 2001b). Chromosome number 2n = 54, FN = 60-62 (Corti et al., 1999; Lavrenchenko et al., 1999).

Distribution

This Ethiopian endemic occurs at high elevations (above 3000 m) east of the Rift Valley, where it inhabits moorland grasslands being most abundant above 3500 m (Yalden et al., 1976; Yalden, 1988; Yalden & Largen, 1992).

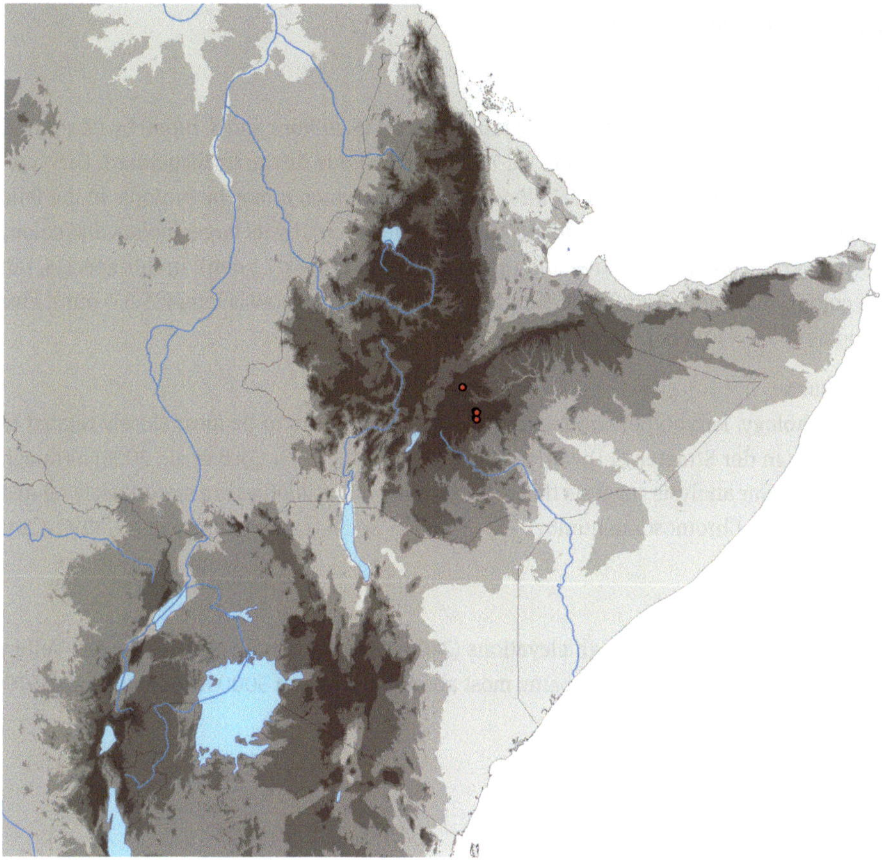

Figure 2.709 Distribution map

Measurements

External and cranial measurements (mm) and mass (g) for *Stenocephalemys albocaudata*, sexes combined

	Mean	Min	Max	N
Mass*	130	83	198	592
HB*	164	106	195	590
Tail*	141	112	175	583
HF*	32	21	36	218
Ear*	27	24	32	218
GLS*	39.4	37.6	41.8	7

*Happold (2013)

Stenocephalomys griseicauda F. Petter, 1972

Diagnosis

This species is readily distinguished from the forest species (*S. albipes* and *S. ruppi*) by its shorter, tail and smaller size (in forest species, tail ca. 120% of HB and is distinctly bicoloured, GLS typically < 34 mm). It is distinguished from *S. albocaudata*, with which it may be syntopic in the Bale Mts (although the two species are typically stratified by elevation), by its larger molars and colour of the tail (in *S. griseicauda*, tail is grey and upper molar toothrow < 7.2 mm). In this species, tail is white in colour, relatively short (ca. 85% of HB), and upper toothrow is long (7.5-8.0 mm). Five pairs of nipples (3 + 2 = 10) are present.

Taxonomic Issues

Based on morphology, karyotype, and habitat this species appears to be most closely related to *S. albocaudata* (Van der Straeten & Dieterlen, 1983; Corti et al., 1999; Corti et al., 2005) although preliminary molecular analysis suggests that *S. griseicauda* and *S. albipes* are more closely related (Fadda et al., 2001b). Chromosome number 2n = 54, FN = 54-58 (Lavrenchenko et al., 1997; Corti et al., 1999).

Distribution

This Ethiopian endemic occurs at high elevations (2400-3900 m) on both sides of the Rift Valley, where it inhabits moorland grasslands being most abundant at about 3000 m (Yalden et al., 1976; Yalden, 1988; Sillero-Zubiri et al., 1995a).

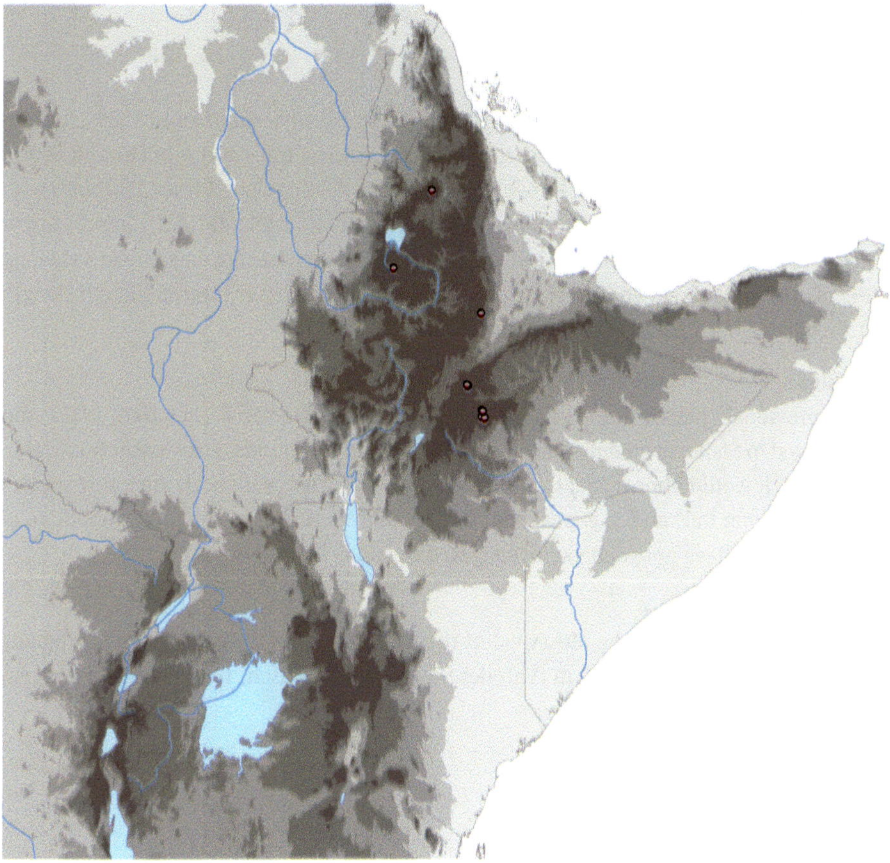

Figure 2.710 Distribution map

Measurements

External and cranial measurements (mm) and mass (g) for *Stenocephalemys griseicauda*, sexes combined

	Mean	Min	Max	N
Mass*	99	70	136	98
HB*	154	130	175	97
Tail*	142	105	160	97
HF*	30	25	33	86
Ear*	24	20	30	85
GLS*	35.9	34.1	37.9	11

*Happold (2013)

Stenocephalemys ruppi (Van der Straeten & Dieterlen, 1983)

Diagnosis

This species is readily distinguished from the moorland species (*S. albocaudata* and *S. griseicauda*) by its longer bicoloured tail and smaller size (in moorland species, tail < 100% of HB and is unicoloured, GLS > 34 mm, upper molar toothrow > 6.5 mm). It is distinguished from *S. albipes* with which it is syntopic by its larger molars and less contrasting bicoloured tail (in *S. albipes*, upper molar toothrow < 5.9 mm; tail black above and white below). In this species, tail is relatively long (ca. 120-125% of HB) grey above and paler below, and upper toothrow is intermediate in length (6.3-7.2 mm). Five pairs of nipples (3 + 2 = 10) are present.

Taxonomic Issues

This taxon was previously placed in the genus *Praomys* (Van der Straeten & Dieterlen, 1983), and subsequently moved to *Stenocephalemys* based on morphological grounds (see Musser & Carleton (2005) for a review). It would appear that this taxon shares characters with both the forest species *S. albipes* (relatively long tail) and *S. griseicauda* (large molars). However, the phylogenetic relationships of this species have not yet been investigated on molecular grounds.

Distribution

This species is endemic to the Gughe Mts, west of the Rift Valley in Ethiopia, where it typically occurs in montane forests at 2700-3200 m (Van der Straeten & Dieterlen, 1983; Yalden & Largen, 1992).

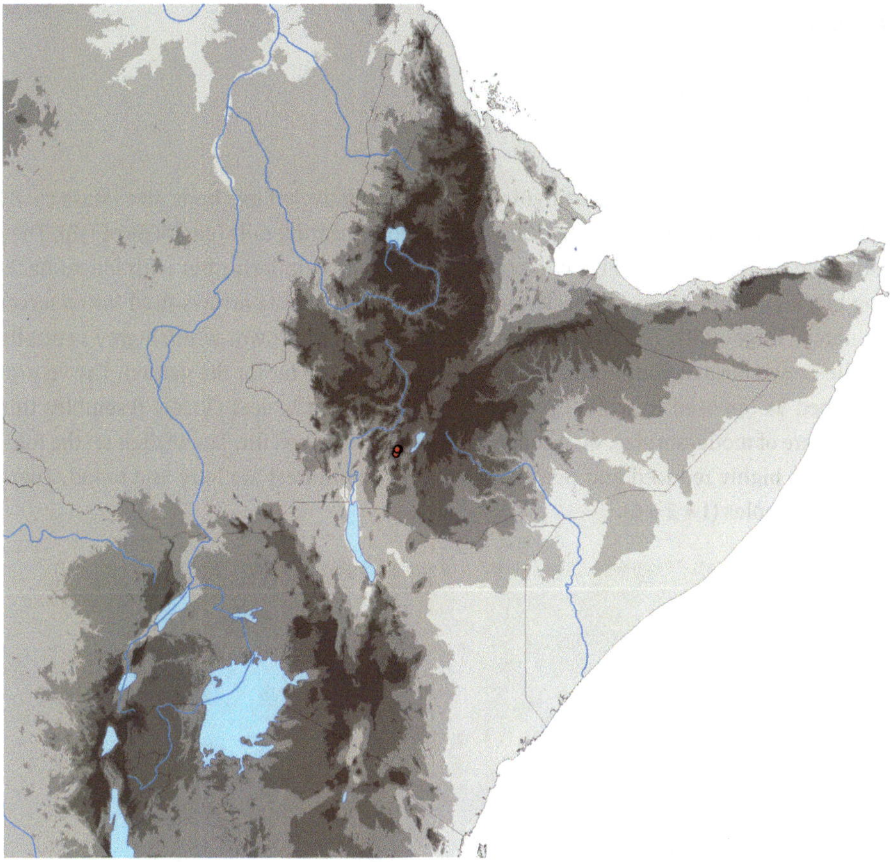

Figure 2.711 Distribution map

Measurements

External and cranial measurements (mm) and mass (g) for *Stenocephalemys ruppi*, sexes combined

	Mean	Min	Max	N
Mass	-	-	-	-
HB*	131.3	124	138	9
Tail*	159.7	143	179	9
HF*	30.7	30	33	8
Ear*	24.1	23	26	8
GLS*	33.5	32.7	34.5	7

*Van der Straeten & Dieterlen (1983)

Genus Account

Stochomys Thomas, 1926 (Target Rat)

Diagnosis

These distinctive rodents are readily distinguished by their relatively large body size (Mass ca. 85 g; HB ca. 140 mm), long whiskers and very long and naked-appearing tails (ca. 146% of HB). They have a distinctive sparse woolly pelage comprising mostly dense underfur but with longer backward-facing black guard hairs on the back and flanks which resemble arrows fired into a target (hence the name, Target Rat). The dorsal colour is a medium red-brown, whitish-grey ventrally with chocolate brown basal colour to the hairs (a diagnostic character for the genus). The ventral fur is grey-tipped as opposed to white-tipped in *Dephomys*, which most closely resembles this genus. The ears are of moderate size, oval shape and possess short red fur. The 1st digit on the forefoot is present but highly reduced and possesses a claw. The hindfeet are long and broad. There are three pairs of nipples (1 + 2 = 6).

Figure 2.712 Photograph of *Stochomys longicaudatus* (Christiane Denys).

Cranial Characters

The skull is large (GLS ca. 38-40 mm) and flattish in profile, with an oval shape having long and deep nasals which hardly taper. The braincase is relatively smoothly contoured but with a well-developed supraorbital crest extending backwards as far as the posterior edge of the parietals; there is also a well-developed supraoccipital crest. The zygomatic arch is quite robust and the zygomatic plate has a rounded antero-dorsal angle and a vertical anterior edge. The palatal foramina are medium in length and extend as far as the root of M^1. The bullae are relatively small (< 5 mm or about 12% of GSL).

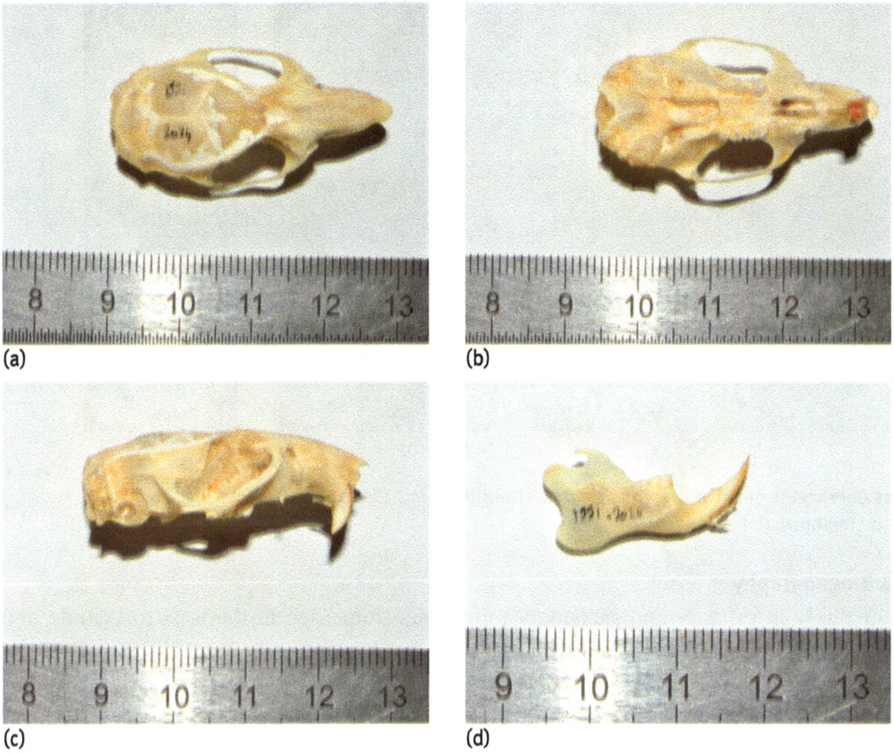

Figure 2.713 a-d Cranium of *Stochomys longicaudatus* (MNHN-ZM-MO 1991-2024) (from top left to bottom right: dorsal view, ventral view, lateral view and lower jaw).

Dentition

The incisors are orthodont to slightly opisthodont and ungrooved. The molars are of normal size (MTR/GSL ± 0.16, Table 10). Cusps are poorly differentiated and aligned in high-crowned, transverse rows of cusps forming backward-facing laminae, much like *Otomys*, especially on the first lobes of M^1 and M^2. On M^1, cusp t9 is small and t3 is small and poorly differentiated from t2. On M^2, t9 is minute and t3 is reduced or absent. Molar M^3 is smaller than M^2, shows traces of three laminae, and lacks t3 (Group 4; Figure 2.439). In the lower molars, there is a well-developed central anterior cusplet on M_1, labial cusplets are present on M_1 and M_2, and posterior cingulum is present on M_1 and M_2, being larger in size on the latter than the former.

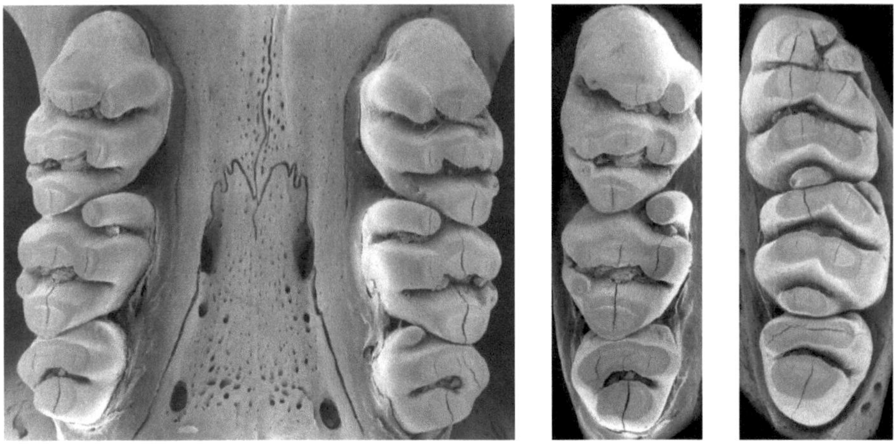

Figure 2.714 Cheek teeth of *Stochomys longicaudatus* (MNHN-ZM-MO 1967-1412) (A: upper cheek teeth; B: upper right toothrow; C: lower right toothrow).

Habitat & Biogeography

They occur rarely in primary and secondary rainforests from Togo in the west to Uganda and Rwanda in the east and southwards to southern DRC.

Fossil Record

None.

Biology & Ecology

They are nocturnal, terrestrial and herbivorous, feeding mostly on fruits and seeds. They seem to breed in most months, have small litters (1-4 embryos) and may live a long time. They appear to be preyed on selectively by small mammalian carnivores (Happold, 2013).

Systematic Notes

Stochomys has been variably associated with the genera *Aethomys*, *Dasymys*, *Mus* and *Rattus* (reviewed by Rosevear, 1969 and Musser & Carleton, 2005) but molar occlusal patterns, supported by immunological and molecular studies show it to be most closely related to *Hybomys* (Misonne, 1969; Watts & Baverstock, 1995; Lecompte et al., 2008; Schenk et al., 2013).

Annotated Species List

One species is currently recognised:
Stochomys longicaudatus (Tullberg, 1893)

Species Accounts

Stochomys longicaudatus (Tullberg, 1893)

Diagnosis

See genus account.

Taxonomic Issues

See genus account.

Distribution

This species is widespread in the lowland forests of tropical Africa from the Volta River in the west to the Albertine Rift in the east.

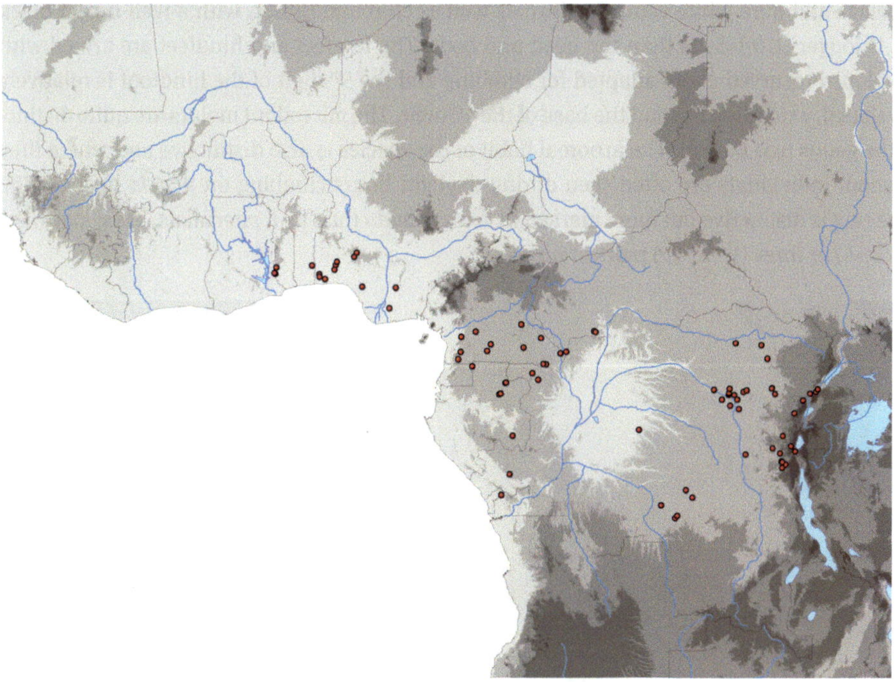

Figure 2.715 Distribution map

Measurements

External and cranial measurements (mm) and mass (g) for *Stochomys longicaudatus*, sexes combined

	Mean	Min	Max	N
Mass	85.2	70	102	5
HB	139.7	100	165	10
Tail	204.1	185	230	10
HF	30.9	25	33	10
Ear	19.8	17	24	9
GLS	38.0	35.3	39.4	10

Genus Account

Thallomys Thomas, 1920 (Acacia Rats or Tree Rats)

Diagnosis

These large (Mass ca. 60-80 g; HB ca. 130-150 mm) rodents have a coarse buffy-brown to grey dorsal pelage and pure white ventral pelage as well as chin and throat, with a well-furred black tail that is longer (110-145 %) than the head and body. The forefeet and hindfeet are armed with well-developed recurved claws adapted for climbing and the 5th digit of the hindfoot is relatively well developed, extending beyond the base of the 4th digit. The microdont molars are quite distinctive for the genus (see below). The arboreal habit of the species is also distinctive and where they are common, individuals are often seen during daylight hours climbing on *Acacia* bushes. The black eye ring is distinctive but more marked in *T. nigricauda* than in *T. paedulcus*. They may have two (0 + 2 =4) or three (1 + 2 = 6) pairs of nipples.

Figure 2.716 Photograph of *Thallomys* sp. (Seth Eiseb).

Cranial Characters

The skull is large (30-37 mm) and quite robust in shape, with flattened braincase, prominent supraorbital and supraoccipital ridges, an average-sized rostrum, robust zygomatic arch and zygomatic plate with quite sharp antero-dorsal angle and vertical anterior edge. The palatal foramina reach to the second row of cusps on M^1. The bullae are enlarged (23% of GSL) although apparently in *T. shortridgei* they are markedly smaller (Musser & Carleton, 2005).

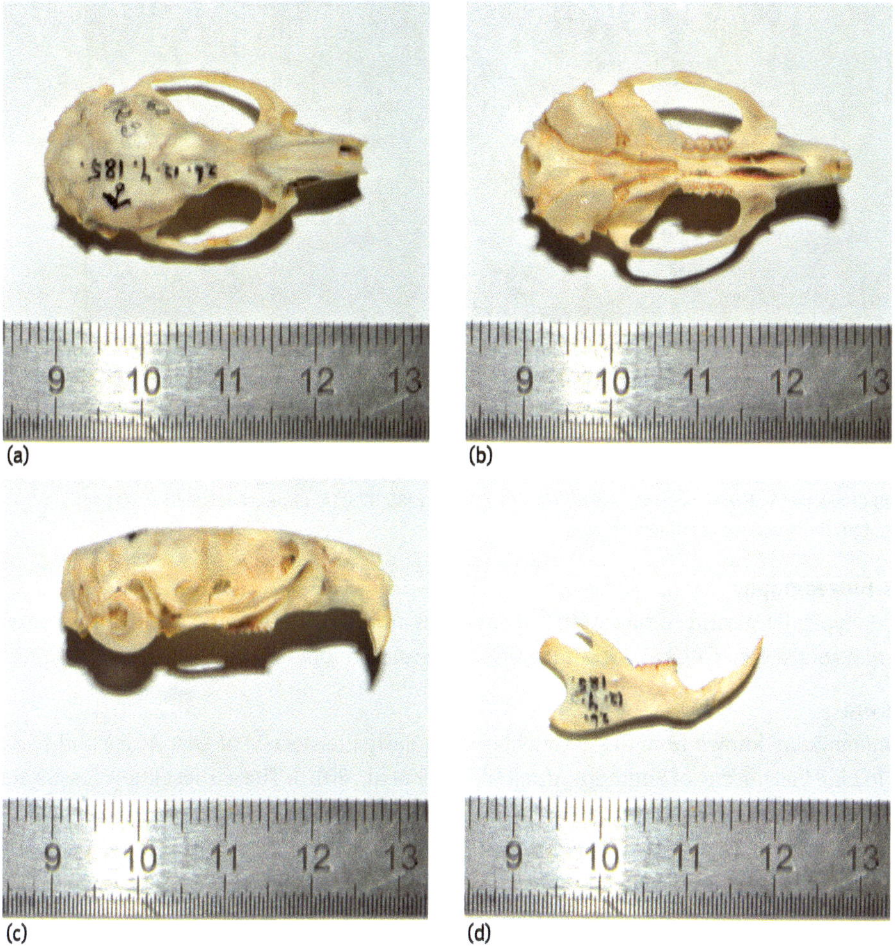

Figure 2.717 a-d Cranium of *Thallomys nigricauda* (MNHN-ZM-MO 1962-1123) (from top left to bottom right: dorsal view, ventral view, lateral view and lower jaw).

Dentition

The upper incisors are orthodont and ungrooved. The molar row is distinctively very narrow and short in proportion to the skull size, showing microdonty (MTR/GSL ± 0.14, Table 10). The cusps are well separated from each other with slight distortion of the transverse rows. On the upper molars, cusp t1 is almost on the same level as t2 with t3 slightly behind; t4 is slightly behind t6. Molars M^2 and M^3 are relatively large. There are traces of stephanodont crests on t1 and t3 and joining t6 and t9. The cusp t3 is unusually present but small on M^3 which has five cusps located on traces of the three laminae (Group 6, Figure 2.439). On the lower molars, the cusps on M_1 are alternated with the first connected to the second one by a longitudinal crest, labial cusplets are often present, the posterior cingulum on M_1 and M_2 are small and equal-sized, and anterior-external cusplets are present on M_2 and M_3.

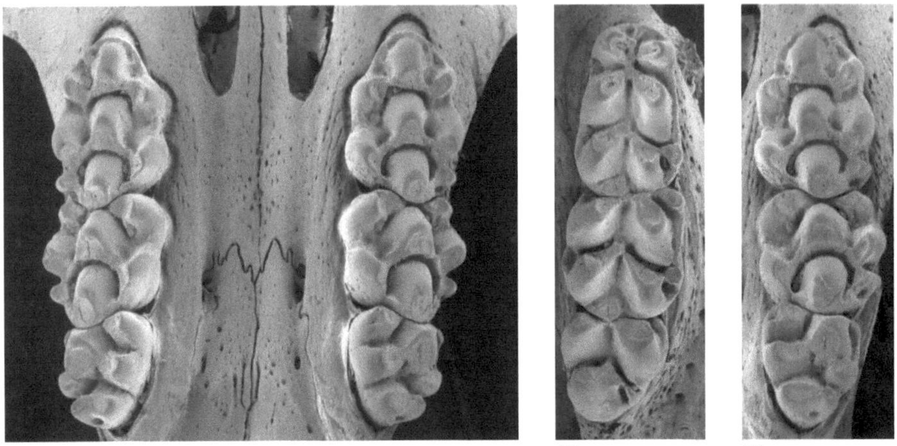

(a) (b) (c)

Figure 2.718 Cheek teeth of *Thallomys nigricauda* (MNHN-ZM-MO 1985-623) (A: upper cheek teeth; B: lower right toothrow; C: upper left toothrow).

Habitat & Biogeography

They occur typically in arid to mesic *Acacia* savannas from northern South Africa, Botswana, Namibia and south-western Angola to Tanzania, Kenya and Ethiopia in the north.

Fossil Record

Fossil *Thallomys* are known from the Early Pliocene to Early Pleistocene of East Africa and Early Pliocene to Late Pleistocene of Southern Africa (Winkler et al., 2010). The earliest know fossils are from 3.7 MYA sediments at Laetoli, East Africa (Denys, 1987a; 2011).
Thallomys laetolilensis Laetoli (Denys, 1987a, 2011)
Thallomys quadrilobatus Olduvai Bed I, Natron (Jaeger, 1976; Denys, 1987b)
Thallomys cf *quadrilobatus* Hadar (Reed, 2011)
Thallomys jaegeri Omo (Wesselman, 1984)
Thallomys sp. Plio-Pleistocene caves of RSA (Pocock, 1987; Senegas, 2000; Denys, 1990c)

Biology & Ecology

They are nocturnal, arboreal and folivorous, foraging and building nests in *Acacia* shrubs (Meyer et al., 2008). Apart from feeding predominantly on the leaves of *Acacia*, but also *Boschia*, *Grewia*, *Terminalia*, *Dichrostachys*, *Ziziphus*, they are known to feed on young *Acacia* pods and occasionally take insects (Meyer et al., 2008). A number of recent studies have elucidated the ecology and physiology of the arid savanna-adapted *T. nigricauda* (e.g., Lovegrove et al., 1991; Eccard et al., 2004; Meyer et al., 2007, 2008, 2010; Coleman & Downs, 2010). They are apparently seasonal breeders, with pregnant females and young being recorded from the summer and autumn months in Southern Africa (Smithers, 1971; De Graaff, 1981). The mean litter size estimated at 30 days was 1.8 (1-4) (Meyer et al., 2008).

Systematic Notes

The affiliations of this genus are unclear and morphological data have not been clear. Immunological data groups it with *Grammomys*, *Pelomys*, *Lemniscomys*, and *Rhabdomys* (Watts & Baverstock, 1995). Musser & Carleton (2005) included it provisionally in the *Oenomys* division, which Lecompte et al. (2008) included in the Arvicanthini tribe based on molecular data. The genus is in desperate need of revision. Many previous treatments considered the genus to be monospecific. However, two species (diagnosable on karyotype) are clearly present in Southern Africa, although their distributional limits are still uncertain (Gordon, 1987; Taylor et al., 1995). The East African population has been named as *T. loringi* on pelage and other characteristics (Musser & Carleton, 2005) but its specific separation from *T. nigricauda* needs to be critically tested. The species *T. shortridgei* from just south of the Orange River is recognised by Musser & Carleton (2005) based on the "acute" description of Thomas & Hinton (1923) and their examination of the type series which reveals distinctively smaller bullae in this species. The species has not been collected in recent times.

Annotated Species List

Four species are currently recognised.
Thallomys loringi (Heller, 1909)
Thallomys nigricauda Thomas, 1882
Thallomys paedulcus (Sundevall, 1846)
Thallomys shortridgei Thomas and Hinton, 1923

Key to Species

It is currently not possible to provide a key to the species of *Thallomys* because the species have been poorly diagnosed; a taxonomic revision is urgently needed.

Species Accounts

***Thallomys loringi* (Heller, 1909)**

Diagnosis

This species apparently has an obvious black facial mask, which may distinguish it from the only other *Thallomys* species known from East Africa, *T. paedulcus* where the face mask is present but rather indistinct (Thomas & Hinton, 1923).

Taxonomic Issues

It is thought that this species is most closely related to *T. nigricauda* (Musser & Carleton, 2005) but a taxonomic analysis has yet to be performed. A molecular study of this species is urgently required to resolve its phylogenetic relationships to other *Thallomys* species. It probably co-occurs in East Africa with *T. paedulcus*.

Distribution

This species has a narrow distribution in southern Kenya and northern Tanzania where it occurs in savanna habitats. Due to possible confusion with *T. paedulcus*, the limits of this species are not yet resolved.

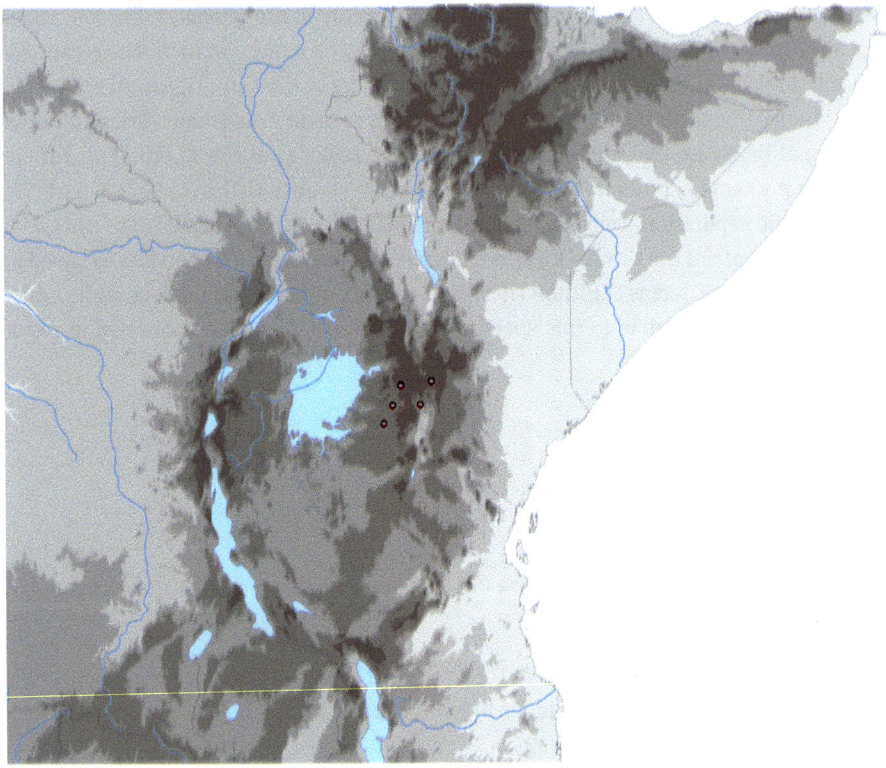

Figure 2.719 Distribution map

Measurements

External and cranial measurements (mm) and mass (g) for *Thallomys loringi*, sexes combined

	Mean	Min	Max	N
Mass	-	-	-	-
HB*	144.9	130	160	11
Tail*	153.3	120	183	11
HF*	26.7	25	28	11
Ear	-	-	-	-
GLS*	33.1	28.4	36.4	9

*Hollister (1919); hindfoot without claw

Thallomys nigricauda Thomas, 1882

Diagnosis

This species cannot be readily distinguished from *T. paedulcus* on morphological grounds, but differs in karyotype (in *T. paedulcus* 2n = 43-46); these two species, however, appear to be parapatric, perhaps overlapping marginally in a narrow contact zone. It differs from *T. shortridgei* by having larger bullae, a more distinct black facial mask, and non-overlapping distribution (*T. shortridgei* is thought to only occur south of the Gariep River, although this has yet to be substantiated). The relationship between this species and *T. loringi* is currently unknown, and the possibility exists that the two species are conspecific; however, they have widely allopatric distributions (*T. loringi* is endemic to East Africa).

Taxonomic Issues

This species has been poorly diagnosed and its relationships with other members of the genus are currently unclear or even confused. For example, based on a morphometric analysis of karyotyped specimens Taylor et al. (1995) and Taylor (2000) reported almost complete overlap in the distributions of *T. nigricauda* and *T. paedulcus* in southern Africa; this has subsequently been shown not to be correct and the two species probably maintain largely non-overlapping distributions. A molecular analysis of the group is urgently needed. Furthermore, the relationship between *T. nigricauda* and the East African *T. loringi* needs investigation. Chromosome number 2n = 47-50 (Gordon, 1987).

Distribution

This species occurs widely in north-western South Africa, western Botswana and Namibia, extending into southern Angola and north along the coast (Hill & Carter, 1941; Crawford-Cabral, 1998), where it inhabits arid *Acacia* savanna. The eastern limits of its distribution are not known and it may extend further east than presented on this map. The records on the Gariep River, in the south, need to be carefully checked to ensure that they do not refer to *T. shortridgei*.

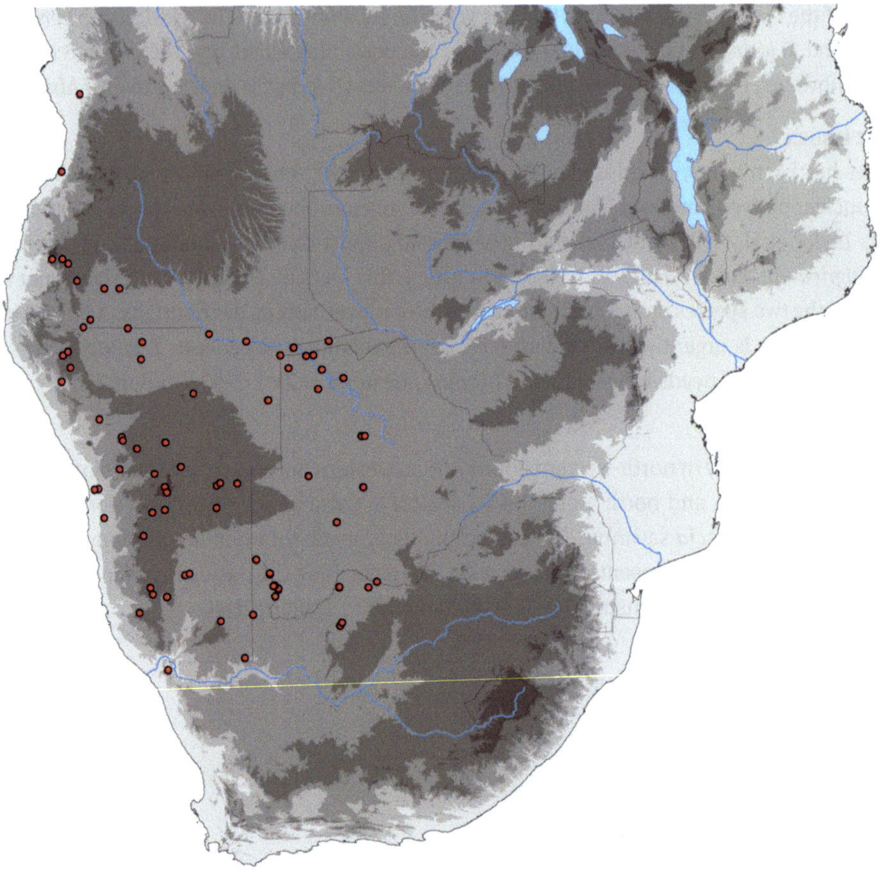

Figure 2.720 Distribution map

Measurements

External and cranial measurements (mm) and mass (g) for *Thallomys nigricauda*, sexes combined

	Mean	Min	Max	N
Mass	62.0	38	85	7
HB	136.8	112	165	15
Tail	149.1	123	195	14
HF*	25.0	23	28	20
Ear	20.9	16	23	15
GLS*	34.1	31.0	37.0	20

*Roberts (1951)

Thallomys paedulcus (Sundevall, 1846)

Diagnosis

This species cannot be readily distinguished from *T. nigricauda* on morphological grounds, but differs in karyotype (in *T. nigricauda* 2n = 47-50); these two species, however, appear to be parapatric, perhaps overlapping marginally in a narrow contact zone (cf. Taylor et al., 1995; and Taylor, 2000). It differs from *T. shortridgei* by having pure white ventral pelage (hairs with white base), and non-overlapping distribution (*T. shortridgei* is thought to only occur south of the Gariep River). The relationship between this species and *T. loringi* is currently not clear, although it has been reported that the black facial mask is distinct in *T. loringi* and poorly developed in *T. paedulcus* (Thomas & Hinton, 1923).

Taxonomic Issues

This species has been poorly diagnosed and its relationships with other members of the genus are currently unclear or even confused (see *T. nigricauda* account for further details). A molecular analysis of the group is urgently needed. Chromosome number 2n = 47-50 (Gordon, 1987).

Distribution

This species is widely distributed in the savannas of Southern Africa, extending patchily into East Africa as far north as Ethiopia and Somalia. The western limits of its distribution in Southern Africa are not known and it may occur more extensively than portrayed in this map.

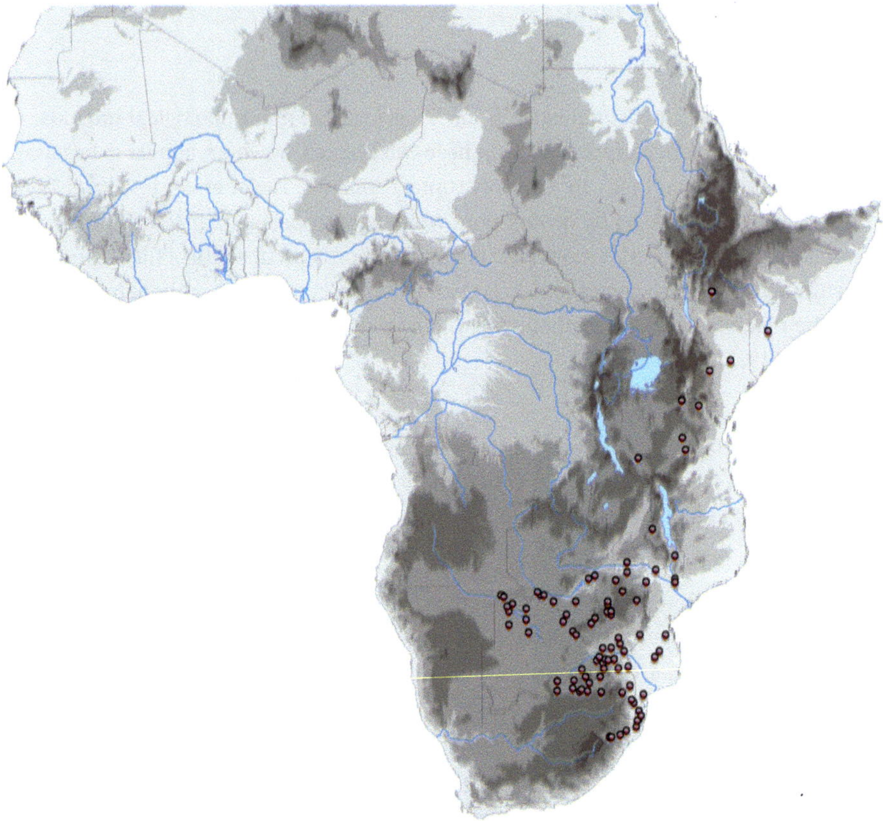

Figure 2.721 Distribution map

Measurements

External and cranial measurements (mm) and mass (g) for *Thallomys paedulcus*, sexes combined

	Mean	Min	Max	N
Mass*	71.6	32	116	33
HB*	131.4	-	-	73
Tail*	141.5	120	184	73
HF*	23.4	20	28	73
Ear*	20.6	16	24	73
GLS	32.3	29.2	35.1	5

*Rautenbach (1982); HB estimated from Total length minus Tail length

Thallomys shortridgei Thomas and Hinton, 1923

Diagnosis

This species can be distinguished from *T. nigricauda* by its smaller bullae (Musser & Carleton, 2005) and indistinct black facial mask (Thomas & Hinton, 1923). It is assumed that the two species are allopatric, separated by the Gariep River, but this needs confirmation.

Taxonomic Issues

This poorly known and diagnosed species requires urgent attention. A molecular study of the species coupled with further collecting of specimens for morphometric analysis is urgently needed, which would elucidate its relationship to other members of the genus.

Distribution

This species is known from just a handful of scattered localities on the south bank of the Gariep River at Upington and Goodhouse, South Africa.

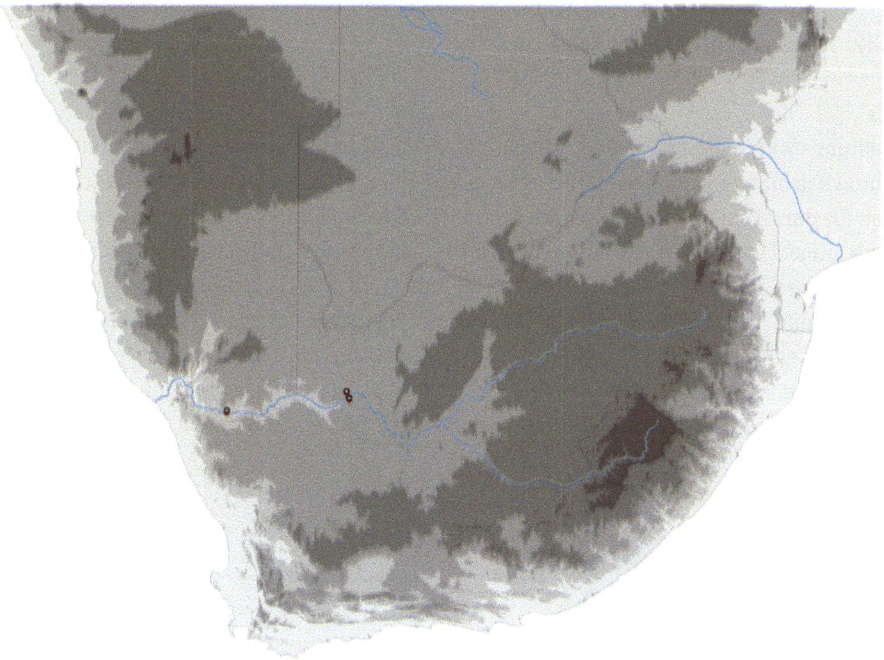

Figure 2.722 Distribution map

Measurements

External and cranial measurements (mm) and mass (g) for *Thallomys shortridgei*, sexes combined

	Mean	Min	Max	N
Mass	-	-	-	-
HB*	-	140	150	4
Tail*	-	183	210	4
HF*	-	27	28	4
Ear*	-	23	25	4
GLS*	-	34.2	37.0	4

*Roberts (1951); measurements of the holotype fall within these ranges except for HF and Ear which are 29 mm and 22 mm, respectively (Thomas & Hinton, 1923)

Genus Account

Thamnomys Thomas, 1907 (Thicket Rats)

Diagnosis

These medium-sized (Mass ca. 40-80 g; HB ca. 120-160 mm) rodents have a soft and long-haired, reddish-brown to orange-rufous dorsal pelage and a whitish ventral pelage which is clearly demarcated from the dorsal colour. There is a dark eye-ring in *T. kempi* but not the other two species. The tail is much longer (127-160 % of HB) than the head and body and usually possesses a terminal tuft of longer fine hairs. The ears are relatively large, rounded and possess reddish brown hairs. There is a white sub-auricular tuft in *T. schoutedeni* but not the other species. The distinctive dental character of this genus is the possession of the cusp t7 in M^1 and M^2. This cusp is only present, but as a reduced ridge, in two other murine genera, *Grammomys* and *Lophuromys*. The four species in the genus can be distinguished on body size. They have two pairs of abdominal nipples (0 + 2 = 4).

Figure 2.723 Photograph of *Thamnomys kempi* museum specimens (FHNH 149343 and 207906) (Julian Peterhans Kerbis).

Cranial Characters

The skull is medium-sized (GSL ca. 34-40 mm) with an unremarkable shape, having a rostrum of average length and breadth, moderately robust zygomatic arch and anterior edge of zygomatic plate which may be tilted forward and convex-concave in shape or vertical and straight. The anterior palatal foramina reach just about to the roots of M^1.

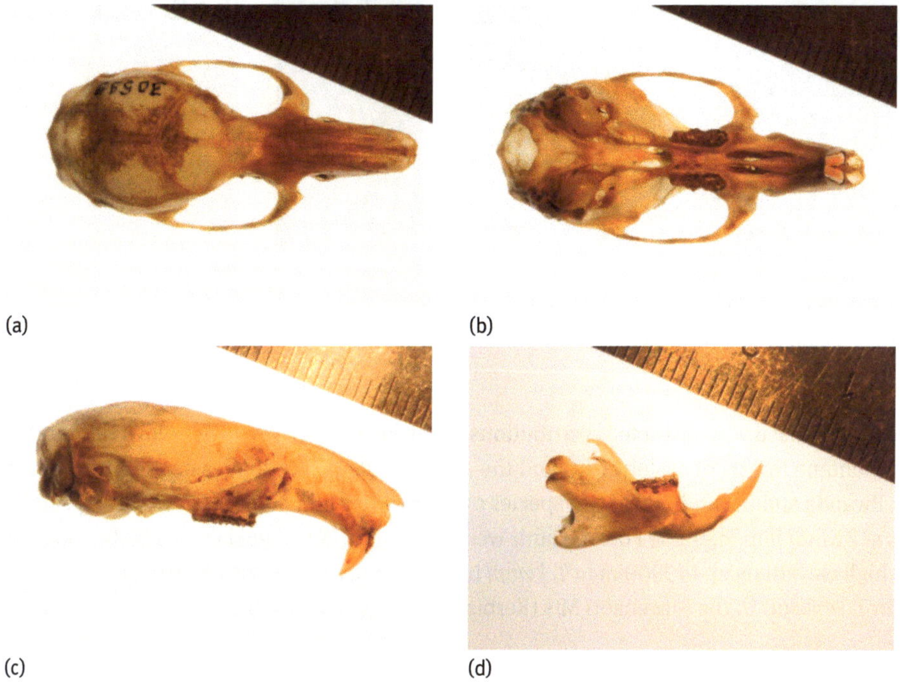

Figure 2.724 a-d Cranium of *Thamnomys kempi* (RMCA RG 30599) (from top left to bottom right: dorsal view, ventral view, lateral view and lower jaw).

Dentition

The upper incisors are orthodont to slightly opisthodont and ungrooved. The molar row is normal-sized (MTR/GSL ± 0.17, Table 10) and unique among murines in retaining all nine cusps in M^1, i.e. including the cusp t7. The cusp t7 is also present in M^2. Unlike in *Grammomys* and *Lophuromys* where t7 may be represented by a ridge, it forms a clear cusp in *Thamnomys* until an advanced age where it may appear reduced by wear. Both M^1 and M^2 have posterior labial accessory cusplets. Molar M^3 has six cusps (or five when the posterior two merge with age) arranged on three laminae and is one of the few genera to retain t3 on this molar (Group 7; Figure 2.439). The median cusps (t2, t5, t8) in the upper molars are enlarged, forming a straight longitudinal line; in this and other respects it is very similar in dentition to *Oenomys*. There are some stephanodont crests in the t1, t3, t6 and t9 but less developed than in *Oenomys* and *Lamottemys*. The lower molars are usually well endowed with supplementary cusplets, including anterior, posterior and up to six labial cusplets on M_1.

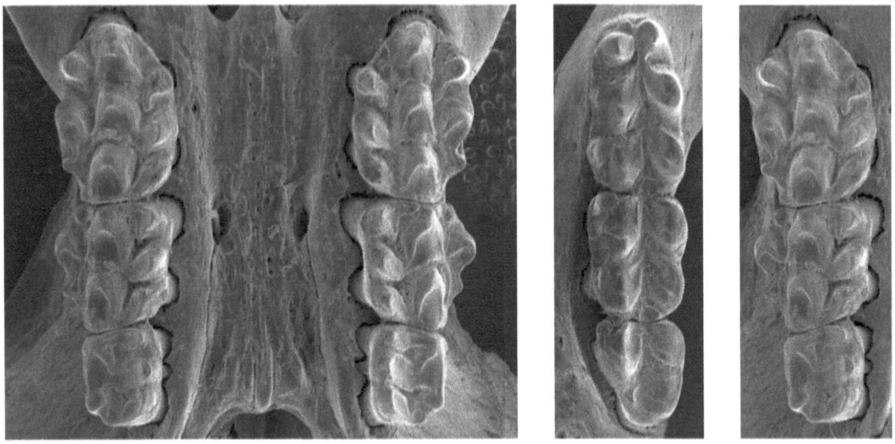

(a) (b) (c)

Figure 2.725 Cheek teeth of *Thamnomys kempi* (RMCA RG 30599) (A: upper cheek teeth; B: lower left toothrow; C: upper right toothrow).

Habitat & Biogeography

All three species have very restricted distributions in thicket habitats within the lowland rainforests and montane forests of eastern DRC and the Albertine Rift extending eastward into western Uganda, Rwanda and Burundi. Up to two species of *Thamnomys* may co-occur as in *T. kempi* and *T. venustus* at Bwindi Impenetrable Forest, south-western Uganda (Kasangaki et al., 2003); both may occur at high elevations up to 3900 m in *T. kempi* in the Virunga Mts (Tuyisingize et al., 2013), and 2600 m in *T. venustus* in the Ruwenzori Mts (Kerbis Peterhans et al., 1998)

Fossil Record

None.

Biology & Ecology

Very little information is available on these rare rodents. They are nocturnal, arboreal and herbivorous, feeding mainly on leaves and fruit. Reproduction occurs throughout the year in *T. kempi* and litter size is small (1-2 embryos) (Happold, 2013).

Systematic Notes

Musser & Carleton (2005) provisionally placed *Thamnomys* in the *Oenomys* division, which Lecompte et al. (2008) included in the Arvicanthini tribe based on molecular data. A close association with *Oenomys* has been supported by both classical and modern cladistic analysis of morphological characters (Hatt, 1940, Lopez-Martinez et al., 1998). Although historically many authors have combined *Thamnomys* and *Grammomys* into a single genus, they are currently recognised to be distinct genera (Musser & Carleton, 2005). The West African taxon, *rutilans*, which was regarded by Rosevear (1969) and others to be a *Thamnomys*, has since been shown to be a synonym of *Grammomys poensis* (Musser & Carleton, 2005). Recent authors have disagreed about the number and accepted names of species in this genus. While Musser & Carleton (2005) recognise *T. kempi*, *T. major* and *T. venustus*, Happold (2013) recognise *T. kempi*, *T. schoutedeni* and *T. venustus*. Musser & Carleton (2005) include *schoutedeni* in *T. venustus*; however, Happold (2013) provides data to show that the two species differ cranially. We also note that *T. venustus* occurs at high altitudes (> 1800 m) in the Albertine Rift (Kasangaki et al., 2003), whereas *schoutedeni* occurs in lower lying rainforests (800-1450 m). Happold (2013) includes *major* within *T. kempi* but Musser & Carleton indicate that the holotype of major is distinctive. Pending a full revision of this genus, we accept both the above-mentioned arguments and thus recognise four species here.

Annotated Species List

Four species are here recognised.
Thamnomys kempi Dollman, 1911
Thamnomys major Hatt, 1934
Thamnomys schoutedeni Hatt, 1934
Thamnomys venustus Thomas, 1907

Key to Species

1	Size larger (GLS > 39 mm, upper toothrow = 7.5 mm)	*T. major*
	Size smaller (GLS < 39 mm, upper toothrow < 7.2 mm)	2
2	Size smaller (GLS < 34.5 mm)	*T. schoutedeni*
	Size larger (GLS > 34.8 mm)	3
3	Size on average smaller (GLS < 35.7 mm); t9 present on M^1 and M^2	*T. venustus*
	Size on average larger (GLS > 35.3 mm); t9 absent on M^1 and M^2	*T. kempi*

Species Accounts

Thamnomys kempi Dollman, 1911

Diagnosis

This species can be distinguished from *T. major* by its smaller size (in *T. major* the upper molar toothrow is 7.5 mm, but < 7.2 mm in *T. kempi*). It is larger than *T. schoutendeni* (GLS < 34.5 mm) and *T. venustus* (GLS < 35.7 mm). Furthermore, the cusp t9 is absent on M^1 and M^2 (t9 is present in *T. venustus*).

Taxonomic Issues

See genus account.

Distribution

This species is restricted to the Lake Kivu region of the Albertine Rift where it occupies thickets and tangles in montane (mostly secondary) forest. The limits of its distribution are not resolved and it may occur more extensively in the Albertine Rift than illustrated in the map.

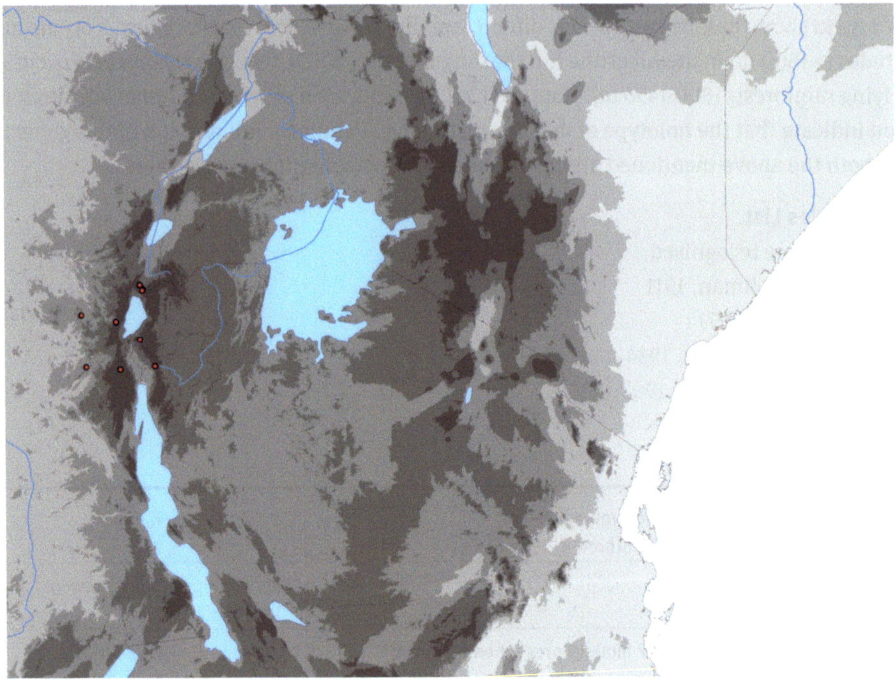

Figure 2.726 Distribution map

Measurements

External and cranial measurements (mm) and mass (g) for *Thamnomys kempi*, sexes combined

	Mean	Min	Max	N
Mass*	84	62	108	48
HB*	155	130	173	48
Tail*	202	170	226	48
HF*	29.3	27	32	48
Ear*	23.6	20	26	48
GLS*	37.5	35.3	39.7	24

*Happold (2013)

Thamnomys major Hatt, 1934

Diagnosis

This species can be distinguished from all other *Thamnomys* species by its large skull size, particularly the length of the upper molar toothrow (in the *T. major* holotype it is 7.5 mm; in all other *Thamnomys* species is it 7.2 mm or shorter).

Taxonomic Issues

This taxon was originally described as a subspecies of *T. kempi* (Hatt, 1934), but elevated to full species status by Musser & Carleton (2005) who point out that the holotype is larger than typical

T. kempi. However, the GLS of the holotype is within the range of *T. kempi* (see the Measurements table for *T. kempi*) and the upper molar row is only slightly longer (in *T. kempi* it is 6.4-7.2 mm). These findings suggest that *T. major* is a large form of *T. kempi*, perhaps as a result of Bergman's rule (the specimens were collected at high elevation, see below). However, we retain this species pending a full revision of the genus.

Distribution

This species is known from the holotype and two paratypes that were collected from *Hagenia* forest on Mt Kirisimbi in the Albertine Rift, eastern DRC at 3900 m (Hatt, 1934; Musser & Carleton, 2005).

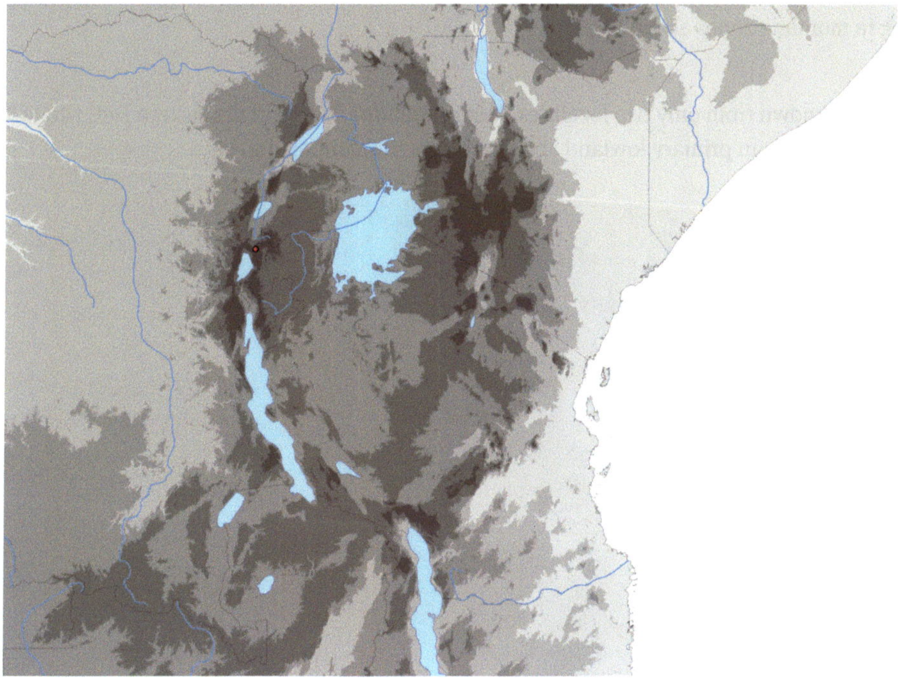

Figure 2.727 Distribution map

Measurements

External and cranial measurements (mm) and mass (g) for *Thamnomys major*, sexes combined

	Mean	Min	Max	N
Mass	-	-	-	-
HB*	156	-	-	1
Tail*	201	-	-	1
HF*	31	-	-	1
Ear	-	-	-	-
GLS*	39.3	-	-	1

*Hatt (1934); measurements of the holotype

Thamnomys schoutedeni Hatt, 1934

Diagnosis

This species does not co-occur with any other member of the genus, and further may be distinguished by its significantly smaller size (in other *Thamnomys* species, GLS > 34.8 mm, HB > 130 mm).

Taxonomic Issues

This taxon was originally described as a subspecies of *T. venustus* (Hatt, 1934), and retained as such by Musser & Carleton (2005). However, Happold (2013) recognise it as a valid species based on its smaller size; it also occupies lower elevational zones than *T. venustus* (the latter typically occurring in montane zones above 1800 m).

Distribution

This species is known from only two localities (Medje and Ituri Forest) in the eastern DRC where it has been collected from primary lowland and secondary submontane forests.

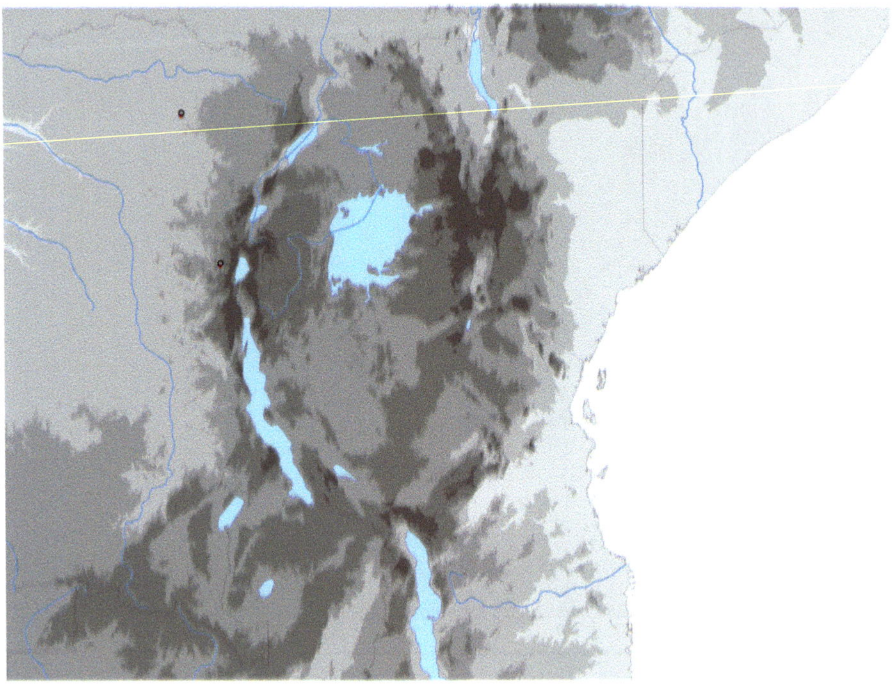

Figure 2.728 Distribution map

Measurements

External and cranial measurements (mm) and mass (g) for *Thamnomys schoutedeni*, sexes combined

	Mean	Min	Max	N
Mass	-	-	-	-
HB*	122	-	-	1
Tail*	203	-	-	1
HF*	25	-	-	1
Ear	-	-	-	-
GLS*	34.3	-	-	1

*Hatt (1934); measurements of the holotype. Happold (2013) present measurements based on 2-4 specimens (sample size in parentheses): Mass = 40-50 (2); HB = 114-122 (4); T = 171-203 (4); HF = 24-26 (4); Ear = 16-20 (4); GLS = 32.8-34.1 (2).

Thamnomys venustus Thomas, 1907

Diagnosis

This species can be distinguished from *T. major* by its smaller size (in *T. major* the upper molar toothrow is 7.5 mm, but < 6.5 mm in *T. venustus*). It is larger than *T. schoutendeni* (GLS < 34.5 mm) but smaller than *T. kempi* (GLS > 35.3 mm). Furthermore, it has a large t9 on M^1 and smaller t9 on M^2 (t9 is absent in *T. kempi*).

Taxonomic Issues

See genus account.

Distribution

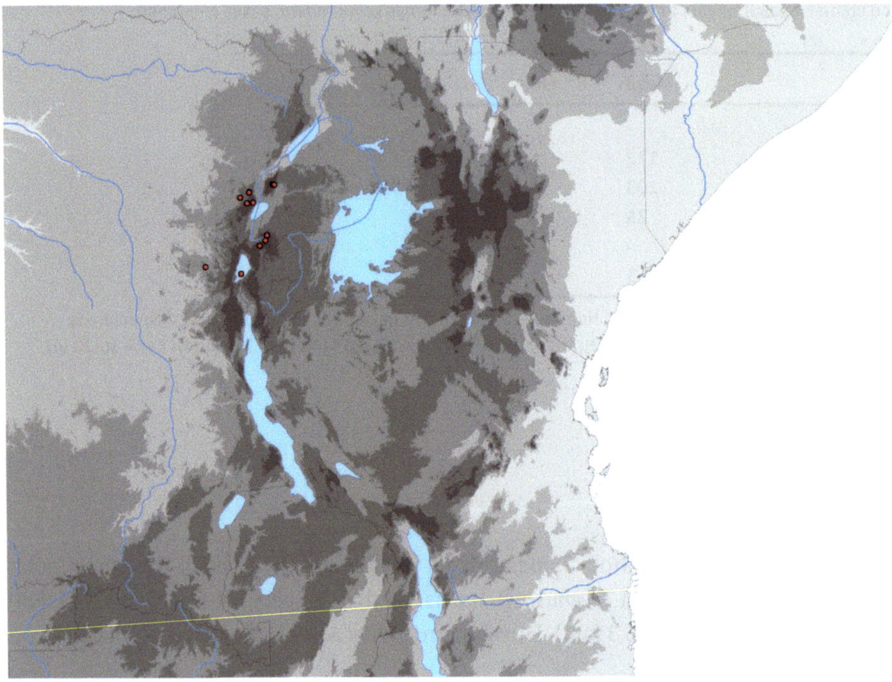

Figure 2.729 Distribution map

Measurements

External and cranial measurements (mm) and mass (g) for *Thamnomys venustus*, sexes combined

	Mean	Min	Max	N
Mass*	64	59	70	6
HB*	146	140	160	6
Tail*	198	183	209	6
HF*	26.3	25	27	6
Ear*	21.3	20	23	6
GLS*	35.2	34.8	35.7	6

*Happold (2013)

Genus Account

Zelotomys Osgood, 1910 (Broad-headed Mice)

Diagnosis

The genus is characterised by its medium size (Mass 45-60 g; 110-130 mm), a scaly white tail which is shorter (70-85 % of HB) than the head and body, and five pairs of nipples (3 + 2 = 10). In *Z. hilde-*

gardae, but not *Z. woosnami*, the head is broadened (due to laterally expanded zygomatic arches), hence the name "Broad-headed Mice". The two species can be distinguished by distribution as well as by dental characters.

Figure 2.730 Photograph of *Zelotomys hildegardeae* (Lauren E. Helgen).

Cranial Characters

The skull is medium in size (GSL ca. 28-32 mm) with a short, broad shape with very short braincase, very gracile zygomatic arch and antero-dorsal angle of zygomatic plate blunt with forward-sloping anterior edge. The palatal foramina are long, extending to the t4 cusp of M^1.

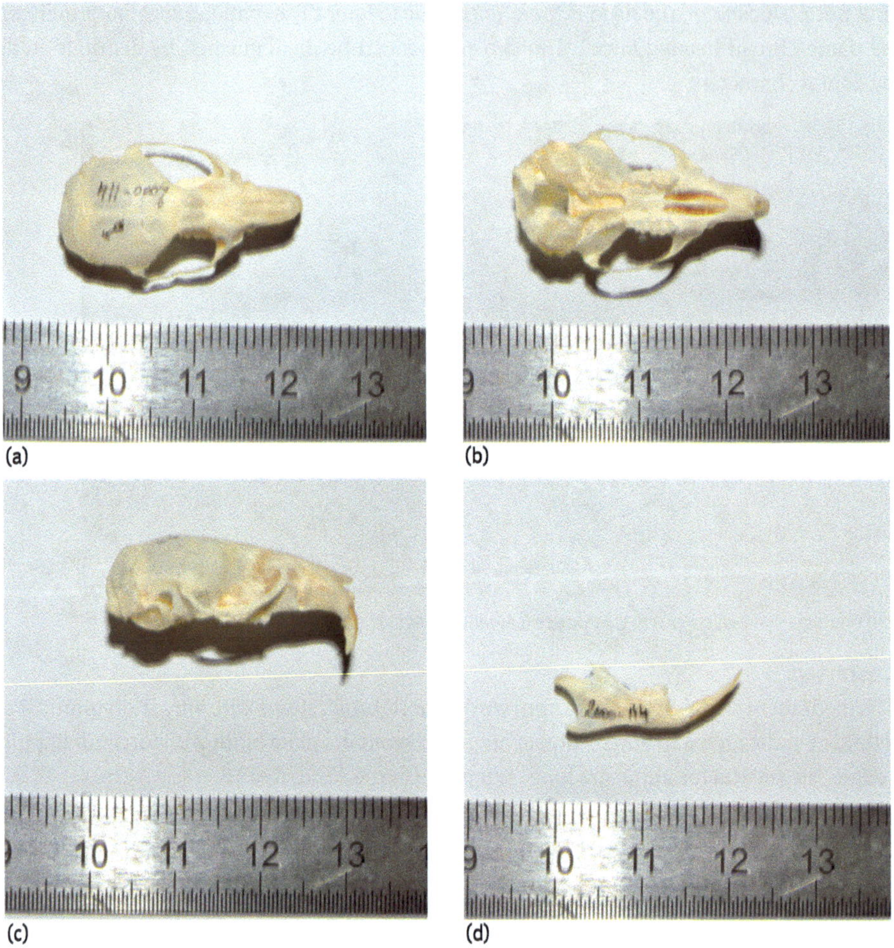

Figure 2.731 a-d Cranium of *Zelotomys hildegardeae* (MNHN-ZM-MO 2000-114) (from top left to bottom right: dorsal view, ventral view, lateral view and lower jaw).

Dentition

The upper incisors are ungrooved and pro-odont in *Z. hildegardae* to opisthodont in *Z. woosnami*. The molars are robust (MTR/GSL ± 0.18, Table 10) and the cusps are well separated and arranged in transverse rows which are not much distorted, the cusp t1 being slightly posterior to t2 and t3, as well as the t4 being posterior to t5 and t6. Molar M^3 is markedly smaller than M^2 and has traces of three laminae (Group 5, Figure 2.439). Molar M_1 has a small posterior cingulum. The first lobe on M_1 is well developed, has two poorly differentiated cusps and is joined by a longitudinal connection to the cusps of the second lobe. There are labial cusplets on M_1 and M_2.

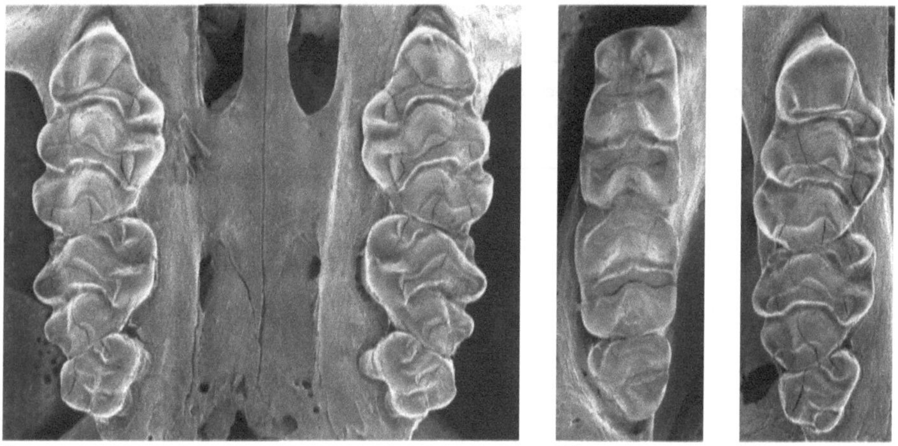

Figure 2.732 Cheek teeth of *Zelotomys hildegardeae* (AMNH 114037) (A: upper cheek teeth; B: lower left toothrow; C: upper right toothrow).

Habitat & Biogeography

This genus extends from South Africa in the south to Kenya and Uganda in the north-east with a single locality in the Central African Republic. The two species inhabit savannas, with *Z. hildegardeae* associated with mesic savannas of South-central and East Africa, and *Z. woosnami* with arid savannas of Southern Africa.

Fossil Record

Fossils are known from the Early Pleistocene of Tanzania and early Plioecene to Late Pleisocene of Southern Africa (Winkler et al., 2010). The earliest fossil occurrence of *Zelotomys* is at 3.3 MYA at Makapansgat, South Africa (Pocock, 1976).
Zelotomys leakeyi Olduvai Bed I (Jaeger, 1976)

Biology & Ecology

Nocturnal, terrestrial murines that occasionally also climb. They are solitary animals with an omnivorous diet (Happold, 2013). Breeding occurs throughout the year and there may be 2-11 embryos (Happold, 2013).

Systematic Notes

Musser & Carleton (2005) grouped *Zelotomys* with *Colomys* based on molar occlusal patterns (Misonne, 1969) and this association is confirmed by phylogenetic analysis of mitochondrial and nuclear sequences (Lecompte et al., 2002b; 2008; Schenk et al., 2013). Together these two genera are included in the Praomyini tribe.

Annotated Species List

Two species are currently recognised:
Zelotomys hildegardeae (Thomas, 1902)
Zelotomys woosnami (Schwann, 1906)

Key to Species

1	Upper incisors pro-odont; occurring in South-central and East Africa	*Z. hildegardeae*
	Upper incisors opisthodont; occurring in Southern Africa, south of the Kunene and Zambezi rivers	*Z. woosnami*

Species Accounts

Zelotomys hildegardeae (Thomas, 1902)

Diagnosis
This species is readily distinguished from *Z. woosnami* by its pro-odont upper incisors and non-overlapping distribution (in *Z. woosnami* the incisors are opisthodont and it is restricted to the arid savannas of Southern Africa).

Taxonomic Issues
Geographic variability in pelage colour has been reported by various authors (see Happold, 2013), but neither molecular nor chromosomal studies have yet been conducted.

Distribution
This species occurs widely in savannas of tropical Africa from Angola and Zambia in the south northwards to East Africa, with an isolated record from the Central African Republic. The species may occur more extensively in the northern savannas, but confirmed records do not currently exist.

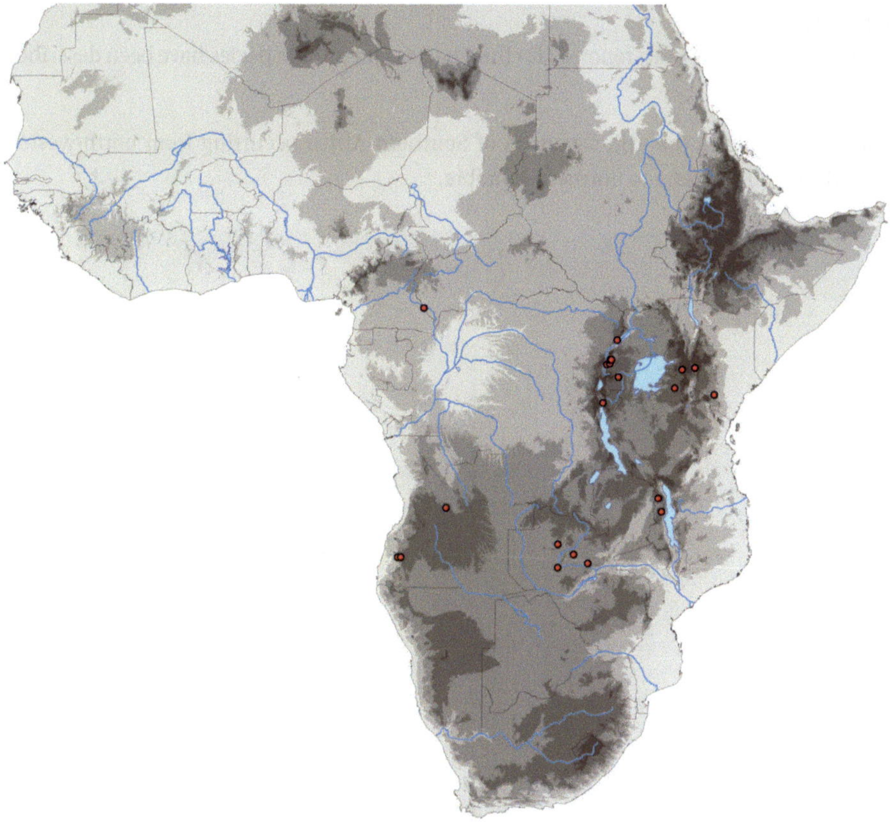

Figure 2.733 Distribution map

Measurements

External and cranial measurements (mm) and mass (g) for *Zelotomys hildegardeae*, sexes combined

	Mean	Min	Max	N
Mass	43	-	-	1
HB	113.0	100	122	5
Tail	91.6	80	103	5
HF	23.6	23	25	5
Ear	16.8	16	17	5
GLS	27.7	-	-	1

Zelotomys woosnami (Schwann, 1906)

Diagnosis

This species is readily distinguished from *Z. hildegardeae* by its opisthodont upper incisors and non-overlapping distribution (in *Z. hildegardeae* the incisors are pro-odont and it occurs in the mesic savannas of tropical Africa).

Taxonomic Issues

No geographic variation has been reported for this species and no subspecies have been described.

Distribution

This species is endemic to the arid savannas of Southern Africa occurring from north-western South Africa through Botswana to northern Namibia.

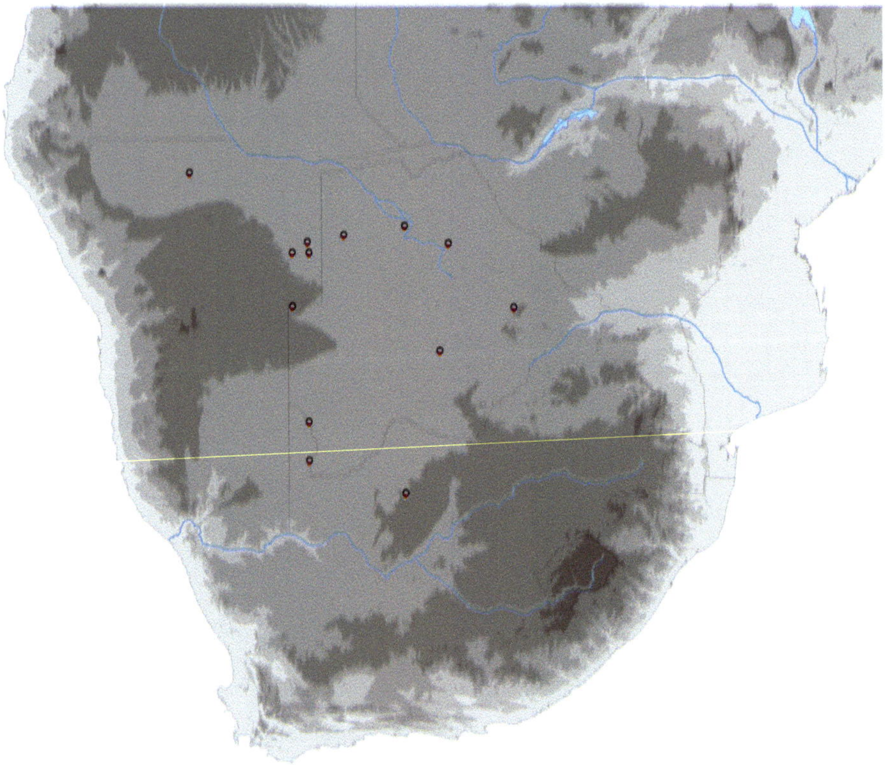

Figure 2.734 Distribution map

Measurements

External and cranial measurements (mm) and mass (g) for *Zelotomys woosnami*, sexes combined

	Mean	Min	Max	N
Mass	58.3	50	75	5
HB	127.5	94	156	11
Tail	102.3	77	134	11
HF	26.5	23	29	11
Ear	18.4	15	21	11
GLS	32.4	27.6	35.3	10

References

Abdel Rahman Ahmed, E. A. 2005. Taxonomic status of *Arvicanthis* Lesson, 1842 from Sudan. PhD thesis, University of KwaZulu Natal, South Africa.

Abdel Rahman Ahmed, E. A., Ducroz, J., Mitchell, A., Lamb, J., Contrafatto, G., Denys, C., Lecompte, E. & Taylor, P. 2008. Phylogeny and historical demography of economically important rodents of the genus *Arvicanthis* (Mammalia: Muridae) from the Nile Valley: of mice and men. Biological Journal of 93: 641-655.

Abdel-Rahman, E. H.,Taylor, P. J.,Contrafatto, G.& Lamb, J. M. 2009. Geometric morphometric and taxonomic status of the genus *Arvicanthis* (Rodentia: Muridae) from Sudan. Mammalian Biology 74: 361-373.

Abiadh, A., Chetoui, M., Lamine-Cheniti, T., Capanna, E. & Colangelo, P. 2010. Molecular phylogenetics of the genus *Gerbillus* (Rodentia, Gerbillinae): implications for systematics, taxonomy and chromosomal evolution. Molecular Phylogenetics and Evolution 56: 513-518.

Aguilar, G. H. 1993. The karyotype and taxonomic status of *Cryptomys hottentotus darlingi* (Rodentia, Bathyergidae). South African Journal of Zoology 28: 201-204.

Aguilar, J. P., Brandy, L. D. & Thaler, L. 1984. Les rongeurs de Salobrena (sud de l'Espagne) et le problème de la migration Messinienne. Paleobiologie Continentale 14: 3-17.

Akpatou, B. K., Nicolas, V., Pires, D., N'Goran, E. & Colyn, M. 2007. Morphometric differentiation between two murid rodents, *Praomys tullbergi* (Thomas, 1894) and *Praomys rostratus* (Miller, 1900), in West Africa. Zootaxa 1607: 21-34.

Ajayi, S. S. 1977. Field observations on the African giant rat *Cricetomys gambianus* Waterhouse in southern Nigeria. East African Wildlife Journal 15: 191-198.

Allen, G.M. 1939. A checklist of African mammals. Bulletin of the Museum of Comparative Zoology at Harvard College 83: 1-763.

Allen, B., Kon, M. & Bar-Yam, Y. 2009. A new phylogenetic diversity measure generalizing the Shannon index and its application to phyllostomid bats. The American Naturalist 174: 236-243.

Ameur, R. 1988. Biochronologie des formations continentales du Neogène et du Quaternaire de l'Algérie. Contribution des micromammifères. Thèse Doctorat Etat Université d'Oran (unpublished).

Amtmann, E. 1965. Zur geographischen farbvariation des Afrikanischen reisenhornchens *Protoxerus stangeri* (Waterhouse 1842) eine quantitative untersuchung zur glogerschen regel. Zeitschrift für Morphologie und Okologie das Tiere 55: 515-529.

Angelici, F. M. & Luiselli, L. 2005. Patterns of specific diversity and population size in small mammals from arboreal and ground-dwelling guilds of a forest area in southern Nigeria. Journal of Zoology, London 265: 9-16.

Aniskin, V. M., Lavrenchenko, L. A., Varshavskii, A. A. & Milishnikov, A. N. 1997a. Karyotypes and chromosomal differentiation of two *Tachyoryctes* species (Rodentia, Tachyoryctinae) from Ethiopia. Russian Journal of Genetics 33: 1079-1084.

Aniskin, V. M., Lavrenchenko, L. A., Varshavskii, A. A. & Milishnikov, A. N. 1997b. Karyotypic differentiation of three harsh-furred mouse species of genus *Lophuromys* (Murinae, Rodentia) from the Bale Mountains National Park, Ethiopia. Russian Journal of Genetics 33: 818-824.

Aniskin, V., Benazzou, T., Biltueva, L., Dobigny, G., Granjon, L. & Volobouev, V. 2005. Unusually extensive karyotype reorganization in four congeneric *Gerbillus* species (Muridae: Gerbillinae). Cytogenetic and Genome Research 112: 1-10.

Ansell, W. F. H. 1957. Some mammals from Northern Rhodesia. Am. Mag. Nat. Hist. ser. 12, 10: 529-551.

Ansell, W. F. H. 1960. The mammals of Northern Rhodesia. Government Printer, Lusaka.

Ansell, W. F. H. 1966. *Thryonomys gregorianus* and *Thryonomys swinderianus* in Zambia. The Puku: 1-15.

Ansell, W. F. H. 1969. Addenda and corrigenda to "Mammals of Northern Rhodesia" No. 3. Puku 5: 1-48.

Ansell, W. F. H. 1973. Addenda and corrigenda to "Mammals of Northern Rhodesia" No. 4. Puku 7: 1-19.

Ansell, W. F. H. 1974. Some mammals from Zambia and adjacent countries. Puku Supplement 1: 1-49.

Ansell, W. F. H. 1978. The mammals of Zambia. National Parks and Wildlife Service of Zambia, Chilanga, Zambia.

Ansell, W. F. H. 1989b. African Mammals 1938-1988. Trendine Press, Zennor, Cornwall.
Ansell, W. F. H. 1989b. Mammals from Malawi, Part II. Nyala 13: 41-56.
Ansell, W. F. H. & Ansell, P. D. H. 1973. The mammals of the north-eastern montane areas of Zambia. Puku 7: 21-69.
Ansell, W. F. H. & Dowsett, R. J. 1988. Mammals of Malawi. Trendine Press, Zennor, Cornwall.
Aplin, K. P., Suzuki, H., Chinen, A. A., Chesser, R. T., Have, ten, J., Donnellan, S. C., Austin, J., Frost, A., Gonzalez, J. P., Herbreteau, V., Catzeflis, F., Soubrier, J., Fang, Y., Robins, J., Matisoo-Smith, E., Bastos, A. D. S., Maryanto, I., Sinaga, M. H., Denys, C., van den Bussche, R. A., Conroy, C., Rowe, K., Cooper, A. & van den Van Den Bussche, R. A. 2011. Multiple Geographic Origins of Commensalism and Complex Dispersal History of Black Rats. PLoS ONE 6: e26357.
Arntzen, L., Wadee, A. & Isaacson, N. 1991. Immune responses of two *Mastomys* sibling species to Yersinia pestis. Infection and Immunity 59: 1966-1971.
Avenant, N. 2011. The potential utility of rodents and other small mammals as indicators of ecosystem 'integrity' of South African grasslands. Wildlife Research 38: 626-639.
Avenant, N. & van der Merwe, M. 2006. Expansion of the greater cane rat, *Thryonomys swinderianus*, into the drier parts of southern Africa. 3 rd International Conference on Rodent Biology and Management: 187-188.
Avery, D.M. 1982. Late Pleistocene coastal environment of the southern Cape Province of South Africa: micromammals from Klasies River Mouth. Journal of Archaeological Science, 14: 405-421.
Avery, D. M. 1995. A preliminary assessment of the micromammalian remains from Gladysvale Cave, South Africa. Palaeontologica Africana 32: 1-10.
Avery, D. M. 1998. An assessment of the lower Pleistocene micromammalian fauna from Swartkrans Members 1-3, Gauteng, South Africa. Geobios 31: 393-414.
Avery, D., Avery, G. & Roberts, A. 2002. A contribution from barn-owl pellets to known micromammalian distribution in KwaZulu-Natal, South Africa. African Zoology: 1-10.
Ba, K., Thiam, M., Dobigny, G., Granjon, L., Mane, Y., Volobouev, V. & Duplantier, J. 2006. Hypothesis on the origin of the invasion of Senegal by *Gerbillus nigeriae* based on chromosomal data. Mammalia 70: 303-305.
Baker, R.J., Qumsiyeh, M.B. & Rautenbach, I.L. 1988. Evidence for eight tandem and five centric fusions in the evolution of the karyotype of *Aethomys namaquensis* A. Smith (Rodentia: Muridae). Genetica 76: 161–169.
Bannikova, A. A., Lavrenchenko, L. A. & Kramerov, D. A. 2005. Phylogenetic relationships between Afrotropical and Palaearctic Crocidura species inferred from Inter-SINE-PCR. Biochemical Systematics and Ecology 33: 45-59.
Barome, P. O., Monnerot, M. & Gautun, J. C. 1998. Intrageneric phylogeny of *Acomys* (Rodentia, Muridae) using mitochondrial gene cytochrome b. Molecular Phylogenetics and Evolution 9: 560-566.
Barome, P. O., Monnerot, M. & Gautun, J. C. 2000. Phylogeny of the genus *Acomys* (Rodentia, Muridae) based on the cytochrome b mitochondrial gene: implications on taxonomy and phylogeography. Mammalia 64: 423-438.
Barome, P. O., Volobouev, V., Monnerot, M., Mfune, J., Chitaukali, W., Gautun, J-C. & Denys, C. 2001. Phylogeny of *Acomys spinosissimus* (Rodentia, Muridae) from north Malawi and Tanzania: evidence from morphological and molecular analysis. Biological Journal of the Linnean Society 73: 321-324.
Barry J.C. 1995. Faunal turnover and diversity in the terrestrial Neogene of Pakistan. Chap. 9. In: Paleoclimate and evolution with emphasis on human origins (E. S. Vrba, H. Denton, T. C. Partridge, H. Burckle Lloyd, eds.).
Baskevich, M. I. & Lavrenchenko, L. A. 2000. Review of karyological studies and the problems of systematics of Ethiopian *Arvicanthis* Lesson 1842 (Rodentia, Muridae). In: Isolated Vertebrate Communities in the Tropics (G. Rheinwald, ed). Bonner Zoologie Monographen 46: 209-215.
Baskevich, M. I., Orlov, N. V., Bekele, A. & Mebrate, A. 1993. Notes on the karyotype of *Tachyoryctes splendens* (Rüppell 1836) (Rodentia Rhizomyidae) from Ethiopia. Tropical Zoology 6: 81-88.

Bastos, A. D., Nair, D., Taylor, P. J., Brettschneider, H., Kirsten, F., Mostert, E., Maltitz, von, E., Lamb, J. M., van Hooft, P., Belmain, S. R., Contrafatto, G., Downs, S. & Chimimba, C. T. 2011. Genetic monitoring detects an overlooked cryptic species and reveals the diversity and distribution of three invasive *Rattus* congeners in South Africa. BMC Genetics 12: 26.

Bates, P. J. J. 1985. Studies of gerbils of the genus *Tatera*: the specific distribution of *Tatera robusta* (Cretzchmar, 1826), *Tatera nigricauda* (Peters, 1878) and *Tatera phillipsi* (de Winton, 1898). Mammalia 49: 37-52.

Bates, P. J. J. 1988. Systematics and zoogeography of *Tatera* (Rodentia; Gerbillinae) of north-east Africa and Asia. Bonner Zoologische Beiträge 39: 265-303.

Begon, M. 2003. Disease: health effects on humans, population effects on rodents. In: Rats, Mice and People: Rodent Biology and Management (G. Singleton, L. A. Hinds, C. J. Krebs, D. M. Spratt, eds). Australian Centre for International Agricultural Research, Canberra.

Bekele, A. 1996a. Rodents of the Menagesha State Forest, Ethiopia, with an emphasis on the endemic *Praomys albipes* Ruppell 1842. Tropical Zoology 9: 201-212.

Bekele, A. 1996b. Population dynamics of the Ethiopian endemic rodent *Praomys albipes* in the Menagesha State Forest. Journal of Zoology, London 238: 1-12.

Bekele, A. & Da Schlitter. 1989. Two new records of rodents from Kenya and Ethiopia. Mammalia: 113-116.

Bekele, A., Corti, M., Marcus, L. F. & Schlitter, D. A. 1993. Systematics and geographic variation of Ethiopian *Arvicanthis* (Rodentia, Muridae). Journal of Zoology, London 230: 117-134.

Bellier, L. 1968. Contribution a l'etude d'*Uranomys ruddi* Dollman. Mammalia 32: 319-329.

Belmain, S. R., Dlamini, N., Eiseb. S., Kirsten, F., Mahlaba, T., Makundi, R., Malebane, P., Maltitz, E. von, Massawe, A., Monadjem. A., Mulungu, L., Siwiya, E., Taylor, P. & Tutjavi, V. 2008. The ECORAT project: developing ecologically-based rodent management for the southern African region. International Pest Control 50: 136-138.

Ben Faleh, A., Ben Othmen, A., Said, K. & Granjon, L. 2010a. Karyotypic variation in two species of jerboas *Jaculus jaculus* and *Jaculus orientalis* (Rodentia, Dipodidae) from Tunisia. Folia Biologica 58: 229-236.

Ben Faleh, A., Cosson, J., Tatard, C., Ben Othmen, A., Said, K. & Granjon, L. 2010b. Are there two cryptic species of the lesser jerboa *Jaculus jaculus* (Rodentia: Dipodidae) in Tunisia? Evidence from molecular, morphometric, and cytogenetic data. Biological Journal of the Linnean Society 99: 673-686.

Ben Faleh, A. R., Granjon, L., Tatard, C., Ben Othmen, A., Said, K. & Cosson, J. F. 2012a. Phylogeography of the Greater Egyptian Jerboa (*Jaculus orientalis*) (Rodentia: Dipodidae) in Mediterranean North Africa. Journal of Zoology, London 286: 208-220.

Ben Faleh, A., Granjon, L., Tatard, C., Boratynski, Z., Cosson, J. & Said, K. 2012b. Phylogeography of two cryptic species of African desert jerboas (Dipodidae: *Jaculus*). Biological Journal of the Linnean Society 107: 27-38.

Benammi, M. 1997. Nouveaux rongeurs du Miocène continental du Jebel Rhassoul (moyenne Moulouya, Maroc). Geobios 30: 713-721.

Benazzou, T., Viegas-Pequignot E., Prod-Homme, M., Lombard, M., Petter, F. & Dutrillaux, B. 1984. Phylogenie chromosomique des Gerbillidae. III. Etude d'espèces des genres *Tatera, Taterillus, Psammomys* et *Pachyuromys*. Annales de Génétique 27: 17-26

Bennett, N. C. & Faulkes, C. G. 2000. African Mole-rats, Ecology and Eusociality. Cambridge University Press, Cambridge, UK.

Bentz, S. & Montgelard, C. 1999. Systematic position of the African dormouse *Graphiurus* (Rodentia, Gliridae) assessed from cytochrome b and 12S rRNA mitochondrial genes. Journal of Mammalian Evolution 6: 67-83.

Beolchini, F. & Corti, M. 2004. The taxonomy of the genus *Tachyoryctes*: a geometric morphometric approach. Italian Journal of Zoology 71: 35-43.

Bernor, R.L., Brunet M., Ginsburg, L., Mein, P., Pickford, M., Rögl, F., Sen, S., Steininger F. & Thomas, H. 1987. A consideration of some major topics concerning old world Miocene mammalian chronology, migrations and paleogeography. Geobios 20: 431-439.

Black, C. C. & Krishtalka, L. 1986. Rodents, bats, and insectivores from Plio-Pleistocene sediments to the east of Lake Turkana, Kenya. Contributions to Science 372: 1-15.
Blanga-Kanfi, S., Miranda, H., Penn, O., Pupko, T., DeBry, R. W. & Huchon, D. 2009. Rodent phylogeny revised: analysis of six nuclear genes from all major rodent clades. BMC Evolutionary Biology 9: 71.
Bohmann, L. 1942. Die Gattung *Dendromus* A. Smith. Versuch einer naturlichen Gruppierung. Zoologischer Anzeiger 139: 33-53.
Bohmann, L. 1952. Die afrikanische Nagergattung *Otomys* F. Cuvier. Zeitschrift fur Säugetierkunde 18: 1-80.
Boratynski, Z., Brito, J. & Mappes, T. 2012. The origin of two cryptic species of African desert jerboas (Dipodidae: *Jaculus*). Biological Journal of the Linnean Society 105: 435-445.
Bouzouggar, A., Barton, N., Vanhaeren, M., d'Errico, F., Collcutt, S., Higham, T., Hodge, E., Parfitt, S., Rhodes, E., Schwenninger, J. L., Stringer, C., Turner, E., Ward, S., Moutmir, A. & Stambouli, A. 2007. 82,000-year-old shell beads from North Africa and implications for the origins of modern human behavior. Proceedings of the National Academy of Science, USA 104: 9964-9969.
Brandy L. D., Sabatier M. & Jaeger J. J. 1980. Implications phylogénétiques et biogéographiques des dernières découvertes de Muridae en Afghanistan, au Pakistan et en Ethiopie. Geobios, Lyon, 13: 639-643.
Breed, W. G. 1995. Variation in sperm head morphology of muroid rodents of Africa: phylogenetic implications. In: Advances in Spermatozoal Phylogeny and Taxonomy (B. G. M. Jamieson, J. Ausio & J-L. Justine, eds). Memoires du Museum national d'Histoire naturelle. Editions du Museum, Paris.
Breed, W. G., Cox, G. A., Leigh, C. M. & Hawkins, P. 1988. Sperm head structure of a murid rodent from southern Africa: the red veld rat *Aethomys chrysophilus*. Gamete Research 19: 191-202
Brito, J. C., Álvares, F., Martínez-Freiría, F., Sierra, P., Sillero, N. & Tarroso, P. 2010. Data on the distribution of mammals from Mauritania, West Africa. Mammalia 74: 449-455.
Britton-Davidian, J., Robinson, T. J. & Veyrunes, F. 2012. Systematics and evolution of the African pygmy mice, subgenus *Nannomys*: A review. Acta Oecologica 42: 41-49.
Brooks, P. M. 1972. Post-natal development of the African bush rat. Zoologica Africana 7: 85-102.
Broom, R. 1934. On the fossil remains associated with *Australopithecus africanus*. South African Journal of Science 31: 471-480.
Broom, R. 1939. The fossil rodents of the Limestone cave at Taungs. Annals of the Transvaal Museum 19: 315-317.
Brouat, C., Tatard, C., Ba, K., Cosson, J., Dobigny, G., Fichet-Calvet, E., Granjon, L., Lecompte, E., Loiseau, A., Mouline, K., Piry, S. & Duplantier, J. 2009. Phylogeography of the Guinea multimammate mouse (*Mastomys erythroleucus*): a case study for Sahelian species in West Africa. Journal of Biogeography 36: 2237-2250.
Brugal, J.P. & Denys, C. 1989. Vertébrés du site acheuléen d'Isenya (Kenya, district de Kajiado). Implications paléoécologiques et paléobiogéographiques. Comptes Rendus de l' Académie des Sciences, Paris, 308, sér..II : 1503-1508.
Brugière, D., Fleury, M-C. & Colyn, M. 2005. Structure of the squirrel community in the Forêt des Abeilles, central Gabon: rediscovery and revalidation of *Funisciurus duchaillui* Sanborn, 1953. Mammalia 69: 223-231.
Bryja, J., Granjon, L., Dobigny, G., Patzenhauerová, H., Konečný, A., Duplantier, J. M., Gauthier, P., Colyn, M., Durnez, L., Lalis, A. & Nicolas, V. 2010. Plio-Pleistocene history of West African Sudanian savanna and the phylogeography of the *Praomys daltoni* complex (Rodentia): the environment/geography/genetic interplay. Molecular Ecology 19: 4783-4799.
Bryja, J., Mikula, O., Patzenhauerová, H., Oguge, N. O., Šumbera, R. & Verheyen, E. 2014. The role of dispersal and vicariance in the Pleistocene history of an East African mountain rodent, *Praomys delectorum*. Journal of Biogeography 41: 196-208.
Bulatova, N., Lavrenchenko, L. A., Orlov, V. & Milishnokov, A. 2002. Notes on chromosomal identification of rodent species in western Ethiopia. Mammalia 66: 128-132.
Burda, H. 2000. Determinants of the distribution and radiation of African mole-rats (Bathyergidae, Rodentia). Ecology or geography? African Small Mammals: 263-277.

Burda, H., Zima, J., Scharff, M., Macholan, M. & Kawalika, M. 1999. The karyotypes of *Cryptomys anselli* sp. nova and *Cryptomys kafuensis* sp. nova: new species of the common mole-rat from Zambia (Rodentia, Bathyergidae). Zeitschrift fur Säugetierkunde 64 36-50.

Burda, H., Chitaukali, W. & Dryden, G. 2005. Taxonomic status and remarks on ecology of the Malawian mole-rat *Cryptomys whytei* (Rodentia, Bathyergidae). Acta Theriologica 50: 529-536.

Butynski, T. M. 1973. Life history and economic value of the springhare (*Pedetes capensis* Forster) in Botswana. Botswana Notes and Records 5: 209-213.

Butynski, T. M. 1979. Reproductive ecology of the Springhaas *Pedetes capensis* in Botswana. Journal of Zoology, London 189: 221-232.

Butynski, T. M. 1982. Pelage and moult in the springhare *Pedetes capensis* in Botswana. African Journal of Ecology 20: 279-287.

Butynski, T. M. 1984. Nocturnal ecology of the springhare, *Pedetes capensis*, in Botswana. African Journal of Ecology 22: 7-22.

Butynski, T. M. & Mattingly, R. 1979. Burrow structure and fossorial ecology of the springhare *Pedetes capensis* in Botswana. African Journal of Ecology 17: 205-215.

Calisher, C. 2007. S393 Bats: important hosts of emerging viruses. International Journal of Antimicrobial Agents 29: S80-S81.

Capanna, E. & Merani, M. S. 1981. Karyotypes of Somali rodent populations. 2. The chromosomes of of *Gerbillus dunni* (Thomas, 1904), *Gerbillus pusillus* Peters, 1878 and *Ammodillus imbellis* (de Winton, 1898) (Cricetidae, Gerbillinae). Monitore Zoologico Italiano 14: 227-240.

Capanna, E., Bekele, A., Capula, M., Castiglia, R., Civitelli, M., Codjia, J., Corti, M. & Fadda, C. 1996. A multidisciplinary approach to the systematics of the genus *Arvicanthis* Lesson, 1842 (Rodentia, Murinae). Mammalia 60: 677-696.

Capanna, E., Codjia, J.T., Chryosostome, C. & Civitelli, M. V. 1997. Les chromosomes des rongeurs du Benin (Afrique de l'Ouest). 3. Murinae. Rendiconti dell'Accademis Nazionale dei Lencei. Scienze Fisiche, Matematiche e Naturali, ser 9, 8: 2-37.

Carleton, M. & Musser, G. 1984. Muroid rodents. In: Orders and Families of Recent Mammals of the World (S. Anderson & J. K. Jones, eds). John Wiley & Sons, New York.

Carleton, M. D. & Robbins, C. B. 1985. On the status and affinities of *Hybomys planifrons* (Miller, 1900) (Rodentia: Muridae). Proceedings of the Biological Society of Washington 98: 956-1003.

Carleton, M. D. & Martinez, C. 1991. Morphometric differentiation among west African populations of the rodent genus *Dasymys* (Muroidea: Murinae), and its taxonomic implications. Proceedings of the Biological Society of Washington 104: 419-436.

Carleton, M. D. & Stanley, W. T. 2005. Review of the *Hylomyscus denniae* complex (Rodentia: Muridae) in Tanzania, with a description of a new species. Proceedings of the Biological Society of Washington 118: 619-646.

Carleton, M. D. & Byrne, E. 2006. The status of *Otomys orestes* dollmani Heller, 1912 (Muridae: Otomyinae), a rodent described from the Mathews Range, central Kenya. Proceedings of the Biological Society of Washington 119: 477-515.

Carleton, M. D. & Stanley, W. T. 2012. Species limits within the *Praomys delectorum* group (Rodentia: Muridae: Murinae) of East Africa: a morphometric reassessment and biogeographical implications. Zoological Journal of the Linnean Society 165: 420-469.

Carleton, M. D. & Van der Straeten, E. 1997. Morphological differentiation among Subsaharan and North African populations of the *Lemniscomys barbarus* complex (Rodentia: Muridae). Proceedings of the Biological Society of Washington 110: 640-680.

Carleton, M. D., Peterhans, J. K. & Stanley, W. T. 2006. Review of the *Hylomyscus denniae* group (Rodentia: Muridae) in eastern Africa, with comments on the generic allocation of *Epimys endorobae* Heller. Proceedings of the Biological Society of Washington 119: 293-3251-34.

Carter, J. & Leonard, B. P. 2002. A review of the literature on the worldwide distribution, spread of, and efforts to eradicate the coypu (*Myocaster coypus*). Wildlife Society Bulletin 30: 162-175.

Castiglia, R., Fadda, C., Corti, M., Scanzani, A., Verheyen, W. & Capanna, E. 2002. Chromosomal evolution in the African Arvicanthine rats (Murinae, Rodentia): comparative cytogenetics of *Lemniscomys* (*L. zebra, L. rosalia, L. striatus*) and *Arvicanthis dembeensis*. Journal of Zoological Systematics and Evolutionary Research 40: 223-231.

Castiglia, R., Corti, M., Colangelo, P., Annesi, F., Capanna, E., Verheyen, W., Sichilima, A. M. & Makundi, R. 2003. Chromosomal and molecular characterization of *Aethomys kaiseri* from Zambia and *Aethomys chrysophilus* from Tanzania (Rodentia, Muridae). Hereditas 139: 81-89.

Castiglia, R., Bekele, A., Makundi, R., Oguge, N. & Corti, M. 2006. Chromosomal diversity in the genus *Arvicanthis* (Rodentia, Muridae) from East Africa: a taxonomic and phylogenetic evaluation. Journal of Zoological Systematics and Evolutionary Research 44: 223-235.

Castiglia, R., Solano, E., Makundi, R., Hulselmans, J., Verheyen, E. & Colangelo, P. 2012. Rapid chromosomal evolution in the mesic four-striped grass rat *Rhabdomys dilectus* (Rodentia, Muridae) revealed by mtDNA phylogeographic analysis. Journal of Zoological Systematics and Evolutionary Research 50: 165-172.

Chaline, J., Mein, P. & Petter, F. 1977. Les grandes lignes d'une classification evolutive des Muroidea. Mammalia 41: 245-252.

Channing, A. 1984. Ecology of the namtap *Graphiurus ocularis* (Rodentia: Gliridae) in the Cedarberg, South Africa. South African Journal of Zoology 19: 144-149.

Chevret, P. 1994. Etude evolutive des Murinae (Rongeurs: Mammiferes) africains par hybridation DNA/DNA. Comparaison avec les approaches morphologiques et paleontologiques. PhD thesis, Universite des Sciences et Techniques du Languedoc, Montpellier-II, France.

Chevret, P. & Dobigny, G. 2005. Systematics and evolution of the subfamily Gerbillinae (Mammalia, Rodentia, Muridae). Molecular Phylogenetics and Evolution 35: 674-688.

Chevret, P., Denys, C., Jaeger, J., Michaux, J. & Catzeflis, F. 1993a. Molecular evidence that the spiny mouse (*Acomys*) is more closely related to gerbils (Gerbillinae) than to true mice (Murinae). Proceedings of the National Academy of Sciences 90: 3433-3436.

Chevret, P., Denys, C., Jaeger, J., Michaux, J. & Catzeflis, F. 1993b. Molecular and paleontological aspects of the tempo and mode of evolution in *Otomys* (Otomyinae: Muridae: Mammalia). Biochemical Systematic and Ecology 21: 123-131.

Chevret, P., Catzeflis, F. & Michaux, J. R. 2001. 'Acomyinae': new molecular evidence for a muroid taxon (Rodentia; Muridae). In: African Small Mammals (C. Denys, L. Granjon & A. Poulet, eds). IRD, Paris.

Chevret, P., Veyrunes, F. & Britton-Davidian, J. 2005. Molecular phylogeny of the genus *Mus* (Rodentia: Murinae) based on mitochondrial and nuclear data. Biological Journal of the Linnean Society 84: 417-427.

Chimimba, C. T. 1998. A taxonomic synthesis of southern African *Aethomys* (Rodentia: Muridae) with a key to species. Mammalia 62: 427-437.

Chimimba, C. T. 2001. Infraspecific morphometric variation in *Aethomys namaquensis* (Rodentia: Muridae) from southern Africa. Journal of Zoology, London 253: 191-210.

Chimimba, C. T. 2005. Phylogenetic relationships in the genus *Aethomys* (Rodentia: Muridae). African Zoology 40: 271-284.

Chimimba, C. T. & Linzey, A. V. 2008. *Aethomys ineptus*. Mammalian Species 809: 1-7

Chimimba, C. T., Dippenaar, N. J. & Robinson, T. J. 1998. Geographic variation in *Aethomys granti* (Rodentia: Muridae) from southern Africa. Annals of the Transvaal Museum 36: 405-412.

Chimimba, C. T., Dippenaar, N. J. & Robinson, T. J. 1999. Morphometric and morphological delineation of southern African species of *Aethomys* (Rodentia: Muridae). Biological Journal of the Linnean Society 67: 501-527.

Chitaukali, W., Burda, H. & Kock, D. 2001. On small mammals of the Nyika Plateau, Malawi. In: African Small Mammals (C. Denys, L. Granjon & A. Poulet, eds), IRD, Paris, pp 415-426.

Cifelli, R. L., Ibui, A., Jacobs, L. L. & Thorington, R. W. 1986. A giant tree squirrel from the late Miocene of Kenya. Journal of Mammalogy 67: 274-283.

Civitelli, M., Codjia, J. & Capanna, E. 1996. The chromosomes of rodents of the Republic of Benin (West Africa): 2. Sciuridae and Thryonomyidae. Rend. Fis. Acc. Lincei 7: 51-56.

Clausnitzer, V. 2000. Ecology of *Otomys barbouri* Lawrence & Loveridge, 1953 (Mammalia, Rodentia): an endemic of the Afro-alpine zone of Mt Elgon, East Africa. Bonner Zoologische Monographien 46: 233-244.

Clausnitzer, V. 2003. Rodents of Mt Elgon, Uganda: ecology, biogeography and the significance of fire. Ecotropical Monographs 3: 1-184.

Clausnitzer, V. & Kityo, R. 2001. Altitudinal distribution of rodents (Muridae and Gliridae) on Mt Elgon, Uganda. Tropical Zoology 14: 95-118.

Coetzee, C. G. 1977. Genus *Steatomys*. In: Mammals of Africa: An Identification Manual (J. Meester & H. W. Setzer, eds). Part 6.8. Smithsonian Institution Press, Washington, DC.

Coetzee, C. G. 2002. The distribution and breeding season of the dassie-rat, *Petromus typicus* (Petromuridae, Rodentia). Folia Zoologica, Brno 51 (Suppl. 1) 23-35.

Coetzee, C. G. & Jackson, P. T. 1999. The comparative behaviour and ecology of *Parotomys brantsii* and *P. littledalei* (Mammalia, Rodentia, Otomyinae). Journal of the Namibia Scientific Society 47: 87-106.

Colangelo, P., Corti, M., Verheyen, E., Annesi, F., Oguge, N., Makundi, R. H. & Verheyen, W. 2005. Mitochondrial phylogeny reveals differential modes of chromosomal evolution in the genus *Tatera* (Rodentia: Gerbillinae) in Africa. Molecular Phylogenetics and Evolution 35: 556-568.

Colangelo, P., Granjon, L., Taylor, P. J. & Corti, M. 2007. Evolutionary systematics in African gerbilline rodents of the genus *Gerbilliscus*: Inference from mitochondrial genes. Molecular Phylogenetics and Evolution 42: 797-806.

Colangelo, P., Castiglia, R., Franchini, P. & Solano, E. 2010a. Pattern of shape variation in the eastern African gerbils of the genus *Gerbilliscus* (Rodentia, Muridae): Environmental correlations and implication for taxonomy and systematics. Mammalian Biology 75: 302-310.

Colangelo, P., Leirs, H., Castiglia, R., Daes, M., Meheretu, Y., Verheyen, E., D'aes, M., Meheretu, Y. & Verheyen, E. 2010b. New data on the distribution and phylogenetic position of *Mastomys awashensis* (Rodentia, Muridae). Mammalian Biology 75: 459-462.

Colangelo, P., Verheyen, E., Leirs, H., Tatard, C., Denys, C., Dobigny, G., Duplantier, J-M., Brouat, C., Granjon, L. & Lecompte, E. 2013. A mitochondrial phylogeographic scenario for the most widespread African rodent, *Mastomys natalensis*. Biological Journal of the Linnean Society 108: 901-916.

Cole, L. 1972. A comparison of *Malacomys longipes* and *Malacomys edwardsi* (Rodentia: Muridae) from a single locality in Ghana. Journal of Mammalogy 53: 1-5.

Colyn, M., Dudu, A. & Mbaelele, M. 1987. Données sur l'exploitation du petit et moyen gibier des forêts ombrophiles du Zaire. Nature et Faune 3: 22-39.

Conroy, G. C., Pickford, M., Senut, B., Van Couvering, J. & Mein, P. 1992. *Otavipithecus namibensis*, first Miocene hominoid from southern Africa. Nature 356: 144-148.

Corbet, G. B. & Jones, L. A. 1965. On the specific characters of the Crested Porcupines, subgenus *Hystrix*. Proceedings of the Zoological Society of London 144: 285-300.

Corbin, C. E. & Cordeiro, N. J. 2006. Gliding characteristics of Lord Derby's anomalure (*Anomalurus derbianus*) in Tanzania. African Journal of Ecology 44: 106-108.

Corti, M., Scanzani, A., Rossi, A. R., Civitelli, M. V., Bekele, A. & Capanna, E. 1999. Karyotypic and genetic divergence in the Ethiopian *Myomys-Stenocephalemys* complex (Mammalia, Rodentia). Italian Journal of Zoology 66: 341-349.

Corti, M., Castiglia, R., Annesi, F. & Verheyen, W. 2004a. Mitochondrial sequences and karyotypes reveal hidden diversity in African pouched mice (subfamily Cricetomyinae, genus *Saccostomus*). Journal of Zoology, London 262: 1-12.

Corti, M., Di Giuliomaria, C. & Verheyen, W. 2004b. Three-dimensional geometric morphometrics of the African genus *Lophuromys* (Rodentia Muridae). Hystrix 11: 145-154.

Corti, M., Castiglia, R., Colangelo, P., Capanna, E., Beolchini, F., Bekele, A., Oguge, N. O., Makundi, R. H., Sichilima, A. M., Leirs, H., Verheyen, W. & Verhagen, R. 2005. Cytotaxonomy of rodent species from Ethiopia, Kenya, Tanzania and Zambia. Belgian Journal of Zoology 135: 197-216.

Coster, P., Benammi, M., Salem, M., Bilal, A. A., Chaimanee, Y., Valentin, X., Brunet, M., Jaeger, J. J. 2012. New hystricognathous rodents from the Early Oligocene of central Libya (Zallah Oasis, Sahara Desert): systematic, phylogenetic, and biochronologic implications. Annals of Carnegie Museum, 80: 239-259.
Coulibaly-N'Golo, D., Allali, B., Kouassi, S. K., Fichet-Calvet, E., Becker-Ziaja, B., Rieger, T., Ölschläger, S., Dosso, H., Denys, C., Meulen, ter, J., Akoua-Koffi, C. & Günther, S. 2011. Novel arenavirus sequences in *Hylomyscus* sp. and *Mus* (*Nannomys*) *setulosus* from Côte d'Ivoire: implications for evolution of arenaviruses in Africa. PLoS ONE 6: e20893.
Cox, C.B. 2000. Plate tectonics, Seaways and Climate in the Historical Biogeography of mammals. Mem. Inst. Oswaldo Cruz, Rio de Janeiro 95: 509-516.
Crawford-Cabral, J. 1998. The Angolan rodents of the superfamily Muroidea. An account of their distribution. Instituto de Investacao Cientifica Tropical, Estudos, Ensaios e Documentos 161: 1-222.
Crawford-Cabral, J. 1999. The short-tailed *Aethomys* of Angola (Mammalia: Rodentia: Muridae). Garcia de Orta, Serie de Zoologica, Lisboa 23: 169-191.
Crespin, L., Papillon, Y., Abdoulaye, D., Granjon, L. & Sicard, B. 2008. Annual flooding, survival and recruitment in a rodent population from the Niger River plain in Mali. Journal of Tropical Ecology 24.
Datiko, D., Bekele, A. & Belay, G. 2007. Species composition, distribution and habitat association of rodents from Arbaminch forest and farmlands, Ethiopia. African Journal of Ecology 45: 651-657.
Daly, M. & Daly, S. 1973. On the feeding ecology of *Psammomys obesus* (Rodentia, Gerbillidae) in the wadi Saoura, Algeria. Mammalia 37: 545-561.
Daly, M. & Daly, S. 1974. Spatial distribution of a leaf-eating Saharan gerbil (*Psammomys obesus*) in relation to its food. Mammalia 38: 591-603.
Davies, C., 1987. Fossil Pedetidae (Rodentia) from Laetoli. In: Laetoli: A Pliocene Site in Northern Tanzania (M. D. Leakey, J. M. Harris, eds.). Clarendon Press, Oxford.
Davis, D. H. S. 1975. Genus *Aethomys* Thomas, 1915. In: The Mammals of Africa: An Identification Manual (J. Meester & H. W. Setzer, eds). Part 6.6. Smithsonian Institution Press, Washington, DC.
De Bruijn, H. & Whybrow, P. J. 1984. A Late Miocene rodent fauna from the Baynunah Formation, Emirate of Abu Dhabi, United Arab Emirates. Proceedings of the Koninklijke Nederlandse Akademie van Wetenschappen, Amsterdam 97: 407-422.
De Graaff, G. 1960. A preliminary investigation of the mammalian microfauna in Pleistocene deposits of caves in the Transvaal. Palaeontologia Africana 7: 59-118.
De Graaff, G. 1981. The Rodents of Southern Africa. Butterworths, Durban.
De Graaff, G. & Nel, J. A. J. 1992. Notes on a single burrow system of the fat mouse *Steatomys pratensis* in the Kruger National Park. Koedoe 35: 123-124.
De Villiers, M. & van Aarde, R. 1994. Aspects of habitat disturbance by Cape porcupines in a savanna ecosystem. South African Journal of Zoology 29: 217-220.
De Winton, W. E. 1896. On collections of rodents made by Mr. J. ffolliot Darling in Mashunaland and Mr. F. C. Selous in Matabeleland, with short field notes by the collectors. Proceedings of the Zoological Society of London 1896: 798-808.
Dean, W. R. J. 1978. Conservation of the white-tailed rat in South Africa. Biological Conservation 13: 133-140.
Decher, J., Kilpatrick, C. W. & Bahian, K. L. 2001. Small mammal zoogeography and diversity in West African forest remnants. Unpublished Progress Report to National Geographic Society.
Delany, M. 1986. Ecology of small rodents in Africa. Mammal Review 16: 1-41.
Delany, M. J. 1975. The Rodents of Uganda. British Museum (Natural History), London.
Dempster, E. R. & Perrin, M. R. 1989. Maternal behaviour and neonatal development in three species of Namib Desert rodents. Journal of Zoology, London 218: 407-419.
Dempster, E. R., Perrin, M. R., Downs, C. T. & Griffin, M. 1998. *Gerbillurus setzeri*. Mammalian Species 598: 1-4.
Dempster, E. R., Perrin, M. R. & Downs, C. T. 1999. *Gerbillurus vallinus*. Mammalian Species 605: 1-4.
Denys, C. 1985. Paleoenvironmental and paleobiogeographical significance of the fossil rodent assemblages of Laetoli (Pliocene, Tanzania). Palaeogeography, Palaeoclimatology and Palaeoecology 52: 77-97.

Denys, C. 1987a. Rodentia and Lagomorpha. 6.1. Fossil rodents (other than Pedetidae) from Laetoli. In: Laetoli, and Pliocene Site in Northern Tanzania (M. D. Leakey & J. M. Harris, eds). Clarendon Press, Oxford, UK.
Denys, C. 1987b. Micromammals from the West Natron Pleistocene deposits (Tanzania). Biostratigraphy and Paleoecology, Sciences Geologiques Bulletin 40: 185-201.
Denys, C. 1988. Apports de l'analyse morphometrique à la détermination des espèces actuelles et fossiles du genre *Saccostomus* (Cricetomyinae, Rodentia). Mammalia 5: 497-532.
Denys, C. 1989a. Two new gerbillids (Rodentia, Mammalia) from Olduvai Bed I (Pleistocene, Tanzania). Neues Jahrbuch für Geologie und Paläontologie Abhandlungen 178: 243-265.
Denys, C. 1989b. A new species of bathyergid rodent from Olduvai Bed I (Tanzania, Lower Pleistocene). Neues Jahrbuch für Geologie und Paläontologie Monatshefte H.5: 257-64.
Denys, C. 1989c. Phylogenetic affinities of the oldest East African *Otomys* (Rodentia, Mammalia) from Olduvai Bed I (Pleistocene, Tanzania). Neues Jachbuch fur Geologie und Paleontologie Monathefie 44: 705-725.
Denys, C. 1990a. First occurrence of *Xerus cf. inauris* (Rodentia, Sciuridae) at Olduvai Bed I (Lower Pleistocene, Tanzania). Paläontologische Zeitschrift 64: 359-365.
Denys, C. 1990b. Deux nouvelles espèces d'*Aethomys* (Rodentia, Muridae) à Langebaanweg (Pliocène, Afrique du Sud): implications phylogénétiques et paléoécologiques. Annales de Paléontologie 76: 41-69.
Denys, 1990c. Implications paléoécologiques et paléobiogéographiques de l'étude de rongeurs Plio-Pléistocènes d'Afrique orientale et australe. Mémoirs Institut Science et Terre, Universite Pierre et Marie Curie, Paris.
Denys, C. 1991. Un nouveau rongeur *Mystromys pocockei* sp. nov. (Cricetinae) du Plioène inférieur de Langebaanweg (Region du Cap, Afrique du Sud). Comptes Rendus de l'Academie des Sciences (Paris), Serie IIa 313: 1335-1345.
Denys, C. 1992. Présence de *Saccostomus* (Rodentia, Mammalia) à Olduvai Bed I (Tanzanie, Pléistocène inférieur). Implications phylétiques et paléobiogéographiques. Geobios 25: 145-154.
Denys, C. 1993.Reéxamen de la dentition de *Lelmacomys buettneri* (Mammalia, Rodentia). Hypothèses sur sa position systematique. Mammalia 57: 613-618.
Denys, C. 1999. Of mice and men. Evolution in East and South Africa during Plio-Pleistocene times. In: African Biogeography, Climate Change & Human Evolution (T. G. Bromage & F. Schrenk, eds). Oxford University Press, Oxford, UK.
Denys, C. 1994. Diet and dental morphology of two coexisting *Aethomys* species (Rodentia) in Mozambique. Implications for diet reconstruction in related extinct species from South Africa. Acta Theriologica 39: 357-364.
Denys, C. 2003. Evolution du genre *Otomys* (Rodentia, Muridae) au Plio-Pléistocène d'Afrique orientale et australe. In: Advances in Paleontology, "Hen to Pantha," Volume in Honor of Constantin Radulesca and Petre Mihai Samson (A. Petculescu & E. Stiuca, eds.). Romanian Academy, "Emil Racovitză," Institute of Speleology, Bucharest, Romania.
Denys, C. 2011. Rodents In: Paleontology and Geology of Laetoli: Human Evolution in Context. Volume 2: -Fossil Hominins and the Associated Fauna (T. Harrison, ed.). Springer.
Denys, C. & Jaeger, J.J. 1992. Rodents of the Miocene site of Fort Ternan (Kenya) first part: Phiomyids, bathyergids, sciurids and anomalurids. Neues Jahrbuch für Geologie und Paläontologie Abhandlungen 185: 63-84.
Denys, C. & Michaux, J. 1992. La troisième molaire supérieure chez les Muridae d'Afrique tropicale et le cas des genres *Acomys, Uranomys* et *Lophuromys*. Bonner Zoologische Beiträge 43: 367-382.
Denys, C. & Tranier, M. 1992. Presence d'*Aethomys* (Mammalia, Rodentia, Muridae) au Tchad et analyse morphometrique préliminaire du complex *A. hindei*. Mammalia 56: 625-656.
Denys, C. & Aniskine, V. 2012. On a new species of *Dendromus* (Rodentia, Nesomyidae) from Mount Nimba, Guinea. Mammalia 76: 295-308.
Denys, C., Michaux, J., Petter, F., Aguilar, J. P. & Jaeger, J. J. 1992. Molar morphology as a clue to the phylogenetic relationships of *Acomys* to the Murinae. Israel Journal of Zoology 38: 253-262.

Denys, C., Gautun, J. C., Tranier, M. & Volobouev, V. 1994. Evolution of the genus *Acomys* (Rodentia, Muridae) from dental and chromosomal patterns. Israel Journal of Zoology 40: 215-246.
Denys, C., Michaux, J., Catzeflis, F., Ducrocq, S. & Chevret, P. 1995. Morphology and molecular data against the monophyly of Dendromurinae (Muridae: Rodentia). Bonner zoologische Beiträge 45: 173-190.
Denys, C., Chitaukali, W., Mfune, J.K., Combrexelle, M. & Cacciani, F. 1999. Diversity of small mammals in owl pellet assemblages of Karonga district, northern Malawi. Acta Zoologica Cracoviensis, 42: 393-396.
Denys, C., Viriot, L., Daams, R., Pelaez-Campomanes, P., Vignaud, P., Andossa, L. & Brunet, M. 2003. A new Pliocene xerine sciurid (Rodentia) from Kossom Bougoudi, Chad. Journal of Vertebrate Paleontology 23: 676-687.
Denys, C., Colyn, M. & Nicolas, V. 2006. First record of the Dollman's tree mouse (*Prionomys batesi*; Mammalia: Nesomyidae) in the Republic of Congo and additional description of this rare Central African rodent. Zootaxa 1318: 59-68.
Denys, C., Lecompte, E., Calvet, E., Camara, M. D., Doré, A., Koulémou, K., Kourouma, F. , Soropogui, B., Sylla, O., Allali B. K., Kouassi-Kan, S., Akoua-Koffi, C., ter Meulen, J. and Koivogui, L. 2005.- Community analysis of Muridae (Mammalia, Rodentia) diversity in Guinea : a special emphasis on Mastomys species and Lassa fever distributions. In: Huber, B.A., Sinclair B. J., Lampe K.H. (eds)., African Biodiversity. Molecules, organisms, Ecosystems. Springer, pp. 339-350.
Denys, C., Lalis, A., Aniskin, V., Kourouma, F., Soropogui, B., Sylla, O., Doré, A., Koulemou, K., Beavogui, Z. B., Sylla, M., Camara, A., Camara, A. B., Camara, A. C., Kouassi Kan, S., Volobouev, V., Camara, C., Koivogui, L., Bernard, A. K. 2009a. New data on the taxonomy and distribution of Rodentia (Mammalia) from the western and coastal regions of Guinea West Africa. Italian Journal of Zoology 76: 111-128.
Denys, C., Missoup, A., Tchiengue, B., Achoundong, G., Ekobo, A., Bilong, C. B., Lembe, D. & Nicolas, V. 2009b. Altitudinal distribution and anthropogenic influence on small mammal assemblages on Mount Kupe, SE Cameroon. Bonner zoologische Beiträge 56: 159-173.
Denys, C., Lalis, A., Kourouma, F., Kouassi, S. K., Nicolas, V., Aniskine, V. & Koivogui, L. 2011a. Discrimination morphologique, genetique et ecologique de deux espèces sympatriques de *Mastomys* (Mammalia: Rodentia) en Guinee maritime (Conakry): implications pour la sante et l'agriculture. Rev. Ecol. (Terre Vie) 66: 1-19.
Denys, C., Lalis, A., Lecompte, É., Cornette, R., Moulin, S., Makundi, R. H., Machang'u, R. S., Volobouev, V. & Aniskine, V. M. 2011b. A faunal survey in Kingu Pira (south Tanzania), with new karyotypes of several small mammals and the description of a new Murid species (Mammalia, Rodentia). Zoosystema 33: 5-47.
Denys, C., Sandrock, O., Kullmer, O., Rozzi, F. R., Bromage, T. G. & Schrenk, F. 2011c. Note on the presence of (Mammalia, Rodentia) in the Malawi Chiwondo Beds (Plio-Pleistocene): taphonomical and palaeoecological implications. Geodiversitas 33: 729-738.
Denys, C., Missoup, A. D., Nicolas, V., Fülling, O., Delapré, A., Bilong Bilong, C. F., Taylor, J. P. & Hutterer, R. 2014. African highlands as mammal diversity hotspots: new records of *Lamottemys okuensis* (Rodentia: Muridae) and other endemic rodents from Mount Oku, Cameroon. Zoostema 36: 647-690.
Deuve, J. L., Bennett, N. C., Britton-Davidian, J. & Robinson, T. J. 2008. Chromosomal phylogeny and evolution of the African mole-rats (Bathyergidae). Chromosome Research 16: 57-74.
Dietrich W. O. 1942 Altestquartäre Säugetiere aus der südlichen Serengeti, Deutsch-Ostafrika. Palaeontographica 94A: 43-133
Dieterlen, F. 1961. Beiträge zur biologie der stachelmaus, *Acomys cahirinus dimidiatus* Cretzschmar. Zeitschrift fur Säugetierkunde 26: 1-13.
Dieterlen, F. 1962. Geburt und geburtshilfe bei der stachelmaus *Acomys cahirinus*. Zeitschrift fur Tierpsychologie 19: 191-222.
Dieterlen, F. 1963. Vergleichende untersuchungen zur ontogenese von stachlemaus (*Acomys*) und wanderratte (*Rattus norvegicus*) (Teil I) – Beträge zum nesthocker-nestfluchter-problem bei nagetieren (Teil II). Zeitschrift für Säugetierkunde 28: 193-227.
Dieterlen, F. 1969. Zur kenntnis von *Delanymys brooksi* Hayman 1962 (Petromyscinae; Cricetidae; Rodentia). Bonner zoologische Beiträge 20: 384-395.

Dieterlen, F. 1971. Beiträge zur systematik, okologie und biologie der gattung *Dendromus* (Dendromurinae, Cricetidae, Rodentia), insbesondere ihrer zentralafrikanischen Formen. Saugetierkundliche Mitteilungen 19: 97-132.
Dieterlen, F. 1975. *Lophuromys medicaudatus* (Muridae; Rodentia). Beschreibung einer neuen art auf grund neuer ergebnisse zur systematischen stellung von *Lophuromys luteogaster* Hatt 1934. Bonner Zoologische Beiträge 26: 293-318.
Dieterlen, F. 1976a. Zweiter fund von *Dendromus kahuziensis* (Dendromurinae; Cricetidae; Rodentia) und weitere Deondromus-Fange im Kivu-Hochland oberhalb 2000 m. Stuttgarter Beiträge zur Naturkunde A 268: 1-5.
Dieterlen, F. 1976b. Die afrikanische Muriden-Gattung *Lophuromys* Peters 1874. Stuttgarter Beiträge zur Naturkunde A 285: 1-96.
Dieterlen, F. 1976c. Bemerkungen uber *Leimacomys buettneri* Matschie, 1893 (Dendromurinae, Cricetidae, Rodentia). Säugetierkunde Mitteilungen 24: 224-228.
Dieterlen, F. 1979. Beiträge zur kenntnis der Gattung *Lophuromys* (Muridae; Rodentia) in Kamerun und Gabun. Bonner Zoologische Beiträge 29: 287-299.
Dieterlen, F. 1983. Zur systematik, verbreitung und okologie von *Colomys goslingi* Thomas & Wroughton, 1907 (Muridae; Rodentia). Bonner zoologische Beiträge 34: 73-106.
Dieterlen, F. 1987. Neue erkenntnise uber afrikanische burstenhaarmause, Gattung *Lophuromys* (Muridae; Rodentia). Bonner Zoologische Beiträge 38: 183-194.
Dieterlen, F. 1991. *Lemniscomys hoogstraali*, a new murid species from Sudan. Bonner Zoologische Beiträge 42: 11-15.
Dieterlen, F. 2005a. Family Anomaluridae. In: Mammal Species of the World: A Taxonomic and Geographic Reference (3rd edn) (D. E. Wilson and D. Reeder, eds). John Hopkins University Press.
Dieterlen, F. 2005b. Family Ctenodactylidae. In: Mammal Species of the World: A Taxonomic and Geographic Reference (3rd edn) (D. E. Wilson and D. Reeder, eds). John Hopkins University Press.
Dieterlen, F. 2005c. Family Pedetidae. In: Mammal Species of the World: A Taxonomic and Geographic Reference (3rd edn) (D. E. Wilson and D. Reeder, eds). John Hopkins University Press, Baltimore.
Dieterlen, F. 2009. Climbing mice of the genus *Dendromus* (Nesomyidae, Dendromurinae) in Sudan and Ethiopia, with the description of a new species. Bonner zoologische Beiträge 56: 185-200.
Dieterlen, F. & Nikolaus, G. 1985. Zur Saugetierfauna des Sudan - weitere erstnachweise und bemerkenswerte funde. Saugetierkundlische Mitteilungen 32: 205-209.
Dieterlen, F. & Rupp, H. 1978. *Megadendromus nikolausi*, Gen nov., sp. nov. (Dendromurinae; Rodentia) ein neuer Nager aus Athiopien. Zeitschrift fur Säugetierkunde 43: 129-145.
Dieterlen, F. & Statzner, B. 1981. The African rodent *Colomys goslingi* Thomas & Wroughton, 1907 – a predator in limnetic ecosystems. Zeitschrift fur Säugetierkunde 46: 369-383.
Dieterlen, F. & Van der Straeten, E. 1984. New specimens of *Malacomys verschureni* from eastern Zaire (Mammalia, Muridae). Revue Zoologique Africaine 98: 861-868.
Dieterlen, F. & Van der Straeten, E. 1988. Deux nouveaux specimens de *Lamottemys okuensis* Petter, 1986 du Cameroun (Muridae: Rodentia). Mammalia 52: 379-385.
Dieterlen, F. & Van der Straeten, E. 1992. Species of the genus *Otomys* from Cameroon and Nigeria and their relationship to East African forms. Bonner Zoologische Beiträge 43: 383-392.
Dippenaar, N. J. & Rautenbach, I. L. 1986. Morphometrics and karyology of the southern African species of the genus *Acomys*. I. Geoffroy Saint-Hillaire, 1838 (Rodentia: Muridae). Annals of the Transvaal Museum 34: 129-183.
Dippenaar, N. J., Meester, J., Rautenbach, I. L. & Wolhuter, D. A. 1983. The status of southern African mammal taxonomy. Annales du Musée Royal de l'Afrique Centrale, Sciences Zoologiques 237: 103-107.
Dippenaar, N. J., Swanepoel, P. & Gordon, D. H. 1993. Diagnostic morphometrics of two medically important southern African rodents, *Mastomys natalensis* and *M. coucha* (Rodentia, Muridae). South African Journal of Science 89: 300-303.

Dobigny, G., Cornette, R., Moulin, S. & Sidiyene, E. 2001a. The mammals of Adrar des Iforas, Mali, with special emphasis on small mammals. Systematics and biogeographical implications. In: African Small Mammals (C. Denys, L. Granjon & A. Poulet, eds). IRD, Paris.

Dobigny, G., Moulin, S., Cornette, R. & Gautun, J-C. 2001b. Rodents from Adrar des Iforas, Mali. Chromosomal data. Mammalia 65: 215-220.

Dobigny, G., Nomao, A. & Gautun, J-C. 2002a. A Cytotaxonomic survey of rodents from Niger: implications for systematics, biodiversity and biogeography. Mammalia 66: 495-523.

Dobigny, G., Baylac, M. & Denys, C. 2002b. Geometric morphometrics, neural networks and diagnosis of sibling *Taterillus* species (Rodentia, Gerbillinae). Biological Journal of the Linnean Society 77: 319-327.

Dobigny, G., Aniskin, V. & Volobouev, V. 2002c. Explosive chromosome evolution and speciation in gerbil genus *Taterillus* (Rodentia, Gerbillinae): a case of two new cryptic species. Cytogenetics and Genome Research 96: 117-124.

Dobigny, G., Granjon, L., Aniskin, V., Ba, K. & Volobouev, V. 2003. A new sibling species of *Taterillus* (Muridae, Gerbillinae) from West Africa. Mammalian Biology 68: 299-316.

Dobginy, G., Aniskin, V., Granjon, L., Cornette, R. & Volobouev, V. 2005. Very recent radiation in West African *Taterillus*: the concerted role of chromosome and climate changes. Heredity 95: 358-368.

Dobigny, G., Lecompte, E., Tatard, C., Gauthier, P., Bâ, K., Denys, C., Duplantier, J. M. & Granjon, L. 2008. An update on the taxonomy and geographic distribution of the cryptic species *Mastomys kollmannspergeri* (Muridae, Murinae) using combined cytogenetic and molecular data. Journal of Zoology, London 276: 368-374.

Dobigny, G., Tatard, C., Kane, M., Gauthier, P., Brouat, C., Ba, K. & Duplantier, J-M. 2011. A cytotaxonomic and DNA-based survey of rodents from Northern Cameroon and Western Chad. Mammalian Biology 76: 417-427.

Dobigny, G., Tatard, C., Gauthier, P., Ba, K., Duplantier, J-M., Granjon, L. & Kergoat, G. J. 2013. Mitochondrial and Nuclear Genes-Based Phylogeography of *Arvicanthis niloticus* (Murinae) and Sub-Saharan Open Habitats Pleistocene History. PlosONE 8(11): e77815.

Downs, S. L. 2012. A molecular phylogenetic study of pygmy mice and house mice (Family Muridae; genus *Mus*) from southern and east Africa. MSc thesis, University of KwaZulu-Natal, Westville.

Downs, C. T. & Perrin, M. R. 1995. The thermal biology of the white-tailed rat *Mystromys albicaudatus*, a cricetine relic in southern temperate African grassland. Comparative Biochemistry and Physiology 110A: 65-69.

Du Plessis, A. & Kerley, G. I. H. 1991. Refuge strategies and habitat segregation in two sympatric rodents, *Otomys unisulcatus* and *Parotomys brantsii*. Journal of Zoology, London 224: 1-10.

Du Plessis, A., Kerley, G. I. H. & Winter, P. E. D. 1992. Refuge microclimates of rodents: a surface nesting *Otomys unisulcatus* and a burrowing *Parotomys brantsii*. Acta Theriologica 37: 351-358.

Du Toit, N., Vuuren, B., Matthee, S. & Matthee, C. 2012. Biome specificity of distinct genetic lineages within the four-striped mouse *Rhabdomys pumilio* (Rodentia: Muridae) from southern Africa with implications for taxonomy. Molecular Phylogenetics and Evolution 65: 75-86.

Dubey, S., Antonin, M., Denys, C. & Vogel, P. 2007. Use of phylogeny to resolve the taxonomy of the widespread and highly polymorphic African giant shrews (*Crocidura olivieri* group, *Crocidurinae, Mammalia*). Zoology 110: 48-57.

Dubois, J-Y., Rakotondravony, D., Hanni, C., Sourouille, P. & Catzeflis, F. M. 1996. Molecular evolutionary relationships of three genera of Nesomyinae, endemic rodent taxa from Madagascar. Journal of Mammalian Evolution 3: 239-259.

Dubois, J-Y. F., Catzeflis, F. M. & Beintema, J. J. 1999. The phylogenetic position of 'Acomyinae' (Rodentia; Mammalia) as sister group of a Murinae + Gerbillinae clade: evidence from the nuclear ribonuclease gene. Molecular phylogenetics and Evolution 13: 181-192.

Ducroz, J-F., Volobouev, V. & Granjon, L. 1998. A molecular perspective on the systematics and evolution of the genus *Arvicanthis* (Rodentia, Muridae): inferences from complete cytochrome b gene sequences. Molecular Phylogenetics and Evolution 10: 104-117.

Ducroz, J-F., Volobouev, V. & Granjon, L. 2001. An assessment of the systematics of Arvicanthine rodents using mitochondrial DNA sequences: evolutionary and biogeographical implications. Journal of Mammalian Evolution 8: 173-206.

Dudu, A., Van der Straeten, E. & Verheyen, W. 1989. Premiere capture de *Hylomyscus parvus* Brosset, Dubost et Heim de Balzac, 1965 au Zaire avec quelques données biometriques (Rodentia, Muridae). Journal of African Zoology 103: 179-182.

Duplantier, J-M. 1989. Les rongeurs myomorphes forestiers du nord-est du Gabon: structure du peuplement, démographie, domaines vitaux. Revue d'Ecologie (La Terre et la Vie) 44: 329-346.

Duplantier, J-M. & Granjon, L. 1990. Rythmes d'activité chez six espèces du muridés du Sénégal appartenant aux genres *Mastomys, Arvicanthis, Myomys* and *Dasymys*. Mammalia 54: 173-182.

Duplantier, J-M., Britton-Davidian, J. & Granjon, L. 1990. Chromosomal characterisation of three species of the genus *Mastomys* in Senegal. Zeitschrift fur Zoologisher Systematik und Evolutionsforschung 28: 289-298.

Duplantier, J-M., Granjon, L. & Bouganaly, H. 1996. Reproductive characteristics of three sympatric species of *Mastomys* in Senegal, as observed in the field and in captivity. Mammalia 60: 629-638.

Durnez, L., Suykerbuyk, P., Nicolas, V., Barriere, P., Verheyen, E., Johnson, C. R., Leirs, H. & Portaels, F. 2010. Terrestrial Small Mammals as Reservoirs of *Mycobacterium ulcerans* in Benin. Applied and Environmental Microbiology 76: 4574-4577.

Eccard, J. A., Meyer, J. & Sundell, J. 2004. Space use, circadian activity pattern, and mating system of the nocturnal tree rat *Thallomys nigricauda*. Journal of Mammalogy 85: 440-445.

Egoscue, H. J. 1972. Breeding the long-tailed pouched rat, *Beamys hindei*, in captivity. Journal of Mammalogy 53: 296-302.

Eiseb, S. J. 2014. Systematics and biogeography of the genus *Mastomys* (Rodentia: Muridae) occurring in Namibia and adjacent countries. PhD thesis

Eisentraut, M. 1968a. Beitrag zur saugetierfauna von Kamerun. Bonner Zoologische Beiträge 19: 1-14.

Eisentraut, M. 1968b. Die tiergeographische bedeutung des Oku-Gebirges im Bamenda-Banso-Hochland (Westkamerun). Bonner zoologische Beiträge 19: 170-175.

Eisentraut, M. 1969. Die verbreitung de Muriden-Gattung *Hylomyscus* auf Fernando Poo und in Westkamerun. Zeitschrift fur Säugetierkunde 34: 296-307

Eisentraut, M. 1970. Die verbreitung der muriden-gattung *Praomys* auf Fernando Poo und in West Kamerun. Zeitschrift fur Säugetierkunde 35: 1-15.

Eisentraut, M. 1975. Weiterer beitrag zur säugetierfauna von Kamerun. Bonner zoologische Beiträge 26: 76-93.

Ellerman, J. R. 1940. The Families and Genera of Living Rodents. Vol. 1: Rodents other than Muridae. British Museum (Natural History), London.

Ellerman, J. R. 1941. The Families and Genera of Living Rodents. Vol. 2: Family Muridae. British Museum (Natural History), London.

Ellerman, J. R., Morris-Scott, T. C. S. & Hayman, R. W. 1953. Southern African Mammals 1758 to 1951: A reclassification. British Museum (Natural History), London.

Ellison, G. T. H. 1995. Thermoregulatory responses of cold-acclimated fat mice (*Steatomys pratensis*). Journal of Mammalogy 76: 240-247.

Emmons, L. H. 1979a. Observations on litter size and development of some African rainforest squirrels. Biotropica 11: 207-213.

Emmons, L. H. 1979b. A note on the forefoot of *Myosciurus pumilio*. Journal of Mammalogy 60: 431-432.

Emmons, L. H. 1980. Ecology and resource partitioning among nine species of African rain forest squirrels. Ecological Monographs 50: 31-54.

Emmons, L. H. 1983. A field study of the African brush-tailed porcupine, *Atherurus africanus*, by radiotelemetry. Mammalia 47: 183-194.

Engelbrecht, A., Taylor, P., Daniels, S. & Rambau, R. 2011. Cryptic speciation in the southern African vlei rat *Otomys irroratus* complex: evidence derived from mitochondrial cyt b and niche modelling. Biological Journal of the Linnean Society 104: 192-206.

Ewer, R. F. 1967. The behaviour of the African Giant Rat (*Cricetomys gambianus* Waterhouse). Zeitschrift fur Tierpsychologie 24: 6-79.

Fa, J. E., Seymour, S., Dupain, J., Amin, R., Albrechtsen, L. & Macdonald, D. 2006. Getting to grips with the magnitude of exploitation: Bushmeat in the Cross–Sanaga rivers region, Nigeria and Cameroon. Biological Conservation 129: 497-510.

Fadda, C. & Corti, M. 1998. Geographic variation of *Arvicanthis* (Rodentia, Muridae) in the Nile Valley. Zeitschrift fur Säugetierkunde 63: 104-113.

Fadda, C. & Corti, M. 2000. Three dimensional geometric study of the Ethiopian *Myomys-Stenocephalemys* complex (Murinae, Rodentia). Hystrix 10: 131-143.

Fadda, C., Castiglia, R., Colangelo, P., Corti, M., Machang'u, R., Makundi, R., Scanzani, A., Tesha, P., Verheyen, W. & Capanna, E. 2001a. The rodent fauna of Tanzania: a cytotaxonomic report from the Maasai Steppe. Rend. Fis. Acc. Lincei: 12: 29-49.

Fadda, C., Corti, M. & Verheyen, E. 2001b. Molecular phylogeny of the *Myomys-Stenocephalemys* complex and its relationships with related African genera. Biochemical Systematics and Ecology 29: 585-596.

Faulkes, C. & Bennett, N. 2013. Plasticity and constraints on social evolution in African mole-rats: ultimate and proximate factors. Philosophical Transactions of the Royal Society B 368: 20120347.

Faulkes, C. G., Verheyen, E., Verheyen, W., Jarvis, J. U. & Bennett, N. C. 2004. Phylogeographical patterns of genetic divergence and speciation in African mole-rats (Family: Bathyergidae). Molecular Ecology 13: 613-629.

Faulkes, C., Mgode, G., Le Comber, S. & Bennett, N. 2010. Cladogenesis and endemism in Tanzanian molerats, genus *Fukomys*: (Rodentia Bathyergidae): a role for tectonics? Biological Journal of the Linnean Society 100: 337-352.

Faulkes, C. G., Bennett, N. C., Cotterill, F. P. D., Stanley, W., Mgode, G. F. & Verheyen, E. 2011. Phylogeography and cryptic diversity of the solitary-dwelling silvery mole-rat, genus *Heliophobius* (family: Bathyergidae). Journal of Zoology, London 285: 324-338.

Fichet-Calvet, E., Jomaa, I., Ben Ismail, R. & Ashford, R. W. 1999. Reproduction and abundance of the fat sand rat (*Psammomys obesus*) in relation to weather conditions in Tunisia. Journal of Zoology, London 248: 15-26.

Fichet-Calvet, E., Jomaa, I., Ben Ismail, R. & Ashford, R. W. 2000. Patterns of infection of haemoparasites in the fat sand rat, *Psammomys obesus*, in Tunisia, and effect on the host. Annals of Tropical Medicine & Parasitology 94: 55-68.

Fichet-Calvet, E., Lecompte, E., Koivogui, L., Soropogui, B., Doré, A., Kourouma, F., Sylla, A., Daffis, S., Koulémou, K. & ter Meulen, J. 2007. Fluctuation of abundance and Lassa virus prevalence in *Mastomys natalensis* in Guinea, West Africa. Vector-Borne and Zoonotic Diseases 7: 119-128.

Fichet-Calvet, E., Lecompte, E., Veyrunes, F., Barriere, P., Nicolas, V. & Koulemou, K. 2012. Diversity and dynamics in a community of small mammals in coastal Guinea, West Africa. Belgian Journal of Zoology 139: 93-102.

Fick, L. G., Kucio, T. A., Fuller, A., Matthee, A. & Mitchell, D. 2009. The relative roles of the parasol-like tail and burrow shuttling in thermoregulation of free-ranging Cape ground squirrels, *Xerus inauris*. Comparative Biochemistry and Physiology Part A 152: 334-340.

Fiedler, L. A. 1988. Rodent problems in Africa. In: Rodent pest management (I. Prakash, ed.), pp. 35-55. CRC Press Inc., Boca Ratos.

Fiedler, L. A. 1994. Rodent Pest Management in Eastern Africa. FAO Plant Protection Paper 123: 1-83.

Fitzgibbon, C. D., Leirs, H. & Verheyen, W. 1995. Distribution, population dynamics and habitat use of the lesser pouched rat, *Beamys hindei*. Journal of Zoology 236: 499-512.

Fjeldsa, J. & Burgess, N. D. 2008.The coincidence of biodiversity patterns and human settlement in Africa. African Journal of Ecology 46: 33-42.

Flynn, L. J. & Winkler, A. 1994. Dispersalist implications of *Paraulacodus indicus*: a South Asian rodent of African affinity. Historical Biology 9: 223-235.

Flynn, L. J., Winkler, A. J., Jacobs, L. L. & Downs W. 2003. Tedford's gerbils from Afghanistan. Bull. Am. Mus.Nat. History, 13, 279: 603-624.

Fülling, O. 1992. Ergahzande angaben uber gaumenfaltenmuster vonhagetierun (Mammalia: Rodentia) aus Kamerun. Bonner Zoologische Beiträge 43: 415-421.

Gautier-Hion, A., Emmons, L. H. & Dubost, G. 1980. A comparison of the diets of three major groups of primary consumers of Gabon (primates, squirrels and ruminants). Oecologia 45: 182-189.

Gautun, J. & Bellier, L. 1970. Contribution a la connaissance de *Hylomyscus baeri* Heim de Balsac et Aellen, 1965. Mammalia 34: 248-251.

Gautun, J. & Petter, F. 1972. Liste préliminaire des rongeurs d'une savane centrale de Côte d'Ivoire: Foro-Foro. Mammalia 36: 634-637.

Gautun, J., Sankhon, I. & Tranier, M. 1986. Nouvelle contribution a la connaissance de rongeurs du massif guineen des monts Nimba (Afrique occidentale). Systématique et apercu quantitatif. Mammalia 50: 205-217.

Ge, D. Y., Liu, X., Lv, X. F., Zhang, Z. Q., Xia, L. & Yang, Q. S. 2013. Historical biogeography and body form evolution of ground squirrels (Sciuridae: Xerinae). Evolutionary Biology doi: 10.1007/s11692-013-9250-7.

Genest, H. & Petter, F. 1973. Les *Taterillus* de Republique Centrafricaine (Rongeurs, Gerbillides). Mammalia 37: 66-75.

Genest-Villard, H. 1967. Revision du genre *Cricetomys* (Rongeurs, Cricetidae). Mammalia 32: 639-656.

Genest-Villard, H. 1972. Contribution a l'écologie et l'éthologie d'un petit rongeur arboricole, *Thamnomys rutilans*, en Republique Centrafricaine. Mammalia 36: 543-578.

Genest-Villard, H. 1973. Données ecologiques sur un 'Leggada' des savanes intra-forestieres du Centre africain: *Mus oubanguii* (Rongeurs, Murides). Mammalia 37: 220-230.

Genest-Villard, H. 1978. Revision systematique du genre *Graphiurus* (Rongeurs, Gliridae). Mammalia 42: 391-426.

Genest-Villard, H. 1980. Regime alimentaire des rongeurs myomorphes de forêt equatoriale (région de M'Baiki, Republique Centrafricaine). Mammalia 44: 423-484.

Gentry, A.W. 1987. Pliocene Bovidae from Laetoli. In: Laetoli: A Pliocene Site in Northern Tanzania (M. D. Leakey & J. M. Harris, eds.). Clarendon Press.

George, W. 1974. Notes on the ecology of gundis (Ctenodactylidae). Symposia of the Zoological Society of London 34: 143-160.

George, W. 1978a. Reproduction in female gundis (Rodentia: Ctenodactylidae). Journal of Zoology, London 185: 57-71.

George, W. 1978b. Combs, fur and coat care related to habitat in the Ctenodactylidae (Rodentia). Zeitschrift fur Säugetierkunde 43: 143-155.

George, W. 1979. The chromosomes of the hystricomorphous family Ctenodactylidae (Rodentia:? Sciuromorpha) and their bearing on the relationships of the four living genera. Zoological Journal of the Linnean Society 65: 261-280.

George, W. 1980. A study in hystricomorph rodent relationships: the karyotypes of *Thryonomys gregorianus, Pedetes capensis* and *Hystrix cristata*. Zoological Journal of the Linnean Society 68: 361-372.

George, W. 1981a. Species-typical calls in the Ctenodactylidae (Rodentia). Journal of Zoology, London 195: 39-52.

George, W. 1981b. The diet of *Petromus typicus* (Petromuridae, Rodentia) in the Augrabies Falls National Park. Koedoe 24: 159-167.

George, W. 1985. Cluster analysis and phylogenetics of five species of Ctenodactylidae (Rodentia). Mammalia 49: 53-63.

George, W. 1988. *Massoutiera mzabi* (Rodentia, Ctenodactylidae) in a climatological trap. Mammalia 52: 331-338.

George, W. & Crowther, G. 1981. Space partitioning between two small mammals in a rocky desert. Biological Journal of the Linnean 15: 195-200.

George, W. & Weir, B. J. 1974. Hystricomorph chromosomes. In: The Biology of Hystricomorph Rodents (I. W. Rowlands & B. J. Weir, eds). Symposia of the Zoological Society of London 34: 79-108.

Geraads, D. 1995. Rongeurs et insectivores (Mammalia) du Pliocene final de Ahl Al Oughlam (Casablanca, Maroc). Geobios 28: 99-115.
Geraads, D., 1998a. Rongeurs du Miocène supérieur de Ch'orora (Ethiopie): Cricetidae, Rhizomyidae, Phiomyidae, Thryonomyidae, Sciuridae. Paleovertebrata 27: 203-216.
Geraads, D. 1998b. Les rongeurs du Mio-Pliocène de Lissasfa (Casablanca, Maroc). Geobios, 31 : 229-245.
Geraads, D. 2001. Rongeurs du Miocène Superieur de Ch'orora, Ethiopie: Murinae, Dendromurinae et conclusion. Paleovertebrata 30: 89-109.
Germain, M. & Petter, F. 2013. Capture a Yaounde (Cameroon) de *Colomys goslingi* (Rodentia, Muridae). Mammalia 34: 722-723.
Gharaibeh, B. M. & Jones, C. 1996. *Myosciurus pumilio*. Mammalian Species 523: 1-3.
Gippoliti, S. & Amori, G. 2011. A new species of mole-rat (Rodentia, Bathyergidae) from the Horn of Africa. Zootaxa 2918: 39-46.
Gomes Rodrigues, H. G., Solé, F., Charles, C., Tafforeau, P., Vianey-Liaud, M. & Viriot, L. 2012. Evolutionary and biological implications of dental mesial drift in rodents: the case of the Ctenodactylidae (Rodentia, Mammalia). PLoS ONE 7: e50197.
Gordon, D. H. 1978. Distribution of sibling species of the *Praomys (Mastomys) natalensis* group in Rhodesia (Mammalia: Rodentia). Journal of Zoology, London 186: 397-401.
Gordon, D. H. 1986. Extensive chromosomal variation in the pouched mouse, *Saccostomus campestris* (Rodentia, Cricetidae) from southern Africa: a preliminary investigation of evolutionary status. Cimbebasia 8: 37-47.
Gordon, D. H. 1987. Discovery of another species of tree rat. Transvaal Museum Bulletin 22: 30-32.
Gordon, D. H. 1991. Chromosomal Variation in the Water Rat *Dasymys incomtus* (Rodentia: Muridae). Journal of Mammalogy 72: 411-414.
Gordon, D. H. & Rautenbach, I. L. 1980. Species complexes in medically important rodents: chromosome studies of *Aethomys, Tatera* and *Saccostomus*. South African Journal of Science 76: 559-561.
Gouat, P., Gouat, J. & Coulon, J. 1984. Répartition et habitat de *Massoutiera mzabi* (Rongeur Cténodactylidé) en Algérie. Mammalia 48: 351-362.
Granjon, L. 2005. Morphological and morphometrical analyses of three cryptic species of *Tatera* Lataste, 1882 (Rodentia: Muridae) from West Africa. Belgian Journal of Zoology 135: 97-102.
Granjon, L. & Dobigny, G. 2003. The importance of cytotaxonomy in understanding the biogeography of African rodents: Lake Chad murids as an example. Mammal Review 33: 77-91.
Granjon, L. & Cosson, J. 2008. Social relationships in *Mastomys huberti* as deduced from field and genetic analyses of multiple capture data. Mammalia 72: 161-168.
Granjon, L. & Duplantier, J-M. 2009. Les rongeurs de l'Afrique sahelo-soudanienne. IRD, Marseille.
Granjon, L. & Duplantier, J. 2011. Guinean biodiversity at the edge: rodents in forest patches of southern Mali. Mammalian Biology 76: 583-591.
Granjon, L. & Traore, M. 2007. Prey selection by barn owls in relation to small-mammal community and population structure in a Sahelian agro-ecosystem. Journal of Tropical Ecology 23: 199.
Granjon, L. & Schlitter, D. 2008. *Ammodillus imbellis*. In: IUCN 2013. IUCN Red List of Threatened Species. Version 2013.2. <www.iucnredlist.org>.
Granjon, L., Duplantier, J., Catalan, J. & Britton-Davidian, J. 1992. Karyotypic data on rodents from Senegal. Israel Journal of Zoology 38: 263-276.
Granjon, L., Duplantier, J., Catalan, J., Britton-Davidian, J. & Bronner, G. 1996. Conspecificity of *Mastomys natalensis* (Rodentia: Muridae) from Senegal and South Africa: evidence from experimental crosses, karyology and biometry. Mammalia 60: 697-706.
Granjon, L., Duplantier, J-M., Catalan, J. & Britton-Davidian, J. 1997. Systematics of the genus *Mastomys* (Thomas, 1915) (Rodentia: Muridae) a review. Belgian Journal of Zoology 127: 7-18.
Granjon, L., Bonnet, A., Hamdine, W. & Volobouev, V. 1999. Reevaluation of the taxonomic status of North African gerbils usually referred to as *Gerbillus pyramidum* (Gerbillinae, Rodentia): chromosomal and biometrical data. Zeitschrift fur Säugetierkunde 64: 298-307.

Granjon, L., Aniskin, V. M., Volobouev, V. & Sicard, B. 2002a. Sand-dwellers in rocky habitats: a new species of *Gerbillus* (Mammalia: Rodentia) from Mali. Journal of Zoology, London 256: 181-190.
Granjon, L., Bruderer, C., Cosson, J-E., Dia, A. T. & Colas, F. 2002b. The small mammal community of a coastal site of south-west Mauritania. African Journal of Ecology 40: 10-17.
Granjon, L., Houssin, C., Lecompte, E., Angaya, M., Cesar, J., Cornette, R., Dobigny, G. & Denys, C. 2004. Community ecology of the terrestrial small mammals of Zakouma National Park, Chad. Acta Theriologica 49: 215-234.
Granjon, L., Cosson, J., Quesseveur, E. & Sicard, B. 2005a. Population dynamics of the multimammate rat *Mastomys huberti* in an annually flooded agricultural region of central Mali. Journal of Mammalogy 86: 997-1008.
Granjon, L., Ba, K., Daouda, I. H. & Duplantier, J-M. 2005b. New data on chromosomes from murid rodents of Benin – the karyotype of *Myomys derooi*. Belgium Journal of Zoology 69: 421-426.
Granjon, L., Colangelo, P., Tatard, C., Colyn, M., Nicolas, G. D. &. V., Dobigny, G. & Nicolas, V. 2012. Intrageneric relationships within *Gerbilliscus* (Rodentia, Muridae, Gerbillinae), with characterization of an additional West African species. Zootaxa 3325: 1-25.
Gratz, N. G. 1997. The burden of rodent-borne diseases in African south of the Sahara. Belgian Journal of Zoology 127: 71-84.
Graur, D., Hide, W. A., Zharkikh, A. & Li, W-H. 1992. The biochemical phylogeny of guinea-pigs and gundis, and the paraphyly of the order rodentia. Comparative Biochemistry and Physiology. 101B: 495-498.
Green, C., Keogh, H., Gordon, D., Pinto, M. & Hartwig, E. 1980. The distribution, identification, and naming of the *Mastomys natalensis* species complex in southern Africa (Rodentia: Muridae). Journal of Zoology, London 192: 17-23.
Greenwood, M. 1958. Fossil Hystricoidea from the Makapan Valley, Transvaal: *Hystrix makapanensis* nom. nov. for *Hystrix* major Greenwood. Annals and Magazine of Natural History 13: 365.
Groves, C. P. 2013. The nature of species: A rejoinder to Zachos et al., Mammalian Biology 78: 7-9.
Groves, C. P. & Grubb, P., 2011. Ungulate Taxonomy. Johns Hopkins University Press, Baltimore.
Grubb, P. 1982. Systematics of sun-squirrels (*Heliosciurus*) in eastern Africa. Bonner zoologische Beiträge 33: 191-204.
Grubb, P., Jones, T. S., Davies, A. G., Edberg, E., Starin, E. D. & Hill, J. E. 1998. Mammals of Ghana, Sierra Leone, and the Gambia. Trendine Press, Zennor, Cornwall.
Habtamu, T. & Bekele, A. 2013. Species composition, relative abundance and habitat association of small mammals along the altitudinal gradient of Jiren Mountain, Jimma, Ethiopia. African Journal of Ecology 51: 37-46.
Haile-Selassie, Y., Wodegabriel, G., White, T. D., Bernor, R. L., Degusta, D., Renne, P. R., Hart, W. K., Vrba, E., Stanley A. & Howell F. C. 2004. Mio-Pliocene mammals from the Middle Awash, Ethiopia. Geobios 37: 536-532.
Hallett, A. F. & Meester, J. 1971. Early postnatal development of the South African hamster *Mystromys albicaudatus*. Zoologica Africana 6: 221-228.
Hamilton, R. & Van Couvering J. 1977. Lower Miocene mammals from South West Africa. Bull. Transvaal Museum 2: 9-11.
Hanney, P. 1965. The Muridae of Malawi (Africa: Nyasaland). Journal of Zoology, London 146: 577-633.
Hanney, P. & Morris, B. 1962. Some observations upon the Pouched Rat in Nyasaland. Journal of Mammalogy 43: 238-248.
Happold, D. C. D. 1966. The mammals of Jebel Marra, Sudan. Journal of Zoology, London 149: 126-136.
Happold, D. C. D. 1967. Biology of the jerboa, *Jaculus jaculus butleri* (Rodentia, Dipodidae), in the Sudan. Journal of Zoology, London 151: 257-275.
Happold, D. C. D. 1969. The mammalian fauna of some jebels in the northern Sudan. Journal of Zoology, London 157: 133-145.
Happold, D. C. D. 1970. Reproduction and development of the Sudanese jerboa, *Jaculus jaculus butleri* (Rodentia, Dipodidae). Journal of Zoology, London 162: 505-515.

Happold, D. C. D. 1977. A population study on small rodents in the tropical rain forest of Nigeria. Revue d'Ecologie (La Terre et la Vie) 31: 385-458.

Happold, D. C. D. 1978. Reproduction, growth and development of a West African forest mouse, *Praomys tullbergi* (Thomas). Mammalia 42: 73-95.

Happold, D. C. D. 1987. The Mammals of Nigeria. Oxford University Press, Oxford, UK.

Happold, D. C. D. (ed.) 2013. Mammals of Africa. Vol. III: Rodents, Hares and Rabbits. Bloomsbury Publishing, London.

Happold, D. C. D. & Happold, M. 1989a. Biogeography of montane small mammals in Malawi, Central Africa. Journal of Biogeography 16: 353-367.

Happold, D. C. D. & Happold, M. 1989b. Demography and habitat selection of small mammals on Zomba Plateau, Malawi. Journal of Zoology, London 219: 581-605.

Happold, D. C. D. & Happold, M. 1997. Conservation of mammals on a tobacco farm on the highlands of Malawi. Biodiversity and Conservation 6: 837-852.

Happold, M. 2005. A new species of *Myotis* (Chiroptera: Vespertilionidae) from central Africa. Acta Chiropterologica: 1-13.

Harper, D. M., Mavuti, K. M. & Muchiri, S. M. 1990. Ecology and management of Lake Naivasha, Kenya, in relation to climatic change, alien species' introductions, and agricultural development. Environmental Conservation 17: 328-336.

Hatt, R. T. 1934. Fourteen hitherto unrecognized African rodents. American Museum Novitates 708: 1-15.

Hatt, R. T. 1940. Lagomorpha and Rodentia other than Sciuridae, Anomaluridae and Idiuridae collected by the American Museum Congo Expedition. Bulletin of the American Museum of Natural History 76: 457-604.

Hautier, L., Michaux, J., Marivaux, L. & Vianey-Liaud, M. 2008. Evolution of the zygomasseteric construction in Rodentia, as revealed by a geometric morphometric analysis of the mandible of *Graphiurus* (Rodentia, Gliridae). Zoological Journal of the Linnean Society 154: 807-821.

Hayman, R. W. 1962. The occurrence of *Delanymys brooksi* (Rodentia, Muridae) in the Congo. Bulletin de l'Institut royal des Sciences naturelles de Belgique 38: 1-4.

Hayman, R. W. 1963. Mammals from Angola, mainly from the Lunda District. Publicacoes Culterais Museu do Dondo 66: 83-139.

Hayman, R. W. 1966. On the affinities of *Nilopegamys plumbeus* Osgood. Annales du Musée Royal de l'Afrique Centrale, Sciences Zoologiques 144: 29-38.

Hayssen, V. 2008. Reproductive effort in squirrels: ecological, phylogenetic, allometric, and latitudinal patterns. Journal of Mammalogy 89: 582-606.

Heaney, L. R. 1985. Systematics of Oriental Pygmy Squirrels of the genera *Exilisciurus* and *Nannosciurus* (Mammalia: Sciuridae). Miscellaneous Publications, Museum of Zoology, University of Michigan 170: 1-58.

Heller, E. 1912. New rodents from British East Africa. Smithsonian Miscellaneous Collections 59: 1-20.

Herron, M. D. & Waterman, J. M. 2004. *Xerus erythropus*. Mammalian Species 748: 1-4.

Herron, M. D., Waterman, J. M. & Parkinson, C. L. 2005. Phylogeny and historical biogeography of African ground squirrels: the role of climate change in the evolution of *Xerus*. Molecular Ecology 14: 2773-2788.

Herzig-Straschil, B. 1978. On the biology of *Xerus inauris* (Zimmermann, 1780) (Rodentia, Sciuridae). Zeitschrift fur Säugetierkunde 43: 262-278.

Herzig-Straschil, B. & Herzig, A. 1989. Biology of *Xerus princeps* (Rodentia, Sciuridae). Madoqua 16: 41-46.

Herzig-Straschil, B., Herzig, A. & Winkler, H. 1991. A morphometric analysis of the skulls of *Xerus inauris* and *Xerus princeps* (Rodentia; Sciuridae). Zeitschrift fur Säugetierkunde 56: 177-187.

Hill, J. E. & Carter, T. D. 1941. The mammals of Angola, Africa. Bulletin of the American Museum of Natural History 78: 1-211.

Hima, K., Thiam, M., Catalan, J., Gauthier, P., Duplantier, J. M., Piry, S., Sembène, M., Britton-Davidian, J., Granjon, L. & Dobigny, G. 2011. Extensive Robertsonian polymorphism in the African rodent *Gerbillus nigeriae*: geographic aspects and meiotic data. Journal of Zoology, London 284: 276-285.

Hlusko, L., 2007. Earliest evidence for *Atherurus* and *Xenohystrix* (Hystricidae, Rodentia) in Africa from the Late Miocene site of Lemudong'o, Kenya. Kirtlandia 56: 86-91.

Hoffman, R. S., Anderson, C., Thorington, R. W. & Heaney, L. R. 1993. Family Sciuridae. In: Mammal Species of the World: A Taxonomic and Geographic Reference (2nd edn) (D.E. Wilson & D. M. Reeder, eds). Smithsonian Institute Press, Washington, DC.
Hoffmann, M., Grubb, P., Groves, C. & Hutterer, R. 2009. A synthesis of African and western Indian Ocean Island mammal taxa (Class: Mammalia) described between 1988 and 2008: an update to Allen (1939) and Ansell (1989). Zootaxa 2205: 1-39.
Holden, M. 1996. Systematic revision of Sub-Saharan African dormice (Rodentia: Myoxidae: *Graphiurus*), Part 1: an introduction to the generic revision, and a revision of *Graphiurus surdus*. American Museum Novitates 3157: 1-44.
Holden, M. 2003. Family Gliridae. In: Mammal Species of the World: A Taxonomic and Geographic Reference (2nd edn) (D. E. Wilson and D. Reeder, eds). John Hopkins University Press, Baltimore.
Holden, M. 2005. Family Gliridae. In: Mammal Species of the World: A Taxonomic and Geographic Reference (3rd edn) (D. E. Wilson and D. Reeder, eds). John Hopkins University Press, Baltimore.
Holden, M. & Musser, G. 2005. Family Dipodidae. In: Mammal Species of the World: A Taxonomic and Geographic Reference (3rd edn) (D. E. Wilson and D. Reeder, eds). John Hopkins University Press, Baltimore.
Holden, M. E. & Levine, R. S. 2009. Chapter 9. Systematic Revision of Sub-Saharan African Dormice (Rodentia: Gliridae: Graphiurus) Part II: Description of a New Species of *Graphiurus* from the Central Congo Basin, Including Morphological and Ecological Niche Comparisons with *G. crassicaudatus* and *G. lorraineus*. Bulletin of the American Museum of Natural History 331: 314-355.
Hollister, N. 1919. East African mammals in the United States National Museum. Part 2, Rodentia, Lagomorpha, and Tubulidentata. Bulletin of the United States National Museum 99: 1-184.
Holroyd, P. A. & Stevens, N. J. 2009. Differentiation of *Phiomys andrewsi* from *Lavocatomys aequatorialis* (n. gen., n. sp.) (Rodentia: Thryonomyoidea) in the Oligo-Miocene Interval on Continental Africa. Journal of Vertebrate Paleontology 29: 1331-1334.
Honacki, J. H., Kinman, K. E. & Koeppl, J. W. 1982. Mammalian Species of the World: A Taxonomic and Geographic Reference. The Association of Systematics Collections, Lawrence, Kansas.
Honeycutt, R. L., Edwards, S. V., Nelson, K. & Nevo, E. 1987. Mitochondrial DNA variation and the phylogeny of African mole-rats (Rodentia: Bathyergidae). Systematic Zoology 36: 280-292.
Honeycutt, R. L., Allard, M. W., Edwards, S. V. & Schlitter, D. A. 1991. Systematics and evolution of the family Bathyergidae. In: The Biology of the Naked Mole-Rat (P. W. Sherman, J. U. M. Jarvis & R. D. Alexander, eds). Princeton University Press, New York.
Horner, D. S., Lefkimmiatis, K., Reyes, A., Gissi, C., Saccone, C. & Pesole, G. 2007. Phylogenetic analyses of complete mitochondrial genome sequences suggest a basal divergence of the enigmatic rodent *Anomalurus*. BMC Evolutionary Biology 7: 16.
Hubbard, C. A. 1970. A new species of *Tatera* from Tanzania with a description of its life history and habits studied in captivity. Zoologica Africana 5: 237-247.
Hubert, B. 1978a. Modern rodent fauna of the lower Omo Valley, Ethiopia. Bulletin of Carnegie Museum of Natural History 6: 109-112.
Hubert, B. 1978b. Revision of the genus *Saccostomus* (Rodentia, Cricetomyinae) with new morphological and chromosomal data from specimens from the lower Omo Valley, Ethiopia. Bulletin of Carnegie Museum of Natural History 6: 48-52.
Hubert, B. 1978c. Caryotype de *Gerbillus pulvinatus* Rhoads, 1896 (Rongeurs, Gerbillides) de la vallée de l'Omo (Ethiopie). Mammalia 42: 225-227.
Hubert, B. & Adam, F. 1983. The regulation of the population dynamics of two Sahelian rodents in Senegal: an hypothesis. Annales du Musée royale d'Afrique centrale, Sciences Zoologique 237: 193-201.
Hubert, B., Adam, F. & Poulet, A. 1973. Liste préliminaire des rongeurs du Senegal. Mammalia 37: 76-87.
Huchon, D., Catzeflis, F. M. & Douzery, E. J. 2000. Variance of molecular datings, evolution of rodents and the phylogenetic affinities between Ctenodactylidae and Hystricognathi. Proceedings of the Royal Society B: Biological Sciences 267: 393-402.

Huchon, D., Madsen, O., Sibbald, M. J. J. B., Ament, K., Stanhope, M. J., Catzeflis, F., de Jong, W. W. & Douzery, E. J. P. 2002. Rodent phylogeny and a timescale for the evolution of Glires: evidence from an extensive taxon sampling using three nuclear genes. Molecular Biology and Evolution 19: 1053-1065.
Huchon D., Chevret P., Jordan U., Kilpatrick C.W., Ranwez V., Jenckins P., Brosius J., Schmitz J. 2007. Multiple molecular evidences for a living mammal fossil. Proceedings of the National Academy of Science 104: 7495-7499.
Hurst, Z. M., McCleery, R. A., Collier, B. A., Fletcher, R. J., Silvy, N. J., Taylor, P. J. & Monadjem, A. 2013. Dynamic Edge Effects in Small Mammal Communities across a Conservation-Agricultural Interface in Swaziland. PLoS ONE 8: e74520.
Hutterer, R. & Dieterlen, F. 1984. Zwei neue arten der gattung *Grammomys* aus Athiopien und Kenia (Mammalia: Muridae). Stuttgarter Beiträge zur Naturkunde 374: 1-18.
Hutterer, R. & Dieterlen, F. 1986. Zur verbreitung und variation von *Desmodilliscus braueri* Wettstein, 1916 (Mammalia: Rodentia). Annalen des Naturhistorischen Museums in Wien 88/89 B: 213-221.
Hutterer, R., Dieterlen, F. & Nikolaus, G. 1992. Small mammals from forest islands of eastern Nigeria and adjacent Cameroon, with systematical and biogeographical notes. Bonner Zoologishe Beiträge 43: 393-414.
Hutterer, R. & Peters, G. 2001. The vocal repertoire of *Graphiurus parvus*, and comparisons with other species of dormice. Trakya University Journal of Scientific Research Series B 2: 69-74.
Ingram, C. M., Burda, H. & Honeycutt, R. L. 2004. Molecular phylogenetics and taxonomy of the African mole-rats, genus *Cryptomys* and the new genus *Coetomys* Gray, 1864. Molecular Phylogenetics and Evolution 31: 997-1014.
Isaacson, M. 1975. The ecology of *Praomys (Mastomys) natalensis* in southern Africa. Bulletin W.H.O. 52: 629-636.
Isaacson, M., Arntzen, L. & Taylor, P. 1981. Susceptibility of members of the *Mastomys natalensis* species complex to experimental infection with *Yersinia pestis*. The Journal of Infectious Diseases 44: 80.
Jackson, T. P. 1998a. The diurnal activity of Brants' Whistling Rat (*Parotomys brantsii*): the effect of seasonal and physical conditions. Transactions of the Royal Society of South Africa 53: 227-236.
Jackson, T. P. 1998b. Seasonal variation in the diet and foraging behaviour of Brants' Whistling Rat, *Parotomys brantsii*, in northern Namaqualand. South African Journal of Zoology 33: 37-41.
Jackson, T. P. 1999. The social organisation and breeding system of Brants' Whitling Rat, *Parotomys brantsii*. Journal of Zoology, London 247: 323-331.
Jackson, T. P. 2000. Adaptation to living in an open arid environment; lessons from the burrow structure of the two southern African whistling rats, *Parotomys brantsii* and *P. littledalei*. Journal of Arid Environments 46: 345-355.
Jackson, P. T. 2001. Factors influencing the food collection behaviour of a central place forager: Brants' Whistling Rat, *Parotomys brantsii*. Journal of Zoology, London 255: 15-23.
Jackson, P. T., Roper, T. J., Conradt, L., Jackson, M. J. & Bennett, N. C. 2002. Alternative refuge strategies and their relation to thermophysiology in two sympatric rodents, *Parotomys brantsii* and *Otomys unisulcatus*. Journal of Arid Environments 51: 21-34.
Jaeger, J. J. 1976. Les Rongeurs (Mammalia, Rodentia) du Pléistocène inférieur d'Olduvai Bed I (Tanzanie). Iè partie: les Muridés. In: Fossil Vertebrates of Africa (R. J. G. Savage, S. C. Coryndon, eds.). Academic Press, London.
Jaeger, J. -J. 1977. Les rongeurs du Miocène moyen et supérieur du Maghreb. Palaeovertebrata 8:1-166.
Jaeger, J. J., Denys, C., Coiffait, B., 1985. New Phiomorpha and Anomaluridae from the late Eocene of North West Africa: phylogenetic implications. In: Evolutionary relationships among rodents: a multidiciplinary analysis (W. P. Luckett & J. L. Hartenberger eds). Plenum Press, New York.
Jaeger, J. J. 1975. Evolution des rongeurs du Miocène à l'actuel en Afrique nord-occidentale. Doctoral thesis, Universite des Sciences et Techniques de Languedoc, Montpellier-II.
Janecek, L. L., Schlitter, D. A. & Rautenbach, I. L. 1991. A genic comparison of spiny mice genus *Acomys*. Journal of Mammalogy 72: 542-552.

Jansa, S. A. & Weksler, M. 2004. Phylogeny of muroid rodents: relationships within and among major lineages as determined by IRBP gene sequences. Molecular Phylogenetics and Evolution 31: 256-276.

Jansa, S. A. Goodman, S. M. & Tucker, P. K. 1999. Molecular phylogeny and biogeography of the native rodents of Madagascar (Muridae: Nesomyidae): a test of the single-origin hypothesis. Cladistics 15: 253-270.

Jansen van Vuuren, B. & Chown, S. L. 2007. Genetic evidence confirms the origin of the house mouse on sub-Antarctic Marion Island. Polar Biology 30: 327-332.

Jarvis, J. U. M. 1973. The structure of a population of mole-rats. *Tachyoryctes splendens* (Rodentia, Rhizomyidae). Journal of Zoology, London 171: 1-14.

Jarvis, J. U. M. 1981. Eusociality in a mammal: cooperative breeding in naked mole-rat colonies. Science 212: 571-573.

Jarvis, J. U. M. & Sherman, P. W. 2002. *Heterocephalus glaber*. Mammalian Species 706: 1-9.

Jones, C. 1971. Notes on the anomalurids of Rio Muni and adjacent areas. Journal of Mammalogy 52: 568-572.

Jori, F., Lopez-Béjar, M. & Houben, P. 1998. The biology and use of the African brush-tailed porcupine (*Atherurus africanus*, Gray, 1842) as a food animal. A review. Biodiversity and Conservation 7: 1417-1426.

Julliot, C., Cajani, S. & Gautier-Hion, A. 1998. Anomalures (Rodentia, Anomaluridae) in Central Gabon: species composition, population densities and ecology. Mammalia 62: 9-21.

Kaleme, P. K., Bates, J. M., Belesi, H. K., Bowie, R. C. K., Gambalemoke, M., Kerbis Peterhans, J., Michaux, J. Mwanga, J. M., Ndara, B. R., Taylor, P. J. & Jansen van Vuuren, B. 2011. Origin and putative colonization routes for invasive rodent taxa in the Democratic Republic of Congo. African Zoology 46: 133-145.

Kalderon, B., Gutman, A., Levy, E., Shafrir, E. & Adler, J.H. 1986. Characterization of stages in development of obesity-diabetes syndrome in sand rat (*Psammomys obesus*). Diabetes 35: 717-724.

Kan Kouassi, S. K., Nicolas, V., Aniskine, V., Lalis, A., Cruaud, C., Couloux, A., Colyn, M., Dosso, M., Koivogui, L., Verheyen, E., Akoua-Koffi, C. & Denys, C. 2008. Taxonomy and biogeography of the African Pygmy mice, Subgenus *Nannomys* (Rodentia, Murinae, Mus) in Ivory Coast and Guinea (West Africa). Mammalia 72: 237-252.

Kasangaki, A., Kityo, R. & Kerbis, J. 2003. Diversity of rodents and shrews along an elevational gradient in Bwindi Impenetrable National Park, south-western Uganda. African Journal of Ecology: 1-9.

Katakweba, A. A. S., Mulungu, L. S., Eiseb, S. J., Mahlaba, T. A., Makundi, R. H., Massawe, A. W., Borremans, B. & Belmain, S. R. 2012. Prevalence of haemoparasites, leptospires and cocobacilli with potential for human infection in the blood of rodents and shrews from selected localities in Tanzania, Namibia and Swaziland. African Zoology. 47: 119-127.

Katuala, P. G. B., Kennis, J., Nicolas, V., Wendelen, W., Hulselmans, J., Verheyen, E., van Houtte, N., Dierckx, T., Dudu, A. M. & Leirs, H. 2008. The presence of *Praomys*, *Lophuromys*, and *Deomys* species (Muridae, Mammalia) in the forest blocks separated by the Congo River and its tributaries (Kisangani region, Democratic Republic of Congo). Mammalia 72: 223-228.

Kawalika, M., Burda, H. & Brüggert, D. 2001. Was Zambia a cradle of the genus *Cryptomys* (Bathyergidae, Rodentia)? A further new ancestral (?) species of *Cryptomys* from Zambia. In: African Small Mammals (C. Denys, L. Granjon & A. Poulet, eds). IRD Editions, Paris.

Keesing, F. 1998. Impacts of ungulates on the demography and diversity of small mammals in central Kenya. Oecologia 116: 381-389.

Keesing, F. & Crawford, T. 2001. Impacts of density and large mammals on space use by pouched mouse (*Saccostomus mearnsi*) in central Kenya. Journal of Tropical Ecology 17: 465-472.

Kennis, J., Nicolas, V., Hulselmans, J., Katuala, P., Wendelen, W., Verheyen, F., Dudu, A. & Leirs, H. 2011. The impact of the Congo River and its tributaries on the rodent genus *Praomys*: speciation orgin or range expansion limit? Zoological Journal of the Linnean Society 163: 983-1002.

Kennis, J., Crespin, L., Amundala, N. D., Dudu, A. M. & Leirs H. 2012. Survival and movement of the Congo forest mouse (*Deomys ferrugineus*): a comparison of primary rainforest and fallow land in Kisangani, Democratic Republic of Congo. African Zoology 47: 147-159.

Kenward, R., Hodder, K., Rose, R. & Walls, C. 1998. Comparative demography of red squirrels (*Sciurus vulgaris*) and grey squirrels (*Sciurus carolinensis*) in deciduous and conifer woodland. Journal of Zoology, London 244: 7-21.
Kerbis Peterhans, J. C. & Patterson, B. D. 1995. The Ethiopian water mouse *Nilopegamys* Osgood, with comments on semi-aquatic adaptations in the African Muridae. Zoological Journal of the Linnean Society 113: 329-349.
Kerbis Peterhans, J. C., Kityo, R. M., Stanley, W. T. & Austin, P. K. 1998. Small mammals along an elevational gradient in Rwenzori Mountains National Park, Uganda. In: Rwenzori Mountains National Park, Uganda: Exploration, Environment & Biology. Conservation, Management and Community Relations (H. Osmaston, J. Tukahirwa, C. Basalirwa & J. Nyakaana, eds). Makerere University, Uganda.
Kerley, G. I. H. 1992. Trophic status of small mammals in the semi-arid Karoo, South Africa. Journal of Zoology, London 226: 563-572.
Kerley, G. I. H., Knight, M. H. & Erasmus, T. 1990. Small mammal microhabitat use and diet in the southern Kalahari, South Africa. South African Journal of Wildlife Research 20: 124-126.
Kingdon, J. 1974. East African Mammals. Vol. II, Part B: Hares and Rodents. Academic Press, London.
Kingdon, J., Agwanda, B., Kinnaird, M., O'Brien, T., Holland, C., Gheysens, T., Boulet-Audet, M. & Vollrath, F. 2012. A poisonous surprise under the coat of the African crested rat. Proceedings of the Royal Society B: Biological Sciences 279: 675-680.
Kirsten, F., & von Maltitz, E. 2005. Technology transfer and promotion of ecologically-based and sustainable rodent control strategies in South Africa R 8190 (ZA 0506). Final technical report. Agricultural Research Council, Pretoria.
Knight, M. H. & Knight-Eloff, A. K. 1987. Digestive tract of the African giant rat *Cricetomys gambianus*. Journal of Zoology, London 213: 7-22.
Knight, M. H. & Skinner, J. D. 1981. Thermoregulatory, reproductive and behavioural adaptations of the big-eared desert mouse, *Malacothrix typica*, to its environment. Journal of Arid Environments 4: 137-145.
Kock, D. 1978. The identity of *Gerbillus bottai* Lastaste, 1882 (Mammalia: Rodentia) from Sennar, Sudan. Bulletin of the Carnegie Museum of Natural History 6: 31-37.
Kock, D., Ingram, C. M., Frabotta, L. J., Honeycutt, R. L. & Burda, H. 2006. On the nomenclature of Bathyergidae and *Fukomys* n. gen. (Mammalia; Rodentia). Zootaxa 1142: 51-55.
Kofron, C. P. 1987. Seasonal reproduction of the springhare (*Pedetes capensis*) in southeastern Zimbabwe. African Journal of Ecology 25: 185-194.
Koprowski, J. L. 1994. *Sciurus carolinensis*. Mammalian Species 480: 1-9.
Korn, H. 1987. Densities and biomasses of non-fossorial southern African savanna rodents during the dry season. Oecologia 72: 410-413.
Koufos, G. D., Kostopoulos, D. S., Vlachou, T. D. 2005. Neogene/Quaternary mammalian migrations in Eastern Mediterranean. Belgian Journal of Zoology 135: 181-190.
Krystufek, B. 2008. Description of a new thicket rat from Kenya: *Grammomys brevirostris* n. sp. Acta Zoologica Lituanica 18: 221-227.
Krystufek, B., Haberl, W., Baxter, R. M. & Zima, J. 2004. Morphology and karyology of two populations of the woodland dormouse *Graphiurus murinus* in the Eastern Cape, South Africa. Folia Zoologica 53: 339-350.
Krystufek, B., Baxter, R. M., Haberl, W., Zima, J. & Buzan, E. V. 2008. Systematics and biogeography of the Mozambique thicket rat, *Grammomys cometes*, in Eastern Cape Province, South Africa. Journal of Mammalogy 89: 325-335.
Kuhn, H-J. 1965. A provisional check-list of the mammals of Liberia. Senckenbergiana Biologica 46: 321-340.
Lalis, A., Evin, A. & Denys, C. 2009. Morphological identification of sibling species: the case of West African *Mastomys* (Rodentia: Muridae) in sympatry. Comptes Rendu Biologies 332: 480-488.
Lalis, A., Leblois, R., Lecompte, E., Denys, C., Meulen, ter, J. & Wirth, T. 2012. The impact of human conflict on the genetics of *Mastomys natalensis* and lassa virus in West Africa. PLoS ONE 7: e37068.
Lamb, J., Abdel-Rahman, E., Ralph, T., Fenton, M., Naidoo, A., Richardson, E., Denys, C., Naidoo, T., Buccas, W., Kajee, H., Hoosen, N., Mallett, D. & Taylor, P. 2006. Phylogeography of southern and northeastern African populations of *Otomops martiensseni* (Chiroptera: Molossidae). Durban Museum Novitates: 1-12.

Lamb, J., Downs, S., Eiseb, S. & Taylor, P. J. 2014. Increased geographic sampling reveals considerable new genetic diversity in the morphologically conservative African Pygmy Mice (genus *Mus*: subgenus *Nannomys*). Mammalian Biology 79: 24-35.

Laurance, W., Croes, B., Guissouegou, N., Buij, R., Dethier, M. & Alonso, A. 2008. Impacts of roads, hunting, and habitat alteration on nocturnal mammals in African rainforests. Conservation Biology 22: 721-732.

Lavocat, R. 1973. Les rongeurs du Miocène d'Afrique Orientale. I. Miocène inférieur. Mémoires et Travaux de l'Ecole Pratique des Hautes Etudes, Institut de Montpellier 1: 1-284.

Lavrenchenko, L. A. 2003. A contribution to the systematics of *Desmomys* Thomas, 1910 (Rodentia, Muridae) with the description of a new species. Bonner Zoologie Beiträge 50: 313-327.

Lavrenchenko, L. A. & Verheyen, E. 2005. An assessment of the systematics of the genus *Desmomys* Thomas, 1910 (Rodentia Muridae) using mitochondrial DNA sequences. African Biodiversity: 363-369.

Lavrenchenko, L. A., Milishnikov, A. N., Aniskin, V. M., Warshavsky, A. A. & Woldegabriel Gebrekidan. 1997. The genetic diversity of small mammals of the Bale Mountains, Ethiopia. Sinet: Ethiopian Journal of Science 20: 213-233.

Lavrenchenko, L., Verheyen, W. & Hulselmans, J. 1998a. Systematic and distributional notes on the *Lophuromys flavopunctatus* Thomas, 1888 species-complex in Ethiopia (Muridae, Rodentia). Bulletin de l'Institut Royal des Sciences Naturelles de Belgique 68: 199-214.

Lavrenchenko, L. A., Likhnova, O. P., Baskevish, M. I. & Bekele, A. 1998b. Systematics and distribution of *Mastomys* (Muridae, Rodentia) from Ethiopia, with the description of a new species. Zeitschrift fur Säugetierkunde 63: 37-51.

Lavrenchenko, L. A., Milishnikov, A. N., Aniskin, V. M. & Warshavsky, A. A. 1999. Systematics and phylogeny of the genus *Stenocephalemys* Frick, 1914 (Rodentia, Muridae): a multidisciplinary approach. Mammalia 63: 475-494.

Lavrenchenko, L. A., Verheyen, W. N., Verheyen, E., Hulselmans, J. & Leirs, H. 2007. Morphometric and genetic study of Ethiopian *Lophuromys flavopunctatus* Thomas, 1888 species complex with description of three new 70-chromosomal species (Muridae, Rodentia). Bulletin de l'Institut Royal des Sciences Naturelles de Belgique 77: 77-117.

Lavrenchenko, L. A., Kruskop, S. V., Bekele, A., Belay, G., Morozov, P. N., Ivlev, Y. F. & Warshavsky, A. A. 2010. Mammals of the Babille Elephant Sanctuary (Eastern Ethiopia). Russian Journal of Theriology 9: 47-60.

Lavrenchenko, L. A., Nadjafova, R.S. & Bulatova, N. S. 2011. Three new karyotypes extend a Robertsonian fan in Ethiopian spiny mice of the genus *Acomys* I. Geoffroy, 1838 (Mammalia, Rodentia). Comparative Cytogenetics 5: 423–431.

Lay, D. M. 1983. Taxonomy of the genus *Gerbillus* (Rodentia: Gerbillinae) with comments on the applications of generic and subgeneric names and an annotated list of species. Zeitschrift fur Säugetierkunde 48: 329-354.

Le Roux, A., Jackson, T. P. & Cherry, M. I. 2002. Differences in alarm vocalizations of sympatric populations of the whistling rats, *Parotomys brantsii* and *P. littledalei*. Journal of Zoology, London 257: 189-194.

Lecompte, E., Van der Straten, E., Petter, F. & Denys, C. 1999. Caracteristiques crânio-dentaires de l'holotype de *Praomys morio* (Rodentia, Muridae). Mammalia 63: 530-534.

Lecompte, E., Denys, C. & Granjon, L. 2001. An identification key for species within the genus *Praomys* (Rodentia: Muridae). In: African Small Mammals (C. Denys, L. Granjon & A. Poulet, eds). IRD, Paris.

Lecompte, E., Granjon, L. & Denys, C. 2002a. The phylogeny of the *Praomys* complex (Rodentia: Muridae) and its phylogeographic implications. Journal of Zoology, Systematic and Evolutionary Research 40: 8-25.

Lecompte, E., Granjon, L., Peterhans, J. K. & Denys, C. 2002b. Cytochrome b-based phylogeny of the *Praomys* group (Rodentia, Murinae): a new African radiation? Comptes Rendu Biologies 325: 827-840.

Lecompte, E., Brouat, C., Duplantier, J-M., Galan, M., Granjon, L., Loiseau, A., Mouline, K. & Cosson, J-F. 2005a. Molecular identification of four cryptic species of *Mastomys* (Rodentia, Murinae). Biochemical Systematics and Ecology 33: 681-689.

Lecompte, E., Denys, C. & Granjon, L. 2005b. Confrontation of morphological and molecular data: The *Praomys* group (Rodentia, Murinae) as a case of adaptive convergences and morphological stasis. Molecular Phylogenetics and Evolution 37: 899-919.

Lecompte, E., Nicolas, V., Colyn, M., Denys, C. & Volobouev, V. 2005c. Description of the karyotype of *Helmyscus fumosus* and of several other murids from the Mount Doudou area (Gabon). Belgian Journal of Zoology 135: 21-25.

Lecompte, E., Fichet-Calvet, E., Daffis, S., Koulémou, K., Sylla, O., Kourouma, F., Doré, A., Soropogui, B., Aniskin, V., Allali, B., Kouassi Kan, S., Lalis, A., Koivogui, L., Günther, S., Denys, C. & ter Meulen, J. 2006. *Mastomys natalensis* and Lassa Fever, West Africa. Emerging Infectious Diseases 12: 1971-1974.

Lecompte, E., Aplin, K., Denys, C., Catzeflis, F., Chades, M. & Chevret, P. 2008. Phylogeny and biogeography of African Murinae based on mitochondrial and nuclear gene sequences, with a new tribal classification of the subfamily. BMC Evolutionary Biology 8: 199.

Leirs, H. 1995. Population Ecology of *Mastomys natalensis* (Smith 1834). Implications for Rodent Control in Africa. Agricultural Edition No. 35. Belgian Administration for Development Cooperation, Brussels.

Leirs, H. 1999. Populations of African rodents: models and the real world. In: Ecologically-based management of rodent pests (G. R. Singleton, L. Hinds, H. Leirs & Z. Zhang, eds), pp. 388-408. ACIAR, Canberra.

Leirs, H. 2003. Management of rodents in crops: the pied piper and his orchestra. In: Rats, Mice and People: Rodent Biology and Management (G. Singleton, L. A. Hinds C. J. Krebs & D. M. Spratt, eds), pp. 180-190. ACIAR Monograph No. 96. Australian Centre for International Agricultural Research, Canberra.

Leirs, H., Verheyen, W., Michiels, M., Verhagen, R. & Stuyck, J. 1989. The relation between rainfall and breeding season of *Mastomys natalensis* (Smith 1834) in Morogoro, Tanzania. Annales de la Societe Royale Zoologique de Belgique 119: 59-64.

Leirs, H., Verhagen, R. & Verheyen, W. 1994. The basis of reproductive seasonality in *Mastomys* rats (Rodentia: Muridae). Journal of Tropical Ecology 10: 55-66.

Leirs, H., Verhagen, R., Verheyen, W., Mwanjabe, P. & Mbise, T. 1996. Forecasting rodent outbreaks in Africa: an ecological basis for *Mastomys* control in Tanzania. Journal of Applied Ecology 33: 937-943.

Lindsay, E. H. 1988. Cricetid rodents from Siwalik deposits near Chinji Village. Part I: Megacricetodontinae, Myocricetodontinae, and Dendromurinae. Palaeovertebrata 18: 95-154.

Linn, I. & Key, G. 1996. Use of space by the African striped ground squirrel *Xerus erythropus*. Mammal Review 26: 9-26.

Linzey, A. & Kesner, M. 1997a. Small mammals of a woodland-savannah ecosystem in Zimbabwe. I. Density and habitat occupancy patterns. Journal of Zoology, London 243: 137-152.

Linzey, A. & Kesner, M. 1997b. Small mammals of a woodland-savannah ecosystem in Zimbabwe. II. Community structure. Journal of Zoology, London. 243.

Linzey, A., Kesner, M., Chimimba, C. & Newbery, C. 2003. Distribution of the veld rat sibling species *Aethomys chrysophilus* and *Aethomys ineptus* (Rodentia: Muridae) in southern Africa. African Zoology 38: 169-174.

Linzey, A. V. & Chimimba, C. T. 2008. *Aethomys chrysophilus* (Rodentia: Muridae). Mammalian Species 808: 1-10.

Long, A. K., Bailey, K., Greene, D. U., Tye, C., Parr, C., Lepage, H. K., Gielow, K. H., Monadjem, A. & McCleery, R. A. 2013. Multi-scale habitat selection of *Mus minutoides* in the Lowveld of Swaziland. African Journal of Ecology 51: 493-500.

López-Antoñanzas, R. & Knoll, F. 2010. The oldest known *Massoutiera* (Rodentia: Ctenodactylinae). Mammalia 74: 461-464.

López-Antoñanzas, R. & Knoll, F. 2011. A comprehensive phylogeny of the gundis (Ctenodactylinae, Ctenodactylidae, Rodentia). Journal of Systematic Palaeontology 9: 379-398.

López-Antoñanzas, R., & Wesselman, H. B. 2013. *Tachyoryctes makooka* (Tachyoryctini, Spalacidae, Rodentia) and its bearing on the phylogeny of the Tachyoryctini. Palaeontology 56: 157-166.

López-Antoñanzas, R., Sen, S. & Mein, P. 2004. Systematics and phylogeny of the cane rats (Rodentia: Thryonomyidae). Zoological Journal of the Linnean Society 142-423-444.

Lopez-Martinez, N., Michaux, J. & Hutterer, R. 1998. The skull of *Stephanomys* and a review of the *Malpaisomys* relationships (Rodentia: Muridae), taxonomic incongruence in murids. Journal of Mammalian Evolution 5: 185-215.

Louchart, A., Wesselman, H., Blumenschine, R. J., Hlusko, L. J., Njau, J. K., Black, M. T., Asnake, M. & White, T.D. 2009. Taphonomic, avian, and small-vertebrate indicators of *Ardipithecus ramidus* Habitat. Science 326: 66, 66e1-66e4.

Lovegrove, B. G., Heldmaier, G. & Knight, M. H. 1991. Seasonal and circadian energetic patterns in an arboreal rodent, *Thallomys paedulcus*, from the Kalahari Desert. Journal of Thermal Biology 16: 199-209.

Lowe, S., Browne, M., Boudjelas, S. & De Porter, M. 2000. 100 of the world's worst invasive alien species: a selection from the Global Invasive Species Database. Published by the Invasive Species Specialist Group (ISSG), a specialist group of the Species Survival Commission (SSC) of the World Conservation Union (IUCN), Gland, Switzerland.

Lunde, D. & Sarmiento, E. 2002. Rodents collected from Kalinzu Forest, Uganda. Mammalian Biology 67: 250-255.

Lynch, C. D. 1983. The mammals of the Orange Free State. Memoirs van die Nasionale Museum, Bloemfontein 18: 1-218.

MacDonald, D. W., Johnson, P. J., Albrechtsen, L., Seymour, S., Dupain, J., Hall, A. & Fa, J. E. 2012. Bushmeat trade in the Cross–Sanaga rivers region: Evidence for the importance of protected areas. Biological Conservation 147: 107-114.

Macholan, M., Scharff, A., Burda, H., Zima, J. & Grùtjen, O. 1998. A new karyotype of a Common mole-rat and taxonomical status of *Cryptomys amatus* from Zambia. Zeitschrift fur Säugetierkunde 63:186-190.

Maddalena, T. E., Van der Straeten, E., Ntahuga, L. & Sparti, A. 1989. Nouvelles données et caryotypes des rongeurs du Burundi. Revue Suisse de Zoologie 96: 939-948.

Maddock, A. H. & Perrin, M. R. 1981. A macroscopical examination of the gastric morphology of *Mystromys albicaudatus* (Smith 1834). South African Journal of Zoology 16: 237-247.

Madikiza, Z. J., Bertolino, S. & Do San, E. L. 2011. Female in space, or female in space and time? First data on the socio-spatial organization and mating system of the woodland dormouse (*Graphiurus murinus*). Journal of Ethology 29: 375-380.

Madikiza, Z. J., Bertolino, S., Baxter, R. M. & Do San, E. L. 2010a. Nest box use by woodland dormice (*Graphiurus murinus*): the influence of life cycle and nest box placement. European Journal of Wildlife Research 56: 735-743.

Madikiza, Z., Bertolino, S., Baxter, R. & Do San, E. L. 2010b. Seasonal, sexual and age related variations in the live trapping success of woodland dormice, *Graphiurus murinus*. Zoological Studies 49: 797-805.

Makundi, R. H., Oguge, N. O. & Mwanjabe, P. S. 1999. Rodent pest management in east Africa – an ecological approach. In: Ecologically-Based Rodent Management (G. R. Singleton, L. Hinds, H. Leirs & Z. Zhang, eds), pp. 460-476. Australian Centre for International Agricultural Research: Canberra.

Makundi, R. H., Massawe, A. W. & Mulungu, L. S. 2005. Rodent population fluctuations in three ecologically heterogeneous locations in north-east, central and south-west Tanzania. Belgian Journal of Zoology 135 (Suppl.): 159–165.

Makundi, R. H., Massawe, A. W., Mulungu, L. S. & Katakweba, A. 2010. Species diversity and population dynamics of rodents in farm-fallow mosaic fields in central Tanzania. African Journal of Ecology 48: 313-320.

Malcolm, J. R. & Ray, J. C. 2000. Influence of timber extraction routes on central African small mammal communities, forest structure, and tree diversity. Conservation Biology 14: 1623-1638.

Manthi, F. K. 2006. The Pliocene micromammalian fauna from Kanapoi, northwestern Kenya, and its contribution to understanding the environment of *Australopithecus anamensis*. Unpublished PhD thesis, University of Cape Town, South Africa.

Manthi F.K., 2007. A preliminary review of the rodent fauna from Lemudong'o, southwestern Kenya, and its implication to the Late Miocene paleoenvironments. Kirtlandia 56: 92-105.

Maree, S. 2002. Phylogenetic relationships and mitochondrial DNA sequences evolution in the African rodent subfamily Otomyinae (Muridae). Unpublished PhD thesis, University of Pretoria, South Africa.

Marivaux, L., Vianey-Liaud, M. & Jaeger, J-J. 2004. High-level phylogeny of early Tertiary rodents: dental evidence. Zoological Journal of the Linnean Society 142: 105-134.

Marivaux , L., Lihoreau, F., Manthi, F.K. & Ducrocq, S. 2012. A New Basal Phiomorph (Rodentia, Hystricognathi) from the Late Oligocene of Lokone (Turkana Basin, Kenya). Journal of Vertebrate Paleontology 32: 646-657.

Maslin, M. A., Christensen, B. 2007. Tectonics, orbital forcing, global climate change, and human evolution in Africa: introduction to the African paleoclimate special volume. Journal of Human Evolution 53: 443-464.

Massawe, A., Mulungu, L. S., Makundi, R. H., Dlamini, N., Eiseb, S. J., Kirsten, F., Mahlaba, T., Malebane, P., Von Maltitz, E., Monadjem, A., Taylor, P., Tutjavi, V. & Belmain, S. R. 2011. Spatial and temporal population dynamics of rodents in three geographically different regions in Africa: Implication for ecologically-based rodent management. African Zoology 46: 393-405.

Mathews, T. & Stynder, D. D. 2011. An analysis of the *Aethomys* (Murinae) community from Langebaanweg (Early Pliocene, South Africa) using geometric morphometrics. Palaeogeography, Palaeoclimatology, Palaeocology 302: 230-242.

Matisoo-Smith, E. & Robins, J. 2009. Mitochondrial DNA evidence for the spread of Pacific rats through Oceania. Biological Invasions 11: 1521-1527.

Matthee, C. & Robinson, T. 1997a. Molecular phylogeny of the springhare, *Pedetes capensis*, based on mitochondrial DNA sequences. Molecular Biology and Evolution 14: 20-29.

Matthee, C. & Robinson, T. 1997b. Mitochondrial DNA phylogeography and comparative cytogenetics of the springhare, *Pedetes capensis* (Mammalia: Rodentia). Journal of Mammalian Evolution 4: 53-73.

Matthey, R. 1957. Cytologie et taxonomie du genre *Meriones* Illiger (Rodentia-Muridae-Gerbillinae). Säugetierkunde Mitteilungen 5: 145-150.

Matthey, R. 1958. Les chromosomes et la position systématique de quelques Murinae africains (Mammalia-Rodentia). Acta Tropica 15: 97-117.

Matthey, R. 1963. La formule chromosomique chez sept espèces et sous-espèces de Murinae Africains. Mammalia 27: 157-176.

Matthey, R. 1965. Etudes de cytogenetique sur des Murinae Africains appartenant aux genres *Arvicanthis*, *Acomys* et *Mastomys* (Rodentia). Mammalia 29: 228-249.

Matthey, R. 1968. Cytogenetique et taxonomie du genre *Acomys*. *A. percivali* Dollman et *A. wilsoni* Thomas, espèces d'Abyssinie. Mammalia 32: 621-627.

Matthey, R. 1970. Caryotypes des Muridés et Dendromuridés originaires de Republique centrafricaine. Mammalia 34: 459-466.

Matthey, R. & Jotterand, M. 1972. L'Analyse du caryotype permet de reconnaitre deux espèces cryptiques confondues sous le nom de *Taterillus gracilis* Th. (Rongeurs, Gerbillidés). Mammalia 36: 193-209.

Matthey, R. & Petter, F. 1970. Etude cytogenetique et taxonomique de 40 *Tatera* et *Taterillus* provenant de Haute-Volta et de Republique Centrafricaine (Rongeurs, Gerbillidae). Mammalia 34: 585-597.

Mboumba, J. F., Deleporte, P., Colyn, M. & Nicolas, V. 2011. Phylogeography of *Mus* (*Nannomys*) **minutoides** (Rodentia, Muridae) in West Central African savannahs: singular vicariance in neighbouring populations. Journal of Zoological Systematics and Evolutionary Research 49: 77-85.

McKechnie, A. E. & Mzilikazi, N. 2011. Heterothermy in Afrotropical mammals and birds: a review. Integrative and Comparative Biology 51: 349-363.

Mdangi, M., Mulungu, L. S., Massawe, A. W., Eiseb, S. J., Tutjavi, V., Kirsten, F., Mahlaba, T., Malebane, P., Maltitz, E., Monadjem, A., Dlamini, N., Makundi, R. H. & Belmain, S. R. 2013. Assessment of rodent damage to stored maize (*Zea mays* L.) on smallholder farms in Tanzania. International Journal of Pest Management: 1-9.

Meerburg, B. G., Singleton, G. R. & Kijlstra, A. 2009. Rodent-borne diseases and their risks for public health. Critical Reviews in Microbiology 35: 221–270.

Meester, J. 1988. Chromosomal speciation in southern African small mammals. South African Journal of Science 84: 721-723.

Meester, J. A. J., Rautenbach, I. L., Dippenaar, N. J. & Baker, C. M. 1986. Classification of southern African mammals. Transvaal Museum Monograph 5: 1-359.

Meheteru, Y., Sumbera, R. & Bryja, J. 2014. Enigmatic Ethiopian endemic rodent *Muriculus imberbis* (Ruppell) represents a separate lineage within genus *Mus*. Mammalia on line : DOI: 10.1515/mammalia-2013-0119, February 2014

Mein, P. 1994. Micromammifères du Miocène supérieur et du Pliocène du rift occidental, Ouganda. In: Geology and Palaeobiology of the Albertine Rift Valley, Uganda-Zaire.Vol. II: Palaeobiology.CIFEG Occas. Publ., Orléans.

Mein, P. & Pickford, M. 2006. Late Miocene micromammals from the Lukeino Formation (6.1 to 5.8 Ma), Kenya. Bulletin mensuel de la Société linnéenne de Lyon 75: 183–223.

Mein P. & Pickford, M. 2010. Vallesian rodents from Sheikh Abdallah, Western Desert, Egypt. Historical Biology, 22, 224-259.

Mein, P., Pickford, M. & Senut, B. 2000a. Late Miocene micromammals from the Harasib karst deposits, Namibia. Part 1- Large muroids and non-muroids rodents. Communications of the Geological Survey of Namibia 12: 375-390.

Mein, P., Pickford, M. & Senut B., 2000b. Late Miocene micromammals from the Harasib karst deposits, Namibia. Part 2A – Myocricetodontinae, Petromyscinae and Namibimyinae (Rodentia, Gerbillinae). Communications of the Geological Survey of Namibia 12: 391-401.

Mein, P., Pickford, M. & Senut, B. 2004. Late Miocene micromammals from the Harasib karst deposits, Namibia. Part 2b. Cricetomyidae, Dendromuridae, and Muridae, with an addendum on the Myocricetodontinae. Communications of the Geological Survey of Namibia 13: 43-61.

Mendelsohn, J. 1982. The feeding ecology of the black-shouldered kite *Elanus caeruleus* (Aves: Accipitridae). Durban Museum Novitates 13: 75-115.

Metz, M. R. & Keesing, F. 2001. Dietary preferences of the pouched mouse (*Saccostomus mearnsi*) in central Kenya. Biotropica 33: 182-187.

Mercer, J. M. & Roth, V. 2003. The effects of Cenozoic global change on squirrel phylogeny. Science 299: 1568-1572.

Mess, A. 2007a. The subplacenta in *Octodon degus* and *Petromus typicus*—two hystricognath rodents without significant placental lobulation. Journal of Experimental Zoology Part B: Molecular and Developmental Evolution 308B: 172-188.

Mess, A. 2007b. Chorioallantoic and yolk sac placentation in the dassie rat *Petromus typicus* and its significance for the evolution of hystricognath rodents. Placenta 28: 1229-1233.

Mess, A. & Ade, M. 2005. Feeding biology of the dassie-rat *Petromus typicus* (Rodentia, Hystricognathi, Petromuridae) in captivity. Belgian Journal of Zoology 135: 45-51.

Meyer, J., Steinhauser, J. & Brandl, R. 2007. Large trees, acacia shrubs, and the density of *Thallomys nigricauda* in the thornveld savannah of South Africa. Journal of Arid Environments 68: 363-370.

Meyer, J., Raudnitschka, D., Steinhauser, J. & Brandl, R. 2008. Biology and ecology of *Thallomys nigricauda* (Rodentia, Muridae) in the thornveld savannah of South Africa. Mammalian Biology 73: 111-118.

Meyer, J., Kohnen, A. & Brandl, R. 2010. Genetic differentiation in an arboreal rodent from African savannas. African Journal of Ecology 48: 831-836.

Michaux, J. & Catzeflis, F. 2000. The bushlike radiation of muroid rodents is exemplified by the molecular phylogeny of the LCAT nuclear gene. Molecular Phylogenetics and Evolution 17: 280-231.

Michaux, J., Reyes, A. & Catzeflis, F. 2001. Evolutionary history of the most speciose mammals: molecular phylogeny of muroid rodents. Molecular Biology and Evolution 18: 2017-2031.

Michaux, J., Hautier, L., Simonin, T. & Vianey-Liaud, M. 2008. Phylogeny, adaptation and mandible shape in Sciuridae (Rodentia, Mammalia). Mammalia 72: 286-296.

Midgley, J. J., Gallaher, K. & Kruger, L. M. 2012. The role of the elephant (*Loxodonta africana*) and the tree squirrel (*Paraxerus cepapi*) in marula (*Sclerocarya birrea*) seed predation, dispersal and germination. Journal of Tropical Ecology 28: 227-231.

Misonne, X. 1965. Presence de *Leggada callewaerti* Thomas au Katanga. Mammalia 29: 426-429.

Misonne, X. 1969. African and Indo-Australian Muridae. Evolutionary trends. Annales du Musée de l'Afrique Centrale, Sciences Zoologiques 172: 1-1-219.

Misonne, X. 1974. Order Rodentia – main text. In: The Mammals of Africa: An Identification Guide (J. Meester & H.W. Setzer, eds). Part 6. Smithsonian Institution Press, Washington.

Missoup A.D. 2010. Systématique et biogéographie des rongeurs des forêts de la ligne volcanique du Cameroun (Afrique ouest centrale). Thèse doctorat Universite Yaoundé 1- MNHN (unpublished).

Missoup, A. D., Nicolas, V., Wendelen, W., Bilong, F. B., Cruaud, C. & Denys, C. 2009. First record of *Hylomyscus walterverheyeni* (Rodentia: Muridae) on the north-western side of the Sanaga River (western Cameroon). Zootaxa 2044: 46-60.

Missoup, A. D., Nicolas, V., Wendelen, W., Keming, E., Bilong Bilong, C. F., Couloux, A., Atanga, E., Hutterer, R. & Denys, C. 2012. Systematics and diversification of *Praomys* species (Rodentia: Muridae) endemic to the Cameroon Volcanic Line (West Central Africa). Zoologica Scripta 41: 327-345.

Monadjem, A. 1997. Stomach contents of 19 species of small mammals from Swaziland. South African Journal of Zoology 32:23-26.

Monadjem, A. 1998a. Distributional patterns and conservation status of mammals of Swaziland, southern Africa. Koedoe 41: 45-59.

Monadjem, A. 1998b. Mammals of Swaziland. The Conservation Trust of Swaziland and Big Game Parks, Mbabane.

Monadjem, A. 1998c. Reproductive biology, age structure and diet of *Mastomys natalensis* (Muridae: Rodentia) and other rodents in a Swaziland grassland. Zeitschrift fur Säugetierkunde 63: 347-356.

Monadjem, A. 1998d. Relative brain size of some southern African myomorph rodents. South African Journal of Zoology 33: 47-49.

Monadjem, A. 1999. Population dynamics of *Mus minutoides* and *Steatomys pratensis* (Muridae: Rodentia) in a subtropical grassland in Swaziland. African Journal of Ecology 37: 202-210.

Monadjem, A. & Perrin, M. R. 1997. Population dynamics of *Lemniscomys rosalia* (Muridae: Rodentia) in a Swaziland grassland: effects of food and fire. South African Journal of Zoology 32: 129-135.

Monadjem, A. & Perrin, M. 1998a. The effect of supplementary food on the home range of the multimammate mouse *Mastomys natalensis*. South African Journal of Wildlife Research.

Monadjem, A. & Perrin, M. R. 1998b. Effects of supplemental food on the habitat selected by *Mastomys natalensis* (Muridae: Rodentia) in a subtropical grassland in Swaziland. South African Journal of Zoology 33: 172-177.

Monadjem, A. & Perrin, M. R. 2003. Population fluctuations and community structure of small mammals in a Swaziland grassland over a three-year period. African Zoology 38: 127-137.

Monadjem, A., Taylor, P. J., Cotterill, F. P. D. & Schoeman, M. C. 2010. Bats of Southern and Central Africa. A Biogeographic and Taxonomic Synthesis. Wits University Press, Johannesburg.

Monadjem, A., Mahlaba, T. A., Dlamini, N., Eiseb, S. J., Belmain, S. R., Mulungu, L. S., Massawe, A. W., Makundi, R. H. & Taylor, P. J. 2011. Impact of crop cycle on movement patterns of pest rodent species between fields and houses in Africa. Wildlife Research 38: 603-609.

Moncrief, N. 1993. Geographic variation in fox squirrels (*Sciurus niger*) and gray squirrels (*S. carolinensis*) of the lower Mississippi River valley. Journal of Mammalogy 74: 547-576.

Monks NJ (1998) Melanism in the tree squirrel, Paraxerus cepapi (A. Smith 1836) in Kyle Recreational Park, Zimbabwe. African J Ecol 36: 102–103.

Monks, N. J. 1988. African Journal of Ecology 36: 102-103.

Montgelard, C. Bentz, S., Tirard, C., Verneau, O. & Catzeflis, F. M. 2002. Molecular systematics of Sciurognathi (Rodentia): the mitochondrial cytochrome b and 12S rRNA genes support the Anomaluroidea (Pedetidae and Anomaluridae): Molecular Phylogenetics and Evolution 22: 220-233.

Montgelard, C., Matthee, C. A. & Robinson, T. J. 2003. Molecular systematics of dormice (Rodentia: Gliridae) and the radiation of *Graphiurus* in Africa. Proceedings of the Royal Society London B 270: 1947-1955.

Moore, J. C. 1959. Relationships among the living squirrels of the Sciurinae. Bulletin of the American Museum of Natural History 118: 153-206.

Moore, J. C. 1961. Geographic variation in some reproductive characteristics of diurnal squirrels. Bulletin of the American Museum of Natural History 122: 1-32.

Morris, B. 1963. Notes on the Giant Rat (*Cricetomys gambianus*) in Nyasaland. African Wild Life 17: 103-108.

Mouline, K., Granjon, L., Galan, M., Tatard, C., Abdoullaye, D., Atteyine, S. A., Duplantier, J-M. & Cosson, J-F. 2008. Phylogeography of a Sahelian rodent species *Mastomys huberti*: a Plio-Pleistocene story of emergence and colonization of humid habitats. Molecular Ecology 17: 1036-1053.

Mugerwa, B., Sheil, D., Ssekiranda, P., van Heist, M. & Ezuma, P. 2013. A camera trap assessment of terrestrial vertebrates in Bwindi Impenetrable National Park, Uganda. African Journal of Ecology 51: 21-31.

Mullin, S. K., Pillay, N., Taylor, P. J. & Campbell, G. 2002. Genetic and morphometric variation in populations of South African *Dasymys incomtus incomtus* (Rodentia, Muridae). Mammalia 66: 381-404.

Mullin, S. K., Taylor, P. J. & Pillay, N. 2004a. Skull size and shape of *Dasymys* (Rodentia, Muridae) from Sub-Saharan Africa. Mammalia 68: 185-220.

Mullin, S. K., Pillay, N. & Taylor, P. J. 2004b. Cranial variation and geographic patterns within the *Dasymys rufulus* complex (Rodentia: Muridae). Journal of Mammalogy 85: 911-923.

Mullin, S. K., Pillay, N. & Taylor, P. J. 2005. The distribution of the water rat *Dasymys* (Muridae) in Africa: a review. South African Journal of Science 101: 117-124.

Mulungu, L. S., Makundi, R. H., Massawe, A. W. & Leirs, H. 2007. Relationship between sampling intensity and precision for estimating damage to maize caused by rodents. Integrative Zoology 2: 131-135.

Mulungu, L. S., Massawe, A. W., Kennis, J., Crauwels, D., Eiseb, S., Mahlaba, T. A., Monadjem, A., Makundi, R. H., Katakweba, A. A. S., Leirs, H. & Belmain, S.R. 2011a. Differences in diet between two rodent species of *Mastomys natalensis* and *Gerbilliscus vicinus* in fallow land habitats in central Tanzania. African Zoology 46: 387-392.

Mulungu, L. S., Mahlaba, T., Massawe, A., Kennis, J., Crauwels, D., Eiseb, E., Monadjem, A., Makundi, R. Katakweba, A. Leirs, H. & Belmain, S. 2011. Dietary preferences of the multimammate mouse (*Mastomys natalensis*, Smith 1832) across different habitats and seasons in Tanzania and Swaziland. Wildlife Research 38: 640-646.

Musser, G. & Carleton, M. D. 1993. Family Muridae. In: Mammal Species of the World: A Taxonomic and Geographic Reference (2rd edn) (D. E. Wilson & D. Reeder, eds). John Hopkins University Press, Baltimore.

Musser, G. & Carleton, M. D. 2005. Superfamily Muroidea. In: Mammal Species of the World: A Taxonomic and Geographic Reference (3rd edn) (D. E. Wilson & D. Reeder, eds). John Hopkins University Press, Baltimore.

Muteka, S. P., Chimimba, C. T. & Bennett, N. C. 2006a. Reproductive seasonality in the Tete veld rat (*Aethomys ineptus*) (Rodentia: Muridae) from southern Africa. Journal of Zoology, London 270: 314-322.

Muteka, S. P., Chimimba, C. T. & Bennett, N. C. 2006b. Reproductive photoresponsiveness in *Aethomys ineptus* and *A. namaquensis* (Rodentia: Muridae) from southern Africa. Journal of Zoology, London 268: 225-231.

Muteka, S. P., Chimimba, C. T. & Bennett, N. C. 2006c. Reproductive seasonality in *Aethomys namaquensis* from southern Africa. Journal of Mammalogy 87: 67-74.

Mzilikazi, N., Madikiza, Z. Oelkrug, R. & Baxter, R. M. 2012. Hibernation in Free-Ranging African Woodland Dormice, *Graphiurus murinus*. In: Living in a Seasonal World (T. Ruf, C. Bieber, W. Arnold & E. Millesi, eds). Springer, Berlin & Heidelberg.

Nadler, C. F. & Hoffmann, R. S. 1974. Chromosomes of the African ground squirrel, *Xerus rutilus* (Rodentia: Sciuridae). Cellular and Molecular Life Sciences 30: 889-891.

Neal, B. R. 1977. Reproduction of the punctuated grass-mouse, *Lemniscomys striatus*, in the Ruwenzori National Park, Uganda (Rodentia: Muridae). Zoologica Africana 12: 419-428.

Neal, B. R. 1982. Reproductive biology of three species of gerbils (Genus *Tatera*) in East Africa. Zeitschrift fur Säugetierkunde 47: 287-296.

Neal, B. R. 1984. Seasonal feeding habits of small mammals in Kenya. Zeitschrift fur Säugetierkunde 49: 226-234.

Nel, J. A. J. & Nolte, H. 1965. Notes on the prey of owls in the Kalahari Gemsbok National Park. Koedoe 8: 75-81.

Nel, J. A. J. & Rautenbach, I. L. 1974. Notes on the activity patterns, food and feeding behaviour of *Parotomys brantsi* (Smith, 1840). Mammalia 38: 7-15.

Nevo, E., Capanna, E., Corti, M., Jarvis, J. U. M. & Hickman, C. G. 1986. Karyotype differentiation in the endemic subterranean mole rats of South Africa (Rodentia: Bathyergidae). Zeitschrift fur Säugetierkunde 51: 36-49.

Nicolas, V. & Colyn, M. 2003. Seasonal variations in population and community structure of small rodents in a tropical forest of Gabon. Canadian Journal of Zoology 81: 1034-1046.

Nicolas, V. & Colyn, M. 2006a. Relative efficiency of three types of small mammal traps in an African rainforest. Belgian Journal of Zoology 136: 107-111.

Nicolas, V. & Colyn, M. 2006b. Swimming ability of 5 species of African rainforest murid rodents in relation to their habitat preferences. Zoological Studies 45: 264-268.

Nicolas, V. & Colyn, M. 2007. Efficiency of fluorescent powder tracking for studying use of space by small mammals in an African rainforest. African Journal of Ecology 45: 577-580.

Nicolas, V., Wendelen, W., Verheyen, W. & Colyn, M. 2003a. Geographical distribution and morphometry of *Heimyscus fumosus* (Brosset et al., 1965). Rev. Ecol. (Terre Vie) 58: 197-208.

Nicolas, V., Wendelen, W., Verheyen, W. & Colyn, M. 2003b. Population structure and reproduction of *Heimyscus fumosus* (Brosset et al., 1965) in south-western Gabon. Rev. Ecol. (Terre Vie) 58: 209-216.

Nicolas, V., Verheyen, E., Verheyen, W., Hulselmans, J., Dillen, M., Akpatou, B., Dudu, A., Wendelen, W. & Colyn, M. 2005. Systematics of African lowland rainforest *Praomys* (Rodentia, Muridae) based on molecular and craniometrical data. Zoological Journal of the Linnean Society 145: 539-553.

Nicolas, V., Quérouil, S., Verheyen, E., Verheyen, W., Mboumba, J. F., Dillen, M. & Colyn, M. 2006. Mitochondrial phylogeny of African wood mice, genus *Hylomyscus* (Rodentia, Muridae): Implications for their taxonomy and biogeography. Molecular Phylogenetics and Evolution 38: 779-793.

Nicolas, V., Mboumba, J., Verheyen, E., Denys, C., Lecompte, E., Olayemi, A., Missoup, A. D., Katuala, P. & Colyn, M. 2008a. Phylogeographic structure and regional history of *Lemniscomys striatus* (Rodentia: Muridae) in tropical Africa. Journal of Biogeography 35: 2074-2089.

Nicolas, V., Bryja, J., Akpatou, B., Konecny, A., Lecompte, E., Colyn, M., Lalis, A., Couloux, A., Denys, C. & Granjon, L. 2008b. Comparative phylogeography of two sibling species of forest-dwelling rodent (*Praomys rostratus* and *P. tullbergi*) in West Africa: different reactions to past forest fragmentation. Molecular Ecology 17: 5118-5134.

Nicolas, V., Wendelen, W., Barriere, P., Dudu, A. & Colyn, M. 2008c. Morphometric variation in *Hylomyscus alleni* and *H. stella* (Rodentia: Muridae), and description of a new species. Journal of Mammalogy 89: 222-231.

Nicolas, V., Granjon, L., Duplantier, J., Cruaud, C. & Dobigny, G. 2009. Phylogeography of spiny mice (genus *Acomys*, Rodentia: Muridae) from the south-western margin of the Sahara with taxonomic implications. Biological Journal of the Linnean Society 98: 29-46.

Nicolas, V., Akpatou, B., Wendelen, W., Peterhans, J. K., Olayemi, A., Decher, J., Missoup, A., Denys, C., Barriere, P., Cruaud, C. & Colyn, M. 2010a. Molecular and morphometric variation in two sibling species of the genus *Praomys* (Rodentia: Muridae): implications for biogeography. Zoological Journal of the Linnean Society 160: 397-419.

Nicolas, V., Natta, A., Barrieré, P., Delapre, A. & Colyn, M. 2010b. Terrestrial small mammal diversity and abundance in central Benin: comparison between habitats, with conservation implications. African Journal of Ecology 48: 1092-1104.

Nicolas, V., Olayemi, A., Wendelen, W. & Colyn, M. 2010c. Mitochondrial DNA and morphometrical identification of a new species of a new species of *Hylomyscus* (Rodentia: Muridae) from West Africa. Zootaxa 2579: 30-44.

Nicolas, V., Missoup, A. D., Denys, C., Kerbis Peterhans, J., Katuala, P., Couloux, A. & Colyn, M. 2011. The roles of rivers and Pleistocene refugia in shaping genetic diversity in *Praomys misonnei* in tropical Africa. Journal of Biogeography 38: 191-207.

Nicolas, V., Missoup, A., Colyn, M., Cruaud, C. & Denys, C. 2012a. West-Central African Pleistocene lowland forest evolution revealed by the phylogeography of Misonne's soft-furred mouse. African Zoology 47: 100-112.

Nicolas, V., Schaeffer, B., Missoup, A. D., Kennis, J., Colyn, M., Denys, C., Tatard, C., Cruaud, C. & Laredo, C. 2012b. Assessment of three mitochondrial genes (16S, Cytb, CO1) for identifying species in the Praomyini tribe (Rodentia: Muridae). PLoS ONE 7: e36586.

Nicolas, V., Ndiaye, A., Benazzou, T., Souttou, K., Delapre, A. & Denys, C. 2014. Phylogeography of the North African dipodil (Rodentia: Muridae) based on cytochrome-b sequences. Journal of Mammalogy 95: 241-253.

Njobe, K. 1997. The distribution and biogeography of the cryptic multimammate mice, *Mastomys natalensis* (A. Smith, 1834) and *Mastomys coucha* (A. Smith, 1836) in southern Africa. Unpublished MSc dissertation, University of Pretoria, South Africa.

Nunes, S., McElhinny, T., Mahoney, M. & Smale, L. 2002. Effects of photoperiod on the reproductive condition of Nile grass rats (*Arvicanthis niloticus*) from an equatorial population. African Journal of Ecology 40: 295-302.

Nunome, M., Yasuda, S. P., Sato, J. J., Vogel, P. & Suzuki, H. 2007. Phylogenetic relationships and divergence times among dormice (Rodentia, Gliridae) based on three nuclear genes. Zoologica Scripta 36: 537-546.

Nyiramana, A., Mendoza, I., Kaplin, B. A. & Forget, P-M. 2011. Evidence for seed dispersal by rodents in tropical montane forest in Africa. Biotropica 43: 654-657.

O'Shea, T. J. 1976. Home range, social behavior, and dominance relationships in the African unstriped ground squirrel, *Xerus rutilus*. Journal of Mammalogy 57: 450-460.

O'Shea, T. J. 1991. *Xerus rutilus*. Mammalian Species 370: 1-5.

Ogada, L. & Keesing, F. 2010. Decline of raptors over a three-year period in Laikipia, central Kenya. Journal of Raptor Research: 1-7.

Olayemi, A., Nicolas, V., Hulselmans, J., Missoup, A., Fichet-Calvet, E., Amundala, D., Dudu, A., Diercx, T., Wendelen, W., Leirs, H. & Verheyen, E. 2012. Taxonomy of the African giant pouched rats (Nesomyidae: *Cricetomys*): molecular and craniometric evidence supports an unexpected high species diversity. Zoological Journal of the Linnean Society 165: 700-719.

Olert, J., Dieterlen, F. & Rupp, H. 1978. Eine neue Muriden-arten aus Sudathiopien. Zeitschrift fur Zoologische Systematik und Evolutionsforschung 16: 297-308.

Osborn, D. J. & Helmy, I. 1980. The contemporary land mammals of Egypt (including Sinai). Fieldiana (Zoology) 6: 1-579.

Padial, J. M. & Tellería, J. L. 2009. Observations of the Felou gundi (*Felovia vae* Lataste, 1886) in central Mauritania (Rodentia: Ctenodactylidae). Mammalia 73: 153-154.

Pavlinov I. Y. 1987. Cladistic analysis of gerbils of the tribe Taterillini (Rodentia, Gerbillinae with notes on cladistic numerical methods. Zool. Zhurn. (Moscow) 66: 903-913. (In Russian with English summary).

Pavlinov, I. J. 2001. Current concepts of gerbillid phylogeny and classification. In: African Small Mammals (C. Denys, L. Granjon & A. Poulet, eds). IRD, Paris.

Pavlinov, I. J. 2008. A review of phylogeny and classification of Gerbillinae (Mammalia: Rodentia). Moscow University Publishing, Moscow.

Pavlinov, I. J. & Potapova, E. G. 2003. Cladistic analysis of the dormouse genus *Graphiurus* Smuts, 1832 (Rodentia, Gliridae), with comments on evolution of its zygomasseteric construction and subgeneric taxonomy. Russian Journal of Theriology 2: 49-58.

Pavlinov, I. J., Dubrowsky, A. Y., Potapova, E.G. & Rossolimo, O. L. 1990. Gerbils of the World. J. Nauka Publications, Moscow. (In Russian).

Pearch, M. J., Bates, P. J. J. & Magin, C. 2001. A review of the small mammal fauna of Djibouti and the results of a recent survey. Mammalia 65: 387-409.

Peinke, D. & Bernard, R. 2005. Life history of the springhare (*Pedetes capensis*) from a strongly seasonal environment in the Eastern Cape Province of South Africa. African Zoology 40: 285-292.

Peinke, D. & Brown, C. 1999. Osmoregulation and water balance in the springhare (*Pedetes capensis*). Journal of Comparative Physiology B 169: 1-10.

Peinke, D. & Brown, C. 2005. Burrow utilization by springhares (*Pedetes capensis*) in the Eastern Cape, South Africa. African Zoology 40: 37-44.

Peinke, D. & Brown, C. 2006. Habitat use by the southern springhare (*Pedetes capensis*) in the Eastern Cape Province, South Africa. South African Journal of Wildlife Research 36: 103-111.
Peinke, D. M. & Brown, C. R. 2003. Metabolism and thermoregulation in the springhare (*Pedetes capensis*). Journal of Comparative Physiology B 173: 347-353.
Pennycuick, C. 2007. Breeding of the lappet-faced and white-headed vultures (*Torgos tracheliotus* Forster and *Trigonoceps occipitalis* Burchell) on the Serengeti Plains, Tanzania. East African Wildlife Journal: 1-18.
Perez del Val, J. P., Juste, J. & Castroviejo, J. J. 1995. A review of *Zenkerella insignis* Matschie, 1899 (Rodentia, Anomaluridae). First records in Bioko island (Equatorial Guinea). Mammalia 59: 441-443.
Perrin, M. R. 1980. The breeding strategies of two coexisting rodents, *Rhabdomys pumilio* and *Otomys irroratus*: with a brief review of some pertinent life history ideas. Acta Oecologia Generalis 1: 383-410.
Perrin, M. R. & Ridgard, B. W. 1999. Thermoregulation and patterns of torpor in the spectacled dormouse, *Graphiurus ocularis* (A. Smith 1829) (Gliridae). Tropical Zoology 12: 253-266.
Perrin, M. R., Dempster, E. R. & Downs, C. T. 1999a. *Gerbillurus paeba*. Mammalian Species 606: 1-6.
Perrin, M. R., Dempster, E. R., Downs, C. T. & Boyer, D. C. 1999b. *Gerbillurus tytonis*. Mammalian Species 607: 1-4.
Petter, F. 1959. Evolution du dessin de la surface d'usure des molaires des gerbilidés. Mammalia 23: 304-315.
Petter, F. 1961. Répartition géographique et écologie des Rongeurs désertiques (du Sahara occidental à l'Iran oriental). Mammalia 25: 1-222.
Petter, F. 1966a. Affinites des genres *Beamys, Saccostomus* et *Cricetomys* (Rongeurs, Cricetomyinae). Annales du Musée Royal de l'Afrique Centrale, Sciences Zoologiques 144: 13-25.
Petter, F. 1966b. *Dendroprionomys rousseloti* gen. nov. sp. nov. Rongeur nouveau du Congo (Cricetidae, Dendromurinae). Mammalia 30: 129-137.
Petter, F. 1966c. La léthargie de *Steatomys opimus* (Rongeurs, Cricetides, Dendromurinés). Mammalia 30: 511-513.
Petter, F. 1967. Particularites dentaires des Petromyscinae Roberts 1951 (Rongeurs, Cricetides). Mammalia 31: 217-224.
Petter, F. 1969. Une souris nouvelle d'Afrique occidentale, *Mus mattheyi* sp. nov. Mammalia 33: 118-123.
Petter, F. 1972. Caractéres morphologiques et répartition géographique de *Mus tenellus* (Thomas, 1903), souris d'Afrique orientale. Mammalia 36: 533-535.
Petter, F. 1973. Tendances évolutives dans le genre *Gerbillus* (Rongeurs, Gerbillides). Mammalia 37: 631-636.
Petter, F. 1975. Subfamily Gerbillinae. In: The Mammals of Africa: An Identification Manual (J. Meester & H. W. Setzer, eds). Part 6.3. Smithsonian Institution Press, Washington, DC.
Petter, F. 1981. Les souris africaines du group *sorella* (Rongeurs, Murides). Mammalia 45: 313-320.
Petter, F. 1982. Les paréntes des *Otomys* du Mont Oku (Cameroun) et des autres formes rapportées à *O. irroratus* (Brants, 1827) (Rodentia, Muridae). Bonner Zoologische Beiträge 33: 215-222.
Petter, F. 1983. Elements d'une révision des *Acomys* africains. Un sous-genre nouveau, *Peracomys* Petter et Roche 1981. Annales du Musée Royal de l'Afrique Centrale, Sciences Zoologiques 237: 109-119.
Petter, F. 1986. Un rongeur nouveau du Mont Oku (Cameroun) *Lamottemys okuensis*, gen. nov., sp. nov. (Rodentia, Muridae). Cimbebasia, ser. A, 8: 97-105.
Petter, F. & Genest, H. 1970. Liste préliminaire des rongeurs myomorphes de Republique centrafricaine. Description de deux souris nouvelles *Mus oubanguii* et *Mus goundae*. Mammalia 34: 451-458.
Petter, F. & Roche, J. 1981. Remarques préliminaires sur la systematique des *Acomys* (Rongeurs, Muridae) *Peracomys*, sous-genre nouveau. Mammalia 45: 381-383.
Petter, F. & Tranier, M. 1975. Contribution a l'etude des *Thamnomys* du groupe *dolichurus* (Rongeurs, Murides). Systématique et caryologie. Mammalia 39: 405-414.
Pickford, M. & Mein, P. 2006. Early Middle Miocene mammals from Morota II, Uganda. Beitr. Paläont. 30: 361-386.
Pickford, M. & Mein, P. 1988. The discovery of fossiliferous Plio-Pleistocene cave fillings in Ngamiland, Botswana. Comptes Rendus de l'Académie des Sciences Paris 307, ser.II: 1681-1686.
Pickford, M. & Mein, P. 2011. New Pedetidae (Rodentia: Mammalia) from the Mio-Pliocene of Africa. Estudios Geologicos 67: 455-469.

Pickford, M., Mein, P. & Senut , B. 1994. Fossiliferous Neogene karst fillings in Angola, Botswana and Namibia. South African Journal of Science 90: 227-230.

Pickford M, Wanas H, Mein P, Soliman H. 2008. Humid conditions in the Western Desert during the Vallesian (Late Miocene). Bull Tethys Geological Soc Cairo. 3:63–79.

Pillay, N. 2003. Reproductive biology of a rare African rodent, the water rat *Dasymys incomtus*. Journal of Mammalogy 84: 505-512.

Pocock, T. N. 1976. Pliocene mammalian microfauna from Laangebaanweg: a few fossil genus linking the Otomyinae with the Murinae. South African Journal of Science 72: 58-60.

Pocock, T. N. 1987. Plio-Pleistocene fossil mammalian microfauna of southern Africa – a preliminary report including description of two new fossil muroid genera (Mammalia: Rodentia). Palaeontologia Africana 26: 69-91.

Poulet, A. 1984. Quelques observations sur la biologie de *Desmodilliscus braueri* Wettstein (Rodentia, Gerbillidae) dans le Sahel du Sénégal. Mammalia 48: 59-64.

Poulet, A. R. 1985. The ecological basis of forecasting rodent outbreaks in Sahelian agrosystem. Acta Zoologica Fennica 173: 107-111.

Primus, A., Harvey, J., Giumondou, S., Mboumba, S., Ngangui, R., Hoffman, F., Baker, R. & Porter, C. A. 2006. Karyology and chromosomal evolution of some small mammals inhabiting the rainforest of the Rabi Oil Field, Gabon. Bulletin of the Biological Society of Washington 12: 371-382.

Qumsiyeh, M. B. 1986. Phylogenetic studies of the rodent family Gerbillidae. I. Chromosomal evolution in the southern African complex. Journal of Mammalogy 67: 680-692.

Qumsiyeh, M. B. & Schlitter, D. A. 1991. Cytogenetic data of the rodent family Gerbillidae. Occasional Papers, Museum Texas Tech University 144: 1-20.

Qumsiyeh, M. B., Hamilton, M. J. & Schlitter, D. A. 1987. Problems of using Robertsonian arrangements in determining monophyly: examples from the genera *Tatera* and *Gerbillurus*. Cytogenetics and Cell Genetics 44: 198-208.

Qumsiyeh, M. B., Hamilton, M. J., Dempster, E. R. & Baker, R. J. 1991. Cytogenetics and systematics of the rodent genus *Gerbillurus*. Journal of Mammalogy 72: 89-96.

Rahm, U. 1962. L'élevage et la reproduction en captivité de l'*Atherurus africanus* (Rongeurs, Hystricidae). Mammalia 26: 1-9.

Rahm, U. 1967. Les Muridés des environs du lac Kivu et des régions voisines (Afrique Centrale) et leur écologie. Revue Suisse Zoologie 74: 439-519.

Rahm, U. 1969. Dokumente uber *Anomalurus* und *Idiurus* des ostlichen Kongo. Zeitschrift fur Säugetierkunde 34: 75-84.

Rahm, H. 1971. Okologie und biologie von *Tachyoryctes ruandae*. Rev. Suisse Zool. 78: 3.

Rahm, H. 1972. Zur Ökologie der Muriden im Regenwaldgebiet des ostlichen Kongo (Zaire). Revue Suisse de Zoologie 79: 1121-1130.

Rahm, H. & Christiaensen, A. 1963. Les mammifères de la région occidentale du Lac Kivu. Annales du Musée Royal de l'Afrique Centrale, Sciences Zoologiques 118: 1-83.

Rambau, R. V., Harrison, W. R., Elder, F. F. B. & Robinson, T. J. 1997. Chromosomes of Brant's whistling rat and genome conservation in the Otomyinae revealed by G-banding and flourescence in situ hybridisation. Cytogenetics and Cell Genetics 78 : 216-220.

Rambau, R. V., Robinson, T. J. & Stanyon, R. 2003. Molecular genetics of *Rhabdomys pumilio* subspecies boundaries: mtDNA phylogeography and karyotypic analysis by fluorescence in situ hybridization. Molecular Phylogenetics and Evolution 28: 564-575.

Ranck, G. L. 1968. The rodents of Libya. Taxonomy, ecology and zoogeographical relationships. Bulletin of the United States National Museum 275: 1-264.

Rathbun, G. & Rathbun, C. 2005. Noki or dassie-rat (*Petromus typicus*) feeding ecology and petrophily. Belgian Journal of Zoology 135: 69-75.

Rathbun, G. B. & Rathbun, C. D. 2006a. Sheltering, basking, and petrophily in the noki or dassie-rat (*Petromus typicus*) in Namibia. Mammalia 70: 269-275.

Rathbun, G. B. & Rathbun, C. D. 2006b. Social monogamy in the noki or dassie-rat (*Petromus typicus*) in Namibia. Mammalian Biology 71: 203-213.

Rautenbach. I. L. 1982. The mammals of the Transvaal. Ecoplan Monograph 1: 111-211.

Rautenbach, I. L. & Schlitter, D. A. 1978. Revision of the genus *Malacomys* of Africa (Mammalia: Muridae). Annals of Carnegie Museum 347: 385-422.

Ray, J. C. & Sunquist, M. E. 2001. Trophic relations in a community of African rainforest carnivores. Oecologia 127: 395-408.

Reed, D. N. 2011. New murid (Mammalia, Rodentia) fossils from a late Pliocene (2.4 MYA) locality, Hadar AL 894, Afar Region, Ethiopia. Journal of Vertebrate Palaeontology 31: 1326-1337.

Renaud, S. 1999. Size and shape variability in relation to species differences and climatic gradients in the African rodent *Oenomys*. Journal of Biogeography 26: 857-865.

Reynolds, J. 1985. Details of the geographic replacement of the red squirrel (*Sciurus vulgaris*) by the grey squirrel (*Sciurus carolinensis*) in eastern England. Journal of Animal Ecology 54: 149-162.

Richardson, E. J. & Perrin, M. R. 1992. Seasonal changes in body mass, torpidity and reproductive activity of captive fat mice, *Steatomys pratensis*. Israel Journal of Zoology 38: 315-322.

Roberts, A. 1951. The mammals of South Africa. Central News Agency, Cape Town, South Africa.

Roberts, A. 1913. The collection of mammals in the Transvaal Museum registered up to the 31st March, 1913, with descriptions of new species. Annals of the Transvaal Museum 67-107.

Robbins, C. B. 1974. Comments on the taxonomy of the West African *Taterillus* (Rodentia, Cricetidae) with the description of a new species. Proceedings of the Biological Society of Washington 87: 395-404.

Robbins, L. W. & Baker, R. J. 1978. Karyotypic data for African mammals, with a description of an in vivo bone marrow technique. Bulletin Carnegie Museum of Natural History 6: 188-210.

Robbins, C. B. & Van der Straeten, E. 1982. A new specimen of *Malacomys verschureni* from Zaire, Central Africa (Rodentia: Muridae). Revue de Zoologie Africaine 96: 216-220.

Robbins, C. B. & Van der Straeten, E. 1989. Comments on the systematics of *Mastomys* Thomas 1815 with the description of a new West African species (Mammalia: Rodentia: Muridae). Senckenbergiana Biologica 69: 1-14.

Robbins, C. B. & Van der Straeten, E. 1996. Small mammals of Togo and Benin. II. Rodentia. Mammalia 60 : 231-242.

Robbins, L. W., Choate, J. R. & Robbins, R. L. 1980. Nongeographic and interspecific variation in four species of *Hylomyscus* (Rodentia, Muridae) in southern Cameroon. Annals of Carnegie Museum 49: 31-48.

Robinson, T. J., Skinner, J. D. & Haim, A. S. 1986. Close chromosomal congruence in two species of ground squirrel: *Xerus inauris* and *X. princeps* (Rodentia: Sciuridae). South African Journal of Zoology 21: 100-105.

Roche, J. & Petter, F. 1968. Faits nouveaux concernant trois Gerbilidés mal connus de Somalie: *Ammodillus imbellis* (de Winton), *Microdillus peeli* (de Winton), Monodia juliani (Saint Leger). Monitore Zoologico Italiano 2 (Suppl): 181-198.

Roche, J. E., Capanna, E., Civitelli, M. V. & Ceraso, A. 1984. Caryotypes des rongeurs de Somalie. 4. Première capture de rongeurs arboricoles du sous-genre *Grammomys* (genre *Thamnomys*, Murides) en Republique de Somalie. Monitore Zoologico Italiano 19: 259-277.

Roper, T. J., Jackson, T. P., Conradt, L. & Bennett, N. C. 2002. Burrow use in Brants' whistling rat *Parotomys brantsii*. Ethology 108: 1-8.

Rosevear, D. R. 1969. The Rodents of West Africa. British Museum (Natural History), London.

Rowe-Rowe, D. T. 1986. Stomach contents of small mammals from the Drakensberg, South Africa. Southern African Journal of Wildlife Research 16: 32-35.

Rupp, H. 1980. Beiträge zur systematik, verbreitung und Ökologie athiopischer nagetiere: ergebnisse mehrerer forschungreisen. Saugetierkundkiche Mitteilungen 28: 81-123.

Russo, I. M. 2003. Molecular systematics of southern African *Aethomys* (Rodentia: Muridae). M.S. thesis, University of Pretoria, Pretoria, South Africa.

Russo, I. M., Chimimba, C. T. & Bloomer, P. 2006. Mitochondrial DNA differentiation between two species of *Aethomys* (rodentia: muridae) from southern Africa. Journal of Mammalogy 87: 545-553.

Russo, I. M., Chimimba, C. T. & Bloomer, P. 2010. Bioregion heterogeneity correlates with extensive mitochondrial DNA diversity in the Namaqua rock mouse, *Micaelamys namaquensis* (Rodentia: Muridae) from southern Africa - evidence for a species complex. BMC Evolutionary Biology 10: 307-16.

Ryan, J. 1989. Evolution of cheek pouches in African pouched rats (Rodentia: Cricetomyinae). Journal of Mammalogy 70: 267-274.

Sabatier, M. 1982. Les rongeurs du site Pliocène à hominidés de Hadar (Ethiopie). Paleovertebrata 12: 1-56.

Sallam, H.M., Seiffert, E.R., Simons, E.L. 2011. Craniodental Morphology and Systematics of a New Family of Hystricognathous Rodents (Gaudeamuridae) from the Late Eocene and Early Oligocene of Egypt. Plos One, 6(2): e16525. doi:10.1371/journal.pone.0016525

Sallam, H. M., Seiffert, E. R., Steiper M. E. & Simons, E. L. 2009. Fossil and molecular evidence constrain scenarios for the early evolutionary and biogeographic history of hystricognathous rodents. Proceedings of the National Academy of Sciences, USA 106: 16722-16727.

Sallam, H. M., Seiffert, E. R., Simons, E. L, Brindley, C. 2010. A large-bodied Anomaluroid rodent from the earliest Late Eocene of Egypt: phylogenetics and biogeographic implications. Journal of Vertebrate Paleontology 30: 1579-1593.

Sanderson, I. 1940. The mammals of the North Cameroons forest area. Being the results of the Percy Sladen Expedition to the Mamfe Division of the British Cameroons. The Transactions of the Zoological Society of London 24: 623-725.

Sandro, B. 2008. Introduction of the American grey squirrel (*Sciurus carolinensis*) in Europe: a case study in biological invasion. Current Science 95: 903-906.

Scantlebury, M., Waterman, J. M. & Bennett, N. 2008. Alternative reproductive tactics in male Cape ground squirrels *Xerus inauris*. Physiology & Behavior 94: 359-367.

Scharff, A., Macholan, M. & Burda, H. 2001. A new karyotype of *Heliophobius argenteocinereus* (Bathyergidae, Rodentia) from Zambia with field notes on the species. Zeitschrift fur Säugetierkunde 66: 376-378.

Schenk, J. J., Rowe, K. C. & Steppan, S. J. 2013. Ecological opportunity and incumbency in the diversification of repeated continental colonizations by muroid rodents. Systematic Biology 6: 837-864.

Schlitter, D. A. 1989. African rodents of special concern: a preliminary assessment. In: Rodents. A World Survey of Species of Conservation Concern (W. Z. Lidicker, ed). Occasional Papers, IUCN Species Survival Commission 4: 33-39. IUCN, Gland.

Schlitter, D. A. & Rautenbach, I.L. 1977. The brush-tailed gerbille *Gerbillurus vallinus* in the Kalahari Germsbok National Park. Koedoe 20: 189-191.

Schlitter, D. A., Rautenbach, I. L. & Coetzee, C. G. 1984. Karyotypes of southern African gerbils, genus *Gerbillurus* Shortridge, 1942 (Rodentia: Cricetidae). Annals of the Carnegie Museum 53: 549-557.

Schlitter, D. A., Robbins, L. & Williams, S. 1985. Taxonomic status of dormice (genus *Graphiurus*) from West and Central Africa. Annals of Carnegie Museum 54: 1-9.

Schradin, C. & Pillay, N. 2005. Intraspecific variation in the spatial and social organisation of the African striped mouse. Journal of Mammalogy 86: 99-107.

Schropter, R. & Klenner-Fringer, B. 1991. Minimizing interspecific competition by different foraging strategies in two North African desert rodents. Acta Theriologica 36: 109-117.

Schunke, A. C. 2005. Systematics and Biogeography of the African Scaly-tailed Squirrels (Mammalia: Rodentia: Anomaluridae). Unpublished PhD thesis. Rheinischen Friedrich-Wilhelms Universitat Bonn, Bonn.

Schunke, A. C. & Hutterer, R. 2004. The variance of variation: Geographic patterns of coat colouration in *Anomalurops* and *Anomalurus* (Mammalia, Rodentia, Anomaluridae). Bonner zoologische Beiträge 53: 169-185.

Schunke, A. C. & Hutterer, R. 2005. Geographic variation in the West African Scaly-tailed Squirrel *Anomalurus pelii* (Schlegel & Muller, 1845) and description of a new subspecies (Rodentia: Anomaluridae). In: African Biodiversity: Molecules, Organisms, Ecosystems (B. A. Huber, B. J. Sinclair & K. H. Lampe, eds). Springer-Verlag, Berlin.

Schunke, A. C. & Hutterer, R. 2007. Geographic Variation of *Idiurus* (Rodentia: Anomaluridae) with Emphasis on Skull Morphometry. American Museum Novitates 3548.
Scopin, A. E., Saveljev, A. P., Suntsova, N. A., Gnophanxay, S., Tikhonov, A. N., Abramov, A.V. 2011. Digestive system of the Laotian rock rat *Laonastes aenigmamus* (Rodentia: Diatomyidae) from the evolutionary viewpoint. Proceedings of the Zoological Institute RAS 315: 3-18.
Searle, J. B., Jones, C. S., Gunduz, I., Scascitelli, M., Jones, E. P., Herman, J. S., Rambau, R. V., Noble, L. R., Berry, R. J., Gimenez, M. D., Johannesdottir, F. 2009a. Of mice and (Viking?) men: phylogeography of British and Irish house mice. Proceedings of the Royal Society, Section B: Biological Sciences 276: 201-207.
Searle, J. B., Jamieson, P. M., Gunduz, I., Stevens, M. I., Jones, E. P., Gemmill, C. E., King, C. M. 2009b. The diverse origins of New Zealand house mice. Proceedings of the Royal Society, Section B: Biological Sciences 276: 209-217.
Seifert, A. W., Kiama, S. G., Seifert M. G., Goheen, J., Palmer, T. M., Maden, M. 2012. Skin shedding and tissue regeneration in African spiny mice (*Acomys*).Nature 489: 561.
Sen, S., Brunet, M. & Heintz, E. 1979. Découverte de rongeurs africains dans le Pliocène d'Afghanistan (bassin de Sarobi) implications paléobiogéographiques et stratigraphiques. Bulletin du Muséum National d'Histoire Naturelle, Section C, Ser 4, 1: 65-75.
Sénégas, F. 2000. Les faunes de rongeurs (Mammalia) plio-pléistocènes de la province de Gauteng (Afrique du Sud) : mises au point et apports systématiques, biochronologiques et précisions paléoenvironnementales. PhD, Université de Montpellier 2, Montpellier, France (unpublished), 2vols, 232 pp.
Sénégas, F. 2001. Interpretation of the dental pattern of the South African fossil *Euryotomys* (Rodentia, Murinae) and origin of otomyine dental morphology. In: African Small Mammals (C. Denys, L. Granjon & A. Poulet, eds). IRD, Paris.
Sénégas, F., 2004. A new species of *Petromus* (Rodentia, Hystricognatha, Petromuridae) from the early Pliocene of South Africa and its palaeoenvironmental implications. Journal of Vertebrate Paleontology 24: 757-763.
Sénégas F. & Avery M. 1998. New evidence for the murine origins of the Otomyinae (Mammalia, Rodentia) and the age of Bolt's Farm (South Africa). South African Journal of Science 94: 503-507.
Senut, B., Pickford, M., Mein, P., Conroy, G. & Van Couvering, J. 1992. Discovery of 12 new Late Cenozoic fossiliferous sites in palaeokarsts of the Otavi Mountains, Namibia. Comptes Rendus des Seances de l'Academie des Sciences (Paris), Serie II 314: 727-733.
Setzer, H. W. 1956. Mammals of Anglo-Egyptian Sudan. Proceedings of the United States National Museum 106: 447-587.
Setzer, H. W. 1975. Genus *Acomys*. In: The Mammals of Africa: An Identification Manual (J. Meester & H. W. Setzer, eds). Part 6.5. Smithsonian Institution Press, Washington, DC.
Sherman, P. W., Braude, S. H. & Jarvis, J. U. M. 1999. Litter sizes and mammary numbers of naked mole-rats: breaking the one half rule. Journal of Mammalogy 80: 720-733.
Shortridge, G. C. 1934. The Mammals of South West Africa, Vols I & II. William Heinemann Ltd, London.
Sicard, B., Tranier, M. & Gautun, J. C. 1988. Un rongeur nouveau du Burkina Faso (ex. Haute Volta): *Taterillus petteri*, sp. nov. (Rodentia, Gerbillidae). Mammalia 52: 187-198.
Sicard, B., Maurel, D., Fuminier, F. & Boissin, J. 1992 Circadian rhythm of photosensitivity and the adaptation of reproductive function to the environment in two populations of *Arvicanthis niloticus* from Mali and Burkina Faso. Journal of Reproduction and Fertility 95: 159-165.
Sillero-Zubiri, C., Tattershall, F. H. & MacDonald, D. W. 1995a. Bale Mountain rodent communities and their relevance to the Ethiopian Wolf (*Canis simensis*). African Journal of Ecology 33: 301-320.
Sillero-Zubiri, C., Tattershall, F. H. & MacDonald, D. W. 1995b. Morphometrics of endemic rodents from the Bale Mountains, Ethiopia. African Journal of Zoology 109: 387-391.
Singleton, G., Hinds, L. A., Leirs, H. & Zhang, Z. (eds.) 1999. Ecologically-Based Management of Rodent Pests. Australian Centre for International Agricultural Research monograph no. 59. ACIAR, Canberra.

Singleton, G., Hinds L. A., Krebs, C. J. & Spratt, D. M. (eds.) 2003. Rats, Mice and People: Rodent Biology and Management. Australian Centre for International Agricultural Research monograph no. 96. ACIAR, Canberra.

Singleton, G., Belmain, S., Brown, P. & Hardy, B. 2010. Rodent outbreaks: ecology and impacts. International Rice Research Institute, Manila, Philippines.

Skinner, J. D. & Chimimba, C. T. 2005. The Mammals of the Southern African Subregion (3rd edn). Cambridge University Press, Cambridge, UK.

Skonhoft, A., Leirs, H., Andreassen, H. P., Mulungu, L. S., Stenseth, N. C. 2006. The bioeconomics of controlling an African rodent pest species. Environment and Development Economics 11: 453-475.

Skurski, D. A. & Waterman, J. M. 2005. *Xerus inauris*. Mammalian Species 781: 1-4.

Sluyds, V., Davis, S., Mercelis, S. & Leirs, H. 2009. Comparison of multimammate mouse (*Mastomys natalensis*) demography in monoculture and mosaic agricultural habitat: implications for pest management. Crop Protection 28: 647-656.

Smithers, R. H. N. 1971. The Mammals of Botswana. Museum Memoir No. 4, National Museums and Monuments of Rhodesia.

Smithers, R. H. N. & Labao Tello, J. L. P. 1976. Checklist and Atlas of the Mammals of Mozambique. Museum Memoir No. 8, National Museums and Monuments of Rhodesia.

Smithers, R. H. N. & Wilson, V. J. 1979. Check List and Atlas of Mammals of Zimbabwe Rhodesia. Museum Memoir No. 9, National Museums and Monuments of Zimbabwe Rhodesia.

Sokolov, E. V., Orlov, N. V., Baskevich, M. I., Bekele, A. & Mebrate, A. 1993. A karyological study of the spiny mouse *Acomys* Geoffroy 1838 (Rodentia Muridae) along the Ethiopian Rift Valley. Tropical Zoology 6: 227-235.

Solano, E., Taylor, P. J., Rautenbach, A., Ropiquet, A. & Castiglia, R. 2014. Cryptic Speciation and Chromosomal Repatterning in the South African Climbing Mice *Dendromus* (Rodentia, Nesomyidae). PlosONE 9(2): e88799. doi:10.1371/journal.pone.0088799.

Stanley, W. T. & Collett, L. 2009. Attack or consumption of *Epomophorus* (Chiroptera) by *Paraxerus* (Rodentia) and *Papio* (Primates) in Tanzania. African Journal of Ecology 47: 792-793.

Stanley, W. & Hutterer, R. 2007. Differences in abundance and species richness between shrews and rodents along an elevational gradient in the Udzungwa Mountains, Tanzania. Acta Theriologica 52: 259-273.

Stanley, W. T. & Foley, C. A. H. 2008. A survey of the small mammals of Minziro Forest, Tanzania, with several additions to the known fauna of the country. mammalia 72.

Stanley, W. & Goodman, S. 2011. Small mammal inventories in the East and West Usambara Mountains, Tanzania. 4. Rodentia. In: Studies of Montane Vertebrates of Tanzania (W. T. Stanley, ed). Fieldiana Life and Earth Sciences, Field Museum of Natural History, Chicago.

Stanley, W., Kihaule, P., Howell, K. & Hutterer, R. 1998. Small mammals of the Eastern Arc Mountains, Tanzania. Journal of East African Natural History 87: 91-100.

Stanley, W. T., Goodman, S. M., Kihaule, P. M. & Howell, K. M. 2000. A survey of the small mammals of the Gonja Forest Reserve, Tanzania. Journal of East African Natural History 89: 73-83.

Stanley, W. T., Gunn, J. & Kihaule, P. M. 2005a. Results of a preliminary small mammal survey of Malundwe Mountain, Mikumi National Park, Tanzania. Journal of East African Natural History 94: 213-222.

Stanley, W. T., Rogers, M. A., Howell, K. M. & Msuya, C. A. 2005b. Results of a survey of small mammals in the Kwamgumi Forest Reserve, East Usambara Mountains, Tanzania. Journal of East African Natural History 94: 223-230.

Stanley, W., Kihaule, P. & Munissi, M. 2007a. Small mammals of two forest reserves in the North Pare mountains, Tanzania. Journal of East African Natural History 96: 215-226.

Stanley, W., Norton, G., Kihaule, P., Collett, L. & McQuaid, K. 2007b. Additional notes on the small mammals of Malundwe Mountain, Mikumi National Park, Tanzania. Journal of East African Natural History 96: 203-214.

Stanley, W. T., Rogers, M. A., Senzota, R. B. M., Mturi, F. A., Kihaule, P. M., Moehlman, P. D. & O'Connor, B. M. 2007c. Surveys of small mammals in Tarangire National Park, Tanzania. Journal of East African Natural History 96: 47-71.

Stehlin, H. G. & Schaub, S. 1951. Die trigonodontie der Simplicidentaten Nager. Schweizerische Paläontologische Abhandlungen 67: 1-384.

Steppan, H. & Dieterlen, F. 1982. Relative brain size in Muridae with special reference to *Colomys goslingi*. Zeitschrift fur Säugetierkunde 47: 38-47.

Steppan, S. J., Adkins, R. M., Spinks, P. Q. & Hale, C. 2005. Multigene phylogeny of the Old World mice, Murinae, reveals distinct geographic lineages and the declining utility of mitochondrial genes compared to nuclear genes. Molecular Phylogenetics and Evolution 37: 370-388.

Steppan, S. J., Storz, B. L. & Hoffmann, R. S. 2004. Nuclear DNA phylogeny of the squirrels (Mammalia: Rodentia) and the evolution of arboreality from c-myc and RAG1. Molecular Phylogenetics and Evolution 30: 703-719.

Stevens, N. J., O'Connor, P. M., Gottfried, M. D., Roberts, E. M., Ngasala S., Dawson, M.R. 2006. *Metaphiomys* (Rodentia: Phiomyidae) from the Paleogene of Sothwestern Tanzania. Journal of Paleontolgy 80: 407-410.

Sumbera, R., Burda, H. & Chitaukali, W. N. 2003a. Reproductive biology of the solitary subterranean bathyergid rodent: the silvery mole-rat (*Heliophobius argenteocinereus*). Journal of Mammalogy 84: 278-287.

Sumbera, R., Burda, H., Chitaukali, W. N. & Kubova, J. 2003b. Silvery mole-rats *Heliophobius argenteocinereus* (Bathyergidae) change their burrow architecture seasonally. Naturwissenchaften 90: 170-173.

Swanepoel, P. & Schlitter, D. A. 1978. Taxonomic review of the fat mice (genus *Steatomys*) of West Africa (Mammalia, Rodentia). Bulletin of Carnegie Museum of Natural History 6: 53-76.

Swynnerton, G. & Hayman, R. 1951. A checklist of the land mammals of the Tanganyika territory and the Zanzibar Protectorate. Journal of East African Natural History 20: 274-392.

Taylor, P. J. 1998. The Smaller Mammals of KwaZulu-Natal. Unversity of Natal Press, Pietermaritzburg.

Taylor, P. J. 2000. Patterns of chromosomal variation in southern African rodents. Journal of Mammalogy 81: 317-331.

Taylor, P. J. & Kumirai, A. 2001. Craniometric relationships between the southern African vlei rat, *Otomys irroratus* (Rodentia, Muridae, Otomyinae), and allied species from north of the Zambezi River. In: African Small Mammals (C. Denys, L. Granjon & A. Poulet, eds). IRD, Paris.

Taylor, P. J., Jarvis, J. U. M. & Crowe, T. 1985. Age determination in the Cape mole-rat *Georychus capensis*. South African Journal of Zoology 20: 261-267.

Taylor, P. J., Rautenbach, I. L., Gordon, D. H., Sink, K. & Lotter, P. 1995. Diagnostic morphometrics and southern African distribution of two sibling species of tree rat, *Thallomys paedulcus* and *Thallomys nigricauda* (Rodentia: Muridae). Durban Museum Novitates 20: 49-62.

Taylor, P. J., Meester, J. & Kearney, T. 1993. The taxonomic status of Saunder's vlei rat, Oto*mys saundersiae* Roberts (Rodentia: Muridae: Otomyinae). Journal of African Zoology 107: 1-26.

Taylor, P. J., Denys, C. & Mukerjee, M. 2004. Phylogeny of the African murid tribe Otomyini (Rodentia), based on morphological and allozyme evidence. Zoologica Scripta 33: 389-402.

Taylor, P. J., Kumirai, A. & Contrafatto, G. 2005. Species with fuzzy borders: the taxonomic status and species limits of Saunders vlei rat, *Otomys saundersiae* Roberts (Rodentia: Muridae: Otomyini). Mammalia 69: 297-322.

Taylor, P. J., Arntzen, L., Hayter, M., Iles, M., Frean, J. & Belmain, S. R. 2008a. Understanding and managing sanitary risks due to rodent zoonoses in an African city: beyond the Boston Model. Integrative Zoology 3: 38-50.

Taylor, P., Rautenbach, A., Schoeman, M. & Combrink, X. 2008b. A winter survey of smaller mammals of uMkuze section of the iSimangaliso Wetland Park, KwaZulu-Natal Province, South Africa. Unpublished report.

Taylor, P. J., Maree, S., Van Sandwyk, J. Baxter, R. & Rambau, R. V. 2009a. When is a species not a species? Uncoupled phenotypic, karyotypic and genotypic divergence in two species of South African laminate-toothed rats (Murinae: Otomyini). Journal of Zoology, London 277: 317-332.

Taylor, P. J., Maree, S., Van Sandwyk, J., Kerbis Peterhans, J., Stanley, W. T., Verheyen, E., Kaliba, P., Verheyen, W., Kaleme, P. & Bennett, N. 2009b. Speciation mirrors geomorphology and palaeoclimatic history in

African laminate-toothed rats (Muridae: Otomyini) of the *Otomys denti* and *Otomys lacustris* species-complexes in the 'Montane Circle' of East Africa. Biological Journal of the Linnean Society 96: 913-941.

Taylor, P. J., Lavrenchenko, L. A., Carleton, M. D., Bennett, N. C., Oosthuizen, C. J. & Maree, S. 2011. Specific limits and emerging diversity patterns in East African populations of laminate-toothed rats, genus *Otomys* (Muridae: Murinae: Otomyini): revision of the *Otomys typus* complex. Zootaxa 3024: 1-66.

Taylor, P. J., Downs, S., Monadjem, A., Eiseb, S. J., Mulungu, L. S., Massawe, A. W., Mahlaba, T. A., Kirsten, F., Von Maltitz, E., Malebane, P., Makundi, R. H., Lamb, J. & Belmain, S. R. 2012. Experimental treatment-control studies of ecologically based rodent management in Africa: balancing conservation and pest management. Wildlife Research 39: 51-61.

Taylor, P. J., Maree, S., Cotterill, F. P. D., Missoup, A. D.,Nicolas, V. Denys, C. 2014. Peripatric speciation across a Neogene volcanic archipelago:molecular and morphological evidence for a Pleistocene radiation of laminate-toothed rats (*Otomys*: Rodentia) across equatorial Africa. Biological Journal of the Linnean Society 113: 320-344.

Tchernov, E. 1968. Rodent faunas, chronostratigraphy and paleobiogeography of the southern Lavent during the Quarternary. Acta Zoologica Cracoviensia 39: 513-530.

ter Meulen, J., Lukashevich, I., Sidibe, K., Inapogui, A., Marx, M., Dorlemann, A., et al., 1996. Hunting of peridomestic rodents and consumption of their meat as possible risk factors for rodent-to-human transmission of Lassa virus in the Republic of Guinea. American Journal of Tropical Medicine and Hygiene 1996 55: 661-6.

Thomas, O. 1909. New African mammals. Annals and Magazine of Natural History, ser. 8, 4: 542-549.

Thomas, O. 1916. Three new African mice of the genus *Dendromus*. Annals and Magazine of Natural History, ser. 8, 18: 241-243.

Thomas, O. & Hinton. 1923. On Mammals collected by Captain Shortridge during the Percy Sladen and Kaffrarian Museum Expedition to the Orange River. Proceedings of the Zoological Society of London 32: 483-499.

Thorington, R. & Hoffmann, R. S. 2005. Family Sciuridae. In: Mammal Species of the World: A Taxonomic and Geographic Reference (3rd edn) (D. E. Wilson and D. Reeder, eds). John Hopkins University Press, Baltimore.

Thorington, R. W., Darrow, K. & Betts, A. D. K. 1997. Comparative myology of the forelimb of squirrels (Sciuridae). Journal of Morphology 234: 155-182.

Tollenaere, C., Brouat, C., Duplantier, J-M., Rahalison, L., Rahelinirina, S., Pascal, M., Moné, H., Mouahid, G., Leirs, H. & Cosson, J-F. 2010. Phylogeography of the introduced species *Rattus rattus* in the western Indian Ocean, with special emphasis on the colonization history of Madagascar. Journal of Biogeography 37:398-410.

Tong, H. 1989. Origine et evolution des Gerbillidae (Mammalia, Rodentia) en Afrique du Nord. Memoires de la Societe Geologique de France 155: 1-120.

Tong, H. & Jaeger, J.-J. 1993. Muroid rodents from the Middle Miocene Fort Ternan locality (Kenya) and their contribution to the phylogeny of muroids. Palaeontographica A, 229: 51-73.

Tranier, M. & Dosso, H. 1979. Recherche caryotypiques sur les rongeurs de Côte d'Ivoire: résultats préliminaires pour les milieux fermés. Mammalia 43: 254-256.

Tranier, M. & Gautun, J. C. 1979. Recherches caryotypiques sur les rongeurs de Côte d'Ivoire: résultats préliminaires pour les milieux ouverts. Le cas d' *Oenomys hypoxanthus ornatus*. Mammalia 43: 252-254.

Tranier, M., Hubert, B. & Petter, F. 1973. *Taterillus* de l'ouest du Tchad et du nord du Cameroun (Rongeurs, Gerbillides). Mammalia 37: 637-641.

Tuyisingize, D., Kerbis Peterhans, J. C., Bronner, G. N. & Stoinski, T. S. 2013. Small mammal community composition in the Volcanoes National Park, Rwanda. Bonn Zoological Bulletin 62: 177-185.

Vale, C. G., Álvares, F. & Brito, J. C. 2012. Distribution, suitable areas and conservation status of the Felou gundi (*Felovia vae* Lataste 1886). Mammalia 76: 201-207.

Van Aarde, R. 1985a. Age determination of Cape porcupines, *Hystrix africaeaustralis*. South African Journal of Zoology 20: 232-236.

Van Aarde, R. 1985b. Reproduction in captive female Cape porcupines (*Hystrix africaeaustralis*). Journal of Reproduction and Fertility 75: 577-582.
Van Aarde, R. 1987a. Demography of a cape porcupine, *Hystrix africaeaustralis*, population. Journal of Zoology, London 213: 205-212.
Van Aarde, R. 1987b. Reproduction in the Cape porcupine *Hystrix africaeaustralis*: an ecological perspective. South African Journal of Science 83: 605-607.
Van Aarde, R. 1987c. Pre-and postnatal growth of the Cape porcupine *Hystrix africaeaustralis*. Journal of Zoology, London 211: 25-33.
Van Daele, P. A. A. G., Dammann, P., Meier, J. L., Kawalika, M., van de Woestijne, C. & Burda, H. 1999. Chromosomal diversity in mole-rats of the genus *Cryptomys* (Rodentia: Bathyergidae) from the Zambezian region: with descriptions of new karyotypes. Journal of Zoology, London 264: 317-326.
Van Daele, P. A. A. G, Faulkes, C. & Verheyen, E. 2007. African mole-rats (Bathyergidae): a complex radiation in tropical soils. In: Subterranean Rodents: News from Underground (S. Begall, H. Burda, C. E. Schleich, eds). Springer-Verlag, Berlin.
Van Daele, P. A. A. G, Blonde, P. & Stjernstedt, R. & Adriaens, D. 2013. A new species of African mole-rat (*Fukomys*, Bathyergidae, Rodentia) from the Zaire-Zambezi watershed. Zootaxa 3636: 171-189.
Van der Merwe, M. 1999. Breeding season and breeding potential of the greater cane rat (*Thryonomys swinderianus*) in captivity in South Africa. South African Journal of Zoology 34: 69-73.
Van der Merwe, M. 2007. Discriminating between *Thryonomys swinderianus* and *Thryonomys gregorianus*. African Zoology 42: 165-171.
Van der Merwe, M. & Avenant, N. L. 2010. The greater canerat, *Thryonomys swinderianus*, is a pest species that is expanding its range in southern Africa. Navorsinge van die Nasionale Museum, Bloemfontein 20: 3-10.
Van der Merwe, M. & van Zyl, A. 2001. Postnatal growth of the greater cane rat *Thryonomys swinderianus* (Thryonomyidae: Rodentia) in Gauteng, South Africa. Mammalia 65: 495-507.
Van der Merwe, M., Skinner, J. D. & Millar, R.P. 1980. Annual reproduction pattern in the springhare *Pedetes capensis*. Journal of Reproduction and Fertility 58: 259-266.
Van der Straeten, E. 1976. *Lemniscomys striatus dieterleni*, a new subspecies of Muridae from Zaire. Revue Zoologique Africaine 90: 431-434.
Van der Straeten, E. 1975. *Lemniscomys bellieri*, a new species of Muridae from the Ivory Coast (Mammalia, Muridae). Revue Zoologique Africaine 89: 906-908.
Van der Straeten, E. 1980a. A new species of *Lemniscomys* (Muridae) from Zambia. Annals of the Cape Province Museum (Natural History) 13: 55-62.
Van der Straeten, E. 1980b. Etude biometrique de *Lemniscomys linulus* (Afrique Occidentale) (Mammalia, Muridae). Revue Zoologique Africaine 94: 185-201.
Van der Straeten, E. 1981. Notes sur *Lemniscomys striatus venustus* (Thomas, 1911). Mammalia 45: 125-128.
Van der Straeten, E. 1984. Etude biometrique de genres *Dephomys* et *Stochomys* avec quelques notes taxonomiques (Mammalia, Muridae). Revue Zoologique Africaine 98: 771-798.
Van der Straeten, E. 1985. Note sur *Hybomys basilii* Eisentraut, 1965. Bonner Zoologische Beiträge 36: 1-8.
Van der Straeten, E. 1999. Notes on *Mastomys pernanus* (Kershaw, 1921). Bonner Zoologische Beiträge 48: 225-230.
Van der Straeten, E. 2008. Notes on the *Praomys* of Angola with the description of a new species (Mammalia: Rodentia: Muridae). Stuttgarter Beiträge zur Naturkunde A 1: 123-131.
Van der Straeten, E. & Dieterlen, F. 1983. Description de *Praomys ruppi*, une nouvelle espèce de Muridae d'Ethiopie. Annale Musée Royal d'Afrique Centrale, Sciences Zoolologique 237: 121-127.
Van der Straeten, E. & Dieterlen, F. 1987. *Praomys misonnei*, a new species of Muridae from eastern Zaire (Mammalia). Stuttgarter Beiträge zur Naturkunde A 402: 1-11.
Van der Straeten, E. & Dudu, A. 1990. Systematics and distribution of *Praomys* from the Masako Forest Reserve (Zaire), with the description of a new species. In: Vertebrates in the Tropics (G. peters & R. Hutterer, eds). Museum Alexander Koenig, Bonn.

Van der Straeten, E. & Hutterer, R. 1986. *Hybomys eisentrauti*, une nouvelle espèce de Muridae du Cameroun (Mammalia, Rodentia). Mammalia 50: 35-42.

Van der Straeten, E. & Kerbis Peterhans, J. 1999. *Praomys degraaffi*, a new species of Muridae (Mammalia) from central Africa. South African Journal of Zoology 34: 80-90.

Van der Straeten, E. & Robbins, C. B. 1997. Further studies on *Mastomys* (Rodentia: Muridae) types and generic distinctions among African Muridae. Mitteilungen aus dem Zoologisches Museum, Berlin 73: 153-163.

Van der Straeten, E. & Verheyen, W. 1978a. Karyological and morphological comparisons of *Lemniscomys striatus* (Linnaeus, 1758) and *Lemniscomys bellieri* van der Straeten, 1975, from Ivory Coast (Mammalia: Muridae). Bulletin Carnegie Museum of Natural History 6: 41-47.

Van der Straeten, E. & Verheyen, W. 1978b. Taxonomical notes on the West-African *Myomys* with the description of *Myomys derooi* (Mammalia-Muridae). Zeitschrift fur Säugetierkunde 43: 31-41.

Van der Straeten, E. & Verheyen, W. 1979a. Notes taxonomiques sur les *Malacomys* de l'Ouest africain avec redescription du patron chromosomique de *Malacomys edwardsi* (Mammalia, Muridae). Revue Zoologique Africaine 93: 10-35.

Van der Straeten, E. & Verheyen, W. 1979b. Note sur la position systematique de *Lemniscomys macculus* (Thomas et Wroughton, 1910) (Mammalia, Muridae). Mammalia 43: 377-389.

Van der Straeten, E. & Verheyen, W. 1980. Relations biometriques dans le groupe specifique *Lemniscomys striatus* (Mammalia, Muridae). Mammalia 44: 73-82.

Van der Straeten, E. & Verheyen, W. 1981. Etude biometrique du genre *Praomys* en Côte d'Ivoire. Bonner Zoologische Beiträge 32: 249-264.

Van der Straeten, E. & Verheyen, W. 1982. Differences biometriques entre *Hybomys univittatus* (Peters) et *Hybomys trivirgatus* (Temminck) de l'Afrique de l'ouest. Bonner Zoologische Beiträge 33: 205-213.

Van der Straeten, E. & Verheyen, W. 1983. Nouvelles captures de *Lophuromys rahmi* et *Delanymys brooksi* en Republique Rwandaise. Mammalia 47: 426-429.

Van der Straeten, E. & Verheyen, W. 1985. Première capture de *Lemniscomys striatus* au Burkina Faso. Mammalia 49: 576-577.

Van der Straeten, E., Verheyen, W. & Harrie, B. 1986. The taxonomic status of *Hybomys univittatus lunaris* Thomas, 1906 (Mammalia: Muridae). Cimbebasia: 1-10.

Van der Straeten, E., Lecompte, E. & Denys, C. 2003. *Praomys petteri*: un nouvelle espèce de Muridae africain (Mammalia, Rodentia). Bonner zoologische Beiträge 50: 329-345.

Van Hensbergen, H. J. & Channing, A. 1989. Habitat preference and use of space by the namtap *Graphiurus ocularis* (Rodentia: Gliridae). Mammalia 53: 25-33.

Van Weers, D. 1993. On the tooth morphology of the long-tailed porcupine *Trichys fasciculata* (Hystricidae: Rodentia), with notes on the genera *Atherurus* and *Hystrix*. Raffles Bulletin of Zoology 41: 251-261.

Venturi, F. P., Chimimba, C. T., Van Aarde, R. J. & Fairall, N. 2004. The distribution of two medically and agriculturally important cryptic rodent species, *Mastomys natalensis* and *M. coucha* (Rodentia: Muridae) in South Africa. African Zoology 39: 235-245.

Verheyen, W. N. 1964. New data on *Lophuromys luteogaster* Hatt 1934. Revue de Zoologie et de Botanique Africaine 70: 341-350.

Verheyen, W. & Van der Straeten, E. 1977. Description of *Malacomys verschureni*, a new Murid-species from Central Africa (Mammalia-Muridae). Revue de Zoologie Africaine 91: 737-744.

Verheyen, W. & Van der Straeten, E. 1980. The caryotype of *Lophuromys nudicaudus* Heller 1911 (Mammalia - Muridae). Revue de Zoologie Africaine 94: 311-316.

Verheyen, W. & Van der Straeten, E. 1985. Karyological comparison of three different species of *Hybomys* (Mammalia: Muridae). In: Proceedings of the International Symposium on African Vertebrates (K-L. Schuchmann, ed). Zoologisches Forschungsinstitut und Museum Alexander Koenig, Bonn 1985.

Verheyen, W. & Verschuren, J. 1966. Rongeurs et Lagomorphes. Exploration du Parc National de la Garamba. Institut des Parcs Nationaux du Congo, Bruxelles 50: 1-71.

Verheyen, E., Colyn, M. & Verheyen, W. 1995. The phylogeny of some African muroids (Rodentia) based upon partial mitochondrial cytochrome b sequences. Belgian Journal of Zoology 125: 403-407.

Verheyen, E., Colyn, M. & Verheyen, W. 1996a. A mitochondrial cytochrome b phylogeny confirms the paraphyly of the Dendromurinae Alston, 1896 (Muridae, Rodentia). Mammalia 60: 780-785.
Verheyen, W., Colyn, M. & Verheyen, W. 1996b. Re-evaluation of the *Lophuromys nudicaudus* Heller, 1911 species-complex with a description of a new species from Zaire (Muridae – Rodentia). . Bulletin de l'Institut Royal des Sciences Naturelles de Belgique, Biologie 66: 241-273.
Verheyen, W., Hulselmans, J., Colyn, M. & Hutterer, R. 1997. Systematics and zoogeography of the small mammal fauna of Cameroun: description of two new *Lophuromys* (Rodentia: Muridae) endemic to Mount Cameroun and Mount Oku. Bulletin de l'Institut Royal des Sciences Naturelles de Belgique, Biologie 67: 163-186.
Verheyen, W., Dierckx, T. & Hulselmans, J. 2000. The brush-furred rats of Angola and southern Congo: description of a new taxon of the *Lophuromys sikapusi* species complex. Bulletin de l'Institut Royal des Sciences Naturelles de Belgique 70: 253-267.
Verheyen, W., Hulselmans, J., Dierckx, T. & Verheyen, E. 2002. The *Lophuromys flavopunctatus* Thomas 1888 s.l. species complex: a craniometric study, with the description and genetic characterization of two new species (Rodentia, Muridae, Africa). Bulletin de l'Institut Royal des Sciences Naturelles de Belgique 72: 141-182.
Verheyen, W., Hulselmans, J., Dierckx, T., Colyn, M., Leirs, H. & Verheyen, E. 2003. A craniometric and genetic approach to the systematics of the genus *Dasymys* Peters, 1875, selection of a neotype and description of three new taxa (Rodentia, Muridae, Africa). Bulletin de l'Institut Royal des Sciences Naturelles de Belgique, Biologie 73: 27-71.
Verheyen, W., Hulselmans, J., Dierckx, T., Mulungu, L., Leirs, H., Corti, M. & Verheyen, E. 2007. The Characterization of the Kilimanjaro *Lophuromys aquilus* True 1892 population and description of five new *Lophuromys* species (Rodentia, Muridae). Bulletin de l'Institut Royal des Sciences Naturelles de Belgique 77: 23-75.
Verheyen, W., Hulselmans, J., Wendelen, W., Leirs, H., Corti, M., Backeljau, T. & Verheyen, E. 2011. Contribution to the systematics and zoogeography of the East-African *Acomys spinosissimus* Peters 1852 species complex and the description of two new species (Rodentia: Muridae). Zootaxa 3059: 1-35.
Vernon, C. J. 1972. An analysis of owl pellets collected in southern Africa. Ostrich 43: 109-124.
Veyrunes, F., Catalan, J., Sicard, B., Robinson, T., Duplantier, J., Granjon, L., Dobigny, G. & Britton-Davidian, J. 2004. Autosome and Sex Chromosome Diversity Among the African Pygmy Mice, Subgenus *Nannomys* (Murinae; Mus). Chromosome Research 12: 369-382.
Veyrunes, F., Britton-Davidian, J., Robinson, T. J., Calvet, E., Denys, C. & Chevret, P. 2005. Molecular phylogeny of the African pygmy mice, subgenus *Nannomys* (Rodentia, Murinae, Mus): Implications for chromosomal evolution. Molecular Phylogenetics and Evolution 36: 358-369.
Veyrunes, F., Dobigny, G., Yang, F., O'Brien, P. C. M., Catalan, J. Robinson, T. J. & Britton-Davidian, J. 2006. Phylogenomics of the genus *Mus* (Rodentia; Muridae): extensive genome repatterning is not restricted to the house mouse. Proceedings of the Royal Society B 273: 2925-2934.
Veyrunes, F., Catalan, J., Tatard, C., Cellier-Holzem, E., Watson, J., Chevret, P., Robinson, T. J. & Britton-Davidian, J. 2010. Mitochondrial and chromosomal insights into karyotypic evolution of the pygmy mouse, *Mus minutoides*, in South Africa. Chromosome Research 18: 563-574.
Vianey-Liaud, M. & Jaeger, J. J. 1996. A new hypothesis for the origin of African Anomaluridae and Graphiuridae (Rodentia). Paleovertebrata 25: 349-358.
Viegas-Pequignot, E., Dutrillaux, B., Prod'Homme, M. & Petter, F. 1983. Chromosomal phylogeny of Muridae: a study of 10 genera. Cytogenetics and Cell Genetics 35: 269-278.
Viljoen, S. 1977a. Behaviour of the bush squirrel, *Paraxerus cepapi cepapi* (A. Smith, 1836). Mammalia 41: 119-166.
Viljoen, S. 1977b. Factors affecting breeding synchronization in an African bush squirrel (*Paraxerus cepapi cepapi*). Journal of Reproduction and Fertility 50: 125-127.
Viljoen, S. 1983a. Feeding habits and comparative feeding rates of three southern African arboreal squirrels. South African Journal of Zoology 18: 378-387.

Viljoen, S. 1983b. Communicatory behaviour of southern African tree squirrels, *Paraxerus palliatus ornatus, P. p. tongensis, P. c. cepapi* and *Funisciurus congicus*. Mammalia 47: 441-461.
Viljoen, S. 1986. Use of space in southern African tree squirrels. Mammalia 50: 293-309.
Viljoen, S. & du Toit, S. 1985. Postnatal development and growth of southern African tree squirrels in the genera *Funisciurus* and *Paraxerus*. Journal of Mammalogy 66: 119-127.
Visser, D. S. & Robinson, T. J. 1986. Cytosystematics of the South African *Aethomys* (Rodentia: Muridae). South African Journal of Zoology 21: 264-268.
Visser, D. S. & Robinson, T. J. 1987. Systematic implications of spermatozoan and bacular morphology for the South African *Aethomys*. Mammalia 51: 447-454.
Volobouev, V., Viegas-Peguignot, E., Petter, F. & Dutrillaux, B. 1987. Karyotypic diversity and taxonomic problems in the genus *Arvicanthis* (Rodentia, Muridae). Genetica 72: 147-150.
Volobouev, V., Viegas-Peguignot, E., Petter, F., Gautun, J. C., Sicard, B. & Dutrillaux, B. 1988. Complex chromosomal polymorphism in *Gerbillus nigeriae* (Rodentia, Gerbillidae). Journal of Mammalogy 69: 131-134.
Volobouev, V., Lombard, M., Tranier, M. & Dutrillaux, B. 1995. Chromosome-banding study in Gerbillinae (Rodentia). I. Comparative analysis of *Gerbillus poecilops, G. henleyi* and *G. nanus*. Journal of Zoological Systematics and Evolution Research 33: 54-61.
Volobouev, V., Sicard, B., Aniskin, V., Gautun, J. & Granjon, L. 2000. Robertsonian polymorphism, B chromosomes variation and sex chromosomes heteromorphism in the African water rat *Dasymys* (Rodentia, Muridae). Chromosome Research 8: 689-697.
Volobouev, V., Hoffmann, A., Sicard, B. & Granjon, L. 2001. Polymorphism and polytypy for pericentric inversions in 38-chromosome *Mastomys* (Rodentia, Murinae) and possible taxonomic implications. Cytogenetics and Cell Genetics 92: 237-242.
Volobouev, V., Aniskin, V. M., Lecompte, E. & Ducroz, J-F. 2002a. Patterns of karyotype evolution in complexes of sibling species within three genera of African murid rodents inferred from the comparison of cytogenetic and molecular data. Cytogenetic and Genome Research 96: 261-275.
Volobouev, V. Ducroz, J-F., Aniskin, V., Britton-Davidian, J., Castiglia, R., Dobigny, G., Granjon, L., Lombard, M., Corti, M., Sicard, B. & Capanna, E. 2002b. Chromosomal characterization of *Arvicanthis* species (Rodentia, Murinae) from Western and Central Africa: implications for taxnomy. Cytogenetic and Genome Research 96: 250-260.
Volobouev , V., Auffray, J. C., Debat , V., Denys C., Gautun, J.C., Tranier, M. 2007a. Species delimitation in the *Acomys dimidiatus - cahirinus* complex inferred from chromosomal and morphological analyses . Biological Journal of the Linnean Society 91: 203-214.
Volobouev, V., Aniskin, V. M., Sicard, B., Dobigny, G. & Granjon, L. 2007b. Systematics and phylogeny of West African gerbils of the genus *Gerbilliscus* (Muridae: Gerbillinae) inferred from comparative G- and C-banding chromosomal analyses. Cytogenetic and Genome Research 116: 269-281.
Vorontsov, N. 1966. Taxonomic position and a survey of the hamsters of the genus *Mystromys* Wagner (Mammalia, Glires). Zoologischeskii Zhurnal 45: 436-4446 (in Russian with English summary).
Wahlert, J. 1984. Relationships of the extinct rodent *Cricetops* to *Lophiomys* and the Cricetinae (Rodentia, Cricetidae). American Museum Novitates 2784: 1-15.
Wahlert, J. H., Sawitzke, S. L. & Holden, M. E. 1993. Cranial anatomy and relationships of dormice (Rodentia, Myoxidae). American Museum Novitates 3061: 1-32.
Waterman, J. M. 1995. The social organization of the Cape ground squirrel (*Xerus inauris*; Rodentia: Sciuridae). Ethology 101: 130-147.
Waterman, J. M. 1996. Reproductive biology of a tropical, non-hibernating ground squirrel. Journal of Mammalogy 77: 134-146.
Waterman, J. M. 2002. Delayed maturity, group fission and the limits of group size in female Cape ground squirrels (Sciuridae: *Xerus inauris*). Journal of Zoology, London 256: 113-120.
Waterman, J. M. & Herron, M. D. 2004. *Xerus princeps*. Mammalian Species 751: 1-3.
Watts, C. H. S. & Baverstock, P. R. 1995. Evolution in some African Murinae (Rodentia) assessed by microcomplement fixation of albumin. Journal of African Zoology 109: 423-433.

Webb, P. I. & Skinner, J. D. 1995. The dormice (Myoxidae) of southern Africa. Hystrix 6: 287-293.
Webb, P. I. & Skinner, J. D. 1996. Summer torpor in African woodland dormice *Graphiurus murinus* (Myoxidae: Graphiurinae). Journal of Comparative Physiology B 166: 325-330.
Wesselman, H. B. 1984. The Omo micromammals. Systematics and paleoecology of Early Man sites from Ethiopia. In: Contributions to Vertebrate Evolution (M. K. Hecht & F. S. Szalay, eds), Basel.
Wesselman, H. B., Black, M. T., Asnake, M. 2009. Small mammals. In: *Ardipithecus kaddaba*: Late Miocene evidence from the Middle Awash, Ethiopia (Y. Haile-Selassie & G Woldegabriel, eds.), University of California Press, Berkeley.
Wessels, W. 1999. Family Gerbillidae. In: The Miocene Land Mammals of Europe (G. E. Rossner & K. Heissig). Dr Friedrich Pfeil, Munich.
Westhuizen, L., Jarvis, J. & Bennett, N. 2013. A case of natural queen succession in a captive colony of naked mole-rats, *Heterocephalus glaber*. African Zoology 48: 56-63.
Williams, S. L., Schlitter, D. A. & Robbins, L. W. 1983. Morphological variation in a natural population of *Cryptomys* (Rodentia, Bathyergidae) from Cameroon. Annales du Musée Royal de l'Afrique Centrale, Sciences Zoologiques 237: 159-172.
Wilson, D. E. & Reeder, D. M. 2005. Mammal Species of the World: A Taxonomic and Geographic Reference. 3rd edition. The John Hopkins University Press, Baltimore.
Winkler, A. J. 1992. Systematics and biogeography of middle Miocene rodents from the Muruyur Beds, Baringo District, Kenya. Journal of Vertebrate Paleontology 12: 236-249.
Winkler, A. J. 1997. Systematic, paleobiogeography and paleoenvironmental significance of rodents from the Ibole Member, Manonga Valley, Tanzania. In: Neogene Paleontology of the Manonga Valley, Tanzania (T. Harrison, ed.). Topics in Geobiology, Plenum Press, New York.
Winkler, A. J. 2002. Neogene paleobiogeography and East African paleoenvironments: contributions from the Tugen Hills rodents and lagomorphs. Journal of Human Evolution 42: 237-256.
Winkler, A. J. 2003. Rodents and lagomorphs from the Miocene and Pliocene of Lothagam, Northern Kenya. In: Lothagam: the Dawn of Humanity in Eastern Africa (M. G. Leakey & J. M. Harris, eds). Columbia University Press, New York.
Winkler, A. J., MacLatchy, L. & Mafabi, M. 2005. Small Rodents and a Lagomorph from the Early Miocene Bukwa locality, Eastern Uganda. Palaeontologia Electronica 8: 1-12.
Winkler, A. J., Denys, C., Avery, M. 2010. Fossil rodents of Africa. In: Fossil Mammals of Africa (L. Werdelin & W. Sanders, eds.). California University Press, Berkeley.
Wirminghaus, J. O. & Perrin, M. R. 1992. Diets of small mammals in a Southern African temperate forest. Israel Journal of Zoology 38: 353-361.
Withers, P. C. 1979. Ecology of a small mammal community on a rocky outcrop of the Namib Desert. Madoqua 11: 229-246.
Withers, P. C. 1983. Seasonal reproduction by small mammals of the Namib desert. Mammalia 47: 105-204.
Wood, A. E. 1955. A revised classification of the rodents. Journal of Mammalogy 36: 165-187.
Wood, A. E. 1965. Unworn teeth and relationships of the African rodent, *Pedetes*. Journal of Mammalogy 46: 419-423.
Woods, C. A. & Kilpatrick, B. 2005. Infraorder Hystricognathi. In: Mammal Species of the World: A Taxonomic and Geographic Reference (3rd edn) (D. E. Wilson and D. Reeder, eds). John Hopkins University Press, Baltimore.
Woods, C. A., Conreras, L., Willner-Chapman, G. & Whidden, H. P. 1992. *Myocaster coypus*. Mammalian Species 398: 1-8.
Yalden, D. W. 1988. Small mammals of the Bale Mountains, Ethiopia. African Journal of Ecology 26: 281-294.
Yalden, D. W. & Largen, M. J. 1992. The endemic mammals of Ethiopia. Mammal Review 22: 115-150.
Yalden, D. W., Largen, M. J. & Kock, D. 1976. Catalogue of the mammals of Ethiopia: 2. Insectivora and Rodentia. Monitore Zoologico Italiano S8: 1-118.
Yalden, D. W., Largen, M. J., Kock, D. & Hillman, J. C. 1996. Catalogue of the mammals of Ethiopia and Eritrea. 7. Revised checklist, zoogeography and conservation. Tropical Zoology 9: 73-164.

Yeboah, S. 1998. Small mammal diversity in the Kakum national park in Ghana. Ghana Journal of Science 38: 25-32.
Yeboah, S. & Dakawa, K. B. 2002. Colony and social structure of the Ghana mole-rat (*Cryptomys zechi*, Matchie) (Rodentia: Bathyergidae). Journal of Zoology, London 256: 85-91.

Appendix

A checklist of the 463 rodent species recorded here as occurring in Sub-Saharan Africa, together with the conservation status (as of January 2014) of each species (IUCN, 2014): CR = Critically Endangered; EN = Endangered; VU = Vulnerable; NT = Near Threatened; LC = Least Concern; DD = Data Deficient; NE = Not Evaluated.

Family	Subfamily	Genus	Species	IUCN status
Anomaluridae	Anomalurinae	*Anomalurus*	*beecrofti*	LC
Anomaluridae	Anomalurinae	*Anomalurus*	*derbianus*	LC
Anomaluridae	Anomalurinae	*Anomalurus*	*pelii*	DD
Anomaluridae	Anomalurinae	*Anomalurus*	*pusillus*	LC
Anomaluridae	Zenkerellinae	*Idiurus*	*macrotis*	LC
Anomaluridae	Zenkerellinae	*Idiurus*	*zenkeri*	LC
Anomaluridae	Zenkerellinae	*Zenkerella*	*insignis*	LC
Bathyergidae		*Bathyergus*	*janetta*	LC
Bathyergidae		*Bathyergus*	*suillus*	LC
Bathyergidae		*Cryptomys*	*hottentotus*	LC
Bathyergidae		*Cryptomys*	*mahali*	NE
Bathyergidae		*Cryptomys*	*natalensis*	NE
Bathyergidae		*Cryptomys*	*nimrodi*	NE
Bathyergidae		*Cryptomys*	*pretoriae*	NE
Bathyergidae		*Fukomys*	*amatus*	NE
Bathyergidae		*Fukomys*	*anselli*	NT
Bathyergidae		*Fukomys*	*bocagei*	LC
Bathyergidae		*Fukomys*	*damarensis*	LC
Bathyergidae		*Fukomys*	*darlingi*	LC
Bathyergidae		*Fukomys*	*foxi*	DD
Bathyergidae		*Fukomys*	*ilariae*	NE
Bathyergidae		*Fukomys*	*kafuensis*	VU
Bathyergidae		*Fukomys*	*mechowi*	LC
Bathyergidae		*Fukomys*	*micklemi*	NE
Bathyergidae		*Fukomys*	*ochraceocinereus*	LC
Bathyergidae		*Fukomys*	*vandewoestijneae*	NE
Bathyergidae		*Fukomys*	*whytei*	NE
Bathyergidae		*Fukomys*	*zechi*	LC
Bathyergidae		*Georychus*	*capensis*	LC
Bathyergidae		*Heliophobius*	*argenteocinereus*	LC
Bathyergidae		*Heliophobius*	*emini*	NE
Bathyergidae		*Heliophobius*	*kapiti*	NE
Bathyergidae		*Heliophobius*	*marungensis*	NE

Bathyergidae		*Heliophobius*	*mottuolei*	NE
Bathyergidae		*Heliophobius*	*spalax*	NE
Bathyergidae		*Heterocephalus*	*glaber*	LC
Ctenodactylidae		*Felovia*	*vae*	DD
Ctenodactylidae		*Massoutiera*	*mzabi*	LC
Ctenodactylidae		*Pectinator*	*spekei*	LC
Dipodidae		*Jaculus*	*deserti*	NE
Dipodidae		*Jaculus*	*jaculus*	LC
Gliridae		*Graphiurus*	*angolensis*	DD
Gliridae		*Graphiurus*	*christyi*	LC
Gliridae		*Graphiurus*	*crassicaudatus*	DD
Gliridae		*Graphiurus*	*johnstoni*	DD
Gliridae		*Graphiurus*	*kelleni*	LC
Gliridae		*Graphiurus*	*lorraineus*	LC
Gliridae		*Graphiurus*	*microtis*	LC
Gliridae		*Graphiurus*	*monardi*	DD
Gliridae		*Graphiurus*	*murinus*	LC
Gliridae		*Graphiurus*	*nagtglasii*	LC
Gliridae		*Graphiurus*	*ocularis*	LC
Gliridae		*Graphiurus*	*platyops*	LC
Gliridae		*Graphiurus*	*rupicola*	LC
Gliridae		*Graphiurus*	*surdus*	LC
Gliridae		*Graphiurus*	*walterverheyeni*	NE
Hystricidae		*Atherurus*	*africanus*	LC
Hystricidae		*Hystrix*	*africaeaustralis*	LC
Hystricidae		*Hystrix*	*cristata*	LC
Muridae	Deomyinae	*Acomys*	*cahirinus*	LC
Muridae	Deomyinae	*Acomys*	*chudeaui*	NE
Muridae	Deomyinae	*Acomys*	*cineraceus*	LC
Muridae	Deomyinae	*Acomys*	*ignitus*	LC
Muridae	Deomyinae	*Acomys*	*johannis*	LC
Muridae	Deomyinae	*Acomys*	*kempi*	LC
Muridae	Deomyinae	*Acomys*	*louisae*	LC
Muridae	Deomyinae	*Acomys*	*mullah*	LC
Muridae	Deomyinae	*Acomys*	*muzei*	NE
Muridae	Deomyinae	*Acomys*	*ngurui*	NE
Muridae	Deomyinae	*Acomys*	*percivali*	LC
Muridae	Deomyinae	*Acomys*	*selousi*	NE
Muridae	Deomyinae	*Acomys*	*spinosissimus*	LC
Muridae	Deomyinae	*Acomys*	*subspinosus*	LC
Muridae	Deomyinae	*Acomys*	*wilsoni*	LC
Muridae	Deomyinae	*Deomys*	*ferrugineus*	LC
Muridae	Deomyinae	*Lophuromys*	*angolensis*	NE

Muridae	Deomyinae	*Lophuromys*	*ansorgei*	NE
Muridae	Deomyinae	*Lophuromys*	*aquilus*	NE
Muridae	Deomyinae	*Lophuromys*	*brevicaudus*	NT
Muridae	Deomyinae	*Lophuromys*	*brunneus*	NE
Muridae	Deomyinae	*Lophuromys*	*chercherensis*	NE
Muridae	Deomyinae	*Lophuromys*	*chrysopus*	LC
Muridae	Deomyinae	*Lophuromys*	*cinereus*	DD
Muridae	Deomyinae	*Lophuromys*	*dieterleni*	EN
Muridae	Deomyinae	*Lophuromys*	*dudi*	NE
Muridae	Deomyinae	*Lophuromys*	*eisentrauti*	EN
Muridae	Deomyinae	*Lophuromys*	*flavopunctatus*	LC
Muridae	Deomyinae	*Lophuromys*	*huttereri*	LC
Muridae	Deomyinae	*Lophuromys*	*kilonzoi*	NE
Muridae	Deomyinae	*Lophuromys*	*laticeps*	NE
Muridae	Deomyinae	*Lophuromys*	*luteogaster*	LC
Muridae	Deomyinae	*Lophuromys*	*machangui*	NE
Muridae	Deomyinae	*Lophuromys*	*makundii*	NE
Muridae	Deomyinae	*Lophuromys*	*margarettae*	NE
Muridae	Deomyinae	*Lophuromys*	*medicaudatus*	VU
Muridae	Deomyinae	*Lophuromys*	*melanonyx*	VU
Muridae	Deomyinae	*Lophuromys*	*menageshae*	NE
Muridae	Deomyinae	*Lophuromys*	*nudicaudus*	LC
Muridae	Deomyinae	*Lophuromys*	*pseudosikapusi*	NE
Muridae	Deomyinae	*Lophuromys*	*rahmi*	EN
Muridae	Deomyinae	*Lophuromys*	*rita*	NE
Muridae	Deomyinae	*Lophuromys*	*roseveari*	LC
Muridae	Deomyinae	*Lophuromys*	*sabunii*	NE
Muridae	Deomyinae	*Lophuromys*	*sikapusi*	LC
Muridae	Deomyinae	*Lophuromys*	*simiensis*	NE
Muridae	Deomyinae	*Lophuromys*	*stanleyi*	NE
Muridae	Deomyinae	*Lophuromys*	*verhageni*	NE
Muridae	Deomyinae	*Lophuromys*	*woosnami*	LC
Muridae	Deomyinae	*Lophuromys*	*zena*	NE
Muridae	Deomyinae	*Uranomys*	*ruddi*	LC
Muridae	Gerbillinae	*Ammodillus*	*imbellis*	DD
Muridae	Gerbillinae	*Desmodilliscus*	*braueri*	LC
Muridae	Gerbillinae	*Desmodillus*	*auricularis*	LC
Muridae	Gerbillinae	*Gerbilliscus*	*afra*	LC
Muridae	Gerbillinae	*Gerbilliscus*	*boehmi*	LC
Muridae	Gerbillinae	*Gerbilliscus*	*brantsii*	LC
Muridae	Gerbillinae	*Gerbilliscus*	*gambianus*	LC
Muridae	Gerbillinae	*Gerbilliscus*	*giffardi*	NE
Muridae	Gerbillinae	*Gerbilliscus*	*guineae*	LC

Muridae	Gerbillinae	*Gerbilliscus*	*inclusus*	LC
Muridae	Gerbillinae	*Gerbilliscus*	*kempi*	LC
Muridae	Gerbillinae	*Gerbilliscus*	*leucogaster*	LC
Muridae	Gerbillinae	*Gerbilliscus*	*nigricaudus*	LC
Muridae	Gerbillinae	*Gerbilliscus*	*paeba*	LC
Muridae	Gerbillinae	*Gerbilliscus*	*phillipsi*	LC
Muridae	Gerbillinae	*Gerbilliscus*	*robustus*	LC
Muridae	Gerbillinae	*Gerbilliscus*	*setzeri*	LC
Muridae	Gerbillinae	*Gerbilliscus*	*tytonis*	LC
Muridae	Gerbillinae	*Gerbilliscus*	*validus*	LC
Muridae	Gerbillinae	*Gerbilliscus*	*vallinus*	LC
Muridae	Gerbillinae	*Gerbilliscus*	*vicinus*	NE
Muridae	Gerbillinae	*Gerbillus*	*acticola*	DD
Muridae	Gerbillinae	*Gerbillus*	*agag*	DD
Muridae	Gerbillinae	*Gerbillus*	*bottai*	DD
Muridae	Gerbillinae	*Gerbillus*	*brockmani*	DD
Muridae	Gerbillinae	*Gerbillus*	*campestris*	LC
Muridae	Gerbillinae	*Gerbillus*	*dunni*	LC
Muridae	Gerbillinae	*Gerbillus*	*gerbillus*	LC
Muridae	Gerbillinae	*Gerbillus*	*harwoodi*	LC
Muridae	Gerbillinae	*Gerbillus*	*henleyi*	LC
Muridae	Gerbillinae	*Gerbillus*	*lowei*	DD
Muridae	Gerbillinae	*Gerbillus*	*mackilligini*	LC
Muridae	Gerbillinae	*Gerbillus*	*nancillus*	DD
Muridae	Gerbillinae	*Gerbillus*	*nanus*	LC
Muridae	Gerbillinae	*Gerbillus*	*nigeriae*	LC
Muridae	Gerbillinae	*Gerbillus*	*pulvinatus*	LC
Muridae	Gerbillinae	*Gerbillus*	*pusillus*	LC
Muridae	Gerbillinae	*Gerbillus*	*pyramidum*	LC
Muridae	Gerbillinae	*Gerbillus*	*rosalinda*	LC
Muridae	Gerbillinae	*Gerbillus*	*rupicola*	LC
Muridae	Gerbillinae	*Gerbillus*	*somalicus*	DD
Muridae	Gerbillinae	*Gerbillus*	*stigmonyx*	DD
Muridae	Gerbillinae	*Gerbillus*	*tarabuli*	LC
Muridae	Gerbillinae	*Gerbillus*	*watersi*	LC
Muridae	Gerbillinae	*Meriones*	*crassus*	LC
Muridae	Gerbillinae	*Microdillus*	*peeli*	LC
Muridae	Gerbillinae	*Pachyuromys*	*duprasi*	LC
Muridae	Gerbillinae	*Psammomys*	*obesus*	LC
Muridae	Gerbillinae	*Taterillus*	*arenarius*	LC
Muridae	Gerbillinae	*Taterillus*	*congicus*	LC
Muridae	Gerbillinae	*Taterillus*	*emini*	LC
Muridae	Gerbillinae	*Taterillus*	*gracilis*	LC

Muridae	Gerbillinae	*Taterillus*	*lacustris*	LC
Muridae	Gerbillinae	*Taterillus*	*petteri*	LC
Muridae	Gerbillinae	*Taterillus*	*pygargus*	LC
Muridae	Gerbillinae	*Taterillus*	*tranieri*	LC
Muridae	Leimacomyinae	*Leimacomys*	*buettneri*	DD
Muridae	Lophiomyinae	*Lophiomys*	*imhausi*	LC
Muridae	Murinae	*Aethomys*	*bocagei*	LC
Muridae	Murinae	*Aethomys*	*chrysophilus*	LC
Muridae	Murinae	*Aethomys*	*hindei*	LC
Muridae	Murinae	*Aethomys*	*ineptus*	LC
Muridae	Murinae	*Aethomys*	*kaiseri*	LC
Muridae	Murinae	*Aethomys*	*nyikae*	LC
Muridae	Murinae	*Aethomys*	*silindensis*	DD
Muridae	Murinae	*Aethomys*	*stannarius*	DD
Muridae	Murinae	*Aethomys*	*thomasi*	LC
Muridae	Murinae	*Arvicanthis*	*abyssinicus*	LC
Muridae	Murinae	*Arvicanthis*	*ansorgei*	LC
Muridae	Murinae	*Arvicanthis*	*blicki*	NT
Muridae	Murinae	*Arvicanthis*	*nairobae*	LC
Muridae	Murinae	*Arvicanthis*	*neumanni*	LC
Muridae	Murinae	*Arvicanthis*	*niloticus*	LC
Muridae	Murinae	*Arvicanthis*	*rufinus*	LC
Muridae	Murinae	*Colomys*	*goslingi*	LC
Muridae	Murinae	*Dasymys*	*alleni*	NE
Muridae	Murinae	*Dasymys*	*cabrali*	NE
Muridae	Murinae	*Dasymys*	*capensis*	NE
Muridae	Murinae	*Dasymys*	cf *incomtus*	NE
Muridae	Murinae	*Dasymys*	*foxi*	DD
Muridae	Murinae	*Dasymys*	*griseifrons*	NE
Muridae	Murinae	*Dasymys*	*incomtus*	LC
Muridae	Murinae	*Dasymys*	*longipilosus*	NE
Muridae	Murinae	*Dasymys*	*medius*	NE
Muridae	Murinae	*Dasymys*	*montanus*	EN
Muridae	Murinae	*Dasymys*	*nudipes*	DD
Muridae	Murinae	*Dasymys*	*robertsi*	NE
Muridae	Murinae	*Dasymys*	*rufulus*	LC
Muridae	Murinae	*Dasymys*	*rwandae*	NE
Muridae	Murinae	*Dasymys*	*sua*	NE
Muridae	Murinae	*Dephomys*	*defua*	LC
Muridae	Murinae	*Dephomys*	*eburneae*	NE
Muridae	Murinae	*Desmomys*	*harringtoni*	LC
Muridae	Murinae	*Desmomys*	*yaldeni*	EN
Muridae	Murinae	*Grammomys*	*aridulus*	DD

Muridae	Murinae	*Grammomys*	*brevirostris*	NE
Muridae	Murinae	*Grammomys*	*buntingi*	DD
Muridae	Murinae	*Grammomys*	*caniceps*	DD
Muridae	Murinae	*Grammomys*	*cometes*	LC
Muridae	Murinae	*Grammomys*	*dolichurus*	LC
Muridae	Murinae	*Grammomys*	*dryas*	NT
Muridae	Murinae	*Grammomys*	*gigas*	EN
Muridae	Murinae	*Grammomys*	*ibeanus*	LC
Muridae	Murinae	*Grammomys*	*kuru*	LC
Muridae	Murinae	*Grammomys*	*macmillani*	LC
Muridae	Murinae	*Grammomys*	*minnae*	VU
Muridae	Murinae	*Grammomys*	*poensis*	NE
Muridae	Murinae	*Grammomys*	*selousi*	NE
Muridae	Murinae	*Heimyscus*	*fumosus*	LC
Muridae	Murinae	*Hybomys*	*badius*	EN
Muridae	Murinae	*Hybomys*	*basilii*	EN
Muridae	Murinae	*Hybomys*	*lunaris*	VU
Muridae	Murinae	*Hybomys*	*planifrons*	LC
Muridae	Murinae	*Hybomys*	*trivirgatus*	LC
Muridae	Murinae	*Hybomys*	*univittatus*	LC
Muridae	Murinae	*Hylomyscus*	*aeta*	LC
Muridae	Murinae	*Hylomyscus*	*alleni*	LC
Muridae	Murinae	*Hylomyscus*	*anselli*	NE
Muridae	Murinae	*Hylomyscus*	*arcimontensis*	NE
Muridae	Murinae	*Hylomyscus*	*baeri*	EN
Muridae	Murinae	*Hylomyscus*	*carillus*	LC
Muridae	Murinae	*Hylomyscus*	*denniae*	LC
Muridae	Murinae	*Hylomyscus*	*endorobae*	NE
Muridae	Murinae	*Hylomyscus*	*grandis*	CR
Muridae	Murinae	*Hylomyscus*	*pamfi*	NE
Muridae	Murinae	*Hylomyscus*	*parvus*	LC
Muridae	Murinae	*Hylomyscus*	*stella*	LC
Muridae	Murinae	*Hylomyscus*	*volcanorum*	NE
Muridae	Murinae	*Hylomyscus*	*walterverheyeni*	NE
Muridae	Murinae	*Lamottemys*	*okuensis*	EN
Muridae	Murinae	*Lemniscomys*	*bellieri*	LC
Muridae	Murinae	*Lemniscomys*	*griselda*	LC
Muridae	Murinae	*Lemniscomys*	*hoogstraali*	DD
Muridae	Murinae	*Lemniscomys*	*linulus*	LC
Muridae	Murinae	*Lemniscomys*	*macculus*	LC
Muridae	Murinae	*Lemniscomys*	*mittendorfi*	VU
Muridae	Murinae	*Lemniscomys*	*rosalia*	LC
Muridae	Murinae	*Lemniscomys*	*roseveari*	DD

Muridae	Murinae	*Lemniscomys*	*striatus*	LC
Muridae	Murinae	*Lemniscomys*	*zebra*	LC
Muridae	Murinae	*Malacomys*	*cansdalei*	LC
Muridae	Murinae	*Malacomys*	*edwardsi*	LC
Muridae	Murinae	*Malacomys*	*longipes*	LC
Muridae	Murinae	*Mastomys*	*awashensis*	VU
Muridae	Murinae	*Mastomys*	*coucha*	LC
Muridae	Murinae	*Mastomys*	*erythroleucus*	LC
Muridae	Murinae	*Mastomys*	*huberti*	LC
Muridae	Murinae	*Mastomys*	*kollmannspergeri*	LC
Muridae	Murinae	*Mastomys*	*natalensis*	LC
Muridae	Murinae	*Mastomys*	*pernanus*	DD
Muridae	Murinae	*Mastomys*	*shortridgei*	LC
Muridae	Murinae	*Micaelamys*	*granti*	LC
Muridae	Murinae	*Micaelamys*	*namaquensis*	LC
Muridae	Murinae	*Mus (Muriculus)*	*imberbis*	LC
Muridae	Murinae	*Mus*	*baoulei*	LC
Muridae	Murinae	*Mus*	*bufo*	LC
Muridae	Murinae	*Mus*	*callewaerti*	LC
Muridae	Murinae	*Mus*	*goundae*	DD
Muridae	Murinae	*Mus*	*haussa*	LC
Muridae	Murinae	*Mus*	*indutus*	LC
Muridae	Murinae	*Mus*	*mahomet*	LC
Muridae	Murinae	*Mus*	*mattheyi*	LC
Muridae	Murinae	*Mus*	*minutoides*	LC
Muridae	Murinae	*Mus*	*musculoides*	LC
Muridae	Murinae	*Mus*	*musculus*	LC
Muridae	Murinae	*Mus*	*neavei*	DD
Muridae	Murinae	*Mus*	*oubanguii*	DD
Muridae	Murinae	*Mus*	*setulosus*	LC
Muridae	Murinae	*Mus*	*setzeri*	LC
Muridae	Murinae	*Mus*	*sorella*	LC
Muridae	Murinae	*Mus*	*tenellus*	LC
Muridae	Murinae	*Mus*	*triton*	LC
Muridae	Murinae	*Mylomys*	*dybowskii*	LC
Muridae	Murinae	*Mylomys*	*rex*	DD
Muridae	Murinae	*Myomyscus*	*angolensis*	LC
Muridae	Murinae	*Myomyscus*	*brockmani*	LC
Muridae	Murinae	*Myomyscus*	*verreauxii*	LC
Muridae	Murinae	*Nilopegamys*	*plumbeus*	CR
Muridae	Murinae	*Oenomys*	*hypoxanthus*	LC
Muridae	Murinae	*Oenomys*	*ornatus*	LC
Muridae	Murinae	*Otomys*	*anchietae*	LC

Muridae	Murinae	*Otomys*	*angoniensis*	LC
Muridae	Murinae	*Otomys*	*auratus*	NE
Muridae	Murinae	*Otomys*	*barbouri*	EN
Muridae	Murinae	*Otomys*	*burtoni*	EN
Muridae	Murinae	*Otomys*	cf *lacustris*	NE
Muridae	Murinae	*Otomys*	cf *sungae*	NE
Muridae	Murinae	*Otomys*	*cheesmani*	NE
Muridae	Murinae	*Otomys*	*cuanzensis*	LC
Muridae	Murinae	*Otomys*	*dartmouthi*	VU
Muridae	Murinae	*Otomys*	*denti*	LC
Muridae	Murinae	*Otomys*	*dollmani*	NE
Muridae	Murinae	*Otomys*	*fortior*	NE
Muridae	Murinae	*Otomys*	*helleri*	NE
Muridae	Murinae	*Otomys*	*irroratus*	LC
Muridae	Murinae	*Otomys*	*jacksoni*	EN
Muridae	Murinae	*Otomys*	*karoensis*	NE
Muridae	Murinae	*Otomys*	*lacustris*	VU
Muridae	Murinae	*Otomys*	*laminatus*	LC
Muridae	Murinae	*Otomys*	*maximus*	NE
Muridae	Murinae	*Otomys*	*occidentalis*	VU
Muridae	Murinae	*Otomys*	*orestes*	DD
Muridae	Murinae	*Otomys*	*simiensis*	NE
Muridae	Murinae	*Otomys*	*sloggetti*	LC
Muridae	Murinae	*Otomys*	*sungae*	NE
Muridae	Murinae	*Otomys*	*thomasi*	NE
Muridae	Murinae	*Otomys*	*tropicalis*	LC
Muridae	Murinae	*Otomys*	*typus*	LC
Muridae	Murinae	*Otomys*	*unisulcatus*	LC
Muridae	Murinae	*Otomys*	*uzungwensis*	EN
Muridae	Murinae	*Otomys*	*yaldeni*	NE
Muridae	Murinae	*Otomys*	*zinki*	NE
Muridae	Murinae	*Parotomys*	*brantsii*	LC
Muridae	Murinae	*Parotomys*	*littledalei*	LC
Muridae	Murinae	*Pelomys*	*campanae*	LC
Muridae	Murinae	*Pelomys*	*fallax*	LC
Muridae	Murinae	*Pelomys*	*hopkinsi*	DD
Muridae	Murinae	*Pelomys*	*isseli*	DD
Muridae	Murinae	*Pelomys*	*minor*	LC
Muridae	Murinae	*Praomys*	*coetzeei*	NE
Muridae	Murinae	*Praomys*	*daltoni*	LC
Muridae	Murinae	*Praomys*	*degraaffii*	VU
Muridae	Murinae	*Praomys*	*delectorum*	LC
Muridae	Murinae	*Praomys*	*hartwigi*	EN

Muridae	Murinae	*Praomys*	*jacksoni*	LC
Muridae	Murinae	*Praomys*	*lukolelae*	LC
Muridae	Murinae	*Praomys*	*minor*	DD
Muridae	Murinae	*Praomys*	*misonnei*	LC
Muridae	Murinae	*Praomys*	*morio*	EN
Muridae	Murinae	*Praomys*	*mutoni*	DD
Muridae	Murinae	*Praomys*	*obscurus*	EN
Muridae	Murinae	*Praomys*	*petteri*	LC
Muridae	Murinae	*Praomys*	*rostratus*	LC
Muridae	Murinae	*Praomys*	*tullbergi*	LC
Muridae	Murinae	*Praomys*	*verschureni*	DD
Muridae	Murinae	*Rattus*	*norvegicus*	LC
Muridae	Murinae	*Rattus*	*rattus*	LC
Muridae	Murinae	*Rattus*	*tanezumi*	LC
Muridae	Murinae	*Rhabdomys*	*bechuanae*	NE
Muridae	Murinae	*Rhabdomys*	*chakae*	NE
Muridae	Murinae	*Rhabdomys*	*dilectus*	NE
Muridae	Murinae	*Rhabdomys*	*intermedius*	NE
Muridae	Murinae	*Rhabdomys*	*pumilio*	LC
Muridae	Murinae	*Stenocephalemys*	*albipes*	LC
Muridae	Murinae	*Stenocephalemys*	*albocaudata*	LC
Muridae	Murinae	*Stenocephalemys*	*griseicauda*	LC
Muridae	Murinae	*Stenocephalemys*	*ruppi*	DD
Muridae	Murinae	*Stochomys*	*longicaudatus*	LC
Muridae	Murinae	*Thallomys*	*loringi*	LC
Muridae	Murinae	*Thallomys*	*nigricauda*	LC
Muridae	Murinae	*Thallomys*	*paedulcus*	LC
Muridae	Murinae	*Thallomys*	*shortridgei*	DD
Muridae	Murinae	*Thamnomys*	*kempi*	VU
Muridae	Murinae	*Thamnomys*	*major*	NE
Muridae	Murinae	*Thamnomys*	*schoutedeni*	DD
Muridae	Murinae	*Thamnomys*	*venustus*	VU
Muridae	Murinae	*Zelotomys*	*hildegardeae*	LC
Muridae	Murinae	*Zelotomys*	*woosnami*	LC
Myocastoridae		*Myocastor*	*coypus*	LC
Nesomyidae	Cricetomyinae	*Beamys*	*hindei*	LC
Nesomyidae	Cricetomyinae	*Beamys*	*major*	NE
Nesomyidae	Cricetomyinae	*Cricetomys*	*ansorgei*	NE
Nesomyidae	Cricetomyinae	*Cricetomys*	*emini*	LC
Nesomyidae	Cricetomyinae	*Cricetomys*	*gambianus*	LC
Nesomyidae	Cricetomyinae	*Cricetomys*	*kivuensis*	NE
Nesomyidae	Cricetomyinae	*Saccostomus*	*campestris*	LC
Nesomyidae	Cricetomyinae	*Saccostomus*	*mearnsi*	LC

Nesomyidae	Delanymyinae	*Delanymys*	*brooksi*	VU
Nesomyidae	Dendromurinae	*Dendromus*	*insignis*	LC
Nesomyidae	Dendromurinae	*Dendromus*	*kahuziensis*	CR
Nesomyidae	Dendromurinae	*Dendromus*	*lachaisei*	NE
Nesomyidae	Dendromurinae	*Dendromus*	*leucostomus*	NE
Nesomyidae	Dendromurinae	*Dendromus*	*lovati*	LC
Nesomyidae	Dendromurinae	*Dendromus*	*melanotis*	LC
Nesomyidae	Dendromurinae	*Dendromus*	*mesomelas*	LC
Nesomyidae	Dendromurinae	*Dendromus*	*messorius*	LC
Nesomyidae	Dendromurinae	*Dendromus*	*mystacalis*	LC
Nesomyidae	Dendromurinae	*Dendromus*	*nyasae*	NE
Nesomyidae	Dendromurinae	*Dendromus*	*nyikae*	LC
Nesomyidae	Dendromurinae	*Dendromus*	*oreas*	VU
Nesomyidae	Dendromurinae	*Dendromus*	*ruppi*	NE
Nesomyidae	Dendromurinae	*Dendromus*	*vernayi*	DD
Nesomyidae	Dendromurinae	*Dendroprionomys*	*rousseloti*	DD
Nesomyidae	Dendromurinae	*Malacothrix*	*typica*	LC
Nesomyidae	Dendromurinae	*Megadendromus*	*nikolausi*	DD
Nesomyidae	Dendromurinae	*Prionomys*	*batesi*	DD
Nesomyidae	Dendromurinae	*Steatomys*	*bocagei*	LC
Nesomyidae	Dendromurinae	*Steatomys*	*caurinus*	LC
Nesomyidae	Dendromurinae	*Steatomys*	*cuppedius*	LC
Nesomyidae	Dendromurinae	*Steatomys*	*jacksoni*	DD
Nesomyidae	Dendromurinae	*Steatomys*	*krebsii*	LC
Nesomyidae	Dendromurinae	*Steatomys*	*opimus*	LC
Nesomyidae	Dendromurinae	*Steatomys*	*parvus*	LC
Nesomyidae	Dendromurinae	*Steatomys*	*pratensis*	LC
Nesomyidae	Mystromyinae	*Mystromys*	*albicaudatus*	EN
Nesomyidae	Petromyscinae	*Petromyscus*	*barbouri*	LC
Nesomyidae	Petromyscinae	*Petromyscus*	*collinus*	LC
Nesomyidae	Petromyscinae	*Petromyscus*	*monticularis*	LC
Nesomyidae	Petromyscinae	*Petromyscus*	*shortridgei*	LC
Pedetidae		*Pedetes*	*capensis*	LC
Pedetidae		*Pedetes*	*surdaster*	LC
Petromuridae		*Petromus*	*typicus*	LC
Sciuridae		*Allosciurus*	*aubinnii*	DD
Sciuridae		*Epixerus*	*ebii*	LC
Sciuridae		*Funisciurus*	*anerythrus*	LC
Sciuridae		*Funisciurus*	*bayonii*	DD
Sciuridae		*Funisciurus*	*carruthersi*	LC
Sciuridae		*Funisciurus*	*congicus*	LC
Sciuridae		*Funisciurus*	*duchaillui*	NE
Sciuridae		*Funisciurus*	*isabella*	LC

Sciuridae		*Funisciurus*	*lemniscatus*	LC
Sciuridae		*Funisciurus*	*leucogenys*	LC
Sciuridae		*Funisciurus*	*pyrropus*	LC
Sciuridae		*Funisciurus*	*substriatus*	DD
Sciuridae		*Heliosciurus*	*gambianus*	LC
Sciuridae		*Heliosciurus*	*mutabilis*	LC
Sciuridae		*Heliosciurus*	*punctatus*	DD
Sciuridae		*Heliosciurus*	*rufobrachium*	LC
Sciuridae		*Heliosciurus*	*ruwenzorii*	LC
Sciuridae		*Heliosciurus*	*undulatus*	LC
Sciuridae		*Myosciurus*	*pumilio*	LC
Sciuridae		*Paraxerus*	*alexandri*	LC
Sciuridae		*Paraxerus*	*boehmi*	LC
Sciuridae		*Paraxerus*	*cepapi*	LC
Sciuridae		*Paraxerus*	*cooperi*	DD
Sciuridae		*Paraxerus*	*flavovittis*	LC
Sciuridae		*Paraxerus*	*lucifer*	DD
Sciuridae		*Paraxerus*	*ochraceus*	LC
Sciuridae		*Paraxerus*	*palliatus*	LC
Sciuridae		*Paraxerus*	*poensis*	LC
Sciuridae		*Paraxerus*	*vexillarius*	NT
Sciuridae		*Paraxerus*	*vincenti*	EN
Sciuridae		*Protoxerus*	*stangeri*	LC
Sciuridae		*Sciurus*	*carolinensis*	LC
Sciuridae		*Xerus*	*erythropus*	LC
Sciuridae		*Xerus*	*inauris*	LC
Sciuridae		*Xerus*	*princeps*	LC
Sciuridae		*Xerus*	*rutilus*	LC
Spalacidae	Tachyoryctinae	*Tachyoryctes*	*ankoliae*	NE
Spalacidae	Tachyoryctinae	*Tachyoryctes*	*annectens*	NE
Spalacidae	Tachyoryctinae	*Tachyoryctes*	*audax*	NE
Spalacidae	Tachyoryctinae	*Tachyoryctes*	*daemon*	NE
Spalacidae	Tachyoryctinae	*Tachyoryctes*	*ibeanus*	NE
Spalacidae	Tachyoryctinae	*Tachyoryctes*	*macrocephalus*	EN
Spalacidae	Tachyoryctinae	*Tachyoryctes*	*naivashae*	NE
Spalacidae	Tachyoryctinae	*Tachyoryctes*	*rex*	NE
Spalacidae	Tachyoryctinae	*Tachyoryctes*	*ruandae*	NE
Spalacidae	Tachyoryctinae	*Tachyoryctes*	*ruddi*	NE
Spalacidae	Tachyoryctinae	*Tachyoryctes*	*spalacinus*	NE
Spalacidae	Tachyoryctinae	*Tachyoryctes*	*splendens*	EN
Spalacidae	Tachyoryctinae	*Tachyoryctes*	*storeyi*	NE
Thryonomyidae		*Thryonomys*	*gregorianus*	LC
Thryonomyidae		*Thryonomys*	*swinderianus*	LC

Index